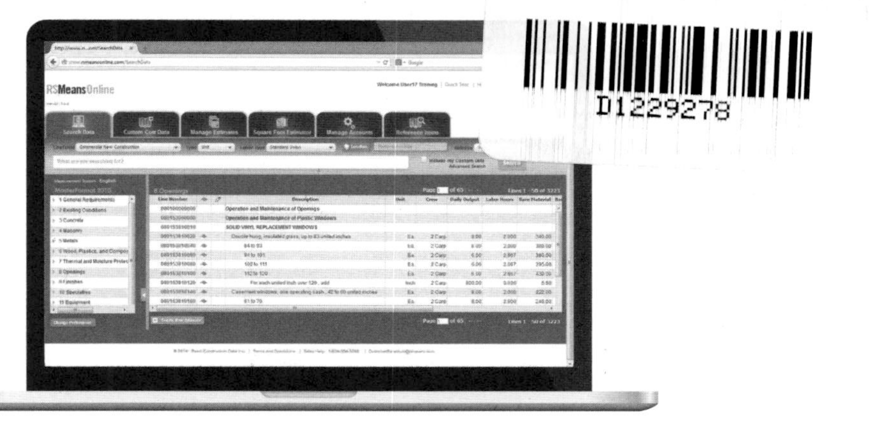

WIN the complete library of RSMeans Online data!

Register your book below to receive your FREE quarterly updates, PLUS you will be entered into a quarterly drawing to win the COMPLETE RSMeans Online library of 2015 data!

Be sure to keep up-to-date in 2015!

Fill out the card below for RSMeans' free quarterly updates, as well as a chance to win the complete RSMeans Online library. Please provide your name, address, and email below and return this card by mail, or to register online:

http://info.thegordiangroup.com/RSMeans.html

Name ______________________________

email ______________________ Title ______________

Company ______________________________

Street ______________________________

City/Town ______________ State/Prov. ______ Zip/Postal Code ______________

RS**Means**

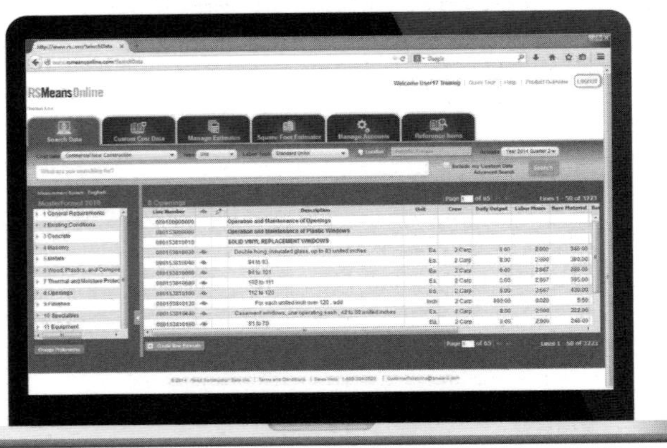

WIN the complete library of RSMeans Online 2015 data!

See other side for details.

NO POSTAGE
NECESSARY
IF MAILED
IN THE
UNITED STATES

BUSINESS REPLY MAIL
FIRST-CLASS MAIL PERMIT NO. 31 NORWELL, MA

POSTAGE WILL BE PAID BY ADDRESSEE

RSMEANS
700 LONGWATER DRIVE
NORWELL MA 02061-1624

RSMeans® Electrical Cost Data

2015 38th annual edition

Adrian C. Charest, PE, Senior Editor

Engineering Director
Bob Mewis, CCC (*1, 14, 41*)

Contributing Editors
Christopher Babbitt
Adrian C. Charest, PE (*26, 27, 28, 48*)
Gary W. Christensen
Cheryl Elsmore
David Fiske
Robert Fortier, PE (*2, 31, 32, 33, 34, 35, 44, 46*)
Robert J. Kuchta (*8*)
Thomas Lane (*6, 7*)
Robert C. McNichols
Genevieve Medeiros
Melville J. Mossman, PE (*21, 22, 23*)
Jeannene D. Murphy
Marilyn Phelan, AIA (*9, 10, 11, 12*)
Stephen C. Plotner (*3, 5*)
Siobhan Soucie (*4, 13*)

**Numbers in italics are the divisional responsibilities for each editor. Please contact the designated editor directly with any questions.*

Vice President Data & Engineering
Chris Anderson

Cover Design
Sara Rutan

Product Manager
Andrea Sillah

Production Manager
Debbie Panarelli

Production
Jill Goodman
Jonathan Forgit
Sara Rutan
Mary Lou Geary

Technical Support
Gary L. Hoitt
Kathryn S. Rodriguez

RSMeans
Construction Publishers & Consultants
700 Longwater Drive
Norwell, MA 02061
United States of America
1-877-763-2526
www.rsmeans.com

Printed in the United States of America
ISSN 1540-6318
ISBN 978-1-940238-53-1

0035 $208.95 per copy (in United States)
Price is subject to change without prior notice.

This book is printed on 50% recycled paper using soy-based ink and is entirely recyclable.

Related RSMeans Products and Services

The engineers at RSMeans suggest the following products and services as companion information resources to *RSMeans Electrical Cost Data:*

Construction Cost Data Books

Electrical Change Order Cost Data 2015
Building Construction Cost Data 2015
Mechanical Cost Data 2015

Reference Books

Electrical Estimating Methods
Estimating Building Costs
RSMeans Estimating Handbook
Green Building: Project Planning & Estimating
How to Estimate with RSMeans Data
Plan Reading & Material Takeoff
Project Scheduling & Management for Construction

Seminars and In-House Training

Unit Price Estimating
RSMeans Online® Training
Practical Project Management for Construction Professionals
Scheduling with MSProject for Construction Professionals
Mechanical & Electrical Estimating

RSMeans Online Bookstore

Visit RSMeans at **www.rsmeans.com** for a one-stop portal for the most reliable and current resources available on the market. Here you can learn about our more than 20 annual cost data books and eBooks, also available on CD and online versions, as well as our libary of reference books and RSMeans seminars.

RSMeans Electronic Data

Get the information found in RSMeans cost books electronically at **RSMeansOnline.com**.

RSMeans Business Solutions

Cost engineers and analysts conduct facility life cycle and benchmark research studies and predictive cost modeling, as well as offer consultation services for conceptual estimating and real property management. Research studies are designed with the application of proprietary cost and project data from RSMeans' extensive North American databases. Analysts offer building product manufacturers qualitative and quantitative market research, as well as time/motion studies for new products, and Web-based dashboards for market opportunity analysis. Clients are from both the public and private sectors, including federal, state, and municipal agencies; corporations; institutions; construction management firms; hospitals; and associations.

Construction Costs for Software Applications

More than 25 unit price and assemblies cost databases are available through a number of leading estimating and facilities management software partners. For more information see the "Other RSMeans Products and Services" pages at the back of this publication.

RSMeans data is also available to federal, state, and local government agencies as multi-year, multi-seat licenses.

For information on our current partners, visit: **www.rsmeans.com/partners** or call 877.763.2526.

Table of Contents

Foreword

Our Mission

Since 1942, RSMeans has been actively engaged in construction cost publishing and consulting throughout North America.

Today, more than 70 years after RSMeans began, our primary objective remains the same: to provide you, the construction and facilities professional, with the most current and comprehensive construction cost data possible.

Whether you are a contractor, owner, architect, engineer, facilities manager, or anyone else who needs a reliable construction cost estimate, you'll find this publication to be a highly useful and necessary tool.

With the constant flow of new construction methods and materials today, it's difficult to find the time to look at and evaluate all the different construction cost possibilities. In addition, because labor and material costs keep changing, last year's cost information is not a reliable basis for today's estimate or budget.

That's why so many construction professionals turn to RSMeans. We keep track of the costs for you, along with a wide range of other key information, from city cost indexes . . . to productivity rates . . . to crew composition . . . to contractor's overhead and profit rates.

RSMeans performs these functions by collecting data from all facets of the industry and organizing it in a format that is instantly accessible to you. From the preliminary budget to the detailed unit price estimate, you'll find the data in this book useful for all phases of construction cost determination.

The Staff, the Organization, and Our Services

When you purchase one of RSMeans' publications, you are, in effect, hiring the services of a full-time staff of construction and engineering professionals.

Our thoroughly experienced and highly qualified staff works daily at collecting, analyzing, and disseminating comprehensive cost information for your needs. These staff members have years of practical construction experience and engineering training prior to joining the firm. As a result, you can count on them not only for accurate cost figures, but also for additional background reference information that will help you create a realistic estimate.

The RSMeans organization is always prepared to help you solve construction problems through its variety of data solutions, including online, CD, and print book formats, as well as cost estimating expertise available via our business solutions, training, and seminars.

Besides a full array of construction cost estimating books, RSMeans also publishes a number of other reference works for the construction industry. Subjects include construction estimating and project and business management, special topics such as green building, and a library of facility management references.

In addition, you can access all of our construction cost data electronically in convenient CD format or on the web. Visit **RSMeansOnline.com** for more information on our 24/7 online cost data.

What's more, you can increase your knowledge and improve your construction estimating and management performance with an RSMeans construction seminar or in-house training program. These two-day seminar programs offer unparalleled opportunities for everyone in your organization to become updated on a wide variety of construction-related issues.

RSMeans is also a worldwide provider of construction cost management and analysis services for commercial and government owners.

In short, RSMeans can provide you with the tools and expertise for constructing accurate and dependable construction estimates and budgets in a variety of ways.

Robert Snow Means Established a Tradition of Quality That Continues Today

Robert Snow Means spent years building RSMeans, making certain he always delivered a quality product.

Today, at RSMeans, we do more than talk about the quality of our data and the usefulness of our books. We stand behind all of our data, from historical cost indexes to construction materials and techniques to current costs.

If you have any questions about our products or services, please call us toll-free at 1-877-763-2526. Our customer service representatives will be happy to assist you. You can also visit our website at **www.rsmeans.com**

How the Book Is Built: An Overview

The Construction Specifications Institute (CSI) and Construction Specifications Canada (CSC) have produced the 2014 edition of MasterFormat®, a system of titles and numbers used extensively to organize construction information.

All unit price data in the RSMeans cost data books is now arranged in the 50-division MasterFormat® 2014 system.

A Powerful Construction Tool

You have in your hands one of the most powerful construction tools available today. A successful project is built on the foundation of an accurate and dependable estimate. This book will enable you to construct just such an estimate.

For the casual user the book is designed to be:

- quickly and easily understood so you can get right to your estimate.
- filled with valuable information so you can understand the necessary factors that go into the cost estimate.

For the regular user, the book is designed to be:

- a handy desk reference that can be quickly referred to for key costs.
- a comprehensive, fully reliable source of current construction costs and productivity rates so you'll be prepared to estimate any project.
- a source book for preliminary project cost, product selections, and alternate materials and methods.

To meet all of these requirements, we have organized the book into the following clearly defined sections.

Quick Start

See our "Quick Start" instructions on the following page to get started right away.

Estimating with RSMeans Unit Price Cost Data

Please refer to these steps for guidance on completing an estimate using RSMeans unit price cost data.

How to Use the Book: The Details

This section contains an in-depth explanation of how the book is arranged . . . and how you can use it to determine a reliable construction cost estimate. It includes information about how we develop our cost figures and how to completely prepare your estimate.

Unit Price Section

All cost data has been divided into the 50 divisions according to the MasterFormat system of classification and numbering. For a listing of these divisions and an outline of their subdivisions, see the Unit Price Section Table of Contents.

Estimating tips are included at the beginning of each division.

Assemblies Section

The cost data in this section has been organized in an "Assemblies" format. These assemblies are the functional elements of a building and are arranged according to the 7 elements of the UNIFORMAT II classification system. For a complete explanation of a typical "Assemblies" page, see "How RSMeans Assemblies Data Works."

Reference Section

This section includes information on Equipment Rental Costs, Crew Listings, Historical Cost Indexes, City Cost Indexes, Location Factors, Reference Tables, Change Orders, Square Foot Costs, and a listing of Abbreviations.

Equipment Rental Costs: This section contains the average costs to rent and operate hundreds of pieces of construction equipment.

Crew Listings: This section lists all of the crews referenced in the book. For the purposes of this book, a crew is composed of more than one trade classification and/or the addition of power equipment to any trade classification. Power equipment is included in the cost of the crew. Costs are shown both with bare labor rates and with the installing contractor's overhead and profit added. For each, the total crew cost per eight-hour day and the composite cost per labor-hour are listed.

Historical Cost Indexes: These indexes provide you with data to adjust construction costs over time.

City Cost Indexes: All costs in this book are U.S. national averages. Costs vary because of the regional economy. You can adjust costs by CSI Division to over 700 locations throughout the U.S. and Canada by using the data in this section.

Location Factors: You can adjust total project costs to over 900 locations throughout the U.S. and Canada by using the data in this section.

Reference Tables: At the beginning of selected major classifications in the Unit Price section are reference numbers shown in a shaded box. These numbers refer you to related information in the Reference Section. In this section, you'll find reference tables, explanations, estimating information that support how we develop the unit price data, technical data, and estimating procedures.

Change Orders: This section includes information on the factors that influence the pricing of change orders.

Square Foot Costs: This section contains costs for 59 different building types that allow you to make a rough estimate for the overall cost of a project or its major components.

Abbreviations: A listing of abbreviations used throughout this book, along with the terms they represent, is included in this section.

Index

A comprehensive listing of all terms and subjects in this book will help you quickly find what you need when you are not sure where it occurs in MasterFormat.

The Scope of This Book

This book is designed to be as comprehensive and as easy to use as possible. To that end we have made certain assumptions and limited its scope in two key ways:

1. We have established material prices based on a national average.
2. We have computed labor costs based on a 30-city national average of union wage rates.

For a more detailed explanation of how the cost data is developed, see "How To Use the Book: The Details."

Project Size/Type

The material prices in RSMeans cost data books are "contractor's prices." They are the prices that contractors can expect to pay at the lumberyards, suppliers'/distributors' warehouses, etc. Small orders of speciality items would be higher than the costs shown, while very large orders, such as truckload lots, would be less. The variation would depend on the size, timing, and negotiating power of the contractor. The labor costs are primarily for new construction or major renovation rather than repairs or minor alterations.

With reasonable exercise of judgment, the figures can be used for any building work.

Absolute Essentials for a Quick Start

If you feel you are ready to use this book and don't think you will need the detailed instructions that begin on the following page, this Absolute Essentials for a Quick Start page is for you. These steps will allow you to get started estimating in a matter of minutes.

1 Scope

Think through the project that you will be estimating, and identify the many individual work tasks that will need to be covered in your estimate.

2 Quantify

Determine the number of units that will be required for each work task that you identified.

3 Pricing

Locate individual Unit Price line items that match the work tasks you identified. The Unit Price Section Table of Contents that begins on page 1 and the Index in the back of the book will help you find these line items.

4 Multiply

Multiply the Total Incl O&P cost for a Unit Price line item in the book by your quantity for that item. The price you calculate will be an estimate for a completed item of work performed by a subcontractor. Keep adding line items in this manner to build your estimate.

5 Project Overhead

Include project overhead items in your estimate. These items are needed to make the job run and are typically, but not always, provided by the General Contractor. They can be found in Division 1. An alternate method of estimating project overhead costs is to apply a percentage of the total project cost.

Include rented tools not included in crews, waste, rubbish handling, and cleanup.

6 Estimate Summary

Include General Contractor's markup on subcontractors, General Contractor's office overhead and profit, and sales tax on materials and equipment.

Adjust your estimate to the project's location by using the City Cost Indexes or Location Factors found in the Reference Section.

Editors' Note: We urge you to spend time reading and understanding the supporting material in the front of this book. An accurate estimate requires experience, knowledge, and careful calculation. The more you know about how we at RSMeans developed the data, the more accurate your estimate will be. In addition, it is important to take into consideration the reference material in the back of the book such as Equipment Listings, Crew Listings, City Cost Indexes, Location Factors, and Reference Tables.

Estimating with RSMeans Unit Price Cost Data

Following these steps will allow you to complete an accurate estimate using RSMeans Unit Price cost data.

1 Scope Out the Project

- Think through the project and identify those CSI Divisions needed in your estimate.
- Identify the individual work tasks that will need to be covered in your estimate.
- The Unit Price data in this book has been divided into 50 Divisions according to the CSI MasterFormat® 2014—their titles are listed on the back cover of your book.
- The Unit Price Section Table of Contents on page 1 may also be helpful when scoping out your project.
- Experienced estimators find it helpful to begin with Division 2 and continue through completion. Division 1 can be estimated after the full project scope is known.

2 Quantify

- Determine the number of units required for each work task that you identified.
- Experienced estimators include an allowance for waste in their quantities. (Waste is not included in RSMeans Unit Price line items unless so stated.)

3 Price the Quantities

- Use the Unit Price Table of Contents, and the Index, to locate individual Unit Price line items for your estimate.
- Reference Numbers indicated within a Unit Price section refer to additional information that you may find useful.
- The crew indicates who is performing the work for that task. Crew codes are expanded in the Crew Listings in the Reference Section to include all trades and equipment that comprise the crew.
- The Daily Output is the amount of work the crew is expected to do in one day.
- The Labor-Hours value is the amount of time it will take for the crew to install one unit of work.
- The abbreviated Unit designation indicates the unit of measure upon which the crew, productivity, and prices are based.
- Bare Costs are shown for materials, labor, and equipment needed to complete the Unit Price line item. Bare costs do not include waste, project overhead, payroll insurance, payroll taxes, main office overhead, or profit.
- The Total Incl O&P cost is the billing rate or invoice amount of the installing contractor or subcontractor who performs the work for the Unit Price line item.

4 Multiply

- Multiply the total number of units needed for your project by the Total Incl O&P cost for each Unit Price line item.
- Be careful that your take off unit of measure matches the unit of measure in the Unit column.
- The price you calculate is an estimate for a completed item of work.
- Keep scoping individual tasks, determining the number of units required for those tasks, matching each task with individual Unit Price line items in the book, and multiply quantities by Total Incl O&P costs.
- An estimate completed in this manner is priced as if a subcontractor, or set of subcontractors, is performing the work. The estimate does not yet include Project Overhead or Estimate Summary components such as General Contractor markups on subcontracted work, General Contractor office overhead and profit, contingency, and location factor.

5 Project Overhead

- Include project overhead items from Division 1 – General Requirements.
- These items are needed to make the job run. They are typically, but not always, provided by the General Contractor. Items include, but are not limited to, field personnel, insurance, performance bond, permits, testing, temporary utilities, field office and storage facilities, temporary scaffolding and platforms, equipment mobilization and demobilization, temporary roads and sidewalks, winter protection, temporary barricades and fencing, temporary security, temporary signs, field engineering and layout, final cleaning and commissioning.
- Each item should be quantified, and matched to individual Unit Price line items in Division 1, and then priced and added to your estimate.
- An alternate method of estimating project overhead costs is to apply a percentage of the total project cost, usually 5% to 15% with an average of 10% (see General Conditions, page ix).
- Include other project related expenses in your estimate such as:
  - Rented equipment not itemized in the Crew Listings
  - Rubbish handling throughout the project (see 02 41 19.19)

6 Estimate Summary

- Include sales tax as required by laws of your state or county.
- Include the General Contractor's markup on self-performed work, usually 5% to 15% with an average of 10%.
- Include the General Contractor's markup on subcontracted work, usually 5% to 15% with an average of 10%.
- Include General Contractor's main office overhead and profit:
  - RSMeans gives general guidelines on the General Contractor's main office overhead (see section 01 31 13.60 and Reference Number R013113-50).
  - RSMeans gives no guidance on the General Contractor's profit.
  - Markups will depend on the size of the General Contractor's operations, his projected annual revenue, the level of risk he is taking on, and on the level of competition in the local area and for this project in particular.
- Include a contingency, usually 3% to 5%, if appropriate.
- Adjust your estimate to the project's location by using the City Cost Indexes or the Location Factors in the Reference Section:
  - Look at the rules on the pages for "How to Use the City Cost Indexes" to see how to apply the Indexes for your location.
  - When the proper Index or Factor has been identified for the project's location, convert it to a multiplier by dividing it by 100, and then multiply that multiplier by your estimated total cost. The original estimated total cost will now be adjusted up or down from the national average to a total that is appropriate for your location.

Editors' Notes:
We urge you to spend time reading and understanding the supporting material in the front of this book. An accurate estimate requires experience, knowledge, and careful calculation. The more you know about how we at RSMeans developed the data, the more accurate your estimate will be. In addition, it is important to take into consideration the reference material in the back of the book such as Equipment Listings, Crew Listings, City Cost Indexes, Location Factors, and Reference Tables.

How to Use the Book: The Details

What's Behind the Numbers? The Development of Cost Data

The staff at RSMeans continually monitors developments in the construction industry in order to ensure reliable, thorough, and up-to-date cost information. While overall construction costs may vary relative to general economic conditions, price fluctuations within the industry are dependent upon many factors. Individual price variations may, in fact, be opposite to overall economic trends. Therefore, costs are constantly tracked and complete updates are published yearly. Also, new items are frequently added in response to changes in materials and methods.

Costs—$ (U.S.)

All costs represent U.S. national averages and are given in U.S. dollars. The RSMeans City Cost Indexes can be used to adjust costs to a particular location. The City Cost Indexes for Canada can be used to adjust U.S. national averages to local costs in Canadian dollars. No exchange rate conversion is necessary.

G The processes or products identified by the green symbol in our publications have been determined to be environmentally responsible and/or resource-efficient solely by the RSMeans engineering staff. The inclusion of the green symbol does not represent compliance with any specific industry association or standard.

Material Costs

The RSMeans staff contacts manufacturers, dealers, distributors, and contractors all across the U.S. and Canada to determine national average material costs. If you have access to current material costs for your specific location, you may wish to make adjustments to reflect differences from the national average. Included within material costs are fasteners for a normal installation. RSMeans engineers use manufacturers' recommendations, written specifications, and/or standard construction practice for size and spacing of fasteners. Adjustments to material costs may be required for your specific application or location. The manufacturer's warranty is assumed. Extended warranties are not included in the material costs. Material costs do not include sales tax.

Labor Costs

Labor costs are based on the average of wage rates from 30 major U.S. cities. Rates are determined from labor union agreements or prevailing wages for construction trades for the current year. Rates, along with overhead and profit markups, are listed on the inside back cover of this book.

- If wage rates in your area vary from those used in this book, or if rate increases are expected within a given year, labor costs should be adjusted accordingly.

Labor costs reflect productivity based on actual working conditions. In addition to actual installation, these figures include time spent during a normal weekday on tasks such as, material receiving and handling, mobilization at site, site movement, breaks, and cleanup.

Productivity data is developed over an extended period so as not to be influenced by abnormal variations and reflects a typical average.

Equipment Costs

Equipment costs include not only rental, but also operating costs for equipment under normal use. The operating costs include parts and labor for routine servicing such as repair and replacement of pumps, filters, and worn lines. Normal operating expendables, such as fuel, lubricants, tires, and electricity (where applicable), are also included. Extraordinary operating expendables with highly variable wear patterns, such as diamond bits and blades, are excluded. These costs are included under materials. Equipment rental rates are obtained from industry sources throughout North America—contractors, suppliers, dealers, manufacturers, and distributors.

Rental rates can also be treated as reimbursement costs for contractor-owned equipment. Owned equipment costs include depreciation, loan payments, interest, taxes, insurance, storage and major repairs.

Equipment costs do not include operators' wages; nor do they include the cost to move equipment to a job site (mobilization) or from a job site (demobilization).

Equipment Cost/Day—The cost of power equipment required for each crew is included in the Crew Listings in the Reference Section (small tools that are considered as essential everyday tools are not listed out separately). The Crew Listings itemize specialized tools and heavy equipment along with labor trades. The daily cost of itemized equipment included in a crew is based on dividing the weekly bare rental rate by 5 (number of working days per week) and then adding the hourly operating cost times 8 (the number of hours per day). This Equipment Cost/Day is shown in the last column of the Equipment Rental Cost pages in the Reference Section.

Mobilization/Demobilization—The cost to move construction equipment from an equipment yard or rental company to the job site and back again is not included in equipment costs. Mobilization (to the site) and demobilization (from the site) costs can be found in the Unit Price Section. If a piece of equipment is already at the job site, it is not appropriate to utilize mob./demob. costs again in an estimate.

Overhead and Profit

Total Cost including O&P for the *Installing Contractor* is shown in the last column on the Unit Price and/or the Assemblies pages of this book. This figure is the sum of the bare material cost plus 10% for profit, the bare labor cost plus total overhead and profit, and the bare equipment cost plus 10% for profit. Details for the calculation of Overhead and Profit on labor are shown on the inside back cover and in the Reference Section of this book. (See "How RSMeans Unit Price Data Works" for an example of this calculation.)

General Conditions

Cost data in this book is presented in two ways: Bare Costs and Total Cost including O&P (Overhead and Profit). General Conditions, or General Requirements, of the contract should also be added to the Total Cost including O&P when applicable. Costs for General Conditions are listed in Division 1 of the Unit Price Section and the Reference

Section of this book. General Conditions for the *Installing Contractor* may range from 0% to 10% of the Total Cost including O&P. For the *General* or *Prime Contractor*, costs for General Conditions may range from 5% to 15% of the Total Cost including O&P, with a figure of 10% as the most typical allowance. If applicable, the Assemblies and Models sections of this book use costs that include the installing contractor's overhead and profit (O&P).

Factors Affecting Costs

Costs can vary depending upon a number of variables. Here's how we have handled the main factors affecting costs.

Quality—The prices for materials and the workmanship upon which productivity is based represent sound construction work. They are also in line with U.S. government specifications.

Overtime—We have made no allowance for overtime. If you anticipate premium time or work beyond normal working hours, be sure to make an appropriate adjustment to your labor costs.

Productivity—The productivity, daily output, and labor-hour figures for each line item are based on working an eight-hour day in daylight hours in moderate temperatures. For work that extends beyond normal work hours or is performed under adverse conditions, productivity may decrease. (See "How RSMeans Unit Price Data Works" for more on productivity.)

Size of Project—The size, scope of work, and type of construction project will have a significant impact on cost. Economies of scale can reduce costs for large projects. Unit costs can often run higher for small projects.

Location—Material prices in this book are for metropolitan areas. However, in dense urban areas, traffic and site storage limitations may increase costs. Beyond a 20-mile radius of large cities, extra trucking or transportation charges may also increase the material costs slightly. On the other hand, lower wage rates may be in effect. Be sure to consider both of these factors when preparing an estimate, particularly if the job site is located in a central city or remote rural location. In addition, highly specialized subcontract items may require travel and per-diem expenses for mechanics.

Other Factors—

- season of year
- contractor management
- weather conditions
- local union restrictions
- building code requirements
- availability of:
  - adequate energy
  - skilled labor
  - building materials
- owner's special requirements/restrictions
- safety requirements
- environmental considerations

Unpredictable Factors—General business conditions influence "in-place" costs of all items. Substitute materials and construction methods may have to be employed. These may affect the installed cost and/or life cycle costs. Such factors may be difficult to evaluate and cannot necessarily be predicted on the basis of the job's location in a particular section of the country. Thus, where these factors apply, you may find significant but unavoidable cost variations for which you will have to apply a measure of judgment to your estimate.

Rounding of Costs

In general, all unit prices in excess of $5.00 have been rounded to make them easier to use and still maintain adequate precision of the results. The rounding rules we have chosen are in the following table.

Prices from . . .	Rounded to the nearest . . .
$.01 to $5.00	$.01
$5.01 to $20.00	$.05
$20.01 to $100.00	$.50
$100.01 to $300.00	$1.00
$300.01 to $1,000.00	$5.00
$1,000.01 to $10,000.00	$25.00
$10,000.01 to $50,000.00	$100.00
$50,000.01 and above	$500.00

How Subcontracted Items Affect Costs

A considerable portion of all large construction jobs is usually subcontracted. In fact, the percentage done by subcontractors is constantly increasing and may run over 90%. Since the workers employed by these companies do nothing else but install their particular product, they soon become expert in that line. The result is, installation by these firms is accomplished so efficiently that the total in-place cost, even adding the general contractor's overhead and profit, is no more, and often less, than if the principal contractor had handled the installation himself/herself. Companies that deal with construction specialties are anxious to have their product perform well and, consequently, the installation will be the best possible.

Contingencies

The allowance for contingencies generally provides for unforeseen construction difficulties. On alterations or repair jobs, 20% is not too much. If drawings are final and only field contingencies are being considered, 2% or 3% is probably sufficient, and often nothing need be added. Contractually, changes in plans will be covered by extras. The contractor should consider inflationary price trends and possible material shortages during the course of the job. These escalation factors are dependent upon both economic conditions and the anticipated time between the estimate and actual construction. If drawings are not complete or approved, or a budget cost is wanted, it is wise to add 5% to 10%. Contingencies, then, are a matter of judgment. Additional allowances for contingencies are shown in Division 1.

Important Estimating Considerations

The productivity or daily output of each craftsman or crew assumes a well-managed job where tradesmen with the proper tools and equipment, along with the appropriate construction materials, are present. Included are daily mobilization and cleanup time, break time and plan layout time. Unless otherwise indicated, time for material movement on site (for items that can be transported by hand) of up to 200' into the building and to the first or second floor is also included. If material has to be transported by other means, over greater distances, or to higher floors, an additional allowance should be considered by the estimator.

While horizontal movement is typically a sole function of distances, vertical transport introduces other variables that can significantly impact productivity. In an occupied building, the use of elevators (assuming access, size, and required protective measures are acceptable) must be understood at the time of the estimate. For new construction, hoist wait and cycle times can easily be 15 minutes and may result in scheduled access extending beyond the normal work day. Finally, all vertical transport will impose strict weight limits likely to preclude the use of any motorized material handling.

The productivity, or daily output, also assumes installation that meets manufacturer/designer/standard specifications. A time allowance for quality control checks, minor adjustments and any task required to ensure the proper function or operation is also included. For items that require connections to services, time is included for positioning, leveling, securing the unit and for making all the necessary connections (and start up where applicable) ensuring a complete installation. Estimating of the services themselves (electrical, plumbing, water, steam, hydraulics, dust collection, etc.) is separate.

In some cases, the estimator must consider the use of a crane and an appropriate crew for the installation of large or heavy items. For those situations where a crane is not included in the assigned crew and as part of the line item cost, equipment rental costs, mobilization and demobilization costs, and operator and support personnel costs must be considered.

Labor-Hours

The labor-hours expressed in this publication are derived by dividing the total daily labor hours for the crew by the daily output. Based on average installation time and the assumptions listed above, the labor-hours include: direct labor, indirect labor, and nonproductive time. A typical day for a craftsman might include, but not be limited to:

- Direct Work
  - □ Measuring and layout
  - □ Preparing materials
  - □ Actual installation
  - □ Quality assurance/quality control
- Indirect Work
  - □ Reading plans or specifications
  - □ Preparing space
  - □ Receiving materials
  - □ Material movement
  - □ Giving or receiving instruction
  - □ Miscellaneous
- Non-Work
  - □ Chatting
  - □ Personal issues
  - □ Breaks
  - □ Interruptions (i.e., sickness, weather, material or equipment shortages, etc.)

If any of the items for a typical day do not apply to the particular work or project situation, the estimator should make any necessary adjustments.

Final Checklist

Estimating can be a straightforward process provided you remember the basics. Here's a checklist of some of the steps you should remember to complete before finalizing your estimate.

Did you remember to . . .

- factor in the City Cost Index for your locale?
- take into consideration which items have been marked up and by how much?
- mark up the entire estimate sufficiently for your purposes?
- read the background information on techniques and technical matters that could impact your project time span and cost?
- include all components of your project in the final estimate?
- double check your figures for accuracy?
- call RSMeans if you have any questions about your estimate or the data you've found in our publications? Remember, RSMeans stands behind its publications. If you have any questions about your estimate . . . about the costs you've used from our books . . . or even about the technical aspects of the job that may affect your estimate, feel free to call the RSMeans editors at 1-877-763-2526.

Free Quarterly Updates

Stay up-to-date throughout 2015 with RSMeans' free cost data updates four times a year. Sign up online to make sure you have access to the newest data. Every quarter we provide the city cost adjustment factors for hundreds of cities and key materials. Register at: **http://info.thegordiangroup.com/RSMeans.html**.

Unit Price Section

Table of Contents

How RSMeans Unit Price Data Works

All RSMeans unit price data is organized in the same way.

It is important to understand the structure, so that you can find information easily and use it correctly.

RSMeans **Line Numbers** consist of 12 characters, which identify a unique location in the database for each task. The first 6 or 8 digits conform to the Construction Specifications Institute MasterFormat® 2014. The remainder of the digits are a further breakdown by RSMeans in order to arrange items in understandable groups of similar tasks. Line numbers are consistent across all RSMeans publications, so a line number in any RSMeans product will always refer to the same unit of work.

RSMeans engineers have created **reference** information to assist you in your estimate. If there is information that applies to a section, it will be indicated at the start of the section. In this case, R033105-10 provides information on the proportionate quantities of formwork, reinforcing, and concrete used in cast-in-place concrete items such as footings, slabs, beams, and columns. The Reference Section is located in the back of the book on the pages with a gray edge.

RSMeans **Descriptions** are shown in a hierarchical structure to make them readable. In order to read a complete description, read up through the indents to the top of the section. Include everything that is above and to the left that is not contradicted by information below. For instance, the complete description for line 03 30 53.40 3550 is "Concrete in place, including forms (4 uses), Grade 60 rebar, concrete (Portland cement Type 1), placement and finishing unless otherwise indicated; Equipment pad (3000 psi), 4' x 4' x 6" thick."

When using **RSMeans data**, it is important to read through an entire section to ensure that you use the data that most closely matches your work. Note that sometimes there is additional information shown in the section that may improve your price. There are frequently lines that further describe, add to, or adjust data for specific situations.

03 30 Cast-In-Place Concrete

03 30 53 – Miscellaneous Cast-In-Place Concrete

03 30 53.40 Concrete In Place

0010	**CONCRETE IN PLACE**	R033105-10
0020	Including forms (4 uses), Grade 60 rebar, concrete (Portland cement	R033105-20
0050	Type I), placement and finishing unless otherwise indicated	R033105-50
0300	Beams (3500 psi), 5 kip per L.F., 10' span	R033105-65
0350	25' span	R033105-70
0500	Chimney foundations (5000 psi), over 5 C.Y.	R033105-85
0510	(3500 psi), under 5 C.Y.	
0700	Columns, square (4000 psi), 12" x 12", less than 2% reinforcing	
3450	Over 10,000 S.F.	
3500	Add per floor for 3 to 6 stories high	
3520	For 7 to 20 stories high	
3540	Equipment pad (3000 psi), 3' x 3' x 6" thick	
3550	4' x 4' x 6" thick	
3560	5' x 5' x 8" thick	
3570	6' x 6' x 8" thick	

The data published in RSMeans print books represents a "national average" cost. This data should be modified to the project location using the **City Cost Indexes** or **Location Factors** tables found in the Reference Section (see pages 492–540). Use the location factors to adjust estimate totals if the project covers multiple trades. Use the city cost indexes (CCI) for single trade projects or projects where a more detailed analysis is required. All figures in the two tables are derived from the same research. The last row of data in the CCI, the weighted average, is the same as the numbers reported for each location in the location factor table.

Crews include labor or labor and equipment necessary to accomplish each task. In this case, Crew C-14H is used. RSMeans selects a crew to represent the workers and equipment that are typically used for that task. In this case, Crew C-14H consists of one carpenter foreman (outside), two carpenters, one rodman, one laborer, one cement finisher, and one gas engine vibrator. Details of all crews can be found in the reference section.

Crews

Crew No.	Bare Costs		Incl. Subs O&P		Cost Per Labor-Hour	
Crew C-14H	Hr.	Daily	Hr.	Daily	Bare Costs	Incl. O&P
1 Carpenter Foreman (outside)	$48.95	$391.60	$75.35	$602.80	$46.33	$71.08
2 Carpenters	46.95	751.20	72.25	1156.00		
1 Rodman (reinf.)	52.55	420.40	82.10	656.80		
1 Laborer	37.60	300.80	57.85	462.80		
1 Cement Finisher	45.00	360.00	66.65	533.20		
1 Gas Engine Vibrator		31.20		34.32	.65	.71
48 L.H., Daily Totals		$2255.20		$3445.92	$46.98	$71.79

The **Daily Output** is the amount of work that the crew can do in a normal 8-hour workday, including mobilization, layout, movement of materials, and cleanup. In this case, crew C-14H can install thirty 4' x 4' x 6" thick concrete pads in a day. Daily output is variable, based on many factors, including the size of the job, location, and environmental conditions. RSMeans data represents work done in daylight (or adequate lighting) and temperate conditions.

Bare Costs are the costs of materials, labor, and equipment that the installing contractor pays. They represent the cost, in U.S. dollars, for one unit of work. They do not include any markups for profit or labor burden.

Crew	Daily Output	Labor-Hours	Unit	Material	Labor	Equipment	Total	Total Incl O&P
				2015 Bare Costs				
C-14A	15.62	12.804	C.Y.	320	605	48	973	1,325
"	18.55	10.782		335	510	40.50	885.50	1,200
C-14C	32.22	3.476		150	157	.97	307.97	405
"	23.71	4.724		177	213	1.32	391.32	525
C-14A	11.96	16.722		365	790	63	1,218	1,700
	2200	.025		.94	1.07	.33	2.34	3.02
	31800	.002			.07	.02	.09	.13
	21200	.003			.11	.03	.14	.21
C-14H	45	1.067	Ea.	47	49.50	.69	97.19	128
	30	1.600		69.50	74	1.04	144.54	191
	18	2.667		122	124	1.73	247.73	325
	14	3.429		164	159	2.23	325.23	425

The **Total Incl O&P column** is the total cost, including overhead and profit, that the installing contractor will charge the customer. This represents the cost of materials plus 10% profit, the cost of labor plus labor burden and 10% profit, and the cost of equipment plus 10% profit. It does not include the general contractor's overhead and profit. Note: See the inside back cover for details of how RSMeans calculates labor burden.

The **Total column** represents the total bare cost for the installing contractor, in U.S. dollars. In this case, the sum of $69.50 for material + $74.00 for labor + $1.04 for equipment is $144.54.

The figure in the **Labor Hours** column is the amount of labor required to perform one unit of work—in this case the amount of labor required to construct one 4' x 4' equipment pad. This figure is calculated by dividing the number of hours of labor in the crew by the daily output (48 labor hours divided by 30 pads = 1.6 hours of labor per pad). Multiply 1.600 times 60 to see the value in minutes: 60 x 1.6 = 96 minutes. Note: the labor hour figure is not dependent on the crew size. A change in crew size will result in a corresponding change in daily output, but the labor hours per unit of work will not change.

All RSMeans unit cost data includes the typical **Unit of Measure** used for estimating that item. For concrete-in-place the typical unit is cubic yards (C.Y.) or each (Ea.). For installing broadloom carpet it is square yard, and for gypsum board it is square foot. The estimator needs to take special care that the unit in the data matches the unit in the take-off. Unit conversions may be found in the Reference Section.

How RSMeans Unit Price Data Works (Continued)

Sample Estimate

This sample demonstrates the elements of an estimate, including a tally of the RSMeans data lines, and a summary of the markups on a contractor's work to arrive at a total cost to the owner. The RSMeans Location Factor is added at the bottom of the estimate to adjust the cost of the work to a specific location.

Work Performed: The body of the estimate shows the RSMeans data selected, including line number, a brief description of each item, its take-off unit and quantity, and the bare costs of materials, labor, and equipment. This estimate also includes a column titled "SubContract." This data is taken from the RSMeans column "Total Incl O&P," and represents the total that a subcontractor would charge a general contractor for the work, including the sub's markup for overhead and profit.

Division 1, General Requirements: This is the first division numerically, but the last division estimated. Division 1 includes project-wide needs provided by the general contractor. These requirements vary by project, but may include temporary facilities and utilities, security, testing, project cleanup, etc. For small projects a percentage can be used, typically between 5% and 15% of project cost. For large projects the costs may be itemized and priced individually.

Bonds: Bond costs should be added to the estimate. The figures here represent a typical performance bond, ensuring the owner that if the general contractor does not complete the obligations in the construction contract the bonding company will pay the cost for completion of the work.

Location Adjustment: RSMeans published data is based on national average costs. If necessary, adjust the total cost of the project using a location factor from the "Location Factor" table or the "City Cost Index" table. Use location factors if the work is general, covering multiple trades. If the work is by a single trade (e.g., masonry) use the more specific data found in the "City Cost Indexes."

This estimate is based on an interactive spreadsheet. A copy of this spreadsheet is located on the RSMeans website at **http:/www.reedconstructiondata.com/rsmeans/extras/546011.** You are free to download it and adjust it to your methodology.

Project Name: Pre-Engineered Steel Building		Architect: As Shown		
Location:	**Anywhere, USA**			
Line Number	**Description**	**Qty**	**Unit**	**Material**
03 30 53.40 3940	Strip footing, 12" x 24", reinforced	34	C.Y.	$4,726.00
03 30 53.40 3950	Strip footing, 12" x 36", reinforced	15	C.Y.	$1,995.00
03 11 13.65 3000	Concrete slab edge forms	500	L.F.	$175.00
03 22 11.10 0200	Welded wire fabric reinforcing	150	C.S.F.	$2,580.00
03 31 13.35 0300	Ready mix concrete, 4000 psi for slab on grade	278	C.Y.	$29,746.00
03 31 13.70 4300	Place, strike off & consolidate concrete slab	278	C.Y.	$0.00
03 35 13.30 0250	Machine float & trowel concrete slab	15,000	S.F.	$0.00
03 15 16.20 0140	Cut control joints in concrete slab	950	L.F.	$57.00
03 39 23.13 0300	Sprayed concrete curing membrane	150	C.S.F.	$1,657.50
Division 03	**Subtotal**			**$40,936.50**
08 36 13.10 2650	Manual 10' x 10' steel sectional overhead door	8	Ea.	$8,800.00
08 36 13.10 2860	Insulation and steel back panel for OH door	800	S.F.	$3,800.00
Division 08	**Subtotal**			**$12,600.00**
13 34 19.50 1100	Pre-Engineered Steel Building, 100' x 150' x 24'	15,000	SF Flr.	$0.00
13 34 19.50 6050	Framing for PESB door opening, 3' x 7'	4	Opng.	$0.00
13 34 19.50 6100	Framing for PESB door opening, 10' x 10'	8	Opng.	$0.00
13 34 19.50 6200	Framing for PESB window opening, 4' x 3'	6	Opng.	$0.00
13 34 19.50 5750	PESB door, 3' x 7', single leaf	4	Opng.	$2,340.00
13 34 19.50 7750	PESB sliding window, 4' x 3' with screen	6	Opng.	$2,280.00
13 34 19.50 6550	PESB gutter, eave type, 26 ga., painted	300	L.F.	$2,085.00
13 34 19.50 8650	PESB roof vent, 12" wide x 10' long	15	Ea.	$540.00
13 34 19.50 6900	PESB insulation, vinyl faced, 4" thick	27,400	S.F.	$11,508.00
Division 13	**Subtotal**			**$18,753.00**
		Subtotal		**$72,289.50**
Division 01	**General Requirements @ 7%**			**5,060.27**
		Estimate Subtotal		**$77,349.77**
		Sales Tax @ 5%		**3,867.49**
		Subtotal A		**81,217.25**
		GC O & P		**8,121.73**
		Subtotal B		**89,338.98**
		Contingency @ 5%		
		Subtotal C		
		Bond @ $12/1000 +10% O&P		
		Subtotal D		
		Location Adjustment Factor		
		Grand Total		

This example shows the cost to construct a pre-engineered steel building. The foundation, doors, windows, and insulation will be installed by the general contractor. A subcontractor will install the structural steel, roofing, and siding.

		01/01/15	STD	
Labor	Equipment	SubContract	Estimate Total	
$3,570.00	$22.10	$0.00		
$1,260.00	$7.80	$0.00		
$1,190.00	$0.00	$0.00		
$4,050.00	$0.00	$0.00		
$0.00	$0.00	$0.00		
$4,753.80	$158.46	$0.00		
$9,000.00	$450.00	$0.00		
$380.00	$85.50	$0.00		
$952.50	$0.00	$0.00		
$25,156.30	**$723.86**	**$0.00**	**$66,816.66**	**Division 03**
$3,320.00	$0.00	$0.00		
$0.00	$0.00	$0.00		
$3,320.00	**$0.00**	**$0.00**	**$15,920.00**	**Division 08**
$0.00	$0.00	$352,500.00		
$0.00	$0.00	$2,280.00		
$0.00	$0.00	$9,200.00		
$0.00	$0.00	$3,330.00		
$672.00	$0.00	$0.00		
$576.00	$67.20	$0.00		
$789.00	$0.00	$0.00		
$3,165.00	$0.00	$0.00		
$9,042.00	$0.00	$0.00		
$14,244.00	**$67.20**	**$367,310.00**	**$400,374.20**	**Division 13**
$42,720.30	**$791.06**	**$367,310.00**	**$483,110.86**	**Subtotal**
2,990.42	**55.37**	**25,711.70**		**Gen. Requirements**
$45,710.72	**$846.43**	**$393,021.70**	**$483,110.86**	**Estimate Subtotal**
	42.32	**9,825.54**		**Sales tax**
45,710.72	**888.76**	**402,847.24**		**Subtotal**
24,912.34	**88.88**	**40,284.72**		**GC O & P**
70,623.06	**977.63**	**443,131.97**	**$604,071.64**	**Subtotal**
			30,203.58	**Contingency**
			$634,275.22	**Subtotal**
			8,372.43	**Bond**
			$642,647.66	**Subtotal**
102.30			**14,780.90**	**Location Adjustment**
			$657,428.55	**Grand Total**

Sales Tax: If the work is subject to state or local sales taxes, the amount must be added to the estimate. Sales tax may be added to material costs, equipment costs, and subcontracted work. In this case, sales tax was added in all three categories. It was assumed that approximately half the subcontracted work would be material cost, so the tax was applied to 50% of the subcontract total.

GC O&P: This entry represents the general contractor's markup on material, labor, equipment, and subcontractor costs. RSMeans' standard markup on materials, equipment, and subcontracted work is 10%. In this estimate, the markup on the labor performed by the GC's workers uses "Skilled Workers Average" shown in Column F on the table "Installing Contractor's Overhead & Profit," which can be found on the inside-back cover of the book.

Contingency: A factor for contingency may be added to any estimate to represent the cost of unknowns that may occur between the time that the estimate is performed and the time the project is constructed. The amount of the allowance will depend on the stage of design at which the estimate is done, and the contractor's assessment of the risk involved. Refer to section 01 21 16.50 for contigency allowances.

Estimating Tips

01 20 00 Price and Payment Procedures

- Allowances that should be added to estimates to cover contingencies and job conditions that are not included in the national average material and labor costs are shown in section 01 21.
- When estimating historic preservation projects (depending on the condition of the existing structure and the owner's requirements), a 15%–20% contingency or allowance is recommended, regardless of the stage of the drawings.

01 30 00 Administrative Requirements

- Before determining a final cost estimate, it is a good practice to review all the items listed in Subdivisions 01 31 and 01 32 to make final adjustments for items that may need customizing to specific job conditions.
- Requirements for initial and periodic submittals can represent a significant cost to the General Requirements of a job. Thoroughly check the submittal specifications when estimating a project to determine any costs that should be included.

01 40 00 Quality Requirements

- All projects will require some degree of Quality Control. This cost is not included in the unit cost of construction listed in each division. Depending upon the terms of the contract, the various costs of inspection and testing can be the responsibility of either the owner or the contractor. Be sure to include the required costs in your estimate.

01 50 00 Temporary Facilities and Controls

- Barricades, access roads, safety nets, scaffolding, security, and many more requirements for the execution of a safe project are elements of direct cost. These costs can easily be overlooked when preparing an estimate. When looking through the major classifications of this subdivision, determine which items apply to each division in your estimate.
- Construction Equipment Rental Costs can be found in the Reference Section in section 01 54 33. Operators' wages are not included in equipment rental costs.
- Equipment mobilization and demobilization costs are not included in equipment rental costs and must be considered separately in section 01 54 36.50.
- The cost of small tools provided by the installing contractor for his workers is covered in the "Overhead" column on the "Installing Contractor's Overhead and Profit" table that lists labor trades, base rates and markups and, therefore, is included in the "Total Incl. O&P" cost of any Unit Price line item.

01 70 00 Execution and Closeout Requirements

- When preparing an estimate, thoroughly read the specifications to determine the requirements for Contract Closeout. Final cleaning, record documentation, operation and maintenance data, warranties and bonds, and spare parts and maintenance materials can all be elements of cost for the completion of a contract. Do not overlook these in your estimate.

Reference Numbers

Reference numbers are shown in shaded boxes at the beginning of some major classifications. These numbers refer to related items in the Reference Section. The reference information may be an estimating procedure, an alternate pricing method, or technical information.

Note: Not all subdivisions listed here necessarily appear in this publication. ■

01 11 Summary of Work

01 11 31 – Professional Consultants

	01 11 31.20 Construction Management Fees	Crew	Daily Output	Labor-Hours	Unit	2015 Bare Costs Material	Labor	Equipment	Total	Total Incl O&P
0010	**CONSTRUCTION MANAGEMENT FEES**									
0020	$1,000,000 job, minimum				Project				4.50%	4.50%
0050	Maximum								7.50%	7.50%
0060	For work to $100,000								10%	10%
0300	$50,000,000 job, minimum								2.50%	2.50%
0350	Maximum				↓				4%	4%

	01 11 31.30 Engineering Fees	Crew	Daily Output	Labor-Hours	Unit	Material	Labor	Equipment	Total	Total Incl O&P
0010	**ENGINEERING FEES** R011110-30									
0020	Educational planning consultant, minimum				Project				.50%	.50%
0100	Maximum				"				2.50%	2.50%
0200	Electrical, minimum				Contrct				4.10%	4.10%
0300	Maximum								10.10%	10.10%
0400	Elevator & conveying systems, minimum								2.50%	2.50%
0500	Maximum								5%	5%
0600	Food service & kitchen equipment, minimum								8%	8%
0700	Maximum								12%	12%
1000	Mechanical (plumbing & HVAC), minimum								4.10%	4.10%
1100	Maximum				↓				10.10%	10.10%

01 21 Allowances

01 21 16 – Contingency Allowances

	01 21 16.50 Contingencies	Crew	Daily Output	Labor-Hours	Unit	Material	Labor	Equipment	Total	Total Incl O&P
0010	**CONTINGENCIES**, Add to estimate									
0020	Conceptual stage				Project				20%	20%
0050	Schematic stage								15%	15%
0100	Preliminary working drawing stage (Design Dev.)								10%	10%
0150	Final working drawing stage				↓				3%	3%

01 21 53 – Factors Allowance

	01 21 53.50 Factors	Crew	Daily Output	Labor-Hours	Unit	Material	Labor	Equipment	Total	Total Incl O&P
0010	**FACTORS** Cost adjustments									
0100	Add to construction costs for particular job requirements									
0500	Cut & patch to match existing construction, add, minimum				Costs	2%	3%			
0550	Maximum					5%	9%			
0800	Dust protection, add, minimum					1%	2%			
0850	Maximum					4%	11%			
1100	Equipment usage curtailment, add, minimum					1%	1%			
1150	Maximum					3%	10%			
1400	Material handling & storage limitation, add, minimum					1%	1%			
1450	Maximum					6%	7%			
1700	Protection of existing work, add, minimum					2%	2%			
1750	Maximum					5%	7%			
2000	Shift work requirements, add, minimum						5%			
2050	Maximum						30%			
2300	Temporary shoring and bracing, add, minimum					2%	5%			
2350	Maximum					5%	12%			
2400	Work inside prisons and high security areas, add, minimum						30%			
2450	Maximum				↓		50%			

01 21 55 – Job Conditions Allowance

01 21 55.50 Job Conditions		Crew	Daily Output	Labor-Hours	Unit	2015 Bare Costs Material	Labor	Equipment	Total	Total Incl O&P
0010	**JOB CONDITIONS** Modifications to applicable									
0020	cost summaries									
0100	Economic conditions, favorable, deduct				Project				2%	2%
0200	Unfavorable, add								5%	5%
0300	Hoisting conditions, favorable, deduct								2%	2%
0400	Unfavorable, add								5%	5%
0700	Labor availability, surplus, deduct								1%	1%
0800	Shortage, add								10%	10%
0900	Material storage area, available, deduct								1%	1%
1000	Not available, add								2%	2%
1100	Subcontractor availability, surplus, deduct								5%	5%
1200	Shortage, add								12%	12%
1300	Work space, available, deduct								2%	2%
1400	Not available, add				↓				5%	5%

01 21 57 – Overtime Allowance

01 21 57.50 Overtime

		Crew	Daily Output	Labor-Hours	Unit	Material	Labor	Equipment	Total	Total Incl O&P
0010	**OVERTIME** for early completion of projects or where R012909-90									
0020	labor shortages exist, add to usual labor, up to				Costs		100%			

01 21 61 – Cost Indexes

01 21 61.10 Construction Cost Index

		Crew	Daily Output	Labor-Hours	Unit	Material	Labor	Equipment	Total	Total Incl O&P
0010	**CONSTRUCTION COST INDEX** (Reference) over 930 zip code locations in									
0020	the U.S. and Canada, total bldg. cost, min. (Fayetteville, AR)				%				75.40%	75.40%
0050	Average								100%	100%
0100	Maximum (New York, NY)				↓				131.80%	131.80%

01 21 61.20 Historical Cost Indexes

		Crew	Daily Output	Labor-Hours	Unit	Material	Labor	Equipment	Total	Total Incl O&P
0010	**HISTORICAL COST INDEXES** (See Reference Section)									

01 21 61.30 Labor Index

		Crew	Daily Output	Labor-Hours	Unit	Material	Labor	Equipment	Total	Total Incl O&P
0010	**LABOR INDEX** (Reference) For over 930 zip code locations in									
0020	the U.S. and Canada, minimum (Beaufort, SC)				%		44.90%			
0050	Average						100%			
0100	Maximum (New York, NY)				↓		168.90%			

01 21 61.50 Material Index

		Crew	Daily Output	Labor-Hours	Unit	Material	Labor	Equipment	Total	Total Incl O&P
0010	**MATERIAL INDEX** (Reference) For over 930 zip code locations in									
0020	the U.S. and Canada, minimum (Oil City, PA)				%	92.40%				
0040	Average					100%				
0060	Maximum (Ketchikan, AK)				↓	132%				

01 21 63 – Taxes

01 21 63.10 Taxes

		Crew	Daily Output	Labor-Hours	Unit	Material	Labor	Equipment	Total	Total Incl O&P
0010	**TAXES** R012909-80									
0020	Sales tax, State, average				%	5.04%				
0050	Maximum R012909-85					7.50%				
0200	Social Security, on first $117,000 of wages						7.65%			
0300	Unemployment, combined Federal and State, minimum						.60%			
0350	Average						7.80%			
0400	Maximum				↓		12.87%			

01 31 Project Management and Coordination

01 31 13 – Project Coordination

01 31 13.20 Field Personnel

	01 31 13.20 Field Personnel	Crew	Daily Output	Labor-Hours	Unit	Material	2015 Bare Costs Labor	Equipment	Total	Total Incl O&P
0010	**FIELD PERSONNEL**									
0020	Clerk, average				Week		450		450	695
0100	Field engineer, minimum						1,075		1,075	1,650
0120	Average						1,400		1,400	2,150
0140	Maximum						1,575		1,575	2,450
0160	General purpose laborer, average						1,500		1,500	2,325
0180	Project manager, minimum						1,975		1,975	3,075
0200	Average						2,275		2,275	3,525
0220	Maximum						2,600		2,600	4,025
0240	Superintendent, minimum						1,925		1,925	2,975
0260	Average						2,125		2,125	3,275
0280	Maximum						2,400		2,400	3,725
0290	Timekeeper, average						1,225		1,225	1,900

01 31 13.30 Insurance

	01 31 13.30 Insurance	Crew	Daily Output	Labor-Hours	Unit	Material	Labor	Equipment	Total	Total Incl O&P
0010	**INSURANCE** R013113-40									
0020	Builders risk, standard, minimum				Job				.24%	.24%
0050	Maximum R013113-60								.64%	.64%
0200	All-risk type, minimum								.25%	.25%
0250	Maximum								.62%	.62%
0400	Contractor's equipment floater, minimum				Value				.50%	.50%
0450	Maximum				"				1.50%	1.50%
0600	Public liability, average				Job				2.02%	2.02%
0800	Workers' compensation & employer's liability, average									
0850	by trade, carpentry, general				Payroll		14.93%			
1000	Electrical						5.76%			
1150	Insulation						11.66%			
1450	Plumbing						6.98%			
1550	Sheet metal work (HVAC)						8.82%			

01 31 13.50 General Contractor's Mark-Up

	01 31 13.50 General Contractor's Mark-Up	Crew	Daily Output	Labor-Hours	Unit	Material	Labor	Equipment	Total	Total Incl O&P
0010	**GENERAL CONTRACTOR'S MARK-UP** on Change Orders									
0200	Extra work, by subcontractors, add				%				10%	10%
0250	By General Contractor, add								15%	15%
0400	Omitted work, by subcontractors, deduct all but								5%	5%
0450	By General Contractor, deduct all but								7.50%	7.50%
0600	Overtime work, by subcontractors, add								15%	15%
0650	By General Contractor, add								10%	10%

01 31 13.60 Installing Contractor's Main Office Overhead

	01 31 13.60 Installing Contractor's Main Office Overhead	Crew	Daily Output	Labor-Hours	Unit	Material	Labor	Equipment	Total	Total Incl O&P
0010	**INSTALLING CONTRACTOR'S MAIN OFFICE OVERHEAD**									
0020	As percent of direct costs, minimum				%				5%	
0050	Average								13%	
0100	Maximum								30%	

01 31 13.80 Overhead and Profit

	01 31 13.80 Overhead and Profit	Crew	Daily Output	Labor-Hours	Unit	Material	Labor	Equipment	Total	Total Incl O&P
0010	**OVERHEAD & PROFIT** Allowance to add to items in this									
0020	book that do not include Subs O&P, average				%				25%	
0100	Allowance to add to items in this book that									
0110	do include Subs O&P, minimum				%				5%	5%
0150	Average								10%	10%
0200	Maximum								15%	15%
0300	Typical, by size of project, under $100,000								30%	
0350	$500,000 project								25%	
0400	$2,000,000 project								20%	
0450	Over $10,000,000 project								15%	

01 31 Project Management and Coordination

01 31 13 – Project Coordination

01 31 13.90 Performance Bond		Crew	Daily Output	Labor-Hours	Unit	Material	Labor	Equipment	Total	Total Incl O&P
0010	PERFORMANCE BOND R013113-80									
0020	For buildings, minimum				Job				.60%	.60%
0100	Maximum				"				2.50%	2.50%

(Material, Labor, Equipment and Total are 2015 Bare Costs)

01 32 Construction Progress Documentation

01 32 33 – Photographic Documentation

01 32 33.50 Photographs

		Crew	Daily Output	Labor-Hours	Unit	Material	Labor	Equipment	Total	Total Incl O&P
0010	PHOTOGRAPHS									
0020	8" x 10", 4 shots, 2 prints ea., std. mounting				Set	475			475	525
0100	Hinged linen mounts					530			530	580
0200	8" x 10", 4 shots, 2 prints each, in color					425			425	470
0300	For I.D. slugs, add to all above				↓	5.30			5.30	5.85
1500	Time lapse equipment, camera and projector, buy				Ea.	2,650			2,650	2,925
1550	Rent per month				"	1,600			1,600	1,750
1700	Cameraman and film, including processing, B.&W.				Day	1,225			1,225	1,350
1720	Color				"	1,425			1,425	1,550

01 41 Regulatory Requirements

01 41 26 – Permit Requirements

01 41 26.50 Permits

		Crew	Daily Output	Labor-Hours	Unit	Material	Labor	Equipment	Total	Total Incl O&P
0010	PERMITS									
0020	Rule of thumb, most cities, minimum				Job				.50%	.50%
0100	Maximum				"				2%	2%

01 51 Temporary Utilities

01 51 13 – Temporary Electricity

01 51 13.50 Temporary Power Equip (Pro-Rated Per Job)

		Crew	Daily Output	Labor-Hours	Unit	Material	Labor	Equipment	Total	Total Incl O&P
0010	TEMPORARY POWER EQUIP (PRO-RATED PER JOB) R015113-65									
0020	Service, overhead feed, 3 use									
0030	100 Amp	1 Elec	1.25	6.400	Ea.	735	350		1,085	1,325
0040	200 Amp		1	8		960	440		1,400	1,700
0050	400 Amp		.75	10.667		1,775	585		2,360	2,825
0060	600 Amp	↓	.50	16	↓	2,600	875		3,475	4,150
0100	Underground feed, 3 use									
0110	100 Amp	1 Elec	2	4	Ea.	700	219		919	1,100
0120	200 Amp		1.15	6.957		955	380		1,335	1,625
0130	400 Amp		1	8		1,775	440		2,215	2,600
0140	600 Amp		.75	10.667		2,225	585		2,810	3,300
0150	800 Amp		.50	16		3,375	875		4,250	5,025
0160	1000 Amp		.35	22.857		3,750	1,250		5,000	6,000
0170	1200 Amp		.25	32		4,225	1,750		5,975	7,275
0180	2000 Amp	↓	.20	40	↓	5,125	2,200		7,325	8,900
0200	Transformers, 3 use									
0210	30 kVA	1 Elec	1	8	Ea.	1,650	440		2,090	2,475
0220	45 kVA		.75	10.667		1,975	585		2,560	3,050
0230	75 kVA		.50	16		3,325	875		4,200	4,950
0240	112.5 kVA	↓	.40	20	↓	3,600	1,100		4,700	5,625

01 51 Temporary Utilities

01 51 13 – Temporary Electricity

	01 51 13.50 Temporary Power Equip (Pro-Rated Per Job)	Crew	Daily Output	Labor-Hours	Unit	2015 Bare Costs Material	Labor	Equipment	Total	Total Incl O&P
0250	Feeder, PVC, CU wire in trench									
0260	60 Amp	1 Elec	96	.083	L.F.	3.76	4.56		8.32	11
0270	100 Amp		85	.094		6.85	5.15		12	15.20
0280	200 Amp		59	.136		16.35	7.40		23.75	29
0290	400 Amp		42	.190		40.50	10.40		50.90	60
0300	Feeder, PVC, aluminum wire in trench									
0310	60 Amp	1 Elec	96	.083	L.F.	4.70	4.56		9.26	12
0320	100 Amp		85	.094		5.30	5.15		10.45	13.55
0330	200 Amp		59	.136		11.65	7.40		19.05	24
0340	400 Amp		42	.190		24	10.40		34.40	42
0350	Feeder, EMT, CU wire									
0360	60 Amp	1 Elec	90	.089	L.F.	3.30	4.86		8.16	10.95
0370	100 Amp		80	.100		6.90	5.45		12.35	15.80
0380	200 Amp		60	.133		15.35	7.30		22.65	28
0390	400 Amp		35	.229		44.50	12.50		57	68
0400	Feeder, EMT, alum. wire									
0410	60 Amp	1 Elec	90	.089	L.F.	5.30	4.86		10.16	13.10
0420	100 Amp		80	.100		6.60	5.45		12.05	15.45
0430	200 Amp		60	.133		16.90	7.30		24.20	29.50
0440	400 Amp		35	.229		30.50	12.50		43	52.50
0500	Equipment, 3 use									
0510	Spider box 50 Amp	1 Elec	8	1	Ea.	950	54.50		1,004.50	1,125
0520	Lighting cord 100′		8	1		118	54.50		172.50	212
0530	Light stanchion		8	1		72	54.50		126.50	161
0540	Temporary cords, 100′, 3 use									
0550	Feeder cord, 50 Amp	1 Elec	16	.500	Ea.	445	27.50		472.50	530
0560	Feeder cord, 100 Amp		12	.667		1,175	36.50		1,211.50	1,350
0570	Tap cord, 50 Amp		12	.667		232	36.50		268.50	310
0580	Tap cord, 100 Amp		6	1.333		1,175	73		1,248	1,400
0590	Temporary cords, 50′, 3 use									
0600	Feeder cord, 50 Amp	1 Elec	16	.500	Ea.	210	27.50		237.50	272
0610	Feeder cord, 100 Amp		12	.667		680	36.50		716.50	805
0620	Tap cord, 50 Amp		12	.667		122	36.50		158.50	190
0630	Tap cord, 100 Amp		6	1.333		680	73		753	860
0700	Connections									
0710	Compressor or pump									
0720	30 Amp	1 Elec	7	1.143	Ea.	21	62.50		83.50	117
0730	60 Amp		5.30	1.509		37.50	82.50		120	166
0740	100 Amp		4	2		85.50	109		194.50	258
0750	Tower crane									
0760	60 Amp	1 Elec	4.50	1.778	Ea.	37.50	97		134.50	188
0770	100 Amp	"	3	2.667	"	85.50	146		231.50	315
0780	Manlift									
0790	Single	1 Elec	3	2.667	Ea.	37	146		183	260
0800	Double	"	2	4	"	67.50	219		286.50	405
0810	Welder with disconnect									
0820	50 Amp	1 Elec	5	1.600	Ea.	234	87.50		321.50	390
0830	100 Amp		3.80	2.105		400	115		515	615
0840	200 Amp		2.50	3.200		725	175		900	1,050
0850	400 Amp		1	8		1,725	440		2,165	2,550
0860	Office trailer									
0870	60 Amp	1 Elec	4.50	1.778	Ea.	115	97		212	273
0880	100 Amp		3	2.667		224	146		370	465

01 51 Temporary Utilities

01 51 13 – Temporary Electricity

01 51 13.50 Temporary Power Equip (Pro-Rated Per Job)		Crew	Daily Output	Labor-Hours	Unit	2015 Bare Costs Material	Labor	Equipment	Total	Total Incl O&P
0890	200 Amp	1 Elec	2	4	Ea.	755	219		974	1,150
0900	Lamping, add per floor				Total				625	690
0910	Maintenance, total temp. power cost				Job				5%	5%

01 51 13.80 Temporary Utilities

Line	Description	Crew	Daily Output	Labor-Hours	Unit	Material	Labor	Equipment	Total	Total Incl O&P
0010	**TEMPORARY UTILITIES**									
0350	Lighting, lamps, wiring, outlets, 40,000 SF building, 8 strings	1 Elec	34	.235	CSF Flr	4.91	12.85		17.76	24.50
0360	16 strings	″	17	.471		9.80	25.50		35.30	49.50
0400	Power for temp lighting only, 6.6 KWH, per month								.92	1.01
0430	11.8 KWH, per month								1.65	1.82
0450	23.6 KWH, per month								3.30	3.63
0600	Power for job duration incl. elevator, etc., minimum								47	51.50
0650	Maximum								110	121
1000	Toilet, portable, see Equip. Rental 01 54 33 in Reference Section									

01 52 Construction Facilities

01 52 13 – Field Offices and Sheds

01 52 13.20 Office and Storage Space

Line	Description	Crew	Daily Output	Labor-Hours	Unit	Material	Labor	Equipment	Total	Total Incl O&P
0010	**OFFICE AND STORAGE SPACE**									
0020	Office trailer, furnished, no hookups, 20′ x 8′, buy	2 Skwk	1	16	Ea.	10,400	780		11,180	12,600
0250	Rent per month					188			188	206
0300	32′ x 8′, buy	2 Skwk	.70	22.857		15,500	1,100		16,600	18,800
0350	Rent per month					239			239	262
0400	50′ x 10′, buy	2 Skwk	.60	26.667		24,100	1,300		25,400	28,500
0450	Rent per month					340			340	375
0500	50′ x 12′, buy	2 Skwk	.50	32		29,400	1,550		30,950	34,700
0550	Rent per month					410			410	455
0700	For air conditioning, rent per month, add					48.50			48.50	53.50
0800	For delivery, add per mile				Mile	11			11	12.10
0890	Delivery each way				Ea.	200			200	220
0900	Bunk house trailer, 8′ x 40′ duplex dorm with kitchen, no hookups, buy	2 Carp	1	16		37,300	750		38,050	42,200
0910	9 man with kitchen and bath, no hookups, buy		1	16		38,000	750		38,750	43,000
0920	18 man sleeper with bath, no hookups, buy		1	16		49,000	750		49,750	55,000
1000	Portable buildings, prefab, on skids, economy, 8′ x 8′		265	.060	S.F.	25	2.83		27.83	32
1100	Deluxe, 8′ x 12′		150	.107	″	20	5		25	29.50
1200	Storage boxes, 20′ x 8′, buy	2 Skwk	1.80	8.889	Ea.	2,775	430		3,205	3,725
1250	Rent per month					82.50			82.50	91
1300	40′ x 8′, buy	2 Skwk	1.40	11.429		3,850	555		4,405	5,075
1350	Rent per month					101			101	111

01 52 13.40 Field Office Expense

Line	Description	Crew	Daily Output	Labor-Hours	Unit	Material	Labor	Equipment	Total	Total Incl O&P
0010	**FIELD OFFICE EXPENSE**									
0100	Office equipment rental average				Month	200			200	220
0120	Office supplies, average				″	80			80	88
0125	Office trailer rental, see Section 01 52 13.20									
0140	Telephone bill; avg. bill/month incl. long dist.				Month	85			85	93.50
0160	Lights & HVAC				″	160			160	176

01 54 Construction Aids

01 54 16 – Temporary Hoists

01 54 16.50 Weekly Forklift Crew		Crew	Daily Output	Labor-Hours	Unit	2015 Bare Costs Material	Labor	Equipment	Total	Total Incl O&P
0010	**WEEKLY FORKLIFT CREW**									
0100	All-terrain forklift, 45′ lift, 35′ reach, 9000 lb. capacity	A-3P	.20	40	Week		1,950	2,625	4,575	5,850

01 54 19 – Temporary Cranes

01 54 19.50 Daily Crane Crews		Crew	Daily Output	Labor-Hours	Unit	Material	Labor	Equipment	Total	Total Incl O&P
0010	**DAILY CRANE CREWS** for small jobs, portal to portal									
0100	12-ton truck-mounted hydraulic crane	A-3H	1	8	Day		415	860	1,275	1,575
0200	25-ton	A-3I	1	8			415	990	1,405	1,725
0300	40-ton	A-3J	1	8			415	1,225	1,640	1,975
0400	55-ton	A-3K	1	16			775	1,625	2,400	2,975
0500	80-ton	A-3L	1	16			775	2,350	3,125	3,750
0600	100-ton	A-3M	1	16			775	2,325	3,100	3,750
0900	If crane is needed on a Saturday, Sunday or Holiday									
0910	At time-and-a-half, add				Day		50%			
0920	At double time, add				"		100%			

01 54 19.60 Monthly Tower Crane Crew		Crew	Daily Output	Labor-Hours	Unit	Material	Labor	Equipment	Total	Total Incl O&P
0010	**MONTHLY TOWER CRANE CREW**, excludes concrete footing									
0100	Static tower crane, 130′ high, 106′ jib, 6200 lb. capacity	A-3N	.05	176	Month		9,100	24,800	33,900	41,100

01 54 23 – Temporary Scaffolding and Platforms

01 54 23.70 Scaffolding		Crew	Daily Output	Labor-Hours	Unit	Material	Labor	Equipment	Total	Total Incl O&P
0010	**SCAFFOLDING** R015423-10									
0015	Steel tube, regular, no plank, labor only to erect & dismantle									
0090	Building exterior, wall face, 1 to 5 stories, 6′-4″ x 5′ frames	3 Carp	8	3	C.S.F.		141		141	217
0200	6 to 12 stories	4 Carp	8	4			188		188	289
0301	13 to 20 stories	5 Clab	8	5			188		188	289
0460	Building interior, wall face area, up to 16′ high	3 Carp	12	2			94		94	145
0560	16′ to 40′ high		10	2.400			113		113	173
0800	Building interior floor area, up to 30′ high		150	.160	C.C.F.		7.50		7.50	11.55
0900	Over 30′ high	4 Carp	160	.200	"		9.40		9.40	14.45
0906	Complete system for face of walls, no plank, material only rent/mo				C.S.F.	35.50			35.50	39.50
0908	Interior spaces, no plank, material only rent/mo				C.C.F.	4.33			4.33	4.76
0910	Steel tubular, heavy duty shoring, buy									
0920	Frames 5′ high 2′ wide				Ea.	81.50			81.50	90
0925	5′ high 4′ wide					93			93	102
0930	6′ high 2′ wide					93.50			93.50	103
0935	6′ high 4′ wide					109			109	120
0940	Accessories									
0945	Cross braces				Ea.	15.50			15.50	17.05
0950	U-head, 8″ x 8″					19.10			19.10	21
0955	J-head, 4″ x 8″					13.90			13.90	15.30
0960	Base plate, 8″ x 8″					15.50			15.50	17.05
0965	Leveling jack					33.50			33.50	36.50
1000	Steel tubular, regular, buy									
1100	Frames 3′ high 5′ wide				Ea.	75			75	82.50
1150	5′ high 5′ wide					86.50			86.50	95
1200	6′-4″ high 5′ wide					118			118	129
1350	7′-6″ high 6′ wide					158			158	173
1500	Accessories cross braces					15.50			15.50	17.05
1550	Guardrail post					16.40			16.40	18.05
1600	Guardrail 7′ section					8.70			8.70	9.60
1650	Screw jacks & plates					21.50			21.50	24
1700	Sidearm brackets					30.50			30.50	33.50

01 54 23.70 Scaffolding		Crew	Daily Output	Labor-Hours	Unit	2015 Bare Costs Material	Labor	Equipment	Total	Total Incl O&P
1750	8″ casters				Ea.	31			31	34
1800	Plank 2″ x 10″ x 16′-0″					57.50			57.50	63.50
1900	Stairway section					286			286	315
1910	Stairway starter bar					32			32	35
1920	Stairway inside handrail					58.50			58.50	64
1930	Stairway outside handrail					86.50			86.50	95
1940	Walk-thru frame guardrail					41.50			41.50	46
2000	Steel tubular, regular, rent/mo.									
2100	Frames 3′ high 5′ wide				Ea.	5			5	5.50
2150	5′ high 5′ wide					5			5	5.50
2200	6′-4″ high 5′ wide					5.50			5.50	6.05
2250	7′-6″ high 6′ wide					10			10	11
2500	Accessories, cross braces					1			1	1.10
2550	Guardrail post					1			1	1.10
2600	Guardrail 7′ section					1			1	1.10
2650	Screw jacks & plates					2			2	2.20
2700	Sidearm brackets					2			2	2.20
2750	8″ casters					8			8	8.80
2800	Outrigger for rolling tower					3			3	3.30
2850	Plank 2″ x 10″ x 16′-0″					10			10	11
2900	Stairway section					35			35	38.50
2940	Walk-thru frame guardrail					2.50			2.50	2.75
3000	Steel tubular, heavy duty shoring, rent/mo.									
3250	5′ high 2′ & 4′ wide				Ea.	8.50			8.50	9.35
3300	6′ high 2′ & 4′ wide					8.50			8.50	9.35
3500	Accessories, cross braces					1			1	1.10
3600	U - head, 8″ x 8″					2.50			2.50	2.75
3650	J - head, 4″ x 8″					2.50			2.50	2.75
3700	Base plate, 8″ x 8″					1			1	1.10
3750	Leveling jack					2.50			2.50	2.75
5700	Planks, 2″ x 10″ x 16′-0″, labor only to erect & remove to 50′ H	3 Carp	72	.333			15.65		15.65	24
5800	Over 50′ high	4 Carp	80	.400			18.80		18.80	29

01 54 23.75 Scaffolding Specialties		Crew	Daily Output	Labor-Hours	Unit	Material	Labor	Equipment	Total	Total Incl O&P
0010	**SCAFFOLDING SPECIALTIES**									
1500	Sidewalk bridge using tubular steel scaffold frames including									
1510	planking (material cost is rent/month)	3 Carp	45	.533	L.F.	8.55	25		33.55	48
1600	For 2 uses per month, deduct from all above					50%				
1700	For 1 use every 2 months, add to all above					100%				
1900	Catwalks, 20″ wide, no guardrails, 7′ span, buy				Ea.	166			166	183
2000	10′ span, buy					233			233	256
3720	Putlog, standard, 8′ span, with hangers, buy					162			162	178
3750	12′ span, buy					203			203	223
3760	Trussed type, 16′ span, buy					365			365	400
3790	22′ span, buy					420			420	465
3795	Rent per month					40			40	44
3800	Rolling ladders with handrails, 30″ wide, buy, 2 step					267			267	294
4000	7 step					760			760	835
4050	10 step					1,075			1,075	1,175
4100	Rolling towers, buy, 5′ wide, 7′ long, 10′ high					1,200			1,200	1,325
4200	For 5′ high added sections, to buy, add					204			204	224
4300	Complete incl. wheels, railings, outriggers,									
4350	21′ high, to buy				Ea.	2,025			2,025	2,225

01 54 Construction Aids

01 54 23 – Temporary Scaffolding and Platforms

01 54 23.75 Scaffolding Specialties	Crew	Daily Output	Labor-Hours	Unit	Material	2015 Bare Costs Labor	Equipment	Total	Total Incl O&P
4400 Rent/month = 5% of purchase cost				Ea.	200			200	220

01 54 26 – Temporary Swing Staging

01 54 26.50 Swing Staging	Crew	Daily Output	Labor-Hours	Unit	Material	Labor	Equipment	Total	Total Incl O&P
0010 **SWING STAGING**, 500 lb. cap., 2' wide to 24' long, hand operated									
0020 steel cable type, with 60' cables, buy				Ea.	4,900			4,900	5,375
0030 Rent per month				"	490			490	540
2200 Move swing staging (setup and remove)	E-4	2	16	Move		850	73	923	1,550

01 54 36 – Equipment Mobilization

01 54 36.50 Mobilization	Crew	Daily Output	Labor-Hours	Unit	Material	Labor	Equipment	Total	Total Incl O&P
0010 **MOBILIZATION** (Use line item again for demobilization) R015436-50									
0015 Up to 25 mi. haul dist. (50 mi. RT for mob/demob crew)									
1200 Small equipment, placed in rear of, or towed by pickup truck	A-3A	4	2	Ea.		97	39	136	191
1300 Equipment hauled on 3-ton capacity towed trailer	A-3Q	2.67	3			146	67	213	295
1400 20-ton capacity	B-34U	2	8			355	237	592	795
1500 40-ton capacity	B-34N	2	8			365	375	740	965
1600 50-ton capacity	B-34V	1	24			1,125	1,050	2,175	2,850
1700 Crane, truck-mounted, up to 75 ton (driver only)	1 Eqhv	4	2			103		103	157
1800 Over 75 ton (with chase vehicle)	A-3E	2.50	6.400			294	62.50	356.50	515
2400 Crane, large lattice boom, requiring assembly	B-34W	.50	144			6,275	7,500	13,775	17,800
2500 For each additional 5 miles haul distance, add						10%	10%		
3000 For large pieces of equipment, allow for assembly/knockdown									
3100 For mob/demob of micro-tunneling equip, see Section 33 05 23.19									

01 55 Vehicular Access and Parking

01 55 23 – Temporary Roads

01 55 23.50 Roads and Sidewalks	Crew	Daily Output	Labor-Hours	Unit	Material	Labor	Equipment	Total	Total Incl O&P
0010 **ROADS AND SIDEWALKS** Temporary									
0050 Roads, gravel fill, no surfacing, 4" gravel depth	B-14	715	.067	S.Y.	4.04	2.67	.51	7.22	9.10
0100 8" gravel depth	"	615	.078	"	8.10	3.10	.59	11.79	14.30
1000 Ramp, 3/4" plywood on 2" x 6" joists, 16" O.C.	2 Carp	300	.053	S.F.	1.58	2.50		4.08	5.60
1100 On 2" x 10" joists, 16" O.C.	"	275	.058	"	2.24	2.73		4.97	6.65

01 56 Temporary Barriers and Enclosures

01 56 13 – Temporary Air Barriers

01 56 13.90 Winter Protection	Crew	Daily Output	Labor-Hours	Unit	Material	Labor	Equipment	Total	Total Incl O&P
0010 **WINTER PROTECTION**									
0100 Framing to close openings	2 Clab	500	.032	S.F.	.46	1.20		1.66	2.35
0200 Tarpaulins hung over scaffolding, 8 uses, not incl. scaffolding		1500	.011		.24	.40		.64	.88
0300 Prefab fiberglass panels, steel frame, 8 uses		1200	.013		2.40	.50		2.90	3.41

01 56 16 – Temporary Dust Barriers

01 56 16.10 Dust Barriers, Temporary	Crew	Daily Output	Labor-Hours	Unit	Material	Labor	Equipment	Total	Total Incl O&P
0010 **DUST BARRIERS, TEMPORARY**									
0020 Spring loaded telescoping pole & head, to 12', erect and dismantle	1 Clab	240	.033	Ea.		1.25		1.25	1.93
0025 Cost per day (based upon 250 days)				Day	.26			.26	.29
0030 To 21', erect and dismantle	1 Clab	240	.033	Ea.		1.25		1.25	1.93
0035 Cost per day (based upon 250 days)				Day	.44			.44	.48
0040 Accessories, caution tape reel, erect and dismantle	1 Clab	480	.017	Ea.		.63		.63	.96
0045 Cost per day (based upon 250 days)				Day	.36			.36	.40

01 56 Temporary Barriers and Enclosures

01 56 16 – Temporary Dust Barriers

01 56 16.10 Dust Barriers, Temporary		Crew	Daily Output	Labor-Hours	Unit	Material	2015 Bare Costs Labor	Equipment	Total	Total Incl O&P
0060	Foam rail and connector, erect and dismantle	1 Clab	240	.033	Ea.		1.25		1.25	1.93
0065	Cost per day (based upon 250 days)				Day	.10			.10	.11
0070	Caution tape	1 Clab	384	.021	C.L.F.	2.70	.78		3.48	4.18
0080	Zipper, standard duty		60	.133	Ea.	8	5		13	16.50
0090	Heavy duty		48	.167	"	9.50	6.25		15.75	20
0100	Polyethylene sheet, 4 mil		37	.216	Sq.	2.90	8.15		11.05	15.70
0110	6 mil	↓	37	.216	"	4.02	8.15		12.17	16.90
1000	Dust partition, 6 mil polyethylene, 1" x 3" frame	2 Carp	2000	.008	S.F.	.30	.38		.68	.90
1080	2" x 4" frame	"	2000	.008	"	.35	.38		.73	.97

01 56 23 – Temporary Barricades

01 56 23.10 Barricades		Crew	Daily Output	Labor-Hours	Unit	Material	Labor	Equipment	Total	Total Incl O&P
0010	**BARRICADES**									
0020	5' high, 3 rail @ 2" x 8", fixed	2 Carp	20	.800	L.F.	6	37.50		43.50	64.50
0150	Movable		30	.533		5	25		30	44
1000	Guardrail, wooden, 3' high, 1" x 6", on 2" x 4" posts		200	.080		1.27	3.76		5.03	7.20
1100	2" x 6", on 4" x 4" posts	↓	165	.097		2.46	4.55		7.01	9.70
1200	Portable metal with base pads, buy					15.55			15.55	17.10
1250	Typical installation, assume 10 reuses	2 Carp	600	.027	↓	2.55	1.25		3.80	4.74
1300	Barricade tape, polyethylene, 7 mil, 3" wide x 500' long roll				Ea.	25			25	27.50
3000	Detour signs, set up and remove									
3010	Reflective aluminum, MUTCD, 24" x 24", post mounted	1 Clab	20	.400	Ea.	2.46	15.05		17.51	25.50
4000	Roof edge portable barrier stands and warning flags, 50 uses	1 Rohe	9100	.001	L.F.	.07	.03		.10	.12
4010	100 uses	"	9100	.001	"	.03	.03		.06	.09

01 56 26 – Temporary Fencing

01 56 26.50 Temporary Fencing		Crew	Daily Output	Labor-Hours	Unit	Material	Labor	Equipment	Total	Total Incl O&P
0010	**TEMPORARY FENCING**									
0020	Chain link, 11 ga., 4' high	2 Clab	400	.040	L.F.	2.95	1.50		4.45	5.55
0100	6' high		300	.053		2.95	2.01		4.96	6.35
0200	Rented chain link, 6' high, to 1000' (up to 12 mo.)		400	.040		4.29	1.50		5.79	7.05
0250	Over 1000' (up to 12 mo.)	↓	300	.053		4.29	2.01		6.30	7.80
0350	Plywood, painted, 2" x 4" frame, 4' high	A-4	135	.178		6.20	7.95		14.15	19
0400	4" x 4" frame, 8' high	"	110	.218		12	9.75		21.75	28
0500	Wire mesh on 4" x 4" posts, 4' high	2 Carp	100	.160		9.85	7.50		17.35	22.50
0550	8' high	"	80	.200	↓	14.90	9.40		24.30	31

01 56 29 – Temporary Protective Walkways

01 56 29.50 Protection		Crew	Daily Output	Labor-Hours	Unit	Material	Labor	Equipment	Total	Total Incl O&P
0010	**PROTECTION**									
0020	Stair tread, 2" x 12" planks, 1 use	1 Carp	75	.107	Tread	5.10	5		10.10	13.30
0100	Exterior plywood, 1/2" thick, 1 use		65	.123		1.97	5.80		7.77	11.05
0200	3/4" thick, 1 use		60	.133	↓	2.78	6.25		9.03	12.70
2200	Sidewalks, 2" x 12" planks, 2 uses		350	.023	S.F.	.85	1.07		1.92	2.59
2300	Exterior plywood, 2 uses, 1/2" thick		750	.011		.33	.50		.83	1.13
2400	5/8" thick		650	.012		.39	.58		.97	1.32
2500	3/4" thick	↓	600	.013	↓	.46	.63		1.09	1.47

01 58 Project Identification

01 58 13 – Temporary Project Signage

01 58 13.50 Signs		Crew	Daily Output	Labor-Hours	Unit	Material	2015 Bare Costs Labor	Equipment	Total	Total Incl O&P
0010	**SIGNS**									
0020	High intensity reflectorized, no posts, buy				Ea.	25			25	27.50

01 74 Cleaning and Waste Management

01 74 13 – Progress Cleaning

01 74 13.20 Cleaning Up

		Crew	Daily Output	Labor-Hours	Unit	Material	Labor	Equipment	Total	Total Incl O&P
0010	**CLEANING UP**									
0020	After job completion, allow, minimum				Job				.30%	.30%
0040	Maximum				"				1%	1%
0050	Cleanup of floor area, continuous, per day, during const.	A-5	24	.750	M.S.F.	2.18	28.50	2.56	33.24	48.50
0100	Final by GC at end of job	"	11.50	1.565	"	2.31	59	5.35	66.66	99.50

01 91 Commissioning

01 91 13 – General Commissioning Requirements

01 91 13.50 Building Commissioning

		Crew	Daily Output	Labor-Hours	Unit	Material	Labor	Equipment	Total	Total Incl O&P
0010	**BUILDING COMMISSIONING**									
0100	Basic building commissioning, minimum				%				.25%	.25%
0150	Maximum								.50%	.50%
0200	Enhanced building commissioning, minimum								.50%	.50%
0250	Maximum				↓				1%	1%

01 93 Facility Maintenance

01 93 13 – Facility Maintenance Procedures

01 93 13.16 Electrical Facilities Maintenance

		Crew	Daily Output	Labor-Hours	Unit	Material	Labor	Equipment	Total	Total Incl O&P
0010	**ELECTRICAL FACILITIES MAINTENANCE**									
0700	Cathodic protection systems									
0720	Check and adjust reading on rectifier	1 Elec	20	.400	Ea.		22		22	33
0730	Check pipe to soil potential		20	.400			22		22	33
0740	Replace lead connection		4	2			109		109	164
0800	Control device, install		5.70	1.404			77		77	115
0810	Disassemble, clean and reinstall		7	1.143			62.50		62.50	93.50
0820	Replace		10.70	.748			41		41	61.50
0830	Trouble shoot	↓	10	.800	↓		44		44	65.50
0900	Demolition, for electrical demolition see Section 26 05 05.10									
1000	Distribution systems and equipment install or repair a breaker									
1010	In power panels up to 200 amps	1 Elec	7	1.143	Ea.		62.50		62.50	93.50
1020	Over 200 amps		2	4			219		219	330
1030	Reset breaker or replace fuse		20	.400			22		22	33
1100	Megger test MCC (each stack)		4	2			109		109	164
1110	MCC vacuum and clean (each stack)		5.30	1.509			82.50		82.50	124
2500	Remove and replace or maint., road fixture & lamp		3	2.667		925	146		1,071	1,250
2510	Fluorescent fixture	↓	7	1.143		67	62.50		129.50	167
2515	Relamp (fluor.) facility area each tube, spot	1 Clab	24	.333		3.38	12.55		15.93	23
2516	Group		100	.080		3.38	3.01		6.39	8.35
2518	Fluorescent fixture, clean (area)	↓	44	.182		.11	6.85		6.96	10.60
2520	Incandescent fixture	1 Elec	11	.727		61.50	40		101.50	128
2530	Lamp (incandescent or fluorescent)	1 Clab	60	.133	↓	3.38	5		8.38	11.40

01 93 13 – Facility Maintenance Procedures

01 93 13.16 Electrical Facilities Maintenance		Crew	Daily Output	Labor-Hours	Unit	2015 Bare Costs Material	Labor	Equipment	Total	Total Incl O&P
2535	Replace cord in socket lamp	1 Elec	13	.615	Ea.	1.66	33.50		35.16	52.50
2540	Ballast electronic type for two tubes		8	1		37	54.50		91.50	123
2541	Starter		30	.267		.37	14.60		14.97	22.50
2545	Replace other lighting parts		11	.727		14.20	40		54.20	75
2550	Switch		11	.727		9.85	40		49.85	70.50
2555	Receptacle		11	.727		7.90	40		47.90	68
2560	Floodlight		4	2		273	109		382	465
2570	Christmas lighting, indoor, per string	1 Clab	16	.500			18.80		18.80	29
2580	Outdoor		13	.615			23		23	35.50
2590	Test battery operated emergency lights		40	.200			7.50		7.50	11.55
2600	Repair and replace component in communication system	1 Elec	6	1.333		53	73		126	167
2700	Repair misc. appliances (incl. clocks, vent fan, blower, etc.)	"	6	1.333			73		73	109
2710	Reset clocks & timers	1 Clab	50	.160			6		6	9.25
2720	Adjust time delay relays	1 Elec	16	.500			27.50		27.50	41
2730	Test specific gravity of lead-acid batteries	1 Clab	80	.100			3.76		3.76	5.80
3000	Motors and generators									
3020	Disassemble, clean and reinstall motor, up to 1/4 HP	1 Elec	4	2	Ea.		109		109	164
3030	Up to 3/4 HP		3	2.667			146		146	219
3040	Up to 10 HP		2	4			219		219	330
3050	Replace part, up to 1/4 HP		6	1.333			73		73	109
3060	Up to 3/4 HP		4	2			109		109	164
3070	Up to 10 HP		3	2.667			146		146	219
3080	Megger test motor windings		5.33	1.501			82		82	123
3082	Motor vibration check		16	.500			27.50		27.50	41
3084	Oil motor bearings		25	.320			17.50		17.50	26
3086	Run test emergency generator for 30 minutes		11	.727			40		40	59.50
3090	Rewind motor, up to 1/4 HP		3	2.667			146		146	219
3100	Up to 3/4 HP		2	4			219		219	330
3110	Up to 10 HP		1.50	5.333			292		292	435
3150	Generator, repair or replace part		4	2			109		109	164
3160	Repair DC generator		2	4			219		219	330
4000	Stub pole, install or remove		3	2.667			146		146	219
4500	Transformer maintenance up to 15 kVA		2.70	2.963			162		162	243

Division Notes

		CREW	DAILY OUTPUT	LABOR-HOURS	UNIT	BARE COSTS MAT.	BARE COSTS LABOR	BARE COSTS EQUIP.	BARE COSTS TOTAL	TOTAL INCL O&P

Estimating Tips

02 30 00 Subsurface Investigation

In preparing estimates on structures involving earthwork or foundations, all information concerning soil characteristics should be obtained. Look particularly for hazardous waste, evidence of prior dumping of debris, and previous stream beds.

02 40 00 Demolition and Structure Moving

The costs shown for selective demolition do not include rubbish handling or disposal. These items should be estimated separately using RSMeans data or other sources.

- Historic preservation often requires that the contractor remove materials from the existing structure, rehab them, and replace them. The estimator must be aware of any related measures and precautions that must be taken when doing selective demolition and cutting and patching. Requirements may include special handling and storage, as well as security.
- In addition to Subdivision 02 41 00, you can find selective demolition items in each division. Example: Roofing demolition is in Division 7.

02 40 00 Building Deconstruction

This section provides costs for the careful dismantling and recycling of most of low-rise building materials.

02 50 00 Containment of Hazardous Waste

This section addresses on-site hazardous waste disposal costs.

02 80 00 Hazardous Material Disposal/ Remediation

This subdivision includes information on hazardous waste handling, asbestos remediation, lead remediation, and mold remediation. See reference R028213-20 and R028319-60 for further guidance in using these unit price lines.

02 90 00 Monitoring Chemical Sampling, Testing Analysis

This section provides costs for on-site sampling and testing hazardous waste.

Reference Numbers

Reference numbers are shown in shaded boxes at the beginning of some major classifications. These numbers refer to related items in the Reference Section. The reference information may be an estimating procedure, an alternate pricing method, or technical information.

Note: Not all subdivisions listed here necessarily appear in this publication. ■

02 41 Demolition

02 41 13 – Selective Site Demolition

02 41 13.17 Demolish, Remove Pavement and Curb		Crew	Daily Output	Labor-Hours	Unit	Material	2015 Bare Costs Labor	Equipment	Total	Total Incl O&P
0010	DEMOLISH, REMOVE PAVEMENT AND CURB R024119-10									
5010	Pavement removal, bituminous roads, up to 3" thick	B-38	690	.058	S.Y.		2.48	1.85	4.33	5.80
5050	4" to 6" thick		420	.095			4.08	3.04	7.12	9.60
5100	Bituminous driveways		640	.063			2.68	1.99	4.67	6.30
5200	Concrete to 6" thick, hydraulic hammer, mesh reinforced		255	.157			6.70	5	11.70	15.75
5300	Rod reinforced		200	.200			8.55	6.40	14.95	20
5400	Concrete, 7" to 24" thick, plain		33	1.212	C.Y.		52	38.50	90.50	122
5500	Reinforced		24	1.667	"		71.50	53	124.50	168
5600	With hand held air equipment, bituminous, to 6" thick	B-39	1900	.025	S.F.		1	.12	1.12	1.67
5700	Concrete to 6" thick, no reinforcing		1600	.030			1.19	.15	1.34	1.99
5800	Mesh reinforced		1400	.034			1.36	.17	1.53	2.27
5900	Rod reinforced		765	.063			2.50	.30	2.80	4.16

02 41 13.23 Utility Line Removal

Line	Description	Crew	Daily Output	Labor-Hours	Unit	Material	Labor	Equipment	Total	Total Incl O&P
0010	UTILITY LINE REMOVAL									
0015	No hauling, abandon catch basin or manhole	B-6	7	3.429	Ea.		142	52	194	274
0020	Remove existing catch basin or manhole, masonry		4	6			248	91	339	480
0030	Catch basin or manhole frames and covers, stored		13	1.846			76	28	104	148
0040	Remove and reset		7	3.429			142	52	194	274

02 41 13.30 Minor Site Demolition

Line	Description	Crew	Daily Output	Labor-Hours	Unit	Material	Labor	Equipment	Total	Total Incl O&P
0010	MINOR SITE DEMOLITION R024119-10									
4000	Sidewalk removal, bituminous, 2" thick	B-6	350	.069	S.Y.		2.83	1.04	3.87	5.45
4010	2-1/2" thick		325	.074			3.05	1.12	4.17	5.90
4050	Brick, set in mortar		185	.130			5.35	1.97	7.32	10.35
4100	Concrete, plain, 4"		160	.150			6.20	2.28	8.48	11.95
4110	Plain, 5"		140	.171			7.05	2.60	9.65	13.70
4120	Plain, 6"		120	.200			8.25	3.04	11.29	16
4200	Mesh reinforced, concrete, 4 "		150	.160			6.60	2.43	9.03	12.75
4210	5 " thick		131	.183			7.55	2.78	10.33	14.60
4220	6 " thick		112	.214			8.85	3.25	12.10	17.15

02 41 19 – Selective Demolition

02 41 19.13 Selective Building Demolition

Line	Description	Crew	Daily Output	Labor-Hours	Unit	Material	Labor	Equipment	Total	Total Incl O&P
0010	SELECTIVE BUILDING DEMOLITION									
0020	Costs related to selective demolition of specific building components									
0025	are included under Common Work Results (XX 05)									
0030	in the component's appropriate division.									

02 41 19.16 Selective Demolition, Cutout

Line	Description	Crew	Daily Output	Labor-Hours	Unit	Material	Labor	Equipment	Total	Total Incl O&P
0010	SELECTIVE DEMOLITION, CUTOUT R024119-10									
0020	Concrete, elev. slab, light reinforcement, under 6 C.F.	B-9	65	.615	C.F.		23.50	3.58	27.08	40
0050	Light reinforcing, over 6 C.F.	"	75	.533	"		20.50	3.10	23.60	34.50
6000	Walls, interior, not including re-framing,									
6010	openings to 5 S.F.									
6100	Drywall to 5/8" thick	1 Clab	24	.333	Ea.		12.55		12.55	19.30
6200	Paneling to 3/4" thick		20	.400			15.05		15.05	23
6300	Plaster, on gypsum lath		20	.400			15.05		15.05	23
6340	On wire lath		14	.571			21.50		21.50	33

02 41 19.19 Selective Demolition

Line	Description	Crew	Daily Output	Labor-Hours	Unit	Material	Labor	Equipment	Total	Total Incl O&P
0010	SELECTIVE DEMOLITION, Rubbish Handling R024119-10									
0020	The following are to be added to the demolition prices									
0600	Dumpster, weekly rental, 1 dump/week, 6 C.Y. capacity (2 Tons)				Week	415			415	455
0700	10 C.Y. capacity (3 Tons)					480			480	530
0725	20 C.Y. capacity (5 Tons) R024119-20					565			565	625

02 41 Demolition

02 41 19 – Selective Demolition

02 41 19.19 Selective Demolition		Crew	Daily Output	Labor-Hours	Unit	Material	2015 Bare Costs Labor	Equipment	Total	Total Incl O&P
0800	30 C.Y. capacity (7 Tons)				Week	730			730	800
0840	40 C.Y. capacity (10 Tons)					775			775	850
2000	Load, haul, dump and return, 0 – 50′ haul, hand carried	2 Clab	24	.667	C.Y.		25		25	38.50
2005	Wheeled		37	.432			16.25		16.25	25
2040	0 – 100′ haul, hand carried		16.50	.970			36.50		36.50	56
2045	Wheeled		25	.640			24		24	37
2080	Haul and return, add per each extra 100′ haul, hand carried		35.50	.451			16.95		16.95	26
2085	Wheeled		54	.296			11.15		11.15	17.15
2120	For travel in elevators, up to 10 floors, add		140	.114			4.30		4.30	6.60
2130	0 – 50′ haul, incl. up to 5 riser stair, hand carried		23	.696			26		26	40
2135	Wheeled		35	.457			17.20		17.20	26.50
2140	6 – 10 riser stairs, hand carried		22	.727			27.50		27.50	42
2145	Wheeled		34	.471			17.70		17.70	27
2150	11 – 20 riser stairs, hand carried		20	.800			30		30	46.50
2155	Wheeled		31	.516			19.40		19.40	30
2160	21 – 40 riser stairs, hand carried		16	1			37.50		37.50	58
2165	Wheeled		24	.667			25		25	38.50
2170	0 – 100′ haul, incl. 5 riser stair, hand carried		15	1.067			40		40	61.50
2175	Wheeled		23	.696			26		26	40
2180	6 – 10 riser stair, hand carried		14	1.143			43		43	66
2185	Wheeled		21	.762			28.50		28.50	44
2190	11 – 20 riser stair, hand carried		12	1.333			50		50	77
2195	Wheeled		18	.889			33.50		33.50	51.50
2200	21 – 40 riser stair, hand carried		8	2			75		75	116
2205	Wheeled		12	1.333			50		50	77
2210	Haul and return, add per each extra 100′ haul, hand carried		35.50	.451			16.95		16.95	26
2215	Wheeled		54	.296			11.15		11.15	17.15
2220	For each additional flight of stairs, up to 5 risers, add		550	.029	Flight		1.09		1.09	1.68
2225	6 – 10 risers, add		275	.058			2.19		2.19	3.37
2230	11 – 20 risers, add		138	.116			4.36		4.36	6.70
2235	21 – 40 risers, add		69	.232			8.70		8.70	13.40
3000	Loading & trucking, including 2 mile haul, chute loaded	B-16	45	.711	C.Y.		27.50	15.35	42.85	59
3040	Hand loading truck, 50′ haul	"	48	.667			26	14.40	40.40	55.50
3080	Machine loading truck	B-17	120	.267			10.90	6.45	17.35	24
5000	Haul, per mile, up to 8 C.Y. truck	B-34B	1165	.007			.28	.59	.87	1.07
5100	Over 8 C.Y. truck	"	1550	.005			.21	.45	.66	.80

02 41 19.20 Selective Demolition, Dump Charges

		Crew	Daily Output	Labor-Hours	Unit	Material	Labor	Equipment	Total	Total Incl O&P
0010	**SELECTIVE DEMOLITION, DUMP CHARGES** R024119-10									
0020	Dump charges, typical urban city, tipping fees only									
0100	Building construction materials				Ton	74			74	81
0200	Trees, brush, lumber					63			63	69.50
0300	Rubbish only					63			63	69.50
0500	Reclamation station, usual charge					74			74	81

02 41 19.27 Selective Demolition, Torch Cutting

		Crew	Daily Output	Labor-Hours	Unit	Material	Labor	Equipment	Total	Total Incl O&P
0010	**SELECTIVE DEMOLITION, TORCH CUTTING** R024119-10									
0020	Steel, 1″ thick plate	E-25	333	.024	L.F.	.84	1.31	.03	2.18	3.24
0040	1″ diameter bar	"	600	.013	Ea.	.14	.73	.02	.89	1.44
1000	Oxygen lance cutting, reinforced concrete walls									
1040	12″ to 16″ thick walls	1 Clab	10	.800	L.F.		30		30	46.50
1080	24″ thick walls	"	6	1.333	"		50		50	77

02 82 13 – Asbestos Abatement

02 82 13.47 Asbestos Waste Pkg., Handling, and Disp.		Crew	Daily Output	Labor-Hours	Unit	2015 Bare Costs Material	Labor	Equipment	Total	Total Incl O&P
0010	ASBESTOS WASTE PACKAGING, HANDLING, AND DISPOSAL									
0100	Collect and bag bulk material, 3 C.F. bags, by hand	A-9	400	.160	Ea.	.82	8.40		9.22	13.95
0200	Large production vacuum loader	A-12	880	.073		.86	3.81	.87	5.54	7.85
1000	Double bag and decontaminate	A-9	960	.067		.82	3.49		4.31	6.35
2000	Containerize bagged material in drums, per 3 C.F. drum	"	800	.080		18.20	4.19		22.39	26.50
3000	Cart bags 50' to dumpster	2 Asbe	400	.040	↓		2.09		2.09	3.26
5000	Disposal charges, not including haul, minimum				C.Y.				61	67
5020	Maximum				"				355	395
9000	For type B (supplied air) respirator equipment, add				%				10%	10%

Estimating Tips

General

- Carefully check all the plans and specifications. Concrete often appears on drawings other than structural drawings, including mechanical and electrical drawings for equipment pads. The cost of cutting and patching is often difficult to estimate. See Subdivision 03 81 for Concrete Cutting, Subdivision 02 41 19.16 for Cutout Demolition, Subdivision 03 05 05.10 for Concrete Demolition, and Subdivision 02 41 19.19 for Rubbish Handling (handling, loading and hauling of debris).
- Always obtain concrete prices from suppliers near the job site. A volume discount can often be negotiated, depending upon competition in the area. Remember to add for waste, particularly for slabs and footings on grade.

03 10 00 Concrete Forming and Accessories

- A primary cost for concrete construction is forming. Most jobs today are constructed with prefabricated forms. The selection of the forms best suited for the job and the total square feet of forms required for efficient concrete forming and placing are key elements in estimating concrete construction. Enough forms must be available for erection to make efficient use of the concrete placing equipment and crew.
- Concrete accessories for forming and placing depend upon the systems used. Study the plans and specifications to ensure that all special accessory requirements have been included in the cost estimate, such as anchor bolts, inserts, and hangers.
- Included within costs for forms-in-place are all necessary bracing and shoring.

03 20 00 Concrete Reinforcing

- Ascertain that the reinforcing steel supplier has included all accessories, cutting, bending, and an allowance for lapping, splicing, and waste. A good rule of thumb is 10% for lapping, splicing, and waste. Also, 10% waste should be allowed for welded wire fabric.
- The unit price items in the subdivisions for Reinforcing In Place, Glass Fiber Reinforcing, and Welded Wire Fabric include the labor to install accessories such as beam and slab bolsters, high chairs, and bar ties and tie wire. The material cost for these accessories is not included; they may be obtained from the Accessories Division.

03 30 00 Cast-In-Place Concrete

- When estimating structural concrete, pay particular attention to requirements for concrete additives, curing methods, and surface treatments. Special consideration for climate, hot or cold, must be included in your estimate. Be sure to include requirements for concrete placing equipment, and concrete finishing.
- For accurate concrete estimating, the estimator must consider each of the following major components individually: forms, reinforcing steel, ready-mix concrete, placement of the concrete, and finishing of the top surface. For faster estimating, Subdivision 03 30 53.40 for Concrete-In-Place can be used; here, various items of concrete work are presented that include the costs of all five major components (unless specifically stated otherwise).

03 40 00 Precast Concrete
03 50 00 Cast Decks and Underlayment

- The cost of hauling precast concrete structural members is often an important factor. For this reason, it is important to get a quote from the nearest supplier. It may become economically feasible to set up precasting beds on the site if the hauling costs are prohibitive.

Reference Numbers

Reference numbers are shown in shaded boxes at the beginning of some major classifications. These numbers refer to related items in the Reference Section. The reference information may be an estimating procedure, an alternate pricing method, or technical information.

Note: Not all subdivisions listed here necessarily appear in this publication. ■

03 01 Maintenance of Concrete

03 01 30 – Maintenance of Cast-In-Place Concrete

03 01 30.62 Concrete Patching		Crew	Daily Output	Labor-Hours	Unit	2015 Bare Costs Material	Labor	Equipment	Total	Total Incl O&P
0010	**CONCRETE PATCHING**									
0100	Floors, 1/4" thick, small areas, regular grout	1 Cefi	170	.047	S.F.	1.46	2.12		3.58	4.74
0150	Epoxy grout	"	100	.080	"	8.05	3.60		11.65	14.25
2000	Walls, including chipping, cleaning and epoxy grout									
2100	1/4" deep	1 Cefi	65	.123	S.F.	7.65	5.55		13.20	16.65
2150	1/2" deep		50	.160		15.35	7.20		22.55	27.50
2200	3/4" deep		40	.200		23	9		32	39

03 11 Concrete Forming

03 11 13 – Structural Cast-In-Place Concrete Forming

03 11 13.40 Forms In Place, Equipment Foundations

		Crew	Daily Output	Labor-Hours	Unit	Material	Labor	Equipment	Total	Total Incl O&P
0010	**FORMS IN PLACE, EQUIPMENT FOUNDATIONS**									
0020	1 use	C-2	160	.300	SFCA	3.45	13.70		17.15	25
0050	2 use		190	.253		1.90	11.55		13.45	19.90
0100	3 use		200	.240		1.38	11		12.38	18.40
0150	4 use		205	.234		1.12	10.70		11.82	17.75

03 11 13.45 Forms In Place, Footings

		Crew	Daily Output	Labor-Hours	Unit	Material	Labor	Equipment	Total	Total Incl O&P
0010	**FORMS IN PLACE, FOOTINGS**									
0020	Continuous wall, plywood, 1 use	C-1	375	.085	SFCA	6.80	3.81		10.61	13.35
0050	2 use		440	.073		3.74	3.24		6.98	9.10
0100	3 use		470	.068		2.72	3.04		5.76	7.65
0150	4 use		485	.066		2.22	2.94		5.16	6.95
5000	Spread footings, job-built lumber, 1 use		305	.105		2.28	4.68		6.96	9.70
5050	2 use		371	.086		1.27	3.85		5.12	7.30
5100	3 use		401	.080		.91	3.56		4.47	6.50
5150	4 use		414	.077		.74	3.45		4.19	6.10

03 11 13.65 Forms In Place, Slab On Grade

		Crew	Daily Output	Labor-Hours	Unit	Material	Labor	Equipment	Total	Total Incl O&P
0010	**FORMS IN PLACE, SLAB ON GRADE**									
3000	Edge forms, wood, 4 use, on grade, to 6" high	C-1	600	.053	L.F.	.35	2.38		2.73	4.04
6000	Trench forms in floor, wood, 1 use		160	.200	SFCA	1.94	8.90		10.84	15.90
6050	2 use		175	.183		1.06	8.15		9.21	13.70
6100	3 use		180	.178		.77	7.95		8.72	13.05
6150	4 use		185	.173		.63	7.70		8.33	12.55

03 15 Concrete Accessories

03 15 05 – Concrete Forming Accessories

03 15 05.75 Sleeves and Chases

		Crew	Daily Output	Labor-Hours	Unit	Material	Labor	Equipment	Total	Total Incl O&P
0010	**SLEEVES AND CHASES**									
0100	Plastic, 1 use, 12" long, 2" diameter	1 Carp	100	.080	Ea.	2.15	3.76		5.91	8.15
0150	4" diameter		90	.089		6.05	4.17		10.22	13.05
0200	6" diameter		75	.107		10.65	5		15.65	19.40
0250	12" diameter		60	.133		31.50	6.25		37.75	44
5000	Sheet metal, 2" diameter G		100	.080		1.36	3.76		5.12	7.30
5100	4" diameter G		90	.089		1.70	4.17		5.87	8.25
5150	6" diameter G		75	.107		1.82	5		6.82	9.70
5200	12" diameter G		60	.133		3.63	6.25		9.88	13.65
6000	Steel pipe, 2" diameter G		100	.080		4.55	3.76		8.31	10.80
6100	4" diameter G		90	.089		16.30	4.17		20.47	24.50
6150	6" diameter G		75	.107		36.50	5		41.50	47.50

03 15 Concrete Accessories

03 15 05 – Concrete Forming Accessories

03 15 05.75 Sleeves and Chases			Crew	Daily Output	Labor-Hours	Unit	2015 Bare Costs Material	Labor	Equipment	Total	Total Incl O&P
6200	12″ diameter	G	1 Carp	60	.133	Ea.	86.50	6.25		92.75	105

03 15 16 – Concrete Construction Joints

03 15 16.20 Control Joints, Saw Cut

			Crew	Daily Output	Labor-Hours	Unit	Material	Labor	Equipment	Total	Total Incl O&P
0010	**CONTROL JOINTS, SAW CUT**										
0100	Sawcut control joints in green concrete										
0120	1″ depth		C-27	2000	.008	L.F.	.04	.36	.08	.48	.66
0140	1-1/2″ depth			1800	.009		.06	.40	.09	.55	.75
0160	2″ depth		↓	1600	.010	↓	.08	.45	.10	.63	.88
0180	Sawcut joint reservoir in cured concrete										
0182	3/8″ wide x 3/4″ deep, with single saw blade		C-27	1000	.016	L.F.	.06	.72	.17	.95	1.31
0184	1/2″ wide x 1″ deep, with double saw blades			900	.018		.11	.80	.19	1.10	1.52
0186	3/4″ wide x 1-1/2″ deep, with double saw blades		↓	800	.020		.23	.90	.21	1.34	1.82
0190	Water blast joint to wash away laitance, 2 passes		C-29	2500	.003			.12	.03	.15	.22
0200	Air blast joint to blow out debris and air dry, 2 passes		C-28	2000	.004	↓		.18	.01	.19	.28
0300	For backer rod, see Section 07 91 23.10										
0342	For joint sealant, see Section 07 92 13.20										

03 15 19 – Cast-In Concrete Anchors

03 15 19.10 Anchor Bolts

			Crew	Daily Output	Labor-Hours	Unit	Material	Labor	Equipment	Total	Total Incl O&P
0010	**ANCHOR BOLTS**										
0015	Made from recycled materials										
0025	Single bolts installed in fresh concrete, no templates										
0030	Hooked w/nut and washer, 1/2″ diameter, 8″ long	G	1 Carp	132	.061	Ea.	1.33	2.85		4.18	5.85
0040	12″ long	G		131	.061		1.48	2.87		4.35	6.05
0050	5/8″ diameter, 8″ long	G		129	.062		2.98	2.91		5.89	7.75
0060	12″ long	G		127	.063		3.67	2.96		6.63	8.60
0070	3/4″ diameter, 8″ long	G		127	.063		3.67	2.96		6.63	8.60
0080	12″ long	G	↓	125	.064	↓	4.59	3		7.59	9.65

03 15 19.45 Machinery Anchors

			Crew	Daily Output	Labor-Hours	Unit	Material	Labor	Equipment	Total	Total Incl O&P
0010	**MACHINERY ANCHORS**, heavy duty, incl. sleeve, floating base nut,										
0020	lower stud & coupling nut, fiber plug, connecting stud, washer & nut.										
0030	For flush mounted embedment in poured concrete heavy equip. pads.										
0200	Stud & bolt, 1/2″ diameter	G	E-16	40	.400	Ea.	72.50	21.50	3.65	97.65	122
0300	5/8″ diameter	G		35	.457		80.50	24.50	4.17	109.17	136
0500	3/4″ diameter	G		30	.533		93	28.50	4.86	126.36	157
0600	7/8″ diameter	G		25	.640		101	34.50	5.85	141.35	177
0800	1″ diameter	G		20	.800		117	43	7.30	167.30	212
0900	1-1/4″ diameter	G	↓	15	1.067	↓	141	57	9.75	207.75	265

03 21 Reinforcement Bars

03 21 11 – Plain Steel Reinforcement Bars

03 21 11.60 Reinforcing In Place

			Crew	Daily Output	Labor-Hours	Unit	Material	Labor	Equipment	Total	Total Incl O&P
0010	**REINFORCING IN PLACE**, 50-60 ton lots, A615 Grade 60										
0020	Includes labor, but not material cost, to install accessories										
0030	Made from recycled materials										
0502	Footings, #4 to #7	G	4 Rodm	4200	.008	Lb.	.48	.40		.88	1.16
0552	#8 to #18	G		7200	.004		.48	.23		.71	.89
0602	Slab on grade, #3 to #7	G	↓	4200	.008	↓	.48	.40		.88	1.16
0900	For other than 50 – 60 ton lots										
1000	Under 10 ton job, #3 to #7, add						25%	10%			
1010	#8 to #18, add						20%	10%			

03 21 Reinforcement Bars

03 21 11 – Plain Steel Reinforcement Bars

03 21 11.60 Reinforcing In Place		Crew	Daily Output	Labor-Hours	Unit	Material	Labor	Equipment	Total	Total Incl O&P
						2015 Bare Costs				
1050	10 – 50 ton job, #3 to #7, add					10%				
1060	#8 to #18, add					5%				
1100	60 – 100 ton job, #3 to #7, deduct					5%				
1110	#8 to #18, deduct					10%				
1150	Over 100 ton job, #3 to #7, deduct					10%				
1160	#8 to #18, deduct					15%				

03 22 Fabric and Grid Reinforcing

03 22 11 – Plain Welded Wire Fabric Reinforcing

03 22 11.10 Plain Welded Wire Fabric		Crew	Daily Output	Labor-Hours	Unit	Material	Labor	Equipment	Total	Total Incl O&P
0010	**PLAIN WELDED WIRE FABRIC** ASTM A185									
0020	Includes labor, but not material cost, to install accessories									
0050	Sheets									
0100	6 x 6 - W1.4 x W1.4 (10 x 10) 21 lb. per C.S.F. G	2 Rodm	35	.457	C.S.F.	14.50	24		38.50	53.50

03 22 13 – Galvanized Welded Wire Fabric Reinforcing

03 22 13.10 Galvanized Welded Wire Fabric		Crew	Daily Output	Labor-Hours	Unit	Material	Labor	Equipment	Total	Total Incl O&P
0010	**GALVANIZED WELDED WIRE FABRIC**									
0100	Add to plain welded wire pricing for galvanized welded wire				Lb.	.23			.23	.25

03 22 16 – Epoxy-Coated Welded Wire Fabric Reinforcing

03 22 16.10 Epoxy-Coated Welded Wire Fabric		Crew	Daily Output	Labor-Hours	Unit	Material	Labor	Equipment	Total	Total Incl O&P
0010	**EPOXY-COATED WELDED WIRE FABRIC**									
0100	Add to plain welded wire pricing for epoxy-coated welded wire				Lb.	.21			.21	.23

03 30 Cast-In-Place Concrete

03 30 53 – Miscellaneous Cast-In-Place Concrete

03 30 53.40 Concrete In Place		Crew	Daily Output	Labor-Hours	Unit	Material	Labor	Equipment	Total	Total Incl O&P
0010	**CONCRETE IN PLACE**									
0020	Including forms (4 uses), Grade 60 rebar, concrete (Portland cement									
0050	Type I), placement and finishing unless otherwise indicated									
3540	Equipment pad (3000 psi), 3′ x 3′ x 6″ thick	C-14H	45	1.067	Ea.	47	49.50	.69	97.19	128
3550	4′ x 4′ x 6″ thick		30	1.600		69.50	74	1.04	144.54	191
3560	5′ x 5′ x 8″ thick		18	2.667		122	124	1.73	247.73	325
3570	6′ x 6′ x 8″ thick		14	3.429		164	159	2.23	325.23	425
3580	8′ x 8′ x 10″ thick		8	6		350	278	3.90	631.90	815
3590	10′ x 10′ x 12″ thick	↓	5	9.600	↓	595	445	6.25	1,046.25	1,350
3800	Footings (3000 psi), spread under 1 C.Y.	C-14C	28	4	C.Y.	166	180	1.12	347.12	460
3825	1 C.Y. to 5 C.Y.		43	2.605		201	117	.73	318.73	405
3850	Over 5 C.Y.	↓	75	1.493		185	67.50	.42	252.92	310
3900	Footings, strip (3000 psi), 18″ x 9″, unreinforced	C-14L	40	2.400		125	105	.79	230.79	300
3920	18″ x 9″, reinforced	C-14C	35	3.200		148	144	.90	292.90	385
3925	20″ x 10″, unreinforced	C-14L	45	2.133		122	93.50	.70	216.20	278
3930	20″ x 10″, reinforced	C-14C	40	2.800		140	126	.78	266.78	350
3935	24″ x 12″, unreinforced	C-14L	55	1.745		120	76.50	.58	197.08	250
3940	24″ x 12″, reinforced	C-14C	48	2.333		139	105	.65	244.65	315
3945	36″ x 12″, unreinforced	C-14L	70	1.371		116	60	.45	176.45	219
3950	36″ x 12″, reinforced	C-14C	60	1.867		133	84	.52	217.52	276
4000	Foundation mat (3000 psi), under 10 C.Y.		38.67	2.896		204	131	.81	335.81	425
4050	Over 20 C.Y.	↓	56.40	1.986	↓	178	89.50	.56	268.06	335

03 30 Cast-In-Place Concrete

03 30 53 – Miscellaneous Cast-In-Place Concrete

03 30 53.40 Concrete In Place		Crew	Daily Output	Labor-Hours	Unit	Material	2015 Bare Costs Labor	Equipment	Total	Total Incl O&P
4650	Slab on grade (3500 psi), not including finish, 4" thick	C-14E	60.75	1.449	C.Y.	124	67.50	.51	192.01	242
4700	6" thick	"	92	.957	↓	119	44.50	.33	163.83	200
4701	Thickened slab edge (3500 psi), for slab on grade poured									
4702	monolithically with slab; depth is in addition to slab thickness;									
4703	formed vertical outside edge, earthen bottom and inside slope									
4705	8" deep x 8" wide bottom, unreinforced	C-14L	2190	.044	L.F.	3.47	1.92	.01	5.40	6.80
4710	8" x 8", reinforced	C-14C	1670	.067		5.75	3.02	.02	8.79	10.95
4715	12" deep x 12" wide bottom, unreinforced	C-14L	1800	.053		7.05	2.34	.02	9.41	11.35
4720	12" x 12", reinforced	C-14C	1310	.086		11.20	3.85	.02	15.07	18.30
4725	16" deep x 16" wide bottom, unreinforced	C-14L	1440	.067		11.85	2.92	.02	14.79	17.50
4730	16" x 16", reinforced	C-14C	1120	.100		16.80	4.51	.03	21.34	25.50
4735	20" deep x 20" wide bottom, unreinforced	C-14L	1150	.083		17.90	3.66	.03	21.59	25.50
4740	20" x 20", reinforced	C-14C	920	.122		24	5.50	.03	29.53	35
4745	24" deep x 24" wide bottom, unreinforced	C-14L	930	.103		25	4.53	.03	29.56	35
4750	24" x 24", reinforced	C-14C	740	.151	↓	33.50	6.80	.04	40.34	47

03 31 Structural Concrete

03 31 13 – Heavyweight Structural Concrete

03 31 13.35 Heavyweight Concrete, Ready Mix		Crew	Daily Output	Labor-Hours	Unit	Material	Labor	Equipment	Total	Total Incl O&P
0010	**HEAVYWEIGHT CONCRETE, READY MIX**, delivered									
0012	Includes local aggregate, sand, Portland cement (Type I) and water									
0015	Excludes all additives and treatments									
0020	2000 psi				C.Y.	97			97	107
0100	2500 psi					99.50			99.50	109
0150	3000 psi					102			102	112
0200	3500 psi					104			104	115
0300	4000 psi					107			107	118
1000	For high early strength (Portland cement Type III), add					10%				
1300	For winter concrete (hot water), add					4.05			4.05	4.46
1410	For mid-range water reducer, add					3.31			3.31	3.64
1420	For high-range water reducer/superplasticizer, add					5.65			5.65	6.20
1430	For retarder, add					3.23			3.23	3.55
1440	For non-Chloride accelerator, add					5.50			5.50	6.05
1450	For Chloride accelerator, per 1%, add					2.90			2.90	3.19
1460	For fiber reinforcing, synthetic (1 lb./C.Y.), add					6.65			6.65	7.30
1500	For Saturday delivery, add				↓	10.80			10.80	11.90
1510	For truck holding/waiting time past 1st hour per load, add				Hr.	94			94	103
1520	For short load (less than 4 C.Y.), add per load				Ea.	60.50			60.50	66.50
2000	For all lightweight aggregate, add				C.Y.	45%				

03 31 13.70 Placing Concrete		Crew	Daily Output	Labor-Hours	Unit	Material	Labor	Equipment	Total	Total Incl O&P
0010	**PLACING CONCRETE**									
0020	Includes labor and equipment to place, level (strike off) and consolidate									
1900	Footings, continuous, shallow, direct chute	C-6	120	.400	C.Y.		15.65	.52	16.17	24.50
1950	Pumped	C-20	150	.427			17.25	5.20	22.45	32.50
2000	With crane and bucket	C-7	90	.800			32.50	13.50	46	65
2100	Footings, continuous, deep, direct chute	C-6	140	.343			13.45	.45	13.90	21
2150	Pumped	C-20	160	.400			16.15	4.89	21.04	30
2200	With crane and bucket	C-7	110	.655			27	11.05	38.05	53
2400	Footings, spread, under 1 C.Y., direct chute	C-6	55	.873			34	1.13	35.13	53.50
2450	Pumped	C-20	65	.985			40	12.05	52.05	74.50
2500	With crane and bucket	C-7	45	1.600	↓		65.50	27	92.50	130

03 31 Structural Concrete

03 31 13 – Heavyweight Structural Concrete

03 31 13.70 Placing Concrete		Crew	Daily Output	Labor-Hours	Unit	Material	2015 Bare Costs Labor	Equipment	Total	Total Incl O&P
2600	Over 5 C.Y., direct chute	C-6	120	.400	C.Y.		15.65	.52	16.17	24.50
2650	Pumped	C-20	150	.427			17.25	5.20	22.45	32.50
2700	With crane and bucket	C-7	100	.720			29.50	12.15	41.65	58.50
2900	Foundation mats, over 20 C.Y., direct chute	C-6	350	.137			5.35	.18	5.53	8.40
2950	Pumped	C-20	400	.160			6.45	1.96	8.41	12
3000	With crane and bucket	C-7	300	.240			9.80	4.05	13.85	19.45

03 35 Concrete Finishing

03 35 13 – High-Tolerance Concrete Floor Finishing

03 35 13.30 Finishing Floors, High Tolerance		Crew	Daily Output	Labor-Hours	Unit	Material	Labor	Equipment	Total	Total Incl O&P
0010	**FINISHING FLOORS, HIGH TOLERANCE**									
0012	Finishing of fresh concrete flatwork requires that concrete									
0013	first be placed, struck off & consolidated									
0015	Basic finishing for various unspecified flatwork									
0100	Bull float only	C-10	4000	.006	S.F.		.26		.26	.38
0125	Bull float & manual float		2000	.012			.51		.51	.76
0150	Bull float, manual float, & broom finish, w/edging & joints		1850	.013			.55		.55	.83
0200	Bull float, manual float & manual steel trowel		1265	.019			.81		.81	1.21
0210	For specified Random Access Floors in ACI Classes 1, 2, 3 and 4 to achieve									
0215	Composite Overall Floor Flatness and Levelness values up to FF35/FL25									
0250	Bull float, machine float & machine trowel (walk-behind)	C-10C	1715	.014	S.F.		.60	.03	.63	.92
0300	Power screed, bull float, machine float & trowel (walk-behind)	C-10D	2400	.010			.43	.05	.48	.69
0350	Power screed, bull float, machine float & trowel (ride-on)	C-10E	4000	.006			.26	.06	.32	.45
0352	For specified Random Access Floors in ACI Classes 5, 6, 7 and 8 to achieve									
0354	Composite Overall Floor Flatness and Levelness values up to FF50/FL50									
0356	Add for two-dimensional restraightening after power float	C-10	6000	.004	S.F.		.17		.17	.25
0358	For specified Random or Defined Access Floors in ACI Class 9 to achieve									
0360	Composite Overall Floor Flatness and Levelness values up to FF100/FL100									
0362	Add for two-dimensional restraightening after bull float & power float	C-10	3000	.008	S.F.		.34		.34	.51
0364	For specified Superflat Defined Access Floors in ACI Class 9 to achieve									
0366	Minimum Floor Flatness and Levelness values of FF100/FL100									
0368	Add for 2-dim'l restraightening after bull float, power float, power trowel	C-10	2000	.012	S.F.		.51		.51	.76

03 54 Cast Underlayment

03 54 16 – Hydraulic Cement Underlayment

03 54 16.50 Cement Underlayment		Crew	Daily Output	Labor-Hours	Unit	Material	Labor	Equipment	Total	Total Incl O&P
0010	**CEMENT UNDERLAYMENT**									
2510	Underlayment, P.C based, self-leveling, 4100 psi, pumped, 1/4" thick	C-8	20000	.003	S.F.	1.55	.12	.04	1.71	1.92
2520	1/2" thick		19000	.003		3.09	.12	.04	3.25	3.63
2530	3/4" thick		18000	.003		4.64	.13	.04	4.81	5.35
2540	1" thick		17000	.003		6.20	.14	.04	6.38	7.05
2550	1-1/2" thick		15000	.004		9.30	.16	.05	9.51	10.50
2560	Hand placed, 1/2" thick	C-18	450	.020		3.09	.76	.13	3.98	4.71
2610	Topping, P.C based, self-leveling, 6100 psi, pumped, 1/4" thick	C-8	20000	.003		2.26	.12	.04	2.42	2.71
2620	1/2" thick		19000	.003		4.52	.12	.04	4.68	5.20
2630	3/4" thick		18000	.003		6.80	.13	.04	6.97	7.70
2660	1" thick		17000	.003		9.05	.14	.04	9.23	10.20
2670	1-1/2" thick		15000	.004		13.55	.16	.05	13.76	15.20
2680	Hand placed, 1/2" thick	C-18	450	.020		4.52	.76	.13	5.41	6.30

03 63 Epoxy Grouting

03 63 05 – Grouting of Dowels and Fasteners

03 63 05.10 Epoxy Only		Crew	Daily Output	Labor-Hours	Unit	2015 Bare Costs Material	Labor	Equipment	Total	Total Incl O&P
0010	**EPOXY ONLY**									
1500	Chemical anchoring, epoxy cartridge, excludes layout, drilling, fastener									
1530	For fastener 3/4" diam. x 6" embedment	2 Skwk	72	.222	Ea.	5.65	10.80		16.45	23
1535	1" diam. x 8" embedment		66	.242		8.45	11.80		20.25	27.50
1540	1-1/4" diam. x 10" embedment		60	.267		16.90	12.95		29.85	38.50
1545	1-3/4" diam. x 12" embedment		54	.296		28	14.40		42.40	53.50
1550	14" embedment		48	.333		34	16.20		50.20	62
1555	2" diam. x 12" embedment		42	.381		45	18.55		63.55	78
1560	18" embedment		32	.500		56.50	24.50		81	99.50

03 81 Concrete Cutting

03 81 13 – Flat Concrete Sawing

03 81 13.50 Concrete Floor/Slab Cutting

		Crew	Daily Output	Labor-Hours	Unit	Material	Labor	Equipment	Total	Total Incl O&P
0010	**CONCRETE FLOOR/SLAB CUTTING**									
0050	Includes blade cost, layout and set-up time									
0300	Saw cut concrete slabs, plain, up to 3" deep	B-89	1060	.015	L.F.	.14	.66	.46	1.26	1.66
0320	Each additional inch of depth		3180	.005		.05	.22	.15	.42	.55
0400	Mesh reinforced, up to 3" deep		980	.016		.16	.72	.50	1.38	1.80
0420	Each additional inch of depth		2940	.005		.05	.24	.17	.46	.60
0500	Rod reinforced, up to 3" deep		800	.020		.19	.88	.61	1.68	2.21
0520	Each additional inch of depth		2400	.007		.06	.29	.20	.55	.73

03 81 13.75 Concrete Saw Blades

		Crew	Daily Output	Labor-Hours	Unit	Material	Labor	Equipment	Total	Total Incl O&P
0010	**CONCRETE SAW BLADES**									
3000	Blades for saw cutting, included in cutting line items									
3020	Diamond, 12" diameter				Ea.	241			241	265
3040	18" diameter					465			465	510
3080	24" diameter					770			770	845
3120	30" diameter					1,100			1,100	1,200
3160	36" diameter					1,425			1,425	1,550
3200	42" diameter					2,625			2,625	2,875

03 82 Concrete Boring

03 82 13 – Concrete Core Drilling

03 82 13.10 Core Drilling

		Crew	Daily Output	Labor-Hours	Unit	Material	Labor	Equipment	Total	Total Incl O&P
0010	**CORE DRILLING**									
0015	Includes bit cost, layout and set-up time									
0020	Reinforced concrete slab, up to 6" thick									
0100	1" diameter core	B-89A	17	.941	Ea.	.18	40.50	6.80	47.48	70
0150	For each additional inch of slab thickness in same hole, add		1440	.011		.03	.48	.08	.59	.86
0200	2" diameter core		16.50	.970		.28	42	7	49.28	72.50
0250	For each additional inch of slab thickness in same hole, add		1080	.015		.05	.64	.11	.80	1.15
0300	3" diameter core		16	1		.38	43	7.25	50.63	75
0350	For each additional inch of slab thickness in same hole, add		720	.022		.06	.96	.16	1.18	1.73
0500	4" diameter core		15	1.067		.49	46	7.70	54.19	80
0550	For each additional inch of slab thickness in same hole, add		480	.033		.08	1.44	.24	1.76	2.58
0700	6" diameter core		14	1.143		.76	49.50	8.25	58.51	86
0750	For each additional inch of slab thickness in same hole, add		360	.044		.13	1.92	.32	2.37	3.45
0900	8" diameter core		13	1.231		1.05	53	8.90	62.95	93
0950	For each additional inch of slab thickness in same hole, add		288	.056		.17	2.40	.40	2.97	4.32

03 82 Concrete Boring

03 82 13 – Concrete Core Drilling

03 82 13.10 Core Drilling		Crew	Daily Output	Labor-Hours	Unit	Material	Labor	Equipment	Total	Total Incl O&P
						2015 Bare Costs				
1100	10″ diameter core	B-89A	12	1.333	Ea.	1.48	57.50	9.65	68.63	101
1150	For each additional inch of slab thickness in same hole, add		240	.067		.25	2.88	.48	3.61	5.25
1300	12″ diameter core		11	1.455		1.78	62.50	10.50	74.78	110
1350	For each additional inch of slab thickness in same hole, add		206	.078		.30	3.35	.56	4.21	6.10
1500	14″ diameter core		10	1.600		2.08	69	11.55	82.63	121
1550	For each additional inch of slab thickness in same hole, add		180	.089		.35	3.83	.64	4.82	7
1700	18″ diameter core		9	1.778		2.82	76.50	12.85	92.17	135
1750	For each additional inch of slab thickness in same hole, add		144	.111		.47	4.79	.80	6.06	8.80
1754	24″ diameter core		8	2		4.02	86.50	14.45	104.97	153
1756	For each additional inch of slab thickness in same hole, add		120	.133		.67	5.75	.96	7.38	10.65
1760	For horizontal holes, add to above						20%	20%		
1770	Prestressed hollow core plank, 8″ thick									
1780	1″ diameter core	B-89A	17.50	.914	Ea.	.24	39.50	6.60	46.34	68.50
1790	For each additional inch of plank thickness in same hole, add		3840	.004		.03	.18	.03	.24	.34
1794	2″ diameter core		17.25	.928		.38	40	6.70	47.08	69.50
1796	For each additional inch of plank thickness in same hole, add		2880	.006		.05	.24	.04	.33	.46
1800	3″ diameter core		17	.941		.51	40.50	6.80	47.81	70.50
1810	For each additional inch of plank thickness in same hole, add		1920	.008		.06	.36	.06	.48	.69
1820	4″ diameter core		16.50	.970		.66	42	7	49.66	73
1830	For each additional inch of plank thickness in same hole, add		1280	.013		.08	.54	.09	.71	1.02
1840	6″ diameter core		15.50	1.032		1.01	44.50	7.45	52.96	78
1850	For each additional inch of plank thickness in same hole, add		960	.017		.13	.72	.12	.97	1.38
1860	8″ diameter core		15	1.067		1.40	46	7.70	55.10	81
1870	For each additional inch of plank thickness in same hole, add		768	.021		.17	.90	.15	1.22	1.75
1880	10″ diameter core		14	1.143		1.98	49.50	8.25	59.73	87.50
1890	For each additional inch of plank thickness in same hole, add		640	.025		.25	1.08	.18	1.51	2.13
1900	12″ diameter core		13.50	1.185		2.37	51	8.55	61.92	91
1910	For each additional inch of plank thickness in same hole, add		548	.029		.30	1.26	.21	1.77	2.50

03 82 16 – Concrete Drilling

03 82 16.10 Concrete Impact Drilling		Crew	Daily Output	Labor-Hours	Unit	Material	Labor	Equipment	Total	Total Incl O&P
0010	**CONCRETE IMPACT DRILLING**									
0020	Includes bit cost, layout and set-up time, no anchors									
0050	Up to 4″ deep in concrete/brick floors/walls									
0100	Holes, 1/4″ diameter	1 Carp	75	.107	Ea.	.07	5		5.07	7.75
0150	For each additional inch of depth in same hole, add		430	.019		.02	.87		.89	1.36
0200	3/8″ diameter		63	.127		.06	5.95		6.01	9.20
0250	For each additional inch of depth in same hole, add		340	.024		.01	1.10		1.11	1.72
0300	1/2″ diameter		50	.160		.06	7.50		7.56	11.60
0350	For each additional inch of depth in same hole, add		250	.032		.01	1.50		1.51	2.33
0400	5/8″ diameter		48	.167		.09	7.85		7.94	12.15
0450	For each additional inch of depth in same hole, add		240	.033		.02	1.56		1.58	2.43
0500	3/4″ diameter		45	.178		.12	8.35		8.47	13
0550	For each additional inch of depth in same hole, add		220	.036		.03	1.71		1.74	2.66
0600	7/8″ diameter		43	.186		.16	8.75		8.91	13.60
0650	For each additional inch of depth in same hole, add		210	.038		.04	1.79		1.83	2.79
0700	1″ diameter		40	.200		.17	9.40		9.57	14.65
0750	For each additional inch of depth in same hole, add		190	.042		.04	1.98		2.02	3.09
0800	1-1/4″ diameter		38	.211		.26	9.90		10.16	15.50
0850	For each additional inch of depth in same hole, add		180	.044		.06	2.09		2.15	3.28
0900	1-1/2″ diameter		35	.229		.39	10.75		11.14	16.95
0950	For each additional inch of depth in same hole, add		165	.048		.10	2.28		2.38	3.61
1000	For ceiling installations, add						40%			

Estimating Tips

05 05 00 Common Work Results for Metals

- Nuts, bolts, washers, connection angles, and plates can add a significant amount to both the tonnage of a structural steel job and the estimated cost. As a rule of thumb, add 10% to the total weight to account for these accessories.
- Type 2 steel construction, commonly referred to as "simple construction," consists generally of field-bolted connections with lateral bracing supplied by other elements of the building, such as masonry walls or x-bracing. The estimator should be aware, however, that shop connections may be accomplished by welding or bolting. The method may be particular to the fabrication shop and may have an impact on the estimated cost.

05 10 00 Structural Steel

- Steel items can be obtained from two sources: a fabrication shop or a metals service center. Fabrication shops can fabricate items under more controlled conditions than can crews in the field. They are also more efficient and can produce items more economically. Metal service centers serve as a source of long mill shapes to both fabrication shops and contractors.
- Most line items in this structural steel subdivision, and most items in 05 50 00 Metal Fabrications, are indicated as being shop fabricated. The bare material cost for these shop fabricated items is the "Invoice Cost" from the shop and includes the mill base price of steel plus mill extras, transportation to the shop, shop drawings and detailing where warranted, shop fabrication and handling, sandblasting and a shop coat of primer paint, all necessary structural bolts, and delivery to the job site. The bare labor cost and bare equipment cost for these shop fabricated items is for field installation or erection.
- Line items in Subdivision 05 12 23.40 Lightweight Framing, and other items scattered in Division 5, are indicated as being field fabricated. The bare material cost for these field fabricated items is the "Invoice Cost" from the metals service center and includes the mill base price of steel plus mill extras, transportation to the metals service center, material handling, and delivery of long lengths of mill shapes to the job site. Material costs for structural bolts and welding rods should be added to the estimate. The bare labor cost and bare equipment cost for these items is for both field fabrication and field installation or erection, and include time for cutting, welding and drilling in the fabricated metal items. Drilling into concrete and fasteners to fasten field fabricated items to other work are not included and should be added to the estimate.

05 20 00 Steel Joist Framing

- In any given project the total weight of open web steel joists is determined by the loads to be supported and the design. However, economies can be realized in minimizing the amount of labor used to place the joists. This is done by maximizing the joist spacing, and therefore minimizing the number of joists required to be installed on the job. Certain spacings and locations may be required by the design, but in other cases maximizing the spacing and keeping it as uniform as possible will keep the costs down.

05 30 00 Steel Decking

- The takeoff and estimating of metal deck involves more than simply the area of the floor or roof and the type of deck specified or shown on the drawings. Many different sizes and types of openings may exist. Small openings for individual pipes or conduits may be drilled after the floor/roof is installed, but larger openings may require special deck lengths as well as reinforcing or structural support. The estimator should determine who will be supplying this reinforcing. Additionally, some deck terminations are part of the deck package, such as screed angles and pour stops, and others will be part of the steel contract, such as angles attached to structural members and cast-in-place angles and plates. The estimator must ensure that all pieces are accounted for in the complete estimate.

05 50 00 Metal Fabrications

- The most economical steel stairs are those that use common materials, standard details, and most importantly, a uniform and relatively simple method of field assembly. Commonly available A36 channels and plates are very good choices for the main stringers of the stairs, as are angles and tees for the carrier members. Risers and treads are usually made by specialty shops, and it is most economical to use a typical detail in as many places as possible. The stairs should be pre-assembled and shipped directly to the site. The field connections should be simple and straightforward to be accomplished efficiently, and with minimum equipment and labor.

Reference Numbers

Reference numbers are shown in shaded boxes at the beginning of some major classifications. These numbers refer to related items in the Reference Section. The reference information may be an estimating procedure, an alternate pricing method, or technical information.

Note: Not all subdivisions listed here necessarily appear in this publication. ■

05 05 Common Work Results for Metals

05 05 19 – Post-Installed Concrete Anchors

05 05 19.10 Chemical Anchors			Crew	Daily Output	Labor-Hours	Unit	2015 Bare Costs Material	Labor	Equipment	Total	Total Incl O&P
0010	CHEMICAL ANCHORS										
0020	Includes layout & drilling										
1430	Chemical anchor, w/rod & epoxy cartridge, 3/4" diam. x 9-1/2" long		B-89A	27	.593	Ea.	9.80	25.50	4.28	39.58	55
1435	1" diameter x 11-3/4" long			24	.667		17.55	29	4.82	51.37	69
1440	1-1/4" diameter x 14" long			21	.762		36.50	33	5.50	75	96.50
1445	1-3/4" diameter x 15" long			20	.800		65	34.50	5.80	105.30	131
1450	18" long			17	.941		78	40.50	6.80	125.30	156
1455	2" diameter x 18" long			16	1		103	43	7.25	153.25	187
1460	24" long			15	1.067		134	46	7.70	187.70	227

05 05 19.20 Expansion Anchors			Crew	Daily Output	Labor-Hours	Unit	Material	Labor	Equipment	Total	Total Incl O&P
0010	EXPANSION ANCHORS										
0100	Anchors for concrete, brick or stone, no layout and drilling										
0200	Expansion shields, zinc, 1/4" diameter, 1-5/16" long, single	G	1 Carp	90	.089	Ea.	.43	4.17		4.60	6.85
0300	1-3/8" long, double	G		85	.094		.54	4.42		4.96	7.40
0400	3/8" diameter, 1-1/2" long, single	G		85	.094		.64	4.42		5.06	7.50
0500	2" long, double	G		80	.100		1.18	4.70		5.88	8.55
0600	1/2" diameter, 2-1/16" long, single	G		80	.100		1.18	4.70		5.88	8.55
0700	2-1/2" long, double	G		75	.107		1.91	5		6.91	9.80
0800	5/8" diameter, 2-5/8" long, single	G		75	.107		2.05	5		7.05	9.95
0900	2-3/4" long, double	G		70	.114		2.70	5.35		8.05	11.20
1000	3/4" diameter, 2-3/4" long, single	G		70	.114		3.09	5.35		8.44	11.65
1100	3-15/16" long, double	G		65	.123		5	5.80		10.80	14.40
2100	Hollow wall anchors for gypsum wall board, plaster or tile										
2300	1/8" diameter, short	G	1 Carp	160	.050	Ea.	.23	2.35		2.58	3.86
2400	Long	G		150	.053		.23	2.50		2.73	4.10
2500	3/16" diameter, short	G		150	.053		.40	2.50		2.90	4.29
2600	Long	G		140	.057		.59	2.68		3.27	4.78
2700	1/4" diameter, short	G		140	.057		.59	2.68		3.27	4.78
2800	Long	G		130	.062		.65	2.89		3.54	5.15
3000	Toggle bolts, bright steel, 1/8" diameter, 2" long	G		85	.094		.19	4.42		4.61	7
3100	4" long	G		80	.100		.25	4.70		4.95	7.55
3400	1/4" diameter, 3" long	G		75	.107		.34	5		5.34	8.05
3500	6" long	G		70	.114		.51	5.35		5.86	8.80
3600	3/8" diameter, 3" long	G		70	.114		.84	5.35		6.19	9.15
3700	6" long	G		60	.133		1.36	6.25		7.61	11.15
3800	1/2" diameter, 4" long	G		60	.133		1.88	6.25		8.13	11.70
3900	6" long	G		50	.160		2.45	7.50		9.95	14.25
4000	Nailing anchors										
4100	Nylon nailing anchor, 1/4" diameter, 1" long		1 Carp	3.20	2.500	C	21	117		138	204
4200	1-1/2" long			2.80	2.857		26	134		160	235
4300	2" long			2.40	3.333		30.50	157		187.50	275
4400	Metal nailing anchor, 1/4" diameter, 1" long	G		3.20	2.500		17.85	117		134.85	201
4500	1-1/2" long	G		2.80	2.857		23.50	134		157.50	232
4600	2" long	G		2.40	3.333		27.50	157		184.50	272
5000	Screw anchors for concrete, masonry,										
5100	stone & tile, no layout or drilling included										
5700	Lag screw shields, 1/4" diameter, short	G	1 Carp	90	.089	Ea.	.34	4.17		4.51	6.75
5900	3/8" diameter, short	G		85	.094		.67	4.42		5.09	7.55
6100	1/2" diameter, short	G		80	.100		.95	4.70		5.65	8.30
6300	5/8" diameter, short	G		70	.114		1.45	5.35		6.80	9.85
6600	Lead, #6 & #8, 3/4" long	G		260	.031		.18	1.44		1.62	2.42
6700	#10 - #14, 1-1/2" long	G		200	.040		.37	1.88		2.25	3.30

05 05 19 – Post-Installed Concrete Anchors

05 05 19.20 Expansion Anchors			Crew	Daily Output	Labor-Hours	Unit	Material	2015 Bare Costs Labor	Equipment	Total	Total Incl O&P
6800	#16 & #18, 1-1/2″ long	G	1 Carp	160	.050	Ea.	.41	2.35		2.76	4.06
6900	Plastic, #6 & #8, 3/4″ long			260	.031		.04	1.44		1.48	2.26
7100	#10 & #12, 1″ long			220	.036		.05	1.71		1.76	2.69
8000	Wedge anchors, not including layout or drilling										
8050	Carbon steel, 1/4″ diameter, 1-3/4″ long	G	1 Carp	150	.053	Ea.	.41	2.50		2.91	4.30
8100	3-1/4″ long	G		140	.057		.54	2.68		3.22	4.72
8150	3/8″ diameter, 2-1/4″ long	G		145	.055		.50	2.59		3.09	4.54
8200	5″ long	G		140	.057		.88	2.68		3.56	5.10
8250	1/2″ diameter, 2-3/4″ long	G		140	.057		.99	2.68		3.67	5.20
8300	7″ long	G		125	.064		1.70	3		4.70	6.50
8350	5/8″ diameter, 3-1/2″ long	G		130	.062		1.86	2.89		4.75	6.50
8400	8-1/2″ long	G		115	.070		3.95	3.27		7.22	9.40
8450	3/4″ diameter, 4-1/4″ long	G		115	.070		2.90	3.27		6.17	8.25
8500	10″ long	G		95	.084		6.60	3.95		10.55	13.35
8550	1″ diameter, 6″ long	G		100	.080		9.30	3.76		13.06	16.05
8575	9″ long	G		85	.094		12.10	4.42		16.52	20
8600	12″ long	G		75	.107		13.10	5		18.10	22
8650	1-1/4″ diameter, 9″ long	G		70	.114		24.50	5.35		29.85	35.50
8700	12″ long	G		60	.133		31.50	6.25		37.75	44
8750	For type 303 stainless steel, add						350%				
8800	For type 316 stainless steel, add						450%				
8950	Self-drilling concrete screw, hex washer head, 3/16″ diam. x 1-3/4″ long	G	1 Carp	300	.027	Ea.	.20	1.25		1.45	2.15
8960	2-1/4″ long	G		250	.032		.24	1.50		1.74	2.57
8970	Phillips flat head, 3/16″ diam. x 1-3/4″ long	G		300	.027		.21	1.25		1.46	2.16
8980	2-1/4″ long	G		250	.032		.23	1.50		1.73	2.56

05 05 21 – Fastening Methods for Metal

05 05 21.15 Drilling Steel		Crew	Daily Output	Labor-Hours	Unit	Material	Labor	Equipment	Total	Total Incl O&P
0010	**DRILLING STEEL**									
1910	Drilling & layout for steel, up to 1/4″ deep, no anchor									
1920	Holes, 1/4″ diameter	1 Sswk	112	.071	Ea.	.09	3.76		3.85	6.65
1925	For each additional 1/4″ depth, add		336	.024		.09	1.25		1.34	2.28
1930	3/8″ diameter		104	.077		.09	4.05		4.14	7.15
1935	For each additional 1/4″ depth, add		312	.026		.09	1.35		1.44	2.44
1940	1/2″ diameter		96	.083		.10	4.39		4.49	7.70
1945	For each additional 1/4″ depth, add		288	.028		.10	1.46		1.56	2.65
1950	5/8″ diameter		88	.091		.14	4.79		4.93	8.45
1955	For each additional 1/4″ depth, add		264	.030		.14	1.60		1.74	2.93
1960	3/4″ diameter		80	.100		.18	5.25		5.43	9.35
1965	For each additional 1/4″ depth, add		240	.033		.18	1.75		1.93	3.25
1970	7/8″ diameter		72	.111		.23	5.85		6.08	10.40
1975	For each additional 1/4″ depth, add		216	.037		.23	1.95		2.18	3.65
1980	1″ diameter		64	.125		.24	6.60		6.84	11.70
1985	For each additional 1/4″ depth, add		192	.042		.24	2.19		2.43	4.07
1990	For drilling up, add						40%			

05 05 23 – Metal Fastenings

05 05 23.10 Bolts and Hex Nuts			Crew	Daily Output	Labor-Hours	Unit	Material	Labor	Equipment	Total	Total Incl O&P
0010	**BOLTS & HEX NUTS**, Steel, A307										
0100	1/4″ diameter, 1/2″ long	G	1 Sswk	140	.057	Ea.	.06	3.01		3.07	5.30
0200	1″ long	G		140	.057		.07	3.01		3.08	5.35
0300	2″ long	G		130	.062		.10	3.24		3.34	5.75
0400	3″ long	G		130	.062		.15	3.24		3.39	5.80
0500	4″ long	G		120	.067		.17	3.51		3.68	6.30

05 05 23.10 Bolts and Hex Nuts

05 05 23.10 Bolts and Hex Nuts			Crew	Daily Output	Labor-Hours	Unit	Material	2015 Bare Costs Labor	Equipment	Total	Total Incl O&P
0600	3/8" diameter, 1" long	G	1 Sswk	130	.062	Ea.	.14	3.24		3.38	5.80
0700	2" long	G		130	.062		.18	3.24		3.42	5.85
0800	3" long	G		120	.067		.24	3.51		3.75	6.35
0900	4" long	G		120	.067		.30	3.51		3.81	6.45
1000	5" long	G		115	.070		.38	3.66		4.04	6.75
1100	1/2" diameter, 1-1/2" long	G		120	.067		.40	3.51		3.91	6.55
1200	2" long	G		120	.067		.46	3.51		3.97	6.60
1300	4" long	G		115	.070		.75	3.66		4.41	7.20
1400	6" long	G		110	.073		1.05	3.83		4.88	7.80
1500	8" long	G		105	.076		1.38	4.01		5.39	8.45
1600	5/8" diameter, 1-1/2" long	G		120	.067		.98	3.51		4.49	7.20
1700	2" long	G		120	.067		1.09	3.51		4.60	7.30
1800	4" long	G		115	.070		1.59	3.66		5.25	8.10
1900	6" long	G		110	.073		2.05	3.83		5.88	8.90
2000	8" long	G		105	.076		3.06	4.01		7.07	10.30
2100	10" long	G		100	.080		3.87	4.21		8.08	11.55
2200	3/4" diameter, 2" long	G		120	.067		1.15	3.51		4.66	7.35
2300	4" long	G		110	.073		1.65	3.83		5.48	8.45
2400	6" long	G		105	.076		2.12	4.01		6.13	9.30
2500	8" long	G		95	.084		3.20	4.43		7.63	11.20
2600	10" long	G		85	.094		4.20	4.96		9.16	13.20
2700	12" long	G		80	.100		4.92	5.25		10.17	14.55
2800	1" diameter, 3" long	G		105	.076		2.69	4.01		6.70	9.90
2900	6" long	G		90	.089		3.94	4.68		8.62	12.50
3000	12" long	G		75	.107		7.10	5.60		12.70	17.55
3100	For galvanized, add						75%				
3200	For stainless, add						350%				

05 05 23.30 Lag Screws

Line	Description		Crew	Daily Output	Labor-Hours	Unit	Material	Labor	Equipment	Total	Total Incl O&P
0010	**LAG SCREWS**										
0020	Steel, 1/4" diameter, 2" long	G	1 Carp	200	.040	Ea.	.09	1.88		1.97	2.99
0100	3/8" diameter, 3" long	G		150	.053		.30	2.50		2.80	4.18
0200	1/2" diameter, 3" long	G		130	.062		.64	2.89		3.53	5.15
0300	5/8" diameter, 3" long	G		120	.067		1.15	3.13		4.28	6.10

05 05 23.35 Machine Screws

Line	Description		Crew	Daily Output	Labor-Hours	Unit	Material	Labor	Equipment	Total	Total Incl O&P
0010	**MACHINE SCREWS**										
0020	Steel, round head, #8 x 1" long	G	1 Carp	4.80	1.667	C	3.93	78.50		82.43	124
0110	#8 x 2" long	G		2.40	3.333		6.20	157		163.20	248
0200	#10 x 1" long	G		4	2		5.20	94		99.20	151
0300	#10 x 2" long	G		2	4		8.65	188		196.65	299

05 05 23.50 Powder Actuated Tools and Fasteners

Line	Description		Crew	Daily Output	Labor-Hours	Unit	Material	Labor	Equipment	Total	Total Incl O&P
0010	**POWDER ACTUATED TOOLS & FASTENERS**										
0020	Stud driver, .22 caliber, single shot					Ea.	148			148	163
0100	.27 caliber, semi automatic, strip					"	440			440	485
0300	Powder load, single shot, .22 cal, power level 2, brown					C	5.35			5.35	5.90
0400	Strip, .27 cal, power level 4, red						7.70			7.70	8.50
0600	Drive pin, .300 x 3/4" long	G	1 Carp	4.80	1.667		4.11	78.50		82.61	125
0700	.300 x 3" long with washer	G	"	4	2		12.65	94		106.65	159

05 05 23.70 Structural Blind Bolts

Line	Description		Crew	Daily Output	Labor-Hours	Unit	Material	Labor	Equipment	Total	Total Incl O&P
0010	**STRUCTURAL BLIND BOLTS**										
0100	1/4" diameter x 1/4" grip	G	1 Sswk	240	.033	Ea.	1.24	1.75		2.99	4.41
0150	1/2" grip	G		216	.037		1.33	1.95		3.28	4.85
0200	3/8" diameter x 1/2" grip	G		232	.034		1.75	1.82		3.57	5.10

05 05 Common Work Results for Metals

05 05 23 – Metal Fastenings

05 05 23.70 Structural Blind Bolts		Crew	Daily Output	Labor-Hours	Unit	Material	2015 Bare Costs Labor	Equipment	Total	Total Incl O&P
0250	3/4" grip	G	1 Sswk	208	.038	Ea.	1.84	2.02		3.86	5.55
0300	1/2" diameter x 1/2" grip	G		224	.036		3.99	1.88		5.87	7.65
0350	3/4" grip	G		200	.040		5.60	2.11		7.71	9.80
0400	5/8" diameter x 3/4" grip	G		216	.037		8.25	1.95		10.20	12.50
0450	1" grip	G	↓	192	.042	↓	9.50	2.19		11.69	14.25

05 12 Structural Steel Framing

05 12 23 – Structural Steel for Buildings

05 12 23.60 Pipe Support Framing			Crew	Daily Output	Labor-Hours	Unit	Material	Labor	Equipment	Total	Total Incl O&P
0010	**PIPE SUPPORT FRAMING**										
0020	Under 10#/L.F., shop fabricated	G	E-4	3900	.008	Lb.	1.78	.44	.04	2.26	2.75
0200	10.1 to 15#/L.F.	G		4300	.007		1.75	.40	.03	2.18	2.65
0400	15.1 to 20#/L.F.	G		4800	.007		1.72	.35	.03	2.10	2.54
0600	Over 20#/L.F.	G	↓	5400	.006	↓	1.70	.32	.03	2.05	2.45

05 35 Raceway Decking Assemblies

05 35 13 – Steel Cellular Decking

05 35 13.50 Cellular Decking			Crew	Daily Output	Labor-Hours	Unit	Material	Labor	Equipment	Total	Total Incl O&P
0010	**CELLULAR DECKING**										
0015	Made from recycled materials										
0200	Cellular units, galv, 1-1/2" deep, Type BC, 20-20 ga., over 15 squares	G	E-4	1460	.022	S.F.	8.10	1.17	.10	9.37	11.05
0250	18-20 ga.	G		1420	.023		9.20	1.20	.10	10.50	12.30
0300	18-18 ga.	G		1390	.023		9.40	1.22	.11	10.73	12.60
0320	16-18 ga.	G		1360	.024		11.25	1.25	.11	12.61	14.65
0340	16-16 ga.	G		1330	.024		12.50	1.28	.11	13.89	16.10
0400	3" deep, Type NC, galvanized, 20-20 ga.	G		1375	.023		8.90	1.24	.11	10.25	12.05
0500	18-20 ga.	G		1350	.024		10.75	1.26	.11	12.12	14.10
0600	18-18 ga.	G		1290	.025		10.70	1.32	.11	12.13	14.15
0700	16-18 ga.	G		1230	.026		12.05	1.38	.12	13.55	15.85
0800	16-16 ga.	G		1150	.028		13.15	1.48	.13	14.76	17.15
1000	4-1/2" deep, Type JC, galvanized, 18-20 ga.	G		1100	.029		12.40	1.55	.13	14.08	16.50
1100	18-18 ga.	G		1040	.031		12.30	1.64	.14	14.08	16.55
1200	16-18 ga.	G		980	.033		13.90	1.74	.15	15.79	18.40
1300	16-16 ga.	G	↓	935	.034	↓	15.10	1.82	.16	17.08	20
1900	For multi-story or congested site, add							50%			

Division Notes

		CREW	DAILY OUTPUT	LABOR-HOURS	UNIT	BARE COSTS MAT.	LABOR	EQUIP.	TOTAL	TOTAL INCL O&P

Estimating Tips

06 05 00 Common Work Results for Wood, Plastics, and Composites

- Common to any wood-framed structure are the accessory connector items such as screws, nails, adhesives, hangers, connector plates, straps, angles, and hold-downs. For typical wood-framed buildings, such as residential projects, the aggregate total for these items can be significant, especially in areas where seismic loading is a concern. For floor and wall framing, the material cost is based on 10 to 25 lbs. per MBF. Hold-downs, hangers, and other connectors should be taken off by the piece.

  Included with material costs are fasteners for a normal installation. RSMeans engineers use manufacturer's recommendations, written specifications, and/or standard construction practice for size and spacing of fasteners. Prices for various fasteners are shown for informational purposes only. Adjustments should be made if unusual fastening conditions exist.

06 10 00 Carpentry

- Lumber is a traded commodity and therefore sensitive to supply and demand in the marketplace. Even in "budgetary" estimating of wood-framed projects, it is advisable to call local suppliers for the latest market pricing.
- Common quantity units for wood-framed projects are "thousand board feet" (MBF). A board foot is a volume of wood, 1" x 1' x 1', or 144 cubic inches. Board-foot quantities are generally calculated using nominal material dimensions— dressed sizes are ignored. Board foot per lineal foot of any stick of lumber can be calculated by dividing the nominal cross-sectional area by 12. As an example, 2,000 lineal feet of 2 x 12 equates to 4 MBF by dividing the nominal area, 2 x 12, by 12, which equals 2, and multiplying by 2,000 to give 4,000 board feet. This simple rule applies to all nominal dimensioned lumber.
- Waste is an issue of concern at the quantity takeoff for any area of construction. Framing lumber is sold in even foot lengths, i.e., 10', 12', 14', 16' and, depending on spans, wall heights, and the grade of lumber, waste is inevitable. A rule of thumb for lumber waste is 5%–10% depending on material quality and the complexity of the framing.
- Wood in various forms and shapes is used in many projects, even where the main structural framing is steel, concrete, or masonry. Plywood as a back-up partition material and 2x boards used as blocking and cant strips around roof edges are two common examples. The estimator should ensure that the costs of all wood materials are included in the final estimate.

06 20 00 Finish Carpentry

- It is necessary to consider the grade of workmanship when estimating labor costs for erecting millwork and interior finish. In practice, there are three grades: premium, custom, and economy. The RSMeans daily output for base and case moldings is in the range of 200 to 250 L.F. per carpenter per day. This is appropriate for most average custom-grade projects. For premium projects, an adjustment to productivity of 25%–50% should be made, depending on the complexity of the job.

Reference Numbers

Reference numbers are shown in shaded boxes at the beginning of some major classifications. These numbers refer to related items in the Reference Section. The reference information may be an estimating procedure, an alternate pricing method, or technical information.

Note: Not all subdivisions listed here necessarily appear in this publication. ■

06 16 Sheathing

06 16 36 – Wood Panel Product Sheathing

06 16 36.10 Sheathing		Crew	Daily Output	Labor-Hours	Unit	2015 Bare Costs Material	Labor	Equipment	Total	Total Incl O&P
0010	**SHEATHING**									
0012	Plywood on roofs, CDX									
0030	5/16" thick	2 Carp	1600	.010	S.F.	.56	.47		1.03	1.33
0035	Pneumatic nailed		1952	.008		.56	.39		.95	1.20
0050	3/8" thick		1525	.010		.60	.49		1.09	1.42
0055	Pneumatic nailed		1860	.009		.60	.40		1	1.28
0100	1/2" thick		1400	.011		.66	.54		1.20	1.55
0105	Pneumatic nailed		1708	.009		.66	.44		1.10	1.40
0200	5/8" thick		1300	.012		.78	.58		1.36	1.75
0205	Pneumatic nailed		1586	.010		.78	.47		1.25	1.59
0300	3/4" thick		1200	.013		.93	.63		1.56	1.98
0305	Pneumatic nailed		1464	.011		.93	.51		1.44	1.81
0500	Plywood on walls, with exterior CDX, 3/8" thick		1200	.013		.60	.63		1.23	1.62
0505	Pneumatic nailed		1488	.011		.60	.50		1.10	1.44
0600	1/2" thick		1125	.014		.66	.67		1.33	1.75
0605	Pneumatic nailed		1395	.011		.66	.54		1.20	1.55
0700	5/8" thick		1050	.015		.78	.72		1.50	1.96
0705	Pneumatic nailed		1302	.012		.78	.58		1.36	1.75
0800	3/4" thick		975	.016		.93	.77		1.70	2.21
0805	Pneumatic nailed		1209	.013		.93	.62		1.55	1.98

Estimating Tips

07 10 00 Dampproofing and Waterproofing

- Be sure of the job specifications before pricing this subdivision. The difference in cost between waterproofing and dampproofing can be great. Waterproofing will hold back standing water. Dampproofing prevents the transmission of water vapor. Also included in this section are vapor retarding membranes.

07 20 00 Thermal Protection

- Insulation and fireproofing products are measured by area, thickness, volume or R-value. Specifications may give only what the specific R-value should be in a certain situation. The estimator may need to choose the type of insulation to meet that R-value.

07 30 00 Steep Slope Roofing
07 40 00 Roofing and Siding Panels

- Many roofing and siding products are bought and sold by the square. One square is equal to an area that measures 100 square feet.

  This simple change in unit of measure could create a large error if the estimator is not observant. Accessories necessary for a complete installation must be figured into any calculations for both material and labor.

07 50 00 Membrane Roofing
07 60 00 Flashing and Sheet Metal
07 70 00 Roofing and Wall Specialties and Accessories

- The items in these subdivisions compose a roofing system. No one component completes the installation, and all must be estimated. Built-up or single-ply membrane roofing systems are made up of many products and installation trades. Wood blocking at roof perimeters or penetrations, parapet coverings, reglets, roof drains, gutters, downspouts, sheet metal flashing, skylights, smoke vents, and roof hatches all need to be considered along with the roofing material. Several different installation trades will need to work together on the roofing system. Inherent difficulties in the scheduling and coordination of various trades must be accounted for when estimating labor costs.

07 90 00 Joint Protection

- To complete the weather-tight shell, the sealants and caulkings must be estimated. Where different materials meet—at expansion joints, at flashing penetrations, and at hundreds of other locations throughout a construction project—they provide another line of defense against water penetration. Often, an entire system is based on the proper location and placement of caulking or sealants. The detailed drawings that are included as part of a set of architectural plans show typical locations for these materials. When caulking or sealants are shown at typical locations, this means the estimator must include them for all the locations where this detail is applicable. Be careful to keep different types of sealants separate, and remember to consider backer rods and primers if necessary.

Reference Numbers

Reference numbers are shown in shaded boxes at the beginning of some major classifications. These numbers refer to related items in the Reference Section. The reference information may be an estimating procedure, an alternate pricing method, or technical information.

Note: Not all subdivisions listed here necessarily appear in this publication. ■

07 84 Firestopping

07 84 13 – Penetration Firestopping

	07 84 13.10 Firestopping	Crew	Daily Output	Labor-Hours	Unit	2015 Bare Costs Material	Labor	Equipment	Total	Total Incl O&P
0010	**FIRESTOPPING** R078413-30									
0100	Metallic piping, non insulated									
0110	Through walls, 2″ diameter	1 Carp	16	.500	Ea.	17.20	23.50		40.70	55
0120	4″ diameter		14	.571		26	27		53	70.50
0130	6″ diameter		12	.667		35.50	31.50		67	87
0140	12″ diameter		10	.800		62.50	37.50		100	127
0150	Through floors, 2″ diameter		32	.250		9.80	11.75		21.55	29
0160	4″ diameter		28	.286		14.20	13.40		27.60	36
0170	6″ diameter		24	.333		18.50	15.65		34.15	44.50
0180	12″ diameter		20	.400		31.50	18.80		50.30	63.50
0190	Metallic piping, insulated									
0200	Through walls, 2″ diameter	1 Carp	16	.500	Ea.	24	23.50		47.50	62.50
0210	4″ diameter		14	.571		33	27		60	77.50
0220	6″ diameter		12	.667		42	31.50		73.50	94
0230	12″ diameter		10	.800		68.50	37.50		106	134
0240	Through floors, 2″ diameter		32	.250		16.60	11.75		28.35	36.50
0250	4″ diameter		28	.286		21	13.40		34.40	43.50
0260	6″ diameter		24	.333		25.50	15.65		41.15	52
0270	12″ diameter		20	.400		31.50	18.80		50.30	63.50
0280	Non metallic piping, non insulated									
0290	Through walls, 2″ diameter	1 Carp	12	.667	Ea.	68.50	31.50		100	124
0300	4″ diameter		10	.800		86	37.50		123.50	153
0310	6″ diameter		8	1		120	47		167	205
0330	Through floors, 2″ diameter		16	.500		54	23.50		77.50	95.50
0340	4″ diameter		6	1.333		66.50	62.50		129	170
0350	6″ diameter		6	1.333		80.50	62.50		143	185
0370	Ductwork, insulated & non insulated, round									
0380	Through walls, 6″ diameter	1 Carp	12	.667	Ea.	35	31.50		66.50	86.50
0390	12″ diameter		10	.800		69.50	37.50		107	135
0400	18″ diameter		8	1		113	47		160	198
0410	Through floors, 6″ diameter		16	.500		18.60	23.50		42.10	56.50
0420	12″ diameter		14	.571		34	27		61	79
0430	18″ diameter		12	.667		59	31.50		90.50	113
0440	Ductwork, insulated & non insulated, rectangular									
0450	With stiffener/closure angle, through walls, 6″ x 12″	1 Carp	8	1	Ea.	28	47		75	104
0460	12″ x 24″		6	1.333		37.50	62.50		100	138
0470	24″ x 48″		4	2		107	94		201	262
0480	With stiffener/closure angle, through floors, 6″ x 12″		10	.800		15.40	37.50		52.90	75
0490	12″ x 24″		8	1		28	47		75	103
0500	24″ x 48″		6	1.333		54	62.50		116.50	156
0510	Multi trade openings									
0520	Through walls, 6″ x 12″	1 Carp	2	4	Ea.	59	188		247	355
0530	12″ x 24″	″	1	8		238	375		613	840
0540	24″ x 48″	2 Carp	1	16		950	750		1,700	2,200
0550	48″ x 96″	″	.75	21.333		3,700	1,000		4,700	5,625
0560	Through floors, 6″ x 12″	1 Carp	2	4		39	188		227	330
0570	12″ x 24″	″	1	8		157	375		532	755
0580	24″ x 48″	2 Carp	.75	21.333		625	1,000		1,625	2,250
0590	48″ x 96″	″	.50	32		2,450	1,500		3,950	4,975
0600	Structural penetrations, through walls									
0610	Steel beams, W8 x 10	1 Carp	8	1	Ea.	37.50	47		84.50	114
0620	W12 x 14		6	1.333		59	62.50		121.50	162
0630	W21 x 44		5	1.600		118	75		193	246

07 84 Firestopping

07 84 13 – Penetration Firestopping

07 84 13.10 Firestopping		Crew	Daily Output	Labor-Hours	Unit	2015 Bare Costs Material	Labor	Equipment	Total	Total Incl O&P
0640	W36 x 135	1 Carp	3	2.667	Ea.	287	125		412	510
0650	Bar joists, 18" deep		6	1.333		54.50	62.50		117	156
0660	24" deep		6	1.333		67.50	62.50		130	171
0670	36" deep		5	1.600		101	75		176	228
0680	48" deep	↓	4	2	↓	118	94		212	275
0690	Construction joints, floor slab at exterior wall									
0700	Precast, brick, block or drywall exterior									
0710	2" wide joint	1 Carp	125	.064	L.F.	8.50	3		11.50	13.95
0720	4" wide joint	"	75	.107	"	17	5		22	26.50
0730	Metal panel, glass or curtain wall exterior									
0740	2" wide joint	1 Carp	40	.200	L.F.	20	9.40		29.40	36.50
0750	4" wide joint	"	25	.320	"	28	15		43	53.50
0760	Floor slab to drywall partition									
0770	Flat joint	1 Carp	100	.080	L.F.	8.40	3.76		12.16	15.05
0780	Fluted joint		50	.160		17.40	7.50		24.90	30.50
0790	Etched fluted joint	↓	75	.107	↓	11.10	5		16.10	19.90
0800	Floor slab to concrete/masonry partition									
0810	Flat joint	1 Carp	75	.107	L.F.	18.75	5		23.75	28
0820	Fluted joint	"	50	.160	"	22.50	7.50		30	36
0830	Concrete/CMU wall joints									
0840	1" wide	1 Carp	100	.080	L.F.	10.25	3.76		14.01	17.10
0850	2" wide		75	.107		18.75	5		23.75	28
0860	4" wide	↓	50	.160	↓	38	7.50		45.50	53
0870	Concrete/CMU floor joints									
0880	1" wide	1 Carp	200	.040	L.F.	5.15	1.88		7.03	8.55
0890	2" wide		150	.053		9.40	2.50		11.90	14.20
0900	4" wide	↓	100	.080	↓	17.90	3.76		21.66	25.50

07 91 Preformed Joint Seals

07 91 13 – Compression Seals

07 91 13.10 Compression Seals		Crew	Daily Output	Labor-Hours	Unit	Material	Labor	Equipment	Total	Total Incl O&P
0010	**COMPRESSION SEALS**									
4900	O-ring type cord, 1/4"	1 Bric	472	.017	L.F.	.37	.78		1.15	1.60
4910	1/2"		440	.018		.95	.84		1.79	2.33
4920	3/4"		424	.019		1.80	.87		2.67	3.31
4930	1"		408	.020		3.40	.91		4.31	5.10
4940	1-1/4"		384	.021		6.30	.96		7.26	8.40
4950	1-1/2"		368	.022		7.95	1		8.95	10.30
4960	1-3/4"		352	.023		13.25	1.05		14.30	16.20
4970	2"	↓	344	.023	↓	18.50	1.07		19.57	22

07 91 23 – Backer Rods

07 91 23.10 Backer Rods		Crew	Daily Output	Labor-Hours	Unit	Material	Labor	Equipment	Total	Total Incl O&P
0010	**BACKER RODS**									
0030	Backer rod, polyethylene, 1/4" diameter	1 Bric	4.60	1.739	C.L.F.	2.13	80.50		82.63	125
0050	1/2" diameter		4.60	1.739		3.40	80.50		83.90	127
0070	3/4" diameter		4.60	1.739		5.55	80.50		86.05	129
0090	1" diameter	↓	4.60	1.739	↓	8.85	80.50		89.35	133

07 91 Preformed Joint Seals

07 91 26 – Joint Fillers

07 91 26.10 Joint Fillers		Crew	Daily Output	Labor-Hours	Unit	2015 Bare Costs Material	Labor	Equipment	Total	Total Incl O&P
0010	JOINT FILLERS									
4360	Butyl rubber filler, 1/4" x 1/4"	1 Bric	290	.028	L.F.	.22	1.27		1.49	2.19
4365	1/2" x 1/2"		250	.032		.89	1.48		2.37	3.23
4370	1/2" x 3/4"		210	.038		1.34	1.76		3.10	4.15
4375	3/4" x 3/4"		230	.035		2.01	1.61		3.62	4.66
4380	1" x 1"		180	.044		2.68	2.05		4.73	6.05
4390	For coloring, add					12%				
4980	Polyethylene joint backing, 1/4" x 2"	1 Bric	2.08	3.846	C.L.F.	12	178		190	284
4990	1/4" x 6"		1.28	6.250	"	28	288		316	470
5600	Silicone, room temp vulcanizing foam seal, 1/4" x 1/2"		1312	.006	L.F.	.33	.28		.61	.80
5610	1/2" x 1/2"		656	.012		.67	.56		1.23	1.59
5620	1/2" x 3/4"		442	.018		1	.84		1.84	2.38
5630	3/4" x 3/4"		328	.024		1.50	1.13		2.63	3.37
5640	1/8" x 1"		1312	.006		.33	.28		.61	.80
5650	1/8" x 3"		442	.018		1	.84		1.84	2.38
5670	1/4" x 3"		295	.027		2	1.25		3.25	4.11
5680	1/4" x 6"		148	.054		3.99	2.49		6.48	8.20
5690	1/2" x 6"		82	.098		8	4.50		12.50	15.65
5700	1/2" x 9"		52.50	.152		12	7.05		19.05	24
5710	1/2" x 12"		33	.242		15.95	11.20		27.15	34.50

07 92 Joint Sealants

07 92 13 – Elastomeric Joint Sealants

07 92 13.20 Caulking and Sealant Options		Crew	Daily Output	Labor-Hours	Unit	Material	Labor	Equipment	Total	Total Incl O&P
0010	CAULKING AND SEALANT OPTIONS									
0050	Latex acrylic based, bulk				Gal.	26.50			26.50	29.50
0055	Bulk in place 1/4" x 1/4" bead	1 Bric	300	.027	L.F.	.08	1.23		1.31	1.97
0060	1/4" x 3/8"		294	.027		.14	1.26		1.40	2.07
0065	1/4" x 1/2"		288	.028		.19	1.28		1.47	2.17
0075	3/8" x 3/8"		284	.028		.21	1.30		1.51	2.21
0080	3/8" x 1/2"		280	.029		.28	1.32		1.60	2.32
0085	3/8" x 5/8"		276	.029		.35	1.34		1.69	2.43
0095	3/8" x 3/4"		272	.029		.42	1.36		1.78	2.53
0100	1/2" x 1/2"		275	.029		.37	1.34		1.71	2.46
0105	1/2" x 5/8"		269	.030		.47	1.37		1.84	2.61
0110	1/2" x 3/4"		263	.030		.56	1.40		1.96	2.76
0115	1/2" x 7/8"		256	.031		.65	1.44		2.09	2.92
0120	1/2" x 1"		250	.032		.75	1.48		2.23	3.07
0125	3/4" x 3/4"		244	.033		.84	1.51		2.35	3.24
0130	3/4" x 1"		225	.036		1.12	1.64		2.76	3.74
0135	1" x 1"		200	.040		1.50	1.85		3.35	4.47
0190	Cartridges				Gal.	32			32	35.50
0200	11 fl. oz. cartridge				Ea.	2.76			2.76	3.04
0600	1/2" x 1/2"	1 Bric	275	.029	L.F.	.45	1.34		1.79	2.55
3200	Polyurethane, 1 or 2 component				Gal.	49			49	54
3300	Cartridges				"	64			64	70.50
3500	Bulk, in place, 1/4" x 1/4"	1 Bric	300	.027	L.F.	.16	1.23		1.39	2.06
3655	1/2" x 1/4"		288	.028		.32	1.28		1.60	2.31
3800	3/4" x 3/8"		272	.029		.72	1.36		2.08	2.87
3900	1" x 1/2"		250	.032		1.28	1.48		2.76	3.66

07 92 Joint Sealants

07 92 19 – Acoustical Joint Sealants

	07 92 19.10 Acoustical Sealant	Crew	Daily Output	Labor-Hours	Unit	Material	Labor	Equipment	Total	Total Incl O&P
						2015 Bare Costs				
0010	**ACOUSTICAL SEALANT**									
0020	Acoustical sealant, elastomeric, cartridges				Ea.	8.50			8.50	9.35
0025	In place, 1/4" x 1/4"	1 Bric	300	.027	L.F.	.35	1.23		1.58	2.26
0030	1/4" x 1/2"		288	.028		.69	1.28		1.97	2.72
0035	1/2" x 1/2"		275	.029		1.39	1.34		2.73	3.57
0040	1/2" x 3/4"		263	.030		2.08	1.40		3.48	4.43
0045	3/4" x 3/4"		244	.033		3.12	1.51		4.63	5.75
0050	1" x 1"	▼	200	.040	▼	5.55	1.85		7.40	8.90

Division Notes

	CREW	DAILY OUTPUT	LABOR-HOURS	UNIT	BARE COSTS MAT.	BARE COSTS LABOR	BARE COSTS EQUIP.	BARE COSTS TOTAL	TOTAL INCL O&P

Division 8 – Openings

Estimating Tips

08 10 00 Doors and Frames

All exterior doors should be addressed for their energy conservation (insulation and seals).

- Most metal doors and frames look alike, but there may be significant differences among them. When estimating these items, be sure to choose the line item that most closely compares to the specification or door schedule requirements regarding:
  - type of metal
  - metal gauge
  - door core material
  - fire rating
  - finish
- Wood and plastic doors vary considerably in price. The primary determinant is the veneer material. Lauan, birch, and oak are the most common veneers. Other variables include the following:
  - hollow or solid core
  - fire rating
  - flush or raised panel
  - finish
- Door pricing includes bore for cylindrical lockset and mortise for hinges.

08 30 00 Specialty Doors and Frames

- There are many varieties of special doors, and they are usually priced per each. Add frames, hardware, or operators required for a complete installation.

08 40 00 Entrances, Storefronts, and Curtain Walls

- Glazed curtain walls consist of the metal tube framing and the glazing material. The cost data in this subdivision is presented for the metal tube framing alone or the composite wall. If your estimate requires a detailed takeoff of the framing, be sure to add the glazing cost and any tints.

08 50 00 Windows

- Most metal windows are delivered preglazed. However, some metal windows are priced without glass. Refer to 08 80 00 Glazing for glass pricing. The grade C indicates commercial grade windows, usually ASTM C-35.
- All wood windows and vinyl are priced preglazed. The glazing is insulating glass. Add the cost of screens and grills if required, and not already included.

08 70 00 Hardware

- Hardware costs add considerably to the cost of a door. The most efficient method to determine the hardware requirements for a project is to review the door and hardware schedule together. One type of door may have different hardware, depending on the door usage.
- Door hinges are priced by the pair, with most doors requiring 1-1/2 pairs per door. The hinge prices do not include installation labor, because it is included in door installation. Hinges are classified according to the frequency of use, base material, and finish.

08 80 00 Glazing

- Different openings require different types of glass. The most common types are:
  - float
  - tempered
  - insulating
  - impact-resistant
  - ballistic-resistant
- Most exterior windows are glazed with insulating glass. Entrance doors and window walls, where the glass is less than 18" from the floor, are generally glazed with tempered glass. Interior windows and some residential windows are glazed with float glass.
- Coastal communities require the use of impact-resistant glass, dependant on wind speed.
- The insulation or 'u' value is a strong consideration, along with solar heat gain, to determine total energy efficiency.

Reference Numbers

Reference numbers are shown in shaded boxes at the beginning of some major classifications. These numbers refer to related items in the Reference Section. The reference information may be an estimating procedure, an alternate pricing method, or technical information.

Note: Not all subdivisions listed here necessarily appear in this publication. ■

08 31 Access Doors and Panels

08 31 13 – Access Doors and Frames

08 31 13.10 Types of Framed Access Doors		Crew	Daily Output	Labor-Hours	Unit	2015 Bare Costs Material	Labor	Equipment	Total	Total Incl O&P
0010	**TYPES OF FRAMED ACCESS DOORS**									
4000	Recessed door for drywall									
4100	Metal 12″ x 12″	1 Carp	6	1.333	Ea.	80	62.50		142.50	185
4150	12″ x 24″		5.50	1.455		114	68.50		182.50	230
4200	24″ x 36″		5	1.600		182	75		257	315

08 74 Access Control Hardware

08 74 13 – Card Key Access Control Hardware

08 74 13.50 Card Key Access

Code	Description	Crew	Daily Output	Labor-Hours	Unit	Material	Labor	Equipment	Total	Total Incl O&P
0010	**CARD KEY ACCESS**									
0020	Computerized system , processor, proximity reader and cards									
0030	Does not inculde door hardware, lockset or wiring									
0040	Card key system for 1 door				Ea.	1,225			1,225	1,350
0060	Card key system for 2 doors					2,125			2,125	2,350
0080	Card key system for 4 doors					2,650			2,650	2,900
0100	Processor for card key access system					850			850	935
0160	Magnetic lock for electric access, 600 Pound holding force					190			190	209
0170	Magnetic lock for electric access, 1200 Pound holding force					190			190	209
0200	Proximity card reader					130			130	143

08 74 13.60 Entrance Card Systems

Code	Description	Crew	Daily Output	Labor-Hours	Unit	Material	Labor	Equipment	Total	Total Incl O&P
0010	**ENTRANCE CARD SYSTEMS**									
0100	Entrance card, barium ferrite				Ea.	4.50			4.50	4.95
0120	Credential					6			6	6.60
0140	Proximity					8.50			8.50	9.35
0160	Weigand					9.75			9.75	10.75
0500	Entrance card reader, barium ferrite	R-19	4	5		295	274		569	735
0520	Credential		4	5		235	274		509	670
0540	Proximity		4	5		315	274		589	760
0560	Weigand		4	5		415	274		689	865
0600	Local processor for card system		4	5		1,750	274		2,024	2,325
0650	Scanner, eye retina		4	5		6,000	274		6,274	7,000
0700	Gate opener, cantilever	R-18	3	8.667		2,350	370		2,720	3,150
0710	Light duty		3	8.667		1,500	370		1,870	2,225
0712	Custom		3	8.667		2,000	370		2,370	2,775
0750	Switch, tamper	R-19	6	3.333		22	183		205	298
0900	Accessories, electric door strike/bolt		5	4		77.50	219		296.50	415
0920	Electromagnetic lock		5	4		169	219		388	515
0940	Keypad for card reader		4	5		580	274		854	1,050

08 74 16 – Keypad Access Control Hardware

08 74 16.50 Keypad Access

Code	Description	Crew	Daily Output	Labor-Hours	Unit	Material	Labor	Equipment	Total	Total Incl O&P
0010	**KEYPAD ACCESS**									
0340	Digital keypad, int/ext, basic, excl. striker/power/wiring	1 Elec	3	2.667	Ea.	109	146		255	340
0350	Lockset, mechanical push-button type, complete, incl hardware	1 Carp	4	2	"	380	94		474	560

08 74 19 – Biometric Identity Access Control Hardware

08 74 19.50 Biometric Identity Access

Code	Description	Crew	Daily Output	Labor-Hours	Unit	Material	Labor	Equipment	Total	Total Incl O&P
0010	**BIOMETRIC IDENTITY ACCESS**									
0200	Fingerprint scanner unit, excl striker/power supply	1 Elec	4	2	Ea.	1,400	109		1,509	1,725
0210	Fingerprint scanner unit, for computer keyboard access		8	1		1,100	54.50		1,154.50	1,275
0220	Hand geometry scanner, mem of 512 users, excl striker/power		3	2.667		2,100	146		2,246	2,525
0230	Memory upgrade for, adds 9,700 user profiles		8	1		300	54.50		354.50	410

08 74 Access Control Hardware

08 74 19 – Biometric Identity Access Control Hardware

08 74 19.50 Biometric Identity Access		Crew	Daily Output	Labor-Hours	Unit	Material	Labor	Equipment	Total	Total Incl O&P
						2015 Bare Costs				
0240	Adds 32,500 user profiles	1 Elec	8	1	Ea.	600	54.50		654.50	740
0250	Prison type, memory of 256 users, excl striker, power		3	2.667		2,600	146		2,746	3,075
0260	Memory upgrade for, adds 3,300 user profiles		8	1		250	54.50		304.50	355
0270	Adds 9,700 user profiles		8	1		460	54.50		514.50	585
0280	Adds 27,900 user profiles		8	1		610	54.50		664.50	750
0290	All weather, mem of 512 users, excl striker/power		3	2.667		3,900	146		4,046	4,525
0300	Facial & fingerprint scanner, combination unit, excl striker/power		3	2.667		4,300	146		4,446	4,950
0310	Access for, for initial setup, excl striker/power		3	2.667		1,100	146		1,246	1,425

08 74 23 – Access Control Accessories

08 74 23.50 Security Access Control Accessories		Crew	Daily Output	Labor-Hours	Unit	Material	Labor	Equipment	Total	Total Incl O&P
0010	**SECURITY ACCESS CONTROL ACCESSORIES**									
0360	Scanner/reader access, power supply/transf, 110 V to 12/24 V	1 Elec	4	2	Ea.	250	109		359	440
0370	Elec/mag strikers, 12/24V		4	2		85	109		194	258
0380	1 hour battery backup power supply		4	2		50	109		159	219
0390	Deadbolt, digital, batt-operated, indoor/outdoor, complete, incl hardware	1 Carp	4	2		250	94		344	420

Division Notes

		CREW	DAILY OUTPUT	LABOR-HOURS	UNIT	BARE COSTS MAT.	LABOR	EQUIP.	TOTAL	TOTAL INCL O&P

Estimating Tips

General

- Room Finish Schedule: A complete set of plans should contain a room finish schedule. If one is not available, it would be well worth the time and effort to obtain one.

09 20 00 Plaster and Gypsum Board

- Lath is estimated by the square yard plus a 5% allowance for waste. Furring, channels, and accessories are measured by the linear foot. An extra foot should be allowed for each accessory miter or stop.
- Plaster is also estimated by the square yard. Deductions for openings vary by preference, from zero deduction to 50% of all openings over 2 feet in width. The estimator should allow one extra square foot for each linear foot of horizontal interior or exterior angle located below the ceiling level. Also, double the areas of small radius work.
- Drywall accessories, studs, track, and acoustical caulking are all measured by the linear foot. Drywall taping is figured by the square foot. Gypsum wallboard is estimated by the square foot. No material deductions should be made for door or window openings under 32 S.F.

09 60 00 Flooring

- Tile and terrazzo areas are taken off on a square foot basis. Trim and base materials are measured by the linear foot. Accent tiles are listed per each. Two basic methods of installation are used. Mud set is approximately 30% more expensive than thin set. In terrazzo work, be sure to include the linear footage of embedded decorative strips, grounds, machine rubbing, and power cleanup.
- Wood flooring is available in strip, parquet, or block configuration. The latter two types are set in adhesives with quantities estimated by the square foot. The laying pattern will influence labor costs and material waste. In addition to the material and labor for laying wood floors, the estimator must make allowances for sanding and finishing these areas, unless the flooring is prefinished.
- Sheet flooring is measured by the square yard. Roll widths vary, so consideration should be given to use the most economical width, as waste must be figured into the total quantity. Consider also the installation methods available, direct glue down or stretched.

09 70 00 Wall Finishes

- Wall coverings are estimated by the square foot. The area to be covered is measured, length by height of wall above baseboards, to calculate the square footage of each wall. This figure is divided by the number of square feet in the single roll which is being used. Deduct, in full, the areas of openings such as doors and windows. Where a pattern match is required allow 25%–30% waste.

09 80 00 Acoustic Treatment

- Acoustical systems fall into several categories. The takeoff of these materials should be by the square foot of area with a 5% allowance for waste. Do not forget about scaffolding, if applicable, when estimating these systems.

09 90 00 Painting and Coating

- A major portion of the work in painting involves surface preparation. Be sure to include cleaning, sanding, filling, and masking costs in the estimate.
- Protection of adjacent surfaces is not included in painting costs. When considering the method of paint application, an important factor is the amount of protection and masking required. These must be estimated separately and may be the determining factor in choosing the method of application.

Reference Numbers

Reference numbers are shown in shaded boxes at the beginning of some major classifications. These numbers refer to related items in the Reference Section. The reference information may be an estimating procedure, an alternate pricing method, or technical information.

Note: Not all subdivisions listed here necessarily appear in this publication. ■

09 22 Supports for Plaster and Gypsum Board

09 22 03 – Fastening Methods for Finishes

09 22 03.20 Drilling Plaster/Drywall		Crew	Daily Output	Labor-Hours	Unit	Material	2015 Bare Costs Labor	Equipment	Total	Total Incl O&P
0010	**DRILLING PLASTER/DRYWALL**									
1100	Drilling & layout for drywall/plaster walls, up to 1" deep, no anchor									
1200	Holes, 1/4" diameter	1 Carp	150	.053	Ea.	.01	2.50		2.51	3.86
1300	3/8" diameter		140	.057		.01	2.68		2.69	4.14
1400	1/2" diameter		130	.062		.01	2.89		2.90	4.46
1500	3/4" diameter		120	.067		.01	3.13		3.14	4.84
1600	1" diameter		110	.073		.02	3.41		3.43	5.25
1700	1-1/4" diameter		100	.080		.03	3.76		3.79	5.85
1800	1-1/2" diameter		90	.089		.05	4.17		4.22	6.45
1900	For ceiling installations, add						40%			

09 69 Access Flooring

09 69 13 – Rigid-Grid Access Flooring

09 69 13.10 Access Floors		Crew	Daily Output	Labor-Hours	Unit	Material	Labor	Equipment	Total	Total Incl O&P
0010	**ACCESS FLOORS**									
0015	Access floor package including panel, pedestal, stringers & laminate cover									
0100	Computer room, greater than 6,000 S.F.	4 Carp	750	.043	S.F.	7.50	2		9.50	11.35
0110	Less than 6,000 S.F.	2 Carp	375	.043		8.15	2		10.15	12.10
0120	Office, greater than 6,000 S.F.	4 Carp	1050	.030		5.35	1.43		6.78	8.10
0250	Panels, particle board or steel, 1250# load, no covering, under 6,000 S.F.	2 Carp	600	.027		3.75	1.25		5	6.05
0300	Over 6,000 S.F.		640	.025		3.21	1.17		4.38	5.35
0400	Aluminum, 24" panels		500	.032		33	1.50		34.50	38.50
0600	For carpet covering, add					8.75			8.75	9.65
0700	For vinyl floor covering, add					9.05			9.05	9.95
0900	For high pressure laminate covering, add					7.50			7.50	8.25
0910	For snap on stringer system, add	2 Carp	1000	.016		1.56	.75		2.31	2.88
0950	Office applications, steel or concrete panels,									
0960	no covering, over 6,000 S.F.	2 Carp	960	.017	S.F.	10.55	.78		11.33	12.80
1000	Machine cutouts after initial installation	1 Carp	50	.160	Ea.	20	7.50		27.50	33.50
1050	Pedestals, 6" to 12"	2 Carp	85	.188		8.40	8.85		17.25	23
1100	Air conditioning grilles, 4" x 12"	1 Carp	17	.471		68.50	22		90.50	109
1150	4" x 18"	"	14	.571		93.50	27		120.50	145
1200	Approach ramps, steel	2 Carp	60	.267	S.F.	24.50	12.50		37	46.50
1300	Aluminum	"	40	.400	"	34	18.80		52.80	66.50
1500	Handrail, 2 rail, aluminum	1 Carp	15	.533	L.F.	109	25		134	158

Estimating Tips

General

- The items in this division are usually priced per square foot or each.
- Many items in Division 10 require some type of support system or special anchors that are not usually furnished with the item. The required anchors must be added to the estimate in the appropriate division.
- Some items in Division 10, such as lockers, may require assembly before installation. Verify the amount of assembly required. Assembly can often exceed installation time.

10 20 00 Interior Specialties

- Support angles and blocking are not included in the installation of toilet compartments, shower/ dressing compartments, or cubicles. Appropriate line items from Divisions 5 or 6 may need to be added to support the installations.
- Toilet partitions are priced by the stall. A stall consists of a side wall, pilaster, and door with hardware. Toilet tissue holders and grab bars are extra.
- The required acoustical rating of a folding partition can have a significant impact on costs. Verify the sound transmission coefficient rating of the panel priced to the specification requirements.
- Grab bar installation does not include supplemental blocking or backing to support the required load. When grab bars are installed at an existing facility, provisions must be made to attach the grab bars to solid structure.

Reference Numbers

Reference numbers are shown in shaded boxes at the beginning of some major classifications. These numbers refer to related items in the Reference Section. The reference information may be an estimating procedure, an alternate pricing method, or technical information.

Note: Not all subdivisions listed here necessarily appear in this publication. ■

10 28 Toilet, Bath, and Laundry Accessories

10 28 16 – Bath Accessories

10 28 16.20 Medicine Cabinets		Crew	Daily Output	Labor-Hours	Unit	2015 Bare Costs Material	Labor	Equipment	Total	Total Incl O&P
0010	**MEDICINE CABINETS**									
0020	With mirror, sst frame, 16" x 22", unlighted	1 Carp	14	.571	Ea.	98	27		125	150
0100	Wood frame		14	.571		128	27		155	183
0300	Sliding mirror doors, 20" x 16" x 4-3/4", unlighted		7	1.143		124	53.50		177.50	219
0400	24" x 19" x 8-1/2", lighted		5	1.600		179	75		254	315
0600	Triple door, 30" x 32", unlighted, plywood body		7	1.143		325	53.50		378.50	445
0700	Steel body		7	1.143		375	53.50		428.50	495
0900	Oak door, wood body, beveled mirror, single door		7	1.143		199	53.50		252.50	300
1000	Double door		6	1.333		380	62.50		442.50	515
1200	Hotel cabinets, stainless, with lower shelf, unlighted		10	.800		200	37.50		237.50	277
1300	Lighted	↓	5	1.600	↓	305	75		380	450

Estimating Tips

General

- The items in this division are usually priced per square foot or each. Many of these items are purchased by the owner for installation by the contractor. Check the specifications for responsibilities and include time for receiving, storage, installation, and mechanical and electrical hookups in the appropriate divisions.
- Many items in Division 11 require some type of support system that is not usually furnished with the item. Examples of these systems include blocking for the attachment of casework and support angles for ceiling-hung projection screens. The required blocking or supports must be added to the estimate in the appropriate division.
- Some items in Division 11 may require assembly or electrical hookups. Verify the amount of assembly required or the need for a hard electrical connection and add the appropriate costs.

Reference Numbers

Reference numbers are shown in shaded boxes at the beginning of some major classifications. These numbers refer to related items in the Reference Section. The reference information may be an estimating procedure, an alternate pricing method, or technical information.

Note: Not all subdivisions listed here necessarily appear in this publication. ■

11 12 Parking Control Equipment

11 12 13 – Parking Key and Card Control Units

11 12 13.10 Parking Control Units		Crew	Daily Output	Labor-Hours	Unit	Material	2015 Bare Costs Labor	Equipment	Total	Total Incl O&P
0010	PARKING CONTROL UNITS									
5100	Card reader	1 Elec	2	4	Ea.	1,950	219		2,169	2,475
5120	Proximity with customer display	2 Elec	1	16		5,575	875		6,450	7,425
6000	Parking control software, basic functionality	1 Elec	.50	16		23,900	875		24,775	27,500
6020	multi-function	"	.20	40	↓	104,500	2,200		106,700	118,500

11 12 16 – Parking Ticket Dispensers

11 12 16.10 Ticket Dispensers		Crew	Daily Output	Labor-Hours	Unit	Material	Labor	Equipment	Total	Total Incl O&P
0010	TICKET DISPENSERS									
5900	Ticket spitter with time/date stamp, standard	2 Elec	2	8	Ea.	6,200	440		6,640	7,475
5920	Mag stripe encoding	"	2	8	"	18,800	440		19,240	21,400

11 12 26 – Parking Fee Collection Equipment

11 12 26.13 Parking Fee Coin Collection Equipment		Crew	Daily Output	Labor-Hours	Unit	Material	Labor	Equipment	Total	Total Incl O&P
0010	PARKING FEE COIN COLLECTION EQUIPMENT									
5200	Cashier booth, average	B-22	1	30	Ea.	10,400	1,325	207	11,932	13,800
5300	Collector station, pay on foot	2 Elec	.20	80		111,000	4,375		115,375	128,500
5320	Credit card only	"	.50	32	↓	20,500	1,750		22,250	25,100

11 12 26.23 Fee Equipment		Crew	Daily Output	Labor-Hours	Unit	Material	Labor	Equipment	Total	Total Incl O&P
0010	FEE EQUIPMENT									
5600	Fee computer	1 Elec	1.50	5.333	Ea.	14,200	292		14,492	16,000

11 12 33 – Parking Gates

11 12 33.13 Lift Arm Parking Gates		Crew	Daily Output	Labor-Hours	Unit	Material	Labor	Equipment	Total	Total Incl O&P
0010	LIFT ARM PARKING GATES									
5000	Barrier gate with programmable controller	2 Elec	3	5.333	Ea.	3,400	292		3,692	4,150
5020	Industrial		3	5.333		5,050	292		5,342	5,975
5500	Exit verifier	↓	1	16		17,600	875		18,475	20,700
5700	Full sign, 4" letters	1 Elec	2	4		1,225	219		1,444	1,675
5800	Inductive loop	2 Elec	4	4		171	219		390	520
5950	Vehicle detector, microprocessor based	1 Elec	3	2.667	↓	425	146		571	690

11 21 Retail and Service Equipment

11 21 73 – Commercial Laundry and Dry Cleaning Equipment

11 21 73.13 Dry Cleaning Equipment		Crew	Daily Output	Labor-Hours	Unit	Material	Labor	Equipment	Total	Total Incl O&P
0010	DRY CLEANING EQUIPMENT									
2000	Dry cleaners, electric, 20 lb. capacity, not incl. rough-in	L-1	.20	80	Ea.	33,700	4,525		38,225	43,800
2050	25 lb. capacity		.17	94.118		48,600	5,325		53,925	61,500
2100	30 lb. capacity		.15	106		51,000	6,050		57,050	65,500
2150	60 lb. capacity	↓	.09	177	↓	79,000	10,100		89,100	102,000

11 21 73.19 Finishing Equipment		Crew	Daily Output	Labor-Hours	Unit	Material	Labor	Equipment	Total	Total Incl O&P
0010	FINISHING EQUIPMENT									
3500	Folders, blankets & sheets, minimum	1 Elec	.17	47.059	Ea.	33,100	2,575		35,675	40,300
3700	King size with automatic stacker		.10	80		60,000	4,375		64,375	72,500
3800	For conveyor delivery, add	↓	.45	17.778	↓	14,500	970		15,470	17,400

11 21 73.23 Commercial Ironing Equipment		Crew	Daily Output	Labor-Hours	Unit	Material	Labor	Equipment	Total	Total Incl O&P
0010	COMMERCIAL IRONING EQUIPMENT									
4500	Ironers, institutional, 110", single roll	1 Elec	.20	40	Ea.	32,100	2,200		34,300	38,600

11 21 Retail and Service Equipment

11 21 73 – Commercial Laundry and Dry Cleaning Equipment

11 21 73.26 Commercial Washers and Extractors		Crew	Daily Output	Labor-Hours	Unit	2015 Bare Costs Material	Labor	Equipment	Total	Total Incl O&P
0010	**COMMERCIAL WASHERS AND EXTRACTORS**, not including rough-in									
6000	Combination washer/extractor, 20 lb. capacity	L-6	1.50	8	Ea.	5,950	460		6,410	7,225
6100	30 lb. capacity		.80	15		9,275	860		10,135	11,500
6200	50 lb. capacity		.68	17.647		10,900	1,000		11,900	13,500
6300	75 lb. capacity		.30	40		20,400	2,300		22,700	26,000
6350	125 lb. capacity		.16	75		27,700	4,300		32,000	36,900
11 21 73.33 Coin-Operated Laundry Equipment										
0010	**COIN-OPERATED LAUNDRY EQUIPMENT**									
5290	Clothes washer									
5300	Commercial, coin operated, average	1 Plum	3	2.667	Ea.	1,250	157		1,407	1,600

11 30 Residential Equipment

11 30 13 – Residential Appliances

11 30 13.15 Cooking Equipment		Crew	Daily Output	Labor-Hours	Unit	Material	Labor	Equipment	Total	Total Incl O&P
0010	**COOKING EQUIPMENT**									
0020	Cooking range, 30" free standing, 1 oven, minimum	2 Clab	10	1.600	Ea.	440	60		500	580
0050	Maximum		4	4		2,025	150		2,175	2,450
0700	Free-standing, 1 oven, 21" wide range, minimum		10	1.600		455	60		515	595
0750	21" wide, maximum		4	4		450	150		600	725
0900	Countertop cooktops, 4 burner, standard, minimum	1 Elec	6	1.333		296	73		369	435
0950	Maximum		3	2.667		1,300	146		1,446	1,675
1050	As above, but with grill and griddle attachment, minimum		6	1.333		1,250	73		1,323	1,475
1100	Maximum		3	2.667		3,700	146		3,846	4,300
1200	Induction cooktop, 30" wide		3	2.667		1,200	146		1,346	1,550
1250	Microwave oven, minimum		4	2		120	109		229	296
1300	Maximum		2	4		465	219		684	840
11 30 13.17 Kitchen Cleaning Equipment										
0010	**KITCHEN CLEANING EQUIPMENT**									
2750	Dishwasher, built-in, 2 cycles, minimum	L-1	4	4	Ea.	238	227		465	600
2800	Maximum		2	8		435	455		890	1,150
2950	4 or more cycles, minimum		4	4		375	227		602	755
2960	Average		4	4		500	227		727	890
3000	Maximum		2	8		1,100	455		1,555	1,875
11 30 13.18 Waste Disposal Equipment										
0010	**WASTE DISPOSAL EQUIPMENT**									
3300	Garbage disposal, sink type, minimum	L-1	10	1.600	Ea.	87.50	90.50		178	232
3350	Maximum	"	10	1.600	"	220	90.50		310.50	380
11 30 13.19 Kitchen Ventilation Equipment										
0010	**KITCHEN VENTILATION EQUIPMENT**									
4150	Hood for range, 2 speed, vented, 30" wide, minimum	L-3	5	3.200	Ea.	70.50	164		234.50	330
4200	Maximum		3	5.333		780	273		1,053	1,275
4300	42" wide, minimum		5	3.200		168	164		332	435
4330	Custom		5	3.200		1,650	164		1,814	2,075
4350	Maximum		3	5.333		2,025	273		2,298	2,650
4500	For ventless hood, 2 speed, add					18.65			18.65	20.50
4650	For vented 1 speed, deduct from maximum					50			50	55
11 30 13.24 Washers										
0010	**WASHERS**									
5000	Residential, 4 cycle, average	1 Plum	3	2.667	Ea.	875	157		1,032	1,200

11 30 Residential Equipment

11 30 15 – Miscellaneous Residential Appliances

11 30 15.23 Water Heaters		Crew	Daily Output	Labor-Hours	Unit	2015 Bare Costs Material	Labor	Equipment	Total	Total Incl O&P
0010	**WATER HEATERS**									
6900	Electric, glass lined, 30 gallon, minimum	L-1	5	3.200	Ea.	430	181		611	745
6950	Maximum		3	5.333		595	300		895	1,100
7100	80 gallon, minimum		2	8		1,225	455		1,680	2,025
7150	Maximum	↓	1	16	↓	1,700	905		2,605	3,225

11 30 15.43 Air Quality

		Crew	Daily Output	Labor-Hours	Unit	Material	Labor	Equipment	Total	Total Incl O&P
0010	**AIR QUALITY**									
2450	Dehumidifier, portable, automatic, 15 pint	1 Elec	4	2	Ea.	152	109		261	330
2550	40 pint		3.75	2.133		209	117		326	405
3550	Heater, electric, built-in, 1250 watt, ceiling type, minimum		4	2		107	109		216	282
3600	Maximum		3	2.667		175	146		321	410
3700	Wall type, minimum		4	2		172	109		281	355
3750	Maximum		3	2.667		185	146		331	420
3900	1500 watt wall type, with blower		4	2		172	109		281	355
3950	3000 watt	↓	3	2.667		350	146		496	605
4850	Humidifier, portable, 8 gallons per day					133			133	146
5000	15 gallons per day				↓	211			211	232

11 32 Unit Kitchens

11 32 13 – Metal Unit Kitchens

11 32 13.10 Commercial Unit Kitchens

		Crew	Daily Output	Labor-Hours	Unit	Material	Labor	Equipment	Total	Total Incl O&P
0010	**COMMERCIAL UNIT KITCHENS**									
1500	Combination range, refrigerator and sink, 30" wide, minimum	L-1	2	8	Ea.	1,100	455		1,555	1,900
1550	Maximum		1	16		1,475	905		2,380	3,000
1570	60" wide, average		1.40	11.429		1,525	650		2,175	2,675
1590	72" wide, average		1.20	13.333		1,625	755		2,380	2,900
1600	Office model, 48" wide		2	8		2,025	455		2,480	2,925
1620	Refrigerator and sink only	↓	2.40	6.667	↓	2,525	380		2,905	3,350
1640	Combination range, refrigerator, sink, microwave									
1660	Oven and ice maker	L-1	.80	20	Ea.	4,550	1,125		5,675	6,700

11 41 Foodservice Storage Equipment

11 41 13 – Refrigerated Food Storage Cases

11 41 13.20 Refrigerated Food Storage Equipment

		Crew	Daily Output	Labor-Hours	Unit	Material	Labor	Equipment	Total	Total Incl O&P
0010	**REFRIGERATED FOOD STORAGE EQUIPMENT**									
2350	Cooler, reach-in, beverage, 6' long	Q-1	6	2.667	Ea.	3,600	141		3,741	4,175

11 42 Food Preparation Equipment

11 42 10 – Commercial Food Preparation Equipment

11 42 10.10 Choppers, Mixers and Misc. Equipment		Crew	Daily Output	Labor-Hours	Unit	Material	2015 Bare Costs Labor	Equipment	Total	Total Incl O&P
0010	**CHOPPERS, MIXERS AND MISC. EQUIPMENT**									
1850	Coffee urn, twin 6 gallon urns	1 Plum	2	4	Ea.	2,400	235		2,635	3,000
3800	Food mixers, bench type, 20 quarts	L-7	7	4		2,775	182		2,957	3,325
3900	60 quarts	"	5	5.600	↓	11,300	254		11,554	12,900

11 44 Food Cooking Equipment

11 44 13 – Commercial Ranges

11 44 13.10 Cooking Equipment

		Crew	Daily Output	Labor-Hours	Unit	Material	Labor	Equipment	Total	Total Incl O&P
0010	**COOKING EQUIPMENT**									
0020	Bake oven, gas, one section	Q-1	8	2	Ea.	5,450	106		5,556	6,150
1300	Broiler, without oven, standard	"	8	2		3,550	106		3,656	4,050
1550	Infrared	L-7	4	7		7,425	320		7,745	8,625
6350	Kettle, w/steam jacket, tilting, w/positive lock, SS, 20 gallons		7	4		8,225	182		8,407	9,325
6600	60 gallons		6	4.667		11,000	212		11,212	12,400
8850	Steamer, electric 27 KW		7	4		10,700	182		10,882	12,100
9100	Electric, 10 KW or gas 100,000 BTU	↓	5	5.600		6,325	254		6,579	7,350
9150	Toaster, conveyor type, 16-22 slices per minute				↓	1,075			1,075	1,200
9200	For deluxe models of above equipment, add					75%				
9400	Rule of thumb: Equipment cost based									
9410	on kitchen work area									
9420	Office buildings, minimum	L-7	77	.364	S.F.	90.50	16.50		107	125
9450	Maximum		58	.483		153	22		175	202
9550	Public eating facilities, minimum		77	.364		119	16.50		135.50	157
9600	Maximum		46	.609		193	27.50		220.50	255
9750	Hospitals, minimum		58	.483		122	22		144	168
9800	Maximum	↓	39	.718	↓	225	32.50		257.50	297

11 46 Food Dispensing Equipment

11 46 16 – Service Line Equipment

11 46 16.10 Commercial Food Dispensing Equipment

		Crew	Daily Output	Labor-Hours	Unit	Material	Labor	Equipment	Total	Total Incl O&P
0010	**COMMERCIAL FOOD DISPENSING EQUIPMENT**									
3300	Food warmer, counter, 1.2 KW				Ea.	665			665	735
3550	1.6 KW					2,150			2,150	2,350
5750	Ice dispenser 567 pound	Q-1	6	2.667	↓	5,200	141		5,341	5,950

11 48 Foodservice Cleaning and Disposal Equipment

11 48 13 – Commercial Dishwashers

11 48 13.10 Dishwashers

		Crew	Daily Output	Labor-Hours	Unit	Material	Labor	Equipment	Total	Total Incl O&P
0010	**DISHWASHERS**									
2700	Dishwasher, commercial, rack type									
2720	10 to 12 racks per hour	Q-1	3.20	5	Ea.	3,525	264		3,789	4,275
2730	Energy star rated, 35 to 40 racks/hour G		1.30	12.308		4,700	650		5,350	6,150
2740	50 to 60 racks/hour G	↓	1.30	12.308	↓	10,200	650		10,850	12,200

11 52 Audio-Visual Equipment

11 52 16 – Projectors

11 52 16.10 Movie Equipment		Crew	Daily Output	Labor-Hours	Unit	Material	2015 Bare Costs Labor	Equipment	Total	Total Incl O&P
0010	**MOVIE EQUIPMENT**									
0020	Changeover, minimum				Ea.	470			470	520
0100	Maximum					915			915	1,000
0800	Lamphouses, incl. rectifiers, xenon, 1,000 watt	1 Elec	2	4		6,725	219		6,944	7,725
0900	1,600 watt		2	4		7,175	219		7,394	8,225
1000	2,000 watt		1.50	5.333		7,700	292		7,992	8,900
1100	4,000 watt		1.50	5.333		9,550	292		9,842	10,900
3700	Sound systems, incl. amplifier, mono, minimum		.90	8.889		3,350	485		3,835	4,400
3800	Dolby/Super Sound, maximum		.40	20		18,300	1,100		19,400	21,800
4100	Dual system, 2 channel, front surround, minimum		.70	11.429		4,675	625		5,300	6,075
4200	Dolby/Super Sound, 4 channel, maximum		.40	20		16,700	1,100		17,800	20,100
5300	Speakers, recessed behind screen, minimum		2	4		1,075	219		1,294	1,500
5400	Maximum		1	8		3,125	440		3,565	4,100
7000	For automation, varying sophistication, minimum		1	8	System	2,425	440		2,865	3,300
7100	Maximum	2 Elec	.30	53.333	"	5,625	2,925		8,550	10,600

11 52 16.20 Movie Equipment- Digital		Crew	Daily Output	Labor-Hours	Unit	Material	Labor	Equipment	Total	Total Incl O&P
0010	**MOVIE EQUIPMENT- DIGITAL**									
1000	Digital 2K projection system, 98" DMD	1 Elec	2	4	Ea.	44,500	219		44,719	49,300
1100	OEM lens		2	4		5,425	219		5,644	6,300
2000	Pedestal with power distribution		2	4		2,075	219		2,294	2,600
3000	Software		2	4		1,750	219		1,969	2,250

11 53 Laboratory Equipment

11 53 19 – Laboratory Sterilizers

11 53 19.13 Sterilizers		Crew	Daily Output	Labor-Hours	Unit	Material	Labor	Equipment	Total	Total Incl O&P
0010	**STERILIZERS**									
0700	Glassware washer, undercounter, minimum	L-1	1.80	8.889	Ea.	6,325	505		6,830	7,700
0710	Maximum	"	1	16	"	13,300	905		14,205	16,000

11 53 33 – Emergency Safety Appliances

11 53 33.13 Emergency Equipment		Crew	Daily Output	Labor-Hours	Unit	Material	Labor	Equipment	Total	Total Incl O&P
0010	**EMERGENCY EQUIPMENT**									
1400	Safety equipment, eye wash, hand held				Ea.	410			410	450
1450	Deluge shower				"	770			770	850

11 53 43 – Service Fittings and Accessories

11 53 43.13 Fittings		Crew	Daily Output	Labor-Hours	Unit	Material	Labor	Equipment	Total	Total Incl O&P
0010	**FITTINGS**									
8000	Alternate pricing method: as percent of lab furniture									
8050	Installation, not incl. plumbing & duct work				% Furn.				22%	22%
8100	Plumbing, final connections, simple system								10%	10%
8110	Moderately complex system								15%	15%
8120	Complex system								20%	20%
8150	Electrical, simple system								10%	10%
8160	Moderately complex system								20%	20%
8170	Complex system								35%	35%

11 61 Broadcast, Theater, and Stage Equipment

11 61 33 – Rigging Systems and Controls

11 61 33.10 Controls		Crew	Daily Output	Labor-Hours	Unit	Material	Labor	Equipment	Total	Total Incl O&P
						2015 Bare Costs				
0010	**CONTROLS**									
0050	Control boards with dimmers and breakers, minimum	1 Elec	1	8	Ea.	12,600	440		13,040	14,600
0100	Average		.50	16		39,700	875		40,575	45,000
0150	Maximum	↓	.20	40	↓	129,000	2,200		131,200	145,500

11 66 Athletic Equipment

11 66 43 – Interior Scoreboards

11 66 43.10 Scoreboards		Crew	Daily Output	Labor-Hours	Unit	Material	Labor	Equipment	Total	Total Incl O&P
0010	**SCOREBOARDS**									
7000	Baseball, minimum	R-3	1.30	15.385	Ea.	4,125	835	106	5,066	5,925
7200	Maximum		.05	400		18,100	21,700	2,750	42,550	55,500
7300	Football, minimum		.86	23.256		5,325	1,275	160	6,760	7,950
7400	Maximum		.20	100		15,900	5,425	690	22,015	26,400
7500	Basketball (one side), minimum		2.07	9.662		2,400	525	66.50	2,991.50	3,525
7600	Maximum		.30	66.667		3,525	3,625	460	7,610	9,800
7700	Hockey-basketball (four sides), minimum		.25	80		5,675	4,350	550	10,575	13,400
7800	Maximum	↓	.15	133	↓	5,750	7,250	920	13,920	18,200

11 71 Medical Sterilizing Equipment

11 71 10 – Medical Sterilizers & Distillers

11 71 10.10 Sterilizers and Distillers		Crew	Daily Output	Labor-Hours	Unit	Material	Labor	Equipment	Total	Total Incl O&P
0010	**STERILIZERS AND DISTILLERS**									
6200	Steam generators, electric 10 kW to 180 kW, freestanding									
6250	Minimum	1 Elec	3	2.667	Ea.	8,775	146		8,921	9,875
6300	Maximum	"	.70	11.429	"	29,200	625		29,825	33,000

11 72 Examination and Treatment Equipment

11 72 53 – Treatment Equipment

11 72 53.13 Medical Treatment Equipment		Crew	Daily Output	Labor-Hours	Unit	Material	Labor	Equipment	Total	Total Incl O&P
0010	**MEDICAL TREATMENT EQUIPMENT**									
6700	Surgical lights, doctor's office, single arm	2 Elec	2	8	Ea.	2,475	440		2,915	3,375
6750	Dual arm	"	1	16	"	4,575	875		5,450	6,325

11 76 Operating Room Equipment

11 76 10 – Operating Room Equipment

11 76 10.10 Surgical Equipment		Crew	Daily Output	Labor-Hours	Unit	Material	Labor	Equipment	Total	Total Incl O&P
0010	**SURGICAL EQUIPMENT**									
6800	Surgical lights, major operating room, dual head, minimum	2 Elec	1	16	Ea.	4,375	875		5,250	6,100
6850	Maximum	"	1	16	"	30,000	875		30,875	34,400

11 78 Mortuary Equipment

11 78 13 – Mortuary Refrigerators

11 78 13.10 Mortuary and Autopsy Equipment		Crew	Daily Output	Labor-Hours	Unit	Material	2015 Bare Costs Labor	Equipment	Total	Total Incl O&P
0010	**MORTUARY AND AUTOPSY EQUIPMENT**									
0015	Autopsy table, standard	1 Plum	1	8	Ea.	9,750	470		10,220	11,400

11 78 16 – Crematorium Equipment

11 78 16.10 Crematory

		Crew	Daily Output	Labor-Hours	Unit	Material	Labor	Equipment	Total	Total Incl O&P
0010	**CREMATORY**									
1500	Crematory, not including building, 1 place	Q-3	.20	160	Ea.	72,500	8,950		81,450	93,000
1750	2 place	"	.10	320	"	103,500	17,900		121,400	141,000

11 82 Facility Solid Waste Handling Equipment

11 82 26 – Facility Waste Compactors

11 82 26.10 Compactors

		Crew	Daily Output	Labor-Hours	Unit	Material	Labor	Equipment	Total	Total Incl O&P
0010	**COMPACTORS**									
0020	Compactors, 115 volt, 250#/hr., chute fed	L-4	1	24	Ea.	12,100	1,050		13,150	15,000
1000	Heavy duty industrial compactor, 0.5 C.Y. capacity		1	24		10,100	1,050		11,150	12,800
1050	1.0 C.Y. capacity		1	24		15,200	1,050		16,250	18,400
1100	3.0 C.Y. capacity		.50	48		25,900	2,125		28,025	31,800
1150	5.0 C.Y. capacity		.50	48		32,600	2,125		34,725	39,100
1200	Combination shredder/compactor (5,000 lb./hr.)	↓	.50	48		63,500	2,125		65,625	73,000
1400	For handling hazardous waste materials, 55 gallon drum packer, std.					19,700			19,700	21,700
1410	55 gallon drum packer w/HEPA filter					24,600			24,600	27,100
1420	55 gallon drum packer w/charcoal & HEPA filter					32,800			32,800	36,100
1430	All of the above made explosion proof, add					1,450			1,450	1,575
5800	Shredder, industrial, minimum					24,000			24,000	26,400
5850	Maximum					128,500			128,500	141,500
5900	Baler, industrial, minimum					9,625			9,625	10,600
5950	Maximum				↓	560,500			560,500	616,500

Estimating Tips

General

- The items in this division are usually priced per square foot or each. Most of these items are purchased by the owner and installed by the contractor. Do not assume the items in Division 12 will be purchased and installed by the contractor. Check the specifications for responsibilities and include receiving, storage, installation, and mechanical and electrical hookups in the appropriate divisions.
- Some items in this division require some type of support system that is not usually furnished with the item. Examples of these systems include blocking for the attachment of casework and heavy drapery rods. The required blocking must be added to the estimate in the appropriate division.

Reference Numbers

Reference numbers are shown in shaded boxes at the beginning of some major classifications. These numbers refer to related items in the Reference Section. The reference information may be an estimating procedure, an alternate pricing method, or technical information.

Note: Not all subdivisions listed here necessarily appear in this publication. ■

Division 12 – Furnishings

12 46 Furnishing Accessories

12 46 19 – Clocks

12 46 19.50 Wall Clocks		Crew	Daily Output	Labor-Hours	Unit	2015 Bare Costs Material	Labor	Equipment	Total	Total Incl O&P
0010	**WALL CLOCKS**									
0080	12" diameter, single face	1 Elec	8	1	Ea.	130	54.50		184.50	225
0100	Double face	"	6.20	1.290	"	300	70.50		370.50	435

Division 13 – Special Construction

Estimating Tips

General

- The items and systems in this division are usually estimated, purchased, supplied, and installed as a unit by one or more subcontractors. The estimator must ensure that all parties are operating from the same set of specifications and assumptions, and that all necessary items are estimated and will be provided. Many times the complex items and systems are covered, but the more common ones, such as excavation or a crane, are overlooked for the very reason that everyone assumes nobody could miss them. The estimator should be the central focus and be able to ensure that all systems are complete.
- Another area where problems can develop in this division is at the interface between systems. The estimator must ensure, for instance, that anchor bolts, nuts, and washers are estimated and included for the air-supported structures and pre-engineered buildings to be bolted to their foundations. Utility supply is a common area where essential items or pieces of equipment can be missed or overlooked, because each subcontractor may feel it is another's responsibility. The estimator should also be aware of certain items which may be supplied as part of a package but installed by others, and ensure that the installing contractor's estimate includes the cost of installation. Conversely, the estimator must also ensure that items are not costed by two different subcontractors, resulting in an inflated overall estimate.

13 30 00 Special Structures

- The foundations and floor slab, as well as rough mechanical and electrical, should be estimated, as this work is required for the assembly and erection of the structure. Generally, as noted in the book, the pre-engineered building comes as a shell. Pricing is based on the size and structural design parameters stated in the reference section. Additional features, such as windows and doors with their related structural framing, must also be included by the estimator. Here again, the estimator must have a clear understanding of the scope of each portion of the work and all the necessary interfaces.

Reference Numbers

Reference numbers are shown in shaded boxes at the beginning of some major classifications. These numbers refer to related items in the Reference Section. The reference information may be an estimating procedure, an alternate pricing method, or technical information.

Note: Not all subdivisions listed here necessarily appear in this publication. ■

13 11 Swimming Pools

13 11 13 – Below-Grade Swimming Pools

13 11 13.50 Swimming Pools		Crew	Daily Output	Labor-Hours	Unit	2015 Bare Costs Material	Labor	Equipment	Total	Total Incl O&P
0010	**SWIMMING POOLS** Residential in-ground, vinyl lined, concrete									
0020	Swimming pools,resi in-ground,vyl lined,conc sides,W/ equip ,sand bot	B-52	300	.187	SF Surf	23.50	8.15	1.98	33.63	40
0100	Metal or polystyrene sides	B-14	410	.117		19.55	4.66	.89	25.10	29.50
0200	Add for vermiculite bottom					1.49			1.49	1.64
0500	Gunite bottom and sides, white plaster finish									
0600	12′ x 30′ pool	B-52	145	.386	SF Surf	43.50	16.85	4.10	64.45	78.50
0720	16′ x 32′ pool		155	.361		39	15.75	3.83	58.58	71
0750	20′ x 40′ pool		250	.224		35	9.75	2.38	47.13	56
0810	Concrete bottom and sides, tile finish									
0820	12′ x 30′ pool	B-52	80	.700	SF Surf	44	30.50	7.45	81.95	103
0830	16′ x 32′ pool		95	.589		36.50	25.50	6.25	68.25	86.50
0840	20′ x 40′ pool		130	.431		29	18.80	4.57	52.37	66
1100	Motel, gunite with plaster finish, incl. medium									
1150	capacity filtration & chlorination	B-52	115	.487	SF Surf	53.50	21	5.15	79.65	97
1200	Municipal, gunite with plaster finish, incl. high									
1250	capacity filtration & chlorination	B-52	100	.560	SF Surf	69.50	24.50	5.95	99.95	120
1350	Add for formed gutters				L.F.	102			102	112
1360	Add for stainless steel gutters				"	300			300	330
1700	Filtration and deck equipment only, as % of total				Total				20%	20%
1800	Deck equipment, rule of thumb, 20′ x 40′ pool				SF Pool				1.18	1.30
1900	5000 S.F. pool				"				1.73	1.90

13 11 46 – Swimming Pool Accessories

13 11 46.50 Swimming Pool Equipment		Crew	Daily Output	Labor-Hours	Unit	Material	Labor	Equipment	Total	Total Incl O&P
0010	**SWIMMING POOL EQUIPMENT**									
0020	Diving stand, stainless steel, 3 meter	2 Carp	.40	40	Ea.	15,100	1,875		16,975	19,500
2100	Lights, underwater, 12 volt, with transformer, 300 watt	1 Elec	1	8		330	440		770	1,025
2200	110 volt, 500 watt, standard		1	8		294	440		734	980
2400	Low water cutoff type		1	8		300	440		740	985
2800	Heaters, see Section 23 52 28.10									

13 21 Controlled Environment Rooms

13 21 13 – Clean Rooms

13 21 13.50 Clean Room Components		Crew	Daily Output	Labor-Hours	Unit	Material	Labor	Equipment	Total	Total Incl O&P
0010	**CLEAN ROOM COMPONENTS**									
1100	Clean room, soft wall, 12′ x 12′, Class 100	1 Carp	.18	44.444	Ea.	18,600	2,075		20,675	23,700
1110	Class 1,000		.18	44.444		15,400	2,075		17,475	20,200
1120	Class 10,000		.21	38.095		13,000	1,800		14,800	17,100
1130	Class 100,000		.21	38.095		12,000	1,800		13,800	16,000
2800	Ceiling grid support, slotted channel struts 4′-0″ O.C., ea. way				S.F.				5.90	6.50
3000	Ceiling panel, vinyl coated foil on mineral substrate									
3020	Sealed, non-perforated				S.F.				1.27	1.40
4000	Ceiling panel seal, silicone sealant, 150 L.F./gal.	1 Carp	150	.053	L.F.	.34	2.50		2.84	4.22
4100	Two sided adhesive tape	"	240	.033	"	.12	1.56		1.68	2.54
4200	Clips, one per panel				Ea.	.99			.99	1.09
6000	HEPA filter, 2′ x 4′, 99.97% eff., 3″ dp beveled frame (silicone seal)					470			470	515
6040	6″ deep skirted frame (channel seal)					440			440	485
6100	99.99% efficient, 3″ deep beveled frame (silicone seal)					525			525	580
6140	6″ deep skirted frame (channel seal)					455			455	500
6200	99.999% efficient, 3″ deep beveled frame (silicone seal)					605			605	665
6240	6″ deep skirted frame (channel seal)					485			485	530

13 21 Controlled Environment Rooms

13 21 13 – Clean Rooms

13 21 13.50 Clean Room Components		Crew	Daily Output	Labor-Hours	Unit	2015 Bare Costs Material	Labor	Equipment	Total	Total Incl O&P
7000	Wall panel systems, including channel strut framing									
7020	Polyester coated aluminum, particle board				S.F.				18.20	20
7100	Porcelain coated aluminum, particle board								32	35
7400	Wall panel support, slotted channel struts, to 12′ high				↓				16.35	18

13 24 Special Activity Rooms

13 24 16 – Saunas

13 24 16.50 Saunas and Heaters		Crew	Daily Output	Labor-Hours	Unit	Material	Labor	Equipment	Total	Total Incl O&P
0010	**SAUNAS AND HEATERS**									
0020	Prefabricated, incl. heater & controls, 7′ high, 6′ x 4′, C/C	L-7	2.20	12.727	Ea.	5,175	580		5,755	6,575
0050	6′ x 4′, C/P		2	14		4,700	635		5,335	6,125
0400	6′ x 5′, C/C		2	14		5,875	635		6,510	7,450
0450	6′ x 5′, C/P		2	14		5,325	635		5,960	6,825
0600	6′ x 6′, C/C		1.80	15.556		6,225	705		6,930	7,925
0650	6′ x 6′, C/P		1.80	15.556		5,675	705		6,380	7,300
0800	6′ x 9′, C/C		1.60	17.500		7,975	795		8,770	10,000
0850	6′ x 9′, C/P		1.60	17.500		7,225	795		8,020	9,150
1000	8′ x 12′, C/C		1.10	25.455		11,700	1,150		12,850	14,700
1050	8′ x 12′, C/P		1.10	25.455		10,500	1,150		11,650	13,400
1200	8′ x 8′, C/C		1.40	20		9,175	910		10,085	11,500
1250	8′ x 8′, C/P		1.40	20		8,450	910		9,360	10,700
1400	8′ x 10′, C/C		1.20	23.333		10,200	1,050		11,250	12,800
1450	8′ x 10′, C/P		1.20	23.333		9,250	1,050		10,300	11,800
1600	10′ x 12′, C/C		1	28		12,200	1,275		13,475	15,500
1650	10′ x 12′, C/P	↓	1	28		11,000	1,275		12,275	14,100
2500	Heaters only (incl. above), wall mounted, to 200 C.F.					685			685	755
2750	To 300 C.F.					930			930	1,025
3000	Floor standing, to 720 C.F., 10,000 watts, w/controls	1 Elec	3	2.667		2,950	146		3,096	3,475
3250	To 1,000 C.F., 16,000 watts	"	3	2.667	↓	3,825	146		3,971	4,425

13 24 26 – Steam Baths

13 24 26.50 Steam Baths and Components		Crew	Daily Output	Labor-Hours	Unit	Material	Labor	Equipment	Total	Total Incl O&P
0010	**STEAM BATHS AND COMPONENTS**									
0020	Heater, timer & head, single, to 140 C.F.	1 Plum	1.20	6.667	Ea.	2,200	390		2,590	3,000
0500	To 300 C.F.		1.10	7.273		2,425	425		2,850	3,325
1000	Commercial size, with blow-down assembly, to 800 C.F.		.90	8.889		5,725	520		6,245	7,100
1500	To 2500 C.F.	↓	.80	10		7,650	585		8,235	9,300
2000	Multiple, motels, apts., 2 baths, w/blow-down assm., 500 C.F.	Q-1	1.30	12.308		6,325	650		6,975	7,950
2500	4 baths	"	.70	22.857	↓	10,200	1,200		11,400	13,000

13 47 Facility Protection

13 47 13 – Cathodic Protection

13 47 13.16 Cathodic Prot. for Underground Storage Tanks		Crew	Daily Output	Labor-Hours	Unit	2015 Bare Costs Material	Labor	Equipment	Total	Total Incl O&P
0010	CATHODIC PROTECTION FOR UNDERGROUND STORAGE TANKS									
1000	Anodes, magnesium type, 9 #	R-15	18.50	2.595	Ea.	41.50	140	16.30	197.80	273
1010	17 #		13	3.692		71.50	199	23	293.50	400
1020	32 #		10	4.800		124	258	30	412	555
1030	48 #		7.20	6.667		164	360	42	566	765
1100	Graphite type w/epoxy cap, 3″ x 60″ (32 #)	R-22	8.40	4.438		128	206		334	455
1110	4″ x 80″ (68 #)		6	6.213		241	289		530	705
1120	6″ x 72″ (80 #)		5.20	7.169		1,500	335		1,835	2,150
1130	6″ x 36″ (45 #)		9.60	3.883		760	181		941	1,100
2000	Rectifiers, silicon type, air cooled, 28 V/10 A	R-19	3.50	5.714		2,300	315		2,615	3,025
2010	20 V/20 A		3.50	5.714		2,375	315		2,690	3,075
2100	Oil immersed, 28 V/10 A		3	6.667		3,175	365		3,540	4,025
2110	20 V/20 A		3	6.667		3,175	365		3,540	4,050
3000	Anode backfill, coke breeze	R-22	3850	.010	Lb.	.22	.45		.67	.92
4000	Cable, HMWPE, No. 8		2.40	15.533	M.L.F.	500	720		1,220	1,650
4010	No. 6		2.40	15.533		745	720		1,465	1,925
4020	No. 4		2.40	15.533		1,125	720		1,845	2,350
4030	No. 2		2.40	15.533		1,750	720		2,470	3,025
4040	No. 1		2.20	16.945		2,400	790		3,190	3,850
4050	No. 1/0		2.20	16.945		2,975	790		3,765	4,475
4060	No. 2/0		2.20	16.945		4,625	790		5,415	6,300
4070	No. 4/0		2	18.640		6,000	865		6,865	7,925
5000	Test station, 7 terminal box, flush curb type w/lockable cover	R-19	12	1.667	Ea.	71.50	91.50		163	216
5010	Reference cell, 2″ dia PVC conduit, cplg., plug, set flush	"	4.80	4.167	"	151	228		379	505

13 49 Radiation Protection

13 49 13 – Integrated X-Ray Shielding Assemblies

13 49 13.50 Lead Sheets		Crew	Daily Output	Labor-Hours	Unit	Material	Labor	Equipment	Total	Total Incl O&P
0010	LEAD SHEETS									
0300	Lead sheets, 1/16″ thick	2 Lath	135	.119	S.F.	10.50	5.10		15.60	19.05
0400	1/8″ thick		120	.133		31	5.75		36.75	42.50
0500	Lead shielding, 1/4″ thick		135	.119		41.50	5.10		46.60	53.50
0550	1/2″ thick		120	.133		73.50	5.75		79.25	89.50
0950	Lead headed nails (average 1 lb. per sheet)				Lb.	8			8	8.80
1000	Butt joints in 1/8″ lead or thicker, 2″ batten strip x 7′ long	2 Lath	240	.067	Ea.	28	2.87		30.87	35
1200	X-ray protection, average radiography or fluoroscopy									
1210	room, up to 300 S.F. floor, 1/16″ lead, economy	2 Lath	.25	64	Total	10,100	2,750		12,850	15,200
1500	7′-0″ walls, deluxe	"	.15	106	"	12,200	4,575		16,775	20,200
1600	Deep therapy X-ray room, 250 kV capacity,									
1800	up to 300 S.F. floor, 1/4″ lead, economy	2 Lath	.08	200	Total	28,300	8,600		36,900	43,800
1900	7′-0″ walls, deluxe	"	.06	266	"	34,900	11,500		46,400	55,500

13 49 19 – Lead-Lined Materials

13 49 19.50 Shielding Lead		Crew	Daily Output	Labor-Hours	Unit	Material	Labor	Equipment	Total	Total Incl O&P
0010	SHIELDING LEAD									
0100	Laminated lead in wood doors, 1/16″ thick, no hardware				S.F.	52.50			52.50	57.50
0200	Lead lined door frame, not incl. hardware,									
0210	1/16″ thick lead, butt prepared for hardware	1 Lath	2.40	3.333	Ea.	810	143		953	1,100
0850	Window frame with 1/16″ lead and voice passage, 36″ x 60″	2 Glaz	2	8		4,200	360		4,560	5,150
0870	24″ x 36″ frame		4	4		2,175	180		2,355	2,675
0900	Lead gypsum board, 5/8″ thick with 1/16″ lead		160	.100	S.F.	10.95	4.51		15.46	18.90

13 49 Radiation Protection

13 49 19 – Lead-Lined Materials

13 49 19.50 Shielding Lead		Crew	Daily Output	Labor-Hours	Unit	2015 Bare Costs Material	Labor	Equipment	Total	Total Incl O&P
0910	1/8″ lead	2 Glaz	140	.114	S.F.	23	5.15		28.15	33.50
0930	1/32″ lead	2 Lath	200	.080		7.95	3.44		11.39	13.80

13 49 21 – Lead Glazing

13 49 21.50 Lead Glazing		Crew	Daily Output	Labor-Hours	Unit	Material	Labor	Equipment	Total	Total Incl O&P
0010	**LEAD GLAZING**									
0600	Lead glass, 1/4″ thick, 2.0 mm LE, 12″ x 16″	2 Glaz	13	1.231	Ea.	380	55.50		435.50	500
0700	24″ x 36″		8	2		1,325	90		1,415	1,575
0800	36″ x 60″		2	8		3,675	360		4,035	4,575
2000	X-ray viewing panels, clear lead plastic									
2010	7 mm thick, 0.3 mm LE, 2.3 lb./S.F.	H-3	139	.115	S.F.	241	4.64		245.64	272
2020	12 mm thick, 0.5 mm LE, 3.9 lb./S.F.		82	.195		355	7.85		362.85	400
2030	18 mm thick, 0.8 mm LE, 5.9 lb./S.F.		54	.296		405	11.95		416.95	465
2040	22 mm thick, 1.0 mm LE, 7.2 lb./S.F.		44	.364		530	14.65		544.65	610
2050	35 mm thick, 1.5 mm LE, 11.5 lb./S.F.		28	.571		815	23		838	930
2060	46 mm thick, 2.0 mm LE, 15.0 lb./S.F.		21	.762		1,050	30.50		1,080.50	1,225
2090	For panels 12 S.F. to 48 S.F., add crating charge				Ea.				50	50

13 49 23 – Integrated RFI/EMI Shielding Assemblies

13 49 23.50 Modular Shielding Partitions		Crew	Daily Output	Labor-Hours	Unit	Material	Labor	Equipment	Total	Total Incl O&P
0010	**MODULAR SHIELDING PARTITIONS**									
4000	X-ray barriers, modular, panels mounted within framework for									
4002	attaching to floor, wall or ceiling, upper portion is clear lead									
4005	plastic window panels 48″H, lower portion is opaque leaded									
4008	steel panels 36″H, structural supports not incl.									
4010	1-section barrier, 36″W x 84″H overall									
4020	0.5 mm LE panels	H-3	6.40	2.500	Ea.	8,100	101		8,201	9,050
4030	0.8 mm LE panels		6.40	2.500		8,725	101		8,826	9,750
4040	1.0 mm LE panels		5.33	3.002		10,200	121		10,321	11,400
4050	1.5 mm LE panels		5.33	3.002		13,600	121		13,721	15,200
4060	2-section barrier, 72″W x 84″H overall									
4070	0.5 mm LE panels	H-3	4	4	Ea.	11,800	161		11,961	13,200
4080	0.8 mm LE panels		4	4		13,100	161		13,261	14,600
4090	1.0 mm LE panels		3.56	4.494		16,000	181		16,181	17,900
5000	1.5 mm LE panels		3.20	5		22,800	201		23,001	25,400
5010	3-section barrier, 108″W x 84″H overall									
5020	0.5 mm LE panels	H-3	3.20	5	Ea.	17,700	201		17,901	19,800
5030	0.8 mm LE panels		3.20	5		19,600	201		19,801	21,800
5040	1.0 mm LE panels		2.67	5.993		24,000	241		24,241	26,800
5050	1.5 mm LE panels		2.46	6.504		34,200	262		34,462	38,000
7000	X-ray barriers, mobile, mounted within framework w/casters on									
7005	bottom, clear lead plastic window panels on upper portion,									
7010	opaque on lower, 30″W x 75″H overall, incl. framework									
7020	24″H upper w/0.5 mm LE, 48″H lower w/0.8 mm LE	1 Carp	16	.500	Ea.	3,800	23.50		3,823.50	4,200
7030	48″W x 75″H overall, incl. framework									
7040	36″H upper w/0.5 mm LE, 36″H lower w/0.8 mm LE	1 Carp	16	.500	Ea.	6,175	23.50		6,198.50	6,825
7050	36″H upper w/1.0 mm LE, 36″H lower w/1.5 mm LE	″	16	.500	″	7,300	23.50		7,323.50	8,075
7060	72″W x 75″H overall, incl. framework									
7070	36″H upper w/0.5 mm LE, 36″H lower w/0.8 mm LE	1 Carp	16	.500	Ea.	7,300	23.50		7,323.50	8,075
7080	36″H upper w/1.0 mm LE, 36″H lower w/1.5 mm LE	″	16	.500	″	9,150	23.50		9,173.50	10,100

13 49 Radiation Protection

13 49 33 – Radio Frequency Shielding

13 49 33.50 Shielding, Radio Frequency		Crew	Daily Output	Labor-Hours	Unit	2015 Bare Costs Material	Labor	Equipment	Total	Total Incl O&P
0010	**SHIELDING, RADIO FREQUENCY**									
0020	Prefabricated, galvanized steel	2 Carp	375	.043	SF Surf	4.46	2		6.46	8
0040	5 oz., copper floor panel		480	.033		3.68	1.56		5.24	6.45
0050	5 oz., copper wall/ceiling panel		155	.103		3.68	4.85		8.53	11.50
0100	12 oz., copper floor panel		470	.034		7.85	1.60		9.45	11.10
0110	12 oz., copper wall/ceiling panel		140	.114		7.85	5.35		13.20	16.90
0150	Door, copper/wood laminate, 4′ x 7′		1.50	10.667	Ea.	7,400	500		7,900	8,900

Estimating Tips

Pipe for fire protection and all uses is located in Subdivisions 21 11 13 and 22 11 13.

The labor adjustment factors listed in Subdivision 22 01 02.20 also apply to Division 21.

Many, but not all, areas in the U.S. require backflow protection in the fire system. It is advisable to check local building codes for specific requirements.

For your reference, the following is a list of the most applicable Fire Codes and Standards which may be purchased from the NFPA, 1 Batterymarch Park, Quincy, MA 02169-7471.

- NFPA 1: Uniform Fire Code
- NFPA 10: Portable Fire Extinguishers
- NFPA 11: Low-, Medium-, and High-Expansion Foam
- NFPA 12: Carbon Dioxide Extinguishing Systems (Also companion 12A)
- NFPA 13: Installation of Sprinkler Systems (Also companion 13D, 13E, and 13R)
- NFPA 14: Installation of Standpipe and Hose Systems
- NFPA 15: Water Spray Fixed Systems for Fire Protection
- NFPA 16: Installation of Foam-Water Sprinkler and Foam-Water Spray Systems
- NFPA 17: Dry Chemical Extinguishing Systems (Also companion 17A)
- NFPA 18: Wetting Agents
- NFPA 20: Installation of Stationary Pumps for Fire Protection
- NFPA 22: Water Tanks for Private Fire Protection
- NFPA 24: Installation of Private Fire Service Mains and their Appurtenances
- NFPA 25: Inspection, Testing and Maintenance of Water-Based Fire Protection

Reference Numbers

Reference numbers are shown in shaded boxes at the beginning of some major classifications. These numbers refer to related items in the Reference Section. The reference information may be an estimating procedure, an alternate pricing method, or technical information.

Note: Not all subdivisions listed here necessarily appear in this publication. ■

21 21 Carbon-Dioxide Fire-Extinguishing Systems

21 21 16 – Carbon-Dioxide Fire-Extinguishing Equipment

21 21 16.50 CO2 Fire Extinguishing System		Crew	Daily Output	Labor-Hours	Unit	2015 Bare Costs Material	Labor	Equipment	Total	Total Incl O&P
0010	**CO_2 FIRE EXTINGUISHING SYSTEM**									
0100	Control panel, single zone with batteries (2 zones det., 1 suppr.)	1 Elec	1	8	Ea.	1,725	440		2,165	2,550
0150	Multizone (4) with batteries (8 zones det., 4 suppr.)	"	.50	16		3,275	875		4,150	4,900
1000	Dispersion nozzle, CO_2, 3" x 5"	1 Plum	18	.444		67	26		93	113
2000	Extinguisher, CO_2 system, high pressure, 75 lb. cylinder	Q-1	6	2.667		1,275	141		1,416	1,625
2100	100 lb. cylinder	"	5	3.200		1,300	169		1,469	1,700
3000	Electro/mechanical release	L-1	4	4		167	227		394	525
3400	Manual pull station	1 Plum	6	1.333		60.50	78.50		139	185
4000	Pneumatic damper release	"	8	1	↓	223	58.50		281.50	335

21 22 Clean-Agent Fire-Extinguishing Systems

21 22 16 – Clean-Agent Fire-Extinguishing Equipment

21 22 16.50 FM200 Fire Extinguishing System		Crew	Daily Output	Labor-Hours	Unit	Material	Labor	Equipment	Total	Total Incl O&P
0010	**FM200 FIRE EXTINGUISHING SYSTEM**									
1100	Dispersion nozzle FM200, 1-1/2"	1 Plum	14	.571	Ea.	67	33.50		100.50	124
2400	Extinguisher, FM200 system, filled, with mounting bracket									
2460	26 lb. container	Q-1	8	2	Ea.	2,300	106		2,406	2,675
2480	44 lb. container		7	2.286		3,050	121		3,171	3,550
2500	63 lb. container		6	2.667		3,575	141		3,716	4,150
2520	101 lb. container		5	3.200		4,775	169		4,944	5,500
2540	196 lb. container	↓	4	4	↓	7,775	211		7,986	8,875
6000	FM200 system, simple nozzle layout, with broad dispersion				C.F.	1.76			1.76	1.94
6020	Complex nozzle layout and/or including underfloor dispersion				"	3.50			3.50	3.85

Estimating Tips

22 10 00 Plumbing Piping and Pumps

This subdivision is primarily basic pipe and related materials. The pipe may be used by any of the mechanical disciplines, i.e., plumbing, fire protection, heating, and air conditioning.

Note: CPVC plastic piping approved for fire protection is located in 21 11 13.

- The labor adjustment factors listed in Subdivision 22 01 02.20 apply throughout Divisions 21, 22, and 23. CAUTION: the correct percentage may vary for the same items. For example, the percentage add for the basic pipe installation should be based on the maximum height that the craftsman must install for that particular section. If the pipe is to be located 14' above the floor but it is suspended on threaded rod from beams, the bottom flange of which is 18' high (4' rods), then the height is actually 18' and the add is 20%. The pipe coverer, however, does not have to go above the 14', and so the add should be 10%.
- Most pipe is priced first as straight pipe with a joint (coupling, weld, etc.) every 10' and a hanger usually every 10'. There are exceptions with hanger spacing such as for cast iron pipe (5') and plastic pipe (3 per 10'). Following each type of pipe there are several lines listing sizes and the amount to be subtracted to delete couplings and hangers. This is for pipe that is to be buried or supported together on trapeze hangers. The reason that the couplings are deleted is that these runs are usually long, and frequently longer lengths of pipe are used. By deleting the couplings, the estimator is expected to look up and add back the correct reduced number of couplings.
- When preparing an estimate, it may be necessary to approximate the fittings. Fittings usually run between 25% and 50% of the cost of the pipe. The lower percentage is for simpler runs, and the higher number is for complex areas, such as mechanical rooms.
- For historic restoration projects, the systems must be as invisible as possible, and pathways must be sought for pipes, conduit, and ductwork. While installations in accessible spaces (such as basements and attics) are relatively straightforward to estimate, labor costs may be more difficult to determine when delivery systems must be concealed.

22 40 00 Plumbing Fixtures

- Plumbing fixture costs usually require two lines: the fixture itself and its "rough-in, supply, and waste."
- In the Assemblies Section (Plumbing D2010) for the desired fixture, the System Components Group at the center of the page shows the fixture on the first line. The rest of the list (fittings, pipe, tubing, etc.) will total up to what we refer to in the Unit Price section as "Rough-in, supply, waste, and vent." Note that for most fixtures we allow a nominal 5' of tubing to reach from the fixture to a main or riser.
- Remember that gas- and oil-fired units need venting.

Reference Numbers

Reference numbers are shown in shaded boxes at the beginning of some major classifications. These numbers refer to related items in the Reference Section. The reference information may be an estimating procedure, an alternate pricing method, or technical information.

Note: Not all subdivisions listed here necessarily appear in this publication. ■

22 01 Operation and Maintenance of Plumbing

22 01 02 – Labor Adjustments

22 01 02.10 Boilers, General		Crew	Daily Output	Labor-Hours	Unit	Material	2015 Bare Costs Labor	Equipment	Total	Total Incl O&P
0010	**BOILERS, GENERAL**, Prices do not include flue piping, elec. wiring,									
0020	gas or oil piping, boiler base, pad, or tankless unless noted									
0100	Boiler H.P.: 10 KW = 34 lb./steam/hr. = 33,475 BTU/hr.									
0150	To convert SFR to BTU rating: Hot water, 150 x SFR;									
0160	Forced hot water, 180 x SFR; steam, 240 x SFR									

22 01 02.20 Labor Adjustment Factors		Crew	Daily Output	Labor-Hours	Unit	Material	Labor	Equipment	Total	Total Incl O&P
0010	**LABOR ADJUSTMENT FACTORS**, (For Div. 21, 22 and 23) R220102-20									
0100	Labor factors, The below are reasonable suggestions, however									
0110	each project must be evaluated for its own peculiarities, and									
0120	the adjustments be increased or decreased depending on the									
0130	severity of the special conditions.									
1000	Add to labor for elevated installation (Above floor level)									
1080	10′ to 14.5′ high						10%			
1100	15′ to 19.5′ high						20%			
1120	20′ to 24.5′ high						25%			
1140	25′ to 29.5′ high						35%			
1160	30′ to 34.5′ high						40%			
1180	35′ to 39.5′ high						50%			
1200	40′ and higher						55%			
2000	Add to labor for crawl space									
2100	3′ high						40%			
2140	4′ high						30%			
3000	Add to labor for multi-story building									
3100	Add per floor for floors 3 thru 19						2%			
3140	Add per floor for floors 20 and up						4%			
4000	Add to labor for working in existing occupied buildings									
4100	Hospital						35%			
4140	Office building						25%			
4180	School						20%			
4220	Factory or warehouse						15%			
4260	Multi dwelling						15%			
5000	Add to labor, miscellaneous									
5100	Cramped shaft						35%			
5140	Congested area						15%			
5180	Excessive heat or cold						30%			
9000	Labor factors, The above are reasonable suggestions, however									
9010	each project should be evaluated for its own peculiarities.									
9100	Other factors to be considered are:									
9140	Movement of material and equipment through finished areas									
9180	Equipment room									
9220	Attic space									
9260	No service road									
9300	Poor unloading/storage area									
9340	Congested site area/heavy traffic									

22 05 29.10 Hangers & Supp. for Plumb'g/HVAC Pipe/Equip.		Crew	Daily Output	Labor-Hours	Unit	2015 Bare Costs Material	Labor	Equipment	Total	Total Incl O&P
0010	**HANGERS AND SUPPORTS FOR PLUMB'G/HVAC PIPE/EQUIP.**									
0011	TYPE numbers per MSS-SP58									
0050	Brackets									
0060	Beam side or wall, malleable iron, TYPE 34									
0070	3/8" threaded rod size	1 Plum	48	.167	Ea.	3.11	9.80		12.91	18.20
0080	1/2" threaded rod size		48	.167		4.39	9.80		14.19	19.65
0090	5/8" threaded rod size		48	.167		8.20	9.80		18	24
0100	3/4" threaded rod size		48	.167		11.30	9.80		21.10	27
0110	7/8" threaded rod size		48	.167		11.70	9.80		21.50	27.50
0120	For concrete installation, add						30%			
0150	Wall, welded steel, medium, TYPE 32									
0160	0 size, 12" wide, 18" deep	1 Plum	34	.235	Ea.	172	13.80		185.80	210
0170	1 size, 18" wide, 24" deep		34	.235		211	13.80		224.80	253
0180	2 size, 24" wide, 30" deep		34	.235		290	13.80		303.80	340
0300	Clamps									
0310	C-clamp, for mounting on steel beam flange, w/locknut, TYPE 23									
0320	3/8" threaded rod size	1 Plum	160	.050	Ea.	2.05	2.94		4.99	6.70
0330	1/2" threaded rod size		160	.050		2.20	2.94		5.14	6.85
0340	5/8" threaded rod size		160	.050		3.70	2.94		6.64	8.50
0350	3/4" threaded rod size		160	.050		4.55	2.94		7.49	9.45
0750	Riser or extension pipe, carbon steel, TYPE 8									
0760	3/4" pipe size	1 Plum	48	.167	Ea.	2.52	9.80		12.32	17.55
0770	1" pipe size		47	.170		2.64	10		12.64	18
0780	1-1/4" pipe size		46	.174		3.21	10.20		13.41	18.95
0790	1-1/2" pipe size		45	.178		3.42	10.45		13.87	19.50
0800	2" pipe size		43	.186		3.54	10.90		14.44	20.50
0810	2-1/2" pipe size		41	.195		3.68	11.45		15.13	21.50
0820	3" pipe size		40	.200		4	11.75		15.75	22
0830	3-1/2" pipe size		39	.205		4.50	12.05		16.55	23
0840	4" pipe size		38	.211		5.45	12.35		17.80	24.50
0850	5" pipe size		37	.216		6.80	12.70		19.50	26.50
0860	6" pipe size		36	.222		8.35	13.05		21.40	29
1150	Insert, concrete									
1160	Wedge type, carbon steel body, malleable iron nut, galvanized									
1170	1/4" threaded rod size	1 Plum	96	.083	Ea.	9.55	4.89		14.44	17.90
1180	3/8" threaded rod size		96	.083		10.30	4.89		15.19	18.75
1190	1/2" threaded rod size		96	.083		10.65	4.89		15.54	19.15
1200	5/8" threaded rod size		96	.083		11.25	4.89		16.14	19.80
1210	3/4" threaded rod size		96	.083		11.95	4.89		16.84	20.50
1220	7/8" threaded rod size		96	.083		12.35	4.89		17.24	21
2650	Rods, carbon steel									
2660	Continuous thread									
2670	1/4" thread size	1 Plum	144	.056	L.F.	1.70	3.26		4.96	6.80
2680	3/8" thread size		144	.056		1.81	3.26		5.07	6.90
2690	1/2" thread size		144	.056		2.86	3.26		6.12	8.10
2700	5/8" thread size		144	.056		4.05	3.26		7.31	9.40
2710	3/4" thread size		144	.056		7.15	3.26		10.41	12.80
2720	7/8" thread size		144	.056		8.95	3.26		12.21	14.80
2725	1/4" thread size, bright finish		144	.056		1.24	3.26		4.50	6.30
2726	1/2" thread size, bright finish		144	.056		4.23	3.26		7.49	9.60
2730	For galvanized, add					40%				
4400	U-bolt, carbon steel									
4410	Standard, with nuts, TYPE 42									

22 05 Common Work Results for Plumbing

22 05 29 – Hangers and Supports for Plumbing Piping and Equipment

22 05 29.10	Hangers & Supp. for Plumb'g/HVAC Pipe/Equip.	Crew	Daily Output	Labor-Hours	Unit	2015 Bare Costs Material	Labor	Equipment	Total	Total Incl O&P
4420	1/2" pipe size	1 Plum	160	.050	Ea.	.98	2.94		3.92	5.50
4430	3/4" pipe size		158	.051		1.03	2.97		4	5.60
4450	1" pipe size		152	.053		1.07	3.09		4.16	5.85
4460	1-1/4" pipe size		148	.054		1.35	3.17		4.52	6.30
4470	1-1/2" pipe size		143	.056		1.43	3.28		4.71	6.55
4480	2" pipe size		139	.058		1.55	3.38		4.93	6.80
4490	2-1/2" pipe size		134	.060		2.53	3.50		6.03	8.10
4500	3" pipe size		128	.063		2.86	3.67		6.53	8.70
4510	3-1/2" pipe size		122	.066		3.08	3.85		6.93	9.20
4520	4" pipe size		117	.068		3.12	4.01		7.13	9.50
4530	5" pipe size		114	.070		3.66	4.12		7.78	10.25
4540	6" pipe size		111	.072		6.95	4.23		11.18	14.05
4580	For plastic coating on 1/2" thru 6" size, add					150%				
8800	Wire cable support system									
8810	Cable with hook terminal and locking device									
8830	2 mm, (.079") dia cable, (100 lb. cap.)									
8840	1 m, (3.3') length, with hook	1 Shee	96	.083	Ea.	3.03	4.66		7.69	10.45
8850	2 m, (6.6') length, with hook		84	.095		3.50	5.35		8.85	12
8860	3 m, (9.9') length, with hook		72	.111		3.97	6.20		10.17	13.85
8870	5 m, (16.4') length, with hook	Q-9	60	.267		4.97	13.45		18.42	26
8880	10 m, (32.8') length, with hook	"	30	.533		7.20	27		34.20	49
8900	3mm, (.118") dia cable, (200 lb. cap.)									
8910	1 m, (3.3') length, with hook	1 Shee	96	.083	Ea.	3.88	4.66		8.54	11.35
8920	2 m, (6.6') length, with hook		84	.095		4.36	5.35		9.71	12.95
8930	3 m, (9.9') length, with hook		72	.111		4.81	6.20		11.01	14.80
8940	5 m, (16.4') length, with hook	Q-9	60	.267		5.95	13.45		19.40	27
8950	10 m, (32.8') length, with hook	"	30	.533		8.55	27		35.55	50.50
9000	Cable system accessories									
9010	Anchor bolt, 3/8", with nut	1 Shee	140	.057	Ea.	1.16	3.20		4.36	6.15
9020	Air duct corner protector		160	.050		.52	2.80		3.32	4.85
9030	Air duct support attachment		140	.057		.99	3.20		4.19	6
9040	Flange clip, hammer-on style									
9044	For flange thickness 3/32" - 9/64", 160 lb. cap.	1 Shee	180	.044	Ea.	.29	2.49		2.78	4.12
9048	For flange thickness 1/8" - 1/4", 200 lb. cap.		160	.050		.29	2.80		3.09	4.60
9052	For flange thickness 5/16" - 1/2", 200 lb. cap.		150	.053		.56	2.98		3.54	5.20
9056	For flange thickness 9/16" - 3/4", 200 lb. cap.		140	.057		.75	3.20		3.95	5.70
9060	Wire insulation protection tube		180	.044	L.F.	.34	2.49		2.83	4.17
9070	Wire cutter				Ea.	41.50			41.50	45.50

22 33 Electric Domestic Water Heaters

22 33 13 – Instantaneous Electric Domestic Water Heaters

22 33 13.10 Hot Water Dispensers

		Crew	Daily Output	Labor-Hours	Unit	Material	Labor	Equipment	Total	Total Incl O&P
0010	**HOT WATER DISPENSERS**									
0160	Commercial, 100 cup, 11.3 amp	1 Plum	14	.571	Ea.	510	33.50		543.50	615
3180	Household, 60 cup	"	14	.571	"	269	33.50		302.50	345

22 33 30 – Residential, Electric Domestic Water Heaters

22 33 30.13 Residential, Small-Capacity Elec. Water Heaters

		Crew	Daily Output	Labor-Hours	Unit	Material	Labor	Equipment	Total	Total Incl O&P
0010	**RESIDENTIAL, SMALL-CAPACITY ELECTRIC DOMESTIC WATER HEATERS**									
1000	Residential, electric, glass lined tank, 5 yr., 10 gal., single element R224000-10	1 Plum	2.30	3.478	Ea.	330	204		534	670
1040	20 gallon, single element R224000-20		2.20	3.636		410	213		623	770

22 33 Electric Domestic Water Heaters

22 33 30 – Residential, Electric Domestic Water Heaters

22 33 30.13 Residential, Small-Capacity Elec. Water Heaters		Crew	Daily Output	Labor-Hours	Unit	2015 Bare Costs Material	Labor	Equipment	Total	Total Incl O&P
1060	30 gallon, double element	1 Plum	2.20	3.636	Ea.	475	213		688	845
1080	40 gallon, double element		2	4		800	235		1,035	1,225
1100	52 gallon, double element		2	4		895	235		1,130	1,350
1120	66 gallon, double element		1.80	4.444		1,200	261		1,461	1,725
1140	80 gallon, double element		1.60	5		1,350	294		1,644	1,925
1180	120 gallon, double element		1.40	5.714		1,900	335		2,235	2,575

22 33 33 – Light-Commercial Electric Domestic Water Heaters

22 33 33.10 Commercial Electric Water Heaters		Crew	Daily Output	Labor-Hours	Unit	2015 Bare Costs Material	Labor	Equipment	Total	Total Incl O&P
0010	**COMMERCIAL ELECTRIC WATER HEATERS**									
4000	Commercial, 100° rise. NOTE: for each size tank, a range of									
4010	heaters between the ones shown are available									
4020	Electric									
4100	5 gal., 3 kW, 12 GPH, 208 volt	1 Plum	2	4	Ea.	2,850	235		3,085	3,500
4120	10 gal., 6 kW, 25 GPH, 208 volt		2	4		3,175	235		3,410	3,825
4130	30 gal., 24 kW, 98 GPH, 208 volt		1.92	4.167		5,200	245		5,445	6,100
4136	40 gal., 36 kW, 148 GPH, 208 volt		1.88	4.255		6,225	250		6,475	7,225
4140	50 gal., 9 kW, 37 GPH, 208 volt		1.80	4.444		4,350	261		4,611	5,175
4160	50 gal., 36 kW, 148 GPH, 208 volt		1.80	4.444		6,650	261		6,911	7,700
4180	80 gal., 12 kW, 49 GPH, 208 volt		1.50	5.333		5,350	315		5,665	6,375
4200	80 gal., 36 kW, 148 GPH, 208 volt		1.50	5.333		7,400	315		7,715	8,625
4220	100 gal., 36 kW, 148 GPH, 208 volt		1.20	6.667		7,725	390		8,115	9,100
4240	120 gal., 36 kW, 148 GPH, 208 volt		1.20	6.667		8,050	390		8,440	9,450
4260	150 gal., 15 kW , 61 GPH, 480 volt		1	8		18,900	470		19,370	21,500
4280	150 gal., 120 kW, 490 GPH, 480 volt		1	8		26,600	470		27,070	29,900
4300	200 gal., 15 kW, 61 GPH, 480 volt	Q-1	1.70	9.412		20,500	495		20,995	23,300
4320	200 gal., 120 kW , 490 GPH, 480 volt		1.70	9.412		28,000	495		28,495	31,600
4340	250 gal., 15 kW, 61 GPH, 480 volt		1.50	10.667		21,100	565		21,665	24,100
4360	250 gal., 150 kW, 615 GPH, 480 volt		1.50	10.667		30,800	565		31,365	34,800
4380	300 gal., 30 kW, 123 GPH, 480 volt		1.30	12.308		23,300	650		23,950	26,600
4400	300 gal., 180 kW, 738 GPH, 480 volt		1.30	12.308		41,700	650		42,350	46,900
4420	350 gal., 30 kW, 123 GPH, 480 volt		1.10	14.545		24,600	770		25,370	28,200
4440	350 gal., 180 kW, 738 GPH, 480 volt		1.10	14.545		34,700	770		35,470	39,300
4460	400 gal., 30 kW, 123 GPH, 480 volt		1	16		27,800	845		28,645	31,900
4480	400 gal., 210 kW, 860 GPH, 480 volt		1	16		40,000	845		40,845	45,300
4500	500 gal., 30 kW, 123 GPH, 480 volt		.80	20		32,500	1,050		33,550	37,300
4520	500 gal., 240 kW, 984 GPH, 480 volt		.80	20		47,800	1,050		48,850	54,000
4540	600 gal., 30 kW, 123 GPH, 480 volt	Q-2	1.20	20		24,700	1,100		25,800	28,900
4560	600 gal., 300 kW, 1230 GPH, 480 volt		1.20	20		37,600	1,100		38,700	43,000
4580	700 gal., 30 kW, 123 GPH, 480 volt		1	24		26,000	1,325		27,325	30,600
4600	700 gal., 300 kW, 1230 GPH, 480 volt		1	24		39,100	1,325		40,425	45,000
4620	800 gal., 60 kW, 245 GPH, 480 volt		.90	26.667		34,800	1,450		36,250	40,500
4640	800 gal., 300 kW, 1230 GPH, 480 volt		.90	26.667		40,000	1,450		41,450	46,200
4660	1000 gal., 60 kW, 245 GPH, 480 volt		.70	34.286		30,800	1,875		32,675	36,700
4680	1000 gal., 480 kW, 1970 GPH, 480 volt		.70	34.286		51,000	1,875		52,875	59,500
4700	1200 gal., 60 kW, 245 GPH, 480 volt		.60	40		52,500	2,200		54,700	61,000
4720	1200 gal., 480 kW, 1970 GPH, 480 volt		.60	40		79,500	2,200		81,700	91,000
4740	1500 gal., 60 kW, 245 GPH, 480 volt		.50	48		69,000	2,625		71,625	80,000
4760	1500 gal., 480 kW, 1970 GPH, 480 volt		.50	48		95,500	2,625		98,125	109,500
5400	Modulating step control for under 90 kW, 2-5 steps	1 Elec	5.30	1.509		810	82.50		892.50	1,025
5440	1 through 5 steps beyond standard		3.20	2.500		221	137		358	450
5460	6 through 10 steps beyond standard		2.70	2.963		455	162		617	745
5480	11 through 18 steps beyond standard		1.60	5		680	274		954	1,150

22 34 Fuel-Fired Domestic Water Heaters

22 34 30 – Residential Gas Domestic Water Heaters

22 34 30.13 Residential, Atmos, Gas Domestic Wtr Heaters		Crew	Daily Output	Labor-Hours	Unit	Material	Labor	Equipment	Total	Total Incl O&P
						2015 Bare Costs				
0010	**RESIDENTIAL, ATMOSPHERIC, GAS DOMESTIC WATER HEATERS**									
2000	Gas fired, foam lined tank, 10 yr., vent not incl.									
2040	30 gallon	1 Plum	2	4	Ea.	895	235		1,130	1,350
2060	40 gallon		1.90	4.211		895	247		1,142	1,350
2080	50 gallon		1.80	4.444		935	261		1,196	1,425
2090	60 gallon		1.70	4.706		1,325	276		1,601	1,875
2100	75 gallon		1.50	5.333		1,350	315		1,665	1,975
2120	100 gallon		1.30	6.154		1,600	360		1,960	2,325
2900	Water heater, safety-drain pan, 26" round		20	.400		37	23.50		60.50	76.50

22 34 46 – Oil-Fired Domestic Water Heaters

22 34 46.10 Residential Oil-Fired Water Heaters		Crew	Daily Output	Labor-Hours	Unit	Material	Labor	Equipment	Total	Total Incl O&P
0010	**RESIDENTIAL OIL-FIRED WATER HEATERS**									
3000	Oil fired, glass lined tank, 5 yr., vent not included, 30 gallon	1 Plum	2	4	Ea.	1,175	235		1,410	1,625
3040	50 gallon		1.80	4.444		1,375	261		1,636	1,925
3060	70 gallon		1.50	5.333		1,975	315		2,290	2,650

Estimating Tips

The labor adjustment factors listed in Subdivision 22 01 02.20 also apply to Division 23.

23 10 00 Facility Fuel Systems

- The prices in this subdivision for above- and below-ground storage tanks do not include foundations or hold-down slabs, unless noted. The estimator should refer to Divisions 3 and 31 for foundation system pricing. In addition to the foundations, required tank accessories, such as tank gauges, leak detection devices, and additional manholes and piping, must be added to the tank prices.

23 50 00 Central Heating Equipment

- When estimating the cost of an HVAC system, check to see who is responsible for providing and installing the temperature control system. It is possible to overlook controls, assuming that they would be included in the electrical estimate.
- When looking up a boiler, be careful on specified capacity. Some manufacturers rate their products on output while others use input.
- Include HVAC insulation for pipe, boiler, and duct (wrap and liner).
- Be careful when looking up mechanical items to get the correct pressure rating and connection type (thread, weld, flange).

23 70 00 Central HVAC Equipment

- Combination heating and cooling units are sized by the air conditioning requirements. (See Reference No. R236000-20 for preliminary sizing guide.)
- A ton of air conditioning is nominally 400 CFM.
- Rectangular duct is taken off by the linear foot for each size, but its cost is usually estimated by the pound. Remember that SMACNA standards now base duct on internal pressure.
- Prefabricated duct is estimated and purchased like pipe: straight sections and fittings.
- Note that cranes or other lifting equipment are not included on any lines in Division 23. For example, if a crane is required to lift a heavy piece of pipe into place high above a gym floor, or to put a rooftop unit on the roof of a four-story building, etc., it must be added. Due to the potential for extreme variation—from nothing additional required to a major crane or helicopter—we feel that including a nominal amount for "lifting contingency" would be useless and detract from the accuracy of the estimate. When using equipment rental cost data from RSMeans, do not forget to include the cost of the operator(s).

Reference Numbers

Reference numbers are shown in shaded boxes at the beginning of some major classifications. These numbers refer to related items in the Reference Section. The reference information may be an estimating procedure, an alternate pricing method, or technical information.

Note: Not all subdivisions listed here necessarily appear in this publication. ■

Note: ***Trade Service****, in part, has been used as a reference source for some of the material prices used in Division 23.*

23 05 Common Work Results for HVAC

23 05 02 – HVAC General

23 05 02.10 Air Conditioning, General		Crew	Daily Output	Labor-Hours	Unit	2015 Bare Costs Material	Labor	Equipment	Total	Total Incl O&P
0010	**AIR CONDITIONING, GENERAL** Prices are for standard efficiencies (SEER 13)									
0020	for upgrade to SEER 14 add					10%				

23 09 Instrumentation and Control for HVAC

23 09 23 – Direct-Digital Control System for HVAC

23 09 23.10 Control Components/DDC Systems		Crew	Daily Output	Labor-Hours	Unit	Material	Labor	Equipment	Total	Total Incl O&P
0010	**CONTROL COMPONENTS/DDC SYSTEMS** (Sub's quote incl. M & L)									
0100	Analog inputs									
0110	Sensors (avg. 50′ run in 1/2″ EMT)									
0120	Duct temperature				Ea.				415	415
0130	Space temperature								665	665
0140	Duct humidity, +/- 3%								695	695
0150	Space humidity, +/- 2%								1,075	1,075
0160	Duct static pressure								565	565
0170	CFM/Transducer								765	765
0180	KW/Transducer								1,350	1,350
0182	KWH totalization (not incl. elec. meter pulse xmtr.)								625	625
0190	Space static pressure				▼				1,075	1,075
1000	Analog outputs (avg. 50′ run in 1/2″ EMT)									
1010	P/I Transducer				Ea.				635	635
1020	Analog output, matl. in MUX								305	305
1030	Pneumatic (not incl. control device)								645	645
1040	Electric (not incl. control device)				▼				380	380
2000	Status (Alarms)									
2100	Digital inputs (avg. 50′ run in 1/2″ EMT)									
2110	Freeze				Ea.				435	435
2120	Fire								395	395
2130	Differential pressure, (air)								595	595
2140	Differential pressure, (water)								975	975
2150	Current sensor								435	435
2160	Duct high temperature thermostat								570	570
2170	Duct smoke detector				▼				705	705
2200	Digital output (avg. 50′ run in 1/2″ EMT)									
2210	Start/stop				Ea.				340	340
2220	On/off (maintained contact)				"				585	585
3000	Controller MUX panel, incl. function boards									
3100	48 point				Ea.				5,275	5,275
3110	128 point				"				7,225	7,225
3200	DDC controller (avg. 50′ run in conduit)									
3210	Mechanical room									
3214	16 point controller (incl. 120 volt/1 phase power supply)				Ea.				3,275	3,275
3229	32 point controller (incl. 120 volt/1 phase power supply)				"				5,425	5,425
3230	Includes software programming and checkout									
3260	Space									
3266	VAV terminal box (incl. space temp. sensor)				Ea.				840	840
3280	Host computer (avg. 50′ run in conduit)									
3281	Package complete with PC, keyboard,									
3282	printer, monitor, basic software				Ea.				3,150	3,150
4000	Front end costs									
4100	Computer (P.C.) with software program				Ea.				6,350	6,350
4200	Color graphics software				▼				3,925	3,925

23 09 23 – Direct-Digital Control System for HVAC

23 09 23.10 Control Components/DDC Systems		Crew	Daily Output	Labor-Hours	Unit	2015 Bare Costs Material	Labor	Equipment	Total	Total Incl O&P
4300	Color graphics slides				Ea.				490	490
4350	Additional printer								980	980
4400	Communications trunk cable				L.F.				3.80	3.80
4500	Engineering labor, (not incl. dftg.)				Point				94	94
4600	Calibration labor								120	120
4700	Start-up, checkout labor								120	120
4800	Programming labor, as req'd									
5000	Communications bus (data transmission cable)									
5010	#18 twisted shielded pair in 1/2" EMT conduit				C.L.F.				380	380
8000	Applications software									
8050	Basic maintenance manager software (not incl. data base entry)				Ea.				1,950	1,950
8100	Time program				Point				6.85	6.85
8120	Duty cycle								13.65	13.65
8140	Optimum start/stop								41.50	41.50
8160	Demand limiting								20.50	20.50
8180	Enthalpy program								41.50	41.50
8200	Boiler optimization				Ea.				1,225	1,225
8220	Chiller optimization				"				1,625	1,625
8240	Custom applications									
8260	Cost varies with complexity									

23 09 43 – Pneumatic Control System for HVAC

23 09 43.10 Pneumatic Control Systems		Crew	Daily Output	Labor-Hours	Unit	Material	Labor	Equipment	Total	Total Incl O&P
0010	**PNEUMATIC CONTROL SYSTEMS**									
0011	Including a nominal 50 ft. of tubing. Add control panelboard if req'd.									
0100	Heating and ventilating, split system									
0200	Mixed air control, economizer cycle, panel readout, tubing									
0220	Up to 10 tons G	Q-19	.68	35.294	Ea.	4,200	1,900		6,100	7,500
0240	For 10 to 20 tons G		.63	37.915		4,500	2,050		6,550	8,050
0260	For over 20 tons G		.58	41.096		4,875	2,225		7,100	8,700
0270	Enthalpy cycle, up to 10 tons		.50	48.387		4,650	2,625		7,275	9,050
0280	For 10 to 20 tons		.46	52.174		5,000	2,825		7,825	9,750
0290	For over 20 tons		.42	56.604		5,425	3,050		8,475	10,600
0300	Heating coil, hot water, 3 way valve,									
0320	Freezestat, limit control on discharge, readout	Q-5	.69	23.088	Ea.	3,125	1,250		4,375	5,300
0500	Cooling coil, chilled water, room									
0520	Thermostat, 3 way valve	Q-5	2	8	Ea.	1,400	430		1,830	2,175
0600	Cooling tower, fan cycle, damper control,									
0620	Control system including water readout in/out at panel	Q-19	.67	35.821	Ea.	5,525	1,925		7,450	9,000
1000	Unit ventilator, day/night operation,									
1100	freezestat, ASHRAE, cycle 2	Q-19	.91	26.374	Ea.	3,050	1,425		4,475	5,525
2000	Compensated hot water from boiler, valve control,									
2100	readout and reset at panel, up to 60 GPM	Q-19	.55	43.956	Ea.	5,725	2,375		8,100	9,875
2120	For 120 GPM		.51	47.059		6,125	2,550		8,675	10,600
2140	For 240 GPM		.49	49.180		6,400	2,650		9,050	11,100
3000	Boiler room combustion air, damper to 5 S.F., controls		1.37	17.582		2,750	950		3,700	4,450
3500	Fan coil, heating and cooling valves, 4 pipe control system		3	8		1,250	435		1,685	2,025
3600	Heat exchanger system controls		.86	27.907		2,675	1,500		4,175	5,225
3900	Multizone control (one per zone), includes thermostat, damper									
3910	motor and reset of discharge temperature	Q-5	.51	31.373	Ea.	2,750	1,675		4,425	5,600
4000	Pneumatic thermostat, including controlling room radiator valve	"	2.43	6.593		830	355		1,185	1,450
4040	Program energy saving optimizer G	Q-19	1.21	19.786		6,900	1,075		7,975	9,200
4060	Pump control system	"	3	8		1,275	435		1,710	2,050

23 09 Instrumentation and Control for HVAC

23 09 43 – Pneumatic Control System for HVAC

23 09 43.10 Pneumatic Control Systems		Crew	Daily Output	Labor-Hours	Unit	2015 Bare Costs Material	Labor	Equipment	Total	Total Incl O&P
4080	Reheat coil control system, not incl. coil	Q-5	2.43	6.593	Ea.	1,075	355		1,430	1,725
4500	Air supply for pneumatic control system									
4600	Tank mounted duplex compressor, starter, alternator,									
4620	piping, dryer, PRV station and filter									
4630	1/2 HP	Q-19	.68	35.139	Ea.	10,300	1,900		12,200	14,200
4640	3/4 HP		.64	37.383		10,800	2,025		12,825	14,900
4650	1 HP		.61	39.539		11,800	2,150		13,950	16,100
4660	1-1/2 HP		.58	41.739		12,500	2,250		14,750	17,200
4680	3 HP		.55	43.956		17,100	2,375		19,475	22,400
4690	5 HP	↓	.42	57.143	↓	29,800	3,100		32,900	37,500
4800	Main air supply, includes 3/8" copper main and labor	Q-5	1.82	8.791	C.L.F.	345	475		820	1,100
4810	If poly tubing used, deduct								30%	30%
7000	Static pressure control for air handling unit, includes pressure									
7010	sensor, receiver controller, readout and damper motors	Q-19	.64	37.383	Ea.	8,175	2,025		10,200	12,000
7020	If return air fan requires control, add								70%	70%
8600	VAV boxes, incl. thermostat, damper motor, reheat coil & tubing	Q-5	1.46	10.989		1,275	590		1,865	2,300
8610	If no reheat coil, deduct				↓				204	204

23 09 53 – Pneumatic and Electric Control System for HVAC

23 09 53.10 Control Components		Crew	Daily Output	Labor-Hours	Unit	Material	Labor	Equipment	Total	Total Incl O&P
0010	**CONTROL COMPONENTS**									
0700	Controller, receiver									
0850	Electric, single snap switch	1 Elec	4	2	Ea.	465	109		574	675
0860	Dual snap switches	"	3	2.667	"	625	146		771	910
3590	Sensor, electric operated									
3620	Humidity	1 Elec	8	1	Ea.	157	54.50		211.50	255
3650	Pressure		8	1		248	54.50		302.50	355
3680	Temperature	↓	10	.800	↓	122	44		166	201
5000	Thermostats									
5200	24 hour, automatic, clock G	1 Shee	8	1	Ea.	165	56		221	268
5220	Electric, low voltage, 2 wire	1 Elec	13	.615		49.50	33.50		83	105
5230	3 wire	"	10	.800	↓	36	44		80	106
7090	Valves, motor controlled, including actuator									
7100	Electric motor actuated									
7200	Brass, two way, screwed									
7210	1/2" pipe size	L-6	36	.333	Ea.	296	19.10		315.10	355
7220	3/4" pipe size		30	.400		525	23		548	610
7230	1" pipe size		28	.429		605	24.50		629.50	700
7240	1-1/2" pipe size		19	.632		665	36		701	785
7250	2" pipe size	↓	16	.750	↓	975	43		1,018	1,150
7350	Brass, three way, screwed									
7360	1/2" pipe size	L-6	33	.364	Ea.	320	21		341	385
7370	3/4" pipe size		27	.444		375	25.50		400.50	450
7380	1" pipe size		25.50	.471		485	27		512	575
7384	1-1/4" pipe size		21	.571		620	33		653	730
7390	1-1/2" pipe size		17	.706		705	40.50		745.50	835
7400	2" pipe size	↓	14	.857	↓	945	49		994	1,100
7550	Iron body, two way, flanged									
7560	2-1/2" pipe size	L-6	4	3	Ea.	1,400	172		1,572	1,800
7570	3" pipe size		3	4		1,475	229		1,704	1,975
7580	4" pipe size	↓	2	6	↓	2,550	345		2,895	3,350
7850	Iron body, three way, flanged									
7860	2-1/2" pipe size	L-6	3	4	Ea.	1,425	229		1,654	1,900

23 09 Instrumentation and Control for HVAC

23 09 53 – Pneumatic and Electric Control System for HVAC

23 09 53.10 Control Components		Crew	Daily Output	Labor-Hours	Unit	2015 Bare Costs Material	Labor	Equipment	Total	Total Incl O&P
7870	3″ pipe size	L-6	2.50	4.800	Ea.	1,650	275		1,925	2,225
7880	4″ pipe size		2	6		2,050	345		2,395	2,775

23 34 HVAC Fans

23 34 13 – Axial HVAC Fans

23 34 13.10 Axial Flow HVAC Fans

Line	Description	Crew	Daily Output	Labor-Hours	Unit	Material	Labor	Equipment	Total	Total Incl O&P
0010	**AXIAL FLOW HVAC FANS**									
0020	Air conditioning and process air handling									
1500	Vaneaxial, low pressure, 2000 CFM, 1/2 HP	Q-20	3.60	5.556	Ea.	2,225	285		2,510	2,875
1520	4,000 CFM, 1 HP		3.20	6.250		2,600	320		2,920	3,325
1540	8,000 CFM, 2 HP		2.80	7.143		3,275	365		3,640	4,175
1560	16,000 CFM, 5 HP		2.40	8.333		4,600	425		5,025	5,725

23 34 14 – Blower HVAC Fans

23 34 14.10 Blower Type HVAC Fans

Line	Description	Crew	Daily Output	Labor-Hours	Unit	Material	Labor	Equipment	Total	Total Incl O&P
0010	**BLOWER TYPE HVAC FANS**									
2000	Blowers, direct drive with motor, complete									
2020	1045 CFM @ .5″ S.P., 1/5 HP	Q-20	18	1.111	Ea.	340	57		397	460
2040	1385 CFM @ .5″ S.P., 1/4 HP		18	1.111		335	57		392	455
2060	1640 CFM @ .5″ S.P., 1/3 HP		18	1.111		305	57		362	420
2080	1760 CFM @ .5″ S.P., 1/2 HP		18	1.111		315	57		372	435
2090	4 speed									
2100	1164 to 1739 CFM @ .5″ S.P., 1/3 HP	Q-20	16	1.250	Ea.	320	64		384	455
2120	1467 to 2218 CFM @ 1.0″ S.P., 3/4 HP	″	14	1.429	″	370	73		443	515
2500	Ceiling fan, right angle, extra quiet, 0.10″ S.P.									
2520	95 CFM	Q-20	20	1	Ea.	300	51		351	410
2540	210 CFM		19	1.053		355	54		409	470
2560	385 CFM		18	1.111		450	57		507	580
2580	885 CFM		16	1.250		890	64		954	1,075
2600	1,650 CFM		13	1.538		1,225	79		1,304	1,475
2620	2,960 CFM		11	1.818		1,650	93		1,743	1,950
2680	For speed control switch, add	1 Elec	16	.500		164	27.50		191.50	222
7500	Utility set, steel construction, pedestal, 1/4″ S.P.									
7520	Direct drive, 150 CFM, 1/8 HP	Q-20	6.40	3.125	Ea.	870	160		1,030	1,200
7540	485 CFM, 1/6 HP		5.80	3.448		1,100	177		1,277	1,475
7560	1950 CFM, 1/2 HP		4.80	4.167		1,275	213		1,488	1,725
7580	2410 CFM, 3/4 HP		4.40	4.545		2,375	233		2,608	2,950
7600	3328 CFM, 1-1/2 HP		3	6.667		2,625	340		2,965	3,425
7680	V-belt drive, drive cover, 3 phase									
7700	800 CFM, 1/4 HP	Q-20	6	3.333	Ea.	980	171		1,151	1,325
7720	1,300 CFM, 1/3 HP		5	4		1,025	205		1,230	1,425
7740	2,000 CFM, 1 HP		4.60	4.348		1,225	223		1,448	1,700
7760	2,900 CFM, 3/4 HP		4.20	4.762		1,650	244		1,894	2,175
7780	3,600 CFM, 3/4 HP		4	5		2,025	256		2,281	2,625
7800	4,800 CFM, 1 HP		3.50	5.714		2,375	293		2,668	3,075
7820	6,700 CFM, 1-1/2 HP		3	6.667		2,950	340		3,290	3,750
7830	7,500 CFM, 2 HP		2.50	8		4,000	410		4,410	5,025
7840	11,000 CFM, 3 HP		2	10		5,325	510		5,835	6,650
7860	13,000 CFM, 3 HP		1.60	12.500		5,425	640		6,065	6,925
7880	15,000 CFM, 5 HP		1	20		5,600	1,025		6,625	7,725
7900	17,000 CFM, 7-1/2 HP		.80	25		6,000	1,275		7,275	8,550

23 34 HVAC Fans

23 34 14 – Blower HVAC Fans

23 34 14.10 Blower Type HVAC Fans		Crew	Daily Output	Labor-Hours	Unit	2015 Bare Costs Material	Labor	Equipment	Total	Total Incl O&P
7920	20,000 CFM, 7-1/2 HP	Q-20	.80	25	Ea.	7,150	1,275		8,425	9,825

23 34 16 – Centrifugal HVAC Fans

23 34 16.10 Centrifugal Type HVAC Fans		Crew	Daily Output	Labor-Hours	Unit	2015 Bare Costs Material	Labor	Equipment	Total	Total Incl O&P
0010	**CENTRIFUGAL TYPE HVAC FANS**									
0200	In-line centrifugal, supply/exhaust booster									
0220	aluminum wheel/hub, disconnect switch, 1/4" S.P.									
0240	500 CFM, 10" diameter connection	Q-20	3	6.667	Ea.	1,300	340		1,640	1,950
0260	1,380 CFM, 12" diameter connection		2	10		1,375	510		1,885	2,300
0280	1,520 CFM, 16" diameter connection		2	10		1,500	510		2,010	2,425
0300	2,560 CFM, 18" diameter connection		1	20		1,625	1,025		2,650	3,350
0320	3,480 CFM, 20" diameter connection		.80	25		1,925	1,275		3,200	4,075
0326	5,080 CFM, 20" diameter connection		.75	26.667		2,100	1,375		3,475	4,400
3500	Centrifugal, airfoil, motor and drive, complete									
3520	1000 CFM, 1/2 HP	Q-20	2.50	8	Ea.	1,900	410		2,310	2,700
3540	2,000 CFM, 1 HP		2	10		2,150	510		2,660	3,125
3560	4,000 CFM, 3 HP		1.80	11.111		2,725	570		3,295	3,875
3580	8,000 CFM, 7-1/2 HP		1.40	14.286		4,100	730		4,830	5,625
3600	12,000 CFM, 10 HP		1	20		5,450	1,025		6,475	7,550
5000	Utility set, centrifugal, V belt drive, motor									
5020	1/4" S.P., 1200 CFM, 1/4 HP	Q-20	6	3.333	Ea.	1,750	171		1,921	2,175
5040	1520 CFM, 1/3 HP		5	4		2,225	205		2,430	2,750
5060	1850 CFM, 1/2 HP		4	5		2,200	256		2,456	2,825
5080	2180 CFM, 3/4 HP		3	6.667		2,600	340		2,940	3,400
5100	1/2" S.P., 3600 CFM, 1 HP		2	10		2,700	510		3,210	3,750
5120	4250 CFM, 1-1/2 HP		1.60	12.500		3,300	640		3,940	4,600
5140	4800 CFM, 2 HP		1.40	14.286		4,000	730		4,730	5,500
5160	6920 CFM, 5 HP		1.30	15.385		4,875	790		5,665	6,575
5180	7700 CFM, 7-1/2 HP		1.20	16.667		5,875	855		6,730	7,775
5200	For explosion proof motor, add					15%				
5500	Fans, industrial exhauster, for air which may contain granular matl.									
5520	1000 CFM, 1-1/2 HP	Q-20	2.50	8	Ea.	2,875	410		3,285	3,800
5540	2000 CFM, 3 HP		2	10		3,500	510		4,010	4,625
5560	4000 CFM, 7-1/2 HP		1.80	11.111		4,875	570		5,445	6,225
5580	8000 CFM, 15 HP		1.40	14.286		6,275	730		7,005	8,025
5600	12,000 CFM, 30 HP		1	20		10,200	1,025		11,225	12,800
7000	Roof exhauster, centrifugal, aluminum housing, 12" galvanized									
7020	curb, bird screen, back draft damper, 1/4" S.P.									
7100	Direct drive, 320 CFM, 11" sq. damper	Q-20	7	2.857	Ea.	705	146		851	1,000
7120	600 CFM, 11" sq. damper		6	3.333		900	171		1,071	1,250
7140	815 CFM, 13" sq. damper		5	4		900	205		1,105	1,300
7160	1450 CFM, 13" sq. damper		4.20	4.762		1,450	244		1,694	1,975
7180	2050 CFM, 16" sq. damper		4	5		1,725	256		1,981	2,300
7200	V-belt drive, 1650 CFM, 12" sq. damper		6	3.333		1,300	171		1,471	1,675
7220	2750 CFM, 21" sq. damper		5	4		1,550	205		1,755	2,000
7230	3500 CFM, 21" sq. damper		4.50	4.444		1,725	228		1,953	2,250
7240	4910 CFM, 23" sq. damper		4	5		2,125	256		2,381	2,725
7260	8525 CFM, 28" sq. damper		3	6.667		2,800	340		3,140	3,600
7280	13,760 CFM, 35" sq. damper		2	10		3,925	510		4,435	5,100
7300	20,558 CFM, 43" sq. damper		1	20		7,875	1,025		8,900	10,200
7320	For 2 speed winding, add					15%				
7340	For explosionproof motor, add					600			600	660
7360	For belt driven, top discharge, add					15%				

23 34 HVAC Fans

23 34 16 – Centrifugal HVAC Fans

23 34 16.10 Centrifugal Type HVAC Fans		Crew	Daily Output	Labor-Hours	Unit	2015 Bare Costs Material	Labor	Equipment	Total	Total Incl O&P
8500	Wall exhausters, centrifugal, auto damper, 1/8" S.P.									
8520	Direct drive, 610 CFM, 1/20 HP	Q-20	14	1.429	Ea.	425	73		498	575
8540	796 CFM, 1/12 HP		13	1.538		880	79		959	1,075
8560	822 CFM, 1/6 HP		12	1.667		1,075	85.50		1,160.50	1,300
8580	1,320 CFM, 1/4 HP		12	1.667		1,250	85.50		1,335.50	1,500
8600	1756 CFM, 1/4 HP		11	1.818		1,250	93		1,343	1,525
8620	1983 CFM, 1/4 HP		10	2		1,300	102		1,402	1,575
8640	2900 CFM, 1/2 HP		9	2.222		1,375	114		1,489	1,675
8660	3307 CFM, 3/4 HP		8	2.500		1,475	128		1,603	1,825
9500	V-belt drive, 3 phase									
9520	2,800 CFM, 1/4 HP	Q-20	9	2.222	Ea.	1,925	114		2,039	2,300
9540	3,740 CFM, 1/2 HP		8	2.500		2,000	128		2,128	2,400
9560	4400 CFM, 3/4 HP		7	2.857		2,025	146		2,171	2,450
9580	5700 CFM, 1-1/2 HP		6	3.333		2,100	171		2,271	2,550

23 34 23 – HVAC Power Ventilators

23 34 23.10 HVAC Power Circulators and Ventilators

		Crew	Daily Output	Labor-Hours	Unit	Material	Labor	Equipment	Total	Total Incl O&P
0010	**HVAC POWER CIRCULATORS AND VENTILATORS**									
3000	Paddle blade air circulator, 3 speed switch									
3020	42", 5,000 CFM high, 3000 CFM low G	1 Elec	2.40	3.333	Ea.	163	182		345	450
3040	52", 6,500 CFM high, 4000 CFM low G	"	2.20	3.636	"	170	199		369	485
3100	For antique white motor, same cost									
3200	For brass plated motor, same cost									
3300	For light adaptor kit, add G				Ea.	41			41	45
6000	Propeller exhaust, wall shutter									
6020	Direct drive, one speed, .075" S.P.									
6100	653 CFM, 1/30 HP	Q-20	10	2	Ea.	192	102		294	365
6120	1033 CFM, 1/20 HP		9	2.222		287	114		401	490
6140	1323 CFM, 1/15 HP		8	2.500		320	128		448	545
6160	2444 CFM, 1/4 HP		7	2.857		400	146		546	665
6300	V-belt drive, 3 phase									
6320	6175 CFM, 3/4 HP	Q-20	5	4	Ea.	2,950	205		3,155	3,550
6340	7500 CFM, 3/4 HP		5	4		3,025	205		3,230	3,625
6360	10,100 CFM, 1 HP		4.50	4.444		3,175	228		3,403	3,850
6380	14,300 CFM, 1-1/2 HP		4	5		3,450	256		3,706	4,175
6400	19,800 CFM, 2 HP		3	6.667		1,950	340		2,290	2,650
6420	26,250 CFM, 3 HP		2.60	7.692		2,150	395		2,545	2,975
6440	38,500 CFM, 5 HP		2.20	9.091		2,550	465		3,015	3,500
6460	46,000 CFM, 7-1/2 HP		2	10		2,725	510		3,235	3,775
6480	51,500 CFM, 10 HP		1.80	11.111		2,825	570		3,395	4,000
6650	Residential, bath exhaust, grille, back draft damper									
6660	50 CFM	Q-20	24	.833	Ea.	63.50	42.50		106	135
6670	110 CFM		22	.909		98	46.50		144.50	179
6680	Light combination, squirrel cage, 100 watt, 70 CFM		24	.833		112	42.50		154.50	188
6700	Light/heater combination, ceiling mounted									
6710	70 CFM, 1450 watt	Q-20	24	.833	Ea.	162	42.50		204.50	243
6800	Heater combination, recessed, 70 CFM		24	.833		67.50	42.50		110	139
6820	With 2 infrared bulbs		23	.870		105	44.50		149.50	184
6900	Kitchen exhaust, grille, complete, 160 CFM		22	.909		108	46.50		154.50	190
6910	180 CFM		20	1		90.50	51		141.50	178
6920	270 CFM		18	1.111		171	57		228	275
6930	350 CFM		16	1.250		129	64		193	240
6940	Residential roof jacks and wall caps									

23 34 HVAC Fans

23 34 23 – HVAC Power Ventilators

23 34 23.10	HVAC Power Circulators and Ventilators	Crew	Daily Output	Labor-Hours	Unit	2015 Bare Costs Material	Labor	Equipment	Total	Total Incl O&P
6944	Wall cap with back draft damper									
6946	3" & 4" diam. round duct	1 Shee	11	.727	Ea.	26	40.50		66.50	90.50
6948	6" diam. round duct	"	11	.727	"	64	40.50		104.50	133
6958	Roof jack with bird screen and back draft damper									
6960	3" & 4" diam. round duct	1 Shee	11	.727	Ea.	26	40.50		66.50	90.50
6962	3-1/4" x 10" rectangular duct	"	10	.800	"	48.50	45		93.50	122
6980	Transition									
6982	3-1/4" x 10" to 6" diam. round	1 Shee	20	.400	Ea.	32	22.50		54.50	69.50
8020	Attic, roof type									
8030	Aluminum dome, damper & curb									
8080	12" diameter, 1000 CFM (gravity)	1 Elec	10	.800	Ea.	575	44		619	695
8090	16" diameter, 1500 CFM (gravity)		9	.889		690	48.50		738.50	835
8100	20" diameter, 2500 CFM (gravity)		8	1		850	54.50		904.50	1,025
8110	26" diameter, 4000 CFM (gravity)		7	1.143		1,025	62.50		1,087.50	1,225
8120	32" diameter, 6500 CFM (gravity)		6	1.333		1,400	73		1,473	1,650
8130	38" diameter, 8000 CFM (gravity)		5	1.600		2,100	87.50		2,187.50	2,425
8140	50" diameter, 13,000 CFM (gravity)		4	2		3,025	109		3,134	3,500
8160	Plastic, ABS dome									
8180	1050 CFM	1 Elec	14	.571	Ea.	168	31.50		199.50	232
8200	1600 CFM	"	12	.667	"	252	36.50		288.50	335
8240	Attic, wall type, with shutter, one speed									
8250	12" diameter, 1000 CFM	1 Elec	14	.571	Ea.	395	31.50		426.50	480
8260	14" diameter, 1500 CFM		12	.667		430	36.50		466.50	530
8270	16" diameter, 2000 CFM		9	.889		485	48.50		533.50	610
8290	Whole house, wall type, with shutter, one speed									
8300	30" diameter, 4800 CFM	1 Elec	7	1.143	Ea.	1,050	62.50		1,112.50	1,250
8310	36" diameter, 7000 CFM		6	1.333		1,125	73		1,198	1,350
8320	42" diameter, 10,000 CFM		5	1.600		1,275	87.50		1,362.50	1,525
8330	48" diameter, 16,000 CFM		4	2		1,575	109		1,684	1,925
8340	For two speed, add					95			95	105
8350	Whole house, lay-down type, with shutter, one speed									
8360	30" diameter, 4500 CFM	1 Elec	8	1	Ea.	1,100	54.50		1,154.50	1,300
8370	36" diameter, 6500 CFM		7	1.143		1,200	62.50		1,262.50	1,400
8380	42" diameter, 9000 CFM		6	1.333		1,300	73		1,373	1,550
8390	48" diameter, 12,000 CFM		5	1.600		1,500	87.50		1,587.50	1,750
8440	For two speed, add					71.50			71.50	78.50
8450	For 12 hour timer switch, add	1 Elec	32	.250		71.50	13.70		85.20	99

23 52 Heating Boilers

23 52 13 – Electric Boilers

23 52 13.10	Electric Boilers, ASME	Crew	Daily Output	Labor-Hours	Unit	Material	Labor	Equipment	Total	Total Incl O&P
0010	**ELECTRIC BOILERS, ASME**, Standard controls and trim									
1000	Steam, 6 KW, 20.5 MBH	Q-19	1.20	20	Ea.	3,950	1,075		5,025	5,975
1040	9 KW, 30.7 MBH		1.20	20		4,025	1,075		5,100	6,050
1060	18 KW, 61.4 MBH		1.20	20		4,150	1,075		5,225	6,200
1080	24 KW, 81.8 MBH		1.10	21.818		4,800	1,175		5,975	7,075
1120	36 KW, 123 MBH		1.10	21.818		5,325	1,175		6,500	7,625
1160	60 KW, 205 MBH		1	24		6,650	1,300		7,950	9,275
1220	112 KW, 382 MBH		.75	32		9,375	1,725		11,100	12,900
1240	148 KW, 505 MBH		.65	36.923		9,600	2,000		11,600	13,600
1260	168 KW, 573 MBH		.60	40		20,400	2,175		22,575	25,800

23 52 13 – Electric Boilers

23 52 13.10 Electric Boilers, ASME		Crew	Daily Output	Labor-Hours	Unit	Material	2015 Bare Costs Labor	Equipment	Total	Total Incl O&P
1280	222 KW, 758 MBH	Q-19	.55	43.636	Ea.	23,800	2,350		26,150	29,800
1300	296 KW, 1010 MBH		.45	53.333		25,000	2,875		27,875	32,000
1320	300 KW, 1023 MBH		.40	60		26,400	3,250		29,650	33,900
1340	370 KW, 1263 MBH		.35	68.571		29,200	3,700		32,900	37,700
1360	444 KW, 1515 MBH		.30	80		31,400	4,325		35,725	41,000
1380	518 KW, 1768 MBH	Q-21	.36	88.889		32,600	4,925		37,525	43,300
1400	592 KW, 2020 MBH		.34	94.118		35,400	5,225		40,625	46,800
1420	666 KW, 2273 MBH		.32	100		36,800	5,550		42,350	48,900
1460	740 KW, 2526 MBH		.28	114		38,300	6,350		44,650	51,500
1480	814 KW, 2778 MBH		.25	128		40,600	7,100		47,700	55,500
1500	962 KW, 3283 MBH		.22	145		44,900	8,075		52,975	61,500
1520	1036 KW, 3536 MBH		.20	160		46,700	8,875		55,575	65,000
1540	1110 KW, 3788 MBH		.19	168		49,600	9,350		58,950	68,500
1560	2070 KW, 7063 MBH		.18	177		68,000	9,875		77,875	90,000
1580	2250 KW, 7677 MBH		.17	188		80,000	10,400		90,400	103,500
1600	2,340 KW, 7984 MBH		.16	200		86,500	11,100		97,600	111,500
2000	Hot water, 7.5 KW, 25.6 MBH	Q-19	1.30	18.462		4,975	1,000		5,975	6,975
2020	15 KW, 51.2 MBH		1.30	18.462		5,000	1,000		6,000	7,000
2040	30 KW, 102 MBH		1.20	20		5,325	1,075		6,400	7,475
2060	45 KW, 164 MBH		1.20	20		5,425	1,075		6,500	7,575
2070	60 KW, 205 MBH		1.20	20		5,525	1,075		6,600	7,700
2080	75 KW, 256 MBH		1.10	21.818		5,900	1,175		7,075	8,275
2100	90 KW, 307 MBH		1.10	21.818		6,000	1,175		7,175	8,375
2120	105 KW, 358 MBH		1	24		6,275	1,300		7,575	8,850
2140	120 KW, 410 MBH		.90	26.667		6,375	1,450		7,825	9,200
2160	135 KW, 461 MBH		.75	32		7,400	1,725		9,125	10,800
2180	150 KW, 512 MBH		.65	36.923		7,500	2,000		9,500	11,200
2200	165 KW, 563 MBH		.60	40		7,600	2,175		9,775	11,600
2220	296 KW, 1010 MBH		.55	43.636		16,100	2,350		18,450	21,300
2280	370 KW, 1263 MBH		.40	60		17,900	3,250		21,150	24,600
2300	444 KW, 1515 MBH		.35	68.571		20,800	3,700		24,500	28,500
2340	518 KW, 1768 MBH	Q-21	.44	72.727		23,100	4,025		27,125	31,500
2360	592 KW, 2020 MBH		.43	74.419		26,100	4,125		30,225	35,000
2400	666 KW, 2273 MBH		.40	80		27,300	4,450		31,750	36,700
2420	740 KW, 2526 MBH		.39	82.051		28,500	4,550		33,050	38,200
2440	814 KW, 2778 MBH		.38	84.211		30,500	4,675		35,175	40,600
2460	888 KW, 3031 MBH		.37	86.486		31,100	4,800		35,900	41,400
2480	962 KW, 3283 MBH		.36	88.889		34,000	4,925		38,925	44,800
2500	1036 KW, 3536 MBH		.34	94.118		35,500	5,225		40,725	47,000
2520	1110 KW, 3788 MBH		.33	96.970		38,800	5,375		44,175	50,500
2540	1440 KW, 4915 MBH		.32	100		43,800	5,550		49,350	56,500
2560	1560 KW, 5323 MBH		.31	103		46,800	5,725		52,525	60,000
2580	1680 KW, 5733 MBH		.30	106		50,000	5,925		55,925	64,000
2600	1800 KW, 6143 MBH		.29	110		53,500	6,125		59,625	67,500
2620	1980 KW, 6757 MBH		.28	114		58,000	6,350		64,350	73,500
2640	2100 KW, 7167 MBH		.27	118		59,500	6,575		66,075	75,500
2660	2220 KW, 7576 MBH		.26	123		65,000	6,825		71,825	82,000
2680	2400 KW, 8191 MBH		.25	128		68,000	7,100		75,100	85,000
2700	2610 KW, 8905 MBH		.24	133		71,000	7,400		78,400	89,000
2720	2790 KW, 9519 MBH		.23	139		75,500	7,725		83,225	94,500
2740	2970 KW, 10133 MBH		.21	152		77,000	8,450		85,450	97,500
2760	3150 KW, 10748 MBH		.19	168		82,500	9,350		91,850	105,000
2780	3240 KW, 11055 MBH		.18	177		84,500	9,875		94,375	108,000

23 52 Heating Boilers

23 52 13 – Electric Boilers

23 52 13.10 Electric Boilers, ASME		Crew	Daily Output	Labor-Hours	Unit	Material	Labor	Equipment	Total	Total Incl O&P
						2015 Bare Costs				
2800	3420 KW, 11669 MBH	Q-21	.17	188	Ea.	90,000	10,400		100,400	114,500
2820	3600 KW, 12,283 MBH		.16	200		94,000	11,100		105,100	120,000

23 52 28 – Swimming Pool Boilers

23 52 28.10 Swimming Pool Heaters		Crew	Daily Output	Labor-Hours	Unit	Material	Labor	Equipment	Total	Total Incl O&P
0010	**SWIMMING POOL HEATERS**, Not including wiring, external									
0020	piping, base or pad,									
2000	Electric, 12 KW, 4,800 gallon pool	Q-19	3	8	Ea.	2,075	435		2,510	2,925
2020	15 KW, 7,200 gallon pool		2.80	8.571		2,100	465		2,565	3,025
2040	24 KW, 9,600 gallon pool		2.40	10		2,425	540		2,965	3,500
2060	30 KW, 12,000 gallon pool		2	12		2,475	650		3,125	3,700
2080	36 KW, 14,400 gallon pool		1.60	15		2,850	810		3,660	4,350
2100	57 KW, 24,000 gallon pool		1.20	20		3,575	1,075		4,650	5,550
9000	To select pool heater: 12 BTUH x S.F. pool area									
9010	X temperature differential = required output									
9050	For electric, KW = gallons x 2.5 divided by 1000									
9100	For family home type pool, double the									
9110	Rated gallon capacity = 1/2°F rise per hour									

23 54 Furnaces

23 54 13 – Electric-Resistance Furnaces

23 54 13.10 Electric Furnaces		Crew	Daily Output	Labor-Hours	Unit	Material	Labor	Equipment	Total	Total Incl O&P
0010	**ELECTRIC FURNACES**, Hot air, blowers, std. controls									
1000	Electric, UL listed									
1100	34.1 MBH	Q-20	4.40	4.545	Ea.	455	233		688	855
1120	51.6 MBH		4.20	4.762		495	244		739	915
1140	68.3 MBH		4	5		515	256		771	955
1160	85.3 MBH		3.80	5.263		520	270		790	985

23 82 Convection Heating and Cooling Units

23 82 16 – Air Coils

23 82 16.20 Duct Heaters		Crew	Daily Output	Labor-Hours	Unit	Material	Labor	Equipment	Total	Total Incl O&P
0010	**DUCT HEATERS**, Electric, 480 V, 3 Ph.									
0020	Finned tubular insert, 500°F									
0100	8" wide x 6" high, 4.0 kW	Q-20	16	1.250	Ea.	765	64		829	945
0120	12" high, 8.0 kW		15	1.333		1,275	68.50		1,343.50	1,500
0140	18" high, 12.0 kW		14	1.429		1,775	73		1,848	2,050
0160	24" high, 16.0 kW		13	1.538		2,300	79		2,379	2,650
0180	30" high, 20.0 kW		12	1.667		2,800	85.50		2,885.50	3,200
0300	12" wide x 6" high, 6.7 kW		15	1.333		815	68.50		883.50	1,000
0320	12" high, 13.3 kW		14	1.429		1,325	73		1,398	1,550
0340	18" high, 20.0 kW		13	1.538		1,850	79		1,929	2,150
0360	24" high, 26.7 kW		12	1.667		2,375	85.50		2,460.50	2,725
0380	30" high, 33.3 kW		11	1.818		2,900	93		2,993	3,325
0500	18" wide x 6" high, 13.3 kW		14	1.429		880	73		953	1,075
0520	12" high, 26.7 kW		13	1.538		1,500	79		1,579	1,775
0540	18" high, 40.0 kW		12	1.667		2,000	85.50		2,085.50	2,325
0560	24" high, 53.3 kW		11	1.818		2,675	93		2,768	3,075
0580	30" high, 66.7 kW		10	2		3,325	102		3,427	3,825
0700	24" wide x 6" high, 17.8 kW		13	1.538		965	79		1,044	1,200

23 82 16.20 Duct Heaters		Crew	Daily Output	Labor-Hours	Unit	Material	2015 Bare Costs Labor	Equipment	Total	Total Incl O&P
0720	12" high, 35.6 kW	Q-20	12	1.667	Ea.	1,625	85.50		1,710.50	1,925
0740	18" high, 53.3 kW		11	1.818		2,300	93		2,393	2,675
0760	24" high, 71.1 kW		10	2		2,950	102		3,052	3,400
0780	30" high, 88.9 kW		9	2.222		3,625	114		3,739	4,175
0900	30" wide x 6" high, 22.2 kW		12	1.667		1,025	85.50		1,110.50	1,250
0920	12" high, 44.4 kW		11	1.818		1,725	93		1,818	2,050
0940	18" high, 66.7 kW		10	2		2,425	102		2,527	2,825
0960	24" high, 88.9 kW		9	2.222		3,125	114		3,239	3,625
0980	30" high, 111.0 kW		8	2.500		3,825	128		3,953	4,425
1400	Note decreased kW available for									
1410	each duct size at same cost									
1420	See line 5000 for modifications and accessories									
2000	Finned tubular flange with insulated									
2020	terminal box, 500°F									
2100	12" wide x 36" high, 54 kW	Q-20	10	2	Ea.	3,775	102		3,877	4,300
2120	40" high, 60 kW		9	2.222		4,350	114		4,464	4,950
2200	24" wide x 36" high, 118.8 kW		9	2.222		4,825	114		4,939	5,500
2220	40" high, 132 kW		8	2.500		4,825	128		4,953	5,500
2400	36" wide x 8" high, 40 kW		11	1.818		1,925	93		2,018	2,275
2420	16" high, 80 kW		10	2		2,575	102		2,677	2,975
2440	24" high, 120 kW		9	2.222		3,350	114		3,464	3,875
2460	32" high, 160 kW		8	2.500		4,300	128		4,428	4,925
2480	36" high, 180 kW		7	2.857		5,300	146		5,446	6,050
2500	40" high, 200 kW		6	3.333		5,875	171		6,046	6,700
2600	40" wide x 8" high, 45 kW		11	1.818		2,000	93		2,093	2,350
2620	16" high, 90 kW		10	2		2,775	102		2,877	3,200
2640	24" high, 135 kW		9	2.222		3,500	114		3,614	4,025
2660	32" high, 180 kW		8	2.500		4,675	128		4,803	5,350
2680	36" high, 202.5 kW		7	2.857		5,375	146		5,521	6,150
2700	40" high, 225 kW		6	3.333		6,125	171		6,296	7,000
2800	48" wide x 8" high, 54.8 kW		10	2		2,100	102		2,202	2,475
2820	16" high, 109.8 kW		9	2.222		2,925	114		3,039	3,375
2840	24" high, 164.4 kW		8	2.500		3,725	128		3,853	4,275
2860	32" high, 219.2 kW		7	2.857		4,950	146		5,096	5,675
2880	36" high, 246.6 kW		6	3.333		5,700	171		5,871	6,525
2900	40" high, 274 kW		5	4		6,450	205		6,655	7,375
3000	56" wide x 8" high, 64 kW		9	2.222		2,400	114		2,514	2,800
3020	16" high, 128 kW		8	2.500		3,325	128		3,453	3,850
3040	24" high, 192 kW		7	2.857		4,025	146		4,171	4,650
3060	32" high, 256 kW		6	3.333		5,625	171		5,796	6,425
3080	36" high, 288kW		5	4		6,450	205		6,655	7,400
3100	40" high, 320kW		4	5		7,125	256		7,381	8,250
3200	64" wide x 8" high, 74kW		8	2.500		2,475	128		2,603	2,900
3220	16" high, 148kW		7	2.857		3,425	146		3,571	3,975
3240	24" high, 222kW		6	3.333		4,350	171		4,521	5,025
3260	32" high, 296kW		5	4		5,825	205		6,030	6,700
3280	36" high, 333kW		4	5		6,950	256		7,206	8,050
3300	40" high, 370kW		3	6.667		7,675	340		8,015	8,950
3800	Note decreased kW available for									
3820	each duct size at same cost									
5000	Duct heater modifications and accessories									
5120	T.C.O. limit auto or manual reset	Q-20	42	.476	Ea.	126	24.50		150.50	175
5140	Thermostat		28	.714		535	36.50		571.50	645

23 82 Convection Heating and Cooling Units

23 82 16 – Air Coils

23 82 16.20 Duct Heaters		Crew	Daily Output	Labor-Hours	Unit	Material (2015 Bare Costs)	Labor (2015 Bare Costs)	Equipment (2015 Bare Costs)	Total (2015 Bare Costs)	Total Incl O&P
5160	Overheat thermocouple (removable)	Q-20	7	2.857	Ea.	760	146		906	1,050
5180	Fan interlock relay		18	1.111		183	57		240	289
5200	Air flow switch		20	1		158	51		209	252
5220	Split terminal box cover		100	.200		52.50	10.25		62.75	73
8000	To obtain BTU multiply kW by 3413									

23 82 27 – Infrared Units

23 82 27.10 Infrared Type Heating Units		Crew	Daily Output	Labor-Hours	Unit	Material	Labor	Equipment	Total	Total Incl O&P
0010	**INFRARED TYPE HEATING UNITS**									
2000	Electric, single or three phase									
2050	6 kW, 20,478 BTU	1 Elec	2.30	3.478	Ea.	705	190		895	1,050
2100	13.5 KW, 40,956 BTU		2.20	3.636		870	199		1,069	1,250
2150	24 KW, 81,912 BTU		2	4		1,750	219		1,969	2,250

23 83 Radiant Heating Units

23 83 33 – Electric Radiant Heaters

23 83 33.10 Electric Heating		Crew	Daily Output	Labor-Hours	Unit	Material	Labor	Equipment	Total	Total Incl O&P
0010	**ELECTRIC HEATING**, not incl. conduit or feed wiring									
1100	Rule of thumb: Baseboard units, including control	1 Elec	4.40	1.818	kW	104	99.50		203.50	263
1300	Baseboard heaters, 2′ long, 350 watt		8	1	Ea.	28.50	54.50		83	114
1400	3′ long, 750 watt		8	1		33	54.50		87.50	119
1600	4′ long, 1000 watt		6.70	1.194		39	65.50		104.50	141
1800	5′ long, 935 watt		5.70	1.404		48	77		125	168
2000	6′ long, 1500 watt		5	1.600		54.50	87.50		142	191
2200	7′ long, 1310 watt		4.40	1.818		63	99.50		162.50	219
2400	8′ long, 2000 watt		4	2		66	109		175	237
2600	9′ long, 1680 watt		3.60	2.222		102	122		224	294
2800	10′ long, 1875 watt		3.30	2.424		190	133		323	410
2950	Wall heaters with fan, 120 to 277 volt									
3160	Recessed, residential, 750 watt	1 Elec	6	1.333	Ea.	134	73		207	256
3170	1000 watt		6	1.333		134	73		207	256
3180	1250 watt		5	1.600		134	87.50		221.50	278
3190	1500 watt		4	2		134	109		243	310
3210	2000 watt		4	2		134	109		243	310
3230	2500 watt		3.50	2.286		335	125		460	555
3240	3000 watt		3	2.667		495	146		641	760
3250	4000 watt		2.70	2.963		495	162		657	790
3260	Commercial, 750 watt		6	1.333		192	73		265	320
3270	1000 watt		6	1.333		192	73		265	320
3280	1250 watt		5	1.600		192	87.50		279.50	345
3290	1500 watt		4	2		192	109		301	375
3300	2000 watt		4	2		192	109		301	375
3310	2500 watt		3.50	2.286		192	125		317	400
3320	3000 watt		3	2.667		345	146		491	600
3330	4000 watt		2.70	2.963		345	162		507	625
3600	Thermostats, integral		16	.500		30	27.50		57.50	74
3800	Line voltage, 1 pole		8	1		19.20	54.50		73.70	103
3810	2 pole		8	1		29.50	54.50		84	115
4000	Heat trace system, 400 degree R238313-10									
4020	115 V, 2.5 watts per L.F.	1 Elec	530	.015	L.F.	7.90	.83		8.73	9.90
4030	5 watts per L.F. R238313-20		530	.015		7.90	.83		8.73	9.90

23 83 33 – Electric Radiant Heaters

23 83 33.10 Electric Heating		Crew	Daily Output	Labor-Hours	Unit	Material	2015 Bare Costs Labor	Equipment	Total	Total Incl O&P
4050	10 watts per L.F.	1 Elec	530	.015	L.F.	7.90	.83		8.73	9.90
4060	208 V, 5 watts per L.F.		530	.015		7.90	.83		8.73	9.90
4080	480 V, 8 watts per L.F.		530	.015		7.90	.83		8.73	9.90
4200	Heater raceway									
4260	Heat transfer cement									
4280	1 gallon				Ea.	60			60	66
4300	5 gallon				"	234			234	257
4320	Snap band, clamp									
4340	3/4" pipe size	1 Elec	470	.017	Ea.		.93		.93	1.39
4360	1" pipe size		444	.018			.99		.99	1.48
4380	1-1/4" pipe size		400	.020			1.09		1.09	1.64
4400	1-1/2" pipe size		355	.023			1.23		1.23	1.85
4420	2" pipe size		320	.025			1.37		1.37	2.05
4440	3" pipe size		160	.050			2.74		2.74	4.10
4460	4" pipe size		100	.080			4.38		4.38	6.55
4480	Thermostat NEMA 3R, 22 amp, 0-150 Deg, 10' cap.		8	1		241	54.50		295.50	345
4500	Thermostat NEMA 4X, 25 amp, 40 Deg, 5-1/2' cap.		7	1.143		241	62.50		303.50	360
4520	Thermostat NEMA 4X, 22 amp, 25-325 Deg, 10' cap.		7	1.143		650	62.50		712.50	810
4540	Thermostat NEMA 4X, 22 amp, 15-140 Deg,		6	1.333		545	73		618	710
4580	Thermostat NEMA 4,7,9, 22 amp, 25-325 Deg, 10' cap.		3.60	2.222		800	122		922	1,050
4600	Thermostat NEMA 4,7,9, 22 amp, 15-140 Deg,		3	2.667		795	146		941	1,100
4720	Fiberglass application tape, 36 yard roll		11	.727		77.50	40		117.50	145
5000	Radiant heating ceiling panels, 2' x 4', 500 watt		16	.500		305	27.50		332.50	375
5050	750 watt		16	.500		310	27.50		337.50	385
5200	For recessed plaster frame, add		32	.250		111	13.70		124.70	143
5300	Infrared quartz heaters, 120 volts, 1000 watts		6.70	1.194		284	65.50		349.50	410
5350	1500 watt		5	1.600		284	87.50		371.50	440
5400	240 volts, 1500 watt		5	1.600		284	87.50		371.50	440
5450	2000 watt		4	2		284	109		393	475
5500	3000 watt		3	2.667		325	146		471	575
5550	4000 watt		2.60	3.077		325	168		493	605
5570	Modulating control		.80	10		146	545		691	980
5600	Unit heaters, heavy duty, with fan & mounting bracket									
5650	Single phase, 208-240-277 volt, 3 kW	1 Elec	6	1.333	Ea.	445	73		518	600
5750	5 kW		5.50	1.455		460	79.50		539.50	625
5800	7 kW		5	1.600		710	87.50		797.50	910
5850	10 kW		4	2		810	109		919	1,050
5950	15 kW		3.80	2.105		1,300	115		1,415	1,600
6000	480 volt, 3 kW		6	1.333		425	73		498	575
6020	4 kW		5.80	1.379		445	75.50		520.50	605
6040	5 kW		5.50	1.455		460	79.50		539.50	625
6060	7 kW		5	1.600		710	87.50		797.50	910
6080	10 kW		4	2		780	109		889	1,025
6100	13 kW		3.80	2.105		1,300	115		1,415	1,600
6120	15 kW		3.70	2.162		1,300	118		1,418	1,600
6140	20 kW		3.50	2.286		1,700	125		1,825	2,025
6300	3 phase, 208-240 volt, 5 kW		5.50	1.455		435	79.50		514.50	600
6320	7 kW		5	1.600		670	87.50		757.50	865
6340	10 kW		4	2		720	109		829	960
6360	15 kW		3.70	2.162		1,200	118		1,318	1,500
6380	20 kW		3.50	2.286		1,725	125		1,850	2,075
6400	25 kW		3.30	2.424		2,025	133		2,158	2,425
6500	480 volt, 5 kW		5.50	1.455		605	79.50		684.50	790

23 83 33 – Electric Radiant Heaters

23 83 33.10 Electric Heating		Crew	Daily Output	Labor-Hours	Unit	Material	2015 Bare Costs Labor	Equipment	Total	Total Incl O&P
6520	7 kW	1 Elec	5	1.600	Ea.	765	87.50		852.50	970
6540	10 kW		4	2		810	109		919	1,050
6560	13 kW		3.80	2.105		1,300	115		1,415	1,600
6580	15 kW		3.70	2.162		1,700	118		1,818	2,025
6600	20 kW		3.50	2.286		1,700	125		1,825	2,025
6620	25 kW		3.30	2.424		2,025	133		2,158	2,425
6630	30 kW		3	2.667		2,350	146		2,496	2,825
6640	40 kW		2	4		3,000	219		3,219	3,625
6650	50 kW		1.60	5		3,625	274		3,899	4,375
6800	Vertical discharge heaters, with fan									
6820	Single phase, 208-240-277 volt, 10 kW	1 Elec	4	2	Ea.	750	109		859	990
6840	15 kW		3.70	2.162		1,250	118		1,368	1,550
6900	3 phase, 208-240 volt, 10 kW		4	2		720	109		829	960
6920	15 kW		3.70	2.162		1,200	118		1,318	1,500
6940	20 kW		3.50	2.286		1,725	125		1,850	2,075
6960	25 kW		3.30	2.424		2,025	133		2,158	2,425
6980	30 kW		3	2.667		2,350	146		2,496	2,825
7000	40 kW		2	4		3,000	219		3,219	3,625
7020	50 kW		1.60	5		3,625	274		3,899	4,375
7100	480 volt, 10 kW		4	2		810	109		919	1,050
7120	15 kW		3.70	2.162		1,300	118		1,418	1,600
7140	20 kW		3.50	2.286		1,700	125		1,825	2,025
7160	25 kW		3.30	2.424		2,025	133		2,158	2,425
7180	30 kW		3	2.667		2,350	146		2,496	2,825
7200	40 kW		2	4		3,000	219		3,219	3,625
7220	50 kW		1.60	5		3,625	274		3,899	4,375
7410	Sill height convector heaters, 5" high x 2' long, 500 watt		6.70	1.194		287	65.50		352.50	415
7420	3' long, 750 watt		6.50	1.231		340	67.50		407.50	470
7430	4' long, 1000 watt		6.20	1.290		390	70.50		460.50	535
7440	5' long, 1250 watt		5.50	1.455		445	79.50		524.50	610
7450	6' long, 1500 watt		4.80	1.667		505	91		596	690
7460	8' long, 2000 watt		3.60	2.222		685	122		807	935
7470	10' long, 2500 watt		3	2.667		850	146		996	1,150
7900	Cabinet convector heaters, 240 volt, three phase,									
7920	3' long, 2000 watt	1 Elec	5.30	1.509	Ea.	1,900	82.50		1,982.50	2,225
7940	3000 watt		5.30	1.509		2,000	82.50		2,082.50	2,300
7960	4000 watt		5.30	1.509		2,050	82.50		2,132.50	2,375
7980	6000 watt		4.60	1.739		2,125	95		2,220	2,475
8000	8000 watt		4.60	1.739		2,200	95		2,295	2,575
8020	4' long, 4000 watt		4.60	1.739		2,075	95		2,170	2,425
8040	6000 watt		4	2		2,150	109		2,259	2,550
8060	8000 watt		4	2		2,225	109		2,334	2,625
8080	10,000 watt		4	2		2,250	109		2,359	2,650
8100	Available also in 208 or 277 volt									
8200	Cabinet unit heaters, 120 to 277 volt, 1 pole,									
8220	wall mounted, 2 kW	1 Elec	4.60	1.739	Ea.	1,875	95		1,970	2,200
8230	3 kW		4.60	1.739		1,950	95		2,045	2,300
8240	4 kW		4.40	1.818		2,000	99.50		2,099.50	2,350
8250	5 kW		4.40	1.818		2,075	99.50		2,174.50	2,450
8260	6 kW		4.20	1.905		2,125	104		2,229	2,475
8270	8 kW		4	2		2,150	109		2,259	2,550
8280	10 kW		3.80	2.105		2,225	115		2,340	2,600
8290	12 kW		3.50	2.286		2,250	125		2,375	2,650

23 83 Radiant Heating Units

23 83 33 – Electric Radiant Heaters

23 83 33.10 Electric Heating		Crew	Daily Output	Labor-Hours	Unit	Material	Labor	Equipment	Total	Total Incl O&P
						2015 Bare Costs				
8300	13.5 kW	1 Elec	2.90	2.759	Ea.	2,700	151		2,851	3,200
8310	16 kW		2.70	2.963		2,725	162		2,887	3,250
8320	20 kW		2.30	3.478		3,475	190		3,665	4,100
8330	24 kW		1.90	4.211		3,525	230		3,755	4,225
8350	Recessed, 2 kW		4.40	1.818		1,875	99.50		1,974.50	2,200
8370	3 kW		4.40	1.818		1,950	99.50		2,049.50	2,300
8380	4 kW		4.20	1.905		2,000	104		2,104	2,350
8390	5 kW		4.20	1.905		2,075	104		2,179	2,425
8400	6 kW		4	2		2,100	109		2,209	2,500
8410	8 kW		3.80	2.105		2,200	115		2,315	2,600
8420	10 kW		3.50	2.286		2,675	125		2,800	3,125
8430	12 kW		2.90	2.759		2,675	151		2,826	3,175
8440	13.5 kW		2.70	2.963		2,700	162		2,862	3,225
8450	16 kW		2.30	3.478		2,875	190		3,065	3,450
8460	20 kW		1.90	4.211		3,275	230		3,505	3,950
8470	24 kW		1.60	5		3,325	274		3,599	4,050
8490	Ceiling mounted, 2 kW		3.20	2.500		1,875	137		2,012	2,250
8510	3 kW		3.20	2.500		1,950	137		2,087	2,350
8520	4 kW		3	2.667		2,000	146		2,146	2,425
8530	5 kW		3	2.667		2,075	146		2,221	2,500
8540	6 kW		2.80	2.857		2,075	156		2,231	2,525
8550	8 kW		2.40	3.333		2,150	182		2,332	2,650
8560	10 kW		2.20	3.636		2,200	199		2,399	2,725
8570	12 kW		2	4		2,250	219		2,469	2,800
8580	13.5 kW		1.50	5.333		2,275	292		2,567	2,950
8590	16 kW		1.30	6.154		2,350	335		2,685	3,075
8600	20 kW		.90	8.889		3,275	485		3,760	4,325
8610	24 kW	↓	.60	13.333	↓	3,325	730		4,055	4,750
8630	208 to 480 V, 3 pole									
8650	Wall mounted, 2 kW	1 Elec	4.60	1.739	Ea.	1,975	95		2,070	2,325
8670	3 kW		4.60	1.739		2,075	95		2,170	2,425
8680	4 kW		4.40	1.818		2,125	99.50		2,224.50	2,475
8690	5 kW		4.40	1.818		2,175	99.50		2,274.50	2,550
8700	6 kW		4.20	1.905		2,200	104		2,304	2,575
8710	8 kW		4	2		2,275	109		2,384	2,675
8720	10 kW		3.80	2.105		2,325	115		2,440	2,725
8730	12 kW		3.50	2.286		2,375	125		2,500	2,775
8740	13.5 kW		2.90	2.759		2,400	151		2,551	2,875
8750	16 kW		2.70	2.963		2,450	162		2,612	2,950
8760	20 kW		2.30	3.478		3,400	190		3,590	4,000
8770	24 kW		1.90	4.211		3,600	230		3,830	4,300
8790	Recessed, 2 kW		4.40	1.818		1,975	99.50		2,074.50	2,325
8810	3 kW		4.40	1.818		2,000	99.50		2,099.50	2,350
8820	4 kW		4.20	1.905		2,075	104		2,179	2,450
8830	5 kW		4.20	1.905		2,100	104		2,204	2,475
8840	6 kW		4	2		2,125	109		2,234	2,500
8850	8 kW		3.80	2.105		2,200	115		2,315	2,600
8860	10 kW		3.50	2.286		2,350	125		2,475	2,750
8870	12 kW		2.90	2.759		2,450	151		2,601	2,925
8880	13.5 kW		2.70	2.963		2,500	162		2,662	3,000
8890	16 kW		2.30	3.478		3,625	190		3,815	4,250
8900	20 kW		1.90	4.211		3,675	230		3,905	4,400
8920	24 kW	↓	1.60	5	↓	3,725	274		3,999	4,500

23 83 Radiant Heating Units

23 83 33 – Electric Radiant Heaters

23 83 33.10 Electric Heating		Crew	Daily Output	Labor-Hours	Unit	2015 Bare Costs Material	Labor	Equipment	Total	Total Incl O&P
8940	Ceiling mount, 2 kW	1 Elec	3.20	2.500	Ea.	2,075	137		2,212	2,475
8950	3 kW		3.20	2.500		2,200	137		2,337	2,625
8960	4 kW		3	2.667		2,200	146		2,346	2,650
8970	5 kW		3	2.667		2,275	146		2,421	2,725
8980	6 kW		2.80	2.857		2,300	156		2,456	2,750
8990	8 kW		2.40	3.333		2,375	182		2,557	2,875
9000	10 kW		2.20	3.636		2,425	199		2,624	2,950
9020	13.5 kW		1.50	5.333		2,500	292		2,792	3,175
9030	16 kW		1.30	6.154		3,625	335		3,960	4,475
9040	20 kW		.90	8.889		3,675	485		4,160	4,775
9060	24 kW		.60	13.333		4,075	730		4,805	5,600

Estimating Tips

26 05 00 Common Work Results for Electrical

- Conduit should be taken off in three main categories—power distribution, branch power, and branch lighting—so the estimator can concentrate on systems and components, therefore making it easier to ensure all items have been accounted for.
- For cost modifications for elevated conduit installation, add the percentages to labor according to the height of installation, and only to the quantities exceeding the different height levels, not to the total conduit quantities.
- Remember that aluminum wiring of equal ampacity is larger in diameter than copper and may require larger conduit.
- If more than three wires at a time are being pulled, deduct percentages from the labor hours of that grouping of wires.
- When taking off grounding systems, identify separately the type and size of wire, and list each unique type of ground connection.
- The estimator should take the weights of materials into consideration when completing a takeoff. Topics to consider include: How will the materials be supported? What methods of support are available? How high will the support structure have to reach? Will the final support structure be able to withstand the total burden? Is the support material included or separate from the fixture, equipment, and material specified?
- Do not overlook the costs for equipment used in the installation. If scaffolding or highlifts are available in the field, contractors may use them in lieu of the proposed ladders and rolling staging.

26 20 00 Low-Voltage Electrical Transmission

- Supports and concrete pads may be shown on drawings for the larger equipment, or the support system may be only a piece of plywood for the back of a panelboard. In either case, it must be included in the costs.

26 40 00 Electrical and Cathodic Protection

- When taking off cathodic protections systems, identify the type and size of cable, and list each unique type of anode connection.

26 50 00 Lighting

- Fixtures should be taken off room by room, using the fixture schedule, specifications, and the ceiling plan. For large concentrations of lighting fixtures in the same area, deduct the percentages from labor hours.

Reference Numbers

Reference numbers are shown in shaded boxes at the beginning of some major classifications. These numbers refer to related items in the Reference Section. The reference information may be an estimating procedure, an alternate pricing method, or technical information.

Note: Not all subdivisions listed here necessarily appear in this publication. ■

Note: ***Trade Service****, in part, has been used as a reference source for some of the material prices used in Division 26.*

26 01 Operation and Maintenance of Electrical Systems

26 01 40 – Operation and Maintenance of Electrical Protection Systems

26 01 40.51 Electrical Systems Repair and Replacement		Crew	Daily Output	Labor-Hours	Unit	Material	2015 Bare Costs Labor	Equipment	Total	Total Incl O&P
0010	**ELECTRICAL SYSTEMS REPAIR AND REPLACEMENT**									
3000	Remove and replace (reinstall), switch cover	1 Elec	60	.133	Ea.		7.30		7.30	10.95
3020	Outlet cover	"	60	.133	"		7.30		7.30	10.95

26 01 50 – Operation and Maintenance of Lighting

26 01 50.81 Luminaire Replacement

Code	Description	Crew	Daily Output	Labor-Hours	Unit	Material	Labor	Equipment	Total	Total Incl O&P
0010	**LUMINAIRE REPLACEMENT**									
0962	Remove and install new replacement lens, 2′ x 2′ troffer	1 Elec	16	.500	Ea.	15.90	27.50		43.40	58.50
0964	2′ x 4′ troffer		16	.500		14.90	27.50		42.40	57.50
3200	Remove and replace (reinstall), lighting fixture		4	2			109		109	164

26 05 Common Work Results for Electrical

26 05 05 – Selective Demolition for Electrical

26 05 05.10 Electrical Demolition

Code	Description	Crew	Daily Output	Labor-Hours	Unit	Material	Labor	Equipment	Total	Total Incl O&P
0010	**ELECTRICAL DEMOLITION** R260105-30									
0020	Conduit to 15′ high, including fittings & hangers									
0100	Rigid galvanized steel, 1/2″ to 1″ diameter	1 Elec	242	.033	L.F.		1.81		1.81	2.71
0120	1-1/4″ to 2″	"	200	.040			2.19		2.19	3.28
0140	2-1/2″ to 3-1/2″	2 Elec	302	.053			2.90		2.90	4.34
0160	4″ to 6″	"	160	.100			5.45		5.45	8.20
0170	PVC #40, 1/2″ to 1″	1 Elec	410	.020			1.07		1.07	1.60
0172	1-1/4″ to 2″		350	.023			1.25		1.25	1.87
0174	2-1/2″		250	.032			1.75		1.75	2.62
0176	3″ to 3-1/2″	2 Elec	340	.047			2.57		2.57	3.86
0178	4″ to 6″	"	230	.070			3.81		3.81	5.70
0200	Electric metallic tubing (EMT), 1/2″ to 1″	1 Elec	394	.020			1.11		1.11	1.66
0220	1-1/4″ to 1-1/2″		326	.025			1.34		1.34	2.01
0240	2″ to 3″		236	.034			1.85		1.85	2.78
0260	3-1/2″ to 4″	2 Elec	310	.052			2.82		2.82	4.23
0270	Armored cable, (BX) avg. 50′ runs									
0280	#14, 2 wire	1 Elec	690	.012	L.F.		.63		.63	.95
0290	#14, 3 wire		571	.014			.77		.77	1.15
0300	#12, 2 wire		605	.013			.72		.72	1.08
0310	#12, 3 wire		514	.016			.85		.85	1.28
0320	#10, 2 wire		514	.016			.85		.85	1.28
0330	#10, 3 wire		425	.019			1.03		1.03	1.54
0340	#8, 3 wire		342	.023			1.28		1.28	1.92
0350	Non metallic sheathed cable (Romex)									
0360	#14, 2 wire	1 Elec	720	.011	L.F.		.61		.61	.91
0370	#14, 3 wire		657	.012			.67		.67	1
0380	#12, 2 wire		629	.013			.70		.70	1.04
0390	#10, 3 wire		450	.018			.97		.97	1.46
0400	Wiremold raceway, including fittings & hangers									
0420	No. 3000	1 Elec	250	.032	L.F.		1.75		1.75	2.62
0440	No. 4000		217	.037			2.02		2.02	3.02
0460	No. 6000		166	.048			2.64		2.64	3.95
0462	Plugmold with receptacle		114	.070			3.84		3.84	5.75
0465	Telephone/power pole		12	.667	Ea.		36.50		36.50	54.50
0470	Non-metallic, straight section		480	.017	L.F.		.91		.91	1.37
0500	Channels, steel, including fittings & hangers									
0520	3/4″ x 1-1/2″	1 Elec	308	.026	L.F.		1.42		1.42	2.13

26 05 Common Work Results for Electrical

26 05 05 – Selective Demolition for Electrical

26 05 05.10 Electrical Demolition		Crew	Daily Output	Labor-Hours	Unit	Material	2015 Bare Costs Labor	Equipment	Total	Total Incl O&P
0540	1-1/2" x 1-1/2"	1 Elec	269	.030	L.F.		1.63		1.63	2.44
0560	1-1/2" x 1-7/8"		229	.035			1.91		1.91	2.86
0600	Copper bus duct, indoor, 3 phase									
0610	Including hangers & supports									
0620	225 amp	2 Elec	135	.119	L.F.		6.50		6.50	9.70
0640	400 amp		106	.151			8.25		8.25	12.35
0660	600 amp		86	.186			10.20		10.20	15.25
0680	1000 amp		60	.267			14.60		14.60	22
0700	1600 amp		40	.400			22		22	33
0720	3000 amp		10	1.600			87.50		87.50	131
0800	Plug-in switches, 600V 3 ph., incl. disconnecting									
0820	wire, conduit terminations, 30 amp	1 Elec	15.50	.516	Ea.		28		28	42.50
0840	60 amp		13.90	.576			31.50		31.50	47
0850	100 amp		10.40	.769			42		42	63
0860	200 amp		6.20	1.290			70.50		70.50	106
0880	400 amp	2 Elec	5.40	2.963			162		162	243
0900	600 amp		3.40	4.706			257		257	385
0920	800 amp		2.60	6.154			335		335	505
0940	1200 amp		2	8			440		440	655
0960	1600 amp		1.70	9.412			515		515	770
1010	Safety switches, 250 or 600V, incl. disconnection									
1050	of wire & conduit terminations									
1100	30 amp	1 Elec	12.30	.650	Ea.		35.50		35.50	53.50
1120	60 amp		8.80	.909			49.50		49.50	74.50
1140	100 amp		7.30	1.096			60		60	90
1160	200 amp		5	1.600			87.50		87.50	131
1180	400 amp	2 Elec	6.80	2.353			129		129	193
1200	600 amp	″	4.60	3.478			190		190	285
1202	Explosion proof, 60 amp	1 Elec	8.80	.909			49.50		49.50	74.50
1204	100 amp		5	1.600			87.50		87.50	131
1206	200 amp		3.33	2.402			131		131	197
1210	Panel boards, incl. removal of all breakers,									
1220	conduit terminations & wire connections									
1230	3 wire, 120/240 V, 100A, to 20 circuits	1 Elec	2.60	3.077	Ea.		168		168	252
1240	200 amps, to 42 circuits	2 Elec	2.60	6.154			335		335	505
1250	400 amps, to 42 circuits	″	2.20	7.273			400		400	595
1260	4 wire, 120/208 V, 125A, to 20 circuits	1 Elec	2.40	3.333			182		182	273
1270	200 amps, to 42 circuits	2 Elec	2.40	6.667			365		365	545
1280	400 amps, to 42 circuits		1.92	8.333			455		455	685
1285	600 amps, to 42 circuits		1.60	10			545		545	820
1300	Transformer, dry type, 1 phase, incl. removal of									
1320	supports, wire & conduit terminations									
1340	1 kVA	1 Elec	7.70	1.039	Ea.		57		57	85
1360	5 kVA		4.70	1.702			93		93	139
1380	10 kVA		3.60	2.222			122		122	182
1400	37.5 kVA	2 Elec	3	5.333			292		292	435
1420	75 kVA	″	2.50	6.400			350		350	525
1440	3 phase to 600V, primary									
1460	3 kVA	1 Elec	3.87	2.067	Ea.		113		113	169
1480	15 kVA	2 Elec	3.67	4.360			238		238	355
1490	25 kVA		3.51	4.558			249		249	375
1500	30 kVA		3.42	4.678			256		256	385
1510	45 kVA		3.18	5.031			275		275	410

26 05 05 – Selective Demolition for Electrical

26 05 05.10 Electrical Demolition		Crew	Daily Output	Labor-Hours	Unit	Material	2015 Bare Costs Labor	Equipment	Total	Total Incl O&P
1520	75 kVA	2 Elec	2.69	5.948	Ea.		325		325	485
1530	112.5 kVA	R-3	2.90	6.897			375	47.50	422.50	610
1540	150 kVA		2.70	7.407			400	51	451	660
1550	300 kVA		1.80	11.111			605	76.50	681.50	990
1560	500 kVA		1.40	14.286			775	98.50	873.50	1,275
1570	750 kVA		1.10	18.182			985	125	1,110	1,625
1600	Pull boxes & cabinets, sheet metal, incl. removal									
1620	of supports and conduit terminations									
1640	6" x 6" x 4"	1 Elec	31.10	.257	Ea.		14.05		14.05	21
1660	12" x 12" x 4"		23.30	.343			18.80		18.80	28
1680	24" x 24" x 6"		12.30	.650			35.50		35.50	53.50
1700	36" x 36" x 8"		7.70	1.039			57		57	85
1720	Junction boxes, 4" sq. & oct.		80	.100			5.45		5.45	8.20
1740	Handy box		107	.075			4.09		4.09	6.15
1760	Switch box		107	.075			4.09		4.09	6.15
1780	Receptacle & switch plates		257	.031			1.70		1.70	2.55
1790	Receptacles & switches, 15 to 30 amp		135	.059			3.24		3.24	4.86
1800	Wire, THW-THWN-THHN, removed from									
1810	in place conduit, to 15' high									
1830	#14	1 Elec	65	.123	C.L.F.		6.75		6.75	10.10
1840	#12		55	.145			7.95		7.95	11.90
1850	#10		45.50	.176			9.60		9.60	14.40
1860	#8		40.40	.198			10.85		10.85	16.25
1870	#6		32.60	.245			13.40		13.40	20
1880	#4	2 Elec	53	.302			16.50		16.50	24.50
1890	#3		50	.320			17.50		17.50	26
1900	#2		44.60	.359			19.60		19.60	29.50
1910	1/0		33.20	.482			26.50		26.50	39.50
1920	2/0		29.20	.548			30		30	45
1930	3/0		25	.640			35		35	52.50
1940	4/0		22	.727			40		40	59.50
1950	250 kcmil		20	.800			44		44	65.50
1960	300 kcmil		19	.842			46		46	69
1970	350 kcmil		18	.889			48.50		48.50	73
1980	400 kcmil		17	.941			51.50		51.50	77
1990	500 kcmil		16.20	.988			54		54	81
2000	Interior fluorescent fixtures, incl. supports									
2010	& whips, to 15' high									
2100	Recessed drop-in 2' x 2', 2 lamp	2 Elec	35	.457	Ea.		25		25	37.50
2120	2' x 4', 2 lamp		33	.485			26.50		26.50	39.50
2140	2' x 4', 4 lamp		30	.533			29		29	43.50
2160	4' x 4', 4 lamp		20	.800			44		44	65.50
2180	Surface mount, acrylic lens & hinged frame									
2200	1' x 4', 2 lamp	2 Elec	44	.364	Ea.		19.90		19.90	30
2220	2' x 2', 2 lamp		44	.364			19.90		19.90	30
2260	2' x 4', 4 lamp		33	.485			26.50		26.50	39.50
2280	4' x 4', 4 lamp		23	.696			38		38	57
2300	Strip fixtures, surface mount									
2320	4' long, 1 lamp	2 Elec	53	.302	Ea.		16.50		16.50	24.50
2340	4' long, 2 lamp		50	.320			17.50		17.50	26
2360	8' long, 1 lamp		42	.381			21		21	31
2380	8' long, 2 lamp		40	.400			22		22	33
2400	Pendant mount, industrial, incl. removal									

26 05 05 – Selective Demolition for Electrical

26 05 05.10 Electrical Demolition		Crew	Daily Output	Labor-Hours	Unit	Material	2015 Bare Costs Labor	Equipment	Total	Total Incl O&P
2410	of chain or rod hangers, to 15′ high									
2420	4′ long, 2 lamp	2 Elec	35	.457	Ea.		25		25	37.50
2440	8′ long, 2 lamp	″	27	.593	″		32.50		32.50	48.50
2460	Interior incandescent, surface, ceiling									
2470	or wall mount, to 12′ high									
2480	Metal cylinder type, 75 Watt	2 Elec	62	.258	Ea.		14.10		14.10	21
2500	150 Watt		62	.258			14.10		14.10	21
2502	300 Watt		53	.302			16.50		16.50	24.50
2520	Metal halide, high bay									
2540	400 Watt	2 Elec	15	1.067	Ea.		58.50		58.50	87.50
2560	1000 Watt		12	1.333			73		73	109
2580	150 Watt, low bay		20	.800			44		44	65.50
2585	Globe type, ceiling mount		62	.258			14.10		14.10	21
2586	Can type, recessed mount		62	.258			14.10		14.10	21
2600	Exterior fixtures, incandescent, wall mount									
2620	100 Watt	2 Elec	50	.320	Ea.		17.50		17.50	26
2640	Quartz, 500 Watt		33	.485			26.50		26.50	39.50
2660	1500 Watt		27	.593			32.50		32.50	48.50
2680	Wall pack, mercury vapor									
2700	175 Watt	2 Elec	25	.640	Ea.		35		35	52.50
2720	250 Watt	″	25	.640			35		35	52.50
7000	Weatherhead/mast, 2″	1 Elec	16	.500			27.50		27.50	41
7002	3″		10	.800			44		44	65.50
7004	3-1/2″		8.50	.941			51.50		51.50	77
7006	4″		8	1			54.50		54.50	82
7100	Service entry cable, #6, +#6 neutral		420	.019	L.F.		1.04		1.04	1.56
7102	#4, +#4 neutral		360	.022			1.22		1.22	1.82
7104	#2, +#4 neutral		345	.023			1.27		1.27	1.90
7106	#2, +#2 neutral		330	.024			1.33		1.33	1.99
9900	Add to labor for higher elevated installation									
9910	15′ to 20′ high, add						10%			
9920	20′ to 25′ high, add						20%			
9930	25′ to 30′ high, add						25%			
9940	30′ to 35′ high, add						30%			
9950	35′ to 40′ high, add						35%			
9960	Over 40′ high, add						40%			

26 05 05.15 Electrical Demolition, Grounding

		Crew	Daily Output	Labor-Hours	Unit	Material	Labor	Equipment	Total	Total Incl O&P
0010	**ELECTRICAL DEMOLITION, GROUNDING** Addition									
0100	Ground clamp, bronze	1 Elec	64	.125	Ea.		6.85		6.85	10.25
0140	Water pipe ground clamp, bronze, heavy duty		24	.333			18.25		18.25	27.50
0150	Ground rod, 8′ to 10′		13	.615			33.50		33.50	50.50
0200	Ground wire, bare armored	2 Elec	900	.018	L.F.		.97		.97	1.46
0240	Bare copper or aluminum	″	2800	.006	″		.31		.31	.47

26 05 05.20 Electrical Demolition, Wiring Methods

		Crew	Daily Output	Labor-Hours	Unit	Material	Labor	Equipment	Total	Total Incl O&P
0010	**ELECTRICAL DEMOLITION, WIRING METHODS** Addition									
0100	Armored cable, w/PVC jacket, in cable tray									
0110	#6 R024119-10	1 Elec	9.30	.860	C.L.F.		47		47	70.50
0120	#4	2 Elec	16.20	.988			54		54	81
0130	#2		13.80	1.159			63.50		63.50	95
0140	#1		12	1.333			73		73	109
0150	1/0		10.80	1.481			81		81	121
0160	2/0		10.20	1.569			86		86	129

26 05 Common Work Results for Electrical

26 05 05 – Selective Demolition for Electrical

26 05 05.20 Electrical Demolition, Wiring Methods		Crew	Daily Output	Labor-Hours	Unit	2015 Bare Costs Material	Labor	Equipment	Total	Total Incl O&P
0170	3/0	2 Elec	9.60	1.667	C.L.F.		91		91	137
0180	4/0		9	1.778			97		97	146
0190	250 kcmil	3 Elec	10.80	2.222			122		122	182
0210	350 kcmil		9.90	2.424			133		133	199
0230	500 kcmil		9	2.667			146		146	219
0240	750 kcmil		8.10	2.963			162		162	243
1100	Control cable, 600 V or less									
1110	3 wires	1 Elec	24	.333	C.L.F.		18.25		18.25	27.50
1120	5 wires		20	.400			22		22	33
1130	7 wires		16.50	.485			26.50		26.50	39.50
1140	9 wires		15	.533			29		29	43.50
1150	12 wires	2 Elec	26	.615			33.50		33.50	50.50
1160	15 wires		22	.727			40		40	59.50
1170	19 wires		19	.842			46		46	69
1180	25 wires		16	1			54.50		54.50	82
1500	Mineral insulated (MI) cable, 600 V									
1510	#10	1 Elec	4.80	1.667	C.L.F.		91		91	137
1520	#8		4.50	1.778			97		97	146
1530	#6		4.20	1.905			104		104	156
1540	#4	2 Elec	7.20	2.222			122		122	182
1550	#2		6.60	2.424			133		133	199
1560	#1		6.30	2.540			139		139	208
1570	1/0		6	2.667			146		146	219
1580	2/0		5.70	2.807			154		154	230
1590	3/0		5.40	2.963			162		162	243
1600	4/0		4.80	3.333			182		182	273
1610	250 kcmil	3 Elec	7.20	3.333			182		182	273
1620	500 kcmil	"	5.90	4.068			223		223	335
2000	Shielded cable, XLP shielding, to 35 kV									
2010	#4	2 Elec	13.20	1.212	C.L.F.		66.50		66.50	99.50
2020	#1		12	1.333			73		73	109
2030	1/0		11.40	1.404			77		77	115
2040	2/0		10.80	1.481			81		81	121
2050	4/0		9.60	1.667			91		91	137
2060	250 kcmil	3 Elec	13.50	1.778			97		97	146
2070	350 kcmil		11.70	2.051			112		112	168
2080	500 kcmil		11.20	2.143			117		117	176
2090	750 kcmil		10.80	2.222			122		122	182
3000	Modular flexible wiring									
3010	Cable set	1 Elec	120	.067	Ea.		3.65		3.65	5.45
3020	Conversion module		48	.167			9.10		9.10	13.65
3030	Switching assembly		96	.083			4.56		4.56	6.85
3200	Undercarpet									
3210	Power or telephone, flat cable	1 Elec	1320	.006	L.F.		.33		.33	.50
3220	Transition block assemblies w/fitting		75	.107	Ea.		5.85		5.85	8.75
3230	Floor box with fitting		60	.133			7.30		7.30	10.95
3300	Data system, cable with connection		50	.160			8.75		8.75	13.10
4000	Cable tray, including fitting & support									
4010	Galvanized steel, 6" wide	2 Elec	310	.052	L.F.		2.82		2.82	4.23
4020	9" wide		295	.054			2.97		2.97	4.45
4030	12" wide		285	.056			3.07		3.07	4.60
4040	18" wide		270	.059			3.24		3.24	4.86
4050	24" wide		260	.062			3.37		3.37	5.05

26 05 05.20 Electrical Demolition, Wiring Methods		Crew	Daily Output	Labor-Hours	Unit	Material	2015 Bare Costs Labor	Equipment	Total	Total Incl O&P
4060	30″ wide	2 Elec	240	.067	L.F.		3.65		3.65	5.45
4070	36″ wide		220	.073			3.98		3.98	5.95
4110	Aluminum, 6″ wide		420	.038			2.08		2.08	3.12
4120	9″ wide		415	.039			2.11		2.11	3.16
4130	12″ wide		390	.041			2.24		2.24	3.36
4140	18″ wide		370	.043			2.37		2.37	3.54
4150	24″ wide		350	.046			2.50		2.50	3.75
4160	30″ wide		325	.049			2.69		2.69	4.03
4170	36″ wide		300	.053			2.92		2.92	4.37
4310	Cable channel, aluminum, 4″ wide straight	1 Elec	240	.033			1.82		1.82	2.73
4320	Cable tray fittings, 6″ to 9″ wide	2 Elec	27	.593	Ea.		32.50		32.50	48.50
4330	12″ to 18″ wide	″	19	.842	″		46		46	69
5000	Conduit nipples, with locknuts and bushings									
5020	1/2″	1 Elec	108	.074	Ea.		4.05		4.05	6.05
5040	3/4″		96	.083			4.56		4.56	6.85
5060	1″		81	.099			5.40		5.40	8.10
5080	1-1/4″		69	.116			6.35		6.35	9.50
5100	1-1/2″		60	.133			7.30		7.30	10.95
5120	2″		54	.148			8.10		8.10	12.15
5140	2-1/2″		45	.178			9.70		9.70	14.55
5160	3″		36	.222			12.15		12.15	18.20
5180	3-1/2″		33	.242			13.25		13.25	19.85
5200	4″		27	.296			16.20		16.20	24.50
5220	5″		21	.381			21		21	31
5240	6″		18	.444			24.50		24.50	36.50
5500	Electric nonmetallic tubing (ENT), flexible, 1/2″ to 1″ diameter		690	.012	L.F.		.63		.63	.95
5510	1-1/4″ to 2″ diameter		300	.027			1.46		1.46	2.19
5600	Flexible metallic tubing, steel, 3/8″ to 3/4″ diameter		600	.013			.73		.73	1.09
5610	1″ to 1-1/4″ diameter		260	.031			1.68		1.68	2.52
5620	1-1/2″ to 2″ diameter		140	.057			3.13		3.13	4.68
5630	2-1/2″ diameter		100	.080			4.38		4.38	6.55
5640	3″ to 3-1/2″ diameter	2 Elec	150	.107			5.85		5.85	8.75
5650	4″ diameter	″	100	.160			8.75		8.75	13.10
5700	Sealtite flexible conduit, 3/8″ to 3/4″ diameter	1 Elec	420	.019			1.04		1.04	1.56
5710	1″ to 1-1/4″ diameter		180	.044			2.43		2.43	3.64
5720	1-1/2″ to 2″ diameter		120	.067			3.65		3.65	5.45
5730	2-1/2″ diameter		80	.100			5.45		5.45	8.20
5740	3″ diameter	2 Elec	150	.107			5.85		5.85	8.75
5750	4″ diameter	″	100	.160			8.75		8.75	13.10
5800	Wiring duct, plastic, 1-1/2″ to 2-1/2″ wide	1 Elec	360	.022			1.22		1.22	1.82
5810	3″ wide		330	.024			1.33		1.33	1.99
5820	4″ wide		300	.027			1.46		1.46	2.19
6000	Floor box and carpet flange		9	.889	Ea.		48.50		48.50	73
6300	Wireway, with fittings and supports, to 15′ high									
6310	2-1/2″ x 2-1/2″	1 Elec	180	.044	L.F.		2.43		2.43	3.64
6320	4″ x 4″	″	160	.050			2.74		2.74	4.10
6330	6″ x 6″	2 Elec	240	.067			3.65		3.65	5.45
6340	8″ x 8″		160	.100			5.45		5.45	8.20
6350	10″ x 10″		120	.133			7.30		7.30	10.95
6360	12″ x 12″		80	.200			10.95		10.95	16.40
6400	Cable reel with receptacle, 120 V/208 V		13	1.231	Ea.		67.50		67.50	101
6500	Equipment connection, to 10 HP	1 Elec	23	.348			19.05		19.05	28.50
6510	To 15 to 30 HP		18	.444			24.50		24.50	36.50

26 05 05 – Selective Demolition for Electrical

26 05 05.20 Electrical Demolition, Wiring Methods

		Crew	Daily Output	Labor-Hours	Unit	Material	2015 Bare Costs Labor	Equipment	Total	Total Incl O&P
6520	To 40 to 60 HP	1 Elec	15	.533	Ea.		29		29	43.50
6530	To 75 to 100 HP		10	.800			44		44	65.50
7008	Receptacle, explosionproof, 120 V, to 30 A		16	.500			27.50		27.50	41
7020	Dimmer switch, 2000 W or less		62	.129			7.05		7.05	10.55

26 05 05.25 Electrical Demolition, Electrical Power

		Crew	Daily Output	Labor-Hours	Unit	Material	2015 Bare Costs Labor	Equipment	Total	Total Incl O&P
0010	**ELECTRICAL DEMOLITION, ELECTRICAL POWER** R024119-10									
0100	Meter centers and sockets									
0120	Meter socket, 4 terminal R260105-30	1 Elec	8.40	.952	Ea.		52		52	78
0140	Trans-socket, 13 terminal, 400 A	"	3.60	2.222			122		122	182
0160	800 A	2 Elec	4.40	3.636			199		199	298
0200	Meter center, 400 A		5.80	2.759			151		151	226
0210	600 A		4	4			219		219	330
0220	800 A		3.30	4.848			265		265	395
0230	1200 A		2.80	5.714			315		315	470
0240	1600 A		2.50	6.400			350		350	525
0300	Base meter devices, 3 meter		3.60	4.444			243		243	365
0310	4 meter		3.30	4.848			265		265	395
0320	5 meter		2.90	5.517			300		300	450
0330	6 meter		2.20	7.273			400		400	595
0340	7 meter		2	8			440		440	655
0350	8 meter		1.90	8.421			460		460	690
0400	Branch meter devices									
0410	Socket w/circuit breaker 200 A, 2 meter	2 Elec	3.30	4.848	Ea.		265		265	395
0420	3 meter		2.90	5.517			300		300	450
0430	4 meter		2.60	6.154			335		335	505
0450	Main circuit breaker, 400 A		5.80	2.759			151		151	226
0460	600 A		4	4			219		219	330
0470	800 A		3.30	4.848			265		265	395
0480	1200 A		2.80	5.714			315		315	470
0490	1600 A		2.50	6.400			350		350	525
0500	Main lug terminal box, 800 A		3.40	4.706			257		257	385
0510	1200 A		2.60	6.154			335		335	505
1000	Motors, 230/460 V, 60 Hz, 3/4 HP	1 Elec	10.70	.748			41		41	61.50
1010	5 HP		9	.889			48.50		48.50	73
1020	10 HP		8	1			54.50		54.50	82
1030	15 HP		6.40	1.250			68.50		68.50	102
1040	20 HP	2 Elec	10.40	1.538			84		84	126
1050	50 HP		9.60	1.667			91		91	137
1060	75 HP		5.60	2.857			156		156	234
1070	100 HP	3 Elec	5.40	4.444			243		243	365
1080	150 HP		3.60	6.667			365		365	545
1090	200 HP		3	8			440		440	655
1200	Variable frequency drive, 460 V, for 5 HP motor size	1 Elec	3.20	2.500			137		137	205
1210	10 HP motor size	"	2.70	2.963			162		162	243
1220	20 HP motor size	2 Elec	3.60	4.444			243		243	365
1230	50 HP motor size	"	2.10	7.619			415		415	625
1240	75 HP motor size	R-3	2.20	9.091			495	62.50	557.50	810
1250	100 HP motor size		2	10			545	69	614	890
1260	150 HP motor size		2	10			545	69	614	890
1270	200 HP motor size		1.70	11.765			640	81	721	1,050
2000	Generator set w/accessories, 3 phase 4 wire, 277/480 V									
2010	7.5 kW	3 Elec	2	12	Ea.		655		655	985

26 05 Common Work Results for Electrical

26 05 05 – Selective Demolition for Electrical

26 05 05.25 Electrical Demolition, Electrical Power		Crew	Daily Output	Labor-Hours	Unit	Material	2015 Bare Costs Labor	Equipment	Total	Total Incl O&P
2020	20 kW	3 Elec	1.70	14.118	Ea.		770		770	1,150
2040	30 kW		1.33	18.045			985		985	1,475
2060	50 kW		1	24			1,325		1,325	1,975
2080	100 kW		.75	32			1,750		1,750	2,625
2100	150 kW		.63	38.095			2,075		2,075	3,125
2120	250 kW		.59	40.678			2,225		2,225	3,325
2140	400 kW		.50	48			2,625		2,625	3,925
2160	500 kW		.44	54.545			2,975		2,975	4,475
2180	750 kW	4 Elec	.56	57.143			3,125		3,125	4,675
2200	1000 kW	"	.46	69.565			3,800		3,800	5,700
3000	Uninterruptible power supply system (UPS)									
3010	Single phase, 120 V, 1 kVA	1 Elec	3.20	2.500	Ea.		137		137	205
3020	2 kVA	2 Elec	3.60	4.444			243		243	365
3030	5 kVA	3 Elec	2.50	9.600			525		525	785
3040	10 kVA		2.30	10.435			570		570	855
3050	15 kVA		1.80	13.333			730		730	1,100
4000	Transformer, incl support, wire & conduit termination									
4010	Buck-boost, single phase, 120/240 V, 0.1 kVA	1 Elec	25	.320	Ea.		17.50		17.50	26
4020	0.5 kVA		12.50	.640			35		35	52.50
4030	1 kVA		6.30	1.270			69.50		69.50	104
4040	5 kVA		3.80	2.105			115		115	173
4800	5 kV or 15 kV primary, 277/480 V second, 112.5 kVA	R-3	3.50	5.714			310	39.50	349.50	510
4810	150 kVA		2.70	7.407			400	51	451	660
4820	225 kVA		2.30	8.696			470	60	530	775
4830	500 kVA		1.45	13.793			750	95	845	1,225
4840	750 kVA		1.33	15.038			815	104	919	1,350
4850	1000 kVA		1.25	16			870	110	980	1,425
4860	2000 kVA		1.05	19.048			1,025	131	1,156	1,700
4870	3000 kVA		.75	26.667			1,450	184	1,634	2,375
6010	Isolation panel, 3 kVA	1 Elec	2.30	3.478			190		190	285
6020	5 kVA		2.20	3.636			199		199	298
6030	7.5 kVA		2.10	3.810			208		208	310
6040	10 kVA		1.80	4.444			243		243	365
6050	15 kVA		1.40	5.714			315		315	470
7000	Power filters & conditioners									
7100	Automatic voltage regulator	2 Elec	6	2.667	Ea.		146		146	219
7200	Capacitor, 1 kVAR	1 Elec	8.60	.930			51		51	76
7210	5 kVAR		5.80	1.379			75.50		75.50	113
7220	10 kVAR		4.80	1.667			91		91	137
7230	15 kVAR		4.20	1.905			104		104	156
7240	20 kVAR		3.50	2.286			125		125	187
7250	30 kVAR		3.40	2.353			129		129	193
7260	50 kVAR		3.20	2.500			137		137	205
7400	Computer isolator transformer									
7410	Single phase, 120/240 V, 0.5 kVAR	1 Elec	12.80	.625	Ea.		34		34	51
7420	1 kVAR		8.50	.941			51.50		51.50	77
7430	2.5 kVAR		6.40	1.250			68.50		68.50	102
7440	5 kVAR		3.70	2.162			118		118	177
7500	Computer regulator transformer									
7510	Single phase, 240 V, 0.5 kVAR	1 Elec	8.50	.941	Ea.		51.50		51.50	77
7520	1 kVAR		6.40	1.250			68.50		68.50	102
7530	2 kVAR		3.20	2.500			137		137	205
7540	Single phase, plug-in unit 120 V, 0.5 kVAR		26	.308			16.85		16.85	25

26 05 05.25 Electrical Demolition, Electrical Power

	26 05 05.25 Electrical Demolition, Electrical Power	Crew	Daily Output	Labor-Hours	Unit	Material	2015 Bare Costs Labor	Equipment	Total	Total Incl O&P
7550	1 kVAR	1 Elec	17	.471	Ea.		25.50		25.50	38.50
7600	Power conditioner transformer									
7610	Single phase 115 V - 240 V, 3 kVA	2 Elec	5.10	3.137	Ea.		172		172	257
7620	5 kVA	″	3.70	4.324			237		237	355
7630	7.5 kVA	3 Elec	4.80	5			274		274	410
7640	10 kVA	″	4.30	5.581			305		305	455
7700	Transient suppressor/voltage regulator									
7710	Single phase 115 or 220 V, 1 kVA	1 Elec	8.50	.941	Ea.		51.50		51.50	77
7720	2 kVA		7.30	1.096			60		60	90
7730	4 kVA		6.80	1.176			64.50		64.50	96.50
7800	Transient voltage suppressor transformer									
7810	Single phase, 115 or 220 V, 3.6 kVA	1 Elec	12.80	.625	Ea.		34		34	51
7820	7.2 kVA		11.50	.696			38		38	57
7830	14.4 kVA		10.20	.784			43		43	64.50
7840	Single phase, plug-in, 120 V, 1.8 kVA		26	.308			16.85		16.85	25
8000	Power measurement & control									
8010	Switchboard instruments, 3 phase 4 wire, indicating unit	1 Elec	25	.320	Ea.		17.50		17.50	26
8020	Recording unit		12	.667			36.50		36.50	54.50
8100	3 current transformers, 3 phase 4 wire, 5 to 800 A		6.40	1.250			68.50		68.50	102
8110	1000 to 1500 A		4.20	1.905			104		104	156
8120	2000 to 4000 A		3.20	2.500			137		137	205

26 05 05.30 Electrical Demolition, Transmission and Distribution

	26 05 05.30 Electrical Demolition, Transmission and Distribution	Crew	Daily Output	Labor-Hours	Unit	Material	Labor	Equipment	Total	Total Incl O&P
0010	**ELECTRICAL DEMOLITION, TRANSMISSION & DISTRIBUTION**									
0100	Load interrupter switch, 600 A, NEMA 1, 4.8 kV	R-3	1.33	15.038	Ea.		815	104	919	1,350
0120	13.8 kV R024119-10	″	1.27	15.748			855	109	964	1,400
0200	Lightning arrester, 4.8 kV	1 Elec	9	.889			48.50		48.50	73
0220	13.8 kV R260105-30		6.67	1.199			65.50		65.50	98.50
0300	Alarm or option items		3.33	2.402			131		131	197

26 05 05.35 Electrical Demolition, L.V. Distribution

	26 05 05.35 Electrical Demolition, L.V. Distribution	Crew	Daily Output	Labor-Hours	Unit	Material	Labor	Equipment	Total	Total Incl O&P
0010	**ELECTRICAL DEMOLITION, L.V. DISTRIBUTION** Addition R024119-10									
0100	Circuit breakers in enclosure									
0120	Enclosed (NEMA 1), 600 V, 3 pole, 30 A R260105-30	1 Elec	12.30	.650	Ea.		35.50		35.50	53.50
0140	60 A		8.80	.909			49.50		49.50	74.50
0160	100 A		7.30	1.096			60		60	90
0180	225 A		5	1.600			87.50		87.50	131
0200	400 A	2 Elec	6.80	2.353			129		129	193
0220	600 A		4.60	3.478			190		190	285
0240	800 A		3.60	4.444			243		243	365
0260	1200 A		3.10	5.161			282		282	425
0280	1600 A		2.80	5.714			315		315	470
0300	2000 A		2.50	6.400			350		350	525
0400	Enclosed (NEMA 7), 600 V, 3 pole, 50 A	1 Elec	8.80	.909			49.50		49.50	74.50
0410	100 A		5.80	1.379			75.50		75.50	113
0420	150 A		3.80	2.105			115		115	173
0430	250 A	2 Elec	6.20	2.581			141		141	211
0440	400 A	″	4.60	3.478			190		190	285
0460	Manual motor starter, NEMA 1	1 Elec	24	.333			18.25		18.25	27.50
0480	NEMA 4 or NEMA 7		15	.533			29		29	43.50
0500	Time switches, single pole single throw		15.40	.519			28.50		28.50	42.50
0520	Photo cell		30	.267			14.60		14.60	22
0540	Load management device, 4 loads		7.70	1.039			57		57	85
0550	8 loads		3.80	2.105			115		115	173

26 05 05 – Selective Demolition for Electrical

26 05 05.35 Electrical Demolition, L.V. Distribution		Crew	Daily Output	Labor-Hours	Unit	Material	2015 Bare Costs Labor	Equipment	Total	Total Incl O&P
0560	Master light control panel	2 Elec	2	8	Ea.		440		440	655
0600	Transfer switches, enclosed, 30 A	1 Elec	12	.667			36.50		36.50	54.50
0610	60 A		9.50	.842			46		46	69
0620	100 A		6.50	1.231			67.50		67.50	101
0630	150 A	2 Elec	12	1.333			73		73	109
0640	260 A		10	1.600			87.50		87.50	131
0660	400 A		8	2			109		109	164
0670	600 A		5	3.200			175		175	262
0680	800 A		4	4			219		219	330
0690	1200 A		3.50	4.571			250		250	375
0700	1600 A		3	5.333			292		292	435
0710	2000 A		2.50	6.400			350		350	525
0800	Air terminal with base and connector	1 Elec	24	.333			18.25		18.25	27.50
1000	Enclosed controller, NEMA 1, 30 A		13.80	.580			31.50		31.50	47.50
1020	60 A		11.50	.696			38		38	57
1040	100 A		9.60	.833			45.50		45.50	68.50
1060	150 A		7.70	1.039			57		57	85
1080	200 A		5.40	1.481			81		81	121
1100	400 A	2 Elec	6.90	2.319			127		127	190
1120	600 A		4.60	3.478			190		190	285
1140	800 A		3.80	4.211			230		230	345
1160	1200 A		3.10	5.161			282		282	425
1200	Control station, NEMA 1	1 Elec	30	.267			14.60		14.60	22
1220	NEMA 7		23	.348			19.05		19.05	28.50
1230	Control switches, push button		69	.116			6.35		6.35	9.50
1240	Indicating light unit		120	.067			3.65		3.65	5.45
1250	Relay		15	.533			29		29	43.50
2000	Motor control center components									
2020	Starter, NEMA 1, size 1	1 Elec	9	.889	Ea.		48.50		48.50	73
2040	Size 2	2 Elec	13.30	1.203			66		66	98.50
2060	Size 3		6.70	2.388			131		131	196
2080	Size 4		5.30	3.019			165		165	247
2100	Size 5		3.30	4.848			265		265	395
2120	NEMA 7, size 1	1 Elec	8.70	.920			50.50		50.50	75.50
2140	Size 2	2 Elec	12.70	1.260			69		69	103
2160	Size 3		6.30	2.540			139		139	208
2180	Size 4		5	3.200			175		175	262
2200	Size 5		3.20	5			274		274	410
2300	Fuse, light contactor, NEMA 1, 30 A	1 Elec	9	.889			48.50		48.50	73
2310	60 A		6.70	1.194			65.50		65.50	98
2320	100 A		3.30	2.424			133		133	199
2330	200 A		2.70	2.963			162		162	243
2350	Motor control center, incoming section	2 Elec	4	4			219		219	330
2352	Structure per section	"	4.20	3.810			208		208	310
2400	Starter & structure, 10 HP	1 Elec	9	.889			48.50		48.50	73
2410	25 HP	2 Elec	13.30	1.203			66		66	98.50
2420	50 HP		6.70	2.388			131		131	196
2430	75 HP		5.30	3.019			165		165	247
2440	100 HP		4.70	3.404			186		186	279
2450	200 HP		3.30	4.848			265		265	395
2460	400 HP		2.70	5.926			325		325	485
2500	Motor starter & control									
2510	Motor starter, NEMA 1, 5 HP	1 Elec	8.90	.899	Ea.		49		49	73.50

26 05 Common Work Results for Electrical

26 05 05 – Selective Demolition for Electrical

26 05 05.35	Electrical Demolition, L.V. Distribution	Crew	Daily Output	Labor-Hours	Unit	Material	2015 Bare Costs Labor	Equipment	Total	Total Incl O&P
2520	10 HP	1 Elec	6.20	1.290	Ea.		70.50		70.50	106
2530	25 HP	2 Elec	8.40	1.905			104		104	156
2540	50 HP		6.90	2.319			127		127	190
2550	100 HP		4.60	3.478			190		190	285
2560	200 HP		3.50	4.571			250		250	375
2570	400 HP		3.10	5.161			282		282	425
2610	Motor starter, NEMA 7, 5 HP	1 Elec	6.20	1.290			70.50		70.50	106
2620	10 HP	"	4.20	1.905			104		104	156
2630	25 HP	2 Elec	6.90	2.319			127		127	190
2640	50 HP		4.60	3.478			190		190	285
2650	100 HP		3.50	4.571			250		250	375
2660	200 HP		1.90	8.421			460		460	690
2710	Combination control unit, NEMA 1, 5 HP	1 Elec	6.90	1.159			63.50		63.50	95
2720	10 HP	"	5	1.600			87.50		87.50	131
2730	25 HP	2 Elec	7.70	2.078			114		114	170
2740	50 HP		5.10	3.137			172		172	257
2750	100 HP		3.10	5.161			282		282	425
2810	NEMA 7, 5 HP	1 Elec	5	1.600			87.50		87.50	131
2820	10 HP	"	3.90	2.051			112		112	168
2830	25 HP	2 Elec	5.10	3.137			172		172	257
2840	50 HP		3.10	5.161			282		282	425
2850	100 HP		2.30	6.957			380		380	570
2860	200 HP		1.50	10.667			585		585	875
3000	Panelboard or load center circuit breaker									
3010	Bolt-on or plug in, 15 A to 50 A	1 Elec	20	.400	Ea.		22		22	33
3020	60 A to 70 A		16	.500			27.50		27.50	41
3030	Bolt-on, 80 A to 100 A		14	.571			31.50		31.50	47
3040	Up to 250 A		7.30	1.096			60		60	90
3050	Motor operated, 30 A		13	.615			33.50		33.50	50.50
3060	60 A		10	.800			44		44	65.50
3070	100 A		8	1			54.50		54.50	82
3200	Switchboard circuit breaker									
3210	15 A - 60 A	1 Elec	19	.421	Ea.		23		23	34.50
3220	70 A - 100 A		14	.571			31.50		31.50	47
3230	125 A - 400 A		10	.800			44		44	65.50
3240	450 A - 600 A		5.30	1.509			82.50		82.50	124
3250	700 A - 800 A		4.30	1.860			102		102	152
3260	1000 A		3.30	2.424			133		133	199
3270	1200 A		2.70	2.963			162		162	243
3500	Switchboard, incoming section, 400 A	2 Elec	3.70	4.324			237		237	355
3510	600 A		3.30	4.848			265		265	395
3520	800 A		2.90	5.517			300		300	450
3530	1200 A		2.40	6.667			365		365	545
3540	1600 A		2.20	7.273			400		400	595
3550	2000 A		2.10	7.619			415		415	625
3560	3000 A		1.90	8.421			460		460	690
3570	4000 A		1.70	9.412			515		515	770
3610	Distribution section, 600 A		4	4			219		219	330
3620	800 A		3.60	4.444			243		243	365
3630	1200 A		3.10	5.161			282		282	425
3640	1600 A		2.90	5.517			300		300	450
3650	2000 A		2.70	5.926			325		325	485
3710	Transition section, 600 A		3.80	4.211			230		230	345

26 05 05.35 Electrical Demolition, L.V. Distribution		Crew	Daily Output	Labor-Hours	Unit	Material	2015 Bare Costs Labor	Equipment	Total	Total Incl O&P
3720	800 A	2 Elec	3.30	4.848	Ea.		265		265	395
3730	1200 A		2.70	5.926			325		325	485
3740	1600 A		2.40	6.667			365		365	545
3750	2000 A		2.20	7.273			400		400	595
3760	2500 A		2.10	7.619			415		415	625
3770	3000 A		1.90	8.421			460		460	690
4000	Bus duct, aluminum or copper, 30 A	1 Elec	200	.040	L.F.		2.19		2.19	3.28
4020	60 A		160	.050			2.74		2.74	4.10
4040	100 A		140	.057			3.13		3.13	4.68
5000	Feedrail, trolley busway, up to 60 A		160	.050			2.74		2.74	4.10
5020	100 A		120	.067			3.65		3.65	5.45
5040	Busway, 50 A		200	.040			2.19		2.19	3.28
6000	Fuse, 30 A		133	.060	Ea.		3.29		3.29	4.93
6010	60 A		133	.060			3.29		3.29	4.93
6020	100 A		105	.076			4.17		4.17	6.25
6030	200 A		95	.084			4.61		4.61	6.90
6040	400 A		80	.100			5.45		5.45	8.20
6050	600 A		53	.151			8.25		8.25	12.35
6060	601 A - 1200 A		42	.190			10.40		10.40	15.60
6070	1500 A - 1600 A		34	.235			12.85		12.85	19.30
6080	1800 A - 2500 A		26	.308			16.85		16.85	25
6090	4000 A		21	.381			21		21	31
6100	4500 A - 5000 A		18	.444			24.50		24.50	36.50
6120	6000 A		15	.533			29		29	43.50
6200	Fuse plug or fustat		100	.080			4.38		4.38	6.55

26 05 05.50 Electrical Demolition, Lighting		Crew	Daily Output	Labor-Hours	Unit	Material	Labor	Equipment	Total	Total Incl O&P
0010	**ELECTRICAL DEMOLITION, LIGHTING** Addition R024119-10									
0100	Fixture hanger, flexible, 1/2" diameter	1 Elec	36	.222	Ea.		12.15		12.15	18.20
0120	3/4" diameter R260105-30	"	30	.267	"		14.60		14.60	22
3000	Light pole, anchor base, excl concrete bases									
3010	Metal light pole, 10'	2 Elec	24	.667	Ea.		36.50		36.50	54.50
3020	16'	"	18	.889			48.50		48.50	73
3030	20'	R-3	8.70	2.299			125	15.85	140.85	204
3040	40'	"	6	3.333			181	23	204	297
3100	Wood light pole, 10'	2 Elec	36	.444			24.50		24.50	36.50
3120	20'	"	24	.667			36.50		36.50	54.50
3140	Bollard light, 42"	1 Elec	9	.889			48.50		48.50	73
3160	Walkway luminaire	"	8.10	.988			54		54	81
4000	Explosionproof									
4010	Metal halide, 175 W	1 Elec	8.70	.920	Ea.		50.50		50.50	75.50
4020	250 W	"	8.10	.988			54		54	81
4030	400 W	2 Elec	14.40	1.111			61		61	91
4050	High pressure sodium, 70 W	1 Elec	9	.889			48.50		48.50	73
4060	100 W		9	.889			48.50		48.50	73
4070	150 W		8.10	.988			54		54	81
4100	Incandescent		8.70	.920			50.50		50.50	75.50
4200	Fluorescent		8.10	.988			54		54	81
5000	Ballast, fluorescent fixture		24	.333			18.25		18.25	27.50
5040	High intensity discharge fixture		24	.333			18.25		18.25	27.50
5300	Exit and emergency lighting									
5310	Exit light	1 Elec	24	.333	Ea.		18.25		18.25	27.50
5320	Emergency battery pack lighting unit		12	.667			36.50		36.50	54.50

26 05 05 – Selective Demolition for Electrical

26 05 05.50 Electrical Demolition, Lighting		Crew	Daily Output	Labor-Hours	Unit	Material	2015 Bare Costs Labor	Equipment	Total	Total Incl O&P
5330	Remote lamp only	1 Elec	80	.100	Ea.		5.45		5.45	8.20
5340	Self-contained fluorescent lamp pack		30	.267			14.60		14.60	22
5500	Track lighting, 8′ section	2 Elec	40	.400			22		22	33
5510	Track lighting fixture	1 Elec	64	.125			6.85		6.85	10.25
5800	Energy saving devices									
5810	Occupancy sensor	1 Elec	21	.381	Ea.		21		21	31
5820	Automatic wall switch		72	.111			6.10		6.10	9.10
5830	Remote power pack		30	.267			14.60		14.60	22
5840	Photoelectric control		24	.333			18.25		18.25	27.50
5850	Fixture whip		100	.080			4.38		4.38	6.55
6000	Lamps									
6010	Fluorescent	1 Elec	200	.040	Ea.		2.19		2.19	3.28
6030	High intensity discharge lamp, up to 400 W		68	.118			6.45		6.45	9.65
6040	Up to 1000 W		45	.178			9.70		9.70	14.55
6050	Quartz		90	.089			4.86		4.86	7.30
6070	Incandescent		360	.022			1.22		1.22	1.82
6080	Exterior, PAR		290	.028			1.51		1.51	2.26
6090	Guards for fluorescent lamp		24	.333			18.25		18.25	27.50

26 05 13 – Medium-Voltage Cables

26 05 13.10 Cable Terminations		Crew	Daily Output	Labor-Hours	Unit	Material	Labor	Equipment	Total	Total Incl O&P
0010	**CABLE TERMINATIONS**, 5 kV to 35 kV									
0100	Indoor, insulation diameter range .64″ to 1.08″									
0300	Padmount, 5 kV	1 Elec	8	1	Ea.	72.50	54.50		127	162
0400	10 kV		6.40	1.250		84.50	68.50		153	195
0500	15 kV		6	1.333		108	73		181	228
0600	25 kV		5.60	1.429		147	78		225	279
0700	insulation diameter range 1.05″ to 1.8″									
0800	Padmount, 5 kV	1 Elec	8	1	Ea.	102	54.50		156.50	194
0900	10 kV		6	1.333		117	73		190	238
1000	15 kV		5.60	1.429		128	78		206	258
1100	25 kV		5.30	1.509		168	82.50		250.50	310
1200	insulation diameter range 1.53″ to 2.32″									
1300	Padmount, 5 KV	1 Elec	7.40	1.081	Ea.	121	59		180	222
1400	10 kV		5.60	1.429		139	78		217	270
1500	15 kV		5.30	1.509		166	82.50		248.50	305
1600	25 kV		5	1.600		238	87.50		325.50	390
1700	Outdoor systems, #4 stranded to 1/0 stranded									
1800	5 kV	1 Elec	7.40	1.081	Ea.	91	59		150	189
1900	15 kV		5.30	1.509		134	82.50		216.50	272
2000	25 kV		5	1.600		180	87.50		267.50	330
2100	35 kV		4.80	1.667		248	91		339	410
2200	#1 solid to 4/0 stranded, 5 kV		6.90	1.159		123	63.50		186.50	231
2300	15 kV		5	1.600		172	87.50		259.50	320
2400	25 kV		4.80	1.667		232	91		323	390
2500	35 kV		4.60	1.739		264	95		359	435
2600	2/0 solid to 350 kcmil stranded, 5 kV		6.40	1.250		164	68.50		232.50	282
2700	15 kV		4.80	1.667		194	91		285	350
2800	25 kV		4.60	1.739		300	95		395	475
2900	35 kV		4.40	1.818		305	99.50		404.50	485
3000	400 kcmil compact to 750 kcmil stranded, 5 kV		6	1.333		196	73		269	325
3100	15 kV		4.60	1.739		234	95		329	400
3200	25 kV		4.40	1.818		340	99.50		439.50	525

26 05 13.10 Cable Terminations		Crew	Daily Output	Labor-Hours	Unit	Material	2015 Bare Costs Labor	Equipment	Total	Total Incl O&P
3300	35 kV	1 Elec	4.20	1.905	Ea.	350	104		454	540
3400	1000 kcmil, 5 kV		5.60	1.429		266	78		344	410
3500	15 kV		4.40	1.818		315	99.50		414.50	500
3600	25 kV		4.20	1.905		350	104		454	540
3700	35 kV		4	2		350	109		459	550

26 05 13.16 Medium-Voltage, Single Cable

		Crew	Daily Output	Labor-Hours	Unit	Material	Labor	Equipment	Total	Total Incl O&P
0010	**MEDIUM-VOLTAGE, SINGLE CABLE** Splicing & terminations not included									
0040	Copper, XLP shielding, 5 kV, #6	2 Elec	4.40	3.636	C.L.F.	160	199		359	475
0050	#4		4.40	3.636		207	199		406	525
0100	#2		4	4		228	219		447	580
0200	#1		4	4		281	219		500	640
0400	1/0		3.80	4.211		315	230		545	690
0600	2/0		3.60	4.444		385	243		628	785
0800	4/0		3.20	5		520	274		794	980
1000	250 kcmil	3 Elec	4.50	5.333		595	292		887	1,100
1200	350 kcmil		3.90	6.154		780	335		1,115	1,375
1400	500 kcmil		3.60	6.667		955	365		1,320	1,600
1600	15 kV, ungrounded neutral, #1	2 Elec	4	4		340	219		559	705
1800	1/0		3.80	4.211		410	230		640	795
2000	2/0		3.60	4.444		465	243		708	875
2200	4/0		3.20	5		620	274		894	1,100
2400	250 kcmil	3 Elec	4.50	5.333		685	292		977	1,200
2600	350 kcmil		3.90	6.154		865	335		1,200	1,450
2800	500 kcmil		3.60	6.667		1,075	365		1,440	1,725
3000	25 kV, grounded neutral, #1/0	2 Elec	3.60	4.444		565	243		808	985
3200	2/0		3.40	4.706		620	257		877	1,075
3400	4/0		3	5.333		780	292		1,072	1,300
3600	250 kcmil	3 Elec	4.20	5.714		970	315		1,285	1,550
3800	350 kcmil		3.60	6.667		1,125	365		1,490	1,800
3900	500 kcmil		3.30	7.273		1,325	400		1,725	2,075
4000	35 kV, grounded neutral, #1/0	2 Elec	3.40	4.706		600	257		857	1,050
4200	2/0		3.20	5		705	274		979	1,175
4400	4/0		2.80	5.714		885	315		1,200	1,450
4600	250 kcmil	3 Elec	3.90	6.154		1,025	335		1,360	1,650
4800	350 kcmil		3.30	7.273		1,250	400		1,650	1,975
5000	500 kcmil		3	8		1,475	440		1,915	2,250
5050	Aluminum, XLP shielding, 5 kV, #2	2 Elec	5	3.200		170	175		345	450
5070	#1		4.40	3.636		176	199		375	490
5090	1/0		4	4		204	219		423	555
5100	2/0		3.80	4.211		229	230		459	595
5150	4/0		3.60	4.444		271	243		514	665
5200	250 kcmil	3 Elec	4.80	5		330	274		604	770
5220	350 kcmil		4.50	5.333		385	292		677	860
5240	500 kcmil		3.90	6.154		495	335		830	1,050
5260	750 kcmil		3.60	6.667		655	365		1,020	1,275
5300	15 kV aluminum, XLP, #1	2 Elec	4.40	3.636		218	199		417	540
5320	1/0		4	4		226	219		445	580
5340	2/0		3.80	4.211		269	230		499	640
5360	4/0		3.60	4.444		299	243		542	695
5380	250 kcmil	3 Elec	4.80	5		355	274		629	800
5400	350 kcmil		4.50	5.333		400	292		692	875
5420	500 kcmil		3.90	6.154		555	335		890	1,125

26 05 Common Work Results for Electrical

26 05 13 – Medium-Voltage Cables

26 05 13.16 Medium-Voltage, Single Cable		Crew	Daily Output	Labor-Hours	Unit	2015 Bare Costs Material	Labor	Equipment	Total	Total Incl O&P
5440	750 kcmil	3 Elec	3.60	6.667	C.L.F.	735	365		1,100	1,350

26 05 19 – Low-Voltage Electrical Power Conductors and Cables

26 05 19.13 Undercarpet Electrical Power Cables

		Crew	Daily Output	Labor-Hours	Unit	Material	Labor	Equipment	Total	Total Incl O&P
0010	UNDERCARPET ELECTRICAL POWER CABLES R260519-80									
0020	Power System									
0100	Cable flat, 3 conductor, #12, w/attached bottom shield	1 Elec	982	.008	L.F.	5.15	.45		5.60	6.30
0200	Shield, top, steel		1768	.005	"	5.50	.25		5.75	6.40
0250	Splice, 3 conductor		48	.167	Ea.	16.75	9.10		25.85	32
0300	Top shield		96	.083		1.49	4.56		6.05	8.50
0350	Tap		40	.200		21.50	10.95		32.45	40
0400	Insulating patch, splice, tap, & end		48	.167		53	9.10		62.10	71.50
0450	Fold		230	.035			1.90		1.90	2.85
0500	Top shield, tap & fold		96	.083		1.49	4.56		6.05	8.50
0700	Transition, block assembly		77	.104		77	5.70		82.70	93.50
0750	Receptacle frame & base		32	.250		42.50	13.70		56.20	67
0800	Cover receptacle		120	.067		3.61	3.65		7.26	9.40
0850	Cover blank		160	.050		4.24	2.74		6.98	8.75
0860	Receptacle, direct connected, single		25	.320		90.50	17.50		108	126
0870	Dual		16	.500		149	27.50		176.50	204
0880	Combination high & low, tension		21	.381		109	21		130	151
0900	Box, floor with cover		20	.400		92	22		114	134
0920	Floor service w/barrier		4	2		259	109		368	450
1000	Wall, surface, with cover		20	.400		60.50	22		82.50	99.50
1100	Wall, flush, with cover		20	.400		42.50	22		64.50	79.50
1450	Cable flat, 5 conductor #12, w/attached bottom shield		800	.010	L.F.	8.45	.55		9	10.05
1550	Shield, top, steel		1768	.005	"	8.35	.25		8.60	9.55
1600	Splice, 5 conductor		48	.167	Ea.	27	9.10		36.10	43.50
1650	Top shield		96	.083		1.49	4.56		6.05	8.50
1700	Tap		48	.167		35.50	9.10		44.60	52.50
1750	Insulating patch, splice tap, & end		83	.096		53	5.25		58.25	66
1800	Transition, block assembly		77	.104		55.50	5.70		61.20	69.50
1850	Box, wall, flush with cover		20	.400		56.50	22		78.50	95
1900	Cable flat, 4 conductor, #12		933	.009	L.F.	6.80	.47		7.27	8.20
1950	3 conductor #10		982	.008		5.90	.45		6.35	7.15
1960	4 conductor #10		933	.009		7.70	.47		8.17	9.20
1970	5 conductor #10		884	.009		9.45	.50		9.95	11.10
2500	Telephone System									
2510	Transition fitting wall box, surface	1 Elec	24	.333	Ea.	45.50	18.25		63.75	77.50
2520	Flush		24	.333		45.50	18.25		63.75	77.50
2530	Flush, for PC board		24	.333		45.50	18.25		63.75	77.50
2540	Floor service box		4	2		235	109		344	425
2550	Cover, surface					14.45			14.45	15.90
2560	Flush					14.45			14.45	15.90
2570	Flush for PC board					14.45			14.45	15.90
2700	Floor fitting w/duplex jack & cover	1 Elec	21	.381		44.50	21		65.50	80
2720	Low profile		53	.151		15.55	8.25		23.80	29.50
2740	Miniature w/duplex jack		53	.151		24	8.25		32.25	39
2760	25 pair kit		21	.381		47	21		68	82.50
2780	Low profile		53	.151		15.90	8.25		24.15	30
2800	Call director kit for 5 cable		19	.421		74.50	23		97.50	117
2820	4 pair kit		19	.421		89	23		112	133
2840	3 pair kit		19	.421		94	23		117	139

26 05 19.13 Undercarpet Electrical Power Cables		Crew	Daily Output	Labor-Hours	Unit	2015 Bare Costs Material	Labor	Equipment	Total	Total Incl O&P
2860	Comb. 25 pair & 3 conductor power	1 Elec	21	.381	Ea.	74.50	21		95.50	113
2880	5 conductor power		21	.381		85.50	21		106.50	125
2900	PC board, 8 per 3 pair		161	.050		64.50	2.72		67.22	75
2920	6 per 4 pair		161	.050		64.50	2.72		67.22	75
2940	3 pair adapter		161	.050		59	2.72		61.72	69
2950	Plug		77	.104		2.74	5.70		8.44	11.50
2960	Couplers		321	.025		7.75	1.36		9.11	10.55
3000	Bottom shield for 25 pair cable		4420	.002	L.F.	.78	.10		.88	1.01
3020	4 pair		4420	.002		.37	.10		.47	.56
3040	Top shield for 25 pair cable		4420	.002		.78	.10		.88	1.01
3100	Cable assembly, double-end, 50′, 25 pair		11.80	.678	Ea.	227	37		264	305
3110	3 pair		23.60	.339		67	18.55		85.55	102
3120	4 pair		23.60	.339		74.50	18.55		93.05	110
3140	Bulk 3 pair		1473	.005	L.F.	1.15	.30		1.45	1.72
3160	4 pair		1473	.005	″	1.43	.30		1.73	2.02
3500	Data System									
3520	Cable 25 conductor w/connection 40′, 75 ohm	1 Elec	14.50	.552	Ea.	67.50	30		97.50	119
3530	Single lead		22	.364		184	19.90		203.90	232
3540	Dual lead		22	.364		229	19.90		248.90	281
3560	Shields same for 25 conductor as 25 pair telephone									
3570	Single & dual, none required									
3590	BNC coax connectors, Plug	1 Elec	40	.200	Ea.	10.05	10.95		21	27.50
3600	TNC coax connectors, Plug	″	40	.200	″	12.90	10.95		23.85	30.50
3700	Cable-bulk									
3710	Single lead	1 Elec	1473	.005	L.F.	2.45	.30		2.75	3.15
3720	Dual lead	″	1473	.005	″	3.55	.30		3.85	4.36
3730	Hand tool crimp				Ea.	510			510	560
3740	Hand tool notch				″	17.55			17.55	19.30
3750	Boxes & floor fitting same as telephone									
3790	Data cable notching, 90°	1 Elec	97	.082	Ea.		4.51		4.51	6.75
3800	180°		60	.133			7.30		7.30	10.95
8100	Drill floor		160	.050		1.96	2.74		4.70	6.25
8200	Marking floor		1600	.005	L.F.		.27		.27	.41
8300	Tape, hold down		6400	.001	″	.16	.07		.23	.28
8350	Tape primer, 500 ft. per can		96	.083	Ea.	35.50	4.56		40.06	46.50
8400	Tool, splicing				″	201			201	221

26 05 19.20 Armored Cable

		Crew	Daily Output	Labor-Hours	Unit	Material	Labor	Equipment	Total	Total Incl O&P
0010	**ARMORED CABLE** R260519-20									
0050	600 volt, copper (BX), #14, 2 conductor, solid	1 Elec	2.40	3.333	C.L.F.	46	182		228	325
0100	3 conductor, solid		2.20	3.636		71	199		270	375
0120	4 conductor, solid		2	4		98.50	219		317.50	440
0150	#12, 2 conductor, solid		2.30	3.478		46	190		236	335
0200	3 conductor, solid		2	4		75.50	219		294.50	415
0220	4 conductor, solid		1.80	4.444		103	243		346	480
0250	#10, 2 conductor, solid		2	4		85	219		304	425
0300	3 conductor, solid		1.60	5		118	274		392	540
0320	4 conductor, solid		1.40	5.714		182	315		497	670
0340	#8, 2 conductor, stranded		1.50	5.333		222	292		514	680
0350	3 conductor, stranded		1.30	6.154		222	335		557	750
0370	4 conductor, stranded		1.10	7.273		315	400		715	945
0380	#6, 2 conductor, stranded		1.30	6.154		233	335		568	760
0390	#4, 3 conductor, stranded		1.40	5.714		555	315		870	1,075

26 05 19.20 Armored Cable		Crew	Daily Output	Labor-Hours	Unit	Material	2015 Bare Costs Labor	Equipment	Total	Total Incl O&P
0400	3 conductor with PVC jacket, in cable tray, #6	1 Elec	3.10	2.581	C.L.F.	540	141		681	800
0450	#4	2 Elec	5.40	2.963		655	162		817	965
0500	#2		4.60	3.478		795	190		985	1,150
0550	#1		4	4		1,025	219		1,244	1,450
0600	1/0		3.60	4.444		1,050	243		1,293	1,525
0650	2/0		3.40	4.706		1,250	257		1,507	1,750
0700	3/0		3.20	5		1,700	274		1,974	2,250
0750	4/0		3	5.333		2,000	292		2,292	2,625
0800	250 kcmil	3 Elec	3.60	6.667		2,325	365		2,690	3,125
0850	350 kcmil		3.30	7.273		3,150	400		3,550	4,050
0900	500 kcmil		3	8		4,350	440		4,790	5,425
0910	4 conductor with PVC jacket, in cable tray, #6	1 Elec	2.70	2.963		670	162		832	980
0920	#4	2 Elec	4.60	3.478		840	190		1,030	1,200
0930	#2		4	4		1,025	219		1,244	1,450
0940	#1		3.60	4.444		1,300	243		1,543	1,825
0950	1/0		3.40	4.706		1,350	257		1,607	1,850
0960	2/0		3.20	5		1,650	274		1,924	2,225
0970	3/0		3	5.333		2,200	292		2,492	2,825
0980	4/0		2.40	6.667		2,550	365		2,915	3,350
0990	250 kcmil	3 Elec	3.30	7.273		3,050	400		3,450	3,975
1000	350 kcmil		3	8		4,125	440		4,565	5,175
1010	500 kcmil		2.70	8.889		5,700	485		6,185	7,000
1050	5 kV, copper, 3 conductor with PVC jacket,									
1060	non-shielded, in cable tray, #4	2 Elec	380	.042	L.F.	7.40	2.30		9.70	11.60
1100	#2		360	.044		9.65	2.43		12.08	14.25
1200	#1		300	.053		12.30	2.92		15.22	17.85
1400	1/0		290	.055		14.20	3.02		17.22	20
1600	2/0		260	.062		16.40	3.37		19.77	23
2000	4/0		240	.067		22	3.65		25.65	29.50
2100	250 kcmil	3 Elec	330	.073		30	3.98		33.98	39
2150	350 kcmil		315	.076		37	4.17		41.17	47
2200	500 kcmil		270	.089		51.50	4.86		56.36	64.50
2400	15 kV, copper, 3 conductor with PVC jacket galv., steel armored									
2500	grounded neutral, in cable tray, #2	2 Elec	300	.053	L.F.	15.60	2.92		18.52	21.50
2600	#1		280	.057		16.60	3.13		19.73	23
2800	1/0		260	.062		19	3.37		22.37	26
2900	2/0		220	.073		25	3.98		28.98	34
3000	4/0		190	.084		28.50	4.61		33.11	38.50
3100	250 kcmil	3 Elec	270	.089		32	4.86		36.86	42.50
3150	350 kcmil		240	.100		37.50	5.45		42.95	49.50
3200	500 kcmil		210	.114		50.50	6.25		56.75	65
3400	15 kV, copper, 3 conductor with PVC jacket,									
3450	ungrounded neutral, in cable tray, #2	2 Elec	260	.062	L.F.	16.80	3.37		20.17	23.50
3500	#1		230	.070		18.60	3.81		22.41	26
3600	1/0		200	.080		21.50	4.38		25.88	30
3700	2/0		190	.084		26	4.61		30.61	36
3800	4/0		160	.100		31.50	5.45		36.95	42.50
4000	250 kcmil	3 Elec	210	.114		37	6.25		43.25	50
4050	350 kcmil		195	.123		48.50	6.75		55.25	63.50
4100	500 kcmil		180	.133		59	7.30		66.30	76
4200	600 volt, aluminum, 3 conductor in cable tray with PVC jacket									
4300	#2	2 Elec	540	.030	L.F.	3.96	1.62		5.58	6.80
4400	#1		460	.035		4.38	1.90		6.28	7.65

26 05 19 – Low-Voltage Electrical Power Conductors and Cables

26 05 19.20 Armored Cable		Crew	Daily Output	Labor-Hours	Unit	Material	Labor	Equipment	Total	Total Incl O&P
						2015 Bare Costs				
4500	#1/0	2 Elec	400	.040	L.F.	5.45	2.19		7.64	9.30
4600	#2/0		360	.044		5.55	2.43		7.98	9.75
4700	#3/0		340	.047		6.50	2.57		9.07	11
4800	#4/0		320	.050		7.85	2.74		10.59	12.75
4900	250 kcmil	3 Elec	450	.053		9.45	2.92		12.37	14.70
5000	350 kcmil		360	.067		11.25	3.65		14.90	17.80
5200	500 kcmil		330	.073		13.95	3.98		17.93	21.50
5300	750 kcmil		285	.084		18	4.61		22.61	26.50
5400	600 volt, aluminum, 4 conductor in cable tray with PVC jacket									
5410	#2	2 Elec	520	.031	L.F.	4.55	1.68		6.23	7.50
5430	#1		440	.036		5.55	1.99		7.54	9.10
5450	1/0		380	.042		6.50	2.30		8.80	10.60
5470	2/0		340	.047		6.60	2.57		9.17	11.15
5480	3/0		320	.050		7.80	2.74		10.54	12.65
5500	4/0		300	.053		9.15	2.92		12.07	14.40
5520	250 kcmil	3 Elec	420	.057		9.80	3.13		12.93	15.50
5540	350 kcmil		330	.073		12.65	3.98		16.63	19.90
5560	500 kcmil		300	.080		15.85	4.38		20.23	24
5580	750 kcmil		270	.089		22	4.86		26.86	31.50
5600	5 kV, aluminum, unshielded in cable tray, #2 with PVC jacket	2 Elec	380	.042		5.55	2.30		7.85	9.55
5700	#1 with PVC jacket		360	.044		6.20	2.43		8.63	10.45
5800	1/0 with PVC jacket		300	.053		6.35	2.92		9.27	11.30
6000	2/0 with PVC jacket		290	.055		6.50	3.02		9.52	11.65
6200	3/0 with PVC jacket		260	.062		7.80	3.37		11.17	13.60
6300	4/0 with PVC jacket		240	.067		9.15	3.65		12.80	15.50
6400	250 kcmil with PVC jacket	3 Elec	330	.073		10	3.98		13.98	16.95
6500	350 kcmil with PVC jacket		315	.076		11.75	4.17		15.92	19.15
6600	500 kcmil with PVC jacket		300	.080		13.90	4.38		18.28	22
6800	750 kcmil with PVC jacket		270	.089		17.15	4.86		22.01	26
6900	15 kV, aluminum, shielded-grounded, #2 with PVC jacket	2 Elec	320	.050		12.60	2.74		15.34	17.95
7000	#1 with PVC jacket		300	.053		12.95	2.92		15.87	18.60
7200	1/0 with PVC jacket		280	.057		13.95	3.13		17.08	20
7300	2/0 with PVC jacket		260	.062		14.20	3.37		17.57	20.50
7400	3/0 with PVC jacket		240	.067		15.90	3.65		19.55	23
7500	4/0 with PVC jacket		220	.073		16.35	3.98		20.33	24
7600	250 kcmil with PVC jacket	3 Elec	300	.080		17.95	4.38		22.33	26.50
7700	350 kcmil with PVC jacket		270	.089		21	4.86		25.86	30.50
7800	500 kcmil with PVC jacket		240	.100		25	5.45		30.45	35.50
8000	750 kcmil with PVC jacket		204	.118		30	6.45		36.45	42.50
8200	15 kV, aluminum, shielded-ungrounded, #1 with PVC jacket	2 Elec	250	.064		15.80	3.50		19.30	22.50
8300	1/0 with PVC jacket		230	.070		16.35	3.81		20.16	23.50
8400	2/0 with PVC jacket		210	.076		17.95	4.17		22.12	26
8500	3/0 with PVC jacket		200	.080		18.20	4.38		22.58	26.50
8600	4/0 with PVC jacket		190	.084		19.80	4.61		24.41	29
8700	250 kcmil with PVC jacket	3 Elec	270	.089		21.50	4.86		26.36	31
8800	350 kcmil with PVC jacket		240	.100		24.50	5.45		29.95	34.50
8900	500 kcmil with PVC jacket		210	.114		29.50	6.25		35.75	41.50
8950	750 kcmil with PVC jacket		174	.138		36.50	7.55		44.05	51.50
9010	600 volt, copper (MC) steel clad, #14, 2 wire	1 Elec	2.40	3.333	C.L.F.	46.50	182		228.50	325
9020	3 wire		2.20	3.636		72	199		271	380
9030	4 wire		2	4		100	219		319	440
9040	#12, 2 wire		2.30	3.478		47	190		237	335
9050	3 wire		2	4		79.50	219		298.50	420

26 05 19 – Low-Voltage Electrical Power Conductors and Cables

26 05 19.20 Armored Cable		Crew	Daily Output	Labor-Hours	Unit	Material	2015 Bare Costs Labor	Equipment	Total	Total Incl O&P
9060	4 wire	1 Elec	1.80	4.444	C.L.F.	107	243		350	485
9070	#10, 2 wire		2	4		98.50	219		317.50	440
9080	3 wire		1.60	5		138	274		412	560
9090	4 wire		1.40	5.714		216	315		531	705
9100	#8, 2 wire, stranded		1.80	4.444		190	243		433	575
9110	3 wire, stranded		1.30	6.154		267	335		602	800
9120	4 wire, stranded		1.10	7.273		360	400		760	990
9130	#6, 2 wire, stranded		1.30	6.154		281	335		616	815
9200	600 volt, copper (MC) aluminum clad, #14, 2 wire		2.65	3.019		45.50	165		210.50	297
9210	3 wire		2.45	3.265		71	179		250	345
9220	4 wire		2.20	3.636		99	199		298	405
9230	#12, 2 wire		2.55	3.137		47	172		219	310
9240	3 wire		2.20	3.636		78.50	199		277.50	385
9250	4 wire		2	4		106	219		325	445
9260	#10, 2 wire		2.20	3.636		97.50	199		296.50	405
9270	3 wire		1.80	4.444		137	243		380	515
9280	4 wire		1.55	5.161		215	282		497	660
9600	Alum (MC) aluminum clad, #6, 3 conductor w/#6 grnd		1.67	4.790		267	262		529	690
9610	4 conductor w/#6 grnd		1.64	4.878		290	267		557	720
9620	#4, 3 conductor w/#6 grnd	2 Elec	2.86	5.594		277	305		582	765
9630	4 conductor w/#6 grnd		2.82	5.674		300	310		610	795
9640	#2, 3 conductor w/#4 grnd		2.50	6.400		380	350		730	945
9650	4 conductor w/#4 grnd		2.47	6.478		430	355		785	1,000
9660	#1, 3 conductor w/#4 grnd		2	8		410	440		850	1,100
9670	4 conductor w/#4 grnd		1.98	8.081		580	440		1,020	1,300
9680	1/0, 3 conductor w/#4 grnd		1.82	8.791		470	480		950	1,225
9690	4 conductor w/#4 grnd		1.79	8.939		570	490		1,060	1,375
9700	2/0, 3 conductor w/#4 grnd		1.75	9.143		590	500		1,090	1,400
9710	4 conductor w/#4 grnd		1.71	9.357		640	510		1,150	1,475
9720	3/0, 3 conductor w/#4 grnd		1.71	9.357		755	510		1,265	1,600
9730	4 conductor w/#4 grnd		1.67	9.581		795	525		1,320	1,650
9740	4/0, 3 conductor w/#2 grnd		1.67	9.581		850	525		1,375	1,725
9750	4 conductor w/#2 grnd		1.61	9.938		975	545		1,520	1,900
9760	250 kcmil, 3 conductor w/#1 grnd	3 Elec	2.42	9.917		1,075	540		1,615	2,025
9770	4 conductor w/#1 grnd		2.36	10.169		1,475	555		2,030	2,450
9775	300 kcmil, 4 conductor w/#1 grnd		2.22	10.811		1,550	590		2,140	2,600
9780	350 kcmil, 3 conductor w/1/0 grnd		2.19	10.959		1,225	600		1,825	2,250
9790	4 conductor w/1/0 grnd		2.10	11.429		1,500	625		2,125	2,600
9800	500 kcmil, 3 conductor w/#1 grnd		2.10	11.429		1,800	625		2,425	2,900
9810	4 conductor w/2/0 grnd		2	12		1,850	655		2,505	3,000
9840	750 kcmil, 3 conductor w/1/0 grnd		1.95	12.308		2,575	675		3,250	3,850
9850	4 conductor w/3/0 grnd		1.89	12.698		2,800	695		3,495	4,150

26 05 19.25 Cable Connectors		Crew	Daily Output	Labor-Hours	Unit	Material	Labor	Equipment	Total	Total Incl O&P
0010	**CABLE CONNECTORS**									
0100	600 volt, nonmetallic, #14-2 wire	1 Elec	160	.050	Ea.	1.44	2.74		4.18	5.70
0200	#14-3 wire to #12-2 wire		133	.060		1.44	3.29		4.73	6.50
0300	#12-3 wire to #10-2 wire		114	.070		1.44	3.84		5.28	7.35
0400	#10-3 wire to #14-4 and #12-4 wire		100	.080		1.44	4.38		5.82	8.15
0500	#8-3 wire to #10-4 wire		80	.100		2.71	5.45		8.16	11.20
0600	#6-3 wire		40	.200		3.65	10.95		14.60	20.50
0800	SER, 3 #8 insulated + 1 #8 ground		32	.250		2.32	13.70		16.02	23
0900	3 #6 + 1 #6 ground		24	.333		2.90	18.25		21.15	30.50

26 05 19 – Low-Voltage Electrical Power Conductors and Cables

26 05 19.25 Cable Connectors		Crew	Daily Output	Labor-Hours	Unit	Material	2015 Bare Costs Labor	Equipment	Total	Total Incl O&P
1000	3 #4 + 1 #6 ground	1 Elec	22	.364	Ea.	3.63	19.90		23.53	34
1100	3 #2 + 1 #4 ground		20	.400		6.60	22		28.60	40.50
1200	3 1/0 + 1 #2 ground		18	.444		17.10	24.50		41.60	55.50
1400	3 2/0 + 1 #1 ground		16	.500		20.50	27.50		48	63.50
1600	3 4/0 + 1 #2/0 ground		14	.571		27	31.50		58.50	76.50
1800	600 volt, armored, #14-2 wire		80	.100		1.01	5.45		6.46	9.30
2200	#14-4, #12-3 and #10-2 wire		40	.200		.91	10.95		11.86	17.40
2400	#12-4, #10-3 and #8-2 wire		32	.250		2.37	13.70		16.07	23
2600	#8-3 and #10-4 wire		26	.308		3.58	16.85		20.43	29
2650	#8-4 wire		22	.364		5.15	19.90		25.05	35.50
2652	Non-PVC jacket connector, #6-3 wire, #6-4 wire		22	.364		44	19.90		63.90	78
2660	1/0-3 wire		11	.727		60	40		100	126
2670	300 kcmil-3 wire		6	1.333		91	73		164	209
2680	700-kcmil-3 wire		4	2		135	109		244	310
2700	PVC jacket connector, #6-3 wire, #6-4 wire		16	.500		8	27.50		35.50	50
2800	#4-3 wire, #4-4 wire		16	.500		8	27.50		35.50	50
2900	#2-3 wire		12	.667		8	36.50		44.50	63.50
3000	#1-3 wire, #2-4 wire		12	.667		13.95	36.50		50.45	70
3200	1/0-3 wire		11	.727		13.95	40		53.95	75
3400	2/0-3 wire, 1/0-4 wire		10	.800		13.95	44		57.95	81
3500	3/0-3 wire, 2/0-4 wire		9	.889		20.50	48.50		69	95.50
3600	4/0-3 wire, 3/0-4 wire		7	1.143		20.50	62.50		83	116
3800	250 kcmil-3 wire, 4/0-4 wire		6	1.333		37.50	73		110.50	150
4000	350 kcmil-3 wire, 250 kcmil-4 wire		5	1.600		37.50	87.50		125	172
4100	350 kcmil-4 wire		4	2		190	109		299	375
4200	500 kcmil-3 wire		4	2		190	109		299	375
4250	500 kcmil-4 wire, 750 kcmil-3 wire		3.50	2.286		259	125		384	470
4300	750 kcmil-4 wire		3	2.667		259	146		405	505
4400	5 kV, armored, #4		8	1		67	54.50		121.50	156
4600	#2		8	1		67	54.50		121.50	156
4800	#1		8	1		86.50	54.50		141	177
5000	1/0		6.40	1.250		107	68.50		175.50	220
5200	2/0		5.30	1.509		107	82.50		189.50	242
5500	4/0		4	2		143	109		252	320
5600	250 kcmil		3.60	2.222		143	122		265	340
5650	350 kcmil		3.20	2.500		172	137		309	395
5700	500 kcmil		2.50	3.200		216	175		391	500
5720	750 kcmil		2.20	3.636		263	199		462	585
5750	1000 kcmil		2	4		325	219		544	685
5800	15 kV, armored, #1		4	2		129	109		238	305
5900	1/0		4	2		161	109		270	340
6000	3/0		3.60	2.222		214	122		336	420
6100	4/0		3.40	2.353		243	129		372	460
6200	250 kcmil		3.20	2.500		274	137		411	505
6300	350 kcmil		2.70	2.963		287	162		449	560
6400	500 kcmil	↓	2	4	↓	325	219		544	685

26 05 19.30 Cable Splicing		Crew	Daily Output	Labor-Hours	Unit	Material	Labor	Equipment	Total	Total Incl O&P
0010	**CABLE SPLICING** URD or similar, ideal conditions									
0100	#6 stranded to #1 stranded, 5 kV	1 Elec	4	2	Ea.	117	109		226	293
0120	15 kV		3.60	2.222		164	122		286	360
0140	25 kV		3.20	2.500		184	137		321	410
0200	#1 stranded to 4/0 stranded, 5 kV	↓	3.60	2.222	↓	127	122		249	320

26 05 19.30 Cable Splicing

	26 05 19.30 Cable Splicing	Crew	Daily Output	Labor-Hours	Unit	2015 Bare Costs Material	Labor	Equipment	Total	Total Incl O&P
0210	15 kV	1 Elec	3.20	2.500	Ea.	174	137		311	395
0220	25 kV		2.80	2.857		184	156		340	435
0300	4/0 stranded to 500 kcmil stranded, 5 kV		3.30	2.424		127	133		260	340
0310	15 kV		2.90	2.759		261	151		412	515
0320	25 kV		2.50	3.200		287	175		462	575
0400	500 kcmil, 5 kV		3.20	2.500		181	137		318	405
0410	15 kV		2.80	2.857		320	156		476	585
0420	25 kV		2.30	3.478		287	190		477	600
0500	600 kcmil, 5 kV		2.90	2.759		181	151		332	425
0510	15 kV		2.40	3.333		330	182		512	635
0520	25 kV		2	4		330	219		549	695
0600	750 kcmil, 5 kV		2.60	3.077		181	168		349	450
0610	15 kV		2.20	3.636		330	199		529	660
0620	25 kV		1.90	4.211		330	230		560	710
0700	1000 kcmil, 5 kV		2.30	3.478		181	190		371	485
0710	15 kV		1.90	4.211		355	230		585	740
0720	25 kV		1.60	5		380	274		654	825

26 05 19.35 Cable Terminations

	26 05 19.35 Cable Terminations	Crew	Daily Output	Labor-Hours	Unit	Material	Labor	Equipment	Total	Total Incl O&P
0010	**CABLE TERMINATIONS**									
0015	Wire connectors, screw type, #22 to #14	1 Elec	260	.031	Ea.	.06	1.68		1.74	2.59
0020	#18 to #12		240	.033		.07	1.82		1.89	2.81
0025	#18 to #10		240	.033		.12	1.82		1.94	2.86
0030	Screw-on connectors, insulated, #18 to #12		240	.033		.25	1.82		2.07	3.01
0035	#16 to #10		230	.035		.27	1.90		2.17	3.15
0040	#14 to #8		210	.038		.30	2.08		2.38	3.45
0045	#12 to #6		180	.044		.62	2.43		3.05	4.32
0050	Terminal lugs, solderless, #16 to #10		50	.160		.37	8.75		9.12	13.50
0100	#8 to #4		30	.267		.73	14.60		15.33	23
0150	#2 to #1		22	.364		1.03	19.90		20.93	31
0200	1/0 to 2/0		16	.500		1.64	27.50		29.14	43
0250	3/0		12	.667		3.30	36.50		39.80	58
0300	4/0		11	.727		4.05	40		44.05	64
0350	250 kcmil		9	.889		3.41	48.50		51.91	77
0400	350 kcmil		7	1.143		4.44	62.50		66.94	98.50
0450	500 kcmil		6	1.333		8.20	73		81.20	118
0500	600 kcmil		5.80	1.379		9.30	75.50		84.80	123
0550	750 kcmil		5.20	1.538		10.95	84		94.95	138
0600	Split bolt connectors, tapped, #6		16	.500		3.78	27.50		31.28	45
0650	#4		14	.571		3.87	31.50		35.37	51.50
0700	#2		12	.667		5.70	36.50		42.20	61
0750	#1		11	.727		7.45	40		47.45	67.50
0800	1/0		10	.800		7.45	44		51.45	73.50
0850	2/0		9	.889		11.70	48.50		60.20	86
0900	3/0		7.20	1.111		16.25	61		77.25	109
1000	4/0		6.40	1.250		19.75	68.50		88.25	124
1100	250 kcmil		5.70	1.404		19.75	77		96.75	137
1200	300 kcmil		5.30	1.509		36	82.50		118.50	164
1400	350 kcmil		4.60	1.739		36	95		131	183
1500	500 kcmil		4	2		62.50	109		171.50	233
1600	Crimp 1 hole lugs, copper or aluminum, 600 volt									
1620	#14	1 Elec	60	.133	Ea.	.56	7.30		7.86	11.55
1630	#12		50	.160		.90	8.75		9.65	14.10

26 05 19.35 Cable Terminations		Crew	Daily Output	Labor-Hours	Unit	Material	2015 Bare Costs Labor	Equipment	Total	Total Incl O&P
1640	#10	1 Elec	45	.178	Ea.	.90	9.70		10.60	15.55
1780	#8		36	.222		1.86	12.15		14.01	20.50
1800	#6		30	.267		2.11	14.60		16.71	24.50
2000	#4		27	.296		2.88	16.20		19.08	27.50
2200	#2		24	.333		4.64	18.25		22.89	32.50
2400	#1		20	.400		4.84	22		26.84	38.50
2500	1/0		17.50	.457		5.20	25		30.20	43
2600	2/0		15	.533		6.30	29		35.30	50.50
2800	3/0		12	.667		7	36.50		43.50	62
3000	4/0		11	.727		7.80	40		47.80	68
3200	250 kcmil		9	.889		9.10	48.50		57.60	83
3400	300 kcmil		8	1		11.05	54.50		65.55	94
3500	350 kcmil		7	1.143		11.50	62.50		74	106
3600	400 kcmil		6.50	1.231		13.75	67.50		81.25	116
3800	500 kcmil		6	1.333		15.70	73		88.70	126
4000	600 kcmil		5.80	1.379		27	75.50		102.50	143
4200	700 kcmil		5.50	1.455		32	79.50		111.50	154
4400	750 kcmil		5.20	1.538		32	84		116	161
4500	Crimp 2-way connectors, copper or alum., 600 volt,									
4510	#14	1 Elec	60	.133	Ea.	2.19	7.30		9.49	13.35
4520	#12		50	.160		2.33	8.75		11.08	15.65
4530	#10		45	.178		2.45	9.70		12.15	17.25
4540	#8		27	.296		2.48	16.20		18.68	27
4600	#6		25	.320		4.17	17.50		21.67	30.50
4800	#4		23	.348		4.76	19.05		23.81	34
5000	#2		20	.400		5.85	22		27.85	39.50
5200	#1		16	.500		5.90	27.50		33.40	47.50
5400	1/0		13	.615		6.50	33.50		40	57.50
5420	2/0		12	.667		8	36.50		44.50	63.50
5440	3/0		11	.727		9.45	40		49.45	70
5460	4/0		10	.800		10.55	44		54.55	77
5480	250 kcmil		9	.889		12.70	48.50		61.20	87
5500	300 kcmil		8.50	.941		14.30	51.50		65.80	93
5520	350 kcmil		8	1		14.75	54.50		69.25	98.50
5540	400 kcmil		7.30	1.096		22	60		82	114
5560	500 kcmil		6.20	1.290		26	70.50		96.50	135
5580	600 kcmil		5.50	1.455		30.50	79.50		110	153
5600	700 kcmil		4.50	1.778		45	97		142	196
5620	750 kcmil		4	2		41.50	109		150.50	210
7000	Compression equipment adapter, aluminum wire, #6		30	.267		10.30	14.60		24.90	33.50
7020	#4		27	.296		9.95	16.20		26.15	35.50
7040	#2		24	.333		10.40	18.25		28.65	39
7060	#1		20	.400		12.30	22		34.30	46.50
7080	1/0		18	.444		13.30	24.50		37.80	51
7100	2/0		15	.533		19.35	29		48.35	65
7140	4/0		11	.727		23	40		63	85
7160	250 kcmil		9	.889		26.50	48.50		75	102
7180	300 kcmil		8	1		26.50	54.50		81	112
7200	350 kcmil		7	1.143		30.50	62.50		93	127
7220	400 kcmil		6.50	1.231		39.50	67.50		107	145
7240	500 kcmil		6	1.333		41	73		114	155
7260	600 kcmil		5.80	1.379		46	75.50		121.50	164
7280	750 kcmil		5.20	1.538		50.50	84		134.50	182

26 05 19 – Low-Voltage Electrical Power Conductors and Cables

26 05 19.35 Cable Terminations		Crew	Daily Output	Labor-Hours	Unit	2015 Bare Costs Material	Labor	Equipment	Total	Total Incl O&P
8000	Compression tool, hand				Ea.	1,200			1,200	1,325
8100	Hydraulic					1,700			1,700	1,875
8500	Hydraulic dies				↓	325			325	360

26 05 19.50 Mineral Insulated Cable		Crew	Daily Output	Labor-Hours	Unit	Material	Labor	Equipment	Total	Total Incl O&P
0010	**MINERAL INSULATED CABLE** 600 volt									
0100	1 conductor, #12	1 Elec	1.60	5	C.L.F.	365	274		639	810
0200	#10		1.60	5		470	274		744	925
0400	#8		1.50	5.333		520	292		812	1,000
0500	#6	↓	1.40	5.714		620	315		935	1,150
0600	#4	2 Elec	2.40	6.667		835	365		1,200	1,475
0800	#2		2.20	7.273		1,175	400		1,575	1,900
0900	#1		2.10	7.619		1,350	415		1,765	2,125
1000	1/0		2	8		1,600	440		2,040	2,400
1100	2/0		1.90	8.421		1,900	460		2,360	2,800
1200	3/0		1.80	8.889		2,275	485		2,760	3,225
1400	4/0	↓	1.60	10		2,625	545		3,170	3,725
1410	250 kcmil	3 Elec	2.40	10		2,975	545		3,520	4,100
1420	350 kcmil		1.95	12.308		3,400	675		4,075	4,725
1430	500 kcmil	↓	1.95	12.308		4,375	675		5,050	5,800
1500	2 conductor, #12	1 Elec	1.40	5.714		760	315		1,075	1,300
1600	#10		1.20	6.667		935	365		1,300	1,575
1800	#8		1.10	7.273		1,075	400		1,475	1,775
2000	#6	↓	1.05	7.619		1,475	415		1,890	2,250
2100	#4	2 Elec	2	8		2,000	440		2,440	2,850
2200	3 conductor, #12	1 Elec	1.20	6.667		950	365		1,315	1,600
2400	#10		1.10	7.273		1,075	400		1,475	1,775
2600	#8		1.05	7.619		1,350	415		1,765	2,100
2800	#6	↓	1	8		1,775	440		2,215	2,600
3000	#4	2 Elec	1.80	8.889		2,225	485		2,710	3,175
3100	4 conductor, #12	1 Elec	1.20	6.667		1,000	365		1,365	1,650
3200	#10		1.10	7.273		1,200	400		1,600	1,925
3400	#8		1	8		1,575	440		2,015	2,375
3600	#6		.90	8.889		2,050	485		2,535	2,975
3620	7 conductor, #12		1.10	7.273		1,275	400		1,675	2,000
3640	#10		1	8	↓	1,675	440		2,115	2,500
3800	Terminations, 600 volt, 1 conductor, #12		8	1	Ea.	16.05	54.50		70.55	99.50
4000	#10		7.60	1.053		16.05	57.50		73.55	104
4100	#8		7.30	1.096		16.05	60		76.05	108
4200	#6		6.70	1.194		16.05	65.50		81.55	116
4400	#4		6.20	1.290		16.05	70.50		86.55	124
4600	#2		5.70	1.404		24	77		101	142
4800	#1		5.30	1.509		24	82.50		106.50	151
5000	1/0		5	1.600		24	87.50		111.50	158
5100	2/0		4.70	1.702		24	93		117	166
5200	3/0		4.30	1.860		24	102		126	179
5400	4/0		4	2		53.50	109		162.50	223
5410	250 kcmil		4	2		53.50	109		162.50	223
5420	350 kcmil		4	2		84.50	109		193.50	257
5430	500 kcmil		4	2		84.50	109		193.50	257
5500	2 conductor, #12		6.70	1.194		16.05	65.50		81.55	116
5600	#10		6.40	1.250		24	68.50		92.50	129
5800	#8	↓	6.20	1.290	↓	24	70.50		94.50	133

26 05 19.50 Mineral Insulated Cable

	26 05 19.50 Mineral Insulated Cable	Crew	Daily Output	Labor-Hours	Unit	Material	Labor	Equipment	Total	Total Incl O&P
						2015 Bare Costs				
6000	#6	1 Elec	5.70	1.404	Ea.	24	77		101	142
6200	#4		5.30	1.509		53.50	82.50		136	183
6400	3 conductor, #12		5.70	1.404		24	77		101	142
6500	#10		5.50	1.455		24	79.50		103.50	146
6600	#8		5.20	1.538		24	84		108	153
6800	#6		4.80	1.667		24	91		115	164
7200	#4		4.60	1.739		53.50	95		148.50	202
7400	4 conductor, #12		4.60	1.739		27	95		122	173
7500	#10		4.40	1.818		27	99.50		126.50	179
7600	#8		4.20	1.905		27	104		131	186
8400	#6		4	2		56.50	109		165.50	226
8500	7 conductor, #12		3.50	2.286		27	125		152	217
8600	#10		3	2.667		58	146		204	283
8800	Crimping tool, plier type					60			60	66
9000	Stripping tool					227			227	249
9200	Hand vise					67.50			67.50	74.50

26 05 19.55 Non-Metallic Sheathed Cable

	26 05 19.55 Non-Metallic Sheathed Cable	Crew	Daily Output	Labor-Hours	Unit	Material	Labor	Equipment	Total	Total Incl O&P
0010	**NON-METALLIC SHEATHED CABLE** 600 volt									
0100	Copper with ground wire, (Romex)									
0150	#14, 2 conductor	1 Elec	2.70	2.963	C.L.F.	24	162		186	270
0200	3 conductor		2.40	3.333		34	182		216	310
0220	4 conductor		2.20	3.636		49	199		248	350
0250	#12, 2 conductor		2.50	3.200		36.50	175		211.50	300
0300	3 conductor		2.20	3.636		52	199		251	355
0320	4 conductor		2	4		76.50	219		295.50	415
0350	#10, 2 conductor		2.20	3.636		56.50	199		255.50	360
0400	3 conductor		1.80	4.444		82.50	243		325.50	455
0420	4 conductor		1.60	5		119	274		393	540
0430	#8, 2 conductor		1.60	5		88.50	274		362.50	510
0450	3 conductor		1.50	5.333		133	292		425	580
0500	#6, 3 conductor		1.40	5.714		215	315		530	705
0520	#4, 3 conductor	2 Elec	2.40	6.667		550	365		915	1,150
0540	#2, 3 conductor	"	2.20	7.273		790	400		1,190	1,475
0550	SE type SER aluminum cable, 3 RHW and									
0600	1 bare neutral, 3 #8 & 1 #8	1 Elec	1.60	5	C.L.F.	158	274		432	585
0650	3 #6 & 1 #6	"	1.40	5.714		179	315		494	665
0700	3 #4 & 1 #6	2 Elec	2.40	6.667		165	365		530	725
0750	3 #2 & 1 #4		2.20	7.273		296	400		696	920
0800	3 #1/0 & 1 #2		2	8		450	440		890	1,150
0850	3 #2/0 & 1 #1		1.80	8.889		530	485		1,015	1,300
0900	3 #4/0 & 1 #2/0		1.60	10		755	545		1,300	1,650
1000	URD - triplex underground distribution cable, alum. 2 #4 + #4 neutral		2.80	5.714		101	315		416	580
1010	2 #2 + #4 neutral		2.65	6.038		126	330		456	635
1020	2 #2 + #2 neutral		2.55	6.275		126	345		471	655
1030	2 1/0 + #2 neutral		2.40	6.667		152	365		517	715
1040	2 1/0 + 1/0 neutral		2.30	6.957		166	380		546	755
1050	2 2/0 + #1 neutral		2.20	7.273		181	400		581	795
1060	2 2/0 + 2/0 neutral		2.10	7.619		197	415		612	840
1070	2 3/0 + 1/0 neutral		1.95	8.205		218	450		668	910
1080	2 3/0 + 3/0 neutral		1.95	8.205		250	450		700	945
1090	2 4/0 + 2/0 neutral		1.85	8.649		261	475		736	995
1100	2 4/0 + 4/0 neutral		1.85	8.649		276	475		751	1,025

26 05 19.55 Non-Metallic Sheathed Cable		Crew	Daily Output	Labor-Hours	Unit	2015 Bare Costs Material	Labor	Equipment	Total	Total Incl O&P
1450	UF underground feeder cable, copper with ground, #14, 2 conductor	1 Elec	4	2	C.L.F.	27	109		136	194
1500	#12, 2 conductor		3.50	2.286		41.50	125		166.50	233
1550	#10, 2 conductor		3	2.667		63.50	146		209.50	289
1600	#14, 3 conductor		3.50	2.286		39.50	125		164.50	231
1650	#12, 3 conductor		3	2.667		59.50	146		205.50	285
1700	#10, 3 conductor		2.50	3.200		93.50	175		268.50	365
1710	#8, 3 conductor		2	4		154	219		373	500
1720	#6, 3 conductor		1.80	4.444		248	243		491	640
2400	SEU service entrance cable, copper 2 conductors, #8 + #8 neutral		1.50	5.333		115	292		407	560
2600	#6 + #8 neutral		1.30	6.154		170	335		505	690
2800	#6 + #6 neutral		1.30	6.154		193	335		528	715
3000	#4 + #6 neutral	2 Elec	2.20	7.273		266	400		666	890
3200	#4 + #4 neutral		2.20	7.273		300	400		700	925
3400	#3 + #5 neutral		2.10	7.619		370	415		785	1,025
3600	#3 + #3 neutral		2.10	7.619		400	415		815	1,075
3800	#2 + #4 neutral		2	8		430	440		870	1,125
4000	#1 + #1 neutral		1.90	8.421		470	460		930	1,200
4200	1/0 + 1/0 neutral		1.80	8.889		750	485		1,235	1,550
4400	2/0 + 2/0 neutral		1.70	9.412		935	515		1,450	1,800
4600	3/0 + 3/0 neutral		1.60	10		1,175	545		1,720	2,125
4620	4/0 + 4/0 neutral		1.45	11.034		1,325	605		1,930	2,350
4800	Aluminum 2 conductors, #8 + #8 neutral	1 Elec	1.60	5		131	274		405	555
5000	#6 + #6 neutral	"	1.40	5.714		132	315		447	615
5100	#4 + #6 neutral	2 Elec	2.50	6.400		155	350		505	695
5200	#4 + #4 neutral		2.40	6.667		169	365		534	730
5300	#2 + #4 neutral		2.30	6.957		207	380		587	800
5400	#2 + #2 neutral		2.20	7.273		226	400		626	845
5450	1/0 + #2 neutral		2.10	7.619		330	415		745	985
5500	1/0 + 1/0 neutral		2	8		345	440		785	1,025
5550	2/0 + #1 neutral		1.90	8.421		370	460		830	1,100
5600	2/0 + 2/0 neutral		1.80	8.889		395	485		880	1,175
5800	3/0 + 1/0 neutral		1.70	9.412		430	515		945	1,250
6000	3/0 + 3/0 neutral		1.70	9.412		470	515		985	1,300
6200	4/0 + 2/0 neutral		1.60	10		510	545		1,055	1,375
6400	4/0 + 4/0 neutral		1.60	10		555	545		1,100	1,425
6500	Service entrance cap for copper SEU									
6600	100 amp	1 Elec	12	.667	Ea.	8.75	36.50		45.25	64
6700	150 amp		10	.800		16.10	44		60.10	83
6800	200 amp		8	1		24.50	54.50		79	109

26 05 19.70 Portable Cord		Crew	Daily Output	Labor-Hours	Unit	Material	Labor	Equipment	Total	Total Incl O&P
0010	**PORTABLE CORD** 600 volt									
0100	Type SO, #18, 2 conductor	1 Elec	980	.008	L.F.	.55	.45		1	1.28
0110	3 conductor		980	.008		.67	.45		1.12	1.41
0120	#16, 2 conductor		840	.010		.65	.52		1.17	1.50
0130	3 conductor		840	.010		.93	.52		1.45	1.80
0140	4 conductor		840	.010		1.21	.52		1.73	2.11
0240	#14, 2 conductor		840	.010		1.07	.52		1.59	1.96
0250	3 conductor		840	.010		1.33	.52		1.85	2.24
0260	4 conductor		840	.010		1.55	.52		2.07	2.49
0280	#12, 2 conductor		840	.010		1.45	.52		1.97	2.38
0290	3 conductor		840	.010		1.95	.52		2.47	2.93
0300	4 conductor		840	.010		2.30	.52		2.82	3.31

26 05 19.70 Portable Cord

26 05 19.70 Portable Cord		Crew	Daily Output	Labor-Hours	Unit	2015 Bare Costs Material	Labor	Equipment	Total	Total Incl O&P
0320	#10, 2 conductor	1 Elec	765	.010	L.F.	2.21	.57		2.78	3.29
0330	3 conductor		765	.010		2.65	.57		3.22	3.78
0340	4 conductor		765	.010		3	.57		3.57	4.16
0360	#8, 2 conductor		555	.014		4.44	.79		5.23	6.05
0370	3 conductor		540	.015		3.89	.81		4.70	5.50
0380	4 conductor		525	.015		4.57	.83		5.40	6.30
0400	#6, 2 conductor		525	.015		3.64	.83		4.47	5.25
0410	3 conductor		490	.016		5.95	.89		6.84	7.90
0420	4 conductor		415	.019		6.50	1.05		7.55	8.75
0440	#4, 2 conductor	2 Elec	830	.019		6.15	1.05		7.20	8.35
0450	3 conductor		700	.023		8.20	1.25		9.45	10.90
0460	4 conductor		660	.024		11.85	1.33		13.18	15
0480	#2, 2 conductor		450	.036		7.30	1.95		9.25	10.95
0490	3 conductor		350	.046		7.30	2.50		9.80	11.80
0500	4 conductor		280	.057		16.15	3.13		19.28	22.50
2000	See 26 27 26.20 for Wiring Devices Elements									

26 05 19.75 Modular Flexible Wiring System

26 05 19.75 Modular Flexible Wiring System		Crew	Daily Output	Labor-Hours	Unit	Material	Labor	Equipment	Total	Total Incl O&P
0010	**MODULAR FLEXIBLE WIRING SYSTEM**									
0020	Commercial system grid ceiling									
0100	Conversion Module	1 Elec	32	.250	Ea.	8.25	13.70		21.95	29.50
0120	Fixture cable, for fixture to fixture, 3 conductor, 15′ long		40	.200		33	10.95		43.95	53
0150	Extender cable, 3 conductor, 15′ long		48	.167		32	9.10		41.10	48.50
0200	Switch drop, 1 level, 9′ long		32	.250		33	13.70		46.70	56.50
0220	2 level, 9′ long		32	.250		30	13.70		43.70	54
0250	Power tee, 9′ long		32	.250		34	13.70		47.70	57.50
1020	Industrial system open ceiling									
1100	Converter, interface between hardwiring and modular wiring	1 Elec	32	.250	Ea.	15.65	13.70		29.35	37.50
1120	Fixture cable, for fixture to fixture, 3 conductor, 21′ long		32	.250		100	13.70		113.70	131
1125	3 conductor, 25′ long		24	.333		116	18.25		134.25	155
1130	3 conductor, 31′ long		19	.421		103	23		126	148
1150	Fixture cord drop, 10′ long		40	.200		27.50	10.95		38.45	47

26 05 19.90 Wire

26 05 19.90 Wire		Crew	Daily Output	Labor-Hours	Unit	Material	Labor	Equipment	Total	Total Incl O&P
0010	**WIRE** R260519-90									
0020	600 volt, copper type THW, solid, #14	1 Elec	13	.615	C.L.F.	7.80	33.50		41.30	59
0030	#12		11	.727		11.95	40		51.95	72.50
0040	#10		10	.800		18.65	44		62.65	86
0050	Stranded, #14		13	.615		9.35	33.50		42.85	61
0100	#12		11	.727		14.35	40		54.35	75.50
0120	#10 R260519-92		10	.800		22.50	44		66.50	90.50
0140	#8		8	1		37.50	54.50		92	123
0160	#6 R260519-93		6.50	1.231		63.50	67.50		131	171
0180	#4	2 Elec	10.60	1.509		100	82.50		182.50	234
0200	#3 R260519-94		10	1.600		126	87.50		213.50	269
0220	#2		9	1.778		158	97		255	320
0240	#1 R260533-22		8	2		200	109		309	385
0260	1/0		6.60	2.424		250	133		383	475
0280	2/0		5.80	2.759		315	151		466	570
0300	3/0		5	3.200		395	175		570	695
0350	4/0		4.40	3.636		500	199		699	850
0400	250 kcmil	3 Elec	6	4		585	219		804	975
0420	300 kcmil		5.70	4.211		700	230		930	1,125
0450	350 kcmil		5.40	4.444		855	243		1,098	1,300

26 05 19.90 Wire		Crew	Daily Output	Labor-Hours	Unit	Material	2015 Bare Costs Labor	Equipment	Total	Total Incl O&P
0480	400 kcmil	3 Elec	5.10	4.706	C.L.F.	985	257		1,242	1,450
0490	500 kcmil		4.80	5		1,150	274		1,424	1,675
0500	600 kcmil		3.90	6.154		1,400	335		1,735	2,025
0510	750 kcmil		3.30	7.273		1,975	400		2,375	2,750
0520	1000 kcmil		2.70	8.889		2,825	485		3,310	3,825
0540	600 volt, aluminum type THHN, stranded, #6	1 Elec	8	1		43	54.50		97.50	130
0560	#4	2 Elec	13	1.231		53.50	67.50		121	160
0580	#2		10.60	1.509		72.50	82.50		155	204
0600	#1		9	1.778		106	97		203	262
0620	1/0		8	2		127	109		236	305
0640	2/0		7.20	2.222		150	122		272	345
0680	3/0		6.60	2.424		186	133		319	405
0700	4/0		6.20	2.581		207	141		348	440
0720	250 kcmil	3 Elec	8.70	2.759		253	151		404	505
0740	300 kcmil		8.10	2.963		350	162		512	630
0760	350 kcmil		7.50	3.200		355	175		530	650
0780	400 kcmil		6.90	3.478		415	190		605	740
0800	500 kcmil		6	4		460	219		679	835
0850	600 kcmil		5.70	4.211		580	230		810	985
0880	700 kcmil		5.10	4.706		670	257		927	1,125
0900	750 kcmil		4.80	5		695	274		969	1,175
0910	1000 kcmil		3.78	6.349		1,025	345		1,370	1,650
0920	600 volt, copper type THWN-THHN, solid, #14	1 Elec	13	.615		7.80	33.50		41.30	59
0940	#12		11	.727		11.95	40		51.95	72.50
0960	#10		10	.800		18.65	44		62.65	86
1000	Stranded, #14		13	.615		8.90	33.50		42.40	60.50
1200	#12		11	.727		13.30	40		53.30	74
1250	#10		10	.800		20.50	44		64.50	88
1300	#8		8	1		33.50	54.50		88	119
1350	#6		6.50	1.231		57.50	67.50		125	165
1400	#4	2 Elec	10.60	1.509		87.50	82.50		170	220
1450	#3		10	1.600		112	87.50		199.50	254
1500	#2		9	1.778		140	97		237	300
1550	#1		8	2		181	109		290	365
1600	1/0		6.60	2.424		217	133		350	435
1650	2/0		5.80	2.759		271	151		422	525
1700	3/0		5	3.200		340	175		515	635
2000	4/0		4.40	3.636		430	199		629	770
2200	250 kcmil	3 Elec	6	4		520	219		739	900
2400	300 kcmil		5.70	4.211		625	230		855	1,025
2600	350 kcmil		5.40	4.444		730	243		973	1,175
2700	400 kcmil		5.10	4.706		825	257		1,082	1,300
2800	500 kcmil		4.80	5		1,025	274		1,299	1,525
2802	600 kcmil		3.90	6.154		1,300	335		1,635	1,925
2804	750 kcmil		3.30	7.273		2,175	400		2,575	2,975
2805	1000 kcmil		1.94	12.371		2,875	675		3,550	4,200
2900	600 volt, copper type XHHW, solid, #14	1 Elec	13	.615		12.60	33.50		46.10	64.50
2920	#12		11	.727		19.90	40		59.90	81.50
2940	#10		10	.800		31	44		75	99.50
3000	Stranded, #14		13	.615		11.05	33.50		44.55	62.50
3020	#12		11	.727		16.60	40		56.60	78
3040	#10		10	.800		25	44		69	93
3060	#8		8	1		38.50	54.50		93	124

26 05 19 – Low-Voltage Electrical Power Conductors and Cables

26 05 19.90 Wire		Crew	Daily Output	Labor-Hours	Unit	2015 Bare Costs Material	Labor	Equipment	Total	Total Incl O&P
3080	#6	1 Elec	6.50	1.231	C.L.F.	63.50	67.50		131	171
3100	#4	2 Elec	10.60	1.509		97.50	82.50		180	231
3120	#2		9	1.778		152	97		249	315
3140	#1		8	2		210	109		319	395
3160	1/0		6.60	2.424		260	133		393	485
3180	2/0		5.80	2.759		325	151		476	585
3200	3/0		5	3.200		405	175		580	710
3220	4/0		4.40	3.636		510	199		709	860
3240	250 kcmil	3 Elec	6	4		565	219		784	955
3260	300 kcmil		5.70	4.211		640	230		870	1,050
3280	350 kcmil		5.40	4.444		745	243		988	1,175
3300	400 kcmil		5.10	4.706		850	257		1,107	1,325
3320	500 kcmil		4.80	5		1,050	274		1,324	1,550
3340	600 kcmil		3.90	6.154		1,350	335		1,685	1,975
3360	750 kcmil		3.30	7.273		2,250	400		2,650	3,075
3380	1000 kcmil		2.40	10		2,925	545		3,470	4,050
5020	600 volt, aluminum type XHHW, stranded, #6	1 Elec	8	1		41	54.50		95.50	127
5040	#4	2 Elec	13	1.231		50.50	67.50		118	157
5060	#2		10.60	1.509		69	82.50		151.50	200
5080	#1		9	1.778		100	97		197	257
5100	1/0		8	2		121	109		230	297
5120	2/0		7.20	2.222		143	122		265	340
5140	3/0		6.60	2.424		177	133		310	395
5160	4/0		6.20	2.581		197	141		338	430
5180	250 kcmil	3 Elec	8.70	2.759		240	151		391	490
5200	300 kcmil		8.10	2.963		330	162		492	610
5220	350 kcmil		7.50	3.200		340	175		515	630
5240	400 kcmil		6.90	3.478		395	190		585	720
5260	500 kcmil		6	4		435	219		654	810
5280	600 kcmil		5.70	4.211		550	230		780	950
5300	700 kcmil		5.40	4.444		635	243		878	1,075
5320	750 kcmil		5.10	4.706		645	257		902	1,100
5340	1000 kcmil		3.60	6.667		955	365		1,320	1,600
5390	600 volt, copper type XLPE-USE (RHW), solid, #14	1 Elec	12	.667		14.40	36.50		50.90	70.50
5400	#12		11	.727		18.85	40		58.85	80.50
5420	#10		10	.800		27.50	44		71.50	95.50
5440	Stranded, #14		13	.615		18.30	33.50		51.80	70.50
5460	#12		11	.727		22	40		62	83.50
5480	#10		10	.800		32	44		76	101
5500	#8		8	1		43	54.50		97.50	130
5520	#6		6.50	1.231		72.50	67.50		140	181
5540	#4	2 Elec	10.60	1.509		110	82.50		192.50	245
5560	#2		9	1.778		178	97		275	340
5580	#1		8	2		241	109		350	430
5600	1/0		6.60	2.424		300	133		433	530
5620	2/0		5.80	2.759		390	151		541	655
5640	3/0		5	3.200		490	175		665	800
5660	4/0		4.40	3.636		540	199		739	890
5680	250 kcmil	3 Elec	6	4		590	219		809	980
5700	300 kcmil		5.70	4.211		765	230		995	1,200
5720	350 kcmil		5.40	4.444		870	243		1,113	1,325
5740	400 kcmil		5.10	4.706		990	257		1,247	1,475
5760	500 kcmil		4.80	5		1,050	274		1,324	1,575

26 05 19 – Low-Voltage Electrical Power Conductors and Cables

26 05 19.90 Wire		Crew	Daily Output	Labor-Hours	Unit	2015 Bare Costs Material	Labor	Equipment	Total	Total Incl O&P
5780	600 kcmil	3 Elec	3.90	6.154	C.L.F.	1,400	335		1,735	2,025
5800	750 kcmil		3.30	7.273		2,500	400		2,900	3,350
5820	1000 kcmil		2.70	8.889		3,300	485		3,785	4,350
5840	600 volt, aluminum type XLPE-USE (RHW), stranded, #6	1 Elec	8	1		49.50	54.50		104	137
5860	#4	2 Elec	13	1.231		57.50	67.50		125	164
5880	#2		10.60	1.509		79.50	82.50		162	212
5900	#1		9	1.778		110	97		207	267
5920	1/0		8	2		136	109		245	315
5940	2/0		7.20	2.222		159	122		281	355
5960	3/0		6.60	2.424		188	133		321	405
5980	4/0		6.20	2.581		206	141		347	440
6000	250 kcmil	3 Elec	8.70	2.759		283	151		434	535
6020	300 kcmil		8.10	2.963		370	162		532	650
6040	350 kcmil		7.50	3.200		375	175		550	675
6060	400 kcmil		6.90	3.478		460	190		650	790
6080	500 kcmil		6	4		505	219		724	885
6100	600 kcmil		5.70	4.211		650	230		880	1,050
6110	700 kcmil		5.40	4.444		745	243		988	1,175
6120	750 kcmil		5.10	4.706		755	257		1,012	1,225

26 05 23 – Control-Voltage Electrical Power Cables

26 05 23.10 Control Cable

		Crew	Daily Output	Labor-Hours	Unit	Material	Labor	Equipment	Total	Total Incl O&P
0010	**CONTROL CABLE**									
0020	600 volt, copper, #14 THWN wire with PVC jacket, 2 wires	1 Elec	9	.889	C.L.F.	30.50	48.50		79	107
0030	3 wires		8	1		42.50	54.50		97	129
0100	4 wires		7	1.143		52	62.50		114.50	151
0150	5 wires		6.50	1.231		65	67.50		132.50	173
0200	6 wires		6	1.333		85.50	73		158.50	203
0300	8 wires		5.30	1.509		106	82.50		188.50	241
0400	10 wires		4.80	1.667		126	91		217	276
0500	12 wires		4.30	1.860		148	102		250	315
0600	14 wires		3.80	2.105		176	115		291	365
0700	16 wires		3.50	2.286		182	125		307	385
0800	18 wires		3.30	2.424		199	133		332	420
0810	19 wires		3.10	2.581		224	141		365	460
0900	20 wires		3	2.667		242	146		388	485
1000	22 wires		2.80	2.857		249	156		405	510

26 05 23.20 Special Wires and Fittings

		Crew	Daily Output	Labor-Hours	Unit	Material	Labor	Equipment	Total	Total Incl O&P
0010	**SPECIAL WIRES & FITTINGS**									
0100	Fixture TFFN 600 volt 90°C stranded, #18	1 Elec	13	.615	C.L.F.	9.50	33.50		43	61
0150	#16		13	.615		12.40	33.50		45.90	64
0500	Thermostat, jacket non-plenum, twisted, #18-2 conductor		8	1		16.45	54.50		70.95	100
0550	#18-3 conductor		7	1.143		21.50	62.50		84	117
0600	#18-4 conductor		6.50	1.231		32	67.50		99.50	136
0650	#18-5 conductor		6	1.333		35.50	73		108.50	148
0700	#18-6 conductor		5.50	1.455		44.50	79.50		124	168
0750	#18-7 conductor		5	1.600		51.50	87.50		139	188
0800	#18-8 conductor		4.80	1.667		58.50	91		149.50	202
2460	Tray cable, type TC, copper, #16-2 conductor		9.40	.851		24	46.50		70.50	96
2464	#16-3 conductor		8.40	.952		31	52		83	112
2468	#16-4 conductor		7.30	1.096		39	60		99	133
2472	#16-5 conductor		6.70	1.194		42.50	65.50		108	145
2476	#16-7 conductor		5.50	1.455		58	79.50		137.50	183

26 05 Common Work Results for Electrical

26 05 23 – Control-Voltage Electrical Power Cables

26 05 23.20 Special Wires and Fittings		Crew	Daily Output	Labor-Hours	Unit	Material	Labor	Equipment	Total	Total Incl O&P
						2015 Bare Costs				
2480	#16-9 conductor	1 Elec	5	1.600	C.L.F.	72.50	87.50		160	211
2484	#16-12 conductor	2 Elec	8.80	1.818		103	99.50		202.50	262
2488	#16-15 conductor		7.20	2.222		126	122		248	320
2492	#16-19 conductor		6.60	2.424		156	133		289	370
2496	#16-25 conductor		6.40	2.500		220	137		357	445
2500	#14-2 conductor	1 Elec	9	.889		29.50	48.50		78	106
2520	#14-3 conductor		8	1		41	54.50		95.50	127
2540	#14-4 conductor		7	1.143		50.50	62.50		113	149
2560	#14-5 conductor		6.50	1.231		62.50	67.50		130	170
2564	#14-7 conductor		5.30	1.509		91.50	82.50		174	225
2568	#14-9 conductor	2 Elec	9.60	1.667		117	91		208	266
2572	#14-12 conductor		8.60	1.860		159	102		261	325
2576	#14-15 conductor		7.20	2.222		173	122		295	375
2578	#14-19 conductor		6.20	2.581		218	141		359	450
2582	#14-25 conductor		5.30	3.019		286	165		451	560
2590	#12-2 conductor	1 Elec	8.40	.952		42	52		94	124
2592	#12-3 conductor		7.60	1.053		49.50	57.50		107	141
2594	#12-4 conductor		6.60	1.212		77.50	66.50		144	185
2596	#12-5 conductor		6.20	1.290		94.50	70.50		165	210
2598	#12-7 conductor	2 Elec	10.40	1.538		139	84		223	279
2602	#12-9 conductor		8.80	1.818		181	99.50		280.50	350
2604	#12-12 conductor		7.80	2.051		236	112		348	425
2606	#12-15 conductor		6.80	2.353		261	129		390	480
2608	#12-19 conductor		6	2.667		320	146		466	570
2610	#12-25 conductor	3 Elec	7.70	3.117		430	170		600	730
2618	#10-2 conductor	1 Elec	8	1		63	54.50		117.50	152
2622	#10-3 conductor	"	7.30	1.096		84	60		144	182
2624	#10-4 conductor	2 Elec	12.80	1.250		120	68.50		188.50	234
2626	#10-5 conductor		11.80	1.356		135	74		209	260
2628	#10-7 conductor		9.40	1.702		195	93		288	355
2630	#10-9 conductor		8.40	1.905		227	104		331	405
2632	#10-12 conductor		7.20	2.222		242	122		364	450
2640	300 V, copper braided shield, PVC jacket									
2650	2 conductor #18 stranded	1 Elec	7	1.143	C.L.F.	47.50	62.50		110	146
2660	3 conductor #18 stranded	"	6	1.333	"	67	73		140	183
3000	Strain relief grip for cable									
3050	Cord, top, #12-3	1 Elec	40	.200	Ea.	10.55	10.95		21.50	28
3060	#12-4		40	.200		10.55	10.95		21.50	28
3070	#12-5		39	.205		11.25	11.20		22.45	29
3100	#10-3		39	.205		11.25	11.20		22.45	29
3110	#10-4		38	.211		11.25	11.50		22.75	29.50
3120	#10-5		38	.211		13.40	11.50		24.90	32
3200	Bottom, #12-3		40	.200		26.50	10.95		37.45	45.50
3210	#12-4		40	.200		26.50	10.95		37.45	45.50
3220	#12-5		39	.205		26.50	11.20		37.70	46.50
3230	#10-3		39	.205		26.50	11.20		37.70	46.50
3300	#10-4		38	.211		26.50	11.50		38	47
3310	#10-5		38	.211		33.50	11.50		45	54
3400	Cable ties, standard, 4" length		190	.042		.17	2.30		2.47	3.64
3410	7" length		160	.050		.19	2.74		2.93	4.31
3420	14.5" length		90	.089		.46	4.86		5.32	7.80
3430	Heavy, 14.5" length		80	.100		.64	5.45		6.09	8.90

26 05 26 – Grounding and Bonding for Electrical Systems

26 05 26.80 Grounding		Crew	Daily Output	Labor-Hours	Unit	Material	Labor (2015 Bare Costs)	Equipment	Total	Total Incl O&P
0010	**GROUNDING** R260526-80									
0030	Rod, copper clad, 8′ long, 1/2″ diameter	1 Elec	5.50	1.455	Ea.	20	79.50		99.50	141
0040	5/8″ diameter		5.50	1.455		19.20	79.50		98.70	140
0050	3/4″ diameter		5.30	1.509		33.50	82.50		116	161
0080	10′ long, 1/2″ diameter		4.80	1.667		22	91		113	162
0090	5/8″ diameter		4.60	1.739		24	95		119	170
0100	3/4″ diameter		4.40	1.818		38.50	99.50		138	192
0130	15′ long, 3/4″ diameter		4	2		54	109		163	224
0150	Coupling, bronze, 1/2″ diameter					4.07			4.07	4.48
0160	5/8″ diameter					5.35			5.35	5.90
0170	3/4″ diameter					13.20			13.20	14.50
0190	Drive studs, 1/2″ diameter					10.70			10.70	11.75
0210	5/8″ diameter					15.55			15.55	17.10
0220	3/4″ diameter					18.15			18.15	20
0230	Clamp, bronze, 1/2″ diameter	1 Elec	32	.250		4.65	13.70		18.35	25.50
0240	5/8″ diameter		32	.250		5.25	13.70		18.95	26.50
0250	3/4″ diameter		32	.250		5.90	13.70		19.60	27
0260	Wire ground bare armored, #8-1 conductor		2	4	C.L.F.	76	219		295	415
0270	#6-1 conductor		1.80	4.444		91.50	243		334.50	465
0280	#4-1 conductor		1.60	5		128	274		402	550
0320	Bare copper wire, #14 solid		14	.571		7.35	31.50		38.85	55
0330	#12		13	.615		12.60	33.50		46.10	64.50
0340	#10		12	.667		18	36.50		54.50	74.50
0350	#8		11	.727		29	40		69	91.50
0360	#6		10	.800		51	44		95	122
0370	#4		8	1		117	54.50		171.50	210
0380	#2		5	1.600		130	87.50		217.50	274
0390	Bare copper wire, stranded, #8		11	.727		29.50	40		69.50	92
0400	#6		10	.800		52	44		96	123
0450	#4	2 Elec	16	1		87.50	54.50		142	179
0600	#2		10	1.600		138	87.50		225.50	283
0650	#1		9	1.778		182	97		279	345
0700	1/0		8	2		197	109		306	380
0750	2/0		7.20	2.222		248	122		370	455
0800	3/0		6.60	2.424		310	133		443	540
1000	4/0		5.70	2.807		395	154		549	665
1200	250 kcmil	3 Elec	7.20	3.333		465	182		647	785
1210	300 kcmil		6.60	3.636		525	199		724	880
1220	350 kcmil		6	4		660	219		879	1,050
1230	400 kcmil		5.70	4.211		770	230		1,000	1,200
1240	500 kcmil		5.10	4.706		945	257		1,202	1,425
1260	750 kcmil		3.60	6.667		1,600	365		1,965	2,325
1270	1000 kcmil		3	8		2,125	440		2,565	3,000
1360	Bare aluminum, stranded, #6	1 Elec	9	.889		14.60	48.50		63.10	89
1370	#4	2 Elec	16	1		25.50	54.50		80	111
1380	#2		13	1.231		36.50	67.50		104	142
1390	#1		10.60	1.509		40.50	82.50		123	169
1400	1/0		9	1.778		46.50	97		143.50	197
1410	2/0		8	2		59	109		168	229
1420	3/0		7.20	2.222		79.50	122		201.50	269
1430	4/0		6.60	2.424		91	133		224	299
1440	250 kcmil	3 Elec	9.30	2.581		119	141		260	340

26 05 26.80 Grounding		Crew	Daily Output	Labor-Hours	Unit	Material	Labor	Equipment	Total	Total Incl O&P
						2015 Bare Costs				
1450	300 kcmil	3 Elec	8.70	2.759	C.L.F.	140	151		291	380
1460	400 kcmil		7.50	3.200		180	175		355	460
1470	500 kcmil		6.90	3.478		214	190		404	520
1480	600 kcmil		6	4		243	219		462	600
1490	700 kcmil		5.70	4.211		267	230		497	640
1500	750 kcmil		5.10	4.706		277	257		534	690
1510	1000 kcmil		4.80	5		350	274		624	795
1800	Water pipe ground clamps, heavy duty									
2000	Bronze, 1/2" to 1" diameter	1 Elec	8	1	Ea.	24	54.50		78.50	108
2100	1-1/4" to 2" diameter		8	1		33	54.50		87.50	119
2200	2-1/2" to 3" diameter		6	1.333		44.50	73		117.50	158
2730	Exothermic weld, 4/0 wire to 1" ground rod		7	1.143		10.70	62.50		73.20	105
2740	4/0 wire to building steel		7	1.143		10.70	62.50		73.20	105
2750	4/0 wire to motor frame		7	1.143		10.70	62.50		73.20	105
2760	4/0 wire to 4/0 wire		7	1.143		10.70	62.50		73.20	105
2770	4/0 wire to #4 wire		7	1.143		10.70	62.50		73.20	105
2780	4/0 wire to #8 wire		7	1.143		10.70	62.50		73.20	105
2790	Mold, reusable, for above					138			138	152
2800	Brazed connections, #6 wire	1 Elec	12	.667		16.15	36.50		52.65	72.50
3000	#2 wire		10	.800		21.50	44		65.50	89.50
3100	3/0 wire		8	1		32.50	54.50		87	118
3200	4/0 wire		7	1.143		37	62.50		99.50	135
3400	250 kcmil wire		5	1.600		43.50	87.50		131	179
3600	500 kcmil wire		4	2		53.50	109		162.50	223
3700	Insulated ground wire, copper #14		13	.615	C.L.F.	8.90	33.50		42.40	60.50
3710	#12		11	.727		13.30	40		53.30	74
3720	#10		10	.800		20.50	44		64.50	88
3730	#8		8	1		33.50	54.50		88	119
3740	#6		6.50	1.231		57.50	67.50		125	165
3750	#4	2 Elec	10.60	1.509		87.50	82.50		170	220
3770	#2		9	1.778		140	97		237	300
3780	#1		8	2		181	109		290	365
3790	1/0		6.60	2.424		217	133		350	435
3800	2/0		5.80	2.759		271	151		422	525
3810	3/0		5	3.200		340	175		515	635
3820	4/0		4.40	3.636		430	199		629	770
3830	250 kcmil	3 Elec	6	4		520	219		739	900
3840	300 kcmil		5.70	4.211		625	230		855	1,025
3850	350 kcmil		5.40	4.444		730	243		973	1,175
3860	400 kcmil		5.10	4.706		825	257		1,082	1,300
3870	500 kcmil		4.80	5		1,025	274		1,299	1,525
3880	600 kcmil		3.90	6.154		1,400	335		1,735	2,025
3890	750 kcmil		3.30	7.273		1,975	400		2,375	2,750
3900	1000 kcmil		2.70	8.889		2,825	485		3,310	3,825
3960	Insulated ground wire, aluminum, #6	1 Elec	8	1		43	54.50		97.50	130
3970	#4	2 Elec	13	1.231		53.50	67.50		121	160
3980	#2		10.60	1.509		72.50	82.50		155	204
3990	#1		9	1.778		106	97		203	262
4000	1/0		8	2		127	109		236	305
4010	2/0		7.20	2.222		150	122		272	345
4020	3/0		6.60	2.424		186	133		319	405
4030	4/0		6.20	2.581		207	141		348	440
4040	250 kcmil	3 Elec	8.70	2.759		253	151		404	505

26 05 26 – Grounding and Bonding for Electrical Systems

26 05 26.80 Grounding		Crew	Daily Output	Labor-Hours	Unit	2015 Bare Costs Material	Labor	Equipment	Total	Total Incl O&P
4050	300 kcmil	3 Elec	8.10	2.963	C.L.F.	350	162		512	630
4060	350 kcmil		7.50	3.200		355	175		530	650
4070	400 kcmil		6.90	3.478		415	190		605	740
4080	500 kcmil		6	4		460	219		679	835
4090	600 kcmil		5.70	4.211		580	230		810	985
4100	700 kcmil		5.10	4.706		670	257		927	1,125
4110	750 kcmil		4.80	5		695	274		969	1,175
5000	Copper Electrolytic ground rod system									
5010	Includes augering hole, mixing bentonite clay,									
5020	Installing rod, and terminating ground wire									
5100	Straight Vertical type, 2″ diam.									
5120	8.5′ long, clamp connection	1 Elec	2.67	2.996	Ea.	770	164		934	1,100
5130	With exothermic weld connection		1.95	4.103		770	224		994	1,175
5140	10′ long		2.35	3.404		900	186		1,086	1,275
5150	With exothermic weld connection		1.78	4.494		900	246		1,146	1,350
5160	12′ long		2.16	3.704		1,025	203		1,228	1,425
5170	With exothermic weld connection		1.67	4.790		1,025	262		1,287	1,525
5180	20′ long		1.74	4.598		1,475	251		1,726	1,975
5190	With exothermic weld connection		1.40	5.714		1,775	315		2,090	2,425
5195	40′ long with exothermic weld connection	2 Elec	2	8		3,075	440		3,515	4,025
5200	L-Shaped, 2″ diam.									
5220	4′ Vert. x 10′ Horz., clamp connection	1 Elec	5.33	1.501	Ea.	1,275	82		1,357	1,550
5230	With exothermic weld connection	″	3.08	2.597	″	1,275	142		1,417	1,650
5300	Protective Box at grade level, with breather slots									
5320	Round 12″ long, fiberlyte	1 Elec	32	.250	Ea.	65	13.70		78.70	92
5330	Concrete	″	16	.500		101	27.50		128.50	152
5400	Bentonite Clay, 50# bag, 1 per 10′ of rod					41.50			41.50	45.50
5500	Equipotential earthing bar	1 Elec	2	4		176	219		395	525

26 05 29 – Hangers and Supports for Electrical Systems

26 05 29.20 Hangers		Crew	Daily Output	Labor-Hours	Unit	Material	Labor	Equipment	Total	Total Incl O&P
0010	**HANGERS** R260533-20									
0015	See section 22 05 29.10 for additional items									
0030	Conduit supports									
0050	Strap w/2 holes, rigid steel conduit									
0100	1/2″ diameter	1 Elec	470	.017	Ea.	.13	.93		1.06	1.53
0150	3/4″ diameter		440	.018		.14	.99		1.13	1.64
0200	1″ diameter		400	.020		.22	1.09		1.31	1.88
0300	1-1/4″ diameter		355	.023		.35	1.23		1.58	2.24
0350	1-1/2″ diameter		320	.025		.40	1.37		1.77	2.49
0400	2″ diameter		266	.030		.44	1.65		2.09	2.95
0500	2-1/2″ diameter		160	.050		.84	2.74		3.58	5
0550	3″ diameter		133	.060		1.08	3.29		4.37	6.10
0600	3-1/2″ diameter		100	.080		2.81	4.38		7.19	9.65
0650	4″ diameter		80	.100		1.46	5.45		6.91	9.80
0700	EMT, 1/2″ diameter		470	.017		.14	.93		1.07	1.54
0800	3/4″ diameter		440	.018		.18	.99		1.17	1.69
0850	1″ diameter		400	.020		.31	1.09		1.40	1.98
0900	1-1/4″ diameter		355	.023		.51	1.23		1.74	2.41
0950	1-1/2″ diameter		320	.025		.55	1.37		1.92	2.66
1000	2″ diameter		266	.030		.80	1.65		2.45	3.35
1100	2-1/2″ diameter		160	.050		1.34	2.74		4.08	5.55
1150	3″ diameter		133	.060		1.88	3.29		5.17	7

26 05 29.20 Hangers		Crew	Daily Output	Labor-Hours	Unit	2015 Bare Costs Material	Labor	Equipment	Total	Total Incl O&P
1200	3-1/2" diameter	1 Elec	100	.080	Ea.	2.10	4.38		6.48	8.85
1250	4" diameter		80	.100		2.44	5.45		7.89	10.90
1400	Hanger, with bolt, 1/2" diameter		200	.040		.53	2.19		2.72	3.86
1450	3/4" diameter		190	.042		.54	2.30		2.84	4.04
1500	1" diameter		176	.045		.94	2.49		3.43	4.75
1550	1-1/4" diameter		160	.050		1.26	2.74		4	5.50
1600	1-1/2" diameter		140	.057		1.57	3.13		4.70	6.40
1650	2" diameter		130	.062		2.04	3.37		5.41	7.30
1700	2-1/2" diameter		100	.080		2.11	4.38		6.49	8.85
1750	3" diameter		64	.125		2.91	6.85		9.76	13.45
1800	3-1/2" diameter		50	.160		3.68	8.75		12.43	17.15
1850	4" diameter		40	.200		8.35	10.95		19.30	25.50
1900	Riser clamps, conduit, 1/2" diameter		40	.200		8.40	10.95		19.35	25.50
1950	3/4" diameter		36	.222		9.10	12.15		21.25	28.50
2000	1" diameter		30	.267		10.15	14.60		24.75	33
2100	1-1/4" diameter		27	.296		11.20	16.20		27.40	37
2150	1-1/2" diameter		27	.296		11.75	16.20		27.95	37.50
2200	2" diameter		20	.400		12.30	22		34.30	46.50
2250	2-1/2" diameter		20	.400		13	22		35	47.50
2300	3" diameter		18	.444		14.15	24.50		38.65	52
2350	3-1/2" diameter		18	.444		16.05	24.50		40.55	54
2400	4" diameter		14	.571		19.20	31.50		50.70	68
2500	Threaded rod, painted, 1/4" diameter		260	.031	L.F.	1.79	1.68		3.47	4.49
2600	3/8" diameter		200	.040		2.89	2.19		5.08	6.45
2700	1/2" diameter		140	.057		3.77	3.13		6.90	8.85
2800	5/8" diameter		100	.080		5.05	4.38		9.43	12.15
2900	3/4" diameter		60	.133		9	7.30		16.30	21
2940	Couplings painted, 1/4" diameter				C	565			565	620
2960	3/8" diameter					655			655	720
2970	1/2" diameter					1,250			1,250	1,375
2980	5/8" diameter					1,950			1,950	2,125
2990	3/4" diameter					2,025			2,025	2,225
3000	Nuts, galvanized, 1/4" diameter					15.15			15.15	16.65
3050	3/8" diameter					22			22	24
3100	1/2" diameter					49			49	53.50
3150	5/8" diameter					125			125	138
3200	3/4" diameter					185			185	203
3250	Washers, galvanized, 1/4" diameter					15.15			15.15	16.65
3300	3/8" diameter					20.50			20.50	23
3350	1/2" diameter					33			33	36.50
3400	5/8" diameter					103			103	113
3450	3/4" diameter					150			150	165
3500	Lock washers, galvanized, 1/4" diameter					15.15			15.15	16.65
3550	3/8" diameter					17.75			17.75	19.50
3600	1/2" diameter					20.50			20.50	22.50
3650	5/8" diameter					42			42	46.50
3700	3/4" diameter					80.50			80.50	88.50
3800	Channels, steel, 3/4" x 1-1/2", 14 Ga	1 Elec	80	.100	L.F.	4.10	5.45		9.55	12.70
3900	1-1/2" x 1-1/2", 12 Ga		70	.114		5.40	6.25		11.65	15.30
4000	1-7/8" x 1-1/2"		60	.133		23	7.30		30.30	36.50
4100	3" x 1-1/2"		50	.160		39.50	8.75		48.25	56.50
4110	1-5/8" x 13/16", 16 Ga.		80	.100		5.20	5.45		10.65	13.90
4114	Trapeze channel support, 12" wide, steel, 12 Ga		8.80	.909	Ea.	29	49.50		78.50	106

26 05 29 – Hangers and Supports for Electrical Systems

26 05 29.20 Hangers		Crew	Daily Output	Labor-Hours	Unit	2015 Bare Costs Material	Labor	Equipment	Total	Total Incl O&P
4116	18" wide, steel	1 Elec	8.30	.964	Ea.	31.50	52.50		84	114
4120	1-5/8" x 1-5/8", 14 Ga.		70	.114	L.F.	6.50	6.25		12.75	16.50
4130	1-5/8" x 7/8", 12 Ga.		70	.114		5.50	6.25		11.75	15.40
4140	1-5/8" x 1-3/8", 12 Ga.		60	.133		7.65	7.30		14.95	19.35
4150	1-5/8" x 1-5/8", 12 Ga.		50	.160		6.90	8.75		15.65	20.50
4160	Flat plate fitting, 2 hole, 3-1/2", material only				Ea.	3.40			3.40	3.74
4170	3 hole, 1-7/16" x 4-1/8"					4.17			4.17	4.59
4200	Spring nuts, long, 1/4"	1 Elec	120	.067		1.17	3.65		4.82	6.75
4250	3/8"		100	.080		1.23	4.38		5.61	7.90
4300	1/2"		80	.100		1.33	5.45		6.78	9.65
4350	Spring nuts, short, 1/4"		120	.067		1.16	3.65		4.81	6.75
4400	3/8"		100	.080		1.20	4.38		5.58	7.85
4450	1/2"		80	.100		1.67	5.45		7.12	10.05
4500	Closure strip		200	.040	L.F.	4.67	2.19		6.86	8.45
4550	End cap		60	.133	Ea.	1.63	7.30		8.93	12.75
4600	End connector 3/4" conduit		40	.200		6.60	10.95		17.55	23.50
4650	Junction box, 1 channel		16	.500		57.50	27.50		85	104
4700	2 channel		14	.571		71.50	31.50		103	126
4750	3 channel		12	.667		78.50	36.50		115	141
4800	4 channel		10	.800		82.50	44		126.50	156
4850	Splice plate		40	.200		9.45	10.95		20.40	27
4900	Continuous concrete insert, 1-1/2" deep, 1' long		16	.500		19.85	27.50		47.35	63
4950	2' long		14	.571		24.50	31.50		56	74
5000	3' long		12	.667		23	36.50		59.50	79.50
5050	4' long		10	.800		34	44		78	103
5100	6' long		8	1		51	54.50		105.50	138
5150	3/4" deep, 1' long		16	.500		14.30	27.50		41.80	57
5200	2' long		14	.571		17.80	31.50		49.30	66.50
5250	3' long		12	.667		18.50	36.50		55	75
5300	4' long		10	.800		24.50	44		68.50	92.50
5350	6' long		8	1		37	54.50		91.50	123
5400	90° angle fitting 2-1/8" x 2-1/8"		60	.133		3.80	7.30		11.10	15.15
5450	Supports, suspension rod type, small		60	.133		24	7.30		31.30	37.50
5500	Large		40	.200		26.50	10.95		37.45	45.50
5550	Beam clamp, small		60	.133		9.90	7.30		17.20	22
5600	Large		40	.200		11.40	10.95		22.35	29
5650	U-support, small		60	.133		6.05	7.30		13.35	17.60
5700	Large		40	.200		12	10.95		22.95	29.50
5750	Concrete insert, cast, for up to 1/2" threaded rod		16	.500		6	27.50		33.50	47.50
5800	Beam clamp, 1/4" clamp, for 1/4" threaded drop rod		32	.250		2.67	13.70		16.37	23.50
5900	3/8" clamp, for 3/8" threaded drop rod		32	.250		5.80	13.70		19.50	27
6000	Strap, rigid conduit, 1/2" diameter		540	.015		1.64	.81		2.45	3.01
6050	3/4" diameter		440	.018		1.26	.99		2.25	2.88
6100	1" diameter		420	.019		2.02	1.04		3.06	3.78
6150	1-1/4" diameter		400	.020		1.58	1.09		2.67	3.38
6200	1-1/2" diameter		400	.020		1.89	1.09		2.98	3.72
6250	2" diameter		267	.030		2.04	1.64		3.68	4.70
6300	2-1/2" diameter		267	.030		2.74	1.64		4.38	5.45
6350	3" diameter		160	.050		2.59	2.74		5.33	6.95
6400	3-1/2" diameter		133	.060		3.49	3.29		6.78	8.75
6450	4" diameter		100	.080		3.61	4.38		7.99	10.50
6500	5" diameter		80	.100		9.95	5.45		15.40	19.10
6550	6" diameter		60	.133		19.40	7.30		26.70	32.50

26 05 29.20 Hangers		Crew	Daily Output	Labor-Hours	Unit	Material	Labor	Equipment	Total	Total Incl O&P
						2015 Bare Costs				
6600	EMT, 1/2" diameter	1 Elec	540	.015	Ea.	1.13	.81		1.94	2.45
6650	3/4" diameter		440	.018		1.21	.99		2.20	2.82
6700	1" diameter		420	.019		1.34	1.04		2.38	3.03
6750	1-1/4" diameter		400	.020		1.51	1.09		2.60	3.30
6800	1-1/2" diameter		400	.020		1.86	1.09		2.95	3.69
6850	2" diameter		267	.030		2.11	1.64		3.75	4.78
6900	2-1/2" diameter		267	.030		2.25	1.64		3.89	4.94
6950	3" diameter		160	.050		2.43	2.74		5.17	6.75
6970	3-1/2" diameter		133	.060		2.72	3.29		6.01	7.90
6990	4" diameter		100	.080		3.14	4.38		7.52	10
7000	Clip, 1 hole for rigid conduit, 1/2" diameter		500	.016		.51	.88		1.39	1.87
7050	3/4" diameter		470	.017		.75	.93		1.68	2.22
7100	1" diameter		440	.018		.93	.99		1.92	2.51
7150	1-1/4" diameter		400	.020		1.97	1.09		3.06	3.81
7200	1-1/2" diameter		355	.023		2.15	1.23		3.38	4.22
7250	2" diameter		320	.025		4.19	1.37		5.56	6.65
7300	2-1/2" diameter		266	.030		9.35	1.65		11	12.70
7350	3" diameter		160	.050		13.35	2.74		16.09	18.75
7400	3-1/2" diameter		133	.060		18.95	3.29		22.24	26
7450	4" diameter		100	.080		42.50	4.38		46.88	53
7500	5" diameter		80	.100		136	5.45		141.45	158
7550	6" diameter		60	.133		143	7.30		150.30	169
7820	Conduit hangers, with bolt & 12" rod, 1/2" diameter		150	.053		3.42	2.92		6.34	8.15
7830	3/4" diameter		145	.055		3.43	3.02		6.45	8.30
7840	1" diameter		135	.059		3.83	3.24		7.07	9.05
7850	1-1/4" diameter		120	.067		4.15	3.65		7.80	10
7860	1-1/2" diameter		110	.073		4.46	3.98		8.44	10.85
7870	2" diameter		100	.080		5.80	4.38		10.18	12.95
7880	2-1/2" diameter		80	.100		5.90	5.45		11.35	14.65
7890	3" diameter		60	.133		6.70	7.30		14	18.30
7900	3-1/2" diameter		45	.178		7.45	9.70		17.15	23
7910	4" diameter		35	.229		12.15	12.50		24.65	32
7920	5" diameter		30	.267		13.45	14.60		28.05	37
7930	6" diameter		25	.320		27.50	17.50		45	56
7950	Jay clamp, 1/2" diameter		32	.250		3.75	13.70		17.45	24.50
7960	3/4" diameter		32	.250		4.85	13.70		18.55	26
7970	1" diameter		32	.250		6.85	13.70		20.55	28
7980	1-1/4" diameter		30	.267		5.65	14.60		20.25	28.50
7990	1-1/2" diameter		30	.267		9.25	14.60		23.85	32
8000	2" diameter		30	.267		13.70	14.60		28.30	37
8010	2-1/2" diameter		28	.286		19.70	15.65		35.35	45
8020	3" diameter		28	.286		19.70	15.65		35.35	45
8030	3-1/2" diameter		25	.320		23.50	17.50		41	52
8040	4" diameter		25	.320		23.50	17.50		41	52
8050	5" diameter		20	.400		99.50	22		121.50	142
8060	6" diameter		16	.500		187	27.50		214.50	247
8070	Channels, 3/4" x 1-1/2" w/12" rods for 1/2" to 1" conduit		30	.267		9.05	14.60		23.65	32
8080	1-1/2" x 1-1/2" w/12" rods for 1-1/4" to 2" conduit		28	.286		9.65	15.65		25.30	34
8090	1-1/2" x 1-1/2" w/12" rods for 2-1/2" to 4" conduit		26	.308		11.80	16.85		28.65	38
8100	1-1/2" x 1-7/8" w/12" rods for 5" to 6" conduit		24	.333		55.50	18.25		73.75	88.50
8110	Beam clamp, conduit, plastic coated steel, 1/2" diam.		30	.267		21	14.60		35.60	45
8120	3/4" diameter		30	.267		22	14.60		36.60	46
8130	1" diameter	↓	30	.267	↓	25	14.60		39.60	49.50

26 05 29.20 Hangers		Crew	Daily Output	Labor-Hours	Unit	Material	2015 Bare Costs Labor	Equipment	Total	Total Incl O&P
8140	1-1/4" diameter	1 Elec	28	.286	Ea.	30.50	15.65		46.15	57
8150	1-1/2" diameter		28	.286		37.50	15.65		53.15	64.50
8160	2" diameter		28	.286		52	15.65		67.65	80.50
8170	2-1/2" diameter		26	.308		53	16.85		69.85	83.50
8180	3" diameter		26	.308		57	16.85		73.85	88
8190	3-1/2" diameter		23	.348		61.50	19.05		80.55	96
8200	4" diameter		23	.348		67	19.05		86.05	102
8210	5" diameter		18	.444		209	24.50		233.50	266
8220	Channels, plastic coated									
8250	3/4" x 1-1/2", w/12" rods for 1/2" to 1" conduit	1 Elec	28	.286	Ea.	35	15.65		50.65	61.50
8260	1-1/2" x 1-1/2", w/12" rods for 1-1/4" to 2" conduit		26	.308		39	16.85		55.85	68
8270	1-1/2" x 1-1/2", w/12" rods for 2-1/2" to 3-1/2" conduit		24	.333		42.50	18.25		60.75	74
8280	1-1/2" x 1-7/8", w/12" rods for 4" to 5" conduit		22	.364		70.50	19.90		90.40	108
8290	1-1/2" x 1-7/8", w/12" rods for 6" conduit		20	.400		77.50	22		99.50	118
8320	Conduit hangers, plastic coated steel, with bolt & 12" rod, 1/2" diam.		140	.057		20	3.13		23.13	26.50
8330	3/4" diameter		135	.059		20.50	3.24		23.74	27.50
8340	1" diameter		125	.064		21	3.50		24.50	28.50
8350	1-1/4" diameter		110	.073		22	3.98		25.98	30.50
8360	1-1/2" diameter		100	.080		25	4.38		29.38	34
8370	2" diameter		90	.089		27.50	4.86		32.36	37.50
8380	2-1/2" diameter		70	.114		32.50	6.25		38.75	45.50
8390	3" diameter		50	.160		39.50	8.75		48.25	56.50
8400	3-1/2" diameter		35	.229		40	12.50		52.50	63
8410	4" diameter		25	.320		59	17.50		76.50	90.50
8420	5" diameter		20	.400		64.50	22		86.50	104
9000	Parallel type, conduit beam clamp, 1/2"		32	.250		3.82	13.70		17.52	24.50
9010	3/4"		32	.250		3.97	13.70		17.67	25
9020	1"		32	.250		4.38	13.70		18.08	25.50
9030	1-1/4"		30	.267		5.70	14.60		20.30	28.50
9040	1-1/2"		30	.267		6.50	14.60		21.10	29
9050	2"		30	.267		9.10	14.60		23.70	32
9060	2-1/2"		28	.286		9.65	15.65		25.30	34
9070	3"		28	.286		13.10	15.65		28.75	38
9090	4"		25	.320		15.70	17.50		33.20	43.50
9110	Right angle, conduit beam clamp, 1/2"		32	.250		2.39	13.70		16.09	23
9120	3/4"		32	.250		2.51	13.70		16.21	23.50
9130	1"		32	.250		2.79	13.70		16.49	23.50
9140	1-1/4"		30	.267		3.20	14.60		17.80	25.50
9150	1-1/2"		30	.267		3.59	14.60		18.19	26
9160	2"		30	.267		5.25	14.60		19.85	28
9170	2-1/2"		28	.286		6.40	15.65		22.05	30.50
9180	3"		28	.286		6.90	15.65		22.55	31
9190	3-1/2"		25	.320		8.35	17.50		25.85	35
9200	4"		25	.320		9	17.50		26.50	36
9230	Adjustable, conduit hanger, 1/2"		32	.250		4.19	13.70		17.89	25
9240	3/4"		32	.250		5.55	13.70		19.25	26.50
9250	1"		32	.250		7	13.70		20.70	28
9260	1-1/4"		30	.267		8.35	14.60		22.95	31
9270	1-1/2"		30	.267		9.70	14.60		24.30	32.50
9280	2"		30	.267		12.45	14.60		27.05	35.50
9290	2-1/2"		28	.286		15.30	15.65		30.95	40.50
9300	3"		28	.286		17.95	15.65		33.60	43.50
9310	3-1/2"		25	.320		20.50	17.50		38	49

26 05 29 – Hangers and Supports for Electrical Systems

26 05 29.20 Hangers		Crew	Daily Output	Labor-Hours	Unit	Material	Labor	Equipment	Total	Total Incl O&P
						2015 Bare Costs				
9320	4"	1 Elec	25	.320	Ea.	23.50	17.50		41	52
9330	5"		20	.400		29	22		51	65
9340	6"		16	.500		34.50	27.50		62	79
9350	Combination conduit hanger, 3/8"		32	.250		11.25	13.70		24.95	33
9360	Adjustable flange 3/8"		32	.250		13.20	13.70		26.90	35

26 05 33 – Raceway and Boxes for Electrical Systems

26 05 33.13 Conduit		Crew	Daily Output	Labor-Hours	Unit	Material	Labor	Equipment	Total	Total Incl O&P
0010	**CONDUIT** To 15' high, includes 2 terminations, 2 elbows, R260533-20									
0020	11 beam clamps, and 11 couplings per 100 L.F.									
0300	Aluminum, 1/2" diameter	1 Elec	100	.080	L.F.	1.75	4.38		6.13	8.45
0500	3/4" diameter R260533-21		90	.089		2.41	4.86		7.27	9.95
0700	1" diameter		80	.100		3.46	5.45		8.91	12
1000	1-1/4" diameter		70	.114		4.27	6.25		10.52	14.05
1030	1-1/2" diameter		65	.123		5.45	6.75		12.20	16.10
1050	2" diameter		60	.133		7.70	7.30		15	19.40
1070	2-1/2" diameter		50	.160		11.50	8.75		20.25	26
1100	3" diameter	2 Elec	90	.178		15.70	9.70		25.40	32
1130	3-1/2" diameter		80	.200		21.50	10.95		32.45	40
1140	4" diameter		70	.229		25	12.50		37.50	46.50
1150	5" diameter		50	.320		55	17.50		72.50	86.50
1160	6" diameter		40	.400		92.50	22		114.50	135
1161	Field bends, 45° to 90°, 1/2" diameter	1 Elec	53	.151	Ea.		8.25		8.25	12.35
1162	3/4" diameter		47	.170			9.30		9.30	13.95
1163	1" diameter		44	.182			9.95		9.95	14.90
1164	1-1/4" diameter		23	.348			19.05		19.05	28.50
1165	1-1/2" diameter		21	.381			21		21	31
1166	2" diameter		16	.500			27.50		27.50	41
1170	Elbows, 1/2" diameter		40	.200		8.60	10.95		19.55	26
1200	3/4" diameter		32	.250		10.95	13.70		24.65	32.50
1230	1" diameter		28	.286		15.75	15.65		31.40	41
1250	1-1/4" diameter		24	.333		26.50	18.25		44.75	56.50
1270	1-1/2" diameter		20	.400		34.50	22		56.50	71
1300	2" diameter		16	.500		50	27.50		77.50	96
1330	2-1/2" diameter		12	.667		94	36.50		130.50	158
1350	3" diameter		8	1		139	54.50		193.50	234
1370	3-1/2" diameter		6	1.333		243	73		316	375
1400	4" diameter		5	1.600		280	87.50		367.50	440
1410	5" diameter		4	2		745	109		854	985
1420	6" diameter		2.50	3.200		1,300	175		1,475	1,675
1430	Couplings, 1/2" diameter		320	.025		2.69	1.37		4.06	5
1450	3/4" diameter		192	.042		4.01	2.28		6.29	7.80
1470	1" diameter		160	.050		5.30	2.74		8.04	9.90
1500	1-1/4" diameter		120	.067		6.80	3.65		10.45	12.95
1530	1-1/2" diameter		96	.083		7.60	4.56		12.16	15.20
1550	2" diameter		80	.100		11.15	5.45		16.60	20.50
1570	2-1/2" diameter		69	.116		26.50	6.35		32.85	38.50
1600	3" diameter		64	.125		33.50	6.85		40.35	47
1630	3-1/2" diameter		56	.143		52.50	7.80		60.30	69
1650	4" diameter		53	.151		56	8.25		64.25	74
1670	5" diameter		48	.167		180	9.10		189.10	212
1690	6" diameter		46	.174		355	9.50		364.50	405
1691	See note on line 26 05 33.13 9995 R260533-30									

26 05 33.13 Conduit		Crew	Daily Output	Labor-Hours	Unit	Material	2015 Bare Costs Labor	Equipment	Total	Total Incl O&P
1750	Rigid galvanized steel, 1/2" diameter	1 Elec	90	.089	L.F.	2.49	4.86		7.35	10.05
1770	3/4" diameter		80	.100		2.74	5.45		8.19	11.20
1800	1" diameter		65	.123		3.95	6.75		10.70	14.45
1830	1-1/4" diameter		60	.133		5.15	7.30		12.45	16.60
1850	1-1/2" diameter		55	.145		6.10	7.95		14.05	18.65
1870	2" diameter		45	.178		7.90	9.70		17.60	23.50
1900	2-1/2" diameter		35	.229		13.75	12.50		26.25	34
1930	3" diameter	2 Elec	50	.320		15.95	17.50		33.45	43.50
1950	3-1/2" diameter		44	.364		20	19.90		39.90	52
1970	4" diameter		40	.400		23	22		45	58.50
1980	5" diameter		30	.533		46	29		75	94
1990	6" diameter		20	.800		66.50	44		110.50	139
1991	Field bends, 45° to 90°, 1/2" diameter	1 Elec	44	.182	Ea.		9.95		9.95	14.90
1992	3/4" diameter		40	.200			10.95		10.95	16.40
1993	1" diameter		36	.222			12.15		12.15	18.20
1994	1-1/4" diameter		19	.421			23		23	34.50
1995	1-1/2" diameter		18	.444			24.50		24.50	36.50
1996	2" diameter		13	.615			33.50		33.50	50.50
2000	Elbows, 1/2" diameter		32	.250		6.65	13.70		20.35	28
2030	3/4" diameter		28	.286		7	15.65		22.65	31
2050	1" diameter		24	.333		10.75	18.25		29	39.50
2070	1-1/4" diameter		18	.444		14.75	24.50		39.25	52.50
2100	1-1/2" diameter		16	.500		18.25	27.50		45.75	61
2130	2" diameter		12	.667		26.50	36.50		63	83.50
2150	2-1/2" diameter		8	1		49	54.50		103.50	136
2170	3" diameter		6	1.333		68	73		141	184
2200	3-1/2" diameter		4.20	1.905		108	104		212	275
2220	4" diameter		4	2		122	109		231	298
2230	5" diameter		3.50	2.286		335	125		460	555
2240	6" diameter		2	4		510	219		729	890
2250	Couplings, 1/2" diameter		267	.030		1.64	1.64		3.28	4.26
2270	3/4" diameter		160	.050		2.01	2.74		4.75	6.30
2300	1" diameter		133	.060		2.96	3.29		6.25	8.20
2330	1-1/4" diameter		100	.080		3.72	4.38		8.10	10.65
2350	1-1/2" diameter		80	.100		4.70	5.45		10.15	13.35
2370	2" diameter		67	.119		6.20	6.55		12.75	16.60
2400	2-1/2" diameter		57	.140		15.35	7.70		23.05	28.50
2430	3" diameter		53	.151		19.90	8.25		28.15	34.50
2450	3-1/2" diameter		47	.170		26.50	9.30		35.80	43.50
2470	4" diameter		44	.182		27	9.95		36.95	44.50
2480	5" diameter		40	.200		62	10.95		72.95	85
2490	6" diameter		38	.211		86.50	11.50		98	112
2491	See note on line 26 05 33.13 9995									
2500	Steel, intermediate conduit (IMC), 1/2" diameter	1 Elec	100	.080	L.F.	1.79	4.38		6.17	8.50
2530	3/4" diameter		90	.089		2.22	4.86		7.08	9.75
2550	1" diameter		70	.114		3.28	6.25		9.53	12.95
2570	1-1/4" diameter		65	.123		4.01	6.75		10.76	14.50
2600	1-1/2" diameter		60	.133		5.35	7.30		12.65	16.85
2630	2" diameter		50	.160		6.40	8.75		15.15	20
2650	2-1/2" diameter		40	.200		11.35	10.95		22.30	29
2670	3" diameter	2 Elec	60	.267		15.40	14.60		30	39
2700	3-1/2" diameter		54	.296		20.50	16.20		36.70	47
2730	4" diameter		50	.320		22	17.50		39.50	50.50

26 05 33.13 Conduit		Crew	Daily Output	Labor-Hours	Unit	2015 Bare Costs Material	Labor	Equipment	Total	Total Incl O&P
2731	Field bends, 45° to 90°, 1/2" diameter	1 Elec	44	.182	Ea.		9.95		9.95	14.90
2732	3/4" diameter		40	.200			10.95		10.95	16.40
2733	1" diameter		36	.222			12.15		12.15	18.20
2734	1-1/4" diameter		19	.421			23		23	34.50
2735	1-1/2" diameter		18	.444			24.50		24.50	36.50
2736	2" diameter		13	.615			33.50		33.50	50.50
2750	Elbows, 1/2" diameter		32	.250		12.10	13.70		25.80	34
2770	3/4" diameter		28	.286		12.90	15.65		28.55	37.50
2800	1" diameter		24	.333		19	18.25		37.25	48.50
2830	1-1/4" diameter		18	.444		29	24.50		53.50	68
2850	1-1/2" diameter		16	.500		35.50	27.50		63	80
2870	2" diameter		12	.667		38	36.50		74.50	96
2900	2-1/2" diameter		8	1		68.50	54.50		123	158
2930	3" diameter		6	1.333		99.50	73		172.50	219
2950	3-1/2" diameter		4.20	1.905		213	104		317	390
2970	4" diameter		4	2		203	109		312	390
3000	Couplings, 1/2" diameter		293	.027		1.64	1.49		3.13	4.04
3030	3/4" diameter		176	.045		2.01	2.49		4.50	5.95
3050	1" diameter		147	.054		2.96	2.98		5.94	7.70
3070	1-1/4" diameter		110	.073		3.72	3.98		7.70	10.05
3100	1-1/2" diameter		88	.091		4.70	4.97		9.67	12.60
3130	2" diameter		73	.110		6.20	6		12.20	15.80
3150	2-1/2" diameter		63	.127		15.35	6.95		22.30	27.50
3170	3" diameter		59	.136		19.90	7.40		27.30	33
3200	3-1/2" diameter		52	.154		26.50	8.40		34.90	42
3230	4" diameter		49	.163		27	8.95		35.95	43
3231	See note on line 26 05 33.13 9995									
4100	Rigid steel, plastic coated, 40 mil thick									
4130	1/2" diameter	1 Elec	80	.100	L.F.	6.95	5.45		12.40	15.85
4150	3/4" diameter		70	.114		7.50	6.25		13.75	17.55
4170	1" diameter		55	.145		9.25	7.95		17.20	22
4200	1-1/4" diameter		50	.160		13.25	8.75		22	27.50
4230	1-1/2" diameter		45	.178		14	9.70		23.70	30
4250	2" diameter		35	.229		18.85	12.50		31.35	39.50
4270	2-1/2" diameter		25	.320		31	17.50		48.50	60
4300	3" diameter	2 Elec	44	.364		35	19.90		54.90	68.50
4330	3-1/2" diameter		40	.400		46.50	22		68.50	84.50
4350	4" diameter		36	.444		48.50	24.50		73	90
4370	5" diameter		30	.533		121	29		150	177
4400	Elbows, 1/2" diameter	1 Elec	28	.286	Ea.	18.25	15.65		33.90	43.50
4430	3/4" diameter		24	.333		15.25	18.25		33.50	44.50
4450	1" diameter		18	.444		17.45	24.50		41.95	55.50
4470	1-1/4" diameter		16	.500		26.50	27.50		54	70
4500	1-1/2" diameter		12	.667		26.50	36.50		63	83.50
4530	2" diameter		8	1		37	54.50		91.50	123
4550	2-1/2" diameter		6	1.333		80	73		153	197
4570	3" diameter		4.20	1.905		132	104		236	300
4600	3-1/2" diameter		4	2		163	109		272	345
4630	4" diameter		3.80	2.105		178	115		293	370
4650	5" diameter		3.50	2.286		440	125		565	670
4680	Couplings, 1/2" diameter		213	.038		5.35	2.05		7.40	8.95
4700	3/4" diameter		128	.063		4.27	3.42		7.69	9.80
4730	1" diameter		107	.075		5.55	4.09		9.64	12.25

26 05 33.13 Conduit		Crew	Daily Output	Labor-Hours	Unit	2015 Bare Costs Material	Labor	Equipment	Total	Total Incl O&P
4750	1-1/4" diameter	1 Elec	80	.100	Ea.	8.30	5.45		13.75	17.35
4770	1-1/2" diameter		64	.125		7.70	6.85		14.55	18.70
4800	2" diameter		53	.151		11.25	8.25		19.50	24.50
4830	2-1/2" diameter		46	.174		33	9.50		42.50	50.50
4850	3" diameter		43	.186		40.50	10.20		50.70	60
4870	3-1/2" diameter		38	.211		50	11.50		61.50	72.50
4900	4" diameter		36	.222		59	12.15		71.15	83
4950	5" diameter		32	.250		188	13.70		201.70	227
4951	See note on line 26 05 33.13 9995									
5000	Electric metallic tubing (EMT), 1/2" diameter	1 Elec	170	.047	L.F.	.67	2.57		3.24	4.60
5020	3/4" diameter		130	.062		.94	3.37		4.31	6.10
5040	1" diameter		115	.070		1.60	3.81		5.41	7.45
5060	1-1/4" diameter		100	.080		2.61	4.38		6.99	9.40
5080	1-1/2" diameter		90	.089		3.35	4.86		8.21	11
5100	2" diameter		80	.100		4.20	5.45		9.65	12.80
5120	2-1/2" diameter		60	.133		9.05	7.30		16.35	21
5140	3" diameter	2 Elec	100	.160		10.60	8.75		19.35	25
5160	3-1/2" diameter		90	.178		13.25	9.70		22.95	29
5180	4" diameter		80	.200		14.50	10.95		25.45	32.50
5200	Field bends, 45° to 90°, 1/2" diameter	1 Elec	89	.090	Ea.		4.92		4.92	7.35
5220	3/4" diameter		80	.100			5.45		5.45	8.20
5240	1" diameter		73	.110			6		6	9
5260	1-1/4" diameter		38	.211			11.50		11.50	17.25
5280	1-1/2" diameter		36	.222			12.15		12.15	18.20
5300	2" diameter		26	.308			16.85		16.85	25
5320	Offsets, 1/2" diameter		65	.123			6.75		6.75	10.10
5340	3/4" diameter		62	.129			7.05		7.05	10.55
5360	1" diameter		53	.151			8.25		8.25	12.35
5380	1-1/4" diameter		30	.267			14.60		14.60	22
5400	1-1/2" diameter		28	.286			15.65		15.65	23.50
5420	2" diameter		20	.400			22		22	33
5700	Elbows, 1" diameter		40	.200		5.35	10.95		16.30	22.50
5720	1-1/4" diameter		32	.250		6.65	13.70		20.35	28
5740	1-1/2" diameter		24	.333		7.75	18.25		26	36
5760	2" diameter		20	.400		11.35	22		33.35	45.50
5780	2-1/2" diameter		12	.667		31	36.50		67.50	88.50
5800	3" diameter		9	.889		41.50	48.50		90	119
5820	3-1/2" diameter		7	1.143		55.50	62.50		118	155
5840	4" diameter		6	1.333		65.50	73		138.50	181
6200	Couplings, set screw, steel, 1/2" diameter		235	.034		1.29	1.86		3.15	4.21
6220	3/4" diameter		188	.043		1.97	2.33		4.30	5.65
6240	1" diameter		157	.051		3.30	2.79		6.09	7.80
6260	1-1/4" diameter		118	.068		6.85	3.71		10.56	13.10
6280	1-1/2" diameter		94	.085		9.60	4.66		14.26	17.50
6300	2" diameter		78	.103		12.85	5.60		18.45	22.50
6320	2-1/2" diameter		67	.119		37	6.55		43.55	50.50
6340	3" diameter		59	.136		40.50	7.40		47.90	55.50
6360	3-1/2" diameter		47	.170		45.50	9.30		54.80	64
6380	4" diameter		43	.186		50	10.20		60.20	70.50
6381	See note on line 26 05 33.13 9995									
6500	Box connectors, set screw, steel, 1/2" diameter	1 Elec	120	.067	Ea.	.99	3.65		4.64	6.55
6520	3/4" diameter		110	.073		1.62	3.98		5.60	7.75
6540	1" diameter		90	.089		3.05	4.86		7.91	10.65

26 05 33.13 Conduit		Crew	Daily Output	Labor-Hours	Unit	2015 Bare Costs Material	Labor	Equipment	Total	Total Incl O&P
6560	1-1/4" diameter	1 Elec	70	.114	Ea.	5.95	6.25		12.20	15.90
6580	1-1/2" diameter		60	.133		8.60	7.30		15.90	20.50
6600	2" diameter		50	.160		11.80	8.75		20.55	26
6620	2-1/2" diameter		36	.222		41.50	12.15		53.65	63.50
6640	3" diameter		27	.296		49.50	16.20		65.70	79
6680	3-1/2" diameter		21	.381		69	21		90	107
6700	4" diameter		16	.500		77	27.50		104.50	126
6740	Insulated box connectors, set screw, steel, 1/2" diameter		120	.067		1.36	3.65		5.01	6.95
6760	3/4" diameter		110	.073		2.14	3.98		6.12	8.30
6780	1" diameter		90	.089		3.94	4.86		8.80	11.65
6800	1-1/4" diameter		70	.114		7.15	6.25		13.40	17.20
6820	1-1/2" diameter		60	.133		10.15	7.30		17.45	22
6840	2" diameter		50	.160		14.75	8.75		23.50	29.50
6860	2-1/2" diameter		36	.222		69.50	12.15		81.65	94.50
6880	3" diameter		27	.296		82.50	16.20		98.70	115
6900	3-1/2" diameter		21	.381		115	21		136	157
6920	4" diameter		16	.500		125	27.50		152.50	179
7000	EMT to conduit adapters, 1/2" diameter (compression)		70	.114		3.85	6.25		10.10	13.60
7020	3/4" diameter		60	.133		5.40	7.30		12.70	16.85
7040	1" diameter		50	.160		8.40	8.75		17.15	22.50
7060	1-1/4" diameter		40	.200		15.65	10.95		26.60	33.50
7080	1-1/2" diameter		30	.267		19.20	14.60		33.80	43
7100	2" diameter		25	.320		28	17.50		45.50	56.50
7200	EMT to Greenfield adapters, 1/2" to 3/8" diameter (compression)		90	.089		2.58	4.86		7.44	10.15
7220	1/2" diameter		90	.089		4.64	4.86		9.50	12.40
7240	3/4" diameter		80	.100		6.05	5.45		11.50	14.85
7260	1" diameter		70	.114		17	6.25		23.25	28
7270	1-1/4" diameter		60	.133		19.50	7.30		26.80	32.50
7280	1-1/2" diameter		50	.160		23	8.75		31.75	38.50
7290	2" diameter		40	.200		33.50	10.95		44.45	53.50
7400	EMT,LB, LR or LL fittings with covers, 1/2" dia, set screw		24	.333		6.15	18.25		24.40	34.50
7420	3/4" diameter		20	.400		8.95	22		30.95	43
7440	1" diameter		16	.500		11.10	27.50		38.60	53
7450	1-1/4" diameter		13	.615		15.15	33.50		48.65	67
7460	1-1/2" diameter		11	.727		20.50	40		60.50	82
7470	2" diameter		9	.889		32.50	48.50		81	109
7600	EMT, "T" fittings with covers, 1/2" diameter, set screw		16	.500		7.90	27.50		35.40	49.50
7620	3/4" diameter		15	.533		10.15	29		39.15	54.50
7640	1" diameter		12	.667		13.90	36.50		50.40	70
7650	1-1/4" diameter		11	.727		20	40		60	81.50
7660	1-1/2" diameter		10	.800		24	44		68	91.50
7670	2" diameter		8	1		31.50	54.50		86	117
8000	EMT, expansion fittings, no jumper, 1/2" diameter		24	.333		70.50	18.25		88.75	105
8020	3/4" diameter		20	.400		82	22		104	124
8040	1" diameter		16	.500		97.50	27.50		125	148
8060	1-1/4" diameter		13	.615		132	33.50		165.50	196
8080	1-1/2" diameter		11	.727		180	40		220	258
8100	2" diameter		9	.889		266	48.50		314.50	365
8110	2-1/2" diameter		7	1.143		425	62.50		487.50	560
8120	3" diameter		6	1.333		520	73		593	680
8140	4" diameter		5	1.600		905	87.50		992.50	1,125
8200	Split adapter, 1/2" diameter		110	.073		2.49	3.98		6.47	8.70
8210	3/4" diameter		90	.089		2.03	4.86		6.89	9.55

26 05 33.13 Conduit		Crew	Daily Output	Labor-Hours	Unit	Material	2015 Bare Costs Labor	Equipment	Total	Total Incl O&P
8220	1" diameter	1 Elec	70	.114	Ea.	2.78	6.25		9.03	12.40
8230	1-1/4" diameter		60	.133		4.49	7.30		11.79	15.90
8240	1-1/2" diameter		50	.160		6.90	8.75		15.65	20.50
8250	2" diameter		36	.222		20.50	12.15		32.65	40.50
8300	1 hole clips, 1/2" diameter		500	.016		.39	.88		1.27	1.74
8320	3/4" diameter		470	.017		.50	.93		1.43	1.94
8340	1" diameter		444	.018		.92	.99		1.91	2.49
8360	1-1/4" diameter		400	.020		1.26	1.09		2.35	3.03
8380	1-1/2" diameter		355	.023		1.90	1.23		3.13	3.94
8400	2" diameter		320	.025		3.01	1.37		4.38	5.35
8420	2-1/2" diameter		266	.030		5.55	1.65		7.20	8.60
8440	3" diameter		160	.050		7	2.74		9.74	11.80
8460	3-1/2" diameter		133	.060		11.05	3.29		14.34	17.15
8480	4" diameter		100	.080		14.35	4.38		18.73	22.50
8500	Clamp back spacers, 1/2" diameter		500	.016		1.18	.88		2.06	2.61
8510	3/4" diameter		470	.017		1.40	.93		2.33	2.93
8520	1" diameter		444	.018		2.44	.99		3.43	4.16
8530	1-1/4" diameter		400	.020		4.36	1.09		5.45	6.45
8540	1-1/2" diameter		355	.023		4.87	1.23		6.10	7.20
8550	2" diameter		320	.025		8.20	1.37		9.57	11.05
8560	2-1/2" diameter		266	.030		15.85	1.65		17.50	19.90
8570	3" diameter		160	.050		22.50	2.74		25.24	29
8580	3-1/2" diameter		133	.060		30.50	3.29		33.79	38.50
8590	4" diameter		100	.080		66.50	4.38		70.88	79.50
8600	Offset connectors, 1/2" diameter		40	.200		4.28	10.95		15.23	21
8610	3/4" diameter		32	.250		6.20	13.70		19.90	27.50
8620	1" diameter		24	.333		6.90	18.25		25.15	35
8650	90° pulling elbows, female, 1/2" diameter, with gasket		24	.333		6.75	18.25		25	35
8660	3/4" diameter		20	.400		11.55	22		33.55	45.50
8700	Couplings, compression, 1/2" diameter, steel		78	.103		3.11	5.60		8.71	11.80
8710	3/4" diameter		67	.119		4.33	6.55		10.88	14.55
8720	1" diameter		59	.136		7.40	7.40		14.80	19.25
8730	1-1/4" diameter		47	.170		14.10	9.30		23.40	29.50
8740	1-1/2" diameter		38	.211		20.50	11.50		32	40
8750	2" diameter		31	.258		28	14.10		42.10	51.50
8760	2-1/2" diameter		24	.333		125	18.25		143.25	166
8770	3" diameter		21	.381		156	21		177	203
8780	3-1/2" diameter		19	.421		254	23		277	315
8790	4" diameter		16	.500		260	27.50		287.50	325
8791	See note on line 26 05 33.13 9995									
8800	Box connectors, compression, 1/2" diam., steel	1 Elec	120	.067	Ea.	2.59	3.65		6.24	8.30
8810	3/4" diameter		110	.073		3.60	3.98		7.58	9.90
8820	1" diameter		90	.089		5.35	4.86		10.21	13.20
8830	1-1/4" diameter		70	.114		13.60	6.25		19.85	24.50
8840	1-1/2" diameter		60	.133		16	7.30		23.30	28.50
8850	2" diameter		50	.160		23	8.75		31.75	38.50
8860	2-1/2" diameter		36	.222		57	12.15		69.15	80.50
8870	3" diameter		27	.296		78.50	16.20		94.70	111
8880	3-1/2" diameter		21	.381		119	21		140	162
8890	4" diameter		16	.500		122	27.50		149.50	175
8900	Box connectors, insulated compression, 1/2" diam., steel		120	.067		2.51	3.65		6.16	8.20
8910	3/4" diameter		110	.073		3.43	3.98		7.41	9.70
8920	1" diameter		90	.089		5.80	4.86		10.66	13.65

26 05 33.13 Conduit		Crew	Daily Output	Labor-Hours	Unit	2015 Bare Costs Material	Labor	Equipment	Total	Total Incl O&P
8930	1-1/4" diameter	1 Elec	70	.114	Ea.	11.65	6.25		17.90	22
8940	1-1/2" diameter		60	.133		17.55	7.30		24.85	30.50
8950	2" diameter		50	.160		25.50	8.75		34.25	41
8960	2-1/2" diameter		36	.222		66.50	12.15		78.65	91
8970	3" diameter		27	.296		87	16.20		103.20	120
8980	3-1/2" diameter		21	.381		127	21		148	170
8990	4" diameter		16	.500		129	27.50		156.50	183
9100	PVC, schedule 40, 1/2" diameter		190	.042	L.F.	.90	2.30		3.20	4.44
9110	3/4" diameter		145	.055		.98	3.02		4	5.60
9120	1" diameter		125	.064		2.11	3.50		5.61	7.55
9130	1-1/4" diameter		110	.073		2.63	3.98		6.61	8.85
9140	1-1/2" diameter		100	.080		2.99	4.38		7.37	9.85
9150	2" diameter		90	.089		3.90	4.86		8.76	11.60
9160	2-1/2" diameter		65	.123		5.45	6.75		12.20	16.10
9170	3" diameter	2 Elec	110	.145		6.30	7.95		14.25	18.80
9180	3-1/2" diameter		100	.160		8.25	8.75		17	22
9190	4" diameter		90	.178		9.65	9.70		19.35	25
9200	5" diameter		70	.229		12.45	12.50		24.95	32.50
9210	6" diameter		60	.267		15.95	14.60		30.55	39.50
9220	Elbows, 1/2" diameter	1 Elec	50	.160	Ea.	.90	8.75		9.65	14.10
9225	3/4" diameter		42	.190		.88	10.40		11.28	16.55
9230	1" diameter		35	.229		1.33	12.50		13.83	20
9235	1-1/4" diameter		28	.286		2.04	15.65		17.69	25.50
9240	1-1/2" diameter		20	.400		2.65	22		24.65	36
9245	2" diameter		16	.500		3.59	27.50		31.09	45
9250	2-1/2" diameter		11	.727		6.45	40		46.45	66.50
9255	3" diameter		9	.889		10.85	48.50		59.35	85
9260	3-1/2" diameter		7	1.143		14	62.50		76.50	109
9265	4" diameter		6	1.333		17.25	73		90.25	128
9270	5" diameter		4	2		28	109		137	195
9275	6" diameter		3	2.667		64.50	146		210.50	290
9312	Couplings, 1/2" diameter		50	.160		.29	8.75		9.04	13.40
9314	3/4" diameter		42	.190		.32	10.40		10.72	15.95
9316	1" diameter		35	.229		.36	12.50		12.86	19.15
9318	1-1/4" diameter		28	.286		.51	15.65		16.16	24
9320	1-1/2" diameter		20	.400		.54	22		22.54	33.50
9322	2" diameter		16	.500		1.07	27.50		28.57	42
9324	2-1/2" diameter		11	.727		1.80	40		41.80	61.50
9326	3" diameter		9	.889		1.97	48.50		50.47	75
9328	3-1/2" diameter		7	1.143		2.51	62.50		65.01	96.50
9330	4" diameter		6	1.333		3.07	73		76.07	112
9332	5" diameter		4	2		7.55	109		116.55	172
9334	6" diameter		3	2.667		10.60	146		156.60	231
9335	See note on line 26 05 33.13 9995									
9340	Field bends, 45° & 90°, 1/2" diameter	1 Elec	45	.178	Ea.		9.70		9.70	14.55
9350	3/4" diameter		40	.200			10.95		10.95	16.40
9360	1" diameter		35	.229			12.50		12.50	18.75
9370	1-1/4" diameter		32	.250			13.70		13.70	20.50
9380	1-1/2" diameter		27	.296			16.20		16.20	24.50
9390	2" diameter		20	.400			22		22	33
9400	2-1/2" diameter		16	.500			27.50		27.50	41
9410	3" diameter		13	.615			33.50		33.50	50.50
9420	3-1/2" diameter		12	.667			36.50		36.50	54.50

26 05 33.13 Conduit		Crew	Daily Output	Labor-Hours	Unit	2015 Bare Costs Material	Labor	Equipment	Total	Total Incl O&P
9430	4″ diameter	1 Elec	10	.800	Ea.		44		44	65.50
9440	5″ diameter		9	.889			48.50		48.50	73
9450	6″ diameter		8	1			54.50		54.50	82
9460	PVC adapters, 1/2″ diameter		50	.160		.25	8.75		9	13.40
9470	3/4″ diameter		42	.190		.41	10.40		10.81	16.05
9480	1″ diameter		38	.211		.56	11.50		12.06	17.85
9490	1-1/4″ diameter		35	.229		.74	12.50		13.24	19.55
9500	1-1/2″ diameter		32	.250		.98	13.70		14.68	21.50
9510	2″ diameter		27	.296		1.22	16.20		17.42	26
9520	2-1/2″ diameter		23	.348		2.14	19.05		21.19	31
9530	3″ diameter		18	.444		2.97	24.50		27.47	40
9540	3-1/2″ diameter		13	.615		4.57	33.50		38.07	55.50
9550	4″ diameter		11	.727		5.30	40		45.30	65.50
9560	5″ diameter		8	1		11.20	54.50		65.70	94.50
9570	6″ diameter		6	1.333		18.70	73		91.70	130
9580	PVC-LB, LR or LL fittings & covers									
9590	1/2″ diameter	1 Elec	20	.400	Ea.	3.04	22		25.04	36.50
9600	3/4″ diameter		16	.500		4.60	27.50		32.10	46
9610	1″ diameter		12	.667		5.05	36.50		41.55	60
9620	1-1/4″ diameter		9	.889		7.80	48.50		56.30	81.50
9630	1-1/2″ diameter		7	1.143		9.30	62.50		71.80	104
9640	2″ diameter		6	1.333		14.75	73		87.75	125
9650	2-1/2″ diameter		6	1.333		52	73		125	167
9660	3″ diameter		5	1.600		50.50	87.50		138	187
9670	3-1/2″ diameter		4	2		53.50	109		162.50	223
9680	4″ diameter		3	2.667		70.50	146		216.50	297
9690	PVC-tee fitting & cover									
9700	1/2″	1 Elec	14	.571	Ea.	5.85	31.50		37.35	53.50
9710	3/4″		13	.615		5.85	33.50		39.35	57
9720	1″		10	.800		6.55	44		50.55	72.50
9730	1-1/4″		9	.889		11.50	48.50		60	85.50
9740	1-1/2″		8	1		13.50	54.50		68	97
9750	2″		7	1.143		20.50	62.50		83	116
9760	PVC-reducers, 3/4″ x 1/2″ diameter					1.41			1.41	1.55
9770	1″ x 1/2″ diameter					3.09			3.09	3.40
9780	1″ x 3/4″ diameter					3.34			3.34	3.67
9790	1-1/4″ x 3/4″ diameter					4.10			4.10	4.51
9800	1-1/4″ x 1″ diameter					4.34			4.34	4.77
9810	1-1/2″ x 1-1/4″ diameter					4.53			4.53	4.98
9820	2″ x 1-1/4″ diameter					5.45			5.45	5.95
9830	2-1/2″ x 2″ diameter					16.85			16.85	18.55
9840	3″ x 2″ diameter					17.40			17.40	19.15
9850	4″ x 3″ diameter					20			20	22
9860	Cement, quart					16.60			16.60	18.30
9870	Gallon					104			104	114
9880	Heat bender, to 6″ diameter					1,850			1,850	2,025
9900	Add to labor for higher elevated installation									
9910	15′ to 20′ high, add						10%			
9920	20′ to 25′ high, add						20%			
9930	25′ to 30′ high, add						25%			
9940	30′ to 35′ high, add						30%			
9950	35′ to 40′ high, add						35%			
9960	Over 40′ high, add						40%			

26 05 33.13 Conduit		Crew	Daily Output	Labor-Hours	Unit	Material	2015 Bare Costs Labor	Equipment	Total	Total Incl O&P
9995	Do not include labor when adding couplings to a fitting installation									

26 05 33.14 Conduit

		Crew	Daily Output	Labor-Hours	Unit	Material	Labor	Equipment	Total	Total Incl O&P
0010	**CONDUIT** To 15′ high, includes 11 couplings per 100′									
0200	Electric metallic tubing, 1/2″diameter	1 Elec	435	.018	L.F.	.48	1.01		1.49	2.04
0220	3/4″ diameter		253	.032		.76	1.73		2.49	3.42
0240	1″ diameter		207	.039		1.26	2.11		3.37	4.56
0260	1-1/4″ diameter		173	.046		2.18	2.53		4.71	6.20
0280	1-1/2″ diameter		153	.052		2.83	2.86		5.69	7.40
0300	2″ diameter		130	.062		3.52	3.37		6.89	8.95
0320	2-1/2″ diameter		92	.087		7.20	4.76		11.96	15.05
0340	3″ diameter	2 Elec	148	.108		8.35	5.90		14.25	18.05
0360	3-1/2″ diameter		134	.119		10.25	6.55		16.80	21
0380	4″ diameter		114	.140		10.90	7.70		18.60	23.50
0500	Steel rigid galvanized, 1/2″ diameter	1 Elec	146	.055		1.93	3		4.93	6.60
0520	3/4″ diameter		125	.064		2.05	3.50		5.55	7.50
0540	1″ diameter		93	.086		2.95	4.71		7.66	10.30
0560	1-1/4″ diameter		88	.091		4.20	4.97		9.17	12.05
0580	1-1/2″ diameter		80	.100		4.67	5.45		10.12	13.35
0600	2″ diameter		65	.123		5.75	6.75		12.50	16.45
0620	2-1/2″ diameter		48	.167		11.20	9.10		20.30	26
0640	3″ diameter	2 Elec	64	.250		12.90	13.70		26.60	34.50
0660	3-1/2″ diameter		60	.267		16.55	14.60		31.15	40
0680	4″ diameter		52	.308		18.15	16.85		35	45
0700	5″ diameter		50	.320		36	17.50		53.50	65.50
0720	6″ diameter		48	.333		47.50	18.25		65.75	80
1000	Steel intermediate conduit (IMC), 1/2 diameter R260533-25	1 Elec	155	.052		1.19	2.82		4.01	5.55
1010	3/4″ diameter		130	.062		1.44	3.37		4.81	6.65
1020	1″ diameter		100	.080		2.17	4.38		6.55	8.95
1030	1-1/4″ diameter		93	.086		2.81	4.71		7.52	10.15
1040	1-1/2″ diameter		85	.094		3.64	5.15		8.79	11.70
1050	2″ diameter		70	.114		4.51	6.25		10.76	14.30
1060	2-1/2″ diameter		53	.151		9.20	8.25		17.45	22.50
1070	3″ diameter	2 Elec	80	.200		12.15	10.95		23.10	30
1080	3-1/2″ diameter		70	.229		14.65	12.50		27.15	35
1090	4″ diameter		60	.267		16.10	14.60		30.70	39.50

26 05 33.15 Conduit Nipples

		Crew	Daily Output	Labor-Hours	Unit	Material	Labor	Equipment	Total	Total Incl O&P
0010	**CONDUIT NIPPLES** With locknuts and bushings									
0100	Aluminum, 1/2″ diameter, close	1 Elec	36	.222	Ea.	8.45	12.15		20.60	27.50
0120	1-1/2″ long		36	.222		8.60	12.15		20.75	27.50
0140	2″ long		36	.222		8.95	12.15		21.10	28
0160	2-1/2″ long		36	.222		9.75	12.15		21.90	29
0180	3″ long		36	.222		9.90	12.15		22.05	29
0200	3-1/2″ long		36	.222		10.65	12.15		22.80	30
0220	4″ long		36	.222		10.85	12.15		23	30
0240	5″ long		36	.222		12.40	12.15		24.55	32
0260	6″ long		36	.222		12.60	12.15		24.75	32
0280	8″ long		36	.222		13.10	12.15		25.25	32.50
0300	10″ long		36	.222		17.20	12.15		29.35	37
0320	12″ long		36	.222		16.60	12.15		28.75	36.50
0340	3/4″ diameter, close		32	.250		10.75	13.70		24.45	32.50
0360	1-1/2″ long		32	.250		10.90	13.70		24.60	32.50
0380	2″ long		32	.250		11.05	13.70		24.75	32.50

26 05 33.15 Conduit Nipples		Crew	Daily Output	Labor-Hours	Unit	2015 Bare Costs Material	Labor	Equipment	Total	Total Incl O&P
0400	2-1/2" long	1 Elec	32	.250	Ea.	11.95	13.70		25.65	33.50
0420	3" long		32	.250		12.20	13.70		25.90	34
0440	3-1/2" long		32	.250		12.90	13.70		26.60	34.50
0460	4" long		32	.250		13.80	13.70		27.50	35.50
0480	5" long		32	.250		14.70	13.70		28.40	36.50
0500	6" long		32	.250		15.40	13.70		29.10	37.50
0520	8" long		32	.250		17.80	13.70		31.50	40
0540	10" long		32	.250		21	13.70		34.70	43.50
0560	12" long		32	.250		29.50	13.70		43.20	53
0580	1" diameter, close		27	.296		16.45	16.20		32.65	42.50
0600	2" long		27	.296		18.75	16.20		34.95	45
0620	2-1/2" long		27	.296		17.80	16.20		34	44
0640	3" long		27	.296		21	16.20		37.20	47.50
0660	3-1/2" long		27	.296		20.50	16.20		36.70	47
0680	4" long		27	.296		23.50	16.20		39.70	50
0700	5" long		27	.296		23	16.20		39.20	49.50
0720	6" long		27	.296		24.50	16.20		40.70	51.50
0740	8" long		27	.296		28.50	16.20		44.70	56
0760	10" long		27	.296		34	16.20		50.20	61.50
0780	12" long		27	.296		37	16.20		53.20	65
0800	1-1/4" diameter, close		23	.348		24	19.05		43.05	55
0820	2" long		23	.348		24.50	19.05		43.55	55.50
0840	2-1/2" long		23	.348		25.50	19.05		44.55	56.50
0860	3" long		23	.348		26.50	19.05		45.55	57.50
0880	3-1/2" long		23	.348		28.50	19.05		47.55	60
0900	4" long		23	.348		29.50	19.05		48.55	61
0920	5" long		23	.348		32.50	19.05		51.55	64.50
0940	6" long		23	.348		35	19.05		54.05	67
0960	8" long		23	.348		44.50	19.05		63.55	77.50
0980	10" long		23	.348		42.50	19.05		61.55	75
1000	12" long		23	.348		62	19.05		81.05	96.50
1020	1-1/2" diameter, close		20	.400		32	22		54	68
1040	2" long		20	.400		33	22		55	69
1060	2-1/2" long		20	.400		33.50	22		55.50	70
1080	3" long		20	.400		35	22		57	71.50
1100	3-1/2" long		20	.400		38	22		60	74.50
1120	4" long		20	.400		38.50	22		60.50	75
1140	5" long		20	.400		42	22		64	79.50
1160	6" long		20	.400		43.50	22		65.50	81
1180	8" long		20	.400		47.50	22		69.50	85.50
1200	10" long		20	.400		62.50	22		84.50	102
1220	12" long		20	.400		75.50	22		97.50	116
1240	2" diameter, close		18	.444		44	24.50		68.50	84.50
1260	2-1/2" long		18	.444		46.50	24.50		71	87.50
1280	3" long		18	.444		48	24.50		72.50	89.50
1300	3-1/2" long		18	.444		50	24.50		74.50	91.50
1320	4" long		18	.444		51	24.50		75.50	93
1340	5" long		18	.444		56.50	24.50		81	98.50
1360	6" long		18	.444		59	24.50		83.50	102
1380	8" long		18	.444		76.50	24.50		101	121
1400	10" long		18	.444		64	24.50		88.50	107
1420	12" long		18	.444		86.50	24.50		111	132
1440	2-1/2" diameter, close	↓	15	.533	↓	107	29		136	161

26 05 33.15 Conduit Nipples		Crew	Daily Output	Labor-Hours	Unit	Material	2015 Bare Costs Labor	Equipment	Total	Total Incl O&P
1460	3″ long	1 Elec	15	.533	Ea.	108	29		137	163
1480	3-1/2″ long		15	.533		110	29		139	165
1500	4″ long		15	.533		113	29		142	169
1520	5″ long		15	.533		119	29		148	175
1540	6″ long		15	.533		118	29		147	174
1560	8″ long		15	.533		129	29		158	186
1580	10″ long		15	.533		140	29		169	197
1600	12″ long		15	.533		149	29		178	207
1620	3″ diameter, close		12	.667		117	36.50		153.50	184
1640	3″ long		12	.667		127	36.50		163.50	194
1660	3-1/2″ long		12	.667		122	36.50		158.50	189
1680	4″ long		12	.667		135	36.50		171.50	204
1700	5″ long		12	.667		130	36.50		166.50	198
1720	6″ long		12	.667		141	36.50		177.50	210
1740	8″ long		12	.667		175	36.50		211.50	248
1760	10″ long		12	.667		173	36.50		209.50	246
1780	12″ long		12	.667		198	36.50		234.50	273
1800	3-1/2″ diameter, close		11	.727		242	40		282	325
1820	4″ long		11	.727		254	40		294	340
1840	5″ long		11	.727		257	40		297	345
1860	6″ long		11	.727		262	40		302	350
1880	8″ long		11	.727		278	40		318	365
1900	10″ long		11	.727		295	40		335	385
1920	12″ long		11	.727		350	40		390	450
1940	4″ diameter, close		9	.889		264	48.50		312.50	365
1960	4″ long		9	.889		291	48.50		339.50	395
1980	5″ long		9	.889		300	48.50		348.50	405
2000	6″ long		9	.889		289	48.50		337.50	395
2020	8″ long		9	.889		345	48.50		393.50	450
2040	10″ long		9	.889		355	48.50		403.50	465
2060	12″ long		9	.889		345	48.50		393.50	455
2080	5″ diameter, close		7	1.143		450	62.50		512.50	590
2100	5″ long		7	1.143		485	62.50		547.50	630
2120	6″ long		7	1.143		495	62.50		557.50	640
2140	8″ long		7	1.143		525	62.50		587.50	670
2160	10″ long		7	1.143		565	62.50		627.50	715
2180	12″ long		7	1.143		585	62.50		647.50	740
2200	6″ diameter, close		6	1.333		780	73		853	965
2220	5″ long		6	1.333		800	73		873	990
2240	6″ long		6	1.333		815	73		888	1,000
2260	8″ long		6	1.333		855	73		928	1,050
2280	10″ long		6	1.333		910	73		983	1,100
2300	12″ long		6	1.333		925	73		998	1,125
2320	Rigid galvanized steel, 1/2″ diameter, close		32	.250		2.98	13.70		16.68	24
2340	1-1/2″ long		32	.250		3.27	13.70		16.97	24
2360	2″ long		32	.250		3.45	13.70		17.15	24.50
2380	2-1/2″ long		32	.250		3.58	13.70		17.28	24.50
2400	3″ long		32	.250		3.76	13.70		17.46	24.50
2420	3-1/2″ long		32	.250		3.94	13.70		17.64	25
2440	4″ long		32	.250		4.12	13.70		17.82	25
2460	5″ long		32	.250		4.41	13.70		18.11	25.50
2480	6″ long		32	.250		4.95	13.70		18.65	26
2500	8″ long		32	.250		7.30	13.70		21	28.50

26 05 33.15 Conduit Nipples		Crew	Daily Output	Labor-Hours	Unit	2015 Bare Costs Material	Labor	Equipment	Total	Total Incl O&P
2520	10″ long	1 Elec	32	.250	Ea.	8.15	13.70		21.85	29.50
2540	12″ long		32	.250		9.15	13.70		22.85	30.50
2560	3/4″ diameter, close		27	.296		3.92	16.20		20.12	29
2580	2″ long		27	.296		4.24	16.20		20.44	29
2600	2-1/2″ long		27	.296		4.47	16.20		20.67	29.50
2620	3″ long		27	.296		4.63	16.20		20.83	29.50
2640	3-1/2″ long		27	.296		4.74	16.20		20.94	29.50
2660	4″ long		27	.296		5.10	16.20		21.30	30
2680	5″ long		27	.296		5.50	16.20		21.70	30.50
2700	6″ long		27	.296		6	16.20		22.20	31
2720	8″ long		27	.296		8.50	16.20		24.70	34
2740	10″ long		27	.296		9.70	16.20		25.90	35
2760	12″ long		27	.296		10.70	16.20		26.90	36.50
2780	1″ diameter, close		23	.348		6.55	19.05		25.60	35.50
2800	2″ long		23	.348		6.80	19.05		25.85	36
2820	2-1/2″ long		23	.348		7	19.05		26.05	36
2840	3″ long		23	.348		7.35	19.05		26.40	36.50
2860	3-1/2″ long		23	.348		7.80	19.05		26.85	37
2880	4″ long		23	.348		8.10	19.05		27.15	37.50
2900	5″ long		23	.348		8.65	19.05		27.70	38
2920	6″ long		23	.348		9.05	19.05		28.10	38.50
2940	8″ long		23	.348		11.90	19.05		30.95	41.50
2960	10″ long		23	.348		14.25	19.05		33.30	44
2980	12″ long		23	.348		15.55	19.05		34.60	45.50
3000	1-1/4″ diameter, close		20	.400		8.85	22		30.85	42.50
3020	2″ long		20	.400		9.05	22		31.05	43
3040	3″ long		20	.400		9.70	22		31.70	43.50
3060	3-1/2″ long		20	.400		10.25	22		32.25	44.50
3080	4″ long		20	.400		10.55	22		32.55	44.50
3100	5″ long		20	.400		11.35	22		33.35	45.50
3120	6″ long		20	.400		12.05	22		34.05	46.50
3140	8″ long		20	.400		16.15	22		38.15	51
3160	10″ long		20	.400		18.90	22		40.90	54
3180	12″ long		20	.400		21	22		43	56
3200	1-1/2″ diameter, close		18	.444		12.65	24.50		37.15	50.50
3220	2″ long		18	.444		12.85	24.50		37.35	50.50
3240	2-1/2″ long		18	.444		13.40	24.50		37.90	51.50
3260	3″ long		18	.444		13.75	24.50		38.25	51.50
3280	3-1/2″ long		18	.444		14.50	24.50		39	52.50
3300	4″ long		18	.444		15	24.50		39.50	53
3320	5″ long		18	.444		15.75	24.50		40.25	54
3340	6″ long		18	.444		17.40	24.50		41.90	55.50
3360	8″ long		18	.444		22	24.50		46.50	60.50
3380	10″ long		18	.444		24.50	24.50		49	63.50
3400	12″ long		18	.444		26	24.50		50.50	65
3420	2″ diameter, close		16	.500		15	27.50		42.50	57.50
3440	2-1/2″ long		16	.500		15.85	27.50		43.35	58.50
3460	3″ long		16	.500		16.75	27.50		44.25	59.50
3480	3-1/2″ long		16	.500		17.65	27.50		45.15	60.50
3500	4″ long		16	.500		18.40	27.50		45.90	61.50
3520	5″ long		16	.500		19.80	27.50		47.30	63
3540	6″ long		16	.500		21	27.50		48.50	64
3560	8″ long	▼	16	.500	▼	26	27.50		53.50	69.50

26 05 33.15 Conduit Nipples		Crew	Daily Output	Labor-Hours	Unit	2015 Bare Costs Material	Labor	Equipment	Total	Total Incl O&P
3580	10″ long	1 Elec	16	.500	Ea.	29	27.50		56.50	73
3600	12″ long		16	.500		31.50	27.50		59	75.50
3620	2-1/2″ diameter, close		13	.615		46	33.50		79.50	101
3640	3″ long		13	.615		45.50	33.50		79	101
3660	3-1/2″ long		13	.615		48.50	33.50		82	104
3680	4″ long		13	.615		49	33.50		82.50	105
3700	5″ long		13	.615		52.50	33.50		86	108
3720	6″ long		13	.615		55	33.50		88.50	111
3740	8″ long		13	.615		62	33.50		95.50	119
3760	10″ long		13	.615		66.50	33.50		100	124
3780	12″ long		13	.615		72.50	33.50		106	130
3800	3″ diameter, close		12	.667		54.50	36.50		91	115
3820	3″ long		12	.667		56.50	36.50		93	117
3900	3-1/2″ long		12	.667		57.50	36.50		94	118
3920	4″ long		12	.667		59	36.50		95.50	120
3940	5″ long		12	.667		62.50	36.50		99	123
3960	6″ long		12	.667		66	36.50		102.50	127
3980	8″ long		12	.667		73.50	36.50		110	136
4000	10″ long		12	.667		80	36.50		116.50	143
4020	12″ long		12	.667		89	36.50		125.50	153
4040	3-1/2″ diameter, close		10	.800		97	44		141	173
4060	4″ long		10	.800		102	44		146	179
4080	5″ long		10	.800		106	44		150	182
4100	6″ long		10	.800		110	44		154	187
4120	8″ long		10	.800		118	44		162	195
4140	10″ long		10	.800		126	44		170	204
4160	12″ long		10	.800		134	44		178	214
4180	4″ diameter, close		8	1		109	54.50		163.50	202
4200	4″ long		8	1		114	54.50		168.50	208
4220	5″ long		8	1		119	54.50		173.50	213
4240	6″ long		8	1		123	54.50		177.50	217
4260	8″ long		8	1		131	54.50		185.50	226
4280	10″ long		8	1		142	54.50		196.50	238
4300	12″ long		8	1		153	54.50		207.50	250
4320	5″ diameter, close		6	1.333		217	73		290	350
4340	5″ long		6	1.333		237	73		310	370
4360	6″ long		6	1.333		242	73		315	375
4380	8″ long		6	1.333		255	73		328	390
4400	10″ long		6	1.333		269	73		342	405
4420	12″ long		6	1.333		290	73		363	430
4440	6″ diameter, close		5	1.600		455	87.50		542.50	630
4460	5″ long		5	1.600		480	87.50		567.50	655
4480	6″ long		5	1.600		485	87.50		572.50	665
4500	8″ long		5	1.600		500	87.50		587.50	680
4520	10″ long		5	1.600		525	87.50		612.50	710
4540	12″ long		5	1.600		540	87.50		627.50	725
4560	Plastic coated, 40 mil thick, 1/2″ diameter, 2″ long		32	.250		17.60	13.70		31.30	40
4580	2-1/2″ long		32	.250		20.50	13.70		34.20	43
4600	3″ long		32	.250		20.50	13.70		34.20	43
4680	3-1/2″ long		32	.250		23.50	13.70		37.20	46.50
4700	4″ long		32	.250		23	13.70		36.70	46
4720	5″ long		32	.250		23	13.70		36.70	46
4740	6″ long	↓	32	.250	↓	24.50	13.70		38.20	47.50

26 05 33.15 Conduit Nipples		Crew	Daily Output	Labor-Hours	Unit	2015 Bare Costs Material	Labor	Equipment	Total	Total Incl O&P
4760	8″ long	1 Elec	32	.250	Ea.	23	13.70		36.70	46
4780	10″ long		32	.250		24	13.70		37.70	47
4800	12″ long		32	.250		25	13.70		38.70	48
4820	3/4″ diameter, 2″ long		26	.308		18.75	16.85		35.60	45.50
4840	2-1/2″ long		26	.308		21.50	16.85		38.35	48.50
4860	3″ long		26	.308		20.50	16.85		37.35	48
4880	3-1/2″ long		26	.308		24.50	16.85		41.35	51.50
4900	4″ long		26	.308		24	16.85		40.85	51.50
4920	5″ long		26	.308		25	16.85		41.85	52.50
4940	6″ long		26	.308		24	16.85		40.85	51.50
4960	8″ long		26	.308		24	16.85		40.85	51.50
4980	10″ long		26	.308		25	16.85		41.85	52.50
5000	12″ long		26	.308		29	16.85		45.85	57
5020	1″ diameter, 2″ long		22	.364		20.50	19.90		40.40	53
5040	2-1/2″ long		22	.364		25	19.90		44.90	57.50
5060	3″ long		22	.364		24	19.90		43.90	56.50
5080	3-1/2″ long		22	.364		25.50	19.90		45.40	58
5100	4″ long		22	.364		28	19.90		47.90	60.50
5120	5″ long		22	.364		28.50	19.90		48.40	61.50
5140	6″ long		22	.364		26.50	19.90		46.40	59.50
5160	8″ long		22	.364		27.50	19.90		47.40	60.50
5180	10″ long		22	.364		29	19.90		48.90	62
5200	12″ long		22	.364		35	19.90		54.90	68.50
5220	1-1/4″ diameter, 2″ long		18	.444		26	24.50		50.50	65
5240	2-1/2″ long		18	.444		28.50	24.50		53	67.50
5260	3″ long		18	.444		30	24.50		54.50	69.50
5280	3-1/2″ long		18	.444		30	24.50		54.50	69.50
5300	4″ long		18	.444		31	24.50		55.50	70.50
5320	5″ long		18	.444		31.50	24.50		56	71.50
5340	6″ long		18	.444		35.50	24.50		60	76
5360	8″ long		18	.444		33	24.50		57.50	73
5380	10″ long		18	.444		37	24.50		61.50	77
5400	12″ long		18	.444		47.50	24.50		72	89
5420	1-1/2″ diameter, 2″ long		16	.500		30	27.50		57.50	74
5440	2-1/2″ long		16	.500		32.50	27.50		60	76.50
5460	3″ long		16	.500		32	27.50		59.50	76
5480	3-1/2″ long		16	.500		34	27.50		61.50	78.50
5500	4″ long		16	.500		38	27.50		65.50	82.50
5520	5″ long		16	.500		39	27.50		66.50	84
5540	6″ long		16	.500		40.50	27.50		68	86
5560	8″ long		16	.500		41.50	27.50		69	86.50
5580	10″ long		16	.500		49	27.50		76.50	95
5600	12″ long		16	.500		62	27.50		89.50	110
5620	2″ diameter, 2-1/2″ long		14	.571		39	31.50		70.50	90
5640	3″ long		14	.571		39	31.50		70.50	89.50
5660	3-1/2″ long		14	.571		40	31.50		71.50	91
5680	4″ long		14	.571		43	31.50		74.50	94.50
5700	5″ long		14	.571		47	31.50		78.50	98.50
5720	6″ long		14	.571		48.50	31.50		80	101
5740	8″ long		14	.571		52	31.50		83.50	104
5760	10″ long		14	.571		62.50	31.50		94	116
5780	12″ long		14	.571		75.50	31.50		107	130
5800	2-1/2″ diameter, 3-1/2″ long		12	.667		80	36.50		116.50	143

26 05 33.15 Conduit Nipples		Crew	Daily Output	Labor-Hours	Unit	Material	2015 Bare Costs Labor	Equipment	Total	Total Incl O&P
5820	4" long	1 Elec	12	.667	Ea.	81.50	36.50		118	144
5840	5" long		12	.667		93.50	36.50		130	158
5860	6" long		12	.667		102	36.50		138.50	168
5880	8" long		12	.667		108	36.50		144.50	173
5900	10" long		12	.667		117	36.50		153.50	184
5920	12" long		12	.667		132	36.50		168.50	200
5940	3" diameter, 3-1/2" long		11	.727		97.50	40		137.50	167
5960	4" long		11	.727		97.50	40		137.50	167
5980	5" long		11	.727		104	40		144	174
6000	6" long		11	.727		113	40		153	184
6020	8" long		11	.727		125	40		165	198
6040	10" long		11	.727		143	40		183	217
6060	12" long		11	.727		163	40		203	239
6080	3-1/2" diameter, 4" long		9	.889		151	48.50		199.50	239
6100	5" long		9	.889		152	48.50		200.50	240
6120	6" long		9	.889		169	48.50		217.50	259
6140	8" long		9	.889		180	48.50		228.50	271
6160	10" long		9	.889		198	48.50		246.50	290
6180	12" long		9	.889		215	48.50		263.50	310
6200	4" diameter, 4" long		7.50	1.067		164	58.50		222.50	269
6220	5" long		7.50	1.067		168	58.50		226.50	273
6240	6" long		7.50	1.067		182	58.50		240.50	288
6260	8" long		7.50	1.067		202	58.50		260.50	310
6280	10" long		7.50	1.067		238	58.50		296.50	350
6300	12" long		7.50	1.067		247	58.50		305.50	360
6320	5" diameter, 5" long		5.50	1.455		262	79.50		341.50	405
6340	6" long		5.50	1.455		276	79.50		355.50	425
6360	8" long		5.50	1.455		286	79.50		365.50	435
6380	10" long		5.50	1.455		305	79.50		384.50	455
6400	12" long		5.50	1.455		310	79.50		389.50	460
6420	6" diameter, 5" long		4.50	1.778		515	97		612	710
6440	6" long		4.50	1.778		525	97		622	725
6460	8" long		4.50	1.778		540	97		637	740
6480	10" long		4.50	1.778		560	97		657	760
6500	12" long		4.50	1.778		580	97		677	785

26 05 33.16 Outlet Boxes		Crew	Daily Output	Labor-Hours	Unit	Material	Labor	Equipment	Total	Total Incl O&P
0010	**OUTLET BOXES** R260533-65									
0020	Pressed steel, octagon, 4"	1 Elec	20	.400	Ea.	2.61	22		24.61	36
0040	For Romex or BX		20	.400		3.91	22		25.91	37.50
0050	For Romex or BX, with bracket		20	.400		6.15	22		28.15	40
0060	Covers, blank		64	.125		1.10	6.85		7.95	11.45
0100	Extension rings		40	.200		4.33	10.95		15.28	21
0150	Square, 4"		20	.400		2.42	22		24.42	35.50
0160	For Romex or BX		20	.400		7	22		29	40.50
0170	For Romex or BX, with bracket		20	.400		7.60	22		29.60	41.50
0200	Extension rings		40	.200		4.37	10.95		15.32	21
0220	2-1/8" deep, 1" KO		20	.400		5.80	22		27.80	39.50
0250	Covers, blank		64	.125		1.18	6.85		8.03	11.55
0260	Raised device		64	.125		2.21	6.85		9.06	12.70
0300	Plaster rings		64	.125		2.40	6.85		9.25	12.90
0350	Square, 4-11/16"		20	.400		6.10	22		28.10	40
0370	2-1/8" deep, 3/4" to 1-1/4" KO		20	.400		7.35	22		29.35	41

26 05 33.16 Outlet Boxes		Crew	Daily Output	Labor-Hours	Unit	2015 Bare Costs Material	Labor	Equipment	Total	Total Incl O&P
0400	Extension rings	1 Elec	40	.200	Ea.	9.85	10.95		20.80	27
0450	Covers, blank		53	.151		2.20	8.25		10.45	14.75
0460	Raised device		53	.151		6.60	8.25		14.85	19.60
0500	Plaster rings		53	.151		6.55	8.25		14.80	19.55
0550	Handy box		27	.296		2.88	16.20		19.08	27.50
0560	Covers, device		64	.125		1.01	6.85		7.86	11.35
0600	Extension rings		54	.148		3.44	8.10		11.54	15.95
0650	Switchbox		27	.296		4.18	16.20		20.38	29
0660	Romex or BX		27	.296		6	16.20		22.20	31
0670	with bracket		27	.296		6.95	16.20		23.15	32
0680	Partition, metal		27	.296		2.62	16.20		18.82	27.50
0700	Masonry, 1 gang, 2-1/2″ deep		27	.296		7.95	16.20		24.15	33.50
0710	3-1/2″ deep		27	.296		8	16.20		24.20	33.50
0750	2 gang, 2-1/2″ deep		20	.400		15	22		37	49.50
0760	3-1/2″ deep		20	.400		12.05	22		34.05	46.50
0800	3 gang, 2-1/2″ deep		13	.615		17.90	33.50		51.40	70
0850	4 gang, 2-1/2″ deep		10	.800		20.50	44		64.50	88.50
0860	5 gang, 2-1/2″ deep		9	.889		22	48.50		70.50	97
0870	6 gang, 2-1/2″ deep		8	1		45	54.50		99.50	132
0880	Masonry thru-the-wall, 1 gang, 4″ block		16	.500		29.50	27.50		57	73.50
0890	6″ block		16	.500		45	27.50		72.50	90.50
0900	8″ block		16	.500		50.50	27.50		78	96.50
0920	2 gang, 6″ block		16	.500		67	27.50		94.50	115
0940	Bar hanger with 3/8″ stud, for wood and masonry boxes		53	.151		5.60	8.25		13.85	18.50
0950	Concrete, set flush, 4″ deep		20	.400		12.30	22		34.30	46.50
1000	Plate with 3/8″ stud		80	.100		8.10	5.45		13.55	17.10
1100	Concrete, floor, 1 gang		5.30	1.509		89.50	82.50		172	223
1150	2 gang		4	2		137	109		246	315
1200	3 gang		2.70	2.963		203	162		365	465
1250	For duplex receptacle, pedestal mounted, add		24	.333		97.50	18.25		115.75	135
1270	Flush mounted, add		27	.296		30.50	16.20		46.70	58
1300	For telephone, pedestal mounted, add		30	.267		113	14.60		127.60	147
1350	Carpet flange, 1 gang		53	.151		50.50	8.25		58.75	68
1400	Cast, 1 gang, FS (2″ deep), 1/2″ hub		12	.667		18.25	36.50		54.75	74.50
1410	3/4″ hub		12	.667		19.30	36.50		55.80	75.50
1420	FD (2-11/16″ deep), 1/2″ hub		12	.667		17.70	36.50		54.20	74
1430	3/4″ hub		12	.667		19.25	36.50		55.75	75.50
1450	2 gang, FS, 1/2″ hub		10	.800		33	44		77	102
1460	3/4″ hub		10	.800		35.50	44		79.50	105
1470	FD, 1/2″ hub		10	.800		39.50	44		83.50	109
1480	3/4″ hub		10	.800		39.50	44		83.50	109
1500	3 gang, FS, 3/4″ hub		9	.889		56	48.50		104.50	135
1510	Switch cover, 1 gang, FS		64	.125		4.83	6.85		11.68	15.55
1520	2 gang		53	.151		7.35	8.25		15.60	20.50
1530	Duplex receptacle cover, 1 gang, FS		64	.125		4.93	6.85		11.78	15.65
1540	2 gang, FS		53	.151		7.50	8.25		15.75	20.50
1542	Weatherproof blank cover, 1 gang		64	.125		1.39	6.85		8.24	11.80
1544	2 gang		53	.151		2.96	8.25		11.21	15.60
1550	Weatherproof switch cover, 1 gang		64	.125		6.40	6.85		13.25	17.30
1554	2 gang		53	.151		12.45	8.25		20.70	26
1600	Weatherproof receptacle cover, 1 gang		64	.125		4.82	6.85		11.67	15.55
1604	2 gang		53	.151		12.50	8.25		20.75	26
1620	Weatherproof receptacle cover, tamper resistant, 1 gang		58	.138		13.50	7.55		21.05	26

26 05 33.16 Outlet Boxes		Crew	Daily Output	Labor-Hours	Unit	2015 Bare Costs Material	Labor	Equipment	Total	Total Incl O&P
1624	2 gang	1 Elec	48	.167	Ea.	27	9.10		36.10	43
1750	FSC, 1 gang, 1/2" hub		11	.727		20	40		60	81.50
1760	3/4" hub		11	.727		22.50	40		62.50	84
1770	2 gang, 1/2" hub		9	.889		36	48.50		84.50	113
1780	3/4" hub		9	.889		39	48.50		87.50	116
1790	FDC, 1 gang, 1/2" hub		11	.727		23	40		63	84.50
1800	3/4" hub		11	.727		25.50	40		65.50	87.50
1810	2 gang, 1/2" hub		9	.889		47.50	48.50		96	125
1820	3/4" hub		9	.889		42	48.50		90.50	119
1850	Weatherproof in-use cover, 1 gang		64	.125		24	6.85		30.85	37
1870	2 gang		53	.151		27	8.25		35.25	42
2000	Poke-thru fitting, fire rated, for 3-3/4" floor		6.80	1.176		131	64.50		195.50	241
2040	For 7" floor		6.80	1.176		165	64.50		229.50	278
2100	Pedestal, 15 amp, duplex receptacle & blank plate		5.25	1.524		138	83.50		221.50	276
2120	Duplex receptacle and telephone plate		5.25	1.524		138	83.50		221.50	276
2140	Pedestal, 20 amp, duplex recept. & phone plate		5	1.600		139	87.50		226.50	283
2160	Telephone plate, both sides		5.25	1.524		131	83.50		214.50	269
2200	Abandonment plate		32	.250		38.50	13.70		52.20	63

26 05 33.17 Outlet Boxes, Plastic		Crew	Daily Output	Labor-Hours	Unit	Material	Labor	Equipment	Total	Total Incl O&P
0010	**OUTLET BOXES, PLASTIC**									
0050	4" diameter, round with 2 mounting nails	1 Elec	25	.320	Ea.	2.39	17.50		19.89	28.50
0100	Bar hanger mounted		25	.320		5.10	17.50		22.60	31.50
0200	4", square with 2 mounting nails		25	.320		4.63	17.50		22.13	31
0300	Plaster ring		64	.125		1.98	6.85		8.83	12.45
0400	Switch box with 2 mounting nails, 1 gang		30	.267		4.39	14.60		18.99	27
0500	2 gang		25	.320		3.33	17.50		20.83	29.50
0600	3 gang		20	.400		4.93	22		26.93	38.50
0700	Old work box		30	.267		5.70	14.60		20.30	28.50
1400	PVC, FSS, 1 gang, 1/2" hub		14	.571		16.35	31.50		47.85	65
1410	3/4" hub		14	.571		13.35	31.50		44.85	61.50
1420	FD, 1 gang for variable terminations		14	.571		11	31.50		42.50	59
1450	FS, 2 gang for variable terminations		12	.667		13	36.50		49.50	69
1480	Weatherproof blank cover, FS, 1 gang		64	.125		4.46	6.85		11.31	15.15
1500	2 gang		53	.151		4.88	8.25		13.13	17.70
1510	Weatherproof switch cover, FS, 1 gang		64	.125		9.85	6.85		16.70	21
1520	2 gang		53	.151		16.20	8.25		24.45	30
1530	Weatherproof duplex receptacle cover, FS, 1 gang		64	.125		13.55	6.85		20.40	25
1540	2 gang		53	.151		14.40	8.25		22.65	28
1750	FSC, 1 gang, 1/2" hub		13	.615		10.85	33.50		44.35	62.50
1760	3/4" hub		13	.615		11.65	33.50		45.15	63.50
1770	FSC, 2 gang, 1/2" hub		11	.727		16.60	40		56.60	78
1780	3/4" hub		11	.727		16.50	40		56.50	77.50
1790	FDC, 1 gang, 1/2" hub		13	.615		11.15	33.50		44.65	63
1800	3/4" hub		13	.615		12.20	33.50		45.70	64
1810	Weatherproof, T box w/3 holes		14	.571		9.90	31.50		41.40	58
1820	4" diameter round w/5 holes		14	.571		9.45	31.50		40.95	57.50
1850	In-use cover, 1 gang		64	.125		8.75	6.85		15.60	19.90
1870	2 gang		53	.151		11	8.25		19.25	24.50

26 05 33.18 Pull Boxes		Crew	Daily Output	Labor-Hours	Unit	Material	Labor	Equipment	Total	Total Incl O&P
0010	**PULL BOXES** R260533-70									
0100	Steel, pull box, NEMA 1, type SC, 6" W x 6" H x 4" D	1 Elec	8	1	Ea.	9.65	54.50		64.15	92.50
0180	8" W x 6" H x 4" D		8	1		13.30	54.50		67.80	96.50

26 05 33 – Raceway and Boxes for Electrical Systems

26 05 33.18 Pull Boxes		Crew	Daily Output	Labor-Hours	Unit	Material	2015 Bare Costs Labor	Equipment	Total	Total Incl O&P
0200	8" W x 8" H x 4" D	1 Elec	8	1	Ea.	14.50	54.50		69	98
0210	10" W x 10" H x 4" D		7	1.143		17.65	62.50		80.15	113
0220	12" W x 12" H x 4" D		6.50	1.231		22.50	67.50		90	126
0230	15" W x 15" H x 4" D		5.20	1.538		42	84		126	173
0240	18" W x 18" H x 4" D		4.40	1.818		42.50	99.50		142	196
0250	6" W x 6" H x 6" D		8	1		12.95	54.50		67.45	96.50
0260	8" W x 8" H x 6" D		7.50	1.067		16.15	58.50		74.65	105
0270	10" W x 10" H x 6" D		5.50	1.455		21	79.50		100.50	142
0300	10" W x 12" H x 6" D		5.30	1.509		24.50	82.50		107	151
0310	12" W x 12" H x 6" D		5.20	1.538		26.50	84		110.50	155
0320	15" W x 15" H x 6" D		4.60	1.739		37	95		132	184
0330	18" W x 18" H x 6" D		4.20	1.905		47.50	104		151.50	209
0340	24" W x 24" H x 6" D		3.20	2.500		95.50	137		232.50	310
0350	12" W x 12" H x 8" D		5	1.600		30.50	87.50		118	165
0360	15" W x 15" H x 8" D		4.50	1.778		45.50	97		142.50	196
0370	18" W x 18" H x 8" D		4	2		61	109		170	231
0380	24" W x 18" H x 6" D		3.70	2.162		87.50	118		205.50	274
0400	16" W x 20" H x 8" D		4	2		84.50	109		193.50	257
0500	20" W x 24" H x 8" D		3.20	2.500		98	137		235	315
0510	24" W x 24" H x 8" D		3	2.667		108	146		254	340
0600	24" W x 36" H x 8" D		2.70	2.963		151	162		313	410
0610	30" W x 30" H x 8" D		2.70	2.963		194	162		356	455
0620	36" W x 36" H x 8" D		2	4		223	219		442	575
0630	24" W x 24" H x 10" D		2.50	3.200		186	175		361	465
0650	Pull box, hinged, NEMA 1, 6" W x 6" H x 4" D		8	1		11.60	54.50		66.10	95
0660	8" W x 8" H x 4" D		8	1		16.45	54.50		70.95	100
0670	10" W x 10" H x 4" D		7	1.143		19.15	62.50		81.65	115
0680	12" W x 12" H x 4" D		6	1.333		27.50	73		100.50	139
0690	15" W x 15" H x 4" D		5.20	1.538		28.50	84		112.50	158
0700	18" W x 18" H x 4" D		4.40	1.818		35	99.50		134.50	188
0710	6" W x 6" H x 6" D		8	1		15.10	54.50		69.60	98.50
0720	8" W x 8" H x 6" D		7.50	1.067		20	58.50		78.50	110
0730	10" W x 10" H x 6" D		5.50	1.455		27	79.50		106.50	149
0740	12" W x 12" H x 6" D		5.20	1.538		32	84		116	162
0800	12" W x 16" H x 6" D		4.70	1.702		49.50	93		142.50	194
0810	15" W x 15" H x 6" D		4.60	1.739		32	95		127	179
0820	18" W x 18" H x 6" D		4.20	1.905		58.50	104		162.50	220
1000	20" W x 20" H x 6" D		3.60	2.222		86	122		208	277
1010	24" W x 24" H x 6" D		3.20	2.500		115	137		252	330
1020	12" W x 12" H x 8" D		5	1.600		69.50	87.50		157	208
1030	15" W x 15" H x 8" D		4.50	1.778		85.50	97		182.50	240
1040	18" W x 18" H x 8" D		4	2		120	109		229	296
1200	20" W x 20" H x 8" D		3.20	2.500		149	137		286	370
1210	24" W x 24" H x 8" D		3	2.667		169	146		315	405
1220	30" W x 30" H x 8" D		2.70	2.963		244	162		406	510
1400	24" W x 36" H x 8" D		2.70	2.963		237	162		399	505
1600	24" W x 42" H x 8" D		2	4		350	219		569	715
1610	36" W x 36" H x 8" D		2	4		350	219		569	715
2100	Pull box, NEMA 3R, type SC, raintight & weatherproof									
2150	6" L x 6" W x 6" D	1 Elec	10	.800	Ea.	14.75	44		58.75	82
2200	8" L x 6" W x 6" D		8	1		20.50	54.50		75	105
2250	10" L x 6" W x 6" D		7	1.143		32	62.50		94.50	129
2300	12" L x 12" W x 6" D		5	1.600		54.50	87.50		142	191

26 05 33.18 Pull Boxes		Crew	Daily Output	Labor-Hours	Unit	2015 Bare Costs Material	Labor	Equipment	Total	Total Incl O&P
2350	16" L x 16" W x 6" D	1 Elec	4.50	1.778	Ea.	77.50	97		174.50	232
2400	20" L x 20" W x 6" D		4	2		101	109		210	275
2450	24" L x 18" W x 8" D		3	2.667		123	146		269	355
2500	24" L x 24" W x 10" D		2.50	3.200		291	175		466	580
2550	30" L x 24" W x 12" D		2	4		385	219		604	750
2600	36" L x 36" W x 12" D	↓	1.50	5.333	↓	430	292		722	910
2800	Cast iron, pull boxes for surface mounting									
3000	NEMA 4, watertight & dust tight									
3050	6" L x 6" W x 6" D	1 Elec	4	2	Ea.	257	109		366	445
3100	8" L x 6" W x 6" D		3.20	2.500		360	137		497	600
3150	10" L x 6" W x 6" D		2.50	3.200		400	175		575	700
3200	12" L x 12" W x 6" D		2.30	3.478		730	190		920	1,075
3250	16" L x 16" W x 6" D		1.30	6.154		960	335		1,295	1,550
3300	20" L x 20" W x 6" D		.80	10		1,800	545		2,345	2,800
3350	24" L x 18" W x 8" D		.70	11.429		2,625	625		3,250	3,825
3400	24" L x 24" W x 10" D		.50	16		4,975	875		5,850	6,775
3450	30" L x 24" W x 12" D		.40	20		5,450	1,100		6,550	7,650
3500	36" L x 36" W x 12" D		.20	40		5,975	2,200		8,175	9,850
3510	NEMA 4 clamp cover, 6" L x 6" W x 4" D		4	2		168	109		277	350
3520	8" L x 6" W x 4" D	↓	4	2	↓	209	109		318	395
4000	NEMA 7, explosionproof									
4050	6" L x 6" W x 6" D	1 Elec	2	4	Ea.	690	219		909	1,100
4100	8" L x 6" W x 6" D		1.80	4.444		955	243		1,198	1,425
4150	10" L x 6" W x 6" D		1.60	5		1,275	274		1,549	1,800
4200	12" L x 12" W x 6" D		1	8		2,225	440		2,665	3,100
4250	16" L x 14" W x 6" D		.60	13.333		3,050	730		3,780	4,475
4300	18" L x 18" W x 8" D		.50	16		5,800	875		6,675	7,675
4350	24" L x 18" W x 8" D		.40	20		7,375	1,100		8,475	9,750
4400	24" L x 24" W x 10" D		.30	26.667		10,000	1,450		11,450	13,300
4450	30" L x 24" W x 12" D		.20	40		14,300	2,200		16,500	19,000
5000	NEMA 9, dust tight 6" L x 6" W x 6" D		3.20	2.500		380	137		517	620
5050	8" L x 6" W x 6" D		2.70	2.963		455	162		617	745
5100	10" L x 6" W x 6" D		2	4		600	219		819	990
5150	12" L x 12" W x 6" D		1.60	5		1,150	274		1,424	1,675
5200	16" L x 16" W x 6" D		1	8		2,000	440		2,440	2,850
5250	18" L x 18" W x 8" D		.70	11.429		3,125	625		3,750	4,350
5300	24" L x 18" W x 8" D		.60	13.333		4,425	730		5,155	5,950
5350	24" L x 24" W x 10" D		.40	20		5,850	1,100		6,950	8,075
5400	30" L x 24" W x 12" D	↓	.30	26.667	↓	8,950	1,450		10,400	12,000
6000	J.I.C. wiring boxes, NEMA 12, dust tight & drip tight									
6050	6" L x 8" W x 4" D	1 Elec	10	.800	Ea.	51.50	44		95.50	122
6100	8" L x 10" W x 4" D		8	1		64	54.50		118.50	153
6150	12" L x 14" W x 6" D		5.30	1.509		124	82.50		206.50	261
6200	14" L x 16" W x 6" D		4.70	1.702		147	93		240	300
6250	16" L x 20" W x 6" D		4.40	1.818		226	99.50		325.50	395
6300	24" L x 30" W x 6" D		3.20	2.500		330	137		467	570
6350	24" L x 30" W x 8" D		2.90	2.759		340	151		491	600
6400	24" L x 36" W x 8" D		2.70	2.963		380	162		542	660
6450	24" L x 42" W x 8" D		2.30	3.478		425	190		615	755
6500	24" L x 48" W x 8" D	↓	2	4	↓	465	219		684	840

	26 05 33.23 Wireway	Crew	Daily Output	Labor-Hours	Unit	Material	2015 Bare Costs Labor	Equipment	Total	Total Incl O&P
0010	**WIREWAY** to 15′ high R260533-60									
0020	For higher elevations, see Section 26 05 36.40									
0100	NEMA 1, Screw cover w/fittings and supports, 2-1/2″ x 2-1/2″	1 Elec	45	.178	L.F.	10.85	9.70		20.55	26.50
0200	4″ x 4″	″	40	.200		11.35	10.95		22.30	29
0400	6″ x 6″	2 Elec	60	.267		19.60	14.60		34.20	43.50
0600	8″ x 8″		40	.400		33.50	22		55.50	70
0620	10″ x 10″		30	.533		48	29		77	96
0640	12″ x 12″		20	.800		61	44		105	133
0800	Elbows, 90°, 2-1/2″	1 Elec	24	.333	Ea.	32.50	18.25		50.75	63
1000	4″		20	.400		36.50	22		58.50	73.50
1200	6″		18	.444		41.50	24.50		66	82
1400	8″		16	.500		66.50	27.50		94	115
1420	10″		12	.667		95	36.50		131.50	159
1440	12″		10	.800		121	44		165	199
1500	Elbows, 45°, 2-1/2″		24	.333		32.50	18.25		50.75	63
1510	4″		20	.400		39.50	22		61.50	76.50
1520	6″		18	.444		41.50	24.50		66	82
1530	8″		16	.500		66.50	27.50		94	115
1540	10″		12	.667		95	36.50		131.50	159
1550	12″		10	.800		169	44		213	251
1600	″T″ box, 2-1/2″		18	.444		37	24.50		61.50	77.50
1800	4″		16	.500		45	27.50		72.50	90.50
2000	6″		14	.571		50.50	31.50		82	103
2200	8″		12	.667		93	36.50		129.50	157
2220	10″		10	.800		124	44		168	202
2240	12″		8	1		175	54.50		229.50	275
2300	Cross, 2-1/2″		16	.500		41.50	27.50		69	86.50
2310	4″		14	.571		50.50	31.50		82	103
2320	6″		12	.667		62.50	36.50		99	123
2400	Panel adapter, 2-1/2″		24	.333		11.15	18.25		29.40	40
2600	4″		20	.400		13.50	22		35.50	48
2800	6″		18	.444		14.45	24.50		38.95	52.50
3000	8″		16	.500		19.90	27.50		47.40	63
3020	10″		14	.571		37	31.50		68.50	87.50
3040	12″		12	.667		46.50	36.50		83	106
3200	Reducer, 4″ to 2-1/2″		24	.333		15.05	18.25		33.30	44
3400	6″ to 4″		20	.400		30	22		52	66
3600	8″ to 6″		18	.444		32.50	24.50		57	72
3620	10″ to 8″		16	.500		41	27.50		68.50	86
3640	12″ to 10″		14	.571		49.50	31.50		81	101
3780	End cap, 2-1/2″		24	.333		4.62	18.25		22.87	32.50
3800	4″		20	.400		5.70	22		27.70	39.50
4000	6″		18	.444		6.85	24.50		31.35	44
4200	8″		16	.500		9.30	27.50		36.80	51.50
4220	10″		14	.571		15.10	31.50		46.60	63.50
4240	12″		12	.667		21	36.50		57.50	77.50
4300	U-connector, 2-1/2″		200	.040		4.62	2.19		6.81	8.40
4320	4″		200	.040		5.70	2.19		7.89	9.55
4340	6″		180	.044		6.85	2.43		9.28	11.15
4360	8″		170	.047		13.85	2.57		16.42	19.10
4380	10″		150	.053		20	2.92		22.92	26.50
4400	12″		130	.062		28	3.37		31.37	35.50

26 05 33.23 Wireway		Crew	Daily Output	Labor-Hours	Unit	2015 Bare Costs Material	Labor	Equipment	Total	Total Incl O&P
4420	Hanger, 2-1/2"	1 Elec	100	.080	Ea.	12.60	4.38		16.98	20.50
4430	4"		100	.080		12.75	4.38		17.13	20.50
4440	6"		80	.100		16.30	5.45		21.75	26
4450	8"		65	.123		24.50	6.75		31.25	37
4460	10"		50	.160		44.50	8.75		53.25	62
4470	12"		40	.200		65	10.95		75.95	88
4475	NEMA 3R, Screw cover w/fittings and supports, 4" x 4"		36	.222	L.F.	16.50	12.15		28.65	36.50
4480	6" x 6"	2 Elec	55	.291		22	15.90		37.90	48.50
4485	8" x 8"		36	.444		34	24.50		58.50	74
4490	12" x 12"		18	.889		54.50	48.50		103	133
4500	Hinged cover, with fittings and supports, 2-1/2" x 2-1/2"	1 Elec	60	.133		15.05	7.30		22.35	27.50
4520	4" x 4"	"	45	.178		19.15	9.70		28.85	35.50
4540	6" x 6"	2 Elec	80	.200		30.50	10.95		41.45	50.50
4560	8" x 8"		60	.267		50	14.60		64.60	77
4580	10" x 10"		50	.320		63.50	17.50		81	95.50
4600	12" x 12"		24	.667		91.50	36.50		128	156
4700	Elbows 90°, 2-1/2" x 2-1/2"	1 Elec	32	.250	Ea.	46.50	13.70		60.20	72
4720	4"		27	.296		62.50	16.20		78.70	93
4730	6"		23	.348		71.50	19.05		90.55	108
4740	8"		18	.444		90.50	24.50		115	136
4750	10"		14	.571		109	31.50		140.50	167
4760	12"		12	.667		175	36.50		211.50	248
4800	Tee box, hinged cover, 2-1/2" x 2-1/2"		23	.348		54	19.05		73.05	88
4810	4"		20	.400		74	22		96	115
4820	6"		18	.444		82.50	24.50		107	127
4830	8"		16	.500		154	27.50		181.50	211
4840	10"		12	.667		157	36.50		193.50	227
4860	12"		10	.800		196	44		240	282
4880	Cross box, hinged cover, 2-1/2" x 2-1/2"		18	.444		60	24.50		84.50	103
4900	4"		16	.500		84	27.50		111.50	134
4920	6"		13	.615		106	33.50		139.50	168
4940	8"		11	.727		154	40		194	229
4960	10"		10	.800		231	44		275	320
4980	12"		9	.889		253	48.50		301.50	350
5000	NEMA 12, Hinged cover, 2-1/2" x 2-1/2"		40	.200	L.F.	37.50	10.95		48.45	57.50
5020	4" x 4"		35	.229		43.50	12.50		56	67
5040	6" x 6"	2 Elec	60	.267		62.50	14.60		77.10	91
5060	8" x 8"	"	50	.320		85.50	17.50		103	120
5120	Elbows 90°, flanged, 2-1/2" x 2-1/2"	1 Elec	23	.348	Ea.	83.50	19.05		102.55	120
5140	4"		20	.400		105	22		127	148
5160	6"		18	.444		133	24.50		157.50	184
5180	8"		15	.533		194	29		223	257
5240	Tee box, flanged, 2-1/2" x 2-1/2"		18	.444		114	24.50		138.50	162
5260	4"		16	.500		133	27.50		160.50	188
5280	6"		15	.533		180	29		209	242
5300	8"		13	.615		262	33.50		295.50	340
5360	Cross box, flanged, 2-1/2" x 2-1/2"		15	.533		154	29		183	213
5380	4"		13	.615		195	33.50		228.50	266
5400	6"		12	.667		254	36.50		290.50	335
5420	8"		10	.800		310	44		354	405
5480	Flange gasket, 2-1/2"		160	.050		4.12	2.74		6.86	8.65
5500	4"		80	.100		5.70	5.45		11.15	14.50
5520	6"		53	.151		7.90	8.25		16.15	21

	26 05 33.23 Wireway	Crew	Daily Output	Labor-Hours	Unit	Material	Labor	Equipment	Total	Total Incl O&P
						2015 Bare Costs				
5530	8″	1 Elec	40	.200	Ea.	10.25	10.95		21.20	27.50

26 05 33.25 Conduit Fittings for Rigid Galvanized Steel

	26 05 33.25 Conduit Fittings for Rigid Galvanized Steel	Crew	Daily Output	Labor-Hours	Unit	Material	Labor	Equipment	Total	Total Incl O&P
0010	**CONDUIT FITTINGS FOR RIGID GALVANIZED STEEL**									
0050	Standard, locknuts, 1/2″ diameter				Ea.	.21			.21	.23
0100	3/4″ diameter					.37			.37	.41
0300	1″ diameter					.63			.63	.69
0500	1-1/4″ diameter					.72			.72	.79
0700	1-1/2″ diameter					1.29			1.29	1.42
1000	2″ diameter					1.80			1.80	1.98
1030	2-1/2″ diameter					4.55			4.55	5
1050	3″ diameter					5.80			5.80	6.40
1070	3-1/2″ diameter					8.85			8.85	9.70
1100	4″ diameter					11.70			11.70	12.90
1110	5″ diameter					25			25	27.50
1120	6″ diameter					67.50			67.50	74.50
1130	Bushings, plastic, 1/2″ diameter	1 Elec	40	.200		.16	10.95		11.11	16.60
1150	3/4″ diameter		32	.250		.27	13.70		13.97	21
1170	1″ diameter		28	.286		.43	15.65		16.08	24
1200	1-1/4″ diameter		24	.333		.65	18.25		18.90	28
1230	1-1/2″ diameter		18	.444		.89	24.50		25.39	37.50
1250	2″ diameter		15	.533		1.64	29		30.64	45.50
1270	2-1/2″ diameter		13	.615		3.80	33.50		37.30	54.50
1300	3″ diameter		12	.667		3.94	36.50		40.44	59
1330	3-1/2″ diameter		11	.727		5.20	40		45.20	65
1350	4″ diameter		9	.889		6.40	48.50		54.90	80
1360	5″ diameter		7	1.143		14.10	62.50		76.60	109
1370	6″ diameter		5	1.600		31	87.50		118.50	165
1390	Steel, 1/2″ diameter		40	.200		.65	10.95		11.60	17.10
1400	3/4″ diameter		32	.250		.77	13.70		14.47	21.50
1430	1″ diameter		28	.286		1.40	15.65		17.05	25
1450	Steel insulated, 1-1/4″ diameter		24	.333		4.66	18.25		22.91	32.50
1470	1-1/2″ diameter		18	.444		5.80	24.50		30.30	43
1500	2″ diameter		15	.533		8.30	29		37.30	52.50
1530	2-1/2″ diameter		13	.615		16.15	33.50		49.65	68.50
1550	3″ diameter		12	.667		22	36.50		58.50	79
1570	3-1/2″ diameter		11	.727		29	40		69	91.50
1600	4″ diameter		9	.889		35	48.50		83.50	112
1610	5″ diameter		7	1.143		73	62.50		135.50	174
1620	6″ diameter		5	1.600		167	87.50		254.50	315
1630	Sealing locknuts, 1/2″ diameter		40	.200		1.31	10.95		12.26	17.85
1650	3/4″ diameter		32	.250		1.41	13.70		15.11	22
1670	1″ diameter		28	.286		2.08	15.65		17.73	26
1700	1-1/4″ diameter		24	.333		3.13	18.25		21.38	31
1730	1-1/2″ diameter		18	.444		3.98	24.50		28.48	41
1750	2″ diameter		15	.533		5.10	29		34.10	49
1760	Grounding bushing, insulated, 1/2″ diameter		32	.250		5.85	13.70		19.55	27
1770	3/4″ diameter		28	.286		6.65	15.65		22.30	31
1780	1″ diameter		20	.400		8.30	22		30.30	42
1800	1-1/4″ diameter		18	.444		10.75	24.50		35.25	48.50
1830	1-1/2″ diameter		16	.500		12.40	27.50		39.90	54.50
1850	2″ diameter		13	.615		17.25	33.50		50.75	69.50
1870	2-1/2″ diameter		12	.667		28	36.50		64.50	85.50

26 05 33.25 Conduit Fittings for Rigid Galvanized Steel		Crew	Daily Output	Labor-Hours	Unit	2015 Bare Costs Material	Labor	Equipment	Total	Total Incl O&P
1900	3" diameter	1 Elec	11	.727	Ea.	29	40		69	91.50
1930	3-1/2" diameter		9	.889		33.50	48.50		82	110
1950	4" diameter		8	1		48.50	54.50		103	136
1960	5" diameter		6	1.333		87.50	73		160.50	206
1970	6" diameter		4	2		123	109		232	299
1990	Coupling, with set screw, 1/2" diameter		50	.160		3.52	8.75		12.27	16.95
2000	3/4" diameter		40	.200		4.54	10.95		15.49	21.50
2030	1" diameter		35	.229		7.35	12.50		19.85	27
2050	1-1/4" diameter		28	.286		12.55	15.65		28.20	37.50
2070	1-1/2" diameter		23	.348		16.05	19.05		35.10	46
2090	2" diameter		20	.400		36	22		58	72.50
2100	2-1/2" diameter		18	.444		78.50	24.50		103	123
2110	3" diameter		15	.533		91.50	29		120.50	145
2120	3-1/2" diameter		12	.667		131	36.50		167.50	199
2130	4" diameter		10	.800		174	44		218	257
2140	5" diameter		9	.889		305	48.50		353.50	415
2150	6" diameter		8	1		540	54.50		594.50	675
2160	Box connector with set screw, plain, 1/2" diameter		70	.114		2.40	6.25		8.65	12
2170	3/4" diameter		60	.133		3.32	7.30		10.62	14.60
2180	1" diameter		50	.160		5.15	8.75		13.90	18.80
2190	Insulated, 1-1/4" diameter		40	.200		13.15	10.95		24.10	31
2200	1-1/2" diameter		30	.267		19.35	14.60		33.95	43.50
2210	2" diameter		20	.400		39	22		61	76
2220	2-1/2" diameter		18	.444		104	24.50		128.50	151
2230	3" diameter		15	.533		150	29		179	208
2240	3-1/2" diameter		12	.667		210	36.50		246.50	286
2250	4" diameter		10	.800		225	44		269	315
2260	5" diameter		9	.889		266	48.50		314.50	365
2270	6" diameter		8	1		280	54.50		334.50	390
2280	LB, LR or LL fittings & covers, 1/2" diameter		16	.500		8.60	27.50		36.10	50.50
2290	3/4" diameter		13	.615		10.65	33.50		44.15	62.50
2300	1" diameter		11	.727		15.55	40		55.55	76.50
2330	1-1/4" diameter		8	1		24.50	54.50		79	109
2350	1-1/2" diameter		6	1.333		32	73		105	145
2370	2" diameter		5	1.600		52.50	87.50		140	189
2380	2-1/2" diameter		4	2		111	109		220	286
2390	3" diameter		3.50	2.286		142	125		267	345
2400	3-1/2" diameter		3	2.667		247	146		393	490
2410	4" diameter		2.50	3.200		270	175		445	560
2420	T fittings, with cover, 1/2" diameter		12	.667		11	36.50		47.50	66.50
2430	3/4" diameter		11	.727		12.70	40		52.70	73.50
2440	1" diameter		9	.889		21.50	48.50		70	97
2450	1-1/4" diameter		6	1.333		28	73		101	140
2470	1-1/2" diameter		5	1.600		38.50	87.50		126	174
2500	2" diameter		4	2		56.50	109		165.50	227
2510	2-1/2" diameter		3.50	2.286		124	125		249	325
2520	3" diameter		3	2.667		173	146		319	410
2530	3-1/2" diameter		2.50	3.200		300	175		475	590
2540	4" diameter		2	4		360	219		579	725
2550	Nipples chase, plain, 1/2" diameter		40	.200		.79	10.95		11.74	17.25
2560	3/4" diameter		32	.250		.93	13.70		14.63	21.50
2570	1" diameter		28	.286		1.90	15.65		17.55	25.50
2600	Insulated, 1-1/4" diameter		24	.333		6.55	18.25		24.80	34.50

26 05 33.25 Conduit Fittings for Rigid Galvanized Steel		Crew	Daily Output	Labor-Hours	Unit	Material	2015 Bare Costs Labor	Equipment	Total	Total Incl O&P
2630	1-1/2" diameter	1 Elec	18	.444	Ea.	8.70	24.50		33.20	46
2650	2" diameter		15	.533		13.05	29		42.05	58
2660	2-1/2" diameter		12	.667		26.50	36.50		63	83.50
2670	3" diameter		10	.800		34	44		78	103
2680	3-1/2" diameter		9	.889		73	48.50		121.50	154
2690	4" diameter		8	1		142	54.50		196.50	239
2700	5" diameter		7	1.143		212	62.50		274.50	325
2710	6" diameter		6	1.333		330	73		403	470
2720	Nipples offset, plain, 1/2" diameter		40	.200		4.42	10.95		15.37	21.50
2730	3/4" diameter		32	.250		4.60	13.70		18.30	25.50
2740	1" diameter		24	.333		6.05	18.25		24.30	34
2750	Insulated, 1-1/4" diameter		20	.400		24.50	22		46.50	60
2760	1-1/2" diameter		18	.444		31	24.50		55.50	70.50
2770	2" diameter		16	.500		48	27.50		75.50	93.50
2780	3" diameter		14	.571		74	31.50		105.50	129
2850	Coupling, expansion, 1/2" diameter		12	.667		43.50	36.50		80	103
2880	3/4" diameter		10	.800		51	44		95	122
2900	1" diameter		8	1		59.50	54.50		114	148
2920	1-1/4" diameter		6.40	1.250		72.50	68.50		141	182
2940	1-1/2" diameter		5.30	1.509		112	82.50		194.50	248
2960	2" diameter		4.60	1.739		169	95		264	330
2980	2-1/2" diameter		3.60	2.222		236	122		358	440
3000	3" diameter		3	2.667		315	146		461	565
3020	3-1/2" diameter		2.80	2.857		385	156		541	660
3040	4" diameter		2.40	3.333		580	182		762	915
3060	5" diameter		2	4		905	219		1,124	1,325
3080	6" diameter		1.80	4.444		1,350	243		1,593	1,850
3100	Expansion deflection, 1/2" diameter		12	.667		249	36.50		285.50	330
3120	3/4" diameter		12	.667		280	36.50		316.50	365
3140	1" diameter		10	.800		298	44		342	395
3160	1-1/4" diameter		6.40	1.250		355	68.50		423.50	490
3180	1-1/2" diameter		5.30	1.509		415	82.50		497.50	580
3200	2" diameter		4.60	1.739		515	95		610	710
3220	2-1/2" diameter		3.60	2.222		700	122		822	950
3240	3" diameter		3	2.667		860	146		1,006	1,175
3260	3-1/2" diameter		2.80	2.857		1,100	156		1,256	1,450
3280	4" diameter		2.40	3.333		1,275	182		1,457	1,675
3300	5" diameter		2	4		1,925	219		2,144	2,450
3320	6" diameter		1.80	4.444		3,350	243		3,593	4,050
3340	Ericson, 1/2" diameter		16	.500		4.95	27.50		32.45	46.50
3360	3/4" diameter		14	.571		6.30	31.50		37.80	54
3380	1" diameter		11	.727		11.70	40		51.70	72.50
3400	1-1/4" diameter		8	1		25	54.50		79.50	110
3420	1-1/2" diameter		7	1.143		30	62.50		92.50	127
3440	2" diameter		5	1.600		60	87.50		147.50	197
3460	2-1/2" diameter		4	2		145	109		254	325
3480	3" diameter		3.50	2.286		221	125		346	430
3500	3-1/2" diameter		3	2.667		370	146		516	625
3520	4" diameter		2.70	2.963		430	162		592	720
3540	5" diameter		2.50	3.200		800	175		975	1,150
3560	6" diameter		2.30	3.478		1,075	190		1,265	1,450
3580	Split, 1/2" diameter		32	.250		3.60	13.70		17.30	24.50
3600	3/4" diameter	↓	27	.296	↓	4.50	16.20		20.70	29.50

26 05 33.25 Conduit Fittings for Rigid Galvanized Steel		Crew	Daily Output	Labor-Hours	Unit	Material	2015 Bare Costs Labor	Equipment	Total	Total Incl O&P
3620	1" diameter	1 Elec	20	.400	Ea.	8.10	22		30.10	42
3640	1-1/4" diameter		16	.500		12.45	27.50		39.95	54.50
3660	1-1/2" diameter		14	.571		14.15	31.50		45.65	62.50
3680	2" diameter		12	.667		28.50	36.50		65	85.50
3700	2-1/2" diameter		10	.800		64	44		108	136
3720	3" diameter		9	.889		96	48.50		144.50	179
3740	3-1/2" diameter		8	1		155	54.50		209.50	252
3760	4" diameter		7	1.143		181	62.50		243.50	293
3780	5" diameter		6	1.333		320	73		393	460
3800	6" diameter		5	1.600		425	87.50		512.50	595
4600	Reducing bushings, 3/4" to 1/2" diameter		54	.148		2.11	8.10		10.21	14.45
4620	1" to 3/4" diameter		46	.174		2.24	9.50		11.74	16.70
4640	1-1/4" to 1" diameter		40	.200		5.20	10.95		16.15	22
4660	1-1/2" to 1-1/4" diameter		36	.222		7.45	12.15		19.60	26.50
4680	2" to 1-1/2" diameter		32	.250		13.55	13.70		27.25	35.50
4740	2-1/2" to 2" diameter		30	.267		15.05	14.60		29.65	38.50
4760	3" to 2-1/2" diameter		28	.286		18.70	15.65		34.35	44
4800	Through-wall seal, 1/2" diameter		8	1		217	54.50		271.50	320
4820	3/4" diameter		7.50	1.067		239	58.50		297.50	350
4840	1" diameter		6.50	1.231		239	67.50		306.50	365
4860	1-1/4" diameter		5.50	1.455		284	79.50		363.50	430
4880	1-1/2" diameter		5	1.600		370	87.50		457.50	535
4900	2" diameter		4.20	1.905		335	104		439	520
4920	2-1/2" diameter		3.50	2.286		450	125		575	680
4940	3" diameter		3	2.667		480	146		626	745
4960	3-1/2" diameter		2.50	3.200		725	175		900	1,050
4980	4" diameter		2	4		735	219		954	1,150
5000	5" diameter		1.50	5.333		950	292		1,242	1,475
5020	6" diameter		1	8		950	440		1,390	1,700
5100	Cable supports, 2 or more wires									
5120	1-1/2" diameter	1 Elec	8	1	Ea.	139	54.50		193.50	235
5140	2" diameter		6	1.333		190	73		263	320
5160	2-1/2" diameter		4	2		204	109		313	390
5180	3" diameter		3.50	2.286		261	125		386	475
5200	3-1/2" diameter		2.60	3.077		340	168		508	625
5220	4" diameter		2	4		430	219		649	800
5240	5" diameter		1.50	5.333		585	292		877	1,075
5260	6" diameter		1	8		950	440		1,390	1,700
5280	Service entrance cap, 1/2" diameter		16	.500		5.65	27.50		33.15	47
5300	3/4" diameter		13	.615		6.50	33.50		40	57.50
5320	1" diameter		10	.800		7.85	44		51.85	74
5340	1-1/4" diameter		8	1		10.55	54.50		65.05	93.50
5360	1-1/2" diameter		6.50	1.231		21	67.50		88.50	124
5380	2" diameter		5.50	1.455		31.50	79.50		111	154
5400	2-1/2" diameter		4	2		84.50	109		193.50	257
5420	3" diameter		3.40	2.353		138	129		267	345
5440	3-1/2" diameter		3	2.667		197	146		343	435
5460	4" diameter		2.70	2.963		256	162		418	525
5750	90° pull elbows steel, female, 1/2" diameter		16	.500		6.80	27.50		34.30	48.50
5760	3/4" diameter		13	.615		7.65	33.50		41.15	59
5780	1" diameter		11	.727		14.05	40		54.05	75
5800	1-1/4" diameter		8	1		19.20	54.50		73.70	103
5820	1-1/2" diameter		6	1.333		27.50	73		100.50	139

26 05 33.25 Conduit Fittings for Rigid Galvanized Steel		Crew	Daily Output	Labor-Hours	Unit	2015 Bare Costs Material	Labor	Equipment	Total	Total Incl O&P
5840	2″ diameter	1 Elec	5	1.600	Ea.	44.50	87.50		132	180
6000	Explosion proof, flexible coupling									
6010	1/2″ diameter, 4″ long	1 Elec	12	.667	Ea.	119	36.50		155.50	186
6020	6″ long		12	.667		107	36.50		143.50	173
6050	12″ long		12	.667		162	36.50		198.50	233
6070	18″ long		12	.667		192	36.50		228.50	266
6090	24″ long		12	.667		239	36.50		275.50	320
6110	30″ long		12	.667		315	36.50		351.50	405
6130	36″ long		12	.667		305	36.50		341.50	390
6140	3/4″ diameter, 4″ long		10	.800		146	44		190	227
6150	6″ long		10	.800		130	44		174	209
6180	12″ long		10	.800		194	44		238	279
6200	18″ long		10	.800		243	44		287	335
6220	24″ long		10	.800		305	44		349	400
6240	30″ long		10	.800		415	44		459	520
6260	36″ long		10	.800		360	44		404	460
6270	1″ diameter, 6″ long		8	1		340	54.50		394.50	450
6300	12″ long		8	1		330	54.50		384.50	440
6320	18″ long		8	1		405	54.50		459.50	525
6340	24″ long		8	1		525	54.50		579.50	655
6360	30″ long		8	1		700	54.50		754.50	850
6380	36″ long		8	1		765	54.50		819.50	925
6390	1-1/4″ diameter, 12″ long		6.40	1.250		590	68.50		658.50	750
6410	18″ long		6.40	1.250		705	68.50		773.50	875
6430	24″ long		6.40	1.250		810	68.50		878.50	995
6450	30″ long		6.40	1.250		855	68.50		923.50	1,050
6470	36″ long		6.40	1.250		1,300	68.50		1,368.50	1,550
6480	1-1/2″ diameter, 12″ long		5.30	1.509		620	82.50		702.50	810
6500	18″ long		5.30	1.509		755	82.50		837.50	955
6520	24″ long		5.30	1.509		1,075	82.50		1,157.50	1,325
6540	30″ long		5.30	1.509		1,225	82.50		1,307.50	1,475
6560	36″ long		5.30	1.509		1,325	82.50		1,407.50	1,575
6570	2″ diameter, 12″ long		4.60	1.739		1,050	95		1,145	1,300
6590	18″ long		4.60	1.739		1,350	95		1,445	1,650
6610	24″ long		4.60	1.739		1,475	95		1,570	1,775
6630	30″ long		4.60	1.739		1,400	95		1,495	1,700
6650	36″ long		4.60	1.739		1,875	95		1,970	2,200
7000	Close up plug, 1/2″ diameter, explosion proof		40	.200		2.24	10.95		13.19	18.85
7010	3/4″ diameter		32	.250		2.63	13.70		16.33	23.50
7020	1″ diameter		28	.286		3.07	15.65		18.72	27
7030	1-1/4″ diameter		24	.333		3.39	18.25		21.64	31
7040	1-1/2″ diameter		18	.444		4.68	24.50		29.18	41.50
7050	2″ diameter		15	.533		8.05	29		37.05	52.50
7060	2-1/2″ diameter		13	.615		13.30	33.50		46.80	65
7070	3″ diameter		12	.667		19.75	36.50		56.25	76
7080	3-1/2″ diameter		11	.727		24.50	40		64.50	86.50
7090	4″ diameter		9	.889		30.50	48.50		79	107
7091	Elbow female, 45°, 1/2″		16	.500		11.15	27.50		38.65	53.50
7092	3/4″		13	.615		13.85	33.50		47.35	66
7093	1″		11	.727		16.55	40		56.55	78
7094	1-1/4″		8	1		25	54.50		79.50	110
7095	1-1/2″		6	1.333		27	73		100	139
7096	2″	↓	5	1.600	↓	31.50	87.50		119	166

26 05 33.25 Conduit Fittings for Rigid Galvanized Steel		Crew	Daily Output	Labor-Hours	Unit	Material	2015 Bare Costs Labor	Equipment	Total	Total Incl O&P
7097	2-1/2"	1 Elec	4.50	1.778	Ea.	92	97		189	247
7098	3"		4.20	1.905		93	104		197	259
7099	3-1/2"		4	2		144	109		253	320
7100	4"		3.80	2.105		158	115		273	345
7101	90°, 1/2"		16	.500		11.25	27.50		38.75	53.50
7102	3/4"		13	.615		12.15	33.50		45.65	64
7103	1"		11	.727		16.60	40		56.60	78
7104	1-1/4"		8	1		26.50	54.50		81	111
7105	1-1/2"		6	1.333		37	73		110	150
7106	2"		5	1.600		58	87.50		145.50	195
7107	2-1/2"		4.50	1.778		136	97		233	295
7110	Elbows 90°, long male & female, 1/2" diameter, explosion proof		16	.500		16.45	27.50		43.95	59
7120	3/4" diameter		13	.615		19	33.50		52.50	71.50
7130	1" diameter		11	.727		24.50	40		64.50	86.50
7140	1-1/4" diameter		8	1		29	54.50		83.50	114
7150	1-1/2" diameter		6	1.333		36.50	73		109.50	149
7160	2" diameter		5	1.600		54	87.50		141.50	191
7170	Capped elbow, 1/2" diameter, explosion proof		11	.727		17.95	40		57.95	79.50
7180	3/4" diameter		8	1		22	54.50		76.50	106
7190	1" diameter		6	1.333		24	73		97	136
7200	1-1/4" diameter		5	1.600		48	87.50		135.50	184
7210	Pulling elbow, 1/2" diameter, explosion proof		11	.727		71	40		111	138
7220	3/4" diameter		8	1		126	54.50		180.50	220
7230	1" diameter		6	1.333		181	73		254	310
7240	1-1/4" diameter		5	1.600		209	87.50		296.50	360
7250	1-1/2" diameter		5	1.600		237	87.50		324.50	390
7260	2" diameter		4	2		293	109		402	485
7270	2-1/2" diameter		3.50	2.286		725	125		850	985
7280	3" diameter		3	2.667		995	146		1,141	1,325
7290	3-1/2" diameter		2.50	3.200		1,250	175		1,425	1,625
7300	4" diameter		2.20	3.636		1,300	199		1,499	1,725
7310	LB conduit body, 1/2" diameter		11	.727		41.50	40		81.50	106
7320	3/4" diameter		8	1		50.50	54.50		105	138
7330	T conduit body, 1/2" diameter		9	.889		45.50	48.50		94	123
7340	3/4" diameter		6	1.333		53.50	73		126.50	168
7350	Explosion proof, round box w/cover, 3 threaded hubs, 1/2" diameter		8	1		45.50	54.50		100	132
7351	3/4" diameter		8	1		65.50	54.50		120	154
7352	1" diameter		7.50	1.067		69.50	58.50		128	164
7353	1-1/4" diameter		7	1.143		126	62.50		188.50	233
7354	1-1/2" diameter		7	1.143		184	62.50		246.50	296
7355	2" diameter		6	1.333		190	73		263	320
7356	Round box w/cover & mtng flange, 3 threaded hubs, 1/2" diameter		8	1		51	54.50		105.50	138
7357	3/4" diameter		8	1		57.50	54.50		112	146
7358	4 threaded hubs, 1" diameter		7	1.143		63	62.50		125.50	163
7400	Unions, 1/2" diameter		20	.400		11.20	22		33.20	45.50
7410	3/4" - 1/2" diameter		16	.500		16.20	27.50		43.70	59
7420	3/4" diameter		16	.500		15.40	27.50		42.90	58
7430	1" diameter		14	.571		28	31.50		59.50	78
7440	1-1/4" diameter		12	.667		41	36.50		77.50	99.50
7450	1-1/2" diameter		10	.800		52.50	44		96.50	123
7460	2" diameter		8.50	.941		67.50	51.50		119	151
7480	2-1/2" diameter		8	1		105	54.50		159.50	198
7490	3" diameter		7	1.143		137	62.50		199.50	245

26 05 33.25 Conduit Fittings for Rigid Galvanized Steel		Crew	Daily Output	Labor-Hours	Unit	Material	2015 Bare Costs Labor	Equipment	Total	Total Incl O&P
7500	3-1/2" diameter	1 Elec	6	1.333	Ea.	254	73		327	390
7510	4" diameter		5	1.600		256	87.50		343.50	415
7680	Reducer, 3/4" to 1/2"		54	.148		2.36	8.10		10.46	14.75
7690	1" to 1/2"		46	.174		4.26	9.50		13.76	18.95
7700	1" to 3/4"		46	.174		4.36	9.50		13.86	19.05
7710	1-1/4" to 3/4"		40	.200		5.35	10.95		16.30	22.50
7720	1-1/4" to 1"		40	.200		6.55	10.95		17.50	23.50
7730	1-1/2" to 1"		36	.222		10.05	12.15		22.20	29.50
7740	1-1/2" to 1-1/4"		36	.222		10.65	12.15		22.80	30
7750	2" to 3/4"		32	.250		9.90	13.70		23.60	31.50
7760	2" to 1-1/4"		32	.250		11.80	13.70		25.50	33.50
7770	2" to 1-1/2"		32	.250		12.55	13.70		26.25	34.50
7780	2-1/2" to 1-1/2"		30	.267		21	14.60		35.60	45
7790	3" to 2"		30	.267		21.50	14.60		36.10	45.50
7800	3-1/2" to 2-1/2"		28	.286		43.50	15.65		59.15	71.50
7810	4" to 3"		28	.286		52.50	15.65		68.15	81.50
7820	Sealing fitting, vertical/horizontal, 1/2" diameter		14.50	.552		15.10	30		45.10	61.50
7830	3/4" diameter		13.30	.602		17.60	33		50.60	69
7840	1" diameter		11.40	.702		22	38.50		60.50	82
7850	1-1/4" diameter		10	.800		28	44		72	96.50
7860	1-1/2" diameter		8.80	.909		42	49.50		91.50	121
7870	2" diameter		8	1		55.50	54.50		110	143
7880	2-1/2" diameter		6.70	1.194		76	65.50		141.50	182
7890	3" diameter		5.70	1.404		98	77		175	223
7900	3-1/2" diameter		4.70	1.702		260	93		353	425
7910	4" diameter		4	2		430	109		539	635
7920	Sealing hubs, 1" by 1-1/2"		12	.667		31.50	36.50		68	89.50
7930	1-1/4" by 2"		10	.800		47	44		91	117
7940	1-1/2" by 2"		9	.889		62	48.50		110.50	141
7950	2" by 2-1/2"		8	1		81	54.50		135.50	172
7960	3" by 4"		7	1.143		137	62.50		199.50	245
7970	4" by 5"		6	1.333		285	73		358	425
7980	Drain, 1/2"		32	.250		83.50	13.70		97.20	113
7990	Breather, 1/2"		32	.250		83.50	13.70		97.20	113
8000	Plastic coated 40 mil thick									
8010	LB, LR or LL conduit body w/cover, 1/2" diameter	1 Elec	13	.615	Ea.	49.50	33.50		83	105
8020	3/4" diameter		11	.727		60.50	40		100.50	126
8030	1" diameter		8	1		72	54.50		126.50	162
8040	1-1/4" diameter		6	1.333		107	73		180	227
8050	1-1/2" diameter		5	1.600		130	87.50		217.50	274
8060	2" diameter		4.50	1.778		190	97		287	355
8070	2-1/2" diameter		4	2		300	109		409	495
8080	3" diameter		3.50	2.286		440	125		565	670
8090	3-1/2" diameter		3	2.667		550	146		696	820
8100	4" diameter		2.50	3.200		615	175		790	935
8150	T conduit body with cover, 1/2" diameter		11	.727		61	40		101	127
8160	3/4" diameter		9	.889		57.50	48.50		106	136
8170	1" diameter		6	1.333		79.50	73		152.50	197
8180	1-1/4" diameter		5	1.600		122	87.50		209.50	266
8190	1-1/2" diameter		4.50	1.778		149	97		246	310
8200	2" diameter		4	2		203	109		312	390
8210	2-1/2" diameter		3.50	2.286		395	125		520	615
8220	3" diameter		3	2.667		510	146		656	785

26 05 33 – Raceway and Boxes for Electrical Systems

26 05 33.25 Conduit Fittings for Rigid Galvanized Steel		Crew	Daily Output	Labor-Hours	Unit	Material	Labor	Equipment	Total	Total Incl O&P
						2015 Bare Costs				
8230	3-1/2″ diameter	1 Elec	2.50	3.200	Ea.	805	175		980	1,150
8240	4″ diameter		2	4		875	219		1,094	1,300
8300	FS conduit body, 1 gang, 3/4″ diameter		11	.727		59	40		99	125
8310	1″ diameter		10	.800		61.50	44		105.50	134
8350	2 gang, 3/4″ diameter		9	.889		107	48.50		155.50	191
8360	1″ diameter		8	1		123	54.50		177.50	217
8400	Duplex receptacle cover		64	.125		43	6.85		49.85	58
8410	Switch cover		64	.125		62.50	6.85		69.35	79.50
8420	Switch, vaportight cover		53	.151		183	8.25		191.25	213
8430	Blank, cover		64	.125		41.50	6.85		48.35	56.50
8520	FSC conduit body, 1 gang, 3/4″ diameter		10	.800		66.50	44		110.50	139
8530	1″ diameter		9	.889		74	48.50		122.50	154
8550	2 gang, 3/4″ diameter		8	1		121	54.50		175.50	215
8560	1″ diameter		7	1.143		116	62.50		178.50	221
8590	Conduit hubs, 1/2″ diameter		18	.444		37	24.50		61.50	77
8600	3/4″ diameter		16	.500		37.50	27.50		65	82.50
8610	1″ diameter		14	.571		47	31.50		78.50	99
8620	1-1/4″ diameter		12	.667		63.50	36.50		100	124
8630	1-1/2″ diameter		10	.800		76	44		120	149
8640	2″ diameter		8.80	.909		89	49.50		138.50	173
8650	2-1/2″ diameter		8.50	.941		149	51.50		200.50	241
8660	3″ diameter		8	1		191	54.50		245.50	292
8670	3-1/2″ diameter		7.50	1.067		288	58.50		346.50	405
8680	4″ diameter		7	1.143		325	62.50		387.50	455
8690	5″ diameter		6	1.333		425	73		498	580
8700	Plastic coated 40 mil thick									
8710	Pipe strap, stamped 1 hole, 1/2″ diameter	1 Elec	470	.017	Ea.	9.80	.93		10.73	12.15
8720	3/4″ diameter		440	.018		10.25	.99		11.24	12.75
8730	1″ diameter		400	.020		10.55	1.09		11.64	13.25
8740	1-1/4″ diameter		355	.023		18.10	1.23		19.33	22
8750	1-1/2″ diameter		320	.025		19.05	1.37		20.42	23
8760	2″ diameter		266	.030		27	1.65		28.65	32
8770	2-1/2″ diameter		200	.040		36.50	2.19		38.69	43.50
8780	3″ diameter		133	.060		48	3.29		51.29	58
8790	3-1/2″ diameter		110	.073		73.50	3.98		77.48	86.50
8800	4″ diameter		90	.089		82	4.86		86.86	97.50
8810	5″ diameter		70	.114		129	6.25		135.25	151
8840	Clamp back spacers, 3/4″ diameter		440	.018		17.45	.99		18.44	20.50
8850	1″ diameter		400	.020		20	1.09		21.09	23.50
8860	1-1/4″ diameter		355	.023		28.50	1.23		29.73	33.50
8870	1-1/2″ diameter		320	.025		39	1.37		40.37	45
8880	2″ diameter		266	.030		60	1.65		61.65	68
8900	3″ diameter		133	.060		82	3.29		85.29	95
8920	4″ diameter		90	.089		103	4.86		107.86	121
8950	Touch-up plastic coating, spray, 12 oz.					52			52	57
8960	Sealing fittings, 1/2″ diameter	1 Elec	11	.727		62	40		102	128
8970	3/4″ diameter		9	.889		63.50	48.50		112	143
8980	1″ diameter		7.50	1.067		74.50	58.50		133	170
8990	1-1/4″ diameter		6.50	1.231		96.50	67.50		164	207
9000	1-1/2″ diameter		5.50	1.455		126	79.50		205.50	258
9010	2″ diameter		4.80	1.667		151	91		242	305
9020	2-1/2″ diameter		4	2		261	109		370	450
9030	3″ diameter		3.50	2.286		279	125		404	490

26 05 33.25 Conduit Fittings for Rigid Galvanized Steel

	26 05 33.25 Conduit Fittings for Rigid Galvanized Steel	Crew	Daily Output	Labor-Hours	Unit	Material	Labor	Equipment	Total	Total Incl O&P
						2015 Bare Costs				
9040	3-1/2" diameter	1 Elec	3	2.667	Ea.	755	146		901	1,050
9050	4" diameter		2.50	3.200		1,025	175		1,200	1,375
9060	5" diameter		1.70	4.706		2,100	257		2,357	2,700
9070	Unions, 1/2" diameter		18	.444		55	24.50		79.50	97
9080	3/4" diameter		15	.533		48.50	29		77.50	97
9090	1" diameter		13	.615		65	33.50		98.50	122
9100	1-1/4" diameter		11	.727		124	40		164	197
9110	1-1/2" diameter		9.50	.842		150	46		196	234
9120	2" diameter		8	1		169	54.50		223.50	268
9130	2-1/2" diameter		7.50	1.067		296	58.50		354.50	415
9140	3" diameter		6.80	1.176		385	64.50		449.50	520
9150	3-1/2" diameter		5.80	1.379		510	75.50		585.50	675
9160	4" diameter		4.80	1.667		615	91		706	810
9170	5" diameter		4	2		945	109		1,054	1,225

26 05 33.30 Electrical Nonmetallic Tubing (ENT)

	26 05 33.30 Electrical Nonmetallic Tubing (ENT)	Crew	Daily Output	Labor-Hours	Unit	Material	Labor	Equipment	Total	Total Incl O&P
0010	**ELECTRICAL NONMETALLIC TUBING (ENT)**									
0050	Flexible, 1/2" diameter	1 Elec	270	.030	L.F.	.74	1.62		2.36	3.24
0100	3/4" diameter		230	.035		1.21	1.90		3.11	4.18
0200	1" diameter		145	.055		2.03	3.02		5.05	6.75
0210	1-1/4" diameter		125	.064		1.98	3.50		5.48	7.45
0220	1-1/2" diameter		100	.080		2.93	4.38		7.31	9.75
0230	2" diameter		75	.107		3.89	5.85		9.74	13.05
0300	Connectors, to outlet box, 1/2" diameter		230	.035	Ea.	1.02	1.90		2.92	3.97
0310	3/4" diameter		210	.038		1.82	2.08		3.90	5.10
0320	1" diameter		200	.040		2.78	2.19		4.97	6.35
0400	Couplings, to conduit, 1/2" diameter		145	.055		1.08	3.02		4.10	5.70
0410	3/4" diameter		130	.062		1.29	3.37		4.66	6.45
0420	1" diameter		125	.064		2.68	3.50		6.18	8.20

26 05 33.35 Flexible Metallic Conduit

	26 05 33.35 Flexible Metallic Conduit	Crew	Daily Output	Labor-Hours	Unit	Material	Labor	Equipment	Total	Total Incl O&P
0010	**FLEXIBLE METALLIC CONDUIT**									
0050	Steel, 3/8" diameter	1 Elec	200	.040	L.F.	.43	2.19		2.62	3.75
0100	1/2" diameter		200	.040		.48	2.19		2.67	3.81
0200	3/4" diameter		160	.050		.67	2.74		3.41	4.84
0250	1" diameter		100	.080		1.21	4.38		5.59	7.90
0300	1-1/4" diameter		70	.114		1.57	6.25		7.82	11.10
0350	1-1/2" diameter		50	.160		2.55	8.75		11.30	15.90
0370	2" diameter		40	.200		3.11	10.95		14.06	19.80
0380	2-1/2" diameter		30	.267		3.77	14.60		18.37	26
0390	3" diameter	2 Elec	50	.320		6.60	17.50		24.10	33.50
0400	3-1/2" diameter		40	.400		7.45	22		29.45	41
0410	4" diameter		30	.533		8.50	29		37.50	53
0420	Connectors, plain, 3/8" diameter	1 Elec	100	.080	Ea.	1.97	4.38		6.35	8.70
0430	1/2" diameter		80	.100		2.28	5.45		7.73	10.70
0440	3/4" diameter		70	.114		2.59	6.25		8.84	12.20
0450	1" diameter		50	.160		5.50	8.75		14.25	19.15
0452	1-1/4" diameter		45	.178		6.75	9.70		16.45	22
0454	1-1/2" diameter		40	.200		10.05	10.95		21	27.50
0456	2" diameter		28	.286		15.05	15.65		30.70	40
0458	2-1/2" diameter		25	.320		25.50	17.50		43	54
0460	3" diameter		20	.400		36.50	22		58.50	73
0462	3-1/2" diameter		16	.500		107	27.50		134.50	159
0464	4" diameter		13	.615		136	33.50		169.50	201

26 05 33.35 Flexible Metallic Conduit		Crew	Daily Output	Labor-Hours	Unit	Material	2015 Bare Costs Labor	Equipment	Total	Total Incl O&P
0490	Insulated, 1" diameter	1 Elec	40	.200	Ea.	6.05	10.95		17	23
0500	1-1/4" diameter		40	.200		13.50	10.95		24.45	31.50
0550	1-1/2" diameter		32	.250		20.50	13.70		34.20	43
0600	2" diameter		23	.348		31	19.05		50.05	62.50
0610	2-1/2" diameter		20	.400		72.50	22		94.50	113
0620	3" diameter		17	.471		96	25.50		121.50	145
0630	3-1/2" diameter		13	.615		195	33.50		228.50	265
0640	4" diameter		10	.800		256	44		300	350
0650	Connectors 90°, plain, 3/8" diameter		80	.100		2.84	5.45		8.29	11.30
0660	1/2" diameter		60	.133		4.77	7.30		12.07	16.20
0700	3/4" diameter		50	.160		7.85	8.75		16.60	22
0750	1" diameter		40	.200		13.40	10.95		24.35	31
0790	Insulated, 1" diameter		40	.200		12.40	10.95		23.35	30
0800	1-1/4" diameter		30	.267		29	14.60		43.60	54
0850	1-1/2" diameter		23	.348		54.50	19.05		73.55	88.50
0900	2" diameter		18	.444		66.50	24.50		91	110
0910	2-1/2" diameter		16	.500		128	27.50		155.50	181
0920	3" diameter		14	.571		163	31.50		194.50	226
0930	3-1/2" diameter		11	.727		495	40		535	605
0940	4" diameter		8	1		690	54.50		744.50	840
0960	Couplings, to flexible conduit, 1/2" diameter		50	.160		1.48	8.75		10.23	14.75
0970	3/4" diameter		40	.200		2.53	10.95		13.48	19.20
0980	1" diameter		35	.229		4.42	12.50		16.92	23.50
0990	1-1/4" diameter		28	.286		9.70	15.65		25.35	34
1000	1-1/2" diameter		23	.348		12.50	19.05		31.55	42.50
1010	2" diameter		20	.400		25.50	22		47.50	61
1020	2-1/2" diameter		18	.444		35	24.50		59.50	75
1030	3" diameter		15	.533		88	29		117	141
1032	Aluminum, 3/8" diameter		210	.038	L.F.	.44	2.08		2.52	3.60
1034	1/2" diameter		210	.038		.51	2.08		2.59	3.68
1036	3/4" diameter		165	.048		.71	2.65		3.36	4.75
1038	1" diameter		105	.076		1.33	4.17		5.50	7.70
1040	1-1/4" diameter		75	.107		1.85	5.85		7.70	10.80
1042	1-1/2" diameter		53	.151		2.65	8.25		10.90	15.25
1044	2" diameter		42	.190		3.38	10.40		13.78	19.30
1046	2-1/2" diameter		32	.250		4.35	13.70		18.05	25.50
1048	3" diameter	2 Elec	53	.302		8.20	16.50		24.70	33.50
1050	3-1/2" diameter		42	.381		8.70	21		29.70	40.50
1052	4" diameter		32	.500		9.15	27.50		36.65	51
1070	Sealtite, 3/8" diameter	1 Elec	140	.057		1.05	3.13		4.18	5.85
1080	1/2" diameter		140	.057		1.13	3.13		4.26	5.90
1090	3/4" diameter		100	.080		1.60	4.38		5.98	8.30
1100	1" diameter		70	.114		2.42	6.25		8.67	12
1200	1-1/4" diameter		50	.160		3.30	8.75		12.05	16.75
1300	1-1/2" diameter		40	.200		3.81	10.95		14.76	20.50
1400	2" diameter		30	.267		4.82	14.60		19.42	27.50
1410	2-1/2" diameter		27	.296		8.20	16.20		24.40	33.50
1420	3" diameter	2 Elec	50	.320		11.35	17.50		28.85	38.50
1440	4" diameter	"	30	.533		17.15	29		46.15	62.50
1490	Connectors, plain, 3/8" diameter	1 Elec	70	.114	Ea.	2.76	6.25		9.01	12.40
1500	1/2" diameter		70	.114		3.77	6.25		10.02	13.50
1700	3/4" diameter		50	.160		5.55	8.75		14.30	19.20
1900	1" diameter		40	.200		9.75	10.95		20.70	27

26 05 33.35 Flexible Metallic Conduit		Crew	Daily Output	Labor-Hours	Unit	Material	2015 Bare Costs Labor	Equipment	Total	Total Incl O&P
1910	Insulated, 1" diameter	1 Elec	40	.200	Ea.	12.40	10.95		23.35	30
2000	1-1/4" diameter		32	.250		19.35	13.70		33.05	42
2100	1-1/2" diameter		27	.296		25.50	16.20		41.70	52.50
2200	2" diameter		20	.400		45	22		67	83
2210	2-1/2" diameter		15	.533		266	29		295	335
2220	3" diameter		12	.667		297	36.50		333.50	380
2240	4" diameter		8	1		370	54.50		424.50	485
2290	Connectors, 90°, 3/8" diameter		70	.114		4.65	6.25		10.90	14.45
2300	1/2" diameter		70	.114		5.70	6.25		11.95	15.60
2400	3/4" diameter		50	.160		6.70	8.75		15.45	20.50
2600	1" diameter		40	.200		13.65	10.95		24.60	31.50
2790	Insulated, 1" diameter		40	.200		22	10.95		32.95	40.50
2800	1-1/4" diameter		32	.250		35.50	13.70		49.20	60
3000	1-1/2" diameter		27	.296		44	16.20		60.20	73
3100	2" diameter		20	.400		61	22		83	101
3110	2-1/2" diameter		14	.571		288	31.50		319.50	360
3120	3" diameter		11	.727		335	40		375	430
3140	4" diameter		7	1.143		430	62.50		492.50	570
4300	Coupling sealtite to rigid, 1/2" diameter		20	.400		3.58	22		25.58	37
4500	3/4" diameter		18	.444		5.25	24.50		29.75	42.50
4800	1" diameter		14	.571		6.90	31.50		38.40	54.50
4900	1-1/4" diameter		12	.667		11.50	36.50		48	67
5000	1-1/2" diameter		11	.727		34.50	40		74.50	97.50
5100	2" diameter		10	.800		35.50	44		79.50	105
5110	2-1/2" diameter		9.50	.842		169	46		215	255
5120	3" diameter		9	.889		187	48.50		235.50	279
5130	3-1/2" diameter		9	.889		209	48.50		257.50	305
5140	4" diameter	↓	8.50	.941	↓	232	51.50		283.50	330

26 05 33.95 Cutting and Drilling		Crew	Daily Output	Labor-Hours	Unit	Material	Labor	Equipment	Total	Total Incl O&P
0010	**CUTTING AND DRILLING**									
0100	Hole drilling to 10' high, concrete wall									
0110	8" thick, 1/2" pipe size	R-31	12	.667	Ea.	.24	36.50	4.59	41.33	60
0120	3/4" pipe size		12	.667		.24	36.50	4.59	41.33	60
0130	1" pipe size		9.50	.842		.38	46	5.80	52.18	76
0140	1-1/4" pipe size		9.50	.842		.38	46	5.80	52.18	76
0150	1-1/2" pipe size		9.50	.842		.38	46	5.80	52.18	76
0160	2" pipe size		4.40	1.818		.51	99.50	12.55	112.56	163
0170	2-1/2" pipe size		4.40	1.818		.51	99.50	12.55	112.56	163
0180	3" pipe size		4.40	1.818		.51	99.50	12.55	112.56	163
0190	3-1/2" pipe size		3.30	2.424		.66	133	16.70	150.36	218
0200	4" pipe size		3.30	2.424		.66	133	16.70	150.36	218
0500	12" thick, 1/2" pipe size		9.40	.851		.35	46.50	5.85	52.70	76.50
0520	3/4" pipe size		9.40	.851		.35	46.50	5.85	52.70	76.50
0540	1" pipe size		7.30	1.096		.56	60	7.55	68.11	99
0560	1-1/4" pipe size		7.30	1.096		.56	60	7.55	68.11	99
0570	1-1/2" pipe size		7.30	1.096		.56	60	7.55	68.11	99
0580	2" pipe size		3.60	2.222		.76	122	15.30	138.06	200
0590	2-1/2" pipe size		3.60	2.222		.76	122	15.30	138.06	200
0600	3" pipe size		3.60	2.222		.76	122	15.30	138.06	200
0610	3-1/2" pipe size		2.80	2.857		.99	156	19.70	176.69	257
0630	4" pipe size		2.50	3.200		.99	175	22	197.99	288
0650	16" thick, 1/2" pipe size	↓	7.60	1.053	↓	.47	57.50	7.25	65.22	95

26 05 33.95 Cutting and Drilling		Crew	Daily Output	Labor-Hours	Unit	Material	2015 Bare Costs Labor	Equipment	Total	Total Incl O&P
0670	3/4" pipe size	R-31	7	1.143	Ea.	.47	62.50	7.85	70.82	103
0690	1" pipe size		6	1.333		.75	73	9.20	82.95	120
0710	1-1/4" pipe size		5.50	1.455		.75	79.50	10	90.25	131
0730	1-1/2" pipe size		5.50	1.455		.75	79.50	10	90.25	131
0750	2" pipe size		3	2.667		1.02	146	18.35	165.37	240
0770	2-1/2" pipe size		2.70	2.963		1.02	162	20.50	183.52	267
0790	3" pipe size		2.50	3.200		1.02	175	22	198.02	288
0810	3-1/2" pipe size		2.30	3.478		1.32	190	24	215.32	315
0830	4" pipe size		2	4		1.32	219	27.50	247.82	360
0850	20" thick, 1/2" pipe size		6.40	1.250		.59	68.50	8.60	77.69	112
0870	3/4" pipe size		6	1.333		.59	73	9.20	82.79	120
0890	1" pipe size		5	1.600		.94	87.50	11	99.44	144
0910	1-1/4" pipe size		4.80	1.667		.94	91	11.50	103.44	151
0930	1-1/2" pipe size		4.60	1.739		.94	95	12	107.94	157
0950	2" pipe size		2.70	2.963		1.27	162	20.50	183.77	267
0970	2-1/2" pipe size		2.40	3.333		1.27	182	23	206.27	300
0990	3" pipe size		2.20	3.636		1.27	199	25	225.27	325
1010	3-1/2" pipe size		2	4		1.64	219	27.50	248.14	360
1030	4" pipe size		1.70	4.706		1.64	257	32.50	291.14	420
1050	24" thick, 1/2" pipe size		5.50	1.455		.71	79.50	10	90.21	131
1070	3/4" pipe size		5.10	1.569		.71	86	10.80	97.51	142
1090	1" pipe size		4.30	1.860		1.13	102	12.80	115.93	167
1110	1-1/4" pipe size		4	2		1.13	109	13.80	123.93	180
1130	1-1/2" pipe size		4	2		1.13	109	13.80	123.93	180
1150	2" pipe size		2.40	3.333		1.52	182	23	206.52	300
1170	2-1/2" pipe size		2.20	3.636		1.52	199	25	225.52	325
1190	3" pipe size		2	4		1.52	219	27.50	248.02	360
1210	3-1/2" pipe size		1.80	4.444		1.97	243	30.50	275.47	400
1230	4" pipe size		1.50	5.333		1.97	292	37	330.97	480
1500	Brick wall, 8" thick, 1/2" pipe size		18	.444		.24	24.50	3.06	27.80	40
1520	3/4" pipe size		18	.444		.24	24.50	3.06	27.80	40
1540	1" pipe size		13.30	.602		.38	33	4.14	37.52	54.50
1560	1-1/4" pipe size		13.30	.602		.38	33	4.14	37.52	54.50
1580	1-1/2" pipe size		13.30	.602		.38	33	4.14	37.52	54.50
1600	2" pipe size		5.70	1.404		.51	77	9.65	87.16	126
1620	2-1/2" pipe size		5.70	1.404		.51	77	9.65	87.16	126
1640	3" pipe size		5.70	1.404		.51	77	9.65	87.16	126
1660	3-1/2" pipe size		4.40	1.818		.66	99.50	12.55	112.71	164
1680	4" pipe size		4	2		.66	109	13.80	123.46	180
1700	12" thick, 1/2" pipe size		14.50	.552		.35	30	3.80	34.15	49.50
1720	3/4" pipe size		14.50	.552		.35	30	3.80	34.15	49.50
1740	1" pipe size		11	.727		.56	40	5	45.56	65.50
1760	1-1/4" pipe size		11	.727		.56	40	5	45.56	65.50
1780	1-1/2" pipe size		11	.727		.56	40	5	45.56	65.50
1800	2" pipe size		5	1.600		.76	87.50	11	99.26	144
1820	2-1/2" pipe size		5	1.600		.76	87.50	11	99.26	144
1840	3" pipe size		5	1.600		.76	87.50	11	99.26	144
1860	3-1/2" pipe size		3.80	2.105		.99	115	14.50	130.49	190
1880	4" pipe size		3.30	2.424		.99	133	16.70	150.69	218
1900	16" thick, 1/2" pipe size		12.30	.650		.47	35.50	4.48	40.45	59
1920	3/4" pipe size		12.30	.650		.47	35.50	4.48	40.45	59
1940	1" pipe size		9.30	.860		.75	47	5.95	53.70	78
1960	1-1/4" pipe size		9.30	.860		.75	47	5.95	53.70	78

26 05 33 – Raceway and Boxes for Electrical Systems

26 05 33.95 Cutting and Drilling		Crew	Daily Output	Labor-Hours	Unit	Material	2015 Bare Costs Labor	Equipment	Total	Total Incl O&P
1980	1-1/2" pipe size	R-31	9.30	.860	Ea.	.75	47	5.95	53.70	78
2000	2" pipe size		4.40	1.818		1.02	99.50	12.55	113.07	164
2010	2-1/2" pipe size		4.40	1.818		1.02	99.50	12.55	113.07	164
2030	3" pipe size		4.40	1.818		1.02	99.50	12.55	113.07	164
2050	3-1/2" pipe size		3.30	2.424		1.32	133	16.70	151.02	219
2070	4" pipe size		3	2.667		1.32	146	18.35	165.67	240
2090	20" thick, 1/2" pipe size		10.70	.748		.59	41	5.15	46.74	68
2110	3/4" pipe size		10.70	.748		.59	41	5.15	46.74	68
2130	1" pipe size		8	1		.94	54.50	6.90	62.34	90.50
2150	1-1/4" pipe size		8	1		.94	54.50	6.90	62.34	90.50
2170	1-1/2" pipe size		8	1		.94	54.50	6.90	62.34	90.50
2190	2" pipe size		4	2		1.27	109	13.80	124.07	181
2210	2-1/2" pipe size		4	2		1.27	109	13.80	124.07	181
2230	3" pipe size		4	2		1.27	109	13.80	124.07	181
2250	3-1/2" pipe size		3	2.667		1.64	146	18.35	165.99	241
2270	4" pipe size		2.70	2.963		1.64	162	20.50	184.14	267
2290	24" thick, 1/2" pipe size		9.40	.851		.71	46.50	5.85	53.06	76.50
2310	3/4" pipe size		9.40	.851		.71	46.50	5.85	53.06	76.50
2330	1" pipe size		7.10	1.127		1.13	61.50	7.75	70.38	102
2350	1-1/4" pipe size		7.10	1.127		1.13	61.50	7.75	70.38	102
2370	1-1/2" pipe size		7.10	1.127		1.13	61.50	7.75	70.38	102
2390	2" pipe size		3.60	2.222		1.52	122	15.30	138.82	201
2410	2-1/2" pipe size		3.60	2.222		1.52	122	15.30	138.82	201
2430	3" pipe size		3.60	2.222		1.52	122	15.30	138.82	201
2450	3-1/2" pipe size		2.80	2.857		1.97	156	19.70	177.67	258
2470	4" pipe size		2.50	3.200		1.97	175	22	198.97	289
3000	Knockouts to 8' high, metal boxes & enclosures									
3020	With hole saw, 1/2" pipe size	1 Elec	53	.151	Ea.		8.25		8.25	12.35
3040	3/4" pipe size		47	.170			9.30		9.30	13.95
3050	1" pipe size		40	.200			10.95		10.95	16.40
3060	1-1/4" pipe size		36	.222			12.15		12.15	18.20
3070	1-1/2" pipe size		32	.250			13.70		13.70	20.50
3080	2" pipe size		27	.296			16.20		16.20	24.50
3090	2-1/2" pipe size		20	.400			22		22	33
4010	3" pipe size		16	.500			27.50		27.50	41
4030	3-1/2" pipe size		13	.615			33.50		33.50	50.50
4050	4" pipe size		11	.727			40		40	59.50

26 05 36 – Cable Trays for Electrical Systems

26 05 36.10 Cable Tray Ladder Type		Crew	Daily Output	Labor-Hours	Unit	Material	Labor	Equipment	Total	Total Incl O&P
0010	**CABLE TRAY LADDER TYPE** w/ftngs. & supports, 4" dp., to 15' elev.									
0100	For higher elevations, see Section 26 05 36.40									
0160	Galvanized steel tray R260536-10									
0170	4" rung spacing, 6" wide	2 Elec	98	.163	L.F.	18.15	8.95		27.10	33.50
0180	9" wide R260536-11		92	.174		19.80	9.50		29.30	36.50
0200	12" wide		86	.186		22	10.20		32.20	39.50
0400	18" wide		82	.195		25.50	10.65		36.15	44
0600	24" wide		78	.205		29	11.20		40.20	49
0650	30" wide		68	.235		37	12.85		49.85	60
0700	36" wide		60	.267		41	14.60		55.60	67
0800	6" rung spacing, 6" wide		100	.160		16.50	8.75		25.25	31.50
0850	9" wide		94	.170		18.15	9.30		27.45	34
0860	12" wide		88	.182		19.05	9.95		29	36

26 05 36.10 Cable Tray Ladder Type		Crew	Daily Output	Labor-Hours	Unit	2015 Bare Costs Material	Labor	Equipment	Total	Total Incl O&P
0870	18″ wide	2 Elec	84	.190	L.F.	21.50	10.40		31.90	39.50
0880	24″ wide		80	.200		24.50	10.95		35.45	43
0890	30″ wide		70	.229		29	12.50		41.50	51
0900	36″ wide		64	.250		31.50	13.70		45.20	55
0910	9″ rung spacing, 6″ wide		102	.157		15.55	8.60		24.15	30
0920	9″ wide		98	.163		15.90	8.95		24.85	31
0930	12″ wide		94	.170		18.30	9.30		27.60	34
0940	18″ wide		90	.178		20.50	9.70		30.20	37
0950	24″ wide		86	.186		23.50	10.20		33.70	41.50
0960	30″ wide		80	.200		25.50	10.95		36.45	45
0970	36″ wide		74	.216		27.50	11.85		39.35	47.50
0980	12″ rung spacing, 6″ wide		106	.151		15.30	8.25		23.55	29
0990	9″ wide		104	.154		15.65	8.40		24.05	30
1000	12″ wide		100	.160		16.50	8.75		25.25	31.50
1010	18″ wide		96	.167		17.75	9.10		26.85	33
1020	24″ wide		94	.170		19.05	9.30		28.35	35
1030	30″ wide		88	.182		21.50	9.95		31.45	39
1040	36″ wide		84	.190		23	10.40		33.40	40.50
1041	18″ rung spacing, 6″ wide		108	.148		15.15	8.10		23.25	29
1042	9″ wide		106	.151		15.40	8.25		23.65	29.50
1043	12″ wide		102	.157		16.40	8.60		25	31
1044	18″ wide		98	.163		17.25	8.95		26.20	32.50
1045	24″ wide		96	.167		17.75	9.10		26.85	33
1046	30″ wide		90	.178		18.70	9.70		28.40	35
1047	36″ wide		86	.186		19.25	10.20		29.45	36.50
1050	Elbows horiz. 9″ rung spacing, 90°, 12″ radius, 6″ wide		9.60	1.667	Ea.	70.50	91		161.50	215
1060	9″ wide		8.40	1.905		75	104		179	239
1070	12″ wide		7.60	2.105		81.50	115		196.50	263
1080	18″ wide		6.20	2.581		107	141		248	330
1090	24″ wide		5.40	2.963		120	162		282	375
1100	30″ wide		4.80	3.333		158	182		340	445
1110	36″ wide		4.20	3.810		184	208		392	510
1120	90° 24″ radius, 6″ wide		9.20	1.739		146	95		241	305
1130	9″ wide		8	2		150	109		259	330
1140	12″ wide		7.20	2.222		154	122		276	350
1150	18″ wide		5.80	2.759		167	151		318	410
1160	24″ wide		5	3.200		180	175		355	460
1170	30″ wide		4.40	3.636		195	199		394	510
1180	36″ wide		3.80	4.211		231	230		461	600
1190	90° 36″ radius, 6″ wide		8.80	1.818		161	99.50		260.50	325
1200	9″ wide		7.60	2.105		169	115		284	360
1210	12″ wide		6.80	2.353		178	129		307	390
1220	18″ wide		5.40	2.963		203	162		365	465
1230	24″ wide		4.60	3.478		223	190		413	530
1240	30″ wide		4	4		268	219		487	625
1250	36″ wide		3.40	4.706		291	257		548	705
1260	45° 12″ radius, 6″ wide		13.20	1.212		54.50	66.50		121	160
1270	9″ wide		11	1.455		56.50	79.50		136	182
1280	12″ wide		9.60	1.667		59	91		150	202
1290	18″ wide		7.60	2.105		62.50	115		177.50	242
1300	24″ wide		6.20	2.581		80	141		221	299
1310	30″ wide		5.40	2.963		102	162		264	355
1320	36″ wide		4.60	3.478		107	190		297	400

26 05 36.10 Cable Tray Ladder Type		Crew	Daily Output	Labor-Hours	Unit	Material	2015 Bare Costs Labor	Equipment	Total	Total Incl O&P
1330	45° 24" radius, 6" wide	2 Elec	12.80	1.250	Ea.	69	68.50		137.50	178
1340	9" wide		10.60	1.509		73.50	82.50		156	205
1350	12" wide		9.20	1.739		82.50	95		177.50	234
1360	18" wide		7.20	2.222		84.50	122		206.50	275
1370	24" wide		5.80	2.759		102	151		253	340
1380	30" wide		5	3.200		122	175		297	395
1390	36" wide		4.20	3.810		140	208		348	465
1400	45° 36" radius, 6" wide		12.40	1.290		99	70.50		169.50	215
1410	9" wide		10.20	1.569		105	86		191	244
1420	12" wide		8.80	1.818		107	99.50		206.50	266
1430	18" wide		6.80	2.353		116	129		245	320
1440	24" wide		5.40	2.963		122	162		284	380
1450	30" wide		4.60	3.478		158	190		348	460
1460	36" wide		3.80	4.211		176	230		406	540
1470	Elbows horizontal, 4" rung spacing, use 9" rung x 1.50									
1480	6" rung spacing use 9" rung x 1.20									
1490	12" rung spacing use 9" rung x .93									
1500	Elbows vert. 9" rung spacing, 90°, 12" radius, 6" wide	2 Elec	9.60	1.667	Ea.	97	91		188	244
1510	9" wide		8.40	1.905		99	104		203	265
1520	12" wide		7.60	2.105		102	115		217	286
1530	18" wide		6.20	2.581		107	141		248	330
1540	24" wide		5.40	2.963		111	162		273	365
1550	30" wide		4.80	3.333		120	182		302	405
1560	36" wide		4.20	3.810		124	208		332	445
1570	24" radius, 6" wide		9.20	1.739		137	95		232	294
1580	9" wide		8	2		141	109		250	320
1590	12" wide		7.20	2.222		146	122		268	340
1600	18" wide		5.80	2.759		154	151		305	395
1610	24" wide		5	3.200		163	175		338	440
1620	30" wide		4.40	3.636		169	199		368	485
1630	36" wide		3.80	4.211		191	230		421	555
1640	36" radius, 6" wide		8.80	1.818		184	99.50		283.50	350
1650	9" wide		7.60	2.105		191	115		306	385
1660	12" wide		6.80	2.353		193	129		322	405
1670	18" wide		5.40	2.963		210	162		372	475
1680	24" wide		4.60	3.478		212	190		402	520
1690	30" wide		4	4		240	219		459	595
1700	36" wide		3.40	4.706		257	257		514	670
1710	Elbows vertical, 4" rung spacing, use 9" rung x 1.25									
1720	6" rung spacing, use 9" rung x 1.15									
1730	12" rung spacing, use 9" rung x .90									
1740	Tee horizontal, 9" rung spacing, 12" radius, 6" wide	2 Elec	5	3.200	Ea.	150	175		325	425
1750	9" wide		4.60	3.478		161	190		351	460
1760	12" wide		4.40	3.636		169	199		368	485
1770	18" wide		4	4		193	219		412	540
1780	24" wide		3.60	4.444		227	243		470	615
1790	30" wide		3.40	4.706		272	257		529	685
1800	36" wide		3	5.333		310	292		602	775
1810	24" radius, 6" wide		4.60	3.478		233	190		423	540
1820	9" wide		4.20	3.810		244	208		452	580
1830	12" wide		4	4		253	219		472	610
1840	18" wide		3.60	4.444		274	243		517	665
1850	24" wide		3.20	5		298	274		572	735

26 05 36.10 Cable Tray Ladder Type		Crew	Daily Output	Labor-Hours	Unit	Material	Labor	Equipment	Total	Total Incl O&P
						2015 Bare Costs				
1860	30" wide	2 Elec	3	5.333	Ea.	325	292		617	790
1870	36" wide		2.60	6.154		415	335		750	960
1880	36" radius, 6" wide		4.20	3.810		350	208		558	695
1890	9" wide		3.80	4.211		365	230		595	745
1900	12" wide		3.60	4.444		385	243		628	790
1910	18" wide		3.20	5		415	274		689	865
1920	24" wide		2.80	5.714		470	315		785	985
1930	30" wide		2.60	6.154		510	335		845	1,075
1940	36" wide		2.20	7.273		550	400		950	1,200
1980	Tee vertical, 9" rung spacing, 12" radius, 6" wide		5.40	2.963		255	162		417	525
1990	9" wide		5.20	3.077		257	168		425	535
2000	12" wide		5	3.200		259	175		434	545
2010	18" wide		4.60	3.478		261	190		451	570
2020	24" wide		4.40	3.636		268	199		467	590
2030	30" wide		4	4		285	219		504	645
2040	36" wide		3.60	4.444		295	243		538	690
2050	24" radius, 6" wide		5	3.200		460	175		635	765
2060	9" wide		4.80	3.333		465	182		647	785
2070	12" wide		4.60	3.478		470	190		660	805
2080	18" wide		4.20	3.810		485	208		693	845
2090	24" wide		4	4		500	219		719	880
2100	30" wide		3.60	4.444		510	243		753	930
2110	36" wide		3.20	5		535	274		809	1,000
2120	36" radius, 6" wide		4.60	3.478		855	190		1,045	1,225
2130	9" wide		4.40	3.636		870	199		1,069	1,250
2140	12" wide		4.20	3.810		880	208		1,088	1,275
2150	18" wide		3.80	4.211		890	230		1,120	1,325
2160	24" wide		3.60	4.444		900	243		1,143	1,350
2170	30" wide		3.20	5		935	274		1,209	1,425
2180	36" wide		2.80	5.714		945	315		1,260	1,525
2190	Tee, 4" rung spacing, use 9" rung x 1.30									
2200	6" rung spacing, use 9" rung x 1.20									
2210	12" rung spacing, use 9" rung x .90									
2220	Cross horizontal, 9" rung spacing, 12" radius, 6" wide	2 Elec	4	4	Ea.	185	219		404	535
2230	9" wide		3.80	4.211		194	230		424	560
2240	12" wide		3.60	4.444		212	243		455	600
2250	18" wide		3.40	4.706		231	257		488	640
2260	24" wide		3	5.333		257	292		549	720
2270	30" wide		2.80	5.714		310	315		625	810
2280	36" wide		2.60	6.154		345	335		680	885
2290	24" radius, 6" wide		3.60	4.444		360	243		603	760
2300	9" wide		3.40	4.706		370	257		627	790
2310	12" wide		3.20	5		380	274		654	830
2320	18" wide		3	5.333		405	292		697	880
2330	24" wide		2.60	6.154		430	335		765	980
2340	30" wide		2.40	6.667		460	365		825	1,050
2350	36" wide		2.20	7.273		535	400		935	1,175
2360	36" radius, 6" wide		3.20	5		500	274		774	960
2370	9" wide		3	5.333		520	292		812	1,000
2380	12" wide		2.80	5.714		545	315		860	1,075
2390	18" wide		2.60	6.154		600	335		935	1,175
2400	24" wide		2.20	7.273		670	400		1,070	1,325
2410	30" wide		2	8		725	440		1,165	1,450

26 05 36.10 Cable Tray Ladder Type		Crew	Daily Output	Labor-Hours	Unit	2015 Bare Costs Material	Labor	Equipment	Total	Total Incl O&P
2420	36" wide	2 Elec	1.80	8.889	Ea.	780	485		1,265	1,575
2430	Cross horizontal, 4" rung spacing, use 9" rung x 1.30									
2440	6" rung spacing, use 9" rung x 1.20									
2450	12" rung spacing, use 9" rung x .90									
2460	Reducer, 9" to 6" wide tray	2 Elec	13	1.231	Ea.	103	67.50		170.50	214
2470	12" to 9" wide tray		12	1.333		105	73		178	224
2480	18" to 12" wide tray		10.40	1.538		105	84		189	241
2490	24" to 18" wide tray		9	1.778		107	97		204	264
2500	30" to 24" wide tray		8	2		109	109		218	284
2510	36" to 30" wide tray		7	2.286		122	125		247	320
2511	Reducer, 18" to 6" wide tray		10.40	1.538		109	84		193	246
2512	24" to 12" wide tray		9	1.778		111	97		208	268
2513	30" to 18" wide tray		8	2		116	109		225	291
2514	30" to 12" wide tray		8	2		120	109		229	296
2515	36" to 24" wide tray		7	2.286		124	125		249	325
2516	36" to 18" wide tray		7	2.286		124	125		249	325
2517	36" to 12" wide tray		7	2.286		124	125		249	325
2520	Dropout or end plate, 6" wide		32	.500		11.45	27.50		38.95	53.50
2530	9" wide		28	.571		13.50	31.50		45	62
2540	12" wide		26	.615		14.75	33.50		48.25	67
2550	18" wide		22	.727		17.35	40		57.35	78.50
2560	24" wide		20	.800		21	44		65	88.50
2570	30" wide		18	.889		23	48.50		71.50	98
2580	36" wide		16	1		26.50	54.50		81	111
2590	Tray connector		48	.333		20.50	18.25		38.75	50
3200	Aluminum tray, 4" deep, 6" rung spacing, 6" wide		134	.119	L.F.	18	6.55		24.55	29.50
3210	9" wide		128	.125		19.10	6.85		25.95	31.50
3220	12" wide		124	.129		20	7.05		27.05	32.50
3230	18" wide		114	.140		22.50	7.70		30.20	36
3240	24" wide		106	.151		26	8.25		34.25	41
3250	30" wide		100	.160		29	8.75		37.75	44.50
3260	36" wide		94	.170		31	9.30		40.30	48.50
3270	9" rung spacing, 6" wide		140	.114		14.65	6.25		20.90	25.50
3280	9" wide		134	.119		15.35	6.55		21.90	26.50
3290	12" wide		130	.123		15.60	6.75		22.35	27.50
3300	18" wide		122	.131		18.20	7.15		25.35	31
3310	24" wide		116	.138		21	7.55		28.55	35
3320	30" wide		108	.148		23.50	8.10		31.60	38
3330	36" wide		100	.160		25	8.75		33.75	40.50
3340	12" rung spacing, 6" wide		146	.110		15.50	6		21.50	26
3350	9" wide		140	.114		15.70	6.25		21.95	26.50
3360	12" wide		134	.119		16.20	6.55		22.75	27.50
3370	18" wide		128	.125		17.15	6.85		24	29
3380	24" wide		124	.129		19.10	7.05		26.15	31.50
3390	30" wide		114	.140		20.50	7.70		28.20	34
3400	36" wide		106	.151		21.50	8.25		29.75	36
3401	18" rung spacing, 6" wide		150	.107		15.40	5.85		21.25	25.50
3402	9" wide tray		144	.111		15.60	6.10		21.70	26.50
3403	12" wide tray		140	.114		16.20	6.25		22.45	27
3404	18" wide tray		134	.119		17.05	6.55		23.60	28.50
3405	24" wide tray		130	.123		18.35	6.75		25.10	30
3406	30" wide tray		120	.133		19.05	7.30		26.35	32
3407	36" wide tray		110	.145		19.90	7.95		27.85	34

26 05 36.10 Cable Tray Ladder Type		Crew	Daily Output	Labor-Hours	Unit	2015 Bare Costs Material	Labor	Equipment	Total	Total Incl O&P
3410	Elbows horiz. 9" rung spacing, 90° 12" radius, 6" wide	2 Elec	9.60	1.667	Ea.	69.50	91		160.50	213
3420	9" wide		8.40	1.905		71	104		175	235
3430	12" wide		7.60	2.105		84.50	115		199.50	266
3440	18" wide		6.20	2.581		95	141		236	315
3450	24" wide		5.40	2.963		118	162		280	375
3460	30" wide		4.80	3.333		125	182		307	410
3470	36" wide		4.20	3.810		150	208		358	475
3480	24" radius, 6" wide		9.20	1.739		123	95		218	279
3490	9" wide		8	2		128	109		237	305
3500	12" wide		7.20	2.222		134	122		256	330
3510	18" wide		5.80	2.759		146	151		297	385
3520	24" wide		5	3.200		161	175		336	440
3530	30" wide		4.40	3.636		174	199		373	490
3540	36" wide		3.80	4.211		198	230		428	565
3550	90°, 36" radius, 6" wide		8.80	1.818		135	99.50		234.50	298
3560	9" wide		7.60	2.105		139	115		254	325
3570	12" wide		6.80	2.353		155	129		284	365
3580	18" wide		5.40	2.963		174	162		336	435
3590	24" wide		4.60	3.478		196	190		386	500
3600	30" wide		4	4		216	219		435	570
3610	36" wide		3.40	4.706		248	257		505	660
3620	45° 12" radius, 6" wide		13.20	1.212		45.50	66.50		112	150
3630	9" wide		11	1.455		47.50	79.50		127	171
3640	12" wide		9.60	1.667		49	91		140	191
3650	18" wide		7.60	2.105		58	115		173	237
3660	24" wide		6.20	2.581		72	141		213	290
3670	30" wide		5.40	2.963		83	162		245	335
3680	36" wide		4.60	3.478		86	190		276	380
3690	45° 24" radius, 6" wide		12.80	1.250		61	68.50		129.50	169
3700	9" wide		10.60	1.509		69.50	82.50		152	200
3710	12" wide		9.20	1.739		71	95		166	221
3720	18" wide		7.20	2.222		76	122		198	266
3730	24" wide		5.80	2.759		88	151		239	325
3740	30" wide		5	3.200		105	175		280	375
3750	36" wide		4.20	3.810		112	208		320	435
3760	45° 36" radius, 6" wide		12.40	1.290		83	70.50		153.50	197
3770	9" wide		10.20	1.569		84.50	86		170.50	222
3780	12" wide		8.80	1.818		86	99.50		185.50	244
3790	18" wide		6.80	2.353		94.50	129		223.50	297
3800	24" wide		5.40	2.963		117	162		279	370
3810	30" wide		4.60	3.478		123	190		313	420
3820	36" wide		3.80	4.211		135	230		365	495
3830	Elbows horizontal, 4" rung spacing, use 9" rung x 1.50									
3840	6" rung spacing, use 9" rung x 1.20									
3850	12" rung spacing, use 9" rung x .93									
3860	Elbows vertical 9" rung spacing, 90° 12" radius, 6" wide	2 Elec	9.60	1.667	Ea.	96.50	91		187.50	243
3870	9" wide		8.40	1.905		101	104		205	268
3880	12" wide		7.60	2.105		103	115		218	286
3890	18" wide		6.20	2.581		106	141		247	330
3900	24" wide		5.40	2.963		113	162		275	370
3910	30" wide		4.80	3.333		118	182		300	405
3920	36" wide		4.20	3.810		122	208		330	445
3930	24" radius, 6" wide		9.20	1.739		105	95		200	258

26 05 36.10 Cable Tray Ladder Type		Crew	Daily Output	Labor-Hours	Unit	Material	Labor	Equipment	Total	Total Incl O&P
3940	9″ wide	2 Elec	8	2	Ea.	108	109		217	283
3950	12″ wide		7.20	2.222		112	122		234	305
3960	18″ wide		5.80	2.759		118	151		269	355
3970	24″ wide		5	3.200		125	175		300	400
3980	30″ wide		4.40	3.636		132	199		331	445
3990	36″ wide		3.80	4.211		140	230		370	500
4000	36″ radius, 6″ wide		8.80	1.818		162	99.50		261.50	325
4010	9″ wide		7.60	2.105		166	115		281	355
4020	12″ wide		6.80	2.353		176	129		305	385
4030	18″ wide		5.40	2.963		183	162		345	445
4040	24″ wide		4.60	3.478		189	190		379	495
4050	30″ wide		4	4		205	219		424	555
4060	36″ wide		3.40	4.706		213	257		470	620
4070	Elbows vertical, 4″ rung spacing, use 9″ rung x 1.25									
4080	6″ rung spacing, use 9″ rung x 1.15									
4090	12″ rung spacing, use 9″ rung x .90									
4100	Tee horizontal 9″ rung spacing, 12″ radius, 6″ wide	2 Elec	5	3.200	Ea.	107	175		282	380
4110	9″ wide		4.60	3.478		115	190		305	410
4120	12″ wide		4.40	3.636		127	199		326	440
4130	18″ wide		4.20	3.810		139	208		347	460
4140	24″ wide		4	4		152	219		371	495
4150	30″ wide		3.60	4.444		177	243		420	560
4160	36″ wide		3.40	4.706		213	257		470	620
4170	24″ radius, 6″ wide		4.60	3.478		183	190		373	485
4180	9″ wide		4.20	3.810		193	208		401	520
4190	12″ wide		4	4		198	219		417	550
4200	18″ wide		3.80	4.211		216	230		446	585
4210	24″ wide		3.60	4.444		237	243		480	625
4220	30″ wide		3.20	5		257	274		531	695
4230	36″ wide		3	5.333		395	292		687	870
4240	36″ radius, 6″ wide		4.20	3.810		259	208		467	595
4250	9″ wide		3.80	4.211		265	230		495	635
4260	12″ wide		3.60	4.444		291	243		534	685
4270	18″ wide		3.40	4.706		305	257		562	720
4280	24″ wide		3.20	5		340	274		614	780
4290	30″ wide		2.80	5.714		430	315		745	940
4300	36″ wide		2.60	6.154		475	335		810	1,025
4310	Tee vertical 9″ rung spacing, 12″ radius, 6″ wide		5.40	2.963		240	162		402	505
4320	9″ wide		5.20	3.077		248	168		416	525
4330	12″ wide		5	3.200		252	175		427	540
4340	18″ wide		4.60	3.478		255	190		445	565
4350	24″ wide		4.40	3.636		275	199		474	605
4360	30″ wide		4.20	3.810		281	208		489	620
4370	36″ wide		4	4		291	219		510	650
4380	24″ radius, 6″ wide		5	3.200		405	175		580	710
4390	9″ wide		4.80	3.333		410	182		592	730
4400	12″ wide		4.60	3.478		420	190		610	745
4410	18″ wide		4.20	3.810		430	208		638	780
4420	24″ wide		4	4		440	219		659	815
4430	30″ wide		3.80	4.211		455	230		685	845
4440	36″ wide		3.60	4.444		490	243		733	905
4450	36″ radius, 6″ wide		4.60	3.478		895	190		1,085	1,275
4460	9″ wide		4.40	3.636		915	199		1,114	1,300

(Material, Labor, Equipment and Total columns are 2015 Bare Costs.)

26 05 36 – Cable Trays for Electrical Systems

26 05 36.10 Cable Tray Ladder Type		Crew	Daily Output	Labor-Hours	Unit	Material	Labor	Equipment	Total	Total Incl O&P
						2015 Bare Costs				
4470	12″ wide	2 Elec	4.20	3.810	Ea.	950	208		1,158	1,350
4480	18″ wide		3.80	4.211		960	230		1,190	1,400
4490	24″ wide		3.60	4.444		985	243		1,228	1,450
4500	30″ wide		3.40	4.706		995	257		1,252	1,475
4510	36″ wide		3.20	5		1,000	274		1,274	1,500
4520	Tees, 4″ rung spacing, use 9″ rung x 1.30									
4530	6″ rung spacing, use 9″ rung x 1.20									
4540	12″ rung spacing, use 9″ rung x .90									
4550	Cross horizontal 9″ rung spacing, 12″ radius, 6″ wide	2 Elec	4.40	3.636	Ea.	149	199		348	460
4560	9″ wide		4.20	3.810		166	208		374	490
4570	12″ wide		4	4		184	219		403	535
4580	18″ wide		3.60	4.444		203	243		446	590
4590	24″ wide		3.40	4.706		210	257		467	615
4600	30″ wide		3	5.333		247	292		539	705
4610	36″ wide		2.80	5.714		310	315		625	810
4620	24″ radius, 6″ wide		4	4		286	219		505	645
4630	9″ wide		3.80	4.211		296	230		526	670
4640	12″ wide		3.60	4.444		305	243		548	700
4650	18″ wide		3.20	5		320	274		594	760
4660	24″ wide		3	5.333		350	292		642	820
4670	30″ wide		2.60	6.154		370	335		705	915
4680	36″ wide		2.40	6.667		420	365		785	1,000
4690	36″ radius, 6″ wide		3.60	4.444		340	243		583	735
4700	9″ wide		3.40	4.706		350	257		607	770
4710	12″ wide		3.20	5		370	274		644	820
4720	18″ wide		2.80	5.714		395	315		710	905
4730	24″ wide		2.60	6.154		490	335		825	1,050
4740	30″ wide		2.20	7.273		545	400		945	1,200
4750	36″ wide		2	8		605	440		1,045	1,325
4760	Cross horizontal, 4″ rung spacing, use 9″ rung x 1.30									
4770	6″ rung spacing, use 9″ rung x 1.20									
4780	12″ rung spacing, use 9″ rung x .90									
4790	Reducer, 9″ to 6″ wide tray	2 Elec	16	1	Ea.	79.50	54.50		134	170
4800	12″ to 9″ wide tray		14	1.143		83	62.50		145.50	185
4810	18″ to 12″ wide tray		12.40	1.290		83	70.50		153.50	197
4820	24″ to 18″ wide tray		10.60	1.509		84.50	82.50		167	217
4830	30″ to 24″ wide tray		9.20	1.739		86	95		181	238
4840	36″ to 30″ wide tray		8	2		91.50	109		200.50	264
4841	Reducer, 18″ to 6″ wide tray		12.40	1.290		83	70.50		153.50	197
4842	24″ to 12″ wide tray		10.60	1.509		84.50	82.50		167	217
4843	30″ to 18″ wide tray		9.20	1.739		86	95		181	238
4844	30″ to 12″ wide tray		9.20	1.739		86	95		181	238
4845	36″ to 24″ wide tray		8	2		91.50	109		200.50	264
4846	36″ to 18″ wide tray		8	2		91.50	109		200.50	264
4847	36″ to 12″ wide tray		8	2		91.50	109		200.50	264
4850	Dropout or end plate, 6″ wide		32	.500		10.45	27.50		37.95	52.50
4860	9″ wide tray		28	.571		11.15	31.50		42.65	59.50
4870	12″ wide tray		26	.615		12.15	33.50		45.65	64
4880	18″ wide tray		22	.727		16.25	40		56.25	77.50
4890	24″ wide tray		20	.800		19.10	44		63.10	86.50
4900	30″ wide tray		18	.889		22.50	48.50		71	98
4910	36″ wide tray		16	1		25	54.50		79.50	110
4920	Tray connector		48	.333		15.20	18.25		33.45	44.50

26 05 Common Work Results for Electrical

26 05 36 – Cable Trays for Electrical Systems

26 05 36.10 Cable Tray Ladder Type		Crew	Daily Output	Labor-Hours	Unit	Material	2015 Bare Costs Labor	Equipment	Total	Total Incl O&P
8000	Elbow 36″ radius horiz., 60°, 6″ wide tray	2 Elec	10.60	1.509	Ea.	115	82.50		197.50	250
8010	9″ wide tray		9	1.778		125	97		222	284
8020	12″ wide tray		7.80	2.051		128	112		240	310
8030	18″ wide tray		6.20	2.581		137	141		278	360
8040	24″ wide tray		5	3.200		161	175		336	440
8050	30″ wide tray		4.40	3.636		164	199		363	480
8060	30°, 6″ wide tray		14	1.143		88	62.50		150.50	190
8070	9″ wide tray		11.40	1.404		93	77		170	217
8080	12″ wide tray		9.80	1.633		94.50	89.50		184	238
8090	18″ wide tray		7.40	2.162		99.50	118		217.50	287
8100	24″ wide tray		5.80	2.759		122	151		273	360
8110	30″ wide tray		4.80	3.333		125	182		307	410
8120	Adjustable, 6″ wide tray		12.40	1.290		117	70.50		187.50	234
8130	9″ wide tray		10.20	1.569		123	86		209	265
8140	12″ wide tray		8.80	1.818		125	99.50		224.50	287
8150	18″ wide tray		6.80	2.353		137	129		266	345
8160	24″ wide tray		5.40	2.963		145	162		307	405
8170	30″ wide tray		4.60	3.478		162	190		352	465
8180	Wye 36″ radius horiz., 45°, 6″ wide tray		4.60	3.478		127	190		317	425
8190	9″ wide tray		4.40	3.636		134	199		333	445
8200	12″ wide tray		4.20	3.810		139	208		347	460
8210	18″ wide tray		3.80	4.211		157	230		387	520
8220	24″ wide tray		3.60	4.444		176	243		419	560
8230	30″ wide tray		3.40	4.706		196	257		453	600
8240	Elbow 36″ radius vert. in/outside, 60°, 6″ wide tray		10.60	1.509		127	82.50		209.50	263
8250	9″ wide tray		9	1.778		132	97		229	291
8260	12″ wide tray		7.80	2.051		134	112		246	315
8270	18″ wide tray		6.20	2.581		139	141		280	365
8280	24″ wide tray		5	3.200		147	175		322	425
8290	30″ wide tray		4.40	3.636		157	199		356	470
8300	45°, 6″ wide tray		12.40	1.290		114	70.50		184.50	232
8310	9″ wide tray		10.20	1.569		121	86		207	262
8320	12″ wide tray		8.80	1.818		123	99.50		222.50	284
8330	18″ wide tray		6.80	2.353		127	129		256	330
8340	24″ wide tray		5.40	2.963		134	162		296	390
8350	30″ wide tray		4.60	3.478		137	190		327	435
8360	30°, 6″ wide tray		14	1.143		100	62.50		162.50	204
8370	9″ wide tray		11.40	1.404		102	77		179	227
8380	12″ wide tray		9.80	1.633		109	89.50		198.50	254
8390	18″ wide tray		7.40	2.162		111	118		229	299
8400	24″ wide tray		5.80	2.759		113	151		264	350
8410	30″ wide tray		4.80	3.333		119	182		301	405
8660	Adjustable, 6″ wide tray		12.40	1.290		114	70.50		184.50	232
8670	9″ wide tray		10.20	1.569		121	86		207	262
8680	12″ wide tray		8.80	1.818		123	99.50		222.50	285
8690	18″ wide tray		6.80	2.353		137	129		266	345
8700	24″ wide tray		5.40	2.963		145	162		307	405
8710	30″ wide tray		4.60	3.478		161	190		351	460
8720	Cross, vertical, 24″ radius, 6″ wide tray		3.60	4.444		780	243		1,023	1,225
8730	9″ wide tray		3.40	4.706		795	257		1,052	1,250
8740	12″ wide tray		3.20	5		820	274		1,094	1,300
8750	18″ wide tray		2.80	5.714		855	315		1,170	1,400
8760	24″ wide tray	↓	2.60	6.154	↓	885	335		1,220	1,475

26 05 36 – Cable Trays for Electrical Systems

26 05 36.10 Cable Tray Ladder Type		Crew	Daily Output	Labor-Hours	Unit	Material	2015 Bare Costs Labor	Equipment	Total	Total Incl O&P
8770	30″ wide tray	2 Elec	2.20	7.273	Ea.	935	400		1,335	1,625
9200	Splice plate	1 Elec	48	.167	Pr.	15.85	9.10		24.95	31
9210	Expansion joint		48	.167		19.10	9.10		28.20	34.50
9220	Horizontal hinged		48	.167		15.85	9.10		24.95	31
9230	Vertical hinged		48	.167		19.10	9.10		28.20	34.50
9240	Ladder hanger, vertical		28	.286	Ea.	4.73	15.65		20.38	28.50
9250	Ladder to channel connector		24	.333		55.50	18.25		73.75	88.50
9260	Ladder to box connector 30″ wide		19	.421		55.50	23		78.50	95.50
9270	24″ wide		20	.400		55.50	22		77.50	94
9280	18″ wide		21	.381		53.50	21		74.50	90
9290	12″ wide		22	.364		50	19.90		69.90	85
9300	9″ wide		23	.348		45	19.05		64.05	78
9310	6″ wide		24	.333		40	18.25		58.25	71.50
9320	Ladder floor flange		24	.333		36	18.25		54.25	67
9330	Cable roller for tray 30″ wide		10	.800		295	44		339	390
9340	24″ wide		11	.727		251	40		291	335
9350	18″ wide		12	.667		239	36.50		275.50	320
9360	12″ wide		13	.615		188	33.50		221.50	258
9370	9″ wide		14	.571		167	31.50		198.50	230
9380	6″ wide		15	.533		135	29		164	193
9390	Pulley, single wheel		12	.667		310	36.50		346.50	395
9400	Triple wheel		10	.800		615	44		659	740
9440	Nylon cable tie, 14″ long		80	.100		.46	5.45		5.91	8.70
9450	Ladder, hold down clamp		60	.133		10.15	7.30		17.45	22
9460	Cable clamp		60	.133		10.30	7.30		17.60	22.50
9470	Wall bracket, 30″ wide tray		19	.421		54.50	23		77.50	94.50
9480	24″ wide tray		20	.400		42.50	22		64.50	79.50
9490	18″ wide tray		21	.381		38	21		59	72.50
9500	12″ wide tray		22	.364		35.50	19.90		55.40	69
9510	9″ wide tray		23	.348		31	19.05		50.05	63
9520	6″ wide tray		24	.333		30.50	18.25		48.75	61

26 05 36.20 Cable Tray Solid Bottom

		Crew	Daily Output	Labor-Hours	Unit	Material	Labor	Equipment	Total	Total Incl O&P
0010	**CABLE TRAY SOLID BOTTOM** w/ftngs. & supports, 3″ dp, to 15′ high									
0200	For higher elevations, see Section 26 05 36.40									
0220	Galvanized steel, tray, 6″ wide R260536-10	2 Elec	120	.133	L.F.	14.65	7.30		21.95	27
0240	12″ wide		100	.160		18.80	8.75		27.55	33.50
0260	18″ wide		70	.229		23	12.50		35.50	44
0280	24″ wide R260536-11		60	.267		27	14.60		41.60	51.50
0300	30″ wide		50	.320		32	17.50		49.50	61
0320	36″ wide		44	.364		45.50	19.90		65.40	80
0340	Elbow horizontal 90°, 12″ radius, 6″ wide		9.60	1.667	Ea.	111	91		202	259
0360	12″ wide		6.80	2.353		125	129		254	330
0370	18″ wide		5.40	2.963		143	162		305	400
0380	24″ wide		4.40	3.636		173	199		372	490
0390	30″ wide		3.80	4.211		208	230		438	575
0400	36″ wide		3.40	4.706		235	257		492	645
0420	24″ radius, 6″ wide		9.20	1.739		158	95		253	315
0440	12″ wide		6.40	2.500		184	137		321	405
0450	18″ wide		5	3.200		214	175		389	495
0460	24″ wide		4	4		248	219		467	605
0470	30″ wide		3.40	4.706		291	257		548	705
0480	36″ wide		3	5.333		325	292		617	795

26 05 36.20 Cable Tray Solid Bottom		Crew	Daily Output	Labor-Hours	Unit	Material	Labor	Equipment	Total	Total Incl O&P
						2015 Bare Costs				
0500	36″ radius, 6″ wide	2 Elec	8.80	1.818	Ea.	231	99.50		330.50	405
0520	12″ wide		6	2.667		261	146		407	505
0530	18″ wide		4.60	3.478		320	190		510	635
0540	24″ wide		3.60	4.444		345	243		588	745
0550	30″ wide		3	5.333		415	292		707	895
0560	36″ wide		2.60	6.154		470	335		805	1,025
0580	Elbow vertical 90°, 12″ radius, 6″ wide		9.60	1.667		135	91		226	285
0600	12″ wide		6.80	2.353		146	129		275	355
0610	18″ wide		5.40	2.963		158	162		320	415
0620	24″ wide		4.40	3.636		169	199		368	485
0630	30″ wide		3.80	4.211		180	230		410	545
0640	36″ wide		3.40	4.706		186	257		443	590
0670	24″ radius, 6″ wide		9.20	1.739		191	95		286	355
0690	12″ wide		6.40	2.500		210	137		347	435
0700	18″ wide		5	3.200		227	175		402	510
0710	24″ wide		4	4		246	219		465	600
0720	30″ wide		3.40	4.706		261	257		518	670
0730	36″ wide		3	5.333		285	292		577	750
0750	36″ radius, 6″ wide		8.80	1.818		261	99.50		360.50	435
0770	12″ wide		6.60	2.424		291	133		424	520
0780	18″ wide		4.60	3.478		320	190		510	635
0790	24″ wide		3.60	4.444		350	243		593	750
0800	30″ wide		3	5.333		380	292		672	855
0810	36″ wide		2.60	6.154		415	335		750	960
0840	Tee horizontal, 12″ radius, 6″ wide		5	3.200		156	175		331	435
0860	12″ wide		4	4		169	219		388	515
0870	18″ wide		3.40	4.706		201	257		458	605
0880	24″ wide		2.80	5.714		225	315		540	715
0890	30″ wide		2.60	6.154		257	335		592	790
0900	36″ wide		2.20	7.273		293	400		693	920
0940	24″ radius, 6″ wide		4.60	3.478		242	190		432	550
0960	12″ wide		3.60	4.444		283	243		526	675
0970	18″ wide		3	5.333		315	292		607	780
0980	24″ wide		2.40	6.667		425	365		790	1,025
0990	30″ wide		2.20	7.273		460	400		860	1,100
1000	36″ wide		1.80	8.889		500	485		985	1,275
1020	36″ radius, 6″ wide		4.20	3.810		380	208		588	730
1040	12″ wide		3.20	5		425	274		699	880
1050	18″ wide		2.60	6.154		460	335		795	1,025
1060	24″ wide		2.20	7.273		595	400		995	1,250
1070	30″ wide		2	8		645	440		1,085	1,375
1080	36″ wide		1.60	10		670	545		1,215	1,550
1100	Tee vertical, 12″ radius, 6″ wide		5	3.200		257	175		432	545
1120	12″ wide		4	4		268	219		487	625
1130	18″ wide		3.60	4.444		272	243		515	665
1140	24″ wide		3.40	4.706		298	257		555	710
1150	30″ wide		3	5.333		320	292		612	785
1160	36″ wide		2.60	6.154		330	335		665	870
1180	24″ radius, 6″ wide		4.60	3.478		380	190		570	700
1200	12″ wide		3.60	4.444		395	243		638	800
1210	18″ wide		3.20	5		415	274		689	865
1220	24″ wide		3	5.333		435	292		727	915
1230	30″ wide		2.60	6.154		480	335		815	1,025

26 05 36.20 Cable Tray Solid Bottom		Crew	Daily Output	Labor-Hours	Unit	2015 Bare Costs Material	Labor	Equipment	Total	Total Incl O&P
1240	36″ wide	2 Elec	2.20	7.273	Ea.	500	400		900	1,150
1260	36″ radius, 6″ wide		4.20	3.810		590	208		798	960
1280	12″ wide		3.20	5		600	274		874	1,075
1290	18″ wide		2.80	5.714		645	315		960	1,175
1300	24″ wide		2.60	6.154		670	335		1,005	1,250
1310	30″ wide		2.20	7.273		780	400		1,180	1,450
1320	36″ wide		2	8		825	440		1,265	1,550
1340	Cross horizontal, 12″ radius, 6″ wide		4	4		186	219		405	535
1360	12″ wide		3.40	4.706		203	257		460	610
1370	18″ wide		2.80	5.714		238	315		553	730
1380	24″ wide		2.40	6.667		268	365		633	840
1390	30″ wide		2	8		298	440		738	980
1400	36″ wide		1.80	8.889		325	485		810	1,100
1420	24″ radius, 6″ wide		3.60	4.444		340	243		583	740
1440	12″ wide		3	5.333		380	292		672	855
1450	18″ wide		2.40	6.667		425	365		790	1,025
1460	24″ wide		2	8		545	440		985	1,250
1470	30″ wide		1.80	8.889		590	485		1,075	1,375
1480	36″ wide		1.60	10		625	545		1,170	1,500
1500	36″ radius, 6″ wide		3.20	5		555	274		829	1,025
1520	12″ wide		2.60	6.154		610	335		945	1,175
1530	18″ wide		2	8		670	440		1,110	1,400
1540	24″ wide		1.80	8.889		790	485		1,275	1,600
1550	30″ wide		1.60	10		855	545		1,400	1,750
1560	36″ wide		1.40	11.429		910	625		1,535	1,925
1580	Drop out or end plate, 6″ wide		32	.500		22.50	27.50		50	66
1600	12″ wide		26	.615		28.50	33.50		62	81.50
1610	18″ wide		22	.727		30.50	40		70.50	93
1620	24″ wide		20	.800		35.50	44		79.50	105
1630	30″ wide		18	.889		40	48.50		88.50	117
1640	36″ wide		16	1		43	54.50		97.50	129
1660	Reducer, 12″ to 6″ wide		12	1.333		109	73		182	229
1680	18″ to 12″ wide		10.60	1.509		111	82.50		193.50	246
1700	18″ to 6″ wide		10.60	1.509		111	82.50		193.50	246
1720	24″ to 18″ wide		9.20	1.739		116	95		211	270
1740	24″ to 12″ wide		9.20	1.739		116	95		211	270
1760	30″ to 24″ wide		8	2		124	109		233	300
1780	30″ to 18″ wide		8	2		124	109		233	300
1800	30″ to 12″ wide		8	2		124	109		233	300
1820	36″ to 30″ wide		7.20	2.222		128	122		250	325
1840	36″ to 24″ wide		7.20	2.222		128	122		250	325
1860	36″ to 18″ wide		7.20	2.222		128	122		250	325
1880	36″ to 12″ wide		7.20	2.222		128	122		250	325
2000	Aluminum tray, 6″ wide		150	.107	L.F.	14.05	5.85		19.90	24
2020	12″ wide		130	.123		18.35	6.75		25.10	30
2030	18″ wide		100	.160		23	8.75		31.75	38.50
2040	24″ wide		90	.178		29	9.70		38.70	46.50
2050	30″ wide		70	.229		34	12.50		46.50	56.50
2060	36″ wide		64	.250		43	13.70		56.70	67.50
2080	Elbow horizontal 90°, 12″ radius, 6″ wide		9.60	1.667	Ea.	113	91		204	261
2100	12″ wide		7.60	2.105		127	115		242	310
2110	18″ wide		6.80	2.353		155	129		284	365
2120	24″ wide		5.80	2.759		177	151		328	420

26 05 36.20 Cable Tray Solid Bottom		Crew	Daily Output	Labor-Hours	Unit	Material	Labor	Equipment	Total	Total Incl O&P
						2015 Bare Costs				
2130	30" wide	2 Elec	5	3.200	Ea.	216	175		391	500
2140	36" wide		4.40	3.636		243	199		442	565
2160	24" radius, 6" wide		9.20	1.739		161	95		256	320
2180	12" wide		7.20	2.222		189	122		311	390
2190	18" wide		6.40	2.500		216	137		353	445
2200	24" wide		5.40	2.963		267	162		429	535
2210	30" wide		4.60	3.478		305	190		495	620
2220	36" wide		4	4		350	219		569	715
2240	36" radius, 6" wide		8.80	1.818		237	99.50		336.50	410
2260	12" wide		6.80	2.353		284	129		413	505
2270	18" wide		6	2.667		340	146		486	590
2280	24" wide		5	3.200		360	175		535	660
2290	30" wide		4.20	3.810		410	208		618	760
2300	36" wide		3.60	4.444		480	243		723	895
2320	Elbow vertical 90°, 12" radius, 6" wide		9.60	1.667		134	91		225	284
2340	12" wide		7.60	2.105		139	115		254	325
2350	18" wide		6.80	2.353		157	129		286	365
2360	24" wide		5.80	2.759		166	151		317	410
2370	30" wide		5	3.200		176	175		351	455
2380	36" wide		4.40	3.636		181	199		380	495
2400	24" radius, 6" wide		9.20	1.739		189	95		284	350
2420	12" wide		7.20	2.222		206	122		328	410
2430	18" wide		6.40	2.500		221	137		358	450
2440	24" wide		5.40	2.963		240	162		402	505
2450	30" wide		4.60	3.478		255	190		445	565
2460	36" wide		4	4		281	219		500	640
2480	36" radius, 6" wide		8.80	1.818		250	99.50		349.50	425
2500	12" wide		6.80	2.353		281	129		410	505
2510	18" wide		6	2.667		305	146		451	555
2520	24" wide		5	3.200		320	175		495	610
2530	30" wide		4.20	3.810		350	208		558	695
2540	36" wide		3.60	4.444		365	243		608	765
2560	Tee horizontal, 12" radius, 6" wide		5	3.200		174	175		349	455
2580	12" wide		4.40	3.636		211	199		410	530
2590	18" wide		4	4		237	219		456	590
2600	24" wide		3.60	4.444		265	243		508	655
2610	30" wide		3	5.333		305	292		597	770
2620	36" wide		2.40	6.667		350	365		715	930
2640	24" radius, 6" wide		4.60	3.478		287	190		477	600
2660	12" wide		4	4		325	219		544	685
2670	18" wide		3.60	4.444		365	243		608	765
2680	24" wide		3	5.333		450	292		742	930
2690	30" wide		2.40	6.667		510	365		875	1,100
2700	36" wide		2.20	7.273		570	400		970	1,225
2720	36" radius, 6" wide		4.20	3.810		465	208		673	825
2740	12" wide		3.60	4.444		510	243		753	925
2750	18" wide		3.20	5		590	274		864	1,050
2760	24" wide		2.60	6.154		685	335		1,020	1,250
2770	30" wide		2	8		755	440		1,195	1,475
2780	36" wide		1.80	8.889		845	485		1,330	1,650
2800	Tee vertical, 12" radius, 6" wide		5	3.200		248	175		423	535
2820	12" wide		4.40	3.636		254	199		453	575
2830	18" wide		4.20	3.810		264	208		472	600

26 05 36.20 Cable Tray Solid Bottom		Crew	Daily Output	Labor-Hours	Unit	Material	2015 Bare Costs Labor	Equipment	Total	Total Incl O&P
2840	24" wide	2 Elec	4	4	Ea.	281	219		500	640
2850	30" wide		3.60	4.444		305	243		548	700
2860	36" wide		3	5.333		325	292		617	790
2880	24" radius, 6" wide		4.60	3.478		365	190		555	685
2900	12" wide		4	4		385	219		604	755
2910	18" wide		3.80	4.211		410	230		640	800
2920	24" wide		3.60	4.444		450	243		693	860
2930	30" wide		3.20	5		490	274		764	950
2940	36" wide		2.60	6.154		520	335		855	1,075
2960	36" radius, 6" wide		4.20	3.810		570	208		778	940
2980	12" wide		3.40	4.706		600	257		857	1,050
2990	18" wide		3.40	4.706		630	257		887	1,075
3000	24" wide		3.20	5		650	274		924	1,125
3010	30" wide		2.80	5.714		705	315		1,020	1,250
3020	36" wide		2.20	7.273		775	400		1,175	1,450
3040	Cross horizontal, 12" radius, 6" wide		4.40	3.636		221	199		420	540
3060	12" wide		4	4		250	219		469	605
3070	18" wide		3.40	4.706		284	257		541	695
3080	24" wide		2.80	5.714		325	315		640	825
3090	30" wide		2.60	6.154		365	335		700	905
3100	36" wide		2.20	7.273		405	400		805	1,050
3120	24" radius, 6" wide		4	4		385	219		604	755
3140	12" wide		3.60	4.444		450	243		693	860
3150	18" wide		3	5.333		485	292		777	970
3160	24" wide		2.40	6.667		590	365		955	1,200
3170	30" wide		2.20	7.273		650	400		1,050	1,300
3180	36" wide		1.80	8.889		675	485		1,160	1,475
3200	36" radius, 6" wide		3.60	4.444		675	243		918	1,100
3220	12" wide		3.20	5		710	274		984	1,200
3230	18" wide		2.60	6.154		790	335		1,125	1,375
3240	24" wide		2	8		860	440		1,300	1,600
3250	30" wide		1.80	8.889		930	485		1,415	1,750
3260	36" wide		1.60	10		1,100	545		1,645	2,025
3280	Dropout, or end plate, 6" wide		32	.500		22.50	27.50		50	65.50
3300	12" wide		26	.615		25	33.50		58.50	78
3310	18" wide		22	.727		30	40		70	92.50
3320	24" wide		20	.800		33	44		77	102
3330	30" wide		18	.889		38.50	48.50		87	116
3340	36" wide		16	1		45	54.50		99.50	132
3380	Reducer, 12" to 6" wide		14	1.143		109	62.50		171.50	214
3400	18" to 12" wide		12	1.333		113	73		186	233
3420	18" to 6" wide		12	1.333		113	73		186	233
3440	24" to 18" wide		10.60	1.509		119	82.50		201.50	255
3460	24" to 12" wide		10.60	1.509		119	82.50		201.50	255
3480	30" to 24" wide		9.20	1.739		130	95		225	286
3500	30" to 18" wide		9.20	1.739		130	95		225	286
3520	30" to 12" wide		9.20	1.739		130	95		225	286
3540	36" to 30" wide		8	2		137	109		246	315
3560	36" to 24" wide		8	2		139	109		248	315
3580	36" to 18" wide		8	2		139	109		248	315
3600	36" to 12" wide	▼	8	2	▼	142	109		251	320

26 05 36.30 Cable Tray Trough		Crew	Daily Output	Labor-Hours	Unit	Material	Labor	Equipment	Total	Total Incl O&P
0010	**CABLE TRAY TROUGH** vented, w/ftngs. & supports, 6″ dp, to 15′ high									
0020	For higher elevations, see Section 26 05 36.40									
0200	Galvanized steel, tray, 6″ wide R260536-10	2 Elec	90	.178	L.F.	16.20	9.70		25.90	32.50
0240	12″ wide		80	.200		16.80	10.95		27.75	35
0260	18″ wide		70	.229		20.50	12.50		33	41.50
0280	24″ wide R260536-11		60	.267		35.50	14.60		50.10	61
0300	30″ wide		50	.320		48	17.50		65.50	78.50
0320	36″ wide		40	.400		59.50	22		81.50	98.50
0340	Elbow horizontal 90°, 12″ radius, 6″ wide		7.60	2.105	Ea.	119	115		234	305
0360	12″ wide		5.60	2.857		141	156		297	390
0370	18″ wide		4.40	3.636		163	199		362	475
0380	24″ wide		3.60	4.444		201	243		444	585
0390	30″ wide		3.20	5		227	274		501	660
0400	36″ wide		2.80	5.714		263	315		578	760
0420	24″ radius, 6″ wide		7.20	2.222		176	122		298	375
0440	12″ wide		5.20	3.077		203	168		371	475
0450	18″ wide		4	4		244	219		463	600
0460	24″ wide		3.20	5		276	274		550	715
0470	30″ wide		2.80	5.714		320	315		635	820
0480	36″ wide		2.40	6.667		370	365		735	950
0500	36″ radius, 6″ wide		6.80	2.353		263	129		392	485
0520	12″ wide		4.80	3.333		300	182		482	605
0530	18″ wide		3.60	4.444		360	243		603	760
0540	24″ wide		2.80	5.714		380	315		695	890
0550	30″ wide		2.40	6.667		430	365		795	1,025
0560	36″ wide		2	8		475	440		915	1,175
0580	Elbow vertical 90°, 12″ radius, 6″ wide		7.60	2.105		154	115		269	345
0600	12″ wide		5.60	2.857		171	156		327	420
0610	18″ wide		4.40	3.636		173	199		372	490
0620	24″ wide		3.60	4.444		199	243		442	585
0630	30″ wide		3.20	5		203	274		477	635
0640	36″ wide		2.80	5.714		214	315		529	705
0660	24″ radius, 6″ wide		7.20	2.222		220	122		342	425
0680	12″ wide		5.20	3.077		233	168		401	510
0690	18″ wide		4	4		259	219		478	615
0700	24″ wide		3.20	5		268	274		542	705
0710	30″ wide		2.80	5.714		300	315		615	800
0720	36″ wide		2.40	6.667		315	365		680	890
0740	36″ radius, 6″ wide		6.80	2.353		298	129		427	520
0760	12″ wide		4.80	3.333		315	182		497	625
0770	18″ wide		3.60	4.444		350	243		593	750
0780	24″ wide		2.80	5.714		375	315		690	885
0790	30″ wide		2.40	6.667		385	365		750	970
0800	36″ wide		2	8		430	440		870	1,125
0820	Tee horizontal, 12″ radius, 6″ wide		4	4		180	219		399	530
0840	12″ wide		3.20	5		201	274		475	630
0850	18″ wide		2.80	5.714		227	315		542	720
0860	24″ wide		2.40	6.667		263	365		628	835
0870	30″ wide		2.20	7.273		300	400		700	925
0880	36″ wide		2	8		330	440		770	1,025
0900	24″ radius, 6″ wide		3.60	4.444		298	243		541	690
0920	12″ wide		2.80	5.714		325	315		640	830

Material, Labor, Equipment and Total are 2015 Bare Costs.

26 05 36.30 Cable Tray Trough		Crew	Daily Output	Labor-Hours	Unit	2015 Bare Costs Material	Labor	Equipment	Total	Total Incl O&P
0930	18″ wide	2 Elec	2.40	6.667	Ea.	365	365		730	945
0940	24″ wide		2	8		470	440		910	1,175
0950	30″ wide		1.80	8.889		510	485		995	1,300
0960	36″ wide		1.60	10		565	545		1,110	1,450
0980	36″ radius, 6″ wide		3.20	5		430	274		704	880
1000	12″ wide		2.40	6.667		490	365		855	1,075
1010	18″ wide		2	8		535	440		975	1,250
1020	24″ wide		1.60	10		640	545		1,185	1,525
1030	30″ wide		1.40	11.429		680	625		1,305	1,675
1040	36″ wide		1.20	13.333		800	730		1,530	1,975
1060	Tee vertical, 12″ radius, 6″ wide		4	4		300	219		519	660
1080	12″ wide		3.20	5		305	274		579	745
1090	18″ wide		3	5.333		315	292		607	780
1100	24″ wide		2.80	5.714		330	315		645	835
1110	30″ wide		2.60	6.154		350	335		685	890
1120	36″ wide		2.20	7.273		355	400		755	985
1140	24″ radius, 6″ wide		3.60	4.444		400	243		643	810
1160	12″ wide		2.80	5.714		425	315		740	935
1170	18″ wide		2.60	6.154		445	335		780	990
1180	24″ wide		2.40	6.667		475	365		840	1,075
1190	30″ wide		2.20	7.273		500	400		900	1,150
1200	36″ wide		1.80	8.889		545	485		1,030	1,325
1220	36″ radius, 6″ wide		3.20	5		640	274		914	1,125
1240	12″ wide		2.40	6.667		655	365		1,020	1,275
1250	18″ wide		2.20	7.273		680	400		1,080	1,350
1260	24″ wide		2	8		720	440		1,160	1,450
1270	30″ wide		1.80	8.889		835	485		1,320	1,650
1280	36″ wide		1.40	11.429		845	625		1,470	1,875
1300	Cross horizontal, 12″ radius, 6″ wide		3.20	5		231	274		505	665
1320	12″ wide		2.80	5.714		233	315		548	725
1330	18″ wide		2.40	6.667		261	365		626	830
1340	24″ wide		2	8		274	440		714	955
1350	30″ wide		1.80	8.889		310	485		795	1,075
1360	36″ wide		1.60	10		345	545		890	1,200
1380	24″ radius, 6″ wide		2.80	5.714		365	315		680	870
1400	12″ wide		2.40	6.667		380	365		745	965
1410	18″ wide		2	8		420	440		860	1,125
1420	24″ wide		1.60	10		535	545		1,080	1,400
1430	30″ wide		1.40	11.429		590	625		1,215	1,575
1440	36″ wide		1.20	13.333		625	730		1,355	1,775
1460	36″ radius, 6″ wide		2.40	6.667		625	365		990	1,225
1480	12″ wide		2	8		645	440		1,085	1,375
1490	18″ wide		1.60	10		655	545		1,200	1,550
1500	24″ wide		1.20	13.333		790	730		1,520	1,975
1510	30″ wide		1	16		870	875		1,745	2,250
1520	36″ wide		.80	20		935	1,100		2,035	2,675
1540	Dropout or end plate, 6″ wide		26	.615		27	33.50		60.50	80
1560	12″ wide		22	.727		33	40		73	95.50
1580	18″ wide		20	.800		36	44		80	105
1600	24″ wide		18	.889		41.50	48.50		90	119
1620	30″ wide		16	1		44.50	54.50		99	131
1640	36″ wide		13.40	1.194		49	65.50		114.50	152
1660	Reducer, 12″ to 6″ wide		9.40	1.702		112	93		205	263

26 05 36.30 Cable Tray Trough		Crew	Daily Output	Labor-Hours	Unit	Material	2015 Bare Costs Labor	Equipment	Total	Total Incl O&P
1680	18″ to 12″ wide	2 Elec	8.40	1.905	Ea.	117	104		221	285
1700	18″ to 6″ wide		8.40	1.905		117	104		221	285
1720	24″ to 18″ wide		7.20	2.222		125	122		247	320
1740	24″ to 12″ wide		7.20	2.222		125	122		247	320
1760	30″ to 24″ wide		6.40	2.500		129	137		266	345
1780	30″ to 18″ wide		6.40	2.500		129	137		266	345
1800	30″ to 12″ wide		6.40	2.500		129	137		266	345
1820	36″ to 30″ wide		5.80	2.759		138	151		289	380
1840	36″ to 24″ wide		5.80	2.759		138	151		289	380
1860	36″ to 18″ wide		5.80	2.759		140	151		291	380
1880	36″ to 12″ wide		5.80	2.759		140	151		291	380
2000	Aluminum, tray, vented, 6″ wide		120	.133	L.F.	18.35	7.30		25.65	31
2010	9″ wide		110	.145		21	7.95		28.95	35
2020	12″ wide		100	.160		23	8.75		31.75	38.50
2030	18″ wide		90	.178		27.50	9.70		37.20	45
2040	24″ wide		80	.200		33	10.95		43.95	52.50
2050	30″ wide		70	.229		43.50	12.50		56	67
2060	36″ wide		60	.267		48	14.60		62.60	74.50
2080	Elbow horiz. 90°, 12″ radius, 6″ wide		7.60	2.105	Ea.	121	115		236	305
2090	9″ wide		7	2.286		132	125		257	330
2100	12″ wide		6.20	2.581		140	141		281	365
2110	18″ wide		5.60	2.857		164	156		320	415
2120	24″ wide		4.60	3.478		189	190		379	495
2130	30″ wide		4	4		233	219		452	585
2140	36″ wide		3.60	4.444		255	243		498	645
2160	24″ radius, 6″ wide		7.20	2.222		176	122		298	375
2180	12″ wide		5.80	2.759		206	151		357	455
2190	18″ wide		5.20	3.077		237	168		405	510
2200	24″ wide		4.20	3.810		270	208		478	605
2210	30″ wide		3.60	4.444		310	243		553	705
2220	36″ wide		3.20	5		335	274		609	780
2240	36″ radius, 6″ wide		6.80	2.353		248	129		377	465
2260	12″ wide		5.40	2.963		282	162		444	555
2270	18″ wide		4.80	3.333		320	182		502	630
2280	24″ wide		3.80	4.211		350	230		580	730
2290	30″ wide		3.40	4.706		410	257		667	840
2300	36″ wide		2.80	5.714		450	315		765	965
2320	Elbow vertical 90°, 12″ radius, 6″ wide		7.60	2.105		147	115		262	335
2330	9″ wide		7	2.286		157	125		282	360
2340	12″ wide		6.20	2.581		159	141		300	385
2350	18″ wide		5.60	2.857		164	156		320	415
2360	24″ wide		4.60	3.478		181	190		371	485
2370	30″ wide		4	4		189	219		408	540
2380	36″ wide		3.60	4.444		191	243		434	575
2400	24″ radius, 6″ wide		7.20	2.222		199	122		321	400
2420	12″ wide		5.80	2.759		216	151		367	465
2430	18″ wide		5.20	3.077		235	168		403	510
2440	24″ wide		4.20	3.810		238	208		446	570
2450	30″ wide		3.60	4.444		252	243		495	640
2460	36″ wide		3.20	5		264	274		538	700
2480	36″ radius, 6″ wide		6.80	2.353		247	129		376	465
2500	12″ wide		5.40	2.963		264	162		426	535
2510	18″ wide		4.80	3.333		296	182		478	600

26 05 36.30 Cable Tray Trough		Crew	Daily Output	Labor-Hours	Unit	Material	Labor	Equipment	Total	Total Incl O&P
						2015 Bare Costs				
2520	24″ wide	2 Elec	3.80	4.211	Ea.	320	230		550	695
2530	30″ wide		3.40	4.706		350	257		607	770
2540	36″ wide		2.80	5.714		360	315		675	870
2560	Tee horizontal, 12″ radius, 6″ wide		4	4		184	219		403	535
2570	9″ wide		3.80	4.211		189	230		419	555
2580	12″ wide		3.60	4.444		206	243		449	590
2590	18″ wide		3.20	5		238	274		512	670
2600	24″ wide		2.80	5.714		274	315		589	770
2610	30″ wide		2.40	6.667		294	365		659	870
2620	36″ wide		2.20	7.273		335	400		735	965
2640	24″ radius, 6″ wide		3.60	4.444		281	243		524	675
2660	12″ wide		3.20	5		320	274		594	760
2670	18″ wide		2.80	5.714		350	315		665	855
2680	24″ wide		2.40	6.667		440	365		805	1,025
2690	30″ wide		2	8		475	440		915	1,175
2700	36″ wide		1.80	8.889		555	485		1,040	1,350
2720	36″ radius, 6″ wide		3.20	5		450	274		724	905
2740	12″ wide		2.80	5.714		520	315		835	1,050
2750	18″ wide		2.40	6.667		590	365		955	1,200
2760	24″ wide		2	8		685	440		1,125	1,400
2770	30″ wide		1.60	10		730	545		1,275	1,625
2780	36″ wide		1.40	11.429		845	625		1,470	1,875
2800	Tee vertical, 12″ radius, 6″ wide		4	4		254	219		473	610
2810	9″ wide		3.80	4.211		255	230		485	625
2820	12″ wide		3.60	4.444		274	243		517	665
2830	18″ wide		3.40	4.706		275	257		532	690
2840	24″ wide		3.20	5		282	274		556	720
2850	30″ wide		3	5.333		296	292		588	760
2860	36″ wide		2.60	6.154		310	335		645	845
2880	24″ radius, 6″ wide		3.60	4.444		360	243		603	760
2900	12″ wide		3.20	5		385	274		659	835
2910	18″ wide		3	5.333		410	292		702	890
2920	24″ wide		2.80	5.714		440	315		755	955
2930	30″ wide		2.60	6.154		475	335		810	1,025
2940	36″ wide		2.20	7.273		485	400		885	1,125
2960	36″ radius, 6″ wide		3.20	5		555	274		829	1,025
2980	12″ wide		2.80	5.714		590	315		905	1,125
2990	18″ wide		2.60	6.154		605	335		940	1,175
3000	24″ wide		2.40	6.667		650	365		1,015	1,250
3010	30″ wide		2.20	7.273		705	400		1,105	1,375
3020	36″ wide		1.80	8.889		740	485		1,225	1,550
3040	Cross horizontal, 12″ radius, 6″ wide		3.60	4.444		230	243		473	620
3050	9″ wide		3.40	4.706		247	257		504	655
3060	12″ wide		3.20	5		254	274		528	690
3070	18″ wide		2.80	5.714		265	315		580	760
3080	24″ wide		2.40	6.667		296	365		661	870
3090	30″ wide		2.20	7.273		360	400		760	990
3100	36″ wide		1.80	8.889		435	485		920	1,200
3120	24″ radius, 6″ wide		3.20	5		410	274		684	865
3140	12″ wide		2.80	5.714		450	315		765	965
3150	18″ wide		2.40	6.667		485	365		850	1,075
3160	24″ wide		2	8		565	440		1,005	1,275
3170	30″ wide	↓	1.80	8.889	↓	605	485		1,090	1,400

26 05 36 – Cable Trays for Electrical Systems

26 05 36.30 Cable Tray Trough		Crew	Daily Output	Labor-Hours	Unit	Material	2015 Bare Costs Labor	Equipment	Total	Total Incl O&P
3180	36" wide	2 Elec	1.40	11.429	Ea.	685	625		1,310	1,700
3200	36" radius, 6" wide		2.80	5.714		685	315		1,000	1,225
3220	12" wide		2.40	6.667		720	365		1,085	1,325
3230	18" wide		2	8		785	440		1,225	1,525
3240	24" wide		1.60	10		905	545		1,450	1,825
3250	30" wide		1.40	11.429		985	625		1,610	2,000
3260	36" wide		1.20	13.333		1,150	730		1,880	2,350
3280	Dropout, or end plate, 6" wide		26	.615		23	33.50		56.50	76
3300	12" wide		22	.727		27	40		67	89.50
3310	18" wide		20	.800		33	44		77	102
3320	24" wide		18	.889		39.50	48.50		88	117
3330	30" wide		16	1		42.50	54.50		97	129
3340	36" wide		14	1.143		48.50	62.50		111	147
3370	Reducer, 9" to 6" wide		12	1.333		113	73		186	233
3380	12" to 6" wide		11.40	1.404		118	77		195	245
3390	12" to 9" wide		11.40	1.404		118	77		195	245
3400	18" to 12" wide		9.60	1.667		127	91		218	276
3420	18" to 6" wide		9.60	1.667		127	91		218	276
3430	18" to 9" wide		9.60	1.667		127	91		218	276
3440	24" to 18" wide		8.40	1.905		135	104		239	305
3460	24" to 12" wide		8.40	1.905		135	104		239	305
3470	24" to 9" wide		8.40	1.905		137	104		241	305
3475	24" to 6" wide		8.40	1.905		137	104		241	305
3480	30" to 24" wide		7.20	2.222		139	122		261	335
3500	30" to 18" wide		7.20	2.222		139	122		261	335
3520	30" to 12" wide		7.20	2.222		142	122		264	340
3540	36" to 30" wide		6.40	2.500		144	137		281	365
3560	36" to 24" wide		6.40	2.500		144	137		281	365
3580	36" to 18" wide		6.40	2.500		144	137		281	365
3600	36" to 12" wide		6.40	2.500		144	137		281	365
3610	Elbow horizontal 60°, 12" radius, 6" wide		7.80	2.051		98	112		210	276
3620	9" wide		7.20	2.222		107	122		229	300
3630	12" wide		6.40	2.500		118	137		255	335
3640	18" wide		5.80	2.759		127	151		278	365
3650	24" wide		4.80	3.333		156	182		338	445
3680	Elbow horizontal 45°, 12" radius, 6" wide		8	2		84.50	109		193.50	257
3690	9" wide		7.40	2.162		89.50	118		207.50	276
3700	12" wide		6.60	2.424		94.50	133		227.50	305
3710	18" wide		6	2.667		106	146		252	335
3720	24" wide		5	3.200		123	175		298	400
3750	Elbow horizontal, 30° 12" radius, 6" wide		8.20	1.951		74	107		181	241
3760	9" wide		7.60	2.105		77.50	115		192.50	258
3770	12" wide		6.80	2.353		83	129		212	284
3780	18" wide		6.20	2.581		89.50	141		230.50	310
3790	24" wide		5.20	3.077		98	168		266	360
3820	Elbow vertical 60° in/outside, 12" radius, 6" wide		7.80	2.051		122	112		234	300
3830	9" wide		7.20	2.222		123	122		245	320
3840	12" wide		6.40	2.500		125	137		262	345
3850	18" wide		5.80	2.759		130	151		281	370
3860	24" wide		4.80	3.333		135	182		317	420
3890	Elbow vertical 45° in/outside, 12" radius, 6" wide		8	2		98	109		207	272
3900	9" wide		7.40	2.162		104	118		222	291
3910	12" wide		6.60	2.424		107	133		240	315

26 05 36 – Cable Trays for Electrical Systems

26 05 36.30 Cable Tray Trough		Crew	Daily Output	Labor-Hours	Unit	Material	Labor	Equipment	Total	Total Incl O&P
						2015 Bare Costs				
3920	18" wide	2 Elec	6	2.667	Ea.	111	146		257	340
3930	24" wide		5	3.200		121	175		296	395
3960	Elbow vertical 30° in/outside, 12" radius, 6" wide		8.20	1.951		84.50	107		191.50	253
3970	9" wide		7.60	2.105		89.50	115		204.50	272
3980	12" wide		6.80	2.353		91.50	129		220.50	293
3990	18" wide		6.20	2.581		95	141		236	315
4000	24" wide		5.20	3.077		96.50	168		264.50	360
4250	Reducer, left or right hand, 24" to 18" wide		8.40	1.905		128	104		232	297
4260	24" to 12" wide		8.40	1.905		128	104		232	297
4270	24" to 9" wide		8.40	1.905		128	104		232	297
4280	24" to 6" wide		8.40	1.905		130	104		234	299
4290	18" to 12" wide		9.60	1.667		120	91		211	269
4300	18" to 9" wide		9.60	1.667		120	91		211	269
4310	18" to 6" wide		9.60	1.667		120	91		211	269
4320	12" to 9" wide		11.40	1.404		115	77		192	241
4330	12" to 6" wide		11.40	1.404		115	77		192	241
4340	9" to 6" wide		12	1.333		112	73		185	232
4350	Splice plate	1 Elec	48	.167		10.65	9.10		19.75	25.50
4360	Splice plate, expansion joint		48	.167		10.90	9.10		20	25.50
4370	Splice plate, hinged, horizontal		48	.167		8.80	9.10		17.90	23.50
4380	Vertical		48	.167		12.50	9.10		21.60	27.50
4390	Trough, hanger, vertical		28	.286		35	15.65		50.65	62
4400	Box connector, 24" wide		20	.400		47.50	22		69.50	85
4410	18" wide		21	.381		39.50	21		60.50	74.50
4420	12" wide		22	.364		38	19.90		57.90	71.50
4430	9" wide		23	.348		36	19.05		55.05	68
4440	6" wide		24	.333		34.50	18.25		52.75	65
4450	Floor flange		24	.333		36	18.25		54.25	67
4460	Hold down clamp		60	.133		4.13	7.30		11.43	15.50
4520	Wall bracket, 24" wide tray		20	.400		36.50	22		58.50	73
4530	18" wide tray		21	.381		35.50	21		56.50	70
4540	12" wide tray		22	.364		18.90	19.90		38.80	51
4550	9" wide tray		23	.348		16.90	19.05		35.95	47
4560	6" wide tray		24	.333		15.35	18.25		33.60	44.50
5000	Cable channel aluminum, vented, 1-1/4" deep, 4" wide, straight		80	.100	L.F.	13.35	5.45		18.80	23
5010	Elbow horizontal, 36" radius, 90°		5	1.600	Ea.	241	87.50		328.50	395
5020	60°		5.50	1.455		184	79.50		263.50	320
5030	45°		6	1.333		150	73		223	274
5040	30°		6.50	1.231		129	67.50		196.50	243
5050	Adjustable		6	1.333		123	73		196	245
5060	Elbow vertical, 36" radius, 90°		5	1.600		259	87.50		346.50	415
5070	60°		5.50	1.455		202	79.50		281.50	340
5080	45°		6	1.333		167	73		240	292
5090	30°		6.50	1.231		147	67.50		214.50	263
5100	Adjustable		6	1.333		123	73		196	245
5110	Splice plate, hinged, horizontal		48	.167		9.20	9.10		18.30	24
5120	Splice plate, hinged, vertical		48	.167		13.10	9.10		22.20	28
5130	Hanger, vertical		28	.286		14.40	15.65		30.05	39.50
5140	Single		28	.286		22	15.65		37.65	47.50
5150	Double		20	.400		22.50	22		44.50	57.50
5160	Channel to box connector		24	.333		29.50	18.25		47.75	60
5170	Hold down clip		80	.100		4.16	5.45		9.61	12.80
5180	Wall bracket, single		28	.286		11.95	15.65		27.60	36.50

26 05 36 – Cable Trays for Electrical Systems

26 05 36.30 Cable Tray Trough		Crew	Daily Output	Labor-Hours	Unit	Material	2015 Bare Costs Labor	Equipment	Total	Total Incl O&P
5190	Double	1 Elec	20	.400	Ea.	15.35	22		37.35	50
5200	Cable roller		16	.500		188	27.50		215.50	248
5210	Splice plate		48	.167		5.80	9.10		14.90	20

26 05 36.40 Cable Tray, Covers and Dividers		Crew	Daily Output	Labor-Hours	Unit	Material	Labor	Equipment	Total	Total Incl O&P
0010	**CABLE TRAY, COVERS AND DIVIDERS** To 15' high R260536-10									
0011	For higher elevations, see lines 9900 – 9960									
0100	Covers, ventilated galv. steel, straight, 6" wide tray size	2 Elec	520	.031	L.F.	5.95	1.68		7.63	9
0200	9" wide tray size		460	.035		8.60	1.90		10.50	12.30
0300	12" wide tray size R260536-11		400	.040		8.75	2.19		10.94	12.95
0400	18" wide tray size		300	.053		11.90	2.92		14.82	17.45
0500	24" wide tray size		220	.073		14.60	3.98		18.58	22
0600	30" wide tray size		180	.089		21	4.86		25.86	30.50
0700	36" wide tray size		160	.100		24	5.45		29.45	34.50
1000	Elbow horizontal 90°, 12" radius, 6" wide tray size		150	.107	Ea.	52	5.85		57.85	66.50
1020	9" wide tray size		128	.125		58	6.85		64.85	74
1040	12" wide tray size		108	.148		61	8.10		69.10	79.50
1060	18" wide tray size		84	.190		84.50	10.40		94.90	109
1080	24" wide tray size		66	.242		102	13.25		115.25	133
1100	30" wide tray size		60	.267		127	14.60		141.60	162
1120	36" wide tray size		50	.320		154	17.50		171.50	195
1160	24" radius, 6" wide tray size		136	.118		87	6.45		93.45	105
1180	9" wide tray size		116	.138		89	7.55		96.55	109
1200	12" wide tray size		96	.167		99	9.10		108.10	123
1220	18" wide tray size		76	.211		122	11.50		133.50	152
1240	24" wide tray size		60	.267		149	14.60		163.60	186
1260	30" wide tray size		52	.308		196	16.85		212.85	240
1280	36" wide tray size		44	.364		227	19.90		246.90	280
1320	36" radius, 6" wide tray size		120	.133		127	7.30		134.30	151
1340	9" wide tray size		104	.154		140	8.40		148.40	167
1360	12" wide tray size		84	.190		150	10.40		160.40	181
1380	18" wide tray size		72	.222		191	12.15		203.15	228
1400	24" wide tray size		52	.308		227	16.85		243.85	275
1420	30" wide tray size		46	.348		270	19.05		289.05	325
1440	36" wide tray size		40	.400		315	22		337	385
1480	Elbow horizontal 45°, 12" radius, 6" wide tray size		150	.107		37	5.85		42.85	50
1500	9" wide tray size		128	.125		44.50	6.85		51.35	59.50
1520	12" wide tray size		108	.148		48.50	8.10		56.60	65
1540	18" wide tray size		88	.182		59	9.95		68.95	80
1560	24" wide tray size		76	.211		69	11.50		80.50	93.50
1580	30" wide tray size		66	.242		81.50	13.25		94.75	109
1600	36" wide tray size		60	.267		91	14.60		105.60	122
1640	24" radius, 6" wide tray size		136	.118		54.50	6.45		60.95	69.50
1660	9" wide tray size		116	.138		61	7.55		68.55	79
1680	12" wide tray size		96	.167		69	9.10		78.10	89.50
1700	18" wide tray size		80	.200		79	10.95		89.95	103
1720	24" wide tray size		70	.229		91	12.50		103.50	119
1740	30" wide tray size		60	.267		116	14.60		130.60	149
1760	36" wide tray size		52	.308		127	16.85		143.85	165
1800	36" radius, 6" wide tray size		120	.133		79	7.30		86.30	98
1820	9" wide tray size		104	.154		89	8.40		97.40	111
1840	12" wide tray size		84	.190		98	10.40		108.40	124
1860	18" wide tray size		76	.211		112	11.50		123.50	141

26 05 36 – Cable Trays for Electrical Systems

26 05 36.40 Cable Tray, Covers and Dividers		Crew	Daily Output	Labor-Hours	Unit	Material	2015 Bare Costs Labor	Equipment	Total	Total Incl O&P
1880	24″ wide tray size	2 Elec	62	.258	Ea.	137	14.10		151.10	172
1900	30″ wide tray size		52	.308		149	16.85		165.85	189
1920	36″ wide tray size		48	.333		180	18.25		198.25	226
1960	Elbow vertical 90°, 12″ radius, 6″ wide tray size		150	.107		43.50	5.85		49.35	57
1980	9″ wide tray size		128	.125		44.50	6.85		51.35	59.50
2000	12″ wide tray size		108	.148		48	8.10		56.10	65
2020	18″ wide tray size		88	.182		54.50	9.95		64.45	75
2040	24″ wide tray size		68	.235		55.50	12.85		68.35	80.50
2060	30″ wide tray size		60	.267		61	14.60		75.60	89.50
2080	36″ wide tray size		50	.320		79	17.50		96.50	113
2120	24″ radius, 6″ wide tray size		136	.118		53.50	6.45		59.95	68.50
2140	9″ wide tray size		116	.138		59	7.55		66.55	76.50
2160	12″ wide tray size		96	.167		61	9.10		70.10	81
2180	18″ wide tray size		80	.200		81.50	10.95		92.45	106
2200	24″ wide tray size		62	.258		91	14.10		105.10	121
2220	30″ wide tray size		52	.308		104	16.85		120.85	139
2240	36″ wide tray size		44	.364		118	19.90		137.90	160
2280	36″ radius, 6″ wide tray size		120	.133		62.50	7.30		69.80	79.50
2300	9″ wide tray size		104	.154		79	8.40		87.40	99.50
2320	12″ wide tray size		84	.190		89	10.40		99.40	114
2340	18″ wide tray size		76	.211		112	11.50		123.50	141
2350	24″ wide tray size		54	.296		125	16.20		141.20	162
2360	30″ wide tray size		46	.348		149	19.05		168.05	193
2370	36″ wide tray size		40	.400		176	22		198	227
2400	Tee horizontal, 12″ radius, 6″ wide tray size		92	.174		79	9.50		88.50	101
2410	9″ wide tray size		80	.200		81.50	10.95		92.45	106
2420	12″ wide tray size		68	.235		92	12.85		104.85	120
2430	18″ wide tray size		60	.267		111	14.60		125.60	144
2440	24″ wide tray size		52	.308		140	16.85		156.85	179
2460	30″ wide tray size		36	.444		165	24.50		189.50	218
2470	36″ wide tray size		30	.533		199	29		228	263
2500	24″ radius, 6″ wide tray size		88	.182		124	9.95		133.95	152
2510	9″ wide tray size		76	.211		143	11.50		154.50	174
2520	12″ wide tray size		64	.250		147	13.70		160.70	183
2530	18″ wide tray size		56	.286		188	15.65		203.65	231
2540	24″ wide tray size		48	.333		289	18.25		307.25	350
2560	30″ wide tray size		32	.500		330	27.50		357.50	405
2570	36″ wide tray size		26	.615		365	33.50		398.50	450
2600	36″ radius, 6″ wide tray size		84	.190		225	10.40		235.40	263
2610	9″ wide tray size		72	.222		229	12.15		241.15	270
2620	12″ wide tray size		60	.267		255	14.60		269.60	300
2630	18″ wide tray size		52	.308		298	16.85		314.85	350
2640	24″ wide tray size		44	.364		390	19.90		409.90	460
2660	30″ wide tray size		28	.571		425	31.50		456.50	515
2670	36″ wide tray size		22	.727		485	40		525	590
2700	Cross horizontal, 12″ radius, 6″ wide tray size		68	.235		118	12.85		130.85	149
2710	9″ wide tray size		64	.250		125	13.70		138.70	158
2720	12″ wide tray size		60	.267		139	14.60		153.60	175
2730	18″ wide tray size		52	.308		165	16.85		181.85	206
2740	24″ wide tray size		36	.444		201	24.50		225.50	258
2760	30″ wide tray size		30	.533		229	29		258	296
2770	36″ wide tray size		28	.571		268	31.50		299.50	340
2800	24″ radius, 6″ wide tray size	↓	64	.250	↓	225	13.70		238.70	268

26 05 36.40 Cable Tray, Covers and Dividers		Crew	Daily Output	Labor-Hours	Unit	Material	2015 Bare Costs Labor	Equipment	Total	Total Incl O&P
2810	9″ wide tray size	2 Elec	60	.267	Ea.	244	14.60		258.60	290
2820	12″ wide tray size		56	.286		263	15.65		278.65	315
2830	18″ wide tray size		48	.333		315	18.25		333.25	375
2840	24″ wide tray size		32	.500		385	27.50		412.50	460
2860	30″ wide tray size		26	.615		430	33.50		463.50	525
2870	36″ wide tray size		24	.667		485	36.50		521.50	590
2900	36″ radius, 6″ wide tray size		60	.267		375	14.60		389.60	435
2910	9″ wide tray size		56	.286		390	15.65		405.65	455
2920	12″ wide tray size		52	.308		410	16.85		426.85	475
2930	18″ wide tray size		44	.364		470	19.90		489.90	550
2940	24″ wide tray size		28	.571		600	31.50		631.50	705
2960	30″ wide tray size		22	.727		645	40		685	770
2970	36″ wide tray size		20	.800		690	44		734	825
3000	Reducer, 9″ to 6″ wide tray size		128	.125		50	6.85		56.85	65.50
3010	12″ to 6″ wide tray size		108	.148		52	8.10		60.10	69.50
3020	12″ to 9″ wide tray size		108	.148		52	8.10		60.10	69.50
3030	18″ to 12″ wide tray size		88	.182		55.50	9.95		65.45	76
3050	18″ to 6″ wide tray size		88	.182		55.50	9.95		65.45	76
3060	24″ to 18″ wide tray size		80	.200		78	10.95		88.95	102
3070	24″ to 12″ wide tray size		80	.200		71.50	10.95		82.45	95
3090	30″ to 24″ wide tray size		70	.229		82.50	12.50		95	109
3100	30″ to 18″ wide tray size		70	.229		82.50	12.50		95	109
3110	30″ to 12″ wide tray size		70	.229		71.50	12.50		84	97.50
3140	36″ to 30″ wide tray size		64	.250		90	13.70		103.70	120
3150	36″ to 24″ wide tray size		64	.250		90	13.70		103.70	120
3160	36″ to 18″ wide tray size		64	.250		90	13.70		103.70	120
3170	36″ to 12″ wide tray size		64	.250		90	13.70		103.70	120
3250	Covers, aluminum, straight, 6″ wide tray size		520	.031	L.F.	5.40	1.68		7.08	8.45
3270	9″ wide tray size		460	.035		6.50	1.90		8.40	10
3290	12″ wide tray size		400	.040		7.80	2.19		9.99	11.85
3310	18″ wide tray size		320	.050		10.30	2.74		13.04	15.45
3330	24″ wide tray size		260	.062		12.85	3.37		16.22	19.20
3350	30″ wide tray size		200	.080		14.20	4.38		18.58	22
3370	36″ wide tray size		180	.089		15.05	4.86		19.91	24
3400	Elbow horizontal 90°, 12″ radius, 6″ wide tray size		150	.107	Ea.	40.50	5.85		46.35	53.50
3410	9″ wide tray size		128	.125		42.50	6.85		49.35	57
3420	12″ wide tray size		108	.148		45.50	8.10		53.60	62
3430	18″ wide tray size		88	.182		60	9.95		69.95	81
3440	24″ wide tray size		70	.229		75.50	12.50		88	102
3460	30″ wide tray size		64	.250		90.50	13.70		104.20	120
3470	36″ wide tray size		54	.296		111	16.20		127.20	147
3500	24″ radius, 6″ wide tray size		136	.118		55.50	6.45		61.95	70.50
3510	9″ wide tray size		116	.138		68.50	7.55		76.05	87
3520	12″ wide tray size		96	.167		75.50	9.10		84.60	96.50
3530	18″ wide tray size		80	.200		89.50	10.95		100.45	115
3540	24″ wide tray size		64	.250		111	13.70		124.70	143
3560	30″ wide tray size		56	.286		135	15.65		150.65	173
3570	36″ wide tray size		48	.333		167	18.25		185.25	212
3600	36″ radius, 6″ wide tray size		120	.133		95	7.30		102.30	116
3610	9″ wide tray size		104	.154		104	8.40		112.40	127
3620	12″ wide tray size		84	.190		118	10.40		128.40	146
3630	18″ wide tray size		76	.211		140	11.50		151.50	171
3640	24″ wide tray size		56	.286		171	15.65		186.65	212

26 05 36 – Cable Trays for Electrical Systems

26 05 36.40 Cable Tray, Covers and Dividers		Crew	Daily Output	Labor-Hours	Unit	Material	2015 Bare Costs Labor	Equipment	Total	Total Incl O&P
3660	30" wide tray size	2 Elec	50	.320	Ea.	196	17.50		213.50	242
3670	36" wide tray size		44	.364		230	19.90		249.90	283
3700	Elbow horizontal 45°, 12" radius, 6" wide tray size		150	.107		30	5.85		35.85	42
3710	9" wide tray size		128	.125		30.50	6.85		37.35	44.50
3720	12" wide tray size		108	.148		34.50	8.10		42.60	50
3730	18" wide tray size		88	.182		39.50	9.95		49.45	58.50
3740	24" wide tray size		80	.200		44	10.95		54.95	65
3760	30" wide tray size		70	.229		55.50	12.50		68	80
3770	36" wide tray size		64	.250		67	13.70		80.70	94
3800	24" radius, 6" wide tray size		136	.118		34.50	6.45		40.95	47.50
3810	9" wide tray size		116	.138		44	7.55		51.55	60
3820	12" wide tray size		96	.167		45.50	9.10		54.60	64
3830	18" wide tray size		80	.200		55.50	10.95		66.45	77.50
3840	24" wide tray size		72	.222		68.50	12.15		80.65	93.50
3860	30" wide tray size		64	.250		79	13.70		92.70	108
3870	36" wide tray size		56	.286		91.50	15.65		107.15	125
3900	36" radius, 6" wide tray size		120	.133		60	7.30		67.30	77
3910	9" wide tray size		104	.154		65	8.40		73.40	84
3920	12" wide tray size		84	.190		68.50	10.40		78.90	91
3930	18" wide tray size		76	.211		82.50	11.50		94	108
3940	24" wide tray size		64	.250		100	13.70		113.70	131
3960	30" wide tray size		56	.286		114	15.65		129.65	150
3970	36" wide tray size		50	.320		130	17.50		147.50	169
4000	Elbow vertical 90°, 12" radius, 6" wide tray size		150	.107		34.50	5.85		40.35	47
4010	9" wide tray size		128	.125		34.50	6.85		41.35	48.50
4020	12" wide tray size		108	.148		38	8.10		46.10	53.50
4030	18" wide tray size		88	.182		44	9.95		53.95	63.50
4040	24" wide tray size		70	.229		45.50	12.50		58	69.50
4060	30" wide tray size		64	.250		46.50	13.70		60.20	72
4070	36" wide tray size		54	.296		56.50	16.20		72.70	86.50
4100	24" radius, 6" wide tray size		136	.118		39.50	6.45		45.95	53
4110	9" wide tray size		116	.138		43	7.55		50.55	59
4120	12" wide tray size		96	.167		46.50	9.10		55.60	65
4130	18" wide tray size		80	.200		56.50	10.95		67.45	78.50
4140	24" wide tray size		64	.250		61.50	13.70		75.20	88
4160	30" wide tray size		56	.286		79	15.65		94.65	111
4170	36" wide tray size		48	.333		86	18.25		104.25	123
4200	36" radius, 6" wide tray size		120	.133		45.50	7.30		52.80	61.50
4210	9" wide tray size		104	.154		56.50	8.40		64.90	74.50
4220	12" wide tray size		84	.190		65	10.40		75.40	87
4230	18" wide tray size		76	.211		79	11.50		90.50	104
4240	24" wide tray size		56	.286		93	15.65		108.65	126
4260	30" wide tray size		50	.320		118	17.50		135.50	156
4270	36" wide tray size		44	.364		127	19.90		146.90	169
4300	Tee horizontal, 12" radius, 6" wide tray size		108	.148		56.50	8.10		64.60	74
4310	9" wide tray size		88	.182		60	9.95		69.95	81
4320	12" wide tray size		80	.200		67	10.95		77.95	90
4330	18" wide tray size		68	.235		79	12.85		91.85	106
4340	24" wide tray size		56	.286		100	15.65		115.65	134
4360	30" wide tray size		44	.364		118	19.90		137.90	160
4370	36" wide tray size		36	.444		142	24.50		166.50	194
4400	24" radius, 6" wide tray size		96	.167		91.50	9.10		100.60	115
4410	9" wide tray size		80	.200		102	10.95		112.95	128

26 05 Common Work Results for Electrical

26 05 36 – Cable Trays for Electrical Systems

26 05 36.40 Cable Tray, Covers and Dividers		Crew	Daily Output	Labor-Hours	Unit	Material	2015 Bare Costs Labor	Equipment	Total	Total Incl O&P
4420	12" wide tray size	2 Elec	72	.222	Ea.	114	12.15		126.15	144
4430	18" wide tray size		60	.267		134	14.60		148.60	169
4440	24" wide tray size		48	.333		213	18.25		231.25	262
4460	30" wide tray size		40	.400		235	22		257	291
4470	36" wide tray size		32	.500		262	27.50		289.50	330
4500	36" radius, 6" wide tray size		88	.182		167	9.95		176.95	199
4510	9" wide tray size		72	.222		171	12.15		183.15	206
4520	12" wide tray size		64	.250		188	13.70		201.70	227
4530	18" wide tray size		56	.286		213	15.65		228.65	258
4540	24" wide tray size		44	.364		270	19.90		289.90	325
4560	30" wide tray size		36	.444		305	24.50		329.50	370
4570	36" wide tray size		28	.571		345	31.50		376.50	420
4600	Cross horizontal, 12" radius, 6" wide tray size		80	.200		86	10.95		96.95	111
4610	9" wide tray size		72	.222		91.50	12.15		103.65	118
4620	12" wide tray size		64	.250		101	13.70		114.70	133
4630	18" wide tray size		56	.286		123	15.65		138.65	160
4640	24" wide tray size		48	.333		142	18.25		160.25	184
4660	30" wide tray size		40	.400		169	22		191	219
4670	36" wide tray size		32	.500		194	27.50		221.50	255
4700	24" radius, 6" wide tray size		72	.222		169	12.15		181.15	204
4710	9" wide tray size		64	.250		181	13.70		194.70	220
4720	12" wide tray size		56	.286		194	15.65		209.65	238
4730	18" wide tray size		48	.333		226	18.25		244.25	277
4740	24" wide tray size		40	.400		270	22		292	330
4760	30" wide tray size		32	.500		315	27.50		342.50	390
4770	36" wide tray size		24	.667		345	36.50		381.50	430
4800	36" radius, 6" wide tray size		64	.250		270	13.70		283.70	320
4810	9" wide tray size		56	.286		282	15.65		297.65	335
4820	12" wide tray size		50	.320		305	17.50		322.50	360
4830	18" wide tray size		44	.364		335	19.90		354.90	400
4840	24" wide tray size		36	.444		410	24.50		434.50	490
4860	30" wide tray size		28	.571		475	31.50		506.50	565
4870	36" wide tray size		22	.727		520	40		560	630
4900	Reducer, 9" to 6" wide tray size		128	.125		43	6.85		49.85	58
4910	12" to 6" wide tray size		108	.148		45	8.10		53.10	61.50
4920	12" to 9" wide tray size		108	.148		45	8.10		53.10	61.50
4930	18" to 12" wide tray size		88	.182		49	9.95		58.95	69
4950	18" to 6" wide tray size		88	.182		49	9.95		58.95	69
4960	24" to 18" wide tray size		80	.200		63	10.95		73.95	86
4970	24" to 12" wide tray size		80	.200		54.50	10.95		65.45	76.50
4990	30" to 24" wide tray size		70	.229		66	12.50		78.50	91.50
5000	30" to 18" wide tray size		70	.229		66	12.50		78.50	91.50
5010	30" to 12" wide tray size		70	.229		66	12.50		78.50	91.50
5040	36" to 30" wide tray size		64	.250		73	13.70		86.70	101
5050	36" to 24" wide tray size		64	.250		73	13.70		86.70	101
5060	36" to 18" wide tray size		64	.250		73	13.70		86.70	101
5070	36" to 12" wide tray size		64	.250		73	13.70		86.70	101
5710	Tray cover hold down clamp	1 Elec	60	.133		11.70	7.30		19	24
8000	Divider strip, straight, galvanized, 3" deep		200	.040	L.F.	6.25	2.19		8.44	10.15
8020	4" deep		180	.044		7.60	2.43		10.03	12.05
8040	6" deep		160	.050		9.95	2.74		12.69	15.05
8060	Aluminum, straight, 3" deep		210	.038		6.25	2.08		8.33	9.95
8080	4" deep		190	.042		7.70	2.30		10	11.95

26 05 36.40 Cable Tray, Covers and Dividers		Crew	Daily Output	Labor-Hours	Unit	Material	Labor	Equipment	Total	Total Incl O&P
						2015 Bare Costs				
8100	6" deep	1 Elec	170	.047	L.F.	9.85	2.57		12.42	14.65
8110	Divider strip vertical fitting 3" deep									
8120	12" radius, galvanized, 30°	1 Elec	28	.286	Ea.	30	15.65		45.65	56.50
8140	45°		27	.296		37.50	16.20		53.70	66
8160	60°		26	.308		40	16.85		56.85	69
8180	90°		25	.320		50	17.50		67.50	81
8200	Aluminum, 30°		29	.276		20	15.10		35.10	44.50
8220	45°		28	.286		23.50	15.65		39.15	49.50
8240	60°		27	.296		27.50	16.20		43.70	55
8260	90°		26	.308		35	16.85		51.85	63.50
8280	24" radius, galvanized, 30°		25	.320		45.50	17.50		63	76
8300	45°		24	.333		51	18.25		69.25	84
8320	60°		23	.348		64.50	19.05		83.55	99.50
8340	90°		22	.364		88	19.90		107.90	127
8360	Aluminum, 30°		26	.308		33.50	16.85		50.35	62
8380	45°		25	.320		38.50	17.50		56	68.50
8400	60°		24	.333		48.50	18.25		66.75	80.50
8420	90°		23	.348		66	19.05		85.05	101
8440	36" radius, galvanized, 30°		22	.364		61	19.90		80.90	97.50
8460	45°		21	.381		71.50	21		92.50	110
8480	60°		20	.400		83.50	22		105.50	125
8500	90°		19	.421		116	23		139	162
8520	Aluminum, 30°		23	.348		51	19.05		70.05	84.50
8540	45°		22	.364		66	19.90		85.90	103
8560	60°		21	.381		85.50	21		106.50	125
8570	90°		20	.400		112	22		134	156
8590	Divider strip vertical fitting 4" deep									
8600	12" radius, galvanized, 30°	1 Elec	27	.296	Ea.	38.50	16.20		54.70	66.50
8610	45°		26	.308		44.50	16.85		61.35	74
8620	60°		25	.320		50	17.50		67.50	81
8630	90°		24	.333		61	18.25		79.25	95
8640	Aluminum, 30°		28	.286		28.50	15.65		44.15	54.50
8650	45°		27	.296		33	16.20		49.20	61
8660	60°		26	.308		38	16.85		54.85	66.50
8670	90°		25	.320		45	17.50		62.50	75.50
8680	24" radius, galvanized, 30°		24	.333		61	18.25		79.25	95
8690	45°		23	.348		77	19.05		96.05	113
8700	60°		22	.364		87	19.90		106.90	126
8710	90°		21	.381		116	21		137	158
8720	Aluminum, 30°		25	.320		45	17.50		62.50	75.50
8730	45°		24	.333		54.50	18.25		72.75	87.50
8740	60°		23	.348		64	19.05		83.05	99
8750	90°		22	.364		88	19.90		107.90	127
8760	36" radius, galvanized, 30°		23	.348		71.50	19.05		90.55	107
8770	45°		22	.364		81.50	19.90		101.40	120
8780	60°		21	.381		102	21		123	144
8790	90°		20	.400		139	22		161	186
8800	Aluminum, 30°		24	.333		68.50	18.25		86.75	103
8810	45°		23	.348		88	19.05		107.05	125
8820	60°		22	.364		107	19.90		126.90	148
8830	90°		21	.381		135	21		156	180
8840	Divider strip vertical fitting 6" deep									
8850	12" radius, galvanized, 30°	1 Elec	24	.333	Ea.	42	18.25		60.25	73.50

26 05 36 – Cable Trays for Electrical Systems

26 05 36.40 Cable Tray, Covers and Dividers		Crew	Daily Output	Labor-Hours	Unit	Material	2015 Bare Costs Labor	Equipment	Total	Total Incl O&P
8860	45°	1 Elec	23	.348	Ea.	46.50	19.05		65.55	80
8870	60°		22	.364		53.50	19.90		73.40	89
8880	90°		21	.381		67	21		88	105
8890	Aluminum, 30°		25	.320		31	17.50		48.50	60
8900	45°		24	.333		37	18.25		55.25	68
8910	60°		23	.348		38.50	19.05		57.55	71
8920	90°		22	.364		45.50	19.90		65.40	80.50
8930	24" radius, galvanized, 30°		23	.348		61	19.05		80.05	96
8940	45°		22	.364		77	19.90		96.90	115
8950	60°		21	.381		88	21		109	128
8960	90°		20	.400		116	22		138	160
8970	Aluminum, 30°		24	.333		45.50	18.25		63.75	78
8980	45°		23	.348		62.50	19.05		81.55	97
8990	60°		22	.364		68.50	19.90		88.40	106
9000	90°		21	.381		93	21		114	133
9010	36" radius, galvanized, 30°		22	.364		71.50	19.90		91.40	109
9020	45°		21	.381		88	21		109	128
9030	60°		20	.400		116	22		138	160
9040	90°		19	.421		149	23		172	199
9050	Aluminum, 30°		23	.348		69.50	19.05		88.55	105
9060	45°		22	.364		93	19.90		112.90	132
9070	60°		21	.381		110	21		131	152
9080	90°		20	.400		132	22		154	178
9120	Divider strip, horizontal fitting, galvanized, 3" deep		33	.242		43	13.25		56.25	67
9130	4" deep		30	.267		48	14.60		62.60	75
9140	6" deep		27	.296		61	16.20		77.20	92
9150	Aluminum, 3" deep		35	.229		32	12.50		44.50	54
9160	4" deep		32	.250		35	13.70		48.70	59
9170	6" deep		29	.276		46.50	15.10		61.60	74
9300	Divider strip protector		300	.027	L.F.	3.82	1.46		5.28	6.40
9310	Fastener, ladder tray				Ea.	.68			.68	.75
9320	Trough or solid bottom tray				"	.46			.46	.51
9900	Add to labor for higher elevated installation									
9910	15' to 20' high add						10%			
9920	20' to 25' high add						20%			
9930	25' to 30' high add						25%			
9940	30' to 35' high add						30%			
9960	Over 40' high add						40%			

26 05 39 – Underfloor Raceways for Electrical Systems

26 05 39.30 Conduit In Concrete Slab

		Crew	Daily Output	Labor-Hours	Unit	Material	Labor	Equipment	Total	Total Incl O&P
0010	**CONDUIT IN CONCRETE SLAB** Including terminations, R260539-30									
0020	fittings and supports									
3230	PVC, schedule 40, 1/2" diameter	1 Elec	270	.030	L.F.	.57	1.62		2.19	3.06
3250	3/4" diameter		230	.035		.66	1.90		2.56	3.57
3270	1" diameter		200	.040		.87	2.19		3.06	4.23
3300	1-1/4" diameter		170	.047		1.17	2.57		3.74	5.15
3330	1-1/2" diameter		140	.057		1.42	3.13		4.55	6.25
3350	2" diameter		120	.067		1.78	3.65		5.43	7.40
3370	2-1/2" diameter		90	.089		3.07	4.86		7.93	10.65
3400	3" diameter	2 Elec	160	.100		3.90	5.45		9.35	12.50
3430	3-1/2" diameter		120	.133		4.94	7.30		12.24	16.40
3440	4" diameter		100	.160		5.40	8.75		14.15	19.05

26 05 39.30 Conduit In Concrete Slab		Crew	Daily Output	Labor-Hours	Unit	2015 Bare Costs Material	Labor	Equipment	Total	Total Incl O&P
3450	5" diameter	2 Elec	80	.200	L.F.	8.35	10.95		19.30	25.50
3460	6" diameter	"	60	.267	"	11.95	14.60		26.55	35
3530	Sweeps, 1" diameter, 30" radius	1 Elec	32	.250	Ea.	41	13.70		54.70	65.50
3550	1-1/4" diameter		24	.333		44.50	18.25		62.75	76.50
3570	1-1/2" diameter		21	.381		47	21		68	82.50
3600	2" diameter		18	.444		52	24.50		76.50	93.50
3630	2-1/2" diameter		14	.571		67.50	31.50		99	122
3650	3" diameter		10	.800		86	44		130	160
3670	3-1/2" diameter		8	1		105	54.50		159.50	198
3700	4" diameter		7	1.143		147	62.50		209.50	256
3710	5" diameter		6	1.333		168	73		241	294
3730	Couplings, 1/2" diameter					.21			.21	.23
3750	3/4" diameter					.24			.24	.26
3770	1" diameter					.36			.36	.40
3800	1-1/4" diameter					.55			.55	.61
3830	1-1/2" diameter					.68			.68	.75
3850	2" diameter					.88			.88	.97
3870	2-1/2" diameter					1.57			1.57	1.73
3900	3" diameter					2.54			2.54	2.79
3930	3-1/2" diameter					3.21			3.21	3.53
3950	4" diameter					3.59			3.59	3.95
3960	5" diameter					9.10			9.10	10
3970	6" diameter					12.35			12.35	13.60
4030	End bells 1" diameter, PVC	1 Elec	60	.133		3.90	7.30		11.20	15.25
4050	1-1/4" diameter		53	.151		4.80	8.25		13.05	17.65
4100	1-1/2" diameter		48	.167		5.95	9.10		15.05	20
4150	2" diameter		34	.235		7.15	12.85		20	27
4170	2-1/2" diameter		27	.296		7.90	16.20		24.10	33
4200	3" diameter		20	.400		8.30	22		30.30	42
4250	3-1/2" diameter		16	.500		9.90	27.50		37.40	52
4300	4" diameter		14	.571		10.75	31.50		42.25	59
4310	5" diameter		12	.667		11.35	36.50		47.85	67
4320	6" diameter		9	.889		13.55	48.50		62.05	88
4350	Rigid galvanized steel, 1/2" diameter		200	.040	L.F.	2.48	2.19		4.67	6
4400	3/4" diameter		170	.047		2.62	2.57		5.19	6.75
4450	1" diameter		130	.062		3.60	3.37		6.97	9
4500	1-1/4" diameter		110	.073		4.94	3.98		8.92	11.40
4600	1-1/2" diameter		100	.080		5.50	4.38		9.88	12.60
4800	2" diameter		90	.089		6.80	4.86		11.66	14.80

26 05 39.40 Conduit In Trench		Crew	Daily Output	Labor-Hours	Unit	Material	Labor	Equipment	Total	Total Incl O&P
0010	**CONDUIT IN TRENCH** Includes terminations and fittings R260539-40									
0020	Does not include excavation or backfill, see Section 31 23 16.00									
0200	Rigid galvanized steel, 2" diameter	1 Elec	150	.053	L.F.	6.40	2.92		9.32	11.40
0400	2-1/2" diameter	"	100	.080		12.45	4.38		16.83	20.50
0600	3" diameter	2 Elec	160	.100		14.55	5.45		20	24
0800	3-1/2" diameter		140	.114		19.20	6.25		25.45	30.50
1000	4" diameter		100	.160		21	8.75		29.75	36.50
1200	5" diameter		80	.200		44	10.95		54.95	64.50
1400	6" diameter		60	.267		61	14.60		75.60	89.50

26 05 43.10 Trench Duct		Crew	Daily Output	Labor-Hours	Unit	Material	2015 Bare Costs Labor	Equipment	Total	Total Incl O&P
0010	**TRENCH DUCT** Steel with cover									
0020	Standard adjustable, depths to 4"									
0100	Straight, single compartment, 9" wide	2 Elec	40	.400	L.F.	112	22		134	156
0200	12" wide		32	.500		136	27.50		163.50	190
0400	18" wide		26	.615		167	33.50		200.50	234
0600	24" wide		22	.727		199	40		239	279
0700	27" wide		21	.762		213	41.50		254.50	297
0800	30" wide		20	.800		234	44		278	325
1000	36" wide		16	1		265	54.50		319.50	375
1020	Two compartment, 9" wide		38	.421		118	23		141	165
1030	12" wide		30	.533		135	29		164	192
1040	18" wide		24	.667		165	36.50		201.50	237
1050	24" wide		20	.800		216	44		260	305
1060	30" wide		18	.889		263	48.50		311.50	365
1070	36" wide		14	1.143		305	62.50		367.50	430
1090	Three compartment, 9" wide		36	.444		135	24.50		159.50	185
1100	12" wide		28	.571		149	31.50		180.50	211
1110	18" wide		22	.727		187	40		227	266
1120	24" wide		18	.889		235	48.50		283.50	330
1130	30" wide		16	1		284	54.50		338.50	390
1140	36" wide		12	1.333		330	73		403	475
1200	Horizontal elbow, 9" wide		5.40	2.963	Ea.	385	162		547	670
1400	12" wide		4.60	3.478		445	190		635	775
1600	18" wide		4	4		570	219		789	960
1800	24" wide		3.20	5		800	274		1,074	1,300
1900	27" wide		3	5.333		910	292		1,202	1,425
2000	30" wide		2.60	6.154		1,075	335		1,410	1,675
2200	36" wide		2.40	6.667		1,400	365		1,765	2,100
2220	Two compartment, 9" wide		3.80	4.211		625	230		855	1,025
2230	12" wide		3	5.333		690	292		982	1,200
2240	18" wide		2.40	6.667		830	365		1,195	1,450
2250	24" wide		2	8		1,050	440		1,490	1,800
2260	30" wide		1.80	8.889		1,400	485		1,885	2,275
2270	36" wide		1.60	10		1,700	545		2,245	2,700
2290	Three compartment, 9" wide		3.60	4.444		655	243		898	1,075
2300	12" wide		2.80	5.714		730	315		1,045	1,275
2310	18" wide		2.20	7.273		890	400		1,290	1,575
2320	24" wide		1.80	8.889		1,125	485		1,610	1,975
2330	30" wide		1.60	10		1,475	545		2,020	2,425
2350	36" wide		1.40	11.429		1,775	625		2,400	2,875
2400	Vertical elbow, 9" wide		5.40	2.963		135	162		297	390
2600	12" wide		4.60	3.478		146	190		336	445
2800	18" wide		4	4		168	219		387	515
3000	24" wide		3.20	5		209	274		483	640
3100	27" wide		3	5.333		218	292		510	675
3200	30" wide		2.60	6.154		230	335		565	760
3400	36" wide		2.40	6.667		253	365		618	825
3600	Cross, 9" wide		4	4		635	219		854	1,025
3800	12" wide		3.20	5		670	274		944	1,150
4000	18" wide		2.60	6.154		800	335		1,135	1,375
4200	24" wide		2.20	7.273		1,025	400		1,425	1,725
4300	27" wide		2.20	7.273		1,200	400		1,600	1,925

26 05 43.10 Trench Duct		Crew	Daily Output	Labor-Hours	Unit	Material	2015 Bare Costs Labor	Equipment	Total	Total Incl O&P
4400	30" wide	2 Elec	2	8	Ea.	1,325	440		1,765	2,100
4600	36" wide		1.80	8.889		1,650	485		2,135	2,550
4620	Two compartment, 9" wide		3.80	4.211		660	230		890	1,075
4630	12" wide		3	5.333		695	292		987	1,200
4640	18" wide		2.40	6.667		840	365		1,205	1,475
4650	24" wide		2	8		1,075	440		1,515	1,825
4660	30" wide		1.80	8.889		1,400	485		1,885	2,275
4670	36" wide		1.60	10		1,700	545		2,245	2,700
4690	Three compartment, 9" wide		3.60	4.444		670	243		913	1,100
4700	12" wide		2.80	5.714		765	315		1,080	1,325
4710	18" wide		2.20	7.273		910	400		1,310	1,600
4720	24" wide		1.80	8.889		1,125	485		1,610	1,975
4730	30" wide		1.60	10		1,475	545		2,020	2,450
4740	36" wide		1.40	11.429		1,825	625		2,450	2,925
4800	End closure, 9" wide		14.40	1.111		39.50	61		100.50	135
5000	12" wide		12	1.333		45.50	73		118.50	159
5200	18" wide		10	1.600		69.50	87.50		157	208
5400	24" wide		8	2		91.50	109		200.50	265
5500	27" wide		7	2.286		106	125		231	305
5600	30" wide		6.60	2.424		115	133		248	325
5800	36" wide		5.80	2.759		136	151		287	375
6000	Tees, 9" wide		4	4		385	219		604	755
6200	12" wide		3.60	4.444		445	243		688	855
6400	18" wide		3.20	5		570	274		844	1,050
6600	24" wide		3	5.333		820	292		1,112	1,350
6700	27" wide		2.80	5.714		910	315		1,225	1,475
6800	30" wide		2.60	6.154		1,075	335		1,410	1,675
7000	36" wide		2	8		1,400	440		1,840	2,200
7020	Two compartment, 9" wide		3.80	4.211		445	230		675	835
7030	12" wide		3.40	4.706		480	257		737	915
7040	18" wide		3	5.333		640	292		932	1,150
7050	24" wide		2.80	5.714		865	315		1,180	1,425
7060	30" wide		2.40	6.667		1,175	365		1,540	1,850
7070	36" wide		1.90	8.421		1,475	460		1,935	2,325
7090	Three compartment, 9" wide		3.60	4.444		510	243		753	925
7100	12" wide		3.20	5		535	274		809	1,000
7110	18" wide		2.80	5.714		680	315		995	1,225
7120	24" wide		2.60	6.154		935	335		1,270	1,525
7130	30" wide		2.20	7.273		1,225	400		1,625	1,950
7140	36" wide		1.80	8.889		1,550	485		2,035	2,450
7200	Riser, and cabinet connector, 9" wide		5.40	2.963		168	162		330	430
7400	12" wide		4.60	3.478		196	190		386	500
7600	18" wide		4	4		241	219		460	595
7800	24" wide		3.20	5		291	274		565	730
7900	27" wide		3	5.333		300	292		592	765
8000	30" wide		2.60	6.154		335	335		670	875
8200	36" wide		2	8		390	440		830	1,075
8400	Insert assembly, cell to conduit adapter, 1-1/4"	1 Elec	16	.500		66.50	27.50		94	114
8500	Adjustable partition	"	320	.025	L.F.	23.50	1.37		24.87	28
8600	Depth of duct over 4", per 1", add					10.30			10.30	11.35
8700	Support post	1 Elec	240	.033		23.50	1.82		25.32	28.50
8800	Cover double tile trim, 2 sides					36.50			36.50	40
8900	4 sides					108			108	119

26 05 43.10 Trench Duct		Crew	Daily Output	Labor-Hours	Unit	Material (2015 Bare Costs)	Labor (2015 Bare Costs)	Equipment (2015 Bare Costs)	Total (2015 Bare Costs)	Total Incl O&P
9160	Trench duct 3-1/2" x 4-1/2", add				L.F.	9.80			9.80	10.80
9170	Trench duct 4" x 5", add					9.80			9.80	10.80
9200	For carpet trim, add					33			33	36.50
9210	For double carpet trim, add				↓	101			101	111

26 05 43.20 Underfloor Duct		Crew	Daily Output	Labor-Hours	Unit	Material	Labor	Equipment	Total	Total Incl O&P
0010	**UNDERFLOOR DUCT** R260543-50									
0100	Duct, 1-3/8" x 3-1/8" blank, standard	2 Elec	160	.100	L.F.	14.25	5.45		19.70	24
0200	1-3/8" x 7-1/4" blank, super duct		120	.133		28.50	7.30		35.80	42.50
0400	7/8" or 1-3/8" insert type, 24" O.C., 1-3/8" x 3-1/8", std.		140	.114		19.05	6.25		25.30	30.50
0600	1-3/8" x 7-1/4", super duct	↓	100	.160	↓	33.50	8.75		42.25	49.50
0800	Junction box, single duct, 1 level, 3-1/8"	1 Elec	4	2	Ea.	430	109		539	635
0820	3-1/8" x 7-1/4"		4	2		500	109		609	715
0840	2 level, 3-1/8" upper & lower		3.20	2.500		500	137		637	755
0860	3-1/8" upper, 7-1/4" lower		2.70	2.963		500	162		662	795
0880	Carpet pan for above		80	.100		355	5.45		360.45	400
0900	Terrazzo pan for above		67	.119		830	6.55		836.55	920
1000	Junction box, single duct, 1 level, 7-1/4"		2.70	2.963		500	162		662	795
1020	2 level, 7-1/4" upper & lower		2.70	2.963		570	162		732	875
1040	2 duct, two 3-1/8" upper & lower		3.20	2.500		755	137		892	1,050
1200	1 level, 2 duct, 3-1/8"		3.20	2.500		570	137		707	835
1220	Carpet pan for above boxes		80	.100		355	5.45		360.45	400
1240	Terrazzo pan for above boxes		67	.119		830	6.55		836.55	920
1260	Junction box, 1 level, two 3-1/8" x one 3-1/8" + one 7-1/4"		2.30	3.478		970	190		1,160	1,350
1280	2 level, two 3-1/8" upper, one 3-1/8" + one 7-1/4" lower		2	4		1,075	219		1,294	1,500
1300	Carpet pan for above boxes		80	.100		355	5.45		360.45	400
1320	Terrazzo pan for above boxes		67	.119		830	6.55		836.55	920
1400	Junction box, 1 level, 2 duct, 7-1/4"		2.30	3.478		1,475	190		1,665	1,900
1420	Two 3-1/8" + one 7-1/4"		2	4		1,475	219		1,694	1,950
1440	Carpet pan for above		80	.100		355	5.45		360.45	400
1460	Terrazzo pan for above		67	.119		830	6.55		836.55	920
1580	Junction box, 1 level, one 3-1/8" + one 7-1/4" x same		2.30	3.478		970	190		1,160	1,350
1600	Triple duct, 3-1/8"		2.30	3.478		970	190		1,160	1,350
1700	Junction box, 1 level, one 3-1/8" + two 7-1/4"		2	4		1,625	219		1,844	2,125
1720	Carpet pan for above		80	.100		355	5.45		360.45	400
1740	Terrazzo pan for above		67	.119		830	6.55		836.55	920
1800	Insert to conduit adapter, 3/4" & 1"		32	.250		35	13.70		48.70	59
2000	Support, single cell		27	.296		52.50	16.20		68.70	82.50
2200	Super duct		16	.500		52.50	27.50		80	99
2400	Double cell		16	.500		52.50	27.50		80	99
2600	Triple cell		11	.727		52.50	40		92.50	118
2800	Vertical elbow, standard duct		10	.800		94.50	44		138.50	170
3000	Super duct		8	1		94.50	54.50		149	186
3200	Cabinet connector, standard duct		32	.250		71	13.70		84.70	98.50
3400	Super duct		27	.296		71	16.20		87.20	103
3600	Conduit adapter, 1" to 1-1/4"		32	.250		71	13.70		84.70	98.50
3800	2" to 1-1/4"		27	.296		85	16.20		101.20	119
4000	Outlet, low tension (tele, computer, etc.)		8	1		99.50	54.50		154	192
4200	High tension, receptacle (120 volt)		8	1		99.50	54.50		154	192
4300	End closure, standard duct		160	.050		3.98	2.74		6.72	8.50
4310	Super duct		160	.050		7.55	2.74		10.29	12.40
4350	Elbow, horiz., standard duct		26	.308		238	16.85		254.85	286
4360	Super duct	↓	26	.308	↓	238	16.85		254.85	286

26 05 43 – Underground Ducts and Raceways for Electrical Systems

26 05 43.20 Underfloor Duct		Crew	Daily Output	Labor-Hours	Unit	2015 Bare Costs Material	Labor	Equipment	Total	Total Incl O&P
4380	Elbow, offset, standard duct	1 Elec	26	.308	Ea.	94.50	16.85		111.35	129
4390	Super duct		26	.308		94.50	16.85		111.35	129
4400	Marker screw assembly for inserts		50	.160		16.80	8.75		25.55	31.50
4410	Y take off, standard duct		26	.308		142	16.85		158.85	181
4420	Super duct		26	.308		142	16.85		158.85	181
4430	Box opening plug, standard duct		160	.050		13.30	2.74		16.04	18.75
4440	Super duct		160	.050		13.30	2.74		16.04	18.75
4450	Sleeve coupling, standard duct		160	.050		35.50	2.74		38.24	43
4460	Super duct		160	.050		43.50	2.74		46.24	52
4470	Conduit adapter, standard duct, 3/4"		32	.250		71	13.70		84.70	98.50
4480	1" or 1-1/4"		32	.250		71	13.70		84.70	98.50
4500	1-1/2"	↓	32	.250	↓	71	13.70		84.70	98.50

26 05 80 – Wiring Connections

26 05 80.10 Motor Connections		Crew	Daily Output	Labor-Hours	Unit	Material	Labor	Equipment	Total	Total Incl O&P
0010	**MOTOR CONNECTIONS**									
0020	Flexible conduit and fittings, 115 volt, 1 phase, up to 1 HP motor	1 Elec	8	1	Ea.	6.35	54.50		60.85	89
0050	2 HP motor R260580-75		6.50	1.231		12.55	67.50		80.05	115
0100	3 HP motor		5.50	1.455		11.05	79.50		90.55	131
0110	230 volt, 3 phase, 3 HP motor		6.78	1.180		7.65	64.50		72.15	105
0112	5 HP motor		5.47	1.463		6.50	80		86.50	127
0114	7-1/2 HP motor		4.61	1.735		9.55	95		104.55	153
0120	10 HP motor		4.20	1.905		21	104		125	180
0150	15 HP motor		3.30	2.424		21	133		154	223
0200	25 HP motor		2.70	2.963		29	162		191	275
0400	50 HP motor		2.20	3.636		59	199		258	365
0600	100 HP motor		1.50	5.333		140	292		432	590
1500	460 volt, 5 HP motor, 3 phase		8	1		6.85	54.50		61.35	89.50
1520	10 HP motor		8	1		6.85	54.50		61.35	89.50
1530	25 HP motor		6	1.333		13.45	73		86.45	124
1540	30 HP motor		6	1.333		13.45	73		86.45	124
1550	40 HP motor		5	1.600		20	87.50		107.50	153
1560	50 HP motor		5	1.600		23	87.50		110.50	156
1570	60 HP motor		3.80	2.105		25	115		140	201
1580	75 HP motor		3.50	2.286		34.50	125		159.50	225
1590	100 HP motor		2.50	3.200		53	175		228	320
1600	125 HP motor		2	4		68	219		287	405
1610	150 HP motor		1.80	4.444		72.50	243		315.50	445
1620	200 HP motor		1.50	5.333		116	292		408	565
2005	460 Volt, 5 HP motor, 3 Phase, w/sealtite		8	1		10.80	54.50		65.30	94
2010	10 HP motor		8	1		10.80	54.50		65.30	94
2015	25 HP motor		6	1.333		21	73		94	132
2020	30 HP motor		6	1.333		21	73		94	132
2025	40 HP motor		5	1.600		37.50	87.50		125	173
2030	50 HP motor		5	1.600		38.50	87.50		126	173
2035	60 HP motor		3.80	2.105		59.50	115		174.50	238
2040	75 HP motor		3.50	2.286		62.50	125		187.50	256
2045	100 HP motor		2.50	3.200		85.50	175		260.50	355
2055	150 HP motor		1.80	4.444		91	243		334	465
2060	200 HP motor	↓	1.50	5.333	↓	605	292		897	1,100

26 05 90.10 Residential Wiring		Crew	Daily Output	Labor-Hours	Unit	Material	2015 Bare Costs Labor	Equipment	Total	Total Incl O&P
0010	**RESIDENTIAL WIRING**									
0020	20' avg. runs and #14/2 wiring incl. unless otherwise noted									
1000	Service & panel, includes 24' SE-AL cable, service eye, meter,									
1010	Socket, panel board, main bkr., ground rod, 15 or 20 amp									
1020	1-pole circuit breakers, and misc. hardware									
1100	100 amp, with 10 branch breakers	1 Elec	1.19	6.723	Ea.	560	370		930	1,175
1110	With PVC conduit and wire		.92	8.696		605	475		1,080	1,375
1120	With RGS conduit and wire		.73	10.959		765	600		1,365	1,750
1150	150 amp, with 14 branch breakers		1.03	7.767		860	425		1,285	1,575
1170	With PVC conduit and wire		.82	9.756		955	535		1,490	1,850
1180	With RGS conduit and wire		.67	11.940		1,275	655		1,930	2,375
1200	200 amp, with 18 branch breakers	2 Elec	1.80	8.889		1,150	485		1,635	2,000
1220	With PVC conduit and wire		1.46	10.959		1,250	600		1,850	2,275
1230	With RGS conduit and wire		1.24	12.903		1,650	705		2,355	2,850
1800	Lightning surge suppressor	1 Elec	32	.250		50.50	13.70		64.20	76.50
2000	Switch devices									
2100	Single pole, 15 amp, Ivory, with a 1-gang box, cover plate,									
2110	Type NM (Romex) cable	1 Elec	17.10	.468	Ea.	12.35	25.50		37.85	52
2120	Type MC (BX) cable		14.30	.559		21.50	30.50		52	69.50
2130	EMT & wire		5.71	1.401		30	76.50		106.50	148
2150	3-way, #14/3, type NM cable		14.55	.550		14.85	30		44.85	61.50
2170	Type MC cable		12.31	.650		27	35.50		62.50	83
2180	EMT & wire		5	1.600		32	87.50		119.50	166
2200	4-way, #14/3, type NM cable		14.55	.550		21	30		51	68
2220	Type MC cable		12.31	.650		33	35.50		68.50	90
2230	EMT & wire		5	1.600		38.50	87.50		126	173
2250	S.P., 20 amp, #12/2, type NM cable		13.33	.600		22	33		55	73
2270	Type MC cable		11.43	.700		28.50	38.50		67	89
2280	EMT & wire		4.85	1.649		41	90		131	180
2290	S.P. rotary dimmer, 600W, no wiring		17	.471		29.50	25.50		55	71
2300	S.P. rotary dimmer, 600W, type NM cable		14.55	.550		34.50	30		64.50	83
2320	Type MC cable		12.31	.650		43.50	35.50		79	102
2330	EMT & wire		5	1.600		53.50	87.50		141	190
2350	3-way rotary dimmer, type NM cable		13.33	.600		28.50	33		61.50	80.50
2370	Type MC cable		11.43	.700		37.50	38.50		76	98.50
2380	EMT & wire		4.85	1.649		47.50	90		137.50	188
2400	Interval timer wall switch, 20 amp, 1-30 min., #12/2									
2410	Type NM cable	1 Elec	14.55	.550	Ea.	56	30		86	107
2420	Type MC cable		12.31	.650		60	35.50		95.50	120
2430	EMT & wire		5	1.600		75	87.50		162.50	214
2500	Decorator style									
2510	S.P., 15 amp, type NM cable	1 Elec	17.10	.468	Ea.	16.25	25.50		41.75	56.50
2520	Type MC cable		14.30	.559		25	30.50		55.50	73.50
2530	EMT & wire		5.71	1.401		34	76.50		110.50	152
2550	3-way, #14/3, type NM cable		14.55	.550		18.75	30		48.75	65.50
2570	Type MC cable		12.31	.650		30.50	35.50		66	87
2580	EMT & wire		5	1.600		36	87.50		123.50	171
2600	4-way, #14/3, type NM cable		14.55	.550		25	30		55	72.50
2620	Type MC cable		12.31	.650		37	35.50		72.50	94
2630	EMT & wire		5	1.600		42	87.50		129.50	178
2650	S.P., 20 amp, #12/2, type NM cable		13.33	.600		26	33		59	77.50
2670	Type MC cable		11.43	.700		32.50	38.50		71	93

26 05 90.10 Residential Wiring		Crew	Daily Output	Labor-Hours	Unit	Material	Labor	Equipment	Total	Total Incl O&P
						2015 Bare Costs				
2680	EMT & wire	1 Elec	4.85	1.649	Ea.	45	90		135	185
2700	S.P., slide dimmer, type NM cable		17.10	.468		28.50	25.50		54	70
2720	Type MC cable		14.30	.559		37.50	30.50		68	87.50
2730	EMT & wire		5.71	1.401		47.50	76.50		124	168
2750	S.P., touch dimmer, type NM cable		17.10	.468		32	25.50		57.50	74
2770	Type MC cable		14.30	.559		41	30.50		71.50	91.50
2780	EMT & wire		5.71	1.401		51.50	76.50		128	172
2800	3-way touch dimmer, type NM cable		13.33	.600		52	33		85	106
2820	Type MC cable		11.43	.700		60.50	38.50		99	125
2830	EMT & wire	↓	4.85	1.649	↓	71	90		161	213
3000	Combination devices									
3100	S.P. switch/15 amp recpt., Ivory, 1-gang box, plate									
3110	Type NM cable	1 Elec	11.43	.700	Ea.	24.50	38.50		63	84.50
3120	Type MC cable		10	.800		33.50	44		77.50	102
3130	EMT & wire		4.40	1.818		43.50	99.50		143	197
3150	S.P. switch/pilot light, type NM cable		11.43	.700		24.50	38.50		63	84.50
3170	Type MC cable		10	.800		33.50	44		77.50	103
3180	EMT & wire		4.43	1.806		43.50	99		142.50	196
3190	2-S.P. switches, 2-#14/2, no wiring		14	.571		6.90	31.50		38.40	54.50
3200	2-S.P. switches, 2-#14/2, type NM cables		10	.800		27.50	44		71.50	95.50
3220	Type MC cable		8.89	.900		41	49		90	119
3230	EMT & wire		4.10	1.951		46.50	107		153.50	211
3250	3-way switch/15 amp recpt., #14/3, type NM cable		10	.800		32	44		76	101
3270	Type MC cable		8.89	.900		44	49		93	122
3280	EMT & wire		4.10	1.951		49	107		156	214
3300	2-3 way switches, 2-#14/3, type NM cables		8.89	.900		41.50	49		90.50	120
3320	Type MC cable		8	1		60.50	54.50		115	149
3330	EMT & wire		4	2		56.50	109		165.50	226
3350	S.P. switch/20 amp recpt., #12/2, type NM cable		10	.800		32	44		76	101
3370	Type MC cable		8.89	.900		36	49		85	114
3380	EMT & wire	↓	4.10	1.951	↓	51	107		158	216
3400	Decorator style									
3410	S.P. switch/15 amp recpt., type NM cable	1 Elec	11.43	.700	Ea.	28.50	38.50		67	88.50
3420	Type MC cable		10	.800		37	44		81	107
3430	EMT & wire		4.40	1.818		47.50	99.50		147	201
3450	S.P. switch/pilot light, type NM cable		11.43	.700		28.50	38.50		67	88.50
3470	Type MC cable		10	.800		37.50	44		81.50	107
3480	EMT & wire		4.40	1.818		47.50	99.50		147	202
3500	2-S.P. switches, 2-#14/2, type NM cables		10	.800		31.50	44		75.50	100
3520	Type MC cable		8.89	.900		44.50	49		93.50	123
3530	EMT & wire		4.10	1.951		50.50	107		157.50	216
3550	3-way/15 amp recpt., #14/3, type NM cable		10	.800		36	44		80	105
3570	Type MC cable		8.89	.900		47.50	49		96.50	127
3580	EMT & wire		4.10	1.951		53	107		160	218
3650	2-3 way switches, 2-#14/3, type NM cables		8.89	.900		45.50	49		94.50	124
3670	Type MC cable		8	1		64.50	54.50		119	153
3680	EMT & wire		4	2		60.50	109		169.50	231
3700	S.P. switch/20 amp recpt., #12/2, type NM cable		10	.800		35.50	44		79.50	105
3720	Type MC cable		8.89	.900		39.50	49		88.50	118
3730	EMT & wire	↓	4.10	1.951	↓	55	107		162	221
4000	Receptacle devices									
4010	Duplex outlet, 15 amp recpt., Ivory, 1-gang box, plate									
4015	Type NM cable	1 Elec	14.55	.550	Ea.	10.80	30		40.80	57

26 05 90.10 Residential Wiring		Crew	Daily Output	Labor-Hours	Unit	2015 Bare Costs Material	Labor	Equipment	Total	Total Incl O&P
4020	Type MC cable	1 Elec	12.31	.650	Ea.	19.75	35.50		55.25	75
4030	EMT & wire		5.33	1.501		28.50	82		110.50	155
4050	With #12/2, type NM cable		12.31	.650		13.30	35.50		48.80	68
4070	Type MC cable		10.67	.750		19.85	41		60.85	83.50
4080	EMT & wire		4.71	1.699		32.50	93		125.50	175
4100	20 amp recpt., #12/2, type NM cable		12.31	.650		19.95	35.50		55.45	75.50
4120	Type MC cable		10.67	.750		26.50	41		67.50	90.50
4130	EMT & wire		4.71	1.699		39	93		132	182
4140	For GFI see Section 26 05 90.10 line 4300 below									
4150	Decorator style, 15 amp recpt., type NM cable	1 Elec	14.55	.550	Ea.	14.70	30		44.70	61
4170	Type MC cable		12.31	.650		23.50	35.50		59	79.50
4180	EMT & wire		5.33	1.501		32.50	82		114.50	159
4200	With #12/2, type NM cable		12.31	.650		17.20	35.50		52.70	72.50
4220	Type MC cable		10.67	.750		23.50	41		64.50	87.50
4230	EMT & wire		4.71	1.699		36.50	93		129.50	179
4250	20 amp recpt. #12/2, type NM cable		12.31	.650		24	35.50		59.50	79.50
4270	Type MC cable		10.67	.750		30.50	41		71.50	95
4280	EMT & wire		4.71	1.699		43	93		136	187
4300	GFI, 15 amp recpt., type NM cable		12.31	.650		41.50	35.50		77	99
4320	Type MC cable		10.67	.750		50.50	41		91.50	117
4330	EMT & wire		4.71	1.699		59	93		152	204
4350	GFI with #12/2, type NM cable		10.67	.750		44	41		85	110
4370	Type MC cable		9.20	.870		50.50	47.50		98	127
4380	EMT & wire		4.21	1.900		63	104		167	226
4400	20 amp recpt., #12/2 type NM cable		10.67	.750		52	41		93	119
4420	Type MC cable		9.20	.870		58.50	47.50		106	136
4430	EMT & wire		4.21	1.900		71.50	104		175.50	235
4500	Weather-proof cover for above receptacles, add		32	.250		3.46	13.70		17.16	24.50
4550	Air conditioner outlet, 20 amp-240 volt recpt.									
4560	30′ of #12/2, 2 pole circuit breaker									
4570	Type NM cable	1 Elec	10	.800	Ea.	63	44		107	135
4580	Type MC cable		9	.889		70.50	48.50		119	151
4590	EMT & wire		4	2		82.50	109		191.50	255
4600	Decorator style, type NM cable		10	.800		68	44		112	141
4620	Type MC cable		9	.889		75.50	48.50		124	156
4630	EMT & wire		4	2		87	109		196	260
4650	Dryer outlet, 30 amp-240 volt recpt., 20′ of #10/3									
4660	2 pole circuit breaker									
4670	Type NM cable	1 Elec	6.41	1.248	Ea.	61.50	68.50		130	170
4680	Type MC cable		5.71	1.401		64.50	76.50		141	186
4690	EMT & wire		3.48	2.299		76	126		202	272
4700	Range outlet, 50 amp-240 volt recpt., 30′ of #8/3									
4710	Type NM cable	1 Elec	4.21	1.900	Ea.	90.50	104		194.50	256
4720	Type MC cable		4	2		124	109		233	300
4730	EMT & wire		2.96	2.703		108	148		256	340
4750	Central vacuum outlet, Type NM cable		6.40	1.250		58.50	68.50		127	167
4770	Type MC cable		5.71	1.401		69.50	76.50		146	192
4780	EMT & wire		3.48	2.299		83.50	126		209.50	280
4800	30 amp-110 volt locking recpt., #10/2 circ. bkr.									
4810	Type NM cable	1 Elec	6.20	1.290	Ea.	67	70.50		137.50	180
4820	Type MC cable		5.40	1.481		80	81		161	209
4830	EMT & wire		3.20	2.500		94	137		231	310
4900	Low voltage outlets									

26 05 90.10 Residential Wiring		Crew	Daily Output	Labor-Hours	Unit	Material	2015 Bare Costs Labor	Equipment	Total	Total Incl O&P
4910	Telephone recpt., 20′ of 4/C phone wire	1 Elec	26	.308	Ea.	9.30	16.85		26.15	35.50
4920	TV recpt., 20′ of RG59U coax wire, F type connector	"	16	.500	"	18.75	27.50		46.25	61.50
4950	Door bell chime, transformer, 2 buttons, 60′ of bellwire									
4970	Economy model	1 Elec	11.50	.696	Ea.	51.50	38		89.50	114
4980	Custom model		11.50	.696		99	38		137	166
4990	Luxury model, 3 buttons		9.50	.842		281	46		327	380
6000	Lighting outlets									
6050	Wire only (for fixture), type NM cable	1 Elec	32	.250	Ea.	6.90	13.70		20.60	28
6070	Type MC cable		24	.333		12.30	18.25		30.55	41
6080	EMT & wire		10	.800		20	44		64	87.50
6100	Box (4″), and wire (for fixture), type NM cable		25	.320		14.45	17.50		31.95	42
6120	Type MC cable		20	.400		19.90	22		41.90	55
6130	EMT & wire		11	.727		27.50	40		67.50	90
6200	Fixtures (use with lines 6050 or 6100 above)									
6210	Canopy style, economy grade	1 Elec	40	.200	Ea.	32	10.95		42.95	51.50
6220	Custom grade		40	.200		53.50	10.95		64.45	75.50
6250	Dining room chandelier, economy grade		19	.421		79.50	23		102.50	122
6260	Custom grade		19	.421		315	23		338	380
6270	Luxury grade		15	.533		715	29		744	830
6310	Kitchen fixture (fluorescent), economy grade		30	.267		71.50	14.60		86.10	101
6320	Custom grade		25	.320		219	17.50		236.50	267
6350	Outdoor, wall mounted, economy grade		30	.267		30	14.60		44.60	55
6360	Custom grade		30	.267		120	14.60		134.60	154
6370	Luxury grade		25	.320		248	17.50		265.50	299
6410	Outdoor PAR floodlights, 1 lamp, 150 watt		20	.400		32.50	22		54.50	68.50
6420	2 lamp, 150 watt each		20	.400		53.50	22		75.50	92
6425	Motion sensing, 2 lamp, 150 watt each		20	.400		87	22		109	129
6430	For infrared security sensor, add		32	.250		132	13.70		145.70	166
6450	Outdoor, quartz-halogen, 300 watt flood		20	.400		39	22		61	75.50
6600	Recessed downlight, round, pre-wired, 50 or 75 watt trim		30	.267		80	14.60		94.60	110
6610	With shower light trim		30	.267		89	14.60		103.60	120
6620	With wall washer trim		28	.286		99.50	15.65		115.15	133
6630	With eye-ball trim		28	.286		99.50	15.65		115.15	133
6700	Porcelain lamp holder		40	.200		2.96	10.95		13.91	19.65
6710	With pull switch		40	.200		6.55	10.95		17.50	23.50
6750	Fluorescent strip, 2-20 watt tube, wrap around diffuser, 24″		24	.333		49.50	18.25		67.75	82
6760	1-34 watt tube, 48″		24	.333		87	18.25		105.25	124
6770	2-34 watt tubes, 48″		20	.400		103	22		125	146
6800	Bathroom heat lamp, 1-250 watt		28	.286		44	15.65		59.65	71.50
6810	2-250 watt lamps		28	.286		70	15.65		85.65	101
6820	For timer switch, see Section 26 05 90.10 line 2400									
6900	Outdoor post lamp, incl. post, fixture, 35′ of #14/2									
6910	Type NMC cable	1 Elec	3.50	2.286	Ea.	258	125		383	470
6920	Photo-eye, add		27	.296		32	16.20		48.20	60
6950	Clock dial time switch, 24 hr., w/enclosure, type NM cable		11.43	.700		73.50	38.50		112	139
6970	Type MC cable		11	.727		82.50	40		122.50	150
6980	EMT & wire		4.85	1.649		91	90		181	235
7000	Alarm systems									
7050	Smoke detectors, box, #14/3, type NM cable	1 Elec	14.55	.550	Ea.	34.50	30		64.50	83
7070	Type MC cable		12.31	.650		44	35.50		79.50	102
7080	EMT & wire		5	1.600		49.50	87.50		137	186
7090	For relay output to security system, add					12			12	13.20
8000	Residential equipment									

26 05 Common Work Results for Electrical

26 05 90 – Residential Applications

26 05 90.10 Residential Wiring		Crew	Daily Output	Labor-Hours	Unit	2015 Bare Costs Material	Labor	Equipment	Total	Total Incl O&P
8050	Disposal hook-up, incl. switch, outlet box, 3′ of flex									
8060	20 amp-1 pole circ. bkr., and 25′ of #12/2									
8070	Type NM cable	1 Elec	10	.800	Ea.	32.50	44		76.50	101
8080	Type MC cable		8	1		39.50	54.50		94	126
8090	EMT & wire		5	1.600		54.50	87.50		142	191
8100	Trash compactor or dishwasher hook-up, incl. outlet box,									
8110	3′ of flex, 15 amp-1 pole circ. bkr., and 25′ of #14/2									
8120	Type NM cable	1 Elec	10	.800	Ea.	24	44		68	91.50
8130	Type MC cable		8	1		34	54.50		88.50	120
8140	EMT & wire		5	1.600		46	87.50		133.50	182
8150	Hot water sink dispensor hook-up, use line 8100									
8200	Vent/exhaust fan hook-up, type NM cable	1 Elec	32	.250	Ea.	6.90	13.70		20.60	28
8220	Type MC cable		24	.333		12.30	18.25		30.55	41
8230	EMT & wire		10	.800		20	44		64	87.50
8250	Bathroom vent fan, 50 CFM (use with above hook-up)									
8260	Economy model	1 Elec	15	.533	Ea.	23	29		52	69
8270	Low noise model		15	.533		41	29		70	88.50
8280	Custom model		12	.667		127	36.50		163.50	195
8300	Bathroom or kitchen vent fan, 110 CFM									
8310	Economy model	1 Elec	15	.533	Ea.	67.50	29		96.50	118
8320	Low noise model	″	15	.533	″	92	29		121	145
8350	Paddle fan, variable speed (w/o lights)									
8360	Economy model (AC motor)	1 Elec	10	.800	Ea.	109	44		153	186
8362	With light kit		10	.800		150	44		194	231
8370	Custom model (AC motor)		10	.800		227	44		271	315
8372	With light kit		10	.800		268	44		312	360
8380	Luxury model (DC motor)		8	1		330	54.50		384.50	445
8382	With light kit		8	1		375	54.50		429.50	490
8390	Remote speed switch for above, add		12	.667		37.50	36.50		74	95.50
8500	Whole house exhaust fan, ceiling mount, 36″, variable speed									
8510	Remote switch, incl. shutters, 20 amp-1 pole circ. bkr.									
8520	30′ of #12/2, type NM cable	1 Elec	4	2	Ea.	1,300	109		1,409	1,600
8530	Type MC cable		3.50	2.286		1,300	125		1,425	1,600
8540	EMT & wire		3	2.667		1,325	146		1,471	1,675
8600	Whirlpool tub hook-up, incl. timer switch, outlet box									
8610	3′ of flex, 20 amp-1 pole GFI circ. bkr.									
8620	30′ of #12/2, type NM cable	1 Elec	5	1.600	Ea.	142	87.50		229.50	287
8630	Type MC cable		4.20	1.905		145	104		249	315
8640	EMT & wire		3.40	2.353		158	129		287	365
8650	Hot water heater hook-up, incl. 1-2 pole circ. bkr., box;									
8660	3′ of flex, 20′ of #10/2, type NM cable	1 Elec	5	1.600	Ea.	33	87.50		120.50	168
8670	Type MC cable		4.20	1.905		43	104		147	204
8680	EMT & wire		3.40	2.353		48.50	129		177.50	246
9000	Heating/air conditioning									
9050	Furnace/boiler hook-up, incl. firestat, local on-off switch									
9060	Emergency switch, and 40′ of type NM cable	1 Elec	4	2	Ea.	52.50	109		161.50	222
9070	Type MC cable		3.50	2.286		65.50	125		190.50	260
9080	EMT & wire		1.50	5.333		83	292		375	525
9100	Air conditioner hook-up, incl. local 60 amp disc. switch									
9110	3′ sealtite, 40 amp, 2 pole circuit breaker									
9130	40′ of #8/2, type NM cable	1 Elec	3.50	2.286	Ea.	163	125		288	365
9140	Type MC cable		3	2.667		216	146		362	455
9150	EMT & wire		1.30	6.154		204	335		539	730

26 05 Common Work Results for Electrical

26 05 90 – Residential Applications

26 05 90.10 Residential Wiring		Crew	Daily Output	Labor-Hours	Unit	Material	2015 Bare Costs Labor	Equipment	Total	Total Incl O&P
9200	Heat pump hook-up, 1-40 & 1-100 amp 2 pole circ. bkr.									
9210	Local disconnect switch, 3′ sealtite									
9220	40′ of #8/2 & 30′ of #3/2									
9230	Type NM cable	1 Elec	1.30	6.154	Ea.	520	335		855	1,075
9240	Type MC cable		1.08	7.407		535	405		940	1,200
9250	EMT & wire		.94	8.511		585	465		1,050	1,350
9500	Thermostat hook-up, using low voltage wire									
9520	Heating only, 25′ of #18-3	1 Elec	24	.333	Ea.	8.80	18.25		27.05	37
9530	Heating/cooling, 25′ of #18-4	"	20	.400	"	11.45	22		33.45	45.50

26 09 Instrumentation and Control for Electrical Systems

26 09 13 – Electrical Power Monitoring

26 09 13.10 Switchboard Instruments

		Crew	Daily Output	Labor-Hours	Unit	Material	Labor	Equipment	Total	Total Incl O&P
0010	**SWITCHBOARD INSTRUMENTS** 3 phase, 4 wire R260913-80									
0100	AC indicating, ammeter & switch	1 Elec	8	1	Ea.	2,500	54.50		2,554.50	2,825
0200	Voltmeter & switch		8	1		2,500	54.50		2,554.50	2,825
0300	Wattmeter		8	1		4,250	54.50		4,304.50	4,750
0400	AC recording, ammeter		4	2		7,550	109		7,659	8,500
0500	Voltmeter		4	2		7,550	109		7,659	8,500
0600	Ground fault protection, zero sequence		2.70	2.963		6,675	162		6,837	7,600
0700	Ground return path		2.70	2.963		6,675	162		6,837	7,600
0800	3 current transformers, 5 to 800 amp		2	4		3,100	219		3,319	3,750
0900	1000 to 1500 amp		1.30	6.154		4,475	335		4,810	5,425
1200	2000 to 4000 amp		1	8		5,275	440		5,715	6,450
1300	Fused potential transformer, maximum 600 volt		8	1		1,175	54.50		1,229.50	1,350

26 09 13.20 Voltage Monitor Systems

		Crew	Daily Output	Labor-Hours	Unit	Material	Labor	Equipment	Total	Total Incl O&P
0010	**VOLTAGE MONITOR SYSTEMS** (test equipment)									
0100	AC voltage monitor system, 120/240 V, one-channel				Ea.	2,750			2,750	3,025
0110	Modem adapter					345			345	380
0120	Add-on detector only					1,450			1,450	1,600
0150	AC voltage remote monitor sys., 3 channel, 120, 230, or 480 V					5,000			5,000	5,500
0160	With internal modem					5,275			5,275	5,825
0170	Combination temperature and humidity probe					775			775	855
0180	Add-on detector only					3,625			3,625	4,000
0190	With internal modem					3,950			3,950	4,350

26 09 13.30 Smart Metering

		Crew	Daily Output	Labor-Hours	Unit	Material	Labor	Equipment	Total	Total Incl O&P
0010	**SMART METERING**, In panel									
0100	Single phase, 120/208 volt, 100 amp G	1 Elec	8.78	.911	Ea.	375	50		425	485
0120	200 amp G		8.78	.911		375	50		425	485
0200	277 volt, 100 amp G		8.78	.911		400	50		450	515
0220	200 amp G		8.78	.911		400	50		450	515
1100	Three phase, 120/208 volt, 100 amp G		4.69	1.706		690	93.50		783.50	900
1120	200 amp G		4.69	1.706		690	93.50		783.50	900
1130	400 amp G		4.69	1.706		690	93.50		783.50	900
1140	800 amp G		4.69	1.706		690	93.50		783.50	900
1150	1600 amp G		4.69	1.706		690	93.50		783.50	900
1200	277/480 volt, 100 amp G		4.69	1.706		775	93.50		868.50	990
1220	200 amp G		4.69	1.706		775	93.50		868.50	990
1230	400 amp G		4.69	1.706		775	93.50		868.50	990
1240	800 amp G		4.69	1.706		775	93.50		868.50	990

26 09 Instrumentation and Control for Electrical Systems

26 09 13 – Electrical Power Monitoring

26 09 13.30 Smart Metering			Crew	Daily Output	Labor-Hours	Unit	2015 Bare Costs Material	Labor	Equipment	Total	Total Incl O&P
1250	1600 amp	G	1 Elec	4.69	1.706	Ea.	785	93.50		878.50	1,000
2000	Data recorder, 8 meters	G		10.97	.729		1,400	40		1,440	1,575
2100	16 meters	G		8.53	.938		1,950	51.50		2,001.50	2,225
3000	Software package, per meter, basic	G					236			236	260
3100	Premium	G					610			610	675

26 09 23 – Lighting Control Devices

26 09 23.10 Energy Saving Lighting Devices			Crew	Daily Output	Labor-Hours	Unit	Material	Labor	Equipment	Total	Total Incl O&P
0010	**ENERGY SAVING LIGHTING DEVICES**										
0100	Occupancy sensors, passive infrared ceiling mounted	G	1 Elec	7	1.143	Ea.	81	62.50		143.50	183
0110	Ultrasonic ceiling mounted	G		7	1.143		89.50	62.50		152	192
0120	Dual technology ceiling mounted	G		6.50	1.231		131	67.50		198.50	245
0150	Automatic wall switches	G		24	.333		64.50	18.25		82.75	98.50
0160	Daylighting sensor, manual control, ceiling mounted	G		7	1.143		111	62.50		173.50	216
0170	Remote and dimming control with remote controller	G		6.50	1.231		152	67.50		219.50	268
0200	Remote power pack	G		10	.800		31	44		75	99.50
0250	Photoelectric control, S.P.S.T. 120 V	G		8	1		19.35	54.50		73.85	104
0300	S.P.S.T. 208 V/277 V	G		8	1		28.50	54.50		83	114
0350	D.P.S.T. 120 V	G		6	1.333		217	73		290	350
0400	D.P.S.T. 208 V/277 V	G		6	1.333		176	73		249	305
0450	S.P.D.T. 208 V/277 V	G		6	1.333		211	73		284	340
0460	Daylight level sensor, wall mounted, on/off or dimming	G		8	1		135	54.50		189.50	231

26 09 36 – Modular Dimming Controls

26 09 36.13 Manual Modular Dimming Controls			Crew	Daily Output	Labor-Hours	Unit	Material	Labor	Equipment	Total	Total Incl O&P
0010	**MANUAL MODULAR DIMMING CONTROLS**										
2000	Lighting control module	G	1 Elec	2	4	Ea.	355	219		574	720

26 12 Medium-Voltage Transformers

26 12 19 – Pad-Mounted, Liquid-Filled, Medium-Voltage Transformers

26 12 19.10 Transformer, Oil-Filled		Crew	Daily Output	Labor-Hours	Unit	Material	Labor	Equipment	Total	Total Incl O&P
0010	**TRANSFORMER, OIL-FILLED** primary delta or Y, R262213-60									
0050	Pad mounted 5 kV or 15 kV, with taps, 277/480 V secondary, 3 phase									
0100	150 kVA	R-3	.65	30.769	Ea.	9,250	1,675	212	11,137	12,900
0110	225 kVA		.55	36.364		10,500	1,975	251	12,726	14,900
0200	300 kVA		.45	44.444		13,200	2,425	305	15,930	18,500
0300	500 kVA		.40	50		18,700	2,725	345	21,770	25,100
0400	750 kVA		.38	52.632		23,700	2,850	365	26,915	30,800
0500	1000 kVA		.26	76.923		28,100	4,175	530	32,805	37,800
0600	1500 kVA		.23	86.957		33,400	4,725	600	38,725	44,500
0700	2000 kVA		.20	100		42,100	5,425	690	48,215	55,500
0710	2500 kVA		.19	105		51,000	5,725	725	57,450	65,500
0720	3000 kVA		.17	117		61,500	6,400	810	68,710	78,000
0800	3750 kVA		.16	125		79,000	6,800	860	86,660	98,000
1990	Pole mounted distribution type, single phase									
2000	13.8 kV primary, 120/240 V secondary, 10 kVA	R-15	7.45	6.443	Ea.	915	345	40.50	1,300.50	1,575
2010	50 kVA		3.70	12.973		1,825	700	81.50	2,606.50	3,175
2020	100 kVA		2.75	17.455		3,200	940	110	4,250	5,050
2030	167 kVA		2.15	22.326		5,675	1,200	140	7,015	8,200
2900	2400 V primary, 120/240 V secondary, 10 kVA		7.45	6.443		965	345	40.50	1,350.50	1,625
2910	15 kVA		6.70	7.164		1,225	385	45	1,655	1,975
2920	25 kVA		6	8		1,525	430	50	2,005	2,375

26 12 Medium-Voltage Transformers

26 12 19 – Pad-Mounted, Liquid-Filled, Medium-Voltage Transformers

26 12 19.10	Transformer, Oil-Filled	Crew	Daily Output	Labor-Hours	Unit	Material	2015 Bare Costs Labor	Equipment	Total	Total Incl O&P
2930	37.5 kVA	R-15	4.30	11.163	Ea.	1,875	600	70	2,545	3,050
2940	50 kVA		3.70	12.973		2,225	700	81.50	3,006.50	3,600
2950	75 kVA		3	16		3,250	860	100	4,210	4,975
2960	100 kVA		2.75	17.455		3,225	940	110	4,275	5,075

26 12 19.20 Transformer, Liquid-Filled

26 12 19.20	Transformer, Liquid-Filled	Crew	Daily Output	Labor-Hours	Unit	Material	Labor	Equipment	Total	Total Incl O&P
0010	**TRANSFORMER, LIQUID-FILLED** Pad mounted									
0020	5 kV or 15 kV primary, 277/480 volt secondary, 3 phase									
0050	225 kVA	R-3	.55	36.364	Ea.	13,600	1,975	251	15,826	18,300
0100	300 kVA		.45	44.444		16,200	2,425	305	18,930	21,800
0200	500 kVA		.40	50		20,400	2,725	345	23,470	27,000
0250	750 kVA		.38	52.632		26,400	2,850	365	29,615	33,700
0300	1000 kVA		.26	76.923		30,600	4,175	530	35,305	40,600
0350	1500 kVA		.23	86.957		35,800	4,725	600	41,125	47,100
0400	2000 kVA		.20	100		44,300	5,425	690	50,415	57,500
0450	2500 kVA		.19	105		53,000	5,725	725	59,450	67,500

26 13 Medium-Voltage Switchgear

26 13 16 – Medium-Voltage Fusible Interrupter Switchgear

26 13 16.10 Switchgear

26 13 16.10	Switchgear	Crew	Daily Output	Labor-Hours	Unit	Material	Labor	Equipment	Total	Total Incl O&P
0010	**SWITCHGEAR**, Incorporate switch with cable connections, transformer,									
0100	& Low Voltage section									
0200	Load interrupter switch, 600 amp, 2 position									
0300	NEMA 1, 4.8 kV, 300 kVA & below w/CLF fuses	R-3	.40	50	Ea.	21,400	2,725	345	24,470	28,100
0400	400 kVA & above w/CLF fuses		.38	52.632		23,900	2,850	365	27,115	31,000
0500	Non fusible		.41	48.780		17,400	2,650	335	20,385	23,400
0600	13.8 kV, 300 kVA & below w/CLF fuses		.38	52.632		26,900	2,850	365	30,115	34,300
0700	400 kVA & above w/CLF fuses		.36	55.556		26,900	3,025	385	30,310	34,500
0800	Non fusible		.40	50		20,400	2,725	345	23,470	27,000
0900	Cable lugs for 2 feeders 4.8 kV or 13.8 kV	1 Elec	8	1		655	54.50		709.50	800
1000	Pothead, one 3 conductor or three 1 conductor		4	2		3,125	109		3,234	3,625
1100	Two 3 conductor or six 1 conductor		2	4		6,200	219		6,419	7,125
1200	Key interlocks		8	1		725	54.50		779.50	875
1300	Lightning arresters, Distribution class (no charge)									
1400	Intermediate class or line type 4.8 kV	1 Elec	2.70	2.963	Ea.	3,525	162		3,687	4,125
1500	13.8 kV		2	4		4,675	219		4,894	5,450
1600	Station class, 4.8 kV		2.70	2.963		6,025	162		6,187	6,875
1700	13.8 kV		2	4		10,400	219		10,619	11,700
1800	Transformers, 4800 volts to 480/277 volts, 75 kVA	R-3	.68	29.412		18,300	1,600	203	20,103	22,800
1900	112.5 kVA		.65	30.769		22,400	1,675	212	24,287	27,300
2000	150 kVA		.57	35.088		25,500	1,900	242	27,642	31,100
2100	225 kVA		.48	41.667		29,300	2,275	287	31,862	35,900
2200	300 kVA		.41	48.780		32,700	2,650	335	35,685	40,300
2300	500 kVA		.36	55.556		43,100	3,025	385	46,510	52,500
2400	750 kVA		.29	68.966		48,900	3,750	475	53,125	59,500
2500	13,800 volts to 480/277 volts, 75 kVA		.61	32.787		25,800	1,775	226	27,801	31,300
2600	112.5 kVA		.55	36.364		34,300	1,975	251	36,526	41,000
2700	150 kVA		.49	40.816		34,600	2,225	281	37,106	41,700
2800	225 kVA		.41	48.780		40,000	2,650	335	42,985	48,300
2900	300 kVA		.37	54.054		40,800	2,925	370	44,095	49,700
3000	500 kVA		.31	64.516		45,100	3,500	445	49,045	55,500
3100	750 kVA		.26	76.923		49,700	4,175	530	54,405	61,500

26 13 Medium-Voltage Switchgear

26 13 16 – Medium-Voltage Fusible Interrupter Switchgear

26 13 16.10 Switchgear		Crew	Daily Output	Labor-Hours	Unit	2015 Bare Costs Material	Labor	Equipment	Total	Total Incl O&P
3200	Forced air cooling & temperature alarm	1 Elec	1	8	Ea.	4,000	440		4,440	5,050
3300	Low voltage components									
3400	Maximum panel height 49-1/2", single or twin row									
3500	Breaker heights, type FA or FH, 6"									
3600	type KA or KH, 8"									
3700	type LA, 11"									
3800	type MA, 14"									
3900	Breakers, 2 pole, 15 to 60 amp, type FA	1 Elec	5.60	1.429	Ea.	350	78		428	500
4000	70 to 100 amp, type FA		4.20	1.905		445	104		549	645
4100	15 to 60 amp, type FH		5.60	1.429		580	78		658	755
4200	70 to 100 amp, type FH		4.20	1.905		675	104		779	900
4300	125 to 225 amp, type KA		3.40	2.353		1,025	129		1,154	1,350
4400	125 to 225 amp, type KH		3.40	2.353		2,525	129		2,654	3,000
4500	125 to 400 amp, type LA		2.50	3.200		1,950	175		2,125	2,375
4600	125 to 600 amp, type MA		1.80	4.444		2,975	243		3,218	3,650
4700	700 & 800 amp, type MA		1.50	5.333		3,850	292		4,142	4,675
4800	3 pole, 15 to 60 amp, type FA		5.30	1.509		455	82.50		537.50	625
4900	70 to 100 amp, type FA		4	2		560	109		669	780
5000	15 to 60 amp, type FH		5.30	1.509		680	82.50		762.50	870
5100	70 to 100 amp, type FH		4	2		770	109		879	1,025
5200	125 to 225 amp, type KA		3.20	2.500		1,300	137		1,437	1,625
5300	125 to 225 amp, type KH		3.20	2.500		2,925	137		3,062	3,425
5400	125 to 400 amp, type LA		2.30	3.478		2,400	190		2,590	2,925
5500	125 to 600 amp, type MA		1.60	5		4,075	274		4,349	4,900
5600	700 & 800 amp, type MA		1.30	6.154		5,300	335		5,635	6,325

26 22 Low-Voltage Transformers

26 22 13 – Low-Voltage Distribution Transformers

26 22 13.10 Transformer, Dry-Type		Crew	Daily Output	Labor-Hours	Unit	Material	Labor	Equipment	Total	Total Incl O&P
0010	**TRANSFORMER, DRY-TYPE** R262213-10									
0050	Single phase, 240/480 volt primary, 120/240 volt secondary									
0100	1 kVA R262213-60	1 Elec	2	4	Ea.	330	219		549	690
0300	2 kVA		1.60	5		490	274		764	950
0500	3 kVA		1.40	5.714		610	315		925	1,150
0700	5 kVA		1.20	6.667		835	365		1,200	1,475
0900	7.5 kVA	2 Elec	2.20	7.273		1,175	400		1,575	1,875
1100	10 kVA		1.60	10		1,450	545		1,995	2,425
1300	15 kVA		1.20	13.333		1,700	730		2,430	2,975
1500	25 kVA		1	16		2,125	875		3,000	3,625
1700	37.5 kVA		.80	20		2,750	1,100		3,850	4,675
1900	50 kVA		.70	22.857		3,250	1,250		4,500	5,475
2100	75 kVA		.65	24.615		4,325	1,350		5,675	6,775
2110	100 kVA	R-3	.90	22.222		5,625	1,200	153	6,978	8,150
2120	167 kVA	"	.80	25		9,325	1,350	172	10,847	12,500
2190	480 V primary 120/240 V secondary, nonvent., 15 kVA	2 Elec	1.20	13.333		1,575	730		2,305	2,825
2200	25 kVA		.90	17.778		2,300	970		3,270	4,000
2210	37 kVA		.75	21.333		2,750	1,175		3,925	4,775
2220	50 kVA		.65	24.615		3,250	1,350		4,600	5,625
2230	75 kVA		.60	26.667		4,325	1,450		5,775	6,925
2240	100 kVA		.50	32		5,625	1,750		7,375	8,800
2250	Low operating temperature(80°C), 25 kVA		1	16		4,200	875		5,075	5,925

26 22 13 – Low-Voltage Distribution Transformers

26 22 13.10 Transformer, Dry-Type		Crew	Daily Output	Labor-Hours	Unit	2015 Bare Costs Material	Labor	Equipment	Total	Total Incl O&P
2260	37 kVA	2 Elec	.80	20	Ea.	4,525	1,100		5,625	6,625
2270	50 kVA		.70	22.857		5,900	1,250		7,150	8,350
2280	75 kVA		.65	24.615		9,425	1,350		10,775	12,400
2290	100 kVA		.55	29.091		9,800	1,600		11,400	13,200
2300	3 phase, 480 volt primary 120/208 volt secondary									
2310	Ventilated, 3 kVA	1 Elec	1	8	Ea.	910	440		1,350	1,650
2700	6 kVA		.80	10		1,025	545		1,570	1,950
2900	9 kVA		.70	11.429		1,075	625		1,700	2,125
3100	15 kVA	2 Elec	1.10	14.545		1,300	795		2,095	2,625
3300	30 kVA		.90	17.778		1,425	970		2,395	3,025
3500	45 kVA		.80	20		1,700	1,100		2,800	3,525
3700	75 kVA		.70	22.857		2,375	1,250		3,625	4,475
3900	112.5 kVA	R-3	.90	22.222		3,400	1,200	153	4,753	5,725
4100	150 kVA		.85	23.529		4,450	1,275	162	5,887	7,000
4300	225 kVA		.65	30.769		6,075	1,675	212	7,962	9,400
4500	300 kVA		.55	36.364		7,600	1,975	251	9,826	11,600
4700	500 kVA		.45	44.444		12,600	2,425	305	15,330	17,900
4800	750 kVA		.35	57.143		21,100	3,100	395	24,595	28,400
4820	1000 kVA		.32	62.500		24,900	3,400	430	28,730	33,000
4850	K-4 rated, 15 kVA	2 Elec	1.10	14.545		3,325	795		4,120	4,850
4855	30 kVA		.90	17.778		5,000	970		5,970	6,925
4860	45 kVA		.80	20		6,000	1,100		7,100	8,250
4865	75 kVA		.70	22.857		9,050	1,250		10,300	11,800
4870	112.5 kVA	R-3	.90	22.222		12,000	1,200	153	13,353	15,300
4875	150 kVA		.85	23.529		15,700	1,275	162	17,137	19,400
4880	225 kVA		.65	30.769		21,900	1,675	212	23,787	26,800
4885	300 kVA		.55	36.364		28,700	1,975	251	30,926	34,800
4890	500 kVA		.45	44.444		40,100	2,425	305	42,830	48,100
4900	K-13 rated, 15 kVA	2 Elec	1.10	14.545		3,775	795		4,570	5,350
4905	30 kVA		.90	17.778		5,675	970		6,645	7,700
4910	45 kVA		.80	20		6,825	1,100		7,925	9,150
4915	75 kVA		.70	22.857		10,300	1,250		11,550	13,200
4920	112.5 kVA	R-3	.90	22.222		13,700	1,200	153	15,053	17,100
4925	150 kVA		.85	23.529		17,900	1,275	162	19,337	21,800
4930	225 kVA		.65	30.769		24,400	1,675	212	26,287	29,500
4935	300 kVA		.55	36.364		32,500	1,975	251	34,726	39,000
4940	500 kVA		.45	44.444		54,000	2,425	305	56,730	63,500
5020	480 volt primary 120/208 volt secondary									
5030	Nonventilated, 15 kVA	2 Elec	1.10	14.545	Ea.	3,225	795		4,020	4,750
5040	30 kVA		.80	20		4,600	1,100		5,700	6,725
5050	45 kVA		.70	22.857		6,650	1,250		7,900	9,175
5060	75 kVA		.65	24.615		8,875	1,350		10,225	11,800
5070	112.5 kVA	R-3	.85	23.529		13,300	1,275	162	14,737	16,700
5081	150 kVA		.85	23.529		16,000	1,275	162	17,437	19,700
5090	225 kVA		.60	33.333		17,900	1,800	230	19,930	22,700
5100	300 kVA		.50	40		20,300	2,175	276	22,751	25,900
5200	Low operating temperature (80°C), 30 kVA	2 Elec	.90	17.778		4,000	970		4,970	5,850
5210	45 kVA		.80	20		5,450	1,100		6,550	7,650
5220	75 kVA		.70	22.857		7,975	1,250		9,225	10,700
5230	112.5 kVA	R-3	.90	22.222		11,000	1,200	153	12,353	14,000
5240	150 kVA		.85	23.529		14,000	1,275	162	15,437	17,500
5250	225 kVA		.65	30.769		19,100	1,675	212	20,987	23,700
5260	300 kVA		.55	36.364		27,000	1,975	251	29,226	33,000

26 22 Low-Voltage Transformers

26 22 13 – Low-Voltage Distribution Transformers

26 22 13.10 Transformer, Dry-Type		Crew	Daily Output	Labor-Hours	Unit	Material	2015 Bare Costs Labor	Equipment	Total	Total Incl O&P
5270	500 kVA	R-3	.45	44.444	Ea.	34,500	2,425	305	37,230	42,000
5380	3 phase, 5 kV primary 277/480 volt secondary									
5400	High voltage, 112.5 kVA	R-3	.85	23.529	Ea.	17,100	1,275	162	18,537	20,900
5410	150 kVA		.65	30.769		18,600	1,675	212	20,487	23,100
5420	225 kVA		.55	36.364		21,900	1,975	251	24,126	27,400
5430	300 kVA		.45	44.444		28,000	2,425	305	30,730	34,800
5440	500 kVA		.35	57.143		36,100	3,100	395	39,595	44,800
5450	750 kVA		.32	62.500		56,500	3,400	430	60,330	67,500
5460	1000 kVA		.30	66.667		66,000	3,625	460	70,085	78,500
5470	1500 kVA		.27	74.074		77,000	4,025	510	81,535	91,000
5480	2000 kVA		.25	80		90,000	4,350	550	94,900	106,000
5490	2500 kVA		.20	100		101,500	5,425	690	107,615	120,500
5500	3000 kVA		.18	111		133,000	6,025	765	139,790	156,500
5590	15 kV primary 277/480 volt secondary									
5600	High voltage, 112.5 kVA	R-3	.85	23.529	Ea.	26,000	1,275	162	27,437	30,700
5610	150 kVA		.65	30.769		30,400	1,675	212	32,287	36,100
5620	225 kVA		.55	36.364		33,600	1,975	251	35,826	40,200
5630	300 kVA		.45	44.444		39,500	2,425	305	42,230	47,500
5640	500 kVA		.35	57.143		50,000	3,100	395	53,495	60,000
5650	750 kVA		.32	62.500		65,500	3,400	430	69,330	77,500
5660	1000 kVA		.30	66.667		74,500	3,625	460	78,585	88,000
5670	1500 kVA		.27	74.074		86,000	4,025	510	90,535	101,000
5680	2000 kVA		.25	80		95,500	4,350	550	100,400	112,000
5690	2500 kVA		.20	100		110,500	5,425	690	116,615	130,500
5700	3000 kVA		.18	111		131,500	6,025	765	138,290	154,500
6000	2400 volt primary, 480 volt secondary, 300 kVA		.45	44.444		28,000	2,425	305	30,730	34,800
6010	500 kVA		.35	57.143		36,100	3,100	395	39,595	44,800
6020	750 kVA		.32	62.500		52,000	3,400	430	55,830	62,500

26 22 13.20 Isolating Panels

		Crew	Daily Output	Labor-Hours	Unit	Material	Labor	Equipment	Total	Total Incl O&P
0010	**ISOLATING PANELS** used with isolating transformers									
0020	For hospital applications									
0100	Critical care area, 8 circuit, 3 kVA	1 Elec	.58	13.793	Ea.	6,750	755		7,505	8,550
0200	5 kVA		.54	14.815		6,850	810		7,660	8,750
0400	7.5 kVA		.52	15.385		7,025	840		7,865	8,975
0600	10 kVA		.44	18.182		7,325	995		8,320	9,550
0800	Operating room power & lighting, 8 circuit, 3 kVA		.58	13.793		4,575	755		5,330	6,150
1000	5 kVA		.54	14.815		5,100	810		5,910	6,825
1200	7.5 kVA		.52	15.385		5,750	840		6,590	7,575
1400	10 kVA		.44	18.182		6,425	995		7,420	8,550
1600	X-ray systems, 15 kVA, 90 amp		.44	18.182		12,200	995		13,195	15,000
1800	25 kVA, 125 amp		.36	22.222		14,800	1,225		16,025	18,100

26 22 13.30 Isolating Transformer

		Crew	Daily Output	Labor-Hours	Unit	Material	Labor	Equipment	Total	Total Incl O&P
0010	**ISOLATING TRANSFORMER**									
0100	Single phase, 120/240 volt primary, 120/240 volt secondary									
0200	0.50 kVA R262213-60	1 Elec	4	2	Ea.	380	109		489	580
0400	1 kVA		2	4		540	219		759	925
0600	2 kVA		1.60	5		805	274		1,079	1,300
0800	3 kVA		1.40	5.714		810	315		1,125	1,350
1000	5 kVA		1.20	6.667		1,075	365		1,440	1,725
1200	7.5 kVA		1.10	7.273		1,350	400		1,750	2,075
1400	10 kVA		.80	10		1,725	545		2,270	2,725
1600	15 kVA		.60	13.333		2,200	730		2,930	3,525

26 22 Low-Voltage Transformers

26 22 13 – Low-Voltage Distribution Transformers

26 22 13.30 Isolating Transformer		Crew	Daily Output	Labor-Hours	Unit	2015 Bare Costs Material	Labor	Equipment	Total	Total Incl O&P
1800	25 kVA	1 Elec	.50	16	Ea.	3,175	875		4,050	4,800
1810	37.5 kVA	2 Elec	.80	20		5,250	1,100		6,350	7,425
1820	75 kVA	"	.65	24.615		7,900	1,350		9,250	10,700
1830	3 phase, 120/240 V primary, 120/240 V secondary, 112.5 kVA	R-3	.90	22.222		9,950	1,200	153	11,303	13,000
1840	150 kVA		.85	23.529		12,700	1,275	162	14,137	16,000
1850	225 kVA		.65	30.769		17,600	1,675	212	19,487	22,100
1860	300 kVA		.55	36.364		23,400	1,975	251	25,626	29,100
1870	500 kVA		.45	44.444		39,000	2,425	305	41,730	47,000
1880	750 kVA		.35	57.143		39,400	3,100	395	42,895	48,500

26 22 13.90 Transformer Handling		Crew	Daily Output	Labor-Hours	Unit	Material	Labor	Equipment	Total	Total Incl O&P
0010	**TRANSFORMER HANDLING** Add to normal labor cost in restricted areas									
5000	Transformers									
5150	15 kVA, approximately 200 pounds	2 Elec	2.70	5.926	Ea.		325		325	485
5160	25 kVA, approximately 300 pounds		2.50	6.400			350		350	525
5170	37.5 kVA, approximately 400 pounds		2.30	6.957			380		380	570
5180	50 kVA, approximately 500 pounds		2	8			440		440	655
5190	75 kVA, approximately 600 pounds		1.80	8.889			485		485	730
5200	100 kVA, approximately 700 pounds		1.60	10			545		545	820
5210	112.5 kVA, approximately 800 pounds	3 Elec	2.20	10.909			595		595	895
5220	125 kVA, approximately 900 pounds		2	12			655		655	985
5230	150 kVA, approximately 1000 pounds		1.80	13.333			730		730	1,100
5240	167 kVA, approximately 1200 pounds		1.60	15			820		820	1,225
5250	200 kVA, approximately 1400 pounds		1.40	17.143			940		940	1,400
5260	225 kVA, approximately 1600 pounds		1.30	18.462			1,000		1,000	1,525
5270	250 kVA, approximately 1800 pounds		1.10	21.818			1,200		1,200	1,800
5280	300 kVA, approximately 2000 pounds		1	24			1,325		1,325	1,975
5290	500 kVA, approximately 3000 pounds		.75	32			1,750		1,750	2,625
5300	600 kVA, approximately 3500 pounds		.67	35.821			1,950		1,950	2,925
5310	750 kVA, approximately 4000 pounds		.60	40			2,200		2,200	3,275
5320	1000 kVA, approximately 5000 pounds		.50	48			2,625		2,625	3,925

26 22 16 – Low-Voltage Buck-Boost Transformers

26 22 16.10 Buck-Boost Transformer		Crew	Daily Output	Labor-Hours	Unit	Material	Labor	Equipment	Total	Total Incl O&P
0010	**BUCK-BOOST TRANSFORMER** R262213-60									
0100	Single phase, 120/240 V primary, 12/24 V secondary									
0200	0.10 kVA	1 Elec	8	1	Ea.	110	54.50		164.50	203
0400	0.25 kVA		5.70	1.404		161	77		238	293
0600	0.50 kVA		4	2		220	109		329	405
0800	0.75 kVA		3.10	2.581		284	141		425	525
1000	1.0 kVA		2	4		355	219		574	720
1200	1.5 kVA		1.80	4.444		430	243		673	840
1400	2.0 kVA		1.60	5		525	274		799	990
1600	3.0 kVA		1.40	5.714		700	315		1,015	1,250
1800	5.0 kVA		1.20	6.667		925	365		1,290	1,575
2000	3 phase, 240 V primary, 208/120 V secondary, 15 kVA	2 Elec	2.40	6.667		2,125	365		2,490	2,900
2200	30 kVA		1.60	10		2,375	545		2,920	3,450
2400	45 kVA		1.40	11.429		2,850	625		3,475	4,075
2600	75 kVA		1.20	13.333		3,475	730		4,205	4,900
2800	112.5 kVA	R-3	1.40	14.286		4,300	775	98.50	5,173.50	6,000
3000	150 kVA		1.10	18.182		5,725	985	125	6,835	7,925
3200	225 kVA		1	20		7,450	1,075	138	8,663	9,975
3400	300 kVA		.90	22.222		10,100	1,200	153	11,453	13,100

26 24 13.10 Incoming Switchboards

	26 24 13.10 Incoming Switchboards	Crew	Daily Output	Labor-Hours	Unit	2015 Bare Costs Material	Labor	Equipment	Total	Total Incl O&P
0010	**INCOMING SWITCHBOARDS** main service section R262419-84									
0100	Aluminum bus bars, not including CT's or PT's									
0200	No main disconnect, includes CT compartment									
0300	120/208 volt, 4 wire, 600 amp	2 Elec	1	16	Ea.	4,225	875		5,100	5,950
0400	800 amp		.88	18.182		4,225	995		5,220	6,150
0500	1000 amp		.80	20		5,075	1,100		6,175	7,225
0600	1200 amp		.72	22.222		5,075	1,225		6,300	7,400
0700	1600 amp		.66	24.242		5,075	1,325		6,400	7,550
0800	2000 amp		.62	25.806		5,450	1,400		6,850	8,125
1000	3000 amp		.56	28.571		7,200	1,575		8,775	10,300
1200	277/480 volt, 4 wire, 600 amp		1	16		4,225	875		5,100	5,950
1300	800 amp		.88	18.182		4,225	995		5,220	6,150
1400	1000 amp		.80	20		5,350	1,100		6,450	7,550
1500	1200 amp		.72	22.222		5,350	1,225		6,575	7,725
1600	1600 amp		.66	24.242		5,350	1,325		6,675	7,875
1700	2000 amp		.62	25.806		5,450	1,400		6,850	8,125
1800	3000 amp		.56	28.571		7,200	1,575		8,775	10,300
1900	4000 amp		.52	30.769		8,925	1,675		10,600	12,400
2000	Fused switch & CT compartment									
2100	120/208 volt, 4 wire, 400 amp	2 Elec	1.12	14.286	Ea.	2,825	780		3,605	4,275
2200	600 amp		.94	17.021		3,350	930		4,280	5,075
2300	800 amp		.84	19.048		11,400	1,050		12,450	14,200
2400	1200 amp		.68	23.529		14,800	1,275		16,075	18,200
2500	277/480 volt, 4 wire, 400 amp		1.14	14.035		3,200	770		3,970	4,675
2600	600 amp		.94	17.021		3,675	930		4,605	5,425
2700	800 amp		.84	19.048		11,400	1,050		12,450	14,200
2800	1200 amp		.68	23.529		14,800	1,275		16,075	18,200
2900	Pressure switch & CT compartment									
3000	120/208 volt, 4 wire, 800 amp	2 Elec	.80	20	Ea.	10,200	1,100		11,300	13,000
3100	1200 amp		.66	24.242		19,800	1,325		21,125	23,800
3200	1600 amp		.62	25.806		21,100	1,400		22,500	25,300
3300	2000 amp		.56	28.571		22,400	1,575		23,975	27,100
3310	2500 amp		.50	32		27,600	1,750		29,350	32,900
3320	3000 amp		.44	36.364		37,200	2,000		39,200	43,900
3330	4000 amp		.40	40		47,700	2,200		49,900	56,000
3340	120/208 volt, 4 wire, 800 amp, with ground fault		.80	20		16,900	1,100		18,000	20,300
3350	1200 amp, with ground fault		.66	24.242		21,900	1,325		23,225	26,100
3360	1600 amp, with ground fault		.62	25.806		23,800	1,400		25,200	28,200
3370	2000 amp, with ground fault		.56	28.571		25,700	1,575		27,275	30,600
3400	277/480 volt, 4 wire, 800 amp, with ground fault		.80	20		16,900	1,100		18,000	20,300
3600	1200 amp, with ground fault		.66	24.242		21,900	1,325		23,225	26,100
4000	1600 amp, with ground fault		.62	25.806		23,800	1,400		25,200	28,200
4200	2000 amp, with ground fault		.56	28.571		25,700	1,575		27,275	30,600
4400	Circuit breaker, molded case & CT compartment									
4600	3 pole, 4 wire, 600 amp	2 Elec	.94	17.021	Ea.	8,825	930		9,755	11,100
4800	800 amp		.84	19.048		10,600	1,050		11,650	13,200
5000	1200 amp		.68	23.529		14,400	1,275		15,675	17,700
5100	Copper bus bars, not incl. CT's or PT's, add, minimum					15%				

26 24 13.20 In Plant Distribution Switchboards

	26 24 13.20 In Plant Distribution Switchboards	Crew	Daily Output	Labor-Hours	Unit	Material	Labor	Equipment	Total	Total Incl O&P
0010	**IN PLANT DISTRIBUTION SWITCHBOARDS**									
0100	Main lugs only, to 600 volt, 3 pole, 3 wire, 200 amp	2 Elec	1.20	13.333	Ea.	1,325	730		2,055	2,550
0110	400 amp		1.20	13.333		1,325	730		2,055	2,550

26 24 13 – Switchboards

26 24 13.20 In Plant Distribution Switchboards		Crew	Daily Output	Labor-Hours	Unit	Material	2015 Bare Costs Labor	Equipment	Total	Total Incl O&P
0120	600 amp	2 Elec	1.20	13.333	Ea.	1,350	730		2,080	2,600
0130	800 amp		1.08	14.815		1,475	810		2,285	2,850
0140	1200 amp		.92	17.391		1,800	950		2,750	3,400
0150	1600 amp		.86	18.605		2,350	1,025		3,375	4,100
0160	2000 amp		.82	19.512		2,600	1,075		3,675	4,475
0250	To 480 volt, 3 pole, 4 wire, 200 amp		1.20	13.333		1,150	730		1,880	2,350
0260	400 amp		1.20	13.333		1,325	730		2,055	2,550
0270	600 amp		1.20	13.333		1,450	730		2,180	2,675
0280	800 amp		1.08	14.815		1,575	810		2,385	2,950
0290	1200 amp		.92	17.391		1,950	950		2,900	3,575
0300	1600 amp		.86	18.605		2,225	1,025		3,250	3,950
0310	2000 amp		.82	19.512		2,575	1,075		3,650	4,450
0400	Main circuit breaker, to 600 volt, 3 pole, 3 wire, 200 amp		1.20	13.333		3,150	730		3,880	4,575
0410	400 amp		1.14	14.035		3,150	770		3,920	4,625
0420	600 amp		1.10	14.545		4,000	795		4,795	5,600
0430	800 amp		1.04	15.385		6,675	840		7,515	8,600
0440	1200 amp		.88	18.182		8,750	995		9,745	11,100
0450	1600 amp		.84	19.048		14,000	1,050		15,050	17,000
0460	2000 amp		.80	20		15,000	1,100		16,100	18,200
0550	277/480 volt, 3 pole, 4 wire, 200 amp		1.20	13.333		3,325	730		4,055	4,750
0560	400 amp		1.14	14.035		3,325	770		4,095	4,800
0570	600 amp		1.10	14.545		4,150	795		4,945	5,775
0580	800 amp		1.04	15.385		7,000	840		7,840	8,950
0590	1200 amp		.88	18.182		9,050	995		10,045	11,500
0600	1600 amp		.84	19.048		14,000	1,050		15,050	17,000
0610	2000 amp		.80	20		15,000	1,100		16,100	18,200
0700	Main fusible switch w/fuse, 208/240 volt, 3 pole, 3 wire, 200 amp		1.20	13.333		3,475	730		4,205	4,925
0710	400 amp		1.14	14.035		3,475	770		4,245	4,975
0720	600 amp		1.10	14.545		4,225	795		5,020	5,850
0730	800 amp		1.04	15.385		8,650	840		9,490	10,800
0740	1200 amp		.88	18.182		10,100	995		11,095	12,600
0800	120/208, 120/240 volt, 3 pole, 4 wire, 200 amp		1.20	13.333		3,075	730		3,805	4,500
0810	400 amp		1.14	14.035		3,075	770		3,845	4,550
0820	600 amp		1.10	14.545		4,025	795		4,820	5,625
0830	800 amp		1.04	15.385		6,425	840		7,265	8,300
0840	1200 amp		.88	18.182		7,450	995		8,445	9,700
0900	480 or 600 volt, 3 pole, 3 wire, 200 amp		1.20	13.333		3,350	730		4,080	4,775
0910	400 amp		1.14	14.035		3,350	770		4,120	4,825
0920	600 amp		1.10	14.545		4,000	795		4,795	5,600
0930	800 amp		1.04	15.385		6,175	840		7,015	8,050
0940	1200 amp		.88	18.182		7,175	995		8,170	9,375
1000	277 or 480 volt, 3 pole, 4 wire, 200 amp		1.20	13.333		3,475	730		4,205	4,925
1010	400 amp		1.14	14.035		3,475	770		4,245	4,975
1020	600 amp		1.10	14.545		4,200	795		4,995	5,800
1030	800 amp		1.04	15.385		6,425	840		7,265	8,325
1040	1200 amp		.88	18.182		7,450	995		8,445	9,700
1120	1600 amp		.76	21.053		13,600	1,150		14,750	16,700
1130	2000 amp		.68	23.529		17,900	1,275		19,175	21,600
1150	Pressure switch, bolted, 3 pole, 208/240 volt, 3 wire, 800 amp		.96	16.667		10,100	910		11,010	12,500
1160	1200 amp		.80	20		12,900	1,100		14,000	15,900
1170	1600 amp		.76	21.053		14,700	1,150		15,850	17,900
1180	2000 amp		.68	23.529		16,800	1,275		18,075	20,300
1200	120/208 or 120/240 volt, 3 pole, 4 wire, 800 amp	↓	.96	16.667	↓	7,975	910		8,885	10,100

26 24 13.20 In Plant Distribution Switchboards		Crew	Daily Output	Labor-Hours	Unit	Material	2015 Bare Costs Labor	Equipment	Total	Total Incl O&P
1210	1200 amp	2 Elec	.80	20	Ea.	9,300	1,100		10,400	11,900
1220	1600 amp		.76	21.053		14,700	1,150		15,850	17,900
1230	2000 amp		.68	23.529		16,800	1,275		18,075	20,300
1300	480 or 600 volt, 3 wire, 800 amp		.96	16.667		10,100	910		11,010	12,500
1310	1200 amp		.80	20		14,100	1,100		15,200	17,200
1320	1600 amp		.76	21.053		15,900	1,150		17,050	19,200
1330	2000 amp		.68	23.529		17,900	1,275		19,175	21,600
1400	277-480 volt, 4 wire, 800 amp		.96	16.667		10,100	910		11,010	12,500
1410	1200 amp		.80	20		14,100	1,100		15,200	17,200
1420	1600 amp		.76	21.053		15,900	1,150		17,050	19,200
1430	2000 amp		.68	23.529		17,900	1,275		19,175	21,600
1500	Main ground fault protector, 1200-2000 amp		5.40	2.963		3,350	162		3,512	3,950
1600	Busway connection, 200 amp		5.40	2.963		505	162		667	800
1610	400 amp		4.60	3.478		505	190		695	840
1620	600 amp		4	4		505	219		724	885
1630	800 amp		3.20	5		505	274		779	965
1640	1200 amp		2.60	6.154		505	335		840	1,050
1650	1600 amp		2.40	6.667		1,050	365		1,415	1,700
1660	2000 amp		2	8		1,050	440		1,490	1,800
1700	Shunt trip for remote operation 200 amp		8	2		670	109		779	900
1710	400 amp		8	2		1,075	109		1,184	1,350
1720	600 amp		8	2		1,250	109		1,359	1,550
1730	800 amp		8	2		1,625	109		1,734	1,950
1740	1200-2000 amp		8	2		3,450	109		3,559	3,950
1800	Motor operated main breaker 200 amp		8	2		3,325	109		3,434	3,825
1810	400 amp		8	2		3,325	109		3,434	3,825
1820	600 amp		8	2		3,400	109		3,509	3,925
1830	800 amp		8	2		3,400	109		3,509	3,925
1840	1200-2000 amp		8	2		3,500	109		3,609	4,025
1900	Current/potential transformer metering compartment 200-800 amp		5.40	2.963		2,900	162		3,062	3,425
1940	1200 amp		5.40	2.963		4,775	162		4,937	5,525
1950	1600-2000 amp		5.40	2.963		6,175	162		6,337	7,025
2000	With watt meter 200-800 amp		4	4		8,525	219		8,744	9,700
2040	1200 amp		4	4		10,500	219		10,719	11,900
2050	1600-2000 amp		4	4		10,500	219		10,719	11,900
2100	Split bus 60-200 amp	1 Elec	5.30	1.509		202	82.50		284.50	345
2130	400 amp	2 Elec	4.60	3.478		350	190		540	670
2140	600 amp		3.60	4.444		430	243		673	835
2150	800 amp		2.60	6.154		545	335		880	1,100
2170	1200 amp		2	8		620	440		1,060	1,350
2250	Contactor control 60 amp	1 Elec	2	4		1,325	219		1,544	1,775
2260	100 amp		1.50	5.333		1,500	292		1,792	2,075
2270	200 amp		1	8		2,250	440		2,690	3,125
2280	400 amp	2 Elec	1	16		6,825	875		7,700	8,800
2290	600 amp		.84	19.048		7,625	1,050		8,675	9,950
2300	800 amp		.72	22.222		9,050	1,225		10,275	11,800
2500	Modifier, two distribution sections, add		.80	20		2,825	1,100		3,925	4,775
2520	Three distribution sections, add		.40	40		6,350	2,200		8,550	10,300
2560	Auxiliary pull section, 20", add		2	8		1,125	440		1,565	1,900
2580	24", add		1.80	8.889		1,125	485		1,610	1,975
2600	30", add		1.60	10		1,125	545		1,670	2,075
2620	36", add		1.40	11.429		1,350	625		1,975	2,425
2640	Dog house, 12", add		2.40	6.667		233	365		598	800

26 24 13.20 In Plant Distribution Switchboards		Crew	Daily Output	Labor-Hours	Unit	Material	Labor	Equipment	Total	Total Incl O&P
2660	18", add	2 Elec	2	8	Ea.	465	440		905	1,175
3000	Transition section between switchboard and transformer									
3050	or motor control center, 4 wire alum. bus, 600 amp	2 Elec	1.14	14.035	Ea.	2,300	770		3,070	3,675
3100	800 amp		1	16		2,575	875		3,450	4,125
3150	1000 amp		.88	18.182		2,900	995		3,895	4,675
3200	1200 amp		.80	20		3,175	1,100		4,275	5,150
3250	1600 amp		.72	22.222		3,750	1,225		4,975	5,975
3300	2000 amp		.66	24.242		4,325	1,325		5,650	6,725
3350	2500 amp		.62	25.806		5,050	1,400		6,450	7,675
3400	3000 amp		.56	28.571		5,800	1,575		7,375	8,725
4000	Weatherproof construction, per vertical section		1.76	9.091		2,650	495		3,145	3,675

Column group "2015 Bare Costs": Material, Labor, Equipment, Total.

26 24 13.30 Distribution Switchboards Section

		Crew	Daily Output	Labor-Hours	Unit	Material	Labor	Equipment	Total	Total Incl O&P
0010	**DISTRIBUTION SWITCHBOARDS SECTION** R262419-80									
0100	Aluminum bus bars, not including breakers									
0160	Subfeed lug-rated at 60 amp	2 Elec	1.30	12.308	Ea.	1,050	675		1,725	2,150
0170	100 amp		1.26	12.698		1,200	695		1,895	2,375
0180	200 amp		1.20	13.333		1,225	730		1,955	2,450
0190	400 amp		1.10	14.545		1,225	795		2,020	2,550
0195	120/208 or 277/480 volt, 4 wire, 400 amp		1.10	14.545		1,300	795		2,095	2,650
0200	600 amp		1	16		1,625	875		2,500	3,075
0300	800 amp		.88	18.182		2,100	995		3,095	3,825
0400	1000 amp		.80	20		2,625	1,100		3,725	4,550
0500	1200 amp		.72	22.222		3,125	1,225		4,350	5,275
0600	1600 amp		.66	24.242		3,575	1,325		4,900	5,925
0700	2000 amp		.62	25.806		4,200	1,400		5,600	6,725
0800	2500 amp		.60	26.667		4,725	1,450		6,175	7,375
0900	3000 amp		.56	28.571		5,725	1,575		7,300	8,650
0950	4000 amp		.52	30.769		8,375	1,675		10,050	11,800

26 24 13.40 Switchboards Feeder Section

		Crew	Daily Output	Labor-Hours	Unit	Material	Labor	Equipment	Total	Total Incl O&P
0010	**SWITCHBOARDS FEEDER SECTION** group mounted devices R262419-82									
0030	Circuit breakers									
0160	FA frame, 15 to 60 amp, 240 volt, 1 pole	1 Elec	8	1	Ea.	120	54.50		174.50	214
0170	2 pole		7	1.143		205	62.50		267.50	320
0180	3 pole		5.30	1.509		305	82.50		387.50	460
0210	480 volt, 1 pole		8	1		151	54.50		205.50	248
0220	2 pole		7	1.143		375	62.50		437.50	505
0230	3 pole		5.30	1.509		485	82.50		567.50	655
0260	600 volt, 2 pole		7	1.143		435	62.50		497.50	570
0270	3 pole		5.30	1.509		560	82.50		642.50	740
0280	FA frame, 70 to 100 amp, 240 volt, 1 pole		7	1.143		191	62.50		253.50	305
0310	2 pole		5	1.600		335	87.50		422.50	495
0320	3 pole		4	2		435	109		544	640
0330	480 volt, 1 pole		7	1.143		228	62.50		290.50	345
0360	2 pole		5	1.600		485	87.50		572.50	665
0370	3 pole		4	2		570	109		679	790
0380	600 volt, 2 pole		5	1.600		550	87.50		637.50	735
0410	3 pole		4	2		685	109		794	920
0420	KA frame, 70 to 225 amp		3.20	2.500		1,250	137		1,387	1,575
0430	LA frame, 125 to 400 amp		2.30	3.478		2,475	190		2,665	3,000
0460	MA frame, 450 to 600 amp		1.60	5		5,000	274		5,274	5,900
0470	700 to 800 amp		1.30	6.154		6,500	335		6,835	7,650
0480	MAL frame, 1000 amp		1	8		6,725	440		7,165	8,050

26 24 13.40 Switchboards Feeder Section		Crew	Daily Output	Labor-Hours	Unit	Material	2015 Bare Costs Labor	Equipment	Total	Total Incl O&P
0490	PA frame, 1200 amp	1 Elec	.80	10	Ea.	13,700	545		14,245	15,900
0500	Branch circuit, fusible switch, 600 volt, double 30/30 amp		4	2		895	109		1,004	1,150
0550	60/60 amp		3.20	2.500		920	137		1,057	1,200
0600	100/100 amp		2.70	2.963		1,150	162		1,312	1,525
0650	Single, 30 amp		5.30	1.509		735	82.50		817.50	935
0700	60 amp		4.70	1.702		815	93		908	1,025
0750	100 amp		4	2		1,075	109		1,184	1,350
0800	200 amp		2.70	2.963		1,425	162		1,587	1,825
0850	400 amp		2.30	3.478		2,625	190		2,815	3,150
0900	600 amp		1.80	4.444		3,200	243		3,443	3,900
0950	800 amp		1.30	6.154		5,375	335		5,710	6,400
1000	1200 amp		.80	10		6,150	545		6,695	7,600
1080	Branch circuit, circuit breakers, high interrupting capacity									
1100	60 amp, 240, 480 or 600 volt, 1 pole	1 Elec	8	1	Ea.	259	54.50		313.50	365
1120	2 pole		7	1.143		665	62.50		727.50	830
1140	3 pole		5.30	1.509		675	82.50		757.50	865
1150	100 amp, 240, 480 or 600 volt, 1 pole		7	1.143		292	62.50		354.50	415
1160	2 pole		5	1.600		775	87.50		862.50	985
1180	3 pole		4	2		730	109		839	970
1200	225 amp, 240, 480 or 600 volt, 2 pole		3.50	2.286		2,150	125		2,275	2,525
1220	3 pole		3.20	2.500		2,275	137		2,412	2,700
1240	400 amp, 240, 480 or 600 volt, 2 pole		2.50	3.200		3,425	175		3,600	4,000
1260	3 pole		2.30	3.478		3,075	190		3,265	3,650
1280	600 amp, 240, 480 or 600 volt, 2 pole		1.80	4.444		3,975	243		4,218	4,750
1300	3 pole		1.60	5		3,750	274		4,024	4,525
1320	800 amp, 240, 480 or 600 volt, 2 pole		1.50	5.333		5,100	292		5,392	6,050
1340	3 pole		1.30	6.154		5,625	335		5,960	6,675
1360	1000 amp, 240, 480 or 600 volt, 2 pole		1.10	7.273		6,675	400		7,075	7,925
1380	3 pole		1	8		7,300	440		7,740	8,675
1400	1200 amp, 240, 480 or 600 volt, 2 pole		.90	8.889		6,925	485		7,410	8,350
1420	3 pole		.80	10		7,925	545		8,470	9,550
1700	Fusible switch, 240 V, 60 amp, 2 pole		3.20	2.500		430	137		567	675
1720	3 pole		3	2.667		590	146		736	870
1740	100 amp, 2 pole		2.70	2.963		435	162		597	725
1760	3 pole		2.50	3.200		660	175		835	985
1780	200 amp, 2 pole		2	4		910	219		1,129	1,325
1800	3 pole		1.90	4.211		1,075	230		1,305	1,525
1820	400 amp, 2 pole		1.50	5.333		1,700	292		1,992	2,300
1840	3 pole		1.30	6.154		2,125	335		2,460	2,850
1860	600 amp, 2 pole		1	8		2,425	440		2,865	3,325
1880	3 pole		.90	8.889		2,975	485		3,460	4,000
1900	240-600 V, 800 amp, 2 pole		.70	11.429		5,400	625		6,025	6,875
1920	3 pole		.60	13.333		6,625	730		7,355	8,375
2000	600 V, 60 amp, 2 pole		3.20	2.500		650	137		787	920
2040	100 amp, 2 pole		2.70	2.963		670	162		832	980
2080	200 amp, 2 pole		2	4		1,150	219		1,369	1,600
2120	400 amp, 2 pole		1.50	5.333		2,300	292		2,592	2,950
2160	600 amp, 2 pole		1	8		2,800	440		3,240	3,725
2500	Branch circuit, circuit breakers, 60 amp, 600 volt, 3 pole		5.30	1.509		525	82.50		607.50	700
2520	240, 480 or 600 volt, 1 pole		8	1		100	54.50		154.50	192
2540	240 volt, 2 pole		7	1.143		195	62.50		257.50	310
2560	480 or 600 volt, 2 pole		7	1.143		365	62.50		427.50	495
2580	240 volt, 3 pole		5.30	1.509		286	82.50		368.50	440

26 24 13 – Switchboards

26 24 13.40 Switchboards Feeder Section		Crew	Daily Output	Labor-Hours	Unit	Material	Labor	Equipment	Total	Total Incl O&P
						2015 Bare Costs				
2600	480 volt, 3 pole	1 Elec	5.30	1.509	Ea.	485	82.50		567.50	660
2620	100 amp, 600 volt, 2 pole		5	1.600		465	87.50		552.50	645
2640	3 pole		4	2		590	109		699	815
2660	480 volt, 2 pole		5	1.600		400	87.50		487.50	570
2680	240 volt, 2 pole		5	1.600		229	87.50		316.50	385
2700	3 pole		4	2		370	109		479	570
2720	480 volt, 3 pole		4	2		540	109		649	760
2740	225 amp, 240, 480 or 600 volt, 2 pole		3.50	2.286		610	125		735	860
2760	3 pole		3.20	2.500		705	137		842	980
2780	400 amp, 240, 480 or 600 volt, 2 pole		2.50	3.200		1,400	175		1,575	1,800
2800	3 pole		2.30	3.478		1,625	190		1,815	2,075
2820	600 amp, 240 or 480 volt, 2 pole		1.80	4.444		2,300	243		2,543	2,900
2840	3 pole		1.60	5		2,825	274		3,099	3,500
2860	800 amp, 240, 480 volt or 600 volt, 2 pole		1.50	5.333		3,450	292		3,742	4,225
2880	3 pole		1.30	6.154		4,025	335		4,360	4,925
2900	1000 amp, 240, 480 or 600 volt, 2 pole		1.10	7.273		4,225	400		4,625	5,250
2920	480 volt, 600 volt, 3 pole		1	8		4,850	440		5,290	6,000
2940	1200 amp, 240, 480 or 600 volt, 2 pole		.90	8.889		5,975	485		6,460	7,300
2960	3 pole		.80	10		6,525	545		7,070	8,000
2980	600 volt, 3 pole	↓	.80	10	↓	6,525	545		7,070	8,000

26 24 16 – Panelboards

26 24 16.10 Load Centers

		Crew	Daily Output	Labor-Hours	Unit	Material	Labor	Equipment	Total	Total Incl O&P
0010	**LOAD CENTERS** (residential type) R262416-50									
0100	3 wire, 120/240 V, 1 phase, including 1 pole plug-in breakers									
0200	100 amp main lugs, indoor, 8 circuits	1 Elec	1.40	5.714	Ea.	178	315		493	665
0300	12 circuits		1.20	6.667		249	365		614	820
0400	Rainproof, 8 circuits		1.40	5.714		214	315		529	705
0500	12 circuits	↓	1.20	6.667		300	365		665	875
0600	200 amp main lugs, indoor, 16 circuits	R-1A	1.80	8.889		330	400		730	975
0700	20 circuits		1.50	10.667		415	480		895	1,175
0800	24 circuits		1.30	12.308		545	555		1,100	1,450
0900	30 circuits		1.20	13.333		630	600		1,230	1,600
1000	40 circuits		.80	20		870	900		1,770	2,325
1200	Rainproof, 16 circuits		1.80	8.889		395	400		795	1,050
1300	20 circuits		1.50	10.667		475	480		955	1,250
1400	24 circuits		1.30	12.308		710	555		1,265	1,625
1500	30 circuits		1.20	13.333		790	600		1,390	1,775
1600	40 circuits		.80	20		1,025	900		1,925	2,525
1800	400 amp main lugs, indoor, 42 circuits		.72	22.222		1,225	1,000		2,225	2,875
1900	Rainproof, 42 circuit	↓	.72	22.222	↓	1,300	1,000		2,300	2,950
2200	Plug in breakers, 20 amp, 1 pole, 4 wire, 120/208 volts									
2210	125 amp main lugs, indoor, 12 circuits	1 Elec	1.20	6.667	Ea.	310	365		675	890
2300	18 circuits		.80	10		440	545		985	1,300
2400	Rainproof, 12 circuits		1.20	6.667		360	365		725	940
2500	18 circuits	↓	.80	10		480	545		1,025	1,350
2600	200 amp main lugs, indoor, 24 circuits	R-1A	1.30	12.308		595	555		1,150	1,500
2700	30 circuits		1.20	13.333		680	600		1,280	1,650
2800	36 circuits		1	16		820	720		1,540	2,000
2900	42 circuits		.80	20		900	900		1,800	2,375
3000	Rainproof, 24 circuits		1.30	12.308		660	555		1,215	1,575
3100	30 circuits		1.20	13.333		745	600		1,345	1,725
3200	36 circuits	↓	1	16	↓	1,025	720		1,745	2,225

26 24 16.10 Load Centers		Crew	Daily Output	Labor-Hours	Unit	Material	2015 Bare Costs Labor	Equipment	Total	Total Incl O&P
3300	42 circuits	R-1A	.80	20	Ea.	1,100	900		2,000	2,600
3500	400 amp main lugs, indoor, 42 circuits		.72	22.222		1,350	1,000		2,350	3,025
3600	Rainproof, 42 circuits		.72	22.222		1,625	1,000		2,625	3,300
3700	Plug-in breakers, 20 amp, 1 pole, 3 wire, 120/240 volts									
3800	100 amp main breaker, indoor, 12 circuits	1 Elec	1.20	6.667	Ea.	335	365		700	910
3900	18 circuits	"	.80	10		450	545		995	1,325
4000	200 amp main breaker, indoor, 20 circuits	R-1A	1.50	10.667		605	480		1,085	1,400
4200	24 circuits		1.30	12.308		730	555		1,285	1,650
4300	30 circuits		1.20	13.333		815	600		1,415	1,800
4400	40 circuits		.90	17.778		1,025	800		1,825	2,350
4500	Rainproof, 20 circuits		1.50	10.667		660	480		1,140	1,450
4600	24 circuits		1.30	12.308		820	555		1,375	1,750
4700	30 circuits		1.20	13.333		900	600		1,500	1,900
4800	40 circuits		.90	17.778		1,100	800		1,900	2,450
5000	400 amp main breaker, indoor, 42 circuits		.72	22.222		2,575	1,000		3,575	4,350
5100	Rainproof, 42 circuits		.72	22.222		3,075	1,000		4,075	4,900
5300	Plug in breakers, 20 amp, 1 pole, 4 wire, 120/208 volts									
5400	200 amp main breaker, indoor, 30 circuits	R-1A	1.20	13.333	Ea.	1,225	600		1,825	2,275
5500	42 circuits		.80	20		1,500	900		2,400	3,025
5600	Rainproof, 30 circuits		1.20	13.333		1,325	600		1,925	2,375
5700	42 circuits		.80	20		1,600	900		2,500	3,125

26 24 16.20 Panelboard and Load Center Circuit Breakers		Crew	Daily Output	Labor-Hours	Unit	Material	Labor	Equipment	Total	Total Incl O&P
0010	**PANELBOARD AND LOAD CENTER CIRCUIT BREAKERS** R262419-82									
0050	Bolt-on, 10,000 amp I.C., 120 volt, 1 pole									
0100	15 to 50 amp	1 Elec	10	.800	Ea.	16.65	44		60.65	84
0200	60 amp		8	1		19.20	54.50		73.70	103
0300	70 amp		8	1		28	54.50		82.50	113
0350	240 volt, 2 pole									
0400	15 to 50 amp	1 Elec	8	1	Ea.	36	54.50		90.50	122
0500	60 amp		7.50	1.067		49	58.50		107.50	141
0600	80 to 100 amp		5	1.600		93.50	87.50		181	234
0700	3 pole, 15 to 60 amp		6.20	1.290		115	70.50		185.50	233
0800	70 amp		5	1.600		146	87.50		233.50	292
0900	80 to 100 amp		3.60	2.222		166	122		288	365
1000	22,000 amp I.C., 240 volt, 2 pole, 70 – 225 amp		2.70	2.963		630	162		792	940
1100	3 pole, 70 – 225 amp		2.30	3.478		700	190		890	1,050
1200	14,000 amp I.C., 277 volts, 1 pole, 15 – 30 amp		8	1		44	54.50		98.50	131
1300	22,000 amp I.C., 480 volts, 2 pole, 70 – 225 amp		2.70	2.963		630	162		792	940
1400	3 pole, 70 – 225 amp		2.30	3.478		780	190		970	1,150
2000	Plug-in panel or load center, 120/240 volt, to 60 amp, 1 pole		12	.667		11.45	36.50		47.95	67
2004	Circuit breaker, 120/240 volt, 20 A, 1 pole with NM cable		6.50	1.231		18.75	67.50		86.25	122
2006	30 A, 1 pole with NM cable		6.50	1.231		18.75	67.50		86.25	122
2010	2 pole		9	.889		26	48.50		74.50	102
2014	50 A, 2 pole with NM cable		5.50	1.455		33	79.50		112.50	156
2020	3 pole		7.50	1.067		89.50	58.50		148	186
2030	100 amp, 2 pole		6	1.333		120	73		193	241
2040	3 pole		4.50	1.778		133	97		230	292
2050	150 to 200 amp, 2 pole		3	2.667		218	146		364	460
2060	Plug-in tandem, 120/240 V, 2-15 A, 1 pole		11	.727		24.50	40		64.50	86.50
2070	1-15 A & 1-20 A		11	.727		24.50	40		64.50	86.50
2080	2-20 A		11	.727		24.50	40		64.50	86.50
2082	Arc fault circuit interrupter, 120/240 V, 1-15 A & 1-20 A, 1 pole		11	.727		65	40		105	131

26 24 16.20 Panelboard and Load Center Circuit Breakers		Crew	Daily Output	Labor-Hours	Unit	Material	2015 Bare Costs Labor	Equipment	Total	Total Incl O&P
2100	High interrupting capacity, 120/240 volt, plug-in, 30 amp, 1 pole	1 Elec	12	.667	Ea.	26	36.50		62.50	83
2110	60 amp, 2 pole		9	.889		61.50	48.50		110	141
2120	3 pole		7.50	1.067		141	58.50		199.50	243
2130	100 amp, 2 pole		6	1.333		153	73		226	278
2140	3 pole		4.50	1.778		223	97		320	390
2150	125 amp, 2 pole		3	2.667		395	146		541	650
2200	Bolt-on, 30 amp, 1 pole		10	.800		33	44		77	102
2210	60 amp, 2 pole		7.50	1.067		70	58.50		128.50	165
2220	3 pole		6.20	1.290		180	70.50		250.50	305
2230	100 amp, 2 pole		5	1.600		196	87.50		283.50	345
2240	3 pole		3.60	2.222		260	122		382	470
2300	Ground fault, 240 volt, 30 amp, 1 pole		7	1.143		102	62.50		164.50	206
2310	2 pole		6	1.333		182	73		255	310
2350	Key operated, 240 volt, 1 pole, 30 amp		7	1.143		84	62.50		146.50	186
2360	Switched neutral, 240 volt, 30 amp, 2 pole		6	1.333		42	73		115	156
2370	3 pole		5.50	1.455		63.50	79.50		143	189
2400	Shunt trip, for 240 volt breaker, 60 amp, 1 pole		4	2		72.50	109		181.50	244
2410	2 pole		3.50	2.286		72.50	125		197.50	267
2420	3 pole		3	2.667		72.50	146		218.50	299
2430	100 amp, 2 pole		3	2.667		72.50	146		218.50	299
2440	3 pole		2.50	3.200		72.50	175		247.50	340
2450	150 amp, 2 pole		2	4		172	219		391	520
2500	Auxiliary switch, for 240 volt breaker, 60 amp, 1 pole		4	2		65.50	109		174.50	237
2510	2 pole		3.50	2.286		65.50	125		190.50	260
2520	3 pole		3	2.667		65.50	146		211.50	292
2530	100 amp, 2 pole		3	2.667		65.50	146		211.50	292
2540	3 pole		2.50	3.200		65.50	175		240.50	335
2550	150 amp, 2 pole		2	4		109	219		328	450
2600	Panel or load center, 277/480 volt, plug-in, 30 amp, 1 pole		12	.667		53	36.50		89.50	113
2610	60 amp, 2 pole		9	.889		159	48.50		207.50	248
2620	3 pole		7.50	1.067		233	58.50		291.50	345
2650	Bolt-on, 60 amp, 2 pole		7.50	1.067		159	58.50		217.50	263
2660	3 pole		6.20	1.290		233	70.50		303.50	365
2700	I-line, 277/480 volt, 30 amp, 1 pole		8	1		61	54.50		115.50	149
2710	60 amp, 2 pole		7.50	1.067		214	58.50		272.50	325
2720	3 pole		6.20	1.290		278	70.50		348.50	410
2730	100 amp, 1 pole		7.50	1.067		117	58.50		175.50	217
2740	2 pole		5	1.600		278	87.50		365.50	435
2750	3 pole		3.50	2.286		330	125		455	545
2800	High interrupting capacity, 277/480 volt, plug-in, 30 amp, 1 pole		12	.667		305	36.50		341.50	395
2810	60 amp, 2 pole		9	.889		475	48.50		523.50	595
2820	3 pole		7	1.143		550	62.50		612.50	700
2830	Bolt-on, 30 amp, 1 pole		8	1		305	54.50		359.50	420
2840	60 amp, 2 pole		7.50	1.067		525	58.50		583.50	670
2850	3 pole		6.20	1.290		610	70.50		680.50	780
2900	I-line, 30 amp, 1 pole		8	1		305	54.50		359.50	420
2910	60 amp, 2 pole		7.50	1.067		440	58.50		498.50	570
2920	3 pole		6.20	1.290		485	70.50		555.50	640
2930	100 amp, 1 pole		7.50	1.067		495	58.50		553.50	635
2940	2 pole		5	1.600		495	87.50		582.50	675
2950	3 pole		3.60	2.222		550	122		672	780
2960	Shunt trip, 277/480 volt breaker, remote oper., 30 amp, 1 pole		4	2		330	109		439	525
2970	60 amp, 2 pole		3.50	2.286		480	125		605	715

26 24 16.20 Panelboard and Load Center Circuit Breakers		Crew	Daily Output	Labor-Hours	Unit	2015 Bare Costs Material	Labor	Equipment	Total	Total Incl O&P
2980	3 pole	1 Elec	3	2.667	Ea.	545	146		691	820
2990	100 amp, 1 pole		3.50	2.286		385	125		510	610
3000	2 pole		3	2.667		545	146		691	820
3010	3 pole		2.50	3.200		595	175		770	915
3050	Under voltage trip, 277/480 volt breaker, 30 amp, 1 pole		4	2		330	109		439	525
3060	60 amp, 2 pole		3.50	2.286		480	125		605	715
3070	3 pole		3	2.667		545	146		691	820
3080	100 amp, 1 pole		3.50	2.286		385	125		510	610
3090	2 pole		3	2.667		545	146		691	820
3100	3 pole		2.50	3.200		595	175		770	915
3150	Motor operated, 277/480 volt breaker, 30 amp, 1 pole		4	2		585	109		694	810
3160	60 amp, 2 pole		3.50	2.286		740	125		865	995
3170	3 pole		3	2.667		800	146		946	1,100
3180	100 amp, 1 pole		3.50	2.286		640	125		765	890
3190	2 pole		3	2.667		800	146		946	1,100
3200	3 pole		2.50	3.200		850	175		1,025	1,200
3250	Panelboard spacers, per pole		40	.200		4.43	10.95		15.38	21.50

26 24 16.30 Panelboards Commercial Applications		Crew	Daily Output	Labor-Hours	Unit	Material	Labor	Equipment	Total	Total Incl O&P
0010	**PANELBOARDS COMMERCIAL APPLICATIONS** R262416-50									
0050	NQOD, w/20 amp 1 pole bolt-on circuit breakers									
0100	3 wire, 120/240 volts, 100 amp main lugs									
0150	10 circuits	1 Elec	1	8	Ea.	545	440		985	1,250
0200	14 circuits		.88	9.091		660	495		1,155	1,475
0250	18 circuits		.75	10.667		720	585		1,305	1,675
0300	20 circuits		.65	12.308		805	675		1,480	1,875
0350	225 amp main lugs, 24 circuits	2 Elec	1.20	13.333		910	730		1,640	2,100
0400	30 circuits		.90	17.778		1,050	970		2,020	2,600
0450	36 circuits		.80	20		1,200	1,100		2,300	2,975
0500	38 circuits		.72	22.222		1,275	1,225		2,500	3,225
0550	42 circuits		.66	24.242		1,350	1,325		2,675	3,450
0600	4 wire, 120/208 volts, 100 amp main lugs, 12 circuits	1 Elec	1	8		640	440		1,080	1,350
0650	16 circuits		.75	10.667		730	585		1,315	1,675
0700	20 circuits		.65	12.308		845	675		1,520	1,925
0750	24 circuits		.60	13.333		875	730		1,605	2,075
0800	30 circuits		.53	15.094		1,050	825		1,875	2,375
0850	225 amp main lugs, 32 circuits	2 Elec	.90	17.778		1,175	970		2,145	2,750
0900	34 circuits		.84	19.048		1,200	1,050		2,250	2,875
0950	36 circuits		.80	20		1,225	1,100		2,325	3,000
1000	42 circuits		.68	23.529		1,375	1,275		2,650	3,425
1040	225 amp main lugs, NEMA 7, 12 circuits		1	16		3,700	875		4,575	5,375
1100	24 circuits		.40	40		4,450	2,200		6,650	8,150
1200	NEHB,w/20 amp, 1 pole bolt-on circuit breakers									
1250	4 wire, 277/480 volts, 100 amp main lugs, 12 circuits	1 Elec	.88	9.091	Ea.	1,225	495		1,720	2,100
1300	20 circuits	"	.60	13.333		1,825	730		2,555	3,100
1350	225 amp main lugs, 24 circuits	2 Elec	.90	17.778		2,075	970		3,045	3,725
1400	30 circuits		.80	20		2,475	1,100		3,575	4,375
1450	36 circuits		.72	22.222		2,875	1,225		4,100	5,000
1500	42 circuits		.60	26.667		3,275	1,450		4,725	5,775
1510	225 amp main lugs, NEMA 7, 12 circuits		.90	17.778		5,500	970		6,470	7,500
1590	24 circuits		.30	53.333		7,425	2,925		10,350	12,500
1600	NQOD panel, w/20 amp, 1 pole, circuit breakers									
1650	3 wire, 120/240 volt with main circuit breaker									

26 24 16.30 Panelboards Commercial Applications		Crew	Daily Output	Labor-Hours	Unit	2015 Bare Costs Material	Labor	Equipment	Total	Total Incl O&P
1700	100 amp main, 12 circuits	1 Elec	.80	10	Ea.	800	545		1,345	1,700
1750	20 circuits	″	.60	13.333		1,025	730		1,755	2,225
1800	225 amp main, 30 circuits	2 Elec	.68	23.529		1,900	1,275		3,175	4,025
1850	42 circuits		.52	30.769		2,200	1,675		3,875	4,950
1900	400 amp main, 30 circuits		.54	29.630		2,625	1,625		4,250	5,325
1950	42 circuits	↓	.50	32	↓	2,925	1,750		4,675	5,850
2000	4 wire, 120/208 volts with main circuit breaker									
2050	100 amp main, 24 circuits	1 Elec	.47	17.021	Ea.	1,175	930		2,105	2,700
2100	30 circuits	″	.40	20		1,325	1,100		2,425	3,100
2200	225 amp main, 32 circuits	2 Elec	.72	22.222		2,225	1,225		3,450	4,250
2250	42 circuits		.56	28.571		2,425	1,575		4,000	5,025
2300	400 amp main, 42 circuits		.48	33.333		3,250	1,825		5,075	6,300
2350	600 amp main, 42 circuits	↓	.40	40	↓	4,825	2,200		7,025	8,575
2400	NEHB, with 20 amp, 1 pole circuit breaker									
2450	4 wire, 277/480 volts with main circuit breaker									
2500	100 amp main, 24 circuits	1 Elec	.42	19.048	Ea.	2,375	1,050		3,425	4,175
2550	30 circuits	″	.38	21.053		2,775	1,150		3,925	4,800
2600	225 amp main, 30 circuits	2 Elec	.72	22.222		3,500	1,225		4,725	5,675
2650	42 circuits		.56	28.571		4,300	1,575		5,875	7,100
2700	400 amp main, 42 circuits		.46	34.783		5,175	1,900		7,075	8,550
2750	600 amp main, 42 circuits	↓	.38	42.105	↓	7,075	2,300		9,375	11,300
2900	Note: the following line items don't include branch circuit breakers									
2910	For branch circuit breakers information, see Section 26 24 16.20									
3010	Main lug, no main breaker, 240 volt, 1 pole, 3 wire, 100 amp	1 Elec	2.30	3.478	Ea.	535	190		725	875
3020	225 amp	2 Elec	2.40	6.667		640	365		1,005	1,250
3030	400 amp	″	1.80	8.889		940	485		1,425	1,750
3060	3 pole, 3 wire, 100 amp	1 Elec	2.30	3.478		575	190		765	920
3070	225 amp	2 Elec	2.40	6.667		680	365		1,045	1,300
3080	400 amp		1.80	8.889		1,025	485		1,510	1,850
3090	600 amp	↓	1.60	10		1,175	545		1,720	2,125
3110	3 pole, 4 wire, 100 amp	1 Elec	2.30	3.478		635	190		825	985
3120	225 amp	2 Elec	2.40	6.667		795	365		1,160	1,425
3130	400 amp		1.80	8.889		1,025	485		1,510	1,850
3140	600 amp	↓	1.60	10		1,175	545		1,720	2,125
3160	480 volt, 3 pole, 3 wire, 100 amp	1 Elec	2.30	3.478		760	190		950	1,125
3170	225 amp	2 Elec	2.40	6.667		920	365		1,285	1,550
3180	400 amp		1.80	8.889		1,275	485		1,760	2,125
3190	600 amp	↓	1.60	10		1,425	545		1,970	2,400
3210	277/480 volt, 3 pole, 4 wire, 100 amp	1 Elec	2.30	3.478		740	190		930	1,100
3220	225 amp	2 Elec	2.40	6.667		900	365		1,265	1,525
3230	400 amp		1.80	8.889		1,250	485		1,735	2,100
3240	600 amp	↓	1.60	10		1,400	545		1,945	2,350
3260	Main circuit breaker, 240 volt, 1 pole, 3 wire, 100 amp	1 Elec	2	4		715	219		934	1,125
3270	225 amp	2 Elec	2	8		1,475	440		1,915	2,250
3280	400 amp	″	1.60	10		2,300	545		2,845	3,350
3310	3 pole, 3 wire, 100 amp	1 Elec	2	4		820	219		1,039	1,225
3320	225 amp	2 Elec	2	8		1,675	440		2,115	2,500
3330	400 amp		1.60	10		2,625	545		3,170	3,725
3360	120/208 volt, 3 pole, 4 wire, 100 amp		4	4		820	219		1,039	1,225
3370	225 amp		2	8		1,675	440		2,115	2,500
3380	400 amp	↓	1.60	10		2,625	545		3,170	3,725
3410	480 volt, 3 pole, 3 wire, 100 amp	1 Elec	2	4		1,150	219		1,369	1,600
3420	225 amp	2 Elec	2	8	↓	1,975	440		2,415	2,825

26 24 Switchboards and Panelboards

26 24 16 – Panelboards

26 24 16.30 Panelboards Commercial Applications		Crew	Daily Output	Labor-Hours	Unit	2015 Bare Costs Material	Labor	Equipment	Total	Total Incl O&P
3430	400 amp	2 Elec	1.60	10	Ea.	2,950	545		3,495	4,075
3460	277/480 volt, 3 pole, 4 wire, 100 amp		4	4		1,125	219		1,344	1,550
3470	225 amp		2	8		1,925	440		2,365	2,775
3480	400 amp		1.60	10		2,975	545		3,520	4,100
3510	Main circuit breaker, HIC, 240 volt, 1 pole, 3 wire, 100 amp	1 Elec	2	4		1,100	219		1,319	1,525
3520	225 amp	2 Elec	2	8		2,875	440		3,315	3,800
3530	400 amp	"	1.60	10		3,850	545		4,395	5,050
3560	3 pole, 3 wire, 100 amp	1 Elec	2	4		1,225	219		1,444	1,675
3570	225 amp	2 Elec	2	8		3,225	440		3,665	4,200
3580	400 amp	"	1.60	10		4,275	545		4,820	5,525
3610	120/208 volt, 3 pole, 4 wire, 100 amp	1 Elec	2	4		1,225	219		1,444	1,675
3620	225 amp	2 Elec	2	8		3,225	440		3,665	4,200
3630	400 amp	"	1.60	10		4,275	545		4,820	5,525
3660	480 volt, 3 pole, 3 wire, 100 amp	1 Elec	2	4		1,825	219		2,044	2,325
3670	225 amp	2 Elec	2	8		3,600	440		4,040	4,600
3680	400 amp	"	1.60	10		4,575	545		5,120	5,850
3710	277/480 volt, 3 pole, 4 wire, 100 amp	1 Elec	2	4		1,725	219		1,944	2,225
3720	225 amp	2 Elec	2	8		3,425	440		3,865	4,425
3730	400 amp	"	1.60	10		4,550	545		5,095	5,825
3760	Main circuit breaker, shunt trip, 100 amp	1 Elec	1.20	6.667		920	365		1,285	1,550
3770	225 amp	2 Elec	1.60	10		2,125	545		2,670	3,150
3780	400 amp	"	1.40	11.429		3,075	625		3,700	4,300

26 24 19 – Motor-Control Centers

26 24 19.20 Motor Control Center Components		Crew	Daily Output	Labor-Hours	Unit	Material	Labor	Equipment	Total	Total Incl O&P
0010	**MOTOR CONTROL CENTER COMPONENTS** R262419-60									
0100	Starter, size 1, FVNR, NEMA 1, type A, fusible	1 Elec	2.70	2.963	Ea.	1,575	162		1,737	2,000
0120	Circuit breaker		2.70	2.963		1,725	162		1,887	2,150
0140	Type B, fusible		2.70	2.963		1,750	162		1,912	2,150
0160	Circuit breaker		2.70	2.963		1,900	162		2,062	2,325
0180	NEMA 12, type A, fusible		2.60	3.077		1,600	168		1,768	2,025
0200	Circuit breaker		2.60	3.077		1,750	168		1,918	2,175
0220	Type B, fusible		2.60	3.077		1,775	168		1,943	2,200
0240	Circuit breaker		2.60	3.077		1,925	168		2,093	2,350
0300	Starter, size 1, FVR, NEMA 1, type A, fusible		2	4		2,275	219		2,494	2,825
0320	Circuit breaker		2	4		2,275	219		2,494	2,825
0340	Type B, fusible		2	4		2,500	219		2,719	3,075
0360	Circuit breaker		2	4		2,500	219		2,719	3,075
0380	NEMA 12, type A, fusible		1.90	4.211		2,300	230		2,530	2,875
0400	Circuit breaker		1.90	4.211		2,300	230		2,530	2,875
0420	Type B, fusible		1.90	4.211		2,525	230		2,755	3,125
0440	Circuit breaker		1.90	4.211		2,525	230		2,755	3,125
0490	Starter size 1, 2 speed, separate winding									
0500	NEMA 1, type A, fusible	1 Elec	2.60	3.077	Ea.	2,975	168		3,143	3,525
0520	Circuit breaker		2.60	3.077		2,975	168		3,143	3,525
0540	Type B, fusible		2.60	3.077		3,275	168		3,443	3,850
0560	Circuit breaker		2.60	3.077		3,275	168		3,443	3,850
0580	NEMA 12, type A, fusible		2.50	3.200		3,050	175		3,225	3,600
0600	Circuit breaker		2.50	3.200		3,050	175		3,225	3,600
0620	Type B, fusible		2.50	3.200		3,350	175		3,525	3,925
0640	Circuit breaker		2.50	3.200		3,350	175		3,525	3,925
0650	Starter size 1, 2 speed, consequent pole									
0660	NEMA 1, type A, fusible	1 Elec	2.60	3.077	Ea.	2,975	168		3,143	3,525

26 24 19.20 Motor Control Center Components		Crew	Daily Output	Labor-Hours	Unit	Material	Labor	Equipment	Total	Total Incl O&P
						2015 Bare Costs				
0680	Circuit breaker	1 Elec	2.60	3.077	Ea.	2,975	168		3,143	3,525
0700	Type B, fusible		2.60	3.077		3,275	168		3,443	3,850
0720	Circuit breaker		2.60	3.077		3,275	168		3,443	3,850
0740	NEMA 12, type A, fusible		2.50	3.200		3,050	175		3,225	3,600
0760	Circuit breaker		2.50	3.200		3,050	175		3,225	3,600
0780	Type B, fusible		2.50	3.200		3,325	175		3,500	3,900
0800	Circuit breaker		2.50	3.200		3,350	175		3,525	3,925
0810	Starter size 1, 2 speed, space only									
0820	NEMA 1, type A, fusible	1 Elec	16	.500	Ea.	670	27.50		697.50	775
0840	Circuit breaker		16	.500		670	27.50		697.50	775
0860	Type B, fusible		16	.500		670	27.50		697.50	775
0880	Circuit breaker		16	.500		670	27.50		697.50	775
0900	NEMA 12, type A, fusible		15	.533		695	29		724	810
0920	Circuit breaker		15	.533		695	29		724	810
0940	Type B, fusible		15	.533		695	29		724	810
0960	Circuit breaker		15	.533		695	29		724	810
1100	Starter size 2, FVNR, NEMA 1, type A, fusible	2 Elec	4	4		1,775	219		1,994	2,300
1120	Circuit breaker		4	4		1,950	219		2,169	2,475
1140	Type B, fusible		4	4		1,975	219		2,194	2,475
1160	Circuit breaker		4	4		2,150	219		2,369	2,675
1180	NEMA 12, type A, fusible		3.80	4.211		1,825	230		2,055	2,350
1200	Circuit breaker		3.80	4.211		1,975	230		2,205	2,525
1220	Type B, fusible		3.80	4.211		1,975	230		2,205	2,525
1240	Circuit breaker		3.80	4.211		2,175	230		2,405	2,725
1300	FVR, NEMA 1, type A, fusible		3.20	5		3,025	274		3,299	3,725
1320	Circuit breaker		3.20	5		3,025	274		3,299	3,725
1340	Type B, fusible		3.20	5		3,325	274		3,599	4,050
1360	Circuit breaker		3.20	5		3,325	274		3,599	4,050
1380	NEMA type 12, type A, fusible		3	5.333		3,075	292		3,367	3,800
1400	Circuit breaker		3	5.333		3,075	292		3,367	3,800
1420	Type B, fusible		3	5.333		3,375	292		3,667	4,150
1440	Circuit breaker		3	5.333		3,375	292		3,667	4,150
1490	Starter size 2, 2 speed, separate winding									
1500	NEMA 1, type A, fusible	2 Elec	3.80	4.211	Ea.	3,375	230		3,605	4,075
1520	Circuit breaker		3.80	4.211		3,375	230		3,605	4,075
1540	Type B, fusible		3.80	4.211		3,700	230		3,930	4,425
1560	Circuit breaker		3.80	4.211		3,700	230		3,930	4,425
1570	NEMA 12, type A, fusible		3.60	4.444		3,425	243		3,668	4,150
1580	Circuit breaker		3.60	4.444		3,425	243		3,668	4,150
1600	Type B, fusible		3.60	4.444		3,775	243		4,018	4,525
1620	Circuit breaker		3.60	4.444		3,775	243		4,018	4,525
1630	Starter size 2, 2 speed, consequent pole									
1640	NEMA 1, type A, fusible	2 Elec	3.80	4.211	Ea.	3,825	230		4,055	4,550
1660	Circuit breaker		3.80	4.211		3,825	230		4,055	4,550
1680	Type B, fusible		3.80	4.211		4,050	230		4,280	4,800
1700	Circuit breaker		3.80	4.211		4,050	230		4,280	4,800
1720	NEMA 12, type A, fusible		3.80	4.211		3,850	230		4,080	4,575
1740	Circuit breaker		3.60	4.444		3,875	243		4,118	4,650
1760	Type B, fusible		3.60	4.444		4,100	243		4,343	4,900
1780	Circuit breaker		3.60	4.444		4,100	243		4,343	4,900
1830	Starter size 2, autotransformer									
1840	NEMA 1, type A, fusible	2 Elec	3.40	4.706	Ea.	5,975	257		6,232	6,950
1860	Circuit breaker		3.40	4.706		6,125	257		6,382	7,100

26 24 19.20 Motor Control Center Components		Crew	Daily Output	Labor-Hours	Unit	Material	Labor	Equipment	Total	Total Incl O&P
						2015 Bare Costs				
1880	Type B, fusible	2 Elec	3.40	4.706	Ea.	6,525	257		6,782	7,550
1900	Circuit breaker		3.40	4.706		6,525	257		6,782	7,550
1920	NEMA 12, type A, fusible		3.20	5		6,075	274		6,349	7,100
1940	Circuit breaker		3.20	5		6,075	274		6,349	7,100
1960	Type B, fusible		3.20	5		6,625	274		6,899	7,700
1980	Circuit breaker		3.20	5		6,625	274		6,899	7,700
2030	Starter size 2, space only									
2040	NEMA 1, type A, fusible	1 Elec	16	.500	Ea.	670	27.50		697.50	775
2060	Circuit breaker		16	.500		670	27.50		697.50	775
2080	Type B, fusible		16	.500		670	27.50		697.50	775
2100	Circuit breaker		16	.500		670	27.50		697.50	775
2120	NEMA 12, type A, fusible		15	.533		695	29		724	810
2140	Circuit breaker		15	.533		695	29		724	810
2160	Type B, fusible		15	.533		695	29		724	810
2180	Circuit breaker		15	.533		695	29		724	810
2300	Starter size 3, FVNR, NEMA 1, type A, fusible	2 Elec	2	8		3,400	440		3,840	4,400
2320	Circuit breaker		2	8		3,025	440		3,465	3,975
2340	Type B, fusible		2	8		3,750	440		4,190	4,775
2360	Circuit breaker		2	8		3,325	440		3,765	4,325
2380	NEMA 12, type A, fusible		1.90	8.421		3,475	460		3,935	4,525
2400	Circuit breaker		1.90	8.421		3,075	460		3,535	4,075
2420	Type B, fusible		1.90	8.421		3,875	460		4,335	4,950
2440	Circuit breaker		1.90	8.421		3,375	460		3,835	4,425
2500	Starter size 3, FVR, NEMA 1, type A, fusible		1.60	10		4,625	545		5,170	5,925
2520	Circuit breaker		1.60	10		4,425	545		4,970	5,700
2540	Type B, fusible		1.60	10		5,050	545		5,595	6,375
2560	Circuit breaker		1.60	10		4,850	545		5,395	6,150
2580	NEMA 12, type A, fusible		1.50	10.667		4,725	585		5,310	6,075
2600	Circuit breaker		1.50	10.667		4,500	585		5,085	5,850
2620	Type B, fusible		1.50	10.667		5,150	585		5,735	6,525
2640	Circuit breaker		1.50	10.667		5,325	585		5,910	6,750
2690	Starter size 3, 2 speed, separate winding									
2700	NEMA 1, type A, fusible	2 Elec	2	8	Ea.	5,325	440		5,765	6,525
2720	Circuit breaker		2	8		4,700	440		5,140	5,825
2740	Type B, fusible		2	8		5,850	440		6,290	7,075
2760	Circuit breaker		2	8		5,150	440		5,590	6,325
2780	NEMA 12, type A, fusible		1.90	8.421		5,450	460		5,910	6,675
2800	Circuit breaker		1.90	8.421		4,800	460		5,260	5,975
2820	Type B, fusible		1.90	8.421		5,950	460		6,410	7,250
2840	Circuit breaker		1.90	8.421		5,250	460		5,710	6,475
2850	Starter size 3, 2 speed, consequent pole									
2860	NEMA 1, type A, fusible	2 Elec	2	8	Ea.	5,950	440		6,390	7,200
2880	Circuit breaker		2	8		5,325	440		5,765	6,500
2900	Type B, fusible		2	8		6,450	440		6,890	7,750
2920	Circuit breaker		2	8		5,775	440		6,215	7,000
2940	NEMA 12, type A, fusible		1.90	8.421		6,050	460		6,510	7,375
2960	Circuit breaker		1.90	8.421		5,425	460		5,885	6,675
2980	Type B, fusible		1.90	8.421		6,575	460		7,035	7,925
3000	Circuit breaker		1.90	8.421		5,875	460		6,335	7,150
3100	Starter size 3, autotransformer, NEMA 1, type A, fusible		1.60	10		7,475	545		8,020	9,050
3120	Circuit breaker		1.60	10		7,400	545		7,945	8,975
3140	Type B, fusible		1.60	10		8,100	545		8,645	9,725
3160	Circuit breaker		1.60	10		8,100	545		8,645	9,725

26 24 19.20 Motor Control Center Components		Crew	Daily Output	Labor-Hours	Unit	Material	Labor	Equipment	Total	Total Incl O&P
						2015 Bare Costs				
3180	NEMA 12, type A, fusible	2 Elec	1.50	10.667	Ea.	7,550	585		8,135	9,175
3200	Circuit breaker		1.50	10.667		7,550	585		8,135	9,175
3220	Type B, fusible		1.50	10.667		8,225	585		8,810	9,925
3240	Circuit breaker		1.50	10.667		8,225	585		8,810	9,925
3260	Starter size 3, space only, NEMA 1, type A, fusible	1 Elec	15	.533		1,150	29		1,179	1,325
3280	Circuit breaker		15	.533		895	29		924	1,025
3300	Type B, fusible		15	.533		1,150	29		1,179	1,325
3320	Circuit breaker		15	.533		895	29		924	1,025
3340	NEMA 12, type A, fusible		14	.571		1,225	31.50		1,256.50	1,400
3360	Circuit breaker		14	.571		950	31.50		981.50	1,100
3380	Type B, fusible		14	.571		1,225	31.50		1,256.50	1,400
3400	Circuit breaker		14	.571		950	31.50		981.50	1,100
3500	Starter size 4, FVNR, NEMA 1, type A, fusible	2 Elec	1.60	10		4,425	545		4,970	5,700
3520	Circuit breaker		1.60	10		4,025	545		4,570	5,250
3540	Type B, fusible		1.60	10		4,850	545		5,395	6,150
3560	Circuit breaker		1.60	10		4,425	545		4,970	5,700
3580	NEMA 12, type A, fusible		1.50	10.667		4,550	585		5,135	5,875
3600	Circuit breaker		1.50	10.667		4,075	585		4,660	5,375
3620	Type B, fusible		1.50	10.667		4,950	585		5,535	6,325
3640	Circuit breaker		1.50	10.667		4,500	585		5,085	5,825
3700	Starter size 4, FVR, NEMA 1, type A, fusible		1.20	13.333		6,050	730		6,780	7,775
3720	Circuit breaker		1.20	13.333		5,450	730		6,180	7,100
3740	Type B, fusible		1.20	13.333		6,600	730		7,330	8,375
3760	Circuit breaker		1.20	13.333		6,000	730		6,730	7,700
3780	NEMA 12, type A, fusible		1.16	13.793		6,200	755		6,955	7,950
3800	Circuit breaker		1.16	13.793		5,550	755		6,305	7,225
3820	Type B, fusible		1.16	13.793		5,600	755		6,355	7,275
3840	Circuit breaker		1.16	13.793		6,100	755		6,855	7,825
3890	Starter size 4, 2 speed, separate windings									
3900	NEMA 1, type A, fusible	2 Elec	1.60	10	Ea.	7,650	545		8,195	9,225
3920	Circuit breaker		1.60	10		5,750	545		6,295	7,150
3940	Type B, fusible		1.60	10		8,400	545		8,945	10,100
3960	Circuit breaker		1.60	10		6,300	545		6,845	7,750
3980	NEMA 12, type A, fusible		1.50	10.667		7,775	585		8,360	9,425
4000	Circuit breaker		1.50	10.667		5,850	585		6,435	7,300
4020	Type B, fusible		1.50	10.667		8,525	585		9,110	10,300
4040	Circuit breaker		1.50	10.667		6,375	585		6,960	7,900
4050	Starter size 4, 2 speed, consequent pole									
4060	NEMA 1, type A, fusible	2 Elec	1.60	10	Ea.	8,925	545		9,470	10,600
4080	Circuit breaker		1.60	10		6,650	545		7,195	8,125
4100	Type B, fusible		1.60	10		9,825	545		10,370	11,600
4120	Circuit breaker		1.60	10		7,275	545		7,820	8,825
4140	NEMA 12, type A, fusible		1.50	10.667		9,075	585		9,660	10,900
4160	Circuit breaker		1.50	10.667		6,750	585		7,335	8,275
4180	Type B, fusible		1.50	10.667		9,950	585		10,535	11,900
4200	Circuit breaker		1.50	10.667		7,375	585		7,960	9,000
4300	Starter size 4, autotransformer, NEMA 1, type A, fusible		1.30	12.308		8,800	675		9,475	10,700
4320	Circuit breaker		1.30	12.308		8,875	675		9,550	10,800
4340	Type B, fusible		1.30	12.308		9,675	675		10,350	11,700
4360	Circuit breaker		1.30	12.308		9,675	675		10,350	11,700
4380	NEMA 12, type A, fusible		1.24	12.903		8,950	705		9,655	10,900
4400	Circuit breaker		1.24	12.903		8,975	705		9,680	10,900
4420	Type B, fusible		1.24	12.903		9,825	705		10,530	11,900

26 24 19.20 Motor Control Center Components		Crew	Daily Output	Labor-Hours	Unit	2015 Bare Costs Material	Labor	Equipment	Total	Total Incl O&P
4440	Circuit breaker	2 Elec	1.24	12.903	Ea.	9,800	705		10,505	11,900
4500	Starter size 4, space only, NEMA 1, type A, fusible	1 Elec	14	.571		1,600	31.50		1,631.50	1,800
4520	Circuit breaker		14	.571		1,150	31.50		1,181.50	1,325
4540	Type B, fusible		14	.571		1,600	31.50		1,631.50	1,800
4560	Circuit breaker		14	.571		1,150	31.50		1,181.50	1,325
4580	NEMA 12, type A, fusible		13	.615		1,700	33.50		1,733.50	1,900
4600	Circuit breaker		13	.615		1,225	33.50		1,258.50	1,400
4620	Type B, fusible		13	.615		1,700	33.50		1,733.50	1,900
4640	Circuit breaker		13	.615		1,225	33.50		1,258.50	1,400
4800	Starter size 5, FVNR, NEMA 1, type A, fusible	2 Elec	1	16		9,150	875		10,025	11,400
4820	Circuit breaker		1	16		6,500	875		7,375	8,450
4840	Type B, fusible		1	16		10,000	875		10,875	12,300
4860	Circuit breaker		1	16		7,150	875		8,025	9,150
4880	NEMA 12, type A, fusible		.96	16.667		9,275	910		10,185	11,600
4900	Circuit breaker		.96	16.667		6,625	910		7,535	8,650
4920	Type B, fusible		.96	16.667		10,200	910		11,110	12,600
4940	Circuit breaker		.96	16.667		7,275	910		8,185	9,375
5000	Starter size 5, FVR, NEMA 1, type A, fusible		.80	20		13,500	1,100		14,600	16,500
5020	Circuit breaker		.80	20		10,600	1,100		11,700	13,300
5040	Type B, fusible		.80	20		14,800	1,100		15,900	17,900
5060	Circuit breaker		.80	20		11,600	1,100		12,700	14,500
5080	NEMA 12, type A, fusible		.76	21.053		13,700	1,150		14,850	16,800
5100	Circuit breaker		.76	21.053		10,700	1,150		11,850	13,500
5120	Type B, fusible		.76	21.053		15,000	1,150		16,150	18,200
5140	Circuit breaker		.76	21.053		11,800	1,150		12,950	14,600
5190	Starter size 5, 2 speed, separate windings									
5200	NEMA 1, type A, fusible	2 Elec	1	16	Ea.	18,600	875		19,475	21,700
5220	Circuit breaker		1	16		13,900	875		14,775	16,500
5240	Type B, fusible		1	16		20,400	875		21,275	23,700
5260	Circuit breaker		1	16		15,200	875		16,075	18,100
5280	NEMA 12, type A, fusible		.96	16.667		18,800	910		19,710	22,100
5300	Circuit breaker		.96	16.667		14,000	910		14,910	16,800
5320	Type B, fusible		.96	16.667		20,700	910		21,610	24,100
5340	Circuit breaker		.96	16.667		15,400	910		16,310	18,300
5400	Starter size 5, autotransformer, NEMA 1, type A, fusible		.70	22.857		15,200	1,250		16,450	18,600
5420	Circuit breaker		.70	22.857		12,600	1,250		13,850	15,800
5440	Type B, fusible		.70	22.857		16,700	1,250		17,950	20,300
5460	Circuit breaker		.70	22.857		13,900	1,250		15,150	17,200
5480	NEMA 12, type A, fusible		.68	23.529		15,400	1,275		16,675	18,800
5500	Circuit breaker		.68	23.529		12,800	1,275		14,075	15,900
5520	Type B, fusible		.68	23.529		16,900	1,275		18,175	20,400
5540	Circuit breakers		.68	23.529		14,000	1,275		15,275	17,300
5600	Starter size 5, space only, NEMA 1, type A, fusible	1 Elec	12	.667		2,050	36.50		2,086.50	2,300
5620	Circuit breaker		12	.667		2,050	36.50		2,086.50	2,300
5640	Type B, fusible		12	.667		2,050	36.50		2,086.50	2,300
5660	Circuit breaker		12	.667		1,350	36.50		1,386.50	1,525
5680	NEMA 12, type A, fusible		11	.727		2,175	40		2,215	2,425
5700	Circuit breaker		11	.727		1,425	40		1,465	1,625
5720	Type B, fusible		11	.727		2,175	40		2,215	2,425
5740	Circuit breaker		11	.727		1,425	40		1,465	1,625
5800	Fuse, light contactor NEMA 1, type A, 30 amp		2.70	2.963		1,575	162		1,737	2,000
5820	60 amp		2	4		1,775	219		1,994	2,300
5840	100 amp		1	8		3,400	440		3,840	4,400

26 24 19.20 Motor Control Center Components		Crew	Daily Output	Labor-Hours	Unit	Material	2015 Bare Costs Labor	Equipment	Total	Total Incl O&P
5860	200 amp	1 Elec	.80	10	Ea.	7,900	545		8,445	9,525
5880	Type B, 30 amp		2.70	2.963		1,725	162		1,887	2,150
5900	60 amp		2	4		1,925	219		2,144	2,450
5920	100 amp		1	8		3,750	440		4,190	4,775
5940	200 amp	2 Elec	1.60	10		8,650	545		9,195	10,300
5960	NEMA 12, type A, 30 amp	1 Elec	2.60	3.077		1,600	168		1,768	2,025
5980	60 amp		1.90	4.211		1,825	230		2,055	2,350
6000	100 amp		.95	8.421		3,475	460		3,935	4,525
6020	200 amp	2 Elec	1.50	10.667		8,025	585		8,610	9,700
6040	Type B, 30 amp	1 Elec	2.60	3.077		1,750	168		1,918	2,175
6060	60 amp		1.90	4.211		1,950	230		2,180	2,500
6080	100 amp		.95	8.421		3,800	460		4,260	4,900
6100	200 amp	2 Elec	1.50	10.667		8,775	585		9,360	10,500
6200	Circuit breaker, light contactor NEMA 1, type A, 30 amp	1 Elec	2.70	2.963		1,725	162		1,887	2,150
6220	60 amp		2	4		1,950	219		2,169	2,475
6240	100 amp		1	8		3,025	440		3,465	3,975
6260	200 amp	2 Elec	1.60	10		6,550	545		7,095	8,025
6280	Type B, 30 amp	1 Elec	2.70	2.963		1,900	162		2,062	2,325
6300	60 amp		2	4		2,175	219		2,394	2,700
6320	100 amp		1	8		3,300	440		3,740	4,275
6340	200 amp	2 Elec	1.60	10		7,150	545		7,695	8,675
6360	NEMA 12, type A, 30 amp	1 Elec	2.60	3.077		1,750	168		1,918	2,175
6380	60 amp		1.90	4.211		1,975	230		2,205	2,525
6400	100 amp		.95	8.421		3,075	460		3,535	4,075
6420	200 amp	2 Elec	1.50	10.667		6,575	585		7,160	8,100
6440	Type B, 30 amp	1 Elec	2.60	3.077		2,050	168		2,218	2,500
6460	60 amp		1.90	4.211		2,200	230		2,430	2,750
6480	100 amp		.95	8.421		3,350	460		3,810	4,375
6500	200 amp	2 Elec	1.50	10.667		7,225	585		7,810	8,825
6600	Fusible switch, NEMA 1, type A, 30 amp	1 Elec	5.30	1.509		1,050	82.50		1,132.50	1,275
6620	60 amp		5	1.600		1,125	87.50		1,212.50	1,350
6640	100 amp		4	2		1,225	109		1,334	1,525
6660	200 amp		3.20	2.500		2,125	137		2,262	2,550
6680	400 amp	2 Elec	4.60	3.478		5,025	190		5,215	5,800
6700	600 amp		3.20	5		5,425	274		5,699	6,375
6720	800 amp		2.60	6.154		14,900	335		15,235	16,900
6740	NEMA 12, type A, 30 amp	1 Elec	5.20	1.538		1,075	84		1,159	1,300
6760	60 amp		4.90	1.633		1,150	89.50		1,239.50	1,375
6780	100 amp		3.90	2.051		1,275	112		1,387	1,575
6800	200 amp		3.10	2.581		2,175	141		2,316	2,600
6820	400 amp	2 Elec	4.40	3.636		5,125	199		5,324	5,925
6840	600 amp		3	5.333		5,575	292		5,867	6,575
6860	800 amp		2.40	6.667		15,000	365		15,365	17,000
6900	Circuit breaker, NEMA 1, type A, 30 amp	1 Elec	5.30	1.509		960	82.50		1,042.50	1,175
6920	60 amp		5	1.600		960	87.50		1,047.50	1,175
6940	100 amp		4	2		960	109		1,069	1,225
6960	225 amp		3.20	2.500		1,725	137		1,862	2,100
6980	400 amp	2 Elec	4.60	3.478		3,250	190		3,440	3,875
7000	600 amp		3.20	5		3,800	274		4,074	4,600
7020	800 amp		2.60	6.154		7,975	335		8,310	9,275
7040	NEMA 12, type A, 30 amp	1 Elec	5.20	1.538		990	84		1,074	1,225
7060	60 amp		4.90	1.633		990	89.50		1,079.50	1,225
7080	100 amp		3.90	2.051		990	112		1,102	1,275

26 24 19.20 Motor Control Center Components		Crew	Daily Output	Labor-Hours	Unit	Material	Labor	Equipment	Total	Total Incl O&P
						2015 Bare Costs				
7100	225 amp	1 Elec	3.10	2.581	Ea.	1,750	141		1,891	2,125
7120	400 amp	2 Elec	4.40	3.636		3,325	199		3,524	3,950
7140	600 amp		3	5.333		3,875	292		4,167	4,675
7160	800 amp		2.40	6.667		8,125	365		8,490	9,475
7300	Incoming line, main lug only, 600 amp, alum., NEMA 1		1.60	10		1,225	545		1,770	2,150
7320	NEMA 12		1.50	10.667		1,250	585		1,835	2,250
7340	Copper, NEMA 1		1.60	10		1,275	545		1,820	2,225
7360	800 amp, alum., NEMA 1		1.50	10.667		3,050	585		3,635	4,225
7380	NEMA 12		1.40	11.429		3,100	625		3,725	4,325
7400	Copper, NEMA 1		1.50	10.667		3,175	585		3,760	4,375
7420	1200 amp, copper, NEMA 1		1.40	11.429		3,275	625		3,900	4,525
7440	Incoming line, fusible switch, 400 amp, alum., NEMA 1		1.20	13.333		4,200	730		4,930	5,700
7460	NEMA 12		1.10	14.545		4,275	795		5,070	5,900
7480	Copper, NEMA 1		1.20	13.333		4,250	730		4,980	5,775
7500	600 amp, alum., NEMA 1		1.10	14.545		5,125	795		5,920	6,825
7520	NEMA 12		1	16		5,200	875		6,075	7,025
7540	Copper, NEMA 1		1.10	14.545		5,175	795		5,970	6,900
7560	Incoming line, circuit breaker, 225 amp, alum., NEMA 1		1.20	13.333		2,175	730		2,905	3,500
7580	NEMA 12		1.10	14.545		2,225	795		3,020	3,650
7600	Copper, NEMA 1		1.20	13.333		2,250	730		2,980	3,575
7620	400 amp, alum., NEMA 1		1.20	13.333		3,250	730		3,980	4,700
7640	NEMA 12		1.10	14.545		3,325	795		4,120	4,850
7660	Copper, NEMA 1		1.20	13.333		3,325	730		4,055	4,750
7680	600 amp, alum., NEMA 1		1.10	14.545		3,800	795		4,595	5,400
7700	NEMA 12		1	16		3,875	875		4,750	5,550
7720	Copper, NEMA 1		1.10	14.545		3,875	795		4,670	5,475
7740	800 amp, copper, NEMA 1		.90	17.778		7,975	970		8,945	10,200
7760	Incoming line, for copper bus, add					120			120	132
7780	For 65000 amp bus bracing, add					179			179	196
7800	For NEMA 3R enclosure, add					5,600			5,600	6,175
7820	For NEMA 12 enclosure, add					155			155	170
7840	For 1/4" x 1" ground bus, add	1 Elec	16	.500		99.50	27.50		127	151
7860	For 1/4" x 2" ground bus, add	"	12	.667		99.50	36.50		136	165
7900	Main rating basic section, alum., NEMA 1, 800 amp	2 Elec	1.40	11.429		325	625		950	1,300
7920	1200 amp	"	1.20	13.333		640	730		1,370	1,800
7940	For copper bus, add					480			480	530
7960	For 65000 amp bus bracing, add					325			325	355
7980	For NEMA 3R enclosure, add					5,600			5,600	6,175
8000	For NEMA 12, enclosure, add					155			155	170
8020	For 1/4" x 1" ground bus, add	1 Elec	16	.500		99.50	27.50		127	151
8040	For 1/4" x 2" ground bus, add		12	.667		99.50	36.50		136	165
8060	Unit devices, pilot light, standard		16	.500		86	27.50		113.50	136
8080	Pilot light, push to test		16	.500		120	27.50		147.50	173
8100	Pilot light, standard, and push button		12	.667		206	36.50		242.50	282
8120	Pilot light, push to test, and push button		12	.667		240	36.50		276.50	320
8140	Pilot light, standard, and select switch		12	.667		206	36.50		242.50	282
8160	Pilot light, push to test, and select switch		12	.667		240	36.50		276.50	320

26 24 19.30 Motor Control Center		Crew	Daily Output	Labor-Hours	Unit	Material	Labor	Equipment	Total	Total Incl O&P
0010	**MOTOR CONTROL CENTER** Consists of starters & structures R262419-60									
0050	Starters, class 1, type B, comb. MCP, FVNR, with									
0100	control transformer, 10 HP, size 1, 12" high	1 Elec	2.70	2.963	Ea.	1,725	162		1,887	2,150
0200	25 HP, size 2, 18" high	2 Elec	4	4		1,950	219		2,169	2,475

26 24 19.30 Motor Control Center		Crew	Daily Output	Labor-Hours	Unit	Material	Labor	Equipment	Total	Total Incl O&P
						2015 Bare Costs				
0300	50 HP, size 3, 24" high	2 Elec	2	8	Ea.	3,025	440		3,465	3,975
0350	75 HP, size 4, 24" high		1.60	10		4,025	545		4,570	5,250
0400	100 HP, size 4, 30" high		1.40	11.429		5,300	625		5,925	6,750
0500	200 HP, size 5, 48" high		1	16		7,900	875		8,775	10,000
0600	400 HP, size 6, 72" high		.80	20		16,300	1,100		17,400	19,600
0800	Structures, 600 amp, 22,000 rms, takes any									
0900	combination of starters up to 72" high	2 Elec	1.60	10	Ea.	1,925	545		2,470	2,950
1000	Back to back, 72" front & 66" back	"	1.20	13.333		2,600	730		3,330	3,950
1100	For copper bus add per structure					268			268	295
1200	For NEMA 12, add per structure					155			155	170
1300	For 42,000 rms, add per structure					201			201	221
1400	For 100,000 rms, size 1 & 2, add					685			685	755
1500	Size 3, add					1,100			1,100	1,225
1600	Size 4, add					895			895	980
1700	For pilot lights, add per starter	1 Elec	16	.500		120	27.50		147.50	173
1800	For push button, add per starter		16	.500		120	27.50		147.50	173
1900	For auxiliary contacts, add per starter		16	.500		179	27.50		206.50	237

26 24 19.40 Motor Starters and Controls		Crew	Daily Output	Labor-Hours	Unit	Material	Labor	Equipment	Total	Total Incl O&P
0010	**MOTOR STARTERS AND CONTROLS** R262419-65									
0050	Magnetic, FVNR, with enclosure and heaters, 480 volt									
0080	2 HP, size 00	1 Elec	3.50	2.286	Ea.	203	125		328	410
0100	5 HP, size 0		2.30	3.478		272	190		462	585
0200	10 HP, size 1		1.60	5		275	274		549	715
0300	25 HP, size 2 R263413-33	2 Elec	2.20	7.273		520	400		920	1,175
0400	50 HP, size 3		1.80	8.889		845	485		1,330	1,650
0500	100 HP, size 4		1.20	13.333		1,875	730		2,605	3,150
0600	200 HP, size 5		.90	17.778		4,375	970		5,345	6,275
0610	400 HP, size 6		.80	20		19,000	1,100		20,100	22,600
0620	NEMA 7, 5 HP, size 0	1 Elec	1.60	5		1,400	274		1,674	1,950
0630	10 HP, size 1	"	1.10	7.273		1,475	400		1,875	2,225
0640	25 HP, size 2	2 Elec	1.80	8.889		2,375	485		2,860	3,350
0650	50 HP, size 3		1.20	13.333		3,575	730		4,305	5,025
0660	100 HP, size 4		.90	17.778		5,775	970		6,745	7,800
0670	200 HP, size 5		.50	32		13,800	1,750		15,550	17,800
0700	Combination, with motor circuit protectors, 5 HP, size 0	1 Elec	1.80	4.444		880	243		1,123	1,325
0800	10 HP, size 1	"	1.30	6.154		915	335		1,250	1,500
0900	25 HP, size 2	2 Elec	2	8		1,275	440		1,715	2,050
1000	50 HP, size 3		1.32	12.121		1,850	665		2,515	3,025
1200	100 HP, size 4		.80	20		4,000	1,100		5,100	6,050
1220	NEMA 7, 5 HP, size 0	1 Elec	1.30	6.154		2,825	335		3,160	3,600
1230	10 HP, size 1	"	1	8		2,900	440		3,340	3,825
1240	25 HP, size 2	2 Elec	1.32	12.121		3,850	665		4,515	5,250
1250	50 HP, size 3		.80	20		6,375	1,100		7,475	8,675
1260	100 HP, size 4		.60	26.667		9,925	1,450		11,375	13,100
1270	200 HP, size 5		.40	40		21,600	2,200		23,800	27,100
1400	Combination, with fused switch, 5 HP, size 0	1 Elec	1.80	4.444		610	243		853	1,025
1600	10 HP, size 1	"	1.30	6.154		650	335		985	1,225
1800	25 HP, size 2	2 Elec	2	8		1,050	440		1,490	1,800
2000	50 HP, size 3		1.32	12.121		1,775	665		2,440	2,975
2200	100 HP, size 4		.80	20		3,125	1,100		4,225	5,075
2610	NEMA 4, with start-stop pushbutton size 1	1 Elec	1.30	6.154		1,625	335		1,960	2,300
2620	Size 2	2 Elec	2	8		2,200	440		2,640	3,050

26 24 19 – Motor-Control Centers

26 24 19.40 Motor Starters and Controls		Crew	Daily Output	Labor-Hours	Unit	2015 Bare Costs Material	Labor	Equipment	Total	Total Incl O&P
2630	Size 3	2 Elec	1.32	12.121	Ea.	3,475	665		4,140	4,800
2640	Size 4		.80	20		5,325	1,100		6,425	7,500
2650	NEMA 4, FVNR, including control transformer									
2660	Size 1	2 Elec	2.60	6.154	Ea.	1,500	335		1,835	2,150
2670	Size 2		2	8		2,150	440		2,590	3,000
2680	Size 3		1.32	12.121		3,425	665		4,090	4,750
2690	Size 4		.80	20		5,275	1,100		6,375	7,450
2710	Magnetic, FVR, control circuit transformer, NEMA 1, size 1	1 Elec	1.30	6.154		885	335		1,220	1,475
2720	Size 2	2 Elec	2	8		1,425	440		1,865	2,200
2730	Size 3		1.32	12.121		2,150	665		2,815	3,375
2740	Size 4		.80	20		4,700	1,100		5,800	6,800
2760	NEMA 4, size 1	1 Elec	1.10	7.273		1,275	400		1,675	2,000
2770	Size 2	2 Elec	1.60	10		2,050	545		2,595	3,075
2780	Size 3		1.20	13.333		3,075	730		3,805	4,500
2790	Size 4		.70	22.857		6,325	1,250		7,575	8,850
2820	NEMA 12, size 1	1 Elec	1.10	7.273		1,050	400		1,450	1,750
2830	Size 2	2 Elec	1.60	10		1,650	545		2,195	2,650
2840	Size 3		1.20	13.333		2,600	730		3,330	3,950
2850	Size 4		.70	22.857		5,350	1,250		6,600	7,775
2870	Combination FVR, fused, w/control XFMR & PB, NEMA 1, size 1	1 Elec	1	8		1,525	440		1,965	2,325
2880	Size 2	2 Elec	1.50	10.667		2,175	585		2,760	3,275
2890	Size 3		1.10	14.545		3,225	795		4,020	4,750
2900	Size 4		.70	22.857		6,550	1,250		7,800	9,100
2910	NEMA 4, size 1	1 Elec	.90	8.889		2,275	485		2,760	3,225
2920	Size 2	2 Elec	1.40	11.429		3,300	625		3,925	4,550
2930	Size 3		1	16		5,200	875		6,075	7,025
2940	Size 4		.60	26.667		8,650	1,450		10,100	11,700
2950	NEMA 12, size 1	1 Elec	1	8		1,725	440		2,165	2,550
2960	Size 2	2 Elec	1.40	11.429		2,450	625		3,075	3,600
2970	Size 3		1	16		3,575	875		4,450	5,250
2980	Size 4		.60	26.667		7,125	1,450		8,575	10,000
3010	Manual, single phase, w/pilot, 1 pole 120 V NEMA 1	1 Elec	6.40	1.250		70	68.50		138.50	179
3020	NEMA 4		4	2		355	109		464	560
3030	2 pole, 120/240 V, NEMA 1		6.40	1.250		81.50	68.50		150	192
3040	NEMA 4		4	2		320	109		429	515
3041	3 phase, 3 pole 600 V, NEMA 1		5.50	1.455		248	79.50		327.50	390
3042	NEMA 4		3.50	2.286		485	125		610	720
3043	NEMA 12		3.50	2.286		273	125		398	485
3070	Auxiliary contact, normally open					84			84	92.50
3500	Magnetic FVNR with NEMA 12, enclosure & heaters, 480 volt									
3600	5 HP, size 0	1 Elec	2.20	3.636	Ea.	231	199		430	550
3700	10 HP, size 1	"	1.50	5.333		350	292		642	820
3800	25 HP, size 2	2 Elec	2	8		650	440		1,090	1,375
3900	50 HP, size 3		1.60	10		1,000	545		1,545	1,925
4000	100 HP, size 4		1	16		2,400	875		3,275	3,925
4100	200 HP, size 5		.80	20		5,750	1,100		6,850	7,975
4200	Combination, with motor circuit protectors, 5 HP, size 0	1 Elec	1.70	4.706		765	257		1,022	1,225
4300	10 HP, size 1	"	1.20	6.667		795	365		1,160	1,425
4400	25 HP, size 2	2 Elec	1.80	8.889		1,200	485		1,685	2,025
4500	50 HP, size 3		1.20	13.333		1,950	730		2,680	3,225
4600	100 HP, size 4		.74	21.622		4,375	1,175		5,550	6,600
4700	Combination, with fused switch, 5 HP, size 0	1 Elec	1.70	4.706		740	257		997	1,200
4800	10 HP, size 1	"	1.20	6.667		770	365		1,135	1,400

26 24 19.40 Motor Starters and Controls		Crew	Daily Output	Labor-Hours	Unit	Material	Labor	Equipment	Total	Total Incl O&P
						2015 Bare Costs				
4900	25 HP, size 2	2 Elec	1.80	8.889	Ea.	1,175	485		1,660	2,025
5000	50 HP, size 3		1.20	13.333		1,875	730		2,605	3,175
5100	100 HP, size 4		.74	21.622		3,825	1,175		5,000	5,975
5200	Factory installed controls, adders to size 0 thru 5									
5300	Start-stop push button	1 Elec	32	.250	Ea.	48.50	13.70		62.20	74
5400	Hand-off-auto-selector switch		32	.250		48.50	13.70		62.20	74
5500	Pilot light		32	.250		91	13.70		104.70	121
5600	Start-stop-pilot		32	.250		139	13.70		152.70	174
5700	Auxiliary contact, NO or NC		32	.250		66.50	13.70		80.20	93.50
5800	NO-NC		32	.250		133	13.70		146.70	167
5810	Magnetic FVR, NEMA 7, w/heaters, size 1		.66	12.121		2,700	665		3,365	3,950
5830	Size 2	2 Elec	1.10	14.545		4,500	795		5,295	6,150
5840	Size 3		.70	22.857		7,225	1,250		8,475	9,825
5850	Size 4		.60	26.667		8,225	1,450		9,675	11,200
5860	Combination w/circuit breakers, heaters, control XFMR PB, size 1	1 Elec	.60	13.333		1,700	730		2,430	2,975
5870	Size 2	2 Elec	.80	20		2,175	1,100		3,275	4,050
5880	Size 3		.50	32		2,900	1,750		4,650	5,825
5890	Size 4		.40	40		5,750	2,200		7,950	9,600
5900	Manual, 240 volt, .75 HP motor	1 Elec	4	2		51.50	109		160.50	221
5910	2 HP motor		4	2		139	109		248	315
6000	Magnetic, 240 volt, 1 or 2 pole, .75 HP motor		4	2		203	109		312	385
6020	2 HP motor		4	2		224	109		333	410
6040	5 HP motor		3	2.667		320	146		466	575
6060	10 HP motor		2.30	3.478		795	190		985	1,150
6100	3 pole, .75 HP motor		3	2.667		203	146		349	440
6120	5 HP motor		2.30	3.478		275	190		465	590
6140	10 HP motor		1.60	5		520	274		794	980
6160	15 HP motor		1.60	5		520	274		794	980
6180	20 HP motor		1.10	7.273		845	400		1,245	1,525
6200	25 HP motor	2 Elec	2.20	7.273		845	400		1,245	1,525
6210	30 HP motor		1.80	8.889		845	485		1,330	1,650
6220	40 HP motor		1.80	8.889		1,875	485		2,360	2,775
6230	50 HP motor		1.80	8.889		1,875	485		2,360	2,775
6240	60 HP motor		1.20	13.333		4,375	730		5,105	5,925
6250	75 HP motor		1.20	13.333		4,375	730		5,105	5,925
6260	100 HP motor		1.20	13.333		4,375	730		5,105	5,925
6270	125 HP motor		.90	17.778		12,300	970		13,270	15,000
6280	150 HP motor		.90	17.778		12,300	970		13,270	15,000
6290	200 HP motor		.90	17.778		12,300	970		13,270	15,000
6400	Starter & nonfused disconnect, 240 volt, 1-2 pole, .75 HP motor	1 Elec	2	4		262	219		481	620
6410	2 HP motor		2	4		283	219		502	640
6420	5 HP motor		1.80	4.444		380	243		623	785
6430	10 HP motor		1.40	5.714		875	315		1,190	1,425
6440	3 pole, .75 HP motor		1.60	5		262	274		536	700
6450	5 HP motor		1.40	5.714		335	315		650	840
6460	10 HP motor		1.10	7.273		595	400		995	1,250
6470	15 HP motor		1	8		595	440		1,035	1,300
6480	20 HP motor	2 Elec	1.50	10.667		1,025	585		1,610	2,000
6490	25 HP motor		1.50	10.667		1,025	585		1,610	2,000
6500	30 HP motor		1.30	12.308		1,025	675		1,700	2,125
6510	40 HP motor		1.24	12.903		2,200	705		2,905	3,475
6520	50 HP motor		1.12	14.286		2,200	780		2,980	3,600
6530	60 HP motor		.90	17.778		4,725	970		5,695	6,650

26 24 19.40 Motor Starters and Controls		Crew	Daily Output	Labor-Hours	Unit	Material	2015 Bare Costs Labor	Equipment	Total	Total Incl O&P
6540	75 HP motor	2 Elec	.76	21.053	Ea.	4,725	1,150		5,875	6,925
6550	100 HP motor		.70	22.857		4,725	1,250		5,975	7,075
6560	125 HP motor		.60	26.667		13,100	1,450		14,550	16,600
6570	150 HP motor		.52	30.769		13,100	1,675		14,775	16,900
6580	200 HP motor		.50	32		13,100	1,750		14,850	17,000
6600	Starter & fused disconnect, 240 volt, 1-2 pole, .75 HP motor	1 Elec	2	4		276	219		495	635
6610	2 HP motor		2	4		297	219		516	655
6620	5 HP motor		1.80	4.444		395	243		638	800
6630	10 HP motor		1.40	5.714		920	315		1,235	1,475
6640	3 pole, .75 HP motor		1.60	5		276	274		550	715
6650	5 HP motor		1.40	5.714		350	315		665	855
6660	10 HP motor		1.10	7.273		640	400		1,040	1,300
6690	15 HP motor		1	8		640	440		1,080	1,350
6700	20 HP motor	2 Elec	1.60	10		1,050	545		1,595	2,000
6710	25 HP motor		1.60	10		1,050	545		1,595	2,000
6720	30 HP motor		1.40	11.429		1,050	625		1,675	2,100
6730	40 HP motor		1.20	13.333		2,325	730		3,055	3,650
6740	50 HP motor		1.20	13.333		2,325	730		3,055	3,650
6750	60 HP motor		.90	17.778		4,825	970		5,795	6,775
6760	75 HP motor		.90	17.778		4,825	970		5,795	6,775
6770	100 HP motor		.70	22.857		4,825	1,250		6,075	7,200
6780	125 HP motor		.54	29.630		13,400	1,625		15,025	17,200
6790	Combination starter & nonfusible disconnect									
6800	240 volt, 1-2 pole, .75 HP motor	1 Elec	2	4	Ea.	640	219		859	1,025
6810	2 HP motor		2	4		640	219		859	1,025
6820	5 HP motor		1.50	5.333		675	292		967	1,175
6830	10 HP motor		1.20	6.667		945	365		1,310	1,600
6840	3 pole, .75 HP motor		1.80	4.444		580	243		823	1,000
6850	5 HP motor		1.30	6.154		610	335		945	1,175
6860	10 HP motor		1	8		945	440		1,385	1,700
6870	15 HP motor		1	8		945	440		1,385	1,700
6880	20 HP motor	2 Elec	1.32	12.121		1,550	665		2,215	2,725
6890	25 HP motor		1.32	12.121		1,550	665		2,215	2,725
6900	30 HP motor		1.32	12.121		1,550	665		2,215	2,725
6910	40 HP motor		.80	20		2,975	1,100		4,075	4,925
6920	50 HP motor		.80	20		2,975	1,100		4,075	4,925
6930	60 HP motor		.70	22.857		6,650	1,250		7,900	9,200
6940	75 HP motor		.70	22.857		6,650	1,250		7,900	9,200
6950	100 HP motor		.70	22.857		6,650	1,250		7,900	9,200
6960	125 HP motor		.60	26.667		17,500	1,450		18,950	21,500
6970	150 HP motor		.60	26.667		17,500	1,450		18,950	21,500
6980	200 HP motor		.60	26.667		17,500	1,450		18,950	21,500
6990	Combination starter and fused disconnect									
7000	240 volt, 1-2 pole, .75 HP motor	1 Elec	2	4	Ea.	595	219		814	985
7010	2 HP motor		2	4		595	219		814	985
7020	5 HP motor		1.50	5.333		625	292		917	1,125
7030	10 HP motor		1.20	6.667		970	365		1,335	1,625
7040	3 pole, .75 HP motor		1.80	4.444		595	243		838	1,025
7050	5 HP motor		1.30	6.154		625	335		960	1,200
7060	10 HP motor		1	8		970	440		1,410	1,725
7070	15 HP motor		1	8		970	440		1,410	1,725
7080	20 HP motor	2 Elec	1.32	12.121		1,625	665		2,290	2,775
7090	25 HP motor		1.32	12.121		1,625	665		2,290	2,775

26 24 19.40 Motor Starters and Controls		Crew	Daily Output	Labor-Hours	Unit	2015 Bare Costs Material	Labor	Equipment	Total	Total Incl O&P
7100	30 HP motor	2 Elec	1.32	12.121	Ea.	1,625	665		2,290	2,775
7110	40 HP motor		.80	20		3,075	1,100		4,175	5,050
7120	50 HP motor		.80	20		3,075	1,100		4,175	5,050
7130	60 HP motor		.80	20		6,875	1,100		7,975	9,225
7140	75 HP motor		.70	22.857		6,875	1,250		8,125	9,450
7150	100 HP motor		.70	22.857		6,875	1,250		8,125	9,450
7160	125 HP motor		.70	22.857		18,200	1,250		19,450	21,900
7170	150 HP motor		.60	26.667		18,200	1,450		19,650	22,200
7180	200 HP motor		.60	26.667		18,200	1,450		19,650	22,200
7190	Combination starter & circuit breaker disconnect									
7200	240 volt, 1-2 pole, .75 HP motor	1 Elec	2	4	Ea.	600	219		819	990
7210	2 HP motor		2	4		600	219		819	990
7220	5 HP motor		1.50	5.333		635	292		927	1,125
7230	10 HP motor		1.20	6.667		965	365		1,330	1,600
7240	3 pole, .75 HP motor		1.80	4.444		620	243		863	1,050
7250	5 HP motor		1.30	6.154		650	335		985	1,225
7260	10 HP motor		1	8		985	440		1,425	1,725
7270	15 HP motor		1	8		985	440		1,425	1,725
7280	20 HP motor	2 Elec	1.32	12.121		1,675	665		2,340	2,850
7290	25 HP motor		1.32	12.121		1,675	665		2,340	2,850
7300	30 HP motor		1.32	12.121		1,675	665		2,340	2,850
7310	40 HP motor		.80	20		3,650	1,100		4,750	5,650
7320	50 HP motor		.80	20		3,650	1,100		4,750	5,650
7330	60 HP motor		.80	20		8,400	1,100		9,500	10,900
7340	75 HP motor		.70	22.857		8,400	1,250		9,650	11,100
7350	100 HP motor		.70	22.857		8,400	1,250		9,650	11,100
7360	125 HP motor		.70	22.857		18,200	1,250		19,450	21,900
7370	150 HP motor		.60	26.667		18,200	1,450		19,650	22,200
7380	200 HP motor		.60	26.667		18,200	1,450		19,650	22,200
7400	Magnetic FVNR with enclosure & heaters, 2 pole,									
7410	230 volt, 1 HP size 00	1 Elec	4	2	Ea.	185	109		294	365
7420	2 HP, size 0		4	2		206	109		315	390
7430	3 HP, size 1		3	2.667		236	146		382	480
7440	5 HP, size 1p		3	2.667		305	146		451	555
7450	115 volt, 1/3 HP, size 00		4	2		185	109		294	365
7460	1 HP, size 0		4	2		206	109		315	390
7470	2 HP, size 1		3	2.667		236	146		382	480
7480	3 HP, size 1P		3	2.667		294	146		440	545
7500	3 pole, 480 volt, 600 HP, size 7	2 Elec	.70	22.857		16,300	1,250		17,550	19,900
7590	Magnetic FVNR with heater, NEMA 1									
7600	600 volt, 3 pole, 5 HP motor	1 Elec	2.30	3.478	Ea.	245	190		435	555
7610	10 HP motor	"	1.60	5		275	274		549	715
7620	25 HP motor	2 Elec	2.20	7.273		520	400		920	1,175
7630	30 HP motor		1.80	8.889		845	485		1,330	1,650
7640	40 HP motor		1.80	8.889		845	485		1,330	1,650
7650	50 HP motor		1.80	8.889		845	485		1,330	1,650
7660	60 HP motor		1.20	13.333		1,875	730		2,605	3,150
7670	75 HP motor		1.20	13.333		2,125	730		2,855	3,450
7680	100 HP motor		1.20	13.333		2,475	730		3,205	3,825
7690	125 HP motor		.90	17.778		4,275	970		5,245	6,175
7700	150 HP motor		.90	17.778		4,500	970		5,470	6,400
7710	200 HP motor		.90	17.778		4,700	970		5,670	6,600
7750	Starter & nonfused disconnect, 600 volt, 3 pole, 5 HP motor	1 Elec	1.40	5.714		360	315		675	865

26 24 19.40 Motor Starters and Controls		Crew	Daily Output	Labor-Hours	Unit	2015 Bare Costs Material	Labor	Equipment	Total	Total Incl O&P
7760	10 HP motor	1 Elec	1.10	7.273	Ea.	390	400		790	1,025
7770	25 HP motor	2 Elec	1.50	10.667		630	585		1,215	1,575
7780	30 HP motor		1.30	12.308		1,025	675		1,700	2,125
7790	40 HP motor		1.30	12.308		1,025	675		1,700	2,125
7800	50 HP motor		1.30	12.308		1,025	675		1,700	2,125
7810	60 HP motor		.92	17.391		2,175	950		3,125	3,800
7820	75 HP motor		.92	17.391		2,425	950		3,375	4,100
7830	100 HP motor		.84	19.048		2,775	1,050		3,825	4,600
7840	125 HP motor		.70	22.857		4,725	1,250		5,975	7,075
7850	150 HP motor		.70	22.857		4,950	1,250		6,200	7,300
7860	200 HP motor		.60	26.667		5,150	1,450		6,600	7,825
7870	Starter & fused disconnect, 600 volt, 3 pole, 5 HP motor	1 Elec	1.40	5.714		445	315		760	960
7880	10 HP motor	"	1.10	7.273		475	400		875	1,125
7890	25 HP motor	2 Elec	1.50	10.667		720	585		1,305	1,675
7900	30 HP motor		1.30	12.308		1,075	675		1,750	2,200
7910	40 HP motor		1.30	12.308		1,075	675		1,750	2,200
7920	50 HP motor		1.30	12.308		1,075	675		1,750	2,200
7930	60 HP motor		.92	17.391		2,325	950		3,275	3,975
7940	75 HP motor		.92	17.391		2,575	950		3,525	4,250
7950	100 HP motor		.84	19.048		2,925	1,050		3,975	4,750
7960	125 HP motor		.70	22.857		4,925	1,250		6,175	7,300
7970	150 HP motor		.70	22.857		5,125	1,250		6,375	7,525
7980	200 HP motor		.60	26.667		5,325	1,450		6,775	8,050
7990	Combination starter and nonfusible disconnect									
8000	600 volt, 3 pole, 5 HP motor	1 Elec	1.80	4.444	Ea.	645	243		888	1,075
8010	10 HP motor	"	1.30	6.154		720	335		1,055	1,300
8020	25 HP motor	2 Elec	2	8		1,050	440		1,490	1,800
8030	30 HP motor		1.32	12.121		1,625	665		2,290	2,775
8040	40 HP motor		1.32	12.121		1,625	665		2,290	2,775
8050	50 HP motor		1.32	12.121		1,925	665		2,590	3,100
8060	60 HP motor		.80	20		3,075	1,100		4,175	5,050
8070	75 HP motor		.80	20		3,075	1,100		4,175	5,050
8080	100 HP motor		.80	20		3,825	1,100		4,925	5,850
8090	125 HP motor		.70	22.857		6,675	1,250		7,925	9,225
8100	150 HP motor		.70	22.857		6,975	1,250		8,225	9,550
8110	200 HP motor		.70	22.857		7,275	1,250		8,525	9,875
8140	Combination starter and fused disconnect									
8150	600 volt, 3 pole, 5 HP motor	1 Elec	1.80	4.444	Ea.	595	243		838	1,025
8160	10 HP motor	"	1.30	6.154		650	335		985	1,225
8170	25 HP motor	2 Elec	2	8		1,050	440		1,490	1,800
8180	30 HP motor		1.32	12.121		1,650	665		2,315	2,825
8190	40 HP motor		1.32	12.121		1,775	665		2,440	2,950
8200	50 HP motor		1.32	12.121		1,775	665		2,440	2,950
8210	60 HP motor		.80	20		3,100	1,100		4,200	5,075
8220	75 HP motor		.80	20		3,100	1,100		4,200	5,075
8230	100 HP motor		.80	20		3,100	1,100		4,200	5,075
8240	125 HP motor		.70	22.857		6,875	1,250		8,125	9,450
8250	150 HP motor		.70	22.857		6,875	1,250		8,125	9,450
8260	200 HP motor		.70	22.857		6,875	1,250		8,125	9,450
8290	Combination starter & circuit breaker disconnect									
8300	600 volt, 3 pole, 5 HP motor	1 Elec	1.80	4.444	Ea.	880	243		1,123	1,325
8310	10 HP motor	"	1.30	6.154		1,075	335		1,410	1,700
8320	25 HP motor	2 Elec	2	8		1,275	440		1,715	2,075

26 24 Switchboards and Panelboards

26 24 19 – Motor-Control Centers

26 24 19.40 Motor Starters and Controls		Crew	Daily Output	Labor-Hours	Unit	2015 Bare Costs Material	Labor	Equipment	Total	Total Incl O&P
8330	30 HP motor	2 Elec	1.32	12.121	Ea.	1,675	665		2,340	2,850
8340	40 HP motor		1.32	12.121		1,675	665		2,340	2,850
8350	50 HP motor		1.32	12.121		1,675	665		2,340	2,850
8360	60 HP motor		.80	20		3,650	1,100		4,750	5,650
8370	75 HP motor		.80	20		3,650	1,100		4,750	5,650
8380	100 HP motor		.80	20		3,650	1,100		4,750	5,650
8390	125 HP motor		.70	22.857		8,400	1,250		9,650	11,100
8400	150 HP motor		.70	22.857		8,400	1,250		9,650	11,100
8410	200 HP motor		.70	22.857		8,400	1,250		9,650	11,100
8430	Starter & circuit breaker disconnect									
8440	600 volt, 3 pole, 5 HP motor	1 Elec	1.40	5.714	Ea.	745	315		1,060	1,300
8450	10 HP motor	"	1.10	7.273		780	400		1,180	1,450
8460	25 HP motor	2 Elec	1.50	10.667		1,025	585		1,610	2,000
8470	30 HP motor		1.30	12.308		1,450	675		2,125	2,600
8480	40 HP motor		1.30	12.308		1,450	675		2,125	2,600
8490	50 HP motor		1.30	12.308		1,450	675		2,125	2,600
8500	60 HP motor		.92	17.391		3,500	950		4,450	5,275
8510	75 HP motor		.92	17.391		3,775	950		4,725	5,575
8520	100 HP motor		.84	19.048		4,100	1,050		5,150	6,075
8530	125 HP motor		.70	22.857		5,925	1,250		7,175	8,375
8540	150 HP motor		.70	22.857		6,125	1,250		7,375	8,600
8550	200 HP motor		.60	26.667		7,475	1,450		8,925	10,400
8900	240 volt, 1-2 pole, .75 HP motor	1 Elec	2	4		705	219		924	1,100
8910	2 HP motor		2	4		725	219		944	1,125
8920	5 HP motor		1.80	4.444		825	243		1,068	1,275
8930	10 HP motor		1.40	5.714		1,425	315		1,740	2,025
8950	3 pole, .75 HP motor		1.60	5		705	274		979	1,175
8970	5 HP motor		1.40	5.714		780	315		1,095	1,325
8980	10 HP motor		1.10	7.273		1,125	400		1,525	1,850
8990	15 HP motor		1	8		1,125	440		1,565	1,900
9100	20 HP motor	2 Elec	1.50	10.667		1,550	585		2,135	2,575
9110	25 HP motor		1.50	10.667		1,550	585		2,135	2,575
9120	30 HP motor		1.30	12.308		1,550	675		2,225	2,700
9130	40 HP motor		1.24	12.903		3,500	705		4,205	4,900
9140	50 HP motor		1.12	14.286		3,500	780		4,280	5,025
9150	60 HP motor		.90	17.778		7,175	970		8,145	9,325
9160	75 HP motor		.76	21.053		7,175	1,150		8,325	9,600
9170	100 HP motor		.70	22.857		7,175	1,250		8,425	9,750
9180	125 HP motor		.60	26.667		16,300	1,450		17,750	20,200
9190	150 HP motor		.52	30.769		16,300	1,675		17,975	20,500
9200	200 HP motor		.50	32		16,300	1,750		18,050	20,600

26 25 13.10 Aluminum Bus Duct		Crew	Daily Output	Labor-Hours	Unit	Material	2015 Bare Costs Labor	Equipment	Total	Total Incl O&P
0010	**ALUMINUM BUS DUCT** 10 ft. long R262513-10									
0050	Indoor 3 pole 4 wire, plug-in, straight section, 225 amp	2 Elec	44	.364	L.F.	125	19.90		144.90	167
0100	400 amp		36	.444		146	24.50		170.50	198
0150	600 amp		32	.500		173	27.50		200.50	231
0200	800 amp		26	.615		198	33.50		231.50	268
0250	1000 amp		24	.667		222	36.50		258.50	300
0270	1200 amp		23	.696		245	38		283	325
0300	1350 amp		22	.727		264	40		304	350
0310	1600 amp		18	.889		297	48.50		345.50	400
0320	2000 amp		16	1		345	54.50		399.50	460
0330	2500 amp		14	1.143		410	62.50		472.50	545
0340	3000 amp		12	1.333		475	73		548	630
0350	Feeder, 600 amp		34	.471		110	25.50		135.50	160
0400	800 amp		28	.571		129	31.50		160.50	189
0450	1000 amp		26	.615		147	33.50		180.50	213
0455	1200 amp		25	.640		147	35		182	215
0500	1350 amp		24	.667		230	36.50		266.50	310
0550	1600 amp		20	.800		267	44		311	360
0600	2000 amp		18	.889		315	48.50		363.50	420
0620	2500 amp		14	1.143		405	62.50		467.50	540
0630	3000 amp		12	1.333		460	73		533	615
0640	4000 amp		10	1.600		665	87.50		752.50	860
0650	Elbow, 225 amp		4.40	3.636	Ea.	735	199		934	1,100
0700	400 amp		3.80	4.211		745	230		975	1,150
0750	600 amp		3.40	4.706		745	257		1,002	1,200
0800	800 amp		3	5.333		775	292		1,067	1,275
0850	1000 amp		2.80	5.714		995	315		1,310	1,575
0870	1200 amp		2.70	5.926		1,250	325		1,575	1,850
0900	1350 amp		2.60	6.154		1,300	335		1,635	1,950
0950	1600 amp		2.40	6.667		1,400	365		1,765	2,075
1000	2000 amp		2	8		1,525	440		1,965	2,325
1020	2500 amp		1.80	8.889		1,800	485		2,285	2,700
1030	3000 amp		1.60	10		2,075	545		2,620	3,125
1040	4000 amp		1.40	11.429		3,350	625		3,975	4,625
1100	Cable tap box end, 225 amp		3.60	4.444		1,450	243		1,693	1,950
1150	400 amp		3.20	5		1,475	274		1,749	2,025
1200	600 amp		2.60	6.154		1,500	335		1,835	2,150
1250	800 amp		2.20	7.273		1,525	400		1,925	2,300
1300	1000 amp		2	8		1,575	440		2,015	2,375
1320	1200 amp		2	8		1,600	440		2,040	2,425
1350	1350 amp		1.60	10		1,625	545		2,170	2,625
1400	1600 amp		1.40	11.429		1,675	625		2,300	2,775
1450	2000 amp		1.20	13.333		1,750	730		2,480	3,025
1460	2500 amp		1	16		1,825	875		2,700	3,300
1470	3000 amp		.80	20		1,925	1,100		3,025	3,750
1480	4000 amp		.60	26.667		2,075	1,450		3,525	4,475
1500	Switchboard stub, 225 amp		5.80	2.759		1,475	151		1,626	1,850
1550	400 amp		5.40	2.963		1,500	162		1,662	1,900
1600	600 amp		4.60	3.478		1,525	190		1,715	1,950
1650	800 amp		4	4		1,550	219		1,769	2,025
1700	1000 amp		3.20	5		1,550	274		1,824	2,125
1720	1200 amp		3.10	5.161		1,600	282		1,882	2,175
1750	1350 amp		3	5.333		1,600	292		1,892	2,200

26 25 13 – Bus Duct/Busway and Fittings

26 25 13.10 Aluminum Bus Duct		Crew	Daily Output	Labor-Hours	Unit	Material	Labor	Equipment	Total	Total Incl O&P
						2015 Bare Costs				
1800	1600 amp	2 Elec	2.60	6.154	Ea.	1,650	335		1,985	2,300
1850	2000 amp		2.40	6.667		1,700	365		2,065	2,400
1860	2500 amp		2.20	7.273		1,750	400		2,150	2,525
1870	3000 amp		2	8		1,825	440		2,265	2,650
1880	4000 amp		1.80	8.889		1,950	485		2,435	2,850
1890	Tee fittings, 225 amp		3.20	5		875	274		1,149	1,375
1900	400 amp		2.80	5.714		875	315		1,190	1,425
1950	600 amp		2.60	6.154		875	335		1,210	1,475
2000	800 amp		2.40	6.667		930	365		1,295	1,575
2050	1000 amp		2.20	7.273		985	400		1,385	1,675
2070	1200 amp		2.10	7.619		1,150	415		1,565	1,900
2100	1350 amp		2	8		1,600	440		2,040	2,400
2150	1600 amp		1.60	10		1,925	545		2,470	2,925
2200	2000 amp		1.20	13.333		2,125	730		2,855	3,425
2220	2500 amp		1	16		2,525	875		3,400	4,100
2230	3000 amp		.80	20		2,900	1,100		4,000	4,825
2240	4000 amp		.60	26.667		4,800	1,450		6,250	7,475
2300	Wall flange, 600 amp		20	.800		196	44		240	281
2310	800 amp		16	1		196	54.50		250.50	297
2320	1000 amp		13	1.231		196	67.50		263.50	315
2325	1200 amp		12	1.333		196	73		269	325
2330	1350 amp		10.80	1.481		196	81		277	335
2340	1600 amp		9	1.778		196	97		293	360
2350	2000 amp		8	2		196	109		305	380
2360	2500 amp		6.60	2.424		196	133		329	415
2370	3000 amp		5.40	2.963		196	162		358	460
2380	4000 amp		4	4		284	219		503	640
2390	5000 amp		3	5.333		284	292		576	745
2400	Vapor barrier		8	2		360	109		469	560
2420	Roof flange kit		4	4		640	219		859	1,025
2600	Expansion fitting, 225 amp		10	1.600		1,250	87.50		1,337.50	1,475
2610	400 amp		8	2		1,250	109		1,359	1,550
2620	600 amp		6	2.667		1,250	146		1,396	1,600
2630	800 amp		4.60	3.478		1,400	190		1,590	1,825
2640	1000 amp		4	4		1,600	219		1,819	2,075
2650	1350 amp		3.60	4.444		2,025	243		2,268	2,600
2660	1600 amp		3.20	5		2,425	274		2,699	3,075
2670	2000 amp		2.80	5.714		2,700	315		3,015	3,450
2680	2500 amp		2.40	6.667		3,275	365		3,640	4,150
2690	3000 amp		2	8		3,775	440		4,215	4,800
2700	4000 amp		1.60	10		4,950	545		5,495	6,275
2800	Reducer nonfused, 400 amp		8	2		875	109		984	1,125
2810	600 amp		6	2.667		875	146		1,021	1,175
2820	800 amp		4.60	3.478		1,050	190		1,240	1,425
2830	1000 amp		4	4		1,225	219		1,444	1,675
2840	1350 amp		3.60	4.444		1,600	243		1,843	2,150
2850	1600 amp		3.20	5		2,175	274		2,449	2,800
2860	2000 amp		2.80	5.714		2,500	315		2,815	3,225
2870	2500 amp		2.40	6.667		3,150	365		3,515	4,000
2880	3000 amp		2	8		3,650	440		4,090	4,650
2890	4000 amp		1.60	10		4,850	545		5,395	6,175
2950	Reducer fuse included, 225 amp		4.40	3.636		2,575	199		2,774	3,125
2960	400 amp		4.20	3.810		2,625	208		2,833	3,175

26 25 13.10 Aluminum Bus Duct		Crew	Daily Output	Labor-Hours	Unit	Material	2015 Bare Costs Labor	Equipment	Total	Total Incl O&P
2970	600 amp	2 Elec	3.60	4.444	Ea.	3,075	243		3,318	3,775
2980	800 amp		3.20	5		4,900	274		5,174	5,800
2990	1000 amp		3	5.333		5,600	292		5,892	6,600
3000	1200 amp		2.80	5.714		5,600	315		5,915	6,650
3010	1600 amp		2.20	7.273		12,800	400		13,200	14,700
3020	2000 amp		1.80	8.889		14,200	485		14,685	16,300
3100	Reducer circuit breaker, 225 amp		4.40	3.636		2,525	199		2,724	3,075
3110	400 amp		4.20	3.810		3,075	208		3,283	3,700
3120	600 amp		3.60	4.444		4,375	243		4,618	5,175
3130	800 amp		3.20	5		5,125	274		5,399	6,050
3140	1000 amp		3	5.333		5,825	292		6,117	6,825
3150	1200 amp		2.80	5.714		7,000	315		7,315	8,175
3160	1600 amp		2.20	7.273		10,300	400		10,700	12,000
3170	2000 amp		1.80	8.889		11,300	485		11,785	13,200
3250	Reducer circuit breaker, 75,000 AIC, 225 amp		4.40	3.636		3,950	199		4,149	4,650
3260	400 amp		4.20	3.810		3,950	208		4,158	4,650
3270	600 amp		3.60	4.444		5,300	243		5,543	6,200
3280	800 amp		3.20	5		5,800	274		6,074	6,800
3290	1000 amp		3	5.333		9,350	292		9,642	10,700
3300	1200 amp		2.80	5.714		9,350	315		9,665	10,800
3310	1600 amp		2.20	7.273		10,300	400		10,700	12,000
3320	2000 amp		1.80	8.889		11,300	485		11,785	13,200
3400	Reducer circuit breaker CLF 225 amp		4.40	3.636		4,075	199		4,274	4,775
3410	400 amp		4.20	3.810		4,825	208		5,033	5,625
3420	600 amp		3.60	4.444		7,225	243		7,468	8,325
3430	800 amp		3.20	5		7,550	274		7,824	8,700
3440	1000 amp		3	5.333		7,875	292		8,167	9,075
3450	1200 amp		2.80	5.714		10,300	315		10,615	11,800
3460	1600 amp		2.20	7.273		10,300	400		10,700	12,000
3470	2000 amp		1.80	8.889		11,300	485		11,785	13,200
3550	Ground bus added to bus duct, 225 amp		320	.050	L.F.	32.50	2.74		35.24	40
3560	400 amp		320	.050		32.50	2.74		35.24	40
3570	600 amp		280	.057		32.50	3.13		35.63	40.50
3580	800 amp		240	.067		32.50	3.65		36.15	41.50
3590	1000 amp		200	.080		32.50	4.38		36.88	42.50
3600	1350 amp		180	.089		32.50	4.86		37.36	43.50
3610	1600 amp		160	.100		32.50	5.45		37.95	44
3620	2000 amp		160	.100		32.50	5.45		37.95	44
3630	2500 amp		140	.114		32.50	6.25		38.75	45.50
3640	3000 amp		120	.133		32.50	7.30		39.80	47
3650	4000 amp		100	.160		32.50	8.75		41.25	49
3810	High short circuit, 400 amp		36	.444		99.50	24.50		124	147
3820	600 amp		32	.500		125	27.50		152.50	178
3830	800 amp		26	.615		144	33.50		177.50	210
3840	1000 amp		24	.667		156	36.50		192.50	227
3850	1350 amp		22	.727		184	40		224	262
3860	1600 amp		18	.889		212	48.50		260.50	305
3870	2000 amp		16	1		248	54.50		302.50	355
3880	2500 amp		14	1.143		415	62.50		477.50	550
3890	3000 amp		12	1.333		470	73		543	625
3920	Cross, 225 amp		5.60	2.857	Ea.	1,300	156		1,456	1,675
3930	400 amp		4.60	3.478		1,300	190		1,490	1,725
3940	600 amp		4	4		1,300	219		1,519	1,775

26 25 13.10 Aluminum Bus Duct		Crew	Daily Output	Labor-Hours	Unit	2015 Bare Costs Material	Labor	Equipment	Total	Total Incl O&P
3950	800 amp	2 Elec	3.40	4.706	Ea.	1,375	257		1,632	1,900
3960	1000 amp		3	5.333		1,450	292		1,742	2,025
3970	1350 amp		2.80	5.714		2,350	315		2,665	3,050
3980	1600 amp		2.20	7.273		2,775	400		3,175	3,650
3990	2000 amp		1.80	8.889		3,050	485		3,535	4,075
4000	2500 amp		1.60	10		3,625	545		4,170	4,800
4010	3000 amp		1.20	13.333		4,150	730		4,880	5,675
4020	4000 amp		1	16		6,350	875		7,225	8,300
4040	Cable tap box center, 225 amp		3.60	4.444		885	243		1,128	1,325
4050	400 amp		3.20	5		945	274		1,219	1,450
4060	600 amp		2.60	6.154		990	335		1,325	1,600
4070	800 amp		2.20	7.273		1,050	400		1,450	1,750
4080	1000 amp		2	8		1,150	440		1,590	1,900
4090	1350 amp		1.60	10		1,425	545		1,970	2,375
4100	1600 amp		1.40	11.429		1,600	625		2,225	2,675
4110	2000 amp		1.20	13.333		1,825	730		2,555	3,100
4120	2500 amp		1	16		2,225	875		3,100	3,725
4130	3000 amp		.80	20		2,450	1,100		3,550	4,350
4140	4000 amp		.60	26.667		3,250	1,450		4,700	5,750
4500	Weatherproof 3 pole 4 wire, feeder, 600 amp		30	.533	L.F.	132	29		161	189
4520	800 amp		24	.667		155	36.50		191.50	225
4540	1000 amp		22	.727		177	40		217	254
4550	1200 amp		21	.762		244	41.50		285.50	330
4560	1350 amp		20	.800		276	44		320	370
4580	1600 amp		17	.941		320	51.50		371.50	425
4600	2000 amp		16	1		375	54.50		429.50	495
4620	2500 amp		12	1.333		485	73		558	645
4640	3000 amp		10	1.600		550	87.50		637.50	735
4660	4000 amp		8	2		795	109		904	1,050
5000	Indoor 3 pole, 3 wire, feeder, 600 amp		40	.400		101	22		123	144
5010	800 amp		32	.500		120	27.50		147.50	173
5020	1000 amp		30	.533		129	29		158	186
5025	1200 amp		29	.552		129	30		159	187
5030	1350 amp		28	.571		175	31.50		206.50	239
5040	1600 amp		24	.667		202	36.50		238.50	278
5050	2000 amp		20	.800		239	44		283	330
5060	2500 amp		16	1		330	54.50		384.50	445
5070	3000 amp		14	1.143		395	62.50		457.50	530
5080	4000 amp		12	1.333		480	73		553	635
5200	Plug-in type, 225 amp		50	.320		109	17.50		126.50	146
5210	400 amp		42	.381		109	21		130	151
5220	600 amp		36	.444		109	24.50		133.50	157
5230	800 amp		30	.533		128	29		157	184
5240	1000 amp		28	.571		137	31.50		168.50	197
5245	1200 amp		27	.593		156	32.50		188.50	221
5250	1350 amp		26	.615		184	33.50		217.50	253
5260	1600 amp		20	.800		212	44		256	299
5270	2000 amp		18	.889		248	48.50		296.50	345
5280	2500 amp		16	1		340	54.50		394.50	455
5290	3000 amp		14	1.143		405	62.50		467.50	540
5300	4000 amp		12	1.333		490	73		563	645
5330	High short circuit, 400 amp		42	.381		109	21		130	151
5340	600 amp		36	.444		109	24.50		133.50	157

26 25 13 – Bus Duct/Busway and Fittings

26 25 13.10	Aluminum Bus Duct	Crew	Daily Output	Labor-Hours	Unit	2015 Bare Costs Material	Labor	Equipment	Total	Total Incl O&P
5350	800 amp	2 Elec	30	.533	L.F.	128	29		157	184
5360	1000 amp		28	.571		137	31.50		168.50	197
5370	1350 amp		26	.615		184	33.50		217.50	253
5380	1600 amp		20	.800		212	44		256	299
5390	2000 amp		18	.889		248	48.50		296.50	345
5400	2500 amp		16	1		390	54.50		444.50	510
5410	3000 amp		14	1.143		405	62.50		467.50	540
5440	Elbow, 225 amp		5	3.200	Ea.	555	175		730	870
5450	400 amp		4.40	3.636		555	199		754	910
5460	600 amp		4	4		555	219		774	940
5470	800 amp		3.40	4.706		595	257		852	1,025
5480	1000 amp		3.20	5		610	274		884	1,075
5485	1200 amp		3.10	5.161		650	282		932	1,150
5490	1350 amp		3	5.333		710	292		1,002	1,225
5500	1600 amp		2.80	5.714		1,100	315		1,415	1,675
5510	2000 amp		2.40	6.667		1,200	365		1,565	1,875
5520	2500 amp		2	8		1,475	440		1,915	2,275
5530	3000 amp		1.80	8.889		1,750	485		2,235	2,650
5540	4000 amp		1.60	10		2,325	545		2,870	3,400
5560	Tee fittings, 225 amp		3.60	4.444		780	243		1,023	1,225
5570	400 amp		3.20	5		780	274		1,054	1,275
5580	600 amp		3	5.333		780	292		1,072	1,300
5590	800 amp		2.80	5.714		835	315		1,150	1,400
5600	1000 amp		2.60	6.154		865	335		1,200	1,450
5605	1200 amp		2.50	6.400		920	350		1,270	1,550
5610	1350 amp		2.40	6.667		1,275	365		1,640	1,950
5620	1600 amp		1.80	8.889		1,500	485		1,985	2,375
5630	2000 amp		1.40	11.429		1,675	625		2,300	2,775
5640	2500 amp		1.20	13.333		2,100	730		2,830	3,400
5650	3000 amp		1	16		2,450	875		3,325	4,000
5660	4000 amp		.70	22.857		3,600	1,250		4,850	5,825
5680	Cross, 225 amp		6.40	2.500		1,275	137		1,412	1,600
5690	400 amp		5.40	2.963		1,275	162		1,437	1,650
5700	600 amp		4.60	3.478		1,275	190		1,465	1,675
5710	800 amp		4	4		1,350	219		1,569	1,800
5720	1000 amp		3.60	4.444		1,375	243		1,618	1,900
5730	1350 amp		3.20	5		2,050	274		2,324	2,650
5740	1600 amp		2.60	6.154		2,400	335		2,735	3,125
5750	2000 amp		2.20	7.273		2,625	400		3,025	3,475
5760	2500 amp		1.80	8.889		3,150	485		3,635	4,200
5770	3000 amp		1.40	11.429		3,775	625		4,400	5,075
5780	4000 amp		1.20	13.333		5,250	730		5,980	6,875
5800	Expansion fitting, 225 amp		11.60	1.379		910	75.50		985.50	1,125
5810	400 amp		9.20	1.739		910	95		1,005	1,150
5820	600 amp		7	2.286		910	125		1,035	1,175
5830	800 amp		5.20	3.077		1,075	168		1,243	1,425
5840	1000 amp		4.60	3.478		1,175	190		1,365	1,575
5850	1350 amp		4.20	3.810		1,450	208		1,658	1,875
5860	1600 amp		3.60	4.444		1,800	243		2,043	2,350
5870	2000 amp		3.20	5		2,125	274		2,399	2,725
5880	2500 amp		2.80	5.714		2,400	315		2,715	3,100
5890	3000 amp		2.40	6.667		2,800	365		3,165	3,625
5900	4000 amp		1.80	8.889		3,825	485		4,310	4,950

26 25 13.10 Aluminum Bus Duct		Crew	Daily Output	Labor-Hours	Unit	Material	Labor	Equipment	Total	Total Incl O&P
						2015 Bare Costs				
5940	Reducer, nonfused, 400 amp	2 Elec	9.20	1.739	Ea.	750	95		845	970
5950	600 amp		7	2.286		750	125		875	1,000
5960	800 amp		5.20	3.077		790	168		958	1,125
5970	1000 amp		4.60	3.478		970	190		1,160	1,350
5980	1350 amp		4.20	3.810		1,450	208		1,658	1,875
5990	1600 amp		3.60	4.444		1,650	243		1,893	2,175
6000	2000 amp		3.20	5		1,925	274		2,199	2,500
6010	2500 amp		2.80	5.714		2,450	315		2,765	3,175
6020	3000 amp		2.20	7.273		2,875	400		3,275	3,750
6030	4000 amp		1.80	8.889		3,925	485		4,410	5,050
6050	Reducer, fuse included, 225 amp		5	3.200		1,975	175		2,150	2,425
6060	400 amp		4.80	3.333		2,625	182		2,807	3,150
6070	600 amp		4.20	3.810		3,350	208		3,558	3,975
6080	800 amp		3.60	4.444		5,175	243		5,418	6,075
6090	1000 amp		3.40	4.706		5,650	257		5,907	6,600
6100	1350 amp		3.20	5		10,300	274		10,574	11,800
6110	1600 amp		2.60	6.154		12,200	335		12,535	14,000
6120	2000 amp		2	8		14,200	440		14,640	16,300
6160	Reducer, circuit breaker, 225 amp		5	3.200		2,425	175		2,600	2,925
6170	400 amp		4.80	3.333		2,975	182		3,157	3,550
6180	600 amp		4.20	3.810		4,250	208		4,458	4,975
6190	800 amp		3.60	4.444		4,975	243		5,218	5,850
6200	1000 amp		3.40	4.706		5,650	257		5,907	6,600
6210	1350 amp		3.20	5		6,825	274		7,099	7,925
6220	1600 amp		2.60	6.154		10,200	335		10,535	11,700
6230	2000 amp		2	8		11,100	440		11,540	13,000
6270	Cable tap box center, 225 amp		4.20	3.810		875	208		1,083	1,275
6280	400 amp		3.60	4.444		875	243		1,118	1,325
6290	600 amp		3	5.333		875	292		1,167	1,400
6300	800 amp		2.60	6.154		955	335		1,290	1,550
6310	1000 amp		2.40	6.667		1,000	365		1,365	1,650
6320	1350 amp		1.80	8.889		1,225	485		1,710	2,075
6330	1600 amp		1.60	10		1,375	545		1,920	2,350
6340	2000 amp		1.40	11.429		1,575	625		2,200	2,675
6350	2500 amp		1.20	13.333		1,975	730		2,705	3,275
6360	3000 amp		1	16		2,225	875		3,100	3,750
6370	4000 amp		.70	22.857		2,650	1,250		3,900	4,775
6390	Cable tap box end, 225 amp		4.20	3.810		535	208		743	895
6400	400 amp		3.60	4.444		535	243		778	950
6410	600 amp		3	5.333		535	292		827	1,025
6420	800 amp		2.60	6.154		585	335		920	1,150
6430	1000 amp		2.40	6.667		630	365		995	1,225
6435	1200 amp		2.10	7.619		685	415		1,100	1,375
6440	1350 amp		1.80	8.889		765	485		1,250	1,575
6450	1600 amp		1.60	10		870	545		1,415	1,775
6460	2000 amp		1.40	11.429		980	625		1,605	2,000
6470	2500 amp		1.20	13.333		1,125	730		1,855	2,350
6480	3000 amp		1	16		1,350	875		2,225	2,775
6490	4000 amp		.70	22.857		1,625	1,250		2,875	3,650
7000	Weatherproof 3 pole 3 wire, feeder, 600 amp		34	.471	L.F.	131	25.50		156.50	183
7020	800 amp		28	.571		144	31.50		175.50	205
7040	1000 amp		26	.615		155	33.50		188.50	221
7050	1200 amp		25	.640		181	35		216	252

26 25 Enclosed Bus Assemblies

26 25 13 – Bus Duct/Busway and Fittings

26 25 13.10 Aluminum Bus Duct	Crew	Daily Output	Labor-Hours	Unit	Material	2015 Bare Costs Labor	Equipment	Total	Total Incl O&P
7060 1350 amp	2 Elec	24	.667	L.F.	210	36.50		246.50	286
7080 1600 amp		20	.800		243	44		287	335
7100 2000 amp		18	.889		287	48.50		335.50	390
7120 2500 amp		14	1.143		395	62.50		457.50	530
7140 3000 amp		12	1.333		475	73		548	635
7160 4000 amp		10	1.600		575	87.50		662.50	760

26 25 13.20 Bus Duct	Crew	Daily Output	Labor-Hours	Unit	Material	Labor	Equipment	Total	Total Incl O&P
0010 **BUS DUCT** 100 amp and less, aluminum or copper, plug-in									
0080 Bus duct, 3 pole 3 wire, 100 amp	1 Elec	42	.190	L.F.	23.50	10.40		33.90	41.50
0110 Elbow		4	2	Ea.	57.50	109		166.50	228
0120 Tee		2	4		83.50	219		302.50	420
0130 Wall flange		8	1		8.35	54.50		62.85	91
0140 Ground kit		16	.500		17.75	27.50		45.25	60.50
0180 3 pole 4 wire, 100 amp		40	.200	L.F.	30.50	10.95		41.45	50
0200 Cable tap box		3.10	2.581	Ea.	78	141		219	297
0300 End closure		16	.500		10.40	27.50		37.90	52.50
0400 Elbow		4	2		65	109		174	236
0500 Tee		2	4		94.50	219		313.50	435
0600 Hangers		10	.800		6.25	44		50.25	72.50
0700 Circuit breakers, 15 to 50 amp, 1 pole		8	1		151	54.50		205.50	249
0800 15 to 60 amp, 2 pole		6.70	1.194		365	65.50		430.50	505
0900 3 pole		5.30	1.509		365	82.50		447.50	525
1000 60 to 100 amp, 1 pole		6.70	1.194		590	65.50		655.50	750
1100 70 to 100 amp, 2 pole		5.30	1.509		590	82.50		672.50	775
1200 3 pole		4.50	1.778		660	97		757	875
1220 Switch, nonfused, 3 pole, 4 wire		8	1		70.50	54.50		125	160
1240 Fused, 3 fuses, 4 wire, 30 amp		8	1		255	54.50		309.50	360
1260 60 amp		5.30	1.509		267	82.50		349.50	420
1280 100 amp		4.50	1.778		445	97		542	635
1300 Plug, fusible, 3 pole 250 volt, 30 amp		5.30	1.509		255	82.50		337.50	405
1310 60 amp		5.30	1.509		300	82.50		382.50	455
1320 100 amp		4.50	1.778		445	97		542	635
1330 3 pole 480 volt, 30 amp		5.30	1.509		291	82.50		373.50	445
1340 60 amp		5.30	1.509		315	82.50		397.50	470
1350 100 amp		4.50	1.778		455	97		552	645
1360 Circuit breaker, 3 pole 250 volt, 60 amp		5.30	1.509		515	82.50		597.50	695
1370 3 pole 480 volt, 100 amp		4.50	1.778		515	97		612	715
2000 Bus duct, 2 wire, 250 volt, 30 amp		60	.133	L.F.	2.82	7.30		10.12	14.05
2100 60 amp		50	.160		2.82	8.75		11.57	16.20
2200 300 volt, 30 amp		60	.133		2.82	7.30		10.12	14.05
2300 60 amp		50	.160		2.82	8.75		11.57	16.20
2400 3 wire, 250 volt, 30 amp		60	.133		2.82	7.30		10.12	14.05
2500 60 amp		50	.160		2.82	8.75		11.57	16.20
2600 480/277 volt, 30 amp		60	.133		2.82	7.30		10.12	14.05
2700 60 amp		50	.160		2.82	8.75		11.57	16.20
2750 End feed, 300 volt 2 wire max. 30 amp		6	1.333	Ea.	42	73		115	155
2800 60 amp		5.50	1.455		42	79.50		121.50	165
2850 30 amp miniature		6	1.333		42	73		115	155
2900 3 wire, 30 amp		6	1.333		53	73		126	167
2950 60 amp		5.50	1.455		53	79.50		132.50	177
3000 30 amp miniature		6	1.333		53	73		126	167
3050 Center feed, 300 volt 2 wire, 30 amp		6	1.333		58	73		131	173

26 25 13.20 Bus Duct		Crew	Daily Output	Labor-Hours	Unit	2015 Bare Costs Material	Labor	Equipment	Total	Total Incl O&P
3100	60 amp	1 Elec	5.50	1.455	Ea.	58	79.50		137.50	183
3150	3 wire, 30 amp		6	1.333		66	73		139	182
3200	60 amp		5.50	1.455		66	79.50		145.50	192
3220	Elbow, 30 amp		6	1.333		40.50	73		113.50	154
3240	60 amp		5.50	1.455		40.50	79.50		120	164
3260	End cap		40	.200		8.10	10.95		19.05	25.50
3280	Strength beam, 10 ft.		15	.533		22.50	29		51.50	68.50
3300	Hanger		24	.333		4.84	18.25		23.09	33
3320	Tap box, nonfusible		6.30	1.270		68.50	69.50		138	180
3340	Fusible switch 30 amp, 1 fuse		6	1.333		299	73		372	440
3360	2 fuse		6	1.333		305	73		378	445
3380	3 fuse		6	1.333		305	73		378	445
3400	Circuit breaker handle on cover, 1 pole		6	1.333		47.50	73		120.50	161
3420	2 pole		6	1.333		65.50	73		138.50	182
3440	3 pole		6	1.333		86	73		159	204
3460	Circuit breaker external operhandle, 1 pole		6	1.333		49	73		122	163
3480	2 pole		6	1.333		68	73		141	184
3500	3 pole		6	1.333		94.50	73		167.50	213
3520	Terminal plug only		16	.500		9.90	27.50		37.40	52
3540	Terminal with receptacle		16	.500		12.80	27.50		40.30	55
3560	Fixture plug		16	.500		8.80	27.50		36.30	50.50
4000	Copper bus duct, lighting, 2 wire 300 volt, 20 amp		70	.114	L.F.	5.20	6.25		11.45	15.05
4020	35 amp		60	.133		5.20	7.30		12.50	16.65
4040	50 amp		55	.145		5.20	7.95		13.15	17.60
4060	60 amp		50	.160		5.20	8.75		13.95	18.80
4080	3 wire 300 volt, 20 amp		70	.114		5.65	6.25		11.90	15.55
4100	35 amp		60	.133		5.65	7.30		12.95	17.15
4120	50 amp		55	.145		5.65	7.95		13.60	18.10
4140	60 amp		50	.160		5.65	8.75		14.40	19.30
4160	Feeder in box, end, 1 circuit		6	1.333	Ea.	78	73		151	195
4180	2 circuit		5.50	1.455		80.50	79.50		160	208
4200	Center, 1 circuit		6	1.333		106	73		179	226
4220	2 circuit		5.50	1.455		110	79.50		189.50	240
4240	End cap		40	.200		12.95	10.95		23.90	30.50
4260	Hanger, surface mount		24	.333		7.75	18.25		26	36
4280	Coupling		40	.200		9.85	10.95		20.80	27.50

26 25 13.30 Copper Bus Duct		Crew	Daily Output	Labor-Hours	Unit	Material	Labor	Equipment	Total	Total Incl O&P
0010	**COPPER BUS DUCT**									
0100	Weatherproof 3 pole 4 wire, feeder duct, 600 amp	2 Elec	24	.667	L.F.	256	36.50		292.50	335
0110	800 amp		18	.889		350	48.50		398.50	460
0120	1000 amp		17	.941		390	51.50		441.50	505
0125	1200 amp		16.50	.970		455	53		508	585
0130	1350 amp		16	1		480	54.50		534.50	605
0140	1600 amp		12	1.333		515	73		588	675
0150	2000 amp		10	1.600		570	87.50		657.50	760
0160	2500 amp		7	2.286		620	125		745	870
0170	3000 amp		5	3.200		860	175		1,035	1,200
0180	4000 amp		3.60	4.444		1,350	243		1,593	1,850
0200	Indoor 3 pole 4 wire, plug-in, bus duct high short circuit, 400 amp		32	.500		246	27.50		273.50	310
0210	600 amp		26	.615		246	33.50		279.50	320
0220	800 amp		20	.800		292	44		336	385
0230	1000 amp		18	.889		325	48.50		373.50	430

26 25 13.30 Copper Bus Duct		Crew	Daily Output	Labor-Hours	Unit	Material	2015 Bare Costs Labor	Equipment	Total	Total Incl O&P
0240	1350 amp	2 Elec	16	1	L.F.	440	54.50		494.50	560
0250	1600 amp		12	1.333		495	73		568	655
0260	2000 amp		10	1.600		630	87.50		717.50	820
0270	2500 amp		8	2		775	109		884	1,025
0280	3000 amp		6	2.667		885	146		1,031	1,200
0310	Cross, 225 amp		3	5.333	Ea.	2,925	292		3,217	3,625
0320	400 amp		2.80	5.714		2,925	315		3,240	3,675
0330	600 amp		2.60	6.154		2,925	335		3,260	3,700
0340	800 amp		2.20	7.273		3,150	400		3,550	4,075
0350	1000 amp		2	8		3,525	440		3,965	4,525
0360	1350 amp		1.80	8.889		3,800	485		4,285	4,900
0370	1600 amp		1.70	9.412		4,225	515		4,740	5,400
0380	2000 amp		1.60	10		6,925	545		7,470	8,450
0390	2500 amp		1.40	11.429		8,375	625		9,000	10,200
0400	3000 amp		1.20	13.333		9,150	730		9,880	11,200
0410	4000 amp		1	16		11,900	875		12,775	14,400
0430	Expansion fitting, 225 amp		5.40	2.963		1,525	162		1,687	1,925
0440	400 amp		4.60	3.478		1,725	190		1,915	2,175
0450	600 amp		4	4		2,150	219		2,369	2,675
0460	800 amp		3.40	4.706		2,525	257		2,782	3,150
0470	1000 amp		3	5.333		2,900	292		3,192	3,600
0480	1350 amp		2.80	5.714		3,450	315		3,765	4,275
0490	1600 amp		2.60	6.154		4,825	335		5,160	5,800
0500	2000 amp		2.20	7.273		5,550	400		5,950	6,700
0510	2500 amp		1.80	8.889		6,750	485		7,235	8,150
0520	3000 amp		1.60	10		7,550	545		8,095	9,125
0530	4000 amp		1.20	13.333		9,725	730		10,455	11,800
0550	Reducer nonfused, 225 amp		5.40	2.963		1,800	162		1,962	2,225
0560	400 amp		4.60	3.478		1,800	190		1,990	2,250
0570	600 amp		4	4		1,800	219		2,019	2,300
0580	800 amp		3.40	4.706		2,175	257		2,432	2,775
0590	1000 amp		3	5.333		2,600	292		2,892	3,300
0600	1350 amp		2.80	5.714		3,750	315		4,065	4,600
0610	1600 amp		2.60	6.154		4,300	335		4,635	5,250
0620	2000 amp		2.20	7.273		5,200	400		5,600	6,300
0630	2500 amp		1.80	8.889		6,550	485		7,035	7,925
0640	3000 amp		1.60	10		7,375	545		7,920	8,925
0650	4000 amp		1.20	13.333		9,600	730		10,330	11,700
0670	Reducer fuse included, 225 amp		4.40	3.636		3,825	199		4,024	4,525
0680	400 amp		4.20	3.810		4,800	208		5,008	5,575
0690	600 amp		3.60	4.444		5,900	243		6,143	6,875
0700	800 amp		3.20	5		8,350	274		8,624	9,575
0710	1000 amp		3	5.333		10,500	292		10,792	11,900
0720	1350 amp		2.80	5.714		16,300	315		16,615	18,400
0730	1600 amp		2.20	7.273		22,000	400		22,400	24,800
0740	2000 amp		1.80	8.889		24,800	485		25,285	28,000
0790	Reducer, circuit breaker, 225 amp		4.40	3.636		4,850	199		5,049	5,625
0800	400 amp		4.20	3.810		5,650	208		5,858	6,525
0810	600 amp		3.60	4.444		8,025	243		8,268	9,225
0820	800 amp		3.20	5		9,425	274		9,699	10,800
0830	1000 amp		3	5.333		10,700	292		10,992	12,200
0840	1350 amp		2.80	5.714		12,200	315		12,515	13,900
0850	1600 amp		2.20	7.273		18,000	400		18,400	20,400

26 25 13.30 Copper Bus Duct		Crew	Daily Output	Labor-Hours	Unit	2015 Bare Costs Material	Labor	Equipment	Total	Total Incl O&P
0860	2000 amp	2 Elec	1.80	8.889	Ea.	19,800	485		20,285	22,500
0910	Cable tap box, center, 225 amp		3.20	5		1,775	274		2,049	2,350
0920	400 amp		2.60	6.154		1,775	335		2,110	2,450
0930	600 amp		2.20	7.273		1,775	400		2,175	2,550
0940	800 amp		2	8		1,950	440		2,390	2,800
0950	1000 amp		1.60	10		2,100	545		2,645	3,150
0960	1350 amp		1.40	11.429		2,525	625		3,150	3,725
0970	1600 amp		1.20	13.333		2,825	730		3,555	4,200
0980	2000 amp		1	16		3,400	875		4,275	5,050
1040	2500 amp		.80	20		4,025	1,100		5,125	6,075
1060	3000 amp		.60	26.667		4,425	1,450		5,875	7,025
1080	4000 amp		.40	40		5,550	2,200		7,750	9,400
1800	Weatherproof 3 pole 3 wire, feeder duct, 600 amp		28	.571	L.F.	258	31.50		289.50	330
1820	800 amp		22	.727		315	40		355	405
1840	1000 amp		20	.800		350	44		394	450
1850	1200 amp		19	.842		365	46		411	470
1860	1350 amp		18	.889		490	48.50		538.50	615
1880	1600 amp		14	1.143		560	62.50		622.50	710
1900	2000 amp		12	1.333		720	73		793	900
1920	2500 amp		8	2		895	109		1,004	1,150
1940	3000 amp		6	2.667		1,025	146		1,171	1,350
1960	4000 amp		4	4		1,350	219		1,569	1,825
2000	Indoor 3 pole 3 wire, feeder duct, 600 amp		32	.500		215	27.50		242.50	278
2010	800 amp		26	.615		261	33.50		294.50	340
2020	1000 amp		24	.667		292	36.50		328.50	375
2025	1200 amp		22	.727		385	40		425	485
2030	1350 amp		20	.800		410	44		454	515
2040	1600 amp		16	1		465	54.50		519.50	595
2050	2000 amp		14	1.143		600	62.50		662.50	755
2060	2500 amp		10	1.600		745	87.50		832.50	950
2070	3000 amp		8	2		860	109		969	1,100
2080	4000 amp		6	2.667		1,125	146		1,271	1,475
2090	5000 amp		5	3.200		1,375	175		1,550	1,775
2200	Indoor 3 pole 3 wire, bus duct plug-in, 225 amp		46	.348		246	19.05		265.05	300
2210	400 amp		36	.444		246	24.50		270.50	310
2220	600 amp		30	.533		246	29		275	315
2230	800 amp		24	.667		292	36.50		328.50	375
2240	1000 amp		20	.800		325	44		369	420
2250	1350 amp		18	.889		440	48.50		488.50	555
2260	1600 amp		14	1.143		495	62.50		557.50	640
2270	2000 amp		12	1.333		630	73		703	800
2280	2500 amp		10	1.600		775	87.50		862.50	980
2290	3000 amp		8	2		885	109		994	1,150
2330	High short circuit, 400 amp		36	.444		246	24.50		270.50	310
2340	600 amp		30	.533		246	29		275	315
2350	800 amp		24	.667		292	36.50		328.50	375
2360	1000 amp		20	.800		325	44		369	420
2370	1350 amp		18	.889		440	48.50		488.50	555
2380	1600 amp		14	1.143		495	62.50		557.50	640
2390	2000 amp		12	1.333		630	73		703	800
2400	2500 amp		10	1.600		775	87.50		862.50	980
2410	3000 amp		8	2		885	109		994	1,150
2440	Elbows, 225 amp		4.60	3.478	Ea.	1,225	190		1,415	1,625

26 25 Enclosed Bus Assemblies

26 25 13 – Bus Duct/Busway and Fittings

26 25 13.30 Copper Bus Duct		Crew	Daily Output	Labor-Hours	Unit	Material	2015 Bare Costs Labor	Equipment	Total	Total Incl O&P
2450	400 amp	2 Elec	4.20	3.810	Ea.	1,225	208		1,433	1,650
2460	600 amp		3.60	4.444		1,225	243		1,468	1,725
2470	800 amp		3.20	5		1,325	274		1,599	1,850
2480	1000 amp		3	5.333		1,375	292		1,667	1,950
2485	1200 amp		2.90	5.517		1,500	300		1,800	2,100
2490	1350 amp		2.80	5.714		1,575	315		1,890	2,200
2500	1600 amp		2.60	6.154		1,700	335		2,035	2,375
2510	2000 amp		2	8		2,050	440		2,490	2,900
2520	2500 amp		1.80	8.889		3,125	485		3,610	4,150
2530	3000 amp		1.60	10		3,425	545		3,970	4,575
2540	4000 amp		1.40	11.429		4,375	625		5,000	5,725
2560	Tee fittings, 225 amp		2.80	5.714		1,450	315		1,765	2,075
2570	400 amp		2.40	6.667		1,450	365		1,815	2,150
2580	600 amp		2	8		1,450	440		1,890	2,250
2590	800 amp		1.80	8.889		1,575	485		2,060	2,475
2600	1000 amp		1.60	10		1,775	545		2,320	2,775
2605	1200 amp		1.50	10.667		1,875	585		2,460	2,950
2610	1350 amp		1.40	11.429		1,975	625		2,600	3,100
2620	1600 amp		1.20	13.333		2,300	730		3,030	3,625
2630	2000 amp		1	16		3,800	875		4,675	5,475
2640	2500 amp		.70	22.857		4,450	1,250		5,700	6,775
2650	3000 amp		.60	26.667		4,900	1,450		6,350	7,575
2660	4000 amp		.50	32		6,350	1,750		8,100	9,600
2680	Cross, 225 amp		3.60	4.444		2,225	243		2,468	2,800
2690	400 amp		3.20	5		2,225	274		2,499	2,825
2700	600 amp		3	5.333		2,225	292		2,517	2,850
2710	800 amp		2.60	6.154		2,650	335		2,985	3,400
2720	1000 amp		2.40	6.667		2,775	365		3,140	3,600
2730	1350 amp		2.20	7.273		3,150	400		3,550	4,075
2740	1600 amp		2	8		3,400	440		3,840	4,375
2750	2000 amp		1.80	8.889		5,375	485		5,860	6,625
2760	2500 amp		1.60	10		6,250	545		6,795	7,700
2770	3000 amp		1.40	11.429		6,825	625		7,450	8,450
2780	4000 amp		1	16		8,725	875		9,600	10,900
2800	Expansion fitting, 225 amp		6.40	2.500		1,775	137		1,912	2,150
2810	400 amp		5.40	2.963		1,775	162		1,937	2,200
2820	600 amp		4.60	3.478		1,775	190		1,965	2,225
2830	800 amp		4	4		2,125	219		2,344	2,650
2840	1000 amp		3.60	4.444		2,325	243		2,568	2,925
2850	1350 amp		3.20	5		2,750	274		3,024	3,425
2860	1600 amp		3	5.333		3,025	292		3,317	3,750
2870	2000 amp		2.60	6.154		3,650	335		3,985	4,500
2880	2500 amp		2.20	7.273		5,075	400		5,475	6,200
2890	3000 amp		1.80	8.889		5,675	485		6,160	6,975
2900	4000 amp		1.40	11.429		7,250	625		7,875	8,900
2920	Reducer nonfused, 225 amp		6.40	2.500		1,475	137		1,612	1,825
2930	400 amp		5.40	2.963		1,475	162		1,637	1,875
2940	600 amp		4.60	3.478		1,475	190		1,665	1,900
2950	800 amp		4	4		1,725	219		1,944	2,200
2960	1000 amp		3.60	4.444		1,925	243		2,168	2,500
2970	1350 amp		3.20	5		2,425	274		2,699	3,075
2980	1600 amp		3	5.333		2,750	292		3,042	3,450
2990	2000 amp		2.60	6.154		3,300	335		3,635	4,125

26 25 13.30 Copper Bus Duct		Crew	Daily Output	Labor-Hours	Unit	Material	2015 Bare Costs Labor	Equipment	Total	Total Incl O&P
3000	2500 amp	2 Elec	2.20	7.273	Ea.	4,825	400		5,225	5,900
3010	3000 amp		1.80	8.889		5,400	485		5,885	6,675
3020	4000 amp		1.40	11.429		7,025	625		7,650	8,650
3040	Reducer fuse included, 225 amp		5	3.200		3,475	175		3,650	4,075
3050	400 amp		4.80	3.333		4,625	182		4,807	5,375
3060	600 amp		4.20	3.810		5,525	208		5,733	6,375
3070	800 amp		3.60	4.444		7,875	243		8,118	9,025
3080	1000 amp		3.40	4.706		9,325	257		9,582	10,600
3090	1350 amp		3.20	5		9,025	274		9,299	10,300
3100	1600 amp		2.60	6.154		21,000	335		21,335	23,600
3110	2000 amp		2	8		23,800	440		24,240	26,900
3160	Reducer circuit breaker, 225 amp		5	3.200		4,450	175		4,625	5,150
3170	400 amp		4.80	3.333		5,450	182		5,632	6,275
3180	600 amp		4.20	3.810		7,825	208		8,033	8,900
3190	800 amp		3.60	4.444		9,150	243		9,393	10,500
3200	1000 amp		3.40	4.706		10,400	257		10,657	11,800
3210	1350 amp		3.20	5		11,900	274		12,174	13,500
3220	1600 amp		2.60	6.154		17,800	335		18,135	20,100
3230	2000 amp		2	8		19,400	440		19,840	22,100
3280	3 pole, 3 wire, cable tap box center, 225 amp		3.60	4.444		2,000	243		2,243	2,575
3290	400 amp		3	5.333		2,000	292		2,292	2,625
3300	600 amp		2.60	6.154		2,000	335		2,335	2,700
3310	800 amp		2.40	6.667		2,250	365		2,615	3,025
3320	1000 amp		1.80	8.889		2,425	485		2,910	3,400
3330	1350 amp		1.60	10		2,975	545		3,520	4,100
3340	1600 amp		1.40	11.429		3,325	625		3,950	4,575
3350	2000 amp		1.20	13.333		4,025	730		4,755	5,525
3360	2500 amp		1	16		4,800	875		5,675	6,575
3370	3000 amp		.70	22.857		5,300	1,250		6,550	7,700
3380	4000 amp		.50	32		6,725	1,750		8,475	10,000
3400	Cable tap box end, 225 amp		3.60	4.444		1,100	243		1,343	1,575
3410	400 amp		3	5.333		1,100	292		1,392	1,625
3420	600 amp		2.60	6.154		1,225	335		1,560	1,850
3430	800 amp		2.40	6.667		1,225	365		1,590	1,900
3440	1000 amp		1.80	8.889		1,350	485		1,835	2,200
3445	1200 amp		1.70	9.412		1,500	515		2,015	2,425
3450	1350 amp		1.60	10		1,600	545		2,145	2,575
3460	1600 amp		1.40	11.429		1,825	625		2,450	2,950
3470	2000 amp		1.20	13.333		2,100	730		2,830	3,400
3480	2500 amp		1	16		2,500	875		3,375	4,050
3490	3000 amp		.70	22.857		2,725	1,250		3,975	4,875
3500	4000 amp		.50	32		3,475	1,750		5,225	6,450
4600	Plug-in, fusible switch w/3 fuses, 3 pole, 250 volt, 30 amp	1 Elec	4	2		435	109		544	645
4610	60 amp		3.60	2.222		570	122		692	805
4620	100 amp		2.70	2.963		815	162		977	1,150
4630	200 amp	2 Elec	3.20	5		1,375	274		1,649	1,900
4640	400 amp		1.40	11.429		3,575	625		4,200	4,850
4650	600 amp		.90	17.778		4,925	970		5,895	6,875
4700	4 pole, 120/208 volt, 30 amp	1 Elec	3.90	2.051		595	112		707	825
4710	60 amp		3.50	2.286		640	125		765	890
4720	100 amp		2.60	3.077		895	168		1,063	1,225
4730	200 amp	2 Elec	3	5.333		1,500	292		1,792	2,075
4740	400 amp		1.30	12.308		3,525	675		4,200	4,875

26 25 13.30 Copper Bus Duct		Crew	Daily Output	Labor-Hours	Unit	Material	Labor	Equipment	Total	Total Incl O&P
						2015 Bare Costs				
4750	600 amp	2 Elec	.80	20	Ea.	4,950	1,100		6,050	7,075
4800	3 pole, 480 volt, 30 amp	1 Elec	4	2		445	109		554	655
4810	60 amp		3.60	2.222		470	122		592	700
4820	100 amp		2.70	2.963		795	162		957	1,125
4830	200 amp	2 Elec	3.20	5		1,375	274		1,649	1,900
4840	400 amp		1.40	11.429		3,200	625		3,825	4,425
4850	600 amp		.90	17.778		4,525	970		5,495	6,425
4860	800 amp		.66	24.242		15,600	1,325		16,925	19,200
4870	1000 amp		.60	26.667		16,600	1,450		18,050	20,400
4880	1200 amp		.50	32		17,400	1,750		19,150	21,800
4890	1600 amp		.44	36.364		19,200	2,000		21,200	24,100
4900	4 pole, 277/480 volt, 30 amp	1 Elec	3.90	2.051		645	112		757	880
4910	60 amp		3.50	2.286		690	125		815	940
4920	100 amp		2.60	3.077		1,000	168		1,168	1,350
4930	200 amp	2 Elec	3	5.333		2,000	292		2,292	2,625
4940	400 amp		1.30	12.308		3,775	675		4,450	5,150
4950	600 amp		.80	20		5,150	1,100		6,250	7,325
5050	800 amp		.60	26.667		17,000	1,450		18,450	20,900
5060	1000 amp		.56	28.571		19,500	1,575		21,075	23,900
5070	1200 amp		.48	33.333		19,700	1,825		21,525	24,400
5080	1600 amp		.42	38.095		21,700	2,075		23,775	27,000
5150	Fusible with starter, 3 pole 250 volt, 30 amp	1 Elec	3.50	2.286		2,400	125		2,525	2,825
5160	60 amp		3.20	2.500		2,550	137		2,687	3,000
5170	100 amp		2.50	3.200		2,875	175		3,050	3,425
5180	200 amp	2 Elec	2.80	5.714		4,750	315		5,065	5,700
5200	3 pole 480 volt, 30 amp	1 Elec	3.50	2.286		2,400	125		2,525	2,825
5210	60 amp		3.20	2.500		2,550	137		2,687	3,000
5220	100 amp		2.50	3.200		2,875	175		3,050	3,425
5230	200 amp	2 Elec	2.80	5.714		4,750	315		5,065	5,700
5300	Fusible with contactor, 3 pole 250 volt, 30 amp	1 Elec	3.50	2.286		2,350	125		2,475	2,750
5310	60 amp		3.20	2.500		2,975	137		3,112	3,475
5320	100 amp		2.50	3.200		4,175	175		4,350	4,850
5330	200 amp	2 Elec	2.80	5.714		4,775	315		5,090	5,725
5400	3 pole 480 volt, 30 amp	1 Elec	3.50	2.286		2,525	125		2,650	2,950
5410	60 amp		3.20	2.500		3,550	137		3,687	4,100
5420	100 amp		2.50	3.200		4,875	175		5,050	5,625
5430	200 amp	2 Elec	2.80	5.714		5,000	315		5,315	5,975
5450	Fusible with capacitor, 3 pole 250 volt, 30 amp	1 Elec	3	2.667		6,075	146		6,221	6,925
5460	60 amp		2	4		7,075	219		7,294	8,125
5500	3 pole 480 volt, 30 amp		3	2.667		5,100	146		5,246	5,825
5510	60 amp		2	4		6,375	219		6,594	7,325
5600	Circuit breaker, 3 pole, 250 volt, 60 amp		4.50	1.778		610	97		707	820
5610	100 amp		3.20	2.500		750	137		887	1,025
5650	4 pole, 120/208 volt, 60 amp		4.40	1.818		690	99.50		789.50	910
5660	100 amp		3.10	2.581		820	141		961	1,100
5700	3 pole, 4 wire 277/480 volt, 60 amp		4.30	1.860		930	102		1,032	1,175
5710	100 amp		3	2.667		1,025	146		1,171	1,375
5720	225 amp	2 Elec	3.20	5		2,300	274		2,574	2,925
5730	400 amp		1.20	13.333		4,775	730		5,505	6,350
5740	600 amp		.96	16.667		6,400	910		7,310	8,425
5750	700 amp		.60	26.667		8,125	1,450		9,575	11,100
5760	800 amp		.60	26.667		8,125	1,450		9,575	11,100
5770	900 amp		.54	29.630		10,700	1,625		12,325	14,200

26 25 13.30 Copper Bus Duct		Crew	Daily Output	Labor-Hours	Unit	2015 Bare Costs Material	Labor	Equipment	Total	Total Incl O&P
5780	1000 amp	2 Elec	.54	29.630	Ea.	10,700	1,625		12,325	14,200
5790	1200 amp		.42	38.095		12,900	2,075		14,975	17,200
5810	Circuit breaker w/HIC fuses, 3 pole 480 volt, 60 amp	1 Elec	4.40	1.818		1,175	99.50		1,274.50	1,450
5820	100 amp	"	3.10	2.581		1,275	141		1,416	1,625
5830	225 amp	2 Elec	3.40	4.706		4,100	257		4,357	4,875
5840	400 amp		1.40	11.429		6,500	625		7,125	8,075
5850	600 amp		1	16		6,600	875		7,475	8,550
5860	700 amp		.64	25		8,750	1,375		10,125	11,700
5870	800 amp		.64	25		8,750	1,375		10,125	11,700
5880	900 amp		.56	28.571		18,900	1,575		20,475	23,200
5890	1000 amp		.56	28.571		18,900	1,575		20,475	23,200
5950	3 pole 4 wire 277/480 volt, 60 amp	1 Elec	4.30	1.860		1,175	102		1,277	1,450
5960	100 amp	"	3	2.667		1,275	146		1,421	1,650
5970	225 amp	2 Elec	3	5.333		4,100	292		4,392	4,925
5980	400 amp		1.10	14.545		6,500	795		7,295	8,350
5990	600 amp		.94	17.021		6,600	930		7,530	8,650
6000	700 amp		.58	27.586		8,750	1,500		10,250	11,900
6010	800 amp		.58	27.586		8,750	1,500		10,250	11,900
6020	900 amp		.52	30.769		18,900	1,675		20,575	23,300
6030	1000 amp		.52	30.769		18,900	1,675		20,575	23,300
6040	1200 amp		.40	40		18,900	2,200		21,100	24,100
6100	Circuit breaker with starter, 3 pole 250 volt, 60 amp	1 Elec	3.20	2.500		1,675	137		1,812	2,025
6110	100 amp	"	2.50	3.200		2,300	175		2,475	2,800
6120	225 amp	2 Elec	3	5.333		2,975	292		3,267	3,700
6130	3 pole 480 volt, 60 amp	1 Elec	3.20	2.500		1,675	137		1,812	2,025
6140	100 amp	"	2.50	3.200		2,300	175		2,475	2,800
6150	225 amp	2 Elec	3	5.333		2,825	292		3,117	3,550
6200	Circuit breaker with contactor, 3 pole 250 volt, 60 amp	1 Elec	3.20	2.500		1,575	137		1,712	1,925
6210	100 amp	"	2.50	3.200		2,150	175		2,325	2,600
6220	225 amp	2 Elec	3	5.333		2,750	292		3,042	3,450
6250	3 pole 480 volt, 60 amp	1 Elec	3.20	2.500		1,575	137		1,712	1,925
6260	100 amp	"	2.50	3.200		2,150	175		2,325	2,600
6270	225 amp	2 Elec	3	5.333		2,450	292		2,742	3,100
6300	Circuit breaker with capacitor, 3 pole 250 volt, 60 amp	1 Elec	2	4		7,900	219		8,119	9,025
6310	3 pole 480 volt, 60 amp		2	4		7,925	219		8,144	9,050
6400	Add control transformer with pilot light to above starter		16	.500		470	27.50		497.50	555
6410	Switch, fusible, mechanically held contactor optional		16	.500		1,250	27.50		1,277.50	1,400
6430	Circuit breaker, mechanically held contactor optional		16	.500		1,250	27.50		1,277.50	1,400
6450	Ground neutralizer, 3 pole		16	.500		57	27.50		84.50	104

26 25 13.40 Copper Bus Duct		Crew	Daily Output	Labor-Hours	Unit	Material	Labor	Equipment	Total	Total Incl O&P
0010	**COPPER BUS DUCT** 10 ft. long									
0050	Indoor 3 pole 4 wire, plug-in, straight section, 225 amp	2 Elec	40	.400	L.F.	230	22		252	286
1000	400 amp		32	.500		230	27.50		257.50	294
1500	600 amp		26	.615		230	33.50		263.50	305
2400	800 amp		20	.800		273	44		317	365
2450	1000 amp		18	.889		300	48.50		348.50	405
2470	1200 amp		17	.941		395	51.50		446.50	505
2500	1350 amp		16	1		415	54.50		469.50	535
2510	1600 amp		12	1.333		470	73		543	625
2520	2000 amp		10	1.600		595	87.50		682.50	785
2530	2500 amp		8	2		730	109		839	970
2540	3000 amp		6	2.667		840	146		986	1,150

26 25 13.40 Copper Bus Duct		Crew	Daily Output	Labor-Hours	Unit	Material	2015 Bare Costs Labor	Equipment	Total	Total Incl O&P
2550	Feeder, 600 amp	2 Elec	28	.571	L.F.	204	31.50		235.50	271
2600	800 amp		22	.727		247	40		287	330
2700	1000 amp		20	.800		276	44		320	370
2750	1200 amp		19	.842		365	46		411	470
2800	1350 amp		18	.889		385	48.50		433.50	500
2900	1600 amp		14	1.143		440	62.50		502.50	580
3000	2000 amp		12	1.333		565	73		638	735
3010	2500 amp		8	2		705	109		814	940
3020	3000 amp		6	2.667		810	146		956	1,100
3030	4000 amp		4	4		1,075	219		1,294	1,500
3040	5000 amp		2	8		1,300	440		1,740	2,100
3100	Elbows, 225 amp		4	4	Ea.	1,375	219		1,594	1,825
3200	400 amp		3.60	4.444		1,375	243		1,618	1,875
3300	600 amp		3.20	5		1,375	274		1,649	1,900
3400	800 amp		2.80	5.714		1,475	315		1,790	2,100
3500	1000 amp		2.60	6.154		1,650	335		1,985	2,325
3550	1200 amp		2.50	6.400		1,850	350		2,200	2,550
3600	1350 amp		2.40	6.667		1,850	365		2,215	2,600
3700	1600 amp		2.20	7.273		2,025	400		2,425	2,825
3800	2000 amp		1.80	8.889		2,500	485		2,985	3,475
3810	2500 amp		1.60	10		3,950	545		4,495	5,175
3820	3000 amp		1.40	11.429		4,325	625		4,950	5,700
3830	4000 amp		1.20	13.333		5,600	730		6,330	7,275
3840	5000 amp		1	16		9,050	875		9,925	11,300
4000	End box, 225 amp		34	.471		165	25.50		190.50	221
4100	400 amp		32	.500		186	27.50		213.50	246
4200	600 amp		28	.571		186	31.50		217.50	252
4300	800 amp		26	.615		186	33.50		219.50	256
4400	1000 amp		24	.667		186	36.50		222.50	260
4410	1200 amp		23	.696		187	38		225	262
4500	1350 amp		22	.727		177	40		217	254
4600	1600 amp		20	.800		177	44		221	260
4700	2000 amp		18	.889		217	48.50		265.50	310
4710	2500 amp		16	1		217	54.50		271.50	320
4720	3000 amp		14	1.143		205	62.50		267.50	320
4730	4000 amp		12	1.333		249	73		322	385
4740	5000 amp		10	1.600		249	87.50		336.50	405
4800	Cable tap box end, 225 amp		3.20	5		1,100	274		1,374	1,600
5000	400 amp		2.60	6.154		1,200	335		1,535	1,825
5100	600 amp		2.20	7.273		1,400	400		1,800	2,150
5200	800 amp		2	8		1,475	440		1,915	2,275
5300	1000 amp		1.60	10		1,500	545		2,045	2,475
5350	1200 amp		1.50	10.667		1,975	585		2,560	3,050
5400	1350 amp		1.40	11.429		2,200	625		2,825	3,350
5500	1600 amp		1.20	13.333		2,475	730		3,205	3,825
5600	2000 amp		1	16		2,750	875		3,625	4,325
5610	2500 amp		.80	20		3,050	1,100		4,150	5,000
5620	3000 amp		.60	26.667		3,550	1,450		5,000	6,100
5630	4000 amp		.40	40		4,100	2,200		6,300	7,800
5640	5000 amp		.20	80		4,850	4,375		9,225	11,900
5700	Switchboard stub, 225 amp		5.40	2.963		1,250	162		1,412	1,650
5800	400 amp		4.60	3.478		1,325	190		1,515	1,725
5900	600 amp		4	4		1,375	219		1,594	1,850

26 25 13.40 Copper Bus Duct		Crew	Daily Output	Labor-Hours	Unit	Material	2015 Bare Costs Labor	Equipment	Total	Total Incl O&P
6000	800 amp	2 Elec	3.20	5	Ea.	1,675	274		1,949	2,225
6100	1000 amp		3	5.333		1,925	292		2,217	2,550
6150	1200 amp		2.80	5.714		2,275	315		2,590	2,975
6200	1350 amp		2.60	6.154		2,400	335		2,735	3,125
6300	1600 amp		2.40	6.667		2,700	365		3,065	3,525
6400	2000 amp		2	8		3,275	440		3,715	4,250
6410	2500 amp		1.80	8.889		3,975	485		4,460	5,100
6420	3000 amp		1.60	10		4,425	545		4,970	5,675
6430	4000 amp		1.40	11.429		5,750	625		6,375	7,250
6440	5000 amp		1.20	13.333		7,050	730		7,780	8,850
6490	Tee fittings, 225 amp		2.40	6.667		1,900	365		2,265	2,625
6500	400 amp		2	8		1,900	440		2,340	2,725
6600	600 amp		1.80	8.889		1,900	485		2,385	2,800
6700	800 amp		1.60	10		2,175	545		2,720	3,225
6750	1000 amp		1.40	11.429		2,525	625		3,150	3,700
6770	1200 amp		1.30	12.308		2,825	675		3,500	4,125
6800	1350 amp		1.20	13.333		3,000	730		3,730	4,400
7000	1600 amp		1	16		3,400	875		4,275	5,050
7100	2000 amp		.80	20		4,025	1,100		5,125	6,100
7110	2500 amp		.60	26.667		4,950	1,450		6,400	7,625
7120	3000 amp		.50	32		5,475	1,750		7,225	8,650
7130	4000 amp		.40	40		7,075	2,200		9,275	11,100
7140	5000 amp		.20	80		8,375	4,375		12,750	15,800
7200	Plug-in fusible switches w/3 fuses, 600 volt, 3 pole, 30 amp	1 Elec	4	2		850	109		959	1,100
7300	60 amp		3.60	2.222		955	122		1,077	1,225
7400	100 amp		2.70	2.963		1,450	162		1,612	1,850
7500	200 amp	2 Elec	3.20	5		2,600	274		2,874	3,275
7600	400 amp		1.40	11.429		7,625	625		8,250	9,300
7700	600 amp		.90	17.778		8,650	970		9,620	11,000
7800	800 amp		.66	24.242		12,100	1,325		13,425	15,300
7900	1200 amp		.50	32		22,800	1,750		24,550	27,600
7910	1600 amp		.44	36.364		21,600	2,000		23,600	26,800
8000	Plug-in circuit breakers, molded case, 15 to 50 amp	1 Elec	4.40	1.818		805	99.50		904.50	1,025
8100	70 to 100 amp	"	3.10	2.581		895	141		1,036	1,200
8200	150 to 225 amp	2 Elec	3.40	4.706		2,425	257		2,682	3,050
8300	250 to 400 amp		1.40	11.429		4,250	625		4,875	5,600
8400	500 to 600 amp		1	16		5,750	875		6,625	7,600
8500	700 to 800 amp		.64	25		7,075	1,375		8,450	9,825
8600	900 to 1000 amp		.56	28.571		10,100	1,575		11,675	13,500
8700	1200 amp		.44	36.364		12,200	2,000		14,200	16,400
8720	1400 amp		.40	40		17,100	2,200		19,300	22,100
8730	1600 amp		.40	40		18,700	2,200		20,900	23,900
8750	Circuit breakers, with current limiting fuse, 15 to 50 amp	1 Elec	4.40	1.818		1,600	99.50		1,699.50	1,925
8760	70 to 100 amp	"	3.10	2.581		1,900	141		2,041	2,300
8770	150 to 225 amp	2 Elec	3.40	4.706		4,100	257		4,357	4,875
8780	250 to 400 amp		1.40	11.429		6,325	625		6,950	7,900
8790	500 to 600 amp		1	16		7,300	875		8,175	9,325
8800	700 to 800 amp		.64	25		12,000	1,375		13,375	15,300
8810	900 to 1000 amp		.56	28.571		13,700	1,575		15,275	17,500
8850	Combination starter FVNR, fusible switch, NEMA size 0, 30 amp	1 Elec	2	4		2,175	219		2,394	2,725
8860	NEMA size 1, 60 amp		1.80	4.444		2,300	243		2,543	2,900
8870	NEMA size 2, 100 amp		1.30	6.154		2,900	335		3,235	3,700
8880	NEMA size 3, 200 amp	2 Elec	2	8		4,625	440		5,065	5,725

26 25 13.40 Copper Bus Duct		Crew	Daily Output	Labor-Hours	Unit	Material	2015 Bare Costs Labor	Equipment	Total	Total Incl O&P
8900	Circuit breaker, NEMA size 0, 30 amp	1 Elec	2	4	Ea.	2,225	219		2,444	2,775
8910	NEMA size 1, 60 amp		1.80	4.444		2,325	243		2,568	2,925
8920	NEMA size 2, 100 amp		1.30	6.154		3,325	335		3,660	4,175
8930	NEMA size 3, 200 amp	2 Elec	2	8		4,225	440		4,665	5,300
8950	Combination contactor, fusible switch, NEMA size 0, 30 amp	1 Elec	2	4		1,275	219		1,494	1,725
8960	NEMA size 1, 60 amp		1.80	4.444		1,300	243		1,543	1,800
8970	NEMA size 2, 100 amp		1.30	6.154		1,925	335		2,260	2,625
8980	NEMA size 3, 200 amp	2 Elec	2	8		2,225	440		2,665	3,100
9000	Circuit breaker, NEMA size 0, 30 amp	1 Elec	2	4		1,450	219		1,669	1,925
9010	NEMA size 1, 60 amp		1.80	4.444		1,500	243		1,743	2,025
9020	NEMA size 2, 100 amp		1.30	6.154		2,300	335		2,635	3,025
9030	NEMA size 3, 200 amp	2 Elec	2	8		2,800	440		3,240	3,725
9050	Control transformer for above, NEMA size 0, 30 amp	1 Elec	8	1		246	54.50		300.50	355
9060	NEMA size 1, 60 amp		8	1		246	54.50		300.50	355
9070	NEMA size 2, 100 amp		7	1.143		345	62.50		407.50	470
9080	NEMA size 3, 200 amp	2 Elec	14	1.143		475	62.50		537.50	620
9100	Comb. fusible switch & lighting control, electrically held, 30 amp	1 Elec	2	4		1,000	219		1,219	1,425
9110	60 amp		1.80	4.444		1,450	243		1,693	1,975
9120	100 amp		1.30	6.154		1,875	335		2,210	2,550
9130	200 amp	2 Elec	2	8		4,575	440		5,015	5,700
9150	Mechanically held, 30 amp	1 Elec	2	4		1,250	219		1,469	1,700
9160	60 amp		1.80	4.444		1,875	243		2,118	2,450
9170	100 amp		1.30	6.154		2,425	335		2,760	3,150
9180	200 amp	2 Elec	2	8		4,925	440		5,365	6,050
9200	Ground bus added to bus duct, 225 amp		320	.050	L.F.	43.50	2.74		46.24	52
9210	400 amp		240	.067		43.50	3.65		47.15	53.50
9220	600 amp		240	.067		43.50	3.65		47.15	53.50
9230	800 amp		160	.100		51	5.45		56.45	64
9240	1000 amp		160	.100		58	5.45		63.45	72
9250	1350 amp		140	.114		83	6.25		89.25	100
9260	1600 amp		120	.133		90	7.30		97.30	110
9270	2000 amp		110	.145		117	7.95		124.95	141
9280	2500 amp		100	.160		145	8.75		153.75	173
9290	3000 amp		90	.178		164	9.70		173.70	195
9300	4000 amp		80	.200		216	10.95		226.95	254
9310	5000 amp		70	.229		262	12.50		274.50	305
9320	High short circuit bracing, add					18.25			18.25	20

26 25 13.60 Copper or Aluminum Bus Duct Fittings		Crew	Daily Output	Labor-Hours	Unit	Material	Labor	Equipment	Total	Total Incl O&P
0010	**COPPER OR ALUMINUM BUS DUCT FITTINGS**									
0100	Flange, wall, with vapor barrier, 225 amp	2 Elec	6.20	2.581	Ea.	795	141		936	1,075
0110	400 amp		6	2.667		795	146		941	1,100
0120	600 amp		5.80	2.759		795	151		946	1,100
0130	800 amp		5.40	2.963		795	162		957	1,125
0140	1000 amp		5	3.200		795	175		970	1,125
0145	1200 amp		4.80	3.333		795	182		977	1,150
0150	1350 amp		4.60	3.478		795	190		985	1,150
0160	1600 amp		4.20	3.810		795	208		1,003	1,175
0170	2000 amp		4	4		795	219		1,014	1,200
0180	2500 amp		3.60	4.444		795	243		1,038	1,250
0190	3000 amp		3.20	5		795	274		1,069	1,275
0200	4000 amp		2.60	6.154		795	335		1,130	1,375
0300	Roof, 225 amp		6.20	2.581		915	141		1,056	1,200

26 25 13.60 Copper or Aluminum Bus Duct Fittings		Crew	Daily Output	Labor-Hours	Unit	Material	2015 Bare Costs Labor	Equipment	Total	Total Incl O&P
0310	400 amp	2 Elec	6	2.667	Ea.	915	146		1,061	1,225
0320	600 amp		5.80	2.759		915	151		1,066	1,225
0330	800 amp		5.40	2.963		915	162		1,077	1,250
0340	1000 amp		5	3.200		915	175		1,090	1,250
0345	1200 amp		4.80	3.333		915	182		1,097	1,275
0350	1350 amp		4.60	3.478		915	190		1,105	1,275
0360	1600 amp		4.20	3.810		915	208		1,123	1,300
0370	2000 amp		4	4		915	219		1,134	1,325
0380	2500 amp		3.60	4.444		915	243		1,158	1,375
0390	3000 amp		3.20	5		915	274		1,189	1,400
0400	4000 amp		2.60	6.154		915	335		1,250	1,500
0420	Support, floor mounted, 225 amp		20	.800		167	44		211	249
0430	400 amp		20	.800		167	44		211	249
0440	600 amp		18	.889		167	48.50		215.50	256
0450	800 amp		16	1		167	54.50		221.50	265
0460	1000 amp		13	1.231		167	67.50		234.50	284
0465	1200 amp		11.80	1.356		167	74		241	294
0470	1350 amp		10.60	1.509		167	82.50		249.50	305
0480	1600 amp		9.20	1.739		167	95		262	325
0490	2000 amp		8	2		167	109		276	345
0500	2500 amp		6.40	2.500		167	137		304	390
0510	3000 amp		5.40	2.963		167	162		329	425
0520	4000 amp		4	4		167	219		386	515
0540	Weather stop, 225 amp		12	1.333		515	73		588	675
0550	400 amp		10	1.600		515	87.50		602.50	695
0560	600 amp		9	1.778		515	97		612	710
0570	800 amp		8	2		515	109		624	730
0580	1000 amp		6.40	2.500		515	137		652	770
0585	1200 amp		5.90	2.712		515	148		663	785
0590	1350 amp		5.40	2.963		515	162		677	810
0600	1600 amp		4.60	3.478		515	190		705	850
0610	2000 amp		4	4		515	219		734	895
0620	2500 amp		3.20	5		515	274		789	975
0630	3000 amp		2.60	6.154		515	335		850	1,075
0640	4000 amp		2	8		515	440		955	1,225
0660	End closure, 225 amp		34	.471		169	25.50		194.50	224
0670	400 amp		32	.500		169	27.50		196.50	226
0680	600 amp		28	.571		169	31.50		200.50	232
0690	800 amp		26	.615		169	33.50		202.50	236
0700	1000 amp		24	.667		169	36.50		205.50	240
0705	1200 amp		23	.696		165	38		203	239
0710	1350 amp		22	.727		165	40		205	242
0720	1600 amp		20	.800		169	44		213	251
0730	2000 amp		18	.889		234	48.50		282.50	330
0740	2500 amp		16	1		245	54.50		299.50	350
0750	3000 amp		14	1.143		245	62.50		307.50	365
0760	4000 amp		12	1.333		245	73		318	380
0780	Switchboard stub, 3 pole 3 wire, 225 amp		6	2.667		995	146		1,141	1,325
0790	400 amp		5.20	3.077		995	168		1,163	1,350
0800	600 amp		4.60	3.478		995	190		1,185	1,375
0810	800 amp		3.60	4.444		1,200	243		1,443	1,700
0820	1000 amp		3.40	4.706		1,400	257		1,657	1,925
0825	1200 amp	↓	3.20	5	↓	1,400	274		1,674	1,950

26 25 Enclosed Bus Assemblies

26 25 13 – Bus Duct/Busway and Fittings

26 25 13.60 Copper or Aluminum Bus Duct Fittings		Crew	Daily Output	Labor-Hours	Unit	Material	2015 Bare Costs Labor	Equipment	Total	Total Incl O&P
0830	1350 amp	2 Elec	3	5.333	Ea.	1,700	292		1,992	2,275
0840	1600 amp		2.80	5.714		1,975	315		2,290	2,650
0850	2000 amp		2.40	6.667		2,325	365		2,690	3,100
0860	2500 amp		2	8		2,825	440		3,265	3,750
0870	3000 amp		1.80	8.889		3,225	485		3,710	4,275
0880	4000 amp		1.60	10		4,175	545		4,720	5,400
0890	5000 amp		1.40	11.429		5,125	625		5,750	6,550
0900	3 pole 4 wire, 225 amp		5.40	2.963		1,250	162		1,412	1,625
0910	400 amp		4.60	3.478		1,250	190		1,440	1,650
0920	600 amp		4	4		1,250	219		1,469	1,700
0930	800 amp		3.20	5		1,500	274		1,774	2,050
0940	1000 amp		3	5.333		1,750	292		2,042	2,325
0950	1350 amp		2.60	6.154		2,225	335		2,560	2,950
0960	1600 amp		2.40	6.667		2,525	365		2,890	3,325
0970	2000 amp		2	8		3,050	440		3,490	4,025
0980	2500 amp		1.80	8.889		3,725	485		4,210	4,825
0990	3000 amp		1.60	10		4,350	545		4,895	5,625
1000	4000 amp		1.40	11.429		5,650	625		6,275	7,150
1050	Service head, weatherproof, 3 pole 3 wire, 225 amp		3	5.333		1,700	292		1,992	2,275
1060	400 amp		2.80	5.714		1,700	315		2,015	2,325
1070	600 amp		2.60	6.154		1,700	335		2,035	2,350
1080	800 amp		2.40	6.667		1,900	365		2,265	2,625
1090	1000 amp		2	8		2,050	440		2,490	2,900
1100	1350 amp		1.80	8.889		2,650	485		3,135	3,650
1110	1600 amp		1.60	10		2,975	545		3,520	4,100
1120	2000 amp		1.40	11.429		3,625	625		4,250	4,900
1130	2500 amp		1.20	13.333		4,325	730		5,055	5,850
1140	3000 amp		.90	17.778		5,050	970		6,020	7,000
1150	4000 amp		.70	22.857		6,400	1,250		7,650	8,925
1200	3 pole 4 wire, 225 amp		2.60	6.154		1,900	335		2,235	2,575
1210	400 amp		2.40	6.667		1,900	365		2,265	2,625
1220	600 amp		2.20	7.273		1,900	400		2,300	2,675
1230	800 amp		2	8		2,200	440		2,640	3,075
1240	1000 amp		1.70	9.412		2,525	515		3,040	3,550
1250	1350 amp		1.50	10.667		3,000	585		3,585	4,175
1260	1600 amp		1.40	11.429		3,275	625		3,900	4,525
1270	2000 amp		1.20	13.333		4,375	730		5,105	5,925
1280	2500 amp		1	16		5,400	875		6,275	7,225
1290	3000 amp		.80	20		6,325	1,100		7,425	8,600
1300	4000 amp		.60	26.667		8,200	1,450		9,650	11,200
1350	Flanged end, 3 pole 3 wire, 225 amp		6	2.667		920	146		1,066	1,250
1360	400 amp		5.20	3.077		920	168		1,088	1,275
1370	600 amp		4.60	3.478		920	190		1,110	1,300
1380	800 amp		3.60	4.444		1,025	243		1,268	1,500
1390	1000 amp		3.40	4.706		1,150	257		1,407	1,625
1395	1200 amp		3.20	5		1,275	274		1,549	1,800
1400	1350 amp		3	5.333		1,375	292		1,667	1,925
1410	1600 amp		2.80	5.714		1,550	315		1,865	2,200
1420	2000 amp		2.40	6.667		1,825	365		2,190	2,575
1430	2500 amp		2	8		2,150	440		2,590	3,000
1440	3000 amp		1.80	8.889		2,450	485		2,935	3,425
1450	4000 amp		1.60	10		3,050	545		3,595	4,175
1500	3 pole 4 wire, 225 amp	↓	5.40	2.963	↓	1,050	162		1,212	1,400

26 25 13.60 Copper or Aluminum Bus Duct Fittings		Crew	Daily Output	Labor-Hours	Unit	Material	2015 Bare Costs Labor	Equipment	Total	Total Incl O&P
1510	400 amp	2 Elec	4.60	3.478	Ea.	1,050	190		1,240	1,425
1520	600 amp		4	4		1,050	219		1,269	1,475
1530	800 amp		3.20	5		1,250	274		1,524	1,775
1540	1000 amp		3	5.333		1,400	292		1,692	1,950
1545	1200 amp		2.80	5.714		1,575	315		1,890	2,200
1550	1350 amp		2.60	6.154		1,700	335		2,035	2,375
1560	1600 amp		2.40	6.667		1,950	365		2,315	2,700
1570	2000 amp		2	8		2,300	440		2,740	3,175
1580	2500 amp		1.80	8.889		2,725	485		3,210	3,725
1590	3000 amp		1.60	10		3,125	545		3,670	4,250
1600	4000 amp		1.40	11.429		4,025	625		4,650	5,375
1650	Hanger, standard, 225 amp		64	.250		22.50	13.70		36.20	45
1660	400 amp		48	.333		22.50	18.25		40.75	52
1670	600 amp		40	.400		22.50	22		44.50	57.50
1680	800 amp		32	.500		22.50	27.50		50	65.50
1690	1000 amp		24	.667		22.50	36.50		59	79
1695	1200 amp		22	.727		22.50	40		62.50	84
1700	1350 amp		20	.800		22.50	44		66.50	90
1710	1600 amp		20	.800		22.50	44		66.50	90
1720	2000 amp		18	.889		22.50	48.50		71	97.50
1730	2500 amp		16	1		22.50	54.50		77	107
1740	3000 amp		16	1		22.50	54.50		77	107
1750	4000 amp		16	1		22.50	54.50		77	107
1800	Spring type, 225 amp		16	1		86.50	54.50		141	177
1810	400 amp		14	1.143		86.50	62.50		149	189
1820	600 amp		14	1.143		86.50	62.50		149	189
1830	800 amp		14	1.143		86.50	62.50		149	189
1840	1000 amp		14	1.143		86.50	62.50		149	189
1845	1200 amp		14	1.143		86.50	62.50		149	189
1850	1350 amp		14	1.143		86.50	62.50		149	189
1860	1600 amp		12	1.333		86.50	73		159.50	204
1870	2000 amp		12	1.333		86.50	73		159.50	204
1880	2500 amp		12	1.333		86.50	73		159.50	204
1890	3000 amp		10	1.600		86.50	87.50		174	226
1900	4000 amp		10	1.600		86.50	87.50		174	226

26 25 13.70 Feedrail		Crew	Daily Output	Labor-Hours	Unit	Material	Labor	Equipment	Total	Total Incl O&P
0010	**FEEDRAIL**, 12 foot mounting									
0050	Trolley busway, 3 pole									
0100	300 volt 60 amp, plain, 10 ft. lengths	1 Elec	50	.160	L.F.	25.50	8.75		34.25	41
0300	Door track		50	.160		41	8.75		49.75	58
0500	Curved track		30	.267		445	14.60		459.60	510
0700	Coupling				Ea.	16.70			16.70	18.35
0900	Center feed	1 Elec	5.30	1.509		78	82.50		160.50	210
1100	End feed		5.30	1.509		68	82.50		150.50	199
1300	Hanger set		24	.333		4.48	18.25		22.73	32.50
3000	600 volt 100 amp, plain, 10 ft. lengths		35	.229	L.F.	58.50	12.50		71	83.50
3300	Door track		35	.229	"	86	12.50		98.50	113
3700	Coupling				Ea.	45.50			45.50	50
4000	End cap	1 Elec	40	.200		44.50	10.95		55.45	65.50
4200	End feed		4	2		199	109		308	385
4500	Trolley, 600 volt, 20 amp		5.30	1.509		425	82.50		507.50	595
4700	30 amp		5.30	1.509		425	82.50		507.50	595

26 25 Enclosed Bus Assemblies

26 25 13 – Bus Duct/Busway and Fittings

26 25 13.70 Feedrail		Crew	Daily Output	Labor-Hours	Unit	2015 Bare Costs Material	Labor	Equipment	Total	Total Incl O&P
4900	Duplex, 40 amp	1 Elec	4	2	Ea.	715	109		824	950
5000	60 amp		4	2		715	109		824	950
5300	Fusible, 20 amp		4	2		995	109		1,104	1,275
5500	30 amp		4	2		995	109		1,104	1,275
5900	300 volt, 20 amp		5.30	1.509		281	82.50		363.50	435
6000	30 amp		5.30	1.509		355	82.50		437.50	515
6300	Fusible, 20 amp		4.70	1.702		345	93		438	520
6500	30 amp		4.70	1.702		480	93		573	670
7300	Busway, 250 volt 50 amp, 2 wire		70	.114	L.F.	20.50	6.25		26.75	32
7330	Coupling				Ea.	42.50			42.50	47
7340	Center feed	1 Elec	6	1.333		495	73		568	655
7350	End feed		6	1.333		109	73		182	229
7360	End cap		40	.200		29	10.95		39.95	48.50
7370	Hanger set		24	.333		2.80	18.25		21.05	30.50
7400	125/250 volt 50 amp, 3 wire		60	.133	L.F.	20.50	7.30		27.80	33.50
7430	Coupling		6	1.333	Ea.	61.50	73		134.50	177
7440	Center feed		6	1.333		495	73		568	655
7450	End feed		6	1.333		109	73		182	229
7460	End cap		40	.200		29	10.95		39.95	48.50
7470	Hanger set		24	.333		3.75	18.25		22	31.50
7480	Trolley, 250 volt, 2 pole, 20 amp		6	1.333		37	73		110	150
7490	30 amp		6	1.333		37	73		110	150
7500	125/250 volt, 3 pole, 20 amp		6	1.333		37	73		110	150
7510	30 amp		6	1.333		37	73		110	150
8000	Cleaning tools, 300 volt, dust remover					107			107	118
8100	Bus bar cleaner					202			202	222
8300	600 volt, dust remover, 60 amp					297			297	325
8400	100 amp					650			650	715
8600	Bus bar cleaner, 60 amp					675			675	745
8700	100 amp					800			800	880

26 27 Low-Voltage Distribution Equipment

26 27 13 – Electricity Metering

26 27 13.10 Meter Centers and Sockets

		Crew	Daily Output	Labor-Hours	Unit	Material	Labor	Equipment	Total	Total Incl O&P
0010	**METER CENTERS AND SOCKETS**									
0100	Sockets, single position, 4 terminal, 100 amp	1 Elec	3.20	2.500	Ea.	43	137		180	253
0200	150 amp		2.30	3.478		48	190		238	340
0300	200 amp		1.90	4.211		92	230		322	445
0400	Transformer rated, 20 amp		3.20	2.500		152	137		289	370
0500	Double position, 4 terminal, 100 amp		2.80	2.857		204	156		360	460
0600	150 amp		2.10	3.810		240	208		448	575
0700	200 amp		1.70	4.706		450	257		707	880
0800	Trans-socket, 13 terminal, 3 CT mounts, 400 amp		1	8		1,325	440		1,765	2,100
0900	800 amp	2 Elec	1.20	13.333		1,550	730		2,280	2,800
1100	Meter centers and sockets, three phase, single pos, 7 terminal, 100 amp	1 Elec	2.80	2.857		164	156		320	415
1200	200 amp		2.10	3.810		295	208		503	635
1400	400 amp		1.70	4.706		720	257		977	1,175
2000	Meter center, main fusible switch, 1P 3W 120/240 volt									
2030	400 amp	2 Elec	1.60	10	Ea.	1,675	545		2,220	2,675
2040	600 amp		1.10	14.545		2,600	795		3,395	4,050
2050	800 amp		.90	17.778		4,325	970		5,295	6,225

26 27 13.10 Meter Centers and Sockets		Crew	Daily Output	Labor-Hours	Unit	Material	2015 Bare Costs Labor	Equipment	Total	Total Incl O&P
2060	Rainproof 1P 3W 120/240 volt, 400 amp	2 Elec	1.60	10	Ea.	1,625	545		2,170	2,600
2070	600 amp		1.10	14.545		2,825	795		3,620	4,300
2080	800 amp		.90	17.778		4,400	970		5,370	6,300
2100	3P 4W 120/208 V, 400 amp		1.60	10		2,125	545		2,670	3,175
2110	600 amp		1.10	14.545		3,350	795		4,145	4,900
2120	800 amp		.90	17.778		5,425	970		6,395	7,425
2130	Rainproof 3P 4W 120/208 V, 400 amp		1.60	10		1,850	545		2,395	2,850
2140	600 amp		1.10	14.545		3,450	795		4,245	5,000
2150	800 amp		.90	17.778		6,425	970		7,395	8,500
2170	Main circuit breaker, 1P 3W 120/240 V									
2180	400 amp	2 Elec	1.60	10	Ea.	2,475	545		3,020	3,550
2190	600 amp		1.10	14.545		3,300	795		4,095	4,850
2200	800 amp		.90	17.778		5,225	970		6,195	7,200
2210	1000 amp		.80	20		6,250	1,100		7,350	8,525
2220	1200 amp		.76	21.053		8,700	1,150		9,850	11,300
2230	1600 amp		.68	23.529		18,900	1,275		20,175	22,700
2240	Rainproof 1P 3W 120/240 V, 400 amp		1.60	10		2,900	545		3,445	4,000
2250	600 amp		1.10	14.545		3,900	795		4,695	5,475
2260	800 amp		.90	17.778		4,550	970		5,520	6,450
2270	1000 amp		.80	20		6,250	1,100		7,350	8,525
2280	1200 amp		.76	21.053		8,450	1,150		9,600	11,000
2300	3P 4W 120/208 V, 400 amp		1.60	10		3,300	545		3,845	4,450
2310	600 amp		1.10	14.545		4,625	795		5,420	6,300
2320	800 amp		.90	17.778		5,500	970		6,470	7,525
2330	1000 amp		.80	20		7,250	1,100		8,350	9,625
2340	1200 amp		.76	21.053		9,250	1,150		10,400	11,900
2350	1600 amp		.68	23.529		18,900	1,275		20,175	22,700
2360	Rainproof 3P 4W 120/208 V, 400 amp		1.60	10		3,300	545		3,845	4,450
2370	600 amp		1.10	14.545		4,625	795		5,420	6,300
2380	800 amp		.90	17.778		5,500	970		6,470	7,525
2390	1000 amp		.76	21.053		7,250	1,150		8,400	9,700
2400	1200 amp		.68	23.529		9,250	1,275		10,525	12,100
2420	Main lugs terminal box, 1P 3W 120/240 V									
2430	800 amp	2 Elec	.94	17.021	Ea.	460	930		1,390	1,900
2440	1200 amp		.72	22.222		925	1,225		2,150	2,850
2450	Rainproof 1P 3W 120/240 V, 225 amp		2.40	6.667		370	365		735	950
2460	800 amp		.94	17.021		545	930		1,475	2,000
2470	1200 amp		.72	22.222		1,100	1,225		2,325	3,025
2500	3P 4W 120/208 V, 800 amp		.94	17.021		605	930		1,535	2,075
2510	1200 amp		.72	22.222		1,225	1,225		2,450	3,150
2520	Rainproof 3P 4W 120/208 V, 225 amp		2.40	6.667		370	365		735	950
2530	800 amp		.94	17.021		605	930		1,535	2,075
2540	1200 amp		.72	22.222		1,225	1,225		2,450	3,150
2590	Basic meter device									
2600	1P 3W 120/240 V 4 jaw 125A sockets, 3 meter	2 Elec	1	16	Ea.	620	875		1,495	1,975
2610	4 meter		.90	17.778		745	970		1,715	2,275
2620	5 meter		.80	20		930	1,100		2,030	2,675
2630	6 meter		.60	26.667		1,075	1,450		2,525	3,350
2640	7 meter		.56	28.571		1,375	1,575		2,950	3,850
2650	8 meter		.52	30.769		1,500	1,675		3,175	4,175
2660	10 meter		.48	33.333		1,875	1,825		3,700	4,775
2680	Rainproof 1P 3W 120/240 V 4 jaw 125A sockets									
2690	3 meter	2 Elec	1	16	Ea.	620	875		1,495	1,975

26 27 13.10 Meter Centers and Sockets		Crew	Daily Output	Labor-Hours	Unit	Material	Labor	Equipment	Total	Total Incl O&P
						2015 Bare Costs				
2700	4 meter	2 Elec	.90	17.778	Ea.	745	970		1,715	2,275
2710	6 meter		.60	26.667		1,075	1,450		2,525	3,350
2720	7 meter		.56	28.571		1,375	1,575		2,950	3,850
2730	8 meter		.52	30.769		1,500	1,675		3,175	4,175
2750	1P 3W 120/240 V 4 jaw sockets									
2760	with 125A circuit breaker, 3 meter	2 Elec	1	16	Ea.	1,175	875		2,050	2,575
2770	4 meter		.90	17.778		1,475	970		2,445	3,075
2780	5 meter		.80	20		1,825	1,100		2,925	3,675
2790	6 meter		.60	26.667		2,150	1,450		3,600	4,550
2800	7 meter		.56	28.571		2,625	1,575		4,200	5,250
2810	8 meter		.52	30.769		2,925	1,675		4,600	5,750
2820	10 meter		.48	33.333		3,675	1,825		5,500	6,775
2830	Rainproof 1P 3W 120/240 V 4 jaw sockets									
2840	with 125A circuit breaker, 3 meter	2 Elec	1	16	Ea.	1,175	875		2,050	2,575
2850	4 meter		.90	17.778		1,475	970		2,445	3,075
2870	6 meter		.60	26.667		2,150	1,450		3,600	4,550
2880	7 meter		.56	28.571		2,625	1,575		4,200	5,250
2890	8 meter		.52	30.769		2,925	1,675		4,600	5,750
2920	1P 3W on 3P 4W 120/208 V system 5 jaw									
2930	125A sockets, 3 meter	2 Elec	1	16	Ea.	620	875		1,495	1,975
2940	4 meter		.90	17.778		745	970		1,715	2,275
2950	5 meter		.80	20		930	1,100		2,030	2,675
2960	6 meter		.60	26.667		1,075	1,450		2,525	3,350
2970	7 meter		.56	28.571		1,375	1,575		2,950	3,850
2980	8 meter		.52	30.769		1,500	1,675		3,175	4,175
2990	10 meter		.48	33.333		1,875	1,825		3,700	4,775
3000	Rainproof 1P 3W on 3P 4W 120/208 V system									
3020	5 jaw 125A sockets, 3 meter	2 Elec	1	16	Ea.	620	875		1,495	1,975
3030	4 meter		.90	17.778		745	970		1,715	2,275
3050	6 meter		.60	26.667		1,075	1,450		2,525	3,350
3060	7 meter		.56	28.571		1,375	1,575		2,950	3,850
3070	8 meter		.52	30.769		1,500	1,675		3,175	4,175
3090	1P 3W on 3P 4W 120/208 V system 5 jaw sockets									
3100	With 125A circuit breaker, 3 meter	2 Elec	1	16	Ea.	1,175	875		2,050	2,575
3110	4 meter		.90	17.778		1,475	970		2,445	3,075
3120	5 meter		.80	20		1,825	1,100		2,925	3,675
3130	6 meter		.60	26.667		2,150	1,450		3,600	4,550
3140	7 meter		.56	28.571		2,625	1,575		4,200	5,250
3150	8 meter		.52	30.769		2,925	1,675		4,600	5,750
3160	10 meter		.48	33.333		3,675	1,825		5,500	6,775
3170	Rainproof 1P 3W on 3P 4W 120/208 V system									
3180	5 jaw sockets w/125A circuit breaker, 3 meter	2 Elec	1	16	Ea.	1,175	875		2,050	2,575
3190	4 meter		.90	17.778		1,475	970		2,445	3,075
3210	6 meter		.60	26.667		2,150	1,450		3,600	4,550
3220	7 meter		.56	28.571		2,625	1,575		4,200	5,250
3230	8 meter		.52	30.769		2,925	1,675		4,600	5,750
3250	1P 3W 120/240 V 4 jaw sockets									
3260	with 200A circuit breaker, 3 meter	2 Elec	1	16	Ea.	1,750	875		2,625	3,225
3270	4 meter		.90	17.778		2,350	970		3,320	4,050
3290	6 meter		.60	26.667		3,500	1,450		4,950	6,025
3300	7 meter		.56	28.571		4,100	1,575		5,675	6,875
3310	8 meter		.56	28.571		4,725	1,575		6,300	7,550
3330	Rainproof 1P 3W 120/240 V 4 jaw sockets									

26 27 13.10 Meter Centers and Sockets		Crew	Daily Output	Labor-Hours	Unit	Material	2015 Bare Costs Labor	Equipment	Total	Total Incl O&P
3350	with 200A circuit breaker, 3 meter	2 Elec	1	16	Ea.	1,750	875		2,625	3,225
3360	4 meter		.90	17.778		2,350	970		3,320	4,050
3380	6 meter		.60	26.667		3,500	1,450		4,950	6,025
3390	7 meter		.56	28.571		4,100	1,575		5,675	6,875
3400	8 meter		.52	30.769		4,725	1,675		6,400	7,725
3420	1P 3W on 3P 4W 120/208 V 5 jaw sockets									
3430	with 200A circuit breaker, 3 meter	2 Elec	1	16	Ea.	1,750	875		2,625	3,225
3440	4 meter		.90	17.778		2,350	970		3,320	4,050
3460	6 meter		.60	26.667		3,500	1,450		4,950	6,025
3470	7 meter		.56	28.571		4,100	1,575		5,675	6,875
3480	8 meter		.52	30.769		4,725	1,675		6,400	7,725
3500	Rainproof 1P 3W on 3P 4W 120/208 V 5 jaw socket									
3510	with 200A circuit breaker, 3 meter	2 Elec	1	16	Ea.	1,750	875		2,625	3,225
3520	4 meter		.90	17.778		2,350	970		3,320	4,050
3540	6 meter		.60	26.667		3,500	1,450		4,950	6,025
3550	7 meter		.56	28.571		4,100	1,575		5,675	6,875
3560	8 meter		.52	30.769		4,725	1,675		6,400	7,725
3600	Automatic circuit closing, add					67.50			67.50	74
3610	Manual circuit closing, add					77			77	84.50
3650	Branch meter device									
3660	3P 4W 208/120 or 240/120 V 7 jaw sockets									
3670	with 200A circuit breaker, 2 meter	2 Elec	.90	17.778	Ea.	2,925	970		3,895	4,650
3680	3 meter		.80	20		4,375	1,100		5,475	6,450
3690	4 meter		.70	22.857		5,825	1,250		7,075	8,275
3700	Main circuit breaker 42,000 rms, 400 amp		1.60	10		2,175	545		2,720	3,200
3710	600 amp		1.10	14.545		4,100	795		4,895	5,725
3720	800 amp		.90	17.778		5,500	970		6,470	7,525
3730	Rainproof main circ. breaker 42,000 rms, 400 amp		1.60	10		2,550	545		3,095	3,625
3740	600 amp		1.10	14.545		4,100	795		4,895	5,725
3750	800 amp		.90	17.778		5,500	970		6,470	7,525
3760	Main circuit breaker 65,000 rms, 400 amp		1.60	10		3,450	545		3,995	4,625
3770	600 amp		1.10	14.545		4,875	795		5,670	6,550
3780	800 amp		.90	17.778		5,500	970		6,470	7,525
3790	1000 amp		.80	20		7,250	1,100		8,350	9,625
3800	1200 amp		.76	21.053		9,250	1,150		10,400	11,900
3810	1600 amp		.68	23.529		18,900	1,275		20,175	22,700
3820	Rainproof main circ. breaker 65,000 rms, 400 amp		1.60	10		3,450	545		3,995	4,625
3830	600 amp		1.10	14.545		4,875	795		5,670	6,550
3840	800 amp		.90	17.778		5,500	970		6,470	7,500
3850	1000 amp		.80	20		7,250	1,100		8,350	9,625
3860	1200 amp		.76	21.053		9,250	1,150		10,400	11,900
3880	Main circuit breaker 100,000 rms, 400 amp		1.60	10		3,450	545		3,995	4,625
3890	600 amp		1.10	14.545		4,875	795		5,670	6,550
3900	800 amp		.90	17.778		5,725	970		6,695	7,725
3910	Rainproof main circ. breaker 100,000 rms, 400 amp		1.60	10		3,450	545		3,995	4,625
3920	600 amp		1.10	14.545		4,875	795		5,670	6,550
3930	800 amp		.90	17.778		5,725	970		6,695	7,725
3940	Main lugs terminal box, 800 amp		.94	17.021		605	930		1,535	2,075
3950	1600 amp		.72	22.222		2,025	1,225		3,250	4,075
3960	Rainproof, 800 amp		.94	17.021		605	930		1,535	2,075
3970	1600 amp		.72	22.222		2,025	1,225		3,250	4,075

26 27 16 – Electrical Cabinets and Enclosures

26 27 16.10 Cabinets		Crew	Daily Output	Labor-Hours	Unit	2015 Bare Costs Material	Labor	Equipment	Total	Total Incl O&P
0010	**CABINETS** R260533-70									
7000	Cabinets, current transformer									
7050	Single door, 24" H x 24" W x 10" D R262716-40	1 Elec	1.60	5	Ea.	152	274		426	575
7100	30" H x 24" W x 10" D		1.30	6.154		165	335		500	685
7150	36" H x 24" W x 10" D		1.10	7.273		177	400		577	790
7200	30" H x 30" W x 10" D		1	8		223	440		663	900
7250	36" H x 30" W x 10" D		.90	8.889		262	485		747	1,025
7300	36" H x 36" W x 10" D		.80	10		270	545		815	1,125
7500	Double door, 48" H x 36" W x 10" D		.60	13.333		590	730		1,320	1,750
7550	24" H x 24" W x 12" D		1	8		173	440		613	845
8000	NEMA 12, double door, floor mounted									
8020	54" H x 42" W x 8" D	2 Elec	6	2.667	Ea.	1,350	146		1,496	1,725
8040	60" H x 48" W x 8" D		5.40	2.963		1,800	162		1,962	2,250
8060	60" H x 48" W x 10" D		5.40	2.963		1,875	162		2,037	2,300
8080	60" H x 60" W x 10" D		5	3.200		2,075	175		2,250	2,550
8100	72" H x 60" W x 10" D		4	4		2,450	219		2,669	3,025
8120	72" H x 72" W x 10" D		3.40	4.706		2,700	257		2,957	3,350
8140	60" H x 48" W x 12" D		3.40	4.706		1,900	257		2,157	2,450
8160	60" H x 60" W x 12" D		3.20	5		2,150	274		2,424	2,775
8180	72" H x 60" W x 12" D		3	5.333		2,425	292		2,717	3,100
8200	72" H x 72" W x 12" D		3	5.333		2,725	292		3,017	3,425
8220	60" H x 48" W x 16" D		3.20	5		1,975	274		2,249	2,575
8240	72" H x 72" W x 16" D		2.60	6.154		2,850	335		3,185	3,650
8260	60" H x 48" W x 20" D		3	5.333		2,175	292		2,467	2,800
8280	72" H x 72" W x 20" D		2.20	7.273		3,075	400		3,475	3,975
8300	60" H x 48" W x 24" D		2.60	6.154		2,300	335		2,635	3,025
8320	72" H x 72" W x 24" D		2	8		3,250	440		3,690	4,225
8340	Pushbutton enclosure, oiltight									
8360	3-1/2" H x 3-1/4" W x 2-3/4" D, for 1 P.B.	1 Elec	12	.667	Ea.	56.50	36.50		93	117
8380	5-3/4" H x 3-1/4" W x 2-3/4" D, for 2 P.B.		11	.727		62	40		102	128
8400	8" H x 3-1/4" W x 2-3/4" D, for 3 P.B.		10.50	.762		67.50	41.50		109	137
8420	10-1/4" H x 3-1/4" W x 2-3/4" D, for 4 P.B.		10.50	.762		74	41.50		115.50	144
8460	12-1/2" H x 3-1/4" W x 3" D, for 5 P.B.		9	.889		88.50	48.50		137	170
8480	9-1/2" H x 6-1/4" W x 3" D, for 6 P.B.		8.50	.941		96.50	51.50		148	183
8500	9-1/2" H x 8-1/2" W x 3" D, for 9 P.B.		8	1		105	54.50		159.50	198
8510	11-3/4" H x 8-1/2" W x 3" D, for 12 P.B.		7	1.143		115	62.50		177.50	221
8520	11-3/4" H x 10-3/4" W x 3" D, for 16 P.B.		6.50	1.231		126	67.50		193.50	239
8540	14" H x 10-3/4" W x 3" D, for 20 P.B.		5	1.600		137	87.50		224.50	282
8560	14" H x 13" W x 3" D, for 25 P.B.		4.50	1.778		150	97		247	310
8580	Sloping front pushbutton enclosures									
8600	3-1/2" H x 7-3/4" W x 4-7/8" D, for 3 P.B.	1 Elec	10	.800	Ea.	88	44		132	163
8620	7-1/4" H x 8-1/2" W x 6-3/4" D, for 6 P.B.		8	1		129	54.50		183.50	224
8640	9-1/2" H x 8-1/2" W x 7-7/8" D, for 9 P.B.		7	1.143		153	62.50		215.50	263
8660	11-1/4" H x 8-1/2" W x 9" D, for 12 P.B.		5	1.600		177	87.50		264.50	325
8680	11-3/4" H x 10" W x 9" D, for 16 P.B.		5	1.600		191	87.50		278.50	340
8700	11-3/4" H x 13" W x 9" D, for 20 P.B.		5	1.600		220	87.50		307.50	375
8720	14" H x 13" W x 10-1/8" D, for 25 P.B.		4.50	1.778		249	97		346	420
8740	Pedestals, not including P.B. enclosure or base									
8760	Straight column 4" x 4"	1 Elec	4.50	1.778	Ea.	265	97		362	435
8780	6" x 6"		4	2		410	109		519	620
8800	Angled column 4" x 4"		4.50	1.778		305	97		402	480
8820	6" x 6"		4	2		455	109		564	670

26 27 16 – Electrical Cabinets and Enclosures

26 27 16.10 Cabinets		Crew	Daily Output	Labor-Hours	Unit	Material	2015 Bare Costs Labor	Equipment	Total	Total Incl O&P
8840	Pedestal, base 18" x 18"	1 Elec	10	.800	Ea.	156	44		200	238
8860	24" x 24"		9	.889		360	48.50		408.50	470
8900	Electronic rack enclosures									
8920	72" H x 19" W x 24" D	1 Elec	1.50	5.333	Ea.	1,925	292		2,217	2,550
8940	72" H x 23" W x 24" D		1.50	5.333		2,175	292		2,467	2,825
8960	72" H x 19" W x 30" D		1.30	6.154		2,300	335		2,635	3,050
8980	72" H x 19" W x 36" D		1.20	6.667		2,650	365		3,015	3,450
9000	72" H x 23" W x 36" D		1.20	6.667		2,925	365		3,290	3,775
9020	NEMA 12 & 4 enclosure panels									
9040	12" x 24"	1 Elec	20	.400	Ea.	36	22		58	73
9060	16" x 12"		20	.400		26.50	22		48.50	62
9080	20" x 16"		20	.400		38.50	22		60.50	75
9100	20" x 20"		19	.421		46	23		69	85
9120	24" x 20"		18	.444		59.50	24.50		84	102
9140	24" x 24"		17	.471		68.50	25.50		94	114
9160	30" x 20"		16	.500		75.50	27.50		103	124
9180	30" x 24"		16	.500		83	27.50		110.50	133
9200	36" x 24"		15	.533		98.50	29		127.50	152
9220	36" x 30"		15	.533		132	29		161	189
9240	42" x 24"		15	.533		112	29		141	167
9260	42" x 30"		14	.571		149	31.50		180.50	211
9280	42" x 36"		14	.571		175	31.50		206.50	239
9300	48" x 24"		14	.571		134	31.50		165.50	194
9320	48" x 30"		14	.571		174	31.50		205.50	238
9340	48" x 36"		13	.615		201	33.50		234.50	272
9360	60" x 36"		12	.667		245	36.50		281.50	325
9400	Wiring trough steel JIC, clamp cover									
9490	4" x 4", 12" long	1 Elec	12	.667	Ea.	94	36.50		130.50	159
9510	24" long		10	.800		119	44		163	197
9530	36" long		8	1		149	54.50		203.50	245
9540	48" long		7	1.143		172	62.50		234.50	283
9550	60" long		6	1.333		202	73		275	330
9560	6" x 6", 12" long		11	.727		125	40		165	197
9580	24" long		9	.889		165	48.50		213.50	255
9600	36" long		7	1.143		202	62.50		264.50	315
9610	48" long		6	1.333		246	73		319	380
9620	60" long		5	1.600		290	87.50		377.50	450

26 27 16.20 Cabinets and Enclosures		Crew	Daily Output	Labor-Hours	Unit	Material	Labor	Equipment	Total	Total Incl O&P
0010	**CABINETS AND ENCLOSURES** Nonmetallic									
0080	Enclosures fiberglass NEMA 4X									
0100	Wall mount, quick release latch door, 20"H x 16"W x 6"D	1 Elec	4.80	1.667	Ea.	615	91		706	810
0110	20"H x 20"W x 6"D		4.50	1.778		735	97		832	955
0120	24"H x 20"W x 6"D		4.20	1.905		785	104		889	1,025
0130	20"H x 16"W x 8"D		4.50	1.778		700	97		797	915
0140	20"H x 20"W x 8"D		4.20	1.905		785	104		889	1,025
0150	24"H x 24"W x 8"D		3.80	2.105		890	115		1,005	1,150
0160	30"H x 24"W x 8"D		3.20	2.500		960	137		1,097	1,250
0170	36"H x 30"W x 8"D		3	2.667		1,350	146		1,496	1,700
0180	20"H x 16"W x 10"D		3.50	2.286		815	125		940	1,075
0190	20"H x 20"W x 10"D		3.20	2.500		870	137		1,007	1,150
0200	24"H x 20"W x 10"D		3	2.667		925	146		1,071	1,250
0210	30"H x 24"W x 10"D		2.80	2.857		1,050	156		1,206	1,375

26 27 16.20 Cabinets and Enclosures		Crew	Daily Output	Labor-Hours	Unit	2015 Bare Costs Material	Labor	Equipment	Total	Total Incl O&P
0220	20"H x 16"W x 12"D	1 Elec	3	2.667	Ea.	865	146		1,011	1,175
0230	20"H x 20"W x 12"D		2.80	2.857		930	156		1,086	1,250
0240	24"H x 24"W x 12"D		2.60	3.077		1,025	168		1,193	1,375
0250	30"H x 24"W x 12"D		2.40	3.333		1,150	182		1,332	1,525
0260	36"H x 30"W x 12"D		2.20	3.636		1,575	199		1,774	2,050
0270	36"H x 36"W x 12"D		2.10	3.810		1,775	208		1,983	2,250
0280	48"H x 36"W x 12"D		2	4		2,000	219		2,219	2,525
0290	60"H x 36"W x 12"D		1.80	4.444		2,275	243		2,518	2,875
0300	30"H x 24"W x 16"D		1.40	5.714		1,325	315		1,640	1,950
0310	48"H x 36"W x 16"D		1.20	6.667		2,225	365		2,590	3,000
0320	60"H x 36"W x 16"D		1	8		2,525	440		2,965	3,425
0480	Freestanding, one door, 72"H x 25"W x 25"D		.80	10		4,275	545		4,820	5,525
0490	Two doors with two panels, 72"H x 49"W x 24"D		.50	16		10,700	875		11,575	13,000
0500	Floor stand kits, for NEMA 4 & 12, 20"W or more, 6"H x 8"D		24	.333		161	18.25		179.25	205
0510	6"H x 10"D		24	.333		176	18.25		194.25	221
0520	6"H x 12"D		24	.333		196	18.25		214.25	244
0530	6"H x 18"D		24	.333		228	18.25		246.25	279
0540	12"H x 8"D		22	.364		203	19.90		222.90	253
0550	12"H x 10"D		22	.364		215	19.90		234.90	266
0560	12"H x 12"D		22	.364		229	19.90		248.90	282
0570	12"H x 16"D		22	.364		267	19.90		286.90	325
0580	12"H x 18"D		22	.364		273	19.90		292.90	330
0590	12"H x 20"D		22	.364		291	19.90		310.90	350
0600	18"H x 8"D		20	.400		242	22		264	299
0610	18"H x 10"D		20	.400		259	22		281	320
0620	18"H x 12"D		20	.400		278	22		300	340
0630	18"H x 16"D		20	.400		310	22		332	375
0640	24"H x 8"D		16	.500		299	27.50		326.50	370
0650	24"H x 10"D		16	.500		310	27.50		337.50	380
0660	24"H x 12"D		16	.500		335	27.50		362.50	405
0670	24"H x 16"D		16	.500		350	27.50		377.50	425
0680	Small, screw cover, 5-1/2"H x 4"W x 4-15/16"D		12	.667		85.50	36.50		122	149
0690	7-1/2"H x 4"W x 4-15/16"D		12	.667		91	36.50		127.50	155
0700	7-1/2"H x 6"W x 5-3/16"D		10	.800		106	44		150	182
0710	9-1/2"H x 6"W x 5-11/16"D		10	.800		108	44		152	185
0720	11-1/2"H x 8"W x 6-11/16"D		8	1		161	54.50		215.50	259
0730	13-1/2"H x 10"W x 7-3/16"D		7	1.143		193	62.50		255.50	305
0740	15-1/2"H x 12"W x 8-3/16"D		6	1.333		253	73		326	390
0750	17-1/2"H x 14"W x 8-11/16"D		5	1.600		300	87.50		387.50	460
0760	Screw cover with window, 6"H x 4"W x 5"D		12	.667		159	36.50		195.50	230
0770	8"H x 4"W x 5"D		11	.727		167	40		207	244
0780	8"H x 6"W x 5"D		11	.727		218	40		258	300
0790	10"H x 6"W x 6"D		10	.800		235	44		279	325
0800	12"H x 8"W x 7"D		8	1		315	54.50		369.50	425
0810	14"H x 10"W x 7"D		7	1.143		345	62.50		407.50	475
0820	16"H x 12"W x 8"D		6	1.333		430	73		503	585
0830	18"H x 14"W x 9"D		5	1.600		520	87.50		607.50	700
0840	Quick-release latch cover, 5-1/2"H x 4"W x 5"D		12	.667		116	36.50		152.50	183
0850	7-1/2"H x 4"W x 5"D		12	.667		124	36.50		160.50	192
0860	7-1/2"H x 6"W x 5-1/4"D		10	.800		134	44		178	214
0870	9-1/2"H x 6"W x 5-3/4"D		10	.800		147	44		191	228
0880	11-1/2"H x 8"W x 6-3/4"D		8	1		210	54.50		264.50	315
0890	13-1/2"H x 10"W x 7-1/4"D		7	1.143		263	62.50		325.50	385

26 27 16.20 Cabinets and Enclosures		Crew	Daily Output	Labor-Hours	Unit	2015 Bare Costs Material	Labor	Equipment	Total	Total Incl O&P
0900	15-1/2"H x 12"W x 8-1/4"D	1 Elec	6	1.333	Ea.	325	73		398	465
0910	17-1/2"H x 14"W x 8-3/4"D		5	1.600		390	87.50		477.50	560
0920	Pushbutton, 1 hole 5-1/2"H x 4"W x 4-15/16"D		12	.667		81	36.50		117.50	144
0930	2 hole 7-1/2"H x 4"W x 4-15/16"D		11	.727		92.50	40		132.50	162
0940	4 hole 7-1/2"H x 6"W x 5-3/16"D		10.50	.762		109	41.50		150.50	183
0950	6 hole 9-1/2"H x 6"W x 5-11/16"D		9	.889		134	48.50		182.50	220
0960	8 hole 11-1/2"H x 8"W x 6-11/16"D		8.50	.941		170	51.50		221.50	264
0970	12 hole 13-1/2"H x 10"W x 7-3/16"D		8	1		219	54.50		273.50	325
0980	20 hole 15-1/2"H x 12"W x 8-3/16"D		5	1.600		299	87.50		386.50	460
0990	30 hole 17-1/2"H x 14"W x 8-11/16"D	↓	4.50	1.778	↓	330	97		427	510
1450	Enclosures polyester NEMA 4X									
1460	Small, screw cover,									
1500	3-15/16"H x 3-15/16"W x 3-1/16"D	1 Elec	12	.667	Ea.	68.50	36.50		105	130
1510	5-3/16"H x 3-5/16"W x 3-1/16"D		12	.667		67.50	36.50		104	129
1520	5-7/8"H x 3-7/8"W x 4-3/16"D		12	.667		71	36.50		107.50	133
1530	5-7/8"H x 5-7/8"W x 4-3/16"D		12	.667		80	36.50		116.50	143
1540	7-5/8"H x 3-5/16"W x 3-1/16"D		12	.667		73.50	36.50		110	136
1550	10-3/16"H x 3-5/16"W x 3-1/16"D		10	.800		87.50	44		131.50	162
1560	Clear cover, 3-15/16"H x 3-15/16"W x 2-7/8"D		12	.667		77.50	36.50		114	140
1570	5-3/16"H x 3-5/16"W x 2-7/8"D		12	.667		82.50	36.50		119	145
1580	5-7/8"H x 3-7/8"W x 4"D		12	.667		102	36.50		138.50	168
1590	5-7/8"H x 5-7/8"W x 4"D		12	.667		126	36.50		162.50	194
1600	7-5/8"H x 3-5/16"W x 2-7/8"D		12	.667		92.50	36.50		129	157
1610	10-3/16"H x 3-5/16"W x 2-7/8"D		10	.800		116	44		160	194
1620	Pushbutton, 1 hole, 5-5/16"H x 3-5/16"W x 3-1/16"D		12	.667		60	36.50		96.50	121
1630	2 hole, 7-5/8"H x 3-5/16"W x 3-1/8"D		11	.727		67.50	40		107.50	134
1640	3 hole, 10-3/16"H x 3-5/16"W x 3-1/16"D		10.50	.762		80	41.50		121.50	151
8000	Wireway fiberglass, straight sect. screwcover, 12" L, 4" W x 4" D		40	.200		194	10.95		204.95	230
8010	6" W x 6" D		30	.267		258	14.60		272.60	305
8020	24" L, 4" W x 4" D		20	.400		236	22		258	292
8030	6" W x 6" D		15	.533		390	29		419	475
8040	36" L, 4" W x 4" D		13.30	.602		295	33		328	375
8050	6" W x 6" D		10	.800		460	44		504	570
8060	48" L, 4" W x 4" D		10	.800		355	44		399	455
8070	6" W x 6" D		7.50	1.067		580	58.50		638.50	730
8080	60" L, 4" W x 4" D		8	1		610	54.50		664.50	750
8090	6" W x 6" D		6	1.333		675	73		748	850
8100	Elbow, 90°, 4" W x 4" D		20	.400		199	22		221	252
8110	6" W x 6" D		18	.444		400	24.50		424.50	475
8120	Elbow, 45°, 4" W x 4" D		20	.400		189	22		211	241
8130	6" W x 6" D		18	.444		375	24.50		399.50	450
8140	Tee, 4" W x 4" D		16	.500		252	27.50		279.50	320
8150	6" W x 6" D		14	.571		460	31.50		491.50	550
8160	Cross, 4" W x 4" D		14	.571		360	31.50		391.50	440
8170	6" W x 6" D		12	.667		775	36.50		811.50	910
8180	Cut-off fitting, w/flange & adhesive, 4" W x 4" D		18	.444		141	24.50		165.50	192
8190	6" W x 6" D		16	.500		305	27.50		332.50	380
8200	Flexible ftng., hvy. neoprene coated nylon, 4" W x 4" D		20	.400		274	22		296	335
8210	6" W x 6" D		18	.444		405	24.50		429.50	480
8220	Closure plate, fiberglass, 4" W x 4" D		20	.400		55	22		77	93.50
8230	6" W x 6" D		18	.444		59.50	24.50		84	102
8240	Box connector, stainless steel type 304, 4" W x 4" D		20	.400		88.50	22		110.50	131
8250	6" W x 6" D	↓	18	.444	↓	104	24.50		128.50	151

26 27 16 – Electrical Cabinets and Enclosures

26 27 16.20 Cabinets and Enclosures		Crew	Daily Output	Labor-Hours	Unit	2015 Bare Costs Material	Labor	Equipment	Total	Total Incl O&P
8260	Hanger, 4" W x 4" D	1 Elec	100	.080	Ea.	23	4.38		27.38	32
8270	6" W x 6" D		80	.100		29	5.45		34.45	40
8280	Straight tube section fiberglass, 4"W x 4"D, 12" long		40	.200		168	10.95		178.95	201
8290	24" long		20	.400		202	22		224	255
8300	36" long		13.30	.602		252	33		285	330
8310	48" long		10	.800		245	44		289	335
8320	60" long		8	1		300	54.50		354.50	410
8330	120" long		4	2		460	109		569	675

26 27 19 – Multi-Outlet Assemblies

26 27 19.10 Wiring Duct		Crew	Daily Output	Labor-Hours	Unit	Material	Labor	Equipment	Total	Total Incl O&P
0010	**WIRING DUCT** Plastic									
1250	PVC, snap-in slots, adhesive backed									
1270	1-1/2"W x 2"H	2 Elec	120	.133	L.F.	4.86	7.30		12.16	16.30
1280	1-1/2"W x 3"H		120	.133		5.85	7.30		13.15	17.35
1290	1-1/2"W x 4"H		120	.133		6.90	7.30		14.20	18.55
1300	2"W x 1"H		120	.133		4.36	7.30		11.66	15.75
1310	2"W x 1-1/2"H		120	.133		4.73	7.30		12.03	16.15
1320	2"W x 2"H		120	.133		4.75	7.30		12.05	16.20
1340	2"W x 3"H		120	.133		6.20	7.30		13.50	17.75
1350	2"W x 4"H		120	.133		7.50	7.30		14.80	19.20
1360	2-1/2"W x 3"H		120	.133		6.65	7.30		13.95	18.30
1370	3"W x 1"H		110	.145		4.95	7.95		12.90	17.35
1390	3"W x 2"H		110	.145		5.90	7.95		13.85	18.40
1400	3"W x 3"H		110	.145		7.15	7.95		15.10	19.75
1410	3"W x 4"H		110	.145		8.95	7.95		16.90	21.50
1420	3"W x 5"H		110	.145		11.75	7.95		19.70	25
1430	4"W x 1-1/2"H		100	.160		5.90	8.75		14.65	19.60
1440	4"W x 2"H		100	.160		6.90	8.75		15.65	20.50
1450	4"W x 3"H		100	.160		7.80	8.75		16.55	21.50
1460	4"W x 4"H		100	.160		9.45	8.75		18.20	23.50
1470	4"W x 5"H		100	.160		13.30	8.75		22.05	27.50
1550	Cover, 1-1/2"W		200	.080		.99	4.38		5.37	7.65
1560	2"W		200	.080		1.22	4.38		5.60	7.90
1570	2-1/2"W		200	.080		1.53	4.38		5.91	8.25
1580	3"W		200	.080		1.86	4.38		6.24	8.60
1590	4"W		200	.080		2.24	4.38		6.62	9

26 27 23 – Indoor Service Poles

26 27 23.40 Surface Raceway		Crew	Daily Output	Labor-Hours	Unit	Material	Labor	Equipment	Total	Total Incl O&P
0010	**SURFACE RACEWAY**									
0090	Metal, straight section									
0100	No. 500	1 Elec	100	.080	L.F.	1.04	4.38		5.42	7.70
0110	No. 700		100	.080		1.17	4.38		5.55	7.85
0400	No. 1500, small pancake		90	.089		2.15	4.86		7.01	9.65
0600	No. 2000, base & cover, blank		90	.089		2.19	4.86		7.05	9.70
0610	Receptacle, 6" O.C.		40	.200		17.55	10.95		28.50	36
0620	12" O.C.		44	.182		11.40	9.95		21.35	27.50
0630	18" O.C.		46	.174		9.45	9.50		18.95	24.50
0650	30" O.C.		50	.160		4.94	8.75		13.69	18.55
0670	No. 2400, base & cover, blank		80	.100		1.89	5.45		7.34	10.30
0680	Receptacle, 6" O.C.		42	.190		34.50	10.40		44.90	53.50
0690	12" O.C.		53	.151		23.50	8.25		31.75	38
0700	18" O.C.		55	.145		16.95	7.95		24.90	30.50

26 27 23.40 Surface Raceway		Crew	Daily Output	Labor-Hours	Unit	2015 Bare Costs Material	Labor	Equipment	Total	Total Incl O&P
0710	24″ O.C.	1 Elec	57	.140	L.F.	9.50	7.70		17.20	22
0720	30″ O.C.		59	.136		6.95	7.40		14.35	18.75
0730	60″ O.C.		61	.131		5.20	7.15		12.35	16.50
0800	No. 3000, base & cover, blank		75	.107		4.18	5.85		10.03	13.35
0810	Receptacle, 6″ O.C.		45	.178		37.50	9.70		47.20	56
0820	12″ O.C.		62	.129		21	7.05		28.05	34
0830	18″ O.C.		64	.125		17.15	6.85		24	29
0840	24″ O.C.		66	.121		12.85	6.65		19.50	24
0850	30″ O.C.		68	.118		11.55	6.45		18	22.50
0860	60″ O.C.		70	.114		8.45	6.25		14.70	18.65
1000	No. 4000, base & cover, blank		65	.123		6.80	6.75		13.55	17.60
1010	Receptacle, 6″ O.C.		41	.195		50.50	10.65		61.15	71.50
1020	12″ O.C.		52	.154		31.50	8.40		39.90	47
1030	18″ O.C.		54	.148		25.50	8.10		33.60	40
1040	24″ O.C.		56	.143		21	7.80		28.80	34.50
1050	30″ O.C.		58	.138		19	7.55		26.55	32.50
1060	60″ O.C.		60	.133		14.60	7.30		21.90	27
1200	No. 6000, base & cover, blank		50	.160		11.40	8.75		20.15	25.50
1210	Receptacle, 6″ O.C.		30	.267		61.50	14.60		76.10	90
1220	12″ O.C.		37	.216		41	11.85		52.85	62.50
1230	18″ O.C.		39	.205		34.50	11.20		45.70	55
1240	24″ O.C.		41	.195		28.50	10.65		39.15	47
1250	30″ O.C.		43	.186		27	10.20		37.20	45
1260	60″ O.C.		45	.178		21	9.70		30.70	37.50
2400	Fittings, elbows, No. 500		40	.200	Ea.	1.90	10.95		12.85	18.50
2800	Elbow cover, No. 2000		40	.200		3.57	10.95		14.52	20.50
2880	Tee, No. 500		42	.190		3.66	10.40		14.06	19.65
2900	No. 2000		27	.296		11.85	16.20		28.05	37.50
3000	Switch box, No. 500		16	.500		11	27.50		38.50	53
3400	Telephone outlet, No. 1500		16	.500		14.10	27.50		41.60	56.50
3600	Junction box, No. 1500		16	.500		9.65	27.50		37.15	51.50
3800	Plugmold wired sections, No. 2000									
4000	1 circuit, 6 outlets, 3 ft. long	1 Elec	8	1	Ea.	36	54.50		90.50	122
4100	2 circuits, 8 outlets, 6 ft. long		5.30	1.509		53	82.50		135.50	183
4110	Tele-power pole, alum, w/2 recept, 10′		4	2		205	109		314	390
4120	12′		3.85	2.078		196	114		310	385
4130	15′		3.70	2.162		310	118		428	515
4140	Steel, w/2 recept, 10′		4	2		130	109		239	305
4150	One phone fitting, 10′		4	2		139	109		248	315
4160	Alum, 4 outlets, 10′		3.70	2.162		262	118		380	465
4300	Overhead distribution systems, 125 volt									
4800	No. 2000, entrance end fitting	1 Elec	20	.400	Ea.	5.50	22		27.50	39
5000	Blank end fitting		40	.200		2.25	10.95		13.20	18.90
5200	Supporting clip		40	.200		1.33	10.95		12.28	17.85
5800	No. 3000, entrance end fitting		20	.400		9.20	22		31.20	43
6000	Blank end fitting		40	.200		2.62	10.95		13.57	19.30
6020	Internal elbow		20	.400		12.10	22		34.10	46.50
6030	External elbow		20	.400		17.10	22		39.10	52
6040	Device bracket		53	.151		4.10	8.25		12.35	16.85
6400	Hanger clamp		32	.250		6.25	13.70		19.95	27.50
7000	No. 4000 Base		90	.089	L.F.	4.40	4.86		9.26	12.15
7200	Divider		100	.080	″	.85	4.38		5.23	7.50
7400	Entrance end fitting		16	.500	Ea.	22.50	27.50		50	65.50

26 27 23.40 Surface Raceway		Crew	Daily Output	Labor-Hours	Unit	Material	2015 Bare Costs Labor	Equipment	Total	Total Incl O&P
7600	Blank end fitting	1 Elec	40	.200	Ea.	6.20	10.95		17.15	23
7610	Recpt. & tele. cover		53	.151		10.55	8.25		18.80	24
7620	External elbow		16	.500		34.50	27.50		62	79
7630	Coupling		53	.151		5.25	8.25		13.50	18.15
7640	Divider clip & coupling		80	.100		1.04	5.45		6.49	9.35
7650	Panel connector		16	.500		22	27.50		49.50	65
7800	Take off connector		16	.500		72.50	27.50		100	121
8000	No. 6000, take off connector		16	.500		86.50	27.50		114	136
8100	Take off fitting		16	.500		64	27.50		91.50	112
8200	Hanger clamp		32	.250		15.55	13.70		29.25	37.50
8230	Coupling					8.50			8.50	9.35
8240	One gang device plate	1 Elec	53	.151		8.30	8.25		16.55	21.50
8250	Two gang device plate		40	.200		10.15	10.95		21.10	27.50
8260	Blank end fitting		40	.200		8.80	10.95		19.75	26
8270	Combination elbow		14	.571		36	31.50		67.50	86.50
8300	Panel connector		16	.500		15.75	27.50		43.25	58.50
8500	Chan-L-Wire system installed in 1-5/8" x 1-5/8" strut. Strut									
8600	not incl., 30 amp, 4 wire, 3 phase	1 Elec	200	.040	L.F.	4.95	2.19		7.14	8.75
8700	Junction box		8	1	Ea.	33	54.50		87.50	119
8800	Insulating end cap		40	.200		8.80	10.95		19.75	26
8900	Strut splice plate		40	.200		11.65	10.95		22.60	29
9000	Tap		40	.200		23	10.95		33.95	42
9100	Fixture hanger		60	.133		10	7.30		17.30	22
9200	Pulling tool					89			89	97.50
9300	Non-metallic, straight section									
9310	7/16" x 7/8", base & cover, blank	1 Elec	160	.050	L.F.	1.72	2.74		4.46	6
9320	Base & cover w/adhesive		160	.050		1.42	2.74		4.16	5.65
9340	7/16" x 1-5/16", base & cover, blank		145	.055		1.88	3.02		4.90	6.60
9350	Base & cover w/adhesive		145	.055		2.17	3.02		5.19	6.90
9370	11/16" x 2-1/4", base & cover, blank		130	.062		2.66	3.37		6.03	8
9380	Base & cover w/adhesive		130	.062		3.01	3.37		6.38	8.35
9385	1-11/16" x 5-1/4", two compartment base & cover w/screws		80	.100		7.50	5.45		12.95	16.45
9400	Fittings, elbows, 7/16" x 7/8"		50	.160	Ea.	1.88	8.75		10.63	15.15
9410	7/16" x 1-5/16"		45	.178		1.95	9.70		11.65	16.70
9420	11/16" x 2-1/4"		40	.200		2.11	10.95		13.06	18.70
9425	1-11/16" x 5-1/4"		28	.286		11.05	15.65		26.70	35.50
9430	Tees, 7/16" x 7/8"		35	.229		2.42	12.50		14.92	21.50
9440	7/16" x 1-5/16"		32	.250		2.49	13.70		16.19	23
9450	11/16" x 2-1/4"		30	.267		2.56	14.60		17.16	25
9455	1-11/16" x 5-1/4"		24	.333		17.75	18.25		36	47
9460	Cover clip, 7/16" x 7/8"		80	.100		.49	5.45		5.94	8.75
9470	7/16" x 1-5/16"		72	.111		.44	6.10		6.54	9.60
9480	11/16" x 2-1/4"		64	.125		.75	6.85		7.60	11.10
9484	1-11/16" x 5-1/4"		42	.190		2.61	10.40		13.01	18.45
9486	Wire clip, 1-11/16" x 5-1/4"		68	.118		.45	6.45		6.90	10.15
9490	Blank end, 7/16" x 7/8"		50	.160		.70	8.75		9.45	13.85
9500	7/16" x 1-5/16"		45	.178		.78	9.70		10.48	15.40
9510	11/16" x 2-1/4"		40	.200		1.18	10.95		12.13	17.70
9515	1-11/16" x 5-1/4"		38	.211		5.45	11.50		16.95	23.50
9520	Round fixture box, 5.5" dia x 1"		25	.320		11.10	17.50		28.60	38
9530	Device box, 1 gang		30	.267		5	14.60		19.60	27.50
9540	2 gang		25	.320		7.30	17.50		24.80	34

26 27 26.10 Low Voltage Switching		Crew	Daily Output	Labor-Hours	Unit	2015 Bare Costs Material	Labor	Equipment	Total	Total Incl O&P
0010	**LOW VOLTAGE SWITCHING**									
3600	Relays, 120 V or 277 V standard	1 Elec	12	.667	Ea.	41.50	36.50		78	101
3800	Flush switch, standard		40	.200		11.50	10.95		22.45	29
4000	Interchangeable		40	.200		15.05	10.95		26	33
4100	Surface switch, standard		40	.200		8.15	10.95		19.10	25.50
4200	Transformer 115 V to 25 V		12	.667		130	36.50		166.50	198
4400	Master control, 12 circuit, manual		4	2		126	109		235	305
4500	25 circuit, motorized		4	2		140	109		249	320
4600	Rectifier, silicon		12	.667		45.50	36.50		82	105
4800	Switchplates, 1 gang, 1, 2 or 3 switch, plastic		80	.100		5	5.45		10.45	13.70
5000	Stainless steel		80	.100		11.35	5.45		16.80	20.50
5400	2 gang, 3 switch, stainless steel		53	.151		23	8.25		31.25	37.50
5500	4 switch, plastic		53	.151		10.35	8.25		18.60	23.50
5600	2 gang, 4 switch, stainless steel		53	.151		21.50	8.25		29.75	36
5700	6 switch, stainless steel		53	.151		43	8.25		51.25	60
5800	3 gang, 9 switch, stainless steel		32	.250		64.50	13.70		78.20	91.50
5900	Receptacle, triple, 1 return, 1 feed		26	.308		42.50	16.85		59.35	72
6000	2 feed		20	.400		42.50	22		64.50	80
6100	Relay gang boxes, flush or surface, 6 gang		5.30	1.509		89.50	82.50		172	223
6200	12 gang		4.70	1.702		101	93		194	250
6400	18 gang		4	2		112	109		221	287
6500	Frame, to hold up to 6 relays		12	.667		83	36.50		119.50	146
7200	Control wire, 2 conductor		6.30	1.270	C.L.F.	29.50	69.50		99	137
7400	3 conductor		5	1.600		41	87.50		128.50	176
7600	19 conductor		2.50	3.200		320	175		495	610
7800	26 conductor		2	4		430	219		649	805
8000	Weatherproof, 3 conductor		5	1.600		86	87.50		173.50	226

26 27 26.20 Wiring Devices Elements

Line	Description	Crew	Daily Output	Labor-Hours	Unit	Material	Labor	Equipment	Total	Total Incl O&P
0010	**WIRING DEVICES ELEMENTS** R262726-90									
0200	Toggle switch, quiet type, single pole, 15 amp	1 Elec	40	.200	Ea.	6.55	10.95		17.50	23.50
0500	20 amp		27	.296		9.85	16.20		26.05	35.50
0510	30 amp		23	.348		19.50	19.05		38.55	50
0530	Lock handle, 20 amp		27	.296		23.50	16.20		39.70	50
0540	Security key, 20 amp		26	.308		116	16.85		132.85	152
0550	Rocker, 15 amp		40	.200		4.10	10.95		15.05	21
0560	20 amp		27	.296		9.50	16.20		25.70	35
0600	3 way, 15 amp		23	.348		5.05	19.05		24.10	34
0800	20 amp		18	.444		9.45	24.50		33.95	47
0810	30 amp		9	.889		24	48.50		72.50	99.50
0830	Lock handle, 20 amp		18	.444		26	24.50		50.50	65.50
0840	Security key, 20 amp		17	.471		110	25.50		135.50	160
0850	Rocker, 15 amp		23	.348		5.75	19.05		24.80	35
0860	20 amp		18	.444		13.65	24.50		38.15	51.50
0900	4 way, 15 amp		15	.533		9.10	29		38.10	53.50
1000	20 amp		11	.727		42.50	40		82.50	107
1020	Lock handle, 20 amp		11	.727		46.50	40		86.50	111
1030	Rocker, 15 amp		15	.533		19.15	29		48.15	64.50
1040	20 amp		11	.727		39	40		79	103
1100	Toggle switch, quiet type, double pole, 15 amp		15	.533		12.30	29		41.30	57
1200	20 amp		11	.727		17.10	40		57.10	78.50
1210	30 amp		9	.889		24.50	48.50		73	100
1230	Lock handle, 20 amp		11	.727		18.45	40		58.45	80

26 27 26.20 Wiring Devices Elements		Crew	Daily Output	Labor-Hours	Unit	Material	Labor	Equipment	Total	Total Incl O&P
						2015 Bare Costs				
1250	Security key, 20 amp	1 Elec	10	.800	Ea.	113	44		157	190
1420	Toggle switch quiet type, 1 pole, 2 throw center off, 15 amp		23	.348		58	19.05		77.05	92
1440	20 amp		18	.444		51	24.50		75.50	92.50
1460	2 pole, 2 throw center off, lock handle, 20 amp		11	.727		60	40		100	126
1480	1 pole, momentary contact, 15 amp		23	.348		17.65	19.05		36.70	48
1500	20 amp		18	.444		29.50	24.50		54	69
1520	Momentary contact, lock handle, 20 amp		18	.444		31	24.50		55.50	70.50
1650	Dimmer switch, 120 volt, incandescent, 600 watt, 1 pole G		16	.500		21	27.50		48.50	64
1700	600 watt, 3 way G		12	.667		15.05	36.50		51.55	71
1750	1000 watt, 1 pole G		16	.500		38	27.50		65.50	82.50
1800	1000 watt, 3 way G		12	.667		73	36.50		109.50	135
2000	1500 watt, 1 pole G		11	.727		84	40		124	152
2100	2000 watt, 1 pole G		8	1		129	54.50		183.50	224
2110	Fluorescent, 600 watt G		15	.533		109	29		138	164
2120	1000 watt G		15	.533		133	29		162	190
2130	1500 watt G		10	.800		246	44		290	335
2160	Explosionproof, toggle switch, wall, single pole 20 amp		5.30	1.509		187	82.50		269.50	330
2180	Receptacle, single outlet, 20 amp		5.30	1.509		330	82.50		412.50	485
2190	30 amp		4	2		685	109		794	915
2290	60 amp		2.50	3.200		775	175		950	1,100
2360	Plug, 20 amp		16	.500		173	27.50		200.50	231
2370	30 amp		12	.667		294	36.50		330.50	380
2380	60 amp		8	1		380	54.50		434.50	500
2410	Furnace, thermal cutoff switch with plate		26	.308		15.75	16.85		32.60	42.50
2460	Receptacle, duplex, 120 volt, grounded, 15 amp		40	.200		1.26	10.95		12.21	17.80
2470	20 amp		27	.296		7.90	16.20		24.10	33
2480	Ground fault interrupting, 15 amp		27	.296		32	16.20		48.20	59.50
2482	20 amp		27	.296		40	16.20		56.20	68.50
2486	Clock receptacle, 15 amp		40	.200		27.50	10.95		38.45	46.50
2490	Dryer, 30 amp		15	.533		4.39	29		33.39	48.50
2500	Range, 50 amp		11	.727		12.15	40		52.15	73
2530	Surge suppresser receptacle, duplex, 20 amp		27	.296		41	16.20		57.20	69.50
2532	Quad, 20 amp		20	.400		79	22		101	120
2540	Isolated ground receptacle, duplex, 20 amp		27	.296		21	16.20		37.20	47.50
2542	Quad, 20 amp		20	.400		32	22		54	68
2550	Simplex, 20 amp		27	.296		23	16.20		39.20	49.50
2560	Simplex, 30 amp		15	.533		27.50	29		56.50	74
2570	Cable reel w/receptacle 50' w/3#12, 120 V, 20 A	2 Elec	2.67	5.999		920	330		1,250	1,500
2600	Wall plates, stainless steel, 1 gang	1 Elec	80	.100		2.56	5.45		8.01	11
2800	2 gang		53	.151		4.33	8.25		12.58	17.10
3000	3 gang		32	.250		8.45	13.70		22.15	30
3100	4 gang		27	.296		11	16.20		27.20	36.50
3110	Brown plastic, 1 gang		80	.100		.38	5.45		5.83	8.60
3120	2 gang		53	.151		.75	8.25		9	13.20
3130	3 gang		32	.250		1.11	13.70		14.81	21.50
3140	4 gang		27	.296		2.86	16.20		19.06	27.50
3150	Brushed brass, 1 gang		80	.100		4.26	5.45		9.71	12.90
3160	Anodized aluminum, 1 gang		80	.100		3.01	5.45		8.46	11.50
3170	Switch cover, weatherproof, 1 gang		60	.133		7	7.30		14.30	18.65
3180	Vandal proof lock, 1 gang		60	.133		14.05	7.30		21.35	26.50
3200	Lampholder, keyless		26	.308		12.65	16.85		29.50	39
3400	Pullchain with receptacle		22	.364		20.50	19.90		40.40	52.50
3500	Pilot light, neon with jewel		27	.296		9.55	16.20		25.75	35

26 27 26.20 Wiring Devices Elements		Crew	Daily Output	Labor-Hours	Unit	2015 Bare Costs Material	Labor	Equipment	Total	Total Incl O&P
3600	Receptacle, 20 amp, 250 volt, NEMA 6	1 Elec	27	.296	Ea.	21.50	16.20		37.70	48
3620	277 volt NEMA 7		27	.296		15.95	16.20		32.15	42
3640	125/250 volt NEMA 10		27	.296		17.95	16.20		34.15	44.50
3680	125/250 volt NEMA 14		25	.320		23	17.50		40.50	51
3700	3 pole, 250 volt NEMA 15		25	.320		23	17.50		40.50	51.50
3720	120/208 volt NEMA 18		25	.320		24	17.50		41.50	52
3740	30 amp, 125 volt NEMA 5		15	.533		17.80	29		46.80	63
3760	250 volt NEMA 6		15	.533		20.50	29		49.50	66
3780	277 volt NEMA 7		15	.533		26.50	29		55.50	73
3820	125/250 volt NEMA 14		14	.571		46.50	31.50		78	98.50
3840	3 pole, 250 volt NEMA 15		14	.571		46	31.50		77.50	97.50
3880	50 amp, 125 volt NEMA 5		11	.727		22.50	40		62.50	84.50
3900	250 volt NEMA 6		11	.727		23	40		63	85
3920	277 volt NEMA 7		11	.727		26.50	40		66.50	88.50
3960	125/250 volt NEMA 14		10	.800		61	44		105	133
3980	3 pole 250 volt NEMA 15		10	.800		62	44		106	134
4020	60 amp, 125/250 volt, NEMA 14		8	1		70.50	54.50		125	160
4040	3 pole, 250 volt NEMA 15		8	1		74.50	54.50		129	164
4060	120/208 volt NEMA 18		8	1		102	54.50		156.50	194
4100	Receptacle locking, 20 amp, 125 volt NEMA L5		27	.296		16.60	16.20		32.80	43
4120	250 volt NEMA L6		27	.296		16.50	16.20		32.70	42.50
4140	277 volt NEMA L7		27	.296		17.15	16.20		33.35	43.50
4150	3 pole, 250 volt, NEMA L11		27	.296		23	16.20		39.20	50
4160	20 amp, 480 volt NEMA L8		27	.296		19.35	16.20		35.55	46
4180	600 volt NEMA L9		27	.296		29	16.20		45.20	56.50
4200	125/250 volt NEMA L10		27	.296		23	16.20		39.20	50
4230	125/250 volt NEMA L14		25	.320		27	17.50		44.50	55.50
4280	250 volt NEMA L15		25	.320		29	17.50		46.50	58
4300	480 volt NEMA L16		25	.320		23.50	17.50		41	51.50
4320	3 phase, 120/208 volt NEMA L18		25	.320		27	17.50		44.50	56
4340	277/480 volt NEMA L19		25	.320		28	17.50		45.50	57
4360	347/600 volt NEMA L20		25	.320		27.50	17.50		45	56.50
4380	120/208 volt NEMA L21		23	.348		29.50	19.05		48.55	61
4400	277/480 volt NEMA L22		23	.348		30	19.05		49.05	61.50
4420	347/600 volt NEMA L23		23	.348		32	19.05		51.05	63.50
4440	30 amp, 125 volt NEMA L5		15	.533		23	29		52	69
4460	250 volt NEMA L6		15	.533		25.50	29		54.50	71.50
4480	277 volt NEMA L7		15	.533		25.50	29		54.50	72
4500	480 volt NEMA L8		15	.533		26.50	29		55.50	73
4520	600 volt NEMA L9		15	.533		27.50	29		56.50	73.50
4540	125/250 volt NEMA L10		15	.533		30	29		59	76.50
4560	3 phase, 250 volt NEMA L11		15	.533		27	29		56	73.50
4620	125/250 volt NEMA L14		14	.571		36.50	31.50		68	87
4640	250 volt NEMA L15		14	.571		37	31.50		68.50	88
4660	480 volt NEMA L16		14	.571		38.50	31.50		70	89
4680	600 volt NEMA L17		14	.571		38	31.50		69.50	88.50
4700	120/208 volt NEMA L18		14	.571		41.50	31.50		73	92.50
4720	277/480 volt NEMA L19		14	.571		42	31.50		73.50	93
4740	347/600 volt NEMA L20		14	.571		43.50	31.50		75	94.50
4760	120/208 volt NEMA L21		13	.615		40	33.50		73.50	94.50
4780	277/480 volt NEMA L22		13	.615		40	33.50		73.50	94.50
4800	347/600 volt NEMA L23		13	.615		53	33.50		86.50	109
4840	Receptacle, corrosion resistant, 15 or 20 amp, 125 volt NEMA L5		27	.296		25.50	16.20		41.70	52.50

26 27 26.20 Wiring Devices Elements		Crew	Daily Output	Labor-Hours	Unit	Material	Labor	Equipment	Total	Total Incl O&P
						2015 Bare Costs				
4860	250 volt NEMA L6	1 Elec	27	.296	Ea.	21.50	16.20		37.70	48
4900	Receptacle, cover plate, phenolic plastic, NEMA 5 & 6		80	.100		.61	5.45		6.06	8.85
4910	NEMA 7-23		80	.100		.75	5.45		6.20	9.05
4920	Stainless steel, NEMA 5 & 6		80	.100		2.71	5.45		8.16	11.20
4930	NEMA 7-23		80	.100		2.59	5.45		8.04	11.05
4940	Brushed brass NEMA 5 & 6		80	.100		5.40	5.45		10.85	14.15
4950	NEMA 7-23		80	.100		5.85	5.45		11.30	14.65
4960	Anodized aluminum, NEMA 5 & 6		80	.100		3.51	5.45		8.96	12.05
4970	NEMA 7-23		80	.100		5.80	5.45		11.25	14.60
4980	Weatherproof NEMA 7-23		60	.133		33	7.30		40.30	47
5100	Plug, 20 amp, 250 volt, NEMA 6		30	.267		15.65	14.60		30.25	39
5110	277 volt NEMA 7		30	.267		19.20	14.60		33.80	43
5120	3 pole, 120/250 volt, NEMA 10		26	.308		18.15	16.85		35	45
5130	125/250 volt NEMA 14		26	.308		36.50	16.85		53.35	65.50
5140	250 volt NEMA 15		26	.308		41	16.85		57.85	70
5150	120/208 volt NEMA 8		26	.308		43.50	16.85		60.35	73
5160	30 amp, 125 volt NEMA 5		13	.615		55	33.50		88.50	111
5170	250 volt NEMA 6		13	.615		52	33.50		85.50	108
5180	277 volt NEMA 7		13	.615		60.50	33.50		94	117
5190	125/250 volt NEMA 14		13	.615		42.50	33.50		76	97
5200	3 pole, 250 volt NEMA 15		12	.667		44	36.50		80.50	103
5210	50 amp, 125 volt NEMA 5		9	.889		61.50	48.50		110	141
5220	250 volt NEMA 6		9	.889		64.50	48.50		113	144
5230	277 volt NEMA 7		9	.889		69	48.50		117.50	149
5240	125/250 volt NEMA 14		9	.889		54	48.50		102.50	133
5250	3 pole, 250 volt NEMA 15		8	1		55	54.50		109.50	143
5260	60 amp, 125/250 volt NEMA 14		7	1.143		61.50	62.50		124	162
5270	3 pole, 250 volt NEMA 15		7	1.143		65	62.50		127.50	165
5280	120/208 volt NEMA 18		7	1.143		82.50	62.50		145	185
5300	Plug angle, 20 amp, 250 volt NEMA 6		30	.267		24	14.60		38.60	48.50
5310	30 amp, 125 volt NEMA 5		13	.615		43	33.50		76.50	98
5320	250 volt NEMA 6		13	.615		44.50	33.50		78	99.50
5330	277 volt NEMA 7		13	.615		51	33.50		84.50	107
5340	125/250 volt NEMA 14		13	.615		47	33.50		80.50	102
5350	3 pole, 250 volt NEMA 15		12	.667		48.50	36.50		85	108
5360	50 amp, 125 volt NEMA 5		9	.889		45	48.50		93.50	123
5370	250 volt NEMA 6		9	.889		46	48.50		94.50	124
5380	277 volt NEMA 7		9	.889		52	48.50		100.50	130
5390	125/250 volt NEMA 14		9	.889		58.50	48.50		107	137
5400	3 pole, 250 volt NEMA 15		8	1		60.50	54.50		115	149
5410	60 amp, 125/250 volt NEMA 14		7	1.143		68	62.50		130.50	169
5420	3 pole, 250 volt NEMA 15		7	1.143		71	62.50		133.50	172
5430	120/208 volt NEMA 18		7	1.143		71.50	62.50		134	172
5500	Plug, locking, 20 amp, 125 volt NEMA L5		30	.267		13.20	14.60		27.80	36.50
5510	250 volt NEMA L6		30	.267		13.15	14.60		27.75	36.50
5520	277 volt NEMA L7		30	.267		13.45	14.60		28.05	37
5530	480 volt NEMA L8		30	.267		14.95	14.60		29.55	38.50
5540	600 volt NEMA L9		30	.267		15	14.60		29.60	38.50
5550	3 pole, 125/250 volt NEMA L10		26	.308		22	16.85		38.85	49
5560	250 volt NEMA L11		26	.308		22	16.85		38.85	49
5570	480 volt NEMA L12		26	.308		25.50	16.85		42.35	53
5580	125/250 volt NEMA L14		26	.308		19.90	16.85		36.75	47
5590	250 volt NEMA L15		26	.308		21	16.85		37.85	48

26 27 26.20 Wiring Devices Elements		Crew	Daily Output	Labor-Hours	Unit	2015 Bare Costs Material	Labor	Equipment	Total	Total Incl O&P
5600	480 volt NEMA L16	1 Elec	26	.308	Ea.	22.50	16.85		39.35	49.50
5610	4 pole, 120/208 volt NEMA L18		24	.333		25	18.25		43.25	55
5620	277/480 volt NEMA L19		24	.333		26	18.25		44.25	56
5630	347/600 volt NEMA L20		24	.333		29	18.25		47.25	59
5640	120/208 volt NEMA L21		24	.333		25.50	18.25		43.75	55.50
5650	277/480 volt NEMA L22		24	.333		27	18.25		45.25	57
5660	347/600 volt NEMA L23		24	.333		28	18.25		46.25	58
5670	30 amp, 125 volt NEMA L5		13	.615		20	33.50		53.50	72.50
5680	250 volt NEMA L6		13	.615		21	33.50		54.50	73.50
5690	277 volt NEMA L7		13	.615		20	33.50		53.50	72.50
5700	480 volt NEMA L8		13	.615		22	33.50		55.50	74.50
5710	600 volt NEMA L9		13	.615		21	33.50		54.50	73.50
5720	3 pole, 125/250 volt NEMA L10		11	.727		24	40		64	85.50
5730	250 volt NEMA L11		11	.727		24	40		64	85.50
5760	125/250 volt NEMA L14		11	.727		27	40		67	89.50
5770	250 volt NEMA L15		11	.727		27	40		67	89.50
5780	480 volt NEMA L16		11	.727		28.50	40		68.50	90.50
5790	600 volt NEMA L17		11	.727		29.50	40		69.50	92
5800	4 pole, 120/208 volt NEMA L18		10	.800		31	44		75	100
5810	120/208 volt NEMA L19		10	.800		31.50	44		75.50	100
5820	347/600 volt NEMA L20		10	.800		32.50	44		76.50	101
5830	120/208 volt NEMA L21		10	.800		30.50	44		74.50	99.50
5840	277/480 volt NEMA L22		10	.800		31	44		75	99.50
5850	347/600 volt NEMA L23		10	.800		35	44		79	104
6000	Connector, 20 amp, 250 volt NEMA 6		30	.267		24	14.60		38.60	48.50
6010	277 volt NEMA 7		30	.267		30.50	14.60		45.10	56
6020	3 pole, 120/250 volt NEMA 10		26	.308		34.50	16.85		51.35	63
6030	125/250 volt NEMA 14		26	.308		34.50	16.85		51.35	63
6040	250 volt NEMA 15		26	.308		35	16.85		51.85	63.50
6050	120/208 volt NEMA 18		26	.308		38	16.85		54.85	67
6060	30 amp, 125 volt NEMA 5		13	.615		60	33.50		93.50	117
6070	250 volt NEMA 6		13	.615		60	33.50		93.50	117
6080	277 volt NEMA 7		13	.615		60	33.50		93.50	117
6110	50 amp, 125 volt NEMA 5		9	.889		83	48.50		131.50	164
6120	250 volt NEMA 6		9	.889		83	48.50		131.50	164
6130	277 volt NEMA 7		9	.889		83	48.50		131.50	164
6200	Connector, locking, 20 amp, 125 volt NEMA L5		30	.267		20	14.60		34.60	44
6210	250 volt NEMA L6		30	.267		20.50	14.60		35.10	44.50
6220	277 volt NEMA L7		30	.267		20.50	14.60		35.10	44.50
6230	480 volt NEMA L8		30	.267		24	14.60		38.60	48
6240	600 volt NEMA L9		30	.267		24.50	14.60		39.10	48.50
6250	3 pole, 125/250 volt NEMA L10		26	.308		33.50	16.85		50.35	62
6260	250 volt NEMA L11		26	.308		33.50	16.85		50.35	62
6280	125/250 volt NEMA L14		26	.308		27.50	16.85		44.35	55.50
6290	250 volt NEMA L15		26	.308		28	16.85		44.85	55.50
6300	480 volt NEMA L16		26	.308		29.50	16.85		46.35	57.50
6310	4 pole, 120/208 volt NEMA L18		24	.333		35.50	18.25		53.75	66.50
6320	277/480 volt NEMA L19		24	.333		40.50	18.25		58.75	72
6330	347/600 volt NEMA L20		24	.333		40	18.25		58.25	71.50
6340	120/208 volt NEMA L21		24	.333		43	18.25		61.25	75
6350	277/480 volt NEMA L22		24	.333		45.50	18.25		63.75	77.50
6360	347/600 volt NEMA L23		24	.333		53	18.25		71.25	86
6370	30 amp, 125 volt NEMA L5		13	.615		41	33.50		74.50	95.50

26 27 26 – Wiring Devices

26 27 26.20 Wiring Devices Elements		Crew	Daily Output	Labor-Hours	Unit	Material	Labor	Equipment	Total	Total Incl O&P
						2015 Bare Costs				
6380	250 volt NEMA L6	1 Elec	13	.615	Ea.	41.50	33.50		75	96
6390	277 volt NEMA L7		13	.615		41.50	33.50		75	96.50
6400	480 volt NEMA L8		13	.615		44	33.50		77.50	99
6410	600 volt NEMA L9		13	.615		43	33.50		76.50	97.50
6420	3 pole, 125/250 volt NEMA L10		11	.727		64.50	40		104.50	131
6430	250 volt NEMA L11		11	.727		64.50	40		104.50	131
6460	125/250 volt NEMA L14		11	.727		56	40		96	121
6470	250 volt NEMA L15		11	.727		56.50	40		96.50	122
6480	480 volt NEMA L16		11	.727		59.50	40		99.50	125
6490	600 volt NEMA L17		11	.727		61.50	40		101.50	127
6500	4 pole, 120/208 volt NEMA L18		10	.800		63	44		107	135
6510	120/208 volt NEMA L19		10	.800		65.50	44		109.50	138
6520	347/600 volt NEMA L20		10	.800		67.50	44		111.50	140
6530	120/208 volt NEMA L21		10	.800		57	44		101	128
6540	277/480 volt NEMA L22		10	.800		58	44		102	129
6550	347/600 volt NEMA L23		10	.800		66	44		110	138
7000	Receptacle computer, 250 volt, 15 amp, 3 pole 4 wire		8	1		82.50	54.50		137	173
7010	20 amp, 2 pole 3 wire		8	1		84	54.50		138.50	175
7020	30 amp, 2 pole 3 wire		6.50	1.231		137	67.50		204.50	251
7030	30 amp, 3 pole 4 wire		6.50	1.231		148	67.50		215.50	263
7040	60 amp, 3 pole 4 wire		4.50	1.778		249	97		346	420
7050	100 amp, 3 pole 4 wire		3	2.667		310	146		456	560
7100	Connector computer, 250 volt, 15 amp, 3 pole 4 wire		27	.296		130	16.20		146.20	168
7110	20 amp, 2 pole 3 wire		27	.296		118	16.20		134.20	154
7120	30 amp, 2 pole 3 wire		15	.533		181	29		210	243
7130	30 amp, 3 pole 4 wire		15	.533		185	29		214	248
7140	60 amp, 3 pole 4 wire		8	1		310	54.50		364.50	420
7150	100 amp, 3 pole 4 wire		4	2		425	109		534	635
7200	Plug, computer, 250 volt, 15 amp, 3 pole 4 wire		27	.296		116	16.20		132.20	153
7210	20 amp, 2 pole, 3 wire		27	.296		109	16.20		125.20	145
7220	30 amp, 2 pole, 3 wire		15	.533		190	29		219	253
7230	30 amp, 3 pole, 4 wire		15	.533		184	29		213	246
7240	60 amp, 3 pole, 4 wire		8	1		276	54.50		330.50	385
7250	100 amp, 3 pole, 4 wire		4	2		345	109		454	545
7300	Connector adapter to flexible conduit, 1/2"		60	.133		2.90	7.30		10.20	14.15
7310	3/4"		50	.160		4.24	8.75		12.99	17.75
7320	1-1/4"		30	.267		12.70	14.60		27.30	36
7330	1-1/2"		23	.348		18.15	19.05		37.20	48.50

26 27 73 – Door Chimes

26 27 73.10 Doorbell System		Crew	Daily Output	Labor-Hours	Unit	Material	Labor	Equipment	Total	Total Incl O&P
0010	**DOORBELL SYSTEM**, incl. transformer, button & signal									
0100	6" bell	1 Elec	4	2	Ea.	129	109		238	305
0200	Buzzer		4	2		106	109		215	281
1000	Door chimes, 2 notes		16	.500		22.50	27.50		50	66
1020	with ambient light		12	.667		109	36.50		145.50	175
1100	Tube type, 3 tube system		12	.667		187	36.50		223.50	261
1180	4 tube system		10	.800		385	44		429	485
1900	For transformer & button, add		5	1.600		15.85	87.50		103.35	148
3000	For push button only		24	.333		2.58	18.25		20.83	30.50
3200	Bell transformer		16	.500		18.05	27.50		45.55	61

26 28 13.10 Fuse Elements		Crew	Daily Output	Labor-Hours	Unit	Material	2015 Bare Costs Labor	Equipment	Total	Total Incl O&P
0010	FUSE ELEMENTS									
0020	Cartridge, nonrenewable									
0050	250 volt, 30 amp	1 Elec	50	.160	Ea.	2.40	8.75		11.15	15.75
0100	60 amp		50	.160		3.68	8.75		12.43	17.15
0150	100 amp		40	.200		16.10	10.95		27.05	34
0200	200 amp		36	.222		37	12.15		49.15	58.50
0250	400 amp		30	.267		74	14.60		88.60	104
0300	600 amp		24	.333		120	18.25		138.25	160
0400	600 volt, 30 amp		40	.200		8.65	10.95		19.60	26
0450	60 amp		40	.200		14.15	10.95		25.10	32
0500	100 amp		36	.222		28.50	12.15		40.65	49
0550	200 amp		30	.267		57.50	14.60		72.10	85.50
0600	400 amp		24	.333		140	18.25		158.25	182
0650	600 amp		20	.400		203	22		225	257
0800	Dual element, time delay, 250 volt, 30 amp		50	.160		9.45	8.75		18.20	23.50
0840	50 amp		50	.160		10.35	8.75		19.10	24.50
0850	60 amp		50	.160		11.60	8.75		20.35	26
0900	100 amp		40	.200		28	10.95		38.95	47.50
0950	200 amp		36	.222		61.50	12.15		73.65	85.50
1000	400 amp		30	.267		110	14.60		124.60	143
1050	600 amp		24	.333		181	18.25		199.25	227
1300	600 volt, 15 to 30 amp		40	.200		18.75	10.95		29.70	37
1350	35 to 60 amp		40	.200		31.50	10.95		42.45	51
1400	70 to 100 amp		36	.222		66.50	12.15		78.65	91
1450	110 to 200 amp		30	.267		129	14.60		143.60	164
1500	225 to 400 amp		24	.333		267	18.25		285.25	320
1550	600 amp		20	.400		355	22		377	430
1800	Class RK1, high capacity, 250 volt, 30 amp		50	.160		8	8.75		16.75	22
1850	60 amp		50	.160		14.75	8.75		23.50	29.50
1900	100 amp		40	.200		34.50	10.95		45.45	54.50
1950	200 amp		36	.222		78.50	12.15		90.65	105
2000	400 amp		30	.267		130	14.60		144.60	165
2050	600 amp		24	.333		182	18.25		200.25	228
2200	600 volt, 30 amp		40	.200		14.90	10.95		25.85	33
2250	60 amp		40	.200		28	10.95		38.95	47.50
2300	100 amp		36	.222		71	12.15		83.15	96.50
2350	200 amp		30	.267		137	14.60		151.60	173
2400	400 amp		24	.333		230	18.25		248.25	281
2450	600 amp		20	.400		263	22		285	320
2700	Class J, current limiting, 250 or 600 volt, 30 amp		40	.200		24.50	10.95		35.45	43.50
2750	60 amp		40	.200		40.50	10.95		51.45	61
2800	100 amp		36	.222		56	12.15		68.15	79.50
2850	200 amp		30	.267		111	14.60		125.60	144
2900	400 amp		24	.333		271	18.25		289.25	325
2950	600 amp		20	.400		405	22		427	480
3100	Class L, current limiting, 250 or 600 volt, 601 to 1200 amp		16	.500		630	27.50		657.50	730
3150	1500-1600 amp		13	.615		900	33.50		933.50	1,050
3200	1800-2000 amp		10	.800		1,025	44		1,069	1,200
3250	2500 amp		10	.800		1,225	44		1,269	1,425
3300	3000 amp		8	1		1,650	54.50		1,704.50	1,875
3350	3500-4000 amp		8	1		1,725	54.50		1,779.50	1,975
3400	4500-5000 amp		6.70	1.194		2,650	65.50		2,715.50	3,025
3450	6000 amp	↓	5.70	1.404	↓	3,550	77		3,627	4,050

26 28 Low-Voltage Circuit Protective Devices

26 28 13 – Fuses

26 28 13.10 Fuse Elements		Crew	Daily Output	Labor-Hours	Unit	Material	Labor	Equipment	Total	Total Incl O&P
						2015 Bare Costs				
3600	Plug, 120 volt, 1 to 10 amp	1 Elec	50	.160	Ea.	3.13	8.75		11.88	16.55
3650	15 to 30 amp		50	.160		4.84	8.75		13.59	18.40
3700	Dual element 0.3 to 14 amp		50	.160		4.59	8.75		13.34	18.15
3750	15 to 30 amp		50	.160		6.65	8.75		15.40	20.50
3800	Fustat, 120 volt, 15 to 30 amp		50	.160		5.75	8.75		14.50	19.45
3850	0.3 to 14 amp		50	.160		7.10	8.75		15.85	21
3900	Adapters 0.3 to 10 amp		50	.160		5.65	8.75		14.40	19.30
3950	15 to 30 amp		50	.160		8.65	8.75		17.40	22.50

26 28 16 – Enclosed Switches and Circuit Breakers

26 28 16.10 Circuit Breakers

		Crew	Daily Output	Labor-Hours	Unit	Material	Labor	Equipment	Total	Total Incl O&P
0010	**CIRCUIT BREAKERS** (in enclosure)									
0100	Enclosed (NEMA 1), 600 volt, 3 pole, 30 amp	1 Elec	3.20	2.500	Ea.	500	137		637	755
0200	60 amp		2.80	2.857		615	156		771	915
0400	100 amp		2.30	3.478		705	190		895	1,050
0500	200 amp		1.50	5.333		1,475	292		1,767	2,050
0600	225 amp		1.50	5.333		1,625	292		1,917	2,225
0700	400 amp	2 Elec	1.60	10		2,775	545		3,320	3,900
0800	600 amp		1.20	13.333		4,025	730		4,755	5,550
1000	800 amp		.94	17.021		5,250	930		6,180	7,175
1200	1000 amp		.84	19.048		6,625	1,050		7,675	8,850
1220	1200 amp		.80	20		8,500	1,100		9,600	11,000
1240	1600 amp		.72	22.222		15,600	1,225		16,825	19,000
1260	2000 amp		.64	25		16,900	1,375		18,275	20,700
1400	1200 amp with ground fault		.80	20		13,500	1,100		14,600	16,600
1600	1600 amp with ground fault		.72	22.222		17,700	1,225		18,925	21,300
1800	2000 amp with ground fault		.64	25		19,000	1,375		20,375	23,000
2000	Disconnect, 240 volt 3 pole, 5 HP motor	1 Elec	3.20	2.500		430	137		567	680
2020	10 HP motor		3.20	2.500		430	137		567	680
2040	15 HP motor		2.80	2.857		430	156		586	710
2060	20 HP motor		2.30	3.478		525	190		715	860
2080	25 HP motor		2.30	3.478		525	190		715	860
2100	30 HP motor		2.30	3.478		525	190		715	860
2120	40 HP motor		2	4		900	219		1,119	1,325
2140	50 HP motor		1.50	5.333		900	292		1,192	1,425
2160	60 HP motor		1.50	5.333		2,075	292		2,367	2,700
2180	75 HP motor	2 Elec	2	8		2,075	440		2,515	2,925
2200	100 HP motor		1.60	10		2,075	545		2,620	3,100
2220	125 HP motor		1.60	10		2,075	545		2,620	3,100
2240	150 HP motor		1.20	13.333		4,025	730		4,755	5,550
2260	200 HP motor		1.20	13.333		5,250	730		5,980	6,875
2300	Enclosed (NEMA 7), explosion proof, 600 volt 3 pole, 50 amp	1 Elec	2.30	3.478		1,525	190		1,715	1,950
2350	100 amp		1.50	5.333		1,575	292		1,867	2,175
2400	150 amp		1	8		3,825	440		4,265	4,850
2450	250 amp	2 Elec	1.60	10		4,775	545		5,320	6,075
2500	400 amp	"	1.20	13.333		5,300	730		6,030	6,925

26 28 16.20 Safety Switches

		Crew	Daily Output	Labor-Hours	Unit	Material	Labor	Equipment	Total	Total Incl O&P
0010	**SAFETY SWITCHES** R262816-80									
0100	General duty 240 volt, 3 pole NEMA 1, fusible, 30 amp	1 Elec	3.20	2.500	Ea.	73.50	137		210.50	286
0200	60 amp		2.30	3.478		124	190		314	420
0300	100 amp		1.90	4.211		213	230		443	580
0400	200 amp		1.30	6.154		455	335		790	1,000
0500	400 amp	2 Elec	1.80	8.889		1,150	485		1,635	2,000

26 28 Low-Voltage Circuit Protective Devices

26 28 16 – Enclosed Switches and Circuit Breakers

26 28 16.20 Safety Switches		Crew	Daily Output	Labor-Hours	Unit	2015 Bare Costs Material	Labor	Equipment	Total	Total Incl O&P
0600	600 amp	2 Elec	1.20	13.333	Ea.	2,150	730		2,880	3,475
0610	Nonfusible, 30 amp	1 Elec	3.20	2.500		59	137		196	270
0650	60 amp		2.30	3.478		78.50	190		268.50	370
0700	100 amp		1.90	4.211		183	230		413	545
0750	200 amp		1.30	6.154		335	335		670	875
0800	400 amp	2 Elec	1.80	8.889		810	485		1,295	1,625
0850	600 amp	"	1.20	13.333		1,575	730		2,305	2,825
1100	Heavy duty, 600 volt, 3 pole NEMA 1 nonfused									
1110	30 amp	1 Elec	3.20	2.500	Ea.	113	137		250	330
1500	60 amp		2.30	3.478		187	190		377	490
1700	100 amp		1.90	4.211		297	230		527	670
1900	200 amp		1.30	6.154		450	335		785	1,000
2100	400 amp	2 Elec	1.80	8.889		1,000	485		1,485	1,825
2300	600 amp		1.20	13.333		1,800	730		2,530	3,075
2500	800 amp		.94	17.021		3,650	930		4,580	5,425
2700	1200 amp		.80	20		4,900	1,100		6,000	7,050
2900	Heavy duty, 240 volt, 3 pole NEMA 1 fusible									
2910	30 amp	1 Elec	3.20	2.500	Ea.	118	137		255	335
3000	60 amp		2.30	3.478		199	190		389	505
3300	100 amp		1.90	4.211		315	230		545	690
3500	200 amp		1.30	6.154		540	335		875	1,100
3700	400 amp	2 Elec	1.80	8.889		1,400	485		1,885	2,250
3900	600 amp		1.20	13.333		2,800	730		3,530	4,175
4100	800 amp		.94	17.021		5,600	930		6,530	7,550
4300	1200 amp		.80	20		7,325	1,100		8,425	9,700
4340	2 pole fusible, 30 amp	1 Elec	3.50	2.286		88.50	125		213.50	285
4350	600 volt, 3 pole, fusible, 30 amp		3.20	2.500		201	137		338	425
4380	60 amp		2.30	3.478		243	190		433	550
4400	100 amp		1.90	4.211		445	230		675	835
4420	200 amp		1.30	6.154		640	335		975	1,200
4440	400 amp	2 Elec	1.80	8.889		1,675	485		2,160	2,550
4450	600 amp		1.20	13.333		2,800	730		3,530	4,175
4460	800 amp		.94	17.021		5,600	930		6,530	7,550
4480	1200 amp		.80	20		7,325	1,100		8,425	9,700
4500	240 volt 3 pole NEMA 3R (no hubs), fusible									
4510	30 amp	1 Elec	3.10	2.581	Ea.	210	141		351	440
4700	60 amp		2.20	3.636		330	199		529	665
4900	100 amp		1.80	4.444		480	243		723	895
5100	200 amp		1.20	6.667		660	365		1,025	1,275
5300	400 amp	2 Elec	1.60	10		1,450	545		1,995	2,425
5500	600 amp	"	1	16		3,025	875		3,900	4,650
5510	Heavy duty, 600 volt, 3 pole 3ph. NEMA 3R fusible, 30 amp	1 Elec	3.10	2.581		335	141		476	580
5520	60 amp		2.20	3.636		390	199		589	730
5530	100 amp		1.80	4.444		615	243		858	1,050
5540	200 amp		1.20	6.667		845	365		1,210	1,475
5550	400 amp	2 Elec	1.60	10		2,000	545		2,545	3,025
5700	600 volt, 3 pole NEMA 3R nonfused									
5710	30 amp	1 Elec	3.10	2.581	Ea.	187	141		328	415
5900	60 amp		2.20	3.636		330	199		529	660
6100	100 amp		1.80	4.444		455	243		698	865
6300	200 amp		1.20	6.667		555	365		920	1,150
6500	400 amp	2 Elec	1.60	10		1,375	545		1,920	2,350
6700	600 amp	"	1	16		2,775	875		3,650	4,350

26 28 16.20 Safety Switches		Crew	Daily Output	Labor-Hours	Unit	Material	2015 Bare Costs Labor	Equipment	Total	Total Incl O&P
6900	600 volt, 6 pole NEMA 3R nonfused, 30 amp	1 Elec	2.70	2.963	Ea.	1,200	162		1,362	1,575
7100	60 amp		2	4		1,375	219		1,594	1,850
7300	100 amp		1.50	5.333		1,700	292		1,992	2,300
7500	200 amp	↓	1.20	6.667	↓	6,550	365		6,915	7,750
7600	600 volt, 3 pole NEMA 7 explosion proof nonfused									
7610	30 amp	1 Elec	2.20	3.636	Ea.	1,325	199		1,524	1,750
7620	60 amp		1.80	4.444		1,575	243		1,818	2,125
7630	100 amp		1.20	6.667		1,925	365		2,290	2,650
7640	200 amp		.80	10		3,975	545		4,520	5,200
7710	600 volt 6 pole, NEMA 3R fusible, 30 amp		2.70	2.963		1,525	162		1,687	1,925
7900	60 amp		2	4		1,600	219		1,819	2,075
8100	100 amp		1.50	5.333		2,250	292		2,542	2,900
8110	240 volt 3 pole, NEMA 12 fusible, 30 amp		3.10	2.581		260	141		401	495
8120	60 amp		2.20	3.636		415	199		614	760
8130	100 amp		1.80	4.444		510	243		753	925
8140	200 amp	↓	1.20	6.667		735	365		1,100	1,350
8150	400 amp	2 Elec	1.60	10		1,675	545		2,220	2,675
8160	600 amp	"	1	16		3,000	875		3,875	4,600
8180	600 volt 3 pole, NEMA 12 fusible, 30 amp	1 Elec	3.10	2.581		340	141		481	585
8190	60 amp		2.20	3.636		405	199		604	745
8200	100 amp		1.80	4.444		655	243		898	1,075
8210	200 amp	↓	1.20	6.667		995	365		1,360	1,650
8220	400 amp	2 Elec	1.60	10		2,025	545		2,570	3,050
8230	600 amp	"	1	16		4,525	875		5,400	6,300
8240	600 volt 3 pole, NEMA 12 nonfused, 30 amp	1 Elec	3.10	2.581		240	141		381	475
8250	60 amp		2.20	3.636		310	199		509	640
8260	100 amp		1.80	4.444		440	243		683	850
8270	200 amp	↓	1.20	6.667		585	365		950	1,175
8280	400 amp	2 Elec	1.60	10		1,450	545		1,995	2,425
8290	600 amp	"	1	16		3,475	875		4,350	5,125
8310	600 volt, 3 pole NEMA 4 fusible, 30 amp	1 Elec	3	2.667		935	146		1,081	1,250
8320	60 amp		2.20	3.636		1,050	199		1,249	1,450
8330	100 amp		1.80	4.444		2,300	243		2,543	2,900
8340	200 amp	↓	1.20	6.667		2,900	365		3,265	3,725
8350	400 amp	2 Elec	1.60	10		5,750	545		6,295	7,150
8360	600 volt 3 pole NEMA 4 nonfused, 30 amp	1 Elec	3	2.667		770	146		916	1,075
8370	60 amp		2.20	3.636		920	199		1,119	1,300
8380	100 amp		1.80	4.444		1,875	243		2,118	2,450
8390	200 amp	↓	1.20	6.667		2,550	365		2,915	3,350
8400	400 amp	2 Elec	1.60	10		5,175	545		5,720	6,525
8490	Motor starters, manual, single phase, NEMA 1	1 Elec	6.40	1.250		65	68.50		133.50	174
8500	NEMA 4		4	2		182	109		291	365
8700	NEMA 7		4	2		200	109		309	385
8900	NEMA 1 with pilot		6.40	1.250		90	68.50		158.50	201
8920	3 pole, NEMA 1, 230/460 volt, 5 HP, size 0		3.50	2.286		207	125		332	415
8940	10 HP, size 1		2	4		245	219		464	600
9010	Disc. switch, 600 V 3 pole fusible, 30 amp, to 10 HP motor		3.20	2.500		355	137		492	595
9050	60 amp, to 30 HP motor		2.30	3.478		815	190		1,005	1,175
9070	100 amp, to 60 HP motor		1.90	4.211		815	230		1,045	1,250
9100	200 amp, to 125 HP motor	↓	1.30	6.154		1,225	335		1,560	1,850
9110	400 amp, to 200 HP motor	2 Elec	1.80	8.889	↓	3,075	485		3,560	4,125

26 28 16.40 Time Switches		Crew	Daily Output	Labor-Hours	Unit	Material	2015 Bare Costs Labor	Equipment	Total	Total Incl O&P
0010	**TIME SWITCHES**									
0100	Single pole, single throw, 24 hour dial	1 Elec	4	2	Ea.	145	109		254	325
0200	24 hour dial with reserve power		3.60	2.222		615	122		737	855
0300	Astronomic dial		3.60	2.222		209	122		331	410
0400	Astronomic dial with reserve power		3.30	2.424		665	133		798	930
0500	7 day calendar dial		3.30	2.424		150	133		283	365
0600	7 day calendar dial with reserve power		3.20	2.500		335	137		472	575
0700	Photo cell 2000 watt		8	1		28.50	54.50		83	114
1080	Load management device, 4 loads		2	4		995	219		1,214	1,425
1100	8 loads		1	8		2,325	440		2,765	3,200

26 29 Low-Voltage Controllers

26 29 13 – Enclosed Controllers

26 29 13.10 Contactors, AC		Crew	Daily Output	Labor-Hours	Unit	Material	2015 Bare Costs Labor	Equipment	Total	Total Incl O&P
0010	**CONTACTORS, AC** Enclosed (NEMA 1)									
0050	Lighting, 600 volt 3 pole, electrically held									
0100	20 amp	1 Elec	4	2	Ea.	315	109		424	510
0200	30 amp		3.60	2.222		360	122		482	575
0300	60 amp		3	2.667		675	146		821	960
0400	100 amp		2.50	3.200		1,000	175		1,175	1,375
0500	200 amp		1.40	5.714		2,525	315		2,840	3,250
0600	300 amp	2 Elec	1.60	10		6,775	545		7,320	8,275
0800	600 volt 3 pole, mechanically held, 30 amp	1 Elec	3.60	2.222		445	122		567	670
0900	60 amp		3	2.667		880	146		1,026	1,200
1000	75 amp		2.80	2.857		1,225	156		1,381	1,575
1100	100 amp		2.50	3.200		1,250	175		1,425	1,625
1200	150 amp		2	4		3,350	219		3,569	4,000
1300	200 amp		1.40	5.714		3,425	315		3,740	4,250
1500	Magnetic with auxiliary contact, size 00, 9 amp		4	2		187	109		296	370
1600	Size 0, 18 amp		4	2		223	109		332	410
1700	Size 1, 27 amp		3.60	2.222		253	122		375	460
1800	Size 2, 45 amp		3	2.667		470	146		616	735
1900	Size 3, 90 amp		2.50	3.200		760	175		935	1,100
2000	Size 4, 135 amp		2.30	3.478		1,725	190		1,915	2,175
2100	Size 5, 270 amp	2 Elec	1.80	8.889		3,650	485		4,135	4,750
2200	Size 6, 540 amp		1.20	13.333		10,700	730		11,430	12,800
2300	Size 7, 810 amp		1	16		14,400	875		15,275	17,100
2310	Size 8, 1215 amp		.80	20		22,300	1,100		23,400	26,300
2500	Magnetic, 240 volt, 1-2 pole, .75 HP motor	1 Elec	4	2		151	109		260	330
2520	2 HP motor		3.60	2.222		169	122		291	370
2540	5 HP motor		2.50	3.200		410	175		585	710
2560	10 HP motor		1.40	5.714		675	315		990	1,200
2600	240 volt or less, 3 pole, .75 HP motor		4	2		151	109		260	330
2620	5 HP motor		3.60	2.222		187	122		309	385
2640	10 HP motor		3.60	2.222		217	122		339	420
2660	15 HP motor		2.50	3.200		435	175		610	735
2700	25 HP motor		2.50	3.200		435	175		610	735
2720	30 HP motor	2 Elec	2.80	5.714		725	315		1,040	1,275
2740	40 HP motor		2.80	5.714		725	315		1,040	1,275
2760	50 HP motor		1.60	10		725	545		1,270	1,625

26 29 13 – Enclosed Controllers

26 29 13.10 Contactors, AC		Crew	Daily Output	Labor-Hours	Unit	2015 Bare Costs Material	Labor	Equipment	Total	Total Incl O&P
2800	75 HP motor	2 Elec	1.60	10	Ea.	1,700	545		2,245	2,700
2820	100 HP motor		1	16		1,700	875		2,575	3,175
2860	150 HP motor		1	16		3,625	875		4,500	5,275
2880	200 HP motor		1	16		3,625	875		4,500	5,275
3000	600 volt, 3 pole, 5 HP motor	1 Elec	4	2		187	109		296	370
3020	10 HP motor		3.60	2.222		217	122		339	420
3040	25 HP motor		3	2.667		435	146		581	695
3100	50 HP motor		2.50	3.200		725	175		900	1,050
3160	100 HP motor	2 Elec	2.80	5.714		1,700	315		2,015	2,350
3220	200 HP motor	"	1.60	10		3,625	545		4,170	4,800

26 29 13.20 Control Stations		Crew	Daily Output	Labor-Hours	Unit	Material	Labor	Equipment	Total	Total Incl O&P
0010	**CONTROL STATIONS**									
0050	NEMA 1, heavy duty, stop/start	1 Elec	8	1	Ea.	135	54.50		189.50	231
0100	Stop/start, pilot light		6.20	1.290		184	70.50		254.50	310
0200	Hand/off/automatic		6.20	1.290		100	70.50		170.50	216
0400	Stop/start/reverse		5.30	1.509		182	82.50		264.50	325
0500	NEMA 7, heavy duty, stop/start		6	1.333		425	73		498	575
0600	Stop/start, pilot light		4	2		515	109		624	735
0700	NEMA 7 or 9, 1 element		6	1.333		345	73		418	490
0800	2 element		6	1.333		445	73		518	600
0900	3 element		4	2		885	109		994	1,150
0910	Selector switch, 2 position		6	1.333		345	73		418	490
0920	3 position		4	2		345	109		454	545
0930	Oiltight, 1 element		8	1		93	54.50		147.50	184
0940	2 element		6.20	1.290		134	70.50		204.50	253
0950	3 element		5.30	1.509		180	82.50		262.50	320
0960	Selector switch, 2 position		6.20	1.290		99	70.50		169.50	215
0970	3 position		5.30	1.509		99	82.50		181.50	233

26 29 13.30 Control Switches		Crew	Daily Output	Labor-Hours	Unit	Material	Labor	Equipment	Total	Total Incl O&P
0010	**CONTROL SWITCHES** Field installed									
6000	Push button 600 V 10A, momentary contact									
6150	Standard operator with colored button	1 Elec	34	.235	Ea.	15.70	12.85		28.55	36.50
6160	With single block 1NO 1NC		18	.444		33	24.50		57.50	73
6170	With double block 2NO 2NC		15	.533		50.50	29		79.50	99
6180	Std operator w/mushroom button 1-9/16" diam.		34	.235		33	12.85		45.85	56
6190	Std operator w/mushroom button 2-1/4" diam.									
6200	With single block 1NO 1NC	1 Elec	18	.444	Ea.	50.50	24.50		75	92
6210	With double block 2NO 2NC		15	.533		68	29		97	119
6500	Maintained contact, selector operator		34	.235		50.50	12.85		63.35	75
6510	With single block 1NO 1NC		18	.444		68	24.50		92.50	112
6520	With double block 2NO 2NC		15	.533		85.50	29		114.50	138
6560	Spring-return selector operator		34	.235		50.50	12.85		63.35	75
6570	With single block 1NO 1NC		18	.444		68	24.50		92.50	112
6580	With double block 2NO 2NC		15	.533		85.50	29		114.50	138
6620	Transformer operator w/illuminated									
6630	button 6 V #12 lamp	1 Elec	32	.250	Ea.	85.50	13.70		99.20	115
6640	With single block 1NO 1NC w/guard		16	.500		103	27.50		130.50	154
6650	With double block 2NO 2NC w/guard		13	.615		120	33.50		153.50	184
6690	Combination operator		34	.235		50.50	12.85		63.35	75
6700	With single block 1NO 1NC		18	.444		68	24.50		92.50	112
6710	With double block 2NO 2NC		15	.533		85.50	29		114.50	138
9000	Indicating light unit, full voltage									

26 29 13 – Enclosed Controllers

26 29 13.30 Control Switches		Crew	Daily Output	Labor-Hours	Unit	2015 Bare Costs Material	Labor	Equipment	Total	Total Incl O&P
9010	110-125 V front mount	1 Elec	32	.250	Ea.	68	13.70		81.70	95.50
9020	130 V resistor type		32	.250		50.50	13.70		64.20	76
9030	6 V transformer type		32	.250		62.50	13.70		76.20	89

26 29 13.40 Relays		Crew	Daily Output	Labor-Hours	Unit	Material	Labor	Equipment	Total	Total Incl O&P
0010	**RELAYS** Enclosed (NEMA 1)									
0050	600 volt AC, 1 pole, 12 amp	1 Elec	5.30	1.509	Ea.	88.50	82.50		171	222
0100	2 pole, 12 amp		5	1.600		88.50	87.50		176	229
0200	4 pole, 10 amp		4.50	1.778		118	97		215	276
0500	250 volt DC, 1 pole, 15 amp		5.30	1.509		116	82.50		198.50	252
0600	2 pole, 10 amp		5	1.600		111	87.50		198.50	253
0700	4 pole, 4 amp		4.50	1.778		147	97		244	305

26 29 23 – Variable-Frequency Motor Controllers

26 29 23.10 Variable Frequency Drives/Adj. Frequency Drives		Crew	Daily Output	Labor-Hours	Unit	Material	Labor	Equipment	Total	Total Incl O&P
0010	**VARIABLE FREQUENCY DRIVES/ADJ. FREQUENCY DRIVES**									
0100	Enclosed (NEMA 1), 460 volt, for 3 HP motor size G	1 Elec	.80	10	Ea.	1,700	545		2,245	2,700
0110	5 HP motor size G		.80	10		1,925	545		2,470	2,925
0120	7.5 HP motor size G		.67	11.940		2,300	655		2,955	3,500
0130	10 HP motor size G		.67	11.940		2,675	655		3,330	3,900
0140	15 HP motor size G	2 Elec	.89	17.978		3,250	985		4,235	5,050
0150	20 HP motor size G		.89	17.978		3,800	985		4,785	5,675
0160	25 HP motor size G		.67	23.881		4,700	1,300		6,000	7,100
0170	30 HP motor size G		.67	23.881		5,900	1,300		7,200	8,425
0180	40 HP motor size G		.67	23.881		7,025	1,300		8,325	9,700
0190	50 HP motor size G		.53	30.189		8,675	1,650		10,325	12,000
0200	60 HP motor size G	R-3	.56	35.714		10,500	1,950	246	12,696	14,700
0210	75 HP motor size G		.56	35.714		12,100	1,950	246	14,296	16,500
0220	100 HP motor size G		.50	40		14,100	2,175	276	16,551	19,100
0230	125 HP motor size G		.50	40		15,700	2,175	276	18,151	20,900
0240	150 HP motor size G		.50	40		17,800	2,175	276	20,251	23,200
0250	200 HP motor size G		.42	47.619		23,400	2,575	330	26,305	30,000
1100	Custom-engineered, 460 volt, for 3 HP motor size G	1 Elec	.56	14.286		2,725	780		3,505	4,175
1110	5 HP motor size G		.56	14.286		2,725	780		3,505	4,175
1120	7.5 HP motor size G		.47	17.021		2,875	930		3,805	4,550
1130	10 HP motor size G		.47	17.021		3,000	930		3,930	4,700
1140	15 HP motor size G	2 Elec	.62	25.806		3,750	1,400		5,150	6,250
1150	20 HP motor size G		.62	25.806		4,225	1,400		5,625	6,775
1160	25 HP motor size G		.47	34.043		4,950	1,850		6,800	8,225
1170	30 HP motor size G		.47	34.043		6,175	1,850		8,025	9,600
1180	40 HP motor size G		.47	34.043		7,300	1,850		9,150	10,800
1190	50 HP motor size G		.37	43.243		8,375	2,375		10,750	12,800
1200	60 HP motor size G	R-3	.39	51.282		12,600	2,775	355	15,730	18,500
1210	75 HP motor size G		.39	51.282		13,400	2,775	355	16,530	19,400
1220	100 HP motor size G		.35	57.143		14,700	3,100	395	18,195	21,300
1230	125 HP motor size G		.35	57.143		15,700	3,100	395	19,195	22,400
1240	150 HP motor size G		.35	57.143		18,000	3,100	395	21,495	24,900
1250	200 HP motor size G		.29	68.966		23,500	3,750	475	27,725	32,000
2000	For complex & special design systems to meet specific									
2010	requirements, obtain quote from vendor.									

26 31 13.50 Solar Energy - Photovoltaics			Crew	Daily Output	Labor-Hours	Unit	2015 Bare Costs Material	Labor	Equipment	Total	Total Incl O&P
0010	SOLAR ENERGY - PHOTOVOLTAICS										
0220	Alt. energy source, photovoltaic module, 6 watt, 15 volts	G	1 Elec	8	1	Ea.	68	54.50		122.50	157
0230	10 watt, 16.3 volts	G		8	1		108	54.50		162.50	201
0240	20 watt, 14.5 volts	G		8	1		163	54.50		217.50	261
0250	36 watt, 17 volts	G		8	1		220	54.50		274.50	325
0260	55 watt, 17 volts	G		8	1		305	54.50		359.50	415
0270	75 watt, 17 volts	G		8	1		390	54.50		444.50	510
0280	130 watt, 33 volts	G		8	1		565	54.50		619.50	700
0290	140 watt, 33 volts	G		8	1		595	54.50		649.50	735
0300	150 watt, 33 volts	G		8	1		655	54.50		709.50	800
0310	DC to AC inverter for, 12 V 2,000 watt	G		4	2		1,175	109		1,284	1,475
0320	12 V, 2,500 watt	G		4	2		1,425	109		1,534	1,725
0330	24 V, 2,500 watt	G		4	2		2,000	109		2,109	2,375
0340	12 V, 3,000 watt	G		3	2.667		1,650	146		1,796	2,050
0350	24 V, 3,000 watt	G		3	2.667		2,425	146		2,571	2,900
0360	24 V, 4,000 watt	G		2	4		3,325	219		3,544	3,975
0370	48 V, 4,000 watt	G		2	4		3,325	219		3,544	3,975
0380	48 V, 5,500 watt	G		2	4		3,775	219		3,994	4,475
0390	PV components, combiner box, 10 lug, NEMA 3R enclosure	G		4	2		191	109		300	375
0400	Fuse, 15 A for combiner box	G		40	.200		18.75	10.95		29.70	37
0410	Battery charger controller w/temperature sensor	G		4	2		490	109		599	705
0420	Digital readout panel, displays hours, volts, amps, etc.	G		4	2		130	109		239	305
0430	Deep cycle solar battery, 6 V, 180 Ah (C/20)	G		8	1		235	54.50		289.50	340
0440	Battery interconn, 15" AWG #2/0, sealed w/copper ring lugs	G		16	.500		12.80	27.50		40.30	55
0442	Battery interconn, 24" AWG #2/0, sealed w/copper ring lugs	G		16	.500		18.35	27.50		45.85	61
0444	Battery interconn, 60" AWG #2/0, sealed w/copper ring lugs	G		16	.500		44	27.50		71.50	89
0446	Batt temp computer probe, RJ11 jack, 15′ cord	G		16	.500		19.70	27.50		47.20	62.50
0450	System disconnect, DC 175 amp circuit breaker	G		8	1		140	54.50		194.50	236
0460	Conduit box for inverter	G		8	1		44.50	54.50		99	131
0470	Low voltage disconnect	G		8	1		40	54.50		94.50	126
0480	Vented battery enclosure, wood	G	1 Carp	2	4		174	188		362	480
0490	PV rack system, roof, non-penetrating ballast, 1 panel	G	R-1A	30.50	.525		895	23.50		918.50	1,025
0500	Penetrating surface mount, on steel framing, 1 panel			5.13	3.122		49	141		190	268
0510	On wood framing, 1 panel			11	1.455		47.50	65.50		113	152
0520	With standoff, 1 panel			11	1.455		55.50	65.50		121	161
0530	Ground, ballast, fixed, 3 panel			20.50	.780		1,150	35		1,185	1,325
0540	4 panel			20.50	.780		1,575	35		1,610	1,775
0550	5 panel			20.50	.780		2,075	35		2,110	2,325
0560	6 panel			20.50	.780		2,375	35		2,410	2,675
0570	Adjustable, 3 panel			20.50	.780		1,225	35		1,260	1,400
0580	4 panel			20.50	.780		1,675	35		1,710	1,875
0590	5 panel			20.50	.780		2,200	35		2,235	2,475
0600	6 panel			20.50	.780		2,525	35		2,560	2,825
0605	Top of Pole, see 32 31 13.30 6710+ for poles										
0610	Passive tracking, 1 panel		R-1A	20.50	.780	Ea.	705	35		740	830
0620	2 panel			20.50	.780		1,425	35		1,460	1,625
0630	3 panel			20.50	.780		2,000	35		2,035	2,250
0640	4 panel			20.50	.780		2,150	35		2,185	2,400
0650	6 panel			20.50	.780		2,425	35		2,460	2,725
0660	8 panel			20.50	.780		3,600	35		3,635	4,000

26 32 Packaged Generator Assemblies

26 32 13 – Engine Generators

26 32 13.13 Diesel-Engine-Driven Generator Sets

26 32 13.13 Diesel-Engine-Driven Generator Sets		Crew	Daily Output	Labor-Hours	Unit	2015 Bare Costs Material	Labor	Equipment	Total	Total Incl O&P
0010	**DIESEL-ENGINE-DRIVEN GENERATOR SETS**									
2000	Diesel engine, including battery, charger,									
2010	muffler, & day tank, 30 kW	R-3	.55	36.364	Ea.	14,000	1,975	251	16,226	18,700
2100	50 kW		.42	47.619		19,800	2,575	330	22,705	26,000
2110	60 kW		.39	51.282		21,100	2,775	355	24,230	27,800
2200	75 kW		.35	57.143		21,300	3,100	395	24,795	28,500
2300	100 kW		.31	64.516		29,900	3,500	445	33,845	38,600
2400	125 kW		.29	68.966		31,700	3,750	475	35,925	40,900
2500	150 kW		.26	76.923		33,800	4,175	530	38,505	44,100
2600	175 kW		.25	80		39,100	4,350	550	44,000	50,000
2700	200 kW		.24	83.333		42,200	4,525	575	47,300	54,000
2800	250 kW		.23	86.957		45,200	4,725	600	50,525	57,500
2850	275 kW		.22	90.909		47,300	4,925	625	52,850	60,000
2900	300 kW		.22	90.909		49,100	4,925	625	54,650	62,000
3000	350 kW		.20	100		55,500	5,425	690	61,615	70,000
3100	400 kW		.19	105		68,500	5,725	725	74,950	85,000
3200	500 kW		.18	111		86,500	6,025	765	93,290	105,000
3220	600 kW		.17	117		112,500	6,400	810	119,710	134,500
3230	650 kW	R-13	.38	110		138,500	5,850	495	144,845	162,000
3240	750 kW		.38	110		140,000	5,850	495	146,345	163,500
3250	800 kW		.36	116		147,500	6,175	520	154,195	172,500
3260	900 kW		.31	135		170,500	7,150	605	178,255	199,000
3270	1000 kW		.27	155		174,500	8,225	695	183,420	205,000

26 32 13.16 Gas-Engine-Driven Generator Sets

26 32 13.16 Gas-Engine-Driven Generator Sets		Crew	Daily Output	Labor-Hours	Unit	Material	Labor	Equipment	Total	Total Incl O&P
0010	**GAS-ENGINE-DRIVEN GENERATOR SETS**									
0020	Gas or gasoline operated, includes battery,									
0050	charger, & muffler									
0200	3 phase 4 wire, 277/480 volt, 7.5 kW	R-3	.83	24.096	Ea.	7,350	1,300	166	8,816	10,200
0300	11.5 kW		.71	28.169		10,400	1,525	194	12,119	14,000
0400	20 kW		.63	31.746		12,300	1,725	219	14,244	16,300
0500	35 kW		.55	36.364		14,600	1,975	251	16,826	19,400
0520	60 kW		.50	40		19,300	2,175	276	21,751	24,800
0600	80 kW		.40	50		24,000	2,725	345	27,070	30,900
0700	100 kW		.33	60.606		26,300	3,300	420	30,020	34,300
0800	125 kW		.28	71.429		54,000	3,875	490	58,365	65,500
0900	185 kW		.25	80		71,000	4,350	550	75,900	85,000

26 33 Battery Equipment

26 33 43 – Battery Chargers

26 33 43.55 Electric Vehicle Charging

26 33 43.55 Electric Vehicle Charging			Crew	Daily Output	Labor-Hours	Unit	Material	Labor	Equipment	Total	Total Incl O&P
0010	**ELECTRIC VEHICLE CHARGING**										
0020	Level 2, wall mounted										
2200	Heavy duty	G	R-1A	15.36	1.042	Ea.	2,525	47		2,572	2,850
2210	with RFID	G		12.29	1.302		2,700	58.50		2,758.50	3,075
2300	Free standing, single connector	G		10.24	1.563		2,900	70.50		2,970.50	3,275
2310	with RFID	G		8.78	1.822		3,625	82		3,707	4,100
2320	Double connector	G		7.68	2.083		4,850	94		4,944	5,475
2330	with RFID	G		6.83	2.343		6,300	106		6,406	7,075

26 33 Battery Equipment

26 33 53 – Static Uninterruptible Power Supply

26 33 53.10 Uninterruptible Power Supply/Conditioner Trans.		Crew	Daily Output	Labor-Hours	Unit	2015 Bare Costs Material	Labor	Equipment	Total	Total Incl O&P
0010	**UNINTERRUPTIBLE POWER SUPPLY/CONDITIONER TRANSFORMERS**									
0100	Volt. regulating, isolating transf., w/invert. & 10 min. battery pack									
0110	Single-phase, 120 V, 0.35 kVA R263353-80	1 Elec	2.29	3.493	Ea.	1,050	191		1,241	1,425
0120	0.5 kVA		2	4		1,100	219		1,319	1,525
0130	For additional 55 min. battery, add to 0.35 kVA		2.29	3.493		620	191		811	965
0140	Add to 0.5 kVA		1.14	7.018		650	385		1,035	1,300
0150	Single-phase, 120 V, 0.75 kVA		.80	10		1,400	545		1,945	2,350
0160	1.0 kVA		.80	10		2,000	545		2,545	3,025
0170	1.5 kVA	2 Elec	1.14	14.035		3,425	770		4,195	4,925
0180	2 kVA	"	.89	17.978		3,725	985		4,710	5,575
0190	3 kVA	R-3	.63	31.746		4,525	1,725	219	6,469	7,825
0200	5 kVA		.42	47.619		6,550	2,575	330	9,455	11,400
0210	7.5 kVA		.33	60.606		8,350	3,300	420	12,070	14,600
0220	10 kVA		.28	71.429		10,800	3,875	490	15,165	18,300
0230	15 kVA		.22	90.909		13,800	4,925	625	19,350	23,200
0240	3 phase, 120/208 V input 120/208 V output, 20 kVA, incl 17 min. battery		.21	95.238		25,300	5,175	655	31,130	36,400
0242	30 kVA, incl 11 min. battery		.20	100		28,400	5,425	690	34,515	40,100
0250	40 kVA, incl 15 min. battery		.20	100		37,800	5,425	690	43,915	50,500
0260	480 V input 277/480 V output, 60 kVA, incl 6 min. battery		.19	105		42,600	5,725	725	49,050	56,000
0262	80 kVA, incl 4 min. battery		.19	105		50,500	5,725	725	56,950	65,500
0400	For additional 34 min./15 min. battery, add to 40 kVA		.71	28.011		13,200	1,525	193	14,918	17,000
0600	For complex & special design systems to meet specific									
0610	requirements, obtain quote from vendor									

26 35 Power Filters and Conditioners

26 35 13 – Capacitors

26 35 13.10 Capacitors Indoor

		Crew	Daily Output	Labor-Hours	Unit	Material	Labor	Equipment	Total	Total Incl O&P
0010	**CAPACITORS INDOOR**									
0020	240 volts, single & 3 phase, 0.5 kVAR	1 Elec	2.70	2.963	Ea.	430	162		592	715
0100	1.0 kVAR		2.70	2.963		515	162		677	815
0150	2.5 kVAR		2	4		580	219		799	970
0200	5.0 kVAR		1.80	4.444		790	243		1,033	1,225
0250	7.5 kVAR		1.60	5		885	274		1,159	1,375
0300	10 kVAR		1.50	5.333		1,025	292		1,317	1,550
0350	15 kVAR		1.30	6.154		1,325	335		1,660	1,950
0400	20 kVAR		1.10	7.273		1,600	400		2,000	2,350
0450	25 kVAR		1	8		1,825	440		2,265	2,675
1000	480 volts, single & 3 phase, 1 kVAR		2.70	2.963		390	162		552	675
1050	2 kVAR		2.70	2.963		450	162		612	740
1100	5 kVAR		2	4		565	219		784	950
1150	7.5 kVAR		2	4		610	219		829	1,000
1200	10 kVAR		2	4		710	219		929	1,100
1250	15 kVAR		2	4		835	219		1,054	1,250
1300	20 kVAR		1.60	5		915	274		1,189	1,400
1350	30 kVAR		1.50	5.333		1,100	292		1,392	1,650
1400	40 kVAR		1.20	6.667		1,375	365		1,740	2,050
1450	50 kVAR		1.10	7.273		1,600	400		2,000	2,375
2000	600 volts, single & 3 phase, 1 kVAR		2.70	2.963		400	162		562	685
2050	2 kVAR		2.70	2.963		460	162		622	755
2100	5 kVAR		2	4		565	219		784	950

26 35 Power Filters and Conditioners

26 35 13 – Capacitors

	26 35 13.10 Capacitors Indoor	Crew	Daily Output	Labor-Hours	Unit	Material	2015 Bare Costs Labor	Equipment	Total	Total Incl O&P
2150	7.5 kVAR	1 Elec	2	4	Ea.	610	219		829	1,000
2200	10 kVAR		2	4		710	219		929	1,100
2250	15 kVAR		1.60	5		840	274		1,114	1,325
2300	20 kVAR		1.60	5		915	274		1,189	1,400
2350	25 kVAR		1.50	5.333		1,000	292		1,292	1,525
2400	35 kVAR		1.40	5.714		1,250	315		1,565	1,850
2450	50 kVAR		1.30	6.154		1,600	335		1,935	2,275

26 35 26 – Harmonic Filters

26 35 26.10 Computer Isolation Transformer

		Crew	Daily Output	Labor-Hours	Unit	Material	Labor	Equipment	Total	Total Incl O&P
0010	**COMPUTER ISOLATION TRANSFORMER**									
0100	Computer grade									
0110	Single-phase, 120/240 V, 0.5 kVA	1 Elec	4	2	Ea.	380	109		489	580
0120	1.0 kVA		2.67	2.996		535	164		699	835
0130	2.5 kVA		2	4		815	219		1,034	1,225
0140	5 kVA		1.14	7.018		930	385		1,315	1,600

26 35 26.20 Computer Regulator Transformer

		Crew	Daily Output	Labor-Hours	Unit	Material	Labor	Equipment	Total	Total Incl O&P
0010	**COMPUTER REGULATOR TRANSFORMER**									
0100	Ferro-resonant, constant voltage, variable transformer									
0110	Single-phase, 240 V, 0.5 kVA	1 Elec	2.67	2.996	Ea.	390	164		554	675
0120	1.0 kVA		2	4		535	219		754	920
0130	2.0 kVA		1	8		915	440		1,355	1,650
0210	Plug-in unit 120 V, 0.14 kVA		8	1		226	54.50		280.50	330
0220	0.25 kVA		8	1		264	54.50		318.50	370
0230	0.5 kVA		8	1		390	54.50		444.50	510
0240	1.0 kVA		5.33	1.501		535	82		617	715
0250	2.0 kVA		4	2		915	109		1,024	1,175

26 35 26.30 Power Conditioner Transformer

		Crew	Daily Output	Labor-Hours	Unit	Material	Labor	Equipment	Total	Total Incl O&P
0010	**POWER CONDITIONER TRANSFORMER**									
0100	Electronic solid state, buck-boost, transformer, w/tap switch									
0110	Single-phase, 115 V, 3.0 kVA, + or - 3% accuracy	2 Elec	1.60	10	Ea.	2,100	545		2,645	3,150
0120	208, 220, 230, or 240 V, 5.0 kVA, + or - 1.5% accuracy	3 Elec	1.60	15		2,725	820		3,545	4,225
0130	5.0 kVA, + or - 6% accuracy	2 Elec	1.14	14.035		2,450	770		3,220	3,850
0140	7.5 kVA, + or - 1.5% accuracy	3 Elec	1.50	16		3,475	875		4,350	5,125
0150	7.5 kVA, + or - 6% accuracy		1.60	15		2,900	820		3,720	4,400
0160	10.0 kVA, + or - 1.5% accuracy		1.33	18.045		4,625	985		5,610	6,550
0170	10.0 kVA, + or - 6% accuracy		1.41	17.021		3,925	930		4,855	5,725

26 35 26.40 Transient Voltage Suppressor Transformer

		Crew	Daily Output	Labor-Hours	Unit	Material	Labor	Equipment	Total	Total Incl O&P
0010	**TRANSIENT VOLTAGE SUPPRESSOR TRANSFORMER**									
0110	Single-phase, 120 V, 1.8 kVA	1 Elec	4	2	Ea.	855	109		964	1,100
0120	3.6 kVA		4	2		1,675	109		1,784	2,025
0130	7.2 kVA		3.20	2.500		2,300	137		2,437	2,725
0150	240 V, 3.6 kVA		4	2		1,675	109		1,784	2,025
0160	7.2 kVA		4	2		2,175	109		2,284	2,575
0170	14.4 kVA		3.20	2.500		3,275	137		3,412	3,800
0210	Plug-in unit, 120 V, 1.8 kVA		8	1		650	54.50		704.50	795

26 35 53 – Voltage Regulators

26 35 53.10 Automatic Voltage Regulators

		Crew	Daily Output	Labor-Hours	Unit	Material	Labor	Equipment	Total	Total Incl O&P
0010	**AUTOMATIC VOLTAGE REGULATORS**									
0100	Computer grade, solid state, variable transf. volt. regulator									
0110	Single-phase, 120 V, 8.6 kVA	2 Elec	1.33	12.030	Ea.	4,900	660		5,560	6,350
0120	17.3 kVA		1.14	14.035		5,775	770		6,545	7,500

26 35 Power Filters and Conditioners

26 35 53 – Voltage Regulators

26 35 53.10 Automatic Voltage Regulators		Crew	Daily Output	Labor-Hours	Unit	Material	Labor	Equipment	Total	Total Incl O&P
						2015 Bare Costs				
0130	208/240 V, 7.5/8.6 kVA	2 Elec	1.33	12.030	Ea.	4,900	660		5,560	6,350
0140	13.5/15.6 kVA		1.33	12.030		5,775	660		6,435	7,325
0150	27.0/31.2 kVA		1.14	14.035		7,300	770		8,070	9,175
0210	Two-phase, single control, 208/240 V, 15.0/17.3 kVA		1.14	14.035		5,775	770		6,545	7,500
0220	Individual phase control, 15.0/17.3 kVA		1.14	14.035		5,775	770		6,545	7,500
0230	30.0/34.6 kVA	3 Elec	1.33	18.045		7,300	985		8,285	9,500
0310	Three-phase single control, 208/240 V, 26/30 kVA	2 Elec	1	16		5,775	875		6,650	7,650
0320	380/480 V, 24/30 kVA	"	1	16		5,775	875		6,650	7,650
0330	43/54 kVA	3 Elec	1.33	18.045		10,500	985		11,485	13,000
0340	Individual phase control, 208 V, 26 kVA	"	1.33	18.045		5,775	985		6,760	7,825
0350	52 kVA	R-3	.91	21.978		7,300	1,200	151	8,651	10,000
0360	340/480 V, 24/30 kVA	2 Elec	1	16		5,775	875		6,650	7,650
0370	43/54 kVA	"	1	16		7,300	875		8,175	9,325
0380	48/60 kVA	3 Elec	1.33	18.045		10,600	985		11,585	13,100
0390	86/108 kVA	R-3	.91	21.978		11,700	1,200	151	13,051	14,900
0500	Standard grade, solid state, variable transformer volt. regulator									
0510	Single-phase, 115 V, 2.3 kVA	1 Elec	2.29	3.493	Ea.	1,875	191		2,066	2,350
0520	4.2 kVA		2	4		3,025	219		3,244	3,650
0530	6.6 kVA		1.14	7.018		3,700	385		4,085	4,650
0540	13.0 kVA		1.14	7.018		6,375	385		6,760	7,600
0550	16.6 kVA	2 Elec	1.23	13.008		7,525	710		8,235	9,375
0610	230 V, 8.3 kVA		1.33	12.030		6,375	660		7,035	8,000
0620	21.4 kVA		1.23	13.008		7,525	710		8,235	9,375
0630	29.9 kVA		1.23	13.008		7,525	710		8,235	9,375
0710	460 V, 9.2 kVA		1.33	12.030		6,375	660		7,035	8,000
0720	20.7 kVA		1.23	13.008		7,525	710		8,235	9,375
0810	Three-phase, 230 V, 13.1 kVA	3 Elec	1.41	17.021		6,375	930		7,305	8,425
0820	19.1 kVA		1.41	17.021		7,525	930		8,455	9,700
0830	25.1 kVA		1.23	19.512		7,525	1,075		8,600	9,900
0840	57.8 kVA	R-3	.95	21.053		13,700	1,150	145	14,995	16,900
0850	74.9 kVA	"	.91	21.978		13,700	1,200	151	15,051	17,000
0910	460 V, 14.3 kVA	3 Elec	1.41	17.021		6,375	930		7,305	8,425
0920	19.1 kVA		1.41	17.021		7,525	930		8,455	9,700
0930	27.9 kVA		1.23	19.512		7,525	1,075		8,600	9,900
0940	59.8 kVA	R-3	1	20		13,700	1,075	138	14,913	16,800
0950	79.7 kVA		.95	21.053		15,300	1,150	145	16,595	18,700
0960	118 kVA		.95	21.053		16,100	1,150	145	17,395	19,600
1000	Laboratory grade, precision, electronic voltage regulator									
1110	Single-phase, 115 V, 0.5 kVA	1 Elec	2.29	3.493	Ea.	1,200	191		1,391	1,600
1120	1.0 kVA		2	4		1,275	219		1,494	1,725
1130	3.0 kVA		.80	10		1,775	545		2,320	2,800
1140	6.0 kVA	2 Elec	1.46	10.959		3,300	600		3,900	4,525
1150	10.0 kVA	3 Elec	1	24		4,300	1,325		5,625	6,700
1160	15.0 kVA	"	1.50	16		4,900	875		5,775	6,700
1210	230 V, 3.0 kVA	1 Elec	.80	10		2,025	545		2,570	3,050
1220	6.0 kVA	2 Elec	1.46	10.959		3,375	600		3,975	4,625
1230	10.0 kVA	3 Elec	1.71	14.035		4,500	770		5,270	6,075
1240	15.0 kVA	"	1.60	15		5,075	820		5,895	6,800

26 35 53.30 Transient Suppressor/Voltage Regulator

		Crew	Daily Output	Labor-Hours	Unit	Material	Labor	Equipment	Total	Total Incl O&P
0010	**TRANSIENT SUPPRESSOR/VOLTAGE REGULATOR** (without isolation)									
0110	Single-phase, 115 V, 1.0 kVA	1 Elec	2.67	2.996	Ea.	1,000	164		1,164	1,350
0120	2.0 kVA		2.29	3.493		1,375	191		1,566	1,800

26 35 Power Filters and Conditioners

26 35 53 – Voltage Regulators

26 35 53.30 Transient Suppressor/Voltage Regulator		Crew	Daily Output	Labor-Hours	Unit	2015 Bare Costs Material	Labor	Equipment	Total	Total Incl O&P
0130	4.0 kVA	1 Elec	2.13	3.756	Ea.	1,700	205		1,905	2,175
0140	220 V, 1.0 kVA		2.67	2.996		1,000	164		1,164	1,350
0150	2.0 kVA		2.29	3.493		1,375	191		1,566	1,800
0160	4.0 kVA		2.13	3.756		1,800	205		2,005	2,275
0210	Plug-in unit, 120 V, 1.0 kVA		8	1		965	54.50		1,019.50	1,125
0220	2.0 kVA		8	1		1,350	54.50		1,404.50	1,575

26 36 Transfer Switches

26 36 13 – Manual Transfer Switches

26 36 13.10 Non-Automatic Transfer Switches		Crew	Daily Output	Labor-Hours	Unit	Material	Labor	Equipment	Total	Total Incl O&P
0010	**NON-AUTOMATIC TRANSFER SWITCHES** enclosed									
0100	Manual operated, 480 volt 3 pole, 30 amp	1 Elec	2.30	3.478	Ea.	1,375	190		1,565	1,800
0150	60 amp		1.90	4.211		1,850	230		2,080	2,375
0200	100 amp		1.30	6.154		2,750	335		3,085	3,525
0250	200 amp	2 Elec	2	8		3,650	440		4,090	4,675
0300	400 amp		1.60	10		5,300	545		5,845	6,675
0350	600 amp		1	16		5,325	875		6,200	7,150
1000	250 volt 3 pole, 30 amp	1 Elec	2.30	3.478		1,000	190		1,190	1,375
1100	60 amp		1.90	4.211		1,450	230		1,680	1,950
1150	100 amp		1.30	6.154		1,825	335		2,160	2,500
1200	200 amp	2 Elec	2	8		3,600	440		4,040	4,625
1300	600 amp	″	1	16		8,100	875		8,975	10,200
1500	Electrically operated, 480 volt 3 pole, 60 amp	1 Elec	1.90	4.211		2,375	230		2,605	2,975
1600	100 amp	″	1.30	6.154		2,375	335		2,710	3,125
1650	200 amp	2 Elec	2	8		3,925	440		4,365	4,975
1700	400 amp		1.60	10		5,500	545		6,045	6,875
1750	600 amp		1	16		7,900	875		8,775	10,000
2000	250 volt 3 pole, 30 amp	1 Elec	2.30	3.478		2,350	190		2,540	2,875
2050	60 amp	″	1.90	4.211		2,350	230		2,580	2,950
2150	200 amp	2 Elec	2	8		3,900	440		4,340	4,925
2200	400 amp		1.60	10		5,425	545		5,970	6,800
2250	600 amp		1	16		7,825	875		8,700	9,925
2500	NEMA 3R, 480 volt 3 pole, 60 amp	1 Elec	1.80	4.444		2,775	243		3,018	3,425
2550	100 amp	″	1.20	6.667		3,575	365		3,940	4,475
2600	200 amp	2 Elec	1.80	8.889		4,375	485		4,860	5,525
2650	400 amp	″	1.40	11.429		6,000	625		6,625	7,525
2800	NEMA 3R, 250 volt 3 pole solid state, 100 amp	1 Elec	1.20	6.667		2,800	365		3,165	3,625
2850	150 amp	2 Elec	1.80	8.889		3,725	485		4,210	4,825
2900	250 volt 2 pole solid state, 100 amp	1 Elec	1.30	6.154		2,725	335		3,060	3,500
2950	150 amp	2 Elec	2	8		3,625	440		4,065	4,650

26 36 23 – Automatic Transfer Switches

26 36 23.10 Automatic Transfer Switch Devices		Crew	Daily Output	Labor-Hours	Unit	Material	Labor	Equipment	Total	Total Incl O&P
0010	**AUTOMATIC TRANSFER SWITCH DEVICES** R263623-60									
0015	Switches, enclosed 120/240 volt, 2 pole, 30 amp	1 Elec	2.40	3.333	Ea.	2,400	182		2,582	2,900
0020	70 amp		2	4		2,400	219		2,619	2,950
0030	100 amp		1.35	5.926		2,400	325		2,725	3,100
0040	225 amp	2 Elec	2.10	7.619		3,525	415		3,940	4,500
0050	400 amp		1.70	9.412		5,375	515		5,890	6,675
0060	600 amp		1.06	15.094		7,875	825		8,700	9,900
0070	800 amp		.84	19.048		9,250	1,050		10,300	11,800

26 36 Transfer Switches

26 36 23 – Automatic Transfer Switches

26 36 23.10 Automatic Transfer Switch Devices		Crew	Daily Output	Labor-Hours	Unit	2015 Bare Costs Material	Labor	Equipment	Total	Total Incl O&P
0100	Switches, enclosed 480 volt, 3 pole, 30 amp	1 Elec	2.30	3.478	Ea.	2,650	190		2,840	3,200
0200	60 amp		1.90	4.211		2,650	230		2,880	3,275
0300	100 amp		1.30	6.154		2,650	335		2,985	3,425
0400	150 amp	2 Elec	2.40	6.667		3,275	365		3,640	4,150
0500	225 amp		2	8		4,200	440		4,640	5,250
0600	260 amp		2	8		4,850	440		5,290	5,975
0700	400 amp		1.60	10		6,325	545		6,870	7,775
0800	600 amp		1	16		8,750	875		9,625	10,900
0900	800 amp		.80	20		10,300	1,100		11,400	13,000
1000	1000 amp		.76	21.053		13,600	1,150		14,750	16,600
1100	1200 amp		.70	22.857		18,600	1,250		19,850	22,400
1200	1600 amp		.60	26.667		21,200	1,450		22,650	25,500
1300	2000 amp		.50	32		23,700	1,750		25,450	28,600
1600	Accessories, time delay on engine starting					208			208	228
1700	Adjustable time delay on retransfer					208			208	228
1800	Shunt trips for customer connections					370			370	405
1900	Maintenance select switch					84.50			84.50	93
2000	Auxiliary contact when normal fails					97.50			97.50	107
2100	Pilot light-emergency					84.50			84.50	93
2200	Pilot light-normal					84.50			84.50	93
2300	Auxiliary contact-closed on normal					97.50			97.50	107
2400	Auxiliary contact-closed on emergency					97.50			97.50	107
2500	Emergency source sensing, frequency relay					430			430	470

26 41 Facility Lightning Protection

26 41 13 – Lightning Protection for Structures

26 41 13.13 Lightning Protection for Buildings

Line	Description	Crew	Daily Output	Labor-Hours	Unit	Material	Labor	Equipment	Total	Total Incl O&P
0010	**LIGHTNING PROTECTION FOR BUILDINGS**									
0200	Air terminals & base, copper									
0400	3/8" diameter x 10" (to 75' high)	1 Elec	8	1	Ea.	22	54.50		76.50	106
0500	1/2" diameter x 12" (over 75' high)		8	1		25.50	54.50		80	110
0520	1/2" diameter x 24"		7.30	1.096		35	60		95	129
0540	1/2" diameter x 60"		6.70	1.194		64	65.50		129.50	168
1000	Aluminum, 1/2" diameter x 12" (to 75' high)		8	1		15.10	54.50		69.60	98.50
1020	1/2" diameter x 24"		7.30	1.096		17.25	60		77.25	109
1040	1/2" diameter x 60"		6.70	1.194		25.50	65.50		91	126
1100	5/8" diameter x 12" (over 75' high)		8	1		16.35	54.50		70.85	100
2000	Cable, copper, 220 lb. per thousand ft. (to 75' high)		320	.025	L.F.	3.03	1.37		4.40	5.40
2100	375 lb. per thousand ft. (over 75' high)		230	.035		5.50	1.90		7.40	8.90
2500	Aluminum, 101 lb. per thousand ft. (to 75' high)		280	.029		.95	1.56		2.51	3.39
2600	199 lb. per thousand ft. (over 75' high)		240	.033		1.40	1.82		3.22	4.27
3000	Arrester, 175 volt AC to ground		8	1	Ea.	103	54.50		157.50	195
3100	650 volt AC to ground		6.70	1.194	"	89.50	65.50		155	197
4000	Air terminals, copper									
4010	3/8" x 10"	1 Elec	50	.160	Ea.	7.10	8.75		15.85	21
4020	1/2" x 12"		38	.211		10.65	11.50		22.15	29
4030	5/8" x 12"		33	.242		16	13.25		29.25	37.50
4040	1/2" x 24"		30	.267		20	14.60		34.60	44
4050	1/2" x 60"		19	.421		49	23		72	88.50
4060	Air terminals, aluminum									
4070	3/8" x 10"	1 Elec	50	.160	Ea.	2.87	8.75		11.62	16.25

26 41 Facility Lightning Protection

26 41 13 – Lightning Protection for Structures

26 41 13.13 Lightning Protection for Buildings		Crew	Daily Output	Labor-Hours	Unit	Material	Labor	Equipment	Total	Total Incl O&P
						2015 Bare Costs				
4080	1/2" x 12"	1 Elec	38	.211	Ea.	3.34	11.50		14.84	21
4090	5/8" x 12"		33	.242		4.62	13.25		17.87	25
4100	1/2" x 24"		30	.267		5.50	14.60		20.10	28
4110	1/2" x 60"		19	.421		13.70	23		36.70	49.50
4200	Air terminal bases, copper									
4210	Adhesive bases, 1/2"	1 Elec	15	.533	Ea.	16.75	29		45.75	62
4215	Adhesive base, lightning air terminal					16.75			16.75	18.45
4220	Bolted bases	1 Elec	9	.889		14.70	48.50		63.20	89
4230	Hinged bases		7	1.143		29.50	62.50		92	126
4240	Side mounted bases		9	.889		15.70	48.50		64.20	90.50
4250	Offset point support		9	.889		29.50	48.50		78	106
4260	Concealed base assembly		5	1.600		29	87.50		116.50	163
4270	Tee connector base		13	.615		37	33.50		70.50	91.50
4280	Tripod base, 36" for 60" air terminal		7	1.143		47.50	62.50		110	146
4290	Intermediate base, 1/2"		15	.533		24	29		53	69.50
4300	Air terminal bases, aluminum									
4310	Adhesive bases, 1/2"	1 Elec	15	.533	Ea.	7.65	29		36.65	52
4320	Bolted bases		9	.889		11.75	48.50		60.25	86
4330	Hinged bases		7	1.143		21.50	62.50		84	117
4340	Side mounted bases		9	.889		11.45	48.50		59.95	85.50
4350	Offset point support		9	.889		17.20	48.50		65.70	92
4360	Concealed base assembly		5	1.600		17.15	87.50		104.65	150
4370	Tee connector base		13	.615		11.60	33.50		45.10	63.50
4380	Tripod base, 36" for 60" air terminal		7	1.143		34	62.50		96.50	131
4390	Intermediate base, 1/2"		15	.533		12	29		41	56.50
4400	Connector cable, copper									
4410	Through wall connector	1 Elec	7	1.143	Ea.	75.50	62.50		138	177
4420	Through roof connector		5	1.600		80.50	87.50		168	220
4430	Beam connector		7	1.143		28.50	62.50		91	125
4440	Double bolt connector		38	.211		11.25	11.50		22.75	29.50
4450	Bar, copper		40	.200		12.40	10.95		23.35	30
4500	Connector cable, aluminum									
4510	Through wall connector	1 Elec	7	1.143	Ea.	30	62.50		92.50	127
4520	Through roof connector		5	1.600		24	87.50		111.50	158
4530	Beam connector		7	1.143		16.50	62.50		79	112
4540	Double bolt connector		38	.211		7.15	11.50		18.65	25
4600	Bonding plates, copper									
4610	I Beam, 8" square	1 Elec	7	1.143	Ea.	19.15	62.50		81.65	115
4620	Purlin, 8" square		7	1.143		28.50	62.50		91	125
4630	Large heavy duty, 16" square		5	1.600		28.50	87.50		116	162
4640	Aluminum, I Beam, 8" square		7	1.143		14.30	62.50		76.80	109
4650	Purlin, 8" square		7	1.143		13.15	62.50		75.65	108
4660	Large heavy duty, 16" square		5	1.600		17.05	87.50		104.55	150
4700	Cable support, copper, loop fastener		38	.211		.77	11.50		12.27	18.10
4710	Adhesive cable holder		38	.211		1.38	11.50		12.88	18.75
4720	Aluminum, loop fastener		38	.211		.27	11.50		11.77	17.55
4730	Adhesive cable holder		38	.211		.49	11.50		11.99	17.80
4740	Bonding strap, copper, 3/4" x 9-1/2"		15	.533		8.55	29		37.55	53
4750	Aluminum, 3/4" x 9-1/2"		15	.533		4.62	29		33.62	48.50
4760	Swivel adapter, copper, 1/2"		73	.110		13.35	6		19.35	23.50
4770	Aluminum, 1/2"		73	.110		9.25	6		15.25	19.15

26 51 Interior Lighting

26 51 13 – Interior Lighting Fixtures, Lamps, and Ballasts

26 51 13.10 Fixture Hangers		Crew	Daily Output	Labor-Hours	Unit	2015 Bare Costs Material	Labor	Equipment	Total	Total Incl O&P
0010	**FIXTURE HANGERS**									
0220	Box hub cover	1 Elec	32	.250	Ea.	4.19	13.70		17.89	25
0240	Canopy		12	.667		8.95	36.50		45.45	64.50
0260	Connecting block		40	.200		2.53	10.95		13.48	19.20
0280	Cushion hanger		16	.500		22.50	27.50		50	66
0300	Box hanger, with mounting strap		8	1		9	54.50		63.50	92
0320	Connecting block		40	.200		2.17	10.95		13.12	18.80
0340	Flexible, 1/2" diameter, 4" long		12	.667		15.60	36.50		52.10	71.50
0360	6" long		12	.667		16.95	36.50		53.45	73
0380	8" long		12	.667		18.75	36.50		55.25	75
0400	10" long		12	.667		20	36.50		56.50	76.50
0420	12" long		12	.667		21	36.50		57.50	78
0440	15" long		12	.667		22	36.50		58.50	79
0460	18" long		12	.667		26	36.50		62.50	83
0480	3/4" diameter, 4" long		10	.800		19.25	44		63.25	86.50
0500	6" long		10	.800		21.50	44		65.50	89
0520	8" long		10	.800		23.50	44		67.50	91
0540	10" long		10	.800		23.50	44		67.50	91.50
0560	12" long		10	.800		25.50	44		69.50	93.50
0580	15" long		10	.800		28.50	44		72.50	97
0600	18" long	▼	10	.800	▼	32	44		76	101

26 51 13.40 Interior HID Fixtures		Crew	Daily Output	Labor-Hours	Unit	Material	Labor	Equipment	Total	Total Incl O&P
0010	**INTERIOR HID FIXTURES** Incl. lamps, and mounting hardware									
0700	High pressure sodium, recessed, round, 70 watt	1 Elec	3.50	2.286	Ea.	430	125		555	660
0720	100 watt		3.50	2.286		460	125		585	690
0740	150 watt		3.20	2.500		515	137		652	770
0760	Square, 70 watt		3.60	2.222		430	122		552	655
0780	100 watt		3.60	2.222		460	122		582	685
0820	250 watt	▼	3	2.667		680	146		826	965
0840	1000 watt	2 Elec	4.80	3.333		1,200	182		1,382	1,600
0860	Surface, round, 70 watt	1 Elec	3	2.667		615	146		761	895
0880	100 watt		3	2.667		630	146		776	915
0900	150 watt		2.70	2.963		660	162		822	970
0920	Square, 70 watt		3	2.667		585	146		731	860
0940	100 watt		3	2.667		615	146		761	895
0980	250 watt		2.50	3.200		665	175		840	995
1040	Pendent, round, 70 watt		3	2.667		640	146		786	920
1060	100 watt		3	2.667		610	146		756	895
1080	150 watt		2.70	2.963		625	162		787	930
1100	Square, 70 watt		3	2.667		665	146		811	950
1120	100 watt		3	2.667		675	146		821	960
1140	150 watt		2.70	2.963		690	162		852	1,000
1160	250 watt		2.50	3.200		935	175		1,110	1,275
1180	400 watt		2.40	3.333		985	182		1,167	1,350
1220	Wall, round, 70 watt		3	2.667		560	146		706	835
1240	100 watt		3	2.667		570	146		716	845
1260	150 watt		2.70	2.963		585	162		747	890
1300	Square, 70 watt		3	2.667		595	146		741	875
1320	100 watt		3	2.667		625	146		771	905
1340	150 watt		2.70	2.963		645	162		807	955
1360	250 watt	▼	2.50	3.200		705	175		880	1,025
1380	400 watt	2 Elec	4.80	3.333	▼	870	182		1,052	1,225

26 51 13.40 Interior HID Fixtures		Crew	Daily Output	Labor-Hours	Unit	2015 Bare Costs Material	Labor	Equipment	Total	Total Incl O&P
1400	1000 watt	2 Elec	3.60	4.444	Ea.	1,250	243		1,493	1,750
1500	Metal halide, recessed, round, 175 watt	1 Elec	3.40	2.353		385	129		514	615
1520	250 watt	"	3.20	2.500		480	137		617	735
1540	400 watt	2 Elec	5.80	2.759		680	151		831	975
1580	Square, 175 watt	1 Elec	3.40	2.353		385	129		514	620
1640	Surface, round, 175 watt		2.90	2.759		535	151		686	815
1660	250 watt		2.70	2.963		850	162		1,012	1,175
1680	400 watt	2 Elec	4.80	3.333		940	182		1,122	1,300
1720	Square, 175 watt	1 Elec	2.90	2.759		580	151		731	865
1800	Pendent, round, 175 watt		2.90	2.759		705	151		856	1,000
1820	250 watt		2.70	2.963		920	162		1,082	1,250
1840	400 watt	2 Elec	4.80	3.333		1,000	182		1,182	1,375
1880	Square, 175 watt	1 Elec	2.90	2.759		505	151		656	780
1900	250 watt	"	2.70	2.963		605	162		767	910
1920	400 watt	2 Elec	4.80	3.333		975	182		1,157	1,350
1980	Wall, round, 175 watt	1 Elec	2.90	2.759		675	151		826	965
2000	250 watt	"	2.70	2.963		845	162		1,007	1,175
2020	400 watt	2 Elec	4.80	3.333		845	182		1,027	1,200
2060	Square, 175 watt	1 Elec	2.90	2.759		575	151		726	855
2080	250 watt	"	2.70	2.963		600	162		762	905
2100	400 watt	2 Elec	4.80	3.333		845	182		1,027	1,200
2800	High pressure sodium, vaporproof, recessed, 70 watt	1 Elec	3.50	2.286		565	125		690	810
2820	100 watt		3.50	2.286		580	125		705	820
2840	150 watt		3.20	2.500		595	137		732	860
2900	Surface, 70 watt		3	2.667		640	146		786	925
2920	100 watt		3	2.667		670	146		816	955
2940	150 watt		2.70	2.963		695	162		857	1,000
3000	Pendent, 70 watt		3	2.667		630	146		776	915
3020	100 watt		3	2.667		650	146		796	935
3040	150 watt		2.70	2.963		690	162		852	1,000
3100	Wall, 70 watt		3	2.667		680	146		826	970
3120	100 watt		3	2.667		710	146		856	1,000
3140	150 watt		2.70	2.963		740	162		902	1,050
3200	Metal halide, vaporproof, recessed, 175 watt		3.40	2.353		505	129		634	750
3220	250 watt		3.20	2.500		585	137		722	845
3240	400 watt	2 Elec	5.80	2.759		730	151		881	1,025
3260	1000 watt	"	4.80	3.333		1,325	182		1,507	1,750
3280	Surface, 175 watt	1 Elec	2.90	2.759		620	151		771	905
3300	250 watt	"	2.70	2.963		830	162		992	1,150
3320	400 watt	2 Elec	4.80	3.333		1,025	182		1,207	1,400
3340	1000 watt	"	3.60	4.444		1,500	243		1,743	2,025
3360	Pendent, 175 watt	1 Elec	2.90	2.759		645	151		796	935
3380	250 watt	"	2.70	2.963		855	162		1,017	1,175
3400	400 watt	2 Elec	4.80	3.333		1,025	182		1,207	1,400
3420	1000 watt	"	3.60	4.444		1,650	243		1,893	2,200
3440	Wall, 175 watt	1 Elec	2.90	2.759		700	151		851	995
3460	250 watt	"	2.70	2.963		905	162		1,067	1,250
3480	400 watt	2 Elec	4.80	3.333		1,075	182		1,257	1,450
3500	1000 watt	"	3.60	4.444		1,725	243		1,968	2,250

26 51 Interior Lighting

26 51 13 – Interior Lighting Fixtures, Lamps, and Ballasts

26 51 13.50 Interior Lighting Fixtures		Crew	Daily Output	Labor-Hours	Unit	2015 Bare Costs Material	Labor	Equipment	Total	Total Incl O&P
0010	**INTERIOR LIGHTING FIXTURES** Including lamps, mounting R265113-40									
0030	hardware and connections									
0100	Fluorescent, C.W. lamps, troffer, recess mounted in grid, RS									
0130	Grid ceiling mount									
0200	Acrylic lens, 1'W x 4'L, two 40 watt	1 Elec	5.70	1.404	Ea.	47.50	77		124.50	167
0210	1'W x 4'L, three 40 watt		5.40	1.481		54	81		135	180
0300	2'W x 2'L, two U40 watt		5.70	1.404		51	77		128	172
0400	2'W x 4'L, two 40 watt		5.30	1.509		50	82.50		132.50	179
0500	2'W x 4'L, three 40 watt		5	1.600		55	87.50		142.50	192
0600	2'W x 4'L, four 40 watt		4.70	1.702		57.50	93		150.50	203
0700	4'W x 4'L, four 40 watt	2 Elec	6.40	2.500		293	137		430	525
0800	4'W x 4'L, six 40 watt		6.20	2.581		305	141		446	545
0900	4'W x 4'L, eight 40 watt		5.80	2.759		315	151		466	570
0910	Acrylic lens, 1'W x 4'L, two 32 watt T8 G	1 Elec	5.70	1.404		59.50	77		136.50	181
0930	2'W x 2'L, two U32 watt T8 G		5.70	1.404		81	77		158	204
0940	2'W x 4'L, two 32 watt T8 G		5.30	1.509		67	82.50		149.50	198
0950	2'W x 4'L, three 32 watt T8 G		5	1.600		68	87.50		155.50	206
0960	2'W x 4'L, four 32 watt T8 G		4.70	1.702		71	93		164	217
1000	Surface mounted, RS									
1030	Acrylic lens with hinged & latched door frame									
1100	1'W x 4'L, two 40 watt	1 Elec	7	1.143	Ea.	64	62.50		126.50	164
1110	1'W x 4'L, three 40 watt		6.70	1.194		66	65.50		131.50	171
1200	2'W x 2'L, two U40 watt		7	1.143		68.50	62.50		131	169
1300	2'W x 4'L, two 40 watt		6.20	1.290		78	70.50		148.50	192
1400	2'W x 4'L, three 40 watt		5.70	1.404		79	77		156	202
1500	2'W x 4'L, four 40 watt		5.30	1.509		81	82.50		163.50	213
1600	4'W x 4'L, four 40 watt	2 Elec	7.20	2.222		400	122		522	620
1700	4'W x 4'L, six 40 watt		6.60	2.424		435	133		568	675
1800	4'W x 4'L, eight 40 watt		6.20	2.581		450	141		591	705
1900	2'W x 8'L, four 40 watt		6.40	2.500		159	137		296	380
2000	2'W x 8'L, eight 40 watt		6.20	2.581		171	141		312	400
2010	Acrylic wrap around lens									
2020	6"W x 4'L, one 40 watt	1 Elec	8	1	Ea.	65	54.50		119.50	154
2030	6"W x 8'L, two 40 watt	2 Elec	8	2		71.50	109		180.50	243
2040	11"W x 4'L, two 40 watt	1 Elec	7	1.143		43.50	62.50		106	141
2050	11"W x 8'L, four 40 watt	2 Elec	6.60	2.424		71.50	133		204.50	278
2060	16"W x 4'L, four 40 watt	1 Elec	5.30	1.509		71.50	82.50		154	203
2070	16"W x 8'L, eight 40 watt	2 Elec	6.40	2.500		156	137		293	375
2080	2'W x 2'L, two U40 watt	1 Elec	7	1.143		87.50	62.50		150	190
2100	Strip fixture									
2130	Surface mounted									
2200	4' long, one 40 watt, RS	1 Elec	8.50	.941	Ea.	27.50	51.50		79	107
2300	4' long, two 40 watt, RS		8	1		38.50	54.50		93	124
2310	4' long, two 32 watt T8, RS G		8	1		77.50	54.50		132	167
2400	4' long, one 40 watt, SL		8	1		46.50	54.50		101	133
2500	4' long, two 40 watt, SL		7	1.143		63	62.50		125.50	163
2580	8' long, one 60 watt T8, SL G	2 Elec	13.40	1.194		83.50	65.50		149	190
2590	8' long, two 60 watt T8, SL G		12.40	1.290		86.50	70.50		157	201
2600	8' long, one 75 watt, SL		13.40	1.194		48	65.50		113.50	151
2700	8' long, two 75 watt, SL		12.40	1.290		58	70.50		128.50	170
2800	4' long, two 60 watt, HO	1 Elec	6.70	1.194		93.50	65.50		159	201
2810	4' long, two 54 watt, T5HO G	"	6.70	1.194		161	65.50		226.50	275

26 51 Interior Lighting

26 51 13 – Interior Lighting Fixtures, Lamps, and Ballasts

26 51 13.50 Interior Lighting Fixtures			Crew	Daily Output	Labor-Hours	Unit	Material	Labor	Equipment	Total	Total Incl O&P
							2015 Bare Costs				
2900	8′ long, two 110 watt, HO		2 Elec	10.60	1.509	Ea.	98.50	82.50		181	232
2910	4′ long, two 115 watt, VHO		1 Elec	6.50	1.231		130	67.50		197.50	244
2920	8′ long, two 215 watt, VHO		2 Elec	10.40	1.538		140	84		224	280
2950	High bay pendent mounted, 16″ W x 4′ L, four 54 watt, T5HO	G		8.90	1.798		229	98.50		327.50	400
2952	2′ W x 4′ L, six 54 watt, T5HO	G		8.50	1.882		305	103		408	490
2954	2′ W x 4′ L, six 32 watt, T8	G		8.50	1.882		179	103		282	350
3000	Strip, pendent mounted, industrial, white porcelain enamel										
3100	4′ long, two 40 watt, RS		1 Elec	5.70	1.404	Ea.	52.50	77		129.50	173
3110	4′ long, two 32 watt T8, RS	G		5.70	1.404		74	77		151	196
3200	4′ long, two 60 watt, HO			5	1.600		82.50	87.50		170	222
3290	8′ long, two 60 watt T8, SL	G	2 Elec	8.80	1.818		115	99.50		214.50	276
3300	8′ long, two 75 watt, SL			8.80	1.818		98	99.50		197.50	257
3400	8′ long, two 110 watt, HO			8	2		125	109		234	300
3410	Acrylic finish, 4′ long, two 40 watt, RS		1 Elec	5.70	1.404		86	77		163	210
3420	4′ long, two 60 watt, HO			5	1.600		160	87.50		247.50	305
3430	4′ long, two 115 watt, VHO			4.80	1.667		207	91		298	365
3440	8′ long, two 75 watt, SL		2 Elec	8.80	1.818		169	99.50		268.50	335
3450	8′ long, two 110 watt, HO			8	2		192	109		301	375
3460	8′ long, two 215 watt, VHO			7.60	2.105		271	115		386	470
3470	Troffer, air handling, 2′W x 4′L with four 32 watt T8	G	1 Elec	4	2		112	109		221	287
3480	2′W x 2′L with two U32 watt T8	G		5.50	1.455		108	79.50		187.50	238
3490	Air connector insulated, 5″ diameter			20	.400		67.50	22		89.50	107
3500	6″ diameter			20	.400		69	22		91	109
3502	Troffer, direct/indirect, 2′W x 4′L with two 32 W T8	G		5.30	1.509		273	82.50		355.50	425
3510	Troffer parabolic lay-in, 1′W x 4′L with one 32 W T8	G		5.70	1.404		114	77		191	241
3520	1′W x 4′L with two 32 W T8	G		5.30	1.509		137	82.50		219.50	274
3525	2′W x 2′L with two U32 W T8	G		5.70	1.404		115	77		192	242
3530	2′W x 4′L with three 32 W T8	G		5	1.600		128	87.50		215.50	271
3535	Downlight, recess mounted	G		8	1		148	54.50		202.50	244
3540	Wall wash reflector, recess mounted	G		8	1		105	54.50		159.50	198
3550	Direct/indirect, 4′ long, stl., pendent mtd.	G		5	1.600		162	87.50		249.50	310
3560	4′ long, alum., pendent mtd.	G		5	1.600		335	87.50		422.50	500
3565	Prefabricated cove, 4′ long, stl. continuous row	G		5	1.600		203	87.50		290.50	355
3570	4′ long, alum. continuous row	G		5	1.600		345	87.50		432.50	510
3580	Wet location, recess mounted, 2′W x 4′L with two 32 watt T8	G		5.30	1.509		237	82.50		319.50	385
3590	Pendent mounted, 2′W x 4′L with two 32 watt T8	G		5.70	1.404		365	77		442	520
4000	Induction lamp, integral ballast, ceiling mounted										
4110	High bay, aluminum reflector, 160 watt		1 Elec	3.20	2.500	Ea.	955	137		1,092	1,250
4120	320 watt		″	3	2.667		1,725	146		1,871	2,100
4130	480 watt		2 Elec	5.80	2.759		2,600	151		2,751	3,100
4150	Low bay, aluminum reflector, 250 watt		1 Elec	3.20	2.500		755	137		892	1,025
4170	Garage, aluminum reflector, 80 watt			3.60	2.222		755	122		877	1,000
4180	Vandalproof, aluminum reflector, 100 watt			3.20	2.500		535	137		672	790
4220	Metal halide, integral ballast, ceiling, recess mounted										
4230	prismatic glass lens, floating door										
4240	2′W x 2′L, 250 watt		1 Elec	3.20	2.500	Ea.	315	137		452	555
4250	2′W x 2′L, 400 watt		2 Elec	5.80	2.759		365	151		516	625
4260	Surface mounted, 2′W x 2′L, 250 watt		1 Elec	2.70	2.963		345	162		507	625
4270	400 watt		2 Elec	4.80	3.333		405	182		587	720
4280	High bay, aluminum reflector,										
4290	Single unit, 400 watt		2 Elec	4.60	3.478	Ea.	410	190		600	735
4300	Single unit, 1000 watt			4	4		590	219		809	980
4310	Twin unit, 400 watt			3.20	5		820	274		1,094	1,325

26 51 13 – Interior Lighting Fixtures, Lamps, and Ballasts

26 51 13.50 Interior Lighting Fixtures		Crew	Daily Output	Labor-Hours	Unit	2015 Bare Costs Material	Labor	Equipment	Total	Total Incl O&P
4320	Low bay, aluminum reflector, 250W DX lamp	1 Elec	3.20	2.500	Ea.	360	137		497	605
4330	400 watt lamp	2 Elec	5	3.200	↓	525	175		700	840
4340	High pressure sodium integral ballast ceiling, recess mounted									
4350	prismatic glass lens, floating door									
4360	2'W x 2'L, 150 watt lamp	1 Elec	3.20	2.500	Ea.	385	137		522	630
4370	2'W x 2'L, 400 watt lamp	2 Elec	5.80	2.759		460	151		611	730
4380	Surface mounted, 2'W x 2'L, 150 watt lamp	1 Elec	2.70	2.963		470	162		632	765
4390	400 watt lamp	2 Elec	4.80	3.333	↓	525	182		707	855
4400	High bay, aluminum reflector,									
4410	Single unit, 400 watt lamp	2 Elec	4.60	3.478	Ea.	380	190		570	700
4430	Single unit, 1000 watt lamp	"	4	4		545	219		764	930
4440	Low bay, aluminum reflector, 150 watt lamp	1 Elec	3.20	2.500		325	137		462	565
4445	High bay H.I.D. quartz restrike	"	16	.500	↓	163	27.50		190.50	221
4450	Incandescent, high hat can, round alzak reflector, prewired									
4470	100 watt	1 Elec	8	1	Ea.	71.50	54.50		126	161
4480	150 watt		8	1		104	54.50		158.50	196
4500	300 watt		6.70	1.194		241	65.50		306.50	365
4520	Round with reflector and baffles, 150 watt		8	1		51	54.50		105.50	138
4540	Round with concentric louver, 150 watt PAR	↓	8	1	↓	78.50	54.50		133	169
4600	Square glass lens with metal trim, prewired									
4630	100 watt	1 Elec	6.70	1.194	Ea.	55	65.50		120.50	159
4700	200 watt		6.70	1.194		97	65.50		162.50	205
4800	300 watt		5.70	1.404		145	77		222	274
4810	500 watt		5	1.600		285	87.50		372.50	445
4900	Ceiling/wall, surface mounted, metal cylinder, 75 watt		10	.800		56	44		100	127
4920	150 watt		10	.800		80.50	44		124.50	154
4930	300 watt		8	1		167	54.50		221.50	266
5000	500 watt		6.70	1.194		360	65.50		425.50	500
5010	Square, 100 watt		8	1		112	54.50		166.50	205
5020	150 watt		8	1		123	54.50		177.50	218
5030	300 watt		7	1.143		335	62.50		397.50	465
5040	500 watt	↓	6	1.333	↓	340	73		413	485
5200	Ceiling, surface mounted, opal glass drum									
5300	8", one 60 watt lamp	1 Elec	10	.800	Ea.	44.50	44		88.50	115
5400	10", two 60 watt lamps		8	1		50	54.50		104.50	137
5500	12", four 60 watt lamps		6.70	1.194		70.50	65.50		136	176
5510	Pendent, round, 100 watt		8	1		112	54.50		166.50	205
5520	150 watt		8	1		122	54.50		176.50	217
5530	300 watt		6.70	1.194		170	65.50		235.50	285
5540	500 watt		5.50	1.455		320	79.50		399.50	475
5550	Square, 100 watt		6.70	1.194		151	65.50		216.50	264
5560	150 watt		6.70	1.194		157	65.50		222.50	271
5570	300 watt		5.70	1.404		230	77		307	365
5580	500 watt		5	1.600		310	87.50		397.50	470
5600	Wall, round, 100 watt		8	1		65.50	54.50		120	154
5620	300 watt		8	1		131	54.50		185.50	226
5630	500 watt		6.70	1.194		375	65.50		440.50	515
5640	Square, 100 watt		8	1		103	54.50		157.50	195
5650	150 watt		8	1		105	54.50		159.50	198
5660	300 watt		7	1.143		165	62.50		227.50	276
5670	500 watt		6	1.333		291	73		364	430
6010	Vapor tight, incandescent, ceiling mounted, 200 watt		6.20	1.290		82.50	70.50		153	197
6020	Recessed, 200 watt	↓	6.70	1.194	↓	123	65.50		188.50	234

26 51 Interior Lighting

26 51 13 – Interior Lighting Fixtures, Lamps, and Ballasts

26 51 13.50 Interior Lighting Fixtures		Crew	Daily Output	Labor-Hours	Unit	2015 Bare Costs Material	Labor	Equipment	Total	Total Incl O&P
6030	Pendent, 200 watt	1 Elec	6.70	1.194	Ea.	82.50	65.50		148	189
6040	Wall, 200 watt		8	1		79.50	54.50		134	170
6100	Fluorescent, surface mounted, 2 lamps, 4'L, RS, 40 watt		3.20	2.500		115	137		252	330
6110	Industrial, 2 lamps 4' long in tandem, 430 MA		2.20	3.636		212	199		411	530
6130	2 lamps 4' long, 800 MA		1.90	4.211		182	230		412	545
6160	Pendent, indust, 2 lamps 4'L in tandem, 430 MA		1.90	4.211		248	230		478	620
6170	2 lamps 4' long, 430 MA		2.30	3.478		166	190		356	470
6180	2 lamps 4' long, 800 MA		1.70	4.706		209	257		466	615
6850	Vandalproof, surface mounted, fluorescent, two 32 watt T8 G		3.20	2.500		252	137		389	485
6860	Incandescent, one 150 watt		8	1		98.50	54.50		153	190
6900	Mirror light, fluorescent, RS, acrylic enclosure, two 40 watt		8	1		128	54.50		182.50	222
6910	One 40 watt		8	1		106	54.50		160.50	198
6920	One 20 watt		12	.667		81.50	36.50		118	145
7000	Low bay, aluminum reflector. 70 watt, high pressure sodium		4	2		275	109		384	470
7010	250 watt		3.20	2.500		365	137		502	610
7020	400 watt	2 Elec	5	3.200		385	175		560	680
7500	Ballast replacement, by weight of ballast, to 15' high									
7520	Indoor fluorescent, less than 2 lb.	1 Elec	10	.800	Ea.	26	44		70	94
7540	Two 40W, watt reducer, 2 to 5 lb.		9.40	.851		41	46.50		87.50	115
7560	Two F96 slimline, over 5 lb.		8	1		77	54.50		131.50	167
7580	Vaportite ballast, less than 2 lb.		9.40	.851		26	46.50		72.50	98
7600	2 lb. to 5 lb.		8.90	.899		41	49		90	119
7620	Over 5 lb.		7.60	1.053		77	57.50		134.50	171
7630	Electronic ballast for two tubes		8	1		37	54.50		91.50	123
7640	Dimmable ballast one lamp G		8	1		106	54.50		160.50	199
7650	Dimmable ballast two-lamp G		7.60	1.053		104	57.50		161.50	201
7690	Emergency ballast (factory installed in fixture)					157			157	173
7990	Decorator									
8000	Pendent RLM in colors, shallow dome, 12" diam. 100 W	1 Elec	8	1	Ea.	80.50	54.50		135	171
8010	Regular dome, 12" diam., 100 watt		8	1		83	54.50		137.50	174
8020	16" diam., 200 watt		7	1.143		84.50	62.50		147	187
8030	18" diam., 300 watt		6	1.333		93	73		166	211
8100	Picture framing light		16	.500		94	27.50		121.50	144
8150	Miniature low voltage, recessed, pinhole		8	1		138	54.50		192.50	234
8160	Star		8	1		134	54.50		188.50	229
8170	Adjustable cone		8	1		168	54.50		222.50	267
8180	Eyeball		8	1		126	54.50		180.50	220
8190	Cone		8	1		130	54.50		184.50	225
8200	Coilex baffle		8	1		127	54.50		181.50	222
8210	Surface mounted, adjustable cylinder		8	1		131	54.50		185.50	226
8250	Chandeliers, incandescent									
8260	24" diam. x 42" high, 6 light candle	1 Elec	6	1.333	Ea.	445	73		518	600
8270	24" diam. x 42" high, 6 light candle w/glass shade		6	1.333		455	73		528	610
8280	17" diam. x 12" high, 8 light w/glass panels		8	1		282	54.50		336.50	390
8300	27" diam. x 29"H, 10 light bohemian lead crystal		4	2		590	109		699	815
8310	21" diam. x 9" high 6 light sculptured ice crystal		8	1		425	54.50		479.50	545
8500	Accent lights, on floor or edge, 0.5 W low volt incandescent									
8520	incl. transformer & fastenings, based on 100' lengths									
8550	Lights in clear tubing, 12" on center	1 Elec	230	.035	L.F.	8.80	1.90		10.70	12.55
8560	6" on center		160	.050		11.50	2.74		14.24	16.75
8570	4" on center		130	.062		17.55	3.37		20.92	24.50
8580	3" on center		125	.064		19.50	3.50		23	27
8590	2" on center		100	.080		28.50	4.38		32.88	37.50

26 51 13 – Interior Lighting Fixtures, Lamps, and Ballasts

26 51 13.50 Interior Lighting Fixtures

	26 51 13.50 Interior Lighting Fixtures	Crew	Daily Output	Labor-Hours	Unit	2015 Bare Costs Material	Labor	Equipment	Total	Total Incl O&P
8600	Carpet, lights both sides 6″ OC, in alum. extrusion	1 Elec	270	.030	L.F.	26.50	1.62		28.12	31.50
8610	In bronze extrusion		270	.030		30	1.62		31.62	35.50
8620	Carpet-bare floor, lights 18″ OC, in alum. extrusion		270	.030		21	1.62		22.62	26
8630	In bronze extrusion		270	.030		25	1.62		26.62	30
8640	Carpet edge-wall, lights 6″ OC in alum. extrusion		270	.030		26.50	1.62		28.12	31.50
8650	In bronze extrusion		270	.030		30	1.62		31.62	35.50
8660	Bare floor, lights 18″ OC, in aluminum extrusion		300	.027		21	1.46		22.46	25.50
8670	In bronze extrusion		300	.027		25	1.46		26.46	29.50
8680	Bare floor conduit, aluminum extrusion		300	.027		7	1.46		8.46	9.90
8690	In bronze extrusion		300	.027		14.05	1.46		15.51	17.65
8700	Step edge to 36″, lights 6″ OC, in alum. extrusion		100	.080	Ea.	71	4.38		75.38	84.50
8710	In bronze extrusion		100	.080		73.50	4.38		77.88	87.50
8720	Step edge to 54″, lights 6″ OC, in alum. extrusion		100	.080		106	4.38		110.38	124
8730	In bronze extrusion		100	.080		112	4.38		116.38	130
8740	Step edge to 72″, lights 6″ OC, in alum. extrusion		100	.080		142	4.38		146.38	163
8750	In bronze extrusion		100	.080		155	4.38		159.38	177
8760	Connector, male		32	.250		2.62	13.70		16.32	23.50
8770	Female with pigtail		32	.250		5.50	13.70		19.20	26.50
8780	Clamps		400	.020		.51	1.09		1.60	2.20
8790	Transformers, 50 watt		8	1		79.50	54.50		134	169
8800	250 watt		4	2		262	109		371	450
8810	1000 watt		2.70	2.963		470	162		632	765

26 51 13.55 Interior LED Fixtures

	26 51 13.55 Interior LED Fixtures		Crew	Daily Output	Labor-Hours	Unit	Material	Labor	Equipment	Total	Total Incl O&P
0010	**INTERIOR LED FIXTURES** Incl. lamps, and mounting hardware										
0100	Downlight, recess mounted, 7.5″ diameter, 25 watt	G	1 Elec	8	1	Ea.	335	54.50		389.50	445
0120	10″ diameter, 36 watt	G		8	1		360	54.50		414.50	475
0160	cylinder, 10 watts	G		8	1		102	54.50		156.50	194
0180	20 watts	G		8	1		585	54.50		639.50	725
1000	Troffer, recess mounted, 2′ x 4′, 3200 Lumens	G		5.30	1.509		138	82.50		220.50	275
1010	4800 Lumens	G		5	1.600		179	87.50		266.50	325
1020	6400 Lumens	G		4.70	1.702		198	93		291	355
1100	Troffer retrofit lamp, 38 watt	G		21	.381		238	21		259	292
1110	60 watt	G		20	.400		340	22		362	410
1120	100 watt	G		18	.444		510	24.50		534.50	595
1200	Troffer, volumetric recess mounted, 2′ x 2′	G		5.70	1.404		251	77		328	390
2000	Strip, surface mounted, one light bar 4′ long, 3500K	G		8.50	.941		299	51.50		350.50	405
2010	5000K	G		8	1		299	54.50		353.50	410
2020	Two light bar 4′ long, 5000K	G		7	1.143		470	62.50		532.50	610
3000	Linear, suspended mounted, one light bar 4′ long, 37 watt	G		6.70	1.194		195	65.50		260.50	315
3010	One light bar 8′ long, 74 watt	G	2 Elec	12.20	1.311		360	71.50		431.50	505
3020	Two light bar 4′ long, 74 watt	G	1 Elec	5.70	1.404		390	77		467	540
3030	Two light bar 8′ long, 148 watt	G	2 Elec	8.80	1.818		450	99.50		549.50	645
4000	High bay, surface mounted, round, 150 watts	G		5.41	2.959		605	162		767	905
4010	2 bars,164 watts	G		5.41	2.959		570	162		732	865
4020	3 bars, 246 watts	G		5.01	3.197		730	175		905	1,075
4030	4 bars, 328 watts	G		4.60	3.478		895	190		1,085	1,275
4040	5 bars, 410 watts	G	3 Elec	4.20	5.716		1,050	315		1,365	1,625
4050	6 bars, 492 watts	G		3.80	6.324		1,200	345		1,545	1,825
4060	7 bars, 574 watts	G		3.39	7.075		1,350	385		1,735	2,075
4070	8 bars, 656 watts	G		2.99	8.029		1,500	440		1,940	2,300
5000	track, lighthead, 6 watt	G	1 Elec	32	.250		54.50	13.70		68.20	80.50
5010	9 watt	G	″	32	.250		61.50	13.70		75.20	88

26 51 Interior Lighting

26 51 13 – Interior Lighting Fixtures, Lamps, and Ballasts

26 51 13.55 Interior LED Fixtures			Crew	Daily Output	Labor-Hours	Unit	2015 Bare Costs Material	Labor	Equipment	Total	Total Incl O&P
6000	Garage, surface mount, 103 watts	G	2 Elec	6.50	2.462	Ea.	970	135		1,105	1,275
6100	pendent mount, 80 watts	G		6.50	2.462		565	135		700	820
6200	95 watts	G		6.50	2.462		635	135		770	900
6300	125 watts	G		6.50	2.462		690	135		825	960

26 51 13.70 Residential Fixtures		Crew	Daily Output	Labor-Hours	Unit	Material	Labor	Equipment	Total	Total Incl O&P
0010	RESIDENTIAL FIXTURES									
0400	Fluorescent, interior, surface, circline, 32 watt & 40 watt	1 Elec	20	.400	Ea.	130	22		152	176
0500	2′ x 2′, two U-tube 32 watt T8		8	1		165	54.50		219.50	263
0700	Shallow under cabinet, two 20 watt		16	.500		62	27.50		89.50	110
0900	Wall mounted, 4′L, two 32 watt T8, with baffle		10	.800		133	44		177	212
2000	Incandescent, exterior lantern, wall mounted, 60 watt		16	.500		55.50	27.50		83	102
2100	Post light, 150W, with 7′ post		4	2		243	109		352	430
2500	Lamp holder, weatherproof with 150W PAR		16	.500		32	27.50		59.50	76
2550	With reflector and guard		12	.667		56	36.50		92.50	117
2600	Interior pendent, globe with shade, 150 watt		20	.400		201	22		223	254

26 51 13.90 Ballast, Replacement HID		Crew	Daily Output	Labor-Hours	Unit	Material	Labor	Equipment	Total	Total Incl O&P
0010	BALLAST, REPLACEMENT HID									
7510	Multi-tap 120/208/240/277 volt									
7550	High pressure sodium, 70 watt	1 Elec	10	.800	Ea.	136	44		180	216
7560	100 watt		9.40	.851		142	46.50		188.50	226
7570	150 watt		9	.889		153	48.50		201.50	241
7580	250 watt		8.50	.941		227	51.50		278.50	325
7590	400 watt		7	1.143		258	62.50		320.50	380
7600	1000 watt		6	1.333		355	73		428	500
7610	Metal halide, 175 watt		8	1		83.50	54.50		138	174
7620	250 watt		8	1		108	54.50		162.50	201
7630	400 watt		7	1.143		135	62.50		197.50	242
7640	1000 watt		6	1.333		231	73		304	365
7650	1500 watt		5	1.600		286	87.50		373.50	445

26 52 Emergency Lighting

26 52 13 – Emergency Lighting Equipments

26 52 13.10 Emergency Lighting and Battery Units		Crew	Daily Output	Labor-Hours	Unit	Material	Labor	Equipment	Total	Total Incl O&P
0010	EMERGENCY LIGHTING AND BATTERY UNITS									
0300	Emergency light units, battery operated									
0350	Twin sealed beam light, 25 watt, 6 volt each									
0500	Lead battery operated	1 Elec	4	2	Ea.	153	109		262	330
0700	Nickel cadmium battery operated		4	2		560	109		669	780
0780	Additional remote mount, sealed beam, 25 W 6 V		26.70	.300		29	16.40		45.40	56.50
0790	Twin sealed beam light, 25 W 6 V each		26.70	.300		54	16.40		70.40	84
0900	Self-contained fluorescent lamp pack		10	.800		156	44		200	237

26 53 Exit Signs

26 53 13 – Exit Lighting

26 53 13.10 Exit Lighting Fixtures		Crew	Daily Output	Labor-Hours	Unit	2015 Bare Costs Material	Labor	Equipment	Total	Total Incl O&P
0010	**EXIT LIGHTING FIXTURES**									
0080	Exit light ceiling or wall mount, incandescent, single face	1 Elec	8	1	Ea.	39	54.50		93.50	125
0100	Double face		6.70	1.194		39	65.50		104.50	141
0120	Explosion proof		3.80	2.105		595	115		710	830
0150	Fluorescent, single face		8	1		67	54.50		121.50	156
0160	Double face		6.70	1.194		75	65.50		140.50	181
0200	LED standard, single face G		8	1		77	54.50		131.50	167
0220	Double face G		6.70	1.194		77	65.50		142.50	183
0230	LED vandal-resistant, single face G		7.27	1.100		212	60		272	325
0240	LED w/battery unit, single face G		4.40	1.818		160	99.50		259.50	325
0260	Double face G		4	2		163	109		272	345
0262	LED w/battery unit, vandal-resistant, single face G		4.40	1.818		245	99.50		344.50	420
0270	Combination emergency light units and exit sign		4	2		179	109		288	360
0290	LED retrofit kits G		60	.133		51	7.30		58.30	67
1780	With emergency battery, explosion proof	R-19	7.70	2.597		3,625	142		3,767	4,200

26 54 Classified Location Lighting

26 54 13 – Classified Lighting

26 54 13.20 Explosionproof		Crew	Daily Output	Labor-Hours	Unit	Material	Labor	Equipment	Total	Total Incl O&P
0010	**EXPLOSIONPROOF**, incl lamps, mounting hardware and connections									
6310	Metal halide with ballast, ceiling, surface mounted, 175 watt	1 Elec	2.90	2.759	Ea.	915	151		1,066	1,225
6320	250 watt	"	2.70	2.963		1,100	162		1,262	1,450
6330	400 watt	2 Elec	4.80	3.333		1,175	182		1,357	1,575
6340	Ceiling, pendent mounted, 175 watt	1 Elec	2.60	3.077		870	168		1,038	1,200
6350	250 watt	"	2.40	3.333		1,050	182		1,232	1,425
6360	400 watt	2 Elec	4.20	3.810		1,125	208		1,333	1,550
6370	Wall, surface mounted, 175 watt	1 Elec	2.90	2.759		980	151		1,131	1,300
6380	250 watt	"	2.70	2.963		1,150	162		1,312	1,525
6390	400 watt	2 Elec	4.80	3.333		1,250	182		1,432	1,650
6400	High pressure sodium, ceiling surface mounted, 70 watt	1 Elec	3	2.667		1,275	146		1,421	1,625
6410	100 watt		3	2.667		1,325	146		1,471	1,675
6420	150 watt		2.70	2.963		1,375	162		1,537	1,750
6430	Pendent mounted, 70 watt		2.70	2.963		1,225	162		1,387	1,600
6440	100 watt		2.70	2.963		1,275	162		1,437	1,650
6450	150 watt		2.40	3.333		1,300	182		1,482	1,725
6460	Wall mounted, 70 watt		3	2.667		1,375	146		1,521	1,725
6470	100 watt		3	2.667		1,425	146		1,571	1,775
6480	150 watt		2.70	2.963		1,450	162		1,612	1,850
6510	Incandescent, ceiling mounted, 200 watt		4	2		1,075	109		1,184	1,350
6520	Pendent mounted, 200 watt		3.50	2.286		925	125		1,050	1,200
6530	Wall mounted, 200 watt		4	2		1,075	109		1,184	1,350
6600	Fluorescent, RS, 4′ long, ceiling mounted, two 40 watt		2.70	2.963		2,850	162		3,012	3,375
6610	Three 40 watt		2.20	3.636		4,125	199		4,324	4,850
6620	Four 40 watt		1.90	4.211		5,300	230		5,530	6,175
6630	Pendent mounted, two 40 watt		2.30	3.478		3,325	190		3,515	3,925
6640	Three 40 watt		1.90	4.211		4,700	230		4,930	5,525
6650	Four 40 watt		1.70	4.706		6,200	257		6,457	7,200

26 55 Special Purpose Lighting

26 55 59 – Display Lighting

26 55 59.10 Track Lighting		Crew	Daily Output	Labor-Hours	Unit	Material	2015 Bare Costs Labor	Equipment	Total	Total Incl O&P
0010	**TRACK LIGHTING**									
0080	Track, 1 circuit, 4′ section	1 Elec	6.70	1.194	Ea.	44.50	65.50		110	147
0100	8′ section	2 Elec	10.60	1.509		69	82.50		151.50	200
0200	12′ section	″	8.80	1.818		104	99.50		203.50	263
0300	3 circuits, 4′ section	1 Elec	6.70	1.194		83	65.50		148.50	190
0400	8′ section	2 Elec	10.60	1.509		109	82.50		191.50	244
0500	12′ section	″	8.80	1.818		159	99.50		258.50	325
1000	Feed kit, surface mounting	1 Elec	16	.500		13	27.50		40.50	55.50
1100	End cover		24	.333		5.55	18.25		23.80	33.50
1200	Feed kit, stem mounting, 1 circuit		16	.500		39	27.50		66.50	83.50
1300	3 circuit		16	.500		39	27.50		66.50	83.50
2000	Electrical joiner, for continuous runs, 1 circuit		32	.250		21	13.70		34.70	43.50
2100	3 circuit		32	.250		59	13.70		72.70	85.50
2200	Fixtures, spotlight, 75W PAR halogen		16	.500		65.50	27.50		93	113
2210	50W MR16 halogen		16	.500		171	27.50		198.50	229
3000	Wall washer, 250 watt tungsten halogen		16	.500		132	27.50		159.50	186
3100	Low voltage, 25/50 watt, 1 circuit		16	.500		133	27.50		160.50	187
3120	3 circuit		16	.500		198	27.50		225.50	259

26 55 61 – Theatrical Lighting

26 55 61.10 Lights		Crew	Daily Output	Labor-Hours	Unit	Material	Labor	Equipment	Total	Total Incl O&P
0010	**LIGHTS**									
2000	Lights, border, quartz, reflector, vented,									
2100	colored or white	1 Elec	20	.400	L.F.	175	22		197	225
2500	Spotlight, follow spot, with transformer, 2,100 watt	″	4	2	Ea.	3,100	109		3,209	3,600
2600	For no transformer, deduct					920			920	1,025
3000	Stationary spot, fresnel quartz, 6″ lens	1 Elec	4	2		133	109		242	310
3100	8″ lens		4	2		234	109		343	420
3500	Ellipsoidal quartz, 1,000W, 6″ lens		4	2		340	109		449	540
3600	12″ lens		4	2		605	109		714	830
4000	Strobe light, 1 to 15 flashes per second, quartz		3	2.667		760	146		906	1,050
4500	Color wheel, portable, five hole, motorized		4	2		197	109		306	380

26 55 63 – Detention Lighting

26 55 63.10 Detention Lighting Fixtures		Crew	Daily Output	Labor-Hours	Unit	Material	Labor	Equipment	Total	Total Incl O&P
0010	**DETENTION LIGHTING FIXTURES**									
1000	Surface mounted, cold rolled steel, 14 ga., 1′ x 4′	1 Elec	5.40	1.481	Ea.	600	81		681	780
1010	2′ x 2′		5.40	1.481		710	81		791	900
1020	2′ x 4′		4.80	1.667		810	91		901	1,025
1100	12 ga., 1′ x 4′		5.40	1.481		845	81		926	1,050
1110	2′ x 2′		5.40	1.481		710	81		791	900
1120	2′ x 4′		4.80	1.667		885	91		976	1,100

26 56 13 – Lighting Poles and Standards

26 56 13.10 Lighting Poles		Crew	Daily Output	Labor-Hours	Unit	2015 Bare Costs Material	Labor	Equipment	Total	Total Incl O&P
0010	**LIGHTING POLES**									
2800	Light poles, anchor base									
2820	not including concrete bases									
2840	Aluminum pole, 8′ high	1 Elec	4	2	Ea.	705	109		814	940
2850	10′ high		4	2		745	109		854	985
2860	12′ high		3.80	2.105		775	115		890	1,025
2870	14′ high		3.40	2.353		805	129		934	1,075
2880	16′ high		3	2.667		880	146		1,026	1,200
3000	20′ high	R-3	2.90	6.897		935	375	47.50	1,357.50	1,625
3200	30′ high		2.60	7.692		1,775	420	53	2,248	2,625
3400	35′ high		2.30	8.696		1,925	470	60	2,455	2,900
3600	40′ high		2	10		2,200	545	69	2,814	3,325
3800	Bracket arms, 1 arm	1 Elec	8	1		121	54.50		175.50	215
4000	2 arms		8	1		243	54.50		297.50	350
4200	3 arms		5.30	1.509		365	82.50		447.50	525
4400	4 arms		5.30	1.509		485	82.50		567.50	660
4500	Steel pole, galvanized, 8′ high		3.80	2.105		610	115		725	845
4510	10′ high		3.70	2.162		635	118		753	875
4520	12′ high		3.40	2.353		690	129		819	950
4530	14′ high		3.10	2.581		730	141		871	1,025
4540	16′ high		2.90	2.759		775	151		926	1,075
4550	18′ high		2.70	2.963		820	162		982	1,150
4600	20′ high	R-3	2.60	7.692		1,100	420	53	1,573	1,875
4800	30′ high		2.30	8.696		1,300	470	60	1,830	2,200
5000	35′ high		2.20	9.091		1,425	495	62.50	1,982.50	2,375
5200	40′ high		1.70	11.765		1,775	640	81	2,496	3,000
5400	Bracket arms, 1 arm	1 Elec	8	1		180	54.50		234.50	280
5600	2 arms		8	1		216	54.50		270.50	320
5800	3 arms		5.30	1.509		254	82.50		336.50	405
6000	4 arms		5.30	1.509		310	82.50		392.50	465
6100	Fiberglass pole, 1 or 2 fixtures, 20′ high	R-3	4	5		675	272	34.50	981.50	1,200
6200	30′ high		3.60	5.556		835	300	38.50	1,173.50	1,425
6300	35′ high		3.20	6.250		1,300	340	43	1,683	1,975
6400	40′ high		2.80	7.143		1,525	390	49	1,964	2,325
6420	Wood pole, 4-1/2″ x 5-1/8″, 8′ high	1 Elec	6	1.333		330	73		403	475
6430	10′ high		6	1.333		375	73		448	525
6440	12′ high		5.70	1.404		475	77		552	640
6450	15′ high		5	1.600		555	87.50		642.50	740
6460	20′ high		4	2		670	109		779	905
7300	Transformer bases, not including concrete bases									
7320	Maximum pole size, steel, 40′ high	1 Elec	2	4	Ea.	1,425	219		1,644	1,875
7340	Cast aluminum, 30′ high		3	2.667		750	146		896	1,050
7350	40′ high		2.50	3.200		1,150	175		1,325	1,500

26 56 16 – Parking Lighting

26 56 16.55 Parking LED Lighting		Crew	Daily Output	Labor-Hours	Unit	Material	Labor	Equipment	Total	Total Incl O&P
0010	**PARKING LED LIGHTING**									
0100	Round pole mounting, 88 lamp watts G	1 Elec	2	4	Ea.	995	219		1,214	1,425
0110	Square pole mounting, 223 lamp watts G	″	2	4	″	1,750	219		1,969	2,250

26 56 19 – Roadway Lighting

26 56 19.20 Roadway Luminaire		Crew	Daily Output	Labor-Hours	Unit	Material	Labor	Equipment	Total	Total Incl O&P
0010	ROADWAY LUMINAIRE									
2650	Roadway area luminaire, low pressure sodium, 135 watt	1 Elec	2	4	Ea.	650	219		869	1,050
2700	180 watt	″	2	4		700	219		919	1,100
2750	Metal halide, 400 watt	2 Elec	4.40	3.636		555	199		754	910
2760	1000 watt		4	4		625	219		844	1,025
2780	High pressure sodium, 400 watt		4.40	3.636		580	199		779	940
2790	1000 watt		4	4		660	219		879	1,050

26 56 19.55 Roadway LED Luminaire			Crew	Daily Output	Labor-Hours	Unit	Material	Labor	Equipment	Total	Total Incl O&P
0010	ROADWAY LED LUMINAIRE										
0100	LED fixture, 72 LEDs, 120 V AC or 12 V DC, equal to 60 watt	G	1 Elec	2.70	2.963	Ea.	595	162		757	895
0110	108 LEDs, 120 V AC or 12 V DC, equal to 90 watt	G		2.70	2.963		695	162		857	1,000
0120	144 LEDs, 120 V AC or 12 V DC, equal to 120 watt	G		2.70	2.963		855	162		1,017	1,175
0130	252 LEDs, 120 V AC or 12 V DC, equal to 210 watt	G	2 Elec	4.40	3.636		1,175	199		1,374	1,600
0140	Replaces high pressure sodium fixture, 75 watt	G	1 Elec	2.70	2.963		720	162		882	1,050
0150	125 watt	G		2.70	2.963		825	162		987	1,150
0160	150 watt	G		2.70	2.963		1,025	162		1,187	1,375
0170	175 watt	G		2.70	2.963		1,225	162		1,387	1,600
0180	200 watt	G		2.70	2.963		1,450	162		1,612	1,825
0190	250 watt	G	2 Elec	4.40	3.636		1,650	199		1,849	2,125
0200	320 watt	G	″	4.40	3.636		1,800	199		1,999	2,275

26 56 23 – Area Lighting

26 56 23.10 Exterior Fixtures		Crew	Daily Output	Labor-Hours	Unit	Material	Labor	Equipment	Total	Total Incl O&P
0010	**EXTERIOR FIXTURES** With lamps									
0200	Wall mounted, incandescent, 100 watt	1 Elec	8	1	Ea.	35	54.50		89.50	121
0400	Quartz, 500 watt		5.30	1.509		50	82.50		132.50	179
0420	1500 watt		4.20	1.905		103	104		207	269
1100	Wall pack, low pressure sodium, 35 watt		4	2		227	109		336	415
1150	55 watt		4	2		270	109		379	460
1160	High pressure sodium, 70 watt		4	2		224	109		333	410
1170	150 watt		4	2		244	109		353	430
1180	Metal Halide, 175 watt		4	2		248	109		357	435
1190	250 watt		4	2		282	109		391	475
1195	400 watt		4	2		335	109		444	530
1250	Induction lamp, 40 watt		4	2		340	109		449	540
1260	80 watt		4	2		580	109		689	805
1278	LED, poly lens, 26 watt		4	2		330	109		439	525
1280	110 watt		4	2		1,300	109		1,409	1,625
1500	LED, glass lens, 13 watt		4	2		330	109		439	525

26 56 23.55 Exterior LED Fixtures			Crew	Daily Output	Labor-Hours	Unit	Material	Labor	Equipment	Total	Total Incl O&P
0010	EXTERIOR LED FIXTURES										
0100	Wall mounted, indoor/outdoor, 12 watt	G	1 Elec	10	.800	Ea.	258	44		302	350
0110	32 watt	G		10	.800		350	44		394	450
0120	66 watt	G		10	.800		500	44		544	615
0200	outdoor, 110 watt	G		10	.800		765	44		809	905
0210	220 watt	G		10	.800		1,350	44		1,394	1,550
0300	modular, type IV, 120 V, 50 lamp watts	G		9	.889		930	48.50		978.50	1,100
0310	101 lamp watts	G		9	.889		1,050	48.50		1,098.50	1,225
0320	126 lamp watts	G		9	.889		1,325	48.50		1,373.50	1,525
0330	202 lamp watts	G		9	.889		1,500	48.50		1,548.50	1,725
0340	240 V, 50 lamp watts	G		8	1		970	54.50		1,024.50	1,150
0350	101 lamp watts	G		8	1		1,100	54.50		1,154.50	1,275

26 56 Exterior Lighting

26 56 23 – Area Lighting

26 56 23.55 Exterior LED Fixtures			Crew	Daily Output	Labor-Hours	Unit	Material	Labor	Equipment	Total	Total Incl O&P
							2015 Bare Costs				
0360	126 lamp watts	G	1 Elec	8	1	Ea.	1,350	54.50		1,404.50	1,575
0370	202 lamp watts	G		8	1		1,550	54.50		1,604.50	1,775
0400	wall pack, glass, 13 lamp watts	G		4	2		360	109		469	560
0410	poly w/photocell, 26 lamp watts	G		4	2		226	109		335	410
0420	50 lamp watts	G		4	2		530	109		639	745
0430	replacement, 40 watts	G		4	2		460	109		569	670
0440	60 watts	G		4	2		580	109		689	805

26 56 26 – Landscape Lighting

26 56 26.20 Landscape Fixtures		Crew	Daily Output	Labor-Hours	Unit	Material	Labor	Equipment	Total	Total Incl O&P
0010	**LANDSCAPE FIXTURES**									
7380	Landscape recessed uplight, incl. housing, ballast, transformer									
7390	& reflector, not incl. conduit, wire, trench									
7420	Incandescent, 250 watt	1 Elec	5	1.600	Ea.	590	87.50		677.50	780
7440	Quartz, 250 watt		5	1.600		560	87.50		647.50	745
7460	500 watt		4	2		575	109		684	800

26 56 33 – Walkway Lighting

26 56 33.10 Walkway Luminaire		Crew	Daily Output	Labor-Hours	Unit	Material	Labor	Equipment	Total	Total Incl O&P
0010	**WALKWAY LUMINAIRE**									
6500	Bollard light, lamp & ballast, 42" high with polycarbonate lens									
6800	Metal halide, 175 watt	1 Elec	3	2.667	Ea.	815	146		961	1,125
6900	High pressure sodium, 70 watt		3	2.667		835	146		981	1,125
7000	100 watt		3	2.667		835	146		981	1,125
7100	150 watt		3	2.667		815	146		961	1,125
7200	Incandescent, 150 watt		3	2.667		595	146		741	875
7810	Walkway luminaire, square 16", metal halide 250 watt		2.70	2.963		635	162		797	945
7820	High pressure sodium, 70 watt		3	2.667		730	146		876	1,025
7830	100 watt		3	2.667		745	146		891	1,025
7840	150 watt		3	2.667		745	146		891	1,025
7850	200 watt		3	2.667		750	146		896	1,050
7910	Round 19", metal halide, 250 watt		2.70	2.963		930	162		1,092	1,275
7920	High pressure sodium, 70 watt		3	2.667		1,025	146		1,171	1,350
7930	100 watt		3	2.667		1,025	146		1,171	1,350
7940	150 watt		3	2.667		1,025	146		1,171	1,350
7950	250 watt		2.70	2.963		1,075	162		1,237	1,425
8000	Sphere 14" opal, incandescent, 200 watt		4	2		295	109		404	490
8020	Sphere 18" opal, incandescent, 300 watt		3.50	2.286		355	125		480	575
8040	Sphere 16" clear, high pressure sodium, 70 watt		3	2.667		615	146		761	900
8050	100 watt		3	2.667		660	146		806	945
8100	Cube 16" opal, incandescent, 300 watt		3.50	2.286		390	125		515	615
8120	High pressure sodium, 70 watt		3	2.667		570	146		716	850
8130	100 watt		3	2.667		585	146		731	865
8230	Lantern, high pressure sodium, 70 watt		3	2.667		510	146		656	785
8240	100 watt		3	2.667		550	146		696	825
8250	150 watt		3	2.667		515	146		661	785
8260	250 watt		2.70	2.963		720	162		882	1,050
8270	Incandescent, 300 watt		3.50	2.286		380	125		505	605
8330	Reflector 22" w/globe, high pressure sodium, 70 watt		3	2.667		490	146		636	760
8340	100 watt		3	2.667		500	146		646	770
8350	150 watt		3	2.667		505	146		651	775
8360	250 watt		2.70	2.963		645	162		807	955

26 56 Exterior Lighting

26 56 33 – Walkway Lighting

26 56 33.55 Walkway LED Luminaire

26 56 33.55 Walkway LED Luminaire		Crew	Daily Output	Labor-Hours	Unit	Material	2015 Bare Costs Labor	Equipment	Total	Total Incl O&P
0010	**WALKWAY LED LUMINAIRE**									
0100	Pole mounted, 86 watts, 4350 lumens G	1 Elec	3	2.667	Ea.	945	146		1,091	1,275
0110	4630 lumens G		3	2.667		1,575	146		1,721	1,975
0120	80 watts, 4000 lumens G		3	2.667		1,500	146		1,646	1,875

26 56 36 – Flood Lighting

26 56 36.20 Floodlights

26 56 36.20 Floodlights		Crew	Daily Output	Labor-Hours	Unit	Material	Labor	Equipment	Total	Total Incl O&P
0010	**FLOODLIGHTS** with ballast and lamp,									
1290	floor mtd, mount with swivel bracket									
1300	Induction lamp, 40 watt	1 Elec	3	2.667	Ea.	450	146		596	715
1310	80 watt		3	2.667		685	146		831	975
1320	150 watt		3	2.667		1,250	146		1,396	1,600
1400	Pole mounted, pole not included									
1950	Metal halide, 175 watt	1 Elec	2.70	2.963	Ea.	340	162		502	615
2000	400 watt	2 Elec	4.40	3.636		420	199		619	765
2200	1000 watt		4	4		580	219		799	965
2210	1500 watt		3.70	4.324		605	237		842	1,025
2250	Low pressure sodium, 55 watt	1 Elec	2.70	2.963		585	162		747	890
2270	90 watt		2	4		645	219		864	1,050
2290	180 watt		2	4		820	219		1,039	1,225
2340	High pressure sodium, 70 watt		2.70	2.963		246	162		408	515
2360	100 watt		2.70	2.963		252	162		414	520
2380	150 watt		2.70	2.963		290	162		452	565
2400	400 watt	2 Elec	4.40	3.636		380	199		579	720
2600	1000 watt	"	4	4		650	219		869	1,050
9005	Solar powered floodlight, w/motion det, incl batt pack for cloudy days G	1 Elec	8	1		119	54.50		173.50	213
9020	Battery pack for G	"	8	1		13.55	54.50		68.05	97

26 56 36.55 LED Floodlights

26 56 36.55 LED Floodlights		Crew	Daily Output	Labor-Hours	Unit	Material	Labor	Equipment	Total	Total Incl O&P
0010	**LED FLOODLIGHTS** with ballast and lamp,									
0020	Pole mounted, pole not included									
0100	11 watt G	1 Elec	4	2	Ea.	345	109		454	545
0110	46 watt G		4	2		1,050	109		1,159	1,325
0120	90 watt G		4	2		1,725	109		1,834	2,075
0130	288 watt G		4	2		2,125	109		2,234	2,500

26 61 Lighting Systems and Accessories

26 61 13 – Lighting Accessories

26 61 13.30 Fixture Whips

26 61 13.30 Fixture Whips		Crew	Daily Output	Labor-Hours	Unit	Material	Labor	Equipment	Total	Total Incl O&P
0010	**FIXTURE WHIPS**									
0080	3/8" Greenfield, 2 connectors, 6' long									
0100	TFFN wire, three #18	1 Elec	32	.250	Ea.	7.30	13.70		21	28.50
0150	Four #18		28	.286		7.75	15.65		23.40	32
0200	Three #16		32	.250		7.35	13.70		21.05	28.50
0250	Four #16		28	.286		7.85	15.65		23.50	32
0300	THHN wire, three #14		32	.250		8.55	13.70		22.25	30
0350	Four #14		28	.286		9.70	15.65		25.35	34
0360	Three #12		32	.250		11.10	13.70		24.80	32.50

26 61 23.10 Lamps		Crew	Daily Output	Labor-Hours	Unit	Material	Labor	Equipment	Total	Total Incl O&P
						2015 Bare Costs				
0010	**LAMPS**									
0080	Fluorescent, rapid start, cool white, 2' long, 20 watt	1 Elec	1	8	C	310	440		750	995
0100	4' long, 40 watt		.90	8.889		236	485		721	990
0120	3' long, 30 watt		.90	8.889		410	485		895	1,175
0125	3' long, 25 watt energy saver R265723-10 G		.90	8.889		1,200	485		1,685	2,025
0150	U-40 watt		.80	10		1,175	545		1,720	2,125
0155	U-34 watt energy saver G		.80	10		1,200	545		1,745	2,150
0170	4' long, 34 watt energy saver G		.90	8.889		430	485		915	1,200
0176	2' long, T8, 17 W energy saver G		1	8		390	440		830	1,075
0178	3' long, T8, 25 W energy saver G		.90	8.889		415	485		900	1,200
0180	4' long, T8, 32 watt energy saver G		.90	8.889		305	485		790	1,075
0200	Slimline, 4' long, 40 watt		.90	8.889		1,250	485		1,735	2,100
0210	4' long, 30 watt energy saver G		.90	8.889		1,250	485		1,735	2,100
0300	8' long, 75 watt		.80	10		1,050	545		1,595	1,975
0350	8' long, 60 watt energy saver G		.80	10		450	545		995	1,325
0400	High output, 4' long, 60 watt		.90	8.889		650	485		1,135	1,450
0410	8' long, 95 watt energy saver G		.80	10		640	545		1,185	1,525
0500	8' long, 110 watt		.80	10		640	545		1,185	1,525
0512	2' long, T5, 14 watt energy saver G		1	8		860	440		1,300	1,600
0514	3' long, T5, 21 watt energy saver G		.90	8.889		1,000	485		1,485	1,850
0516	4' long, T5, 28 watt energy saver G		.90	8.889		1,300	485		1,785	2,150
0517	4' long, T5, 54 watt energy saver G		.90	8.889		1,575	485		2,060	2,450
0520	Very high output, 4' long, 110 watt		.90	8.889		2,175	485		2,660	3,125
0525	8' long, 195 watt energy saver G		.70	11.429		1,775	625		2,400	2,875
0550	8' long, 215 watt		.70	11.429		1,575	625		2,200	2,650
0554	Full spectrum, 4' long, 60 watt		.90	8.889		1,250	485		1,735	2,100
0556	6' long, 85 watt		.90	8.889		1,300	485		1,785	2,150
0558	8' long, 110 watt		.80	10		1,825	545		2,370	2,825
0560	Twin tube compact lamp G		.90	8.889		360	485		845	1,125
0570	Double twin tube compact lamp G		.80	10		805	545		1,350	1,700
0600	Mercury vapor, mogul base, deluxe white, 100 watt		.30	26.667		3,650	1,450		5,100	6,200
0650	175 watt		.30	26.667		1,725	1,450		3,175	4,050
0700	250 watt		.30	26.667		3,000	1,450		4,450	5,475
0800	400 watt		.30	26.667		4,350	1,450		5,800	6,950
0900	1000 watt		.20	40		7,350	2,200		9,550	11,400
1000	Metal halide, mogul base, 175 watt		.30	26.667		2,550	1,450		4,000	5,000
1100	250 watt		.30	26.667		2,900	1,450		4,350	5,350
1200	400 watt		.30	26.667		4,750	1,450		6,200	7,400
1300	1000 watt		.20	40		6,625	2,200		8,825	10,600
1320	1000 watt, 125,000 initial lumens		.20	40		22,100	2,200		24,300	27,600
1330	1500 watt		.20	40		27,900	2,200		30,100	34,000
1350	High pressure sodium, 70 watt		.30	26.667		3,125	1,450		4,575	5,600
1360	100 watt		.30	26.667		3,250	1,450		4,700	5,750
1370	150 watt		.30	26.667		3,325	1,450		4,775	5,850
1380	250 watt		.30	26.667		4,225	1,450		5,675	6,825
1400	400 watt		.30	26.667		5,525	1,450		6,975	8,250
1450	1000 watt		.20	40		10,000	2,200		12,200	14,300
1500	Low pressure sodium, 35 watt		.30	26.667		12,000	1,450		13,450	15,400
1550	55 watt		.30	26.667		11,900	1,450		13,350	15,300
1600	90 watt		.30	26.667		13,400	1,450		14,850	17,000
1650	135 watt		.20	40		20,000	2,200		22,200	25,300
1700	180 watt		.20	40		23,100	2,200		25,300	28,700

26 61 Lighting Systems and Accessories

26 61 23 – Lamps Applications

26 61 23.10 Lamps		Crew	Daily Output	Labor-Hours	Unit	2015 Bare Costs Material	Labor	Equipment	Total	Total Incl O&P
1750	Quartz line, clear, 500 watt	1 Elec	1.10	7.273	C	1,375	400		1,775	2,125
1760	1500 watt		.20	40		3,050	2,200		5,250	6,625
1762	Spot, MR 16, 50 watt		1.30	6.154		1,075	335		1,410	1,675
1770	Tungsten halogen, T4, 400 watt		1.10	7.273		3,975	400		4,375	4,975
1775	T3, 1200 watt		.30	26.667		4,900	1,450		6,350	7,575
1778	PAR 30, 50 watt		1.30	6.154		1,150	335		1,485	1,750
1780	PAR 38, 90 watt		1.30	6.154		960	335		1,295	1,550
1800	Incandescent, interior, A21, 100 watt		1.60	5		233	274		507	665
1900	A21, 150 watt		1.60	5		166	274		440	595
2000	A23, 200 watt		1.60	5		330	274		604	775
2200	PS 35, 300 watt		1.60	5		860	274		1,134	1,350
2210	PS 35, 500 watt		1.60	5		1,425	274		1,699	1,950
2230	PS 52, 1000 watt		1.30	6.154		2,550	335		2,885	3,300
2240	PS 52, 1500 watt		1.30	6.154		6,825	335		7,160	8,025
2300	R30, 75 watt		1.30	6.154		620	335		955	1,175
2400	R40, 100 watt		1.30	6.154		620	335		955	1,175
2500	Exterior, PAR 38, 75 watt		1.30	6.154		1,850	335		2,185	2,550
2600	PAR 38, 150 watt		1.30	6.154		2,000	335		2,335	2,700
2700	PAR 46, 200 watt		1.10	7.273		3,875	400		4,275	4,850
2800	PAR 56, 300 watt		1.10	7.273		4,425	400		4,825	5,475
3000	Guards, fluorescent lamp, 4' long		1	8		1,400	440		1,840	2,200
3200	8' long		.90	8.889		2,800	485		3,285	3,825

26 61 23.55 LED Lamps		Crew	Daily Output	Labor-Hours	Unit	Material	Labor	Equipment	Total	Total Incl O&P
0010	**LED LAMPS**									
0100	LED lamp, interior, shape A60, equal to 60 watt G	1 Elec	160	.050	Ea.	22.50	2.74		25.24	28.50
0200	Globe frosted A60, equal to 60 watt G		160	.050		22.50	2.74		25.24	28.50
0300	Globe earth, equal to 100 watt G		160	.050		74	2.74		76.74	85.50
1100	MR16, 3 watt, replacement of halogen lamp 25 watt G		130	.062		21.50	3.37		24.87	28.50
1200	6 watt replacement of halogen lamp 45 watt G		130	.062		47	3.37		50.37	56.50
2100	10 watt, PAR20, equal to 60 watt G		130	.062		47.50	3.37		50.87	57.50
2200	15 watt, PAR30, equal to 100 watt G		130	.062		79	3.37		82.37	92

26 71 Electrical Machines

26 71 13 – Motors Applications

26 71 13.10 Handling		Crew	Daily Output	Labor-Hours	Unit	Material	Labor	Equipment	Total	Total Incl O&P
0010	**HANDLING** Add to normal labor cost for restricted areas									
5000	Motors									
5100	1/2 HP, 23 pounds	1 Elec	4	2	Ea.		109		109	164
5110	3/4 HP, 28 pounds		4	2			109		109	164
5120	1 HP, 33 pounds		4	2			109		109	164
5130	1-1/2 HP, 44 pounds		3.20	2.500			137		137	205
5140	2 HP, 56 pounds		3	2.667			146		146	219
5150	3 HP, 71 pounds		2.30	3.478			190		190	285
5160	5 HP, 82 pounds		1.90	4.211			230		230	345
5170	7-1/2 HP, 124 pounds		1.50	5.333			292		292	435
5180	10 HP, 144 pounds		1.20	6.667			365		365	545
5190	15 HP, 185 pounds		1	8			440		440	655
5200	20 HP, 214 pounds	2 Elec	1.50	10.667			585		585	875
5210	25 HP, 266 pounds		1.40	11.429			625		625	935
5220	30 HP, 310 pounds		1.20	13.333			730		730	1,100
5230	40 HP, 400 pounds		1	16			875		875	1,300

26 71 13 – Motors Applications

26 71 13.10 Handling		Crew	Daily Output	Labor-Hours	Unit	Material	2015 Bare Costs Labor	Equipment	Total	Total Incl O&P
5240	50 HP, 450 pounds	2 Elec	.90	17.778	Ea.		970		970	1,450
5250	75 HP, 680 pounds		.80	20			1,100		1,100	1,650
5260	100 HP, 870 pounds	3 Elec	1	24			1,325		1,325	1,975
5270	125 HP, 940 pounds		.80	30			1,650		1,650	2,450
5280	150 HP, 1200 pounds		.70	34.286			1,875		1,875	2,800
5290	175 HP, 1300 pounds		.60	40			2,200		2,200	3,275
5300	200 HP, 1400 pounds		.50	48			2,625		2,625	3,925

26 71 13.20 Motors		Crew	Daily Output	Labor-Hours	Unit	Material	Labor	Equipment	Total	Total Incl O&P
0010	**MOTORS** 230/460 volts, 60 HZ R263413-30									
0050	Dripproof, premium efficiency, 1.15 service factor									
0060	1800 RPM, 1/4 HP	1 Elec	5.33	1.501	Ea.	190	82		272	330
0070	1/3 HP		5.33	1.501		195	82		277	340
0080	1/2 HP		5.33	1.501		218	82		300	365
0090	3/4 HP		5.33	1.501		248	82		330	395
0100	1 HP		4.50	1.778		279	97		376	450
0150	2 HP		4.50	1.778		335	97		432	515
0200	3 HP		4.50	1.778		560	97		657	760
0250	5 HP		4.50	1.778		610	97		707	815
0300	7.5 HP		4.20	1.905		810	104		914	1,050
0350	10 HP		4	2		990	109		1,099	1,275
0400	15 HP		3.20	2.500		1,375	137		1,512	1,725
0450	20 HP	2 Elec	5.20	3.077		1,750	168		1,918	2,175
0500	25 HP		5	3.200		2,050	175		2,225	2,525
0550	30 HP		4.80	3.333		2,250	182		2,432	2,750
0600	40 HP		4	4		2,925	219		3,144	3,550
0650	50 HP		3.20	5		3,200	274		3,474	3,925
0700	60 HP		2.80	5.714		4,075	315		4,390	4,950
0750	75 HP		2.40	6.667		4,150	365		4,515	5,125
0800	100 HP	3 Elec	2.70	8.889		5,350	485		5,835	6,600
0850	125 HP		2.10	11.429		6,100	625		6,725	7,650
0900	150 HP		1.80	13.333		7,925	730		8,655	9,825
0950	200 HP		1.50	16		9,650	875		10,525	11,900
1000	1200 RPM, 1 HP	1 Elec	4.50	1.778		475	97		572	665
1050	2 HP		4.50	1.778		560	97		657	760
1100	3 HP		4.50	1.778		720	97		817	935
1150	5 HP		4.50	1.778		945	97		1,042	1,175
1200	3600 RPM, 2 HP		4.50	1.778		465	97		562	660
1250	3 HP		4.50	1.778		475	97		572	665
1300	5 HP		4.50	1.778		515	97		612	715
1350	Totally enclosed, premium efficiency 1.15 service factor									
1360	1800 RPM, 1/4 HP	1 Elec	5.33	1.501	Ea.	263	82		345	415
1370	1/3 HP		5.33	1.501		291	82		373	445
1380	1/2 HP		5.33	1.501		340	82		422	500
1390	3/4 HP		5.33	1.501		375	82		457	535
1400	1 HP		4.50	1.778		485	97		582	675
1450	2 HP		4.50	1.778		575	97		672	775
1500	3 HP		4.50	1.778		615	97		712	820
1550	5 HP		4.50	1.778		730	97		827	950
1600	7.5 HP		4.20	1.905		960	104		1,064	1,200
1650	10 HP		4	2		1,100	109		1,209	1,375
1700	15 HP		3.20	2.500		1,625	137		1,762	2,000
1750	20 HP	2 Elec	5.20	3.077		2,225	168		2,393	2,675

26 71 Electrical Machines

26 71 13 – Motors Applications

26 71 13.20 Motors		Crew	Daily Output	Labor-Hours	Unit	2015 Bare Costs Material	Labor	Equipment	Total	Total Incl O&P
1800	25 HP	2 Elec	5	3.200	Ea.	2,575	175		2,750	3,075
1850	30 HP		4.80	3.333		2,725	182		2,907	3,250
1900	40 HP		4	4		3,325	219		3,544	3,975
1950	50 HP		3.20	5		3,575	274		3,849	4,350
2000	60 HP		2.80	5.714		5,275	315		5,590	6,275
2050	75 HP		2.40	6.667		6,425	365		6,790	7,625
2100	100 HP	3 Elec	2.70	8.889		8,200	485		8,685	9,725
2150	125 HP		2.10	11.429		11,500	625		12,125	13,500
2200	150 HP		1.80	13.333		12,600	730		13,330	14,900
2250	200 HP		1.50	16		15,900	875		16,775	18,800
2300	1200 RPM, 1 HP	1 Elec	4.50	1.778		410	97		507	595
2350	2 HP		4.50	1.778		470	97		567	665
2400	3 HP		4.50	1.778		640	97		737	850
2450	5 HP		4.50	1.778		870	97		967	1,100
2500	3600 RPM, 2 HP		4.50	1.778		365	97		462	550
2550	3 HP		4.50	1.778		455	97		552	645
2600	5 HP		4.50	1.778		570	97		667	770

Division Notes

		CREW	DAILY OUTPUT	LABOR-HOURS	UNIT	BARE COSTS MAT.	LABOR	EQUIP.	TOTAL	TOTAL INCL O&P

Estimating Tips

27 20 00 Data Communications
27 30 00 Voice Communications
27 40 00 Audio-Video Communications

When estimating material costs for special systems, it is always prudent to obtain manufacturers' quotations for equipment prices and special installation requirements which will affect the total costs.

Reference Numbers

Reference numbers are shown in shaded boxes at the beginning of some major classifications. These numbers refer to related items in the Reference Section. The reference information may be an estimating procedure, an alternate pricing method, or technical information.

Note: Not all subdivisions listed here necessarily appear in this publication. ■

Note: ***Trade Service****, in part, has been used as a reference source for some of the material prices used in Division 27.*

27 01 Operation and Maintenance of Communications Systems

27 01 30 – Operation and Maintenance of Voice Communications

27 01 30.51 Operation and Maintenance of Voice Equipment	Crew	Daily Output	Labor-Hours	Unit	Material	2015 Bare Costs Labor	Equipment	Total	Total Incl O&P
0010 **OPERATION AND MAINTENANCE OF VOICE EQUIPMENT**									
3400 Remove and replace (reinstall), speaker	1 Elec	6	1.333	Ea.		73		73	109

27 05 Common Work Results for Communications

27 05 05 – Selective Demolition for Communications

27 05 05.20 Electrical Demolition, Communications	Crew	Daily Output	Labor-Hours	Unit	Material	2015 Bare Costs Labor	Equipment	Total	Total Incl O&P
0010 **ELECTRICAL DEMOLITION, COMMUNICATIONS** R024119-10									
0100 Fiber optics									
0120 Cable R260105-30	1 Elec	2400	.003	L.F.		.18		.18	.27
0160 Multi-channel rack enclosure		6	1.333	Ea.		73		73	109
0180 Patch panel	↓	18	.444	"		24.50		24.50	36.50
0200 Communication cables & fittings									
0220 Voice/data outlet	1 Elec	140	.057	Ea.		3.13		3.13	4.68
0240 Telephone cable		2800	.003	L.F.		.16		.16	.23
0260 Phone jack		135	.059	Ea.		3.24		3.24	4.86
0300 High performance cable, 2 pair		3000	.003	L.F.		.15		.15	.22
0320 4 pair		2100	.004			.21		.21	.31
0340 25 pair		900	.009	↓		.49		.49	.73
0400 Terminal cabinet	↓	5	1.600	Ea.		87.50		87.50	131
1000 Nurse call system									
1020 Station	1 Elec	24	.333	Ea.		18.25		18.25	27.50
1040 Standard call button		24	.333			18.25		18.25	27.50
1060 Corridor dome light or zone indictor	↓	24	.333			18.25		18.25	27.50
1080 Master control station	2 Elec	2	8	↓		440		440	655

27 05 05.30 Electrical Demolition, Sound and Video	Crew	Daily Output	Labor-Hours	Unit	Material	2015 Bare Costs Labor	Equipment	Total	Total Incl O&P
0010 **ELECTRICAL DEMOLITION, SOUND & VIDEO** R024119-10									
0100 Cables									
0120 TV antenna lead-in cable R260105-30	1 Elec	2100	.004	L.F.		.21		.21	.31
0140 Sound cable		2400	.003			.18		.18	.27
0160 Microphone cable		2400	.003			.18		.18	.27
0180 Coaxial cable		2400	.003	↓		.18		.18	.27
0200 Doorbell system, not including wires, cables, and conduit		16	.500	Ea.		27.50		27.50	41
0220 Door chime or devices	↓	36	.222	"		12.15		12.15	18.20
0300 Public address system, not including wires, cables, and conduit									
0320 Conventional office	1 Elec	16	.500	Speaker		27.50		27.50	41
0340 Conventional industrial	"	8	1	"		54.50		54.50	82
0352 PA cabinet/panel	2 Elec	5	3.200	Ea.		175		175	262
0400 Sound system, not including wires, cables, and conduit									
0410 Components	1 Elec	24	.333	Ea.		18.25		18.25	27.50
0412 Speaker		24	.333			18.25		18.25	27.50
0416 Volume control		24	.333			18.25		18.25	27.50
0420 Intercom, master station		6	1.333			73		73	109
0440 Remote station		24	.333			18.25		18.25	27.50
0460 Intercom outlets		24	.333			18.25		18.25	27.50
0480 Handset	↓	12	.667	↓		36.50		36.50	54.50
0500 Emergency call system, not including wires, cables, and conduit									
0520 Annunciator	1 Elec	4	2	Ea.		109		109	164
0540 Devices	"	16	.500			27.50		27.50	41
0600 Master door, buzzer type unit	2 Elec	1.60	10	↓		545		545	820
0800 TV System, not including wires, cables, and conduit									
0820 Master TV antenna system, per outlet	1 Elec	39	.205	Outlet		11.20		11.20	16.80

27 05 Common Work Results for Communications

27 05 05 – Selective Demolition for Communications

27 05 05.30	Electrical Demolition, Sound and Video	Crew	Daily Output	Labor-Hours	Unit	Material	2015 Bare Costs Labor	Equipment	Total	Total Incl O&P
0840	School application, per outlet	1 Elec	16	.500	Outlet		27.50		27.50	41
0860	Amplifier		12	.667	Ea.		36.50		36.50	54.50
0880	Antenna	↓	6	1.333	"		73		73	109
0900	One camera & one monitor	2 Elec	7.80	2.051	Total		112		112	168
0920	One camera	1 Elec	8	1	Ea.		54.50		54.50	82

27 11 Communications Equipment Room Fittings

27 11 16 – Communications Cabinets, Racks, Frames and Enclosures

27 11 16.10	Public Phone	Crew	Daily Output	Labor-Hours	Unit	Material	Labor	Equipment	Total	Total Incl O&P
0010	**PUBLIC PHONE**									
7600	Telephone with wood backboard									
7620	Single door, 12" H x 12" W x 4" D	1 Elec	5.30	1.509	Ea.	88	82.50		170.50	221
7650	18" H x 12" W x 4" D		4.70	1.702		103	93		196	253
7700	24" H x 12" W x 4" D		4.20	1.905		113	104		217	280
7720	18" H x 18" W x 4" D		4.20	1.905		123	104		227	292
7750	24" H x 18" W x 4" D		4	2		175	109		284	355
7780	36" H x 36" W x 4" D		3.60	2.222		199	122		321	400
7800	24" H x 24" W x 6" D		3.60	2.222		218	122		340	420
7820	30" H x 24" W x 6" D		3.20	2.500		258	137		395	490
7850	30" H x 30" W x 6" D		2.70	2.963		300	162		462	580
7880	36" H x 30" W x 6" D		2.50	3.200		370	175		545	665
7900	48" H x 36" W x 6" D		2.20	3.636		610	199		809	970
7920	Double door, 48" H x 36" W x 6" D	↓	2	4	↓	850	219		1,069	1,275

27 11 19 – Communications Termination Blocks and Patch Panels

27 11 19.10	Termination Blocks and Patch Panels	Crew	Daily Output	Labor-Hours	Unit	Material	Labor	Equipment	Total	Total Incl O&P
0010	**TERMINATION BLOCKS AND PATCH PANELS**									
2960	Patch panel, RJ-45/110 type, 24 ports	2 Elec	6	2.667	Ea.	192	146		338	430
3000	48 ports	3 Elec	6	4		320	219		539	685
3040	96 ports	"	4	6		535	330		865	1,075
3100	Punch down termination per port	1 Elec	107	.075	↓		4.09		4.09	6.15

27 13 Communications Backbone Cabling

27 13 23 – Communications Optical Fiber Backbone Cabling

27 13 23.13	Communications Optical Fiber	Crew	Daily Output	Labor-Hours	Unit	Material	Labor	Equipment	Total	Total Incl O&P
0010	**COMMUNICATIONS OPTICAL FIBER**									
0040	Specialized tools & techniques cause installation costs to vary.									
0070	Fiber optic, cable, bulk simplex, single mode R271323-40	1 Elec	8	1	C.L.F.	22.50	54.50		77	107
0080	Multi mode		8	1		42	54.50		96.50	129
0090	4 strand, single mode		7.34	1.090		38	59.50		97.50	131
0095	Multi mode		7.34	1.090		50.50	59.50		110	145
0100	12 strand, single mode		6.67	1.199		90	65.50		155.50	198
0105	Multi mode	↓	6.67	1.199	↓	96.50	65.50		162	205
0150	Jumper				Ea.	33			33	36.50
0200	Pigtail					33.50			33.50	37
0300	Connector	1 Elec	24	.333		23.50	18.25		41.75	53
0350	Finger splice		32	.250		32.50	13.70		46.20	56.50
0400	Transceiver (low cost bi-directional)		8	1		420	54.50		474.50	540
0450	Rack housing, 4 rack spaces, 12 panels (144 fibers)		2	4		500	219		719	880
0500	Patch panel, 12 ports	↓	6	1.333	↓	247	73		320	380

27 15 Communications Horizontal Cabling

27 15 10 – Special Communications Cabling

27 15 10.23 Sound and Video Cables and Fittings		Crew	Daily Output	Labor-Hours	Unit	2015 Bare Costs Material	Labor	Equipment	Total	Total Incl O&P
0010	SOUND AND VIDEO CABLES & FITTINGS									
0900	TV antenna lead-in, 300 ohm, #20-2 conductor	1 Elec	7	1.143	C.L.F.	19.50	62.50		82	115
0950	Coaxial, feeder outlet		7	1.143		17.35	62.50		79.85	113
1000	Coaxial, main riser		6	1.333		25	73		98	136
1100	Sound, shielded with drain, #22-2 conductor		8	1		13.95	54.50		68.45	97.50
1150	#22-3 conductor		7.50	1.067		17.55	58.50		76.05	107
1200	#22-4 conductor		6.50	1.231		22	67.50		89.50	126
1250	Nonshielded, #22-2 conductor		10	.800		14.25	44		58.25	81
1300	#22-3 conductor		9	.889		21.50	48.50		70	96.50
1350	#22-4 conductor		8	1		28.50	54.50		83	114
1400	Microphone cable		8	1		49.50	54.50		104	137

27 15 13 – Communications Copper Horizontal Cabling

27 15 13.13 Communication Cables		Crew	Daily Output	Labor-Hours	Unit	Material	Labor	Equipment	Total	Total Incl O&P
0010	COMMUNICATION CABLES									
2200	Telephone twisted, PVC insulation, #22-2 conductor	1 Elec	10	.800	C.L.F.	10.40	44		54.40	77
2250	#22-3 conductor		9	.889		12.45	48.50		60.95	86.50
2300	#22-4 conductor		8	1		15.90	54.50		70.40	99.50
2350	#18-2 conductor		9	.889		14.55	48.50		63.05	89
2370	Telephone jack, eight pins		32	.250	Ea.	4.12	13.70		17.82	25
5000	High performance unshielded twisted pair (UTP)									
5100	Cable, category 3, #24, 2 pair solid, PVC jacket R271513-75	1 Elec	10	.800	C.L.F.	9.25	44		53.25	75.50
5200	4 pair solid, PVC jacket		7	1.143		13.90	62.50		76.40	109
5300	25 pair solid, PVC jacket		3	2.667		78.50	146		224.50	305
5400	2 pair solid, plenum		10	.800		13.05	44		57.05	80
5500	4 pair solid, plenum		7	1.143		14.15	62.50		76.65	109
5600	25 pair solid, plenum		3	2.667		111	146		257	340
5700	4 pair stranded, PVC jacket		7	1.143		30.50	62.50		93	127
7000	Category 5, #24, 4 pair solid, PVC jacket		7	1.143		14.95	62.50		77.45	110
7100	4 pair solid, plenum		7	1.143		45.50	62.50		108	144
7200	4 pair stranded, PVC jacket		7	1.143		22.50	62.50		85	118
7210	Category 5e, #24, 4 pair solid, PVC jacket		7	1.143		14.30	62.50		76.80	109
7212	4 pair solid, plenum		7	1.143		23.50	62.50		86	120
7214	4 pair stranded, PVC jacket		7	1.143		25.50	62.50		88	122
7240	Category 6, #24, 4 pair solid, PVC jacket		7	1.143		20.50	62.50		83	116
7242	4 pair solid, plenum		7	1.143		24.50	62.50		87	121
7244	4 pair stranded, PVC jacket		7	1.143		25.50	62.50		88	122
7300	Connector, RJ-45, category 5		80	.100	Ea.	1.33	5.45		6.78	9.65
7302	Shielded RJ-45, category 5		72	.111		3.39	6.10		9.49	12.85
7310	Jack, UTP RJ-45, category 3		72	.111		2.35	6.10		8.45	11.70
7312	Category 5		65	.123		5.05	6.75		11.80	15.65
7314	Category 5e		65	.123		3.29	6.75		10.04	13.70
7316	Category 6		65	.123		3.34	6.75		10.09	13.75
7322	Jack, shielded RJ-45, category 5		60	.133		6.10	7.30		13.40	17.65
7324	Category 5e		60	.133		6.10	7.30		13.40	17.65
7326	Category 6		60	.133		6.10	7.30		13.40	17.65
7400	Voice/data expansion module, category 5e		8	1		46.50	54.50		101	133
8000	Multipair unshielded non-plenum cable, 150 V PVC jacket									
8002	#22, 2 pair	1 Elec	8.40	.952	C.L.F.	16.75	52		68.75	96.50
8003	3 pair		8	1		22	54.50		76.50	107
8004	4 pair		7.30	1.096		42	60		102	136
8006	6 pair		6.20	1.290		42.50	70.50		113	153
8008	8 pair		5.70	1.404		55.50	77		132.50	176

27 15 13 – Communications Copper Horizontal Cabling

27 15 13.13 Communication Cables		Crew	Daily Output	Labor-Hours	Unit	Material	2015 Bare Costs Labor	Equipment	Total	Total Incl O&P
8010	10 pair	1 Elec	5.30	1.509	C.L.F.	64.50	82.50		147	195
8012	12 pair		4.20	1.905		140	104		244	310
8015	15 pair		3.80	2.105		190	115		305	380
8020	20 pair	2 Elec	6.60	2.424		195	133		328	415
8025	25 pair		6	2.667		226	146		372	470
8030	30 pair		5.60	2.857		375	156		531	650
8040	40 pair		5.20	3.077		495	168		663	790
8050	50 pair		4.90	3.265		615	179		794	950
8100	Multipair unshielded non-plenum cable, 300 V PVC jacket									
8102	#20, 2 pair	1 Elec	7.30	1.096	C.L.F.	26	60		86	119
8103	3 pair		6.70	1.194		34	65.50		99.50	135
8104	4 pair		6	1.333		38	73		111	151
8106	6 pair		5.30	1.509		59	82.50		141.50	189
8108	8 pair		4.40	1.818		97	99.50		196.50	255
8110	10 pair	2 Elec	7.30	2.192		107	120		227	297
8112	12 pair		6.20	2.581		135	141		276	360
8115	15 pair		5.90	2.712		169	148		317	410
8201	#18, 1 pair	1 Elec	8	1		63	54.50		117.50	152
8202	2 pair		6.50	1.231		80	67.50		147.50	190
8203	3 pair		5.60	1.429		121	78		199	250
8204	4 pair		5	1.600		158	87.50		245.50	305
8206	6 pair		4.40	1.818		266	99.50		365.50	440
8208	8 pair		4	2		245	109		354	435
8215	15 pair	2 Elec	5	3.200		650	175		825	980
8300	Multipair shielded non-plenum cable, 300 V PVC jacket									
8303	#22, 3 pair	1 Elec	6.70	1.194	C.L.F.	62.50	65.50		128	167
8306	6 pair		5.70	1.404		125	77		202	252
8309	9 pair		5	1.600		195	87.50		282.50	345
8312	12 pair	2 Elec	8	2		490	109		599	705
8315	15 pair		7.30	2.192		505	120		625	740
8317	17 pair		6.80	2.353		690	129		819	950
8319	19 pair		6.40	2.500		690	137		827	965
8327	27 pair		5.90	2.712		1,250	148		1,398	1,600
8402	#20, 2 pair	1 Elec	6.70	1.194		77.50	65.50		143	184
8403	3 pair		6.20	1.290		97	70.50		167.50	212
8406	6 pair		4.40	1.818		244	99.50		343.50	415
8409	9 pair		3.60	2.222		280	122		402	490
8412	12 pair	2 Elec	5.90	2.712		565	148		713	840
8415	15 pair	"	5.70	2.807		475	154		629	750
8502	#18, 2 pair	1 Elec	6.20	1.290		133	70.50		203.50	252
8503	3 pair		5.30	1.509		208	82.50		290.50	350
8504	4 pair		4.70	1.702		247	93		340	410
8506	6 pair		4	2		385	109		494	590
8509	9 pair		3.40	2.353		590	129		719	845
8515	15 pair	2 Elec	4.80	3.333		985	182		1,167	1,350

27 15 33 – Communications Coaxial Horizontal Cabling

27 15 33.10 Coaxial Cable and Fittings

		Crew	Daily Output	Labor-Hours	Unit	Material	Labor	Equipment	Total	Total Incl O&P
0010	COAXIAL CABLE & FITTINGS									
3500	Coaxial connectors, 50 ohm impedance quick disconnect									
3540	BNC plug, for RG A/U #58 cable	1 Elec	42	.190	Ea.	4.65	10.40		15.05	20.50
3550	RG A/U #59 cable		42	.190		4.65	10.40		15.05	20.50
3560	RG A/U #62 cable		42	.190		4.65	10.40		15.05	20.50

27 15 Communications Horizontal Cabling

27 15 33 – Communications Coaxial Horizontal Cabling

27 15 33.10 Coaxial Cable and Fittings		Crew	Daily Output	Labor-Hours	Unit	2015 Bare Costs Material	Labor	Equipment	Total	Total Incl O&P
3600	BNC jack, for RG A/U #58 cable	1 Elec	42	.190	Ea.	4.90	10.40		15.30	21
3610	RG A/U #59 cable		42	.190		4.90	10.40		15.30	21
3620	RG A/U #62 cable		42	.190		4.90	10.40		15.30	21
3660	BNC panel jack, for RG A/U #58 cable		40	.200		7.60	10.95		18.55	25
3670	RG A/U #59 cable		40	.200		7.60	10.95		18.55	25
3680	RG A/U #62 cable		40	.200		7.60	10.95		18.55	25
3720	BNC bulkhead jack, for RG A/U #58 cable		40	.200		7.85	10.95		18.80	25
3730	RG A/U #59 cable		40	.200		7.85	10.95		18.80	25
3740	RG A/U #62 cable		40	.200		7.85	10.95		18.80	25
3850	Coaxial cable, RG A/U 58, 50 ohm		8	1	C.L.F.	49.50	54.50		104	137
3860	RG A/U 59, 75 ohm		8	1		39	54.50		93.50	125
3870	RG A/U 62, 93 ohm		8	1		46.50	54.50		101	133
3875	RG 6/U, 75 ohm		8	1		28.50	54.50		83	114
3950	Fire rated, RG A/U 58, 50 ohm		8	1		94.50	54.50		149	186
3960	RG A/U 59, 75 ohm		8	1		125	54.50		179.50	219
3970	RG A/U 62, 93 ohm		8	1		110	54.50		164.50	203

27 15 43 – Communications Faceplates and Connectors

27 15 43.13 Communication Outlets		Crew	Daily Output	Labor-Hours	Unit	Material	Labor	Equipment	Total	Total Incl O&P
0010	COMMUNICATION OUTLETS									
0100	Voice/data devices not included									
0120	Voice/Data outlets, single opening	1 Elec	48	.167	Ea.	7.85	9.10		16.95	22.50
0140	Two jack openings		48	.167		2.56	9.10		11.66	16.45
0160	One jack & one 3/4" round opening		48	.167		6.95	9.10		16.05	21.50
0180	One jack & one twinaxial opening		48	.167		7.60	9.10		16.70	22
0200	One jack & one connector cabling opening		48	.167		6.95	9.10		16.05	21.50
0220	Two 3/8" coaxial openings		48	.167		6.95	9.10		16.05	21.50
0300	Data outlets, single opening		48	.167		6.95	9.10		16.05	21.50
0320	One 25-pin subminiature opening		48	.167		6.95	9.10		16.05	21.50
1000	Voice/Data wall plate plastic, 1 gang, 1-port		72	.111		2.08	6.10		8.18	11.40
1020	2-port		72	.111		2.08	6.10		8.18	11.40
1040	3-port		72	.111		2.08	6.10		8.18	11.40
1060	4-port		72	.111		2.08	6.10		8.18	11.40
1080	6-port		72	.111		2.08	6.10		8.18	11.40
1100	2 gang, 6-port		48	.167		6.25	9.10		15.35	20.50
1120	Voice/Data wall plate stainless steel, 1 gang, 1-port		72	.111		6.30	6.10		12.40	16.05
1140	2-port		72	.111		6.45	6.10		12.55	16.15
1160	3-port		72	.111		6.45	6.10		12.55	16.20
1180	4-port		72	.111		6.45	6.10		12.55	16.15
1200	2 gang, 6-port		48	.167		11.20	9.10		20.30	26

27 21 Data Communications Network Equipment

27 21 23 – Data Communications Switches and Hubs

27 21 23.10 Switching and Routing Equipment		Crew	Daily Output	Labor-Hours	Unit	Material	Labor	Equipment	Total	Total Incl O&P
0010	SWITCHING AND ROUTING EQUIPMENT									
1100	Network hub, dual speed, 24 ports, includes cabinet	3 Elec	.66	36.364	Ea.	1,725	2,000		3,725	4,850
2000	Network switch, 10/100/1000 Mbps, 24 ports		.75	32		1,875	1,750		3,625	4,675
2040	48 ports		.66	36.364		2,350	2,000		4,350	5,575

27 32 Voice Communications Terminal Equipment

27 32 36 – TTY Equipment

27 32 36.10 TTY Telephone Equipment		Crew	Daily Output	Labor-Hours	Unit	2015 Bare Costs Material	Labor	Equipment	Total	Total Incl O&P
0010	**TTY TELEPHONE EQUIPMENT**									
1620	Telephone, TTY, compact, pocket type				Ea.	253			253	278
1630	Advanced, desk type	2 Elec	20	.800		510	44		554	625
1640	Full-featured public, wall type	"	4	4		725	219		944	1,125

27 41 Audio-Video Systems

27 41 33 – Master Antenna Television Systems

27 41 33.10 T.V. Systems

27 41 33.10 T.V. Systems		Crew	Daily Output	Labor-Hours	Unit	Material	Labor	Equipment	Total	Total Incl O&P
0010	**T.V. SYSTEMS**, not including rough-in wires, cables & conduits									
0100	Master TV antenna system									
0200	VHF reception & distribution, 12 outlets	1 Elec	6	1.333	Outlet	133	73		206	255
0400	30 outlets		10	.800		141	44		185	221
0600	100 outlets		13	.615		144	33.50		177.50	209
0800	VHF & UHF reception & distribution, 12 outlets		6	1.333		213	73		286	345
1000	30 outlets		10	.800		141	44		185	221
1200	100 outlets		13	.615		144	33.50		177.50	209
1400	School and deluxe systems, 12 outlets		2.40	3.333		281	182		463	585
1600	30 outlets		4	2		246	109		355	435
1800	80 outlets		5.30	1.509		237	82.50		319.50	385
1900	Amplifier		4	2	Ea.	705	109		814	940
5000	Antenna, small		6	1.333		47	73		120	161
5100	Large		4	2		198	109		307	380
5110	Rotor unit		8	1		77.50	54.50		132	168
5120	Single booster		8	1		38	54.50		92.50	124
5130	Antenna pole, 10′		3.20	2.500		28	137		165	236
6100	Satellite TV system	2 Elec	1	16		2,150	875		3,025	3,650
6110	Dish, mesh, 10′ diam.	"	2.40	6.667		1,175	365		1,540	1,850
6111	Two way RF/IF tapeoff	1 Elec	36	.222		3.56	12.15		15.71	22
6112	Two way RF/IF splitter		24	.333		14.75	18.25		33	43.50
6113	Line amplifier		24	.333		14.75	18.25		33	43.50
6114	Line splitters		36	.222		4.69	12.15		16.84	23.50
6115	Line multi switches		8	1		680	54.50		734.50	830
6120	Motor unit		2.40	3.333		370	182		552	685
7000	Home theater, widescreen, 42″, high definition, TV		10.24	.781		510	42.50		552.50	625
7050	Flat wall mount bracket		10.24	.781		129	42.50		171.50	206
7100	7 channel home theater receiver		10.24	.781		380	42.50		422.50	485
7200	Home theater speakers		10.24	.781	Set	191	42.50		233.50	274
7300	Home theater programmable remote		10.24	.781	Ea.	259	42.50		301.50	350
8000	Main video splitter		4	2		2,450	109		2,559	2,875
8010	Video distribution units		4	2		124	109		233	300

27 51 Distributed Audio-Video Communications Systems

27 51 16 – Public Address and Mass Notification Systems

27 51 16.10 Public Address System		Crew	Daily Output	Labor-Hours	Unit	2015 Bare Costs Material	Labor	Equipment	Total	Total Incl O&P
0010	**PUBLIC ADDRESS SYSTEM**									
0100	Conventional, office	1 Elec	5.33	1.501	Speaker	135	82		217	272
0200	Industrial	"	2.70	2.963	"	261	162		423	530

27 51 19 – Sound Masking Systems

27 51 19.10 Sound System		Crew	Daily Output	Labor-Hours	Unit	2015 Bare Costs Material	Labor	Equipment	Total	Total Incl O&P
0010	**SOUND SYSTEM**, not including rough-in wires, cables & conduits									
0100	Components, projector outlet	1 Elec	8	1	Ea.	46	54.50		100.50	133
0200	Microphone		4	2		81.50	109		190.50	254
0400	Speakers, ceiling or wall		8	1		119	54.50		173.50	213
0600	Trumpets		4	2		222	109		331	410
0800	Privacy switch		8	1		88.50	54.50		143	180
1000	Monitor panel		4	2		395	109		504	600
1200	Antenna, AM/FM		4	2		138	109		247	315
1400	Volume control		8	1		90	54.50		144.50	181
1600	Amplifier, 250 watts		1	8		1,275	440		1,715	2,050
1800	Cabinets		1	8		860	440		1,300	1,600
2000	Intercom, 30 station capacity, master station	2 Elec	2	8		2,350	440		2,790	3,250
2020	10 station capacity	"	4	4		1,300	219		1,519	1,775
2200	Remote station	1 Elec	8	1		166	54.50		220.50	264
2400	Intercom outlets		8	1		97.50	54.50		152	189
2600	Handset		4	2		320	109		429	520
2800	Emergency call system, 12 zones, annunciator		1.30	6.154		970	335		1,305	1,575
3000	Bell		5.30	1.509		100	82.50		182.50	234
3200	Light or relay		8	1		50	54.50		104.50	137
3400	Transformer		4	2		220	109		329	405
3600	House telephone, talking station		1.60	5		475	274		749	930
3800	Press to talk, release to listen		5.30	1.509		110	82.50		192.50	245
4000	System-on button					66			66	72.50
4200	Door release	1 Elec	4	2		118	109		227	294
4400	Combination speaker and microphone		8	1		201	54.50		255.50	305
4600	Termination box		3.20	2.500		63	137		200	275
4800	Amplifier or power supply		5.30	1.509		725	82.50		807.50	925
5000	Vestibule door unit		16	.500	Name	133	27.50		160.50	188
5200	Strip cabinet		27	.296	Ea.	252	16.20		268.20	300
5400	Directory		16	.500		119	27.50		146.50	172
6000	Master door, button buzzer type, 100 unit	2 Elec	.54	29.630		1,200	1,625		2,825	3,750
6020	200 unit		.30	53.333		2,275	2,925		5,200	6,875
6040	300 unit		.20	80		3,475	4,375		7,850	10,400
6060	Transformer	1 Elec	8	1		31	54.50		85.50	116
6080	Door opener		5.30	1.509		44	82.50		126.50	173
6100	Buzzer with door release and plate		4	2		44	109		153	213
6200	Intercom type, 100 unit	2 Elec	.54	29.630		1,500	1,625		3,125	4,075
6220	200 unit		.30	53.333		2,925	2,925		5,850	7,600
6240	300 unit		.20	80		4,425	4,375		8,800	11,400
6260	Amplifier	1 Elec	2	4		220	219		439	575
6280	Speaker with door release	"	4	2		66	109		175	237

27 52 Healthcare Communications and Monitoring Systems

27 52 23 – Nurse Call/Code Blue Systems

27 52 23.10 Nurse Call Systems		Crew	Daily Output	Labor-Hours	Unit	2015 Bare Costs Material	Labor	Equipment	Total	Total Incl O&P
0010	**NURSE CALL SYSTEMS**									
0100	Single bedside call station	1 Elec	8	1	Ea.	231	54.50		285.50	335
0200	Ceiling speaker station		8	1		67.50	54.50		122	156
0400	Emergency call station		8	1		72	54.50		126.50	162
0600	Pillow speaker		8	1		177	54.50		231.50	277
0800	Double bedside call station		4	2		142	109		251	320
1000	Duty station		4	2		148	109		257	325
1200	Standard call button		8	1		87	54.50		141.50	178
1400	Lights, corridor, dome or zone indicator	↓	8	1	↓	49	54.50		103.50	136
1600	Master control station for 20 stations	2 Elec	.65	24.615	Total	3,975	1,350		5,325	6,400

27 53 Distributed Systems

27 53 13 – Clock Systems

27 53 13.50 Clock Equipments		Crew	Daily Output	Labor-Hours	Unit	Material	Labor	Equipment	Total	Total Incl O&P
0010	**CLOCK EQUIPMENTS**, not including wires & conduits									
0100	Time system components, master controller	1 Elec	.33	24.242	Ea.	1,850	1,325		3,175	4,025
0200	Program bell		8	1		86.50	54.50		141	178
0400	Combination clock & speaker		3.20	2.500		214	137		351	440
0600	Frequency generator		2	4		2,400	219		2,619	2,975
0800	Job time automatic stamp recorder	↓	4	2		540	109		649	760
1600	Master time clock system, clocks & bells, 20 room	4 Elec	.20	160		6,200	8,750		14,950	19,900
1800	50 room	"	.08	400		12,700	21,900		34,600	46,800
1900	Time clock	1 Elec	3.20	2.500		445	137		582	695
2000	100 cards in & out, 1 color					9.15			9.15	10.10
2200	2 colors					9.15			9.15	10.10
2800	Metal rack for 25 cards	1 Elec	7	1.143		44	62.50		106.50	142
4000	Wireless time systems component, master controller	"	2	4	↓	605	219		824	1,000
4010	For transceiver and antenna, see Section 28 39 10.10									
4100	Wireless analog clock w/battery operated	1 Elec	8	1	Ea.	113	54.50		167.50	207
4200	Wireless digital clock w/battery operated	"	8	1	"	370	54.50		424.50	485

Division Notes

		CREW	DAILY OUTPUT	LABOR-HOURS	UNIT	BARE COSTS MAT.	BARE COSTS LABOR	BARE COSTS EQUIP.	BARE COSTS TOTAL	TOTAL INCL O&P

Estimating Tips

- When estimating material costs for electronic safety and security systems, it is always prudent to obtain manufacturers' quotations for equipment prices and special installation requirements that affect the total cost.
- Fire alarm systems consist of control panels, annunciator panels, battery with rack, charger, and fire alarm actuating and indicating devices. Some fire alarm systems include speakers, telephone lines, door closer controls, and other components. Be careful not to overlook the costs related to installation for these items. Also be aware of costs for integrated automation instrumentation and terminal devices, control equipment, control wiring, and programming.
- Security equipment includes items such as CCTV, access control, and other detection and identification systems to perform alert and alarm functions. Be sure to consider the costs related to installation for this security equipment, such as for integrated automation instrumentation and terminal devices, control equipment, control wiring, and programming.

Reference Numbers

Reference numbers are shown in shaded boxes at the beginning of some major classifications. These numbers refer to related items in the Reference Section. The reference information may be an estimating procedure, an alternate pricing method, or technical information.

Note: Not all subdivisions listed here necessarily appear in this publication. ■

Division 28 – Electronic Safety & Security

28 01 Operation and Maint. of Electronic Safety and Security

28 01 30 – Operation and Maint. of Electronic Detection and Alarm

28 01 30.51 Maint. and Admin. of Elec. Detection and Alarm		Crew	Daily Output	Labor-Hours	Unit	2015 Bare Costs Material	Labor	Equipment	Total	Total Incl O&P
0010	**MAINT. AND ADMIN. OF ELEC. DETECTION AND ALARM**									
3300	Remove and replace (reinstall), fire alarm device	1 Elec	5.33	1.501	Ea.		82		82	123

28 05 Common Work Results for Electronic Safety and Security

28 05 05 – Selective Demolition for Electronic Safety and Security

28 05 05.10 Safety and Security Demolition

		Crew	Daily Output	Labor-Hours	Unit	Material	Labor	Equipment	Total	Total Incl O&P
0010	**SAFETY AND SECURITY DEMOLITION**									
1050	Finger print/card reader	1 Elec	8	1	Ea.		54.50		54.50	82
1060	Video camera		8	1			54.50		54.50	82
1070	Motion detector, multi-channel		7	1.143			62.50		62.50	93.50
1090	Infrared detector		7	1.143			62.50		62.50	93.50
1130	Video monitor, 19"		8	1			54.50		54.50	82
1210	Fire alarm horn and strobe light		16	.500			27.50		27.50	41
1220	Flame detector		23	.348			19.05		19.05	28.50
1230	Duct detector		10	.800			44		44	65.50
1240	Smoke detector		23	.348			19.05		19.05	28.50
1250	Fire alarm pull station		23	.348			19.05		19.05	28.50
1260	Fire alarm control panel, 4 to 8 zone	2 Elec	4	4			219		219	330
1270	12 to 16 zone	"	2.67	5.993			330		330	490
1280	Fire alarm annunciation panel, 4 to 8 zone	1 Elec	6	1.333			73		73	109
1290	12 to 16 zone	2 Elec	6	2.667			146		146	219

28 05 13 – Conductors and Cables for Electronic Safety and Security

28 05 13.23 Fire Alarm Communications Conductors and Cables

		Crew	Daily Output	Labor-Hours	Unit	Material	Labor	Equipment	Total	Total Incl O&P
0010	**FIRE ALARM COMMUNICATIONS CONDUCTORS AND CABLES**									
1500	Fire alarm FEP teflon 150 volt to 200°C									
1550	#22, 1 pair	1 Elec	10	.800	C.L.F.	63	44		107	135
1600	2 pair		8	1		103	54.50		157.50	196
1650	4 pair		7	1.143		160	62.50		222.50	270
1700	6 pair		6	1.333		208	73		281	335
1750	8 pair		5.50	1.455		261	79.50		340.50	405
1800	10 pair		5	1.600		315	87.50		402.50	475
1850	#18, 1 pair		8	1		72	54.50		126.50	161
1900	2 pair		6.50	1.231		137	67.50		204.50	252
1950	4 pair		4.80	1.667		208	91		299	365
2000	6 pair		4	2		273	109		382	465
2050	8 pair		3.50	2.286		355	125		480	580
2100	10 pair		3	2.667		410	146		556	670

28 13 Access Control

28 13 53 – Security Access Detection

28 13 53.13 Security Access Metal Detectors

		Crew	Daily Output	Labor-Hours	Unit	Material	Labor	Equipment	Total	Total Incl O&P
0010	**SECURITY ACCESS METAL DETECTORS**									
0240	Metal detector, hand-held, wand type, unit only				Ea.	89			89	98
0250	Metal detector, walk through portal type, single zone	1 Elec	2	4		3,700	219		3,919	4,400
0260	Multi-zone	"	2	4		4,700	219		4,919	5,500

28 13 Access Control

28 13 53 – Security Access Detection

28 13 53.16 Security Access X-Ray Equipment		Crew	Daily Output	Labor-Hours	Unit	2015 Bare Costs Material	Labor	Equipment	Total	Total Incl O&P
0010	**SECURITY ACCESS X-RAY EQUIPMENT**									
0290	X-ray machine, desk top, for mail/small packages/letters	1 Elec	4	2	Ea.	3,425	109		3,534	3,950
0300	Conveyor type, incl monitor		2	4		16,000	219		16,219	17,900
0310	Includes additional features		2	4		28,600	219		28,819	31,700
0320	X-ray machine, large unit, for airports, incl monitor	2 Elec	1	16		40,000	875		40,875	45,300
0330	Full console	"	.50	32		68,500	1,750		70,250	78,000

28 13 53.23 Security Access Explosive Detection Equipment		Crew	Daily Output	Labor-Hours	Unit	Material	Labor	Equipment	Total	Total Incl O&P
0010	**SECURITY ACCESS EXPLOSIVE DETECTION EQUIPMENT**									
0270	Explosives detector, walk through portal type	1 Elec	2	4	Ea.	44,200	219		44,419	48,900
0280	Hand-held, battery operated				"				25,500	28,100

28 16 Intrusion Detection

28 16 16 – Intrusion Detection Systems Infrastructure

28 16 16.50 Intrusion Detection		Crew	Daily Output	Labor-Hours	Unit	Material	Labor	Equipment	Total	Total Incl O&P
0010	**INTRUSION DETECTION**, not including wires & conduits									
0100	Burglar alarm, battery operated, mechanical trigger	1 Elec	4	2	Ea.	278	109		387	470
0200	Electrical trigger		4	2		330	109		439	530
0400	For outside key control, add		8	1		84	54.50		138.50	175
0600	For remote signaling circuitry, add		8	1		125	54.50		179.50	219
0800	Card reader, flush type, standard		2.70	2.963		930	162		1,092	1,275
1000	Multi-code		2.70	2.963		1,200	162		1,362	1,575
1010	Card reader, proximity type		2.70	2.963		315	162		477	595
1200	Door switches, hinge switch		5.30	1.509		58.50	82.50		141	189
1400	Magnetic switch		5.30	1.509		69	82.50		151.50	200
1600	Exit control locks, horn alarm		4	2		277	109		386	470
1800	Flashing light alarm		4	2		305	109		414	500
2000	Indicating panels, 1 channel		2.70	2.963		370	162		532	650
2200	10 channel	2 Elec	3.20	5		1,050	274		1,324	1,550
2400	20 channel		2	8		2,450	440		2,890	3,350
2600	40 channel		1.14	14.035		4,450	770		5,220	6,050
2800	Ultrasonic motion detector, 12 volt	1 Elec	2.30	3.478		230	190		420	540
3000	Infrared photoelectric detector		4	2		189	109		298	370
3200	Passive infrared detector		4	2		283	109		392	475
3400	Glass break alarm switch		8	1		93	54.50		147.50	184
3420	Switchmats, 30" x 5'		5.30	1.509		84.50	82.50		167	217
3440	30" x 25'		4	2		203	109		312	385
3460	Police connect panel		4	2		244	109		353	435
3480	Telephone dialer		5.30	1.509		385	82.50		467.50	550
3500	Alarm bell		4	2		104	109		213	279
3520	Siren		4	2		146	109		255	325
3540	Microwave detector, 10' to 200'		2	4		670	219		889	1,075
3560	10' to 350'		2	4		1,950	219		2,169	2,475

28 23 Video Surveillance

28 23 13 – Video Surveillance Control and Management Systems

28 23 13.10 Closed Circuit Television System		Crew	Daily Output	Labor-Hours	Unit	Material	2015 Bare Costs Labor	Equipment	Total	Total Incl O&P
0010	**CLOSED CIRCUIT TELEVISION SYSTEM**									
2000	Surveillance, one station (camera & monitor)	2 Elec	2.60	6.154	Total	1,350	335		1,685	1,975
2200	For additional camera stations, add	1 Elec	2.70	2.963	Ea.	755	162		917	1,075
2400	Industrial quality, one station (camera & monitor)	2 Elec	2.60	6.154	Total	2,800	335		3,135	3,575
2600	For additional camera stations, add	1 Elec	2.70	2.963	Ea.	1,725	162		1,887	2,125
2610	For low light, add		2.70	2.963		1,375	162		1,537	1,775
2620	For very low light, add		2.70	2.963		10,200	162		10,362	11,400
2800	For weatherproof camera station, add		1.30	6.154		1,050	335		1,385	1,675
3000	For pan and tilt, add		1.30	6.154		2,725	335		3,060	3,500
3200	For zoom lens - remote control, add		2	4		2,525	219		2,744	3,100
3400	Extended zoom lens		2	4		9,200	219		9,419	10,400
3410	For automatic iris for low light, add		2	4		2,200	219		2,419	2,750
3600	Educational T.V. studio, basic 3 camera system, black & white,									
3800	electrical & electronic equip. only	4 Elec	.80	40	Total	13,100	2,200		15,300	17,700
4000	Full console		.28	114		56,000	6,250		62,250	71,000
4100	As above, but color system		.28	114		74,000	6,250		80,250	90,500
4120	Full console		.12	266		321,000	14,600		335,600	375,000
4200	For film chain, black & white, add	1 Elec	1	8	Ea.	15,000	440		15,440	17,200
4250	Color, add		.25	32		18,200	1,750		19,950	22,600
4400	For video recorders, add		1	8		3,150	440		3,590	4,125
4600	Premium	4 Elec	.40	80		26,200	4,375		30,575	35,400

28 23 19 – Digital Video Recorders and Analog Recording Devices

28 23 19.10 Digital Video Recorder (DVR)		Crew	Daily Output	Labor-Hours	Unit	Material	Labor	Equipment	Total	Total Incl O&P
0010	**DIGITAL VIDEO RECORDER (DVR)**									
0100	Pentaplex hybrid, internet protocol, and hard drive									
0200	4 channel	1 Elec	1.33	6.015	Ea.	1,350	330		1,680	1,975
0300	8 channel		1	8		2,650	440		3,090	3,575
0400	16 channel		1	8		2,925	440		3,365	3,875

28 23 23 – Video Surveillance Systems Infrastructure

28 23 23.50 Video Surveillance Equipments		Crew	Daily Output	Labor-Hours	Unit	Material	Labor	Equipment	Total	Total Incl O&P
0010	**VIDEO SURVEILLANCE EQUIPMENTS**									
0200	Video cameras, wireless, hidden in exit signs, clocks, etc., incl. receiver	1 Elec	3	2.667	Ea.	108	146		254	340
0210	Accessories for video recorder, single camera		3	2.667		183	146		329	420
0220	For multiple cameras		3	2.667		1,750	146		1,896	2,150
0230	Video cameras, wireless, for under vehicle searching, complete		2	4		10,200	219		10,419	11,500
0400	Internet protocol network camera, day/night, color & power supply		2.60	3.077		1,025	168		1,193	1,375
0500	Monitor, color flat screen, liquid crystal display (LCD), 15"		2.70	2.963		695	162		857	1,000
0520	17"		2.70	2.963		820	162		982	1,150
0540	19"		2.70	2.963		920	162		1,082	1,275

28 31 Fire Detection and Alarm

28 31 23 – Fire Detection and Alarm Annunciation Panels and Fire Stations

28 31 23.50 Alarm Panels and Devices		Crew	Daily Output	Labor-Hours	Unit	2015 Bare Costs Material	Labor	Equipment	Total	Total Incl O&P
0010	**ALARM PANELS AND DEVICES**, not including wires & conduits									
3594	Fire, alarm control panel									
3600	4 zone	2 Elec	2	8	Ea.	400	440		840	1,100
3800	8 zone		1	16		780	875		1,655	2,150
4000	12 zone		.67	23.988		2,400	1,300		3,700	4,600
4020	Alarm device	1 Elec	8	1		238	54.50		292.50	345
4050	Actuating device	"	8	1		335	54.50		389.50	450
4160	Alarm control panel, addressable w/o voice, up to 200 points	2 Elec	1.14	13.998		4,475	765		5,240	6,075
4170	addressable w/voice, up to 400 points	"	.73	22.008		9,425	1,200		10,625	12,200
4175	Addressable interface device	1 Elec	7.25	1.103		135	60.50		195.50	239
4200	Battery and rack		4	2		410	109		519	620
4400	Automatic charger		8	1		585	54.50		639.50	720
4600	Signal bell		8	1		78.50	54.50		133	169
4800	Trouble buzzer or manual station		8	1		83.50	54.50		138	174
5600	Strobe and horn		5.30	1.509		152	82.50		234.50	291
5610	Strobe and horn (ADA type)		5.30	1.509		152	82.50		234.50	291
5620	Visual alarm (ADA type)		6.70	1.194		103	65.50		168.50	211
5800	Fire alarm horn		6.70	1.194		61	65.50		126.50	165
6000	Door holder, electro-magnetic		4	2		103	109		212	277
6200	Combination holder and closer		3.20	2.500		123	137		260	340
6600	Drill switch		8	1		370	54.50		424.50	490
6800	Master box		2.70	2.963		6,400	162		6,562	7,300
7000	Break glass station		8	1		55.50	54.50		110	143
7010	Break glass station, addressable		7.25	1.103		148	60.50		208.50	253
7800	Remote annunciator, 8 zone lamp		1.80	4.444		209	243		452	595
8000	12 zone lamp	2 Elec	2.60	6.154		335	335		670	875
8200	16 zone lamp	"	2.20	7.273		420	400		820	1,050

28 31 43 – Fire Detection Sensors

28 31 43.50 Fire and Heat Detectors		Crew	Daily Output	Labor-Hours	Unit	Material	Labor	Equipment	Total	Total Incl O&P
0010	**FIRE & HEAT DETECTORS**									
5000	Detector, rate of rise	1 Elec	8	1	Ea.	51	54.50		105.50	138
5010	Heat addressable type		7.25	1.103		257	60.50		317.50	375
5100	Fixed temperature		8	1		51	54.50		105.50	138

28 31 46 – Smoke Detection Sensors

28 31 46.50 Smoke Detectors		Crew	Daily Output	Labor-Hours	Unit	Material	Labor	Equipment	Total	Total Incl O&P
0010	**SMOKE DETECTORS**									
5200	Smoke detector, ceiling type	1 Elec	6.20	1.290	Ea.	110	70.50		180.50	227
5240	Smoke detector addressable type		6	1.333		221	73		294	355
5400	Duct type		3.20	2.500		325	137		462	565
5420	Duct addressable type		3.20	2.500		510	137		647	765

28 39 Mass Notification Systems

28 39 10 – Notification Systems

28 39 10.10 Mass Notification System		Crew	Daily Output	Labor-Hours	Unit	2015 Bare Costs Material	Labor	Equipment	Total	Total Incl O&P
0010	**MASS NOTIFICATION SYSTEM**									
0100	Wireless command center, 10,000 devices	2 Elec	1.33	12.030	Ea.	3,600	660		4,260	4,925
0200	Option, email notification					1,750			1,750	1,925
0210	Remote device supervision & monitor					2,450			2,450	2,675
0300	Antenna VHF or UHF, for medium range	1 Elec	4	2		129	109		238	305
0310	For high-power transmitter		2	4		670	219		889	1,075
0400	Transmitter, 25 watt		4	2		2,025	109		2,134	2,400
0410	40 watt		2.66	3.008		2,700	165		2,865	3,225
0420	100 watt		1.33	6.015		6,775	330		7,105	7,950
0500	Wireless receiver/control module for speaker		8	1		265	54.50		319.50	375
0600	Desktop paging controller, stand alone		4	2		370	109		479	575

Estimating Tips

31 05 00 Common Work Results for Earthwork

- Estimating the actual cost of performing earthwork requires careful consideration of the variables involved. This includes items such as type of soil, whether water will be encountered, dewatering, whether banks need bracing, disposal of excavated earth, and length of haul to fill or spoil sites, etc. If the project has large quantities of cut or fill, consider raising or lowering the site to reduce costs, while paying close attention to the effect on site drainage and utilities.
- If the project has large quantities of fill, creating a borrow pit on the site can significantly lower the costs.
- It is very important to consider what time of year the project is scheduled for completion. Bad weather can create large cost overruns from dewatering, site repair, and lost productivity from cold weather.

Reference Numbers

Reference numbers are shown in shaded boxes at the beginning of some major classifications. These numbers refer to related items in the Reference Section. The reference information may be an estimating procedure, an alternate pricing method, or technical information.

Note: Not all subdivisions listed here necessarily appear in this publication. ■

31 23 16.13 Excavating, Trench

31 23 16.13 Excavating, Trench		Crew	Daily Output	Labor-Hours	Unit	Material	2015 Bare Costs Labor	Equipment	Total	Total Incl O&P
0010	**EXCAVATING, TRENCH**									
0011	Or continuous footing									
0020	Common earth with no sheeting or dewatering included									
0050	1′ to 4′ deep, 3/8 C.Y. excavator	B-11C	150	.107	B.C.Y.		4.70	2.43	7.13	9.85
0060	1/2 C.Y. excavator	B-11M	200	.080			3.53	1.96	5.49	7.55
0062	3/4 C.Y. excavator	B-12F	270	.059			2.65	2.43	5.08	6.70
1400	By hand with pick and shovel 2′ to 6′ deep, light soil	1 Clab	8	1			37.50		37.50	58
1500	Heavy soil	"	4	2	↓		75		75	116
3000	Backfill trench, F.E. loader, wheel mtd., 1 C.Y. bucket									
3020	Minimal haul	B-10R	400	.030	L.C.Y.		1.39	.75	2.14	2.93
5020	Loam & Sandy clay with no sheeting or dewatering included									
5050	1′ to 4′ deep, 3/8 C.Y. tractor loader/backhoe	B-11C	162	.099	B.C.Y.		4.36	2.25	6.61	9.10
5060	1/2 C.Y. excavator	B-11M	216	.074			3.27	1.81	5.08	7
5070	3/4 C.Y. excavator	B-12F	292	.055	↓		2.45	2.24	4.69	6.20
6020	Sand & gravel with no sheeting or dewatering included									
6050	1′ to 4′ deep, 3/8 C.Y. excavator	B-11C	165	.097	B.C.Y.		4.28	2.21	6.49	9
6060	1/2 C.Y. excavator	B-11M	220	.073			3.21	1.78	4.99	6.85
6070	3/4 C.Y. excavator	B-12F	297	.054	↓		2.41	2.20	4.61	6.10
7020	Dense hard clay with no sheeting or dewatering included									
7050	1′ to 4′ deep, 3/8 C.Y. excavator	B-11C	132	.121	B.C.Y.		5.35	2.76	8.11	11.20
7060	1/2 C.Y. excavator	B-11M	176	.091			4.01	2.23	6.24	8.55
7070	3/4 C.Y. excavator	B-12F	238	.067	↓		3	2.75	5.75	7.60

31 23 16.14 Excavating, Utility Trench

31 23 16.14 Excavating, Utility Trench		Crew	Daily Output	Labor-Hours	Unit	Material	2015 Bare Costs Labor	Equipment	Total	Total Incl O&P
0010	**EXCAVATING, UTILITY TRENCH**									
0011	Common earth									
0050	Trenching with chain trencher, 12 H.P., operator walking									
0100	4″ wide trench, 12″ deep	B-53	800	.010	L.F.		.49	.09	.58	.83
0150	18″ deep		750	.011			.52	.09	.61	.89
0200	24″ deep		700	.011			.56	.10	.66	.95
0300	6″ wide trench, 12″ deep		650	.012			.60	.10	.70	1.03
0350	18″ deep		600	.013			.65	.11	.76	1.10
0400	24″ deep		550	.015			.71	.12	.83	1.21
0450	36″ deep		450	.018			.86	.15	1.01	1.48
0600	8″ wide trench, 12″ deep		475	.017			.82	.14	.96	1.40
0650	18″ deep		400	.020			.97	.17	1.14	1.67
0700	24″ deep		350	.023			1.11	.19	1.30	1.90
0750	36″ deep	↓	300	.027	↓		1.30	.23	1.53	2.22
1000	Backfill by hand including compaction, add									
1050	4″ wide trench, 12″ deep	A-1G	800	.010	L.F.		.38	.07	.45	.66
1100	18″ deep		530	.015			.57	.11	.68	.99
1150	24″ deep		400	.020			.75	.14	.89	1.32
1300	6″ wide trench, 12″ deep		540	.015			.56	.10	.66	.97
1350	18″ deep		405	.020			.74	.14	.88	1.29
1400	24″ deep		270	.030			1.11	.21	1.32	1.94
1450	36″ deep		180	.044			1.67	.31	1.98	2.91
1600	8″ wide trench, 12″ deep		400	.020			.75	.14	.89	1.32
1650	18″ deep		265	.030			1.14	.21	1.35	1.98
1700	24″ deep		200	.040			1.50	.28	1.78	2.62
1750	36″ deep	↓	135	.059	↓		2.23	.42	2.65	3.89
2000	Chain trencher, 40 H.P. operator riding									
2050	6″ wide trench and backfill, 12″ deep	B-54	1200	.007	L.F.		.32	.28	.60	.80
2100	18″ deep	↓	1000	.008	↓		.39	.34	.73	.96

31 23 16.14 Excavating, Utility Trench		Crew	Daily Output	Labor-Hours	Unit	Material	2015 Bare Costs Labor	Equipment	Total	Total Incl O&P
2150	24″ deep	B-54	975	.008	L.F.		.40	.34	.74	.99
2200	36″ deep		900	.009			.43	.37	.80	1.07
2250	48″ deep		750	.011			.52	.45	.97	1.28
2300	60″ deep		650	.012			.60	.52	1.12	1.48
2400	8″ wide trench and backfill, 12″ deep		1000	.008			.39	.34	.73	.96
2450	18″ deep		950	.008			.41	.35	.76	1.01
2500	24″ deep		900	.009			.43	.37	.80	1.07
2550	36″ deep		800	.010			.49	.42	.91	1.20
2600	48″ deep		650	.012			.60	.52	1.12	1.48
2700	12″ wide trench and backfill, 12″ deep		975	.008			.40	.34	.74	.99
2750	18″ deep		860	.009			.45	.39	.84	1.12
2800	24″ deep		800	.010			.49	.42	.91	1.20
2850	36″ deep		725	.011			.54	.46	1	1.32
3000	16″ wide trench and backfill, 12″ deep		835	.010			.47	.40	.87	1.15
3050	18″ deep		750	.011			.52	.45	.97	1.28
3100	24″ deep		700	.011			.56	.48	1.04	1.37
3200	Compaction with vibratory plate, add								35%	35%
5100	Hand excavate and trim for pipe bells after trench excavation									
5200	8″ pipe	1 Clab	155	.052	L.F.		1.94		1.94	2.99
5300	18″ pipe	″	130	.062	″		2.31		2.31	3.56
31 23 16.15 Excavating, Utility Trench, Plow										
0010	**EXCAVATING, UTILITY TRENCH, PLOW**									
0100	Single cable, plowed into fine material	B-54C	3800	.004	L.F.		.19	.31	.50	.63
0200	Two cable		3200	.005			.22	.37	.59	.75
0300	Single cable, plowed into coarse material		2000	.008			.35	.60	.95	1.20
31 23 16.16 Structural Excavation for Minor Structures										
0010	**STRUCTURAL EXCAVATION FOR MINOR STRUCTURES**									
0015	Hand, pits to 6′ deep, sandy soil	1 Clab	8	1	B.C.Y.		37.50		37.50	58
0100	Heavy soil or clay		4	2			75		75	116
0300	Pits 6′ to 12′ deep, sandy soil		5	1.600			60		60	92.50
0500	Heavy soil or clay		3	2.667			100		100	154
0700	Pits 12′ to 18′ deep, sandy soil		4	2			75		75	116
0900	Heavy soil or clay		2	4			150		150	231
6030	Common earth, hydraulic backhoe, 1/2 C.Y. bucket	B-12E	55	.291			13	8.15	21.15	29
6035	3/4 C.Y. bucket	B-12F	90	.178			7.95	7.30	15.25	20
6040	1 C.Y. bucket	B-12A	108	.148			6.60	7.50	14.10	18.40
6050	1-1/2 C.Y. bucket	B-12B	144	.111			4.96	7.15	12.11	15.40
6060	2 C.Y. bucket	B-12C	200	.080			3.57	5.90	9.47	11.90
6070	Sand and gravel, 3/4 C.Y. bucket	B-12F	100	.160			7.15	6.55	13.70	18.10
6080	1 C.Y. bucket	B-12A	120	.133			5.95	6.75	12.70	16.55
6090	1-1/2 C.Y. bucket	B-12B	160	.100			4.47	6.45	10.92	13.90
6100	2 C.Y. bucket	B-12C	220	.073			3.25	5.35	8.60	10.85
6110	Clay, till, or blasted rock, 3/4 C.Y. bucket	B-12F	80	.200			8.95	8.20	17.15	22.50
6120	1 C.Y. bucket	B-12A	95	.168			7.50	8.55	16.05	21
6130	1-1/2 C.Y. bucket	B-12B	130	.123			5.50	7.90	13.40	17.10
6140	2 C.Y. bucket	B-12C	175	.091			4.08	6.70	10.78	13.65
6230	Sandy clay & loam, hydraulic backhoe, 1/2 C.Y. bucket	B-12E	60	.267			11.90	7.45	19.35	26.50
6235	3/4 C.Y. bucket	B-12F	98	.163			7.30	6.70	14	18.50
6240	1 C.Y. bucket	B-12A	116	.138			6.15	7	13.15	17.10
6250	1-1/2 C.Y. bucket	B-12B	156	.103			4.58	6.60	11.18	14.25

31 23 Excavation and Fill

31 23 19 – Dewatering

31 23 19.20 Dewatering Systems		Crew	Daily Output	Labor-Hours	Unit	Material	2015 Bare Costs Labor	Equipment	Total	Total Incl O&P
0010	**DEWATERING SYSTEMS**									
0020	Excavate drainage trench, 2′ wide, 2′ deep	B-11C	90	.178	C.Y.		7.85	4.05	11.90	16.40
0100	2′ wide, 3′ deep, with backhoe loader	″	135	.119			5.25	2.70	7.95	10.95
0200	Excavate sump pits by hand, light soil	1 Clab	7.10	1.127			42.50		42.50	65
0300	Heavy soil	″	3.50	2.286			86		86	132
0500	Pumping 8 hr., attended 2 hrs. per day, including 20 L.F.									
0550	of suction hose & 100 L.F. discharge hose									
0600	2″ diaphragm pump used for 8 hours	B-10H	4	3	Day		139	18.70	157.70	232
0650	4″ diaphragm pump used for 8 hours	B-10I	4	3			139	30.50	169.50	245
0800	8 hrs. attended, 2″ diaphragm pump	B-10H	1	12			555	75	630	925
0900	3″ centrifugal pump	B-10J	1	12			555	84	639	940
1000	4″ diaphragm pump	B-10I	1	12			555	122	677	980
1100	6″ centrifugal pump	B-10K	1	12			555	370	925	1,250

31 23 23 – Fill

31 23 23.13 Backfill

		Crew	Daily Output	Labor-Hours	Unit	Material	Labor	Equipment	Total	Total Incl O&P
0010	**BACKFILL**									
0015	By hand, no compaction, light soil	1 Clab	14	.571	L.C.Y.		21.50		21.50	33
0100	Heavy soil		11	.727	″		27.50		27.50	42
0300	Compaction in 6″ layers, hand tamp, add to above		20.60	.388	E.C.Y.		14.60		14.60	22.50
0400	Roller compaction operator walking, add	B-10A	100	.120			5.55	1.80	7.35	10.45
0500	Air tamp, add	B-9D	190	.211			8	1.40	9.40	13.85
0600	Vibrating plate, add	A-1D	60	.133			5	.60	5.60	8.35
0800	Compaction in 12″ layers, hand tamp, add to above	1 Clab	34	.235			8.85		8.85	13.60
0900	Roller compaction operator walking, add	B-10A	150	.080			3.70	1.20	4.90	6.95
1000	Air tamp, add	B-9	285	.140			5.35	.82	6.17	9.10
1100	Vibrating plate, add	A-1E	90	.089			3.34	.52	3.86	5.70
1300	Dozer backfilling, bulk, up to 300′ haul, no compaction	B-10B	1200	.010	L.C.Y.		.46	1.16	1.62	1.97
1400	Air tamped, add	B-11B	80	.200	E.C.Y.		8.60	3.77	12.37	17.30
1600	Compacting backfill, 6″ to 12″ lifts, vibrating roller	B-10C	800	.015			.69	2.23	2.92	3.51
1700	Sheepsfoot roller	B-10D	750	.016			.74	2.42	3.16	3.79
1900	Dozer backfilling, trench, up to 300′ haul, no compaction	B-10B	900	.013	L.C.Y.		.62	1.54	2.16	2.63
2000	Air tamped, add	B-11B	80	.200	E.C.Y.		8.60	3.77	12.37	17.30
2200	Compacting backfill, 6″ to 12″ lifts, vibrating roller	B-10C	700	.017			.79	2.55	3.34	4.01
2300	Sheepsfoot roller	B-10D	650	.018			.85	2.79	3.64	4.37

31 23 23.16 Fill By Borrow and Utility Bedding

		Crew	Daily Output	Labor-Hours	Unit	Material	Labor	Equipment	Total	Total Incl O&P
0010	**FILL BY BORROW AND UTILITY BEDDING**									
0049	Utility bedding, for pipe & conduit, not incl. compaction									
0050	Crushed or screened bank run gravel	B-6	150	.160	L.C.Y.	25.50	6.60	2.43	34.53	41
0100	Crushed stone 3/4″ to 1/2″		150	.160		23.50	6.60	2.43	32.53	39
0200	Sand, dead or bank		150	.160		17.85	6.60	2.43	26.88	32.50
0500	Compacting bedding in trench	A-1D	90	.089	E.C.Y.		3.34	.40	3.74	5.60
0610	See Section 31 23 23.20 for hauling mileage add.									

31 23 23.20 Hauling

		Crew	Daily Output	Labor-Hours	Unit	Material	Labor	Equipment	Total	Total Incl O&P
0010	**HAULING**									
0011	Excavated or borrow, loose cubic yards									
0012	no loading equipment, including hauling,waiting, loading/dumping									
0013	time per cycle (wait, load, travel, unload or dump & return)									
0014	8 C.Y. truck, 15 MPH ave, cycle 0.5 miles, 10 min. wait/Ld./Uld.	B-34A	320	.025	L.C.Y.		1	1.29	2.29	2.92
0016	cycle 1 mile		272	.029			1.18	1.51	2.69	3.44
0018	cycle 2 miles		208	.038			1.54	1.98	3.52	4.50
0020	cycle 4 miles		144	.056			2.23	2.86	5.09	6.50

31 23 23 – Fill

31 23 23.20 Hauling		Crew	Daily Output	Labor-Hours	Unit	Material	2015 Bare Costs Labor	Equipment	Total	Total Incl O&P
0022	cycle 6 miles	B-34A	112	.071	L.C.Y.		2.86	3.67	6.53	8.35
0024	cycle 8 miles		88	.091			3.64	4.67	8.31	10.65
0026	20 MPH ave,cycle 0.5 mile		336	.024			.95	1.22	2.17	2.79
0028	cycle 1 mile		296	.027			1.08	1.39	2.47	3.17
0030	cycle 2 miles		240	.033			1.33	1.71	3.04	3.90
0032	cycle 4 miles		176	.045			1.82	2.34	4.16	5.30
0034	cycle 6 miles		136	.059			2.36	3.02	5.38	6.90
0036	cycle 8 miles		112	.071			2.86	3.67	6.53	8.35
0044	25 MPH ave, cycle 4 miles		192	.042			1.67	2.14	3.81	4.88
0046	cycle 6 miles		160	.050			2	2.57	4.57	5.85
0048	cycle 8 miles		128	.063			2.50	3.21	5.71	7.30
0050	30 MPH ave, cycle 4 miles		216	.037			1.48	1.90	3.38	4.33
0052	cycle 6 miles		176	.045			1.82	2.34	4.16	5.30
0054	cycle 8 miles		144	.056			2.23	2.86	5.09	6.50
0114	15 MPH ave, cycle 0.5 mile, 15 min. wait/Ld./Uld.		224	.036			1.43	1.84	3.27	4.18
0116	cycle 1 mile		200	.040			1.60	2.06	3.66	4.68
0118	cycle 2 miles		168	.048			1.91	2.45	4.36	5.55
0120	cycle 4 miles		120	.067			2.67	3.43	6.10	7.80
0122	cycle 6 miles		96	.083			3.34	4.28	7.62	9.75
0124	cycle 8 miles		80	.100			4.01	5.15	9.16	11.70
0126	20 MPH ave, cycle 0.5 mile		232	.034			1.38	1.77	3.15	4.04
0128	cycle 1 mile		208	.038			1.54	1.98	3.52	4.50
0130	cycle 2 miles		184	.043			1.74	2.23	3.97	5.10
0132	cycle 4 miles		144	.056			2.23	2.86	5.09	6.50
0134	cycle 6 miles		112	.071			2.86	3.67	6.53	8.35
0136	cycle 8 miles		96	.083			3.34	4.28	7.62	9.75
0144	25 MPH ave, cycle 4 miles		152	.053			2.11	2.71	4.82	6.15
0146	cycle 6 miles		128	.063			2.50	3.21	5.71	7.30
0148	cycle 8 miles		112	.071			2.86	3.67	6.53	8.35
0150	30 MPH ave, cycle 4 miles		168	.048			1.91	2.45	4.36	5.55
0152	cycle 6 miles		144	.056			2.23	2.86	5.09	6.50
0154	cycle 8 miles		120	.067			2.67	3.43	6.10	7.80
0214	15 MPH ave, cycle 0.5 mile, 20 min wait/Ld./Uld.		176	.045			1.82	2.34	4.16	5.30
0216	cycle 1 mile		160	.050			2	2.57	4.57	5.85
0218	cycle 2 miles		136	.059			2.36	3.02	5.38	6.90
0220	cycle 4 miles		104	.077			3.08	3.95	7.03	9
0222	cycle 6 miles		88	.091			3.64	4.67	8.31	10.65
0224	cycle 8 miles		72	.111			4.45	5.70	10.15	13.05
0226	20 MPH ave, cycle 0.5 mile		176	.045			1.82	2.34	4.16	5.30
0228	cycle 1 mile		168	.048			1.91	2.45	4.36	5.55
0230	cycle 2 miles		144	.056			2.23	2.86	5.09	6.50
0232	cycle 4 miles		120	.067			2.67	3.43	6.10	7.80
0234	cycle 6 miles		96	.083			3.34	4.28	7.62	9.75
0236	cycle 8 miles		88	.091			3.64	4.67	8.31	10.65
0244	25 MPH ave, cycle 4 miles		128	.063			2.50	3.21	5.71	7.30
0246	cycle 6 miles		112	.071			2.86	3.67	6.53	8.35
0248	cycle 8 miles		96	.083			3.34	4.28	7.62	9.75
0250	30 MPH ave, cycle 4 miles		136	.059			2.36	3.02	5.38	6.90
0252	cycle 6 miles		120	.067			2.67	3.43	6.10	7.80
0254	cycle 8 miles		104	.077			3.08	3.95	7.03	9
0314	15 MPH ave, cycle 0.5 mile, 25 min wait/Ld./Uld.		144	.056			2.23	2.86	5.09	6.50
0316	cycle 1 mile		128	.063			2.50	3.21	5.71	7.30
0318	cycle 2 miles		112	.071			2.86	3.67	6.53	8.35

31 23 23.20 Hauling		Crew	Daily Output	Labor-Hours	Unit	Material	2015 Bare Costs Labor	Equipment	Total	Total Incl O&P
0320	cycle 4 miles	B-34A	96	.083	L.C.Y.		3.34	4.28	7.62	9.75
0322	cycle 6 miles		80	.100			4.01	5.15	9.16	11.70
0324	cycle 8 miles		64	.125			5	6.45	11.45	14.60
0326	20 MPH ave, cycle 0.5 mile		144	.056			2.23	2.86	5.09	6.50
0328	cycle 1 mile		136	.059			2.36	3.02	5.38	6.90
0330	cycle 2 miles		120	.067			2.67	3.43	6.10	7.80
0332	cycle 4 miles		104	.077			3.08	3.95	7.03	9
0334	cycle 6 miles		88	.091			3.64	4.67	8.31	10.65
0336	cycle 8 miles		80	.100			4.01	5.15	9.16	11.70
0344	25 MPH ave, cycle 4 miles		112	.071			2.86	3.67	6.53	8.35
0346	cycle 6 miles		96	.083			3.34	4.28	7.62	9.75
0348	cycle 8 miles		88	.091			3.64	4.67	8.31	10.65
0350	30 MPH ave, cycle 4 miles		112	.071			2.86	3.67	6.53	8.35
0352	cycle 6 miles		104	.077			3.08	3.95	7.03	9
0354	cycle 8 miles		96	.083			3.34	4.28	7.62	9.75
0414	15 MPH ave, cycle 0.5 mile, 30 min wait/Ld./Uld.		120	.067			2.67	3.43	6.10	7.80
0416	cycle 1 mile		112	.071			2.86	3.67	6.53	8.35
0418	cycle 2 miles		96	.083			3.34	4.28	7.62	9.75
0420	cycle 4 miles		80	.100			4.01	5.15	9.16	11.70
0422	cycle 6 miles		72	.111			4.45	5.70	10.15	13.05
0424	cycle 8 miles		64	.125			5	6.45	11.45	14.60
0426	20 MPH ave, cycle 0.5 mile		120	.067			2.67	3.43	6.10	7.80
0428	cycle 1 mile		112	.071			2.86	3.67	6.53	8.35
0430	cycle 2 miles		104	.077			3.08	3.95	7.03	9
0432	cycle 4 miles		88	.091			3.64	4.67	8.31	10.65
0434	cycle 6 miles		80	.100			4.01	5.15	9.16	11.70
0436	cycle 8 miles		72	.111			4.45	5.70	10.15	13.05
0444	25 MPH ave, cycle 4 miles		96	.083			3.34	4.28	7.62	9.75
0446	cycle 6 miles		88	.091			3.64	4.67	8.31	10.65
0448	cycle 8 miles		80	.100			4.01	5.15	9.16	11.70
0450	30 MPH ave, cycle 4 miles		96	.083			3.34	4.28	7.62	9.75
0452	cycle 6 miles		88	.091			3.64	4.67	8.31	10.65
0454	cycle 8 miles		80	.100			4.01	5.15	9.16	11.70
0514	15 MPH ave, cycle 0.5 mile, 35 min wait/Ld./Uld.		104	.077			3.08	3.95	7.03	9
0516	cycle 1 mile		96	.083			3.34	4.28	7.62	9.75
0518	cycle 2 miles		88	.091			3.64	4.67	8.31	10.65
0520	cycle 4 miles		72	.111			4.45	5.70	10.15	13.05
0522	cycle 6 miles		64	.125			5	6.45	11.45	14.60
0524	cycle 8 miles		56	.143			5.70	7.35	13.05	16.75
0526	20 MPH ave, cycle 0.5 mile		104	.077			3.08	3.95	7.03	9
0528	cycle 1 mile		96	.083			3.34	4.28	7.62	9.75
0530	cycle 2 miles		96	.083			3.34	4.28	7.62	9.75
0532	cycle 4 miles		80	.100			4.01	5.15	9.16	11.70
0534	cycle 6 miles		72	.111			4.45	5.70	10.15	13.05
0536	cycle 8 miles		64	.125			5	6.45	11.45	14.60
0544	25 MPH ave, cycle 4 miles		88	.091			3.64	4.67	8.31	10.65
0546	cycle 6 miles		80	.100			4.01	5.15	9.16	11.70
0548	cycle 8 miles		72	.111			4.45	5.70	10.15	13.05
0550	30 MPH ave, cycle 4 miles		88	.091			3.64	4.67	8.31	10.65
0552	cycle 6 miles		80	.100			4.01	5.15	9.16	11.70
0554	cycle 8 miles		72	.111			4.45	5.70	10.15	13.05
1014	12 C.Y. truck, cycle 0.5 mile, 15 MPH ave, 15 min. wait/Ld./Uld.	B-34B	336	.024			.95	2.06	3.01	3.70
1016	cycle 1 mile		300	.027			1.07	2.30	3.37	4.14

31 23 Excavation and Fill

31 23 23 – Fill

31 23 23.20 Hauling		Crew	Daily Output	Labor-Hours	Unit	Material	2015 Bare Costs Labor	Equipment	Total	Total Incl O&P
1018	cycle 2 miles	B-34B	252	.032	L.C.Y.		1.27	2.74	4.01	4.94
1020	cycle 4 miles		180	.044			1.78	3.84	5.62	6.90
1022	cycle 6 miles		144	.056			2.23	4.80	7.03	8.65
1024	cycle 8 miles		120	.067			2.67	5.75	8.42	10.40
1025	cycle 10 miles		96	.083			3.34	7.20	10.54	12.95
1026	20 MPH ave, cycle 0.5 mile		348	.023			.92	1.99	2.91	3.57
1028	cycle 1 mile		312	.026			1.03	2.21	3.24	3.99
1030	cycle 2 miles		276	.029			1.16	2.50	3.66	4.51
1032	cycle 4 miles		216	.037			1.48	3.20	4.68	5.75
1034	cycle 6 miles		168	.048			1.91	4.11	6.02	7.40
1036	cycle 8 miles		144	.056			2.23	4.80	7.03	8.65
1038	cycle 10 miles		120	.067			2.67	5.75	8.42	10.40
1040	25 MPH ave, cycle 4 miles		228	.035			1.41	3.03	4.44	5.45
1042	cycle 6 miles		192	.042			1.67	3.60	5.27	6.50
1044	cycle 8 miles		168	.048			1.91	4.11	6.02	7.40
1046	cycle 10 miles		144	.056			2.23	4.80	7.03	8.65
1050	30 MPH ave, cycle 4 miles		252	.032			1.27	2.74	4.01	4.94
1052	cycle 6 miles		216	.037			1.48	3.20	4.68	5.75
1054	cycle 8 miles		180	.044			1.78	3.84	5.62	6.90
1056	cycle 10 miles		156	.051			2.05	4.43	6.48	8
1060	35 MPH ave, cycle 4 miles		264	.030			1.21	2.62	3.83	4.71
1062	cycle 6 miles		228	.035			1.41	3.03	4.44	5.45
1064	cycle 8 miles		204	.039			1.57	3.39	4.96	6.10
1066	cycle 10 miles		180	.044			1.78	3.84	5.62	6.90
1068	cycle 20 miles		120	.067			2.67	5.75	8.42	10.40
1069	cycle 30 miles		84	.095			3.81	8.25	12.06	14.80
1070	cycle 40 miles		72	.111			4.45	9.60	14.05	17.30
1072	40 MPH ave, cycle 6 miles		240	.033			1.33	2.88	4.21	5.20
1074	cycle 8 miles		216	.037			1.48	3.20	4.68	5.75
1076	cycle 10 miles		192	.042			1.67	3.60	5.27	6.50
1078	cycle 20 miles		120	.067			2.67	5.75	8.42	10.40
1080	cycle 30 miles		96	.083			3.34	7.20	10.54	12.95
1082	cycle 40 miles		72	.111			4.45	9.60	14.05	17.30
1084	cycle 50 miles		60	.133			5.35	11.50	16.85	20.50
1094	45 MPH ave, cycle 8 miles		216	.037			1.48	3.20	4.68	5.75
1096	cycle 10 miles		204	.039			1.57	3.39	4.96	6.10
1098	cycle 20 miles		132	.061			2.43	5.25	7.68	9.40
1100	cycle 30 miles		108	.074			2.97	6.40	9.37	11.55
1102	cycle 40 miles		84	.095			3.81	8.25	12.06	14.80
1104	cycle 50 miles		72	.111			4.45	9.60	14.05	17.30
1106	50 MPH ave, cycle 10 miles		216	.037			1.48	3.20	4.68	5.75
1108	cycle 20 miles		144	.056			2.23	4.80	7.03	8.65
1110	cycle 30 miles		108	.074			2.97	6.40	9.37	11.55
1112	cycle 40 miles		84	.095			3.81	8.25	12.06	14.80
1114	cycle 50 miles		72	.111			4.45	9.60	14.05	17.30
1214	15 MPH ave, cycle 0.5 mile, 20 min. wait/Ld./Uld.		264	.030			1.21	2.62	3.83	4.71
1216	cycle 1 mile		240	.033			1.33	2.88	4.21	5.20
1218	cycle 2 miles		204	.039			1.57	3.39	4.96	6.10
1220	cycle 4 miles		156	.051			2.05	4.43	6.48	8
1222	cycle 6 miles		132	.061			2.43	5.25	7.68	9.40
1224	cycle 8 miles		108	.074			2.97	6.40	9.37	11.55
1225	cycle 10 miles		96	.083			3.34	7.20	10.54	12.95
1226	20 MPH ave, cycle 0.5 mile		264	.030			1.21	2.62	3.83	4.71

31 23 23.20 Hauling		Crew	Daily Output	Labor-Hours	Unit	Material	2015 Bare Costs Labor	Equipment	Total	Total Incl O&P
1228	cycle 1 mile	B-34B	252	.032	L.C.Y.		1.27	2.74	4.01	4.94
1230	cycle 2 miles		216	.037			1.48	3.20	4.68	5.75
1232	cycle 4 miles		180	.044			1.78	3.84	5.62	6.90
1234	cycle 6 miles		144	.056			2.23	4.80	7.03	8.65
1236	cycle 8 miles		132	.061			2.43	5.25	7.68	9.40
1238	cycle 10 miles		108	.074			2.97	6.40	9.37	11.55
1240	25 MPH ave, cycle 4 miles		192	.042			1.67	3.60	5.27	6.50
1242	cycle 6 miles		168	.048			1.91	4.11	6.02	7.40
1244	cycle 8 miles		144	.056			2.23	4.80	7.03	8.65
1246	cycle 10 miles		132	.061			2.43	5.25	7.68	9.40
1250	30 MPH ave, cycle 4 miles		204	.039			1.57	3.39	4.96	6.10
1252	cycle 6 miles		180	.044			1.78	3.84	5.62	6.90
1254	cycle 8 miles		156	.051			2.05	4.43	6.48	8
1256	cycle 10 miles		144	.056			2.23	4.80	7.03	8.65
1260	35 MPH ave, cycle 4 miles		216	.037			1.48	3.20	4.68	5.75
1262	cycle 6 miles		192	.042			1.67	3.60	5.27	6.50
1264	cycle 8 miles		168	.048			1.91	4.11	6.02	7.40
1266	cycle 10 miles		156	.051			2.05	4.43	6.48	8
1268	cycle 20 miles		108	.074			2.97	6.40	9.37	11.55
1269	cycle 30 miles		72	.111			4.45	9.60	14.05	17.30
1270	cycle 40 miles		60	.133			5.35	11.50	16.85	20.50
1272	40 MPH ave, cycle 6 miles		192	.042			1.67	3.60	5.27	6.50
1274	cycle 8 miles		180	.044			1.78	3.84	5.62	6.90
1276	cycle 10 miles		156	.051			2.05	4.43	6.48	8
1278	cycle 20 miles		108	.074			2.97	6.40	9.37	11.55
1280	cycle 30 miles		84	.095			3.81	8.25	12.06	14.80
1282	cycle 40 miles		72	.111			4.45	9.60	14.05	17.30
1284	cycle 50 miles		60	.133			5.35	11.50	16.85	20.50
1294	45 MPH ave, cycle 8 miles		180	.044			1.78	3.84	5.62	6.90
1296	cycle 10 miles		168	.048			1.91	4.11	6.02	7.40
1298	cycle 20 miles		120	.067			2.67	5.75	8.42	10.40
1300	cycle 30 miles		96	.083			3.34	7.20	10.54	12.95
1302	cycle 40 miles		72	.111			4.45	9.60	14.05	17.30
1304	cycle 50 miles		60	.133			5.35	11.50	16.85	20.50
1306	50 MPH ave, cycle 10 miles		180	.044			1.78	3.84	5.62	6.90
1308	cycle 20 miles		132	.061			2.43	5.25	7.68	9.40
1310	cycle 30 miles		96	.083			3.34	7.20	10.54	12.95
1312	cycle 40 miles		84	.095			3.81	8.25	12.06	14.80
1314	cycle 50 miles		72	.111			4.45	9.60	14.05	17.30
1414	15 MPH ave, cycle 0.5 mile, 25 min. wait/Ld./Uld.		204	.039			1.57	3.39	4.96	6.10
1416	cycle 1 mile		192	.042			1.67	3.60	5.27	6.50
1418	cycle 2 miles		168	.048			1.91	4.11	6.02	7.40
1420	cycle 4 miles		132	.061			2.43	5.25	7.68	9.40
1422	cycle 6 miles		120	.067			2.67	5.75	8.42	10.40
1424	cycle 8 miles		96	.083			3.34	7.20	10.54	12.95
1425	cycle 10 miles		84	.095			3.81	8.25	12.06	14.80
1426	20 MPH ave, cycle 0.5 mile		216	.037			1.48	3.20	4.68	5.75
1428	cycle 1 mile		204	.039			1.57	3.39	4.96	6.10
1430	cycle 2 miles		180	.044			1.78	3.84	5.62	6.90
1432	cycle 4 miles		156	.051			2.05	4.43	6.48	8
1434	cycle 6 miles		132	.061			2.43	5.25	7.68	9.40
1436	cycle 8 miles		120	.067			2.67	5.75	8.42	10.40
1438	cycle 10 miles		96	.083			3.34	7.20	10.54	12.95

31 23 23.20 Hauling		Crew	Daily Output	Labor-Hours	Unit	Material	2015 Bare Costs Labor	Equipment	Total	Total Incl O&P
1440	25 MPH ave, cycle 4 miles	B-34B	168	.048	L.C.Y.		1.91	4.11	6.02	7.40
1442	cycle 6 miles		144	.056			2.23	4.80	7.03	8.65
1444	cycle 8 miles		132	.061			2.43	5.25	7.68	9.40
1446	cycle 10 miles		108	.074			2.97	6.40	9.37	11.55
1450	30 MPH ave, cycle 4 miles		168	.048			1.91	4.11	6.02	7.40
1452	cycle 6 miles		156	.051			2.05	4.43	6.48	8
1454	cycle 8 miles		132	.061			2.43	5.25	7.68	9.40
1456	cycle 10 miles		120	.067			2.67	5.75	8.42	10.40
1460	35 MPH ave, cycle 4 miles		180	.044			1.78	3.84	5.62	6.90
1462	cycle 6 miles		156	.051			2.05	4.43	6.48	8
1464	cycle 8 miles		144	.056			2.23	4.80	7.03	8.65
1466	cycle 10 miles		132	.061			2.43	5.25	7.68	9.40
1468	cycle 20 miles		96	.083			3.34	7.20	10.54	12.95
1469	cycle 30 miles		72	.111			4.45	9.60	14.05	17.30
1470	cycle 40 miles		60	.133			5.35	11.50	16.85	20.50
1472	40 MPH ave, cycle 6 miles		168	.048			1.91	4.11	6.02	7.40
1474	cycle 8 miles		156	.051			2.05	4.43	6.48	8
1476	cycle 10 miles		144	.056			2.23	4.80	7.03	8.65
1478	cycle 20 miles		96	.083			3.34	7.20	10.54	12.95
1480	cycle 30 miles		84	.095			3.81	8.25	12.06	14.80
1482	cycle 40 miles		60	.133			5.35	11.50	16.85	20.50
1484	cycle 50 miles		60	.133			5.35	11.50	16.85	20.50
1494	45 MPH ave, cycle 8 miles		156	.051			2.05	4.43	6.48	8
1496	cycle 10 miles		144	.056			2.23	4.80	7.03	8.65
1498	cycle 20 miles		108	.074			2.97	6.40	9.37	11.55
1500	cycle 30 miles		84	.095			3.81	8.25	12.06	14.80
1502	cycle 40 miles		72	.111			4.45	9.60	14.05	17.30
1504	cycle 50 miles		60	.133			5.35	11.50	16.85	20.50
1506	50 MPH ave, cycle 10 miles		156	.051			2.05	4.43	6.48	8
1508	cycle 20 miles		120	.067			2.67	5.75	8.42	10.40
1510	cycle 30 miles		96	.083			3.34	7.20	10.54	12.95
1512	cycle 40 miles		72	.111			4.45	9.60	14.05	17.30
1514	cycle 50 miles		60	.133			5.35	11.50	16.85	20.50
1614	15 MPH, cycle 0.5 mile, 30 min. wait/Ld./Uld.		180	.044			1.78	3.84	5.62	6.90
1616	cycle 1 mile		168	.048			1.91	4.11	6.02	7.40
1618	cycle 2 miles		144	.056			2.23	4.80	7.03	8.65
1620	cycle 4 miles		120	.067			2.67	5.75	8.42	10.40
1622	cycle 6 miles		108	.074			2.97	6.40	9.37	11.55
1624	cycle 8 miles		84	.095			3.81	8.25	12.06	14.80
1625	cycle 10 miles		84	.095			3.81	8.25	12.06	14.80
1626	20 MPH ave, cycle 0.5 mile		180	.044			1.78	3.84	5.62	6.90
1628	cycle 1 mile		168	.048			1.91	4.11	6.02	7.40
1630	cycle 2 miles		156	.051			2.05	4.43	6.48	8
1632	cycle 4 miles		132	.061			2.43	5.25	7.68	9.40
1634	cycle 6 miles		120	.067			2.67	5.75	8.42	10.40
1636	cycle 8 miles		108	.074			2.97	6.40	9.37	11.55
1638	cycle 10 miles		96	.083			3.34	7.20	10.54	12.95
1640	25 MPH ave, cycle 4 miles		144	.056			2.23	4.80	7.03	8.65
1642	cycle 6 miles		132	.061			2.43	5.25	7.68	9.40
1644	cycle 8 miles		108	.074			2.97	6.40	9.37	11.55
1646	cycle 10 miles		108	.074			2.97	6.40	9.37	11.55
1650	30 MPH ave, cycle 4 miles		144	.056			2.23	4.80	7.03	8.65
1652	cycle 6 miles		132	.061			2.43	5.25	7.68	9.40

31 23 23.20 Hauling		Crew	Daily Output	Labor-Hours	Unit	Material	2015 Bare Costs Labor	Equipment	Total	Total Incl O&P
1654	cycle 8 miles	B-34B	120	.067	L.C.Y.		2.67	5.75	8.42	10.40
1656	cycle 10 miles		108	.074			2.97	6.40	9.37	11.55
1660	35 MPH ave, cycle 4 miles		156	.051			2.05	4.43	6.48	8
1662	cycle 6 miles		144	.056			2.23	4.80	7.03	8.65
1664	cycle 8 miles		132	.061			2.43	5.25	7.68	9.40
1666	cycle 10 miles		120	.067			2.67	5.75	8.42	10.40
1668	cycle 20 miles		84	.095			3.81	8.25	12.06	14.80
1669	cycle 30 miles		72	.111			4.45	9.60	14.05	17.30
1670	cycle 40 miles		60	.133			5.35	11.50	16.85	20.50
1672	40 MPH, cycle 6 miles		144	.056			2.23	4.80	7.03	8.65
1674	cycle 8 miles		132	.061			2.43	5.25	7.68	9.40
1676	cycle 10 miles		120	.067			2.67	5.75	8.42	10.40
1678	cycle 20 miles		96	.083			3.34	7.20	10.54	12.95
1680	cycle 30 miles		72	.111			4.45	9.60	14.05	17.30
1682	cycle 40 miles		60	.133			5.35	11.50	16.85	20.50
1684	cycle 50 miles		48	.167			6.70	14.40	21.10	26
1694	45 MPH ave, cycle 8 miles		144	.056			2.23	4.80	7.03	8.65
1696	cycle 10 miles		132	.061			2.43	5.25	7.68	9.40
1698	cycle 20 miles		96	.083			3.34	7.20	10.54	12.95
1700	cycle 30 miles		84	.095			3.81	8.25	12.06	14.80
1702	cycle 40 miles		60	.133			5.35	11.50	16.85	20.50
1704	cycle 50 miles		60	.133			5.35	11.50	16.85	20.50
1706	50 MPH ave, cycle 10 miles		132	.061			2.43	5.25	7.68	9.40
1708	cycle 20 miles		108	.074			2.97	6.40	9.37	11.55
1710	cycle 30 miles		84	.095			3.81	8.25	12.06	14.80
1712	cycle 40 miles		72	.111			4.45	9.60	14.05	17.30
1714	cycle 50 miles		60	.133			5.35	11.50	16.85	20.50
2000	Hauling, 8 C.Y. truck, small project cost per hour	B-34A	8	1	Hr.		40	51.50	91.50	117
2100	12 C.Y. Truck	B-34B	8	1			40	86.50	126.50	156
2150	16.5 C.Y. Truck	B-34C	8	1			40	91	131	161
2175	18 C.Y. 8 wheel Truck	B-34I	8	1			40	109	149	180
2200	20 C.Y. Truck	B-34D	8	1			40	92.50	132.50	163
2300	Grading at dump, or embankment if required, by dozer	B-10B	1000	.012	L.C.Y.		.56	1.39	1.95	2.38
2310	Spotter at fill or cut, if required	1 Clab	8	1	Hr.		37.50		37.50	58
9014	18 C.Y. truck, 8 wheels,15 min. wait/Ld./Uld.,15 MPH, cycle 0.5 mi.	B-34I	504	.016	L.C.Y.		.64	1.72	2.36	2.86
9016	cycle 1 mile		450	.018			.71	1.93	2.64	3.20
9018	cycle 2 miles		378	.021			.85	2.30	3.15	3.81
9020	cycle 4 miles		270	.030			1.19	3.22	4.41	5.35
9022	cycle 6 miles		216	.037			1.48	4.02	5.50	6.65
9024	cycle 8 miles		180	.044			1.78	4.83	6.61	8
9025	cycle 10 miles		144	.056			2.23	6.05	8.28	10
9026	20 MPH ave, cycle 0.5 mile		522	.015			.61	1.67	2.28	2.76
9028	cycle 1 mile		468	.017			.68	1.86	2.54	3.07
9030	cycle 2 miles		414	.019			.77	2.10	2.87	3.48
9032	cycle 4 miles		324	.025			.99	2.68	3.67	4.45
9034	cycle 6 miles		252	.032			1.27	3.45	4.72	5.70
9036	cycle 8 miles		216	.037			1.48	4.02	5.50	6.65
9038	cycle 10 miles		180	.044			1.78	4.83	6.61	8
9040	25 MPH ave, cycle 4 miles		342	.023			.94	2.54	3.48	4.21
9042	cycle 6 miles		288	.028			1.11	3.02	4.13	5
9044	cycle 8 miles		252	.032			1.27	3.45	4.72	5.70
9046	cycle 10 miles		216	.037			1.48	4.02	5.50	6.65
9050	30 MPH ave, cycle 4 miles		378	.021			.85	2.30	3.15	3.81

31 23 23.20 Hauling		Crew	Daily Output	Labor-Hours	Unit	Material	2015 Bare Costs Labor	Equipment	Total	Total Incl O&P
9052	cycle 6 miles	B-34I	324	.025	L.C.Y.		.99	2.68	3.67	4.45
9054	cycle 8 miles		270	.030			1.19	3.22	4.41	5.35
9056	cycle 10 miles		234	.034			1.37	3.71	5.08	6.15
9060	35 MPH ave, cycle 4 miles		396	.020			.81	2.19	3	3.63
9062	cycle 6 miles		342	.023			.94	2.54	3.48	4.21
9064	cycle 8 miles		288	.028			1.11	3.02	4.13	5
9066	cycle 10 miles		270	.030			1.19	3.22	4.41	5.35
9068	cycle 20 miles		162	.049			1.98	5.35	7.33	8.90
9070	cycle 30 miles		126	.063			2.54	6.90	9.44	11.45
9072	cycle 40 miles		90	.089			3.56	9.65	13.21	16
9074	40 MPH ave, cycle 6 miles		360	.022			.89	2.41	3.30	4.01
9076	cycle 8 miles		324	.025			.99	2.68	3.67	4.45
9078	cycle 10 miles		288	.028			1.11	3.02	4.13	5
9080	cycle 20 miles		180	.044			1.78	4.83	6.61	8
9082	cycle 30 miles		144	.056			2.23	6.05	8.28	10
9084	cycle 40 miles		108	.074			2.97	8.05	11.02	13.35
9086	cycle 50 miles		90	.089			3.56	9.65	13.21	16
9094	45 MPH ave, cycle 8 miles		324	.025			.99	2.68	3.67	4.45
9096	cycle 10 miles		306	.026			1.05	2.84	3.89	4.70
9098	cycle 20 miles		198	.040			1.62	4.39	6.01	7.30
9100	cycle 30 miles		144	.056			2.23	6.05	8.28	10
9102	cycle 40 miles		126	.063			2.54	6.90	9.44	11.45
9104	cycle 50 miles		108	.074			2.97	8.05	11.02	13.35
9106	50 MPH ave, cycle 10 miles		324	.025			.99	2.68	3.67	4.45
9108	cycle 20 miles		216	.037			1.48	4.02	5.50	6.65
9110	cycle 30 miles		162	.049			1.98	5.35	7.33	8.90
9112	cycle 40 miles		126	.063			2.54	6.90	9.44	11.45
9114	cycle 50 miles		108	.074			2.97	8.05	11.02	13.35
9214	20 min. wait/Ld./Uld.,15 MPH, cycle 0.5 mi.		396	.020			.81	2.19	3	3.63
9216	cycle 1 mile		360	.022			.89	2.41	3.30	4.01
9218	cycle 2 miles		306	.026			1.05	2.84	3.89	4.70
9220	cycle 4 miles		234	.034			1.37	3.71	5.08	6.15
9222	cycle 6 miles		198	.040			1.62	4.39	6.01	7.30
9224	cycle 8 miles		162	.049			1.98	5.35	7.33	8.90
9225	cycle 10 miles		144	.056			2.23	6.05	8.28	10
9226	20 MPH ave, cycle 0.5 mile		396	.020			.81	2.19	3	3.63
9228	cycle 1 mile		378	.021			.85	2.30	3.15	3.81
9230	cycle 2 miles		324	.025			.99	2.68	3.67	4.45
9232	cycle 4 miles		270	.030			1.19	3.22	4.41	5.35
9234	cycle 6 miles		216	.037			1.48	4.02	5.50	6.65
9236	cycle 8 miles		198	.040			1.62	4.39	6.01	7.30
9238	cycle 10 miles		162	.049			1.98	5.35	7.33	8.90
9240	25 MPH ave, cycle 4 miles		288	.028			1.11	3.02	4.13	5
9242	cycle 6 miles		252	.032			1.27	3.45	4.72	5.70
9244	cycle 8 miles		216	.037			1.48	4.02	5.50	6.65
9246	cycle 10 miles		198	.040			1.62	4.39	6.01	7.30
9250	30 MPH ave, cycle 4 miles		306	.026			1.05	2.84	3.89	4.70
9252	cycle 6 miles		270	.030			1.19	3.22	4.41	5.35
9254	cycle 8 miles		234	.034			1.37	3.71	5.08	6.15
9256	cycle 10 miles		216	.037			1.48	4.02	5.50	6.65
9260	35 MPH ave, cycle 4 miles		324	.025			.99	2.68	3.67	4.45
9262	cycle 6 miles		288	.028			1.11	3.02	4.13	5
9264	cycle 8 miles		252	.032			1.27	3.45	4.72	5.70

31 23 23.20 Hauling		Crew	Daily Output	Labor-Hours	Unit	Material	2015 Bare Costs Labor	Equipment	Total	Total Incl O&P
9266	cycle 10 miles	B-34I	234	.034	L.C.Y.		1.37	3.71	5.08	6.15
9268	cycle 20 miles		162	.049			1.98	5.35	7.33	8.90
9270	cycle 30 miles		108	.074			2.97	8.05	11.02	13.35
9272	cycle 40 miles		90	.089			3.56	9.65	13.21	16
9274	40 MPH ave, cycle 6 miles		288	.028			1.11	3.02	4.13	5
9276	cycle 8 miles		270	.030			1.19	3.22	4.41	5.35
9278	cycle 10 miles		234	.034			1.37	3.71	5.08	6.15
9280	cycle 20 miles		162	.049			1.98	5.35	7.33	8.90
9282	cycle 30 miles		126	.063			2.54	6.90	9.44	11.45
9284	cycle 40 miles		108	.074			2.97	8.05	11.02	13.35
9286	cycle 50 miles		90	.089			3.56	9.65	13.21	16
9294	45 MPH ave, cycle 8 miles		270	.030			1.19	3.22	4.41	5.35
9296	cycle 10 miles		252	.032			1.27	3.45	4.72	5.70
9298	cycle 20 miles		180	.044			1.78	4.83	6.61	8
9300	cycle 30 miles		144	.056			2.23	6.05	8.28	10
9302	cycle 40 miles		108	.074			2.97	8.05	11.02	13.35
9304	cycle 50 miles		90	.089			3.56	9.65	13.21	16
9306	50 MPH ave, cycle 10 miles		270	.030			1.19	3.22	4.41	5.35
9308	cycle 20 miles		198	.040			1.62	4.39	6.01	7.30
9310	cycle 30 miles		144	.056			2.23	6.05	8.28	10
9312	cycle 40 miles		126	.063			2.54	6.90	9.44	11.45
9314	cycle 50 miles		108	.074			2.97	8.05	11.02	13.35
9414	25 min. wait/Ld./Uld.,15 MPH, cycle 0.5 mi.		306	.026			1.05	2.84	3.89	4.70
9416	cycle 1 mile		288	.028			1.11	3.02	4.13	5
9418	cycle 2 miles		252	.032			1.27	3.45	4.72	5.70
9420	cycle 4 miles		198	.040			1.62	4.39	6.01	7.30
9422	cycle 6 miles		180	.044			1.78	4.83	6.61	8
9424	cycle 8 miles		144	.056			2.23	6.05	8.28	10
9425	cycle 10 miles		126	.063			2.54	6.90	9.44	11.45
9426	20 MPH ave, cycle 0.5 mile		324	.025			.99	2.68	3.67	4.45
9428	cycle 1 mile		306	.026			1.05	2.84	3.89	4.70
9430	cycle 2 miles		270	.030			1.19	3.22	4.41	5.35
9432	cycle 4 miles		234	.034			1.37	3.71	5.08	6.15
9434	cycle 6 miles		198	.040			1.62	4.39	6.01	7.30
9436	cycle 8 miles		180	.044			1.78	4.83	6.61	8
9438	cycle 10 miles		144	.056			2.23	6.05	8.28	10
9440	25 MPH ave, cycle 4 miles		252	.032			1.27	3.45	4.72	5.70
9442	cycle 6 miles		216	.037			1.48	4.02	5.50	6.65
9444	cycle 8 miles		198	.040			1.62	4.39	6.01	7.30
9446	cycle 10 miles		180	.044			1.78	4.83	6.61	8
9450	30 MPH ave, cycle 4 miles		252	.032			1.27	3.45	4.72	5.70
9452	cycle 6 miles		234	.034			1.37	3.71	5.08	6.15
9454	cycle 8 miles		198	.040			1.62	4.39	6.01	7.30
9456	cycle 10 miles		180	.044			1.78	4.83	6.61	8
9460	35 MPH ave, cycle 4 miles		270	.030			1.19	3.22	4.41	5.35
9462	cycle 6 miles		234	.034			1.37	3.71	5.08	6.15
9464	cycle 8 miles		216	.037			1.48	4.02	5.50	6.65
9466	cycle 10 miles		198	.040			1.62	4.39	6.01	7.30
9468	cycle 20 miles		144	.056			2.23	6.05	8.28	10
9470	cycle 30 miles		108	.074			2.97	8.05	11.02	13.35
9472	cycle 40 miles		90	.089			3.56	9.65	13.21	16
9474	40 MPH ave, cycle 6 miles		252	.032			1.27	3.45	4.72	5.70
9476	cycle 8 miles		234	.034			1.37	3.71	5.08	6.15

31 23 23.20 Hauling		Crew	Daily Output	Labor-Hours	Unit	Material	2015 Bare Costs Labor	Equipment	Total	Total Incl O&P
9478	cycle 10 miles	B-34I	216	.037	L.C.Y.		1.48	4.02	5.50	6.65
9480	cycle 20 miles		144	.056			2.23	6.05	8.28	10
9482	cycle 30 miles		126	.063			2.54	6.90	9.44	11.45
9484	cycle 40 miles		90	.089			3.56	9.65	13.21	16
9486	cycle 50 miles		90	.089			3.56	9.65	13.21	16
9494	45 MPH ave, cycle 8 miles		234	.034			1.37	3.71	5.08	6.15
9496	cycle 10 miles		216	.037			1.48	4.02	5.50	6.65
9498	cycle 20 miles		162	.049			1.98	5.35	7.33	8.90
9500	cycle 30 miles		126	.063			2.54	6.90	9.44	11.45
9502	cycle 40 miles		108	.074			2.97	8.05	11.02	13.35
9504	cycle 50 miles		90	.089			3.56	9.65	13.21	16
9506	50 MPH ave, cycle 10 miles		234	.034			1.37	3.71	5.08	6.15
9508	cycle 20 miles		180	.044			1.78	4.83	6.61	8
9510	cycle 30 miles		144	.056			2.23	6.05	8.28	10
9512	cycle 40 miles		108	.074			2.97	8.05	11.02	13.35
9514	cycle 50 miles		90	.089			3.56	9.65	13.21	16
9614	30 min. wait/Ld./Uld.,15 MPH, cycle 0.5 mi.		270	.030			1.19	3.22	4.41	5.35
9616	cycle 1 mile		252	.032			1.27	3.45	4.72	5.70
9618	cycle 2 miles		216	.037			1.48	4.02	5.50	6.65
9620	cycle 4 miles		180	.044			1.78	4.83	6.61	8
9622	cycle 6 miles		162	.049			1.98	5.35	7.33	8.90
9624	cycle 8 miles		126	.063			2.54	6.90	9.44	11.45
9625	cycle 10 miles		126	.063			2.54	6.90	9.44	11.45
9626	20 MPH ave, cycle 0.5 mile		270	.030			1.19	3.22	4.41	5.35
9628	cycle 1 mile		252	.032			1.27	3.45	4.72	5.70
9630	cycle 2 miles		234	.034			1.37	3.71	5.08	6.15
9632	cycle 4 miles		198	.040			1.62	4.39	6.01	7.30
9634	cycle 6 miles		180	.044			1.78	4.83	6.61	8
9636	cycle 8 miles		162	.049			1.98	5.35	7.33	8.90
9638	cycle 10 miles		144	.056			2.23	6.05	8.28	10
9640	25 MPH ave, cycle 4 miles		216	.037			1.48	4.02	5.50	6.65
9642	cycle 6 miles		198	.040			1.62	4.39	6.01	7.30
9644	cycle 8 miles		180	.044			1.78	4.83	6.61	8
9646	cycle 10 miles		162	.049			1.98	5.35	7.33	8.90
9650	30 MPH ave, cycle 4 miles		216	.037			1.48	4.02	5.50	6.65
9652	cycle 6 miles		198	.040			1.62	4.39	6.01	7.30
9654	cycle 8 miles		180	.044			1.78	4.83	6.61	8
9656	cycle 10 miles		162	.049			1.98	5.35	7.33	8.90
9660	35 MPH ave, cycle 4 miles		234	.034			1.37	3.71	5.08	6.15
9662	cycle 6 miles		216	.037			1.48	4.02	5.50	6.65
9664	cycle 8 miles		198	.040			1.62	4.39	6.01	7.30
9666	cycle 10 miles		180	.044			1.78	4.83	6.61	8
9668	cycle 20 miles		126	.063			2.54	6.90	9.44	11.45
9670	cycle 30 miles		108	.074			2.97	8.05	11.02	13.35
9672	cycle 40 miles		90	.089			3.56	9.65	13.21	16
9674	40 MPH ave, cycle 6 miles		216	.037			1.48	4.02	5.50	6.65
9676	cycle 8 miles		198	.040			1.62	4.39	6.01	7.30
9678	cycle 10 miles		180	.044			1.78	4.83	6.61	8
9680	cycle 20 miles		144	.056			2.23	6.05	8.28	10
9682	cycle 30 miles		108	.074			2.97	8.05	11.02	13.35
9684	cycle 40 miles		90	.089			3.56	9.65	13.21	16
9686	cycle 50 miles		72	.111			4.45	12.05	16.50	20
9694	45 MPH ave, cycle 8 miles		216	.037			1.48	4.02	5.50	6.65

31 23 23.20 Hauling		Crew	Daily Output	Labor-Hours	Unit	Material	2015 Bare Costs Labor	Equipment	Total	Total Incl O&P
9696	cycle 10 miles	B-34I	198	.040	L.C.Y.		1.62	4.39	6.01	7.30
9698	cycle 20 miles		144	.056			2.23	6.05	8.28	10
9700	cycle 30 miles		126	.063			2.54	6.90	9.44	11.45
9702	cycle 40 miles		108	.074			2.97	8.05	11.02	13.35
9704	cycle 50 miles		90	.089			3.56	9.65	13.21	16
9706	50 MPH ave, cycle 10 miles		198	.040			1.62	4.39	6.01	7.30
9708	cycle 20 miles		162	.049			1.98	5.35	7.33	8.90
9710	cycle 30 miles		126	.063			2.54	6.90	9.44	11.45
9712	cycle 40 miles		108	.074			2.97	8.05	11.02	13.35
9714	cycle 50 miles	↓	90	.089	↓		3.56	9.65	13.21	16
31 23 23.23 Compaction										
0010	**COMPACTION**									
7000	Walk behind, vibrating plate 18" wide, 6" lifts, 2 passes	A-1D	200	.040	E.C.Y.		1.50	.18	1.68	2.51
7020	3 passes		185	.043			1.63	.20	1.83	2.72
7040	4 passes	↓	140	.057	↓		2.15	.26	2.41	3.59

Estimating Tips

32 01 00 Operations and Maintenance of Exterior Improvements

- Recycling of asphalt pavement is becoming very popular and is an alternative to removal and replacement. It can be a good value engineering proposal if removed pavement can be recycled, either at the project site or at another site that is reasonably close to the project site. Sections on repair of flexible and rigid pavement are included.

32 10 00 Bases, Ballasts, and Paving

- When estimating paving, keep in mind the project schedule. Also note that prices for asphalt and concrete are generally higher in the cold seasons. Lines for pavement markings, including tactile warning systems and fence lines, are included.

32 90 00 Planting

- The timing of planting and guarantee specifications often dictate the costs for establishing tree and shrub growth and a stand of grass or ground cover. Establish the work performance schedule to coincide with the local planting season. Maintenance and growth guarantees can add from 20%–100% to the total landscaping cost and can be contractually cumbersome. The cost to replace trees and shrubs can be as high as 5% of the total cost, depending on the planting zone, soil conditions, and time of year.

Reference Numbers

Reference numbers are shown in shaded boxes at the beginning of some major classifications. These numbers refer to related items in the Reference Section. The reference information may be an estimating procedure, an alternate pricing method, or technical information.

Note: Not all subdivisions listed here necessarily appear in this publication. ■

32 31 Fences and Gates

32 31 13 – Chain Link Fences and Gates

32 31 13.30 Fence, Chain Link, Gates and Posts		Crew	Daily Output	Labor-Hours	Unit	Material	2015 Bare Costs Labor	Equipment	Total	Total Incl O&P
0010	**FENCE, CHAIN LINK, GATES & POSTS**									
0011	(1/3 post length in ground)									
6710	Corner post, galv. steel, 4" OD, set in conc., 4'	B-80	65	.492	Ea.	97	20.50	11.10	128.60	150
6715	6'		63	.508		106	21	11.45	138.45	162
6720	7'		61	.525		131	21.50	11.80	164.30	190
6725	8'		65	.492		141	20.50	11.10	172.60	198
7900	Auger fence post hole, 3' deep, medium soil, by hand	1 Clab	30	.267			10.05		10.05	15.45
7925	By machine	B-80	175	.183			7.55	4.12	11.67	16.05
7950	Rock, with jackhammer	B-9	32	1.250			47.50	7.30	54.80	81
7975	With rock drill	B-47C	65	.246			10.60	25	35.60	43.50

Estimating Tips

33 10 00 Water Utilities
33 30 00 Sanitary Sewerage Utilities
33 40 00 Storm Drainage Utilities

- Never assume that the water, sewer, and drainage lines will go in at the early stages of the project. Consider the site access needs before dividing the site in half with open trenches, loose pipe, and machinery obstructions. Always inspect the site to establish that the site drawings are complete. Check off all existing utilities on your drawings as you locate them. Be especially careful with underground utilities because appurtenances are sometimes buried during regrading or repaving operations. If you find any discrepancies, mark up the site plan for further research. Differing site conditions can be very costly if discovered later in the project.
- See also Section 33 01 00 for restoration of pipe where removal/replacement may be undesirable. Use of new types of piping materials can reduce the overall project cost. Owners/design engineers should consider the installing contractor as a valuable source of current information on utility products and local conditions that could lead to significant cost savings.

Reference Numbers

Reference numbers are shown in shaded boxes at the beginning of some major classifications. These numbers refer to related items in the Reference Section. The reference information may be an estimating procedure, an alternate pricing method, or technical information.

Note: Not all subdivisions listed here necessarily appear in this publication. ■

Note: ***Trade Service****, in part, has been used as a reference source for some of the material prices used in Division 33.*

33 05 16 – Utility Structures

	33 05 16.13 Precast Concrete Utility Boxes	Crew	Daily Output	Labor-Hours	Unit	2015 Bare Costs Material	Labor	Equipment	Total	Total Incl O&P
0010	**PRECAST CONCRETE UTILITY BOXES**, 6″ thick									
0050	5′ x 10′ x 6′ high, I.D.	B-13	2	28	Ea.	3,750	1,150	370	5,270	6,275
0100	6′ x 10′ x 6′ high, I.D.		2	28		3,900	1,150	370	5,420	6,425
0150	5′ x 12′ x 6′ high, I.D.		2	28		4,125	1,150	370	5,645	6,675
0200	6′ x 12′ x 6′ high, I.D.		1.80	31.111		4,600	1,275	410	6,285	7,475
0250	6′ x 13′ x 6′ high, I.D.		1.50	37.333		6,050	1,525	490	8,065	9,550
0300	8′ x 14′ x 7′ high, I.D.		1	56		6,525	2,300	735	9,560	11,500

33 05 23 – Trenchless Utility Installation

	33 05 23.19 Microtunneling	Crew	Daily Output	Labor-Hours	Unit	Material	Labor	Equipment	Total	Total Incl O&P
0010	**MICROTUNNELING**									
0011	Not including excavation, backfill, shoring,									
0020	or dewatering, average 50′/day, slurry method									
0100	24″ to 48″ outside diameter, minimum				L.F.				875	965
0110	Adverse conditions, add				%				50%	50%
1000	Rent microtunneling machine, average monthly lease				Month				97,500	107,000
1010	Operating technician				Day				630	705
1100	Mobilization and demobilization, minimum				Job				41,200	45,900
1110	Maximum				″				445,500	490,500

	33 05 23.20 Horizontal Boring	Crew	Daily Output	Labor-Hours	Unit	Material	Labor	Equipment	Total	Total Incl O&P
0010	**HORIZONTAL BORING**									
0011	Casing only, 100′ minimum,									
0020	not incl. jacking pits or dewatering									
0100	Roadwork, 1/2″ thick wall, 24″ diameter casing	B-42	20	3.200	L.F.	121	136	68	325	420
0200	36″ diameter		16	4		223	170	85	478	605
0300	48″ diameter		15	4.267		310	181	90.50	581.50	725
0500	Railroad work, 24″ diameter		15	4.267		121	181	90.50	392.50	515
0600	36″ diameter		14	4.571		223	194	97	514	655
0700	48″ diameter		12	5.333		310	226	113	649	820
0900	For ledge, add								20%	20%
1000	Small diameter boring, 3″, sandy soil	B-82	900	.018		22	.77	.09	22.86	25.50
1040	Rocky soil	″	500	.032		22	1.38	.17	23.55	26.50
1100	Prepare jacking pits, incl. mobilization & demobilization, minimum				Ea.				3,225	3,700
1101	Maximum				″				22,000	25,500

	33 05 23.22 Directional Drilling	Crew	Daily Output	Labor-Hours	Unit	Material	Labor	Equipment	Total	Total Incl O&P
0010	**DIRECTIONAL DRILLING**									
0011	Excluding access and splice pits (if required) and conduit (required)									
0012	Drilled Hole diameters shown should be 50% larger than conduit									
0013	Assume access to H20 & removal of spoil/drilling mud as non-hazard material									
0014	Actual production rates adj. to account for risk factors for each soil type									
0100	Sand, silt, clay, common earth									
0110	Mobilization or demobilization	B-82A	1	32	Ea.		1,375	2,425	3,800	4,775
0120	6″ diameter		480	.067	L.F.		2.87	5.05	7.92	10
0130	12″ diameter		270	.119			5.10	9	14.10	17.70
0140	18″ diameter		180	.178			7.65	13.50	21.15	26.50
0150	24″ diameter		135	.237			10.20	18.05	28.25	35.50
0200	Hard clay, cobble, random boulders									
0210	Mobilization or demobilization	B-82B	1	32	Ea.		1,375	2,700	4,075	5,075
0220	6″ diameter		400	.080	L.F.		3.45	6.80	10.25	12.70
0230	12″ diameter		220	.145			6.25	12.30	18.55	23
0240	18″ diameter		145	.221			9.50	18.70	28.20	35
0250	24″ diameter		105	.305			13.15	26	39.15	48.50
0260	30″ diameter		80	.400			17.25	34	51.25	64

33 05 Common Work Results for Utilities

33 05 23 – Trenchless Utility Installation

33 05 23.22 Directional Drilling		Crew	Daily Output	Labor-Hours	Unit	Material	2015 Bare Costs Labor	Equipment	Total	Total Incl O&P
0270	36″ diameter	B-82B	60	.533	L.F.		23	45	68	84.50
0300	Hard rock (solid bed, 24,000+ psi)									
0310	Mobilization or Demobilization	B-82C	1	32	Ea.		1,375	3,025	4,400	5,425
0320	6″ diameter		75	.427	L.F.		18.40	40	58.40	72.50
0330	12″ diameter		40	.800			34.50	75.50	110	136
0340	18″ diameter		25	1.280			55	121	176	217
0350	24″ diameter		20	1.600			69	151	220	271
0360	30″ diameter		15	2.133			92	201	293	360
0370	36″ diameter		12	2.667			115	251	366	450

33 05 26 – Utility Identification

33 05 26.05 Utility Connection		Crew	Daily Output	Labor-Hours	Unit	Material	Labor	Equipment	Total	Total Incl O&P
0010	**UTILITY CONNECTION**									
0020	Water, sanitary, stormwater, gas, single connection	B-14	1	48	Ea.	3,225	1,900	365	5,490	6,875
0030	Telecommunication	″	1	48	″	395	1,900	365	2,660	3,750

33 05 26.10 Utility Accessories		Crew	Daily Output	Labor-Hours	Unit	Material	Labor	Equipment	Total	Total Incl O&P
0010	**UTILITY ACCESSORIES**									
0400	Underground tape, detectable, reinforced, alum. foil core, 2″	1 Clab	150	.053	C.L.F.	6.50	2.01		8.51	10.25
0500	6″	″	140	.057	″	27.50	2.15		29.65	33.50

33 44 Storm Utility Water Drains

33 44 13 – Utility Area Drains

33 44 13.13 Catchbasins		Crew	Daily Output	Labor-Hours	Unit	Material	Labor	Equipment	Total	Total Incl O&P
0010	**CATCHBASINS**									
0011	Not including footing & excavation									
1600	Frames & grates, C.I., 24″ square, 500 lb.	B-6	7.80	3.077	Ea.	340	127	46.50	513.50	620
1700	26″ D shape, 600 lb.		7	3.429		505	142	52	699	830
1800	Light traffic, 18″ diameter, 100 lb.		10	2.400		123	99	36.50	258.50	325
1900	24″ diameter, 300 lb.		8.70	2.759		196	114	42	352	435
2000	36″ diameter, 900 lb.		5.80	4.138		570	171	63	804	955
2100	Heavy traffic, 24″ diameter, 400 lb.		7.80	3.077		244	127	46.50	417.50	515
2200	36″ diameter, 1150 lb.		3	8		795	330	121	1,246	1,525
2300	Mass. State standard, 26″ diameter, 475 lb.		7	3.429		266	142	52	460	565
2400	30″ diameter, 620 lb.		7	3.429		345	142	52	539	655
2500	Watertight, 24″ diameter, 350 lb.		7.80	3.077		320	127	46.50	493.50	595
2600	26″ diameter, 500 lb.		7	3.429		425	142	52	619	745
2700	32″ diameter, 575 lb.		6	4		850	165	60.50	1,075.50	1,250
2800	3 piece cover & frame, 10″ deep,									
2900	1200 lb., for heavy equipment	B-6	3	8	Ea.	1,050	330	121	1,501	1,825
3000	Raised for paving 1-1/4″ to 2″ high									
3100	4 piece expansion ring									
3200	20″ to 26″ diameter	1 Clab	3	2.667	Ea.	158	100		258	330
3300	30″ to 36″ diameter	″	3	2.667	″	217	100		317	395
3320	Frames and covers, existing, raised for paving, 2″, including									
3340	row of brick, concrete collar, up to 12″ wide frame	B-6	18	1.333	Ea.	45	55	20	120	156
3360	20″ to 26″ wide frame		11	2.182		67.50	90	33	190.50	249
3380	30″ to 36″ wide frame		9	2.667		83.50	110	40.50	234	305
3400	Inverts, single channel brick	D-1	3	5.333		97	225		322	450
3500	Concrete		5	3.200		104	135		239	320
3600	Triple channel, brick		2	8		148	335		483	680
3700	Concrete		3	5.333		139	225		364	500

33 49 Storm Drainage Structures

33 49 13 – Storm Drainage Manholes, Frames, and Covers

33 49 13.10 Storm Drainage Manholes, Frames and Covers		Crew	Daily Output	Labor-Hours	Unit	2015 Bare Costs Material	Labor	Equipment	Total	Total Incl O&P
0010	**STORM DRAINAGE MANHOLES, FRAMES & COVERS**									
0020	Excludes footing, excavation, backfill (See line items for frame & cover)									
0050	Brick, 4′ inside diameter, 4′ deep	D-1	1	16	Ea.	520	675		1,195	1,600
0100	6′ deep		.70	22.857		740	960		1,700	2,300
0150	8′ deep		.50	32		955	1,350		2,305	3,100
0200	For depths over 8′, add		4	4	V.L.F.	83.50	168		251.50	350
0400	Concrete blocks (radial), 4′ I.D., 4′ deep		1.50	10.667	Ea.	415	450		865	1,150
0500	6′ deep		1	16		565	675		1,240	1,650
0600	8′ deep		.70	22.857		710	960		1,670	2,250
0700	For depths over 8′, add		5.50	2.909	V.L.F.	77	122		199	272
0800	Concrete, cast in place, 4′ x 4′, 8″ thick, 4′ deep	C-14H	2	24	Ea.	505	1,100	15.60	1,620.60	2,275
0900	6′ deep		1.50	32		730	1,475	21	2,226	3,100
1000	8′ deep		1	48		1,050	2,225	31	3,306	4,575
1100	For depths over 8′, add		8	6	V.L.F.	119	278	3.90	400.90	560
1110	Precast, 4′ I.D., 4′ deep	B-22	4.10	7.317	Ea.	725	320	50.50	1,095.50	1,350
1120	6′ deep		3	10		925	440	69	1,434	1,775
1130	8′ deep		2	15		1,075	660	103	1,838	2,300
1140	For depths over 8′, add		16	1.875	V.L.F.	127	82.50	12.90	222.40	280
1150	5′ I.D., 4′ deep	B-6	3	8	Ea.	1,650	330	121	2,101	2,450
1160	6′ deep		2	12		1,925	495	182	2,602	3,075
1170	8′ deep		1.50	16		2,400	660	243	3,303	3,925
1180	For depths over 8′, add		12	2	V.L.F.	280	82.50	30.50	393	470
1190	6′ I.D., 4′ deep		2	12	Ea.	2,150	495	182	2,827	3,325
1200	6′ deep		1.50	16		2,600	660	243	3,503	4,125
1210	8′ deep		1	24		3,200	990	365	4,555	5,425
1220	For depths over 8′, add		8	3	V.L.F.	380	124	45.50	549.50	655
1250	Slab tops, precast, 8″ thick									
1300	4′ diameter manhole	B-6	8	3	Ea.	252	124	45.50	421.50	515
1400	5′ diameter manhole		7.50	3.200		410	132	48.50	590.50	710
1500	6′ diameter manhole		7	3.429		635	142	52	829	975
3800	Steps, heavyweight cast iron, 7″ x 9″	1 Bric	40	.200		19.25	9.25		28.50	35
3900	8″ x 9″		40	.200		23	9.25		32.25	39.50
3928	12″ x 10-1/2″		40	.200		27	9.25		36.25	43.50
4000	Standard sizes, galvanized steel		40	.200		22	9.25		31.25	38
4100	Aluminum		40	.200		24	9.25		33.25	40.50
4150	Polyethylene		40	.200		26	9.25		35.25	42.50

33 71 Electrical Utility Transmission and Distribution

33 71 13 – Electrical Utility Towers

33 71 13.23 Steel Electrical Utility Towers		Crew	Daily Output	Labor-Hours	Unit	Material	Labor	Equipment	Total	Total Incl O&P
0010	**STEEL ELECTRICAL UTILITY TOWERS**									
0100	Excavation and backfill, earth	R-5	135.38	.650	C.Y.		31	10.85	41.85	59
0105	Rock		21.46	4.101	″		196	68.50	264.50	370
0200	Steel footings (grillage) in earth		3.91	22.506	Ton	2,175	1,075	375	3,625	4,450
0205	In rock		3.20	27.500	″	2,175	1,325	460	3,960	4,875
0290	See also Section 33 05 23.19									
0300	Rock anchors	R-5	5.87	14.991	Ea.	575	715	250	1,540	1,975
0400	Concrete foundations	″	12.85	6.848	C.Y.	134	325	114	573	770
0490	See also Section 03 30 53.40									
0500	Towers-material handling and spotting	R-7	22.56	2.128	Ton		82.50	9.15	91.65	137
0540	Steel tower erection	R-5	7.65	11.503		2,125	550	192	2,867	3,375

33 71 13 – Electrical Utility Towers

33 71 13.23 Steel Electrical Utility Towers		Crew	Daily Output	Labor-Hours	Unit	Material	Labor	Equipment	Total	Total Incl O&P
						2015 Bare Costs				
0550	Lace and box	R-5	7.10	12.394	Ton	2,125	590	206	2,921	3,450
0560	Painting total structure		1.47	59.864	Ea.	460	2,850	995	4,305	5,925
0570	Disposal of surplus material	R-7	20.87	2.300	Mile		89	9.90	98.90	148
0600	Special towers-material handling and spotting	"	12.31	3.899	Ton		151	16.75	167.75	250
0640	Special steel structure erection	R-6	6.52	13.497		2,650	645	525	3,820	4,475
0650	Special steel lace and box	"	6.29	13.990		2,650	670	545	3,865	4,525
0670	Disposal of surplus material	R-7	7.87	6.099	Mile		236	26	262	395

33 71 13.80 Transmission Line Right of Way

Line	Description	Crew	Daily Output	Labor-Hours	Unit	Material	Labor	Equipment	Total	Total Incl O&P
0010	**TRANSMISSION LINE RIGHT OF WAY**									
0100	Clearing right of way	B-87	6.67	5.997	Acre		288	480	768	970
0200	Restoration & seeding	B-10D	4	3	"	440	139	455	1,034	1,200

33 71 16 – Electrical Utility Poles

33 71 16.23 Steel Electrical Utility Poles

Line	Description	Crew	Daily Output	Labor-Hours	Unit	Material	Labor	Equipment	Total	Total Incl O&P
0010	**STEEL ELECTRICAL UTILITY POLES**									
0100	Steel, square, tapered shaft, 20′	R-15A	6	8	Ea.	365	385	50	800	1,025
0120	25′		5.35	8.972		385	430	56.50	871.50	1,125
0140	30′		4.80	10		455	480	63	998	1,300
0160	35′		4.35	11.034		470	530	69.50	1,069.50	1,400
0180	40′		4	12		665	575	75.50	1,315.50	1,675
0300	Galvanized, round, tapered shaft, 40′		4	12		1,975	575	75.50	2,625.50	3,125
0320	45′		3.70	12.973		2,200	625	81.50	2,906.50	3,425
0340	50′		3.45	13.913		2,425	670	87.50	3,182.50	3,750
0360	55′		3	16		2,550	770	100	3,420	4,075
0380	60′		2.65	18.113		2,850	870	114	3,834	4,575
0400	65′	R-13	2.20	19.091		3,525	1,000	85	4,610	5,500
0420	70′		1.90	22.105		3,850	1,175	98.50	5,123.50	6,075
0440	75′		1.70	24.706		5,050	1,300	110	6,460	7,625
0460	80′		1.50	28		4,575	1,475	125	6,175	7,375
0480	85′		1.40	30		4,950	1,575	134	6,659	7,975
0500	90′		1.25	33.600		5,300	1,775	150	7,225	8,700
0520	95′		1.15	36.522		5,725	1,925	163	7,813	9,375
0540	100′		1	42		6,175	2,225	187	8,587	10,300
0700	12 sided, tapered shaft, 40′	R-15A	4	12		1,100	575	75.50	1,750.50	2,150
0720	45′		3.70	12.973		1,225	625	81.50	1,931.50	2,375
0740	50′		3.45	13.913		1,400	670	87.50	2,157.50	2,650
0760	55′		3	16		1,600	770	100	2,470	3,050
0780	60′		2.65	18.113		1,675	870	114	2,659	3,300
0800	65′	R-13	2.20	19.091		1,975	1,000	85	3,060	3,800
0820	70′		1.90	22.105		2,150	1,175	98.50	3,423.50	4,200
0840	75′		1.70	24.706		2,325	1,300	110	3,735	4,625
0860	80′		1.50	28		3,950	1,475	125	5,550	6,700
0880	85′		1.40	30		6,400	1,575	134	8,109	9,575
0900	90′		1.25	33.600		8,375	1,775	150	10,300	12,000
0920	95′		1.15	36.522		10,500	1,925	163	12,588	14,700
0940	100′		1	42		12,800	2,225	187	15,212	17,600
0960	105′		.90	46.667		14,200	2,475	208	16,883	19,500
0980	Ladder clips					225			225	248
1000	Galvanized steel, round, tapered, w/one 6′ arm, 20′	R-15A	9.20	5.217		900	251	33	1,184	1,400
1020	25′		7.65	6.275		980	300	39.50	1,319.50	1,575
1040	30′		6.55	7.328		1,075	350	46	1,471	1,750
1060	35′		5.75	8.348		1,175	400	52.50	1,627.50	1,950
1080	40′		5.15	9.320		1,250	450	58.50	1,758.50	2,125

33 71 16 – Electrical Utility Poles

33 71 16.23 Steel Electrical Utility Poles		Crew	Daily Output	Labor-Hours	Unit	2015 Bare Costs Material	Labor	Equipment	Total	Total Incl O&P
1200	Two 6′ arms, 20′	R-15A	6.55	7.328	Ea.	1,150	350	46	1,546	1,850
1220	25′		5.75	8.348		1,175	400	52.50	1,627.50	1,975
1240	30′		5.15	9.320		1,300	450	58.50	1,808.50	2,200
1260	35′		4.60	10.435		1,400	500	65.50	1,965.50	2,375
1280	40′		4.20	11.429		1,625	550	72	2,247	2,675
1400	One 12′ truss arm, 25′		7.65	6.275		1,125	300	39.50	1,464.50	1,725
1420	30′		6.55	7.328		1,150	350	46	1,546	1,850
1440	35′		5.75	8.348		1,250	400	52.50	1,702.50	2,050
1460	40′		5.15	9.320		1,400	450	58.50	1,908.50	2,300
1480	45′		4.60	10.435		1,625	500	65.50	2,190.50	2,625
1600	Two 12′ truss arms, 25′		5.75	8.348		1,325	400	52.50	1,777.50	2,125
1620	30′		5.15	9.320		1,375	450	58.50	1,883.50	2,250
1640	35′		4.60	10.435		1,475	500	65.50	2,040.50	2,425
1660	40′		4.20	11.429		1,625	550	72	2,247	2,675
1680	45′		3.85	12.468		1,825	600	78.50	2,503.50	3,025
3400	Galvanized steel, tapered, 10′		15.50	3.097		279	149	19.45	447.45	550
3420	12′		11.50	4.174		410	201	26	637	785
3440	14′		9.20	5.217		300	251	33	584	745
3460	16′		8.45	5.680		315	273	35.50	623.50	800
3480	18′		7.65	6.275		325	300	39.50	664.50	855
3500	20′		7.10	6.761		305	325	42.50	672.50	870
5200	Galvanized steel, bracket type, 2″, 6′ long		18.50	2.595		139	125	16.30	280.30	360
5220	8′ long		15.50	3.097		173	149	19.45	341.45	435
5240	10′ long		12.50	3.840		290	185	24	499	625
6000	Digging holes in earth, average	R-5	25.14	3.500			167	58.50	225.50	315
6010	In rock, average	″	4.51	19.512			930	325	1,255	1,750
6020	Formed plate pole structure									
6030	Material handling and spotting	R-7	2.40	20	Ea.		775	86	861	1,300
6040	Erect steel plate pole	R-5	1.95	45.128		9,150	2,150	750	12,050	14,200
6050	Guys, anchors and hardware for pole, in earth		7.04	12.500		555	595	208	1,358	1,750
6060	In rock		17.96	4.900		665	234	81.50	980.50	1,175
6070	Foundations for line poles									
6080	Excavation, in earth	R-5	135.38	.650	C.Y.		31	10.85	41.85	59
6090	In rock		20	4.400			210	73.50	283.50	400
6110	Concrete foundations		11	8		134	380	133	647	875

33 71 16.33 Wood Electrical Utility Poles

		Crew	Daily Output	Labor-Hours	Unit	Material	Labor	Equipment	Total	Total Incl O&P
0010	**WOOD ELECTRICAL UTILITY POLES**									
0011	Excludes excavation, backfill and cast-in-place concrete									
1020	12″ Ponderosa Pine Poles treated 0.40 ACQ, 16′	R-3	3.20	6.250	Ea.	880	340	43	1,263	1,525
5000	Wood, class 3 yellow pine, penta-treated, 25′	R-15A	8.60	5.581		256	268	35	559	725
5020	30′		7.70	6.234		292	300	39	631	820
5040	35′		5.80	8.276		365	400	52	817	1,050
5060	40′		5.30	9.057		440	435	57	932	1,200
5080	45′		4.70	10.213		475	490	64	1,029	1,325
5100	50′		4.20	11.429		585	550	72	1,207	1,550
5120	55′		3.80	12.632		685	605	79.50	1,369.50	1,775
5140	60′		3.50	13.714		915	660	86	1,661	2,100
5160	65′		3.20	15		1,200	720	94	2,014	2,500
5180	70′		3	16		1,675	770	100	2,545	3,125
5200	75′		2.80	17.143		2,175	825	108	3,108	3,750
6000	Wood, class 1 type C, CCA/ACA-treated, 25′		8.60	5.581		256	268	35	559	725
6020	30′		7.70	6.234		315	300	39	654	845

33 71 Electrical Utility Transmission and Distribution

33 71 16 – Electrical Utility Poles

33 71 16.33 Wood Electrical Utility Poles		Crew	Daily Output	Labor-Hours	Unit	2015 Bare Costs Material	Labor	Equipment	Total	Total Incl O&P
6040	35′	R-15A	5.80	8.276	Ea.	440	400	52	892	1,125
6060	40′		5.30	9.057		510	435	57	1,002	1,275
6080	45′		4.70	10.213		620	490	64	1,174	1,500
6100	50′		4.20	11.429		680	550	72	1,302	1,650
6120	55′		3.80	12.632		765	605	79.50	1,449.50	1,850
6200	Electric & tel sitework, 20′ high, treated wd., see Section 26 56 13.10	R-3	3.10	6.452		224	350	44.50	618.50	820
6400	25′ high R337119-30		2.90	6.897		265	375	47.50	687.50	905
6600	30′ high		2.60	7.692		435	420	53	908	1,150
6800	35′ high		2.40	8.333		495	455	57.50	1,007.50	1,300
7000	40′ high		2.30	8.696		710	470	60	1,240	1,550
7200	45′ high		1.70	11.765		865	640	81	1,586	2,000
7400	Cross arms with hardware & insulators									
7600	4′ long	1 Elec	2.50	3.200	Ea.	150	175		325	425
7800	5′ long		2.40	3.333		172	182		354	460
8000	6′ long		2.20	3.636		166	199		365	480
9000	Disposal of pole & hardware surplus material	R-7	20.87	2.300	Mile		89	9.90	98.90	148
9100	Disposal of crossarms & hardware surplus material	"	40	1.200	"		46.50	5.15	51.65	77

33 71 19 – Electrical Underground Ducts and Manholes

33 71 19.15 Underground Ducts and Manholes		Crew	Daily Output	Labor-Hours	Unit	Material	Labor	Equipment	Total	Total Incl O&P
0010	**UNDERGROUND DUCTS AND MANHOLES**									
0011	Not incl. excavation, backfill and concrete, in slab or duct bank									
1000	Direct burial									
1010	PVC, schedule 40, w/coupling, 1/2″ diameter	1 Elec	340	.024	L.F.	.35	1.29		1.64	2.32
1020	3/4″ diameter		290	.028		.47	1.51		1.98	2.78
1030	1″ diameter		260	.031		.75	1.68		2.43	3.35
1040	1-1/2″ diameter		210	.038		1.14	2.08		3.22	4.37
1050	2″ diameter		180	.044		1.48	2.43		3.91	5.25
1060	3″ diameter	2 Elec	240	.067		2.82	3.65		6.47	8.55
1070	4″ diameter		160	.100		3.95	5.45		9.40	12.55
1080	5″ diameter		120	.133		5.75	7.30		13.05	17.25
1090	6″ diameter		90	.178		7.60	9.70		17.30	23
1110	Elbows, 1/2″ diameter	1 Elec	48	.167	Ea.	.90	9.10		10	14.65
1120	3/4″ diameter		38	.211		.88	11.50		12.38	18.20
1130	1″ diameter		32	.250		1.33	13.70		15.03	22
1140	1-1/2″ diameter		21	.381		2.65	21		23.65	34
1150	2″ diameter		16	.500		3.59	27.50		31.09	45
1160	3″ diameter		12	.667		10.85	36.50		47.35	66.50
1170	4″ diameter		9	.889		17.25	48.50		65.75	92
1180	5″ diameter		8	1		28	54.50		82.50	113
1190	6″ diameter		5	1.600		64.50	87.50		152	202
1210	Adapters, 1/2″ diameter		52	.154		.25	8.40		8.65	12.90
1220	3/4″ diameter		43	.186		.41	10.20		10.61	15.70
1230	1″ diameter		39	.205		.56	11.20		11.76	17.40
1240	1-1/2″ diameter		35	.229		.98	12.50		13.48	19.85
1250	2″ diameter		26	.308		1.22	16.85		18.07	26.50
1260	3″ diameter		20	.400		2.97	22		24.97	36.50
1270	4″ diameter		14	.571		5.30	31.50		36.80	53
1280	5″ diameter		12	.667		11.20	36.50		47.70	67
1290	6″ diameter		9	.889		18.70	48.50		67.20	93.50
1340	Bell end & cap, 1-1/2″ diameter		35	.229		8.55	12.50		21.05	28
1350	Bell end & plug, 2″ diameter		26	.308		9.65	16.85		26.50	35.50
1360	3″ diameter		20	.400		11.70	22		33.70	46

33 71 19 – Electrical Underground Ducts and Manholes

33 71 19.15 Underground Ducts and Manholes		Crew	Daily Output	Labor-Hours	Unit	Material	2015 Bare Costs Labor	Equipment	Total	Total Incl O&P
1370	4" diameter	1 Elec	14	.571	Ea.	14.75	31.50		46.25	63.50
1380	5" diameter		12	.667		16.75	36.50		53.25	73
1390	6" diameter		9	.889		20	48.50		68.50	95
1450	Base spacer, 2" diameter		56	.143		1.55	7.80		9.35	13.40
1460	3" diameter		46	.174		1.58	9.50		11.08	16
1470	4" diameter		41	.195		1.64	10.65		12.29	17.80
1480	5" diameter		37	.216		2.02	11.85		13.87	19.90
1490	6" diameter		34	.235		2.75	12.85		15.60	22.50
1550	Intermediate spacer, 2" diameter		60	.133		1.50	7.30		8.80	12.60
1560	3" diameter		46	.174		1.69	9.50		11.19	16.10
1570	4" diameter		41	.195		1.55	10.65		12.20	17.70
1580	5" diameter		37	.216		1.85	11.85		13.70	19.75
1590	6" diameter		34	.235		2.70	12.85		15.55	22.50
4010	PVC, schedule 80, w/coupling, 1/2" diameter		215	.037	L.F.	.92	2.04		2.96	4.06
4020	3/4" diameter		180	.044		1.24	2.43		3.67	5
4030	1" diameter		145	.055		1.76	3.02		4.78	6.45
4040	1-1/2" diameter		120	.067		2.93	3.65		6.58	8.65
4050	2" diameter		100	.080		4.05	4.38		8.43	11
4060	3" diameter	2 Elec	130	.123		8.25	6.75		15	19.20
4070	4" diameter		90	.178		12	9.70		21.70	28
4080	5" diameter		70	.229		17.25	12.50		29.75	37.50
4090	6" diameter		50	.320		21.50	17.50		39	50
4110	Elbows, 1/2" diameter	1 Elec	29	.276	Ea.	1.33	15.10		16.43	24
4120	3/4" diameter		23	.348		2.57	19.05		21.62	31.50
4130	1" diameter		20	.400		3.18	22		25.18	36.50
4140	1-1/2" diameter		16	.500		6.95	27.50		34.45	48.50
4150	2" diameter		12	.667		8.25	36.50		44.75	63.50
4160	3" diameter		9	.889		23	48.50		71.50	98.50
4170	4" diameter		7	1.143		46.50	62.50		109	145
4180	5" diameter		6	1.333		195	73		268	325
4190	6" diameter		4	2		231	109		340	420
4210	Adapter, 1/2" diameter		39	.205		.25	11.20		11.45	17.10
4220	3/4" diameter		33	.242		.41	13.25		13.66	20.50
4230	1" diameter		29	.276		.56	15.10		15.66	23
4240	1-1/2" diameter		26	.308		.98	16.85		17.83	26
4250	2" diameter		23	.348		1.22	19.05		20.27	30
4260	3" diameter		18	.444		2.97	24.50		27.47	40
4270	4" diameter		13	.615		5.30	33.50		38.80	56.50
4280	5" diameter		11	.727		11.20	40		51.20	72
4290	6" diameter		8	1		18.70	54.50		73.20	103
4310	Bell end & cap, 1-1/2" diameter		26	.308		8.55	16.85		25.40	34.50
4320	Bell end & plug, 2" diameter		23	.348		9.65	19.05		28.70	39
4330	3" diameter		18	.444		11.70	24.50		36.20	49.50
4340	4" diameter		13	.615		14.75	33.50		48.25	67
4350	5" diameter		11	.727		16.75	40		56.75	78
4360	6" diameter		8	1		20	54.50		74.50	104
4370	Base spacer, 2" diameter		42	.190		1.55	10.40		11.95	17.30
4380	3" diameter		33	.242		1.58	13.25		14.83	21.50
4390	4" diameter		29	.276		1.64	15.10		16.74	24.50
4400	5" diameter		26	.308		2.02	16.85		18.87	27
4410	6" diameter		25	.320		2.75	17.50		20.25	29
4420	Intermediate spacer, 2" diameter		45	.178		1.50	9.70		11.20	16.20
4430	3" diameter		34	.235		1.69	12.85		14.54	21

33 71 19.15 Underground Ducts and Manholes		Crew	Daily Output	Labor-Hours	Unit	2015 Bare Costs Material	Labor	Equipment	Total	Total Incl O&P
4440	4" diameter	1 Elec	31	.258	Ea.	1.55	14.10		15.65	22.50
4450	5" diameter		28	.286		1.85	15.65		17.50	25.50
4460	6" diameter		25	.320		2.70	17.50		20.20	29

33 71 19.17 Electric and Telephone Underground

		Crew	Daily Output	Labor-Hours	Unit	Material	Labor	Equipment	Total	Total Incl O&P
0010	**ELECTRIC AND TELEPHONE UNDERGROUND**									
0011	Not including excavation									
0200	backfill and cast in place concrete									
0250	For bedding, see Section 31 23 23.16 R337119-30									
0400	Hand holes, precast concrete, with concrete cover									
0600	2' x 2' x 3' deep	R-3	2.40	8.333	Ea.	405	455	57.50	917.50	1,200
0800	3' x 3' x 3' deep		1.90	10.526		525	570	72.50	1,167.50	1,525
1000	4' x 4' x 4' deep		1.40	14.286		1,400	775	98.50	2,273.50	2,825
1200	Manholes, precast with iron racks & pulling irons, C.I. frame									
1400	and cover, 4' x 6' x 7' deep	B-13	2	28	Ea.	6,050	1,150	370	7,570	8,800
1600	6' x 8' x 7' deep		1.90	29.474		6,800	1,200	390	8,390	9,750
1800	6' x 10' x 7' deep		1.80	31.111		7,625	1,275	410	9,310	10,800
4200	Underground duct, banks ready for concrete fill, min. of 7.5"									
4400	between conduits, center to center									
4580	PVC, type EB, 1 @ 2" diameter	2 Elec	480	.033	L.F.	.71	1.82		2.53	3.51
4600	2 @ 2" diameter		240	.067		1.41	3.65		5.06	7
4800	4 @ 2" diameter		120	.133		2.82	7.30		10.12	14.05
4900	1 @ 3" diameter		400	.040		.98	2.19		3.17	4.36
5000	2 @ 3" diameter		200	.080		1.96	4.38		6.34	8.70
5200	4 @ 3" diameter		100	.160		3.91	8.75		12.66	17.40
5300	1 @ 4" diameter		320	.050		1.52	2.74		4.26	5.80
5400	2 @ 4" diameter		160	.100		3.05	5.45		8.50	11.55
5600	4 @ 4" diameter		80	.200		6.10	10.95		17.05	23
5800	6 @ 4" diameter		54	.296		9.15	16.20		25.35	34.50
5810	1 @ 5" diameter		260	.062		2.24	3.37		5.61	7.50
5820	2 @ 5" diameter		130	.123		4.48	6.75		11.23	15.05
5840	4 @ 5" diameter		70	.229		8.95	12.50		21.45	28.50
5860	6 @ 5" diameter		50	.320		13.45	17.50		30.95	41
5870	1 @ 6" diameter		200	.080		3.21	4.38		7.59	10.10
5880	2 @ 6" diameter		100	.160		6.40	8.75		15.15	20
5900	4 @ 6" diameter		50	.320		12.80	17.50		30.30	40
5920	6 @ 6" diameter		30	.533		19.25	29		48.25	64.50
6200	Rigid galvanized steel, 2 @ 2" diameter		180	.089		12.90	4.86		17.76	21.50
6400	4 @ 2" diameter		90	.178		26	9.70		35.70	43
6800	2 @ 3" diameter		100	.160		29	8.75		37.75	44.50
7000	4 @ 3" diameter		50	.320		57.50	17.50		75	89.50
7200	2 @ 4" diameter		70	.229		41.50	12.50		54	64.50
7400	4 @ 4" diameter		34	.471		83	25.50		108.50	130
7600	6 @ 4" diameter		22	.727		124	40		164	197
7620	2 @ 5" diameter		60	.267		85.50	14.60		100.10	116
7640	4 @ 5" diameter		30	.533		171	29		200	232
7660	6 @ 5" diameter		18	.889		257	48.50		305.50	355
7680	2 @ 6" diameter		40	.400		116	22		138	161
7700	4 @ 6" diameter		20	.800		232	44		276	320
7720	6 @ 6" diameter		14	1.143		350	62.50		412.50	480
7800	For cast-in-place concrete - Add									
7810	Under 1 C.Y.	C-6	16	3	C.Y.	163	118	3.90	284.90	360
7820	1 C.Y. - 5 C.Y.		19.20	2.500		147	98	3.25	248.25	315

33 71 19 – Electrical Underground Ducts and Manholes

33 71 19.17 Electric and Telephone Underground		Crew	Daily Output	Labor-Hours	Unit	2015 Bare Costs Material	Labor	Equipment	Total	Total Incl O&P
7830	Over 5 C.Y.	C-6	24	2	C.Y.	122	78.50	2.60	203.10	257
7850	For reinforcing rods - Add									
7860	#4 to #7	2 Rodm	1.10	14.545	Ton	970	765		1,735	2,275
7870	#8 to #14	"	1.50	10.667	"	970	560		1,530	1,950
8000	Fittings, PVC type EB, elbow, 2″ diameter	1 Elec	16	.500	Ea.	15.90	27.50		43.40	58.50
8200	3″ diameter		14	.571		20.50	31.50		52	69.50
8400	4″ diameter		12	.667		28.50	36.50		65	86
8420	5″ diameter		10	.800		152	44		196	233
8440	6″ diameter		9	.889		173	48.50		221.50	263
8500	Coupling, 2″ diameter					.72			.72	.79
8600	3″ diameter					3.23			3.23	3.55
8700	4″ diameter					3.53			3.53	3.88
8720	5″ diameter					6.65			6.65	7.30
8740	6″ diameter					22.50			22.50	24.50
8800	Adapter, 2″ diameter	1 Elec	26	.308		1.02	16.85		17.87	26
9000	3″ diameter		20	.400		2.90	22		24.90	36
9200	4″ diameter		16	.500		4.23	27.50		31.73	45.50
9220	5″ diameter		13	.615		10.25	33.50		43.75	62
9240	6″ diameter		10	.800		13.30	44		57.30	80
9400	End bell, 2″ diameter		16	.500		1.30	27.50		28.80	42.50
9600	3″ diameter		14	.571		6.65	31.50		38.15	54.50
9800	4″ diameter		12	.667		4.26	36.50		40.76	59
9810	5″ diameter		10	.800		13	44		57	80
9820	6″ diameter		8	1		66	54.50		120.50	155
9830	5° angle coupling, 2″ diameter		26	.308		19.60	16.85		36.45	46.50
9840	3″ diameter		20	.400		29.50	22		51.50	65.50
9850	4″ diameter		16	.500		15.25	27.50		42.75	58
9860	5″ diameter		13	.615		14.75	33.50		48.25	66.50
9870	6″ diameter		10	.800		23	44		67	90.50
9880	Expansion joint, 2″ diameter		16	.500		33.50	27.50		61	78
9890	3″ diameter		18	.444		64	24.50		88.50	107
9900	4″ diameter		12	.667		85	36.50		121.50	148
9910	5″ diameter		10	.800		114	44		158	192
9920	6″ diameter		8	1		143	54.50		197.50	239
9930	Heat bender, 2″ diameter					545			545	600
9940	6″ diameter					1,850			1,850	2,025
9950	Cement, quart					16.60			16.60	18.30
9960	Nylon polyethylene pull rope, 1/4″	2 Elec	2000	.008	L.F.	.16	.44		.60	.84

33 71 23 – Insulators and Fittings

33 71 23.16 Post Insulators		Crew	Daily Output	Labor-Hours	Unit	Material	Labor	Equipment	Total	Total Incl O&P
0010	**POST INSULATORS**									
7400	Insulators, pedestal type	R-11	112	.500	Ea.		26	7.70	33.70	47.50
7490	See also line 33 71 39.13 1000									

33 71 26 – Transmission and Distribution Equipment

33 71 26.13 Capacitor Banks		Crew	Daily Output	Labor-Hours	Unit	Material	Labor	Equipment	Total	Total Incl O&P
0010	**CAPACITOR BANKS**									
1300	Station capacitors									
1350	Synchronous, 13 to 26 kV	R-11	3.11	18.006	MVAR	6,650	935	277	7,862	9,025
1360	46 kV		3.33	16.817		8,500	875	258	9,633	11,000
1370	69 kV		3.81	14.698		8,350	765	226	9,341	10,600
1380	161 kV		6.51	8.602		7,825	445	132	8,402	9,425
1390	500 kV		10.37	5.400		6,800	280	83	7,163	7,975

33 71 26 – Transmission and Distribution Equipment

33 71 26.13 Capacitor Banks		Crew	Daily Output	Labor-Hours	Unit	2015 Bare Costs Material	Labor	Equipment	Total	Total Incl O&P
1450	Static, 13 to 26 kV	R-11	3.11	18.006	MVAR	5,650	935	277	6,862	7,900
1460	46 kV		3.01	18.605		7,125	965	286	8,376	9,625
1470	69 kV		3.81	14.698		6,925	765	226	7,916	9,025
1480	161 kV		6.51	8.602		6,425	445	132	7,002	7,900
1490	500 kV		10.37	5.400		5,875	280	83	6,238	6,950
1600	Voltage regulators, 13 to 26 kV		.75	74.667	Ea.	252,500	3,875	1,150	257,525	285,000

33 71 26.23 Current Transformers

		Crew	Daily Output	Labor-Hours	Unit	Material	Labor	Equipment	Total	Total Incl O&P
0010	**CURRENT TRANSFORMERS**									
4050	Current transformers, 13 to 26 kV	R-11	14	4	Ea.	3,100	208	61.50	3,369.50	3,800
4060	46 kV		9.33	6.002		9,075	310	92	9,477	10,500
4070	69 kV		7	8		9,400	415	123	9,938	11,100
4080	161 kV		1.87	29.947		30,500	1,550	460	32,510	36,400

33 71 26.26 Potential Transformers

		Crew	Daily Output	Labor-Hours	Unit	Material	Labor	Equipment	Total	Total Incl O&P
0010	**POTENTIAL TRANSFORMERS**									
4100	Potential transformers, 13 to 26 kV	R-11	11.20	5	Ea.	4,450	260	77	4,787	5,350
4110	46 kV		8	7		9,125	365	108	9,598	10,700
4120	69 kV		6.22	9.003		9,675	465	138	10,278	11,500
4130	161 kV		2.24	25		20,900	1,300	385	22,585	25,400
4140	500 kV		1.40	40		62,500	2,075	615	65,190	72,500

33 71 39 – High-Voltage Wiring

33 71 39.13 Overhead High-Voltage Wiring

		Crew	Daily Output	Labor-Hours	Unit	Material	Labor	Equipment	Total	Total Incl O&P
0010	**OVERHEAD HIGH-VOLTAGE WIRING** R337116-60									
0100	Conductors, primary circuits									
0110	Material handling and spotting	R-5	9.78	8.998	W.Mile		430	150	580	815
0120	For river crossing, add		11	8			380	133	513	725
0150	Conductors, per wire, 210 to 636 kcmil		1.96	44.898		10,400	2,150	750	13,300	15,500
0160	795 to 954 kcmil		1.87	47.059		20,700	2,250	785	23,735	27,100
0170	1000 to 1600 kcmil		1.47	59.864		35,500	2,850	995	39,345	44,400
0180	Over 1600 kcmil		1.35	65.185		49,600	3,125	1,075	53,800	60,500
0200	For river crossing, add, 210 to 636 kcmil		1.24	70.968			3,400	1,175	4,575	6,425
0220	795 to 954 kcmil		1.09	80.734			3,850	1,350	5,200	7,300
0230	1000 to 1600 kcmil		.97	90.722			4,325	1,500	5,825	8,225
0240	Over 1600 kcmil		.87	101			4,825	1,675	6,500	9,150
0300	Joints and dead ends	R-8	6	8	Ea.	1,425	385	58.50	1,868.50	2,225
0400	Sagging	R-5	7.33	12.001	W.Mile		575	200	775	1,075
0500	Clipping, per structure, 69 kV	R-10	9.60	5	Ea.		258	64.50	322.50	460
0510	161 kV		5.33	9.006			465	116	581	830
0520	345 to 500 kV		2.53	18.972			980	245	1,225	1,750
0600	Make and install jumpers, per structure, 69 kV	R-8	3.20	15		385	725	110	1,220	1,650
0620	161 kV		1.20	40		770	1,925	292	2,987	4,100
0640	345 to 500 kV		.32	150		1,300	7,250	1,100	9,650	13,600
0700	Spacers	R-10	68.57	.700		77	36	9	122	149
0720	For river crossings, add	"	60	.800			41.50	10.30	51.80	73.50
0800	Installing pulling line (500 kV only)	R-9	1.45	44.138	W.Mile	765	2,000	242	3,007	4,125
0810	Disposal of surplus material, high voltage conductors	R-7	6.96	6.897	Mile		267	29.50	296.50	445
0820	With trailer mounted reel stands	"	13.71	3.501	"		136	15.05	151.05	226
0900	Insulators and hardware, primary circuits									
0920	Material handling and spotting, 69 kV	R-7	480	.100	Ea.		3.87	.43	4.30	6.40
0930	161 kV		685.71	.070			2.71	.30	3.01	4.50
0950	345 to 500 kV		960	.050			1.94	.22	2.16	3.22
1000	Disk insulators, 69 kV	R-5	880	.100		78.50	4.78	1.67	84.95	95.50
1020	161 kV		977.78	.090		89.50	4.30	1.50	95.30	107

33 71 39.13 Overhead High-Voltage Wiring		Crew	Daily Output	Labor-Hours	Unit	2015 Bare Costs Material	Labor	Equipment	Total	Total Incl O&P
1040	345 to 500 kV	R-5	1100	.080	Ea.	89.50	3.82	1.33	94.65	106
1060	See Section 33 71 23.16 for pin or pedestal insulator									
1100	Install disk insulator at river crossing, add									
1110	69 kV	R-5	586.67	.150	Ea.		7.15	2.50	9.65	13.60
1120	161 kV		880	.100			4.78	1.67	6.45	9.05
1140	345 to 500 kV		880	.100			4.78	1.67	6.45	9.05
1150	Disposal of surplus material, high voltage insulators	R-7	41.74	1.150	Mile		44.50	4.94	49.44	74
1300	Overhead ground wire installation									
1320	Material handling and spotting	R-7	5.65	8.496	W.Mile		330	36.50	366.50	545
1340	Overhead ground wire	R-5	1.76	50		3,000	2,400	835	6,235	7,825
1350	At river crossing, add		1.17	75.214			3,600	1,250	4,850	6,800
1360	Disposal of surplus material, grounding wire		41.74	2.108	Mile		101	35	136	191
1400	Installing conductors, underbuilt circuits									
1420	Material handling and spotting	R-7	5.65	8.496	W.Mile		330	36.50	366.50	545
1440	Conductors, per wire, 210 to 636 kcmil	R-5	1.96	44.898		10,400	2,150	750	13,300	15,500
1450	795 to 954 kcmil		1.87	47.059		20,700	2,250	785	23,735	27,100
1460	1000 to 1600 kcmil		1.47	59.864		35,500	2,850	995	39,345	44,400
1470	Over 1600 kcmil		1.35	65.185		49,600	3,125	1,075	53,800	60,500
1500	Joints and dead ends	R-8	6	8	Ea.	1,425	385	58.50	1,868.50	2,225
1550	Sagging	R-5	8.80	10	W.Mile		480	167	647	905
1600	Clipping, per structure, 69 kV	R-10	9.60	5	Ea.		258	64.50	322.50	460
1620	161 kV		5.33	9.006			465	116	581	830
1640	345 to 500 kV		2.53	18.972			980	245	1,225	1,750
1700	Making and installing jumpers, per structure, 69 kV	R-8	5.87	8.177		385	395	59.50	839.50	1,100
1720	161 kV		.96	50		770	2,425	365	3,560	4,900
1740	345 to 500 kV		.32	150		1,300	7,250	1,100	9,650	13,600
1800	Spacers	R-10	96	.500		77	26	6.45	109.45	131
1810	Disposal of surplus material, conductors & hardware	R-7	6.96	6.897	Mile		267	29.50	296.50	445
2000	Insulators and hardware for underbuilt circuits									
2100	Material handling and spotting	R-7	1200	.040	Ea.		1.55	.17	1.72	2.57
2150	Disk insulators, 69 kV	R-8	600	.080		78.50	3.87	.58	82.95	93
2160	161 kV		686	.070		89.50	3.38	.51	93.39	104
2170	345 to 500 kV		800	.060		89.50	2.90	.44	92.84	103
2180	Disposal of surplus material, insulators & hardware	R-7	41.74	1.150	Mile		44.50	4.94	49.44	74
2300	Sectionalizing switches, 69 kV	R-5	1.26	69.841	Ea.	20,200	3,325	1,175	24,700	28,500
2310	161 kV		.80	110		22,900	5,250	1,825	29,975	35,200
2500	Protective devices		5.50	16		6,600	765	267	7,632	8,725
2600	Clearance poles, 8 poles per mile									
2650	In earth, 69 kV	R-5	1.16	75.862	Mile	5,450	3,625	1,275	10,350	12,900
2660	161 kV	"	.64	137		8,950	6,575	2,300	17,825	22,300
2670	345 to 500 kV	R-6	.48	183		10,700	8,750	7,125	26,575	32,900
2800	In rock, 69 kV	R-5	.69	127		5,450	6,100	2,125	13,675	17,500
2820	161 kV	"	.35	251		8,950	12,000	4,200	25,150	32,700
2840	345 to 500 kV	R-6	.24	366		10,700	17,500	14,300	42,500	54,000

33 72 Utility Substations

33 72 26 – Substation Bus Assemblies

33 72 26.13 Aluminum Substation Bus Assemblies		Crew	Daily Output	Labor-Hours	Unit	2015 Bare Costs Material	Labor	Equipment	Total	Total Incl O&P
0010	**ALUMINUM SUBSTATION BUS ASSEMBLIES**									
7300	Bus	R-11	590	.095	Lb.	2.75	4.93	1.46	9.14	12.05

33 72 33 – Control House Equipment

33 72 33.33 Raceway/Boxes for Utility Substations		Crew	Daily Output	Labor-Hours	Unit	Material	Labor	Equipment	Total	Total Incl O&P
0010	**RACEWAY/BOXES FOR UTILITY SUBSTATIONS**									
7000	Conduit, conductors, and insulators									
7100	Conduit, metallic	R-11	560	.100	Lb.	2.52	5.20	1.54	9.26	12.25
7110	Non-metallic	"	800	.070	"	7.75	3.63	1.08	12.46	15.20
7190	See Section 26 05 33									
7200	Wire and cable	R-11	700	.080	Lb.	8.20	4.15	1.23	13.58	16.60
7290	See Section 26 05 19									

33 72 33.36 Cable Trays for Utility Substations		Crew	Daily Output	Labor-Hours	Unit	Material	Labor	Equipment	Total	Total Incl O&P
0010	**CABLE TRAYS FOR UTILITY SUBSTATIONS**									
7700	Cable tray	R-11	40	1.400	L.F.	23.50	72.50	21.50	117.50	159
7790	See Section 26 05 36									

33 72 33.43 Substation Backup Batteries		Crew	Daily Output	Labor-Hours	Unit	Material	Labor	Equipment	Total	Total Incl O&P
0010	**SUBSTATION BACKUP BATTERIES**									
9120	Battery chargers	R-11	11.20	5	Ea.	3,800	260	77	4,137	4,650
9200	Control batteries	"	14	4	K.A.H.	87	208	61.50	356.50	475

33 72 33.46 Substation Converter Stations		Crew	Daily Output	Labor-Hours	Unit	Material	Labor	Equipment	Total	Total Incl O&P
0010	**SUBSTATION CONVERTER STATIONS**									
9000	Station service equipment									
9100	Conversion equipment									
9110	Station service transformers	R-11	5.60	10	Ea.	91,500	520	154	92,174	101,500

33 73 Utility Transformers

33 73 23 – Dry-Type Utility Transformers

33 73 23.20 Primary Transformers		Crew	Daily Output	Labor-Hours	Unit	Material	Labor	Equipment	Total	Total Incl O&P
0010	**PRIMARY TRANSFORMERS**									
1000	Main conversion equipment									
1050	Power transformers, 13 to 26 kV	R-11	1.72	32.558	MVA	23,900	1,700	500	26,100	29,400
1060	46 kV		3.50	16		22,400	830	246	23,476	26,100
1070	69 kV		3.11	18.006		19,100	935	277	20,312	22,700
1080	110 kV		3.29	17.021		18,100	885	261	19,246	21,600
1090	161 kV		4.31	12.993		16,800	675	200	17,675	19,700
1100	500 kV		7	8	↓	16,800	415	123	17,338	19,200
1200	Grounding transformers	↓	3.11	18.006	Ea.	105,500	935	277	106,712	117,500

33 75 High-Voltage Switchgear and Protection Devices

33 75 13 – Air High-Voltage Circuit Breaker

33 75 13.13 High-Voltage Circuit Breaker, Air		Crew	Daily Output	Labor-Hours	Unit	Material	2015 Bare Costs Labor	Equipment	Total	Total Incl O&P
0010	**HIGH-VOLTAGE CIRCUIT BREAKER, AIR**									
2100	Air circuit breakers, 13 to 26 kV	R-11	.56	100	Ea.	59,000	5,200	1,525	65,725	74,500
2110	161 kV	"	.24	233	"	254,000	12,100	3,575	269,675	301,000

33 75 16 – Oil High-Voltage Circuit Breaker

33 75 16.13 Oil Circuit Breaker

		Crew	Daily Output	Labor-Hours	Unit	Material	Labor	Equipment	Total	Total Incl O&P
0010	**OIL CIRCUIT BREAKER**									
2000	Power circuit breakers									
2050	Oil circuit breakers, 13 to 26 kV	R-11	1.12	50	Ea.	57,000	2,600	770	60,370	67,000
2060	46 kV		.75	74.667		83,000	3,875	1,150	88,025	98,000
2070	69 kV		.45	124		192,500	6,450	1,900	200,850	223,500
2080	161 kV		.16	350		293,000	18,200	5,375	316,575	355,500
2090	500 kV	↓	.06	933	↓	1,095,000	48,400	14,300	1,157,700	1,293,500

33 75 19 – Gas High-Voltage Circuit Breaker

33 75 19.13 Gas Circuit Breaker

		Crew	Daily Output	Labor-Hours	Unit	Material	Labor	Equipment	Total	Total Incl O&P
0010	**GAS CIRCUIT BREAKER**									
2000	Power circuit breakers									
2150	Gas circuit breakers, 13 to 26 kV	R-11	.56	100	Ea.	222,000	5,200	1,525	228,725	253,500
2160	161 kV		.08	700		291,500	36,300	10,800	338,600	387,000
2170	500 kV	↓	.04	1400	↓	1,025,000	72,500	21,500	1,119,000	1,260,500

33 75 23 – Vacuum High-Voltage Circuit Breaker

33 75 23.13 Vacuum Circuit Breaker

		Crew	Daily Output	Labor-Hours	Unit	Material	Labor	Equipment	Total	Total Incl O&P
0010	**VACUUM CIRCUIT BREAKER**									
2000	Power circuit breakers									
2200	Vacuum circuit breakers, 13 to 26 kV	R-11	.56	100	Ea.	50,500	5,200	1,525	57,225	65,000

33 75 36 – High-Voltage Utility Fuses

33 75 36.13 Fuses

		Crew	Daily Output	Labor-Hours	Unit	Material	Labor	Equipment	Total	Total Incl O&P
0010	**FUSES**									
8000	Protective equipment									
8250	Fuses, 13 to 26 kV	R-11	18.67	2.999	Ea.	2,075	156	46	2,277	2,550
8260	46 kV		11.20	5		2,375	260	77	2,712	3,075
8270	69 kV		8	7		2,525	365	108	2,998	3,450
8280	161 kV	↓	4.67	11.991	↓	3,100	620	184	3,904	4,575

33 75 39 – High-Voltage Surge Arresters

33 75 39.13 Surge Arresters

		Crew	Daily Output	Labor-Hours	Unit	Material	Labor	Equipment	Total	Total Incl O&P
0010	**SURGE ARRESTERS**									
8000	Protective equipment									
8050	Lightning arresters, 13 to 26 kV	R-11	18.67	2.999	Ea.	1,550	156	46	1,752	1,975
8060	46 kV		14	4		4,150	208	61.50	4,419.50	4,925
8070	69 kV		11.20	5		5,325	260	77	5,662	6,325
8080	161 kV		5.60	10		7,400	520	154	8,074	9,100
8090	500 kV	↓	1.40	40	↓	25,200	2,075	615	27,890	31,500

33 75 43 – Shunt Reactors

33 75 43.13 Reactors

		Crew	Daily Output	Labor-Hours	Unit	Material	Labor	Equipment	Total	Total Incl O&P
0010	**REACTORS**									
8000	Protective equipment									
8150	Reactors and resistors, 13 to 26 kV	R-11	28	2	Ea.	3,700	104	30.50	3,834.50	4,275
8160	46 kV		4.31	12.993		11,100	675	200	11,975	13,400
8170	69 kV		2.80	20		17,800	1,050	305	19,155	21,400
8180	161 kV	↓	2.24	25	↓	20,000	1,300	385	21,685	24,400

33 75 High-Voltage Switchgear and Protection Devices

33 75 43 – Shunt Reactors

33 75 43.13 Reactors		Crew	Daily Output	Labor-Hours	Unit	Material	Labor	Equipment	Total	Total Incl O&P
						2015 Bare Costs				
8190	500 kV	R-11	.08	700	Ea.	78,500	36,300	10,800	125,600	153,000

33 75 53 – High-Voltage Switches

33 75 53.13 Switches		Crew	Daily Output	Labor-Hours	Unit	Material	Labor	Equipment	Total	Total Incl O&P
0010	**SWITCHES**									
3000	Disconnecting switches									
3050	Gang operated switches									
3060	Manual operation, 13 to 26 kV	R-11	1.65	33.939	Ea.	14,100	1,750	520	16,370	18,700
3070	46 kV		1.12	50		23,700	2,600	770	27,070	30,700
3080	69 kV		.80	70		25,900	3,625	1,075	30,600	35,200
3090	161 kV		.56	100		31,100	5,200	1,525	37,825	43,700
3100	500 kV		.14	400		83,500	20,800	6,150	110,450	130,000
3110	Motor operation, 161 kV		.51	109		46,800	5,700	1,675	54,175	62,000
3120	500 kV		.28	200		127,500	10,400	3,075	140,975	159,000
3250	Circuit switches, 161 kV		.41	136		92,000	7,100	2,100	101,200	114,000
3300	Single pole switches									
3350	Disconnecting switches, 13 to 26 kV	R-11	28	2	Ea.	12,600	104	30.50	12,734.50	14,000
3360	46 kV		8	7		22,200	365	108	22,673	25,100
3370	69 kV		5.60	10		24,000	520	154	24,674	27,300
3380	161 kV		2.80	20		89,500	1,050	305	90,855	100,500
3390	500 kV		.22	254		254,500	13,200	3,900	271,600	304,000
3450	Grounding switches, 46 kV		5.60	10		33,300	520	154	33,974	37,500
3460	69 kV		3.73	15.013		34,000	780	231	35,011	38,800
3470	161 kV		2.24	25		36,300	1,300	385	37,985	42,300
3480	500 kV		.62	90.323		45,400	4,700	1,375	51,475	58,500

33 79 Site Grounding

33 79 83 – Site Grounding Conductors

33 79 83.13 Grounding Wire, Bar, and Rod		Crew	Daily Output	Labor-Hours	Unit	Material	Labor	Equipment	Total	Total Incl O&P
0010	**GROUNDING WIRE, BAR, AND ROD**									
7500	Grounding systems	R-11	280	.200	Lb.	10.80	10.40	3.07	24.27	31

33 81 Communications Structures

33 81 13 – Communications Transmission Towers

33 81 13.10 Radio Towers		Crew	Daily Output	Labor-Hours	Unit	Material	Labor	Equipment	Total	Total Incl O&P
0010	**RADIO TOWERS**									
0020	Guyed, 50′ H, 40 lb. sect., 70 MPH basic wind spd.	2 Sswk	1	16	Ea.	2,750	840		3,590	4,500
0100	Wind load 90 MPH basic wind speed	"	1	16		3,700	840		4,540	5,550
0300	190′ high, 40 lb. section, wind load 70 MPH basic wind speed	K-2	.33	72.727		9,550	3,550	920	14,020	17,500
0400	200′ high, 70 lb. section, wind load 90 MPH basic wind speed		.33	72.727		15,900	3,550	920	20,370	24,500
0600	300′ high, 70 lb. section, wind load 70 MPH basic wind speed		.20	120		25,500	5,850	1,525	32,875	39,500
0700	270′ high, 90 lb. section, wind load 90 MPH basic wind speed		.20	120		27,600	5,850	1,525	34,975	41,800
0800	400′ high, 100 lb. section, wind load 70 MPH basic wind speed		.14	171		38,100	8,375	2,175	48,650	58,500
0900	Self-supporting, 60′ high, wind load 70 MPH basic wind speed		.80	30		4,500	1,475	380	6,355	7,825
0910	60′ high, wind load 90 MPH basic wind speed		.45	53.333		5,275	2,600	675	8,550	10,900
1000	120′ high, wind load 70 MPH basic wind speed		.40	60		10,000	2,925	760	13,685	16,700
1200	190′ high, wind load 90 MPH basic wind speed		.20	120		28,300	5,850	1,525	35,675	42,600
2000	For states west of Rocky Mountains, add for shipping					10%				

Division Notes

I		CREW	DAILY OUTPUT	LABOR-HOURS	UNIT	BARE COSTS MAT.	LABOR	EQUIP.	TOTAL	TOTAL INCL O&P

Estimating Tips

34 11 00 Rail Tracks

This subdivision includes items that may involve either repair of existing, or construction of new, railroad tracks. Additional preparation work, such as the roadbed earthwork, would be found in Division 31. Additional new construction siding and turnouts are found in Subdivision 34 72. Maintenance of railroads is found under 34 01 23 Operation and Maintenance of Railways.

34 40 00 Traffic Signals

This subdivision includes traffic signal systems. Other traffic control devices such as traffic signs are found in Subdivision 10 14 53 Traffic Signage.

34 70 00 Vehicle Barriers

This subdivision includes security vehicle barriers, guide and guard rails, crash barriers, and delineators. The actual maintenance and construction of concrete and asphalt pavement is found in Division 32.

Reference Numbers

Reference numbers are shown in shaded boxes at the beginning of some major classifications. These numbers refer to related items in the Reference Section. The reference information may be an estimating procedure, an alternate pricing method, or technical information.

Note: Not all subdivisions listed here necessarily appear in this publication. ■

34 43 Airfield Signaling and Control Equipment

34 43 13 – Airfield Signals and Lighting

34 43 13.16 Airport Lighting		Crew	Daily Output	Labor-Hours	Unit	Material	2015 Bare Costs Labor	Equipment	Total	Total Incl O&P
0010	**AIRPORT LIGHTING**									
0100	Runway centerline, bidir., semi-flush, 200 W, w/shallow insert base	R-22	12.40	3.006	Ea.	1,725	140		1,865	2,100
0110	for mounting in base housing		18.60	2.004		865	93		958	1,100
0120	Flush, 200 W, w/shallow insert base		12.40	3.006		1,700	140		1,840	2,075
0130	for mounting in base housing		18.64	2		885	93		978	1,125
0140	45 W, for mounting in base housing		18.64	2		885	93		978	1,125
0150	Touchdown zone light, unidirectional, 200 W, w/shallow insert base		12.40	3.006		1,575	140		1,715	1,925
0160	115 W		12.40	3.006		1,525	140		1,665	1,875
0170	Bi-directional, 62 W		12.40	3.006		1,600	140		1,740	1,950
0180	Unidirectional, 200 W, for mounting in base housing		18.60	2.004		700	93		793	915
0190	115 W		18.60	2.004		755	93		848	970
0200	Bi-directional, 62 W		18.60	2.004		840	93		933	1,075
0210	Runway edge & threshold light, bidir., 200 W, for base housing		9.36	3.983		940	185		1,125	1,300
0220	300 W		9.36	3.983		1,325	185		1,510	1,725
0230	499 W		7.44	5.011		2,200	233		2,433	2,775
0240	Threshold & approach light, unidir., 200 W, for base housing		9.36	3.983		545	185		730	880
0250	499 W		7.44	5.011		665	233		898	1,075
0260	Runway edge, bi-directional, 2-115 W, for base housing		12.40	3.006		1,450	140		1,590	1,775
0270	2-185 W		12.40	3.006		1,600	140		1,740	1,950
0280	Runway threshold & end, bidir., 2-115 W, for base housing		12.40	3.006		1,600	140		1,740	1,950
0370	45 W, flush, for mounting in base housing		18.64	2		780	93		873	995
0380	115 W		18.64	2		795	93		888	1,025
0400	Transformer, 45 W		37.20	1.002		171	46.50		217.50	259
0410	65 W		37.20	1.002		175	46.50		221.50	264
0420	100 W		37.20	1.002		217	46.50		263.50	310
0430	200 W, 6.6/6.6 A		37.20	1.002		213	46.50		259.50	305
0440	20/6.6 A		24.80	1.503		228	70		298	355
0450	500 W		24.80	1.503		298	70		368	435
0460	Base housing, 24" deep, 12" diameter		18.64	2		480	93		573	670
0470	15" diameter		18.64	2		1,050	93		1,143	1,300
0480	Connection kit for 1 conductor cable		24.80	1.503		34.50	70		104.50	144
0500	Elevated runway marker, high intensity quartz, edge, 115 W		14.88	2.505		385	117		502	600
0510	175 W		14.88	2.505		395	117		512	610
0520	Threshold, 115 W		14.88	2.505		430	117		547	650
0530	Edge, 200 W, w/base and transformer		11.85	3.146		1,075	146		1,221	1,400
0540	Elevated runway light, quartz, runway edge, 45 W		14.88	2.505		226	117		343	425
0550	Threshold, 115 W		14.88	2.505		375	117		492	585
0560	Taxiway, 30 W		14.88	2.505		207	117		324	405
0570	Fixture, 200 W, w/base and transformer		11.85	3.146		1,075	146		1,221	1,400
0580	Elevated threshold marker, quartz, 115 W		14.88	2.505		475	117		592	695
0590	175 W		14.88	2.505		475	117		592	695

34 43 23 – Weather Observation Equipment

34 43 23.16 Airfield Wind Cones		Crew	Daily Output	Labor-Hours	Unit	Material	Labor	Equipment	Total	Total Incl O&P
0010	**AIRFIELD WIND CONES**									
1200	Wind cone, 12' lighted assembly, rigid, w/obstruction light	R-21	1.36	24.118	Ea.	9,200	1,325	44	10,569	12,100
1210	Without obstruction light		1.52	21.579		7,725	1,175	39.50	8,939.50	10,300
1220	Unlighted assembly, w/obstruction light		1.68	19.524		8,425	1,075	35.50	9,535.50	10,900
1230	Without obstruction light		1.84	17.826		7,725	975	32.50	8,732.50	9,975
1240	Wind cone slip fitter, 2-1/2" pipe		21.84	1.502		102	82	2.75	186.75	238
1250	Wind cone sock, 12' x 3', cotton		6.56	5		560	274	9.15	843.15	1,025
1260	Nylon		6.56	5		565	274	9.15	848.15	1,050

Estimating Tips

- When estimating costs for the installation of electrical power generation equipment, factors to review include access to the job site, access and setting at the installation site, required connections, uncrating pads, anchors, leveling, final assembly of the components, and temporary protection from physical damage, including from exposure to the environment.
- Be aware of the cost of equipment supports, concrete pads, and vibration isolators, and cross-reference to other trades' specifications. Also, review site and structural drawings for items that must be included in the estimates.
- It is important to include items that are not documented in the plans and specifications but must be priced. These items include, but are not limited to, testing, dust protection, roof penetration, core drilling concrete floors and walls, patching, cleanup, and final adjustments. Add a contingency or allowance for utility company fees for power hookups, if needed.
- The project size and scope of electrical power generation equipment will have a significant impact on cost. The intent of RSMeans cost data is to provide a benchmark cost so that owners, engineers, and electrical contractors will have a comfortable number with which to start a project. Additionally, there are many websites available to use for research and to obtain a vendor's quote to finalize costs.

Reference Numbers

Reference numbers are shown in shaded boxes at the beginning of some major classifications. These numbers refer to related items in the Reference Section. The reference information may be an estimating procedure, an alternate pricing method, or technical information.

Note: Not all subdivisions listed here necessarily appear in this publication. ■

48 15 Wind Energy Electrical Power Generation Equipment

48 15 13 – Wind Turbines

48 15 13.50 Wind Turbines and Components		Crew	Daily Output	Labor-Hours	Unit	2015 Bare Costs Material	Labor	Equipment	Total	Total Incl O&P
0010	**WIND TURBINES & COMPONENTS**									
0500	Complete system, grid connected									
1000	20 kW, 31′ dia, incl. labor & material G				System				49,900	49,900
1010	Enhanced G								92,000	92,000
1500	10 kW, 23′ dia, incl. labor & material G								74,000	74,000
2000	2.4 kW, 12′ dia, incl. labor & material G								18,000	18,000
2900	Component system									
3000	Turbine, 400 watt, 3′ dia G	1 Elec	3.41	2.346	Ea.	650	128		778	905
3100	600 watt, 3′ dia G		2.56	3.125		880	171		1,051	1,225
3200	1000 watt, 9′ dia G		2.05	3.902		3,650	213		3,863	4,350
3400	Mounting hardware									
3500	30′ guyed tower kit G	2 Clab	5.12	3.125	Ea.	415	118		533	640
3505	3′ galvanized helical earth screw G	1 Clab	8	1		49.50	37.50		87	113
3510	Attic mount kit G	1 Rofc	2.56	3.125		154	125		279	385
3520	Roof mount kit G	1 Clab	3.41	2.346		136	88		224	286
8900	Equipment									
9100	DC to AC inverter for, 48 V, 4,000 watt G	1 Elec	2	4	Ea.	2,325	219		2,544	2,900

48 18 Fuel Cell Electrical Power Generation Equipment

48 18 13 – Electrical Power Generation Fuel Cells

48 18 13.10 Fuel Cells		Crew	Daily Output	Labor-Hours	Unit	Material	Labor	Equipment	Total	Total Incl O&P
0010	**FUEL CELLS**									
2001	Complete system, natural gas, installed, 200 kW G				System				1,150,000	1,150,000
2002	400 kW G								2,168,000	2,168,000
2003	600 kW G								3,060,000	3,060,000
2004	800 kW G								4,000,000	4,000,000
2005	1000 kW G								5,005,000	5,005,000
2010	Comm type, for battery charging, uses hydrogen, 100 watt, 12 V G				Ea.	2,100			2,100	2,325
2020	500 watt, 12 V G					4,300			4,300	4,750
2030	1000 watt, 48 V G					7,050			7,050	7,775
2040	Small, for demonstration or battery charging units, 12 V G					190			190	209
2050	Spare anode set for G					43			43	47
2060	3 watts, uses hydrogen gas G					440			440	480
2070	6 watts, uses hydrogen gas G					790			790	870
2080	10 watts, uses hydrogen gas G					1,150			1,150	1,250

Assemblies Section

Table of Contents

How RSMeans Assemblies Data Works

Assemblies estimating provides a fast and reasonably accurate way to develop construction costs. An assembly is the grouping of individual work items, with appropriate quantities, to provide a cost for a major construction component in a convenient unit of measure.

An assemblies estimate is often used during early stages of design development to compare the cost impact of various design alternatives on total building cost.

Assemblies estimates are also used as an efficient tool to verify construction estimates.

Assemblies estimates do not require a completed design or detailed drawings. Instead, they are based on the general size of the structure and other known parameters of the project. The degree of accuracy of an assemblies estimate is generally within +/- 15%.

RSMeans assemblies are identified by a **unique 12-character identifier**. The assemblies are numbered using UNIFORMAT II, ASTM Standard E1557. The first 6 characters represent this system to Level 4. The last 6 characters represent further breakdown by RSMeans in order to arrange items in understandable groups of similar tasks. Line numbers are consistent across all RSMeans publications, so a line number in any RSMeans assemblies data set will always refer to the same work.

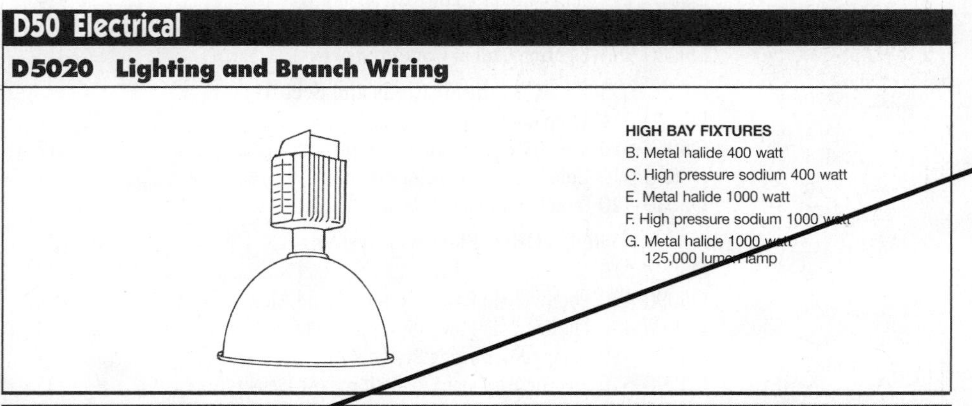

System Components	QUANTITY	UNIT	COST PER S.F. MAT.	INST.	TOTAL
SYSTEM D5020 222 0240					
HIGH INTENSITY DISCHARGE FIXTURE, 8'-10' ABOVE WORK PLANE					
1 WATT/S.F., TYPE B, 29 FC, 2 FIXTURES/1000 S.F.					
Steel intermediate conduit, (IMC) 1/2" diam	.100	L.F.	.20	.66	.86
Wire, 600V, type THWN-THHN, copper, solid, #10	.002	C.L.F.	.04	.13	.17
Steel outlet box 4" concrete	.002	Ea.	.03	.07	.10
Steel outlet box plate with stud, 4" concrete	.002	Ea.	.02	.02	.04
Metal halide, hi bay, aluminum reflector, 400 W lamp	.002	Ea.	.90	.57	1.47
TOTAL			1.19	1.45	2.64

D5020 222	H.I.D. Fixture, High Bay, 8'-10' (by Wattage)	COST PER S.F. MAT.	INST.	TOTAL
0190	High intensity discharge fixture, 8'-10' above work plane			
0240	1 watt/S.F., type B, 29 FC, 2 fixtures/1000 S.F.	1.19	1.45	2.64
0280	Type C, 54 FC, 2 fixtures/1000 S.F.	1.01	1.10	2.11
0400	2 watt/S.F., type B, 59 FC, 4 fixtures/1000 S.F.	2.34	2.78	5.12
0440	Type C, 108 FC, 4 fixtures/1000 S.F.	2	2.12	4.12
0560	3 watt/S.F., type B, 103 FC, 7 fixtures/1000 S.F.	3.95	4.45	8.40
0600	Type C, 189 FC, 6 fixtures/1000 S.F.	3.20	2.75	5.95
0720	4 watt/S.F., type B, 133 FC, 9 fixtures/1000 S.F.	5.10	5.85	10.95
0760	Type C, 243 FC, 9 fixtures/1000 S.F.	4.48	4.73	9.21
0880	5 watt/S.F., type B, 162 FC, 11 fixtures/1000 S.F.	6.30	7.25	13.55
0920	Type C, 297 FC, 11 fixtures/1000 S.F.	5.50	5.85	11.35

Information is available in the Reference Section to assist the estimator with estimating procedures, alternate pricing methods, and additional technical information.

The **Reference Box** indicates the exact location of this information in the Reference Section at the back of the book. The "R" stands for "reference," and the remaining characters are the RSMeans line numbers.

RSMeans assemblies descriptions appear in two formats: narrative and table. **Narrative descriptions** are shown in a hierarchical structure to make them readable. In order to read a complete description, read up through the indents to the top of the section. Include everything that is above and to the left that is not contradicted by information below.

For supplemental customizable square foot estimating forms, visit: **http://www.reedconstructiondata.com/rsmeans/assemblies**

Most assemblies consist of three major elements: a graphic, the system components, and the cost data itself. The **Graphic** is a visual representation showing the typical appearance of the assembly in question, frequently accompanied by additional explanatory technical information describing the class of items. The **System Components** is a listing of the individual tasks that make up the assembly, including the quantity and unit of measure for each item, along with the cost of material and installation. The **Assemblies Data** below lists prices for other similar systems with dimensional and/or size variations.

All RSMeans assemblies costs represent the cost for the installing contractor. An allowance for profit has been added to all material, labor, and equipment rental costs.
A markup for labor burdens, including workers' compensation, fixed overhead, and business overhead, is included with installation costs.

The data published in RSMeans print books represents a "national average" cost. This data should be modified to the project location using the **City Cost Indexes** or **Location Factors** tables found in the Reference Section in the back of the book.

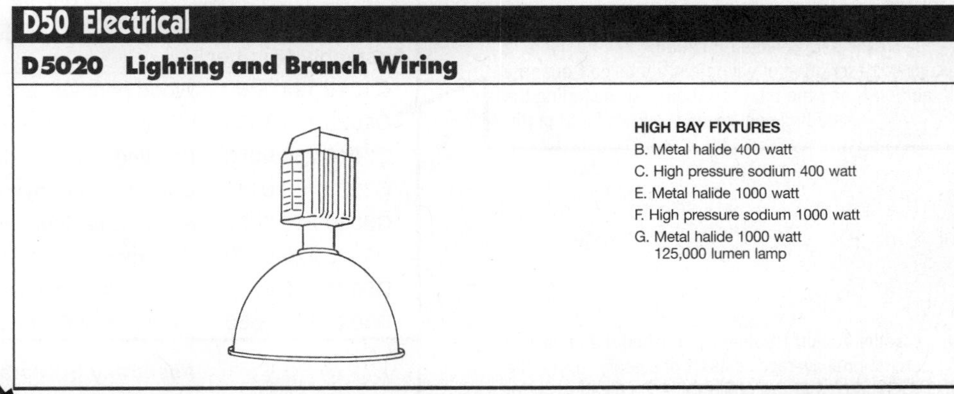

System components are listed separately to detail what is included in the development of the total system price.

System Components	QUANTITY	UNIT	COST PER S.F. MAT.	COST PER S.F. INST.	COST PER S.F. TOTAL
SYSTEM D5020 222 0240					
HIGH INTENSITY DISCHARGE FIXTURE, 8'-10' ABOVE WORK PLANE					
1 WATT/S.F., TYPE B, 29 FC, 2 FIXTURES/1000 S.F.					
Steel intermediate conduit, (IMC) 1/2" diam	.100	L.F.	.20	.66	.86
Wire, 600V, type THWN-THHN, copper, solid, #10	.002	C.L.F.	.04	.13	.17
Steel outlet box 4" concrete	.002	Ea.	.03	.07	.10
Steel outlet box plate with stud, 4" concrete	.002	Ea.	.02	.02	.04
Metal halide, hi bay, aluminum reflector, 400 W lamp	.002	Ea.	.90	.57	1.47
TOTAL			1.19	1.45	2.64

All RSMeans assemblies data includes a typical **Unit of Measure** used for estimating that item. For instance, while the unit of measure for the graphic shown here is Each, some A/C systems are estimated by the square foot (S.F.). The estimator needs to take special care that the unit in the data matches the unit in the takeoff. Abbreviations can be found in the Reference Section.

D5020 222	H.I.D. Fixture, High Bay, 8'-10' (by Wattage)	COST PER S.F. MAT.	COST PER S.F. INST.	COST PER S.F. TOTAL
0190	High intensity discharge fixture, 8'-10' above work plane			
0240	1 watt/S.F., type B, 29 FC, 2 fixtures/1000 S.F.	1.19	1.45	2.64
0280	Type C, 54 FC, 2 fixtures/1000 S.F.	1.01	1.10	2.11
0400	2 watt/S.F., type B, 59 FC, 4 fixtures/1000 S.F.	2.34	2.78	5.12
0440	Type C, 108 FC, 4 fixtures/1000 S.F.	2	2.12	4.12
0560	3 watt/S.F., type B, 103 FC, 7 fixtures/1000 S.F.	3.95	4.45	8.40
0600	Type C, 189 FC, 6 fixtures/1000 S.F.	3.20	2.75	5.95
0720	4 watt/S.F., type B, 133 FC, 9 fixtures/1000 S.F.	5.10	5.85	10.95
0760	Type C, 243 FC, 9 fixtures/1000 S.F.	4.48	4.73	9.21
0880	5 watt/S.F., type B, 162 FC, 11 fixtures/1000 S.F.	6.30	7.25	13.55
0920	Type C, 297 FC, 11 fixtures/1000 S.F.	5.50	5.85	11.35

How RSMeans Assemblies Data Works (Continued)

Sample Estimate

This sample demonstrates the elements of an estimate, including a tally of the RSMeans data lines. Published assemblies costs include all markups for labor burden and profit for the installing contractor. This estimate adds a summary of the markups applied by a general contractor on the installing contractors' work. These figures represent the total cost to the owner. The RSMeans location factor is added at the bottom of the estimate to adjust the cost of the work to a specific location

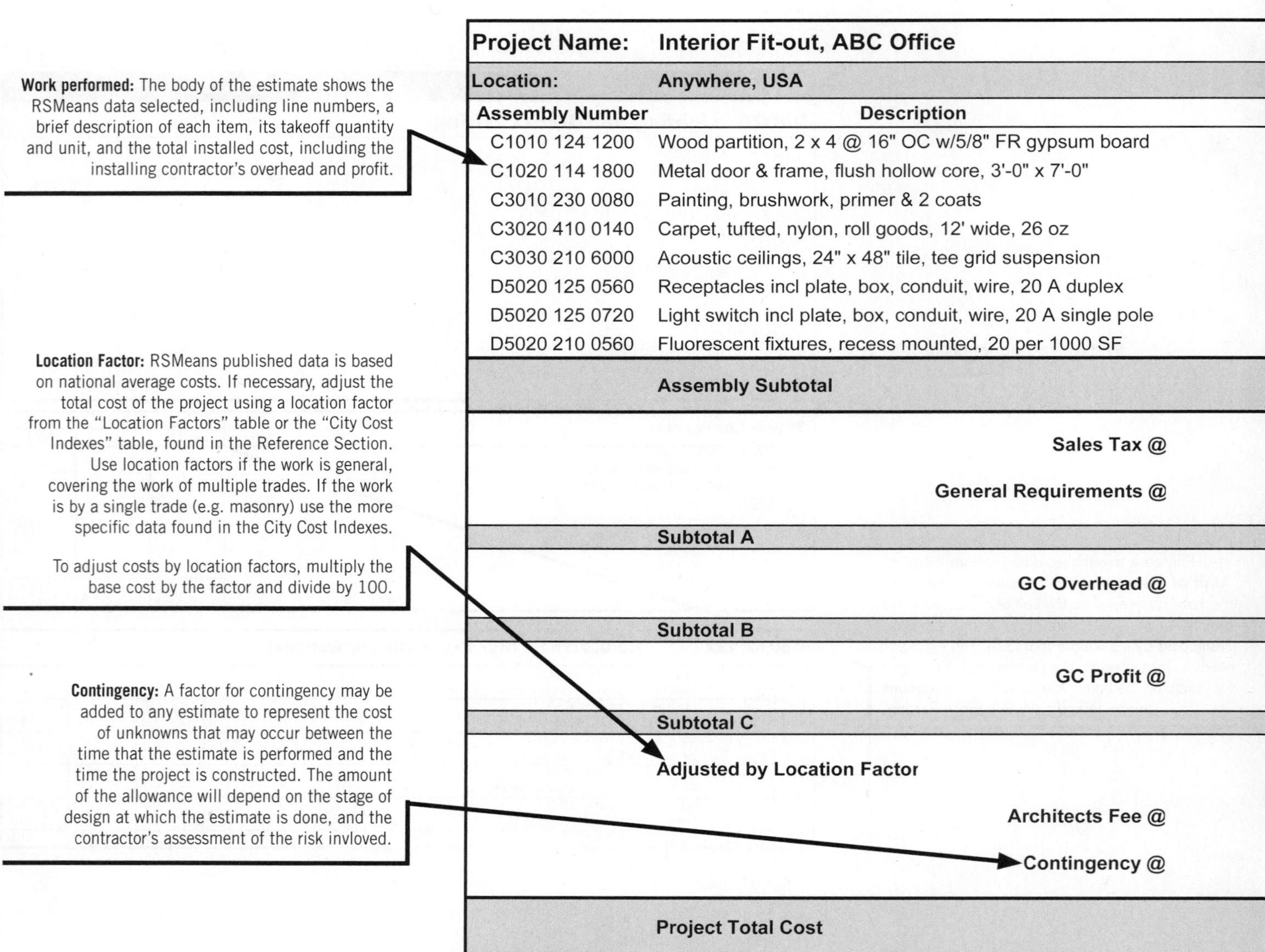

Work performed: The body of the estimate shows the RSMeans data selected, including line numbers, a brief description of each item, its takeoff quantity and unit, and the total installed cost, including the installing contractor's overhead and profit.

Location Factor: RSMeans published data is based on national average costs. If necessary, adjust the total cost of the project using a location factor from the "Location Factors" table or the "City Cost Indexes" table, found in the Reference Section. Use location factors if the work is general, covering the work of multiple trades. If the work is by a single trade (e.g. masonry) use the more specific data found in the City Cost Indexes.

To adjust costs by location factors, multiply the base cost by the factor and divide by 100.

Contingency: A factor for contingency may be added to any estimate to represent the cost of unknowns that may occur between the time that the estimate is performed and the time the project is constructed. The amount of the allowance will depend on the stage of design at which the estimate is done, and the contractor's assessment of the risk invloved.

Project Name:	**Interior Fit-out, ABC Office**
Location:	**Anywhere, USA**
Assembly Number	**Description**
C1010 124 1200	Wood partition, 2 x 4 @ 16" OC w/5/8" FR gypsum board
C1020 114 1800	Metal door & frame, flush hollow core, 3'-0" x 7'-0"
C3010 230 0080	Painting, brushwork, primer & 2 coats
C3020 410 0140	Carpet, tufted, nylon, roll goods, 12' wide, 26 oz
C3030 210 6000	Acoustic ceilings, 24" x 48" tile, tee grid suspension
D5020 125 0560	Receptacles incl plate, box, conduit, wire, 20 A duplex
D5020 125 0720	Light switch incl plate, box, conduit, wire, 20 A single pole
D5020 210 0560	Fluorescent fixtures, recess mounted, 20 per 1000 SF
	Assembly Subtotal
	Sales Tax @
	General Requirements @
	Subtotal A
	GC Overhead @
	Subtotal B
	GC Profit @
	Subtotal C
	Adjusted by Location Factor
	Architects Fee @
	Contingency @
	Project Total Cost

This estimate is based on an interactive spreadsheet. A copy of this spreadsheet is located on the RSMeans website at **http:/www.reedconstructiondata.com/rsmeans/extras/546011**. You are free to download it and adjust it to your methodology.

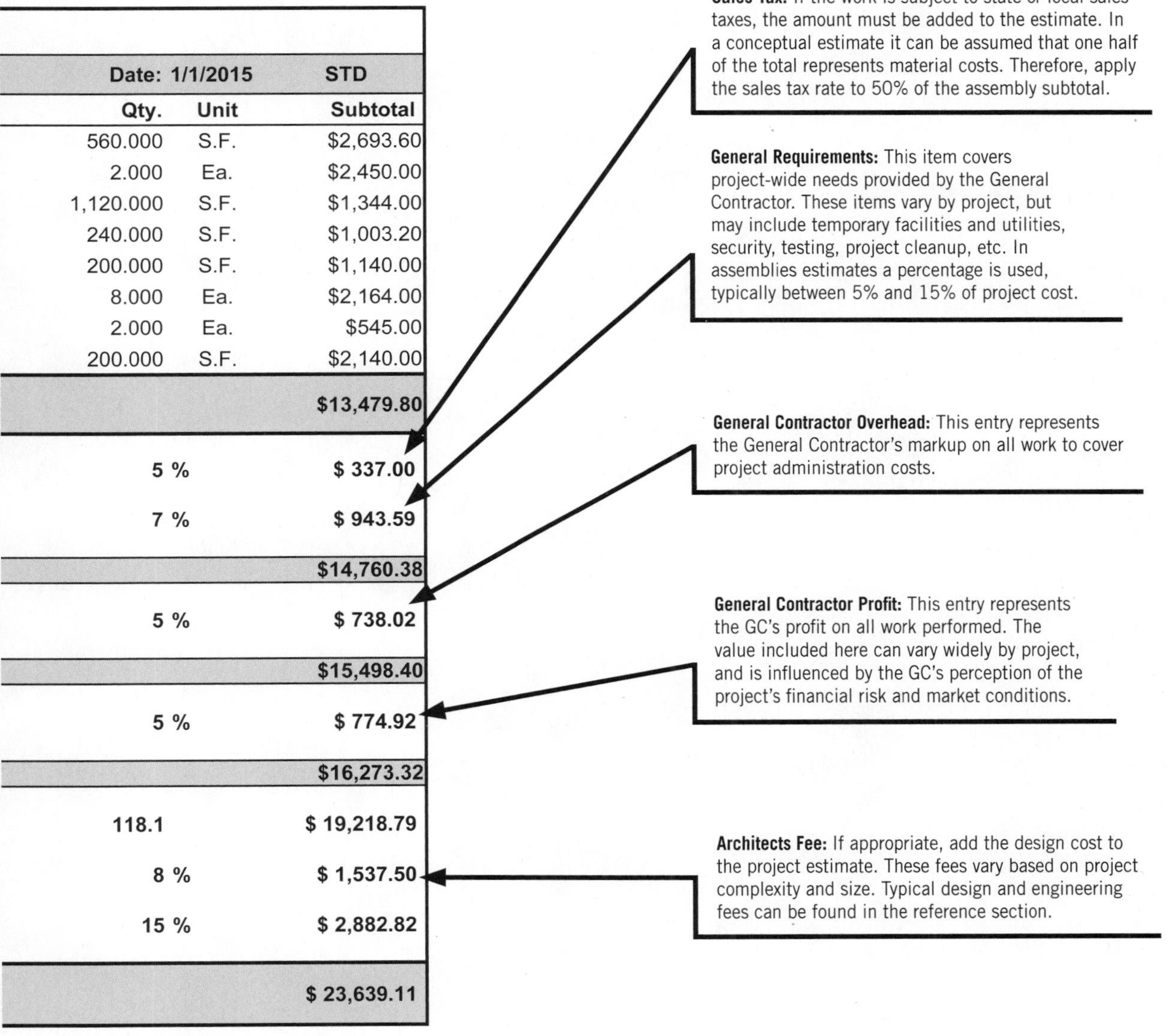

Date: 1/1/2015		STD
Qty.	Unit	Subtotal
560.000	S.F.	$2,693.60
2.000	Ea.	$2,450.00
1,120.000	S.F.	$1,344.00
240.000	S.F.	$1,003.20
200.000	S.F.	$1,140.00
8.000	Ea.	$2,164.00
2.000	Ea.	$545.00
200.000	S.F.	$2,140.00
		$13,479.80
5 %		$ 337.00
7 %		$ 943.59
		$14,760.38
5 %		$ 738.02
		$15,498.40
5 %		$ 774.92
		$16,273.32
118.1		$ 19,218.79
8 %		$ 1,537.50
15 %		$ 2,882.82
		$ 23,639.11

Sales Tax: If the work is subject to state or local sales taxes, the amount must be added to the estimate. In a conceptual estimate it can be assumed that one half of the total represents material costs. Therefore, apply the sales tax rate to 50% of the assembly subtotal.

General Requirements: This item covers project-wide needs provided by the General Contractor. These items vary by project, but may include temporary facilities and utilities, security, testing, project cleanup, etc. In assemblies estimates a percentage is used, typically between 5% and 15% of project cost.

General Contractor Overhead: This entry represents the General Contractor's markup on all work to cover project administration costs.

General Contractor Profit: This entry represents the GC's profit on all work performed. The value included here can vary widely by project, and is influenced by the GC's perception of the project's financial risk and market conditions.

Architects Fee: If appropriate, add the design cost to the project estimate. These fees vary based on project complexity and size. Typical design and engineering fees can be found in the reference section.

D2020 Domestic Water Distribution

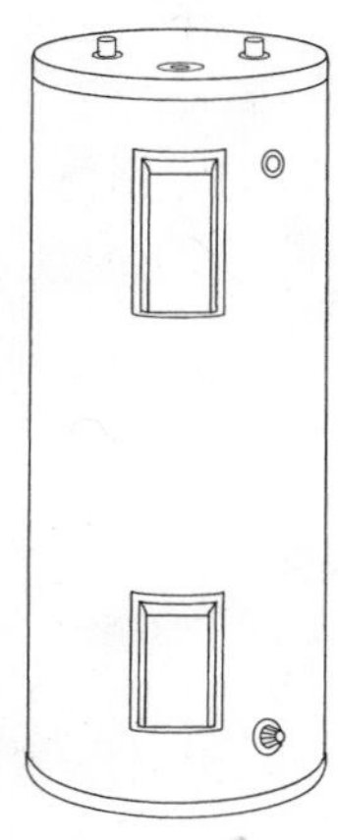

Installation includes piping and fittings within 10′ of heater. Electric water heaters do not require venting.

1 Kilowatt hour will raise:			
Gallons of Water	Degrees F	Gallons of Water	Degrees F
4.1	100°	6.8	60°
4.5	90°	8.2	50°
5.1	80°	10.0	40°
5.9	70°		

System Components	QUANTITY	UNIT	COST EACH MAT.	COST EACH INST.	COST EACH TOTAL
SYSTEM D2020 210 1780					
ELECTRIC WATER HEATER, RESIDENTIAL, 100°F RISE					
10 GALLON TANK, 7 GPH					
Water heater, residential electric, glass lined tank, 10 gal.	1.000	Ea.	360	310	670
Copper tubing, type L, solder joint, hanger 10′ OC 1/2″ diam.	30.000	L.F.	122.10	262.50	384.60
Wrought copper 90° elbow for solder joints 1/2″ diam.	4.000	Ea.	4.84	142	146.84
Wrought copper Tee for solder joints, 1/2″ diam.	2.000	Ea.	4.12	109	113.12
Union, wrought copper, 1/2″ diam.	2.000	Ea.	30.90	75	105.90
Valve, gate, bronze, 125 lb, NRS, soldered 1/2″ diam.	2.000	Ea.	119	59	178
Relief valve, bronze, press & temp, self-close, 3/4″ IPS	1.000	Ea.	278	25.50	303.50
Wrought copper adapter, CTS to MPT 3/4″ IPS	1.000	Ea.	4.22	41.50	45.72
Copper tubing, type L, solder joints, 3/4″ diam.	1.000	L.F.	5.70	9.35	15.05
Wrought copper 90° elbow for solder joints 3/4″ diam.	1.000	Ea.	2.71	37.50	40.21
TOTAL			931.59	1,071.35	2,002.94

D2020 210	Electric Water Heaters - Residential Systems	COST EACH MAT.	COST EACH INST.	COST EACH TOTAL
1760	Electric water heater, residential, 100°F rise			
1780	10 gallon tank, 7 GPH	930	1,075	2,005
1820	20 gallon tank, 7 GPH (R224000 -20)	1,125	1,150	2,275
1860	30 gallon tank, 7 GPH	1,400	1,200	2,600
1900	40 gallon tank, 8 GPH	1,925	1,325	3,250
1940	52 gallon tank, 10 GPH	2,050	1,325	3,375
1980	66 gallon tank, 13 GPH	3,200	1,525	4,725
2020	80 gallon tank, 16 GPH	3,375	1,600	4,975
2060	120 gallon tank, 23 GPH	4,675	1,850	6,525

D2020 Domestic Water Distribution

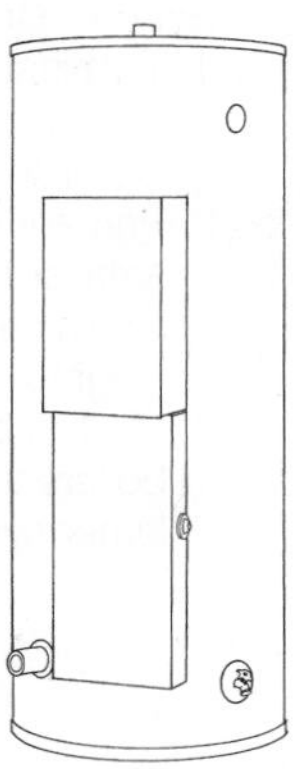

Systems below include piping and fittings within 10′ of heater. Electric water heaters do not require venting.

System Components	QUANTITY	UNIT	COST EACH MAT.	COST EACH INST.	COST EACH TOTAL
SYSTEM D2020 240 1820					
ELECTRIC WATER HEATER, COMMERCIAL, 100°F RISE					
50 GALLON TANK, 9 KW, 37 GPH					
Water heater, commercial, electric, 50 Gal, 9 KW, 37 GPH	1.000	Ea.	4,775	395	5,170
Copper tubing, type L, solder joint, hanger 10′ OC, 3/4″ diam	34.000	L.F.	193.80	317.90	511.70
Wrought copper 90° elbow for solder joints 3/4″ diam	5.000	Ea.	13.55	187.50	201.05
Wrought copper Tee for solder joints, 3/4″ diam	2.000	Ea.	9.92	118	127.92
Wrought copper union for soldered joints, 3/4″ diam.	2.000	Ea.	38.70	79	117.70
Valve, gate, bronze, 125 lb, NRS, soldered 3/4″ diam	2.000	Ea.	140	71	211
Relief valve, bronze, press & temp, self-close, 3/4″ IPS	1.000	Ea.	278	25.50	303.50
Wrought copper adapter, copper tubing to male, 3/4″ IPS	1.000	Ea.	4.22	41.50	45.72
TOTAL			5,453.19	1,235.40	6,688.59

D2020 240	Electric Water Heaters - Commercial Systems		COST EACH MAT.	COST EACH INST.	COST EACH TOTAL
1800	Electric water heater, commercial, 100°F rise				
1820	50 gallon tank, 9 KW 37 GPH		5,450	1,225	6,675
1860	80 gal, 12 KW 49 GPH	R224000 -10	7,025	1,525	8,550
1900	36 KW 147 GPH		9,475	1,650	11,125
1940	120 gal, 36 KW 147 GPH	R224000 -20	10,200	1,775	11,975
1980	150 gal, 120 KW 490 GPH		30,500	1,900	32,400
2020	200 gal, 120 KW 490 GPH		32,100	1,950	34,050
2060	250 gal, 150 KW 615 GPH		35,900	2,275	38,175
2100	300 gal, 180 KW 738 GPH		47,900	2,400	50,300
2140	350 gal, 30 KW 123 GPH		29,000	2,575	31,575
2180	180 KW 738 GPH		40,100	2,575	42,675
2220	500 gal, 30 KW 123 GPH		37,700	3,050	40,750
2260	240 KW 984 GPH		54,500	3,050	57,550
2300	700 gal, 30 KW 123 GPH		31,200	3,475	34,675
2340	300 KW 1230 GPH		45,600	3,475	49,075
2380	1000 gal, 60 KW 245 GPH		38,400	4,825	43,225
2420	480 KW 1970 GPH		61,000	4,850	65,850
2460	1500 gal, 60 KW 245 GPH		80,500	6,000	86,500
2500	480 KW 1970 GPH		110,000	6,000	116,000

D3020 Heat Generating Systems

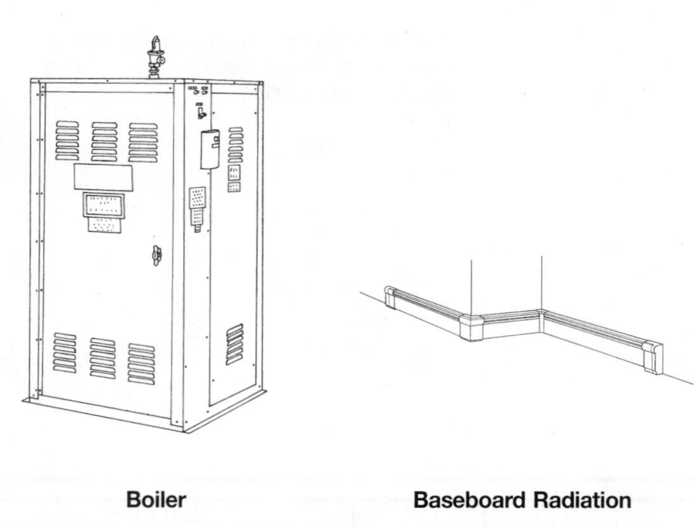

Small Electric Boiler System Considerations:

1. Terminal units are fin tube baseboard radiation rated at 720 BTU/hr with 200° water temperature or 820 BTU/hr steam.
2. Primary use being for residential or smaller supplementary areas, the floor levels are based on 7-1/2′ ceiling heights.
3. All distribution piping is copper for boilers through 205 MBH. All piping for larger systems is steel pipe.

System Components	QUANTITY	UNIT	COST EACH MAT.	COST EACH INST.	COST EACH TOTAL
SYSTEM D3020 102 1120					
SMALL HEATING SYSTEM, HYDRONIC, ELECTRIC BOILER					
1,480 S.F., 61 MBH, STEAM, 1 FLOOR					
Boiler, electric steam, std cntrls, trim, ftngs and valves, 18 KW, 61.4 MBH	1.000	Ea.	5,032.50	1,787.50	6,820
Copper tubing type L, solder joint, hanger 10′OC, 1-1/4″ diam	160.000	L.F.	1,816	1,960	3,776
Radiation, 3/4″ copper tube w/alum fin baseboard pkg 7″ high	60.000	L.F.	486	1,350	1,836
Rough in baseboard panel or fin tube with valves & traps	10.000	Set	2,630	6,800	9,430
Pipe covering, calcium silicate w/cover, 1″ wall 1-1/4″ diam	160.000	L.F.	716.80	1,136	1,852.80
Low water cut-off, quick hookup, in gage glass tappings	1.000	Ea.	281	45	326
TOTAL			10,962.30	13,078.50	24,040.80
COST PER S.F.			7.41	8.84	16.25

D3020 102	Small Heating Systems, Hydronic, Electric Boilers	COST PER S.F. MAT.	COST PER S.F. INST.	COST PER S.F. TOTAL
1100	Small heating systems, hydronic, electric boilers			
1120	Steam, 1 floor, 1480 S.F., 61 M.B.H.	7.43	8.85	16.28
1160	3,000 S.F., 123 M.B.H.	5.65	7.75	13.40
1200	5,000 S.F., 205 M.B.H.	4.93	7.15	12.08
1240	2 floors, 12,400 S.F., 512 M.B.H.	3.92	7.10	11.02
1280	3 floors, 24,800 S.F., 1023 M.B.H.	4.41	7.05	11.46
1320	34,750 S.F., 1,433 M.B.H.	4.04	6.85	10.89
1360	Hot water, 1 floor, 1,000 S.F., 41 M.B.H.	12.55	4.91	17.46
1400	2,500 S.F., 103 M.B.H.	8.80	8.85	17.65
1440	2 floors, 4,850 S.F., 205 M.B.H.	8.50	10.60	19.10
1480	3 floors, 9,700 S.F., 410 M.B.H.	8.80	10.95	19.75

D3020 Heat Generating Systems

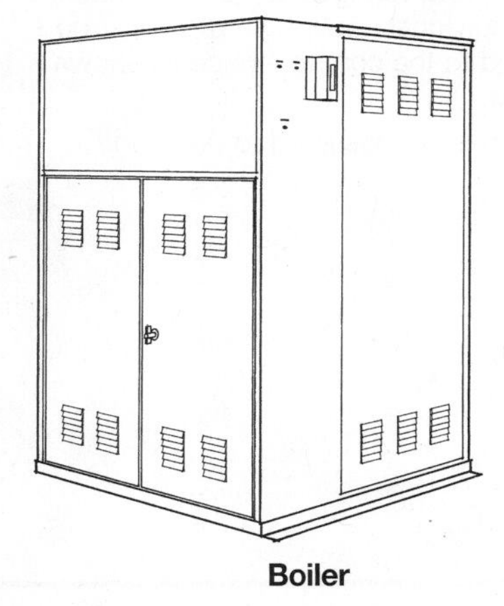
Boiler

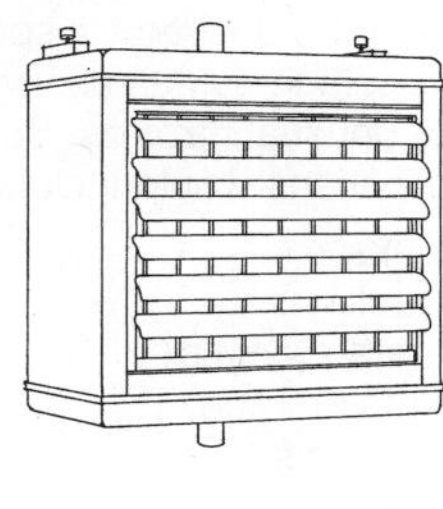
Unit Heater

Large Electric Boiler System Considerations:

1. Terminal units are all unit heaters of the same size. Quantities are varied to accommodate total requirements.
2. All air is circulated through the heaters a minimum of three times per hour.
3. As the capacities are adequate for commercial use, floor levels are based on 10′ ceiling heights.
4. All distribution piping is black steel pipe.

System Components	QUANTITY	UNIT	COST EACH MAT.	COST EACH INST.	COST EACH TOTAL
SYSTEM D3020 104 1240					
LARGE HEATING SYSTEM, HYDRONIC, ELECTRIC BOILER					
9,280 S.F., 135 KW, 461 MBH, 1 FLOOR					
Boiler, electric hot water, std ctrls, trim, ftngs, valves, 135 KW, 461 MBH	1.000	Ea.	12,225	3,900	16,125
Expansion tank, painted steel, 60 Gal capacity ASME	1.000	Ea.	4,175	217	4,392
Circulating pump, CI, close cpld, 50 GPM, 2 HP, 2″ pipe conn	1.000	Ea.	2,925	425	3,350
Unit heater, 1 speed propeller, horizontal, 200° EWT, 72.7 MBH	7.000	Ea.	6,125	1,652	7,777
Unit heater piping hookup with controls	7.000	Set	5,530	10,675	16,205
Pipe, steel, black, schedule 40, welded, 2-1/2″ diam	380.000	L.F.	5,339	10,773	16,112
Pipe covering, calcium silicate w/cover, 1″ wall, 2-1/2″ diam	380.000	L.F.	2,109	2,793	4,902
TOTAL			38,428	30,435	68,863
COST PER S.F.			4.14	3.28	7.42

D3020 104	Large Heating Systems, Hydronic, Electric Boilers	COST PER S.F. MAT.	COST PER S.F. INST.	COST PER S.F. TOTAL
1230	Large heating systems, hydronic, electric boilers			
1240	9,280 S.F., 135 K.W., 461 M.B.H., 1 floor	4.14	3.28	7.42
1280	14,900 S.F., 240 K.W., 820 M.B.H., 2 floors	5.55	5.35	10.90
1320	18,600 S.F., 296 K.W., 1,010 M.B.H., 3 floors	5.45	5.75	11.20
1360	26,100 S.F., 420 K.W., 1,432 M.B.H., 4 floors	5.35	5.65	11
1400	39,100 S.F., 666 K.W., 2,273 M.B.H., 4 floors	4.59	4.73	9.32
1440	57,700 S.F., 900 K.W., 3,071 M.B.H., 5 floors	4.39	4.67	9.06
1480	111,700 S.F., 1,800 K.W., 6,148 M.B.H., 6 floors	3.89	3.99	7.88
1520	149,000 S.F., 2,400 K.W., 8,191 M.B.H., 8 floors	3.84	4	7.84
1560	223,300 S.F., 3,600 K.W., 12,283 M.B.H., 14 floors	4.10	4.53	8.63

D4090 Other Fire Protection Systems

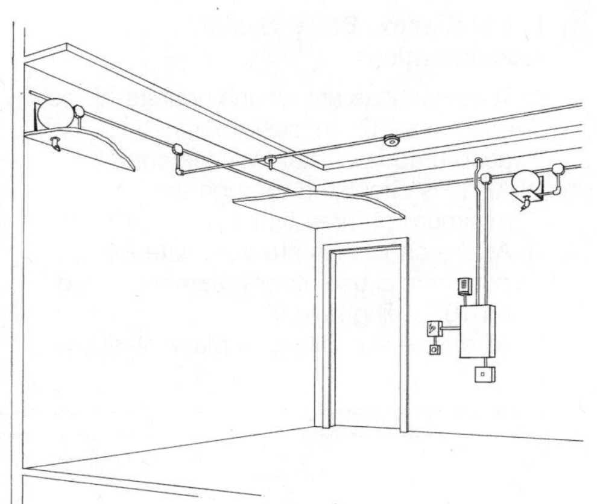

General: Automatic fire protection (suppression) systems other than water sprinklers may be desired for special environments, high risk areas, isolated locations or unusual hazards. Some typical applications would include:

Paint dip tanks
Securities vaults
Electronic data processing
Tape and data storage
Transformer rooms
Spray booths
Petroleum storage
High rack storage

Piping and wiring costs are dependent on the individual application and must be added to the component costs shown below.

All areas are assumed to be open.

D4090 910	Fire Suppression Unit Components	COST EACH		
		MAT.	INST.	TOTAL
0020	Detectors with brackets			
0040	Fixed temperature heat detector	56	82	138
0060	Rate of temperature rise detector	56	82	138
0080	Ion detector (smoke) detector	121	106	227
0200	Extinguisher agent			
0240	200 lb FM200, container	8,550	320	8,870
0280	75 lb carbon dioxide cylinder	1,400	213	1,613
0320	Dispersion nozzle			
0340	FM200 1-1/2" dispersion nozzle	73.50	50.50	124
0380	Carbon dioxide 3" x 5" dispersion nozzle	73.50	39.50	113
0420	Control station			
0440	Single zone control station with batteries	1,900	655	2,555
0470	Multizone (4) control station with batteries	3,600	1,300	4,900
0490				
0500	Electric mechanical release	184	340	524
0520				
0550	Manual pull station	66.50	118	184.50
0570				
0640	Battery standby power 10" x 10" x 17"	455	164	619
0700				
0740	Bell signalling device	86.50	82	168.50

D4090 920	FM200 Systems	COST PER C.F.		
		MAT.	INST.	TOTAL
0820	Average FM200 system, minimum			1.94
0840	Maximum			3.85

D5010 Electrical Service/Distribution

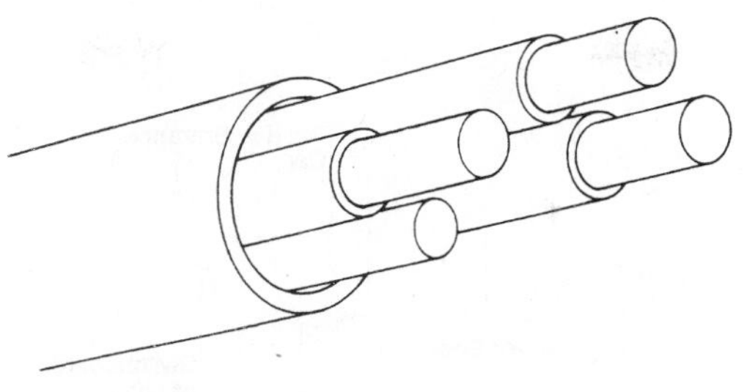

System Components	QUANTITY	UNIT	COST PER L.F. MAT.	COST PER L.F. INST.	COST PER L.F. TOTAL
SYSTEM D5010 110 0200					
HIGH VOLTAGE CABLE, NEUTRAL AND CONDUIT INCLUDED, COPPER #2, 5 kV					
Shielded cable, no splice/termn, copper, XLP shielding, 5 kV, #2	.030	C.L.F.	7.53	9.90	17.43
Wire, 600 volt, type THW, copper, stranded, #8	.010	C.L.F.	.41	.82	1.23
Rigid galv steel conduit to 15′ H, 2″ diam, w/term, ftng & support	1.000	L.F.	8.70	14.55	23.25
TOTAL			16.64	25.27	41.91

D5010 110	High Voltage Shielded Conductors	COST PER L.F. MAT.	COST PER L.F. INST.	COST PER L.F. TOTAL
0200	High voltage cable, neutral & conduit included, copper #2, 5 kV	16.65	25.50	42.15
0240	Copper #1, 5 kV	25	29.50	54.50
0280	15 kV	27	29.50	56.50
0320	Copper 1/0, 5 kV	28.50	37.50	66
0360	15 kV	31.50	37.50	69
0400	25 kV	37	38	75
0440	35 kV	38	38.50	76.50
0480	Copper 2/0, 5 kV	31	38	69
0520	15 kV	34	38	72
0560	25 kV	39	39	78
0600	35 kV	46.50	43.50	90
0640	Copper 4/0, 5 kV	41	44	85
0680	15 kV	44	44	88
0720	25 kV	49.50	44.50	94
0760	35 kV	53	45.50	98.50
0800	Copper 250 kcmil, 5 kV	43.50	44.50	88
0840	15 kV	46.50	44.50	91
0880	25 kV	56	45.50	101.50
0920	35 kV	61.50	49.50	111
0960	Copper 350 kcmil, 5 kV	53	49.50	102.50
1000	15 kV	55.50	49.50	105
1040	25 kV	64.50	51	115.50
1080	35 kV	68.50	52.50	121
1120	Copper 500 kcmil, 5 kV	85	62	147
1160	15 kV	88.50	62	150.50
1200	25 kV	97.50	63.50	161
1240	35 kV	101	65	166

D5010 Electrical Service/Distribution

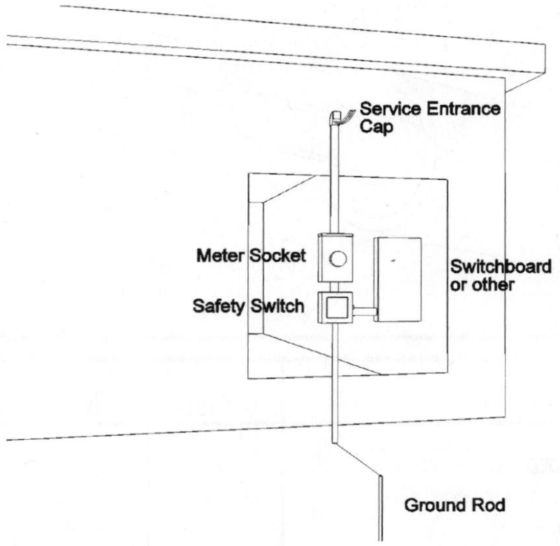

Service Entrance Cap
Meter Socket
Switchboard or other
Safety Switch
Ground Rod

System Components	QUANTITY	UNIT	COST EACH MAT.	COST EACH INST.	COST EACH TOTAL
SYSTEM D5010 120 0220					
SERVICE INSTALLATION, INCLUDES BREAKERS, METERING, 20′ CONDUIT & WIRE					
3 PHASE, 4 WIRE, 60 A					
Wire, 600 volt, copper type XHHW, stranded #6	.600	C.L.F.	41.70	60.60	102.30
Rigid galvanized steel conduit, 3/4″ including fittings	20.000	L.F.	60.40	164	224.40
Service entrance cap, 3/4″ diameter	1.000	Ea.	7.15	50.50	57.65
Conduit LB fitting with cover, 3/4″ diameter	1.000	Ea.	11.75	50.50	62.25
Meter socket, three phase, 100 A	1.000	Ea.	181	234	415
Safety switches, heavy duty, 240 volt, 3 pole NEMA 1 fusible, 60 amp	1.000	Ea.	219	285	504
Grounding, wire ground bare armored, #8-1 conductor	.200	C.L.F.	16.80	66	82.80
Grounding, clamp, bronze, 3/4″ diameter	1.000	Ea.	6.50	20.50	27
Grounding, rod, copper clad, 8′ long, 3/4″ diameter	1.000	Ea.	37	124	161
Wireway w/fittings, 2-1/2″ x 2-1/2″	1.000	L.F.	11.95	14.55	26.50
TOTAL			593.25	1,069.65	1,662.90

D5010 120	Overhead Electric Service, 3 Phase - 4 Wire	COST EACH MAT.	COST EACH INST.	COST EACH TOTAL
0200	Service installation, includes breakers, metering, 20′ conduit & wire			
0220	3 phase, 4 wire, 120/208 volts, 60 A	595	1,075	1,670
0240	3 phase, 4 wire, 120/208 volts, 100 A	790	1,200	1,990
0245	100 A w/circuit breaker	1,575	1,500	3,075
0280	200 A	1,475	1,650	3,125
0285	200 A w/circuit breaker	3,100	2,100	5,200
0320	400 A	2,925	3,325	6,250
0325	400 A w/circuit breaker	6,000	4,150	10,150
0360	600 A	5,025	4,900	9,925
0365	600 A, w/switchboard	9,675	6,200	15,875
0400	800 A	6,950	5,875	12,825
0405	800 A, w/switchboard	11,600	7,375	18,975
0440	1000 A	8,625	7,250	15,875
0445	1000 A, w/switchboard	14,200	8,900	23,100
0480	1200 A	11,500	8,275	19,775
0485	1200 A, w/groundfault switchboard	35,600	10,200	45,800
0520	1600 A	14,200	10,800	25,000
0525	1600 A, w/groundfault switchboard	40,300	12,900	53,200

D5010 Electrical Service/Distribution

D5010 120	Overhead Electric Service, 3 Phase - 4 Wire	COST EACH		
		MAT.	INST.	TOTAL
0560	2000 A	19,000	13,100	32,100
0565	2000 A, w/groundfault switchboard	48,700	16,200	64,900
0610	1 phase, 3 wire, 120/240 volts, 100 A (no safety switch)	190	590	780
0615	100 A w/load center	555	1,125	1,680
0620	200 A	445	815	1,260
0625	200 A w/load center	1,250	1,650	2,900

D5010 Electrical Service/Distribution

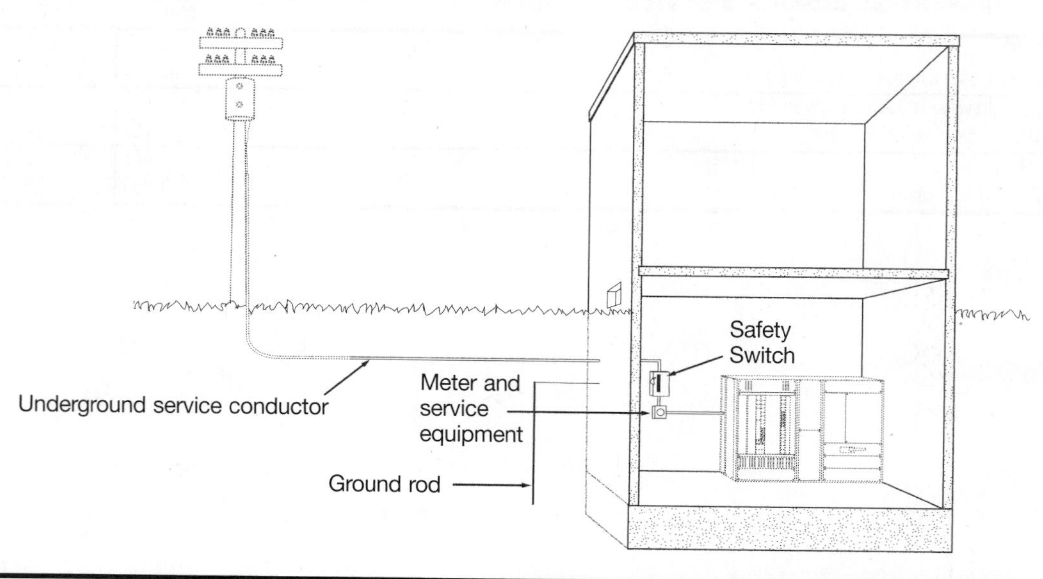

System Components	QUANTITY	UNIT	COST EACH MAT.	COST EACH INST.	COST EACH TOTAL
SYSTEM D5010 130 1000					
2000 AMP UNDERGROUND ELECTRIC SERVICE WITH GROUNDFAULT SWITCHBOARD INCLUDING EXCAVATION, BACKFILL, AND COMPACTION					
Excavate Trench	44.440	B.C.Y.		298.19	298.19
4 inch conduit bank	108.000	L.F.	1,085.40	2,646	3,731.40
4 inch fitting	8.000	Ea.	252	436	688
4 inch bells	4.000	Ea.	18.76	218	236.76
Concrete material	16.580	C.Y.	1,856.96		1,856.96
Concrete placement	16.580	C.Y.		407.37	407.37
Backfill trench	33.710	L.C.Y.		98.77	98.77
Compact fill material in trench	25.930	E.C.Y.		128.87	128.87
Dispose of excess fill material on-site	24.070	L.C.Y.		282.58	282.58
500 kcmil power cable	18.000	C.L.F.	22,950	7,380	30,330
Wire, 600 volt, type THW, copper, stranded, 1/0	6.000	C.L.F.	1,650	1,194	2,844
Saw cutting, concrete walls, plain, per inch of depth	64.000	L.F.	3.20	513.92	517.12
Meter centers and sockets, single pos, 4 terminal, 400 amp	5.000	Ea.	3,975	1,925	5,900
Safety switch, 400 amp	5.000	Ea.	7,625	3,650	11,275
Ground rod clamp	1.000	Ea.	6.50	20.50	27
Wireway	1.000	L.F.	11.95	14.55	26.50
600 volt stranded copper wire, 1/0	1.200	C.L.F.	330	238.80	568.80
Flexible metallic conduit	120.000	L.F.	88.80	492	580.80
Ground rod	6.000	Ea.	357	984	1,341
TOTAL			40,210.57	20,928.55	61,139.12

D5010 130	Underground Electric Service	COST EACH MAT.	COST EACH INST.	COST EACH TOTAL
0950	Underground electric service including excavation,backfill, and compaction			
1000	3 phase, 4 wire, 277/480 volts, 2000 A	40,200	20,900	61,100
1050	2000 A w/groundfault switch	68,500	23,400	91,900
1100	1600 A	31,900	16,900	48,800
1150	1600 A w/groundfault switchboard	58,000	19,000	77,000
1200	1200 A	24,400	13,400	37,800
1250	1200 A w/groundfault switchboard	48,500	15,400	63,900
1400	800 A	17,700	11,200	28,900

D5010 Electrical Service/Distribution

D5010 130	Underground Electric Service	COST EACH		
		MAT.	INST.	TOTAL
1450	800 A w/ switchboard	22,300	12,700	35,000
1500	600 A	11,400	10,300	21,700
1550	600 A w/ switchboard	16,000	11,500	27,500
1600	1 phase, 3 wire, 120/240 volts, 200 A	3,225	3,550	6,775
1650	200 A w/load center	4,050	4,400	8,450
1700	100 A	2,200	2,900	5,100
1750	100 A w/load center	3,000	3,750	6,750

D5010 Electrical Service/Distribution

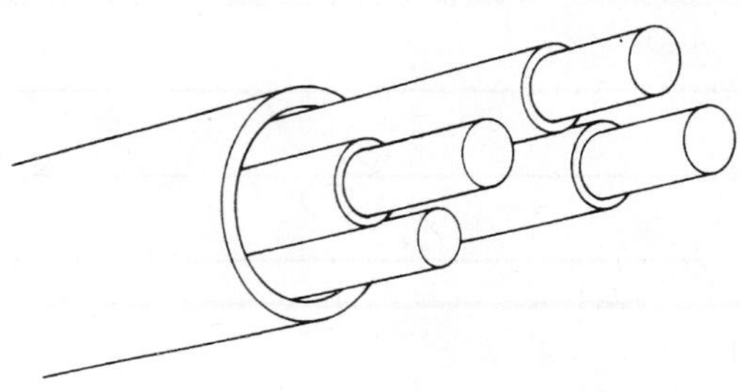

System Components	QUANTITY	UNIT	COST PER L.F. MAT.	INST.	TOTAL
SYSTEM D5010 230 0200					
FEEDERS, INCLUDING STEEL CONDUIT & WIRE, 60 A					
Rigid galvanized steel conduit, 3/4", including fittings	1.000	L.F.	3.02	8.20	11.22
Wire 600 volt, type XHHW copper stranded #6	.030	C.L.F.	2.09	3.03	5.12
Wire 600 volt, type XHHW copper stranded #8	.010	C.L.F.	.42	.82	1.24
TOTAL			5.53	12.05	17.58

D5010 230	Feeder Installation	COST PER L.F. MAT.	INST.	TOTAL
0200	Feeder installation 600 V, including RGS conduit and XHHW wire, 60 A	5.55	12.05	17.60
0240	100 A	10.05	15.70	25.75
0280	200 A	23.50	23.50	47
0320	400 A	46.50	47.50	94
0360	600 A	87.50	77	164.50
0400	800 A	113	91.50	204.50
0440	1000 A	142	120	262
0480	1200 A	175	154	329
0520	1600 A	227	183	410
0560	2000 A	284	241	525
1200	Branch installation 600 V, including EMT conduit and THW wire, 15 A	1.15	5.90	7.05
1240	20 A	1.37	6.25	7.62
1280	30 A	1.74	6.50	8.24
1320	50 A	3.40	9	12.40
1360	65 A	4.27	9.55	13.82
1400	85 A	6.60	11.10	17.70
1440	100 A	7.40	11.30	18.70
1480	130 A	11	13.25	24.25
1520	150 A	13.55	15.20	28.75
1560	200 A	18.75	17.30	36.05

D5010 Electrical Service/Distribution

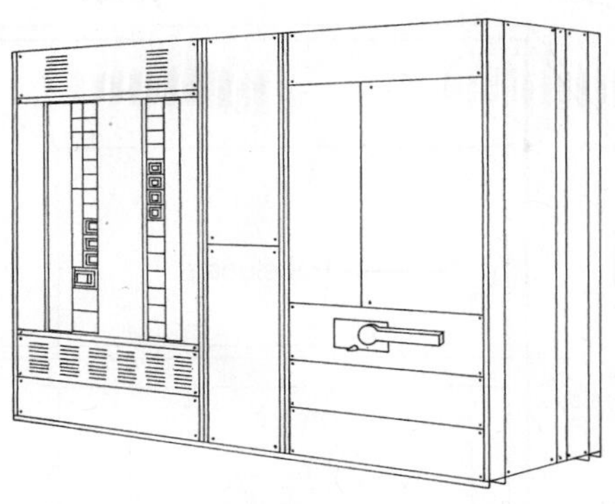

System Components	QUANTITY	UNIT	COST EACH MAT.	INST.	TOTAL
SYSTEM D5010 240 0240					
SWITCHGEAR INSTALLATION, INCL SWBD, PANELS & CIRC BREAKERS, 600 A					
Switchboards, 120/208 V, 600 amp	1.000	Ea.	4,650	1,300	5,950
Aluminum bus bars, 120/208 V, 600 amp	1.000	Ea.	1,775	1,300	3,075
Feeder section, circuit breakers, KA frame, 70 to 225 amp	1.000	Ea.	1,375	205	1,580
Feeder section, circuit breakers, LA frame, 125 to 400 amp	1.000	Ea.	2,725	285	3,010
TOTAL			10,525	3,090	13,615

D5010 240	Switchgear	COST EACH MAT.	INST.	TOTAL
0190	Switchgear installation, including switchboard, panels, & circuit breaker			
0200	120/208 V, 1 phase, 400 A	8,375	2,775	11,150
0240	600 A	10,500	3,100	13,600
0280	800 A	13,900	3,625	17,525
0300	1000 A	17,000	4,000	21,000
0320	1200 A	17,800	4,500	22,300
0360	1600 A	27,400	5,175	32,575
0400	2000 A	32,600	5,675	38,275
0500	277/480 V, 3 phase, 400 A	11,300	5,175	16,475
0520	600 A	15,500	5,900	21,400
0540	800 A	18,500	6,625	25,125
0560	1000 A	23,100	7,300	30,400
0580	1200 A	25,000	8,150	33,150
0600	1600 A	34,200	8,925	43,125
0620	2000 A	41,800	9,925	51,725

D5010 Electrical Service/Distribution

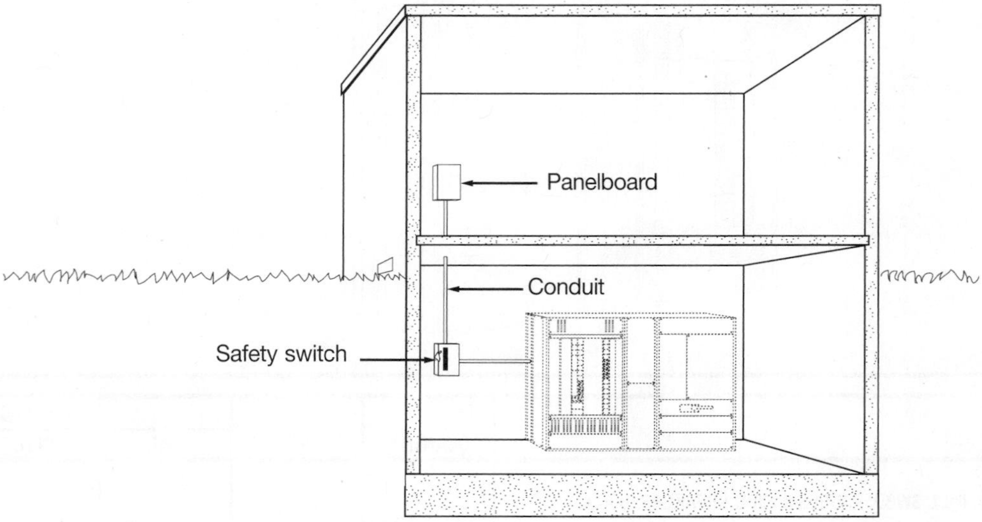

Panelboard
Conduit
Safety switch

System Components	QUANTITY	UNIT	COST EACH MAT.	COST EACH INST.	COST EACH TOTAL
SYSTEM D5010 250 1020					
PANELBOARD INSTALLATION, INCLUDES PANELBOARD, CONDUCTOR, & CONDUIT					
Conduit, galvanized steel, 1-1/4"dia.	37.000	L.F.	209.05	405.15	614.20
Panelboards, NQOD, 4 wire, 120/208 volts, 100 amp main, 24 circuits	1.000	Ea.	1,300	1,400	2,700
Wire, 600 volt, type THW, copper, stranded, #3	1.480	C.L.F.	204.24	193.88	398.12
TOTAL			1,713.29	1,999.03	3,712.32

D5010 250	Panelboard	COST EACH MAT.	COST EACH INST.	COST EACH TOTAL
0900	Panelboards, NQOD, 4 wire, 120/208 volts w/conductor & conduit			
1000	100 A, 0 stories, 0′ horizontal	1,300	1,400	2,700
1020	1 stories, 25′ horizontal	1,725	2,000	3,725
1040	5 stories, 50′ horizontal	2,525	3,175	5,700
1060	10 stories, 75′ horizontal	3,475	4,550	8,025
1080	225A, 0 stories, 0′ horizontal	2,425	1,825	4,250
2000	1 stories, 25′ horizontal	3,650	2,900	6,550
2020	5 stories, 50′ horizontal	6,100	5,025	11,125
2040	10 stories, 75′ horizontal	8,925	7,500	16,425
2060	400A, 0 stories, 0′ horizontal	3,575	2,725	6,300
2080	1 stories, 25′ horizontal	5,275	4,475	9,750
3000	5 stories, 50′ horizontal	8,600	7,925	16,525
3020	10 stories, 75′ horizontal	18,000	15,500	33,500
3040	600 A, 0 stories, 0′ horizontal	5,300	3,275	8,575
3060	1 stories, 25′ horizontal	8,800	6,125	14,925
3080	5 stories, 50′ horizontal	15,700	11,700	27,400
4000	10 stories, 75′ horizontal	35,500	23,100	58,600
4010	Panelboards, NEHB, 4 wire, 277/480 volts w/conductor, conduit, & safety switch			
4020	100 A, 0 stories, 0′ horizontal, includes safety switch	2,975	1,900	4,875
4040	1 stories, 25′ horizontal	3,375	2,500	5,875
4060	5 stories, 50′ horizontal	4,200	3,675	7,875
4080	10 stories, 75′ horizontal	5,150	5,050	10,200
5000	225 A, 0 stories, 0′ horizontal	4,450	2,325	6,775
5020	1 stories, 25′ horizontal	5,675	3,400	9,075
5040	5 stories, 50′ horizontal	8,100	5,525	13,625
5060	10 stories, 75′ horizontal	10,900	8,025	18,925

D5010 Electrical Service/Distribution

D5010 250	Panelboard	COST EACH MAT.	COST EACH INST.	COST EACH TOTAL
5080	400 A, 0 stories, 0' horizontal	7,225	3,575	10,800
6000	1 stories, 25' horizontal	8,925	5,325	14,250
6020	5 stories, 50' horizontal	12,300	8,775	21,075
6040	10 stories, 75' horizontal	21,700	16,400	38,100
6060	600 A, 0 stories, 0' horizontal	10,900	4,550	15,450
6080	1 stories, 25' horizontal	14,400	7,400	21,800
7000	5 stories, 50' horizontal	21,300	13,000	34,300
7020	10 stories, 75' horizontal	41,100	24,400	65,500

D5020 Lighting and Branch Wiring

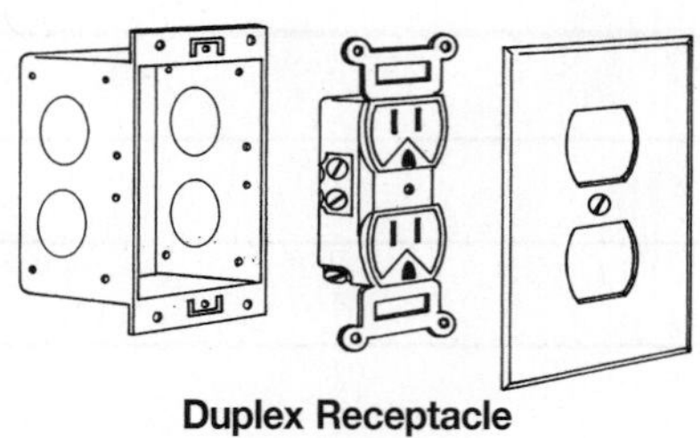

Duplex Receptacle

System Components	QUANTITY	UNIT	COST PER S.F. MAT.	COST PER S.F. INST.	COST PER S.F. TOTAL
SYSTEM D5020 110 0200					
RECEPTACLES INCL. PLATE, BOX, CONDUIT, WIRE & TRANS. WHEN REQUIRED					
2.5 PER 1000 S.F., .3 WATTS PER S.F.					
Steel intermediate conduit, (IMC) 1/2" diam	167.000	L.F.	.33	1.09	1.42
Wire 600V type THWN-THHN, copper solid #12	3.340	C.L.F.	.04	.20	.24
Wiring device, receptacle, duplex, 120V grounded, 15 amp	2.500	Ea.		.04	.04
Wall plate, 1 gang, brown plastic	2.500	Ea.		.02	.02
Steel outlet box 4" square	2.500	Ea.	.01	.08	.09
Steel outlet box 4" plaster rings	2.500	Ea.	.01	.03	.04
TOTAL			.39	1.46	1.85

D5020 110	Receptacle (by Wattage)	COST PER S.F. MAT.	COST PER S.F. INST.	COST PER S.F. TOTAL
0190	Receptacles include plate, box, conduit, wire & transformer when required			
0200	2.5 per 1000 S.F., .3 watts per S.F.	.39	1.46	1.85
0240	With transformer	.47	1.53	2
0280	4 per 1000 S.F., .5 watts per S.F.	.44	1.70	2.14
0320	With transformer	.57	1.82	2.39
0360	5 per 1000 S.F., .6 watts per S.F.	.52	2.01	2.53
0400	With transformer	.68	2.16	2.84
0440	8 per 1000 S.F., .9 watts per S.F.	.54	2.23	2.77
0480	With transformer	.80	2.46	3.26
0520	10 per 1000 S.F., 1.2 watts per S.F.	.57	2.41	2.98
0560	With transformer	.89	2.70	3.59
0600	16.5 per 1000 S.F., 2.0 watts per S.F.	.65	3.02	3.67
0640	With transformer	1.18	3.50	4.68
0680	20 per 1000 S.F., 2.4 watts per S.F.	.70	3.30	4
0720	With transformer	1.34	3.89	5.23

D5020 Lighting and Branch Wiring

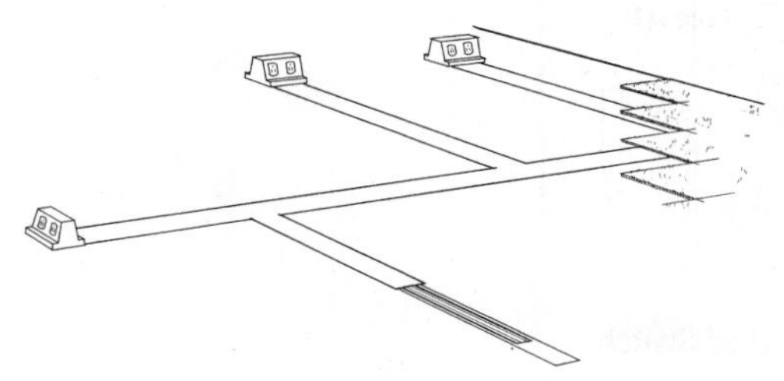

Underfloor Receptacle System

Description: Table D5020 115 includes installed costs of raceways and copper wire from panel to and including receptacle.

National Electrical Code prohibits use of undercarpet system in residential, school or hospital buildings. Can only be used with carpet squares.

Low density = (1) Outlet per 259 S.F. of floor area.

High density = (1) Outlet per 127 S.F. of floor area.

System Components	QUANTITY	UNIT	COST PER S.F. MAT.	COST PER S.F. INST.	COST PER S.F. TOTAL
SYSTEM D5020 115 0200					
RECEPTACLE SYSTEMS, UNDERFLOOR DUCT, 5′ ON CENTER, LOW DENSITY					
Underfloor duct 3-1/8″ x 7/8″ w/insert 24″ on center	.190	L.F.	3.99	1.78	5.77
Underfloor duct junction box, single duct, 3-1/8″	.003	Ea.	1.41	.49	1.90
Underfloor junction box carpet pan	.003	Ea.	1.17	.02	1.19
Underfloor duct outlet, high tension receptacle	.004	Ea.	.44	.33	.77
Wire 600V type THWN-THHN copper solid #12	.010	C.L.F.	.13	.60	.73
Vertical elbow for underfloor duct, 3-1/8″, included					
Underfloor duct conduit adapter, 2″ x 1-1/4″, included					
TOTAL			7.14	3.22	10.36

D5020 115	Receptacles, Floor	COST PER S.F. MAT.	COST PER S.F. INST.	COST PER S.F. TOTAL
0200	Receptacle systems, underfloor duct, 5′ on center, low density	7.15	3.22	10.37
0240	High density	7.70	4.14	11.84
0280	7′ on center, low density	5.65	2.78	8.43
0320	High density	6.20	3.70	9.90
0400	Poke thru fittings, low density	1.26	1.58	2.84
0440	High density	2.50	3.14	5.64
0520	Telepoles, using Romex, low density	1.19	.97	2.16
0560	High density	2.38	1.94	4.32
0600	Using EMT, low density	1.26	1.28	2.54
0640	High density	2.50	2.56	5.06
0720	Conduit system with floor boxes, low density	1.17	1.11	2.28
0760	High density	2.34	2.21	4.55
0840	Undercarpet power system, 3 conductor with 5 conductor feeder, low density	1.55	.39	1.94
0880	High density	3.04	.81	3.85

D5020 Lighting and Branch Wiring

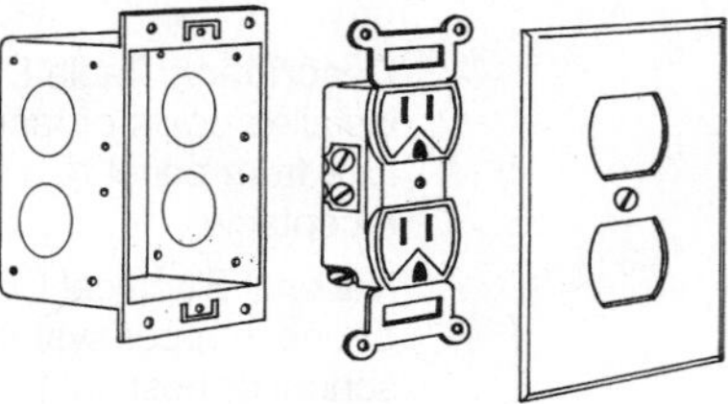

Duplex Receptacle

Wall Switch

System Components	QUANTITY	UNIT	COST PER S.F. MAT.	COST PER S.F. INST.	COST PER S.F. TOTAL
SYSTEM D5020 120 0520					
RECEPTACLES AND WALL SWITCHES					
4 RECEPTACLES PER 400 S.F.					
Steel intermediate conduit, (IMC), 1/2" diam	.220	L.F.	.43	1.44	1.87
Wire, 600 volt, type THWN-THHN, copper, solid #12	.005	C.L.F.	.07	.30	.37
Steel outlet box 4" square	.010	Ea.	.03	.33	.36
Steel outlet box, 4" square, plaster rings	.010	Ea.	.03	.10	.13
Receptacle, duplex, 120 volt grounded, 15 amp	.010	Ea.	.01	.16	.17
Wall plate, 1 gang, brown plastic	.010	Ea.		.08	.08
TOTAL			.57	2.41	2.98

D5020 120	Receptacles & Wall Switches	COST PER S.F. MAT.	COST PER S.F. INST.	COST PER S.F. TOTAL
0520	Receptacles and wall switches, 400 S.F., 4 receptacles	.57	2.41	2.98
0560	6 receptacles	.63	2.83	3.46
0600	8 receptacles	.68	3.29	3.97
0640	1 switch	.13	.54	.67
0680	600 S.F., 6 receptacles	.57	2.41	2.98
0720	8 receptacles	.61	2.74	3.35
0760	10 receptacles	.66	3.08	3.74
0800	2 switches	.19	.71	.90
0840	1000 S.F., 10 receptacles	.56	2.41	2.97
0880	12 receptacles	.57	2.55	3.12
0920	14 receptacles	.62	2.74	3.36
0960	2 switches	.10	.40	.50
1000	1600 S.F., 12 receptacles	.50	2.14	2.64
1040	14 receptacles	.53	2.34	2.87
1080	16 receptacles	.56	2.41	2.97
1120	4 switches	.12	.46	.58
1160	2000 S.F., 14 receptacles	.53	2.20	2.73
1200	16 receptacles	.52	2.21	2.73
1240	18 receptacles	.53	2.34	2.87
1280	4 switches	.10	.40	.50
1320	3000 S.F., 12 receptacles	.43	1.68	2.11
1360	18 receptacles	.52	2.08	2.60
1400	24 receptacles	.52	2.21	2.73
1440	6 switches	.10	.40	.50
1480	3600 S.F., 20 receptacles	.50	2.01	2.51
1520	24 receptacles	.50	2.07	2.57
1560	28 receptacles	.52	2.14	2.66
1600	8 switches	.12	.46	.58
1640	4000 S.F., 16 receptacles	.43	1.68	2.11
1680	24 receptacles	.52	2.08	2.60
1720	30 receptacles	.50	2.14	2.64
1760	8 switches	.10	.40	.50

D5020 Lighting and Branch Wiring

D5020 120	Receptacles & Wall Switches	COST PER S.F.		
		MAT.	INST.	TOTAL
1800	5000 S.F., 20 receptacles	.43	1.68	2.11
1840	26 receptacles	.50	2.01	2.51
1880	30 receptacles	.52	2.08	2.60
1920	10 switches	.10	.40	.50

D5020 Lighting and Branch Wiring

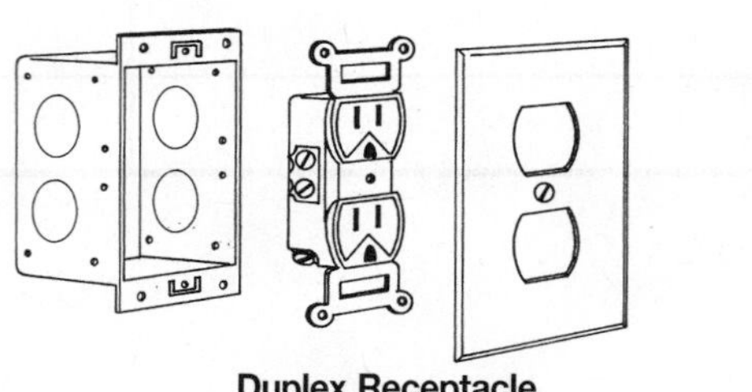

System Components	QUANTITY	UNIT	COST PER EACH MAT.	INST.	TOTAL
SYSTEM D5020 125 0520					
RECEPTACLES AND WALL SWITCHES, RECEPTICLE DUPLEX 120 V GROUNDED, 15 A					
Electric metallic tubing conduit, (EMT), 3/4" diam	22.000	L.F.	22.88	111.10	133.98
Wire, 600 volt, type THWN-THHN, copper, solid #12	.630	C.L.F.	8.25	37.49	45.74
Steel outlet box 4" square	1.000	Ea.	2.66	33	35.66
Steel outlet box, 4" square, plaster rings	1.000	Ea.	2.64	10.25	12.89
Receptacle, duplex, 120 volt grounded, 15 amp	1.000	Ea.	1.39	16.40	17.79
Wall plate, 1 gang, brown plastic	1.000	Ea.	.42	8.20	8.62
TOTAL			38.24	216.44	254.68

D5020 125	Receptacles & Switches by Each	COST PER EACH MAT.	INST.	TOTAL
0460	Receptacles & Switches, with box, plate, 3/4" EMT conduit & wire			
0520	Receptacle duplex 120 V grounded, 15 A	38	216	254
0560	20 A	45.50	225	270.50
0600	Receptacle duplex ground fault interrupting, 15 A	72	225	297
0640	20 A	81	225	306
0680	Toggle switch single, 15 A	44	216	260
0720	20 A	47.50	225	272.50
0760	3 way switch, 15 A	42.50	229	271.50
0800	20 A	47.50	237	284.50
0840	4 way switch, 15 A	47	244	291
0880	20 A	84	260	344

D5020 Lighting and Branch Wiring

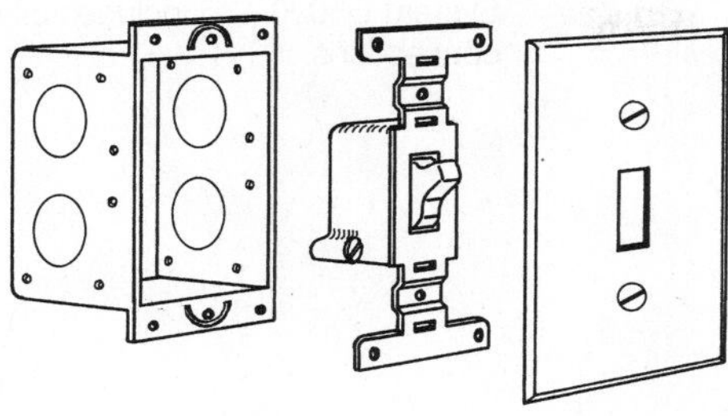

Description: Table D5020 130 includes the cost for switch, plate, box, conduit in slab or EMT exposed and copper wire. Add 20% for exposed conduit.

No power required for switches.

Federal energy guidelines recommend the maximum lighting area controlled per switch shall not exceed 1000 S.F. and that areas over 500 S.F. shall be so controlled that total illumination can be reduced by at least 50%.

System Components	QUANTITY	UNIT	COST PER S.F. MAT.	COST PER S.F. INST.	COST PER S.F. TOTAL
SYSTEM D5020 130 0360					
WALL SWITCHES, 5.0 PER 1000 S.F.					
Steel, intermediate conduit (IMC), 1/2" diameter	88.000	L.F.	.17	.58	.75
Wire, 600V type THWN-THHN, copper solid #12	1.710	C.L.F.	.02	.10	.12
Toggle switch, single pole, 15 amp	5.000	Ea.	.04	.08	.12
Wall plate, 1 gang, brown plastic	5.000	Ea.		.04	.04
Steel outlet box 4" plaster rings	5.000	Ea.	.01	.17	.18
Plaster rings	5.000	Ea.	.01	.05	.06
TOTAL			.25	1.02	1.27

D5020 130	Wall Switch by Sq. Ft.	COST PER S.F. MAT.	COST PER S.F. INST.	COST PER S.F. TOTAL
0200	Wall switches, 1.0 per 1000 S.F.	.06	.23	.29
0240	1.2 per 1000 S.F.	.07	.26	.33
0280	2.0 per 1000 S.F.	.10	.38	.48
0320	2.5 per 1000 S.F.	.13	.48	.61
0360	5.0 per 1000 S.F.	.25	1.02	1.27
0400	10.0 per 1000 S.F.	.53	2.05	2.58

D5020 Lighting and Branch Wiring

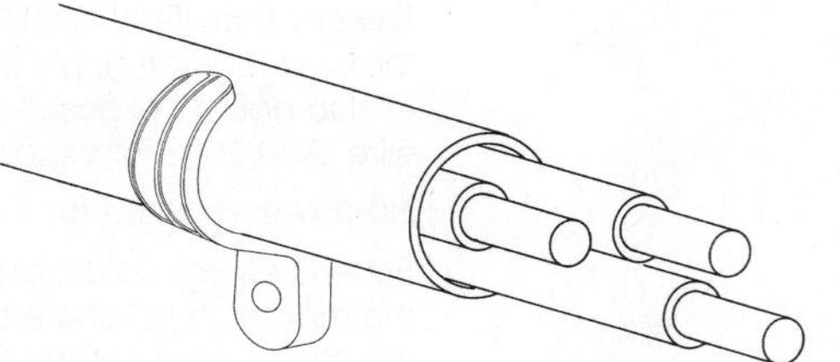

System D5020 135 includes all wiring and connections.

System Components	QUANTITY	UNIT	COST PER S.F. MAT.	COST PER S.F. INST.	COST PER S.F. TOTAL
SYSTEM D5020 135 0200					
MISCELLANEOUS POWER, TO .5 WATTS					
Steel intermediate conduit, (IMC) 1/2" diam	15.000	L.F.	.03	.10	.13
Wire 600V type THWN-THHN, copper solid #12	.325	C.L.F.		.02	.02
TOTAL			.03	.12	.15

D5020 135	Miscellaneous Power	COST PER S.F. MAT.	COST PER S.F. INST.	COST PER S.F. TOTAL
0200	Miscellaneous power, to .5 watts	.03	.12	.15
0240	.8 watts	.05	.17	.22
0280	1 watt	.07	.22	.29
0320	1.2 watts	.08	.26	.34
0360	1.5 watts	.09	.30	.39
0400	1.8 watts	.10	.35	.45
0440	2 watts	.12	.42	.54
0480	2.5 watts	.15	.52	.67
0520	3 watts	.18	.61	.79

D5020 Lighting and Branch Wiring

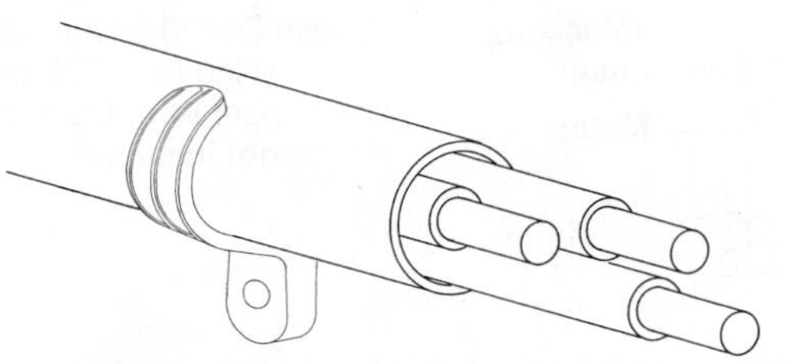

System D5020 140 includes all wiring and connections for central air conditioning units.

System Components	QUANTITY	UNIT	COST PER S.F. MAT.	COST PER S.F. INST.	COST PER S.F. TOTAL
SYSTEM D5020 140 0200					
CENTRAL AIR CONDITIONING POWER, 1 WATT					
Steel intermediate conduit, 1/2" diam.	.030	L.F.	.06	.20	.26
Wire 600V type THWN-THHN, copper solid #12	.001	C.L.F.	.01	.06	.07
TOTAL			.07	.26	.33

D5020 140	Central A. C. Power (by Wattage)	COST PER S.F. MAT.	COST PER S.F. INST.	COST PER S.F. TOTAL
0200	Central air conditioning power, 1 watt	.07	.26	.33
0220	2 watts	.08	.30	.38
0240	3 watts	.11	.33	.44
0280	4 watts	.18	.44	.62
0320	6 watts	.34	.59	.93
0360	8 watts	.43	.63	1.06
0400	10 watts	.63	.74	1.37

D5020 Lighting and Branch Wiring

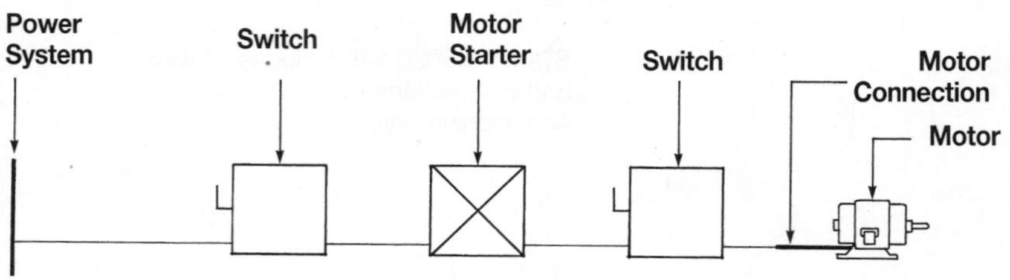

System D5020 145 installed cost of motor wiring using 50′ of rigid conduit and copper wire. **Cost and setting of motor not included.**

System Components	QUANTITY	UNIT	COST EACH MAT.	COST EACH INST.	COST EACH TOTAL
SYSTEM D5020 145 0200					
MOTOR INST., SINGLE PHASE, 115V, TO AND INCLUDING 1/3 HP MOTOR SIZE					
Wire 600V type THWN-THHN, copper solid #12	1.250	C.L.F.	16.38	74.38	90.76
Steel intermediate conduit, (IMC) 1/2″ diam.	50.000	L.F.	98.50	327.50	426
Magnetic FVNR, 115V, 1/3 HP, size 00 starter	1.000	Ea.	203	164	367
Safety switch, fused, heavy duty, 240V 2P 30 amp	1.000	Ea.	97.50	187	284.50
Safety switch, non fused, heavy duty, 600V, 3 phase, 30 A	1.000	Ea.	125	205	330
Flexible metallic conduit, Greenfield 1/2″ diam.	1.500	L.F.	.80	4.92	5.72
Connectors for flexible metallic conduit Greenfield 1/2″ diam.	1.000	Ea.	2.51	8.20	10.71
Coupling for Greenfield to conduit 1/2″ diam. flexible metalic conduit	1.000	Ea.	1.63	13.10	14.73
Fuse cartridge nonrenewable, 250V 30 amp	1.000	Ea.	2.64	13.10	15.74
TOTAL			547.96	997.20	1,545.16

D5020 145	Motor Installation	COST EACH MAT.	COST EACH INST.	COST EACH TOTAL
0200	Motor installation, single phase, 115V, 1/3 HP motor size	550	995	1,545
0240	1 HP motor size	570	995	1,565
0280	2 HP motor size	620	1,050	1,670
0320	3 HP motor size	710	1,075	1,785
0360	230V, 1 HP motor size	550	1,000	1,550
0400	2 HP motor size	590	1,000	1,590
0440	3 HP motor size	660	1,075	1,735
0520	Three phase, 200V, 1-1/2 HP motor size	640	1,100	1,740
0560	3 HP motor size	720	1,200	1,920
0600	5 HP motor size	760	1,350	2,110
0640	7-1/2 HP motor size	790	1,375	2,165
0680	10 HP motor size	1,200	1,725	2,925
0720	15 HP motor size	1,625	1,925	3,550
0760	20 HP motor size	2,075	2,200	4,275
0800	25 HP motor size	2,125	2,225	4,350
0840	30 HP motor size	3,300	2,625	5,925
0880	40 HP motor size	4,125	3,100	7,225
0920	50 HP motor size	7,250	3,600	10,850
0960	60 HP motor size	7,550	3,800	11,350
1000	75 HP motor size	9,500	4,350	13,850
1040	100 HP motor size	27,200	5,125	32,325
1080	125 HP motor size	27,500	5,650	33,150
1120	150 HP motor size	31,800	6,650	38,450
1160	200 HP motor size	32,800	7,850	40,650
1240	230V, 1-1/2 HP motor size	615	1,100	1,715
1280	3 HP motor size	690	1,200	1,890
1320	5 HP motor size	730	1,325	2,055
1360	7-1/2 HP motor size	730	1,325	2,055
1400	10 HP motor size	1,125	1,625	2,750
1440	15 HP motor size	1,275	1,775	3,050
1480	20 HP motor size	1,975	2,150	4,125
1520	25 HP motor size	2,075	2,200	4,275

D5020 Lighting and Branch Wiring

D5020 145	Motor Installation	COST EACH		
		MAT.	INST.	TOTAL
1560	30 HP motor size	2,100	2,225	4,325
1600	40 HP motor size	3,975	3,025	7,000
1640	50 HP motor size	4,300	3,200	7,500
1680	60 HP motor size	7,275	3,600	10,875
1720	75 HP motor size	8,750	4,075	12,825
1760	100 HP motor size	10,300	4,550	14,850
1800	125 HP motor size	27,700	5,300	33,000
1840	150 HP motor size	29,300	6,025	35,325
1880	200 HP motor size	31,500	6,700	38,200
1960	460V, 2 HP motor size	760	1,100	1,860
2000	5 HP motor size	835	1,200	2,035
2040	10 HP motor size	840	1,325	2,165
2080	15 HP motor size	1,125	1,525	2,650
2120	20 HP motor size	1,200	1,625	2,825
2160	25 HP motor size	1,275	1,725	3,000
2200	30 HP motor size	1,625	1,850	3,475
2240	40 HP motor size	2,100	2,000	4,100
2280	50 HP motor size	2,350	2,200	4,550
2320	60 HP motor size	3,500	2,600	6,100
2360	75 HP motor size	4,150	2,875	7,025
2400	100 HP motor size	4,625	3,200	7,825
2440	125 HP motor size	7,600	3,625	11,225
2480	150 HP motor size	9,425	4,050	13,475
2520	200 HP motor size	11,100	4,575	15,675
2600	575V, 2 HP motor size	760	1,100	1,860
2640	5 HP motor size	835	1,200	2,035
2680	10 HP motor size	840	1,325	2,165
2720	20 HP motor size	1,125	1,525	2,650
2760	25 HP motor size	1,200	1,625	2,825
2800	30 HP motor size	1,625	1,850	3,475
2840	50 HP motor size	1,750	1,925	3,675
2880	60 HP motor size	3,475	2,575	6,050
2920	75 HP motor size	3,500	2,600	6,100
2960	100 HP motor size	4,150	2,875	7,025
3000	125 HP motor size	7,400	3,550	10,950
3040	150 HP motor size	7,600	3,625	11,225
3080	200 HP motor size	9,575	4,100	13,675

D5020 Lighting and Branch Wiring

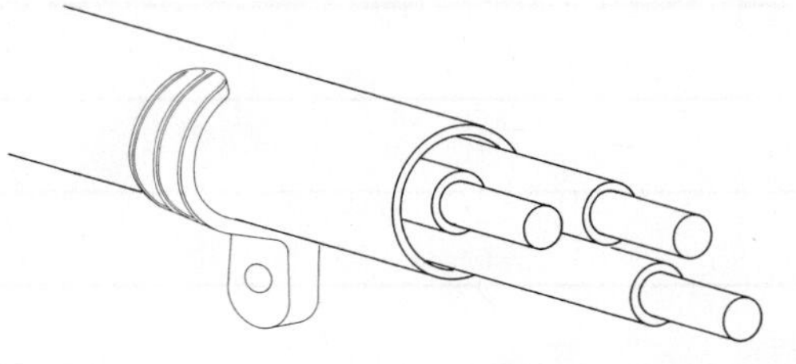

System Components	QUANTITY	UNIT	COST PER L.F. MAT.	INST.	TOTAL
SYSTEM D5020 155 0200					
MOTOR FEEDER SYSTEMS, SINGLE PHASE, UP TO 115V, 1HP OR 230V, 2HP					
Steel intermediate conduit, (IMC) 1/2" diam	1.000	L.F.	1.97	6.55	8.52
Wire 600V type THWN-THHN, copper solid #12	.020	C.L.F.	.26	1.19	1.45
TOTAL			2.23	7.74	9.97

D5020 155	Motor Feeder	COST PER L.F. MAT.	INST.	TOTAL
0200	Motor feeder systems, single phase, feed up to 115V 1HP or 230V 2 HP	2.23	7.75	9.98
0240	115V 2HP, 230V 3HP	2.38	7.85	10.23
0280	115V 3HP	2.71	8.20	10.91
0360	Three phase, feed to 200V 3HP, 230V 5HP, 460V 10HP, 575V 10HP	2.36	8.35	10.71
0440	200V 5HP, 230V 7.5HP, 460V 15HP, 575V 20HP	2.59	8.50	11.09
0520	200V 10HP, 230V 10HP, 460V 30HP, 575V 30HP	3.08	9	12.08
0600	200V 15HP, 230V 15HP, 460V 40HP, 575V 50HP	4.35	10.35	14.70
0680	200V 20HP, 230V 25HP, 460V 50HP, 575V 60HP	6.50	13.05	19.55
0760	200V 25HP, 230V 30HP, 460V 60HP, 575V 75HP	7.30	13.30	20.60
0840	200V 30HP	8.25	13.75	22
0920	230V 40HP, 460V 75HP, 575V 100HP	10.40	15	25.40
1000	200V 40HP	11.55	16.05	27.60
1080	230V 50HP, 460V 100HP, 575V 125HP	14.85	17.75	32.60
1160	200V 50HP, 230V 60HP, 460V 125HP, 575V 150HP	17.15	18.80	35.95
1240	200V 60HP, 460V 150HP	21	22	43
1320	230V 75HP, 575V 200HP	24	23	47
1400	200V 75HP	28	23.50	51.50
1480	230V 100HP, 460V 200HP	36.50	27.50	64
1560	200V 100HP	50.50	34.50	85
1640	230V 125HP	50.50	34.50	85
1720	200V 125HP, 230V 150HP	58.50	42.50	101
1800	200V 150HP	71	46.50	117.50
1880	200V 200HP	90	49	139
1960	230V 200HP	92	50.50	142.50

D5020 Lighting and Branch Wiring

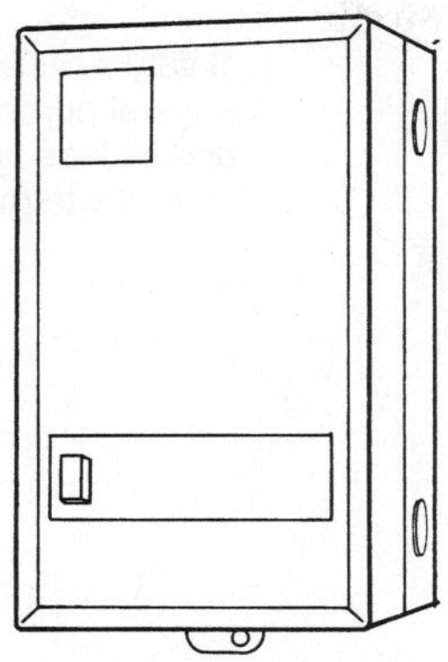

Starters are full voltage, type NEMA 1 for general purpose indoor application with motor overload protection and include mounting and wire connections.

System Components	QUANTITY	UNIT	COST EACH MAT.	COST EACH INST.	COST EACH TOTAL
SYSTEM D5020 160 0200					
MAGNETIC STARTER, SIZE 00 TO 1/3 HP, 1 PHASE 115V OR 1 HP 230V					
Magnetic starter, size 00, to 1/3 HP, 1 phase, 115V or 1 HP 230V	1.000	Ea.	203	164	367
TOTAL			203	164	367

D5020 160	Magnetic Starter	COST EACH MAT.	COST EACH INST.	COST EACH TOTAL
0200	Magnetic starter, size 00, to 1/3 HP, 1 phase, 115V or 1 HP 230V	203	164	367
0280	Size 00, to 1-1/2 HP, 3 phase, 200-230V or 2 HP 460-575V	223	187	410
0360	Size 0, to 1 HP, 1 phase, 115V or 2 HP 230V	226	164	390
0440	Size 0, to 3 HP, 3 phase, 200-230V or 5 HP 460-575V	300	285	585
0520	Size 1, to 2 HP, 1 phase, 115V or 3 HP 230V	260	219	479
0600	Size 1, to 7-1/2 HP, 3 phase, 200-230V or 10 HP 460-575V	305	410	715
0680	Size 2, to 10 HP, 3 phase, 200V, 15 HP-230V or 25 HP 460-575V	570	595	1,165
0760	Size 3, to 25 HP, 3 phase, 200V, 30 HP-230V or 50 HP 460-575V	930	730	1,660
0840	Size 4, to 40 HP, 3 phase, 200V, 50 HP-230V or 100 HP 460-575V	2,050	1,100	3,150
0920	Size 5, to 75 HP, 3 phase, 200V, 100 HP-230V or 200 HP 460-575V	4,825	1,450	6,275
1000	Size 6, to 150 HP, 3 phase, 200V, 200 HP-230V or 400 HP 460-575V	20,900	1,650	22,550

D5020 Lighting and Branch Wiring

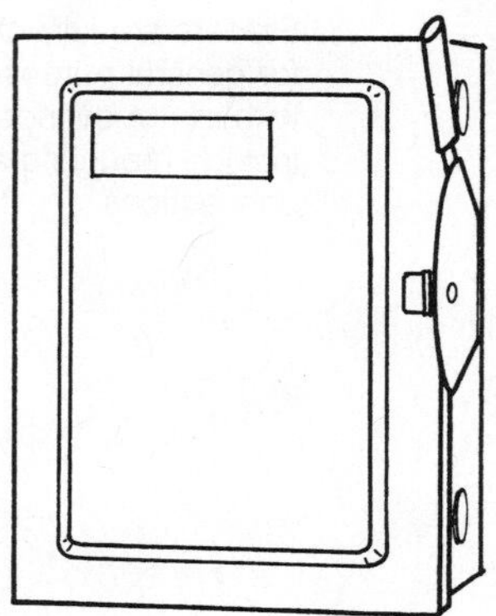

Safety switches are type NEMA 1 for general purpose indoor application, and include time delay fuses, insulation and wire terminations.

System Components	QUANTITY	UNIT	COST EACH MAT.	COST EACH INST.	COST EACH TOTAL
SYSTEM D5020 165 0200					
SAFETY SWITCH, 30A FUSED, 1 PHASE, 115V OR 230V					
Safety switch fused, hvy duty, 240V 2p 30 amp	1.000	Ea.	97.50	187	284.50
Fuse, dual element time delay 250V, 30 amp	2.000	Ea.	20.80	26.20	47
TOTAL			118.30	213.20	331.50

D5020 165	Safety Switches	COST EACH MAT.	COST EACH INST.	COST EACH TOTAL
0200	Safety switch, 30 A fused, 1 phase, 2 HP 115 V or 3 HP, 230 V	118	213	331
0280	3 phase, 5 HP, 200 V or 7 1/2 HP, 230 V	161	244	405
0360	15 HP, 460 V or 20 HP, 575 V	283	254	537
0440	60 A fused, 3 phase, 15 HP 200 V or 15 HP 230 V	257	325	582
0520	30 HP 460 V or 40 HP 575 V	370	335	705
0600	100 A fused, 3 phase, 20 HP 200 V or 25 HP 230 V	440	395	835
0680	50 HP 460 V or 60 HP 575 V	710	400	1,110
0760	200 A fused, 3 phase, 50 HP 200 V or 60 HP 230 V	800	560	1,360
0840	125 HP 460 V or 150 HP 575 V	1,125	570	1,695
0920	400 A fused, 3 phase, 100 HP 200 V or 125 HP 230 V	1,900	795	2,695
1000	250 HP 460 V or 350 HP 575 V	2,700	815	3,515
1020	600 A fused, 3 phase, 150 HP 200 V or 200 HP 230 V	3,675	1,175	4,850
1040	400 HP 460 V	4,250	1,200	5,450

D5020 Lighting and Branch Wiring

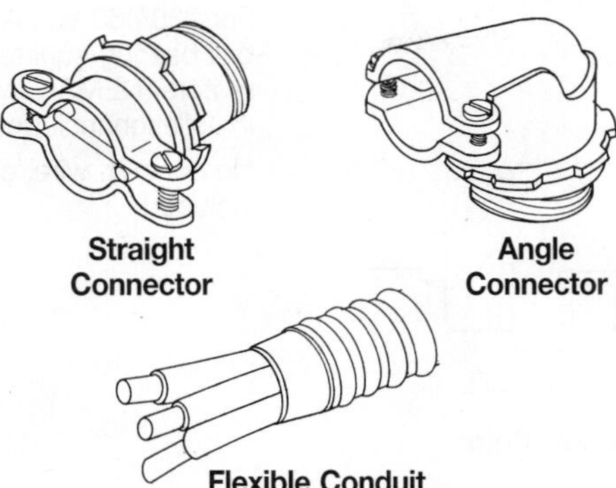

Table below includes costs for the flexible conduit. Not included are wire terminations and testing motor for correct rotation.

System Components	QUANTITY	UNIT	COST EACH MAT.	COST EACH INST.	COST EACH TOTAL
SYSTEM D5020 170 0200					
MOTOR CONNECTIONS, SINGLE PHASE, 115V/230V UP TO 1 HP					
Motor connection, flexible conduit & fittings, 1 HP motor 115V	1.000	Ea.	7	82	89
TOTAL			7	82	89

D5020 170	Motor Connections	COST EACH MAT.	COST EACH INST.	COST EACH TOTAL
0200	Motor connections, single phase, 115/230V, up to 1 HP	7	82	89
0240	Up to 3 HP	13.80	101	114.80
0280	Three phase, 200/230/460/575V, up to 3 HP	8.40	96.50	104.90
0320	Up to 5 HP	7.15	120	127.15
0360	Up to 7-1/2 HP	10.50	142	152.50
0400	Up to 10 HP	23.50	156	179.50
0440	Up to 15 HP	23.50	199	222.50
0480	Up to 25 HP	32	243	275
0520	Up to 50 HP	65	298	363
0560	Up to 100 HP	154	435	589

D5020 Lighting and Branch Wiring

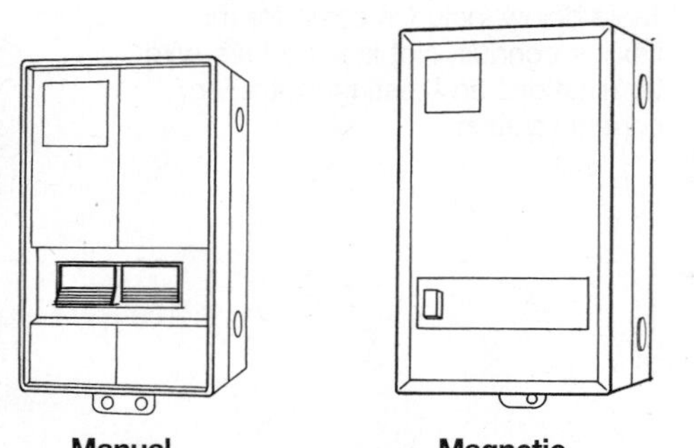

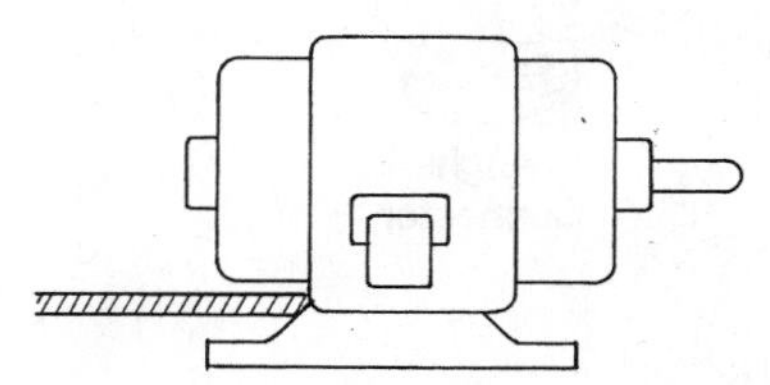

For 230/460 Volt A.C., 3 phase, 60 cycle ball bearing squirrel cage induction motors, NEMA Class B standard line. Installation included.

No conduit, wire, or terminations included.

System Components	QUANTITY	UNIT	COST EACH MAT.	COST EACH INST.	COST EACH TOTAL
SYSTEM D5020 175 0220					
MOTOR, DRIPPROOF CLASS B INSULATION, 1.15 SERVICE FACTOR, WITH STARTER					
1 H.P., 1200 RPM WITH MANUAL STARTER					
Motor, dripproof, class B insul, 1.15 serv fact, 1200 RPM, 1 HP	1.000	Ea.	520	146	666
Motor starter, manual, 3 phase, 1 HP motor	1.000	Ea.	228	187	415
TOTAL			748	333	1,081

D5020 175	Motor & Starter	COST EACH MAT.	COST EACH INST.	COST EACH TOTAL
0190	Motor, dripproof, premium efficient, 1.15 service factor			
0200	1 HP, 1200 RPM, motor only	520	146	666
0220	With manual starter	750	335	1,085
0240	With magnetic starter	745	335	1,080
0260	1800 RPM, motor only	305	146	451
0280	With manual starter	535	335	870
0300	With magnetic starter	530	335	865
0320	2 HP, 1200 RPM, motor only	615	146	761
0340	With manual starter	845	335	1,180
0360	With magnetic starter	915	430	1,345
0380	1800 RPM, motor only	370	146	516
0400	With manual starter	600	335	935
0420	With magnetic starter	670	430	1,100
0440	3600 RPM, motor only	515	146	661
0460	With manual starter	745	335	1,080
0480	With magnetic starter	815	430	1,245
0500	3 HP, 1200 RPM, motor only	790	146	936
0520	With manual starter	1,025	335	1,360
0540	With magnetic starter	1,100	430	1,530
0560	1800 RPM, motor only	615	146	761
0580	With manual starter	845	335	1,180
0600	With magnetic starter	915	430	1,345
0620	3600 RPM, motor only	520	146	666
0640	With manual starter	750	335	1,085
0660	With magnetic starter	820	430	1,250
0680	5 HP, 1200 RPM, motor only	1,025	146	1,171
0700	With manual starter	1,300	475	1,775
0720	With magnetic starter	1,325	555	1,880
0740	1800 RPM, motor only	670	146	816
0760	With manual starter	940	475	1,415
0780	With magnetic starter	975	555	1,530
0800	3600 RPM, motor only	570	146	716

D5020 Lighting and Branch Wiring

D5020 175	Motor & Starter	MAT.	INST.	TOTAL
		COST EACH		
0820	With manual starter	840	475	1,315
0840	With magnetic starter	875	555	1,430
0860	7.5 HP, 1800 RPM, motor only	890	156	1,046
0880	With manual starter	1,150	485	1,635
0900	With magnetic starter	1,450	750	2,200
0920	10 HP, 1800 RPM, motor only	1,100	164	1,264
0940	With manual starter	1,375	495	1,870
0960	With magnetic starter	1,675	760	2,435
0980	15 HP, 1800 RPM, motor only	1,525	205	1,730
1000	With magnetic starter	2,100	800	2,900
1040	20 HP, 1800 RPM, motor only	1,925	252	2,177
1060	With magnetic starter	2,850	980	3,830
1100	25 HP, 1800 RPM, motor only	2,275	262	2,537
1120	With magnetic starter	3,200	990	4,190
1160	30 HP, 1800 RPM, motor only	2,475	273	2,748
1180	With magnetic starter	3,400	1,000	4,400
1220	40 HP, 1800 RPM, motor only	3,225	330	3,555
1240	With magnetic starter	5,275	1,425	6,700
1280	50 HP, 1800 RPM, motor only	3,525	410	3,935
1300	With magnetic starter	5,575	1,500	7,075
1340	60 HP, 1800 RPM, motor only	4,475	470	4,945
1360	With magnetic starter	9,300	1,925	11,225
1400	75 HP, 1800 RPM, motor only	4,575	545	5,120
1420	With magnetic starter	9,400	2,000	11,400
1460	100 HP, 1800 RPM, motor only	5,875	730	6,605
1480	With magnetic starter	10,700	2,175	12,875
1520	125 HP, 1800 RPM, motor only	6,725	935	7,660
1540	With magnetic starter	27,600	2,575	30,175
1580	150 HP, 1800 RPM, motor only	8,725	1,100	9,825
1600	With magnetic starter	29,600	2,750	32,350
1640	200 HP, 1800 RPM, motor only	10,600	1,300	11,900
1660	With magnetic starter	31,500	2,950	34,450
1680	Totally encl, premium efficient, 1.0 ser. fac., 1HP, 1200 RPM, motor only	450	146	596
1700	With manual starter	680	335	1,015
1720	With magnetic starter	675	335	1,010
1740	1800 RPM, motor only	530	146	676
1760	With manual starter	760	335	1,095
1780	With magnetic starter	755	335	1,090
1800	2 HP, 1200 RPM, motor only	520	146	666
1820	With manual starter	750	335	1,085
1840	With magnetic starter	820	430	1,250
1860	1800 RPM, motor only	630	146	776
1880	With manual starter	860	335	1,195
1900	With magnetic starter	930	430	1,360
1920	3600 RPM, motor only	405	146	551
1940	With manual starter	635	335	970
1960	With magnetic starter	705	430	1,135
1980	3 HP, 1200 RPM, motor only	705	146	851
2000	With manual starter	935	335	1,270
2020	With magnetic starter	1,000	430	1,430
2040	1800 RPM, motor only	675	146	821
2060	With manual starter	905	335	1,240
2080	With magnetic starter	975	430	1,405
2100	3600 RPM, motor only	500	146	646
2120	With manual starter	730	335	1,065
2140	With magnetic starter	800	430	1,230
2160	5 HP, 1200 RPM, motor only	960	146	1,106
2180	With manual starter	1,225	475	1,700

D5020 Lighting and Branch Wiring

D5020 175	Motor & Starter	COST EACH		
		MAT.	INST.	TOTAL
2200	With magnetic starter	1,275	555	1,830
2220	1800 RPM, motor only	805	146	951
2240	With manual starter	1,075	475	1,550
2260	With magnetic starter	1,100	555	1,655
2280	3600 RPM, motor only	625	146	771
2300	With manual starter	895	475	1,370
2320	With magnetic starter	930	555	1,485
2340	7.5 HP, 1800 RPM, motor only	1,050	156	1,206
2360	With manual starter	1,325	485	1,810
2380	With magnetic starter	1,625	750	2,375
2400	10 HP, 1800 RPM, motor only	1,200	164	1,364
2420	With manual starter	1,475	495	1,970
2440	With magnetic starter	1,775	760	2,535
2460	15 HP, 1800 RPM, motor only	1,800	205	2,005
2480	With magnetic starter	2,375	800	3,175
2500	20 HP, 1800 RPM, motor only	2,425	252	2,677
2520	With magnetic starter	3,350	980	4,330
2540	25 HP, 1800 RPM, motor only	2,825	262	3,087
2560	With magnetic starter	3,750	990	4,740
2580	30 HP, 1800 RPM, motor only	2,975	273	3,248
2600	With magnetic starter	3,900	1,000	4,900
2620	40 HP, 1800 RPM, motor only	3,650	330	3,980
2640	With magnetic starter	5,700	1,425	7,125
2660	50 HP, 1800 RPM, motor only	3,950	410	4,360
2680	With magnetic starter	6,000	1,500	7,500
2700	60 HP, 1800 RPM, motor only	5,800	470	6,270
2720	With magnetic starter	10,600	1,925	12,525
2740	75 HP, 1800 RPM, motor only	7,075	545	7,620
2760	With magnetic starter	11,900	2,000	13,900
2780	100 HP, 1800 RPM, motor only	9,000	730	9,730
2800	With magnetic starter	13,800	2,175	15,975
2820	125 HP, 1800 RPM, motor only	12,600	935	13,535
2840	With magnetic starter	33,500	2,575	36,075
2860	150 HP, 1800 RPM, motor only	13,800	1,100	14,900
2880	With magnetic starter	34,700	2,750	37,450
2900	200 HP, 1800 RPM, motor only	17,500	1,300	18,800
2920	With magnetic starter	38,400	2,950	41,350

General: The cost of the lighting portion of the electrical costs is dependent upon:

1. The footcandle requirement of the proposed building.
2. The type of fixtures required.
3. The ceiling heights of the building.
4. Reflectance value of ceilings, walls and floors.
5. Fixture efficiencies and spacing vs. mounting height ratios.

Footcandle Requirements: See Table D5020-204 for Footcandle and Watts per S.F. determination.

Table D5020-201 IESNA* Recommended Illumination Levels in Footcandles

Commercial Buildings			Industrial Buildings		
Type	**Description**	**Footcandles**	**Type**	**Description**	**Footcandles**
Bank	Lobby	50	Assembly Areas	Rough bench & machine work	50
	Customer Areas	70		Medium bench & machine work	100
	Teller Stations	150		Fine bench & machine work	500
	Accounting Areas	150	Inspection Areas	Ordinary	50
Offices	Routine Work	100		Difficult	100
	Accounting	150		Highly Difficult	200
	Drafting	200	Material Handling	Loading	20
	Corridors, Halls, Washrooms	30		Stock Picking	30
Schools	Reading or Writing	70		Packing, Wrapping	50
	Drafting, Labs, Shops	100	Stairways	Service Areas	20
	Libraries	70	Washrooms	Service Areas	20
	Auditoriums, Assembly	15	Storage Areas	Inactive	5
	Auditoriums, Exhibition	30		Active, Rough, Bulky	10
Stores	Circulation Areas	30		Active, Medium	20
	Stock Rooms	30		Active, Fine	50
	Merchandise Areas, Service	100	Garages	Active Traffic Areas	20
	Self-Service Areas	200		Service & Repair	100

*IESNA - Illuminating Engineering Society of North America

Table D5020-202 General Lighting Loads by Occupancies

Type of Occupancy	Unit Load per S.F. (Watts)
Armories and Auditoriums	1
Banks	5
Barber Shops and Beauty Parlors	3
Churches	1
Clubs	2
Court Rooms	2
*Dwelling Units	3
Garages — Commercial (storage)	½
Hospitals	2
*Hotels and Motels, including apartment houses without provisions for cooking by tenants	2
Industrial Commercial (Loft) Buildings	2
Lodge Rooms	1½
Office Buildings	5
Restaurants	2
Schools	3
Stores	3
Warehouses (storage)	¼
*In any of the above occupancies except one-family dwellings and individual dwelling units of multi-family dwellings:	
Assembly Halls and Auditoriums	1
Halls, Corridors, Closets	½
Storage Spaces	¼

Table D5020-203 Lighting Limit (Connected Load) for Listed Occupancies: New Building Proposed Energy Conservation Guideline

Type of Use	Maximum Watts per S.F.
Interior	
Category A: Classrooms, office areas, automotive mechanical areas, museums, conference rooms, drafting rooms, clerical areas, laboratories, merchandising areas, kitchens, examining rooms, book stacks, athletic facilities.	3.00
Category B: Auditoriums, waiting areas, spectator areas, restrooms, dining areas, transportation terminals, working corridors in prisons and hospitals, book storage areas, active inventory storage, hospital bedrooms, hotel and motel bedrooms, enclosed shopping mall concourse areas, stairways.	1.00
Category C: Corridors, lobbies, elevators, inactive storage areas.	0.50
Category D: Indoor parking.	0.25
Exterior	
Category E: Building perimeter: wall-wash, facade, canopy.	5.00 (per linear foot)
Category F: Outdoor parking.	0.10

Table D5020-204 Procedure for Calculating Footcandles and Watts Per Square Foot

1. Initial footcandles = No. of fixtures × lamps per fixture × lumens per lamp × coefficient of utilization ÷ square feet
2. Maintained footcandles = initial footcandles × maintenance factor
3. Watts per square foot = No. of fixtures × lamps × (lamp watts + ballast watts) ÷ square feet

Example: To find footcandles and watts per S.F. for an office 20′ x 20′ with 11 fluorescent fixtures each having 4–40 watt C.W. lamps.

Based on good reflectance and clean conditions:

Lumens per lamp = 40 watt cool white at 3150 lumens per lamp RD5020-250, Table RD5020-251

Coefficient of utilization = .42 (varies from .62 for light colored areas to .27 for dark)

Maintenance factor = .75 (varies from .80 for clean areas with good maintenance to .50 for poor)

Ballast loss = 8 watts per lamp. (Varies with manufacturer. See manufacturers' catalog.)

1. Initial footcandles:

$$\frac{11 \times 4 \times 3150 \times .42}{400} = \frac{58{,}212}{400} = 145 \text{ footcandles}$$

2. Maintained footcandles:

   145 × .75 = 109 footcandles

3. Watts per S.F.

$$\frac{11 \times 4\ (40 + 8)}{400} = \frac{2{,}112}{400} = 5.3 \text{ watts per S.F.}$$

Table D5020-205 Approximate Watts Per Square Foot for Popular Fixture Types

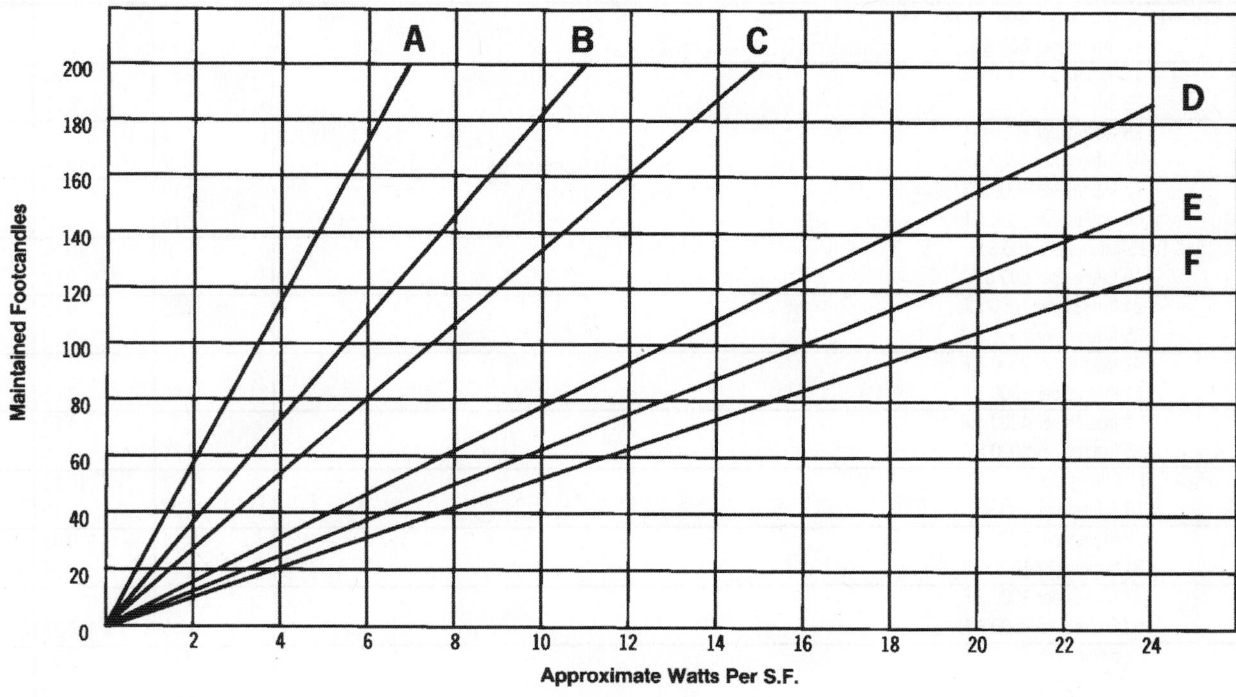

Due to the many variables involved, use for preliminary estimating only:

a. Fluorescent – industrial System D5020 208
b. Fluorescent – lens unit System D5020 208 Fixture types B & C
c. Fluorescent – louvered unit
d. Incandescent – open reflector System D5020 214, Type D
e. Incandescent – lens unit System D5020 214, Type A
f. Incandescent – down light System D5020 214, Type B

D5020 Lighting and Branch Wiring

A. Strip Fixture

B. Surface Mounted

C. Recessed

D. Pendent Mounted

Design Assumptions:

1. A 100 footcandle average maintained level of illumination.
2. Ceiling heights range from 9′ to 11′.
3. Average reflectance values are assumed for ceilings, walls and floors.
4. Cool white (CW) fluorescent lamps with 3150 lumens for 40 watt lamps and 6300 lumens for 8′ slimline lamps.
5. Four 40 watt lamps per 4′ fixture and two 8′ lamps per 8′ fixture.
6. Average fixture efficiency values and spacing to mounting height ratios.
7. Installation labor is average U.S. rate as of January 1.

System Components	QUANTITY	UNIT	COST PER S.F. MAT.	INST.	TOTAL
SYSTEM D5020 208 0520					
FLUORESCENT FIXTURES MOUNTED 9′-11″ ABOVE FLOOR, 100 FC					
TYPE A, 8 FIXTURES PER 400 S.F.					
Conduit, steel intermediate, 1/2″ diam.	.185	L.F.	.36	1.21	1.57
Wire, 600V, type THWN-THHN, copper, solid, #12	.004	C.L.F.	.05	.22	.27
Fluorescent strip fixture 8′ long, surface mounted, two 75W SL	.020	Ea.	1.27	2.12	3.39
Steel outlet box 4″ concrete	.020	Ea.	.27	.66	.93
Steel outlet box plate with stud, 4″ concrete	.020	Ea.	.18	.16	.34
Fixture hangers, flexible, 1/2″ diameter, 4″ long	.040	Ea.	.69	2.18	2.87
Fixture whip, THHN wire, three #12, 3/8″ Greenfield	.040	Ea.	.49	.82	1.31
TOTAL			3.31	7.37	10.68

D5020 208	Fluorescent Fixtures (by Type)	COST PER S.F. MAT.	INST.	TOTAL
0520	Fluorescent fixtures, type A, 8 fixtures per 400 S.F.	3.31	7.35	10.66
0560	11 fixtures per 600 S.F.	2.99	6.75	9.74
0600	17 fixtures per 1000 S.F.	2.80	6.25	9.05
0640	23 fixtures per 1600 S.F.	2.35	5.25	7.60
0680	28 fixtures per 2000 S.F.	2.31	5.15	7.46
0720	41 fixtures per 3000 S.F.	2.27	5.05	7.32
0800	53 fixtures per 4000 S.F.	2.19	4.90	7.09
0840	64 fixtures per 5000 S.F.	2.16	4.78	6.94
0880	Type B, 11 fixtures per 400 S.F.	5.25	10.50	15.75
0920	15 fixtures per 600 S.F.	4.71	9.50	14.21
0960	24 fixtures per 1000 S.F.	4.53	9.15	13.68
1000	35 fixtures per 1600 S.F.	4.15	8.30	12.45
1040	42 fixtures per 2000 S.F.	3.95	7.95	11.90
1080	61 fixtures per 3000 S.F.	3.80	7.65	11.45
1160	80 fixtures per 4000 S.F.	2.71	4.87	7.58
1200	98 fixtures per 5000 S.F.	3.74	7.45	11.19
1240	Type C, 11 fixtures per 400 S.F.	4.52	10.90	15.42
1280	14 fixtures per 600 S.F.	3.78	9.15	12.93
1320	23 fixtures per 1000 S.F.	3.74	9.05	12.79
1360	34 fixtures per 1600 S.F.	3.45	8.35	11.80
1400	43 fixtures per 2000 S.F.	3.56	8.55	12.11
1440	63 fixtures per 3000 S.F.	3.41	8.25	11.66
1520	81 fixtures per 4000 S.F.	3.29	7.95	11.24
1560	101 fixtures per 5000 S.F.	3.27	7.90	11.17
1600	Type D, 8 fixtures per 400 S.F.	4.20	8.25	12.45
1640	12 fixtures per 600 S.F.	4.20	8.25	12.45
1680	19 fixtures per 1000 S.F.	3.99	7.85	11.84
1720	27 fixtures per 1600 S.F.	3.56	7	10.56
1760	34 fixtures per 2000 S.F.	3.56	7	10.56
1800	48 fixtures per 3000 S.F.	3.36	6.60	9.96
1880	64 fixtures per 4000 S.F.	3.36	6.60	9.96
1920	79 fixtures per 5000 S.F.	3.36	6.55	9.91

D5020 Lighting and Branch Wiring

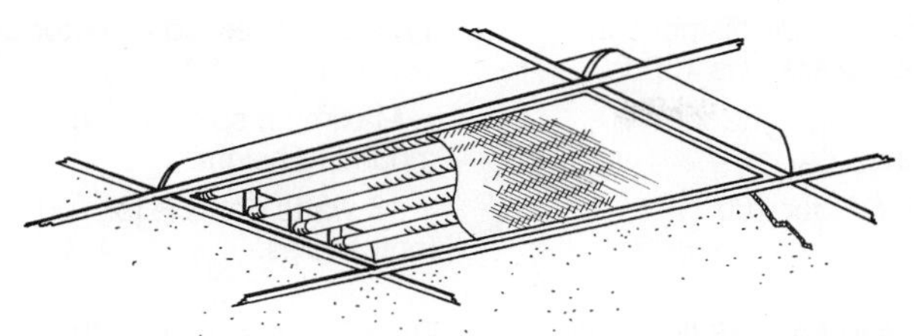

Type C. Recessed, mounted on grid ceiling suspension system, 2′ x 4′, four 40 watt lamps, acrylic prismatic diffusers.

5.3 watts per S.F. for 100 footcandles.
3 watts per S.F. for 57 footcandles.

System Components	QUANTITY	UNIT	COST PER S.F. MAT.	INST.	TOTAL
SYSTEM D5020 210 0200					
FLUORESCENT FIXTURES RECESS MOUNTED IN CEILING					
1 WATT PER S.F., 20 FC, 5 FIXTURES PER 1000 S.F.					
Steel intermediate conduit, (IMC) 1/2″ diam.	.128	L.F.	.25	.84	1.09
Wire, 600 volt, type THW, copper, solid, #12	.003	C.L.F.	.04	.18	.22
Fluorescent fixture, recessed, 2′x 4′, four 40W, w/lens, for grid ceiling	.005	Ea.	.32	.70	1.02
Steel outlet box 4″ square	.005	Ea.	.07	.17	.24
Fixture whip, Greenfield w/#12 THHN wire	.005	Ea.	.04	.04	.08
TOTAL			.72	1.93	2.65

D5020 210	Fluorescent Fixtures (by Wattage)	COST PER S.F. MAT.	INST.	TOTAL
0190	Fluorescent fixtures recess mounted in ceiling			
0195	T12, standard 40 watt lamps			
0200	1 watt per S.F., 20 FC, 5 fixtures @40 watts per 1000 S.F.	.72	1.93	2.65
0240	2 watt per S.F., 40 FC, 10 fixtures @40 watt per 1000 S.F.	1.44	3.78	5.22
0280	3 watt per S.F., 60 FC, 15 fixtures @40 watt per 1000 S.F	2.14	5.70	7.84
0320	4 watt per S.F., 80 FC, 20 fixtures @40 watt per 1000 S.F.	2.86	7.55	10.41
0400	5 watt per S.F., 100 FC, 25 fixtures @40 watt per 1000 S.F.	3.58	9.50	13.08
0450	T8, energy saver 32 watt lamps			
0500	0.8 watt per S.F., 20 FC, 5 fixtures @32 watt per 1000 S.F.	.79	1.93	2.72
0520	1.6 watt per S.F., 40 FC, 10 fixtures @32 watt per 1000 S.F.	1.58	3.78	5.36
0540	2.4 watt per S.F., 60 FC, 15 fixtures @ 32 watt per 1000 S.F	2.36	5.70	8.06
0560	3.2 watt per S.F., 80 FC, 20 fixtures @32 watt per 1000 S.F.	3.15	7.55	10.70
0580	4 watt per S.F., 100 FC, 25 fixtures @32 watt per 1000 S.F.	3.94	9.50	13.44

D5020 Lighting and Branch Wiring

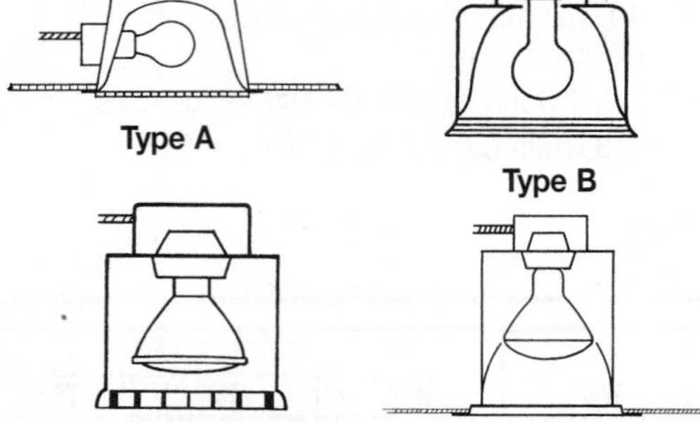

Type A. Recessed wide distribution reflector with flat glass lens 150 W.

Maximum spacing = 1.2 x mounting height.

13 watts per S.F. for 100 footcandles.

Type C. Recessed PAR–38 flood lamp with concentric louver 150 W.

Maximum spacing = 0.5 x mounting height.

19 watts per S.F. for 100 footcandles.

Type B. Recessed reflector down light with baffles 150 W.

Maximum spacing = 0.8 x mounting height.

18 watts per S.F. for 100 footcandles.

Type D. Recessed R–40 flood lamp with reflector skirt.

Maximum spacing = 0.7 x mounting height.

15 watts per S.F. for 100 footcandles.

System Components	QUANTITY	UNIT	COST PER S.F. MAT.	COST PER S.F. INST.	COST PER S.F. TOTAL
SYSTEM D5020 214 0400					
INCANDESCENT FIXTURE RECESS MOUNTED, 100 FC					
TYPE A, 34 FIXTURES PER 400 S.F.					
Steel intermediate conduit, (IMC) 1/2″ diam	1.060	L.F.	2.09	6.94	9.03
Wire, 600V, type THWN-THHN, copper, solid, #12	.033	C.L.F.	.43	1.96	2.39
Steel outlet box 4″ square	.085	Ea.	1.15	2.81	3.96
Fixture whip, Greenfield w/#12 THHN wire	.085	Ea.	.76	.70	1.46
Incandescent fixture, recessed, w/lens, prewired, square trim, 200W	.085	Ea.	9.10	8.33	17.43
TOTAL			13.53	20.74	34.27

D5020 214	Incandescent Fixture (by Type)	COST PER S.F. MAT.	COST PER S.F. INST.	COST PER S.F. TOTAL
0380	Incandescent fixture recess mounted, 100 FC			
0400	Type A, 34 fixtures per 400 S.F.	13.55	20.50	34.05
0440	49 fixtures per 600 S.F.	13.15	20.50	33.65
0480	63 fixtures per 800 S.F.	12.80	20	32.80
0520	90 fixtures per 1200 S.F.	12.35	19.75	32.10
0560	116 fixtures per 1600 S.F.	12.10	19.60	31.70
0600	143 fixtures per 2000 S.F.	12	19.50	31.50
0640	Type B, 47 fixtures per 400 S.F.	12.55	26	38.55
0680	66 fixtures per 600 S.F.	12.05	25.50	37.55
0720	88 fixtures per 800 S.F.	12.05	25.50	37.55
0760	127 fixtures per 1200 S.F.	11.80	25.50	37.30
0800	160 fixtures per 1600 S.F.	11.60	25	36.60
0840	206 fixtures per 2000 S.F.	11.55	25	36.55
0880	Type C, 51 fixtures per 400 S.F.	17.20	27.50	44.70
0920	74 fixtures per 600 S.F.	16.70	27	43.70
0960	97 fixtures per 800 S.F.	16.50	26.50	43
1000	142 fixtures per 1200 S.F.	16.25	26.50	42.75
1040	186 fixtures per 1600 S.F.	16.05	26.50	42.55
1080	230 fixtures per 2000 S.F.	15.95	26.50	42.45
1120	Type D, 39 fixtures per 400 S.F.	16.05	21.50	37.55
1160	57 fixtures per 600 S.F.	15.65	21.50	37.15
1200	75 fixtures per 800 S.F.	15.55	21.50	37.05
1240	109 fixtures per 1200 S.F.	15.20	21	36.20
1280	143 fixtures per 1600 S.F.	14.90	21	35.90
1320	176 fixtures per 2000 S.F.	14.80	20.50	35.30

D5020 Lighting and Branch Wiring

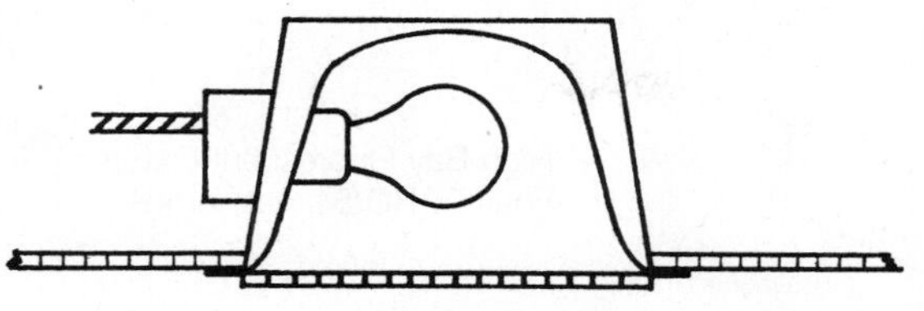

Type A. Recessed, wide distribution reflector with flat glass lens.

150 watt inside frost—2500 lumens per lamp.

PS–25 extended service lamp.

Maximum spacing = 1.2 x mounting height.

13 watts per S.F. for 100 footcandles.

System Components	QUANTITY	UNIT	COST PER S.F. MAT.	COST PER S.F. INST.	COST PER S.F. TOTAL
SYSTEM D5020 216 0200					
INCANDESCENT FIXTURE RECESS MOUNTED, TYPE A					
1 WATT PER S.F., 8 FC, 6 FIXT PER 1000 S.F.					
Steel intermediate conduit, (IMC) 1/2" diam	.091	L.F.	.18	.60	.78
Wire, 600V, type THWN-THHN, copper, solid, #12	.002	C.L.F.	.03	.12	.15
Incandescent fixture, recessed, w/lens, prewired, square trim, 200W	.006	Ea.	.64	.59	1.23
Steel outlet box 4" square	.006	Ea.	.08	.20	.28
Fixture whip, Greenfield w/#12 THHN wire	.006	Ea.	.05	.05	.10
TOTAL			.98	1.56	2.54

D5020 216	Incandescent Fixture (by Wattage)	COST PER S.F. MAT.	COST PER S.F. INST.	COST PER S.F. TOTAL
0190	Incandescent fixture recess mounted, type A			
0200	1 watt per S.F., 8 FC, 6 fixtures per 1000 S.F.	.98	1.56	2.54
0240	2 watt per S.F., 16 FC, 12 fixtures per 1000 S.F.	1.96	3.11	5.07
0280	3 watt per S.F., 24 FC, 18 fixtures, per 1000 S.F.	2.94	4.59	7.53
0320	4 watt per S.F., 32 FC, 24 fixtures per 1000 S.F.	3.91	6.15	10.06
0400	5 watt per S.F., 40 FC, 30 fixtures per 1000 S.F.	4.91	7.70	12.61

D5020 Lighting and Branch Wiring

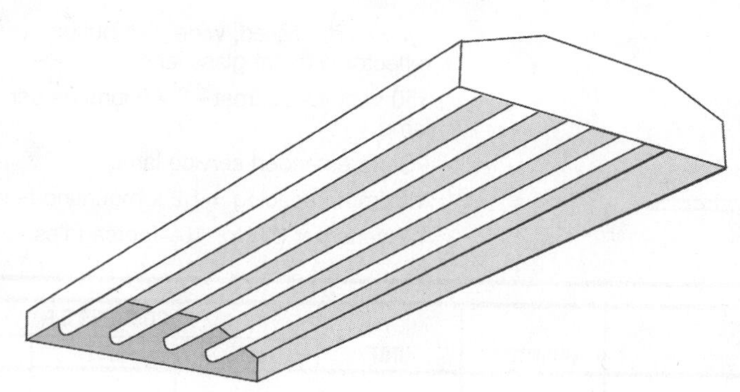

High Bay Fluorescent Fixtures
Four T5 HO/54 watt lamps

System Components

System Components	QUANTITY	UNIT	COST PER S.F. MAT.	COST PER S.F. INST.	COST PER S.F. TOTAL
SYSTEM D5020 218 0200					
FLUORESCENT HIGH BAY FIXTURE, 4 LAMP, 8'-10' ABOVE WORK PLANE					
0.5 WATT/S.F., 29 FC, 2 FIXTURES/1000 S.F.					
Steel intermediate conduit, (IMC) 1/2" diam	.100	L.F.	.20	.66	.86
Wire, 600V, type THWN-THHN, copper, solid, #10	.002	C.L.F.	.04	.13	.17
Steel outlet box 4" concrete	.002	Ea.	.03	.07	.10
Steel outlet box plate with stud, 4" concrete	.002	Ea.	.02	.02	.04
Flourescent , hi bay, 4 lamp, T-5 fixture	.002	Ea.	.50	.29	.79
TOTAL			.79	1.17	1.96

D5020 218 Fluorescent Fixture, High Bay, 8'-10' (by Wattage)

	D5020 218 Fluorescent Fixture, High Bay, 8'-10' (by Wattage)	COST PER S.F. MAT.	COST PER S.F. INST.	COST PER S.F. TOTAL
0190	Fluorescent high bay-4 lamp fixture, 8'-10' above work plane			
0200	.5 watt/SF, 29 FC, 2 fixtures per 1000 S.F.	.79	1.17	1.96
0400	1 watt/SF, 59 FC, 4 fixtures per 1000 S.F.	1.54	2.23	3.77
0600	1.5 watt/SF, 103 FC, 7 fixtures per 1000 S.F.	2.56	3.48	6.04
0800	2 watt/SF, 133 FC, 9 fixtures per 1000 S.F.	3.33	4.59	7.92
1000	2.5 watt/SF, 162 FC, 11 fixtures per 1000 S.F.	4.11	5.70	9.81

D5020 Lighting and Branch Wiring

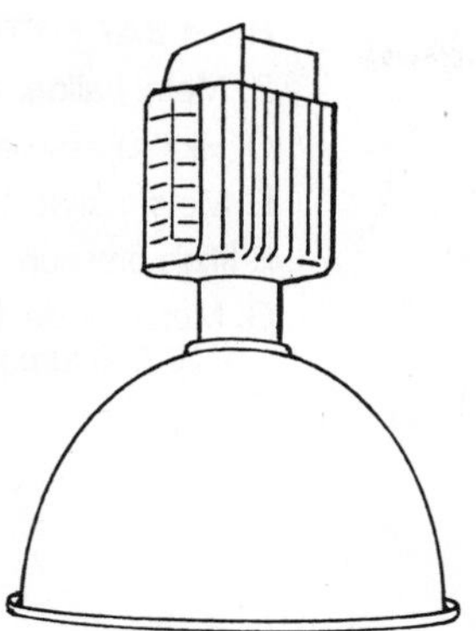

HIGH BAY FIXTURES
B. Metal halide 400 watt
C. High pressure sodium 400 watt
E. Metal halide 1000 watt
F. High pressure sodium 1000 watt
G. Metal halide 1000 watt
125,000 lumen lamp

System Components	QUANTITY	UNIT	COST PER S.F. MAT.	COST PER S.F. INST.	COST PER S.F. TOTAL
SYSTEM D5020 220 0880					
HIGH INTENSITY DISCHARGE FIXTURE, 8'-10' ABOVE WORK PLANE, 100 FC					
TYPE B, 8 FIXTURES PER 900 S.F.					
Steel intermediate conduit, (IMC) 1/2" diam	.460	L.F.	.91	3.01	3.92
Wire, 600V, type THWN-THHN, copper, solid, #10	.009	C.L.F.	.18	.59	.77
Steel outlet box 4" concrete	.009	Ea.	.12	.30	.42
Steel outlet box plate with stud 4" concrete	.009	Ea.	.08	.07	.15
Metal halide, hi bay, aluminum reflector, 400 W lamp	.009	Ea.	4.05	2.57	6.62
TOTAL			5.34	6.54	11.88

D5020 220	H.I.D. Fixture, High Bay, 8'-10' (by Type)	COST PER S.F. MAT.	COST PER S.F. INST.	COST PER S.F. TOTAL
0500	High intensity discharge fixture, 8'-10' above work plane, 100 FC			
0880	Type B, 8 fixtures per 900 S.F.	5.35	6.55	11.90
0920	15 fixtures per 1800 S.F.	4.90	6.30	11.20
0960	24 fixtures per 3000 S.F.	4.90	6.30	11.20
1000	31 fixtures per 4000 S.F.	4.89	6.30	11.19
1040	38 fixtures per 5000 S.F.	4.89	6.30	11.19
1080	60 fixtures per 8000 S.F.	4.89	6.30	11.19
1120	72 fixtures per 10000 S.F.	4.37	5.85	10.22
1160	115 fixtures per 16000 S.F.	4.37	5.85	10.22
1200	230 fixtures per 32000 S.F.	4.37	5.85	10.22
1240	Type C, 4 fixtures per 900 S.F.	2.56	3.99	6.55
1280	8 fixtures per 1800 S.F.	2.57	4.02	6.59
1320	13 fixtures per 3000 S.F.	2.56	3.99	6.55
1360	17 fixtures per 4000 S.F.	2.56	3.99	6.55
1400	21 fixtures per 5000 S.F.	2.46	3.66	6.12
1440	33 fixtures per 8000 S.F.	2.44	3.60	6.04
1480	40 fixtures per 10000 S.F.	2.38	3.40	5.78
1520	63 fixtures per 16000 S.F.	2.38	3.40	5.78
1560	126 fixtures per 32000 S.F.	2.38	3.40	5.78

D5020 Lighting and Branch Wiring

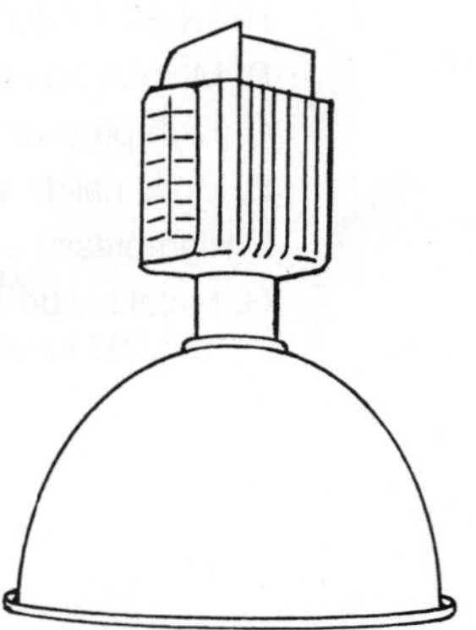

HIGH BAY FIXTURES

B. Metal halide 400 watt
C. High pressure sodium 400 watt
E. Metal halide 1000 watt
F. High pressure sodium 1000 watt
G. Metal halide 1000 watt
125,000 lumen lamp

System Components	QUANTITY	UNIT	COST PER S.F. MAT.	COST PER S.F. INST.	COST PER S.F. TOTAL
SYSTEM D5020 222 0240 HIGH INTENSITY DISCHARGE FIXTURE, 8′-10′ ABOVE WORK PLANE 1 WATT/S.F., TYPE B, 29 FC, 2 FIXTURES/1000 S.F.					
Steel intermediate conduit, (IMC) 1/2″ diam	.100	L.F.	.20	.66	.86
Wire, 600V, type THWN-THHN, copper, solid, #10	.002	C.L.F.	.04	.13	.17
Steel outlet box 4″ concrete	.002	Ea.	.03	.07	.10
Steel outlet box plate with stud, 4″ concrete	.002	Ea.	.02	.02	.04
Metal halide, hi bay, aluminum reflector, 400 W lamp	.002	Ea.	.90	.57	1.47
TOTAL			1.19	1.45	2.64

D5020 222	H.I.D. Fixture, High Bay, 8′-10′ (by Wattage)	COST PER S.F. MAT.	COST PER S.F. INST.	COST PER S.F. TOTAL
0190	High intensity discharge fixture, 8′-10′ above work plane			
0240	1 watt/S.F., type B, 29 FC, 2 fixtures/1000 S.F.	1.19	1.45	2.64
0280	Type C, 54 FC, 2 fixtures/1000 S.F.	1.01	1.10	2.11
0400	2 watt/S.F., type B, 59 FC, 4 fixtures/1000 S.F.	2.34	2.78	5.12
0440	Type C, 108 FC, 4 fixtures/1000 S.F.	2	2.12	4.12
0560	3 watt/S.F., type B, 103 FC, 7 fixtures/1000 S.F.	3.95	4.45	8.40
0600	Type C, 189 FC, 6 fixtures/1000 S.F.	3.20	2.75	5.95
0720	4 watt/S.F., type B, 133 FC, 9 fixtures/1000 S.F.	5.10	5.85	10.95
0760	Type C, 243 FC, 9 fixtures/1000 S.F.	4.48	4.73	9.21
0880	5 watt/S.F., type B, 162 FC, 11 fixtures/1000 S.F.	6.30	7.25	13.55
0920	Type C, 297 FC, 11 fixtures/1000 S.F.	5.50	5.85	11.35

D5020 Lighting and Branch Wiring

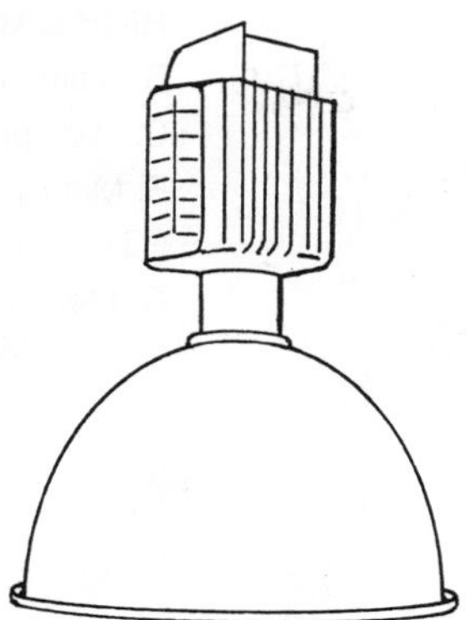

HIGH BAY FIXTURES
B. Metal halide 400 watt
C. High pressure sodium 400 watt
E. Metal halide 1000 watt
F. High pressure sodium 1000 watt
G. Metal halide 1000 watt
125,000 lumen lamp

System Components	QUANTITY	UNIT	COST PER S.F. MAT.	COST PER S.F. INST.	COST PER S.F. TOTAL
SYSTEM D5020 224 1240					
HIGH INTENSITY DISCHARGE FIXTURE, 16′ ABOVE WORK PLANE, 100 FC					
TYPE C, 5 FIXTURES PER 900 S.F.					
Steel intermediate conduit, (IMC) 1/2″ diam	.260	L.F.	.51	1.70	2.21
Wire, 600V, type THWN-THHN, copper, solid, #10	.007	C.L.F.	.14	.46	.60
Steel outlet box 4″ concrete	.006	Ea.	.08	.20	.28
Steel outlet box plate with stud, 4″ concrete	.006	Ea.	.05	.05	.10
High pressure sodium, hi bay, aluminum reflector, 400 W lamp	.006	Ea.	2.49	1.71	4.20
TOTAL			3.27	4.12	7.39

D5020 224	H.I.D. Fixture, High Bay, 16′ (by Type)	COST PER S.F. MAT.	COST PER S.F. INST.	COST PER S.F. TOTAL
0510	High intensity discharge fixture, 16′ above work plane, 100 FC			
1240	Type C, 5 fixtures per 900 S.F.	3.27	4.12	7.39
1280	9 fixtures per 1800 S.F.	3	4.32	7.32
1320	15 fixtures per 3000 S.F.	3	4.32	7.32
1360	18 fixtures per 4000 S.F.	2.86	3.86	6.72
1400	22 fixtures per 5000 S.F.	2.86	3.86	6.72
1440	36 fixtures per 8000 S.F.	2.86	3.86	6.72
1480	42 fixtures per 10,000 S.F.	2.56	3.98	6.54
1520	65 fixtures per 16,000 S.F.	2.56	3.98	6.54
1600	Type G, 4 fixtures per 900 S.F.	4.22	6.50	10.72
1640	6 fixtures per 1800 S.F.	3.53	6.10	9.63
1720	9 fixtures per 4000 S.F.	2.92	5.90	8.82
1760	11 fixtures per 5000 S.F.	2.92	5.90	8.82
1840	21 fixtures per 10,000 S.F.	2.74	5.35	8.09
1880	33 fixtures per 16,000 S.F.	2.74	5.35	8.09

D5020 Lighting and Branch Wiring

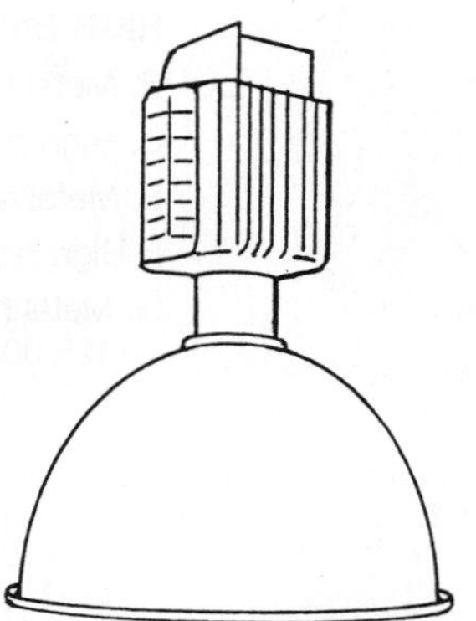

HIGH BAY FIXTURES

B. Metal halide 400 watt
C. High pressure sodium 400 watt
E. Metal halide 1000 watt
F. High pressure sodium 1000 watt
G. Metal halide 1000 watt
125,000 lumen lamp

System Components	QUANTITY	UNIT	COST PER S.F. MAT.	COST PER S.F. INST.	COST PER S.F. TOTAL
SYSTEM D5020 226 0240					
HIGH INTENSITY DISCHARGE FIXTURE, 16′ ABOVE WORK PLANE					
1 WATT/S.F., TYPE E, 42 FC, 1 FIXTURE/1000 S.F.					
Steel intermediate conduit, (IMC) 1/2″ diam	.160	L.F.	.32	1.05	1.37
Wire, 600V, type THWN-THHN, copper, solid, #10	.003	C.L.F.	.06	.20	.26
Steel outlet box 4″ concrete	.001	Ea.	.01	.03	.04
Steel outlet box plate with stud, 4″ concrete	.001	Ea.	.01	.01	.02
Metal halide, hi bay, aluminum reflector, 1000 W lamp	.001	Ea.	.65	.33	.98
TOTAL			1.05	1.62	2.67

D5020 226	H.I.D. Fixture, High Bay, 16′ (by Wattage)	COST PER S.F. MAT.	COST PER S.F. INST.	COST PER S.F. TOTAL
0190	High intensity discharge fixture, 16′ above work plane			
0240	1 watt/S.F., type E, 42 FC, 1 fixture/1000 S.F.	1.05	1.62	2.67
0280	Type G, 52 FC, 1 fixture/1000 S.F.	1.05	1.62	2.67
0320	Type C, 54 FC, 2 fixture/1000 S.F.	1.23	1.82	3.05
0440	2 watt/S.F., type E, 84 FC, 2 fixture/1000 S.F.	2.12	3.31	5.43
0480	Type G, 105 FC, 2 fixture/1000 S.F.	2.12	3.31	5.43
0520	Type C, 108 FC, 4 fixture/1000 S.F.	2.45	3.60	6.05
0640	3 watt/S.F., type E, 126 FC, 3 fixture/1000 S.F.	3.18	4.91	8.09
0680	Type G, 157 FC, 3 fixture/1000 S.F.	3.18	4.91	8.09
0720	Type C, 162 FC, 6 fixture/1000 S.F.	3.66	5.40	9.06
0840	4 watt/S.F., type E, 168 FC, 4 fixture/1000 S.F.	4.24	6.60	10.84
0880	Type G, 210 FC, 4 fixture/1000 S.F.	4.24	6.60	10.84
0920	Type C, 243 FC, 9 fixture/1000 S.F.	5.35	7.55	12.90
1040	5 watt/S.F., type E, 210 FC, 5 fixture/1000 S.F.	5.30	8.20	13.50
1080	Type G, 262 FC, 5 fixture/1000 S.F.	5.30	8.20	13.50
1120	Type C, 297 FC, 11 fixture/1000 S.F.	6.55	9.35	15.90

D5020 Lighting and Branch Wiring

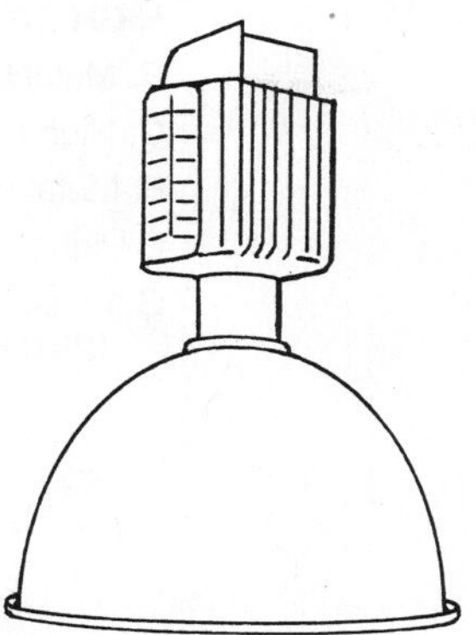

HIGH BAY FIXTURES
B. Metal halide 400 watt
C. High pressure sodium 400 watt
E. Metal halide 1000 watt
F. High pressure sodium 1000 watt
G. Metal halide 1000 watt
125,000 lumen lamp

System Components	QUANTITY	UNIT	COST PER S.F. MAT.	COST PER S.F. INST.	COST PER S.F. TOTAL
SYSTEM D5020 228 1240					
HIGH INTENSITY DISCHARGE FIXTURE, 20′ ABOVE WORK PLANE, 100 FC					
TYPE C, 6 FIXTURES PER 900 S.F.					
Steel intermediate conduit, (IMC) 1/2″ diam	.350	L.F.	.69	2.29	2.98
Wire, 600V, type THWN-THHN, copper, solid, #10.	.011	C.L.F.	.23	.72	.95
Steel outlet box 4″ concrete	.007	Ea.	.09	.23	.32
Steel outlet box plate with stud, 4″ concrete	.007	Ea.	.06	.06	.12
High pressure sodium, hi bay, aluminum reflector, 400 W lamp	.007	Ea.	2.91	2	4.91
TOTAL			3.98	5.30	9.28

D5020 228	H.I.D. Fixture, High Bay, 20′ (by Type)	COST PER S.F. MAT.	COST PER S.F. INST.	COST PER S.F. TOTAL
0510	High intensity discharge fixture 20′ above work plane, 100 FC			
1240	Type C, 6 fixtures per 900 S.F.	3.98	5.30	9.28
1280	10 fixtures per 1800 S.F.	3.54	4.97	8.51
1320	16 fixtures per 3000 S.F.	3.15	4.79	7.94
1360	20 fixtures per 4000 S.F.	3.15	4.79	7.94
1400	24 fixtures per 5000 S.F.	3.15	4.79	7.94
1440	38 fixtures per 8000 S.F.	3.09	4.59	7.68
1520	68 fixtures per 16000 S.F.	2.89	5.05	7.94
1560	132 fixtures per 32000 S.F.	2.89	5.05	7.94
1600	Type G, 4 fixtures per 900 S.F.	4.18	6.40	10.58
1640	6 fixtures per 1800 S.F.	3.53	6.10	9.63
1680	7 fixtures per 3000 S.F.	3.47	5.90	9.37
1720	10 fixtures per 4000 S.F.	3.45	5.80	9.25
1760	11 fixtures per 5000 S.F.	3.02	6.25	9.27
1800	18 fixtures per 8000 S.F.	3.02	6.25	9.27
1840	22 fixtures per 10000 S.F.	3.02	6.25	9.27
1880	34 fixtures per 16000 S.F.	2.96	6.05	9.01
1920	66 fixtures per 32000 S.F.	2.80	5.55	8.35

D5020 Lighting and Branch Wiring

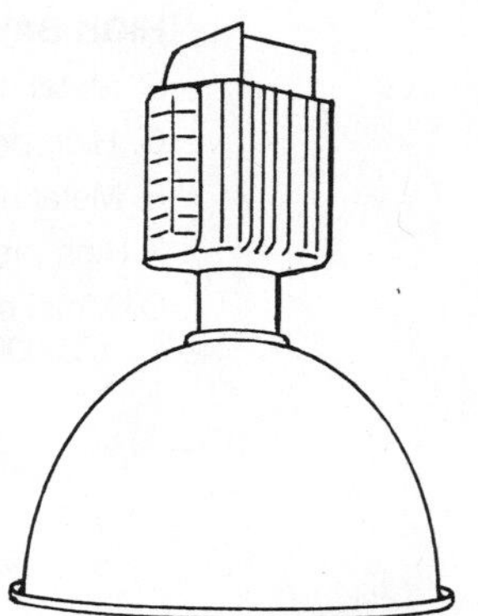

HIGH BAY FIXTURES

B. Metal halide 400 watt
C. High pressure sodium 400 watt
E. Metal halide 1000 watt
F. High pressure sodium 1000 watt
G. Metal halide 1000 watt
125,000 lumen lamp

System Components	QUANTITY	UNIT	COST PER S.F. MAT.	COST PER S.F. INST.	COST PER S.F. TOTAL
SYSTEM D5020 230 0240					
HIGH INTENSITY DISCHARGE FIXTURE, 20′ ABOVE WORK PLANE					
1 WATT/S.F., TYPE E, 40 FC, 1 FIXTURE 1000 S.F.					
Steel intermediate conduit, (IMC) 1/2″ diam	.160	L.F.	.32	1.05	1.37
Wire, 600V, type THWN-THHN, copper, solid, #10	.005	C.L.F.	.10	.33	.43
Steel outlet box 4″ concrete	.001	Ea.	.01	.03	.04
Steel outlet box plate with stud, 4″ concrete	.001	Ea.	.01	.01	.02
Metal halide, hi bay, aluminum reflector, 1000 W lamp	.001	Ea.	.65	.33	.98
TOTAL			1.09	1.75	2.84

D5020 230	H.I.D. Fixture, High Bay, 20′ (by Wattage)	COST PER S.F. MAT.	COST PER S.F. INST.	COST PER S.F. TOTAL
0190	High intensity discharge fixture, 20′ above work plane			
0240	1 watt/S.F., type E, 40 FC, 1 fixture/1000 S.F.	1.09	1.75	2.84
0280	Type G, 50 FC, 1 fixture/1000 S.F.	1.09	1.75	2.84
0320	Type C, 52 FC, 2 fixtures/1000 S.F.	1.29	2.02	3.31
0440	2 watt/S.F., type E, 81 FC, 2 fixtures/1000 S.F.	2.19	3.51	5.70
0480	Type G, 101 FC, 2 fixtures/1000 S.F.	2.19	3.51	5.70
0520	Type C, 104 FC, 4 fixtures/1000 S.F.	2.55	3.96	6.51
0640	3 watt/S.F., type E, 121 FC, 3 fixtures/1000 S.F.	3.28	5.25	8.53
0680	Type G, 151 FC, 3 fixtures/1000 S.F.	3.28	5.25	8.53
0720	Type C, 155 FC, 6 fixtures/1000 S.F.	3.84	6	9.84
0840	4 watt/S.F., type E, 161 FC, 4 fixtures/1000 S.F.	4.36	7	11.36
0880	Type G, 202 FC, 4 fixtures/1000 S.F.	4.36	7	11.36
0920	Type C, 233 FC, 9 fixtures/1000 S.F.	5.55	8.25	13.80
1040	5 watt/S.F., type E, 202 FC, 5 fixtures/1000 S.F.	5.45	8.75	14.20
1080	Type G, 252 FC, 5 fixtures/1000 S.F.	5.45	8.75	14.20
1120	Type C, 285 FC, 11 fixtures/1000 S.F.	6.85	10.25	17.10

D5020 Lighting and Branch Wiring

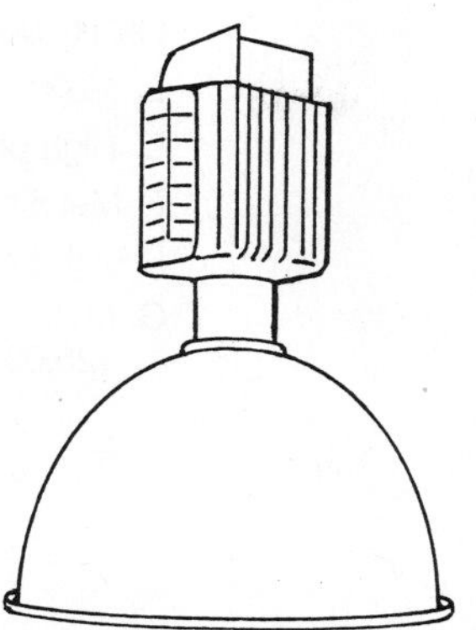

HIGH BAY FIXTURES
B. Metal halide 400 watt
C. High pressure sodium 400 watt
E. Metal halide 1000 watt
F. High pressure sodium 1000 watt
G. Metal halide 1000 watt
125,000 lumen lamp

System Components	QUANTITY	UNIT	COST PER S.F. MAT.	COST PER S.F. INST.	COST PER S.F. TOTAL
SYSTEM D5020 232 1240					
HIGH INTENSITY DISCHARGE FIXTURE, 30′ ABOVE WORK PLANE, 100 FC					
TYPE F, 4 FIXTURES PER 900 S.F.					
Steel intermediate conduit, (IMC) 1/2″ diam	.580	L.F.	1.14	3.80	4.94
Wire, 600V, type THWN-THHN, copper, solid, #10	.018	C.L.F.	.37	1.18	1.55
Steel outlet box 4″ concrete	.004	Ea.	.05	.13	.18
Steel outlet box plate with stud, 4″ concrete	.004	Ea.	.04	.03	.07
High pressure sodium, hi bay, aluminum, 1000 W lamp	.004	Ea.	2.40	1.32	3.72
TOTAL			4	6.46	10.46

D5020 232	H.I.D. Fixture, High Bay, 30′ (by Type)	COST PER S.F. MAT.	COST PER S.F. INST.	COST PER S.F. TOTAL
0510	High intensity discharge fixture, 30′ above work plane, 100 FC			
1240	Type F, 4 fixtures per 900 S.F.	4	6.45	10.45
1280	6 fixtures per 1800 S.F.	3.40	6.15	9.55
1320	8 fixtures per 3000 S.F.	3.40	6.15	9.55
1360	9 fixtures per 4000 S.F.	2.80	5.85	8.65
1400	10 fixtures per 5000 S.F.	2.80	5.85	8.65
1440	17 fixtures per 8000 S.F.	2.80	5.85	8.65
1480	18 fixtures per 10,000 S.F.	2.66	5.40	8.06
1520	27 fixtures per 16,000 S.F.	2.66	5.40	8.06
1560	52 fixtures per 32000 S.F.	2.64	5.35	7.99
1600	Type G, 4 fixtures per 900 S.F.	4.32	6.85	11.17
1640	6 fixtures per 1800 S.F.	3.57	6.20	9.77
1680	9 fixtures per 3000 S.F.	3.51	6	9.51
1720	11 fixtures per 4000 S.F.	3.45	5.80	9.25
1760	13 fixtures per 5000 S.F.	3.45	5.80	9.25
1800	21 fixtures per 8000 S.F.	3.45	5.80	9.25
1840	23 fixtures per 10,000 S.F.	3	6.20	9.20
1880	36 fixtures per 16,000 S.F.	3	6.20	9.20
1920	70 fixtures per 32,000 S.F.	3	6.20	9.20

D5020 Lighting and Branch Wiring

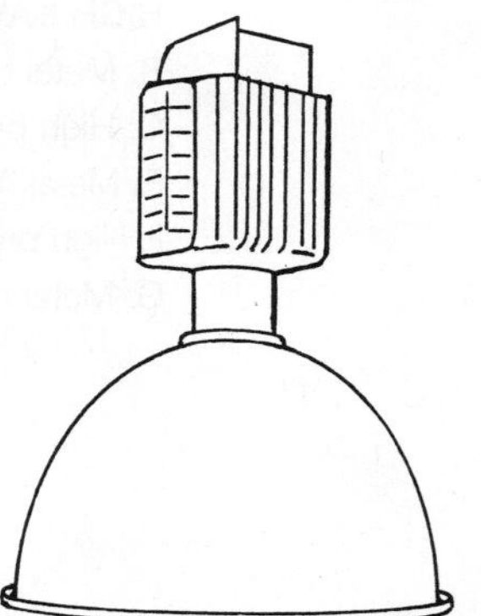

HIGH BAY FIXTURES
B. Metal halide 400 watt
C. High pressure sodium 400 watt
E. Metal halide 1000 watt
F. High pressure sodium 1000 watt
G. Metal halide 1000 watt
125,000 lumen lamp

System Components	QUANTITY	UNIT	COST PER S.F. MAT.	COST PER S.F. INST.	COST PER S.F. TOTAL
SYSTEM D5020 234 0240					
HIGH INTENSITY DISCHARGE FIXTURE, 30′ ABOVE WORK PLANE					
1 WATT/S.F., TYPE E, 37 FC, 1 FIXTURE/1000 S.F.					
Steel intermediate conduit, (IMC) 1/2″ diam	.196	L.F.	.39	1.28	1.67
Wire, 600V type THWN-THHN, copper, solid, #10	.006	C.L.F.	.12	.39	.51
Steel outlet box 4″ concrete	.001	Ea.	.01	.03	.04
Steel outlet box plate with stud, 4″ concrete	.001	Ea.	.01	.01	.02
Metal halide, hi bay, aluminum reflector, 1000 W lamp	.001	Ea.	.65	.33	.98
TOTAL			1.18	2.04	3.22

D5020 234	H.I.D. Fixture, High Bay, 30′ (by Wattage)	COST PER S.F. MAT.	COST PER S.F. INST.	COST PER S.F. TOTAL
0190	High intensity discharge fixture, 30′ above work plane			
0240	1 watt/S.F., type E, 37 FC, 1 fixture/1000 S.F.	1.18	2.04	3.22
0280	Type G, 45 FC., 1 fixture/1000 S.F.	1.18	2.04	3.22
0320	Type F, 50 FC, 1 fixture/1000 S.F.	.99	1.60	2.59
0440	2 watt/S.F., type E, 74 FC, 2 fixtures/1000 S.F.	2.37	4.11	6.48
0480	Type G, 92 FC, 2 fixtures/1000 S.F.	2.37	4.11	6.48
0520	Type F, 100 FC, 2 fixtures/1000 S.F.	2.01	3.28	5.29
0640	3 watt/S.F., type E, 110 FC, 3 fixtures/1000 S.F.	3.57	6.20	9.77
0680	Type G, 138 FC, 3 fixtures/1000 S.F.	3.57	6.20	9.77
0720	Type F, 150 FC, 3 fixtures/1000 S.F.	3.01	4.87	7.88
0840	4 watt/S.F., type E, 148 FC, 4 fixtures/1000 S.F.	4.74	8.25	12.99
0880	Type G, 185 FC, 4 fixtures/1000 S.F.	4.74	8.25	12.99
0920	Type F, 200 FC, 4 fixtures/1000 S.F.	4.03	6.55	10.58
1040	5 watt/S.F., type E, 185 FC, 5 fixtures/1000 S.F.	5.95	10.30	16.25
1080	Type G, 230 FC, 5 fixtures/1000 S.F.	5.95	10.30	16.25
1120	Type F, 250 FC, 5 fixtures/1000 S.F.	5	8.15	13.15

D5020 Lighting and Branch Wiring

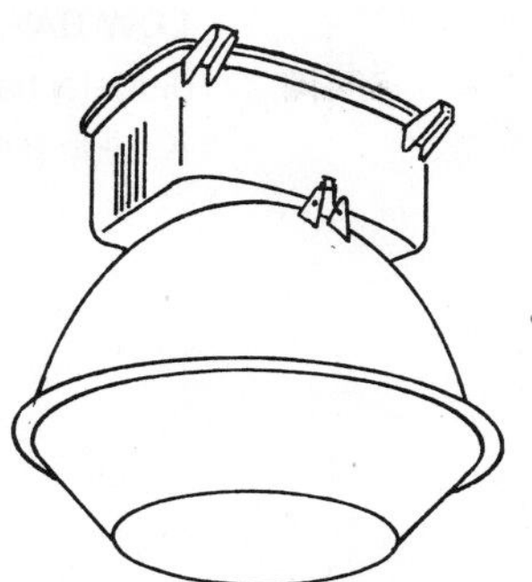

LOW BAY FIXTURES
J. Metal halide 250 watt
K. High pressure sodium 150 watt

System Components	QUANTITY	UNIT	COST PER S.F. MAT.	COST PER S.F. INST.	COST PER S.F. TOTAL
SYSTEM D5020 236 0920					
HIGH INTENSITY DISCHARGE FIXTURE, 8′-10′ ABOVE WORK PLANE, 50 FC					
TYPE J, 13 FIXTURES PER 1800 S.F.					
Steel intermediate conduit, (IMC) 1/2″ diam	.550	L.F.	1.08	3.60	4.68
Wire, 600V, type THWN-THHN, copper, solid, #10	.012	C.L.F.	.25	.79	1.04
Steel outlet box 4″ concrete	.007	Ea.	.09	.23	.32
Steel outlet box plate with stud, 4″ concrete	.007	Ea.	.06	.06	.12
Metal halide, lo bay, aluminum reflector, 250 W DX lamp	.007	Ea.	2.80	1.44	4.24
TOTAL			4.28	6.12	10.40

D5020 236	H.I.D. Fixture, Low Bay, 8′-10′ (by Type)	COST PER S.F. MAT.	COST PER S.F. INST.	COST PER S.F. TOTAL
0510	High intensity discharge fixture, 8′-10′ above work plane, 50 FC			
0880	Type J, 7 fixtures per 900 S.F.	4.65	6.15	10.80
0920	13 fixtures per 1800 S.F.	4.28	6.10	10.38
0960	21 fixtures per 3000 S.F.	4.28	6.10	10.38
1000	28 fixtures per 4000 S.F.	4.28	6.10	10.38
1040	35 fixtures per 5000 S.F.	4.28	6.10	10.38
1120	62 fixtures per 10,000 S.F.	3.86	5.85	9.71
1160	99 fixtures per 16,000 S.F.	3.86	5.85	9.71
1200	199 fixtures per 32,000 S.F.	3.86	5.85	9.71
1240	Type K, 9 fixtures per 900 S.F.	4.68	5.25	9.93
1280	16 fixtures per 1800 S.F.	4.31	5.10	9.41
1320	26 fixtures per 3000 S.F.	4.29	5.05	9.34
1360	31 fixtures per 4000 S.F.	3.95	4.91	8.86
1400	39 fixtures per 5000 S.F.	3.95	4.91	8.86
1440	62 fixtures per 8000 S.F.	3.95	4.91	8.86
1480	78 fixtures per 10,000 S.F.	3.95	4.91	8.86
1520	124 fixtures per 16,000 S.F.	3.91	4.79	8.70
1560	248 fixtures per 32,000 S.F.	3.76	5.35	9.11

D5020 Lighting and Branch Wiring

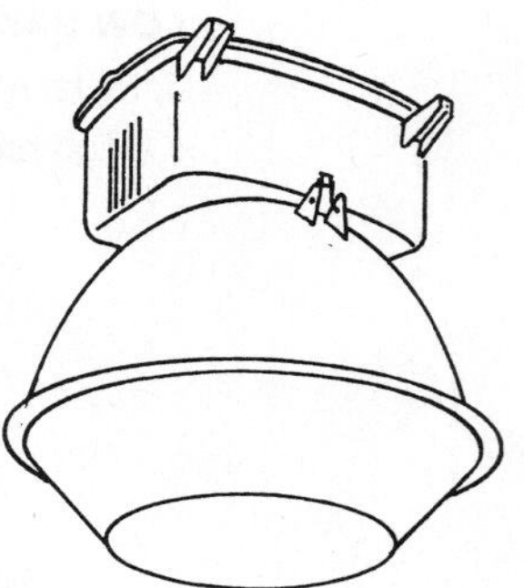

LOW BAY FIXTURES

J. Metal halide 250 watt

K. High pressure sodium 150 watt

System Components	QUANTITY	UNIT	COST PER S.F. MAT.	COST PER S.F. INST.	COST PER S.F. TOTAL
SYSTEM D5020 238 0240					
HIGH INTENSITY DISCHARGE FIXTURE, 8′-10′ ABOVE WORK PLANE					
1 WATT/S.F., TYPE J, 30 FC, 4 FIXTURES/1000 S.F.					
Steel intermediate conduit, (IMC) 1/2″ diam	.280	L.F.	.55	1.83	2.38
Wire, 600V, type THWN-THHN, copper, solid, #10	.008	C.L.F.	.16	.52	.68
Steel outlet box 4″ concrete	.004	Ea.	.05	.13	.18
Steel outlet box plate with stud, 4″ concrete	.004	Ea.	.04	.03	.07
Metal halide, lo bay, aluminum reflector, 250 W DX lamp	.004	Ea.	1.60	.82	2.42
TOTAL			2.40	3.33	5.73

D5020 238	H.I.D. Fixture, Low Bay, 8′-10′ (by Wattage)	COST PER S.F. MAT.	COST PER S.F. INST.	COST PER S.F. TOTAL
0190	High intensity discharge fixture, 8′-10′ above work plane			
0240	1 watt/S.F., type J, 30 FC, 4 fixtures/1000 S.F.	2.40	3.33	5.73
0280	Type K, 29 FC, 5 fixtures/1000 S.F.	2.44	3.01	5.45
0400	2 watt/S.F., type J, 52 FC, 7 fixtures/1000 S.F.	4.29	6.15	10.44
0440	Type K, 63 FC, 11 fixtures/1000 S.F.	5.30	6.30	11.60
0560	3 watt/S.F., type J, 81 FC, 11 fixtures/1000 S.F.	6.65	9.25	15.90
0600	Type K, 92 FC, 16 fixtures/1000 S.F.	7.75	9.30	17.05
0720	4 watt/S.F., type J, 103 FC, 14 fixtures/1000 S.F.	8.55	12.20	20.75
0760	Type K, 127 FC, 22 fixtures/1000 S.F.	10.60	12.60	23.20
0880	5 watt/S.F., type J, 133 FC, 18 fixtures/1000 S.F.	10.90	15.35	26.25
0920	Type K, 155 FC, 27 fixtures/1000 S.F.	13.05	15.65	28.70

D5020 Lighting and Branch Wiring

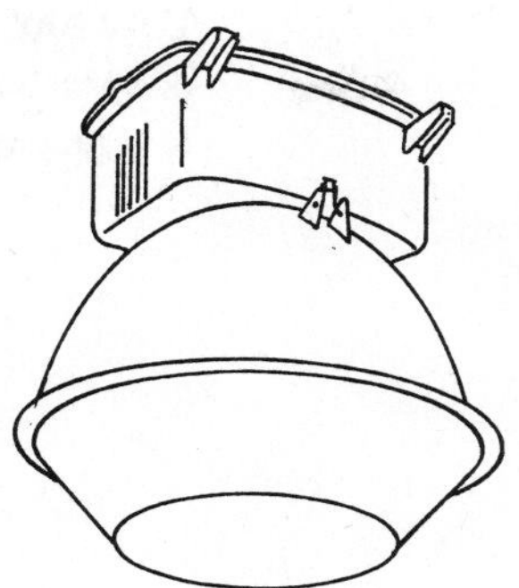

LOW BAY FIXTURES

J. Metal halide 250 watt

K. High pressure sodium 150 watt

System Components	QUANTITY	UNIT	COST PER S.F. MAT.	COST PER S.F. INST.	COST PER S.F. TOTAL
SYSTEM D5020 240 0880					
HIGH INTENSITY DISCHARGE FIXTURE, 16′ ABOVE WORK PLANE, 50 FC					
TYPE J, 9 FIXTURES PER 900 S.F.					
Steel intermediate conduit, (IMC) 1/2″ diam	.630	L.F.	1.24	4.13	5.37
Wire, 600V type, THWN-THHN, copper, solid, #10	.012	C.L.F.	.25	.79	1.04
Steel outlet box 4″ concrete	.010	Ea.	.14	.33	.47
Steel outlet box plate with stud, 4″ concrete	.010	Ea.	.09	.08	.17
Metal halide, lo bay, aluminum reflector, 250 W DX lamp	.010	Ea.	4	2.05	6.05
TOTAL			5.72	7.38	13.10

D5020 240	H.I.D. Fixture, Low Bay, 16′ (by Type)	COST PER S.F. MAT.	COST PER S.F. INST.	COST PER S.F. TOTAL
0510	High intensity discharge fixture, 16′ above work plane, 50 FC			
0880	Type J, 9 fixtures per 900 S.F.	5.70	7.40	13.10
0920	14 fixtures per 1800 S.F.	4.89	6.95	11.84
0960	24 fixtures per 3000 S.F.	4.93	7.10	12.03
1000	32 fixtures per 4000 S.F.	4.93	7.10	12.03
1040	35 fixtures per 5000 S.F.	4.52	6.90	11.42
1080	56 fixtures per 8000 S.F.	4.52	6.90	11.42
1120	70 fixtures per 10,000 S.F.	4.52	6.90	11.42
1160	111 fixtures per 16,000 S.F.	4.52	6.90	11.42
1200	222 fixtures per 32,000 S.F.	4.52	6.90	11.42
1240	Type K, 11 fixtures per 900 S.F.	5.80	6.90	12.70
1280	20 fixtures per 1800 S.F.	5.30	6.40	11.70
1320	29 fixtures per 3000 S.F.	4.99	6.25	11.24
1360	39 fixtures per 4000 S.F.	4.97	6.20	11.17
1400	44 fixtures per 5000 S.F.	4.62	6.10	10.72
1440	62 fixtures per 8000 S.F.	4.62	6.10	10.72
1480	87 fixtures per 10,000 S.F.	4.62	6.10	10.72
1520	138 fixtures per 16,000 S.F.	4.62	6.10	10.72

D5020 Lighting and Branch Wiring

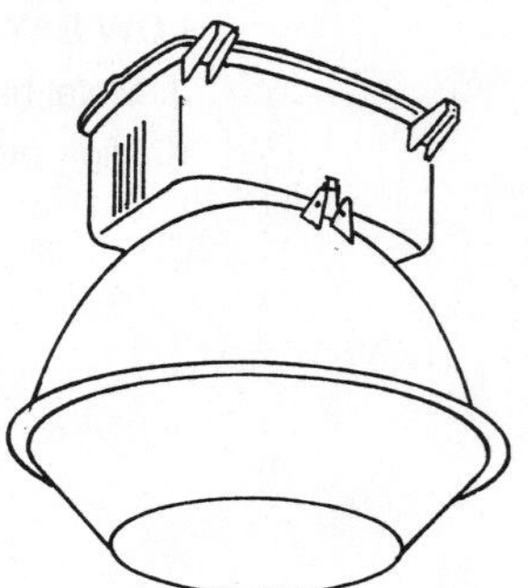

LOW BAY FIXTURES

J. Metal halide 250 watt

K. High pressure sodium 150 watt

System Components	QUANTITY	UNIT	COST PER S.F. MAT.	COST PER S.F. INST.	COST PER S.F. TOTAL
SYSTEM D5020 242 0240					
HIGH INTENSITY DISCHARGE FIXTURE, 16′ ABOVE WORK PLANE					
1 WATT/S.F., TYPE J, 28 FC, 4 FIXTURES/1000 S.F.					
Steel intermediate conduit, (IMC) 1/2″ diam	.328	L.F.	.65	2.15	2.80
Wire, 600V, type THWN-THHN, copper, solid, #10	.010	C.L.F.	.21	.66	.87
Steel outlet box 4″ concrete	.004	Ea.	.05	.13	.18
Steel outlet box plate with stud, 4″ concrete	.004	Ea.	.04	.03	.07
Metal halide, lo bay, aluminum reflector, 250 W DX lamp	.004	Ea.	1.60	.82	2.42
TOTAL			2.55	3.79	6.34

D5020 242	H.I.D. Fixture, Low Bay, 16′ (by Wattage)	COST PER S.F. MAT.	COST PER S.F. INST.	COST PER S.F. TOTAL
0190	High intensity discharge fixture, mounted 16′ above work plane			
0240	1 watt/S.F., type J, 28 FC, 4 fixt./1000 S.F.	2.55	3.79	6.34
0280	Type K, 27 FC, 5 fixt./1000 S.F.	2.83	4.25	7.08
0400	2 watt/S.F., type J, 48 FC, 7 fixt/1000 S.F.	4.65	7.35	12
0440	Type K, 58 FC, 11 fixt/1000 S.F.	6	8.70	14.70
0560	3 watt/S.F., type J, 75 FC, 11 fixt/1000 S.F.	7.20	11.15	18.35
0600	Type K, 85 FC, 16 fixt/1000 S.F.	8.85	12.90	21.75
0720	4 watt/S.F., type J, 95 FC, 14 fixt/1000 S.F.	9.30	14.65	23.95
0760	Type K, 117 FC, 22 fixt/1000 S.F.	12.05	17.35	29.40
0880	5 watt/S.F., type J, 122 FC, 18 fixt/1000 S.F.	11.85	18.45	30.30
0920	Type K, 143 FC, 27 fixt/1000 S.F.	14.85	21.50	36.35

D5020 Lighting and Branch Wiring

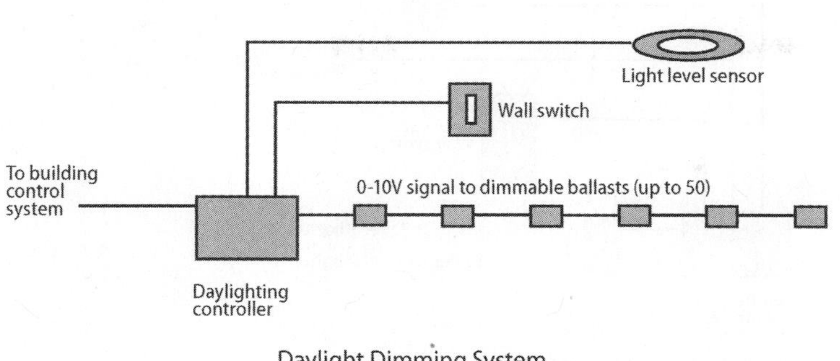

Daylight Dimming System

System Components	QUANTITY	UNIT	COST PER S.F. MAT.	COST PER S.F. INST.	COST PER S.F. TOTAL
SYSTEM D5020 290 0800					
DAYLIGHT DIMMING CONTROL SYSTEM					
5 FIXTURES PER 1000 M.S.F.					
Tray cable, type TC, copper #16-4 conductor	.300	C.L.F.	.01	.03	.04
Wire, 600 volt, type THWN-THHN, copper, solid, #12	.320	C.L.F.		.02	.02
Conduit (EMT) , to 15′ H, incl 2 termn,2 elb&11 bm clp per 100′, 3/4″	10.000	L.F.	.01	.05	.06
Cabinet, hinged, steel, NEMA 1, 12″W x 12″H x 4″D	.200	Ea.	.01	.02	.03
Lighting control module	.200	Ea.	.08	.07	.15
Dimmable ballast three-lamp	5.000	Ea.	.68	.48	1.16
Daylight level sensor, wall mounted, on/off or dimming	.200	Ea.	.03	.02	.05
Automatic wall switches	.200	Ea.	.01	.01	.02
Remote power pack	.100	Ea.		.01	.01
TOTAL			.83	.71	1.54

D5020 290	Daylight Dimming System	COST PER S.F. MAT.	COST PER S.F. INST.	COST PER S.F. TOTAL
0500	Daylight Dimming System (no fixtures or fixture power)			
0800	5 fixtures per 1000 S.F.	.83	.71	1.54
1000	10 fixtures per 1000 S.F.	1.67	1.29	2.96
2000	12 fixtures per 1000 S.F.	1.98	1.55	3.53
3000	15 fixtures per 1000 S.F.	2.48	1.94	4.42
4000	20 fixtures per 1000 S.F.	3.30	2.55	5.85
5000	25 fixtures per 1000 S.F.	4.13	3.17	7.30
6000	50 fixtures per 1000 S.F.	8.25	6.25	14.50

D5020 Lighting and Branch Wiring

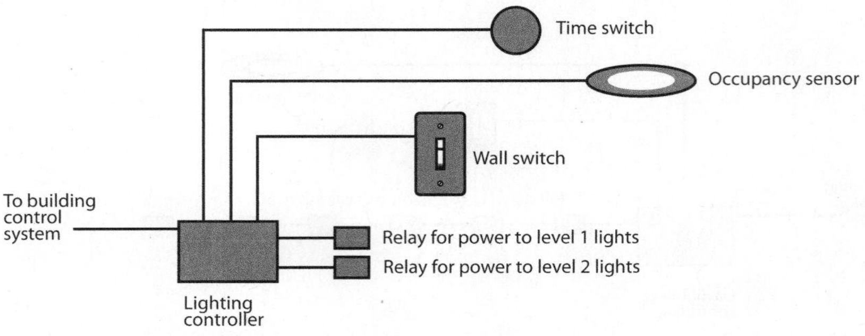

Time switch
Occupancy sensor
Wall switch
To building control system
Relay for power to level 1 lights
Relay for power to level 2 lights
Lighting controller

Lighting Control System

System Components	QUANTITY	UNIT	COST PER S.F. MAT.	COST PER S.F. INST.	COST PER S.F. TOTAL
SYSTEM D5020 295 0800					
LIGHTING ON/OFF CONTROL SYSTEM					
5 FIXTURES PER 1000 S.F.					
Tray cable, type TC, copper #16-4 conductor	.600	C.L.F.	.03	.05	.08
Wire, 600 volt, type THWN-THHN, copper, solid, #12	.320	C.L.F.		.02	.02
Conduit (EMT) , to 15′ H, incl 2 termn,2 elb&11 bm clp per 100′, 3/4″	10.000	L.F.	.01	.05	.06
Cabinet, hinged, steel, NEMA 1, 12″W x 12″H x 4″D	.400	Ea.	.01	.04	.05
Relays, 120 V or 277 V standard	.800	Ea.	.04	.04	.08
24 hour dial with reserve power	.400	Ea.	.27	.07	.34
Lighting control module	.200	Ea.	.08	.07	.15
Occupancy sensors, passive infrared ceiling mounted	.400	Ea.	.04	.04	.08
Automatic wall switches	.400	Ea.	.03	.01	.04
Remote power pack	.200	Ea.	.01	.01	.02
TOTAL			.52	.40	.92

D5020 295	Lighting On/Off Control System	COST PER S.F. MAT.	COST PER S.F. INST.	COST PER S.F. TOTAL
0500	Includes occupancy and time switching (no fixtures or fixture power)			
0800	5 fixtures per 1000 SF	.52	.40	.92
1000	10 fixtures per 1000 S.F.	.60	.46	1.06
2000	12 fixtures per 1000 S.F.	.68	.55	1.23
3000	15 fixtures per 1000 S.F.	.87	.70	1.57
4000	20 fixtures per 1000 S.F.	1.14	.89	2.03
5000	25 fixtures per 1000 S.F.	1.44	1.09	2.53
6000	50 fixtures per 1000 S.F.	2.86	2.12	4.98

D5030 Communications and Security

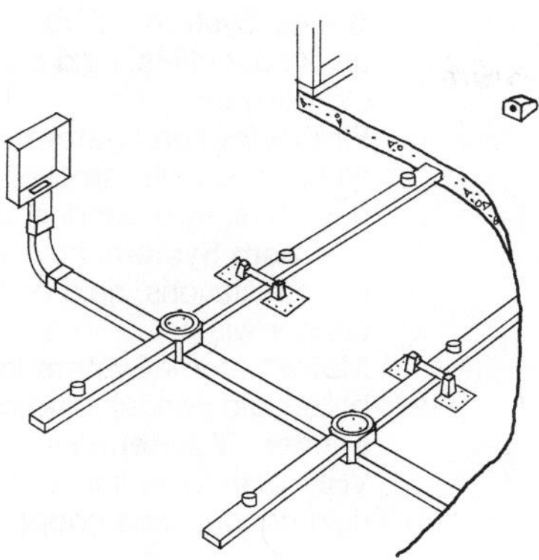

Description: System below includes telephone fitting installed. Does not include cable.

When poke thru fittings and telepoles are used for power, they can also be used for telephones at a negligible additional cost.

System Components	QUANTITY	UNIT	COST PER S.F. MAT.	COST PER S.F. INST.	COST PER S.F. TOTAL
SYSTEM D5030 310 0200					
TELEPHONE SYSTEMS, UNDERFLOOR DUCT, 5′ ON CENTER, LOW DENSITY					
Underfloor duct 7-1/4″ w/insert 2′ O.C. 1-3/8″ x 7-1/4″ super duct	.190	L.F.	6.94	2.49	9.43
Vertical elbow for underfloor superduct, 7-1/4″, included					
Underfloor duct conduit adapter, 2″ x 1-1/4″, included					
Underfloor duct junction box, single duct, 7-1/4″ x 3 1/8″	.003	Ea.	1.65	.49	2.14
Underfloor junction box carpet pan	.003	Ea.	1.17	.02	1.19
Underfloor duct outlet, low tension	.004	Ea.	.44	.33	.77
TOTAL			10.20	3.33	13.53

D5030 310	Telephone Systems	COST PER S.F. MAT.	COST PER S.F. INST.	COST PER S.F. TOTAL
0200	Telephone systems, underfloor duct, 5′ on center, low density	10.20	3.33	13.53
0240	5′ on center, high density	10.65	3.66	14.31
0280	7′ on center, low density	8.15	2.78	10.93
0320	7′ on center, high density	8.60	3.11	11.71
0400	Poke thru fittings, low density	1.18	1.19	2.37
0440	High density	2.36	2.36	4.72
0520	Telepoles, low density	1.15	.71	1.86
0560	High density	2.30	1.42	3.72
0640	Conduit system with floor boxes, low density	1.34	1.18	2.52
0680	High density	2.68	2.31	4.99
1020	Telephone wiring for offices & laboratories, 8 jacks/M.S.F.	.39	1.80	2.19

D5030 810	Security & Detection Systems	COST EACH MAT.	COST EACH INST.	COST EACH TOTAL
0200	Security system, head end equipment	29,500	4,375	33,875
0240	Security system, door/window contact biased, box, conduit & cable	265	495	760
0280	Security system, door/window contact balanced, box, conduit & cable	265	495	760
0440	Security system, proximity card reader, box, conduit & cable	470	655	1,125
1600	Card control entrance system, to 6 zone, including hardware for 100 doors	42,900	30,400	73,300

D5030 Communications and Security

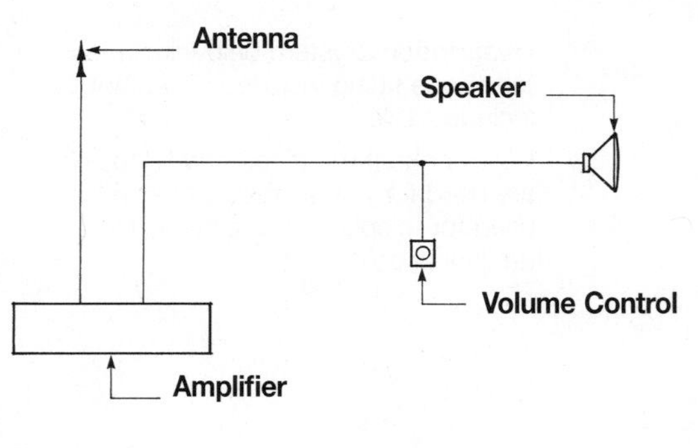

Sound System Includes AM–FM antenna, outlets, rigid conduit, and copper wire.
Fire Detection System Includes pull stations, signals, smoke and heat detectors, rigid conduit, and copper wire.
Intercom System Includes master and remote stations, rigid conduit, and copper wire.
Master Clock System Includes clocks, bells, rigid conduit, and copper wire.
Master TV Antenna Includes antenna, VHF–UHF reception and distribution, rigid conduit, and copper wire.

System Components	QUANTITY	UNIT	COST EACH MAT.	INST.	TOTAL
SYSTEM D5030 910 0220					
SOUND SYSTEM, INCLUDES OUTLETS, BOXES, CONDUIT & WIRE					
Steel intermediate conduit, (IMC) 1/2″ diam	1200.000	L.F.	2,364	7,860	10,224
Wire sound shielded w/drain, #22-2 conductor	15.500	C.L.F.	237.93	1,271	1,508.93
Sound system speakers ceiling or wall	12.000	Ea.	1,572	984	2,556
Sound system volume control	12.000	Ea.	1,188	984	2,172
Sound system amplifier, 250 Watts	1.000	Ea.	1,400	655	2,055
Sound system antenna, AM FM	1.000	Ea.	152	164	316
Sound system monitor panel	1.000	Ea.	435	164	599
Sound system cabinet	1.000	Ea.	945	655	1,600
Steel outlet box 4″ square	12.000	Ea.	31.92	396	427.92
Steel outlet box 4″ plaster rings	12.000	Ea.	31.68	123	154.68
TOTAL			8,357.53	13,256	21,613.53

D5030 910	Communication & Alarm Systems	COST EACH MAT.	INST.	TOTAL
0200	Communication & alarm systems, includes outlets, boxes, conduit & wire			
0210	Sound system, 6 outlets	5,800	8,300	14,100
0220	12 outlets	8,350	13,300	21,650
0240	30 outlets	14,900	25,100	40,000
0280	100 outlets	45,300	84,000	129,300
0320	Fire detection systems, non-addressable, 12 detectors	3,300	6,875	10,175
0360	25 detectors	5,800	11,600	17,400
0400	50 detectors	11,400	23,100	34,500
0440	100 detectors	21,200	41,900	63,100
0450	Addressable type, 12 detectors	4,975	6,925	11,900
0452	25 detectors	9,075	11,700	20,775
0454	50 detectors	17,500	23,000	40,500
0456	100 detectors	33,900	42,300	76,200
0458	Fire alarm control panel, 8 zone, excluding wire and conduit	860	1,300	2,160
0459	12 zone	2,625	1,975	4,600
0460	Fire alarm command center, addressable without voice, excl. wire & conduit	4,925	1,150	6,075
0462	Addressable with voice	10,400	1,800	12,200
0480	Intercom systems, 6 stations	4,000	5,675	9,675
0520	12 stations	7,700	11,300	19,000
0560	25 stations	12,900	21,700	34,600
0600	50 stations	24,900	40,800	65,700
0640	100 stations	49,200	79,500	128,700
0680	Master clock systems, 6 rooms	5,175	9,250	14,425
0720	12 rooms	8,100	15,800	23,900
0760	20 rooms	11,200	22,300	33,500
0800	30 rooms	18,300	41,100	59,400

D5030 Communications and Security

D5030 910	Communication & Alarm Systems	COST EACH		
		MAT.	INST.	TOTAL
0840	50 rooms	29,700	69,500	99,200
0880	100 rooms	57,500	137,500	195,000
0920	Master TV antenna systems, 6 outlets	2,425	5,850	8,275
0960	12 outlets	4,500	10,900	15,400
1000	30 outlets	11,200	25,200	36,400
1040	100 outlets	37,300	82,000	119,300

D5030 920	Data Communication	COST PER M.S.F.		
		MAT.	INST.	TOTAL
0100	Data communication system, incl data/voice outlets, boxes, conduit, & cable			
0102	Data and voice system, 2 data/voice outlets per 1000 S.F.	166	485	651
0104	4 data/voice outlets per 1000 S.F.	315	955	1,270
0106	6 data/voice outlets per 1000 S.F.	450	1,400	1,850
0110	8 data/voice outlets per 1000 S.F.	575	1,875	2,450

D5090 Other Electrical Systems

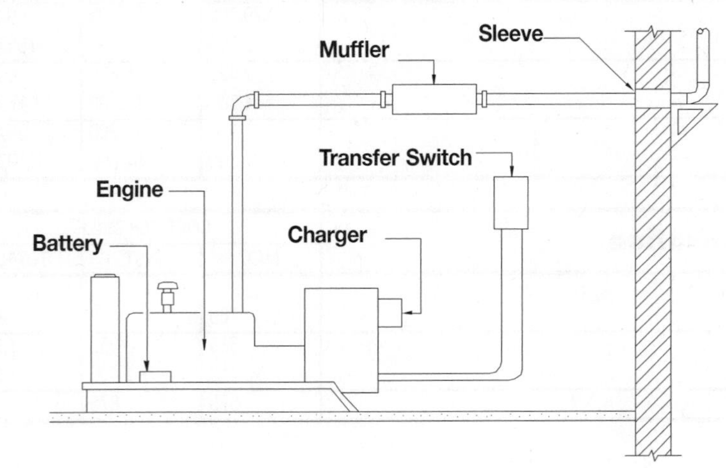

Description: System below tabulates the installed cost for generators by kW. Included in costs are battery, charger, muffler, and transfer switch.

No conduit, wire, or terminations included.

System Components	QUANTITY	UNIT	COST PER kW MAT.	COST PER kW INST.	COST PER kW TOTAL
SYSTEM D5090 210 0200					
GENERATOR SET, INCL. BATTERY, CHARGER, MUFFLER & TRANSFER SWITCH					
GAS/GASOLINE OPER., 3 PHASE, 4 WIRE, 277/480V, 7.5 kW					
Generator set, gas or gasoline operated, 3 ph 4 W, 277/480 V, 7.5 kW	.133	Ea.	1,076.67	287.60	1,364.27
TOTAL			1,076.67	287.60	1,364.27

D5090 210	Generators (by kW)	COST PER kW MAT.	COST PER kW INST.	COST PER kW TOTAL
0190	Generator sets, include battery, charger, muffler & transfer switch			
0200	Gas/gasoline operated, 3 phase, 4 wire, 277/480 volt, 7.5 kW	1,075	288	1,363
0240	11.5 kW	1,000	219	1,219
0280	20 kW	675	142	817
0320	35 kW	460	93	553
0360	80 kW	330	56	386
0400	100 kW	289	54	343
0440	125 kW	470	51	521
0480	185 kW	420	39	459
0560	Diesel engine with fuel tank, 30 kW	515	108	623
0600	50 kW	435	84.50	519.50
0640	75 kW	310	68	378
0680	100 kW	330	57.50	387.50
0720	125 kW	278	49	327
0760	150 kW	248	46	294
0800	175 kW	246	41	287
0840	200 kW	233	37	270
0880	250 kW	199	31	230
0920	300 kW	180	27.50	207.50
0960	350 kW	174	25.50	199.50
1000	400 kW	189	23.50	212.50
1040	500 kW	190	19.80	209.80
1200	750 kW	205	12.45	217.45
1400	1000 kW	192	13.05	205.05

D5090 Other Electrical Systems

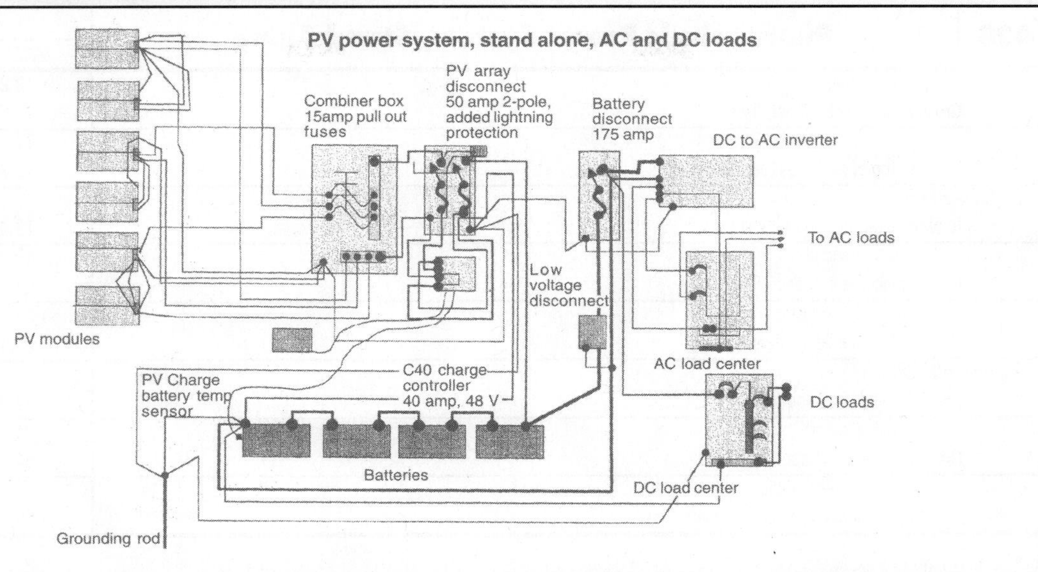

System Components	QUANTITY	UNIT	COST EACH MAT.	COST EACH INST.	COST EACH TOTAL
SYSTEM D5090 420 0100					
PHOTOVOLTAIC POWER SYSTEM, STAND ALONE					
Alternative energy sources, photovoltaic module, 75 watt, 17 volts	12.000	Ea.	5,160	984	6,144
PV rack system, roof, penetrating surface mount, on wood framing, 1 panel	12.000	Ea.	624	1,194	1,818
DC to AC inverter for, 24 V, 2,500 watt	1.000	Ea.	2,200	164	2,364
PV components, combiner box, 10 lug, NEMA 3R enclosure	1.000	Ea.	210	164	374
Alternative energy sources, PV components, fuse, 15 A for combiner box	10.000	Ea.	205	164	369
Battery charger controller w/temperature sensor	1.000	Ea.	540	164	704
Digital readout panel, displays hours, volts, amps, etc.	1.000	Ea.	143	164	307
Deep cycle solar battery, 6 V, 180 Ah (C/20)	4.000	Ea.	1,032	328	1,360
Battery interconnection, 15″ AWG #2/0, sealed w/copper ring lugs	3.000	Ea.	42.30	123	165.30
Battery interconn, 24″ AWG #2/0, sealed w/copper ring lugs	2.000	Ea.	40	82	122
Battery interconn, 60″ AWG #2/0, sealed w/copper ring lugs	2.000	Ea.	96	82	178
Batt temp computer probe, RJ11 jack, 15′ cord	1.000	Ea.	21.50	41	62.50
System disconnect, DC 175 amp circuit breaker	1.000	Ea.	154	82	236
Conduit box for inverter	1.000	Ea.	48.50	82	130.50
Low voltage disconnect	1.000	Ea.	44	82	126
Vented battery enclosure, wood	1.000	Ea.	191	289	480
Grounding, rod, copper clad, 8′ long, 5/8″ diameter	1.000	Ea.	21	119	140
Grounding, clamp, bronze, 5/8″ dia	1.000	Ea.	5.80	20.50	26.30
Bare copper wire, stranded, #8	1.000	C.L.F.	32.50	59.50	92
Wire, 600 volt, type THW, copper, stranded, #12	3.600	C.L.F.	56.88	214.20	271.08
Wire, 600 volt, type THW, copper, stranded, #10	1.050	C.L.F.	26.25	68.78	95.03
Cond to 15′ H,incl 2 termn,2 elb&11 bm CLP per 100′,galv stl,1/2″ dia	120.000	L.F.	328.80	876	1,204.80
Conduit,to 15′ H,incl 2 termn,2 elb&11 bm clp per 100′,(EMT), 1″ dia	30.000	L.F.	52.80	171	223.80
Lightning surge suppressor	1.000	Ea.	56	20.50	76.50
General duty 240 volt, 2 pole, nonfusible, NEMA 3R, 60 amp	1.000	Ea.	255	285	540
Load centers, 3 wire, 120/240V, 100 amp main lugs, indoor, 8 circuits	2.000	Ea.	392	940	1,332
Circuit breaker, Plug-in, 120/240 volt, to 60 amp, 1 pole	5.000	Ea.	63	272.50	335.50
Fuses, dual element, time delay, 250 volt, 50 amp	2.000	Ea.	22.80	26.20	49
TOTAL			12,064.13	7,262.18	19,326.31

D5090 420	Photovoltaic Power System, Stand Alone	COST EACH MAT.	COST EACH INST.	COST EACH TOTAL
0050	24 V capacity, 900 W (~102 SF)			
0100	Roof mounted, on wood framing	12,100	7,250	19,350

D5090 Other Electrical Systems

D5090 420	Photovoltaic Power System, Stand Alone	COST EACH		
		MAT.	INST.	TOTAL
0150	with standoff	12,200	7,250	19,450
0200	Ground placement, ballast, fixed	17,600	7,875	25,475
0250	Adjustable	17,900	7,875	25,775
0300	Top of pole, passive tracking	17,700	7,875	25,575
0350	1.8 kW (~204 SF)			
0400	Roof mounted, on wood framing	19,000	11,500	30,500
0450	with standoff	19,200	11,500	30,700
0500	Ground placement, ballast, fixed	29,200	11,000	40,200
0550	Adjustable	29,800	11,000	40,800
0600	Top of pole, passive tracking	30,500	11,000	41,500
0650	3.2 kW (~408 SF)			
0700	Roof mounted, on wood framing	34,300	19,600	53,900
0750	with standoff	34,700	19,600	54,300
0800	Ground placement, ballast, fixed	54,000	17,000	71,000
0850	Adjustable	55,000	17,000	72,000
0900	Top of pole, passive tracking	56,500	16,900	73,400

D5090 Other Electrical Systems

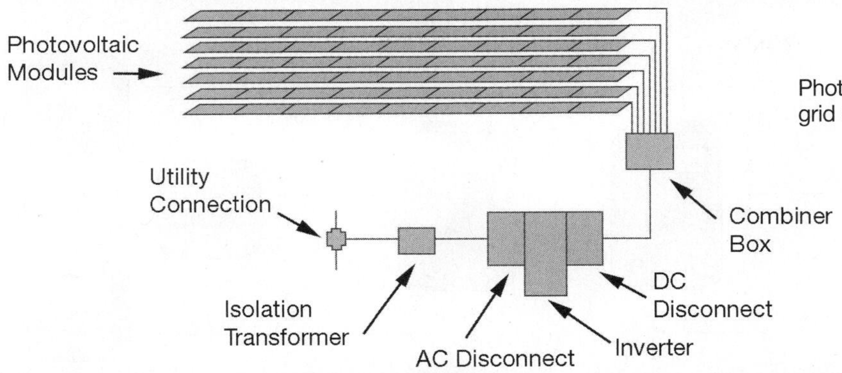

Photovoltaic Modules
Utility Connection
Combiner Box
DC Disconnect
Isolation Transformer
AC Disconnect
Inverter

Photovoltaic power system, grid connected 10 kW

System Components	QUANTITY	UNIT	COST EACH MAT.	COST EACH INST.	COST EACH TOTAL
SYSTEM D5090 430 0100					
PHOTOVOLTAIC POWER SYSTEM, GRID CONNECTED 10 kW					
Photovoltaic module, 150 watt, 33 volts	60.000	Ea.	43,200	4,920	48,120
PV rack system, roof, non-penetrating ballast, 1 panel	60.000	Ea.	59,100	2,160	61,260
DC to AC inverter for, 24 V, 4,000 watt	3.000	Ea.	10,950	990	11,940
Combiner box 10 lug, NEMA 3R	1.000	Ea.	210	164	374
15 amp fuses	10.000	Ea.	205	164	369
Safety switch, 60 amp	1.000	Ea.	365	298	663
Safety switch, 100 amp	1.000	Ea.	680	365	1,045
Utility connection, 3 pole breaker	1.000	Ea.	257	106	363
Fuse, 60 A	3.000	Ea.	38.40	39.30	77.70
Fuse, 100 A	3.000	Ea.	93	49.20	142.20
10 kVA isolation tranformer	1.000	Ea.	1,900	820	2,720
EMT Conduit w/fittings & support	1.000	L.F.	316.80	1,026	1,342.80
RGS Conduit w/fittings & support	267.000	L.F.	1,158.78	2,696.70	3,855.48
Enclosure 24" x 24" x 10", NEMA 4	1.000	Ea.	5,475	1,300	6,775
Wire, 600 volt, type THWN-THHN, copper, stranded, #6	8.000	C.L.F.	508	808	1,316
Wire, 600 volt, copper type XLPE-USE(RHW), stranded, #12	18.000	C.L.F.	432	1,071	1,503
Grounding, bare copper wire stranded, 4/0	2.000	C.L.F.	870	460	1,330
Grounding, exothermic weld, 4/0 wire to building steel	4.000	Ea.	47.20	374	421.20
Grounding, brazed connections, #6 wire	2.000	Ea.	35.50	109	144.50
Insulated ground wire, copper, #6	.400	C.L.F.	25.40	40.40	65.80
TOTAL			125,867.08	17,960.60	143,827.68

D5090 430	Photovoltaic Power System, Grid Connected	COST EACH MAT.	COST EACH INST.	COST EACH TOTAL
0050	10 kW			
0100	Roof mounted, non-penetrating ballast	126,000	18,000	144,000
0200	Penetrating surface mount, on steel framing	70,000	28,600	98,600
0300	on wood framing	70,000	21,800	91,800
0400	with standoff	70,500	21,800	92,300
0500	Ground placement, ballast, fixed	94,000	18,000	112,000
0600	adjustable	95,500	18,000	113,500
0700	Top of pole, passive tracking	96,000	18,500	114,500
1050	20 kW			
1100	Roof mounted, non-penetrating ballast	245,000	30,000	275,000
1200	Penetrating surface mount, on steel framing	133,500	51,500	185,000
1300	on wood framing	133,000	37,600	170,600
1400	with standoff	134,000	37,600	171,600
1500	Ground placement, ballast, fixed	181,000	29,100	210,100
1600	adjustable	184,000	29,100	213,100
1700	Top of pole, passive tracking	185,000	30,000	215,000

D5090 Other Electrical Systems

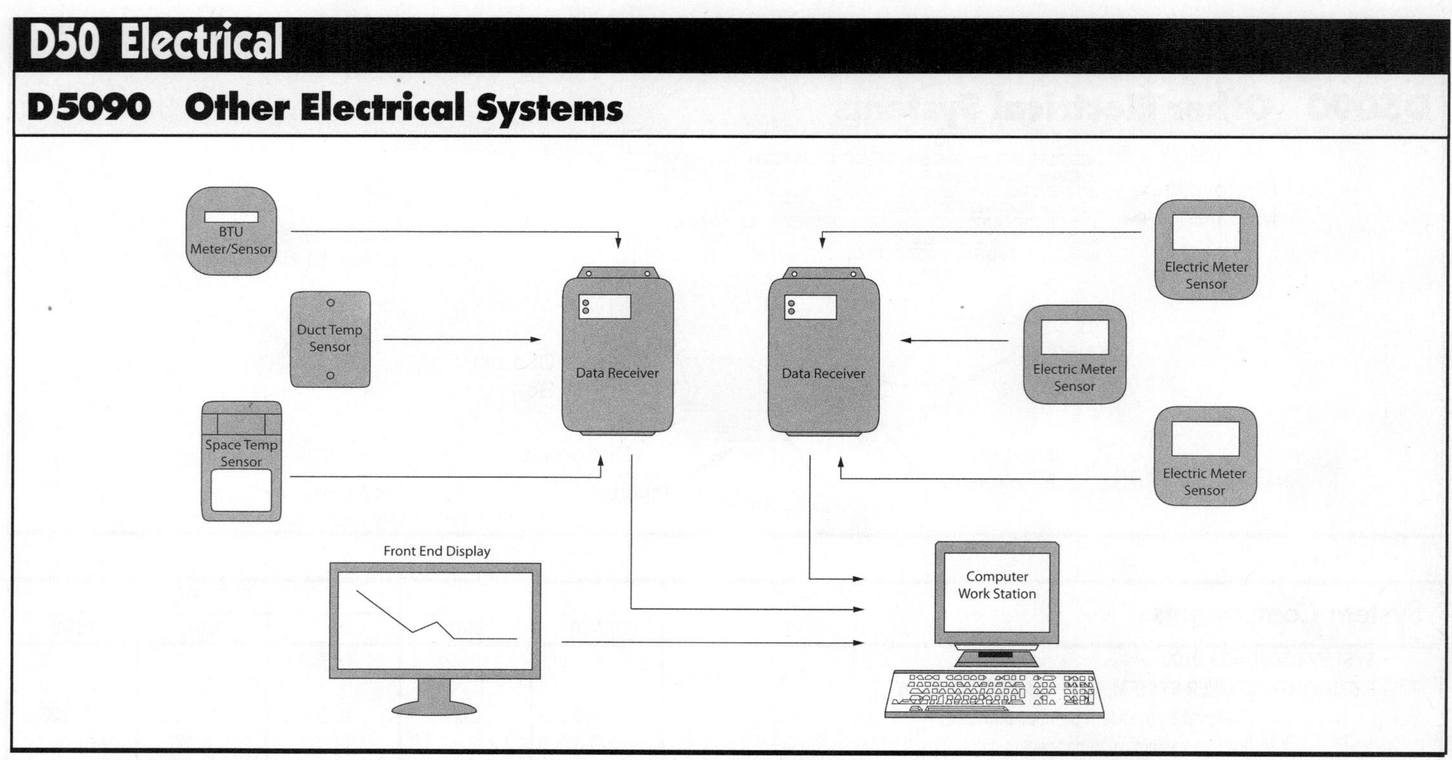

System Components	QUANTITY	UNIT	COST EACH MAT.	INST.	TOTAL
SYSTEM D5090 480 1100					
ELECTRICAL, SINGLE PHASE, 1 METER					
#18 twisted shielded pair in 1/2" EMT conduit	.050	C.L.F.	8.10	10.90	19
Wire, 600 volt, type THW, copper, solid, #12	.150	C.L.F.	1.97	8.93	10.90
Conduit (EMT) , to 15' H, incl 2 termn,2 elb&11 bm clp per 100', 3/4"	5.000	L.F.	5.20	25.25	30.45
Outlet boxes, pressed steel, handy box	1.000	Ea.	3.17	24.50	27.67
Outlet boxes, pressed steel, handy box, covers, device	1.000	Ea.	1.11	10.25	11.36
Wiring devices, receptacle, duplex, 120 volt, ground, 20 amp	1.000	Ea.	8.70	24.50	33.20
Single phase, 277 volt, 200 amp	1.000	Ea.	440	74.50	514.50
Data recorder, 8 meters	1.000	Ea.	1,525	60	1,585
Software package, per meter, premium	1.000	Ea.	675		675
TOTAL			2,668.25	238.83	2,907.08

D5090 480	Energy Monitoring Systems	COST EACH MAT.	INST.	TOTAL
1000	Electrical			
1100	Single phase, 1 meter	2,675	239	2,914
1110	4 meters	6,800	1,800	8,600
1120	8 meters	13,700	3,700	17,400
1200	Three phase, 1 meter	3,000	305	3,305
1210	5 meters	10,000	2,825	12,825
1220	10 meters	20,500	5,775	26,275
1230	25 meters	46,700	11,500	58,200
2000	Mechanical			
2100	BTU, 1 meter	3,725	875	4,600
2110	w/1 duct sensor	3,925	1,075	5,000
2120	& 1 space sensor	4,450	1,225	5,675
2130	& 5 space sensors	6,525	1,800	8,325
2140	& 10 space sensors	9,750	2,550	12,300
2200	BTU, 3 meters	8,100	2,500	10,600
2210	w/3 duct sensors	8,725	3,125	11,850
2220	& 3 space sensors	10,300	3,550	13,850
2230	& 15 space sensors	16,500	5,325	21,825
2240	& 30 space sensors	25,000	7,525	32,525
9000	Front end display	700	128	828
9100	Computer workstation	1,450	1,700	3,150

D5090 Other Electrical Systems

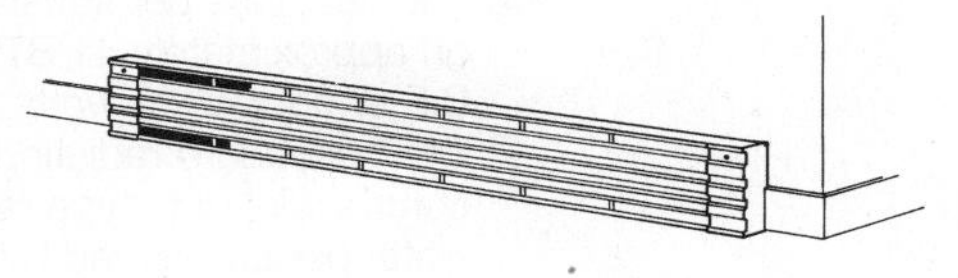

Low Density and Medium Density Baseboard Radiation

The costs shown in Table below are based on the following system considerations:

1. The heat loss per square foot is based on approximately 34 BTU/hr. per S.F. of floor or 10 watts per S.F. of floor.
2. Baseboard radiation is based on the low watt density type rated 187 watts per L.F. and the medium density type rated 250 watts per L.F.
3. Thermostat is not included.
4. Wiring costs include branch circuit wiring.

System Components	QUANTITY	UNIT	COST PER S.F. MAT.	INST.	TOTAL
SYSTEM D5090 510 1000					
ELECTRIC BASEBOARD RADIATION, LOW DENSITY, 900 S.F., 31 MBH, 9 kW					
Electric baseboard radiator, 5′ long 935 watt	.011	Ea.	.58	1.27	1.85
Steel intermediate conduit, (IMC) 1/2″ diam	.170	L.F.	.33	1.11	1.44
Wire 600 volt, type THW, copper, solid, #12	.005	C.L.F.	.07	.30	.37
TOTAL			.98	2.68	3.66

D5090 510	Electric Baseboard Radiation (Low Density)	COST PER S.F. MAT.	INST.	TOTAL
1000	Electric baseboard radiation, low density, 900 S.F., 31 MBH, 9 kW	.98	2.68	3.66
1200	1500 S.F., 51 MBH, 15 kW	.97	2.62	3.59
1400	2100 S.F., 72 MBH, 21 kW	.87	2.34	3.21
1600	3000 S.F., 102 MBH, 30 kW	.78	2.14	2.92
2000	Medium density, 900 S.F., 31 MBH, 9 kW	.87	2.45	3.32
2200	1500 S.F., 51 MBH, 15 kW	.86	2.39	3.25
2400	2100 S.F., 72 MBH, 21 kW	.81	2.23	3.04
2600	3000 S.F., 102 MBH, 30 kW	.73	2.02	2.75

D5090 Other Electrical Systems

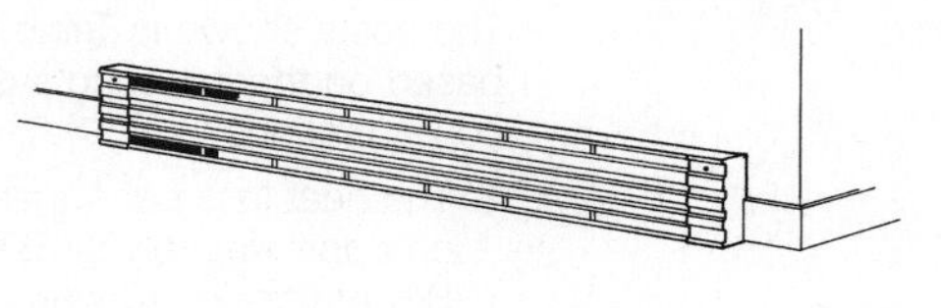

Commercial Duty Baseboard Radiation
The costs shown in Table below are based on the following system considerations:

1. The heat loss per square foot is based on approximately 41 BTU/hr. per S.F. of floor or 12 watts per S.F.
2. The baseboard radiation is of the commercial duty type rated 250 watts per L.F. served by 277 volt, single phase power.
3. Thermostat is not included.
4. Wiring costs include branch circuit wiring.

System Components	QUANTITY	UNIT	COST PER S.F. MAT.	COST PER S.F. INST.	COST PER S.F. TOTAL
SYSTEM D5090 520 1000					
ELECTRIC BASEBOARD RADIATION, MEDIUM DENSITY, 1230 S.F., 51 MBH, 15 kW					
Electric baseboard radiator, 5′ long	.013	Ea.	.68	1.50	2.18
Steel intermediate conduit, (IMC) 1/2″ diam	.154	L.F.	.30	1.01	1.31
Wire 600 volt, type THW, copper, solid, #12	.004	C.L.F.	.05	.24	.29
TOTAL			1.03	2.75	3.78

D5090 520	Electric Baseboard Radiation (Medium Density)	COST PER S.F. MAT.	COST PER S.F. INST.	COST PER S.F. TOTAL
1000	Electric baseboard radiation, medium density, 1230 SF, 51 MBH, 15 kW	1.03	2.75	3.78
1200	2500 S.F. floor area, 106 MBH, 31 kW	.97	2.57	3.54
1400	3700 S.F. floor area, 157 MBH, 46 kW	.94	2.50	3.44
1600	4800 S.F. floor area, 201 MBH, 59 kW	.87	2.32	3.19
1800	11,300 S.F. floor area, 464 MBH, 136 kW	.82	2.12	2.94
2000	30,000 S.F. floor area, 1229 MBH, 360 kW	.81	2.07	2.88

G Building Sitework

G1030 Site Earthwork

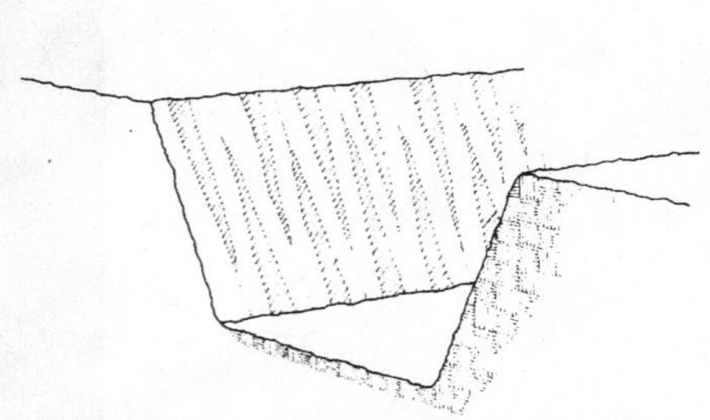

Trenching Systems are shown on a cost per linear foot basis. The systems include: excavation; backfill and removal of spoil; and compaction for various depths and trench bottom widths. The backfill has been reduced to accommodate a pipe of suitable diameter and bedding.

The slope for trench sides varies from none to 1:1.

The Expanded System Listing shows Trenching Systems that range from 2′ to 12′ in width. Depths range from 2′ to 25′.

System Components	QUANTITY	UNIT	COST PER L.F. EQUIP.	LABOR	TOTAL
SYSTEM G1030 805 1310					
TRENCHING COMMON EARTH, NO SLOPE, 2′ WIDE, 2′ DP, 3/8 C.Y. BUCKET					
Excavation, trench, hyd. backhoe, track mtd., 3/8 C.Y. bucket	.148	B.C.Y.	.40	1.07	1.47
Backfill and load spoil, from stockpile	.153	L.C.Y.	.13	.32	.45
Compaction by vibrating plate, 6″ lifts, 4 passes	.118	E.C.Y.	.03	.39	.42
Remove excess spoil, 8 C.Y. dump truck, 2 mile roundtrip	.040	L.C.Y.	.15	.16	.31
TOTAL			.71	1.94	2.65

G1030 805	Trenching Common Earth	COST PER L.F. EQUIP.	LABOR	TOTAL
1310	Trenching, common earth, no slope, 2′ wide, 2′ deep, 3/8 C.Y. bucket	.71	1.94	2.65
1320	3′ deep, 3/8 C.Y. bucket	1	2.93	3.93
1330	4′ deep, 3/8 C.Y. bucket	1.29	3.90	5.19
1340	6′ deep, 3/8 C.Y. bucket	1.67	5.05	6.72
1350	8′ deep, 1/2 C.Y. bucket	2.21	6.70	8.91
1360	10′ deep, 1 C.Y. bucket	3.46	7.95	11.41
1400	4′ wide, 2′ deep, 3/8 C.Y. bucket	1.60	3.83	5.43
1410	3′ deep, 3/8 C.Y. bucket	2.20	5.80	8
1420	4′ deep, 1/2 C.Y. bucket	2.56	6.50	9.06
1430	6′ deep, 1/2 C.Y. bucket	4.12	10.45	14.57
1440	8′ deep, 1/2 C.Y. bucket	6.60	13.45	20.05
1450	10′ deep, 1 C.Y. bucket	7.95	16.70	24.65
1460	12′ deep, 1 C.Y. bucket	10.25	21.50	31.75
1470	15′ deep, 1-1/2 C.Y. bucket	9.40	19.05	28.45
1480	18′ deep, 2-1/2 C.Y. bucket	13	26.50	39.50
1520	6′ wide, 6′ deep, 5/8 C.Y. bucket w/trench box	8.50	15.45	23.95
1530	8′ deep, 3/4 C.Y. bucket	11.30	20.50	31.80
1540	10′ deep, 1 C.Y. bucket	11.50	21	32.50
1550	12′ deep, 1-1/2 C.Y. bucket	12.35	22.50	34.85
1560	16′ deep, 2-1/2 C.Y. bucket	16.90	28.50	45.40
1570	20′ deep, 3-1/2 C.Y. bucket	22	34	56
1580	24′ deep, 3-1/2 C.Y. bucket	26	40.50	66.50
1640	8′ wide, 12′ deep, 1-1/2 C.Y. bucket w/trench box	17.30	28.50	45.80
1650	15′ deep, 1-1/2 C.Y. bucket	22.50	37.50	60
1660	18′ deep, 2-1/2 C.Y. bucket	24.50	38	62.50
1680	24′ deep, 3-1/2 C.Y. bucket	35.50	52.50	88
1730	10′ wide, 20′ deep, 3-1/2 C.Y. bucket w/trench box	28.50	50	78.50
1740	24′ deep, 3-1/2 C.Y. bucket	42.50	60	102.50
1780	12′ wide, 20′ deep, 3-1/2 C.Y. bucket w/trench box	45	63.50	108.50
1790	25′ deep, bucket	55.50	81	136.50
1800	1/2 to 1 slope, 2′ wide, 2′ deep, 3/8 C.Y. bucket	1	2.93	3.93
1810	3′ deep, 3/8 C.Y. bucket	1.67	5.15	6.82

G1030 Site Earthwork

G1030 805	Trenching Common Earth	COST PER L.F.		
		EQUIP.	LABOR	TOTAL
1820	4' deep, 3/8 C.Y. bucket	2.49	7.85	10.34
1840	6' deep, 3/8 C.Y. bucket	4	12.70	16.70
1860	8' deep, 1/2 C.Y. bucket	6.35	20.50	26.85
1880	10' deep, 1 C.Y. bucket	11.90	28	39.90
2300	4' wide, 2' deep, 3/8 C.Y. bucket	1.90	4.81	6.71
2310	3' deep, 3/8 C.Y. bucket	2.87	8	10.87
2320	4' deep, 1/2 C.Y. bucket	3.61	9.90	13.51
2340	6' deep, 1/2 C.Y. bucket	6.90	18.55	25.45
2360	8' deep, 1/2 C.Y. bucket	12.75	27	39.75
2380	10' deep, 1 C.Y. bucket	17.60	38.50	56.10
2400	12' deep, 1 C.Y. bucket	24	52	76
2430	15' deep, 1-1/2 C.Y. bucket	26.50	55.50	82
2460	18' deep, 2-1/2 C.Y. bucket	46	87	133
2840	6' wide, 6' deep, 5/8 C.Y. bucket w/trench box	12.55	23	35.55
2860	8' deep, 3/4 C.Y. bucket	18.20	34.50	52.70
2880	10' deep, 1 C.Y. bucket	18.55	35	53.55
2900	12' deep, 1-1/2 C.Y. bucket	23.50	46	69.50
2940	16' deep, 2-1/2 C.Y. bucket	38.50	67.50	106
2980	20' deep, 3-1/2 C.Y. bucket	55	91	146
3020	24' deep, 3-1/2 C.Y. bucket	77.50	125	202.50
3100	8' wide, 12' deep, 1-1/2 C.Y. bucket w/trench box	29	52.50	81.50
3120	15' deep, 1-1/2 C.Y. bucket	42	76	118
3140	18' deep, 2-1/2 C.Y. bucket	53	90.50	143.50
3180	24' deep, 3-1/2 C.Y. bucket	86.50	137	223.50
3270	10' wide, 20' deep, 3-1/2 C.Y. bucket w/trench box	55.50	105	160.50
3280	24' deep, 3-1/2 C.Y. bucket	96	149	245
3370	12' wide, 20' deep, 3-1/2 C.Y. bucket w/trench box	80.50	122	202.50
3380	25' deep, 3-1/2 C.Y. bucket	112	174	286
3500	1 to 1 slope, 2' wide, 2' deep, 3/8 C.Y. bucket	1.29	3.90	5.19
3520	3' deep, 3/8 C.Y. bucket	3.70	8.80	12.50
3540	4' deep, 3/8 C.Y. bucket	3.68	11.75	15.43
3560	6' deep, 1/2 C.Y. bucket	4	12.70	16.70
3580	8' deep, 1/2 C.Y. bucket	7.90	25.50	33.40
3600	10' deep, 1 C.Y. bucket	20.50	48.50	69
3800	4' wide, 2' deep, 3/8 C.Y. bucket	2.20	5.80	8
3820	3' deep, 3/8 C.Y. bucket	3.54	10.20	13.74
3840	4' deep, 1/2 C.Y. bucket	4.64	13.30	17.94
3860	6' deep, 1/2 C.Y. bucket	9.65	26.50	36.15
3880	8' deep, 1/2 C.Y. bucket	18.95	41	59.95
3900	10' deep, 1 C.Y. bucket	27.50	60	87.50
3920	12' deep, 1 C.Y. bucket	40	86.50	126.50
3940	15' deep, 1-1/2 C.Y. bucket	43.50	92.50	136
3960	18' deep, 2-1/2 C.Y. bucket	63.50	120	183.50
4030	6' wide, 6' deep, 5/8 C.Y. bucket w/trench box	16.50	31	47.50
4040	8' deep, 3/4 C.Y. bucket	23.50	42.50	66
4050	10' deep, 1 C.Y. bucket	27	52	79
4060	12' deep, 1-1/2 C.Y. bucket	36	70.50	106.50
4070	16' deep, 2-1/2 C.Y. bucket	60.50	107	167.50
4080	20' deep, 3-1/2 C.Y. bucket	88.50	149	237.50
4090	24' deep, 3-1/2 C.Y. bucket	129	209	338
4500	8' wide, 12' deep, 1-1/2 C.Y. bucket w/trench box	41	76	117
4550	15' deep, 1-1/2 C.Y. bucket	62	114	176
4600	18' deep, 2-1/2 C.Y. bucket	80.50	140	220.50
4650	24' deep, 3-1/2 C.Y. bucket	138	221	359
4800	10' wide, 20' deep, 3-1/2 C.Y. bucket w/trench box	82.50	160	242.50
4850	24' deep, 3-1/2 C.Y. bucket	147	233	380
4950	12' wide, 20' deep, 3-1/2 C.Y. bucket w/trench box	116	180	296
4980	25' deep, 3-1/2 C.Y. bucket	167	263	430

G1030 Site Earthwork

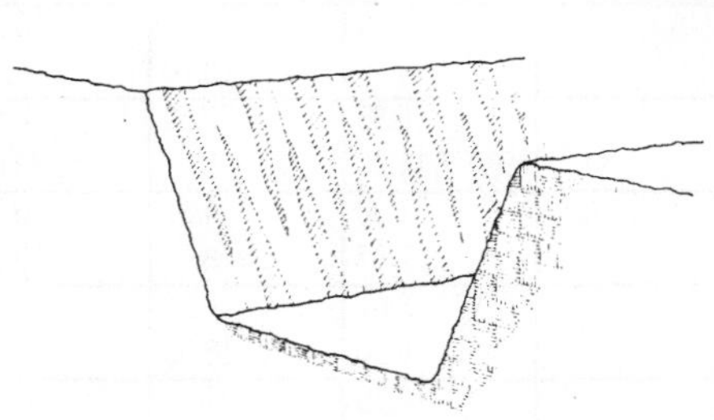

Trenching Systems are shown on a cost per linear foot basis. The systems include: excavation; backfill and removal of spoil; and compaction for various depths and trench bottom widths. The backfill has been reduced to accommodate a pipe of suitable diameter and bedding.

The slope for trench sides varies from none to 1:1.

The Expanded System Listing shows Trenching Systems that range from 2′ to 12′ in width. Depths range from 2′ to 25′.

System Components	QUANTITY	UNIT	COST PER L.F. EQUIP.	LABOR	TOTAL
SYSTEM G1030 806 1310					
TRENCHING LOAM & SANDY CLAY, NO SLOPE, 2′ WIDE, 2′ DP, 3/8 C.Y. BUCKET					
Excavation, trench, hyd. backhoe, track mtd., 3/8 C.Y. bucket	.148	B.C.Y.	.37	.98	1.35
Backfill and load spoil, from stockpile	.165	L.C.Y.	.14	.35	.49
Compaction by vibrating plate 18″ wide, 6″ lifts, 4 passes	.118	E.C.Y.	.03	.39	.42
Remove excess spoil, 8 C.Y. dump truck, 2 mile roundtrip	.042	L.C.Y.	.16	.17	.33
TOTAL			.70	1.89	2.59

G1030 806	Trenching Loam & Sandy Clay	COST PER L.F. EQUIP.	LABOR	TOTAL
1310	Trenching, loam & sandy clay, no slope, 2′ wide, 2′ deep, 3/8 C.Y. bucket	.70	1.89	2.59
1320	3′ deep, 3/8 C.Y. bucket	1.08	3.13	4.21
1330	4′ deep, 3/8 C.Y. bucket	1.27	3.81	5.10
1340	6′ deep, 3/8 C.Y. bucket	1.80	4.51	6.30
1350	8′ deep, 1/2 C.Y. bucket	2.38	5.95	8.35
1360	10′ deep, 1 C.Y. bucket	2.69	6.40	9.10
1400	4′ wide, 2′ deep, 3/8 C.Y. bucket	1.60	3.75	5.35
1410	3′ deep, 3/8 C.Y. bucket	2.18	5.65	7.85
1420	4′ deep, 1/2 C.Y. bucket	2.55	6.40	8.95
1430	6′ deep, 1/2 C.Y. bucket	4.39	9.35	13.75
1440	8′ deep, 1/2 C.Y. bucket	6.45	13.25	19.70
1450	10′ deep, 1 C.Y. bucket	6.45	13.60	20
1460	12′ deep, 1 C.Y. bucket	8.10	16.90	25
1470	15′ deep, 1-1/2 C.Y. bucket	9.80	19.70	29.50
1480	18′ deep, 2-1/2 C.Y. bucket	11.50	21.50	33
1520	6′ wide, 6′ deep, 5/8 C.Y. bucket w/trench box	8.25	15.20	23.50
1530	8′ deep, 3/4 C.Y. bucket	10.90	19.95	31
1540	10′ deep, 1 C.Y. bucket	10.55	20.50	31
1550	12′ deep, 1-1/2 C.Y. bucket	12.05	23	35
1560	16′ deep, 2-1/2 C.Y. bucket	16.50	28.50	45
1570	20′ deep, 3-1/2 C.Y. bucket	20	34	54
1580	24′ deep, 3-1/2 C.Y. bucket	25.50	41.50	67
1640	8′ wide, 12′ deep, 1-1/4 C.Y. bucket w/trench box	17.05	29	46
1650	15′ deep, 1-1/2 C.Y. bucket	21	36.50	57.50
1660	18′ deep, 2-1/2 C.Y. bucket	25	41.50	66.50
1680	24′ deep, 3-1/2 C.Y. bucket	34.50	53.50	88
1730	10′ wide, 20′ deep, 3-1/2 C.Y. bucket w/trench box	34.50	53.50	88
1740	24′ deep, 3-1/2 C.Y. bucket	43.50	66	110
1780	12′ wide, 20′ deep, 3-1/2 C.Y. bucket w/trench box	42	63.50	106
1790	25′ deep, 3-1/2 C.Y. bucket	54.50	82	137
1800	1/2:1 slope, 2′ wide, 2′ deep, 3/8 C.Y. bucket	.98	2.86	3.84
1810	3′ deep, 3/8 C.Y. bucket	1.63	5	6.65
1820	4′ deep, 3/8 C.Y. bucket	2.43	7.65	10.10
1840	6′ deep, 3/8 C.Y. bucket	4.31	11.30	15.60

G1030 Site Earthwork

G1030 806	Trenching Loam & Sandy Clay	COST PER L.F.		
		EQUIP.	LABOR	TOTAL
1860	8' deep, 1/2 C.Y. bucket	6.85	18.10	25
1880	10' deep, 1 C.Y. bucket	9.15	22.50	31.50
2300	4' wide, 2' deep, 3/8 C.Y. bucket	1.88	4.70	6.60
2310	3' deep, 3/8 C.Y. bucket	2.83	7.80	10.65
2320	4' deep, 1/2 C.Y. bucket	3.57	9.70	13.25
2340	6' deep, 1/2 C.Y. bucket	7.35	16.65	24
2360	8' deep, 1/2 C.Y. bucket	12.40	27	39.50
2380	10' deep, 1 C.Y. bucket	14.10	31	45
2400	12' deep, 1 C.Y. bucket	23.50	51.50	75
2430	15' deep, 1-1/2 C.Y. bucket	27.50	57.50	85
2460	18' deep, 2-1/2 C.Y. bucket	45.50	88	134
2840	6' wide, 6' deep, 5/8 C.Y. bucket w/trench box	11.95	23	35
2860	8' deep, 3/4 C.Y. bucket	17.50	34	51.50
2880	10' deep, 1 C.Y. bucket	18.90	38	57
2900	12' deep, 1-1/2 C.Y. bucket	23.50	46.50	70
2940	16' deep, 2-1/2 C.Y. bucket	37.50	68.50	106
2980	20' deep, 3-1/2 C.Y. bucket	53	92.50	146
3020	24' deep, 3-1/2 C.Y. bucket	75	127	202
3100	8' wide, 12' deep, 1-1/2 C.Y. bucket w/trench box	28.50	52.50	81
3120	15' deep, 1-1/2 C.Y. bucket	39	74	113
3140	18' deep, 2-1/2 C.Y. bucket	52	91.50	144
3180	24' deep, 3-1/2 C.Y. bucket	84	139	223
3270	10' wide, 20' deep, 3-1/2 C.Y. bucket w/trench box	67.50	112	180
3280	24' deep, 3-1/2 C.Y. bucket	93	151	244
3320	12' wide, 20' deep, 3-1/2 C.Y. bucket w/trench box	75	122	197
3380	25' deep, 3-1/2 C.Y. bucket w/trench box	102	164	266
3500	1:1 slope, 2' wide, 2' deep, 3/8 C.Y. bucket	1.27	3.81	5.10
3520	3' deep, 3/8 C.Y. bucket	2.28	7.15	9.45
3540	4' deep, 3/8 C.Y. bucket	3.59	11.45	15.05
3560	6' deep, 1/2 C.Y. bucket	4.31	11.30	15.60
3580	8' deep, 1/2 C.Y. bucket	11.35	30	41.50
3600	10' deep, 1 C.Y. bucket	15.65	39	54.50
3800	4' wide, 2' deep, 3/8 C.Y. bucket	2.18	5.65	7.85
3820	3' deep, 1/2 C.Y. bucket	3.48	9.95	13.45
3840	4' deep, 1/2 C.Y. bucket	4.58	13.05	17.65
3860	6' deep, 1/2 C.Y. bucket	10.30	24	34.50
3880	8' deep, 1/2 C.Y. bucket	18.40	40.50	59
3900	10' deep, 1 C.Y. bucket	22	49	71
3920	12' deep, 1 C.Y. bucket	31.50	69	101
3940	15' deep, 1-1/2 C.Y. bucket	45.50	95.50	141
3960	18' deep, 2-1/2 C.Y. bucket	62	121	183
4030	6' wide, 6' deep, 5/8 C.Y. bucket w/trench box	15.70	31	46.50
4040	8' deep, 3/4 C.Y. bucket	24	47.50	71.50
4050	10' deep, 1 C.Y. bucket	27	56	83
4060	12' deep, 1-1/2 C.Y. bucket	35	70.50	106
4070	16' deep, 2-1/2 C.Y. bucket	59	108	167
4080	20' deep, 3-1/2 C.Y. bucket	86	151	237
4090	24' deep, 3-1/2 C.Y. bucket	125	212	335
4500	8' wide, 12' deep, 1-1/4 C.Y. bucket w/trench box	40	76.50	117
4550	15' deep, 1-1/2 C.Y. bucket	56.50	111	168
4600	18' deep, 2-1/2 C.Y. bucket	78.50	142	221
4650	24' deep, 3-1/2 C.Y. bucket	134	224	360
4800	10' wide, 20' deep, 3-1/2 C.Y. bucket w/trench box	100	171	271
4850	24' deep, 3-1/2 C.Y. bucket	143	237	380
4950	12' wide, 20' deep, 3-1/2 C.Y. bucket w/trench box	108	181	289
4980	25' deep, 3-1/2 C.Y. bucket	162	267	430

G1030 Site Earthwork

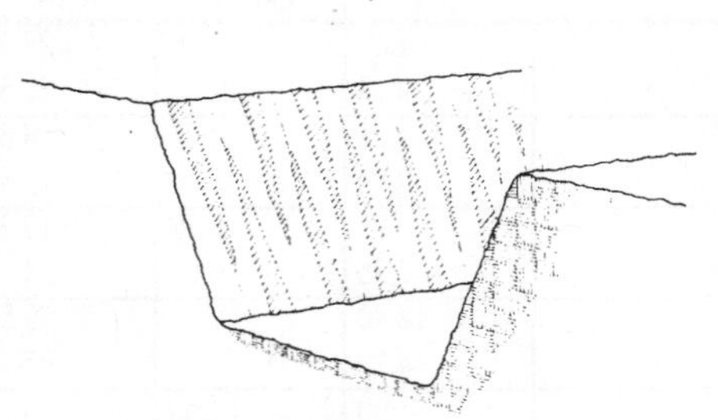

Trenching Systems are shown on a cost per linear foot basis. The systems include: excavation; backfill and removal of spoil; and compaction for various depths and trench bottom widths. The backfill has been reduced to accommodate a pipe of suitable diameter and bedding.

The slope for trench sides varies from none to 1:1.

The Expanded System Listing shows Trenching Systems that range from 2′ to 12′ in width. Depths range from 2′ to 25′.

System Components	QUANTITY	UNIT	COST PER L.F. EQUIP.	COST PER L.F. LABOR	COST PER L.F. TOTAL
SYSTEM G1030 807 1310					
TRENCHING SAND & GRAVEL, NO SLOPE, 2′ WIDE, 2′ DEEP, 3/8 C.Y. BUCKET					
Excavation, trench, hyd. backhoe, track mtd., 3/8 C.Y. bucket	.148	B.C.Y.	.36	.97	1.33
Backfill and load spoil, from stockpile	.140	L.C.Y.	.11	.30	.41
Compaction by vibrating plate 18″ wide, 6″ lifts, 4 passes	.118	E.C.Y.	.03	.39	.42
Remove excess spoil, 8 C.Y. dump truck, 2 mile roundtrip	.035	L.C.Y.	.13	.14	.27
TOTAL			.63	1.80	2.43

G1030 807	Trenching Sand & Gravel	COST PER L.F. EQUIP.	COST PER L.F. LABOR	COST PER L.F. TOTAL
1310	Trenching, sand & gravel, no slope, 2′ wide, 2′ deep, 3/8 C.Y. bucket	.63	1.80	2.43
1320	3′ deep, 3/8 C.Y. bucket	1.01	3.01	4.02
1330	4′ deep, 3/8 C.Y. bucket	1.18	3.62	4.80
1340	6′ deep, 3/8 C.Y. bucket	1.70	4.30	6
1350	8′ deep, 1/2 C.Y. bucket	2.25	5.70	7.95
1360	10′ deep, 1 C.Y. bucket	2.50	6.05	8.55
1400	4′ wide, 2′ deep, 3/8 C.Y. bucket	1.46	3.54	5
1410	3′ deep, 3/8 C.Y. bucket	2.01	5.35	7.35
1420	4′ deep, 1/2 C.Y. bucket	2.33	6.05	8.40
1430	6′ deep, 1/2 C.Y. bucket	4.16	8.95	13.10
1440	8′ deep, 1/2 C.Y. bucket	6.05	12.60	18.65
1450	10′ deep, 1 C.Y. bucket	6.05	12.85	18.90
1460	12′ deep, 1 C.Y. bucket	7.65	16	23.50
1470	15′ deep, 1-1/2 C.Y. bucket	9.25	18.55	28
1480	18′ deep, 2-1/2 C.Y. bucket	10.85	20.50	31.50
1520	6′ wide, 6′ deep, 5/8 C.Y. bucket w/trench box	7.75	14.35	22
1530	8′ deep, 3/4 C.Y. bucket	10.30	18.95	29.50
1540	10′ deep, 1 C.Y. bucket	9.95	19.15	29
1550	12′ deep, 1-1/2 C.Y. bucket	11.40	21.50	33
1560	16′ deep, 2 C.Y. bucket	15.75	27	43
1570	20′ deep, 3-1/2 C.Y. bucket	19.10	32	51
1580	24′ deep, 3-1/2 C.Y. bucket	24	39	63
1640	8′ wide, 12′ deep, 1-1/2 C.Y. bucket w/trench box	16.05	27	43
1650	15′ deep, 1-1/2 C.Y. bucket	19.50	34.50	54
1660	18′ deep, 2-1/2 C.Y. bucket	23.50	39	62.50
1680	24′ deep, 3-1/2 C.Y. bucket	32.50	50	82.50
1730	10′ wide, 20′ deep, 3-1/2 C.Y. bucket w/trench box	32.50	50	82.50
1740	24′ deep, 3-1/2 C.Y. bucket	41	62	103
1780	12′ wide, 20′ deep, 3-1/2 C.Y. bucket w/trench box	39.50	59.50	99
1790	25′ deep, 3-1/2 C.Y. bucket	51	77	128
1800	1/2:1 slope, 2′ wide, 2′ deep, 3/8 C.Y. bucket	.91	2.72	3.63
1810	3′ deep, 3/8 C.Y. bucket	1.53	4.77	6.30
1820	4′ deep, 3/8 C.Y. bucket	2.27	7.30	9.55
1840	6′ deep, 3/8 C.Y. bucket	4.10	10.80	14.90

G1030 Site Earthwork

G1030 807	Trenching Sand & Gravel	COST PER L.F.		
		EQUIP.	LABOR	TOTAL
1860	8′ deep, 1/2 C.Y. bucket	6.55	17.30	24
1880	10′ deep, 1 C.Y. bucket	8.55	21.50	30
2300	4′ wide, 2′ deep, 3/8 C.Y. bucket	1.74	4.45	6.20
2310	3′ deep, 3/8 C.Y. bucket	2.62	7.40	10
2320	4′ deep, 1/2 C.Y. bucket	3.29	9.20	12.50
2340	6′ deep, 1/2 C.Y. bucket	7	15.95	23
2360	8′ deep, 1/2 C.Y. bucket	11.75	25.50	37.50
2380	10′ deep, 1 C.Y. bucket	13.35	29.50	43
2400	12′ deep, 1 C.Y. bucket	22.50	49	71.50
2430	15′ deep, 1-1/2 C.Y. bucket	26	54.50	80.50
2460	18′ deep, 2-1/2 C.Y. bucket	43	83	126
2840	6′ wide, 6′ deep, 5/8 C.Y. bucket w/trench box	11.35	22	33.50
2860	8′ deep, 3/4 C.Y. bucket	16.60	32	48.50
2880	10′ deep, 1 C.Y. bucket	17.85	36	54
2900	12′ deep, 1-1/2 C.Y. bucket	22.50	44	66.50
2940	16′ deep, 2 C.Y. bucket	36	64.50	101
2980	20′ deep, 3-1/2 C.Y. bucket	50.50	87	138
3020	24′ deep, 3-1/2 C.Y. bucket	71.50	119	191
3100	8′ wide, 12′ deep, 1-1/4 C.Y. bucket w/trench box	27	50	77
3120	15′ deep, 1-1/2 C.Y. bucket	36.50	69.50	106
3140	18′ deep, 2-1/2 C.Y. bucket	49	86	135
3180	24′ deep, 3-1/2 C.Y. bucket	80	130	210
3270	10′ wide, 20′ deep, 3-1/2 C.Y. bucket w/trench box	64	105	169
3280	24′ deep, 3-1/2 C.Y. bucket	88	142	230
3370	12′ wide, 20′ deep, 3-1/2 C.Y. bucket w/trench box	71.50	117	189
3380	25′ deep, 3-1/2 C.Y. bucket	103	164	267
3500	1:1 slope, 2′ wide, 2′ deep, 3/8 C.Y. bucket	2.05	4.46	6.50
3520	3′ deep, 3/8 C.Y. bucket	2.14	6.85	9
3540	4′ deep, 3/8 C.Y. bucket	3.36	10.95	14.30
3560	6′ deep, 3/8 C.Y. bucket	4.10	10.80	14.90
3580	8′ deep, 1/2 C.Y. bucket	10.80	29	40
3600	10′ deep, 1 C.Y. bucket	14.65	36.50	51
3800	4′ wide, 2′ deep, 3/8 C.Y. bucket	2.01	5.35	7.35
3820	3′ deep, 3/8 C.Y. bucket	3.24	9.50	12.75
3840	4′ deep, 1/2 C.Y. bucket	4.23	12.35	16.60
3860	6′ deep, 1/2 C.Y. bucket	9.85	23	33
3880	8′ deep, 1/2 C.Y. bucket	17.45	38.50	56
3900	10′ deep, 1 C.Y. bucket	20.50	46	66.50
3920	12′ deep, 1 C.Y. bucket	30	65.50	95.50
3940	15′ deep, 1-1/2 C.Y. bucket	43	90	133
3960	18′ deep, 2-1/2 C.Y. bucket	59	114	173
4030	6′ wide, 6′ deep, 5/8 C.Y. bucket w/trench box	14.90	29.50	44.50
4040	8′ deep, 3/4 C.Y. bucket	23	45.50	68.50
4050	10′ deep, 1 C.Y. bucket	25.50	53	78.50
4060	12′ deep, 1-1/2 C.Y. bucket	33	66.50	99.50
4070	16′ deep, 2 C.Y. bucket	56.50	102	159
4080	20′ deep, 3-1/2 C.Y. bucket	81.50	142	224
4090	24′ deep, 3-1/2 C.Y. bucket	119	199	320
4500	8′ wide, 12′ deep, 1-1/2 C.Y. bucket w/trench box	38	72.50	111
4550	15′ deep, 1-1/2 C.Y. bucket	53.50	105	159
4600	18′ deep, 2-1/2 C.Y. bucket	74.50	134	209
4650	24′ deep, 3-1/2 C.Y. bucket	127	210	335
4800	10′ wide, 20′ deep, 3-1/2 C.Y. bucket w/trench box	95	160	255
4850	24′ deep, 3-1/2 C.Y. bucket	135	222	355
4950	12′ wide, 20′ deep, 3-1/2 C.Y. bucket w/trench box	102	170	272
4980	25′ deep, 3-1/2 C.Y. bucket	154	251	405

G4010 Electrical Distribution

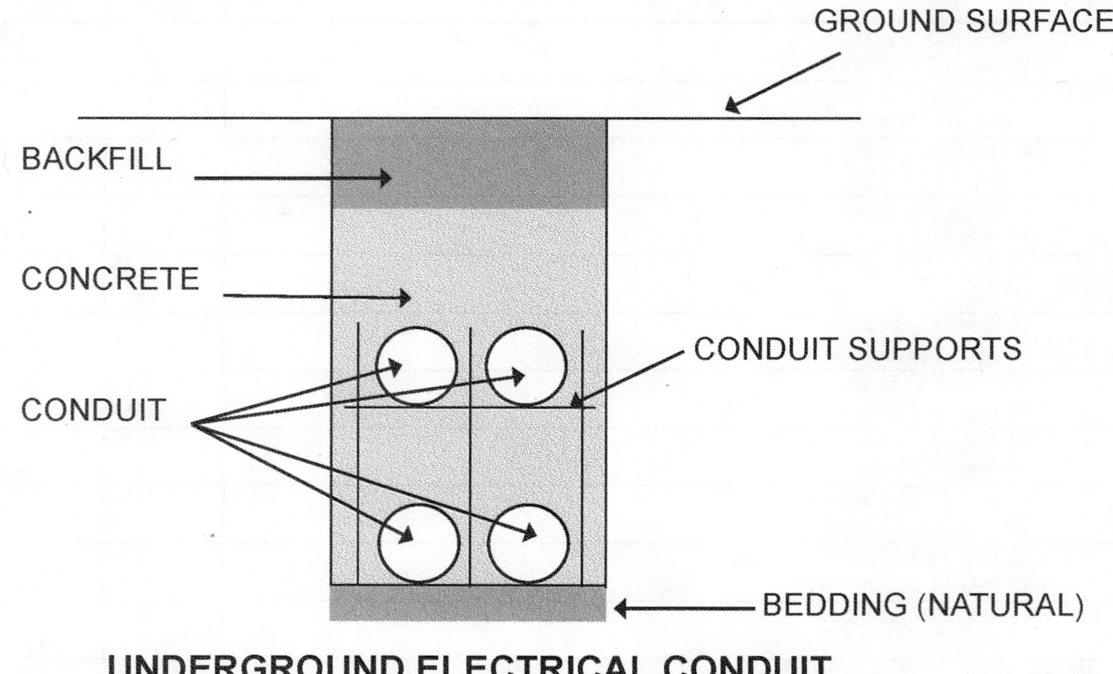

System Components	QUANTITY	UNIT	COST EACH MAT.	COST EACH INST.	COST EACH TOTAL
SYSTEM G4010 312 1000					
2000 AMP UNDERGROUND SERVICE INCLUDING COMMON EARTH EXCAVATION, CONCRETE, BACKFILL AND COMPACTION					
Excavate Trench	44.440	B.C.Y.		298.19	298.19
4 inch conduit bank	108.000	L.F.	1,085.40	2,646	3,731.40
4 inch fitting	8.000	Ea.	252	436	688
4 inch bells	4.000	Ea.	18.76	218	236.76
Concrete material	16.580	C.Y.	1,856.96		1,856.96
Concrete placement	16.580	C.Y.		407.37	407.37
Backfill trench	33.710	L.C.Y.		98.77	98.77
Compact fill material in trench	25.930	E.C.Y.		128.87	128.87
Dispose of excess fill material on-site	24.070	L.C.Y.		282.58	282.58
500 kcmil power cable	18.000	C.L.F.	22,950	7,380	30,330
Wire, 600 volt stranded copper wire, type THW, 1/0	6.000	C.L.F.	1,650	1,194	2,844
TOTAL			27,813.12	13,089.78	40,902.90

G4010 312	Underground Power Feed	COST EACH MAT.	COST EACH INST.	COST EACH TOTAL
0950	Underground electrical including common earth excavation,concrete, backfill and compaction			
1000	2000 Amp service, 100′ length, 4′ depth	27,800	13,100	40,900
1100	1600 Amp service	21,800	10,100	31,900
1200	1200 Amp service	16,900	8,375	25,275
1300	1000 Amp service	14,500	7,650	22,150
1400	800 Amp service	12,500	7,200	19,700
1500	600 Amp service	8,350	6,675	15,025

G4010 Electrical Distribution

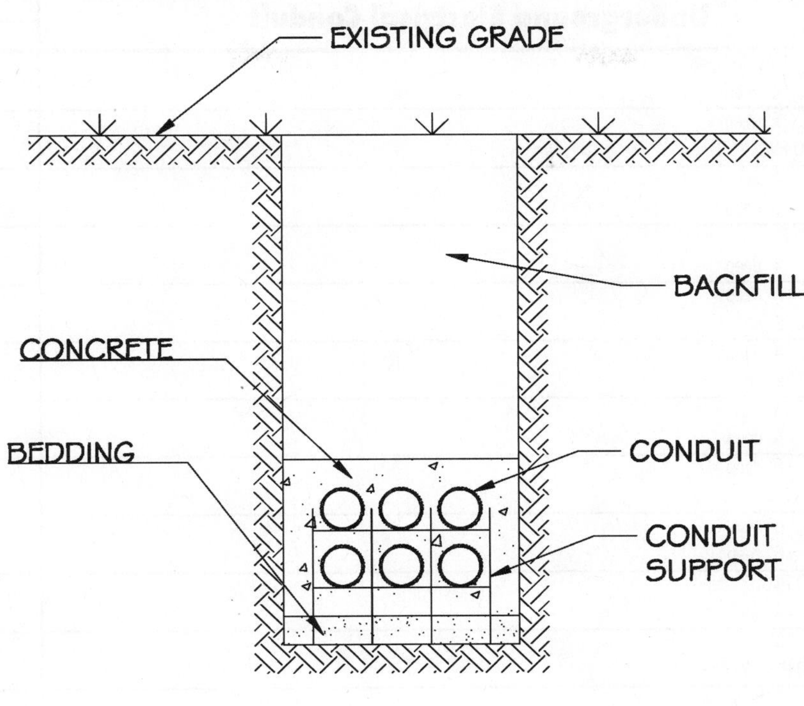

System Components	QUANTITY	UNIT	COST L.F. MAT.	COST L.F. INST.	COST L.F. TOTAL
SYSTEM G4010 320 1012 UNDERGROUND ELECTRICAL CONDUIT, 2″ DIAMETER, INCLUDING EXCAVATION, CONCRETE, BEDDING, BACKFILL, AND COMPACTION					
Excavate Trench	.150	B.C.Y.		1.01	1.01
Base spacer	.200	Ea.	.34	2.34	2.68
Conduit	1.000	L.F.	1.63	3.64	5.27
Concrete material	.050	C.Y.	5.60		5.60
Concrete placement	.050	C.Y.		1.23	1.23
Backfill trench	.120	L.C.Y.		.35	.35
Compact fill material in trench	.100	E.C.Y.		.49	.49
Dispose of excess fill material on-site	.120	L.C.Y.		1.40	1.40
Marking tape	.010	C.L.F.	.30	.03	.33
TOTAL			7.87	10.49	18.36

G4010 320	Underground Electrical Conduit	COST L.F. MAT.	COST L.F. INST.	COST L.F. TOTAL
0950	Underground electrical conduit, including common earth excavation,			
0952	Concrete, bedding, backfill and compaction			
1012	2″ dia. Schedule 40 PVC, 2′ deep	7.85	10.50	18.35
1014	4′ deep	7.85	15	22.85
1016	6′ deep	7.85	19.35	27.20
1022	2 @ 2″ dia. Schedule 40 PVC, 2′ deep	9.85	16.45	26.30
1024	4′ deep	9.85	21	30.85
1026	6′ deep	9.85	25	34.85
1032	3 @ 2″ dia. Schedule 40 PVC, 2′ deep	11.80	22.50	34.30
1034	4′ deep	11.80	27	38.80
1036	6′ deep	11.80	31	42.80
1042	4 @ 2″ dia. Schedule 40 PVC, 2′ deep	16	28	44
1044	4′ deep	16	32.50	48.50
1046	6′ deep	16	37	53

G4010 Electrical Distribution

G4010 320	Underground Electrical Conduit	COST L.F.		
		MAT.	INST.	TOTAL
1062	6 @ 2" dia. Schedule 40 PVC, 2′ deep	19.95	40	59.95
1064	4′ deep	19.95	44	63.95
1066	6′ deep	19.95	49	68.95
1082	8 @ 2" dia. Schedule 40 PVC, 2′ deep	24	52	76
1084	4′ deep	24	56	80
1086	6′ deep	24	60.50	84.50
1092	9 @ 2" dia. Schedule 40 PVC, 2′ deep	26	57.50	83.50
1094	4′ deep	28	62	90
1096	6′ deep	28	66.50	94.50
1212	3" dia. Schedule 40 PVC, 2′ deep	10.45	12.85	23.30
1214	4′ deep	10.45	17.40	27.85
1216	6′ deep	10.45	21.50	31.95
1222	2 @ 3" dia. Schedule 40 PVC, 2′ deep	12.80	21	33.80
1224	4′ deep	12.80	25.50	38.30
1226	6′ deep	12.80	30	42.80
1232	3 @ 3" dia. Schedule 40 PVC, 2′ deep	16.25	29.50	45.75
1234	4′ deep	16.25	34	50.25
1236	6′ deep	16.25	38	54.25
1242	4 @ 3" dia. Schedule 40 PVC, 2′ deep	23	37.50	60.50
1244	4′ deep	23	42	65
1246	6′ deep	23	46.50	69.50
1262	6 @ 3" dia. Schedule 40 PVC, 2′ deep	29	54	83
1264	4′ deep	29	58.50	87.50
1266	6′ deep	29	63	92
1282	8 @ 3" dia. Schedule 40 PVC, 2′ deep	36	71	107
1284	4′ deep	36	75	111
1286	6′ deep	36	80	116
1292	9 @ 3" dia. Schedule 40 PVC, 2′ deep	41.50	80	121.50
1294	4′ deep	42.50	83.50	126
1296	6′ deep	42.50	87.50	130
1312	4" dia. Schedule 40 PVC, 2′ deep	11.75	15.75	27.50
1314	4′ deep	11.75	20.50	32.25
1316	6′ deep	11.75	24.50	36.25
1322	2 @ 4" dia. Schedule 40 PVC, 2′ deep	16.45	27.50	43.95
1324	4′ deep	16.45	31.50	47.95
1326	6′ deep	16.45	36	52.45
1332	3 @ 4" dia. Schedule 40 PVC, 2′ deep	20	38.50	58.50
1334	4′ deep	20	42.50	62.50
1336	6′ deep	20	47	67
1342	4 @ 4" dia. Schedule 40 PVC, 2′ deep	29	50	79
1344	4′ deep	29	54.50	83.50
1346	6′ deep	29	59	88
1362	6 @ 4" dia. Schedule 40 PVC, 2′ deep	37.50	73	110.50
1364	4′ deep	37.50	77	114.50
1366	6′ deep	37.50	81.50	119
1382	8 @ 4" dia. Schedule 40 PVC, 2′ deep	48	96	144
1384	4′ deep	48	101	149
1386	6′ deep	48	107	155
1392	9 @ 4" dia. Schedule 40 PVC, 2′ deep	54	108	162
1394	4′ deep	54	111	165
1396	6′ deep	54	115	169
1412	5" dia. Schedule 40 PVC, 2′ deep	13.75	18.85	32.60
1414	4′ deep	13.75	23	36.75
1416	6′ deep	13.75	27.50	41.25
1422	2 @ 5" dia. Schedule 40 PVC, 2′ deep	20.50	33.50	54
1424	4′ deep	20.50	37.50	58
1426	6′ deep	20.50	42	62.50
1432	3 @ 5" dia. Schedule 40 PVC, 2′ deep	27.50	48	75.50

G4010 Electrical Distribution

G4010 320	Underground Electrical Conduit	COST L.F.		
		MAT.	INST.	TOTAL
1434	4′ deep	27.50	53	80.50
1436	6′ deep	27.50	57	84.50
1442	4 @ 5″ dia. Schedule 40 PVC, 2′ deep	37.50	62	99.50
1444	4′ deep	37.50	66.50	104
1446	6′ deep	37.50	70.50	108
1462	6 @ 5″ dia. Schedule 40 PVC, 2′ deep	49.50	90.50	140
1464	4′ deep	49.50	95.50	145
1466	6′ deep	49.50	100	149.50
1482	8 @ 5″ dia. Schedule 40 PVC, 2′ deep	65.50	121	186.50
1484	4′ deep	65.50	127	192.50
1486	6′ deep	65.50	132	197.50
1492	9 @ 5″ dia. Schedule 40 PVC, 2′ deep	73	136	209
1494	4′ deep	73	139	212
1496	6′ deep	73	143	216
1512	6″ dia. Schedule 40 PVC, 2′ deep	17.10	23	40.10
1514	4′ deep	17.10	27	44.10
1516	6′ deep	17.10	32	49.10
1522	2 @ 6″ dia. Schedule 40 PVC, 2′ deep	25	41	66
1524	4′ deep	25	45	70
1526	6′ deep	25	50	75
1532	3 @ 6″ dia. Schedule 40 PVC, 2′ deep	35	60	95
1534	4′ deep	35	65.50	100.50
1536	6′ deep	35	71	106
1542	4 @ 6″ dia. Schedule 40 PVC, 2′ deep	46	77	123
1544	4′ deep	46	82	128
1546	6′ deep	46	86	132
1562	6 @ 6″ dia. Schedule 40 PVC, 2′ deep	65	115	180
1564	4′ deep	65	120	185
1566	6′ deep	65	126	191
1582	8 @ 6″ dia. Schedule 40 PVC, 2′ deep	86.50	153	239.50
1584	4′ deep	86.50	160	246.50
1586	6′ deep	86.50	167	253.50
1592	9 @ 6″ dia. Schedule 40 PVC, 2′ deep	96.50	172	268.50
1594	4′ deep	96.50	175	271.50
1596	6′ deep	96.50	180	276.50

G4020 Site Lighting

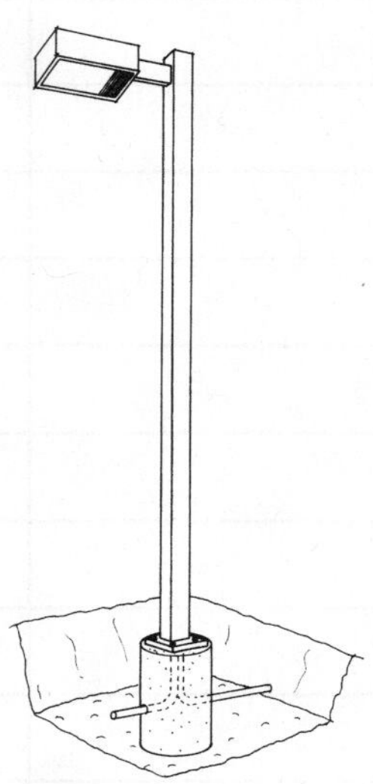

The Site Lighting System includes the complete unit from foundation to electrical fixtures. Each system includes: excavation; concrete base; backfill by hand; compaction with a plate compacter; pole of specified material; all fixtures; and lamps.

The Expanded System Listing shows Site Lighting Systems that use one of two types of lamps: high pressure sodium; and metal halide. Systems are listed for 400-watt and 1000-watt lamps. Pole height varies from 20′ to 40′. There are four types of poles possibly listed: aluminum, fiberglass, steel and wood.

System Components	QUANTITY	UNIT	COST EACH MAT.	COST EACH INST.	COST EACH TOTAL
SYSTEM G4020 110 5820					
SITE LIGHTING, METAL HALIDE, 400 WATT, ALUMINUM POLE, 20′ HIGH					
Excavating, by hand, pits to 6′ deep, heavy soil	2.368	C.Y.		274.69	274.69
Concrete in place incl. forms and reinf. stl. spread footings under 1 CY	.465	C.Y.	85.10	129.39	214.49
Backfill	1.903	C.Y.		79.93	79.93
Compact, vibrating plate	1.903	C.Y.		10.88	10.88
Aluminum light pole, 20′, no concrete base	1.000	Ea.	1,025	612	1,637
Road area luminaire wire	.300	C.L.F.	11.10	24.60	35.70
Roadway area luminaire, metal halide 400 W	1.000	Ea.	610	298	908
TOTAL			1,731.20	1,429.49	3,160.69

G4020 110	Site Lighting	COST EACH MAT.	COST EACH INST.	COST EACH TOTAL
2320	Site lighting, high pressure sodium, 400 watt, aluminum pole, 20′ high	1,750	1,425	3,175
2340	30′ high	2,700	1,750	4,450
2360	40′ high	3,200	2,250	5,450
2520	Fiberglass pole, 20′ high	1,475	1,275	2,750
2540	30′ high	1,675	1,575	3,250
2560	40′ high	2,450	2,000	4,450
2720	Steel pole, 20′ high	1,925	1,500	3,425
2740	30′ high	2,175	1,850	4,025
2760	40′ high	2,725	2,425	5,150
2920	Wood pole, 20′ high	985	1,400	2,385
2940	30′ high	1,225	1,750	2,975
2960	40′ high	1,575	2,150	3,725
3120	1000 watt, aluminum pole, 20′ high	1,850	1,450	3,300
3140	30′ high	2,800	1,775	4,575
3160	40′ high	3,300	2,300	5,600
3320	Fiberglass pole, 20′ high	1,550	1,300	2,850
3340	30′ high	1,775	1,600	3,375
3360	40′ high	2,550	2,025	4,575
3420	Steel pole, 20′ high	2,025	1,525	3,550
3440	30′ high	2,275	1,875	4,150

G4020 Site Lighting

G4020 110	Site Lighting	COST EACH		
		MAT.	INST.	TOTAL
3460	40' high	2,825	2,450	5,275
3520	Wood pole, 20' high	1,075	1,425	2,500
3540	30' high	1,325	1,775	3,100
3560	40' high	1,650	2,175	3,825
5820	Metal halide, 400 watt, aluminum pole, 20' high	1,725	1,425	3,150
5840	30' high	2,675	1,750	4,425
5860	40' high	3,175	2,250	5,425
6120	Fiberglass pole, 20' high	1,450	1,275	2,725
6140	30' high	1,650	1,575	3,225
6160	40' high	2,425	2,000	4,425
6320	Steel pole, 20' high	1,900	1,500	3,400
6340	30' high	2,150	1,850	4,000
6360	40' high	2,700	2,425	5,125
7620	1000 watt, aluminum pole, 20' high	1,800	1,450	3,250
7640	30' high	2,750	1,775	4,525
7660	40' high	3,250	2,300	5,550
7900	Fiberglass pole, 20' high	1,525	1,300	2,825
7920	30' high	1,725	1,600	3,325
7940	40' high	2,500	2,025	4,525
8120	Steel pole, 20' high	1,975	1,525	3,500
8140	30' high	2,225	1,875	4,100
8160	40' high	2,775	2,450	5,225

G4020 Site Lighting

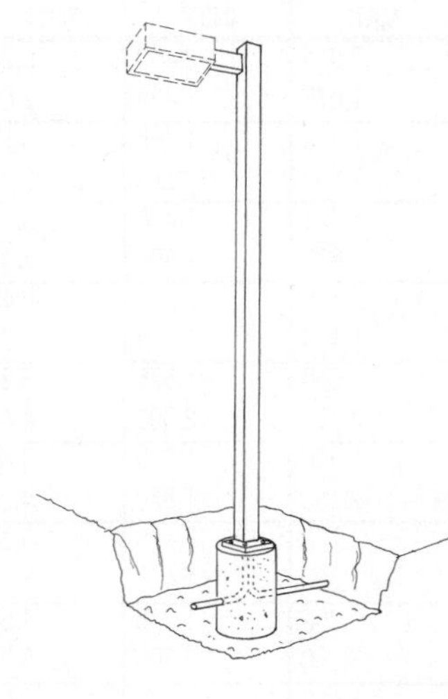

Table G4020 210 Procedure for Calculating Floodlights Required for Various Footcandles
Poles should not be spaced more than 4 times the fixture mounting height for good light distribution.

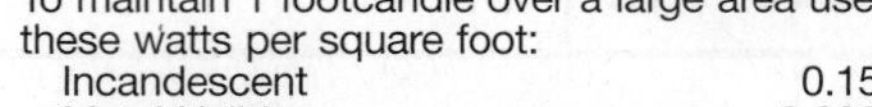

To maintain 1 footcandle over a large area use these watts per square foot:

Incandescent	0.15
Metal Halide	0.032
Mercury Vapor	0.05
High Pressure Sodium	0.024

Estimating Chart
Select Lamp type.

Determine total square feet.

Chart will show quantity of fixtures to provide 1 footcandle initial, at intersection of lines. Multiply fixture quantity by desired footcandle level.

Chart based on use of wide beam luminaires in an area whose dimensions are large compared to mounting height and is approximate only.

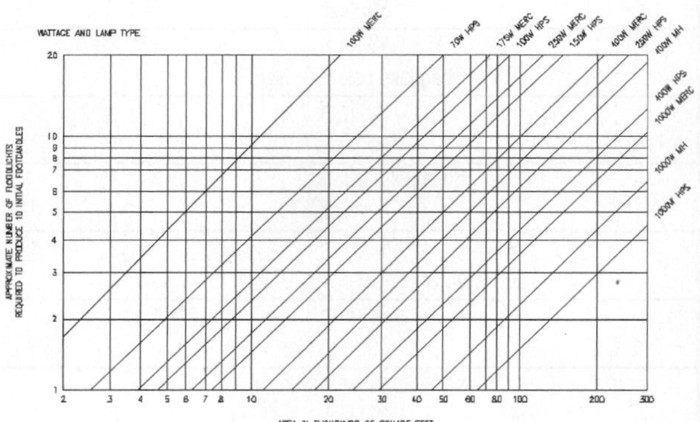

System Components	QUANTITY	UNIT	COST EACH MAT.	INST.	TOTAL
SYSTEM G4020 210 0200					
LIGHT POLES, ALUMINUM, 20′ HIGH, 1 ARM BRACKET					
Aluminum light pole, 20′, no concrete base	1.000	Ea.	1,025	612	1,637
Bracket arm for Aluminum light pole	1.000	Ea.	133	82	215
Excavation by hand, pits to 6′ deep, heavy soil or clay	2.368	C.Y.		274.69	274.69
Footing, concrete incl forms, reinforcing, spread, under 1 C.Y.	.465	C.Y.	85.10	129.39	214.49
Backfill by hand	1.903	C.Y.		79.93	79.93
Compaction vibrating plate	1.903	C.Y.		10.88	10.88
TOTAL			1,243.10	1,188.89	2,431.99

G4020 210	Light Pole (Installed)	COST EACH MAT.	INST.	TOTAL
0200	Light pole, aluminum, 20′ high, 1 arm bracket	1,250	1,175	2,425
0240	2 arm brackets	1,375	1,175	2,550
0280	3 arm brackets	1,500	1,225	2,725
0320	4 arm brackets	1,650	1,225	2,875
0360	30′ high, 1 arm bracket	2,200	1,500	3,700
0400	2 arm brackets	2,325	1,500	3,825
0440	3 arm brackets	2,450	1,525	3,975
0480	4 arm brackets	2,600	1,525	4,125
0680	40′ high, 1 arm bracket	2,675	2,000	4,675
0720	2 arm brackets	2,825	2,000	4,825
0760	3 arm brackets	2,950	2,050	5,000
0800	4 arm brackets	3,100	2,050	5,150
0840	Steel, 20′ high, 1 arm bracket	1,475	1,250	2,725
0880	2 arm brackets	1,525	1,250	2,775
0920	3 arm brackets	1,575	1,300	2,875
0960	4 arm brackets	1,625	1,300	2,925
1000	30′ high, 1 arm bracket	1,725	1,600	3,325
1040	2 arm brackets	1,775	1,600	3,375
1080	3 arm brackets	1,800	1,650	3,450
1120	4 arm brackets	1,875	1,650	3,525
1320	40′ high, 1 arm bracket	2,275	2,175	4,450
1360	2 arm brackets	2,325	2,175	4,500
1400	3 arm brackets	2,350	2,225	4,575
1440	4 arm brackets	2,425	2,225	4,650

Reference Section

All the reference information is in one section, making it easy to find what you need to know . . . and easy to use the book on a daily basis. This section is visually identified by a vertical black bar on the page edges.

In this Reference Section, we've included Equipment Rental Costs, a listing of rental and operating costs; Crew Listings, a full listing of all crews, equipment, and their costs; Historical Cost Indexes for cost comparisons over time; City Cost Indexes and Location Factors for adjusting costs to the region you are in; Reference Tables, where you will find explanations, estimating information and procedures, or technical data; Square Foot Costs that allow you to make a rough estimate for the overall cost of a project; and an explanation of all the Abbreviations in the book.

Table of Contents

Estimating Tips

- This section contains the average costs to rent and operate hundreds of pieces of construction equipment. This is useful information when estimating the time and material requirements of any particular operation in order to establish a unit or total cost. Equipment costs include not only rental, but also operating costs for equipment under normal use.

Rental Costs

- Equipment rental rates are obtained from the following industry sources throughout North America: contractors, suppliers, dealers, manufacturers, and distributors.
- Rental rates vary throughout the country, with larger cities generally having lower rates. Lease plans for new equipment are available for periods in excess of six months, with a percentage of payments applying toward purchase.
- Monthly rental rates vary from 2% to 5% of the purchase price of the equipment depending on the anticipated life of the equipment and its wearing parts.
- Weekly rental rates are about 1/3 the monthly rates, and daily rental rates are about 1/3 the weekly rate.
- Rental rates can also be treated as reimbursement costs for contractor-owned equipment. Owned equipment costs include depreciation, loan payments, interest, taxes, insurance, storage, and major repairs.

Operating Costs

- The operating costs include parts and labor for routine servicing, such as repair and replacement of pumps, filters and worn lines. Normal operating expendables, such as fuel, lubricants, tires and electricity (where applicable), are also included.
- Extraordinary operating expendables with highly variable wear patterns, such as diamond bits and blades, are excluded. These costs can be found as material costs in the Unit Price section.
- The hourly operating costs listed do not include the operator's wages.

Equipment Cost/Day

- Any power equipment required by a crew is shown in the Crew Listings with a daily cost.
- The daily cost of equipment needed by a crew is based on dividing the weekly rental rate by 5 (number of working days in the week), and then adding the hourly operating cost times 8 (the number of hours in a day). This "Equipment Cost/Day" is shown in the far right column of the Equipment Rental pages.
- If equipment is needed for only one or two days, it is best to develop your own cost by including components for daily rent and hourly operating cost. This is important when the listed Crew for a task does not contain the equipment needed, such as a crane for lifting mechanical heating/cooling equipment up onto a roof.
- If the quantity of work is less than the crew's Daily Output shown for a Unit Price line item that includes a bare unit equipment cost, it is recommended to estimate one day's rental cost and operating cost for equipment shown in the Crew Listing for that line item.

Mobilization/ Demobilization

- The cost to move construction equipment from an equipment yard or rental company to the job site and back again is not included in equipment rental costs listed in the Reference Section, nor in the bare equipment cost of any unit price line item, nor in any equipment costs shown in the Crew listings.
- Mobilization (to the site) and demobilization (from the site) costs can be found in the Unit Price Section.
- If a piece of equipment is already at the job site, it is not appropriate to utilize mobilization/demobilization. costs again in an estimate. ■

01 54 33 | Equipment Rental

			UNIT	HOURLY OPER. COST	RENT PER DAY	RENT PER WEEK	RENT PER MONTH	EQUIPMENT COST/DAY	
10	0010	**CONCRETE EQUIPMENT RENTAL** without operators R015433-10							10
	0200	Bucket, concrete lightweight, 1/2 C.Y.	Ea.	.80	23.50	70	210	20.40	
	0300	1 C.Y.		.90	27.50	83	249	23.80	
	0400	1-1/2 C.Y.		1.10	36.50	110	330	30.80	
	0500	2 C.Y.		1.25	45	135	405	37	
	0580	8 C.Y.		6.10	257	770	2,300	202.80	
	0600	Cart, concrete, self-propelled, operator walking, 10 C.F.		3.25	56.50	170	510	60	
	0700	Operator riding, 18 C.F.		5.35	95	285	855	99.80	
	0800	Conveyer for concrete, portable, gas, 16″ wide, 26′ long		12.60	123	370	1,100	174.80	
	0900	46′ long		13.00	148	445	1,325	193	
	1000	56′ long		13.10	157	470	1,400	198.80	
	1100	Core drill, electric, 2-1/2 H.P., 1″ to 8″ bit diameter		1.77	68.50	205	615	55.15	
	1150	11 H.P., 8″ to 18″ cores		5.95	113	340	1,025	115.60	
	1200	Finisher, concrete floor, gas, riding trowel, 96″ wide		12.75	145	435	1,300	189	
	1300	Gas, walk-behind, 3 blade, 36″ trowel		2.15	20	60	180	29.20	
	1400	4 blade, 48″ trowel		4.20	27.50	83	249	50.20	
	1500	Float, hand-operated (Bull float) 48″ wide		.08	13.35	40	120	8.65	
	1570	Curb builder, 14 H.P., gas, single screw		14.65	248	745	2,225	266.20	
	1590	Double screw		15.35	290	870	2,600	296.80	
	1600	Floor grinder, concrete and terrazzo, electric, 22″ path		2.57	158	475	1,425	115.55	
	1700	Edger, concrete, electric, 7″ path		1.04	51.50	155	465	39.30	
	1750	Vacuum pick-up system for floor grinders, wet/dry		1.49	81.50	245	735	60.90	
	1800	Mixer, powered, mortar and concrete, gas, 6 C.F., 18 H.P.		8.65	118	355	1,075	140.20	
	1900	10 C.F., 25 H.P.		10.80	143	430	1,300	172.40	
	2000	16 C.F.		11.15	165	495	1,475	188.20	
	2100	Concrete, stationary, tilt drum, 2 C.Y.		6.90	232	695	2,075	194.20	
	2120	Pump, concrete, truck mounted 4″ line 80′ boom		25.00	865	2,600	7,800	720	
	2140	5″ line, 110′ boom		32.40	1,150	3,435	10,300	946.20	
	2160	Mud jack, 50 C.F. per hr.		7.30	125	375	1,125	133.40	
	2180	225 C.F. per hr.		9.50	145	435	1,300	163	
	2190	Shotcrete pump rig, 12 C.Y./hr.		14.85	220	660	1,975	250.80	
	2200	35 C.Y./hr.		17.50	235	705	2,125	281	
	2600	Saw, concrete, manual, gas, 18 H.P.		6.75	45	135	405	81	
	2650	Self-propelled, gas, 30 H.P.		13.30	102	305	915	167.40	
	2675	V-groove crack chaser, manual, gas, 6 H.P.		2.35	17.35	52	156	29.20	
	2700	Vibrators, concrete, electric, 60 cycle, 2 H.P.		.46	8.65	26	78	8.90	
	2800	3 H.P.		.60	11.65	35	105	11.80	
	2900	Gas engine, 5 H.P.		2.00	16	48	144	25.60	
	3000	8 H.P.		2.75	15.35	46	138	31.20	
	3050	Vibrating screed, gas engine, 8 H.P.		2.91	71.50	215	645	66.30	
	3120	Concrete transit mixer, 6 x 4, 250 H.P., 8 C.Y., rear discharge		61.55	570	1,715	5,150	835.40	
	3200	Front discharge		72.45	700	2,095	6,275	998.60	
	3300	6 x 6, 285 H.P., 12 C.Y., rear discharge		71.50	660	1,980	5,950	968	
	3400	Front discharge	↓	74.00	705	2,120	6,350	1,016	
20	0010	**EARTHWORK EQUIPMENT RENTAL** without operators R015433-10							20
	0040	Aggregate spreader, push type 8′ to 12′ wide	Ea.	3.05	25.50	76	228	39.60	
	0045	Tailgate type, 8′ wide		2.90	32.50	98	294	42.80	
	0055	Earth auger, truck-mounted, for fence & sign posts, utility poles		19.15	440	1,320	3,950	417.20	
	0060	For borings and monitoring wells		46.65	675	2,030	6,100	779.20	
	0070	Portable, trailer mounted		3.30	32.50	98	294	46	
	0075	Truck-mounted, for caissons, water wells		100.15	2,900	8,705	26,100	2,542	
	0080	Horizontal boring machine, 12″ to 36″ diameter, 45 H.P.		24.20	192	575	1,725	308.60	
	0090	12″ to 48″ diameter, 65 H.P.		34.35	335	1,000	3,000	474.80	
	0095	Auger, for fence posts, gas engine, hand held		.60	6	18	54	8.40	
	0100	Excavator, diesel hydraulic, crawler mounted, 1/2 C.Y. cap.		24.65	420	1,255	3,775	448.20	
	0120	5/8 C.Y. capacity		32.55	550	1,645	4,925	589.40	
	0140	3/4 C.Y. capacity		35.60	615	1,850	5,550	654.80	
	0150	1 C.Y. capacity	↓	49.70	690	2,075	6,225	812.60	

01 54 33 | Equipment Rental

			UNIT	HOURLY OPER. COST	RENT PER DAY	RENT PER WEEK	RENT PER MONTH	EQUIPMENT COST/DAY	
20	0200	1-1/2 C.Y. capacity	Ea.	59.80	920	2,760	8,275	1,030	20
	0300	2 C.Y. capacity		67.35	1,050	3,180	9,550	1,175	
	0320	2-1/2 C.Y. capacity		102.55	1,300	3,925	11,800	1,605	
	0325	3-1/2 C.Y. capacity		144.45	2,150	6,420	19,300	2,440	
	0330	4-1/2 C.Y. capacity		173.40	2,625	7,880	23,600	2,963	
	0335	6 C.Y. capacity		219.80	2,900	8,680	26,000	3,494	
	0340	7 C.Y. capacity		222.05	3,025	9,070	27,200	3,590	
	0342	Excavator attachments, bucket thumbs		3.20	245	735	2,200	172.60	
	0345	Grapples		2.75	193	580	1,750	138	
	0346	Hydraulic hammer for boom mounting, 4000 ft lb.		12.40	350	1,045	3,125	308.20	
	0347	5000 ft lb.		14.40	425	1,280	3,850	371.20	
	0348	8000 ft lb.		21.30	625	1,875	5,625	545.40	
	0349	12,000 ft lb.		23.35	750	2,245	6,725	635.80	
	0350	Gradall type, truck mounted, 3 ton @ 15′ radius, 5/8 C.Y.		43.65	890	2,665	8,000	882.20	
	0370	1 C.Y. capacity		47.85	1,025	3,095	9,275	1,002	
	0400	Backhoe-loader, 40 to 45 H.P., 5/8 C.Y. capacity		14.90	240	720	2,150	263.20	
	0450	45 H.P. to 60 H.P., 3/4 C.Y. capacity		23.40	295	885	2,650	364.20	
	0460	80 H.P., 1-1/4 C.Y. capacity		25.60	310	935	2,800	391.80	
	0470	112 H.P., 1-1/2 C.Y. capacity		40.95	645	1,930	5,800	713.60	
	0482	Backhoe-loader attachment, compactor, 20,000 lb.		5.80	142	425	1,275	131.40	
	0485	Hydraulic hammer, 750 ft lb.		3.30	96.50	290	870	84.40	
	0486	Hydraulic hammer, 1200 ft lb.		6.30	217	650	1,950	180.40	
	0500	Brush chipper, gas engine, 6″ cutter head, 35 H.P.		11.10	105	315	945	151.80	
	0550	Diesel engine, 12″ cutter head, 130 H.P.		27.55	285	855	2,575	391.40	
	0600	15″ cutter head, 165 H.P.		33.10	335	1,005	3,025	465.80	
	0750	Bucket, clamshell, general purpose, 3/8 C.Y.		1.30	38.50	115	345	33.40	
	0800	1/2 C.Y.		1.40	45	135	405	38.20	
	0850	3/4 C.Y.		1.55	55	165	495	45.40	
	0900	1 C.Y.		1.60	58.50	175	525	47.80	
	0950	1-1/2 C.Y.		2.55	81.50	245	735	69.40	
	1000	2 C.Y.		2.70	90	270	810	75.60	
	1010	Bucket, dragline, medium duty, 1/2 C.Y.		.75	23.50	70	210	20	
	1020	3/4 C.Y.		.75	24.50	73	219	20.60	
	1030	1 C.Y.		.80	26	78	234	22	
	1040	1-1/2 C.Y.		1.20	40	120	360	33.60	
	1050	2 C.Y.		1.30	45	135	405	37.40	
	1070	3 C.Y.		1.95	61.50	185	555	52.60	
	1200	Compactor, manually guided 2-drum vibratory smooth roller, 7.5 H.P.		7.25	203	610	1,825	180	
	1250	Rammer/tamper, gas, 8″		2.75	46.50	140	420	50	
	1260	15″		3.05	53.50	160	480	56.40	
	1300	Vibratory plate, gas, 18″ plate, 3000 lb. blow		2.70	24.50	73	219	36.20	
	1350	21″ plate, 5000 lb. blow		3.40	32.50	98	294	46.80	
	1370	Curb builder/extruder, 14 H.P., gas, single screw		14.65	248	745	2,225	266.20	
	1390	Double screw		15.35	290	870	2,600	296.80	
	1500	Disc harrow attachment, for tractor		.44	73.50	221	665	47.70	
	1810	Feller buncher, shearing & accumulating trees, 100 H.P.		48.30	755	2,260	6,775	838.40	
	1860	Grader, self-propelled, 25,000 lb.		39.50	640	1,925	5,775	701	
	1910	30,000 lb.		43.15	650	1,955	5,875	736.20	
	1920	40,000 lb.		64.90	1,100	3,275	9,825	1,174	
	1930	55,000 lb.		84.15	1,700	5,130	15,400	1,699	
	1950	Hammer, pavement breaker, self-propelled, diesel, 1000 to 1250 lb.		29.90	350	1,055	3,175	450.20	
	2000	1300 to 1500 lb.		44.85	705	2,110	6,325	780.80	
	2050	Pile driving hammer, steam or air, 4150 ft lb. @ 225 bpm		10.40	475	1,420	4,250	367.20	
	2100	8750 ft lb. @ 145 bpm		12.40	660	1,975	5,925	494.20	
	2150	15,000 ft lb. @ 60 bpm		14.00	790	2,370	7,100	586	
	2200	24,450 ft lb. @ 111 bpm		15.00	875	2,630	7,900	646	
	2250	Leads, 60′ high for pile driving hammers up to 20,000 ft lb.		3.30	80.50	242	725	74.80	
	2300	90′ high for hammers over 20,000 ft lb.	↓	4.95	141	424	1,275	124.40	

01 54 33 | Equipment Rental

			UNIT	HOURLY OPER. COST	RENT PER DAY	RENT PER WEEK	RENT PER MONTH	EQUIPMENT COST/DAY	
20	2350	Diesel type hammer, 22,400 ft lb.	Ea.	22.55	460	1,380	4,150	456.40	20
	2400	41,300 ft lb.		32.00	540	1,625	4,875	581	
	2450	141,000 ft lb.		51.65	960	2,875	8,625	988.20	
	2500	Vib. elec. hammer/extractor, 200 kW diesel generator, 34 H.P.		54.90	635	1,900	5,700	819.20	
	2550	80 H.P.		100.65	925	2,775	8,325	1,360	
	2600	150 H.P.		190.90	1,775	5,300	15,900	2,587	
	2800	Log chipper, up to 22" diameter, 600 H.P.		63.30	640	1,915	5,750	889.40	
	2850	Logger, for skidding & stacking logs, 150 H.P.		52.75	815	2,445	7,325	911	
	2860	Mulcher, diesel powered, trailer mounted		24.05	212	635	1,900	319.40	
	2900	Rake, spring tooth, with tractor		18.23	345	1,042	3,125	354.25	
	3000	Roller, vibratory, tandem, smooth drum, 20 H.P.		8.80	145	435	1,300	157.40	
	3050	35 H.P.		11.10	247	740	2,225	236.80	
	3100	Towed type vibratory compactor, smooth drum, 50 H.P.		26.00	315	940	2,825	396	
	3150	Sheepsfoot, 50 H.P.		27.35	345	1,035	3,100	425.80	
	3170	Landfill compactor, 220 H.P.		91.00	1,475	4,440	13,300	1,616	
	3200	Pneumatic tire roller, 80 H.P.		16.80	350	1,050	3,150	344.40	
	3250	120 H.P.		25.20	600	1,800	5,400	561.60	
	3300	Sheepsfoot vibratory roller, 240 H.P.		68.95	1,100	3,270	9,800	1,206	
	3320	340 H.P.		100.85	1,575	4,730	14,200	1,753	
	3350	Smooth drum vibratory roller, 75 H.P.		24.85	585	1,755	5,275	549.80	
	3400	125 H.P.		32.55	710	2,135	6,400	687.40	
	3410	Rotary mower, brush, 60", with tractor		22.60	315	950	2,850	370.80	
	3420	Rototiller, walk-behind, gas, 5 H.P.		1.96	51.50	155	465	46.70	
	3422	8 H.P.		3.08	80	240	720	72.65	
	3440	Scrapers, towed type, 7 C.Y. capacity		5.80	113	340	1,025	114.40	
	3450	10 C.Y. capacity		6.60	157	470	1,400	146.80	
	3500	15 C.Y. capacity		7.10	180	540	1,625	164.80	
	3525	Self-propelled, single engine, 14 C.Y. capacity		114.76	1,600	4,830	14,500	1,884	
	3550	Dual engine, 21 C.Y. capacity		173.80	2,175	6,490	19,500	2,688	
	3600	31 C.Y. capacity		231.70	3,075	9,195	27,600	3,693	
	3640	44 C.Y. capacity		287.80	3,950	11,885	35,700	4,679	
	3650	Elevating type, single engine, 11 C.Y. capacity		71.45	985	2,955	8,875	1,163	
	3700	22 C.Y. capacity		139.55	2,350	7,025	21,100	2,521	
	3710	Screening plant 110 H.P. w/5' x 10' screen		25.35	400	1,200	3,600	442.80	
	3720	5' x 16' screen		29.65	505	1,515	4,550	540.20	
	3850	Shovel, crawler-mounted, front-loading, 7 C.Y. capacity		258.85	3,200	9,610	28,800	3,993	
	3855	12 C.Y. capacity		386.45	4,400	13,220	39,700	5,736	
	3860	Shovel/backhoe bucket, 1/2 C.Y.		2.50	66.50	200	600	60	
	3870	3/4 C.Y.		2.55	75	225	675	65.40	
	3880	1 C.Y.		2.65	83.50	250	750	71.20	
	3890	1-1/2 C.Y.		2.75	96.50	290	870	80	
	3910	3 C.Y.		3.15	130	390	1,175	103.20	
	3950	Stump chipper, 18" deep, 30 H.P.		7.82	205	615	1,850	185.55	
	4110	Dozer, crawler, torque converter, diesel 80 H.P.		29.05	400	1,200	3,600	472.40	
	4150	105 H.P.		35.60	530	1,590	4,775	602.80	
	4200	140 H.P.		51.25	795	2,380	7,150	886	
	4260	200 H.P.		77.25	1,275	3,845	11,500	1,387	
	4310	300 H.P.		100.65	1,825	5,460	16,400	1,897	
	4360	410 H.P.		132.60	2,250	6,720	20,200	2,405	
	4370	500 H.P.		171.15	3,225	9,650	29,000	3,299	
	4380	700 H.P.		280.90	4,550	13,660	41,000	4,979	
	4400	Loader, crawler, torque conv., diesel, 1-1/2 C.Y., 80 H.P.		31.25	455	1,360	4,075	522	
	4450	1-1/2 to 1-3/4 C.Y., 95 H.P.		34.60	590	1,765	5,300	629.80	
	4510	1-3/4 to 2-1/4 C.Y., 130 H.P.		53.80	875	2,630	7,900	956.40	
	4530	2-1/2 to 3-1/4 C.Y., 190 H.P.		65.95	1,100	3,300	9,900	1,188	
	4560	3-1/2 to 5 C.Y., 275 H.P.		87.80	1,425	4,295	12,900	1,561	
	4610	Front end loader, 4WD, articulated frame, diesel, 1 to 1-1/4 C.Y., 70 H.P.		19.75	235	705	2,125	299	
20	4620	1-1/2 to 1-3/4 C.Y., 95 H.P.	↓	24.80	300	900	2,700	378.40	20

01 54 33 | Equipment Rental

			UNIT	HOURLY OPER. COST	RENT PER DAY	RENT PER WEEK	RENT PER MONTH	EQUIPMENT COST/DAY	
20	4650	1-3/4 to 2 C.Y., 130 H.P.	Ea.	29.15	380	1,140	3,425	461.20	20
	4710	2-1/2 to 3-1/2 C.Y., 145 H.P.		33.10	425	1,275	3,825	519.80	
	4730	3 to 4-1/2 C.Y., 185 H.P.		42.40	550	1,650	4,950	669.20	
	4760	5-1/4 to 5-3/4 C.Y., 270 H.P.		67.50	905	2,710	8,125	1,082	
	4810	7 to 9 C.Y., 475 H.P.		114.05	1,700	5,125	15,400	1,937	
	4870	9 - 11 C.Y., 620 H.P.		160.55	2,775	8,290	24,900	2,942	
	4880	Skid steer loader, wheeled, 10 C.F., 30 H.P. gas		10.45	150	450	1,350	173.60	
	4890	1 C.Y., 78 H.P., diesel		20.20	262	785	2,350	318.60	
	4892	Skid-steer attachment, auger		.48	80.50	242	725	52.25	
	4893	Backhoe		.66	111	332	995	71.70	
	4894	Broom		.74	124	372	1,125	80.30	
	4895	Forks		.22	36	108	325	23.35	
	4896	Grapple		.56	92.50	278	835	60.10	
	4897	Concrete hammer		1.06	177	531	1,600	114.70	
	4898	Tree spade		1.03	172	515	1,550	111.25	
	4899	Trencher		.54	90	270	810	58.30	
	4900	Trencher, chain, boom type, gas, operator walking, 12 H.P.		5.00	46.50	140	420	68	
	4910	Operator riding, 40 H.P.		20.15	290	870	2,600	335.20	
	5000	Wheel type, diesel, 4′ deep, 12″ wide		85.95	845	2,535	7,600	1,195	
	5100	6′ deep, 20″ wide		92.75	1,925	5,740	17,200	1,890	
	5150	Chain type, diesel, 5′ deep, 8″ wide		38.35	560	1,675	5,025	641.80	
	5200	Diesel, 8′ deep, 16″ wide		155.95	3,600	10,770	32,300	3,402	
	5202	Rock trencher, wheel type, 6″ wide x 18″ deep		21.50	335	1,000	3,000	372	
	5206	Chain type, 18″ wide x 7′ deep		112.00	2,800	8,395	25,200	2,575	
	5210	Tree spade, self-propelled		14.38	267	800	2,400	275.05	
	5250	Truck, dump, 2-axle, 12 ton, 8 C.Y. payload, 220 H.P.		34.40	227	680	2,050	411.20	
	5300	Three axle dump, 16 ton, 12 C.Y. payload, 400 H.P.		61.00	340	1,015	3,050	691	
	5310	Four axle dump, 25 ton, 18 C.Y. payload, 450 H.P.		71.75	490	1,475	4,425	869	
	5350	Dump trailer only, rear dump, 16-1/2 C.Y.		5.45	138	415	1,250	126.60	
	5400	20 C.Y.		5.90	157	470	1,400	141.20	
	5450	Flatbed, single axle, 1-1/2 ton rating		25.55	68.50	205	615	245.40	
	5500	3 ton rating		30.65	96.50	290	870	303.20	
	5550	Off highway rear dump, 25 ton capacity		73.90	1,275	3,800	11,400	1,351	
	5600	35 ton capacity		82.50	1,425	4,260	12,800	1,512	
	5610	50 ton capacity		102.65	1,700	5,080	15,200	1,837	
	5620	65 ton capacity		105.65	1,700	5,090	15,300	1,863	
	5630	100 ton capacity		151.40	2,850	8,570	25,700	2,925	
	6000	Vibratory plow, 25 H.P., walking	↓	8.45	60	180	540	103.60	
40	0010	**GENERAL EQUIPMENT RENTAL** without operators R015433 -10							40
	0150	Aerial lift, scissor type, to 15′ high, 1000 lb. cap., electric	Ea.	3.05	51.50	155	465	55.40	
	0160	To 25′ high, 2000 lb. capacity		3.45	66.50	200	600	67.60	
	0170	Telescoping boom to 40′ high, 500 lb. capacity, diesel		13.55	320	965	2,900	301.40	
	0180	To 45′ high, 500 lb. capacity		14.75	350	1,055	3,175	329	
	0190	To 60′ high, 600 lb. capacity		17.20	495	1,490	4,475	435.60	
	0195	Air compressor, portable, 6.5 CFM, electric		.67	12.65	38	114	12.95	
	0196	Gasoline		.81	19	57	171	17.90	
	0200	Towed type, gas engine, 60 CFM		13.55	48.50	145	435	137.40	
	0300	160 CFM		15.80	50	150	450	156.40	
	0400	Diesel engine, rotary screw, 250 CFM		17.05	108	325	975	201.40	
	0500	365 CFM		23.10	132	395	1,175	263.80	
	0550	450 CFM		29.35	165	495	1,475	333.80	
	0600	600 CFM		51.70	228	685	2,050	550.60	
	0700	750 CFM	↓	51.90	237	710	2,125	557.20	
	0800	For silenced models, small sizes, add to rent		3%	5%	5%	5%		
	0900	Large sizes, add to rent		5%	7%	7%	7%		
	0930	Air tools, breaker, pavement, 60 lb.	Ea.	.50	9.65	29	87	9.80	
	0940	80 lb.		.50	10.35	31	93	10.20	
	0950	Drills, hand (jackhammer) 65 lb.	↓	.60	17.35	52	156	15.20	

01 54 33 | Equipment Rental

			UNIT	HOURLY OPER. COST	RENT PER DAY	RENT PER WEEK	RENT PER MONTH	EQUIPMENT COST/DAY	
40	0960	Track or wagon, swing boom, 4″ drifter	Ea.	62.50	880	2,640	7,925	1,028	40
	0970	5″ drifter		74.45	1,075	3,235	9,700	1,243	
	0975	Track mounted quarry drill, 6″ diameter drill		128.10	1,600	4,765	14,300	1,978	
	0980	Dust control per drill		1.02	23.50	71	213	22.35	
	0990	Hammer, chipping, 12 lb.		.55	26	78	234	20	
	1000	Hose, air with couplings, 50′ long, 3/4″ diameter		.03	5	15	45	3.25	
	1100	1″ diameter		.04	6.35	19	57	4.10	
	1200	1-1/2″ diameter		.05	9	27	81	5.80	
	1300	2″ diameter		.07	12	36	108	7.75	
	1400	2-1/2″ diameter		.11	19	57	171	12.30	
	1410	3″ diameter		.14	23	69	207	14.90	
	1450	Drill, steel, 7/8″ x 2′		.05	8.65	26	78	5.60	
	1460	7/8″ x 6′		.05	9	27	81	5.80	
	1520	Moil points		.02	3.33	10	30	2.15	
	1525	Pneumatic nailer w/accessories		.58	38.50	115	345	27.65	
	1530	Sheeting driver for 60 lb. breaker		.04	6	18	54	3.90	
	1540	For 90 lb. breaker		.12	8	24	72	5.75	
	1550	Spade, 25 lb.		.45	6.65	20	60	7.60	
	1560	Tamper, single, 35 lb.		.55	36.50	109	325	26.20	
	1570	Triple, 140 lb.		.82	54.50	164	490	39.35	
	1580	Wrenches, impact, air powered, up to 3/4″ bolt		.40	12.65	38	114	10.80	
	1590	Up to 1-1/4″ bolt		.50	23.50	70	210	18	
	1600	Barricades, barrels, reflectorized, 1 to 99 barrels		.03	4.60	13.80	41.50	3	
	1610	100 to 200 barrels		.02	3.53	10.60	32	2.30	
	1620	Barrels with flashers, 1 to 99 barrels		.03	5.25	15.80	47.50	3.40	
	1630	100 to 200 barrels		.03	4.20	12.60	38	2.75	
	1640	Barrels with steady burn type C lights		.04	7	21	63	4.50	
	1650	Illuminated board, trailer mounted, with generator		3.50	130	390	1,175	106	
	1670	Portable barricade, stock, with flashers, 1 to 6 units		.03	5.25	15.80	47.50	3.40	
	1680	25 to 50 units		.03	4.90	14.70	44	3.20	
	1685	Butt fusion machine, wheeled, 1.5 HP electric, 2″ - 8″ diameter pipe		2.63	167	500	1,500	121.05	
	1690	Tracked, 20 HP diesel, 4″-12″ diameter pipe		11.14	490	1,465	4,400	382.10	
	1695	83 HP diesel, 8″ - 24″ diameter pipe		30.46	975	2,930	8,800	829.70	
	1700	Carts, brick, hand powered, 1000 lb. capacity		.50	83.50	251	755	54.20	
	1800	Gas engine, 1500 lb., 7-1/2′ lift		4.17	115	345	1,025	102.35	
	1822	Dehumidifier, medium, 6 lb./hr., 150 CFM		1.00	62	186	560	45.20	
	1824	Large, 18 lb./hr., 600 CFM		2.02	126	378	1,125	91.75	
	1830	Distributor, asphalt, trailer mounted, 2000 gal., 38 H.P. diesel		9.95	325	980	2,950	275.60	
	1840	3000 gal., 38 H.P. diesel		11.45	355	1,070	3,200	305.60	
	1850	Drill, rotary hammer, electric		1.02	26	78	234	23.75	
	1860	Carbide bit, 1-1/2″ diameter, add to electric rotary hammer		.02	3.61	10.84	32.50	2.35	
	1865	Rotary, crawler, 250 H.P.		148.50	2,050	6,185	18,600	2,425	
	1870	Emulsion sprayer, 65 gal., 5 H.P. gas engine		2.91	97	291	875	81.50	
	1880	200 gal., 5 H.P. engine		7.85	162	485	1,450	159.80	
	1900	Floor auto-scrubbing machine, walk-behind, 28″ path		4.98	325	970	2,900	233.85	
	1930	Floodlight, mercury vapor, or quartz, on tripod, 1000 watt		.44	20.50	62	186	15.90	
	1940	2000 watt		.83	41	123	370	31.25	
	1950	Floodlights, trailer mounted with generator, 1 - 300 watt light		3.65	71.50	215	645	72.20	
	1960	2 - 1000 watt lights		4.85	96.50	290	870	96.80	
	2000	4 - 300 watt lights		4.55	93.50	280	840	92.40	
	2005	Foam spray rig, incl. box trailer, compressor, generator, proportioner		35.03	490	1,465	4,400	573.25	
	2020	Forklift, straight mast, 12′ lift, 5000 lb., 2 wheel drive, gas		26.35	202	605	1,825	331.80	
	2040	21′ lift, 5000 lb., 4 wheel drive, diesel		20.90	240	720	2,150	311.20	
	2050	For rough terrain, 42′ lift, 35′ reach, 9000 lb., 110 H.P.		29.55	485	1,450	4,350	526.40	
	2060	For plant, 4 ton capacity, 80 H.P., 2 wheel drive, gas		16.15	93.50	280	840	185.20	
	2080	10 ton capacity, 120 H.P., 2 wheel drive, diesel		24.10	162	485	1,450	289.80	
	2100	Generator, electric, gas engine, 1.5 kW to 3 kW		3.75	11.35	34	102	36.80	
	2200	5 kW	↓	4.80	15.35	46	138	47.60	

01 54 33 | Equipment Rental

	Line	Description	UNIT	HOURLY OPER. COST	RENT PER DAY	RENT PER WEEK	RENT PER MONTH	EQUIPMENT COST/DAY	
40	2300	10 kW	Ea.	9.10	36.50	110	330	94.80	40
	2400	25 kW		10.60	85	255	765	135.80	
	2500	Diesel engine, 20 kW		12.10	66.50	200	600	136.80	
	2600	50 kW		23.35	103	310	930	248.80	
	2700	100 kW		43.30	128	385	1,150	423.40	
	2800	250 kW		85.15	235	705	2,125	822.20	
	2850	Hammer, hydraulic, for mounting on boom, to 500 ft lb.		2.55	75	225	675	65.40	
	2860	1000 ft lb.		4.35	127	380	1,150	110.80	
	2900	Heaters, space, oil or electric, 50 MBH		2.01	7.65	23	69	20.70	
	3000	100 MBH		3.76	10.65	32	96	36.50	
	3100	300 MBH		11.03	38.50	115	345	111.25	
	3150	500 MBH		18.15	45	135	405	172.20	
	3200	Hose, water, suction with coupling, 20′ long, 2″ diameter		.02	3	9	27	1.95	
	3210	3″ diameter		.03	4.33	13	39	2.85	
	3220	4″ diameter		.03	5	15	45	3.25	
	3230	6″ diameter		.11	17.65	53	159	11.50	
	3240	8″ diameter		.27	44.50	133	400	28.75	
	3250	Discharge hose with coupling, 50′ long, 2″ diameter		.01	1.33	4	12	.90	
	3260	3″ diameter		.01	2.33	7	21	1.50	
	3270	4″ diameter		.02	3.67	11	33	2.35	
	3280	6″ diameter		.06	9.35	28	84	6.10	
	3290	8″ diameter		.18	30	90	270	19.45	
	3295	Insulation blower		.78	6	18	54	9.85	
	3300	Ladders, extension type, 16′ to 36′ long		.14	22.50	68	204	14.70	
	3400	40′ to 60′ long		.19	31	93	279	20.10	
	3405	Lance for cutting concrete		2.35	71.50	215	645	61.80	
	3407	Lawn mower, rotary, 22″, 5 H.P.		1.80	49	147	440	43.80	
	3408	48″ self propelled		2.98	75	225	675	68.85	
	3410	Level, electronic, automatic, with tripod and leveling rod		1.20	75.50	226	680	54.80	
	3430	Laser type, for pipe and sewer line and grade		.73	48.50	145	435	34.85	
	3440	Rotating beam for interior control		.78	52	156	470	37.45	
	3460	Builder's optical transit, with tripod and rod		.09	14.65	44	132	9.50	
	3500	Light towers, towable, with diesel generator, 2000 watt		4.55	93.50	280	840	92.40	
	3600	4000 watt		4.85	96.50	290	870	96.80	
	3700	Mixer, powered, plaster and mortar, 6 C.F., 7 H.P.		2.70	20	60	180	33.60	
	3800	10 C.F., 9 H.P.		2.80	31.50	94	282	41.20	
	3850	Nailer, pneumatic		.58	38.50	115	345	27.65	
	3900	Paint sprayers complete, 8 CFM		1.04	69.50	208	625	49.90	
	4000	17 CFM		1.90	127	380	1,150	91.20	
	4020	Pavers, bituminous, rubber tires, 8′ wide, 50 H.P., diesel		31.15	490	1,465	4,400	542.20	
	4030	10′ wide, 150 H.P.		106.10	1,800	5,370	16,100	1,923	
	4050	Crawler, 8′ wide, 100 H.P., diesel		88.65	1,800	5,385	16,200	1,786	
	4060	10′ wide, 150 H.P.		113.15	2,200	6,610	19,800	2,227	
	4070	Concrete paver, 12′ to 24′ wide, 250 H.P.		101.40	1,550	4,675	14,000	1,746	
	4080	Placer-spreader-trimmer, 24′ wide, 300 H.P.		150.25	2,650	7,960	23,900	2,794	
	4100	Pump, centrifugal gas pump, 1-1/2″ diam., 65 GPM		3.90	48.50	145	435	60.20	
	4200	2″ diameter, 130 GPM		5.35	53.50	160	480	74.80	
	4300	3″ diameter, 250 GPM		5.65	55	165	495	78.20	
	4400	6″ diameter, 1500 GPM		30.45	172	515	1,550	346.60	
	4500	Submersible electric pump, 1-1/4″ diameter, 55 GPM		.39	16.35	49	147	12.90	
	4600	1-1/2″ diameter, 83 GPM		.43	18.65	56	168	14.65	
	4700	2″ diameter, 120 GPM		1.50	23.50	70	210	26	
	4800	3″ diameter, 300 GPM		2.69	41.50	125	375	46.50	
	4900	4″ diameter, 560 GPM		12.34	158	475	1,425	193.70	
	5000	6″ diameter, 1590 GPM		18.40	212	635	1,900	274.20	
	5100	Diaphragm pump, gas, single, 1-1/2″ diameter		1.18	49.50	148	445	39.05	
	5200	2″ diameter		4.25	61.50	185	555	71	
	5300	3″ diameter	↓	4.25	61.50	185	555	71	

01 54 33 | Equipment Rental

			UNIT	HOURLY OPER. COST	RENT PER DAY	RENT PER WEEK	RENT PER MONTH	EQUIPMENT COST/DAY	
40	5400	Double, 4″ diameter	Ea.	6.40	105	315	945	114.20	40
	5450	Pressure washer 5 GPM, 3000 psi		4.90	51.50	155	465	70.20	
	5460	7 GPM, 3000 psi		6.45	60	180	540	87.60	
	5500	Trash pump, self-priming, gas, 2″ diameter		4.60	21	63	189	49.40	
	5600	Diesel, 4″ diameter		9.15	88.50	265	795	126.20	
	5650	Diesel, 6″ diameter		25.25	147	440	1,325	290	
	5655	Grout Pump		26.35	268	805	2,425	371.80	
	5700	Salamanders, L.P. gas fired, 100,000 Btu		3.96	13.65	41	123	39.90	
	5705	50,000 Btu		2.21	10.35	31	93	23.90	
	5720	Sandblaster, portable, open top, 3 C.F. capacity		.55	26.50	80	240	20.40	
	5730	6 C.F. capacity		.95	40	120	360	31.60	
	5740	Accessories for above		.13	21.50	65	195	14.05	
	5750	Sander, floor		.70	14.35	43	129	14.20	
	5760	Edger		.50	14.35	43	129	12.60	
	5800	Saw, chain, gas engine, 18″ long		2.30	21.50	64	192	31.20	
	5900	Hydraulic powered, 36″ long		.75	65	195	585	45	
	5950	60″ long		.75	66.50	200	600	46	
	6000	Masonry, table mounted, 14″ diameter, 5 H.P.		1.32	56.50	170	510	44.55	
	6050	Portable cut-off, 8 H.P.		2.50	33.50	100	300	40	
	6100	Circular, hand held, electric, 7-1/4″ diameter		.23	4.67	14	42	4.65	
	6200	12″ diameter		.23	8	24	72	6.65	
	6250	Wall saw, w/hydraulic power, 10 H.P.		9.70	61.50	185	555	114.60	
	6275	Shot blaster, walk-behind, 20″ wide		4.75	293	880	2,650	214	
	6280	Sidewalk broom, walk-behind		2.52	78.50	235	705	67.15	
	6300	Steam cleaner, 100 gallons per hour		3.70	76.50	230	690	75.60	
	6310	200 gallons per hour		5.35	95	285	855	99.80	
	6340	Tar Kettle/Pot, 400 gallons		15.60	75	225	675	169.80	
	6350	Torch, cutting, acetylene-oxygen, 150′ hose, excludes gases		.30	15	45	135	11.40	
	6360	Hourly operating cost includes tips and gas		19.00				152	
	6410	Toilet, portable chemical		.13	21	63	189	13.65	
	6420	Recycle flush type		.15	25	75	225	16.20	
	6430	Toilet, fresh water flush, garden hose,		.18	30.50	91	273	19.65	
	6440	Hoisted, non-flush, for high rise		.15	24.50	74	222	16	
	6465	Tractor, farm with attachment		21.50	297	890	2,675	350	
	6480	Trailers, platform, flush deck, 2 axle, 3 ton capacity		1.45	20	60	180	23.60	
	6500	25 ton capacity		5.45	117	350	1,050	113.60	
	6600	40 ton capacity		7.00	163	490	1,475	154	
	6700	3 axle, 50 ton capacity		7.55	180	540	1,625	168.40	
	6800	75 ton capacity		9.40	235	705	2,125	216.20	
	6810	Trailer mounted cable reel for high voltage line work		5.45	260	779	2,325	199.40	
	6820	Trailer mounted cable tensioning rig		10.85	515	1,550	4,650	396.80	
	6830	Cable pulling rig		73.05	2,900	8,680	26,000	2,320	
	6900	Water tank trailer, engine driven discharge, 5000 gallons		7.00	142	425	1,275	141	
	6925	10,000 gallons		9.50	197	590	1,775	194	
	6950	Water truck, off highway, 6000 gallons		88.15	770	2,315	6,950	1,168	
	7010	Tram car for high voltage line work, powered, 2 conductor		6.60	141	423	1,275	137.40	
	7020	Transit (builder's level) with tripod		.09	14.65	44	132	9.50	
	7030	Trench box, 3000 lb., 6′ x 8′		.56	93.50	280	840	60.50	
	7040	7200 lb., 6′ x 20′		.75	125	375	1,125	81	
	7050	8000 lb., 8′ x 16′		1.08	180	540	1,625	116.65	
	7060	9500 lb., 8′ x 20′		1.21	201	603	1,800	130.30	
	7065	11,000 lb., 8′ x 24′		1.27	211	633	1,900	136.75	
	7070	12,000 lb., 10′ x 20′		1.50	249	748	2,250	161.60	
	7100	Truck, pickup, 3/4 ton, 2 wheel drive		13.65	58.50	175	525	144.20	
	7200	4 wheel drive		13.95	73.50	220	660	155.60	
	7250	Crew carrier, 9 passenger		19.30	86.50	260	780	206.40	
	7290	Flat bed truck, 20,000 lb. GVW		21.70	125	375	1,125	248.60	
	7300	Tractor, 4 x 2, 220 H.P.	↓	30.20	197	590	1,775	359.60	

01 54 33 | Equipment Rental

		01 54 33 | Equipment Rental		UNIT	HOURLY OPER. COST	RENT PER DAY	RENT PER WEEK	RENT PER MONTH	EQUIPMENT COST/DAY
40	7410	330 H.P.		Ea.	44.75	270	810	2,425	520
	7500	6 x 4, 380 H.P.			51.35	315	945	2,825	599.80
	7600	450 H.P.			62.30	380	1,145	3,425	727.40
	7610	Tractor, with A frame, boom and winch, 225 H.P.			33.00	272	815	2,450	427
	7620	Vacuum truck, hazardous material, 2500 gallons			12.05	290	870	2,600	270.40
	7625	5,000 gallons			14.26	405	1,220	3,650	358.10
	7650	Vacuum, HEPA, 16 gallon, wet/dry			.85	18	54	162	17.60
	7655	55 gallon, wet/dry			.80	27	81	243	22.60
	7660	Water tank, portable			.16	26.50	80	240	17.30
	7690	Sewer/catch basin vacuum, 14 C.Y., 1500 gallons			18.53	610	1,830	5,500	514.25
	7700	Welder, electric, 200 amp			3.88	16.35	49	147	40.85
	7800	300 amp			5.71	20	60	180	57.70
	7900	Gas engine, 200 amp			14.65	23.50	70	210	131.20
	8000	300 amp			16.38	24.50	74	222	145.85
	8100	Wheelbarrow, any size			.08	13	39	117	8.45
	8200	Wrecking ball, 4000 lb.		↓	2.35	68.50	205	615	59.80
50	0010	**HIGHWAY EQUIPMENT RENTAL** without operators	R015433 -10						
	0050	Asphalt batch plant, portable drum mixer, 100 ton/hr.		Ea.	78.06	1,425	4,270	12,800	1,478
	0060	200 ton/hr.			89.18	1,500	4,530	13,600	1,619
	0070	300 ton/hr.			106.34	1,775	5,325	16,000	1,916
	0100	Backhoe attachment, long stick, up to 185 H.P., 10.5′ long			.35	23.50	70	210	16.80
	0140	Up to 250 H.P., 12′ long			.38	25	75	225	18.05
	0180	Over 250 H.P., 15′ long			.53	35	105	315	25.25
	0200	Special dipper arm, up to 100 H.P., 32′ long			1.08	71.50	215	645	51.65
	0240	Over 100 H.P., 33′ long			1.34	89.50	268	805	64.30
	0280	Catch basin/sewer cleaning truck, 3 ton, 9 C.Y., 1000 gal.			42.70	385	1,160	3,475	573.60
	0300	Concrete batch plant, portable, electric, 200 C.Y./hr.			22.87	515	1,550	4,650	492.95
	0520	Grader/dozer attachment, ripper/scarifier, rear mounted, up to 135 H.P.			2.90	56.50	170	510	57.20
	0540	Up to 180 H.P.			3.90	88.50	265	795	84.20
	0580	Up to 250 H.P.			5.00	117	350	1,050	110
	0700	Pvmt. removal bucket, for hyd. excavator, up to 90 H.P.			1.90	53.50	160	480	47.20
	0740	Up to 200 H.P.			2.15	73.50	220	660	61.20
	0780	Over 200 H.P.			2.25	85	255	765	69
	0900	Aggregate spreader, self-propelled, 187 H.P.			53.00	685	2,050	6,150	834
	1000	Chemical spreader, 3 C.Y.			3.30	43.50	130	390	52.40
	1900	Hammermill, traveling, 250 H.P.			78.83	2,075	6,240	18,700	1,879
	2000	Horizontal borer, 3″ diameter, 13 H.P. gas driven			6.35	55	165	495	83.80
	2150	Horizontal directional drill, 20,000 lb. thrust, 78 H.P. diesel			31.15	675	2,025	6,075	654.20
	2160	30,000 lb. thrust, 115 H.P.			38.90	1,025	3,100	9,300	931.20
	2170	50,000 lb. thrust, 170 H.P.			55.65	1,325	3,960	11,900	1,237
	2190	Mud trailer for HDD, 1500 gallons, 175 H.P., gas			31.90	152	455	1,375	346.20
	2200	Hydromulcher, diesel, 3000 gallon, for truck mounting			23.40	245	735	2,200	334.20
	2300	Gas, 600 gallon			8.60	98.50	295	885	127.80
	2400	Joint & crack cleaner, walk behind, 25 H.P.			3.95	50	150	450	61.60
	2500	Filler, trailer mounted, 400 gallons, 20 H.P.			9.50	210	630	1,900	202
	3000	Paint striper, self-propelled, 40 gallon, 22 H.P.			7.30	155	465	1,400	151.40
	3100	120 gallon, 120 H.P.			23.00	395	1,185	3,550	421
	3200	Post drivers, 6″ I-Beam frame, for truck mounting			19.05	380	1,135	3,400	379.40
	3400	Road sweeper, self-propelled, 8′ wide, 90 H.P.			39.80	575	1,720	5,150	662.40
	3450	Road sweeper, vacuum assisted, 4 C.Y., 220 gallons			75.75	635	1,900	5,700	986
	4000	Road mixer, self-propelled, 130 H.P.			46.85	765	2,295	6,875	833.80
	4100	310 H.P.			83.25	2,100	6,315	18,900	1,929
	4220	Cold mix paver, incl. pug mill and bitumen tank, 165 H.P.			97.00	2,275	6,855	20,600	2,147
	4240	Pavement brush, towed			3.20	93.50	280	840	81.60
	4250	Paver, asphalt, wheel or crawler, 130 H.P., diesel			96.65	2,275	6,795	20,400	2,132
	4300	Paver, road widener, gas 1′ to 6′, 67 H.P.			48.50	895	2,680	8,050	924
	4400	Diesel, 2′ to 14′, 88 H.P.			62.70	1,050	3,135	9,400	1,129
	4600	Slipform pavers, curb and gutter, 2 track, 75 H.P.		↓	57.20	900	2,700	8,100	997.60

01 54 33 | Equipment Rental

				UNIT	HOURLY OPER. COST	RENT PER DAY	RENT PER WEEK	RENT PER MONTH	EQUIPMENT COST/DAY	
50	4700	4 track, 165 H.P.		Ea.	41.50	740	2,220	6,650	776	50
	4800	Median barrier, 215 H.P.			60.50	1,050	3,150	9,450	1,114	
	4901	Trailer, low bed, 75 ton capacity			10.10	235	705	2,125	221.80	
	5000	Road planer, walk behind, 10″ cutting width, 10 H.P.			3.70	33.50	100	300	49.60	
	5100	Self-propelled, 12″ cutting width, 64 H.P.			10.40	113	340	1,025	151.20	
	5120	Traffic line remover, metal ball blaster, truck mounted, 115 H.P.			47.95	730	2,190	6,575	821.60	
	5140	Grinder, truck mounted, 115 H.P.			54.60	790	2,375	7,125	911.80	
	5160	Walk-behind, 11 H.P.			4.20	53.50	160	480	65.60	
	5200	Pavement profiler, 4′ to 6′ wide, 450 H.P.			254.50	3,325	9,960	29,900	4,028	
	5300	8′ to 10′ wide, 750 H.P.			399.20	4,350	13,055	39,200	5,805	
	5400	Roadway plate, steel, 1″ x 8′ x 20′			.08	13.35	40	120	8.65	
	5600	Stabilizer, self-propelled, 150 H.P.			49.30	600	1,795	5,375	753.40	
	5700	310 H.P.			99.05	1,675	5,030	15,100	1,798	
	5800	Striper, truck mounted, 120 gallon paint, 460 H.P.			64.20	485	1,455	4,375	804.60	
	5900	Thermal paint heating kettle, 115 gallons			8.20	25.50	77	231	81	
	6000	Tar kettle, 330 gallon, trailer mounted			12.25	56.50	170	510	132	
	7000	Tunnel locomotive, diesel, 8 to 12 ton			31.85	585	1,750	5,250	604.80	
	7005	Electric, 10 ton			26.00	665	1,995	5,975	607	
	7010	Muck cars, 1/2 C.Y. capacity			2.00	24.50	73	219	30.60	
	7020	1 C.Y. capacity			2.25	32.50	97	291	37.40	
	7030	2 C.Y. capacity			2.35	36.50	110	330	40.80	
	7040	Side dump, 2 C.Y. capacity			2.60	45	135	405	47.80	
	7050	3 C.Y. capacity			3.45	50	150	450	57.60	
	7060	5 C.Y. capacity			4.95	63.50	190	570	77.60	
	7100	Ventilating blower for tunnel, 7-1/2 H.P.			2.05	51.50	155	465	47.40	
	7110	10 H.P.			2.28	51.50	155	465	49.25	
	7120	20 H.P.			3.48	67.50	202	605	68.25	
	7140	40 H.P.			5.76	96.50	290	870	104.10	
	7160	60 H.P.			8.77	152	455	1,375	161.15	
	7175	75 H.P.			11.66	207	620	1,850	217.30	
	7180	200 H.P.			23.65	305	910	2,725	371.20	
	7800	Windrow loader, elevating		↓	55.00	1,300	3,935	11,800	1,227	
60	0010	**LIFTING AND HOISTING EQUIPMENT RENTAL** without operators	R015433 -10							60
	0120	Aerial lift truck, 2 person, to 80′		Ea.	26.30	705	2,120	6,350	634.40	
	0140	Boom work platform, 40′ snorkel	R015433 -15		12.00	275	825	2,475	261	
	0150	Crane, flatbed mounted, 3 ton capacity			16.35	188	565	1,700	243.80	
	0200	Crane, climbing, 106′ jib, 6000 lb. capacity, 410 fpm	R312316 -45		39.35	1,650	4,960	14,900	1,307	
	0300	101′ jib, 10,250 lb. capacity, 270 fpm			45.95	2,100	6,280	18,800	1,624	
	0500	Tower, static, 130′ high, 106′ jib, 6200 lb. capacity at 400 fpm			43.20	1,900	5,730	17,200	1,492	
	0600	Crawler mounted, lattice boom, 1/2 C.Y., 15 tons at 12′ radius			36.98	625	1,870	5,600	669.85	
	0700	3/4 C.Y., 20 tons at 12′ radius			49.31	780	2,340	7,025	862.50	
	0800	1 C.Y., 25 tons at 12′ radius			65.75	1,050	3,120	9,350	1,150	
	0900	1-1/2 C.Y., 40 tons at 12′ radius			65.75	1,050	3,150	9,450	1,156	
	1000	2 C.Y., 50 tons at 12′ radius			69.70	1,225	3,680	11,000	1,294	
	1100	3 C.Y., 75 tons at 12′ radius			74.70	1,425	4,305	12,900	1,459	
	1200	100 ton capacity, 60′ boom			84.60	1,650	4,950	14,900	1,667	
	1300	165 ton capacity, 60′ boom			107.75	1,925	5,785	17,400	2,019	
	1400	200 ton capacity, 70′ boom			130.55	2,400	7,210	21,600	2,486	
	1500	350 ton capacity, 80′ boom			182.90	3,625	10,845	32,500	3,632	
	1600	Truck mounted, lattice boom, 6 x 4, 20 tons at 10′ radius			37.11	1,075	3,260	9,775	948.90	
	1700	25 tons at 10′ radius			40.09	1,175	3,550	10,700	1,031	
	1800	8 x 4, 30 tons at 10′ radius			43.61	1,250	3,780	11,300	1,105	
	1900	40 tons at 12′ radius			46.70	1,325	3,950	11,900	1,164	
	2000	60 tons at 15′ radius			53.07	1,400	4,180	12,500	1,261	
	2050	82 tons at 15′ radius			59.80	1,475	4,460	13,400	1,370	
	2100	90 tons at 15′ radius			67.36	1,625	4,860	14,600	1,511	
	2200	115 tons at 15′ radius			76.11	1,800	5,430	16,300	1,695	
	2300	150 tons at 18′ radius		↓	83.65	1,900	5,720	17,200	1,813	

01 54 | Construction Aids

	01 54 33 | Equipment Rental		UNIT	HOURLY OPER. COST	RENT PER DAY	RENT PER WEEK	RENT PER MONTH	EQUIPMENT COST/DAY	
60	2350	165 tons at 18′ radius	Ea.	90.05	2,025	6,060	18,200	1,932	60
	2400	Truck mounted, hydraulic, 12 ton capacity		42.60	520	1,565	4,700	653.80	
	2500	25 ton capacity		44.95	630	1,885	5,650	736.60	
	2550	33 ton capacity		45.50	645	1,930	5,800	750	
	2560	40 ton capacity		58.85	750	2,250	6,750	920.80	
	2600	55 ton capacity		76.80	855	2,570	7,700	1,128	
	2700	80 ton capacity		100.45	1,375	4,105	12,300	1,625	
	2720	100 ton capacity		94.20	1,425	4,250	12,800	1,604	
	2740	120 ton capacity		100.05	1,525	4,575	13,700	1,715	
	2760	150 ton capacity		127.60	2,000	5,995	18,000	2,220	
	2800	Self-propelled, 4 x 4, with telescoping boom, 5 ton		17.55	225	675	2,025	275.40	
	2900	12-1/2 ton capacity		32.30	360	1,075	3,225	473.40	
	3000	15 ton capacity		33.00	375	1,130	3,400	490	
	3050	20 ton capacity		35.95	445	1,335	4,000	554.60	
	3100	25 ton capacity		37.55	500	1,495	4,475	599.40	
	3150	40 ton capacity		46.00	560	1,675	5,025	703	
	3200	Derricks, guy, 20 ton capacity, 60′ boom, 75′ mast		27.55	405	1,214	3,650	463.20	
	3300	100′ boom, 115′ mast		43.21	695	2,090	6,275	763.70	
	3400	Stiffleg, 20 ton capacity, 70′ boom, 37′ mast		30.11	525	1,580	4,750	556.90	
	3500	100′ boom, 47′ mast		46.29	845	2,530	7,600	876.30	
	3550	Helicopter, small, lift to 1250 lb. maximum, w/pilot		103.70	3,275	9,800	29,400	2,790	
	3600	Hoists, chain type, overhead, manual, 3/4 ton		.10	.33	1	3	1	
	3900	10 ton		.70	6	18	54	9.20	
	4000	Hoist and tower, 5000 lb. cap., portable electric, 40′ high		4.95	233	699	2,100	179.40	
	4100	For each added 10′ section, add		.11	18.35	55	165	11.90	
	4200	Hoist and single tubular tower, 5000 lb. electric, 100′ high		6.70	325	976	2,925	248.80	
	4300	For each added 6′-6″ section, add		.19	31.50	95	285	20.50	
	4400	Hoist and double tubular tower, 5000 lb., 100′ high		7.19	360	1,075	3,225	272.50	
	4500	For each added 6′-6″ section, add		.21	35	105	315	22.70	
	4550	Hoist and tower, mast type, 6000 lb., 100′ high		7.76	370	1,115	3,350	285.10	
	4570	For each added 10′ section, add		.13	21.50	65	195	14.05	
	4600	Hoist and tower, personnel, electric, 2000 lb., 100′ @ 125 fpm		16.31	990	2,970	8,900	724.50	
	4700	3000 lb., 100′ @ 200 fpm		18.62	1,125	3,360	10,100	820.95	
	4800	3000 lb., 150′ @ 300 fpm		20.62	1,250	3,760	11,300	916.95	
	4900	4000 lb., 100′ @ 300 fpm		21.38	1,275	3,840	11,500	939.05	
	5000	6000 lb., 100′ @ 275 fpm		23.06	1,350	4,030	12,100	990.50	
	5100	For added heights up to 500′, add	L.F.	.01	1.67	5	15	1.10	
	5200	Jacks, hydraulic, 20 ton	Ea.	.05	2	6	18	1.60	
	5500	100 ton		.40	11.65	35	105	10.20	
	6100	Jacks, hydraulic, climbing w/50′ jackrods, control console, 30 ton cap.		2.01	134	402	1,200	96.50	
	6150	For each added 10′ jackrod section, add		.05	3.33	10	30	2.40	
	6300	50 ton capacity		3.23	215	646	1,950	155.05	
	6350	For each added 10′ jackrod section, add		.06	4	12	36	2.90	
	6500	125 ton capacity		8.45	565	1,690	5,075	405.60	
	6550	For each added 10′ jackrod section, add		.58	38.50	115	345	27.65	
	6600	Cable jack, 10 ton capacity with 200′ cable		1.68	112	336	1,000	80.65	
	6650	For each added 50′ of cable, add		.20	13.35	40	120	9.60	
70	0010	**WELLPOINT EQUIPMENT RENTAL** without operators	R015433 -10						70
	0020	Based on 2 months rental							
	0100	Combination jetting & wellpoint pump, 60 H.P. diesel	Ea.	18.40	330	996	3,000	346.40	
	0200	High pressure gas jet pump, 200 H.P., 300 psi	″	44.06	284	851	2,550	522.70	
	0300	Discharge pipe, 8″ diameter	L.F.	.01	.54	1.62	4.86	.40	
	0350	12″ diameter		.01	.79	2.38	7.15	.55	
	0400	Header pipe, flows up to 150 GPM, 4″ diameter		.01	.49	1.47	4.41	.35	
	0500	400 GPM, 6″ diameter		.01	.57	1.72	5.15	.40	
	0600	800 GPM, 8″ diameter		.01	.79	2.38	7.15	.55	
	0700	1500 GPM, 10″ diameter		.01	.84	2.51	7.55	.60	

01 54 33 | Equipment Rental

			UNIT	HOURLY OPER. COST	RENT PER DAY	RENT PER WEEK	RENT PER MONTH	EQUIPMENT COST/DAY	
70	0800	2500 GPM, 12″ diameter	L.F.	.02	1.58	4.74	14.20	1.10	70
	0900	4500 GPM, 16″ diameter		.03	2.02	6.07	18.20	1.45	
	0950	For quick coupling aluminum and plastic pipe, add	↓	.03	2.09	6.28	18.85	1.50	
	1100	Wellpoint, 25′ long, with fittings & riser pipe, 1-1/2″ or 2″ diameter	Ea.	.06	4.18	12.54	37.50	3	
	1200	Wellpoint pump, diesel powered, 4″ suction, 20 H.P.		7.83	191	574	1,725	177.45	
	1300	6″ suction, 30 H.P.		10.70	237	712	2,125	228	
	1400	8″ suction, 40 H.P.		14.45	325	976	2,925	310.80	
	1500	10″ suction, 75 H.P.		22.27	380	1,141	3,425	406.35	
	1600	12″ suction, 100 H.P.		31.72	605	1,810	5,425	615.75	
	1700	12″ suction, 175 H.P.	↓	47.41	670	2,010	6,025	781.30	
80	0010	**MARINE EQUIPMENT RENTAL** without operators R015433-10							80
	0200	Barge, 400 Ton, 30′ wide x 90′ long	Ea.	17.50	1,050	3,180	9,550	776	
	0240	800 Ton, 45′ wide x 90′ long		21.25	1,300	3,870	11,600	944	
	2000	Tugboat, diesel, 100 H.P.		38.20	217	650	1,950	435.60	
	2040	250 H.P.		79.25	395	1,180	3,550	870	
	2080	380 H.P.		156.80	1,175	3,535	10,600	1,961	
	3000	Small work boat, gas, 16-foot, 50 H.P.		17.15	60	180	540	173.20	
	4000	Large, diesel, 48-foot, 200 H.P.	↓	89.40	1,250	3,740	11,200	1,463	

Crews

Crew No.	Bare Costs		Incl. Subs O&P		Cost Per Labor-Hour	
Crew A-1	**Hr.**	**Daily**	**Hr.**	**Daily**	**Bare Costs**	**Incl. O&P**
1 Building Laborer	$37.60	$300.80	$57.85	$462.80	$37.60	$57.85
1 Concrete Saw, Gas Manual		81.00		89.10	10.13	11.14
8 L.H., Daily Totals		$381.80		$551.90	$47.73	$68.99
Crew A-1A	**Hr.**	**Daily**	**Hr.**	**Daily**	**Bare Costs**	**Incl. O&P**
1 Skilled Worker	$48.65	$389.20	$75.15	$601.20	$48.65	$75.15
1 Shot Blaster, 20"		214.00		235.40	26.75	29.43
8 L.H., Daily Totals		$603.20		$836.60	$75.40	$104.58
Crew A-1B	**Hr.**	**Daily**	**Hr.**	**Daily**	**Bare Costs**	**Incl. O&P**
1 Building Laborer	$37.60	$300.80	$57.85	$462.80	$37.60	$57.85
1 Concrete Saw		167.40		184.14	20.93	23.02
8 L.H., Daily Totals		$468.20		$646.94	$58.52	$80.87
Crew A-1C	**Hr.**	**Daily**	**Hr.**	**Daily**	**Bare Costs**	**Incl. O&P**
1 Building Laborer	$37.60	$300.80	$57.85	$462.80	$37.60	$57.85
1 Chain Saw, Gas, 18"		31.20		34.32	3.90	4.29
8 L.H., Daily Totals		$332.00		$497.12	$41.50	$62.14
Crew A-1D	**Hr.**	**Daily**	**Hr.**	**Daily**	**Bare Costs**	**Incl. O&P**
1 Building Laborer	$37.60	$300.80	$57.85	$462.80	$37.60	$57.85
1 Vibrating Plate, Gas, 18"		36.20		39.82	4.53	4.98
8 L.H., Daily Totals		$337.00		$502.62	$42.13	$62.83
Crew A-1E	**Hr.**	**Daily**	**Hr.**	**Daily**	**Bare Costs**	**Incl. O&P**
1 Building Laborer	$37.60	$300.80	$57.85	$462.80	$37.60	$57.85
1 Vibrating Plate, Gas, 21"		46.80		51.48	5.85	6.43
8 L.H., Daily Totals		$347.60		$514.28	$43.45	$64.28
Crew A-1F	**Hr.**	**Daily**	**Hr.**	**Daily**	**Bare Costs**	**Incl. O&P**
1 Building Laborer	$37.60	$300.80	$57.85	$462.80	$37.60	$57.85
1 Rammer/Tamper, Gas, 8"		50.00		55.00	6.25	6.88
8 L.H., Daily Totals		$350.80		$517.80	$43.85	$64.72
Crew A-1G	**Hr.**	**Daily**	**Hr.**	**Daily**	**Bare Costs**	**Incl. O&P**
1 Building Laborer	$37.60	$300.80	$57.85	$462.80	$37.60	$57.85
1 Rammer/Tamper, Gas, 15"		56.40		62.04	7.05	7.75
8 L.H., Daily Totals		$357.20		$524.84	$44.65	$65.61
Crew A-1H	**Hr.**	**Daily**	**Hr.**	**Daily**	**Bare Costs**	**Incl. O&P**
1 Building Laborer	$37.60	$300.80	$57.85	$462.80	$37.60	$57.85
1 Exterior Steam Cleaner		75.60		83.16	9.45	10.40
8 L.H., Daily Totals		$376.40		$545.96	$47.05	$68.25
Crew A-1J	**Hr.**	**Daily**	**Hr.**	**Daily**	**Bare Costs**	**Incl. O&P**
1 Building Laborer	$37.60	$300.80	$57.85	$462.80	$37.60	$57.85
1 Cultivator, Walk-Behind, 5 H.P.		46.70		51.37	5.84	6.42
8 L.H., Daily Totals		$347.50		$514.17	$43.44	$64.27
Crew A-1K	**Hr.**	**Daily**	**Hr.**	**Daily**	**Bare Costs**	**Incl. O&P**
1 Building Laborer	$37.60	$300.80	$57.85	$462.80	$37.60	$57.85
1 Cultivator, Walk-Behind, 8 H.P.		72.65		79.92	9.08	9.99
8 L.H., Daily Totals		$373.45		$542.72	$46.68	$67.84
Crew A-1M	**Hr.**	**Daily**	**Hr.**	**Daily**	**Bare Costs**	**Incl. O&P**
1 Building Laborer	$37.60	$300.80	$57.85	$462.80	$37.60	$57.85
1 Snow Blower, Walk-Behind		67.15		73.86	8.39	9.23
8 L.H., Daily Totals		$367.95		$536.66	$45.99	$67.08

Crew No.	Bare Costs		Incl. Subs O&P		Cost Per Labor-Hour	
Crew A-2	**Hr.**	**Daily**	**Hr.**	**Daily**	**Bare Costs**	**Incl. O&P**
2 Laborers	$37.60	$601.60	$57.85	$925.60	$38.10	$58.27
1 Truck Driver (light)	39.10	312.80	59.10	472.80		
1 Flatbed Truck, Gas, 1.5 Ton		245.40		269.94	10.23	11.25
24 L.H., Daily Totals		$1159.80		$1668.34	$48.33	$69.51
Crew A-2A	**Hr.**	**Daily**	**Hr.**	**Daily**	**Bare Costs**	**Incl. O&P**
2 Laborers	$37.60	$601.60	$57.85	$925.60	$38.10	$58.27
1 Truck Driver (light)	39.10	312.80	59.10	472.80		
1 Flatbed Truck, Gas, 1.5 Ton		245.40		269.94		
1 Concrete Saw		167.40		184.14	17.20	18.92
24 L.H., Daily Totals		$1327.20		$1852.48	$55.30	$77.19
Crew A-2B	**Hr.**	**Daily**	**Hr.**	**Daily**	**Bare Costs**	**Incl. O&P**
1 Truck Driver (light)	$39.10	$312.80	$59.10	$472.80	$39.10	$59.10
1 Flatbed Truck, Gas, 1.5 Ton		245.40		269.94	30.68	33.74
8 L.H., Daily Totals		$558.20		$742.74	$69.78	$92.84
Crew A-3A	**Hr.**	**Daily**	**Hr.**	**Daily**	**Bare Costs**	**Incl. O&P**
1 Equip. Oper. (light)	$48.60	$388.80	$73.75	$590.00	$48.60	$73.75
1 Pickup Truck, 4 x 4, 3/4 Ton		155.60		171.16	19.45	21.40
8 L.H., Daily Totals		$544.40		$761.16	$68.05	$95.14
Crew A-3B	**Hr.**	**Daily**	**Hr.**	**Daily**	**Bare Costs**	**Incl. O&P**
1 Equip. Oper. (medium)	$50.60	$404.80	$76.75	$614.00	$45.33	$68.65
1 Truck Driver (heavy)	40.05	320.40	60.55	484.40		
1 Dump Truck, 12 C.Y., 400 H.P.		691.00		760.10		
1 F.E. Loader, W.M.,2.5 C.Y.		519.80		571.78	75.67	83.24
16 L.H., Daily Totals		$1936.00		$2430.28	$121.00	$151.89
Crew A-3C	**Hr.**	**Daily**	**Hr.**	**Daily**	**Bare Costs**	**Incl. O&P**
1 Equip. Oper. (light)	$48.60	$388.80	$73.75	$590.00	$48.60	$73.75
1 Loader, Skid Steer, 78 H.P.		318.60		350.46	39.83	43.81
8 L.H., Daily Totals		$707.40		$940.46	$88.42	$117.56
Crew A-3D	**Hr.**	**Daily**	**Hr.**	**Daily**	**Bare Costs**	**Incl. O&P**
1 Truck Driver (light)	$39.10	$312.80	$59.10	$472.80	$39.10	$59.10
1 Pickup Truck, 4 x 4, 3/4 Ton		155.60		171.16		
1 Flatbed Trailer, 25 Ton		113.60		124.96	33.65	37.02
8 L.H., Daily Totals		$582.00		$768.92	$72.75	$96.11
Crew A-3E	**Hr.**	**Daily**	**Hr.**	**Daily**	**Bare Costs**	**Incl. O&P**
1 Equip. Oper. (crane)	$51.70	$413.60	$78.45	$627.60	$45.88	$69.50
1 Truck Driver (heavy)	40.05	320.40	60.55	484.40		
1 Pickup Truck, 4 x 4, 3/4 Ton		155.60		171.16	9.72	10.70
16 L.H., Daily Totals		$889.60		$1283.16	$55.60	$80.20
Crew A-3F	**Hr.**	**Daily**	**Hr.**	**Daily**	**Bare Costs**	**Incl. O&P**
1 Equip. Oper. (crane)	$51.70	$413.60	$78.45	$627.60	$45.88	$69.50
1 Truck Driver (heavy)	40.05	320.40	60.55	484.40		
1 Pickup Truck, 4 x 4, 3/4 Ton		155.60		171.16		
1 Truck Tractor, 6x4, 380 H.P.		599.80		659.78		
1 Lowbed Trailer, 75 Ton		221.80		243.98	61.08	67.18
16 L.H., Daily Totals		$1711.20		$2186.92	$106.95	$136.68

Crew No.	Bare Costs		Incl. Subs O&P		Cost Per Labor-Hour	
Crew A-3G	**Hr.**	**Daily**	**Hr.**	**Daily**	**Bare Costs**	**Incl. O&P**
1 Equip. Oper. (crane)	$51.70	$413.60	$78.45	$627.60	$45.88	$69.50
1 Truck Driver (heavy)	40.05	320.40	60.55	484.40		
1 Pickup Truck, 4 x 4, 3/4 Ton		155.60		171.16		
1 Truck Tractor, 6x4, 450 H.P.		727.40		800.14		
1 Lowbed Trailer, 75 Ton		221.80		243.98	69.05	75.95
16 L.H., Daily Totals		$1838.80		$2327.28	$114.93	$145.46
Crew A-3H	**Hr.**	**Daily**	**Hr.**	**Daily**	**Bare Costs**	**Incl. O&P**
1 Equip. Oper. (crane)	$51.70	$413.60	$78.45	$627.60	$51.70	$78.45
1 Hyd. Crane, 12 Ton (Daily)		860.80		946.88	107.60	118.36
8 L.H., Daily Totals		$1274.40		$1574.48	$159.30	$196.81
Crew A-3I	**Hr.**	**Daily**	**Hr.**	**Daily**	**Bare Costs**	**Incl. O&P**
1 Equip. Oper. (crane)	$51.70	$413.60	$78.45	$627.60	$51.70	$78.45
1 Hyd. Crane, 25 Ton (Daily)		989.60		1088.56	123.70	136.07
8 L.H., Daily Totals		$1403.20		$1716.16	$175.40	$214.52
Crew A-3J	**Hr.**	**Daily**	**Hr.**	**Daily**	**Bare Costs**	**Incl. O&P**
1 Equip. Oper. (crane)	$51.70	$413.60	$78.45	$627.60	$51.70	$78.45
1 Hyd. Crane, 40 Ton (Daily)		1221.00		1343.10	152.63	167.89
8 L.H., Daily Totals		$1634.60		$1970.70	$204.32	$246.34
Crew A-3K	**Hr.**	**Daily**	**Hr.**	**Daily**	**Bare Costs**	**Incl. O&P**
1 Equip. Oper. (crane)	$51.70	$413.60	$78.45	$627.60	$48.45	$73.50
1 Equip. Oper. (oiler)	45.20	361.60	68.55	548.40		
1 Hyd. Crane, 55 Ton (Daily)		1469.00		1615.90		
1 P/U Truck, 3/4 Ton (Daily)		167.20		183.92	102.26	112.49
16 L.H., Daily Totals		$2411.40		$2975.82	$150.71	$185.99
Crew A-3L	**Hr.**	**Daily**	**Hr.**	**Daily**	**Bare Costs**	**Incl. O&P**
1 Equip. Oper. (crane)	$51.70	$413.60	$78.45	$627.60	$48.45	$73.50
1 Equip. Oper. (oiler)	45.20	361.60	68.55	548.40		
1 Hyd. Crane, 80 Ton (Daily)		2174.00		2391.40		
1 P/U Truck, 3/4 Ton (Daily)		167.20		183.92	146.32	160.96
16 L.H., Daily Totals		$3116.40		$3751.32	$194.78	$234.46
Crew A-3M	**Hr.**	**Daily**	**Hr.**	**Daily**	**Bare Costs**	**Incl. O&P**
1 Equip. Oper. (crane)	$51.70	$413.60	$78.45	$627.60	$48.45	$73.50
1 Equip. Oper. (oiler)	45.20	361.60	68.55	548.40		
1 Hyd. Crane, 100 Ton (Daily)		2169.00		2385.90		
1 P/U Truck, 3/4 Ton (Daily)		167.20		183.92	146.01	160.61
16 L.H., Daily Totals		$3111.40		$3745.82	$194.46	$234.11
Crew A-3N	**Hr.**	**Daily**	**Hr.**	**Daily**	**Bare Costs**	**Incl. O&P**
1 Equip. Oper. (crane)	$51.70	$413.60	$78.45	$627.60	$51.70	$78.45
1 Tower Cane (monthly)		1128.00		1240.80	141.00	155.10
8 L.H., Daily Totals		$1541.60		$1868.40	$192.70	$233.55
Crew A-3P	**Hr.**	**Daily**	**Hr.**	**Daily**	**Bare Costs**	**Incl. O&P**
1 Equip. Oper. (light)	$48.60	$388.80	$73.75	$590.00	$48.60	$73.75
1 A.T. Forklift, 42' lift		526.40		579.04	65.80	72.38
8 L.H., Daily Totals		$915.20		$1169.04	$114.40	$146.13
Crew A-3Q	**Hr.**	**Daily**	**Hr.**	**Daily**	**Bare Costs**	**Incl. O&P**
1 Equip. Oper. (light)	$48.60	$388.80	$73.75	$590.00	$48.60	$73.75
1 Pickup Truck, 4 x 4, 3/4 Ton		155.60		171.16		
1 Flatbed trailer, 3 Ton		23.60		25.96	22.40	24.64
8 L.H., Daily Totals		$568.00		$787.12	$71.00	$98.39

Crew No.	Bare Costs		Incl. Subs O&P		Cost Per Labor-Hour	
Crew A-4	**Hr.**	**Daily**	**Hr.**	**Daily**	**Bare Costs**	**Incl. O&P**
2 Carpenters	$46.95	$751.20	$72.25	$1156.00	$44.75	$68.43
1 Painter, Ordinary	40.35	322.80	60.80	486.40		
24 L.H., Daily Totals		$1074.00		$1642.40	$44.75	$68.43
Crew A-5	**Hr.**	**Daily**	**Hr.**	**Daily**	**Bare Costs**	**Incl. O&P**
2 Laborers	$37.60	$601.60	$57.85	$925.60	$37.77	$57.99
.25 Truck Driver (light)	39.10	78.20	59.10	118.20		
.25 Flatbed Truck, Gas, 1.5 Ton		61.35		67.48	3.41	3.75
18 L.H., Daily Totals		$741.15		$1111.29	$41.17	$61.74
Crew A-6	**Hr.**	**Daily**	**Hr.**	**Daily**	**Bare Costs**	**Incl. O&P**
1 Instrument Man	$48.65	$389.20	$75.15	$601.20	$46.88	$71.92
1 Rodman/Chainman	45.10	360.80	68.70	549.60		
1 Level, Electronic		54.80		60.28	3.42	3.77
16 L.H., Daily Totals		$804.80		$1211.08	$50.30	$75.69
Crew A-7	**Hr.**	**Daily**	**Hr.**	**Daily**	**Bare Costs**	**Incl. O&P**
1 Chief of Party	$60.90	$487.20	$93.20	$745.60	$51.55	$79.02
1 Instrument Man	48.65	389.20	75.15	601.20		
1 Rodman/Chainman	45.10	360.80	68.70	549.60		
1 Level, Electronic		54.80		60.28	2.28	2.51
24 L.H., Daily Totals		$1292.00		$1956.68	$53.83	$81.53
Crew A-8	**Hr.**	**Daily**	**Hr.**	**Daily**	**Bare Costs**	**Incl. O&P**
1 Chief of Party	$60.90	$487.20	$93.20	$745.60	$49.94	$76.44
1 Instrument Man	48.65	389.20	75.15	601.20		
2 Rodmen/Chainmen	45.10	721.60	68.70	1099.20		
1 Level, Electronic		54.80		60.28	1.71	1.88
32 L.H., Daily Totals		$1652.80		$2506.28	$51.65	$78.32
Crew A-9	**Hr.**	**Daily**	**Hr.**	**Daily**	**Bare Costs**	**Incl. O&P**
1 Asbestos Foreman	$52.85	$422.80	$82.30	$658.40	$52.41	$81.60
7 Asbestos Workers	52.35	2931.60	81.50	4564.00		
64 L.H., Daily Totals		$3354.40		$5222.40	$52.41	$81.60
Crew A-10A	**Hr.**	**Daily**	**Hr.**	**Daily**	**Bare Costs**	**Incl. O&P**
1 Asbestos Foreman	$52.85	$422.80	$82.30	$658.40	$52.52	$81.77
2 Asbestos Workers	52.35	837.60	81.50	1304.00		
24 L.H., Daily Totals		$1260.40		$1962.40	$52.52	$81.77
Crew A-10B	**Hr.**	**Daily**	**Hr.**	**Daily**	**Bare Costs**	**Incl. O&P**
1 Asbestos Foreman	$52.85	$422.80	$82.30	$658.40	$52.48	$81.70
3 Asbestos Workers	52.35	1256.40	81.50	1956.00		
32 L.H., Daily Totals		$1679.20		$2614.40	$52.48	$81.70
Crew A-10C	**Hr.**	**Daily**	**Hr.**	**Daily**	**Bare Costs**	**Incl. O&P**
3 Asbestos Workers	$52.35	$1256.40	$81.50	$1956.00	$52.35	$81.50
1 Flatbed Truck, Gas, 1.5 Ton		245.40		269.94	10.23	11.25
24 L.H., Daily Totals		$1501.80		$2225.94	$62.58	$92.75
Crew A-10D	**Hr.**	**Daily**	**Hr.**	**Daily**	**Bare Costs**	**Incl. O&P**
2 Asbestos Workers	$52.35	$837.60	$81.50	$1304.00	$50.40	$77.50
1 Equip. Oper. (crane)	51.70	413.60	78.45	627.60		
1 Equip. Oper. (oiler)	45.20	361.60	68.55	548.40		
1 Hydraulic Crane, 33 Ton		750.00		825.00	23.44	25.78
32 L.H., Daily Totals		$2362.80		$3305.00	$73.84	$103.28

Crew No.	Bare Costs		Incl. Subs O&P		Cost Per Labor-Hour	
Crew A-11	**Hr.**	**Daily**	**Hr.**	**Daily**	**Bare Costs**	**Incl. O&P**
1 Asbestos Foreman	$52.85	$422.80	$82.30	$658.40	$52.41	$81.60
7 Asbestos Workers	52.35	2931.60	81.50	4564.00		
2 Chip. Hammers, 12 Lb., Elec.		40.00		44.00	.63	.69
64 L.H., Daily Totals		$3394.40		$5266.40	$53.04	$82.29

Crew A-12	Hr.	Daily	Hr.	Daily	Bare Costs	Incl. O&P
1 Asbestos Foreman	$52.85	$422.80	$82.30	$658.40	$52.41	$81.60
7 Asbestos Workers	52.35	2931.60	81.50	4564.00		
1 Trk-Mtd Vac, 14 CY, 1500 Gal.		514.25		565.67		
1 Flatbed Truck, 20,000 GVW		248.60		273.46	11.92	13.11
64 L.H., Daily Totals		$4117.25		$6061.53	$64.33	$94.71

Crew A-13	Hr.	Daily	Hr.	Daily	Bare Costs	Incl. O&P
1 Equip. Oper. (light)	$48.60	$388.80	$73.75	$590.00	$48.60	$73.75
1 Trk-Mtd Vac, 14 CY, 1500 Gal.		514.25		565.67		
1 Flatbed Truck, 20,000 GVW		248.60		273.46	95.36	104.89
8 L.H., Daily Totals		$1151.65		$1429.14	$143.96	$178.64

Crew B-1	Hr.	Daily	Hr.	Daily	Bare Costs	Incl. O&P
1 Labor Foreman (outside)	$39.60	$316.80	$60.95	$487.60	$38.27	$58.88
2 Laborers	37.60	601.60	57.85	925.60		
24 L.H., Daily Totals		$918.40		$1413.20	$38.27	$58.88

Crew B-1A	Hr.	Daily	Hr.	Daily	Bare Costs	Incl. O&P
1 Labor Foreman (outside)	$39.60	$316.80	$60.95	$487.60	$38.27	$58.88
2 Laborers	37.60	601.60	57.85	925.60		
2 Cutting Torches		22.80		25.08		
2 Sets of Gases		304.00		334.40	13.62	14.98
24 L.H., Daily Totals		$1245.20		$1772.68	$51.88	$73.86

Crew B-1B	Hr.	Daily	Hr.	Daily	Bare Costs	Incl. O&P
1 Labor Foreman (outside)	$39.60	$316.80	$60.95	$487.60	$41.63	$63.77
2 Laborers	37.60	601.60	57.85	925.60		
1 Equip. Oper. (crane)	51.70	413.60	78.45	627.60		
2 Cutting Torches		22.80		25.08		
2 Sets of Gases		304.00		334.40		
1 Hyd. Crane, 12 Ton		653.80		719.18	30.64	33.71
32 L.H., Daily Totals		$2312.60		$3119.46	$72.27	$97.48

Crew B-1C	Hr.	Daily	Hr.	Daily	Bare Costs	Incl. O&P
1 Labor Foreman (outside)	$39.60	$316.80	$60.95	$487.60	$38.27	$58.88
2 Laborers	37.60	601.60	57.85	925.60		
1 Aerial Lift Truck, 60' Boom		435.60		479.16	18.15	19.97
24 L.H., Daily Totals		$1354.00		$1892.36	$56.42	$78.85

Crew B-1D	Hr.	Daily	Hr.	Daily	Bare Costs	Incl. O&P
2 Laborers	$37.60	$601.60	$57.85	$925.60	$37.60	$57.85
1 Small Work Boat, Gas, 50 H.P.		173.20		190.52		
1 Pressure Washer, 7 GPM		87.60		96.36	16.30	17.93
16 L.H., Daily Totals		$862.40		$1212.48	$53.90	$75.78

Crew B-1E	Hr.	Daily	Hr.	Daily	Bare Costs	Incl. O&P
1 Labor Foreman (outside)	$39.60	$316.80	$60.95	$487.60	$38.10	$58.63
3 Laborers	37.60	902.40	57.85	1388.40		
1 Work Boat, Diesel, 200 H.P.		1463.00		1609.30		
2 Pressure Washer, 7 GPM		175.20		192.72	51.19	56.31
32 L.H., Daily Totals		$2857.40		$3678.02	$89.29	$114.94

Crew No.	Bare Costs		Incl. Subs O&P		Cost Per Labor-Hour	
Crew B-1F	**Hr.**	**Daily**	**Hr.**	**Daily**	**Bare Costs**	**Incl. O&P**
2 Skilled Workers	$48.65	$778.40	$75.15	$1202.40	$44.97	$69.38
1 Laborer	37.60	300.80	57.85	462.80		
1 Small Work Boat, Gas, 50 H.P.		173.20		190.52		
1 Pressure Washer, 7 GPM		87.60		96.36	10.87	11.95
24 L.H., Daily Totals		$1340.00		$1952.08	$55.83	$81.34

Crew B-1G	Hr.	Daily	Hr.	Daily	Bare Costs	Incl. O&P
2 Laborers	$37.60	$601.60	$57.85	$925.60	$37.60	$57.85
1 Small Work Boat, Gas, 50 H.P.		173.20		190.52	10.82	11.91
16 L.H., Daily Totals		$774.80		$1116.12	$48.42	$69.76

Crew B-1H	Hr.	Daily	Hr.	Daily	Bare Costs	Incl. O&P
2 Skilled Workers	$48.65	$778.40	$75.15	$1202.40	$44.97	$69.38
1 Laborer	37.60	300.80	57.85	462.80		
1 Small Work Boat, Gas, 50 H.P.		173.20		190.52	7.22	7.94
24 L.H., Daily Totals		$1252.40		$1855.72	$52.18	$77.32

Crew B-1J	Hr.	Daily	Hr.	Daily	Bare Costs	Incl. O&P
1 Labor Foreman (inside)	$38.10	$304.80	$58.65	$469.20	$37.85	$58.25
1 Laborer	37.60	300.80	57.85	462.80		
16 L.H., Daily Totals		$605.60		$932.00	$37.85	$58.25

Crew B-1K	Hr.	Daily	Hr.	Daily	Bare Costs	Incl. O&P
1 Carpenter Foreman (inside)	$47.45	$379.60	$73.05	$584.40	$47.20	$72.65
1 Carpenter	46.95	375.60	72.25	578.00		
16 L.H., Daily Totals		$755.20		$1162.40	$47.20	$72.65

Crew B-2	Hr.	Daily	Hr.	Daily	Bare Costs	Incl. O&P
1 Labor Foreman (outside)	$39.60	$316.80	$60.95	$487.60	$38.00	$58.47
4 Laborers	37.60	1203.20	57.85	1851.20		
40 L.H., Daily Totals		$1520.00		$2338.80	$38.00	$58.47

Crew B-2A	Hr.	Daily	Hr.	Daily	Bare Costs	Incl. O&P
1 Labor Foreman (outside)	$39.60	$316.80	$60.95	$487.60	$38.27	$58.88
2 Laborers	37.60	601.60	57.85	925.60		
1 Aerial Lift Truck, 60' Boom		435.60		479.16	18.15	19.97
24 L.H., Daily Totals		$1354.00		$1892.36	$56.42	$78.85

Crew B-3	Hr.	Daily	Hr.	Daily	Bare Costs	Incl. O&P
1 Labor Foreman (outside)	$39.60	$316.80	$60.95	$487.60	$40.92	$62.42
2 Laborers	37.60	601.60	57.85	925.60		
1 Equip. Oper. (medium)	50.60	404.80	76.75	614.00		
2 Truck Drivers (heavy)	40.05	640.80	60.55	968.80		
1 Crawler Loader, 3 C.Y.		1188.00		1306.80		
2 Dump Trucks, 12 C.Y., 400 H.P.		1382.00		1520.20	53.54	58.90
48 L.H., Daily Totals		$4534.00		$5823.00	$94.46	$121.31

Crew B-3A	Hr.	Daily	Hr.	Daily	Bare Costs	Incl. O&P
4 Laborers	$37.60	$1203.20	$57.85	$1851.20	$40.20	$61.63
1 Equip. Oper. (medium)	50.60	404.80	76.75	614.00		
1 Hyd. Excavator, 1.5 C.Y.		1030.00		1133.00	25.75	28.32
40 L.H., Daily Totals		$2638.00		$3598.20	$65.95	$89.95

Crew B-3B	Hr.	Daily	Hr.	Daily	Bare Costs	Incl. O&P
2 Laborers	$37.60	$601.60	$57.85	$925.60	$41.46	$63.25
1 Equip. Oper. (medium)	50.60	404.80	76.75	614.00		
1 Truck Driver (heavy)	40.05	320.40	60.55	484.40		
1 Backhoe Loader, 80 H.P.		391.80		430.98		
1 Dump Truck, 12 C.Y., 400 H.P.		691.00		760.10	33.84	37.22
32 L.H., Daily Totals		$2409.60		$3215.08	$75.30	$100.47

Crew No.	Bare Costs		Incl. Subs O&P		Cost Per Labor-Hour	
Crew B-3C	**Hr.**	**Daily**	**Hr.**	**Daily**	**Bare Costs**	**Incl. O&P**
3 Laborers	$37.60	$902.40	$57.85	$1388.40	$40.85	$62.58
1 Equip. Oper. (medium)	50.60	404.80	76.75	614.00		
1 Crawler Loader, 4 C.Y.		1561.00		1717.10	48.78	53.66
32 L.H., Daily Totals		$2868.20		$3719.50	$89.63	$116.23
Crew B-4	**Hr.**	**Daily**	**Hr.**	**Daily**	**Bare Costs**	**Incl. O&P**
1 Labor Foreman (outside)	$39.60	$316.80	$60.95	$487.60	$38.34	$58.82
4 Laborers	37.60	1203.20	57.85	1851.20		
1 Truck Driver (heavy)	40.05	320.40	60.55	484.40		
1 Truck Tractor, 220 H.P.		359.60		395.56		
1 Flatbed Trailer, 40 Ton		154.00		169.40	10.70	11.77
48 L.H., Daily Totals		$2354.00		$3388.16	$49.04	$70.59
Crew B-5	**Hr.**	**Daily**	**Hr.**	**Daily**	**Bare Costs**	**Incl. O&P**
1 Labor Foreman (outside)	$39.60	$316.80	$60.95	$487.60	$41.60	$63.69
4 Laborers	37.60	1203.20	57.85	1851.20		
2 Equip. Oper. (medium)	50.60	809.60	76.75	1228.00		
1 Air Compressor, 250 cfm		201.40		221.54		
2 Breakers, Pavement, 60 lb.		19.60		21.56		
2 -50' Air Hoses, 1.5"		11.60		12.76		
1 Crawler Loader, 3 C.Y.		1188.00		1306.80	25.37	27.90
56 L.H., Daily Totals		$3750.20		$5129.46	$66.97	$91.60
Crew B-5A	**Hr.**	**Daily**	**Hr.**	**Daily**	**Bare Costs**	**Incl. O&P**
1 Labor Foreman (outside)	$39.60	$316.80	$60.95	$487.60	$41.26	$63.03
6 Laborers	37.60	1804.80	57.85	2776.80		
2 Equip. Oper. (medium)	50.60	809.60	76.75	1228.00		
1 Equip. Oper. (light)	48.60	388.80	73.75	590.00		
2 Truck Drivers (heavy)	40.05	640.80	60.55	968.80		
1 Air Compressor, 365 cfm		263.80		290.18		
2 Breakers, Pavement, 60 lb.		19.60		21.56		
8 -50' Air Hoses, 1"		32.80		36.08		
2 Dump Trucks, 8 C.Y., 220 H.P.		822.40		904.64	11.86	13.05
96 L.H., Daily Totals		$5099.40		$7303.66	$53.12	$76.08
Crew B-5B	**Hr.**	**Daily**	**Hr.**	**Daily**	**Bare Costs**	**Incl. O&P**
1 Powderman	$48.65	$389.20	$75.15	$601.20	$45.00	$68.38
2 Equip. Oper. (medium)	50.60	809.60	76.75	1228.00		
3 Truck Drivers (heavy)	40.05	961.20	60.55	1453.20		
1 F.E. Loader, W.M.,2.5 C.Y.		519.80		571.78		
3 Dump Trucks, 12 C.Y., 400 H.P.		2073.00		2280.30		
1 Air Compressor, 365 CFM		263.80		290.18	59.51	65.46
48 L.H., Daily Totals		$5016.60		$6424.66	$104.51	$133.85
Crew B-5C	**Hr.**	**Daily**	**Hr.**	**Daily**	**Bare Costs**	**Incl. O&P**
3 Laborers	$37.60	$902.40	$57.85	$1388.40	$42.55	$64.80
1 Equip. Oper. (medium)	50.60	404.80	76.75	614.00		
2 Truck Drivers (heavy)	40.05	640.80	60.55	968.80		
1 Equip. Oper. (crane)	51.70	413.60	78.45	627.60		
1 Equip. Oper. (oiler)	45.20	361.60	68.55	548.40		
2 Dump Trucks, 12 C.Y., 400 H.P.		1382.00		1520.20		
1 Crawler Loader, 4 C.Y.		1561.00		1717.10		
1 S.P. Crane, 4x4, 25 Ton		599.40		659.34	55.35	60.88
64 L.H., Daily Totals		$6265.60		$8043.84	$97.90	$125.69

Crew No.	Bare Costs		Incl. Subs O&P		Cost Per Labor-Hour	
Crew B-5D	**Hr.**	**Daily**	**Hr.**	**Daily**	**Bare Costs**	**Incl. O&P**
1 Labor Foreman (outside)	$39.60	$316.80	$60.95	$487.60	$41.41	$63.30
4 Laborers	37.60	1203.20	57.85	1851.20		
2 Equip. Oper. (medium)	50.60	809.60	76.75	1228.00		
1 Truck Driver (heavy)	40.05	320.40	60.55	484.40		
1 Air Compressor, 250 cfm		201.40		221.54		
2 Breakers, Pavement, 60 lb.		19.60		21.56		
2 -50' Air Hoses, 1.5"		11.60		12.76		
1 Crawler Loader, 3 C.Y.		1188.00		1306.80		
1 Dump Truck, 12 C.Y., 400 H.P.		691.00		760.10	32.99	36.29
64 L.H., Daily Totals		$4761.60		$6373.96	$74.40	$99.59
Crew B-6	**Hr.**	**Daily**	**Hr.**	**Daily**	**Bare Costs**	**Incl. O&P**
2 Laborers	$37.60	$601.60	$57.85	$925.60	$41.27	$63.15
1 Equip. Oper. (light)	48.60	388.80	73.75	590.00		
1 Backhoe Loader, 48 H.P.		364.20		400.62	15.18	16.69
24 L.H., Daily Totals		$1354.60		$1916.22	$56.44	$79.84
Crew B-6A	**Hr.**	**Daily**	**Hr.**	**Daily**	**Bare Costs**	**Incl. O&P**
.5 Labor Foreman (outside)	$39.60	$158.40	$60.95	$243.80	$43.20	$66.03
1 Laborer	37.60	300.80	57.85	462.80		
1 Equip. Oper. (medium)	50.60	404.80	76.75	614.00		
1 Vacuum Truck, 5000 Gal.		358.10		393.91	17.91	19.70
20 L.H., Daily Totals		$1222.10		$1714.51	$61.10	$85.73
Crew B-6B	**Hr.**	**Daily**	**Hr.**	**Daily**	**Bare Costs**	**Incl. O&P**
2 Labor Foremen (outside)	$39.60	$633.60	$60.95	$975.20	$38.27	$58.88
4 Laborers	37.60	1203.20	57.85	1851.20		
1 S.P. Crane, 4x4, 5 Ton		275.40		302.94		
1 Flatbed Truck, Gas, 1.5 Ton		245.40		269.94		
1 Butt Fusion Mach., 4"-12" diam.		382.10		420.31	18.81	20.69
48 L.H., Daily Totals		$2739.70		$3819.59	$57.08	$79.57
Crew B-6C	**Hr.**	**Daily**	**Hr.**	**Daily**	**Bare Costs**	**Incl. O&P**
2 Labor Foremen (outside)	$39.60	$633.60	$60.95	$975.20	$38.27	$58.88
4 Laborers	37.60	1203.20	57.85	1851.20		
1 S.P. Crane, 4x4, 12 Ton		473.40		520.74		
1 Flatbed Truck, Gas, 3 Ton		303.20		333.52		
1 Butt Fusion Mach., 8"-24" diam.		829.70		912.67	33.46	36.81
48 L.H., Daily Totals		$3443.10		$4593.33	$71.73	$95.69
Crew B-7	**Hr.**	**Daily**	**Hr.**	**Daily**	**Bare Costs**	**Incl. O&P**
1 Labor Foreman (outside)	$39.60	$316.80	$60.95	$487.60	$40.10	$61.52
4 Laborers	37.60	1203.20	57.85	1851.20		
1 Equip. Oper. (medium)	50.60	404.80	76.75	614.00		
1 Brush Chipper, 12", 130 H.P.		391.40		430.54		
1 Crawler Loader, 3 C.Y.		1188.00		1306.80		
2 Chain Saws, Gas, 36" Long		90.00		99.00	34.78	38.26
48 L.H., Daily Totals		$3594.20		$4789.14	$74.88	$99.77
Crew B-7A	**Hr.**	**Daily**	**Hr.**	**Daily**	**Bare Costs**	**Incl. O&P**
2 Laborers	$37.60	$601.60	$57.85	$925.60	$41.27	$63.15
1 Equip. Oper. (light)	48.60	388.80	73.75	590.00		
1 Rake w/Tractor		354.25		389.68		
2 Chain Saw, Gas, 18"		62.40		68.64	17.36	19.10
24 L.H., Daily Totals		$1407.05		$1973.92	$58.63	$82.25

Crew No.	Bare Costs		Incl. Subs O&P		Cost Per Labor-Hour	
Crew B-7B	**Hr.**	**Daily**	**Hr.**	**Daily**	**Bare Costs**	**Incl. O&P**
1 Labor Foreman (outside)	$39.60	$316.80	$60.95	$487.60	$40.09	$61.38
4 Laborers	37.60	1203.20	57.85	1851.20		
1 Equip. Oper. (medium)	50.60	404.80	76.75	614.00		
1 Truck Driver (heavy)	40.05	320.40	60.55	484.40		
1 Brush Chipper, 12", 130 H.P.		391.40		430.54		
1 Crawler Loader, 3 C.Y.		1188.00		1306.80		
2 Chain Saws, Gas, 36" Long		90.00		99.00		
1 Dump Truck, 8 C.Y., 220 H.P.		411.20		452.32	37.15	40.87
56 L.H., Daily Totals		$4325.80		$5725.86	$77.25	$102.25

Crew B-7C	Hr.	Daily	Hr.	Daily	Bare Costs	Incl. O&P
1 Labor Foreman (outside)	$39.60	$316.80	$60.95	$487.60	$40.09	$61.38
4 Laborers	37.60	1203.20	57.85	1851.20		
1 Equip. Oper. (medium)	50.60	404.80	76.75	614.00		
1 Truck Driver (heavy)	40.05	320.40	60.55	484.40		
1 Brush Chipper, 12", 130 H.P.		391.40		430.54		
1 Crawler Loader, 3 C.Y.		1188.00		1306.80		
2 Chain Saws, Gas, 36" Long		90.00		99.00		
1 Dump Truck, 12 C.Y., 400 H.P.		691.00		760.10	42.15	46.37
56 L.H., Daily Totals		$4605.60		$6033.64	$82.24	$107.74

Crew B-8	Hr.	Daily	Hr.	Daily	Bare Costs	Incl. O&P
1 Labor Foreman (outside)	$39.60	$316.80	$60.95	$487.60	$42.66	$64.97
2 Laborers	37.60	601.60	57.85	925.60		
2 Equip. Oper. (medium)	50.60	809.60	76.75	1228.00		
1 Equip. Oper. (oiler)	45.20	361.60	68.55	548.40		
2 Truck Drivers (heavy)	40.05	640.80	60.55	968.80		
1 Hyd. Crane, 25 Ton		736.60		810.26		
1 Crawler Loader, 3 C.Y.		1188.00		1306.80		
2 Dump Trucks, 12 C.Y., 400 H.P.		1382.00		1520.20	51.67	56.83
64 L.H., Daily Totals		$6037.00		$7795.66	$94.33	$121.81

Crew B-9	Hr.	Daily	Hr.	Daily	Bare Costs	Incl. O&P
1 Labor Foreman (outside)	$39.60	$316.80	$60.95	$487.60	$38.00	$58.47
4 Laborers	37.60	1203.20	57.85	1851.20		
1 Air Compressor, 250 cfm		201.40		221.54		
2 Breakers, Pavement, 60 lb.		19.60		21.56		
2 -50' Air Hoses, 1.5"		11.60		12.76	5.82	6.40
40 L.H., Daily Totals		$1752.60		$2594.66	$43.81	$64.87

Crew B-9A	Hr.	Daily	Hr.	Daily	Bare Costs	Incl. O&P
2 Laborers	$37.60	$601.60	$57.85	$925.60	$38.42	$58.75
1 Truck Driver (heavy)	40.05	320.40	60.55	484.40		
1 Water Tank Trailer, 5000 Gal.		141.00		155.10		
1 Truck Tractor, 220 H.P.		359.60		395.56		
2 -50' Discharge Hoses, 3"		3.00		3.30	20.98	23.08
24 L.H., Daily Totals		$1425.60		$1963.96	$59.40	$81.83

Crew B-9B	Hr.	Daily	Hr.	Daily	Bare Costs	Incl. O&P
2 Laborers	$37.60	$601.60	$57.85	$925.60	$38.42	$58.75
1 Truck Driver (heavy)	40.05	320.40	60.55	484.40		
2 -50' Discharge Hoses, 3"		3.00		3.30		
1 Water Tank Trailer, 5000 Gal.		141.00		155.10		
1 Truck Tractor, 220 H.P.		359.60		395.56		
1 Pressure Washer		70.20		77.22	23.91	26.30
24 L.H., Daily Totals		$1495.80		$2041.18	$62.33	$85.05

Crew B-9D	Hr.	Daily	Hr.	Daily	Bare Costs	Incl. O&P
1 Labor Foreman (outside)	$39.60	$316.80	$60.95	$487.60	$38.00	$58.47
4 Common Laborers	37.60	1203.20	57.85	1851.20		
1 Air Compressor, 250 cfm		201.40		221.54		
2 -50' Air Hoses, 1.5"		11.60		12.76		
2 Air Powered Tampers		52.40		57.64	6.63	7.30
40 L.H., Daily Totals		$1785.40		$2630.74	$44.63	$65.77

Crew B-10	Hr.	Daily	Hr.	Daily	Bare Costs	Incl. O&P
1 Equip. Oper. (medium)	$50.60	$404.80	$76.75	$614.00	$46.27	$70.45
.5 Laborer	37.60	150.40	57.85	231.40		
12 L.H., Daily Totals		$555.20		$845.40	$46.27	$70.45

Crew B-10A	Hr.	Daily	Hr.	Daily	Bare Costs	Incl. O&P
1 Equip. Oper. (medium)	$50.60	$404.80	$76.75	$614.00	$46.27	$70.45
.5 Laborer	37.60	150.40	57.85	231.40		
1 Roller, 2-Drum, W.B., 7.5 H.P.		180.00		198.00	15.00	16.50
12 L.H., Daily Totals		$735.20		$1043.40	$61.27	$86.95

Crew B-10B	Hr.	Daily	Hr.	Daily	Bare Costs	Incl. O&P
1 Equip. Oper. (medium)	$50.60	$404.80	$76.75	$614.00	$46.27	$70.45
.5 Laborer	37.60	150.40	57.85	231.40		
1 Dozer, 200 H.P.		1387.00		1525.70	115.58	127.14
12 L.H., Daily Totals		$1942.20		$2371.10	$161.85	$197.59

Crew B-10C	Hr.	Daily	Hr.	Daily	Bare Costs	Incl. O&P
1 Equip. Oper. (medium)	$50.60	$404.80	$76.75	$614.00	$46.27	$70.45
.5 Laborer	37.60	150.40	57.85	231.40		
1 Dozer, 200 H.P.		1387.00		1525.70		
1 Vibratory Roller, Towed, 23 Ton		396.00		435.60	148.58	163.44
12 L.H., Daily Totals		$2338.20		$2806.70	$194.85	$233.89

Crew B-10D	Hr.	Daily	Hr.	Daily	Bare Costs	Incl. O&P
1 Equip. Oper. (medium)	$50.60	$404.80	$76.75	$614.00	$46.27	$70.45
.5 Laborer	37.60	150.40	57.85	231.40		
1 Dozer, 200 H.P.		1387.00		1525.70		
1 Sheepsft. Roller, Towed		425.80		468.38	151.07	166.17
12 L.H., Daily Totals		$2368.00		$2839.48	$197.33	$236.62

Crew B-10E	Hr.	Daily	Hr.	Daily	Bare Costs	Incl. O&P
1 Equip. Oper. (medium)	$50.60	$404.80	$76.75	$614.00	$46.27	$70.45
.5 Laborer	37.60	150.40	57.85	231.40		
1 Tandem Roller, 5 Ton		157.40		173.14	13.12	14.43
12 L.H., Daily Totals		$712.60		$1018.54	$59.38	$84.88

Crew B-10F	Hr.	Daily	Hr.	Daily	Bare Costs	Incl. O&P
1 Equip. Oper. (medium)	$50.60	$404.80	$76.75	$614.00	$46.27	$70.45
.5 Laborer	37.60	150.40	57.85	231.40		
1 Tandem Roller, 10 Ton		236.80		260.48	19.73	21.71
12 L.H., Daily Totals		$792.00		$1105.88	$66.00	$92.16

Crew B-10G	Hr.	Daily	Hr.	Daily	Bare Costs	Incl. O&P
1 Equip. Oper. (medium)	$50.60	$404.80	$76.75	$614.00	$46.27	$70.45
.5 Laborer	37.60	150.40	57.85	231.40		
1 Sheepsfoot Roller, 240 H.P.		1206.00		1326.60	100.50	110.55
12 L.H., Daily Totals		$1761.20		$2172.00	$146.77	$181.00

Crew B-10H	Hr.	Daily	Hr.	Daily	Bare Costs	Incl. O&P
1 Equip. Oper. (medium)	$50.60	$404.80	$76.75	$614.00	$46.27	$70.45
.5 Laborer	37.60	150.40	57.85	231.40		
1 Diaphragm Water Pump, 2"		71.00		78.10		
1 -20' Suction Hose, 2"		1.95		2.15		
2 -50' Discharge Hoses, 2"		1.80		1.98	6.23	6.85
12 L.H., Daily Totals		$629.95		$927.63	$52.50	$77.30

Crew B-10I	Hr.	Daily	Hr.	Daily	Bare Costs	Incl. O&P
1 Equip. Oper. (medium)	$50.60	$404.80	$76.75	$614.00	$46.27	$70.45
.5 Laborer	37.60	150.40	57.85	231.40		
1 Diaphragm Water Pump, 4"		114.20		125.62		
1 -20' Suction Hose, 4"		3.25		3.58		
2 -50' Discharge Hoses, 4"		4.70		5.17	10.18	11.20
12 L.H., Daily Totals		$677.35		$979.76	$56.45	$81.65

Crew B-10J	Hr.	Daily	Hr.	Daily	Bare Costs	Incl. O&P
1 Equip. Oper. (medium)	$50.60	$404.80	$76.75	$614.00	$46.27	$70.45
.5 Laborer	37.60	150.40	57.85	231.40		
1 Centrifugal Water Pump, 3"		78.20		86.02		
1 -20' Suction Hose, 3"		2.85		3.13		
2 -50' Discharge Hoses, 3"		3.00		3.30	7.00	7.70
12 L.H., Daily Totals		$639.25		$937.86	$53.27	$78.15

Crew B-10K	Hr.	Daily	Hr.	Daily	Bare Costs	Incl. O&P
1 Equip. Oper. (medium)	$50.60	$404.80	$76.75	$614.00	$46.27	$70.45
.5 Laborer	37.60	150.40	57.85	231.40		
1 Centr. Water Pump, 6"		346.60		381.26		
1 -20' Suction Hose, 6"		11.50		12.65		
2 -50' Discharge Hoses, 6"		12.20		13.42	30.86	33.94
12 L.H., Daily Totals		$925.50		$1252.73	$77.13	$104.39

Crew B-10L	Hr.	Daily	Hr.	Daily	Bare Costs	Incl. O&P
1 Equip. Oper. (medium)	$50.60	$404.80	$76.75	$614.00	$46.27	$70.45
.5 Laborer	37.60	150.40	57.85	231.40		
1 Dozer, 80 H.P.		472.40		519.64	39.37	43.30
12 L.H., Daily Totals		$1027.60		$1365.04	$85.63	$113.75

Crew B-10M	Hr.	Daily	Hr.	Daily	Bare Costs	Incl. O&P
1 Equip. Oper. (medium)	$50.60	$404.80	$76.75	$614.00	$46.27	$70.45
.5 Laborer	37.60	150.40	57.85	231.40		
1 Dozer, 300 H.P.		1897.00		2086.70	158.08	173.89
12 L.H., Daily Totals		$2452.20		$2932.10	$204.35	$244.34

Crew B-10N	Hr.	Daily	Hr.	Daily	Bare Costs	Incl. O&P
1 Equip. Oper. (medium)	$50.60	$404.80	$76.75	$614.00	$46.27	$70.45
.5 Laborer	37.60	150.40	57.85	231.40		
1 F.E. Loader, T.M., 1.5 C.Y		522.00		574.20	43.50	47.85
12 L.H., Daily Totals		$1077.20		$1419.60	$89.77	$118.30

Crew B-10O	Hr.	Daily	Hr.	Daily	Bare Costs	Incl. O&P
1 Equip. Oper. (medium)	$50.60	$404.80	$76.75	$614.00	$46.27	$70.45
.5 Laborer	37.60	150.40	57.85	231.40		
1 F.E. Loader, T.M., 2.25 C.Y.		956.40		1052.04	79.70	87.67
12 L.H., Daily Totals		$1511.60		$1897.44	$125.97	$158.12

Crew B-10P	Hr.	Daily	Hr.	Daily	Bare Costs	Incl. O&P
1 Equip. Oper. (medium)	$50.60	$404.80	$76.75	$614.00	$46.27	$70.45
.5 Laborer	37.60	150.40	57.85	231.40		
1 Crawler Loader, 3 C.Y.		1188.00		1306.80	99.00	108.90
12 L.H., Daily Totals		$1743.20		$2152.20	$145.27	$179.35

Crew B-10Q	Hr.	Daily	Hr.	Daily	Bare Costs	Incl. O&P
1 Equip. Oper. (medium)	$50.60	$404.80	$76.75	$614.00	$46.27	$70.45
.5 Laborer	37.60	150.40	57.85	231.40		
1 Crawler Loader, 4 C.Y.		1561.00		1717.10	130.08	143.09
12 L.H., Daily Totals		$2116.20		$2562.50	$176.35	$213.54

Crew B-10R	Hr.	Daily	Hr.	Daily	Bare Costs	Incl. O&P
1 Equip. Oper. (medium)	$50.60	$404.80	$76.75	$614.00	$46.27	$70.45
.5 Laborer	37.60	150.40	57.85	231.40		
1 F.E. Loader, W.M., 1 C.Y.		299.00		328.90	24.92	27.41
12 L.H., Daily Totals		$854.20		$1174.30	$71.18	$97.86

Crew B-10S	Hr.	Daily	Hr.	Daily	Bare Costs	Incl. O&P
1 Equip. Oper. (medium)	$50.60	$404.80	$76.75	$614.00	$46.27	$70.45
.5 Laborer	37.60	150.40	57.85	231.40		
1 F.E. Loader, W.M., 1.5 C.Y.		378.40		416.24	31.53	34.69
12 L.H., Daily Totals		$933.60		$1261.64	$77.80	$105.14

Crew B-10T	Hr.	Daily	Hr.	Daily	Bare Costs	Incl. O&P
1 Equip. Oper. (medium)	$50.60	$404.80	$76.75	$614.00	$46.27	$70.45
.5 Laborer	37.60	150.40	57.85	231.40		
1 F.E. Loader, W.M.,2.5 C.Y.		519.80		571.78	43.32	47.65
12 L.H., Daily Totals		$1075.00		$1417.18	$89.58	$118.10

Crew B-10U	Hr.	Daily	Hr.	Daily	Bare Costs	Incl. O&P
1 Equip. Oper. (medium)	$50.60	$404.80	$76.75	$614.00	$46.27	$70.45
.5 Laborer	37.60	150.40	57.85	231.40		
1 F.E. Loader, W.M., 5.5 C.Y.		1082.00		1190.20	90.17	99.18
12 L.H., Daily Totals		$1637.20		$2035.60	$136.43	$169.63

Crew B-10V	Hr.	Daily	Hr.	Daily	Bare Costs	Incl. O&P
1 Equip. Oper. (medium)	$50.60	$404.80	$76.75	$614.00	$46.27	$70.45
.5 Laborer	37.60	150.40	57.85	231.40		
1 Dozer, 700 H.P.		4979.00		5476.90	414.92	456.41
12 L.H., Daily Totals		$5534.20		$6322.30	$461.18	$526.86

Crew B-10W	Hr.	Daily	Hr.	Daily	Bare Costs	Incl. O&P
1 Equip. Oper. (medium)	$50.60	$404.80	$76.75	$614.00	$46.27	$70.45
.5 Laborer	37.60	150.40	57.85	231.40		
1 Dozer, 105 H.P.		602.80		663.08	50.23	55.26
12 L.H., Daily Totals		$1158.00		$1508.48	$96.50	$125.71

Crew B-10X	Hr.	Daily	Hr.	Daily	Bare Costs	Incl. O&P
1 Equip. Oper. (medium)	$50.60	$404.80	$76.75	$614.00	$46.27	$70.45
.5 Laborer	37.60	150.40	57.85	231.40		
1 Dozer, 410 H.P.		2405.00		2645.50	200.42	220.46
12 L.H., Daily Totals		$2960.20		$3490.90	$246.68	$290.91

Crew B-10Y	Hr.	Daily	Hr.	Daily	Bare Costs	Incl. O&P
1 Equip. Oper. (medium)	$50.60	$404.80	$76.75	$614.00	$46.27	$70.45
.5 Laborer	37.60	150.40	57.85	231.40		
1 Vibr. Roller, Towed, 12 Ton		549.80		604.78	45.82	50.40
12 L.H., Daily Totals		$1105.00		$1450.18	$92.08	$120.85

Crew B-11A	Hr.	Daily	Hr.	Daily	Bare Costs	Incl. O&P
1 Equipment Oper. (med.)	$50.60	$404.80	$76.75	$614.00	$44.10	$67.30
1 Laborer	37.60	300.80	57.85	462.80		
1 Dozer, 200 H.P.		1387.00		1525.70	86.69	95.36
16 L.H., Daily Totals		$2092.60		$2602.50	$130.79	$162.66

Crew No.	Bare Costs		Incl. Subs O&P		Cost Per Labor-Hour	
Crew B-11B	**Hr.**	**Daily**	**Hr.**	**Daily**	**Bare Costs**	**Incl. O&P**
1 Equipment Oper. (light)	$48.60	$388.80	$73.75	$590.00	$43.10	$65.80
1 Laborer	37.60	300.80	57.85	462.80		
1 Air Powered Tamper		26.20		28.82		
1 Air Compressor, 365 cfm		263.80		290.18		
2 -50' Air Hoses, 1.5"		11.60		12.76	18.85	20.73
16 L.H., Daily Totals		$991.20		$1384.56	$61.95	$86.53

Crew B-11C	Hr.	Daily	Hr.	Daily	Bare Costs	Incl. O&P
1 Equipment Oper. (med.)	$50.60	$404.80	$76.75	$614.00	$44.10	$67.30
1 Laborer	37.60	300.80	57.85	462.80		
1 Backhoe Loader, 48 H.P.		364.20		400.62	22.76	25.04
16 L.H., Daily Totals		$1069.80		$1477.42	$66.86	$92.34

Crew B-11J	Hr.	Daily	Hr.	Daily	Bare Costs	Incl. O&P
1 Equipment Oper. (med.)	$50.60	$404.80	$76.75	$614.00	$44.10	$67.30
1 Laborer	37.60	300.80	57.85	462.80		
1 Grader, 30,000 Lbs.		736.20		809.82		
1 Ripper, Beam & 1 Shank		84.20		92.62	51.27	56.40
16 L.H., Daily Totals		$1526.00		$1979.24	$95.38	$123.70

Crew B-11K	Hr.	Daily	Hr.	Daily	Bare Costs	Incl. O&P
1 Equipment Oper. (med.)	$50.60	$404.80	$76.75	$614.00	$44.10	$67.30
1 Laborer	37.60	300.80	57.85	462.80		
1 Trencher, Chain Type, 8' D		3402.00		3742.20	212.63	233.89
16 L.H., Daily Totals		$4107.60		$4819.00	$256.73	$301.19

Crew B-11L	Hr.	Daily	Hr.	Daily	Bare Costs	Incl. O&P
1 Equipment Oper. (med.)	$50.60	$404.80	$76.75	$614.00	$44.10	$67.30
1 Laborer	37.60	300.80	57.85	462.80		
1 Grader, 30,000 Lbs.		736.20		809.82	46.01	50.61
16 L.H., Daily Totals		$1441.80		$1886.62	$90.11	$117.91

Crew B-11M	Hr.	Daily	Hr.	Daily	Bare Costs	Incl. O&P
1 Equipment Oper. (med.)	$50.60	$404.80	$76.75	$614.00	$44.10	$67.30
1 Laborer	37.60	300.80	57.85	462.80		
1 Backhoe Loader, 80 H.P.		391.80		430.98	24.49	26.94
16 L.H., Daily Totals		$1097.40		$1507.78	$68.59	$94.24

Crew B-11N	Hr.	Daily	Hr.	Daily	Bare Costs	Incl. O&P
1 Labor Foreman (outside)	$39.60	$316.80	$60.95	$487.60	$42.34	$64.19
2 Equipment Operators (med.)	50.60	809.60	76.75	1228.00		
6 Truck Drivers (heavy)	40.05	1922.40	60.55	2906.40		
1 F.E. Loader, W.M., 5.5 C.Y.		1082.00		1190.20		
1 Dozer, 410 H.P.		2405.00		2645.50		
6 Dump Trucks, Off Hwy., 50 Ton		11022.00		12124.20	201.51	221.67
72 L.H., Daily Totals		$17557.80		$20581.90	$243.86	$285.86

Crew B-11Q	Hr.	Daily	Hr.	Daily	Bare Costs	Incl. O&P
1 Equipment Operator (med.)	$50.60	$404.80	$76.75	$614.00	$46.27	$70.45
.5 Laborer	37.60	150.40	57.85	231.40		
1 Dozer, 140 H.P.		886.00		974.60	73.83	81.22
12 L.H., Daily Totals		$1441.20		$1820.00	$120.10	$151.67

Crew B-11R	Hr.	Daily	Hr.	Daily	Bare Costs	Incl. O&P
1 Equipment Operator (med.)	$50.60	$404.80	$76.75	$614.00	$46.27	$70.45
.5 Laborer	37.60	150.40	57.85	231.40		
1 Dozer, 200 H.P.		1387.00		1525.70	115.58	127.14
12 L.H., Daily Totals		$1942.20		$2371.10	$161.85	$197.59

Crew B-11S	Hr.	Daily	Hr.	Daily	Bare Costs	Incl. O&P
1 Equipment Operator (med.)	$50.60	$404.80	$76.75	$614.00	$46.27	$70.45
.5 Laborer	37.60	150.40	57.85	231.40		
1 Dozer, 300 H.P.		1897.00		2086.70		
1 Ripper, Beam & 1 Shank		84.20		92.62	165.10	181.61
12 L.H., Daily Totals		$2536.40		$3024.72	$211.37	$252.06

Crew B-11T	Hr.	Daily	Hr.	Daily	Bare Costs	Incl. O&P
1 Equipment Operator (med.)	$50.60	$404.80	$76.75	$614.00	$46.27	$70.45
.5 Laborer	37.60	150.40	57.85	231.40		
1 Dozer, 410 H.P.		2405.00		2645.50		
1 Ripper, Beam & 2 Shanks		110.00		121.00	209.58	230.54
12 L.H., Daily Totals		$3070.20		$3611.90	$255.85	$300.99

Crew B-11U	Hr.	Daily	Hr.	Daily	Bare Costs	Incl. O&P
1 Equipment Operator (med.)	$50.60	$404.80	$76.75	$614.00	$46.27	$70.45
.5 Laborer	37.60	150.40	57.85	231.40		
1 Dozer, 520 H.P.		3299.00		3628.90	274.92	302.41
12 L.H., Daily Totals		$3854.20		$4474.30	$321.18	$372.86

Crew B-11V	Hr.	Daily	Hr.	Daily	Bare Costs	Incl. O&P
3 Laborers	$37.60	$902.40	$57.85	$1388.40	$37.60	$57.85
1 Roller, 2-Drum, W.B., 7.5 H.P.		180.00		198.00	7.50	8.25
24 L.H., Daily Totals		$1082.40		$1586.40	$45.10	$66.10

Crew B-11W	Hr.	Daily	Hr.	Daily	Bare Costs	Incl. O&P
1 Equipment Operator (med.)	$50.60	$404.80	$76.75	$614.00	$40.73	$61.67
1 Common Laborer	37.60	300.80	57.85	462.80		
10 Truck Drivers (heavy)	40.05	3204.00	60.55	4844.00		
1 Dozer, 200 H.P.		1387.00		1525.70		
1 Vibratory Roller, Towed, 23 Ton		396.00		435.60		
10 Dump Trucks, 8 C.Y., 220 H.P.		4112.00		4523.20	61.41	67.55
96 L.H., Daily Totals		$9804.60		$12405.30	$102.13	$129.22

Crew B-11Y	Hr.	Daily	Hr.	Daily	Bare Costs	Incl. O&P
1 Labor Foreman (outside)	$39.60	$316.80	$60.95	$487.60	$42.16	$64.49
5 Common Laborers	37.60	1504.00	57.85	2314.00		
3 Equipment Operators (med.)	50.60	1214.40	76.75	1842.00		
1 Dozer, 80 H.P.		472.40		519.64		
2 Roller, 2-Drum, W.B., 7.5 H.P.		360.00		396.00		
4 Vibrating Plate, Gas, 21"		187.20		205.92	14.16	15.58
72 L.H., Daily Totals		$4054.80		$5765.16	$56.32	$80.07

Crew B-12A	Hr.	Daily	Hr.	Daily	Bare Costs	Incl. O&P
1 Equip. Oper. (crane)	$51.70	$413.60	$78.45	$627.60	$44.65	$68.15
1 Laborer	37.60	300.80	57.85	462.80		
1 Hyd. Excavator, 1 C.Y.		812.60		893.86	50.79	55.87
16 L.H., Daily Totals		$1527.00		$1984.26	$95.44	$124.02

Crew B-12B	Hr.	Daily	Hr.	Daily	Bare Costs	Incl. O&P
1 Equip. Oper. (crane)	$51.70	$413.60	$78.45	$627.60	$44.65	$68.15
1 Laborer	37.60	300.80	57.85	462.80		
1 Hyd. Excavator, 1.5 C.Y.		1030.00		1133.00	64.38	70.81
16 L.H., Daily Totals		$1744.40		$2223.40	$109.03	$138.96

Crew B-12C	Hr.	Daily	Hr.	Daily	Bare Costs	Incl. O&P
1 Equip. Oper. (crane)	$51.70	$413.60	$78.45	$627.60	$44.65	$68.15
1 Laborer	37.60	300.80	57.85	462.80		
1 Hyd. Excavator, 2 C.Y.		1175.00		1292.50	73.44	80.78
16 L.H., Daily Totals		$1889.40		$2382.90	$118.09	$148.93

Crew No.	Bare Costs		Incl. Subs O&P		Cost Per Labor-Hour	
Crew B-12D	**Hr.**	**Daily**	**Hr.**	**Daily**	**Bare Costs**	**Incl. O&P**
1 Equip. Oper. (crane)	$51.70	$413.60	$78.45	$627.60	$44.65	$68.15
1 Laborer	37.60	300.80	57.85	462.80		
1 Hyd. Excavator, 3.5 C.Y.		2440.00		2684.00	152.50	167.75
16 L.H., Daily Totals		$3154.40		$3774.40	$197.15	$235.90
Crew B-12E	**Hr.**	**Daily**	**Hr.**	**Daily**	**Bare Costs**	**Incl. O&P**
1 Equip. Oper. (crane)	$51.70	$413.60	$78.45	$627.60	$44.65	$68.15
1 Laborer	37.60	300.80	57.85	462.80		
1 Hyd. Excavator, .5 C.Y.		448.20		493.02	28.01	30.81
16 L.H., Daily Totals		$1162.60		$1583.42	$72.66	$98.96
Crew B-12F	**Hr.**	**Daily**	**Hr.**	**Daily**	**Bare Costs**	**Incl. O&P**
1 Equip. Oper. (crane)	$51.70	$413.60	$78.45	$627.60	$44.65	$68.15
1 Laborer	37.60	300.80	57.85	462.80		
1 Hyd. Excavator, .75 C.Y.		654.80		720.28	40.92	45.02
16 L.H., Daily Totals		$1369.20		$1810.68	$85.58	$113.17
Crew B-12G	**Hr.**	**Daily**	**Hr.**	**Daily**	**Bare Costs**	**Incl. O&P**
1 Equip. Oper. (crane)	$51.70	$413.60	$78.45	$627.60	$44.65	$68.15
1 Laborer	37.60	300.80	57.85	462.80		
1 Crawler Crane, 15 Ton		669.85		736.84		
1 Clamshell Bucket, .5 C.Y.		38.20		42.02	44.25	48.68
16 L.H., Daily Totals		$1422.45		$1869.26	$88.90	$116.83
Crew B-12H	**Hr.**	**Daily**	**Hr.**	**Daily**	**Bare Costs**	**Incl. O&P**
1 Equip. Oper. (crane)	$51.70	$413.60	$78.45	$627.60	$44.65	$68.15
1 Laborer	37.60	300.80	57.85	462.80		
1 Crawler Crane, 25 Ton		1150.00		1265.00		
1 Clamshell Bucket, 1 C.Y.		47.80		52.58	74.86	82.35
16 L.H., Daily Totals		$1912.20		$2407.98	$119.51	$150.50
Crew B-12I	**Hr.**	**Daily**	**Hr.**	**Daily**	**Bare Costs**	**Incl. O&P**
1 Equip. Oper. (crane)	$51.70	$413.60	$78.45	$627.60	$44.65	$68.15
1 Laborer	37.60	300.80	57.85	462.80		
1 Crawler Crane, 20 Ton		862.50		948.75		
1 Dragline Bucket, .75 C.Y.		20.60		22.66	55.19	60.71
16 L.H., Daily Totals		$1597.50		$2061.81	$99.84	$128.86
Crew B-12J	**Hr.**	**Daily**	**Hr.**	**Daily**	**Bare Costs**	**Incl. O&P**
1 Equip. Oper. (crane)	$51.70	$413.60	$78.45	$627.60	$44.65	$68.15
1 Laborer	37.60	300.80	57.85	462.80		
1 Gradall, 5/8 C.Y.		882.20		970.42	55.14	60.65
16 L.H., Daily Totals		$1596.60		$2060.82	$99.79	$128.80
Crew B-12K	**Hr.**	**Daily**	**Hr.**	**Daily**	**Bare Costs**	**Incl. O&P**
1 Equip. Oper. (crane)	$51.70	$413.60	$78.45	$627.60	$44.65	$68.15
1 Laborer	37.60	300.80	57.85	462.80		
1 Gradall, 3 Ton, 1 C.Y.		1002.00		1102.20	62.63	68.89
16 L.H., Daily Totals		$1716.40		$2192.60	$107.28	$137.04
Crew B-12L	**Hr.**	**Daily**	**Hr.**	**Daily**	**Bare Costs**	**Incl. O&P**
1 Equip. Oper. (crane)	$51.70	$413.60	$78.45	$627.60	$44.65	$68.15
1 Laborer	37.60	300.80	57.85	462.80		
1 Crawler Crane, 15 Ton		669.85		736.84		
1 F.E. Attachment, .5 C.Y.		60.00		66.00	45.62	50.18
16 L.H., Daily Totals		$1444.25		$1893.23	$90.27	$118.33

Crew No.	Bare Costs		Incl. Subs O&P		Cost Per Labor-Hour	
Crew B-12M	**Hr.**	**Daily**	**Hr.**	**Daily**	**Bare Costs**	**Incl. O&P**
1 Equip. Oper. (crane)	$51.70	$413.60	$78.45	$627.60	$44.65	$68.15
1 Laborer	37.60	300.80	57.85	462.80		
1 Crawler Crane, 20 Ton		862.50		948.75		
1 F.E. Attachment, .75 C.Y.		65.40		71.94	57.99	63.79
16 L.H., Daily Totals		$1642.30		$2111.09	$102.64	$131.94
Crew B-12N	**Hr.**	**Daily**	**Hr.**	**Daily**	**Bare Costs**	**Incl. O&P**
1 Equip. Oper. (crane)	$51.70	$413.60	$78.45	$627.60	$44.65	$68.15
1 Laborer	37.60	300.80	57.85	462.80		
1 Crawler Crane, 25 Ton		1150.00		1265.00		
1 F.E. Attachment, 1 C.Y.		71.20		78.32	76.33	83.96
16 L.H., Daily Totals		$1935.60		$2433.72	$120.97	$152.11
Crew B-12O	**Hr.**	**Daily**	**Hr.**	**Daily**	**Bare Costs**	**Incl. O&P**
1 Equip. Oper. (crane)	$51.70	$413.60	$78.45	$627.60	$44.65	$68.15
1 Laborer	37.60	300.80	57.85	462.80		
1 Crawler Crane, 40 Ton		1156.00		1271.60		
1 F.E. Attachment, 1.5 C.Y.		80.00		88.00	77.25	84.97
16 L.H., Daily Totals		$1950.40		$2450.00	$121.90	$153.13
Crew B-12P	**Hr.**	**Daily**	**Hr.**	**Daily**	**Bare Costs**	**Incl. O&P**
1 Equip. Oper. (crane)	$51.70	$413.60	$78.45	$627.60	$44.65	$68.15
1 Laborer	37.60	300.80	57.85	462.80		
1 Crawler Crane, 40 Ton		1156.00		1271.60		
1 Dragline Bucket, 1.5 C.Y.		33.60		36.96	74.35	81.78
16 L.H., Daily Totals		$1904.00		$2398.96	$119.00	$149.94
Crew B-12Q	**Hr.**	**Daily**	**Hr.**	**Daily**	**Bare Costs**	**Incl. O&P**
1 Equip. Oper. (crane)	$51.70	$413.60	$78.45	$627.60	$44.65	$68.15
1 Laborer	37.60	300.80	57.85	462.80		
1 Hyd. Excavator, 5/8 C.Y.		589.40		648.34	36.84	40.52
16 L.H., Daily Totals		$1303.80		$1738.74	$81.49	$108.67
Crew B-12S	**Hr.**	**Daily**	**Hr.**	**Daily**	**Bare Costs**	**Incl. O&P**
1 Equip. Oper. (crane)	$51.70	$413.60	$78.45	$627.60	$44.65	$68.15
1 Laborer	37.60	300.80	57.85	462.80		
1 Hyd. Excavator, 2.5 C.Y.		1605.00		1765.50	100.31	110.34
16 L.H., Daily Totals		$2319.40		$2855.90	$144.96	$178.49
Crew B-12T	**Hr.**	**Daily**	**Hr.**	**Daily**	**Bare Costs**	**Incl. O&P**
1 Equip. Oper. (crane)	$51.70	$413.60	$78.45	$627.60	$44.65	$68.15
1 Laborer	37.60	300.80	57.85	462.80		
1 Crawler Crane, 75 Ton		1459.00		1604.90		
1 F.E. Attachment, 3 C.Y.		103.20		113.52	97.64	107.40
16 L.H., Daily Totals		$2276.60		$2808.82	$142.29	$175.55
Crew B-12V	**Hr.**	**Daily**	**Hr.**	**Daily**	**Bare Costs**	**Incl. O&P**
1 Equip. Oper. (crane)	$51.70	$413.60	$78.45	$627.60	$44.65	$68.15
1 Laborer	37.60	300.80	57.85	462.80		
1 Crawler Crane, 75 Ton		1459.00		1604.90		
1 Dragline Bucket, 3 C.Y.		52.60		57.86	94.47	103.92
16 L.H., Daily Totals		$2226.00		$2753.16	$139.13	$172.07
Crew B-12Y	**Hr.**	**Daily**	**Hr.**	**Daily**	**Bare Costs**	**Incl. O&P**
1 Equip. Oper. (crane)	$51.70	$413.60	$78.45	$627.60	$42.30	$64.72
2 Laborers	37.60	601.60	57.85	925.60		
1 Hyd. Excavator, 3.5 C.Y.		2440.00		2684.00	101.67	111.83
24 L.H., Daily Totals		$3455.20		$4237.20	$143.97	$176.55

Crew B-12Z	Hr.	Daily	Hr.	Daily	Bare Costs	Incl. O&P
1 Equip. Oper. (crane)	$51.70	$413.60	$78.45	$627.60	$42.30	$64.72
2 Laborers	37.60	601.60	57.85	925.60		
1 Hyd. Excavator, 2.5 C.Y.		1605.00		1765.50	66.88	73.56
24 L.H., Daily Totals		$2620.20		$3318.70	$109.18	$138.28

Crew B-13	Hr.	Daily	Hr.	Daily	Bare Costs	Incl. O&P
1 Labor Foreman (outside)	$39.60	$316.80	$60.95	$487.60	$40.99	$62.76
4 Laborers	37.60	1203.20	57.85	1851.20		
1 Equip. Oper. (crane)	51.70	413.60	78.45	627.60		
1 Equip. Oper. (oiler)	45.20	361.60	68.55	548.40		
1 Hyd. Crane, 25 Ton		736.60		810.26	13.15	14.47
56 L.H., Daily Totals		$3031.80		$4325.06	$54.14	$77.23

Crew B-13A	Hr.	Daily	Hr.	Daily	Bare Costs	Incl. O&P
1 Labor Foreman (outside)	$39.60	$316.80	$60.95	$487.60	$42.30	$64.46
2 Laborers	37.60	601.60	57.85	925.60		
2 Equipment Operators (med.)	50.60	809.60	76.75	1228.00		
2 Truck Drivers (heavy)	40.05	640.80	60.55	968.80		
1 Crawler Crane, 75 Ton		1459.00		1604.90		
1 Crawler Loader, 4 C.Y.		1561.00		1717.10		
2 Dump Trucks, 8 C.Y., 220 H.P.		822.40		904.64	68.61	75.48
56 L.H., Daily Totals		$6211.20		$7836.64	$110.91	$139.94

Crew B-13B	Hr.	Daily	Hr.	Daily	Bare Costs	Incl. O&P
1 Labor Foreman (outside)	$39.60	$316.80	$60.95	$487.60	$40.99	$62.76
4 Laborers	37.60	1203.20	57.85	1851.20		
1 Equip. Oper. (crane)	51.70	413.60	78.45	627.60		
1 Equip. Oper. (oiler)	45.20	361.60	68.55	548.40		
1 Hyd. Crane, 55 Ton		1128.00		1240.80	20.14	22.16
56 L.H., Daily Totals		$3423.20		$4755.60	$61.13	$84.92

Crew B-13C	Hr.	Daily	Hr.	Daily	Bare Costs	Incl. O&P
1 Labor Foreman (outside)	$39.60	$316.80	$60.95	$487.60	$40.99	$62.76
4 Laborers	37.60	1203.20	57.85	1851.20		
1 Equip. Oper. (crane)	51.70	413.60	78.45	627.60		
1 Equip. Oper. (oiler)	45.20	361.60	68.55	548.40		
1 Crawler Crane, 100 Ton		1667.00		1833.70	29.77	32.74
56 L.H., Daily Totals		$3962.20		$5348.50	$70.75	$95.51

Crew B-13D	Hr.	Daily	Hr.	Daily	Bare Costs	Incl. O&P
1 Laborer	$37.60	$300.80	$57.85	$462.80	$44.65	$68.15
1 Equip. Oper. (crane)	51.70	413.60	78.45	627.60		
1 Hyd. Excavator, 1 C.Y.		812.60		893.86		
1 Trench Box		81.00		89.10	55.85	61.44
16 L.H., Daily Totals		$1608.00		$2073.36	$100.50	$129.59

Crew B-13E	Hr.	Daily	Hr.	Daily	Bare Costs	Incl. O&P
1 Laborer	$37.60	$300.80	$57.85	$462.80	$44.65	$68.15
1 Equip. Oper. (crane)	51.70	413.60	78.45	627.60		
1 Hyd. Excavator, 1.5 C.Y.		1030.00		1133.00		
1 Trench Box		81.00		89.10	69.44	76.38
16 L.H., Daily Totals		$1825.40		$2312.50	$114.09	$144.53

Crew B-13F	Hr.	Daily	Hr.	Daily	Bare Costs	Incl. O&P
1 Laborer	$37.60	$300.80	$57.85	$462.80	$44.65	$68.15
1 Equip. Oper. (crane)	51.70	413.60	78.45	627.60		
1 Hyd. Excavator, 3.5 C.Y.		2440.00		2684.00		
1 Trench Box		81.00		89.10	157.56	173.32
16 L.H., Daily Totals		$3235.40		$3863.50	$202.21	$241.47

Crew B-13G	Hr.	Daily	Hr.	Daily	Bare Costs	Incl. O&P
1 Laborer	$37.60	$300.80	$57.85	$462.80	$44.65	$68.15
1 Equip. Oper. (crane)	51.70	413.60	78.45	627.60		
1 Hyd. Excavator, .75 C.Y.		654.80		720.28		
1 Trench Box		81.00		89.10	45.99	50.59
16 L.H., Daily Totals		$1450.20		$1899.78	$90.64	$118.74

Crew B-13H	Hr.	Daily	Hr.	Daily	Bare Costs	Incl. O&P
1 Laborer	$37.60	$300.80	$57.85	$462.80	$44.65	$68.15
1 Equip. Oper. (crane)	51.70	413.60	78.45	627.60		
1 Gradall, 5/8 C.Y.		882.20		970.42		
1 Trench Box		81.00		89.10	60.20	66.22
16 L.H., Daily Totals		$1677.60		$2149.92	$104.85	$134.37

Crew B-13I	Hr.	Daily	Hr.	Daily	Bare Costs	Incl. O&P
1 Laborer	$37.60	$300.80	$57.85	$462.80	$44.65	$68.15
1 Equip. Oper. (crane)	51.70	413.60	78.45	627.60		
1 Gradall, 3 Ton, 1 C.Y.		1002.00		1102.20		
1 Trench Box		81.00		89.10	67.69	74.46
16 L.H., Daily Totals		$1797.40		$2281.70	$112.34	$142.61

Crew B-13J	Hr.	Daily	Hr.	Daily	Bare Costs	Incl. O&P
1 Laborer	$37.60	$300.80	$57.85	$462.80	$44.65	$68.15
1 Equip. Oper. (crane)	51.70	413.60	78.45	627.60		
1 Hyd. Excavator, 2.5 C.Y.		1605.00		1765.50		
1 Trench Box		81.00		89.10	105.38	115.91
16 L.H., Daily Totals		$2400.40		$2945.00	$150.03	$184.06

Crew B-13K	Hr.	Daily	Hr.	Daily	Bare Costs	Incl. O&P
2 Equip. Opers. (crane)	$51.70	$827.20	$78.45	$1255.20	$51.70	$78.45
1 Hyd. Excavator, .75 C.Y.		654.80		720.28		
1 Hyd. Hammer, 4000 ft-lb		308.20		339.02		
1 Hyd. Excavator, .75 C.Y.		654.80		720.28	101.11	111.22
16 L.H., Daily Totals		$2445.00		$3034.78	$152.81	$189.67

Crew B-13L	Hr.	Daily	Hr.	Daily	Bare Costs	Incl. O&P
2 Equip. Opers. (crane)	$51.70	$827.20	$78.45	$1255.20	$51.70	$78.45
1 Hyd. Excavator, 1.5 C.Y.		1030.00		1133.00		
1 Hyd. Hammer, 5000 ft-lb		371.20		408.32		
1 Hyd. Excavator, .75 C.Y.		654.80		720.28	128.50	141.35
16 L.H., Daily Totals		$2883.20		$3516.80	$180.20	$219.80

Crew B-13M	Hr.	Daily	Hr.	Daily	Bare Costs	Incl. O&P
2 Equip. Opers. (crane)	$51.70	$827.20	$78.45	$1255.20	$51.70	$78.45
1 Hyd. Excavator, 2.5 C.Y.		1605.00		1765.50		
1 Hyd. Hammer, 8000 ft-lb		545.40		599.94		
1 Hyd. Excavator, 1.5 C.Y.		1030.00		1133.00	198.78	218.65
16 L.H., Daily Totals		$4007.60		$4753.64	$250.47	$297.10

Crew B-13N	Hr.	Daily	Hr.	Daily	Bare Costs	Incl. O&P
2 Equip. Opers. (crane)	$51.70	$827.20	$78.45	$1255.20	$51.70	$78.45
1 Hyd. Excavator, 3.5 C.Y.		2440.00		2684.00		
1 Hyd. Hammer, 12,000 ft-lb		635.80		699.38		
1 Hyd. Excavator, 1.5 C.Y.		1030.00		1133.00	256.61	282.27
16 L.H., Daily Totals		$4933.00		$5771.58	$308.31	$360.72

Crew No.	Bare Costs		Incl. Subs O&P		Cost Per Labor-Hour	
Crew B-14	**Hr.**	**Daily**	**Hr.**	**Daily**	**Bare Costs**	**Incl. O&P**
1 Labor Foreman (outside)	$39.60	$316.80	$60.95	$487.60	$39.77	$61.02
4 Laborers	37.60	1203.20	57.85	1851.20		
1 Equip. Oper. (light)	48.60	388.80	73.75	590.00		
1 Backhoe Loader, 48 H.P.		364.20		400.62	7.59	8.35
48 L.H., Daily Totals		$2273.00		$3329.42	$47.35	$69.36
Crew B-14A	**Hr.**	**Daily**	**Hr.**	**Daily**	**Bare Costs**	**Incl. O&P**
1 Equip. Oper. (crane)	$51.70	$413.60	$78.45	$627.60	$47.00	$71.58
.5 Laborer	37.60	150.40	57.85	231.40		
1 Hyd. Excavator, 4.5 C.Y.		2963.00		3259.30	246.92	271.61
12 L.H., Daily Totals		$3527.00		$4118.30	$293.92	$343.19
Crew B-14B	**Hr.**	**Daily**	**Hr.**	**Daily**	**Bare Costs**	**Incl. O&P**
1 Equip. Oper. (crane)	$51.70	$413.60	$78.45	$627.60	$47.00	$71.58
.5 Laborer	37.60	150.40	57.85	231.40		
1 Hyd. Excavator, 6 C.Y.		3494.00		3843.40	291.17	320.28
12 L.H., Daily Totals		$4058.00		$4702.40	$338.17	$391.87
Crew B-14C	**Hr.**	**Daily**	**Hr.**	**Daily**	**Bare Costs**	**Incl. O&P**
1 Equip. Oper. (crane)	$51.70	$413.60	$78.45	$627.60	$47.00	$71.58
.5 Laborer	37.60	150.40	57.85	231.40		
1 Hyd. Excavator, 7 C.Y.		3590.00		3949.00	299.17	329.08
12 L.H., Daily Totals		$4154.00		$4808.00	$346.17	$400.67
Crew B-14F	**Hr.**	**Daily**	**Hr.**	**Daily**	**Bare Costs**	**Incl. O&P**
1 Equip. Oper. (crane)	$51.70	$413.60	$78.45	$627.60	$47.00	$71.58
.5 Laborer	37.60	150.40	57.85	231.40		
1 Hyd. Shovel, 7 C.Y.		3993.00		4392.30	332.75	366.02
12 L.H., Daily Totals		$4557.00		$5251.30	$379.75	$437.61
Crew B-14G	**Hr.**	**Daily**	**Hr.**	**Daily**	**Bare Costs**	**Incl. O&P**
1 Equip. Oper. (crane)	$51.70	$413.60	$78.45	$627.60	$47.00	$71.58
.5 Laborer	37.60	150.40	57.85	231.40		
1 Hyd. Shovel, 12 C.Y.		5736.00		6309.60	478.00	525.80
12 L.H., Daily Totals		$6300.00		$7168.60	$525.00	$597.38
Crew B-14J	**Hr.**	**Daily**	**Hr.**	**Daily**	**Bare Costs**	**Incl. O&P**
1 Equip. Oper. (medium)	$50.60	$404.80	$76.75	$614.00	$46.27	$70.45
.5 Laborer	37.60	150.40	57.85	231.40		
1 F.E. Loader, 8 C.Y.		1937.00		2130.70	161.42	177.56
12 L.H., Daily Totals		$2492.20		$2976.10	$207.68	$248.01
Crew B-14K	**Hr.**	**Daily**	**Hr.**	**Daily**	**Bare Costs**	**Incl. O&P**
1 Equip. Oper. (medium)	$50.60	$404.80	$76.75	$614.00	$46.27	$70.45
.5 Laborer	37.60	150.40	57.85	231.40		
1 F.E. Loader, 10 C.Y.		2942.00		3236.20	245.17	269.68
12 L.H., Daily Totals		$3497.20		$4081.60	$291.43	$340.13
Crew B-15	**Hr.**	**Daily**	**Hr.**	**Daily**	**Bare Costs**	**Incl. O&P**
1 Equipment Oper. (med.)	$50.60	$404.80	$76.75	$614.00	$42.71	$64.79
.5 Laborer	37.60	150.40	57.85	231.40		
2 Truck Drivers (heavy)	40.05	640.80	60.55	968.80		
2 Dump Trucks, 12 C.Y., 400 H.P.		1382.00		1520.20		
1 Dozer, 200 H.P.		1387.00		1525.70	98.89	108.78
28 L.H., Daily Totals		$3965.00		$4860.10	$141.61	$173.57

Crew No.	Bare Costs		Incl. Subs O&P		Cost Per Labor-Hour	
Crew B-16	**Hr.**	**Daily**	**Hr.**	**Daily**	**Bare Costs**	**Incl. O&P**
1 Labor Foreman (outside)	$39.60	$316.80	$60.95	$487.60	$38.71	$59.30
2 Laborers	37.60	601.60	57.85	925.60		
1 Truck Driver (heavy)	40.05	320.40	60.55	484.40		
1 Dump Truck, 12 C.Y., 400 H.P.		691.00		760.10	21.59	23.75
32 L.H., Daily Totals		$1929.80		$2657.70	$60.31	$83.05
Crew B-17	**Hr.**	**Daily**	**Hr.**	**Daily**	**Bare Costs**	**Incl. O&P**
2 Laborers	$37.60	$601.60	$57.85	$925.60	$40.96	$62.50
1 Equip. Oper. (light)	48.60	388.80	73.75	590.00		
1 Truck Driver (heavy)	40.05	320.40	60.55	484.40		
1 Backhoe Loader, 48 H.P.		364.20		400.62		
1 Dump Truck, 8 C.Y., 220 H.P.		411.20		452.32	24.23	26.65
32 L.H., Daily Totals		$2086.20		$2852.94	$65.19	$89.15
Crew B-17A	**Hr.**	**Daily**	**Hr.**	**Daily**	**Bare Costs**	**Incl. O&P**
2 Labor Foremen (outside)	$39.60	$633.60	$60.95	$975.20	$40.41	$62.24
6 Laborers	37.60	1804.80	57.85	2776.80		
1 Skilled Worker Foreman (out)	50.65	405.20	78.25	626.00		
1 Skilled Worker	48.65	389.20	75.15	601.20		
80 L.H., Daily Totals		$3232.80		$4979.20	$40.41	$62.24
Crew B-17B	**Hr.**	**Daily**	**Hr.**	**Daily**	**Bare Costs**	**Incl. O&P**
2 Laborers	$37.60	$601.60	$57.85	$925.60	$40.96	$62.50
1 Equip. Oper. (light)	48.60	388.80	73.75	590.00		
1 Truck Driver (heavy)	40.05	320.40	60.55	484.40		
1 Backhoe Loader, 48 H.P.		364.20		400.62		
1 Dump Truck, 12 C.Y., 400 H.P.		691.00		760.10	32.98	36.27
32 L.H., Daily Totals		$2366.00		$3160.72	$73.94	$98.77
Crew B-18	**Hr.**	**Daily**	**Hr.**	**Daily**	**Bare Costs**	**Incl. O&P**
1 Labor Foreman (outside)	$39.60	$316.80	$60.95	$487.60	$38.27	$58.88
2 Laborers	37.60	601.60	57.85	925.60		
1 Vibrating Plate, Gas, 21"		46.80		51.48	1.95	2.15
24 L.H., Daily Totals		$965.20		$1464.68	$40.22	$61.03
Crew B-19	**Hr.**	**Daily**	**Hr.**	**Daily**	**Bare Costs**	**Incl. O&P**
1 Pile Driver Foreman (outside)	$48.10	$384.80	$76.15	$609.20	$47.64	$74.20
4 Pile Drivers	46.10	1475.20	73.00	2336.00		
2 Equip. Oper. (crane)	51.70	827.20	78.45	1255.20		
1 Equip. Oper. (oiler)	45.20	361.60	68.55	548.40		
1 Crawler Crane, 40 Ton		1156.00		1271.60		
1 Lead, 90' High		124.40		136.84		
1 Hammer, Diesel, 22k ft-lb		456.40		502.04	27.14	29.85
64 L.H., Daily Totals		$4785.60		$6659.28	$74.78	$104.05
Crew B-19A	**Hr.**	**Daily**	**Hr.**	**Daily**	**Bare Costs**	**Incl. O&P**
1 Pile Driver Foreman (outside)	$48.10	$384.80	$76.15	$609.20	$47.64	$74.20
4 Pile Drivers	46.10	1475.20	73.00	2336.00		
2 Equip. Oper. (crane)	51.70	827.20	78.45	1255.20		
1 Equip. Oper. (oiler)	45.20	361.60	68.55	548.40		
1 Crawler Crane, 75 Ton		1459.00		1604.90		
1 Lead, 90' high		124.40		136.84		
1 Hammer, Diesel, 41k ft-lb		581.00		639.10	33.82	37.20
64 L.H., Daily Totals		$5213.20		$7129.64	$81.46	$111.40

Crew No.	Bare Costs		Incl. Subs O&P		Cost Per Labor-Hour	
Crew B-19B	**Hr.**	**Daily**	**Hr.**	**Daily**	**Bare Costs**	**Incl. O&P**
1 Pile Driver Foreman (outside)	$48.10	$384.80	$76.15	$609.20	$47.64	$74.20
4 Pile Drivers	46.10	1475.20	73.00	2336.00		
2 Equip. Oper. (crane)	51.70	827.20	78.45	1255.20		
1 Equip. Oper. (oiler)	45.20	361.60	68.55	548.40		
1 Crawler Crane, 40 Ton		1156.00		1271.60		
1 Lead, 90' High		124.40		136.84		
1 Hammer, Diesel, 22k ft-lb		456.40		502.04		
1 Barge, 400 Ton		776.00		853.60	39.26	43.19
64 L.H., Daily Totals		$5561.60		$7512.88	$86.90	$117.39

Crew B-19C	**Hr.**	**Daily**	**Hr.**	**Daily**	**Bare Costs**	**Incl. O&P**
1 Pile Driver Foreman (outside)	$48.10	$384.80	$76.15	$609.20	$47.64	$74.20
4 Pile Drivers	46.10	1475.20	73.00	2336.00		
2 Equip. Oper. (crane)	51.70	827.20	78.45	1255.20		
1 Equip. Oper. (oiler)	45.20	361.60	68.55	548.40		
1 Crawler Crane, 75 Ton		1459.00		1604.90		
1 Lead, 90' High		124.40		136.84		
1 Hammer, Diesel, 41k ft-lb		581.00		639.10		
1 Barge, 400 Ton		776.00		853.60	45.94	50.54
64 L.H., Daily Totals		$5989.20		$7983.24	$93.58	$124.74

Crew B-20	**Hr.**	**Daily**	**Hr.**	**Daily**	**Bare Costs**	**Incl. O&P**
1 Labor Foreman (outside)	$39.60	$316.80	$60.95	$487.60	$41.95	$64.65
1 Skilled Worker	48.65	389.20	75.15	601.20		
1 Laborer	37.60	300.80	57.85	462.80		
24 L.H., Daily Totals		$1006.80		$1551.60	$41.95	$64.65

Crew B-20A	**Hr.**	**Daily**	**Hr.**	**Daily**	**Bare Costs**	**Incl. O&P**
1 Labor Foreman (outside)	$39.60	$316.80	$60.95	$487.60	$45.71	$69.59
1 Laborer	37.60	300.80	57.85	462.80		
1 Plumber	58.70	469.60	88.65	709.20		
1 Plumber Apprentice	46.95	375.60	70.90	567.20		
32 L.H., Daily Totals		$1462.80		$2226.80	$45.71	$69.59

Crew B-21	**Hr.**	**Daily**	**Hr.**	**Daily**	**Bare Costs**	**Incl. O&P**
1 Labor Foreman (outside)	$39.60	$316.80	$60.95	$487.60	$43.34	$66.62
1 Skilled Worker	48.65	389.20	75.15	601.20		
1 Laborer	37.60	300.80	57.85	462.80		
.5 Equip. Oper. (crane)	51.70	206.80	78.45	313.80		
.5 S.P. Crane, 4x4, 5 Ton		137.70		151.47	4.92	5.41
28 L.H., Daily Totals		$1351.30		$2016.87	$48.26	$72.03

Crew B-21A	**Hr.**	**Daily**	**Hr.**	**Daily**	**Bare Costs**	**Incl. O&P**
1 Labor Foreman (outside)	$39.60	$316.80	$60.95	$487.60	$46.91	$71.36
1 Laborer	37.60	300.80	57.85	462.80		
1 Plumber	58.70	469.60	88.65	709.20		
1 Plumber Apprentice	46.95	375.60	70.90	567.20		
1 Equip. Oper. (crane)	51.70	413.60	78.45	627.60		
1 S.P. Crane, 4x4, 12 Ton		473.40		520.74	11.84	13.02
40 L.H., Daily Totals		$2349.80		$3375.14	$58.74	$84.38

Crew B-21B	**Hr.**	**Daily**	**Hr.**	**Daily**	**Bare Costs**	**Incl. O&P**
1 Labor Foreman (outside)	$39.60	$316.80	$60.95	$487.60	$40.82	$62.59
3 Laborers	37.60	902.40	57.85	1388.40		
1 Equip. Oper. (crane)	51.70	413.60	78.45	627.60		
1 Hyd. Crane, 12 Ton		653.80		719.18	16.34	17.98
40 L.H., Daily Totals		$2286.60		$3222.78	$57.16	$80.57

Crew No.	Bare Costs		Incl. Subs O&P		Cost Per Labor-Hour	
Crew B-21C	**Hr.**	**Daily**	**Hr.**	**Daily**	**Bare Costs**	**Incl. O&P**
1 Labor Foreman (outside)	$39.60	$316.80	$60.95	$487.60	$40.99	$62.76
4 Laborers	37.60	1203.20	57.85	1851.20		
1 Equip. Oper. (crane)	51.70	413.60	78.45	627.60		
1 Equip. Oper. (oiler)	45.20	361.60	68.55	548.40		
2 Cutting Torches		22.80		25.08		
2 Sets of Gases		304.00		334.40		
1 Lattice Boom Crane, 90 Ton		1511.00		1662.10	32.82	36.10
56 L.H., Daily Totals		$4133.00		$5536.38	$73.80	$98.86

Crew B-22	**Hr.**	**Daily**	**Hr.**	**Daily**	**Bare Costs**	**Incl. O&P**
1 Labor Foreman (outside)	$39.60	$316.80	$60.95	$487.60	$43.90	$67.41
1 Skilled Worker	48.65	389.20	75.15	601.20		
1 Laborer	37.60	300.80	57.85	462.80		
.75 Equip. Oper. (crane)	51.70	310.20	78.45	470.70		
.75 S.P. Crane, 4x4, 5 Ton		206.55		227.21	6.88	7.57
30 L.H., Daily Totals		$1523.55		$2249.51	$50.78	$74.98

Crew B-22A	**Hr.**	**Daily**	**Hr.**	**Daily**	**Bare Costs**	**Incl. O&P**
1 Labor Foreman (outside)	$39.60	$316.80	$60.95	$487.60	$43.03	$66.05
1 Skilled Worker	48.65	389.20	75.15	601.20		
2 Laborers	37.60	601.60	57.85	925.60		
1 Equipment Operator, Crane	51.70	413.60	78.45	627.60		
1 S.P. Crane, 4x4, 5 Ton		275.40		302.94		
1 Butt Fusion Mach., 4"-12" diam.		382.10		420.31	16.44	18.08
40 L.H., Daily Totals		$2378.70		$3365.25	$59.47	$84.13

Crew B-22B	**Hr.**	**Daily**	**Hr.**	**Daily**	**Bare Costs**	**Incl. O&P**
1 Labor Foreman (outside)	$39.60	$316.80	$60.95	$487.60	$43.03	$66.05
1 Skilled Worker	48.65	389.20	75.15	601.20		
2 Laborers	37.60	601.60	57.85	925.60		
1 Equip. Oper. (crane)	51.70	413.60	78.45	627.60		
1 S.P. Crane, 4x4, 5 Ton		275.40		302.94		
1 Butt Fusion Mach., 8"-24" diam.		829.70		912.67	27.63	30.39
40 L.H., Daily Totals		$2826.30		$3857.61	$70.66	$96.44

Crew B-22C	**Hr.**	**Daily**	**Hr.**	**Daily**	**Bare Costs**	**Incl. O&P**
1 Skilled Worker	$48.65	$389.20	$75.15	$601.20	$43.13	$66.50
1 Laborer	37.60	300.80	57.85	462.80		
1 Butt Fusion Mach., 2"-8" diam.		121.05		133.16	7.57	8.32
16 L.H., Daily Totals		$811.05		$1197.16	$50.69	$74.82

Crew B-23	**Hr.**	**Daily**	**Hr.**	**Daily**	**Bare Costs**	**Incl. O&P**
1 Labor Foreman (outside)	$39.60	$316.80	$60.95	$487.60	$38.00	$58.47
4 Laborers	37.60	1203.20	57.85	1851.20		
1 Drill Rig, Truck-Mounted		2542.00		2796.20		
1 Flatbed Truck, Gas, 3 Ton		303.20		333.52	71.13	78.24
40 L.H., Daily Totals		$4365.20		$5468.52	$109.13	$136.71

Crew B-23A	**Hr.**	**Daily**	**Hr.**	**Daily**	**Bare Costs**	**Incl. O&P**
1 Labor Foreman (outside)	$39.60	$316.80	$60.95	$487.60	$42.60	$65.18
1 Laborer	37.60	300.80	57.85	462.80		
1 Equip. Oper. (medium)	50.60	404.80	76.75	614.00		
1 Drill Rig, Truck-Mounted		2542.00		2796.20		
1 Pickup Truck, 3/4 Ton		144.20		158.62	111.93	123.12
24 L.H., Daily Totals		$3708.60		$4519.22	$154.53	$188.30

Crew B-23B	Hr.	Daily	Hr.	Daily	Bare Costs	Incl. O&P
1 Labor Foreman (outside)	$39.60	$316.80	$60.95	$487.60	$42.60	$65.18
1 Laborer	37.60	300.80	57.85	462.80		
1 Equip. Oper. (medium)	50.60	404.80	76.75	614.00		
1 Drill Rig, Truck-Mounted		2542.00		2796.20		
1 Pickup Truck, 3/4 Ton		144.20		158.62		
1 Centr. Water Pump, 6″		346.60		381.26	126.37	139.00
24 L.H., Daily Totals		$4055.20		$4900.48	$168.97	$204.19

Crew B-24	Hr.	Daily	Hr.	Daily	Bare Costs	Incl. O&P
1 Cement Finisher	$45.00	$360.00	$66.65	$533.20	$43.18	$65.58
1 Laborer	37.60	300.80	57.85	462.80		
1 Carpenter	46.95	375.60	72.25	578.00		
24 L.H., Daily Totals		$1036.40		$1574.00	$43.18	$65.58

Crew B-25	Hr.	Daily	Hr.	Daily	Bare Costs	Incl. O&P
1 Labor Foreman (outside)	$39.60	$316.80	$60.95	$487.60	$41.33	$63.29
7 Laborers	37.60	2105.60	57.85	3239.60		
3 Equip. Oper. (medium)	50.60	1214.40	76.75	1842.00		
1 Asphalt Paver, 130 H.P.		2132.00		2345.20		
1 Tandem Roller, 10 Ton		236.80		260.48		
1 Roller, Pneum. Whl., 12 Ton		344.40		378.84	30.83	33.91
88 L.H., Daily Totals		$6350.00		$8553.72	$72.16	$97.20

Crew B-25B	Hr.	Daily	Hr.	Daily	Bare Costs	Incl. O&P
1 Labor Foreman (outside)	$39.60	$316.80	$60.95	$487.60	$42.10	$64.41
7 Laborers	37.60	2105.60	57.85	3239.60		
4 Equip. Oper. (medium)	50.60	1619.20	76.75	2456.00		
1 Asphalt Paver, 130 H.P.		2132.00		2345.20		
2 Tandem Rollers, 10 Ton		473.60		520.96		
1 Roller, Pneum. Whl., 12 Ton		344.40		378.84	30.73	33.80
96 L.H., Daily Totals		$6991.60		$9428.20	$72.83	$98.21

Crew B-25C	Hr.	Daily	Hr.	Daily	Bare Costs	Incl. O&P
1 Labor Foreman (outside)	$39.60	$316.80	$60.95	$487.60	$42.27	$64.67
3 Laborers	37.60	902.40	57.85	1388.40		
2 Equip. Oper. (medium)	50.60	809.60	76.75	1228.00		
1 Asphalt Paver, 130 H.P.		2132.00		2345.20		
1 Tandem Roller, 10 Ton		236.80		260.48	49.35	54.28
48 L.H., Daily Totals		$4397.60		$5709.68	$91.62	$118.95

Crew B-25D	Hr.	Daily	Hr.	Daily	Bare Costs	Incl. O&P
1 Labor Foreman (outside)	$39.60	$316.80	$60.95	$487.60	$42.39	$64.83
3 Laborers	37.60	902.40	57.85	1388.40		
2.125 Equip. Oper. (medium)	50.60	860.20	76.75	1304.75		
.125 Truck Driver (heavy)	40.05	40.05	60.55	60.55		
.125 Truck Tractor, 6x4, 380 H.P.		74.97		82.47		
.125 Dist. Tanker, 3000 Gallon		38.20		42.02		
1 Asphalt Paver, 130 H.P.		2132.00		2345.20		
1 Tandem Roller, 10 Ton		236.80		260.48	49.64	54.60
50 L.H., Daily Totals		$4601.43		$5971.47	$92.03	$119.43

Crew B-25E	Hr.	Daily	Hr.	Daily	Bare Costs	Incl. O&P
1 Labor Foreman (outside)	$39.60	$316.80	$60.95	$487.60	$42.50	$64.97
3 Laborers	37.60	902.40	57.85	1388.40		
2.250 Equip. Oper. (medium)	50.60	910.80	76.75	1381.50		
.25 Truck Driver (heavy)	40.05	80.10	60.55	121.10		
.25 Truck Tractor, 6x4, 380 H.P.		149.95		164.94		
.25 Dist. Tanker, 3000 Gallon		76.40		84.04		
1 Asphalt Paver, 130 H.P.		2132.00		2345.20		
1 Tandem Roller, 10 Ton		236.80		260.48	49.91	54.90
52 L.H., Daily Totals		$4805.25		$6233.27	$92.41	$119.87

Crew B-26	Hr.	Daily	Hr.	Daily	Bare Costs	Incl. O&P
1 Labor Foreman (outside)	$39.60	$316.80	$60.95	$487.60	$42.18	$64.57
6 Laborers	37.60	1804.80	57.85	2776.80		
2 Equip. Oper. (medium)	50.60	809.60	76.75	1228.00		
1 Rodman (reinf.)	52.55	420.40	82.10	656.80		
1 Cement Finisher	45.00	360.00	66.65	533.20		
1 Grader, 30,000 Lbs.		736.20		809.82		
1 Paving Mach. & Equip.		2794.00		3073.40	40.12	44.13
88 L.H., Daily Totals		$7241.80		$9565.62	$82.29	$108.70

Crew B-26A	Hr.	Daily	Hr.	Daily	Bare Costs	Incl. O&P
1 Labor Foreman (outside)	$39.60	$316.80	$60.95	$487.60	$42.18	$64.57
6 Laborers	37.60	1804.80	57.85	2776.80		
2 Equip. Oper. (medium)	50.60	809.60	76.75	1228.00		
1 Rodman (reinf.)	52.55	420.40	82.10	656.80		
1 Cement Finisher	45.00	360.00	66.65	533.20		
1 Grader, 30,000 Lbs.		736.20		809.82		
1 Paving Mach. & Equip.		2794.00		3073.40		
1 Concrete Saw		167.40		184.14	42.02	46.22
88 L.H., Daily Totals		$7409.20		$9749.76	$84.20	$110.79

Crew B-26B	Hr.	Daily	Hr.	Daily	Bare Costs	Incl. O&P
1 Labor Foreman (outside)	$39.60	$316.80	$60.95	$487.60	$42.88	$65.59
6 Laborers	37.60	1804.80	57.85	2776.80		
3 Equip. Oper. (medium)	50.60	1214.40	76.75	1842.00		
1 Rodman (reinf.)	52.55	420.40	82.10	656.80		
1 Cement Finisher	45.00	360.00	66.65	533.20		
1 Grader, 30,000 Lbs.		736.20		809.82		
1 Paving Mach. & Equip.		2794.00		3073.40		
1 Concrete Pump, 110′ Boom		946.20		1040.82	46.63	51.29
96 L.H., Daily Totals		$8592.80		$11220.44	$89.51	$116.88

Crew B-26C	Hr.	Daily	Hr.	Daily	Bare Costs	Incl. O&P
1 Labor Foreman (outside)	$39.60	$316.80	$60.95	$487.60	$41.34	$63.35
6 Laborers	37.60	1804.80	57.85	2776.80		
1 Equip. Oper. (medium)	50.60	404.80	76.75	614.00		
1 Rodman (reinf.)	52.55	420.40	82.10	656.80		
1 Cement Finisher	45.00	360.00	66.65	533.20		
1 Paving Mach. & Equip.		2794.00		3073.40		
1 Concrete Saw		167.40		184.14	37.02	40.72
80 L.H., Daily Totals		$6268.20		$8325.94	$78.35	$104.07

Crew B-27	Hr.	Daily	Hr.	Daily	Bare Costs	Incl. O&P
1 Labor Foreman (outside)	$39.60	$316.80	$60.95	$487.60	$38.10	$58.63
3 Laborers	37.60	902.40	57.85	1388.40		
1 Berm Machine		296.80		326.48	9.28	10.20
32 L.H., Daily Totals		$1516.00		$2202.48	$47.38	$68.83

Crew B-28	Hr.	Daily	Hr.	Daily	Bare Costs	Incl. O&P
2 Carpenters	$46.95	$751.20	$72.25	$1156.00	$43.83	$67.45
1 Laborer	37.60	300.80	57.85	462.80		
24 L.H., Daily Totals		$1052.00		$1618.80	$43.83	$67.45

Crew B-29	Hr.	Daily	Hr.	Daily	Bare Costs	Incl. O&P
1 Labor Foreman (outside)	$39.60	$316.80	$60.95	$487.60	$40.99	$62.76
4 Laborers	37.60	1203.20	57.85	1851.20		
1 Equip. Oper. (crane)	51.70	413.60	78.45	627.60		
1 Equip. Oper. (oiler)	45.20	361.60	68.55	548.40		
1 Gradall, 5/8 C.Y.		882.20		970.42	15.75	17.33
56 L.H., Daily Totals		$3177.40		$4485.22	$56.74	$80.09

Crew B-30	Bare Costs Hr.	Bare Costs Daily	Incl. Subs O&P Hr.	Incl. Subs O&P Daily	Cost Per Labor-Hour Bare Costs	Cost Per Labor-Hour Incl. O&P
1 Equip. Oper. (medium)	$50.60	$404.80	$76.75	$614.00	$43.57	$65.95
2 Truck Drivers (heavy)	40.05	640.80	60.55	968.80		
1 Hyd. Excavator, 1.5 C.Y.		1030.00		1133.00		
2 Dump Trucks, 12 C.Y., 400 H.P.		1382.00		1520.20	100.50	110.55
24 L.H., Daily Totals		$3457.60		$4236.00	$144.07	$176.50

Crew B-31	Hr.	Daily	Hr.	Daily	Bare Costs	Incl. O&P
1 Labor Foreman (outside)	$39.60	$316.80	$60.95	$487.60	$39.87	$61.35
3 Laborers	37.60	902.40	57.85	1388.40		
1 Carpenter	46.95	375.60	72.25	578.00		
1 Air Compressor, 250 cfm		201.40		221.54		
1 Sheeting Driver		5.75		6.33		
2 -50' Air Hoses, 1.5"		11.60		12.76	5.47	6.02
40 L.H., Daily Totals		$1813.55		$2694.63	$45.34	$67.37

Crew B-32	Hr.	Daily	Hr.	Daily	Bare Costs	Incl. O&P
1 Laborer	$37.60	$300.80	$57.85	$462.80	$47.35	$72.03
3 Equip. Oper. (medium)	50.60	1214.40	76.75	1842.00		
1 Grader, 30,000 Lbs.		736.20		809.82		
1 Tandem Roller, 10 Ton		236.80		260.48		
1 Dozer, 200 H.P.		1387.00		1525.70	73.75	81.13
32 L.H., Daily Totals		$3875.20		$4900.80	$121.10	$153.15

Crew B-32A	Hr.	Daily	Hr.	Daily	Bare Costs	Incl. O&P
1 Laborer	$37.60	$300.80	$57.85	$462.80	$46.27	$70.45
2 Equip. Oper. (medium)	50.60	809.60	76.75	1228.00		
1 Grader, 30,000 Lbs.		736.20		809.82		
1 Roller, Vibratory, 25 Ton		687.40		756.14	59.32	65.25
24 L.H., Daily Totals		$2534.00		$3256.76	$105.58	$135.70

Crew B-32B	Hr.	Daily	Hr.	Daily	Bare Costs	Incl. O&P
1 Laborer	$37.60	$300.80	$57.85	$462.80	$46.27	$70.45
2 Equip. Oper. (medium)	50.60	809.60	76.75	1228.00		
1 Dozer, 200 H.P.		1387.00		1525.70		
1 Roller, Vibratory, 25 Ton		687.40		756.14	86.43	95.08
24 L.H., Daily Totals		$3184.80		$3972.64	$132.70	$165.53

Crew B-32C	Hr.	Daily	Hr.	Daily	Bare Costs	Incl. O&P
1 Labor Foreman (outside)	$39.60	$316.80	$60.95	$487.60	$44.43	$67.82
2 Laborers	37.60	601.60	57.85	925.60		
3 Equip. Oper. (medium)	50.60	1214.40	76.75	1842.00		
1 Grader, 30,000 Lbs.		736.20		809.82		
1 Tandem Roller, 10 Ton		236.80		260.48		
1 Dozer, 200 H.P.		1387.00		1525.70	49.17	54.08
48 L.H., Daily Totals		$4492.80		$5851.20	$93.60	$121.90

Crew B-33A	Hr.	Daily	Hr.	Daily	Bare Costs	Incl. O&P
1 Equip. Oper. (medium)	$50.60	$404.80	$76.75	$614.00	$46.89	$71.35
.5 Laborer	37.60	150.40	57.85	231.40		
.25 Equip. Oper. (medium)	50.60	101.20	76.75	153.50		
1 Scraper, Towed, 7 C.Y.		114.40		125.84		
1.25 Dozers, 300 H.P.		2371.25		2608.38	177.55	195.30
14 L.H., Daily Totals		$3142.05		$3733.11	$224.43	$266.65

Crew B-33B	Hr.	Daily	Hr.	Daily	Bare Costs	Incl. O&P
1 Equip. Oper. (medium)	$50.60	$404.80	$76.75	$614.00	$46.89	$71.35
.5 Laborer	37.60	150.40	57.85	231.40		
.25 Equip. Oper. (medium)	50.60	101.20	76.75	153.50		
1 Scraper, Towed, 10 C.Y.		146.80		161.48		
1.25 Dozers, 300 H.P.		2371.25		2608.38	179.86	197.85
14 L.H., Daily Totals		$3174.45		$3768.76	$226.75	$269.20

Crew B-33C	Hr.	Daily	Hr.	Daily	Bare Costs	Incl. O&P
1 Equip. Oper. (medium)	$50.60	$404.80	$76.75	$614.00	$46.89	$71.35
.5 Laborer	37.60	150.40	57.85	231.40		
.25 Equip. Oper. (medium)	50.60	101.20	76.75	153.50		
1 Scraper, Towed, 15 C.Y.		164.80		181.28		
1.25 Dozers, 300 H.P.		2371.25		2608.38	181.15	199.26
14 L.H., Daily Totals		$3192.45		$3788.55	$228.03	$270.61

Crew B-33D	Hr.	Daily	Hr.	Daily	Bare Costs	Incl. O&P
1 Equip. Oper. (medium)	$50.60	$404.80	$76.75	$614.00	$46.89	$71.35
.5 Laborer	37.60	150.40	57.85	231.40		
.25 Equip. Oper. (medium)	50.60	101.20	76.75	153.50		
1 S.P. Scraper, 14 C.Y.		1884.00		2072.40		
.25 Dozer, 300 H.P.		474.25		521.67	168.45	185.29
14 L.H., Daily Totals		$3014.65		$3592.97	$215.33	$256.64

Crew B-33E	Hr.	Daily	Hr.	Daily	Bare Costs	Incl. O&P
1 Equip. Oper. (medium)	$50.60	$404.80	$76.75	$614.00	$46.89	$71.35
.5 Laborer	37.60	150.40	57.85	231.40		
.25 Equip. Oper. (medium)	50.60	101.20	76.75	153.50		
1 S.P. Scraper, 21 C.Y.		2688.00		2956.80		
.25 Dozer, 300 H.P.		474.25		521.67	225.88	248.46
14 L.H., Daily Totals		$3818.65		$4477.38	$272.76	$319.81

Crew B-33F	Hr.	Daily	Hr.	Daily	Bare Costs	Incl. O&P
1 Equip. Oper. (medium)	$50.60	$404.80	$76.75	$614.00	$46.89	$71.35
.5 Laborer	37.60	150.40	57.85	231.40		
.25 Equip. Oper. (medium)	50.60	101.20	76.75	153.50		
1 Elev. Scraper, 11 C.Y.		1163.00		1279.30		
.25 Dozer, 300 H.P.		474.25		521.67	116.95	128.64
14 L.H., Daily Totals		$2293.65		$2799.88	$163.83	$199.99

Crew B-33G	Hr.	Daily	Hr.	Daily	Bare Costs	Incl. O&P
1 Equip. Oper. (medium)	$50.60	$404.80	$76.75	$614.00	$46.89	$71.35
.5 Laborer	37.60	150.40	57.85	231.40		
.25 Equip. Oper. (medium)	50.60	101.20	76.75	153.50		
1 Elev. Scraper, 22 C.Y.		2521.00		2773.10		
.25 Dozer, 300 H.P.		474.25		521.67	213.95	235.34
14 L.H., Daily Totals		$3651.65		$4293.68	$260.83	$306.69

Crew B-33H	Hr.	Daily	Hr.	Daily	Bare Costs	Incl. O&P
.5 Laborer	$37.60	$150.40	$57.85	$231.40	$46.89	$71.35
1 Equipment Operator (med.)	50.60	404.80	76.75	614.00		
.25 Equipment Operator (med.)	50.60	101.20	76.75	153.50		
1 S.P. Scraper, 44 C.Y.		4679.00		5146.90		
.25 Dozer, 410 H.P.		601.25		661.38	377.16	414.88
14 L.H., Daily Totals		$5936.65		$6807.18	$424.05	$486.23

Crew B-33J	Hr.	Daily	Hr.	Daily	Bare Costs	Incl. O&P
1 Equipment Operator (med.)	$50.60	$404.80	$76.75	$614.00	$50.60	$76.75
1 S.P. Scraper, 14 C.Y.		1884.00		2072.40	235.50	259.05
8 L.H., Daily Totals		$2288.80		$2686.40	$286.10	$335.80

Crew B-33K	Hr.	Daily	Hr.	Daily	Bare Costs	Incl. O&P
1 Equipment Operator (med.)	$50.60	$404.80	$76.75	$614.00	$46.89	$71.35
.25 Equipment Operator (med.)	50.60	101.20	76.75	153.50		
.5 Laborer	37.60	150.40	57.85	231.40		
1 S.P. Scraper, 31 C.Y.		3693.00		4062.30		
.25 Dozer, 410 H.P.		601.25		661.38	306.73	337.41
14 L.H., Daily Totals		$4950.65		$5722.57	$353.62	$408.76

Crew No.	Bare Costs		Incl. Subs O&P		Cost Per Labor-Hour	
Crew B-34A	**Hr.**	**Daily**	**Hr.**	**Daily**	**Bare Costs**	**Incl. O&P**
1 Truck Driver (heavy)	$40.05	$320.40	$60.55	$484.40	$40.05	$60.55
1 Dump Truck, 8 C.Y., 220 H.P.		411.20		452.32	51.40	56.54
8 L.H., Daily Totals		$731.60		$936.72	$91.45	$117.09

Crew B-34B	Hr.	Daily	Hr.	Daily	Bare Costs	Incl. O&P
1 Truck Driver (heavy)	$40.05	$320.40	$60.55	$484.40	$40.05	$60.55
1 Dump Truck, 12 C.Y., 400 H.P.		691.00		760.10	86.38	95.01
8 L.H., Daily Totals		$1011.40		$1244.50	$126.43	$155.56

Crew B-34C	Hr.	Daily	Hr.	Daily	Bare Costs	Incl. O&P
1 Truck Driver (heavy)	$40.05	$320.40	$60.55	$484.40	$40.05	$60.55
1 Truck Tractor, 6x4, 380 H.P.		599.80		659.78		
1 Dump Trailer, 16.5 C.Y.		126.60		139.26	90.80	99.88
8 L.H., Daily Totals		$1046.80		$1283.44	$130.85	$160.43

Crew B-34D	Hr.	Daily	Hr.	Daily	Bare Costs	Incl. O&P
1 Truck Driver (heavy)	$40.05	$320.40	$60.55	$484.40	$40.05	$60.55
1 Truck Tractor, 6x4, 380 H.P.		599.80		659.78		
1 Dump Trailer, 20 C.Y.		141.20		155.32	92.63	101.89
8 L.H., Daily Totals		$1061.40		$1299.50	$132.68	$162.44

Crew B-34E	Hr.	Daily	Hr.	Daily	Bare Costs	Incl. O&P
1 Truck Driver (heavy)	$40.05	$320.40	$60.55	$484.40	$40.05	$60.55
1 Dump Truck, Off Hwy., 25 Ton		1351.00		1486.10	168.88	185.76
8 L.H., Daily Totals		$1671.40		$1970.50	$208.93	$246.31

Crew B-34F	Hr.	Daily	Hr.	Daily	Bare Costs	Incl. O&P
1 Truck Driver (heavy)	$40.05	$320.40	$60.55	$484.40	$40.05	$60.55
1 Dump Truck, Off Hwy., 35 Ton		1512.00		1663.20	189.00	207.90
8 L.H., Daily Totals		$1832.40		$2147.60	$229.05	$268.45

Crew B-34G	Hr.	Daily	Hr.	Daily	Bare Costs	Incl. O&P
1 Truck Driver (heavy)	$40.05	$320.40	$60.55	$484.40	$40.05	$60.55
1 Dump Truck, Off Hwy., 50 Ton		1837.00		2020.70	229.63	252.59
8 L.H., Daily Totals		$2157.40		$2505.10	$269.68	$313.14

Crew B-34H	Hr.	Daily	Hr.	Daily	Bare Costs	Incl. O&P
1 Truck Driver (heavy)	$40.05	$320.40	$60.55	$484.40	$40.05	$60.55
1 Dump Truck, Off Hwy., 65 Ton		1863.00		2049.30	232.88	256.16
8 L.H., Daily Totals		$2183.40		$2533.70	$272.93	$316.71

Crew B-34I	Hr.	Daily	Hr.	Daily	Bare Costs	Incl. O&P
1 Truck Driver (heavy)	$40.05	$320.40	$60.55	$484.40	$40.05	$60.55
1 Dump Truck, 18 C.Y., 450 H.P.		869.00		955.90	108.63	119.49
8 L.H., Daily Totals		$1189.40		$1440.30	$148.68	$180.04

Crew B-34J	Hr.	Daily	Hr.	Daily	Bare Costs	Incl. O&P
1 Truck Driver (heavy)	$40.05	$320.40	$60.55	$484.40	$40.05	$60.55
1 Dump Truck, Off Hwy., 100 Ton		2925.00		3217.50	365.63	402.19
8 L.H., Daily Totals		$3245.40		$3701.90	$405.68	$462.74

Crew B-34K	Hr.	Daily	Hr.	Daily	Bare Costs	Incl. O&P
1 Truck Driver (heavy)	$40.05	$320.40	$60.55	$484.40	$40.05	$60.55
1 Truck Tractor, 6x4, 450 H.P.		727.40		800.14		
1 Lowbed Trailer, 75 Ton		221.80		243.98	118.65	130.51
8 L.H., Daily Totals		$1269.60		$1528.52	$158.70	$191.07

Crew B-34L	Hr.	Daily	Hr.	Daily	Bare Costs	Incl. O&P
1 Equip. Oper. (light)	$48.60	$388.80	$73.75	$590.00	$48.60	$73.75
1 Flatbed Truck, Gas, 1.5 Ton		245.40		269.94	30.68	33.74
8 L.H., Daily Totals		$634.20		$859.94	$79.28	$107.49

Crew B-34M	Hr.	Daily	Hr.	Daily	Bare Costs	Incl. O&P
1 Equip. Oper. (light)	$48.60	$388.80	$73.75	$590.00	$48.60	$73.75
1 Flatbed Truck, Gas, 3 Ton		303.20		333.52	37.90	41.69
8 L.H., Daily Totals		$692.00		$923.52	$86.50	$115.44

Crew B-34N	Hr.	Daily	Hr.	Daily	Bare Costs	Incl. O&P
1 Truck Driver (heavy)	$40.05	$320.40	$60.55	$484.40	$45.33	$68.65
1 Equip. Oper. (medium)	50.60	404.80	76.75	614.00		
1 Truck Tractor, 6x4, 380 H.P.		599.80		659.78		
1 Flatbed Trailer, 40 Ton		154.00		169.40	47.11	51.82
16 L.H., Daily Totals		$1479.00		$1927.58	$92.44	$120.47

Crew B-34P	Hr.	Daily	Hr.	Daily	Bare Costs	Incl. O&P
1 Pipe Fitter	$59.75	$478.00	$90.20	$721.60	$49.82	$75.35
1 Truck Driver (light)	39.10	312.80	59.10	472.80		
1 Equip. Oper. (medium)	50.60	404.80	76.75	614.00		
1 Flatbed Truck, Gas, 3 Ton		303.20		333.52		
1 Backhoe Loader, 48 H.P.		364.20		400.62	27.81	30.59
24 L.H., Daily Totals		$1863.00		$2542.54	$77.63	$105.94

Crew B-34Q	Hr.	Daily	Hr.	Daily	Bare Costs	Incl. O&P
1 Pipe Fitter	$59.75	$478.00	$90.20	$721.60	$50.18	$75.92
1 Truck Driver (light)	39.10	312.80	59.10	472.80		
1 Equip. Oper. (crane)	51.70	413.60	78.45	627.60		
1 Flatbed Trailer, 25 Ton		113.60		124.96		
1 Dump Truck, 8 C.Y., 220 H.P.		411.20		452.32		
1 Hyd. Crane, 25 Ton		736.60		810.26	52.56	57.81
24 L.H., Daily Totals		$2465.80		$3209.54	$102.74	$133.73

Crew B-34R	Hr.	Daily	Hr.	Daily	Bare Costs	Incl. O&P
1 Pipe Fitter	$59.75	$478.00	$90.20	$721.60	$50.18	$75.92
1 Truck Driver (light)	39.10	312.80	59.10	472.80		
1 Equip. Oper. (crane)	51.70	413.60	78.45	627.60		
1 Flatbed Trailer, 25 Ton		113.60		124.96		
1 Dump Truck, 8 C.Y., 220 H.P.		411.20		452.32		
1 Hyd. Crane, 25 Ton		736.60		810.26		
1 Hyd. Excavator, 1 C.Y.		812.60		893.86	86.42	95.06
24 L.H., Daily Totals		$3278.40		$4103.40	$136.60	$170.97

Crew B-34S	Hr.	Daily	Hr.	Daily	Bare Costs	Incl. O&P
2 Pipe Fitters	$59.75	$956.00	$90.20	$1443.20	$52.81	$79.85
1 Truck Driver (heavy)	40.05	320.40	60.55	484.40		
1 Equip. Oper. (crane)	51.70	413.60	78.45	627.60		
1 Flatbed Trailer, 40 Ton		154.00		169.40		
1 Truck Tractor, 6x4, 380 H.P.		599.80		659.78		
1 Hyd. Crane, 80 Ton		1625.00		1787.50		
1 Hyd. Excavator, 2 C.Y.		1175.00		1292.50	111.06	122.16
32 L.H., Daily Totals		$5243.80		$6464.38	$163.87	$202.01

Crew B-34T	Hr.	Daily	Hr.	Daily	Bare Costs	Incl. O&P
2 Pipe Fitters	$59.75	$956.00	$90.20	$1443.20	$52.81	$79.85
1 Truck Driver (heavy)	40.05	320.40	60.55	484.40		
1 Equip. Oper. (crane)	51.70	413.60	78.45	627.60		
1 Flatbed Trailer, 40 Ton		154.00		169.40		
1 Truck Tractor, 6x4, 380 H.P.		599.80		659.78		
1 Hyd. Crane, 80 Ton		1625.00		1787.50	74.34	81.77
32 L.H., Daily Totals		$4068.80		$5171.88	$127.15	$161.62

Crew B-34U	Bare Costs Hr.	Bare Costs Daily	Incl. Subs O&P Hr.	Incl. Subs O&P Daily	Cost Per Labor-Hour Bare Costs	Cost Per Labor-Hour Incl. O&P
1 Truck Driver (heavy)	$40.05	$320.40	$60.55	$484.40	$44.33	$67.15
1 Equip. Oper. (light)	48.60	388.80	73.75	590.00		
1 Truck Tractor, 220 H.P.		359.60		395.56		
1 Flatbed Trailer, 25 Ton		113.60		124.96	29.57	32.53
16 L.H., Daily Totals		$1182.40		$1594.92	$73.90	$99.68

Crew B-34V	Hr.	Daily	Hr.	Daily	Bare Costs	Incl. O&P
1 Truck Driver (heavy)	$40.05	$320.40	$60.55	$484.40	$46.78	$70.92
1 Equip. Oper. (crane)	51.70	413.60	78.45	627.60		
1 Equip. Oper. (light)	48.60	388.80	73.75	590.00		
1 Truck Tractor, 6x4, 450 H.P.		727.40		800.14		
1 Equipment Trailer, 50 Ton		168.40		185.24		
1 Pickup Truck, 4 x 4, 3/4 Ton		155.60		171.16	43.81	48.19
24 L.H., Daily Totals		$2174.20		$2858.54	$90.59	$119.11

Crew B-34W	Hr.	Daily	Hr.	Daily	Bare Costs	Incl. O&P
5 Truck Drivers (heavy)	$40.05	$1602.00	$60.55	$2422.00	$43.66	$66.21
2 Equip. Opers. (crane)	51.70	827.20	78.45	1255.20		
1 Equip. Oper. (mechanic)	51.65	413.20	78.35	626.80		
1 Laborer	37.60	300.80	57.85	462.80		
4 Truck Tractor, 6x4, 380 H.P.		2399.20		2639.12		
2 Equipment Trailer, 50 Ton		336.80		370.48		
2 Flatbed Trailer, 40 Ton		308.00		338.80		
1 Pickup Truck, 4 x 4, 3/4 Ton		155.60		171.16		
1 S.P. Crane, 4x4, 20 Ton		554.60		610.06	52.14	57.36
72 L.H., Daily Totals		$6897.40		$8896.42	$95.80	$123.56

Crew B-35	Hr.	Daily	Hr.	Daily	Bare Costs	Incl. O&P
1 Labor Foreman (outside)	$39.60	$316.80	$60.95	$487.60	$46.91	$71.60
1 Skilled Worker	48.65	389.20	75.15	601.20		
1 Welder (plumber)	58.70	469.60	88.65	709.20		
1 Laborer	37.60	300.80	57.85	462.80		
1 Equip. Oper. (crane)	51.70	413.60	78.45	627.60		
1 Equip. Oper. (oiler)	45.20	361.60	68.55	548.40		
1 Welder, Electric, 300 amp		57.70		63.47		
1 Hyd. Excavator, .75 C.Y.		654.80		720.28	14.84	16.33
48 L.H., Daily Totals		$2964.10		$4220.55	$61.75	$87.93

Crew B-35A	Hr.	Daily	Hr.	Daily	Bare Costs	Incl. O&P
1 Labor Foreman (outside)	$39.60	$316.80	$60.95	$487.60	$45.58	$69.64
2 Laborers	37.60	601.60	57.85	925.60		
1 Skilled Worker	48.65	389.20	75.15	601.20		
1 Welder (plumber)	58.70	469.60	88.65	709.20		
1 Equip. Oper. (crane)	51.70	413.60	78.45	627.60		
1 Equip. Oper. (oiler)	45.20	361.60	68.55	548.40		
1 Welder, Gas Engine, 300 amp		145.85		160.44		
1 Crawler Crane, 75 Ton		1459.00		1604.90	28.66	31.52
56 L.H., Daily Totals		$4157.25		$5664.94	$74.24	$101.16

Crew B-36	Hr.	Daily	Hr.	Daily	Bare Costs	Incl. O&P
1 Labor Foreman (outside)	$39.60	$316.80	$60.95	$487.60	$43.20	$66.03
2 Laborers	37.60	601.60	57.85	925.60		
2 Equip. Oper. (medium)	50.60	809.60	76.75	1228.00		
1 Dozer, 200 H.P.		1387.00		1525.70		
1 Aggregate Spreader		39.60		43.56		
1 Tandem Roller, 10 Ton		236.80		260.48	41.59	45.74
40 L.H., Daily Totals		$3391.40		$4470.94	$84.78	$111.77

Crew B-36A	Hr.	Daily	Hr.	Daily	Bare Costs	Incl. O&P
1 Labor Foreman (outside)	$39.60	$316.80	$60.95	$487.60	$45.31	$69.09
2 Laborers	37.60	601.60	57.85	925.60		
4 Equip. Oper. (medium)	50.60	1619.20	76.75	2456.00		
1 Dozer, 200 H.P.		1387.00		1525.70		
1 Aggregate Spreader		39.60		43.56		
1 Tandem Roller, 10 Ton		236.80		260.48		
1 Roller, Pneum. Whl., 12 Ton		344.40		378.84	35.85	39.44
56 L.H., Daily Totals		$4545.40		$6077.78	$81.17	$108.53

Crew B-36B	Hr.	Daily	Hr.	Daily	Bare Costs	Incl. O&P
1 Labor Foreman (outside)	$39.60	$316.80	$60.95	$487.60	$44.66	$68.03
2 Laborers	37.60	601.60	57.85	925.60		
4 Equip. Oper. (medium)	50.60	1619.20	76.75	2456.00		
1 Truck Driver (heavy)	40.05	320.40	60.55	484.40		
1 Grader, 30,000 Lbs.		736.20		809.82		
1 F.E. Loader, Crl, 1.5 C.Y.		629.80		692.78		
1 Dozer, 300 H.P.		1897.00		2086.70		
1 Roller, Vibratory, 25 Ton		687.40		756.14		
1 Truck Tractor, 6x4, 450 H.P.		727.40		800.14		
1 Water Tank Trailer, 5000 Gal.		141.00		155.10	75.29	82.82
64 L.H., Daily Totals		$7676.80		$9654.28	$119.95	$150.85

Crew B-36C	Hr.	Daily	Hr.	Daily	Bare Costs	Incl. O&P
1 Labor Foreman (outside)	$39.60	$316.80	$60.95	$487.60	$46.29	$70.35
3 Equip. Oper. (medium)	50.60	1214.40	76.75	1842.00		
1 Truck Driver (heavy)	40.05	320.40	60.55	484.40		
1 Grader, 30,000 Lbs.		736.20		809.82		
1 Dozer, 300 H.P.		1897.00		2086.70		
1 Roller, Vibratory, 25 Ton		687.40		756.14		
1 Truck Tractor, 6x4, 450 H.P.		727.40		800.14		
1 Water Tank Trailer, 5000 Gal.		141.00		155.10	104.72	115.20
40 L.H., Daily Totals		$6040.60		$7421.90	$151.01	$185.55

Crew B-36D	Hr.	Daily	Hr.	Daily	Bare Costs	Incl. O&P
1 Labor Foreman (outside)	$39.60	$316.80	$60.95	$487.60	$47.85	$72.80
3 Equip. Oper. (medium)	50.60	1214.40	76.75	1842.00		
1 Grader, 30,000 Lbs.		736.20		809.82		
1 Dozer, 300 H.P.		1897.00		2086.70		
1 Roller, Vibratory, 25 Ton		687.40		756.14	103.77	114.15
32 L.H., Daily Totals		$4851.80		$5982.26	$151.62	$186.95

Crew B-37	Hr.	Daily	Hr.	Daily	Bare Costs	Incl. O&P
1 Labor Foreman (outside)	$39.60	$316.80	$60.95	$487.60	$39.77	$61.02
4 Laborers	37.60	1203.20	57.85	1851.20		
1 Equip. Oper. (light)	48.60	388.80	73.75	590.00		
1 Tandem Roller, 5 Ton		157.40		173.14	3.28	3.61
48 L.H., Daily Totals		$2066.20		$3101.94	$43.05	$64.62

Crew B-37A	Hr.	Daily	Hr.	Daily	Bare Costs	Incl. O&P
2 Laborers	$37.60	$601.60	$57.85	$925.60	$38.10	$58.27
1 Truck Driver (light)	39.10	312.80	59.10	472.80		
1 Flatbed Truck, Gas, 1.5 Ton		245.40		269.94		
1 Tar Kettle, T.M.		132.00		145.20	15.73	17.30
24 L.H., Daily Totals		$1291.80		$1813.54	$53.83	$75.56

Crew B-37B	Hr.	Daily	Hr.	Daily	Bare Costs	Incl. O&P
3 Laborers	$37.60	$902.40	$57.85	$1388.40	$37.98	$58.16
1 Truck Driver (light)	39.10	312.80	59.10	472.80		
1 Flatbed Truck, Gas, 1.5 Ton		245.40		269.94		
1 Tar Kettle, T.M.		132.00		145.20	11.79	12.97
32 L.H., Daily Totals		$1592.60		$2276.34	$49.77	$71.14

Crew No.	Bare Costs		Incl. Subs O&P		Cost Per Labor-Hour	
Crew B-37C	**Hr.**	**Daily**	**Hr.**	**Daily**	**Bare Costs**	**Incl. O&P**
2 Laborers	$37.60	$601.60	$57.85	$925.60	$38.35	$58.48
2 Truck Drivers (light)	39.10	625.60	59.10	945.60		
2 Flatbed Trucks, Gas, 1.5 Ton		490.80		539.88		
1 Tar Kettle, T.M.		132.00		145.20	19.46	21.41
32 L.H., Daily Totals		$1850.00		$2556.28	$57.81	$79.88
Crew B-37D	**Hr.**	**Daily**	**Hr.**	**Daily**	**Bare Costs**	**Incl. O&P**
1 Laborer	$37.60	$300.80	$57.85	$462.80	$38.35	$58.48
1 Truck Driver (light)	39.10	312.80	59.10	472.80		
1 Pickup Truck, 3/4 Ton		144.20		158.62	9.01	9.91
16 L.H., Daily Totals		$757.80		$1094.22	$47.36	$68.39
Crew B-37E	**Hr.**	**Daily**	**Hr.**	**Daily**	**Bare Costs**	**Incl. O&P**
3 Laborers	$37.60	$902.40	$57.85	$1388.40	$41.46	$63.18
1 Equip. Oper. (light)	48.60	388.80	73.75	590.00		
1 Equip. Oper. (medium)	50.60	404.80	76.75	614.00		
2 Truck Drivers (light)	39.10	625.60	59.10	945.60		
4 Barrels w/ Flasher		13.60		14.96		
1 Concrete Saw		167.40		184.14		
1 Rotary Hammer Drill		23.75		26.13		
1 Hammer Drill Bit		2.35		2.59		
1 Loader, Skid Steer, 30 H.P.		173.60		190.96		
1 Conc. Hammer Attach.		114.70		126.17		
1 Vibrating Plate, Gas, 18"		36.20		39.82		
2 Flatbed Trucks, Gas, 1.5 Ton		490.80		539.88	18.26	20.08
56 L.H., Daily Totals		$3344.00		$4662.64	$59.71	$83.26
Crew B-37F	**Hr.**	**Daily**	**Hr.**	**Daily**	**Bare Costs**	**Incl. O&P**
3 Laborers	$37.60	$902.40	$57.85	$1388.40	$37.98	$58.16
1 Truck Driver (light)	39.10	312.80	59.10	472.80		
4 Barrels w/ Flasher		13.60		14.96		
1 Concrete Mixer, 10 C.F.		172.40		189.64		
1 Air Compressor, 60 cfm		137.40		151.14		
1 -50' Air Hose, 3/4"		3.25		3.58		
1 Spade (Chipper)		7.60		8.36		
1 Flatbed Truck, Gas, 1.5 Ton		245.40		269.94	18.11	19.93
32 L.H., Daily Totals		$1794.85		$2498.82	$56.09	$78.09
Crew B-37G	**Hr.**	**Daily**	**Hr.**	**Daily**	**Bare Costs**	**Incl. O&P**
1 Labor Foreman (outside)	$39.60	$316.80	$60.95	$487.60	$39.77	$61.02
4 Laborers	37.60	1203.20	57.85	1851.20		
1 Equip. Oper. (light)	48.60	388.80	73.75	590.00		
1 Berm Machine		296.80		326.48		
1 Tandem Roller, 5 Ton		157.40		173.14	9.46	10.41
48 L.H., Daily Totals		$2363.00		$3428.42	$49.23	$71.43
Crew B-37H	**Hr.**	**Daily**	**Hr.**	**Daily**	**Bare Costs**	**Incl. O&P**
1 Labor Foreman (outside)	$39.60	$316.80	$60.95	$487.60	$39.77	$61.02
4 Laborers	37.60	1203.20	57.85	1851.20		
1 Equip. Oper. (light)	48.60	388.80	73.75	590.00		
1 Tandem Roller, 5 Ton		157.40		173.14		
1 Flatbed Trucks, Gas, 1.5 Ton		245.40		269.94		
1 Tar Kettle, T.M.		132.00		145.20	11.14	12.26
48 L.H., Daily Totals		$2443.60		$3517.08	$50.91	$73.27

Crew No.	Bare Costs		Incl. Subs O&P		Cost Per Labor-Hour	
Crew B-37I	**Hr.**	**Daily**	**Hr.**	**Daily**	**Bare Costs**	**Incl. O&P**
3 Laborers	$37.60	$902.40	$57.85	$1388.40	$41.46	$63.18
1 Equip. Oper. (light)	48.60	388.80	73.75	590.00		
1 Equip. Oper. (medium)	50.60	404.80	76.75	614.00		
2 Truck Drivers (light)	39.10	625.60	59.10	945.60		
4 Barrels w/ Flasher		13.60		14.96		
1 Concrete Saw		167.40		184.14		
1 Rotary Hammer Drill		23.75		26.13		
1 Hammer Drill Bit		2.35		2.59		
1 Air Compressor, 60 cfm		137.40		151.14		
1 -50' Air Hose, 3/4"		3.25		3.58		
1 Spade (Chipper)		7.60		8.36		
1 Loader, Skid Steer, 30 H.P.		173.60		190.96		
1 Conc. Hammer Attach.		114.70		126.17		
1 Concrete Mixer, 10 C.F.		172.40		189.64		
1 Vibrating Plate, Gas, 18"		36.20		39.82		
2 Flatbed Trucks, Gas, 1.5 Ton		490.80		539.88	23.98	26.38
56 L.H., Daily Totals		$3664.65		$5015.35	$65.44	$89.56
Crew B-37J	**Hr.**	**Daily**	**Hr.**	**Daily**	**Bare Costs**	**Incl. O&P**
1 Labor Foreman (outside)	$39.60	$316.80	$60.95	$487.60	$39.77	$61.02
4 Laborers	37.60	1203.20	57.85	1851.20		
1 Equip. Oper. (light)	48.60	388.80	73.75	590.00		
1 Air Compressor, 60 cfm		137.40		151.14		
1 -50' Air Hose, 3/4"		3.25		3.58		
2 Concrete Mixer, 10 C.F.		344.80		379.28		
2 Flatbed Trucks, Gas, 1.5 Ton		490.80		539.88		
1 Shot Blaster, 20"		214.00		235.40	24.80	27.28
48 L.H., Daily Totals		$3099.05		$4238.07	$64.56	$88.29
Crew B-37K	**Hr.**	**Daily**	**Hr.**	**Daily**	**Bare Costs**	**Incl. O&P**
1 Labor Foreman (outside)	$39.60	$316.80	$60.95	$487.60	$39.77	$61.02
4 Laborers	37.60	1203.20	57.85	1851.20		
1 Equip. Oper. (light)	48.60	388.80	73.75	590.00		
1 Air Compressor, 60 cfm		137.40		151.14		
1 -50' Air Hose, 3/4"		3.25		3.58		
2 Flatbed Trucks, Gas, 1.5 Ton		490.80		539.88		
1 Shot Blaster, 20"		214.00		235.40	17.61	19.37
48 L.H., Daily Totals		$2754.25		$3858.80	$57.38	$80.39
Crew B-38	**Hr.**	**Daily**	**Hr.**	**Daily**	**Bare Costs**	**Incl. O&P**
1 Labor Foreman (outside)	$39.60	$316.80	$60.95	$487.60	$42.80	$65.43
2 Laborers	37.60	601.60	57.85	925.60		
1 Equip. Oper. (light)	48.60	388.80	73.75	590.00		
1 Equip. Oper. (medium)	50.60	404.80	76.75	614.00		
1 Backhoe Loader, 48 H.P.		364.20		400.62		
1 Hyd. Hammer, (1200 lb.)		180.40		198.44		
1 F.E. Loader, W.M., 4 C.Y.		669.20		736.12		
1 Pvmt. Rem. Bucket		61.20		67.32	31.88	35.06
40 L.H., Daily Totals		$2987.00		$4019.70	$74.67	$100.49
Crew B-39	**Hr.**	**Daily**	**Hr.**	**Daily**	**Bare Costs**	**Incl. O&P**
1 Labor Foreman (outside)	$39.60	$316.80	$60.95	$487.60	$39.77	$61.02
4 Laborers	37.60	1203.20	57.85	1851.20		
1 Equip. Oper. (light)	48.60	388.80	73.75	590.00		
1 Air Compressor, 250 cfm		201.40		221.54		
2 Breakers, Pavement, 60 lb.		19.60		21.56		
2 -50' Air Hoses, 1.5"		11.60		12.76	4.85	5.33
48 L.H., Daily Totals		$2141.40		$3184.66	$44.61	$66.35

Crew B-40	Bare Costs Hr.	Bare Costs Daily	Incl. Subs O&P Hr.	Incl. Subs O&P Daily	Bare Costs	Incl. O&P
1 Pile Driver Foreman (outside)	$48.10	$384.80	$76.15	$609.20	$47.64	$74.20
4 Pile Drivers	46.10	1475.20	73.00	2336.00		
2 Equip. Oper. (crane)	51.70	827.20	78.45	1255.20		
1 Equip. Oper. (oiler)	45.20	361.60	68.55	548.40		
1 Crawler Crane, 40 Ton		1156.00		1271.60		
1 Vibratory Hammer & Gen.		2587.00		2845.70	58.48	64.33
64 L.H., Daily Totals		$6791.80		$8866.10	$106.12	$138.53

Crew B-40B	Hr.	Daily	Hr.	Daily	Bare Costs	Incl. O&P
1 Labor Foreman (outside)	$39.60	$316.80	$60.95	$487.60	$41.55	$63.58
3 Laborers	37.60	902.40	57.85	1388.40		
1 Equip. Oper. (crane)	51.70	413.60	78.45	627.60		
1 Equip. Oper. (oiler)	45.20	361.60	68.55	548.40		
1 Lattice Boom Crane, 40 Ton		1164.00		1280.40	24.25	26.68
48 L.H., Daily Totals		$3158.40		$4332.40	$65.80	$90.26

Crew B-41	Hr.	Daily	Hr.	Daily	Bare Costs	Incl. O&P
1 Labor Foreman (outside)	$39.60	$316.80	$60.95	$487.60	$38.95	$59.84
4 Laborers	37.60	1203.20	57.85	1851.20		
.25 Equip. Oper. (crane)	51.70	103.40	78.45	156.90		
.25 Equip. Oper. (oiler)	45.20	90.40	68.55	137.10		
.25 Crawler Crane, 40 Ton		289.00		317.90	6.57	7.22
44 L.H., Daily Totals		$2002.80		$2950.70	$45.52	$67.06

Crew B-42	Hr.	Daily	Hr.	Daily	Bare Costs	Incl. O&P
1 Labor Foreman (outside)	$39.60	$316.80	$60.95	$487.60	$42.44	$66.35
4 Laborers	37.60	1203.20	57.85	1851.20		
1 Equip. Oper. (crane)	51.70	413.60	78.45	627.60		
1 Equip. Oper. (oiler)	45.20	361.60	68.55	548.40		
1 Welder	52.65	421.20	91.45	731.60		
1 Hyd. Crane, 25 Ton		736.60		810.26		
1 Welder, Gas Engine, 300 amp		145.85		160.44		
1 Horz. Boring Csg. Mch.		474.80		522.28	21.21	23.33
64 L.H., Daily Totals		$4073.65		$5739.38	$63.65	$89.68

Crew B-43	Hr.	Daily	Hr.	Daily	Bare Costs	Incl. O&P
1 Labor Foreman (outside)	$39.60	$316.80	$60.95	$487.60	$41.55	$63.58
3 Laborers	37.60	902.40	57.85	1388.40		
1 Equip. Oper. (crane)	51.70	413.60	78.45	627.60		
1 Equip. Oper. (oiler)	45.20	361.60	68.55	548.40		
1 Drill Rig, Truck-Mounted		2542.00		2796.20	52.96	58.25
48 L.H., Daily Totals		$4536.40		$5848.20	$94.51	$121.84

Crew B-44	Hr.	Daily	Hr.	Daily	Bare Costs	Incl. O&P
1 Pile Driver Foreman (outside)	$48.10	$384.80	$76.15	$609.20	$46.69	$72.86
4 Pile Drivers	46.10	1475.20	73.00	2336.00		
2 Equip. Oper. (crane)	51.70	827.20	78.45	1255.20		
1 Laborer	37.60	300.80	57.85	462.80		
1 Crawler Crane, 40 Ton		1156.00		1271.60		
1 Lead, 60' High		74.80		82.28		
1 Hammer, Diesel, 15K ft.-lbs.		586.00		644.60	28.39	31.23
64 L.H., Daily Totals		$4804.80		$6661.68	$75.08	$104.09

Crew B-45	Hr.	Daily	Hr.	Daily	Bare Costs	Incl. O&P
1 Equip. Oper. (medium)	$50.60	$404.80	$76.75	$614.00	$45.33	$68.65
1 Truck Driver (heavy)	40.05	320.40	60.55	484.40		
1 Dist. Tanker, 3000 Gallon		305.60		336.16		
1 Truck Tractor, 6x4, 380 H.P.		599.80		659.78	56.59	62.25
16 L.H., Daily Totals		$1630.60		$2094.34	$101.91	$130.90

Crew B-46	Hr.	Daily	Hr.	Daily	Bare Costs	Incl. O&P
1 Pile Driver Foreman (outside)	$48.10	$384.80	$76.15	$609.20	$42.18	$65.95
2 Pile Drivers	46.10	737.60	73.00	1168.00		
3 Laborers	37.60	902.40	57.85	1388.40		
1 Chain Saw, Gas, 36" Long		45.00		49.50	.94	1.03
48 L.H., Daily Totals		$2069.80		$3215.10	$43.12	$66.98

Crew B-47	Hr.	Daily	Hr.	Daily	Bare Costs	Incl. O&P
1 Blast Foreman (outside)	$39.60	$316.80	$60.95	$487.60	$41.93	$64.18
1 Driller	37.60	300.80	57.85	462.80		
1 Equip. Oper. (light)	48.60	388.80	73.75	590.00		
1 Air Track Drill, 4"		1028.00		1130.80		
1 Air Compressor, 600 cfm		550.60		605.66		
2 -50' Air Hoses, 3"		29.80		32.78	67.02	73.72
24 L.H., Daily Totals		$2614.80		$3309.64	$108.95	$137.90

Crew B-47A	Hr.	Daily	Hr.	Daily	Bare Costs	Incl. O&P
1 Drilling Foreman (outside)	$39.60	$316.80	$60.95	$487.60	$45.50	$69.32
1 Equip. Oper. (heavy)	51.70	413.60	78.45	627.60		
1 Equip. Oper. (oiler)	45.20	361.60	68.55	548.40		
1 Air Track Drill, 5"		1243.00		1367.30	51.79	56.97
24 L.H., Daily Totals		$2335.00		$3030.90	$97.29	$126.29

Crew B-47C	Hr.	Daily	Hr.	Daily	Bare Costs	Incl. O&P
1 Laborer	$37.60	$300.80	$57.85	$462.80	$43.10	$65.80
1 Equip. Oper. (light)	48.60	388.80	73.75	590.00		
1 Air Compressor, 750 cfm		557.20		612.92		
2 -50' Air Hoses, 3"		29.80		32.78		
1 Air Track Drill, 4"		1028.00		1130.80	100.94	111.03
16 L.H., Daily Totals		$2304.60		$2829.30	$144.04	$176.83

Crew B-47E	Hr.	Daily	Hr.	Daily	Bare Costs	Incl. O&P
1 Labor Foreman (outside)	$39.60	$316.80	$60.95	$487.60	$38.10	$58.63
3 Laborers	37.60	902.40	57.85	1388.40		
1 Flatbed Truck, Gas, 3 Ton		303.20		333.52	9.47	10.42
32 L.H., Daily Totals		$1522.40		$2209.52	$47.58	$69.05

Crew B-47G	Hr.	Daily	Hr.	Daily	Bare Costs	Incl. O&P
1 Labor Foreman (outside)	$39.60	$316.80	$60.95	$487.60	$40.85	$62.60
2 Laborers	37.60	601.60	57.85	925.60		
1 Equip. Oper. (light)	48.60	388.80	73.75	590.00		
1 Air Track Drill, 4"		1028.00		1130.80		
1 Air Compressor, 600 cfm		550.60		605.66		
2 -50' Air Hoses, 3"		29.80		32.78		
1 Gunite Pump Rig		371.80		408.98	61.88	68.07
32 L.H., Daily Totals		$3287.40		$4181.42	$102.73	$130.67

Crew B-47H	Hr.	Daily	Hr.	Daily	Bare Costs	Incl. O&P
1 Skilled Worker Foreman (out)	$50.65	$405.20	$78.25	$626.00	$49.15	$75.92
3 Skilled Workers	48.65	1167.60	75.15	1803.60		
1 Flatbed Truck, Gas, 3 Ton		303.20		333.52	9.47	10.42
32 L.H., Daily Totals		$1876.00		$2763.12	$58.63	$86.35

Crew B-48	Bare Costs Hr.	Bare Costs Daily	Incl. Subs O&P Hr.	Incl. Subs O&P Daily	Cost Per Labor-Hour Bare Costs	Cost Per Labor-Hour Incl. O&P
1 Labor Foreman (outside)	$39.60	$316.80	$60.95	$487.60	$42.56	$65.04
3 Laborers	37.60	902.40	57.85	1388.40		
1 Equip. Oper. (crane)	51.70	413.60	78.45	627.60		
1 Equip. Oper. (oiler)	45.20	361.60	68.55	548.40		
1 Equip. Oper. (light)	48.60	388.80	73.75	590.00		
1 Centr. Water Pump, 6"		346.60		381.26		
1 -20' Suction Hose, 6"		11.50		12.65		
1 -50' Discharge Hose, 6"		6.10		6.71		
1 Drill Rig, Truck-Mounted		2542.00		2796.20	51.90	57.09
56 L.H., Daily Totals		$5289.40		$6838.82	$94.45	$122.12

Crew B-49	Hr.	Daily	Hr.	Daily	Bare Costs	Incl. O&P
1 Labor Foreman (outside)	$39.60	$316.80	$60.95	$487.60	$44.27	$68.02
3 Laborers	37.60	902.40	57.85	1388.40		
2 Equip. Oper. (crane)	51.70	827.20	78.45	1255.20		
2 Equip. Oper. (oilers)	45.20	723.20	68.55	1096.80		
1 Equip. Oper. (light)	48.60	388.80	73.75	590.00		
2 Pile Drivers	46.10	737.60	73.00	1168.00		
1 Hyd. Crane, 25 Ton		736.60		810.26		
1 Centr. Water Pump, 6"		346.60		381.26		
1 -20' Suction Hose, 6"		11.50		12.65		
1 -50' Discharge Hose, 6"		6.10		6.71		
1 Drill Rig, Truck-Mounted		2542.00		2796.20	41.40	45.53
88 L.H., Daily Totals		$7538.80		$9993.08	$85.67	$113.56

Crew B-50	Hr.	Daily	Hr.	Daily	Bare Costs	Incl. O&P
2 Pile Driver Foremen (outside)	$48.10	$769.60	$76.15	$1218.40	$45.30	$70.66
6 Pile Drivers	46.10	2212.80	73.00	3504.00		
2 Equip. Oper. (crane)	51.70	827.20	78.45	1255.20		
1 Equip. Oper. (oiler)	45.20	361.60	68.55	548.40		
3 Laborers	37.60	902.40	57.85	1388.40		
1 Crawler Crane, 40 Ton		1156.00		1271.60		
1 Lead, 60' High		74.80		82.28		
1 Hammer, Diesel, 15K ft.-lbs.		586.00		644.60		
1 Air Compressor, 600 cfm		550.60		605.66		
2 -50' Air Hoses, 3"		29.80		32.78		
1 Chain Saw, Gas, 36" Long		45.00		49.50	21.81	23.99
112 L.H., Daily Totals		$7515.80		$10600.82	$67.11	$94.65

Crew B-51	Hr.	Daily	Hr.	Daily	Bare Costs	Incl. O&P
1 Labor Foreman (outside)	$39.60	$316.80	$60.95	$487.60	$38.18	$58.58
4 Laborers	37.60	1203.20	57.85	1851.20		
1 Truck Driver (light)	39.10	312.80	59.10	472.80		
1 Flatbed Truck, Gas, 1.5 Ton		245.40		269.94	5.11	5.62
48 L.H., Daily Totals		$2078.20		$3081.54	$43.30	$64.20

Crew B-52	Hr.	Daily	Hr.	Daily	Bare Costs	Incl. O&P
1 Carpenter Foreman (outside)	$48.95	$391.60	$75.35	$602.80	$43.61	$66.75
1 Carpenter	46.95	375.60	72.25	578.00		
3 Laborers	37.60	902.40	57.85	1388.40		
1 Cement Finisher	45.00	360.00	66.65	533.20		
.5 Rodman (reinf.)	52.55	210.20	82.10	328.40		
.5 Equip. Oper. (medium)	50.60	202.40	76.75	307.00		
.5 Crawler Loader, 3 C.Y.		594.00		653.40	10.61	11.67
56 L.H., Daily Totals		$3036.20		$4391.20	$54.22	$78.41

Crew B-53	Hr.	Daily	Hr.	Daily	Bare Costs	Incl. O&P
1 Equip. Oper. (light)	$48.60	$388.80	$73.75	$590.00	$48.60	$73.75
1 Trencher, Chain, 12 H.P.		68.00		74.80	8.50	9.35
8 L.H., Daily Totals		$456.80		$664.80	$57.10	$83.10

Crew B-54	Hr.	Daily	Hr.	Daily	Bare Costs	Incl. O&P
1 Equip. Oper. (light)	$48.60	$388.80	$73.75	$590.00	$48.60	$73.75
1 Trencher, Chain, 40 H.P.		335.20		368.72	41.90	46.09
8 L.H., Daily Totals		$724.00		$958.72	$90.50	$119.84

Crew B-54A	Hr.	Daily	Hr.	Daily	Bare Costs	Incl. O&P
.17 Labor Foreman (outside)	$39.60	$53.86	$60.95	$82.89	$49.00	$74.45
1 Equipment Operator (med.)	50.60	404.80	76.75	614.00		
1 Wheel Trencher, 67 H.P.		1195.00		1314.50	127.67	140.44
9.36 L.H., Daily Totals		$1653.66		$2011.39	$176.67	$214.89

Crew B-54B	Hr.	Daily	Hr.	Daily	Bare Costs	Incl. O&P
.25 Labor Foreman (outside)	$39.60	$79.20	$60.95	$121.90	$48.40	$73.59
1 Equipment Operator (med.)	50.60	404.80	76.75	614.00		
1 Wheel Trencher, 150 H.P.		1890.00		2079.00	189.00	207.90
10 L.H., Daily Totals		$2374.00		$2814.90	$237.40	$281.49

Crew B-54C	Hr.	Daily	Hr.	Daily	Bare Costs	Incl. O&P
1 Laborer	$37.60	$300.80	$57.85	$462.80	$44.10	$67.30
1 Equipment Operator (med.)	50.60	404.80	76.75	614.00		
1 Wheel Trencher, 67 H.P.		1195.00		1314.50	74.69	82.16
16 L.H., Daily Totals		$1900.60		$2391.30	$118.79	$149.46

Crew B-54D	Hr.	Daily	Hr.	Daily	Bare Costs	Incl. O&P
1 Laborer	$37.60	$300.80	$57.85	$462.80	$44.10	$67.30
1 Equipment Operator (med.)	50.60	404.80	76.75	614.00		
1 Rock Trencher, 6" Width		372.00		409.20	23.25	25.57
16 L.H., Daily Totals		$1077.60		$1486.00	$67.35	$92.88

Crew B-54E	Hr.	Daily	Hr.	Daily	Bare Costs	Incl. O&P
1 Laborer	$37.60	$300.80	$57.85	$462.80	$44.10	$67.30
1 Equipment Operator (med.)	50.60	404.80	76.75	614.00		
1 Rock Trencher, 18" Width		2575.00		2832.50	160.94	177.03
16 L.H., Daily Totals		$3280.60		$3909.30	$205.04	$244.33

Crew B-55	Hr.	Daily	Hr.	Daily	Bare Costs	Incl. O&P
2 Laborers	$37.60	$601.60	$57.85	$925.60	$38.10	$58.27
1 Truck Driver (light)	39.10	312.80	59.10	472.80		
1 Truck-Mounted Earth Auger		779.20		857.12		
1 Flatbed Truck, Gas, 3 Ton		303.20		333.52	45.10	49.61
24 L.H., Daily Totals		$1996.80		$2589.04	$83.20	$107.88

Crew B-56	Hr.	Daily	Hr.	Daily	Bare Costs	Incl. O&P
1 Laborer	$37.60	$300.80	$57.85	$462.80	$43.10	$65.80
1 Equip. Oper. (light)	48.60	388.80	73.75	590.00		
1 Air Track Drill, 4"		1028.00		1130.80		
1 Air Compressor, 600 cfm		550.60		605.66		
1 -50' Air Hose, 3"		14.90		16.39	99.59	109.55
16 L.H., Daily Totals		$2283.10		$2805.65	$142.69	$175.35

Crew No.	Bare Costs		Incl. Subs O&P		Cost Per Labor-Hour	
Crew B-57	**Hr.**	**Daily**	**Hr.**	**Daily**	**Bare Costs**	**Incl. O&P**
1 Labor Foreman (outside)	$39.60	$316.80	$60.95	$487.60	$43.38	$66.23
2 Laborers	37.60	601.60	57.85	925.60		
1 Equip. Oper. (crane)	51.70	413.60	78.45	627.60		
1 Equip. Oper. (light)	48.60	388.80	73.75	590.00		
1 Equip. Oper. (oiler)	45.20	361.60	68.55	548.40		
1 Crawler Crane, 25 Ton		1150.00		1265.00		
1 Clamshell Bucket, 1 C.Y.		47.80		52.58		
1 Centr. Water Pump, 6"		346.60		381.26		
1 -20' Suction Hose, 6"		11.50		12.65		
20 -50' Discharge Hoses, 6"		122.00		134.20	34.96	38.45
48 L.H., Daily Totals		$3760.30		$5024.89	$78.34	$104.69
Crew B-58	**Hr.**	**Daily**	**Hr.**	**Daily**	**Bare Costs**	**Incl. O&P**
2 Laborers	$37.60	$601.60	$57.85	$925.60	$41.27	$63.15
1 Equip. Oper. (light)	48.60	388.80	73.75	590.00		
1 Backhoe Loader, 48 H.P.		364.20		400.62		
1 Small Helicopter, w/ Pilot		2790.00		3069.00	131.43	144.57
24 L.H., Daily Totals		$4144.60		$4985.22	$172.69	$207.72
Crew B-59	**Hr.**	**Daily**	**Hr.**	**Daily**	**Bare Costs**	**Incl. O&P**
1 Truck Driver (heavy)	$40.05	$320.40	$60.55	$484.40	$40.05	$60.55
1 Truck Tractor, 220 H.P.		359.60		395.56		
1 Water Tank Trailer, 5000 Gal.		141.00		155.10	62.58	68.83
8 L.H., Daily Totals		$821.00		$1035.06	$102.63	$129.38
Crew B-59A	**Hr.**	**Daily**	**Hr.**	**Daily**	**Bare Costs**	**Incl. O&P**
2 Laborers	$37.60	$601.60	$57.85	$925.60	$38.42	$58.75
1 Truck Driver (heavy)	40.05	320.40	60.55	484.40		
1 Water Tank Trailer, 5000 Gal.		141.00		155.10		
1 Truck Tractor, 220 H.P.		359.60		395.56	20.86	22.94
24 L.H., Daily Totals		$1422.60		$1960.66	$59.27	$81.69
Crew B-60	**Hr.**	**Daily**	**Hr.**	**Daily**	**Bare Costs**	**Incl. O&P**
1 Labor Foreman (outside)	$39.60	$316.80	$60.95	$487.60	$44.13	$67.31
2 Laborers	37.60	601.60	57.85	925.60		
1 Equip. Oper. (crane)	51.70	413.60	78.45	627.60		
2 Equip. Oper. (light)	48.60	777.60	73.75	1180.00		
1 Equip. Oper. (oiler)	45.20	361.60	68.55	548.40		
1 Crawler Crane, 40 Ton		1156.00		1271.60		
1 Lead, 60' High		74.80		82.28		
1 Hammer, Diesel, 15K ft.-lbs.		586.00		644.60		
1 Backhoe Loader, 48 H.P.		364.20		400.62	38.95	42.84
56 L.H., Daily Totals		$4652.20		$6168.30	$83.08	$110.15
Crew B-61	**Hr.**	**Daily**	**Hr.**	**Daily**	**Bare Costs**	**Incl. O&P**
1 Labor Foreman (outside)	$39.60	$316.80	$60.95	$487.60	$40.20	$61.65
3 Laborers	37.60	902.40	57.85	1388.40		
1 Equip. Oper. (light)	48.60	388.80	73.75	590.00		
1 Cement Mixer, 2 C.Y.		194.20		213.62		
1 Air Compressor, 160 cfm		156.40		172.04	8.77	9.64
40 L.H., Daily Totals		$1958.60		$2851.66	$48.97	$71.29
Crew B-62	**Hr.**	**Daily**	**Hr.**	**Daily**	**Bare Costs**	**Incl. O&P**
2 Laborers	$37.60	$601.60	$57.85	$925.60	$41.27	$63.15
1 Equip. Oper. (light)	48.60	388.80	73.75	590.00		
1 Loader, Skid Steer, 30 H.P.		173.60		190.96	7.23	7.96
24 L.H., Daily Totals		$1164.00		$1706.56	$48.50	$71.11

Crew No.	Bare Costs		Incl. Subs O&P		Cost Per Labor-Hour	
Crew B-62A	**Hr.**	**Daily**	**Hr.**	**Daily**	**Bare Costs**	**Incl. O&P**
2 Laborers	$37.60	$601.60	$57.85	$925.60	$41.27	$63.15
1 Equip. Oper. (light)	48.60	388.80	73.75	590.00		
1 Loader, Skid Steer, 30 H.P.		173.60		190.96		
1 Trencher Attachment		58.30		64.13	9.66	10.63
24 L.H., Daily Totals		$1222.30		$1770.69	$50.93	$73.78
Crew B-63	**Hr.**	**Daily**	**Hr.**	**Daily**	**Bare Costs**	**Incl. O&P**
4 Laborers	$37.60	$1203.20	$57.85	$1851.20	$39.80	$61.03
1 Equip. Oper. (light)	48.60	388.80	73.75	590.00		
1 Loader, Skid Steer, 30 H.P.		173.60		190.96	4.34	4.77
40 L.H., Daily Totals		$1765.60		$2632.16	$44.14	$65.80
Crew B-63B	**Hr.**	**Daily**	**Hr.**	**Daily**	**Bare Costs**	**Incl. O&P**
1 Labor Foreman (inside)	$38.10	$304.80	$58.65	$469.20	$40.48	$62.02
2 Laborers	37.60	601.60	57.85	925.60		
1 Equip. Oper. (light)	48.60	388.80	73.75	590.00		
1 Loader, Skid Steer, 78 H.P.		318.60		350.46	9.96	10.95
32 L.H., Daily Totals		$1613.80		$2335.26	$50.43	$72.98
Crew B-64	**Hr.**	**Daily**	**Hr.**	**Daily**	**Bare Costs**	**Incl. O&P**
1 Laborer	$37.60	$300.80	$57.85	$462.80	$38.35	$58.48
1 Truck Driver (light)	39.10	312.80	59.10	472.80		
1 Power Mulcher (small)		151.80		166.98		
1 Flatbed Truck, Gas, 1.5 Ton		245.40		269.94	24.82	27.31
16 L.H., Daily Totals		$1010.80		$1372.52	$63.17	$85.78
Crew B-65	**Hr.**	**Daily**	**Hr.**	**Daily**	**Bare Costs**	**Incl. O&P**
1 Laborer	$37.60	$300.80	$57.85	$462.80	$38.35	$58.48
1 Truck Driver (light)	39.10	312.80	59.10	472.80		
1 Power Mulcher (Large)		319.40		351.34		
1 Flatbed Truck, Gas, 1.5 Ton		245.40		269.94	35.30	38.83
16 L.H., Daily Totals		$1178.40		$1556.88	$73.65	$97.31
Crew B-66	**Hr.**	**Daily**	**Hr.**	**Daily**	**Bare Costs**	**Incl. O&P**
1 Equip. Oper. (light)	$48.60	$388.80	$73.75	$590.00	$48.60	$73.75
1 Loader-Backhoe, 40 H.P.		263.20		289.52	32.90	36.19
8 L.H., Daily Totals		$652.00		$879.52	$81.50	$109.94
Crew B-67	**Hr.**	**Daily**	**Hr.**	**Daily**	**Bare Costs**	**Incl. O&P**
1 Millwright	$49.15	$393.20	$72.35	$578.80	$48.88	$73.05
1 Equip. Oper. (light)	48.60	388.80	73.75	590.00		
1 Forklift, R/T, 4,000 Lb.		311.20		342.32	19.45	21.40
16 L.H., Daily Totals		$1093.20		$1511.12	$68.33	$94.44
Crew B-67B	**Hr.**	**Daily**	**Hr.**	**Daily**	**Bare Costs**	**Incl. O&P**
1 Millwright Foreman (inside)	$49.65	$397.20	$73.10	$584.80	$49.40	$72.72
1 Millwright	49.15	393.20	72.35	578.80		
16 L.H., Daily Totals		$790.40		$1163.60	$49.40	$72.72
Crew B-68	**Hr.**	**Daily**	**Hr.**	**Daily**	**Bare Costs**	**Incl. O&P**
2 Millwrights	$49.15	$786.40	$72.35	$1157.60	$48.97	$72.82
1 Equip. Oper. (light)	48.60	388.80	73.75	590.00		
1 Forklift, R/T, 4,000 Lb.		311.20		342.32	12.97	14.26
24 L.H., Daily Totals		$1486.40		$2089.92	$61.93	$87.08
Crew B-68A	**Hr.**	**Daily**	**Hr.**	**Daily**	**Bare Costs**	**Incl. O&P**
1 Millwright Foreman (inside)	$49.65	$397.20	$73.10	$584.80	$49.32	$72.60
2 Millwrights	49.15	786.40	72.35	1157.60		
1 Forklift, 8,000 Lb.		185.20		203.72	7.72	8.49
24 L.H., Daily Totals		$1368.80		$1946.12	$57.03	$81.09

Crew No.	Bare Costs		Incl. Subs O&P		Cost Per Labor-Hour	
Crew B-68B	**Hr.**	**Daily**	**Hr.**	**Daily**	**Bare Costs**	**Incl. O&P**
1 Millwright Foreman (inside)	$49.65	$397.20	$73.10	$584.80	$53.54	$79.86
2 Millwrights	49.15	786.40	72.35	1157.60		
2 Electricians	54.70	875.20	81.95	1311.20		
2 Plumbers	58.70	939.20	88.65	1418.40		
1 Forklift, 5,000 Lb.		331.80		364.98	5.92	6.52
56 L.H., Daily Totals		$3329.80		$4836.98	$59.46	$86.37
Crew B-68C	**Hr.**	**Daily**	**Hr.**	**Daily**	**Bare Costs**	**Incl. O&P**
1 Millwright Foreman (inside)	$49.65	$397.20	$73.10	$584.80	$53.05	$79.01
1 Millwright	49.15	393.20	72.35	578.80		
1 Electrician	54.70	437.60	81.95	655.60		
1 Plumber	58.70	469.60	88.65	709.20		
1 Forklift, 5,000 Lb.		331.80		364.98	10.37	11.41
32 L.H., Daily Totals		$2029.40		$2893.38	$63.42	$90.42
Crew B-68D	**Hr.**	**Daily**	**Hr.**	**Daily**	**Bare Costs**	**Incl. O&P**
1 Labor Foreman (inside)	$38.10	$304.80	$58.65	$469.20	$41.43	$63.42
1 Laborer	37.60	300.80	57.85	462.80		
1 Equip. Oper. (light)	48.60	388.80	73.75	590.00		
1 Forklift, 5,000 Lb.		331.80		364.98	13.82	15.21
24 L.H., Daily Totals		$1326.20		$1886.98	$55.26	$78.62
Crew B-68E	**Hr.**	**Daily**	**Hr.**	**Daily**	**Bare Costs**	**Incl. O&P**
1 Struc. Steel Foreman (inside)	$53.15	$425.20	$92.30	$738.40	$52.75	$91.62
3 Struc. Steel Workers	52.65	1263.60	91.45	2194.80		
1 Welder	52.65	421.20	91.45	731.60		
1 Forklift, 8,000 Lb.		185.20		203.72	4.63	5.09
40 L.H., Daily Totals		$2295.20		$3868.52	$57.38	$96.71
Crew B-68F	**Hr.**	**Daily**	**Hr.**	**Daily**	**Bare Costs**	**Incl. O&P**
1 Skilled Worker Foreman (out)	$50.65	$405.20	$78.25	$626.00	$49.32	$76.18
2 Skilled Workers	48.65	778.40	75.15	1202.40		
1 Forklift, 5,000 Lb.		331.80		364.98	13.82	15.21
24 L.H., Daily Totals		$1515.40		$2193.38	$63.14	$91.39
Crew B-68G	**Hr.**	**Daily**	**Hr.**	**Daily**	**Bare Costs**	**Incl. O&P**
2 Structural Steel Workers	$52.65	$842.40	$91.45	$1463.20	$52.65	$91.45
1 Forklift, 5,000 Lb.		331.80		364.98	20.74	22.81
16 L.H., Daily Totals		$1174.20		$1828.18	$73.39	$114.26
Crew B-69	**Hr.**	**Daily**	**Hr.**	**Daily**	**Bare Costs**	**Incl. O&P**
1 Labor Foreman (outside)	$39.60	$316.80	$60.95	$487.60	$41.55	$63.58
3 Laborers	37.60	902.40	57.85	1388.40		
1 Equip. Oper. (crane)	51.70	413.60	78.45	627.60		
1 Equip. Oper. (oiler)	45.20	361.60	68.55	548.40		
1 Hyd. Crane, 80 Ton		1625.00		1787.50	33.85	37.24
48 L.H., Daily Totals		$3619.40		$4839.50	$75.40	$100.82
Crew B-69A	**Hr.**	**Daily**	**Hr.**	**Daily**	**Bare Costs**	**Incl. O&P**
1 Labor Foreman (outside)	$39.60	$316.80	$60.95	$487.60	$41.33	$62.98
3 Laborers	37.60	902.40	57.85	1388.40		
1 Equip. Oper. (medium)	50.60	404.80	76.75	614.00		
1 Concrete Finisher	45.00	360.00	66.65	533.20		
1 Curb/Gutter Paver, 2-Track		997.60		1097.36	20.78	22.86
48 L.H., Daily Totals		$2981.60		$4120.56	$62.12	$85.84

Crew No.	Bare Costs		Incl. Subs O&P		Cost Per Labor-Hour	
Crew B-69B	**Hr.**	**Daily**	**Hr.**	**Daily**	**Bare Costs**	**Incl. O&P**
1 Labor Foreman (outside)	$39.60	$316.80	$60.95	$487.60	$41.33	$62.98
3 Laborers	37.60	902.40	57.85	1388.40		
1 Equip. Oper. (medium)	50.60	404.80	76.75	614.00		
1 Cement Finisher	45.00	360.00	66.65	533.20		
1 Curb/Gutter Paver, 4-Track		776.00		853.60	16.17	17.78
48 L.H., Daily Totals		$2760.00		$3876.80	$57.50	$80.77
Crew B-70	**Hr.**	**Daily**	**Hr.**	**Daily**	**Bare Costs**	**Incl. O&P**
1 Labor Foreman (outside)	$39.60	$316.80	$60.95	$487.60	$43.46	$66.39
3 Laborers	37.60	902.40	57.85	1388.40		
3 Equip. Oper. (medium)	50.60	1214.40	76.75	1842.00		
1 Grader, 30,000 Lbs.		736.20		809.82		
1 Ripper, Beam & 1 Shank		84.20		92.62		
1 Road Sweeper, S.P., 8' wide		662.40		728.64		
1 F.E. Loader, W.M., 1.5 C.Y.		378.40		416.24	33.24	36.56
56 L.H., Daily Totals		$4294.80		$5765.32	$76.69	$102.95
Crew B-70A	**Hr.**	**Daily**	**Hr.**	**Daily**	**Bare Costs**	**Incl. O&P**
1 Laborer	$37.60	$300.80	$57.85	$462.80	$48.00	$72.97
4 Equip. Oper. (medium)	50.60	1619.20	76.75	2456.00		
1 Grader, 40,000 Lbs.		1174.00		1291.40		
1 F.E. Loader, W.M.,2.5 C.Y.		519.80		571.78		
1 Dozer, 80 H.P.		472.40		519.64		
1 Roller, Pneum. Whl., 12 Ton		344.40		378.84	62.77	69.04
40 L.H., Daily Totals		$4430.60		$5680.46	$110.77	$142.01
Crew B-71	**Hr.**	**Daily**	**Hr.**	**Daily**	**Bare Costs**	**Incl. O&P**
1 Labor Foreman (outside)	$39.60	$316.80	$60.95	$487.60	$43.46	$66.39
3 Laborers	37.60	902.40	57.85	1388.40		
3 Equip. Oper. (medium)	50.60	1214.40	76.75	1842.00		
1 Pvmt. Profiler, 750 H.P.		5805.00		6385.50		
1 Road Sweeper, S.P., 8' wide		662.40		728.64		
1 F.E. Loader, W.M., 1.5 C.Y.		378.40		416.24	122.25	134.47
56 L.H., Daily Totals		$9279.40		$11248.38	$165.70	$200.86
Crew B-72	**Hr.**	**Daily**	**Hr.**	**Daily**	**Bare Costs**	**Incl. O&P**
1 Labor Foreman (outside)	$39.60	$316.80	$60.95	$487.60	$44.35	$67.69
3 Laborers	37.60	902.40	57.85	1388.40		
4 Equip. Oper. (medium)	50.60	1619.20	76.75	2456.00		
1 Pvmt. Profiler, 750 H.P.		5805.00		6385.50		
1 Hammermill, 250 H.P.		1879.00		2066.90		
1 Windrow Loader		1227.00		1349.70		
1 Mix Paver 165 H.P.		2147.00		2361.70		
1 Roller, Pneum. Whl., 12 Ton		344.40		378.84	178.16	195.98
64 L.H., Daily Totals		$14240.80		$16874.64	$222.51	$263.67
Crew B-73	**Hr.**	**Daily**	**Hr.**	**Daily**	**Bare Costs**	**Incl. O&P**
1 Labor Foreman (outside)	$39.60	$316.80	$60.95	$487.60	$45.98	$70.05
2 Laborers	37.60	601.60	57.85	925.60		
5 Equip. Oper. (medium)	50.60	2024.00	76.75	3070.00		
1 Road Mixer, 310 H.P.		1929.00		2121.90		
1 Tandem Roller, 10 Ton		236.80		260.48		
1 Hammermill, 250 H.P.		1879.00		2066.90		
1 Grader, 30,000 Lbs.		736.20		809.82		
.5 F.E. Loader, W.M., 1.5 C.Y.		189.20		208.12		
.5 Truck Tractor, 220 H.P.		179.80		197.78		
.5 Water Tank Trailer, 5000 Gal.		70.50		77.55	81.57	89.73
64 L.H., Daily Totals		$8162.90		$10225.75	$127.55	$159.78

Crews

Crew No.	Bare Costs		Incl. Subs O&P		Cost Per Labor-Hour	
Crew B-74	**Hr.**	**Daily**	**Hr.**	**Daily**	**Bare Costs**	**Incl. O&P**
1 Labor Foreman (outside)	$39.60	$316.80	$60.95	$487.60	$44.96	$68.36
1 Laborer	37.60	300.80	57.85	462.80		
4 Equip. Oper. (medium)	50.60	1619.20	76.75	2456.00		
2 Truck Drivers (heavy)	40.05	640.80	60.55	968.80		
1 Grader, 30,000 Lbs.		736.20		809.82		
1 Ripper, Beam & 1 Shank		84.20		92.62		
2 Stabilizers, 310 H.P.		3596.00		3955.60		
1 Flatbed Truck, Gas, 3 Ton		303.20		333.52		
1 Chem. Spreader, Towed		52.40		57.64		
1 Roller, Vibratory, 25 Ton		687.40		756.14		
1 Water Tank Trailer, 5000 Gal.		141.00		155.10		
1 Truck Tractor, 220 H.P.		359.60		395.56	93.13	102.44
64 L.H., Daily Totals		$8837.60		$10931.20	$138.09	$170.80
Crew B-75	**Hr.**	**Daily**	**Hr.**	**Daily**	**Bare Costs**	**Incl. O&P**
1 Labor Foreman (outside)	$39.60	$316.80	$60.95	$487.60	$45.66	$69.48
1 Laborer	37.60	300.80	57.85	462.80		
4 Equip. Oper. (medium)	50.60	1619.20	76.75	2456.00		
1 Truck Driver (heavy)	40.05	320.40	60.55	484.40		
1 Grader, 30,000 Lbs.		736.20		809.82		
1 Ripper, Beam & 1 Shank		84.20		92.62		
2 Stabilizers, 310 H.P.		3596.00		3955.60		
1 Dist. Tanker, 3000 Gallon		305.60		336.16		
1 Truck Tractor, 6x4, 380 H.P.		599.80		659.78		
1 Roller, Vibratory, 25 Ton		687.40		756.14	107.31	118.04
56 L.H., Daily Totals		$8566.40		$10500.92	$152.97	$187.52
Crew B-76	**Hr.**	**Daily**	**Hr.**	**Daily**	**Bare Costs**	**Incl. O&P**
1 Dock Builder Foreman (outside)	$48.10	$384.80	$76.15	$609.20	$47.47	$74.07
5 Dock Builders	46.10	1844.00	73.00	2920.00		
2 Equip. Oper. (crane)	51.70	827.20	78.45	1255.20		
1 Equip. Oper. (oiler)	45.20	361.60	68.55	548.40		
1 Crawler Crane, 50 Ton		1294.00		1423.40		
1 Barge, 400 Ton		776.00		853.60		
1 Hammer, Diesel, 15K ft.-lbs.		586.00		644.60		
1 Lead, 60' High		74.80		82.28		
1 Air Compressor, 600 cfm		550.60		605.66		
2 -50' Air Hoses, 3"		29.80		32.78	45.99	50.59
72 L.H., Daily Totals		$6728.80		$8975.12	$93.46	$124.65
Crew B-76A	**Hr.**	**Daily**	**Hr.**	**Daily**	**Bare Costs**	**Incl. O&P**
1 Labor Foreman (outside)	$39.60	$316.80	$60.95	$487.60	$40.56	$62.15
5 Laborers	37.60	1504.00	57.85	2314.00		
1 Equip. Oper. (crane)	51.70	413.60	78.45	627.60		
1 Equip. Oper. (oiler)	45.20	361.60	68.55	548.40		
1 Crawler Crane, 50 Ton		1294.00		1423.40		
1 Barge, 400 Ton		776.00		853.60	32.34	35.58
64 L.H., Daily Totals		$4666.00		$6254.60	$72.91	$97.73
Crew B-77	**Hr.**	**Daily**	**Hr.**	**Daily**	**Bare Costs**	**Incl. O&P**
1 Labor Foreman (outside)	$39.60	$316.80	$60.95	$487.60	$38.30	$58.72
3 Laborers	37.60	902.40	57.85	1388.40		
1 Truck Driver (light)	39.10	312.80	59.10	472.80		
1 Crack Cleaner, 25 H.P.		61.60		67.76		
1 Crack Filler, Trailer Mtd.		202.00		222.20		
1 Flatbed Truck, Gas, 3 Ton		303.20		333.52	14.17	15.59
40 L.H., Daily Totals		$2098.80		$2972.28	$52.47	$74.31

Crew No.	Bare Costs		Incl. Subs O&P		Cost Per Labor-Hour	
Crew B-78	**Hr.**	**Daily**	**Hr.**	**Daily**	**Bare Costs**	**Incl. O&P**
1 Labor Foreman (outside)	$39.60	$316.80	$60.95	$487.60	$38.18	$58.58
4 Laborers	37.60	1203.20	57.85	1851.20		
1 Truck Driver (light)	39.10	312.80	59.10	472.80		
1 Paint Striper, S.P., 40 Gallon		151.40		166.54		
1 Flatbed Truck, Gas, 3 Ton		303.20		333.52		
1 Pickup Truck, 3/4 Ton		144.20		158.62	12.48	13.72
48 L.H., Daily Totals		$2431.60		$3470.28	$50.66	$72.30
Crew B-78A	**Hr.**	**Daily**	**Hr.**	**Daily**	**Bare Costs**	**Incl. O&P**
1 Equip. Oper. (light)	$48.60	$388.80	$73.75	$590.00	$48.60	$73.75
1 Line Rem. (Metal Balls) 115 H.P.		821.60		903.76	102.70	112.97
8 L.H., Daily Totals		$1210.40		$1493.76	$151.30	$186.72
Crew B-78B	**Hr.**	**Daily**	**Hr.**	**Daily**	**Bare Costs**	**Incl. O&P**
2 Laborers	$37.60	$601.60	$57.85	$925.60	$38.82	$59.62
.25 Equip. Oper. (light)	48.60	97.20	73.75	147.50		
1 Pickup Truck, 3/4 Ton		144.20		158.62		
1 Line Rem.,11 H.P.,Walk Behind		65.60		72.16		
.25 Road Sweeper, S.P., 8' wide		165.60		182.16	20.86	22.94
18 L.H., Daily Totals		$1074.20		$1486.04	$59.68	$82.56
Crew B-78C	**Hr.**	**Daily**	**Hr.**	**Daily**	**Bare Costs**	**Incl. O&P**
1 Labor Foreman (outside)	$39.60	$316.80	$60.95	$487.60	$38.18	$58.58
4 Laborers	37.60	1203.20	57.85	1851.20		
1 Truck Driver (light)	39.10	312.80	59.10	472.80		
1 Paint Striper, T.M., 120 Gal.		804.60		885.06		
1 Flatbed Truck, Gas, 3 Ton		303.20		333.52		
1 Pickup Truck, 3/4 Ton		144.20		158.62	26.08	28.69
48 L.H., Daily Totals		$3084.80		$4188.80	$64.27	$87.27
Crew B-78D	**Hr.**	**Daily**	**Hr.**	**Daily**	**Bare Costs**	**Incl. O&P**
2 Labor Foremen (outside)	$39.60	$633.60	$60.95	$975.20	$38.15	$58.59
7 Laborers	37.60	2105.60	57.85	3239.60		
1 Truck Driver (light)	39.10	312.80	59.10	472.80		
1 Paint Striper, T.M., 120 Gal.		804.60		885.06		
1 Flatbed Truck, Gas, 3 Ton		303.20		333.52		
3 Pickup Trucks, 3/4 Ton		432.60		475.86		
1 Air Compressor, 60 cfm		137.40		151.14		
1 -50' Air Hose, 3/4"		3.25		3.58		
1 Breakers, Pavement, 60 lb.		9.80		10.78	21.14	23.25
80 L.H., Daily Totals		$4742.85		$6547.53	$59.29	$81.84
Crew B-78E	**Hr.**	**Daily**	**Hr.**	**Daily**	**Bare Costs**	**Incl. O&P**
2 Labor Foremen (outside)	$39.60	$633.60	$60.95	$975.20	$38.06	$58.47
9 Laborers	37.60	2707.20	57.85	4165.20		
1 Truck Driver (light)	39.10	312.80	59.10	472.80		
1 Paint Striper, T.M., 120 Gal.		804.60		885.06		
1 Flatbed Truck, Gas, 3 Ton		303.20		333.52		
4 Pickup Trucks, 3/4 Ton		576.80		634.48		
2 Air Compressor, 60 cfm		274.80		302.28		
2 -50' Air Hose, 3/4"		6.50		7.15		
2 Breakers, Pavement, 60 lb.		19.60		21.56	20.68	22.75
96 L.H., Daily Totals		$5639.10		$7797.25	$58.74	$81.22

Crew No.	Bare Costs		Incl. Subs O&P		Cost Per Labor-Hour	
Crew B-78F	**Hr.**	**Daily**	**Hr.**	**Daily**	**Bare Costs**	**Incl. O&P**
2 Labor Foremen (outside)	$39.60	$633.60	$60.95	$975.20	$37.99	$58.38
11 Laborers	37.60	3308.80	57.85	5090.80		
1 Truck Driver (light)	39.10	312.80	59.10	472.80		
1 Paint Striper, T.M., 120 Gal.		804.60		885.06		
1 Flatbed Truck, Gas, 3 Ton		303.20		333.52		
7 Pickup Trucks, 3/4 Ton		1009.40		1110.34		
3 Air Compressor, 60 cfm		412.20		453.42		
3 -50' Air Hose, 3/4"		9.75		10.73		
3 Breakers, Pavement, 60 lb.		29.40		32.34	22.93	25.23
112 L.H., Daily Totals		$6823.75		$9364.20	$60.93	$83.61

Crew B-79	Hr.	Daily	Hr.	Daily	Bare Costs	Incl. O&P
1 Labor Foreman (outside)	$39.60	$316.80	$60.95	$487.60	$38.30	$58.72
3 Laborers	37.60	902.40	57.85	1388.40		
1 Truck Driver (light)	39.10	312.80	59.10	472.80		
1 Paint Striper, T.M., 120 Gal.		804.60		885.06		
1 Heating Kettle, 115 Gallon		81.00		89.10		
1 Flatbed Truck, Gas, 3 Ton		303.20		333.52		
2 Pickup Trucks, 3/4 Ton		288.40		317.24	36.93	40.62
40 L.H., Daily Totals		$3009.20		$3973.72	$75.23	$99.34

Crew B-79A	Hr.	Daily	Hr.	Daily	Bare Costs	Incl. O&P
1.5 Equip. Oper. (light)	$48.60	$583.20	$73.75	$885.00	$48.60	$73.75
.5 Line Remov. (Grinder) 115 H.P.		455.90		501.49		
1 Line Rem. (Metal Balls) 115 H.P.		821.60		903.76	106.46	117.10
12 L.H., Daily Totals		$1860.70		$2290.25	$155.06	$190.85

Crew B-79B	Hr.	Daily	Hr.	Daily	Bare Costs	Incl. O&P
1 Laborer	$37.60	$300.80	$57.85	$462.80	$37.60	$57.85
1 Set of Gases		152.00		167.20	19.00	20.90
8 L.H., Daily Totals		$452.80		$630.00	$56.60	$78.75

Crew B-79C	Hr.	Daily	Hr.	Daily	Bare Costs	Incl. O&P
1 Labor Foreman (outside)	$39.60	$316.80	$60.95	$487.60	$38.10	$58.47
5 Laborers	37.60	1504.00	57.85	2314.00		
1 Truck Driver (light)	39.10	312.80	59.10	472.80		
1 Paint Striper, T.M., 120 Gal.		804.60		885.06		
1 Heating Kettle, 115 Gallon		81.00		89.10		
1 Flatbed Truck, Gas, 3 Ton		303.20		333.52		
3 Pickup Trucks, 3/4 Ton		432.60		475.86		
1 Air Compressor, 60 cfm		137.40		151.14		
1 -50' Air Hose, 3/4"		3.25		3.58		
1 Breakers, Pavement, 60 lb.		9.80		10.78	31.64	34.80
56 L.H., Daily Totals		$3905.45		$5223.44	$69.74	$93.28

Crew B-79D	Hr.	Daily	Hr.	Daily	Bare Costs	Incl. O&P
2 Labor Foremen (outside)	$39.60	$633.60	$60.95	$975.20	$38.29	$58.78
5 Laborers	37.60	1504.00	57.85	2314.00		
1 Truck Driver (light)	39.10	312.80	59.10	472.80		
1 Paint Striper, T.M., 120 Gal.		804.60		885.06		
1 Heating Kettle, 115 Gallon		81.00		89.10		
1 Flatbed Truck, Gas, 3 Ton		303.20		333.52		
4 Pickup Trucks, 3/4 Ton		576.80		634.48		
1 Air Compressor, 60 cfm		137.40		151.14		
1 -50' Air Hose, 3/4"		3.25		3.58		
1 Breakers, Pavement, 60 lb.		9.80		10.78	29.94	32.93
64 L.H., Daily Totals		$4366.45		$5869.65	$68.23	$91.71

Crew B-79E	Hr.	Daily	Hr.	Daily	Bare Costs	Incl. O&P
2 Labor Foremen (outside)	$39.60	$633.60	$60.95	$975.20	$38.15	$58.59
7 Laborers	37.60	2105.60	57.85	3239.60		
1 Truck Driver (light)	39.10	312.80	59.10	472.80		
1 Paint Striper, T.M., 120 Gal.		804.60		885.06		
1 Heating Kettle, 115 Gallon		81.00		89.10		
1 Flatbed Truck, Gas, 3 Ton		303.20		333.52		
5 Pickup Trucks, 3/4 Ton		721.00		793.10		
2 Air Compressors, 60 cfm		274.80		302.28		
2 -50' Air Hoses, 3/4"		6.50		7.15		
2 Breakers, Pavement, 60 lb.		19.60		21.56	27.63	30.40
80 L.H., Daily Totals		$5262.70		$7119.37	$65.78	$88.99

Crew B-80	Hr.	Daily	Hr.	Daily	Bare Costs	Incl. O&P
1 Labor Foreman (outside)	$39.60	$316.80	$60.95	$487.60	$41.23	$62.91
1 Laborer	37.60	300.80	57.85	462.80		
1 Truck Driver (light)	39.10	312.80	59.10	472.80		
1 Equip. Oper. (light)	48.60	388.80	73.75	590.00		
1 Flatbed Truck, Gas, 3 Ton		303.20		333.52		
1 Earth Auger, Truck-Mtd.		417.20		458.92	22.51	24.76
32 L.H., Daily Totals		$2039.60		$2805.64	$63.74	$87.68

Crew B-80A	Hr.	Daily	Hr.	Daily	Bare Costs	Incl. O&P
3 Laborers	$37.60	$902.40	$57.85	$1388.40	$37.60	$57.85
1 Flatbed Truck, Gas, 3 Ton		303.20		333.52	12.63	13.90
24 L.H., Daily Totals		$1205.60		$1721.92	$50.23	$71.75

Crew B-80B	Hr.	Daily	Hr.	Daily	Bare Costs	Incl. O&P
3 Laborers	$37.60	$902.40	$57.85	$1388.40	$40.35	$61.83
1 Equip. Oper. (light)	48.60	388.80	73.75	590.00		
1 Crane, Flatbed Mounted, 3 Ton		243.80		268.18	7.62	8.38
32 L.H., Daily Totals		$1535.00		$2246.58	$47.97	$70.21

Crew B-80C	Hr.	Daily	Hr.	Daily	Bare Costs	Incl. O&P
2 Laborers	$37.60	$601.60	$57.85	$925.60	$38.10	$58.27
1 Truck Driver (light)	39.10	312.80	59.10	472.80		
1 Flatbed Truck, Gas, 1.5 Ton		245.40		269.94		
1 Manual Fence Post Auger, Gas		8.40		9.24	10.57	11.63
24 L.H., Daily Totals		$1168.20		$1677.58	$48.67	$69.90

Crew B-81	Hr.	Daily	Hr.	Daily	Bare Costs	Incl. O&P
1 Laborer	$37.60	$300.80	$57.85	$462.80	$42.75	$65.05
1 Equip. Oper. (medium)	50.60	404.80	76.75	614.00		
1 Truck Driver (heavy)	40.05	320.40	60.55	484.40		
1 Hydromulcher, T.M., 3000 Gal.		334.20		367.62		
1 Truck Tractor, 220 H.P.		359.60		395.56	28.91	31.80
24 L.H., Daily Totals		$1719.80		$2324.38	$71.66	$96.85

Crew B-81A	Hr.	Daily	Hr.	Daily	Bare Costs	Incl. O&P
1 Laborer	$37.60	$300.80	$57.85	$462.80	$38.35	$58.48
1 Truck Driver (light)	39.10	312.80	59.10	472.80		
1 Hydromulcher, T.M., 600 Gal.		127.80		140.58		
1 Flatbed Truck, Gas, 3 Ton		303.20		333.52	26.94	29.63
16 L.H., Daily Totals		$1044.60		$1409.70	$65.29	$88.11

Crew B-82	Hr.	Daily	Hr.	Daily	Bare Costs	Incl. O&P
1 Laborer	$37.60	$300.80	$57.85	$462.80	$43.10	$65.80
1 Equip. Oper. (light)	48.60	388.80	73.75	590.00		
1 Horiz. Borer, 6 H.P.		83.80		92.18	5.24	5.76
16 L.H., Daily Totals		$773.40		$1144.98	$48.34	$71.56

Crew No.	Bare Costs		Incl. Subs O&P		Cost Per Labor-Hour	
Crew B-82A	**Hr.**	**Daily**	**Hr.**	**Daily**	**Bare Costs**	**Incl. O&P**
2 Laborers	$37.60	$601.60	$57.85	$925.60	$43.10	$65.80
2 Equip. Opers. (light)	48.60	777.60	73.75	1180.00		
2 Dump Truck, 8 C.Y., 220 H.P.		822.40		904.64		
1 Flatbed Trailer, 25 Ton		113.60		124.96		
1 Horiz. Dir. Drill, 20k lb. Thrust		654.20		719.62		
1 Mud Trailer for HDD, 1500 Gal.		346.20		380.82		
1 Pickup Truck, 4 x 4, 3/4 Ton		155.60		171.16		
1 Flatbed trailer, 3 Ton		23.60		25.96		
1 Loader, Skid Steer, 78 H.P.		318.60		350.46	76.07	83.68
32 L.H., Daily Totals		$3813.40		$4783.22	$119.17	$149.48
Crew B-82B	**Hr.**	**Daily**	**Hr.**	**Daily**	**Bare Costs**	**Incl. O&P**
2 Laborers	$37.60	$601.60	$57.85	$925.60	$43.10	$65.80
2 Equip. Opers. (light)	48.60	777.60	73.75	1180.00		
2 Dump Truck, 8 C.Y., 220 H.P.		822.40		904.64		
1 Flatbed Trailer, 25 Ton		113.60		124.96		
1 Horiz. Dir. Drill, 30k lb. Thrust		931.20		1024.32		
1 Mud Trailer for HDD, 1500 Gal.		346.20		380.82		
1 Pickup Truck, 4 x 4, 3/4 Ton		155.60		171.16		
1 Flatbed trailer, 3 Ton		23.60		25.96		
1 Loader, Skid Steer, 78 H.P.		318.60		350.46	84.72	93.20
32 L.H., Daily Totals		$4090.40		$5087.92	$127.83	$159.00
Crew B-82C	**Hr.**	**Daily**	**Hr.**	**Daily**	**Bare Costs**	**Incl. O&P**
2 Laborers	$37.60	$601.60	$57.85	$925.60	$43.10	$65.80
2 Equip. Opers. (light)	48.60	777.60	73.75	1180.00		
2 Dump Truck, 8 C.Y., 220 H.P.		822.40		904.64		
1 Flatbed Trailer, 25 Ton		113.60		124.96		
1 Horiz. Dir. Drill, 50k lb. Thrust		1237.00		1360.70		
1 Mud Trailer for HDD, 1500 Gal.		346.20		380.82		
1 Pickup Truck, 4 x 4, 3/4 Ton		155.60		171.16		
1 Flatbed trailer, 3 Ton		23.60		25.96		
1 Loader, Skid Steer, 78 H.P.		318.60		350.46	94.28	103.71
32 L.H., Daily Totals		$4396.20		$5424.30	$137.38	$169.51
Crew B-82D	**Hr.**	**Daily**	**Hr.**	**Daily**	**Bare Costs**	**Incl. O&P**
1 Equip. Oper. (light)	$48.60	$388.80	$73.75	$590.00	$48.60	$73.75
1 Mud Trailer for HDD, 1500 Gal.		346.20		380.82	43.27	47.60
8 L.H., Daily Totals		$735.00		$970.82	$91.88	$121.35
Crew B-83	**Hr.**	**Daily**	**Hr.**	**Daily**	**Bare Costs**	**Incl. O&P**
1 Tugboat Captain	$50.60	$404.80	$76.75	$614.00	$44.10	$67.30
1 Tugboat Hand	37.60	300.80	57.85	462.80		
1 Tugboat, 250 H.P.		870.00		957.00	54.38	59.81
16 L.H., Daily Totals		$1575.60		$2033.80	$98.47	$127.11
Crew B-84	**Hr.**	**Daily**	**Hr.**	**Daily**	**Bare Costs**	**Incl. O&P**
1 Equip. Oper. (medium)	$50.60	$404.80	$76.75	$614.00	$50.60	$76.75
1 Rotary Mower/Tractor		370.80		407.88	46.35	50.98
8 L.H., Daily Totals		$775.60		$1021.88	$96.95	$127.74
Crew B-85	**Hr.**	**Daily**	**Hr.**	**Daily**	**Bare Costs**	**Incl. O&P**
3 Laborers	$37.60	$902.40	$57.85	$1388.40	$40.69	$62.17
1 Equip. Oper. (medium)	50.60	404.80	76.75	614.00		
1 Truck Driver (heavy)	40.05	320.40	60.55	484.40		
1 Aerial Lift Truck, 80'		634.40		697.84		
1 Brush Chipper, 12", 130 H.P.		391.40		430.54		
1 Pruning Saw, Rotary		6.65		7.32	25.81	28.39
40 L.H., Daily Totals		$2660.05		$3622.49	$66.50	$90.56

Crew No.	Bare Costs		Incl. Subs O&P		Cost Per Labor-Hour	
Crew B-86	**Hr.**	**Daily**	**Hr.**	**Daily**	**Bare Costs**	**Incl. O&P**
1 Equip. Oper. (medium)	$50.60	$404.80	$76.75	$614.00	$50.60	$76.75
1 Stump Chipper, S.P.		185.55		204.10	23.19	25.51
8 L.H., Daily Totals		$590.35		$818.11	$73.79	$102.26
Crew B-86A	**Hr.**	**Daily**	**Hr.**	**Daily**	**Bare Costs**	**Incl. O&P**
1 Equip. Oper. (medium)	$50.60	$404.80	$76.75	$614.00	$50.60	$76.75
1 Grader, 30,000 Lbs.		736.20		809.82	92.03	101.23
8 L.H., Daily Totals		$1141.00		$1423.82	$142.63	$177.98
Crew B-86B	**Hr.**	**Daily**	**Hr.**	**Daily**	**Bare Costs**	**Incl. O&P**
1 Equip. Oper. (medium)	$50.60	$404.80	$76.75	$614.00	$50.60	$76.75
1 Dozer, 200 H.P.		1387.00		1525.70	173.38	190.71
8 L.H., Daily Totals		$1791.80		$2139.70	$223.97	$267.46
Crew B-87	**Hr.**	**Daily**	**Hr.**	**Daily**	**Bare Costs**	**Incl. O&P**
1 Laborer	$37.60	$300.80	$57.85	$462.80	$48.00	$72.97
4 Equip. Oper. (medium)	50.60	1619.20	76.75	2456.00		
2 Feller Bunchers, 100 H.P.		1676.80		1844.48		
1 Log Chipper, 22" Tree		889.40		978.34		
1 Dozer, 105 H.P.		602.80		663.08		
1 Chain Saw, Gas, 36" Long		45.00		49.50	80.35	88.39
40 L.H., Daily Totals		$5134.00		$6454.20	$128.35	$161.35
Crew B-88	**Hr.**	**Daily**	**Hr.**	**Daily**	**Bare Costs**	**Incl. O&P**
1 Laborer	$37.60	$300.80	$57.85	$462.80	$48.74	$74.05
6 Equip. Oper. (medium)	50.60	2428.80	76.75	3684.00		
2 Feller Bunchers, 100 H.P.		1676.80		1844.48		
1 Log Chipper, 22" Tree		889.40		978.34		
2 Log Skidders, 50 H.P.		1822.00		2004.20		
1 Dozer, 105 H.P.		602.80		663.08		
1 Chain Saw, Gas, 36" Long		45.00		49.50	89.93	98.92
56 L.H., Daily Totals		$7765.60		$9686.40	$138.67	$172.97
Crew B-89	**Hr.**	**Daily**	**Hr.**	**Daily**	**Bare Costs**	**Incl. O&P**
1 Equip. Oper. (light)	$48.60	$388.80	$73.75	$590.00	$43.85	$66.42
1 Truck Driver (light)	39.10	312.80	59.10	472.80		
1 Flatbed Truck, Gas, 3 Ton		303.20		333.52		
1 Concrete Saw		167.40		184.14		
1 Water Tank, 65 Gal.		17.30		19.03	30.49	33.54
16 L.H., Daily Totals		$1189.50		$1599.49	$74.34	$99.97
Crew B-89A	**Hr.**	**Daily**	**Hr.**	**Daily**	**Bare Costs**	**Incl. O&P**
1 Skilled Worker	$48.65	$389.20	$75.15	$601.20	$43.13	$66.50
1 Laborer	37.60	300.80	57.85	462.80		
1 Core Drill (Large)		115.60		127.16	7.22	7.95
16 L.H., Daily Totals		$805.60		$1191.16	$50.35	$74.45
Crew B-89B	**Hr.**	**Daily**	**Hr.**	**Daily**	**Bare Costs**	**Incl. O&P**
1 Equip. Oper. (light)	$48.60	$388.80	$73.75	$590.00	$43.85	$66.42
1 Truck Driver (light)	39.10	312.80	59.10	472.80		
1 Wall Saw, Hydraulic, 10 H.P.		114.60		126.06		
1 Generator, Diesel, 100 kW		423.40		465.74		
1 Water Tank, 65 Gal.		17.30		19.03		
1 Flatbed Truck, Gas, 3 Ton		303.20		333.52	53.66	59.02
16 L.H., Daily Totals		$1560.10		$2007.15	$97.51	$125.45

Crew No.	Bare Costs		Incl. Subs O&P		Cost Per Labor-Hour	
Crew B-90	**Hr.**	**Daily**	**Hr.**	**Daily**	**Bare Costs**	**Incl. O&P**
1 Labor Foreman (outside)	$39.60	$316.80	$60.95	$487.60	$41.21	$62.89
3 Laborers	37.60	902.40	57.85	1388.40		
2 Equip. Oper. (light)	48.60	777.60	73.75	1180.00		
2 Truck Drivers (heavy)	40.05	640.80	60.55	968.80		
1 Road Mixer, 310 H.P.		1929.00		2121.90		
1 Dist. Truck, 2000 Gal.		275.60		303.16	34.45	37.89
64 L.H., Daily Totals		$4842.20		$6449.86	$75.66	$100.78
Crew B-90A	**Hr.**	**Daily**	**Hr.**	**Daily**	**Bare Costs**	**Incl. O&P**
1 Labor Foreman (outside)	$39.60	$316.80	$60.95	$487.60	$45.31	$69.09
2 Laborers	37.60	601.60	57.85	925.60		
4 Equip. Oper. (medium)	50.60	1619.20	76.75	2456.00		
2 Graders, 30,000 Lbs.		1472.40		1619.64		
1 Tandem Roller, 10 Ton		236.80		260.48		
1 Roller, Pneum. Whl., 12 Ton		344.40		378.84	36.67	40.34
56 L.H., Daily Totals		$4591.20		$6128.16	$81.99	$109.43
Crew B-90B	**Hr.**	**Daily**	**Hr.**	**Daily**	**Bare Costs**	**Incl. O&P**
1 Labor Foreman (outside)	$39.60	$316.80	$60.95	$487.60	$44.43	$67.82
2 Laborers	37.60	601.60	57.85	925.60		
3 Equip. Oper. (medium)	50.60	1214.40	76.75	1842.00		
1 Roller, Pneum. Whl., 12 Ton		344.40		378.84		
1 Road Mixer, 310 H.P.		1929.00		2121.90	47.36	52.10
48 L.H., Daily Totals		$4406.20		$5755.94	$91.80	$119.92
Crew B-90C	**Hr.**	**Daily**	**Hr.**	**Daily**	**Bare Costs**	**Incl. O&P**
1 Labor Foreman (outside)	$39.60	$316.80	$60.95	$487.60	$42.00	$64.02
4 Laborers	37.60	1203.20	57.85	1851.20		
3 Equip. Oper. (medium)	50.60	1214.40	76.75	1842.00		
3 Truck Drivers (heavy)	40.05	961.20	60.55	1453.20		
3 Road Mixers, 310 H.P.		5787.00		6365.70	65.76	72.34
88 L.H., Daily Totals		$9482.60		$11999.70	$107.76	$136.36
Crew B-90D	**Hr.**	**Daily**	**Hr.**	**Daily**	**Bare Costs**	**Incl. O&P**
1 Labor Foreman (outside)	$39.60	$316.80	$60.95	$487.60	$41.32	$63.07
6 Laborers	37.60	1804.80	57.85	2776.80		
3 Equip. Oper. (medium)	50.60	1214.40	76.75	1842.00		
3 Truck Drivers (heavy)	40.05	961.20	60.55	1453.20		
3 Road Mixers, 310 H.P.		5787.00		6365.70	55.64	61.21
104 L.H., Daily Totals		$10084.20		$12925.30	$96.96	$124.28
Crew B-90E	**Hr.**	**Daily**	**Hr.**	**Daily**	**Bare Costs**	**Incl. O&P**
1 Labor Foreman (outside)	$39.60	$316.80	$60.95	$487.60	$42.43	$64.79
4 Laborers	37.60	1203.20	57.85	1851.20		
3 Equip. Oper. (medium)	50.60	1214.40	76.75	1842.00		
1 Truck Driver (heavy)	40.05	320.40	60.55	484.40		
1 Road Mixers, 310 H.P.		1929.00		2121.90	26.79	29.47
72 L.H., Daily Totals		$4983.80		$6787.10	$69.22	$94.27
Crew B-91	**Hr.**	**Daily**	**Hr.**	**Daily**	**Bare Costs**	**Incl. O&P**
1 Labor Foreman (outside)	$39.60	$316.80	$60.95	$487.60	$44.66	$68.03
2 Laborers	37.60	601.60	57.85	925.60		
4 Equip. Oper. (medium)	50.60	1619.20	76.75	2456.00		
1 Truck Driver (heavy)	40.05	320.40	60.55	484.40		
1 Dist. Tanker, 3000 Gallon		305.60		336.16		
1 Truck Tractor, 6x4, 380 H.P.		599.80		659.78		
1 Aggreg. Spreader, S.P.		834.00		917.40		
1 Roller, Pneum. Whl., 12 Ton		344.40		378.84		
1 Tandem Roller, 10 Ton		236.80		260.48	36.26	39.89
64 L.H., Daily Totals		$5178.60		$6906.26	$80.92	$107.91

Crew No.	Bare Costs		Incl. Subs O&P		Cost Per Labor-Hour	
Crew B-91B	**Hr.**	**Daily**	**Hr.**	**Daily**	**Bare Costs**	**Incl. O&P**
1 Laborer	$37.60	$300.80	$57.85	$462.80	$44.10	$67.30
1 Equipment Oper. (med.)	50.60	404.80	76.75	614.00		
1 Road Sweeper, Vac. Assist.		986.00		1084.60	61.63	67.79
16 L.H., Daily Totals		$1691.60		$2161.40	$105.72	$135.09
Crew B-91C	**Hr.**	**Daily**	**Hr.**	**Daily**	**Bare Costs**	**Incl. O&P**
1 Laborer	$37.60	$300.80	$57.85	$462.80	$38.35	$58.48
1 Truck Driver (light)	39.10	312.80	59.10	472.80		
1 Catch Basin Cleaning Truck		573.60		630.96	35.85	39.44
16 L.H., Daily Totals		$1187.20		$1566.56	$74.20	$97.91
Crew B-91D	**Hr.**	**Daily**	**Hr.**	**Daily**	**Bare Costs**	**Incl. O&P**
1 Labor Foreman (outside)	$39.60	$316.80	$60.95	$487.60	$43.13	$65.77
5 Laborers	37.60	1504.00	57.85	2314.00		
5 Equip. Oper. (medium)	50.60	2024.00	76.75	3070.00		
2 Truck Drivers (heavy)	40.05	640.80	60.55	968.80		
1 Aggreg. Spreader, S.P.		834.00		917.40		
2 Truck Tractor, 6x4, 380 H.P.		1199.60		1319.56		
2 Dist. Tanker, 3000 Gallon		611.20		672.32		
2 Pavement Brush, Towed		163.20		179.52		
2 Roller, Pneum. Whl., 12 Ton		688.80		757.68	33.62	36.99
104 L.H., Daily Totals		$7982.40		$10686.88	$76.75	$102.76
Crew B-92	**Hr.**	**Daily**	**Hr.**	**Daily**	**Bare Costs**	**Incl. O&P**
1 Labor Foreman (outside)	$39.60	$316.80	$60.95	$487.60	$38.10	$58.63
3 Laborers	37.60	902.40	57.85	1388.40		
1 Crack Cleaner, 25 H.P.		61.60		67.76		
1 Air Compressor, 60 cfm		137.40		151.14		
1 Tar Kettle, T.M.		132.00		145.20		
1 Flatbed Truck, Gas, 3 Ton		303.20		333.52	19.82	21.80
32 L.H., Daily Totals		$1853.40		$2573.62	$57.92	$80.43
Crew B-93	**Hr.**	**Daily**	**Hr.**	**Daily**	**Bare Costs**	**Incl. O&P**
1 Equip. Oper. (medium)	$50.60	$404.80	$76.75	$614.00	$50.60	$76.75
1 Feller Buncher, 100 H.P.		838.40		922.24	104.80	115.28
8 L.H., Daily Totals		$1243.20		$1536.24	$155.40	$192.03
Crew B-94A	**Hr.**	**Daily**	**Hr.**	**Daily**	**Bare Costs**	**Incl. O&P**
1 Laborer	$37.60	$300.80	$57.85	$462.80	$37.60	$57.85
1 Diaphragm Water Pump, 2"		71.00		78.10		
1 -20' Suction Hose, 2"		1.95		2.15		
2 -50' Discharge Hoses, 2"		1.80		1.98	9.34	10.28
8 L.H., Daily Totals		$375.55		$545.02	$46.94	$68.13
Crew B-94B	**Hr.**	**Daily**	**Hr.**	**Daily**	**Bare Costs**	**Incl. O&P**
1 Laborer	$37.60	$300.80	$57.85	$462.80	$37.60	$57.85
1 Diaphragm Water Pump, 4"		114.20		125.62		
1 -20' Suction Hose, 4"		3.25		3.58		
2 -50' Discharge Hoses, 4"		4.70		5.17	15.27	16.80
8 L.H., Daily Totals		$422.95		$597.16	$52.87	$74.65
Crew B-94C	**Hr.**	**Daily**	**Hr.**	**Daily**	**Bare Costs**	**Incl. O&P**
1 Laborer	$37.60	$300.80	$57.85	$462.80	$37.60	$57.85
1 Centrifugal Water Pump, 3"		78.20		86.02		
1 -20' Suction Hose, 3"		2.85		3.13		
2 -50' Discharge Hoses, 3"		3.00		3.30	10.51	11.56
8 L.H., Daily Totals		$384.85		$555.26	$48.11	$69.41

Crew No.	Bare Costs		Incl. Subs O&P		Cost Per Labor-Hour	
Crew B-94D	**Hr.**	**Daily**	**Hr.**	**Daily**	**Bare Costs**	**Incl. O&P**
1 Laborer	$37.60	$300.80	$57.85	$462.80	$37.60	$57.85
1 Centr. Water Pump, 6"		346.60		381.26		
1 -20' Suction Hose, 6"		11.50		12.65		
2 -50' Discharge Hoses, 6"		12.20		13.42	46.29	50.92
8 L.H., Daily Totals		$671.10		$870.13	$83.89	$108.77
Crew C-1	**Hr.**	**Daily**	**Hr.**	**Daily**	**Bare Costs**	**Incl. O&P**
3 Carpenters	$46.95	$1126.80	$72.25	$1734.00	$44.61	$68.65
1 Laborer	37.60	300.80	57.85	462.80		
32 L.H., Daily Totals		$1427.60		$2196.80	$44.61	$68.65
Crew C-2	**Hr.**	**Daily**	**Hr.**	**Daily**	**Bare Costs**	**Incl. O&P**
1 Carpenter Foreman (outside)	$48.95	$391.60	$75.35	$602.80	$45.73	$70.37
4 Carpenters	46.95	1502.40	72.25	2312.00		
1 Laborer	37.60	300.80	57.85	462.80		
48 L.H., Daily Totals		$2194.80		$3377.60	$45.73	$70.37
Crew C-2A	**Hr.**	**Daily**	**Hr.**	**Daily**	**Bare Costs**	**Incl. O&P**
1 Carpenter Foreman (outside)	$48.95	$391.60	$75.35	$602.80	$45.40	$69.43
3 Carpenters	46.95	1126.80	72.25	1734.00		
1 Cement Finisher	45.00	360.00	66.65	533.20		
1 Laborer	37.60	300.80	57.85	462.80		
48 L.H., Daily Totals		$2179.20		$3332.80	$45.40	$69.43
Crew C-3	**Hr.**	**Daily**	**Hr.**	**Daily**	**Bare Costs**	**Incl. O&P**
1 Rodman Foreman (outside)	$54.55	$436.40	$85.20	$681.60	$48.57	$75.38
4 Rodmen (reinf.)	52.55	1681.60	82.10	2627.20		
1 Equip. Oper. (light)	48.60	388.80	73.75	590.00		
2 Laborers	37.60	601.60	57.85	925.60		
3 Stressing Equipment		30.60		33.66		
.5 Grouting Equipment		81.50		89.65	1.75	1.93
64 L.H., Daily Totals		$3220.50		$4947.71	$50.32	$77.31
Crew C-4	**Hr.**	**Daily**	**Hr.**	**Daily**	**Bare Costs**	**Incl. O&P**
1 Rodman Foreman (outside)	$54.55	$436.40	$85.20	$681.60	$53.05	$82.88
3 Rodmen (reinf.)	52.55	1261.20	82.10	1970.40		
3 Stressing Equipment		30.60		33.66	.96	1.05
32 L.H., Daily Totals		$1728.20		$2685.66	$54.01	$83.93
Crew C-4A	**Hr.**	**Daily**	**Hr.**	**Daily**	**Bare Costs**	**Incl. O&P**
2 Rodmen (reinf.)	$52.55	$840.80	$82.10	$1313.60	$52.55	$82.10
4 Stressing Equipment		40.80		44.88	2.55	2.81
16 L.H., Daily Totals		$881.60		$1358.48	$55.10	$84.91
Crew C-5	**Hr.**	**Daily**	**Hr.**	**Daily**	**Bare Costs**	**Incl. O&P**
1 Rodman Foreman (outside)	$54.55	$436.40	$85.20	$681.60	$51.66	$80.09
4 Rodmen (reinf.)	52.55	1681.60	82.10	2627.20		
1 Equip. Oper. (crane)	51.70	413.60	78.45	627.60		
1 Equip. Oper. (oiler)	45.20	361.60	68.55	548.40		
1 Hyd. Crane, 25 Ton		736.60		810.26	13.15	14.47
56 L.H., Daily Totals		$3629.80		$5295.06	$64.82	$94.55
Crew C-6	**Hr.**	**Daily**	**Hr.**	**Daily**	**Bare Costs**	**Incl. O&P**
1 Labor Foreman (outside)	$39.60	$316.80	$60.95	$487.60	$39.17	$59.83
4 Laborers	37.60	1203.20	57.85	1851.20		
1 Cement Finisher	45.00	360.00	66.65	533.20		
2 Gas Engine Vibrators		62.40		68.64	1.30	1.43
48 L.H., Daily Totals		$1942.40		$2940.64	$40.47	$61.26

Crew No.	Bare Costs		Incl. Subs O&P		Cost Per Labor-Hour	
Crew C-7	**Hr.**	**Daily**	**Hr.**	**Daily**	**Bare Costs**	**Incl. O&P**
1 Labor Foreman (outside)	$39.60	$316.80	$60.95	$487.60	$40.93	$62.46
5 Laborers	37.60	1504.00	57.85	2314.00		
1 Cement Finisher	45.00	360.00	66.65	533.20		
1 Equip. Oper. (medium)	50.60	404.80	76.75	614.00		
1 Equip. Oper. (oiler)	45.20	361.60	68.55	548.40		
2 Gas Engine Vibrators		62.40		68.64		
1 Concrete Bucket, 1 C.Y.		23.80		26.18		
1 Hyd. Crane, 55 Ton		1128.00		1240.80	16.86	18.55
72 L.H., Daily Totals		$4161.40		$5832.82	$57.80	$81.01
Crew C-7A	**Hr.**	**Daily**	**Hr.**	**Daily**	**Bare Costs**	**Incl. O&P**
1 Labor Foreman (outside)	$39.60	$316.80	$60.95	$487.60	$38.46	$58.91
5 Laborers	37.60	1504.00	57.85	2314.00		
2 Truck Drivers (heavy)	40.05	640.80	60.55	968.80		
2 Conc. Transit Mixers		2032.00		2235.20	31.75	34.92
64 L.H., Daily Totals		$4493.60		$6005.60	$70.21	$93.84
Crew C-7B	**Hr.**	**Daily**	**Hr.**	**Daily**	**Bare Costs**	**Incl. O&P**
1 Labor Foreman (outside)	$39.60	$316.80	$60.95	$487.60	$40.56	$62.15
5 Laborers	37.60	1504.00	57.85	2314.00		
1 Equipment Operator, Crane	51.70	413.60	78.45	627.60		
1 Equipment Oiler	45.20	361.60	68.55	548.40		
1 Conc. Bucket, 2 C.Y.		37.00		40.70		
1 Lattice Boom Crane, 165 Ton		1932.00		2125.20	30.77	33.84
64 L.H., Daily Totals		$4565.00		$6143.50	$71.33	$95.99
Crew C-7C	**Hr.**	**Daily**	**Hr.**	**Daily**	**Bare Costs**	**Incl. O&P**
1 Labor Foreman (outside)	$39.60	$316.80	$60.95	$487.60	$41.10	$62.96
5 Laborers	37.60	1504.00	57.85	2314.00		
2 Equipment Operators (med.)	50.60	809.60	76.75	1228.00		
2 F.E. Loaders, W.M., 4 C.Y.		1338.40		1472.24	20.91	23.00
64 L.H., Daily Totals		$3968.80		$5501.84	$62.01	$85.97
Crew C-7D	**Hr.**	**Daily**	**Hr.**	**Daily**	**Bare Costs**	**Incl. O&P**
1 Labor Foreman (outside)	$39.60	$316.80	$60.95	$487.60	$39.74	$60.99
5 Laborers	37.60	1504.00	57.85	2314.00		
1 Equip. Oper. (medium)	50.60	404.80	76.75	614.00		
1 Concrete Conveyer		198.80		218.68	3.55	3.90
56 L.H., Daily Totals		$2424.40		$3634.28	$43.29	$64.90
Crew C-8	**Hr.**	**Daily**	**Hr.**	**Daily**	**Bare Costs**	**Incl. O&P**
1 Labor Foreman (outside)	$39.60	$316.80	$60.95	$487.60	$41.86	$63.51
3 Laborers	37.60	902.40	57.85	1388.40		
2 Cement Finishers	45.00	720.00	66.65	1066.40		
1 Equip. Oper. (medium)	50.60	404.80	76.75	614.00		
1 Concrete Pump (Small)		720.00		792.00	12.86	14.14
56 L.H., Daily Totals		$3064.00		$4348.40	$54.71	$77.65
Crew C-8A	**Hr.**	**Daily**	**Hr.**	**Daily**	**Bare Costs**	**Incl. O&P**
1 Labor Foreman (outside)	$39.60	$316.80	$60.95	$487.60	$40.40	$61.30
3 Laborers	37.60	902.40	57.85	1388.40		
2 Cement Finishers	45.00	720.00	66.65	1066.40		
48 L.H., Daily Totals		$1939.20		$2942.40	$40.40	$61.30

Crew C-8B	Bare Costs Hr.	Bare Costs Daily	Incl. Subs O&P Hr.	Incl. Subs O&P Daily	Cost Per Labor-Hour Bare Costs	Cost Per Labor-Hour Incl. O&P
1 Labor Foreman (outside)	$39.60	$316.80	$60.95	$487.60	$40.60	$62.25
3 Laborers	37.60	902.40	57.85	1388.40		
1 Equip. Oper. (medium)	50.60	404.80	76.75	614.00		
1 Vibrating Power Screed		66.30		72.93		
1 Roller, Vibratory, 25 Ton		687.40		756.14		
1 Dozer, 200 H.P.		1387.00		1525.70	53.52	58.87
40 L.H., Daily Totals		$3764.70		$4844.77	$94.12	$121.12

Crew C-8C	Hr.	Daily	Hr.	Daily	Bare Costs	Incl. O&P
1 Labor Foreman (outside)	$39.60	$316.80	$60.95	$487.60	$41.33	$62.98
3 Laborers	37.60	902.40	57.85	1388.40		
1 Cement Finisher	45.00	360.00	66.65	533.20		
1 Equip. Oper. (medium)	50.60	404.80	76.75	614.00		
1 Shotcrete Rig, 12 C.Y./hr		250.80		275.88		
1 Air Compressor, 160 cfm		156.40		172.04		
4 -50' Air Hoses, 1"		16.40		18.04		
4 -50' Air Hoses, 2"		31.00		34.10	9.47	10.42
48 L.H., Daily Totals		$2438.60		$3523.26	$50.80	$73.40

Crew C-8D	Hr.	Daily	Hr.	Daily	Bare Costs	Incl. O&P
1 Labor Foreman (outside)	$39.60	$316.80	$60.95	$487.60	$42.70	$64.80
1 Laborer	37.60	300.80	57.85	462.80		
1 Cement Finisher	45.00	360.00	66.65	533.20		
1 Equipment Oper. (light)	48.60	388.80	73.75	590.00		
1 Air Compressor, 250 cfm		201.40		221.54		
2 -50' Air Hoses, 1"		8.20		9.02	6.55	7.21
32 L.H., Daily Totals		$1576.00		$2304.16	$49.25	$72.00

Crew C-8E	Hr.	Daily	Hr.	Daily	Bare Costs	Incl. O&P
1 Labor Foreman (outside)	$39.60	$316.80	$60.95	$487.60	$41.00	$62.48
3 Laborers	37.60	902.40	57.85	1388.40		
1 Cement Finisher	45.00	360.00	66.65	533.20		
1 Equipment Oper. (light)	48.60	388.80	73.75	590.00		
1 Shotcrete Rig, 35 C.Y./hr		281.00		309.10		
1 Air Compressor, 250 cfm		201.40		221.54		
4 -50' Air Hoses, 1"		16.40		18.04		
4 -50' Air Hoses, 2"		31.00		34.10	11.04	12.14
48 L.H., Daily Totals		$2497.80		$3581.98	$52.04	$74.62

Crew C-10	Hr.	Daily	Hr.	Daily	Bare Costs	Incl. O&P
1 Laborer	$37.60	$300.80	$57.85	$462.80	$42.53	$63.72
2 Cement Finishers	45.00	720.00	66.65	1066.40		
24 L.H., Daily Totals		$1020.80		$1529.20	$42.53	$63.72

Crew C-10B	Hr.	Daily	Hr.	Daily	Bare Costs	Incl. O&P
3 Laborers	$37.60	$902.40	$57.85	$1388.40	$40.56	$61.37
2 Cement Finishers	45.00	720.00	66.65	1066.40		
1 Concrete Mixer, 10 C.F.		172.40		189.64		
2 Trowels, 48" Walk-Behind		100.40		110.44	6.82	7.50
40 L.H., Daily Totals		$1895.20		$2754.88	$47.38	$68.87

Crew C-10C	Hr.	Daily	Hr.	Daily	Bare Costs	Incl. O&P
1 Laborer	$37.60	$300.80	$57.85	$462.80	$42.53	$63.72
2 Cement Finishers	45.00	720.00	66.65	1066.40		
1 Trowel, 48" Walk-Behind		50.20		55.22	2.09	2.30
24 L.H., Daily Totals		$1071.00		$1584.42	$44.63	$66.02

Crew C-10D	Hr.	Daily	Hr.	Daily	Bare Costs	Incl. O&P
1 Laborer	$37.60	$300.80	$57.85	$462.80	$42.53	$63.72
2 Cement Finishers	45.00	720.00	66.65	1066.40		
1 Vibrating Power Screed		66.30		72.93		
1 Trowel, 48" Walk-Behind		50.20		55.22	4.85	5.34
24 L.H., Daily Totals		$1137.30		$1657.35	$47.39	$69.06

Crew C-10E	Hr.	Daily	Hr.	Daily	Bare Costs	Incl. O&P
1 Laborer	$37.60	$300.80	$57.85	$462.80	$42.53	$63.72
2 Cement Finishers	45.00	720.00	66.65	1066.40		
1 Vibrating Power Screed		66.30		72.93		
1 Cement Trowel, 96" Ride-On		189.00		207.90	10.64	11.70
24 L.H., Daily Totals		$1276.10		$1810.03	$53.17	$75.42

Crew C-10F	Hr.	Daily	Hr.	Daily	Bare Costs	Incl. O&P
1 Laborer	$37.60	$300.80	$57.85	$462.80	$42.53	$63.72
2 Cement Finishers	45.00	720.00	66.65	1066.40		
1 Aerial Lift Truck, 60' Boom		435.60		479.16	18.15	19.97
24 L.H., Daily Totals		$1456.40		$2008.36	$60.68	$83.68

Crew C-11	Hr.	Daily	Hr.	Daily	Bare Costs	Incl. O&P
1 Struc. Steel Foreman (outside)	$54.65	$437.20	$94.95	$759.60	$51.94	$87.85
6 Struc. Steel Workers	52.65	2527.20	91.45	4389.60		
1 Equip. Oper. (crane)	51.70	413.60	78.45	627.60		
1 Equip. Oper. (oiler)	45.20	361.60	68.55	548.40		
1 Lattice Boom Crane, 150 Ton		1813.00		1994.30	25.18	27.70
72 L.H., Daily Totals		$5552.60		$8319.50	$77.12	$115.55

Crew C-12	Hr.	Daily	Hr.	Daily	Bare Costs	Incl. O&P
1 Carpenter Foreman (outside)	$48.95	$391.60	$75.35	$602.80	$46.52	$71.40
3 Carpenters	46.95	1126.80	72.25	1734.00		
1 Laborer	37.60	300.80	57.85	462.80		
1 Equip. Oper. (crane)	51.70	413.60	78.45	627.60		
1 Hyd. Crane, 12 Ton		653.80		719.18	13.62	14.98
48 L.H., Daily Totals		$2886.60		$4146.38	$60.14	$86.38

Crew C-13	Hr.	Daily	Hr.	Daily	Bare Costs	Incl. O&P
1 Struc. Steel Worker	$52.65	$421.20	$91.45	$731.60	$50.75	$85.05
1 Welder	52.65	421.20	91.45	731.60		
1 Carpenter	46.95	375.60	72.25	578.00		
1 Welder, Gas Engine, 300 amp		145.85		160.44	6.08	6.68
24 L.H., Daily Totals		$1363.85		$2201.64	$56.83	$91.73

Crew C-14	Hr.	Daily	Hr.	Daily	Bare Costs	Incl. O&P
1 Carpenter Foreman (outside)	$48.95	$391.60	$75.35	$602.80	$46.18	$70.93
5 Carpenters	46.95	1878.00	72.25	2890.00		
4 Laborers	37.60	1203.20	57.85	1851.20		
4 Rodmen (reinf.)	52.55	1681.60	82.10	2627.20		
2 Cement Finishers	45.00	720.00	66.65	1066.40		
1 Equip. Oper. (crane)	51.70	413.60	78.45	627.60		
1 Equip. Oper. (oiler)	45.20	361.60	68.55	548.40		
1 Hyd. Crane, 80 Ton		1625.00		1787.50	11.28	12.41
144 L.H., Daily Totals		$8274.60		$12001.10	$57.46	$83.34

Crew No.	Bare Costs		Incl. Subs O&P		Cost Per Labor-Hour	
Crew C-14A	**Hr.**	**Daily**	**Hr.**	**Daily**	**Bare Costs**	**Incl. O&P**
1 Carpenter Foreman (outside)	$48.95	$391.60	$75.35	$602.80	$47.25	$72.75
16 Carpenters	46.95	6009.60	72.25	9248.00		
4 Rodmen (reinf.)	52.55	1681.60	82.10	2627.20		
2 Laborers	37.60	601.60	57.85	925.60		
1 Cement Finisher	45.00	360.00	66.65	533.20		
1 Equip. Oper. (medium)	50.60	404.80	76.75	614.00		
1 Gas Engine Vibrator		31.20		34.32		
1 Concrete Pump (Small)		720.00		792.00	3.76	4.13
200 L.H., Daily Totals		$10200.40		$15377.12	$51.00	$76.89

Crew No.	Bare Costs		Incl. Subs O&P		Cost Per Labor-Hour	
Crew C-14B	**Hr.**	**Daily**	**Hr.**	**Daily**	**Bare Costs**	**Incl. O&P**
1 Carpenter Foreman (outside)	$48.95	$391.60	$75.35	$602.80	$47.16	$72.52
16 Carpenters	46.95	6009.60	72.25	9248.00		
4 Rodmen (reinf.)	52.55	1681.60	82.10	2627.20		
2 Laborers	37.60	601.60	57.85	925.60		
2 Cement Finishers	45.00	720.00	66.65	1066.40		
1 Equip. Oper. (medium)	50.60	404.80	76.75	614.00		
1 Gas Engine Vibrator		31.20		34.32		
1 Concrete Pump (Small)		720.00		792.00	3.61	3.97
208 L.H., Daily Totals		$10560.40		$15910.32	$50.77	$76.49

Crew No.	Bare Costs		Incl. Subs O&P		Cost Per Labor-Hour	
Crew C-14C	**Hr.**	**Daily**	**Hr.**	**Daily**	**Bare Costs**	**Incl. O&P**
1 Carpenter Foreman (outside)	$48.95	$391.60	$75.35	$602.80	$45.08	$69.36
6 Carpenters	46.95	2253.60	72.25	3468.00		
2 Rodmen (reinf.)	52.55	840.80	82.10	1313.60		
4 Laborers	37.60	1203.20	57.85	1851.20		
1 Cement Finisher	45.00	360.00	66.65	533.20		
1 Gas Engine Vibrator		31.20		34.32	.28	.31
112 L.H., Daily Totals		$5080.40		$7803.12	$45.36	$69.67

Crew No.	Bare Costs		Incl. Subs O&P		Cost Per Labor-Hour	
Crew C-14D	**Hr.**	**Daily**	**Hr.**	**Daily**	**Bare Costs**	**Incl. O&P**
1 Carpenter Foreman (outside)	$48.95	$391.60	$75.35	$602.80	$46.80	$71.97
18 Carpenters	46.95	6760.80	72.25	10404.00		
2 Rodmen (reinf.)	52.55	840.80	82.10	1313.60		
2 Laborers	37.60	601.60	57.85	925.60		
1 Cement Finisher	45.00	360.00	66.65	533.20		
1 Equip. Oper. (medium)	50.60	404.80	76.75	614.00		
1 Gas Engine Vibrator		31.20		34.32		
1 Concrete Pump (Small)		720.00		792.00	3.76	4.13
200 L.H., Daily Totals		$10110.80		$15219.52	$50.55	$76.10

Crew No.	Bare Costs		Incl. Subs O&P		Cost Per Labor-Hour	
Crew C-14E	**Hr.**	**Daily**	**Hr.**	**Daily**	**Bare Costs**	**Incl. O&P**
1 Carpenter Foreman (outside)	$48.95	$391.60	$75.35	$602.80	$46.44	$71.68
2 Carpenters	46.95	751.20	72.25	1156.00		
4 Rodmen (reinf.)	52.55	1681.60	82.10	2627.20		
3 Laborers	37.60	902.40	57.85	1388.40		
1 Cement Finisher	45.00	360.00	66.65	533.20		
1 Gas Engine Vibrator		31.20		34.32	.35	.39
88 L.H., Daily Totals		$4118.00		$6341.92	$46.80	$72.07

Crew No.	Bare Costs		Incl. Subs O&P		Cost Per Labor-Hour	
Crew C-14F	**Hr.**	**Daily**	**Hr.**	**Daily**	**Bare Costs**	**Incl. O&P**
1 Labor Foreman (outside)	$39.60	$316.80	$60.95	$487.60	$42.76	$64.06
2 Laborers	37.60	601.60	57.85	925.60		
6 Cement Finishers	45.00	2160.00	66.65	3199.20		
1 Gas Engine Vibrator		31.20		34.32	.43	.48
72 L.H., Daily Totals		$3109.60		$4646.72	$43.19	$64.54

Crew No.	Bare Costs		Incl. Subs O&P		Cost Per Labor-Hour	
Crew C-14G	**Hr.**	**Daily**	**Hr.**	**Daily**	**Bare Costs**	**Incl. O&P**
1 Labor Foreman (outside)	$39.60	$316.80	$60.95	$487.60	$42.11	$63.32
2 Laborers	37.60	601.60	57.85	925.60		
4 Cement Finishers	45.00	1440.00	66.65	2132.80		
1 Gas Engine Vibrator		31.20		34.32	.56	.61
56 L.H., Daily Totals		$2389.60		$3580.32	$42.67	$63.93

Crew No.	Bare Costs		Incl. Subs O&P		Cost Per Labor-Hour	
Crew C-14H	**Hr.**	**Daily**	**Hr.**	**Daily**	**Bare Costs**	**Incl. O&P**
1 Carpenter Foreman (outside)	$48.95	$391.60	$75.35	$602.80	$46.33	$71.08
2 Carpenters	46.95	751.20	72.25	1156.00		
1 Rodman (reinf.)	52.55	420.40	82.10	656.80		
1 Laborer	37.60	300.80	57.85	462.80		
1 Cement Finisher	45.00	360.00	66.65	533.20		
1 Gas Engine Vibrator		31.20		34.32	.65	.71
48 L.H., Daily Totals		$2255.20		$3445.92	$46.98	$71.79

Crew No.	Bare Costs		Incl. Subs O&P		Cost Per Labor-Hour	
Crew C-14L	**Hr.**	**Daily**	**Hr.**	**Daily**	**Bare Costs**	**Incl. O&P**
1 Carpenter Foreman (outside)	$48.95	$391.60	$75.35	$602.80	$43.84	$67.24
6 Carpenters	46.95	2253.60	72.25	3468.00		
4 Laborers	37.60	1203.20	57.85	1851.20		
1 Cement Finisher	45.00	360.00	66.65	533.20		
1 Gas Engine Vibrator		31.20		34.32	.33	.36
96 L.H., Daily Totals		$4239.60		$6489.52	$44.16	$67.60

Crew No.	Bare Costs		Incl. Subs O&P		Cost Per Labor-Hour	
Crew C-14M	**Hr.**	**Daily**	**Hr.**	**Daily**	**Bare Costs**	**Incl. O&P**
1 Carpenter Foreman (outside)	$48.95	$391.60	$75.35	$602.80	$45.77	$70.13
2 Carpenters	46.95	751.20	72.25	1156.00		
1 Rodman (reinf.)	52.55	420.40	82.10	656.80		
2 Laborers	37.60	601.60	57.85	925.60		
1 Cement Finisher	45.00	360.00	66.65	533.20		
1 Equip. Oper. (medium)	50.60	404.80	76.75	614.00		
1 Gas Engine Vibrator		31.20		34.32		
1 Concrete Pump (Small)		720.00		792.00	11.74	12.91
64 L.H., Daily Totals		$3680.80		$5314.72	$57.51	$83.04

Crew No.	Bare Costs		Incl. Subs O&P		Cost Per Labor-Hour	
Crew C-15	**Hr.**	**Daily**	**Hr.**	**Daily**	**Bare Costs**	**Incl. O&P**
1 Carpenter Foreman (outside)	$48.95	$391.60	$75.35	$602.80	$44.24	$67.64
2 Carpenters	46.95	751.20	72.25	1156.00		
3 Laborers	37.60	902.40	57.85	1388.40		
2 Cement Finishers	45.00	720.00	66.65	1066.40		
1 Rodman (reinf.)	52.55	420.40	82.10	656.80		
72 L.H., Daily Totals		$3185.60		$4870.40	$44.24	$67.64

Crew No.	Bare Costs		Incl. Subs O&P		Cost Per Labor-Hour	
Crew C-16	**Hr.**	**Daily**	**Hr.**	**Daily**	**Bare Costs**	**Incl. O&P**
1 Labor Foreman (outside)	$39.60	$316.80	$60.95	$487.60	$41.86	$63.51
3 Laborers	37.60	902.40	57.85	1388.40		
2 Cement Finishers	45.00	720.00	66.65	1066.40		
1 Equip. Oper. (medium)	50.60	404.80	76.75	614.00		
1 Gunite Pump Rig		371.80		408.98		
2 -50' Air Hoses, 3/4"		6.50		7.15		
2 -50' Air Hoses, 2"		15.50		17.05	7.03	7.74
56 L.H., Daily Totals		$2737.80		$3989.58	$48.89	$71.24

Crew No.	Bare Costs		Incl. Subs O&P		Cost Per Labor-Hour	
Crew C-16A	**Hr.**	**Daily**	**Hr.**	**Daily**	**Bare Costs**	**Incl. O&P**
1 Laborer	$37.60	$300.80	$57.85	$462.80	$44.55	$66.97
2 Cement Finishers	45.00	720.00	66.65	1066.40		
1 Equip. Oper. (medium)	50.60	404.80	76.75	614.00		
1 Gunite Pump Rig		371.80		408.98		
2 -50' Air Hoses, 3/4"		6.50		7.15		
2 -50' Air Hoses, 2"		15.50		17.05		
1 Aerial Lift Truck, 60' Boom		435.60		479.16	25.92	28.51
32 L.H., Daily Totals		$2255.00		$3055.54	$70.47	$95.49

Crew No.	Bare Costs		Incl. Subs O&P		Cost Per Labor-Hour	
Crew C-17	**Hr.**	**Daily**	**Hr.**	**Daily**	**Bare Costs**	**Incl. O&P**
2 Skilled Worker Foremen (out)	$50.65	$810.40	$78.25	$1252.00	$49.05	$75.77
8 Skilled Workers	48.65	3113.60	75.15	4809.60		
80 L.H., Daily Totals		$3924.00		$6061.60	$49.05	$75.77
Crew C-17A	**Hr.**	**Daily**	**Hr.**	**Daily**	**Bare Costs**	**Incl. O&P**
2 Skilled Worker Foremen (out)	$50.65	$810.40	$78.25	$1252.00	$49.08	$75.80
8 Skilled Workers	48.65	3113.60	75.15	4809.60		
.125 Equip. Oper. (crane)	51.70	51.70	78.45	78.45		
.125 Hyd. Crane, 80 Ton		203.13		223.44	2.51	2.76
81 L.H., Daily Totals		$4178.82		$6363.49	$51.59	$78.56
Crew C-17B	**Hr.**	**Daily**	**Hr.**	**Daily**	**Bare Costs**	**Incl. O&P**
2 Skilled Worker Foremen (out)	$50.65	$810.40	$78.25	$1252.00	$49.11	$75.84
8 Skilled Workers	48.65	3113.60	75.15	4809.60		
.25 Equip. Oper. (crane)	51.70	103.40	78.45	156.90		
.25 Hyd. Crane, 80 Ton		406.25		446.88		
.25 Trowel, 48" Walk-Behind		12.55		13.81	5.11	5.62
82 L.H., Daily Totals		$4446.20		$6679.18	$54.22	$81.45
Crew C-17C	**Hr.**	**Daily**	**Hr.**	**Daily**	**Bare Costs**	**Incl. O&P**
2 Skilled Worker Foremen (out)	$50.65	$810.40	$78.25	$1252.00	$49.15	$75.87
8 Skilled Workers	48.65	3113.60	75.15	4809.60		
.375 Equip. Oper. (crane)	51.70	155.10	78.45	235.35		
.375 Hyd. Crane, 80 Ton		609.38		670.31	7.34	8.08
83 L.H., Daily Totals		$4688.48		$6967.26	$56.49	$83.94
Crew C-17D	**Hr.**	**Daily**	**Hr.**	**Daily**	**Bare Costs**	**Incl. O&P**
2 Skilled Worker Foremen (out)	$50.65	$810.40	$78.25	$1252.00	$49.18	$75.90
8 Skilled Workers	48.65	3113.60	75.15	4809.60		
.5 Equip. Oper. (crane)	51.70	206.80	78.45	313.80		
.5 Hyd. Crane, 80 Ton		812.50		893.75	9.67	10.64
84 L.H., Daily Totals		$4943.30		$7269.15	$58.85	$86.54
Crew C-17E	**Hr.**	**Daily**	**Hr.**	**Daily**	**Bare Costs**	**Incl. O&P**
2 Skilled Worker Foremen (out)	$50.65	$810.40	$78.25	$1252.00	$49.05	$75.77
8 Skilled Workers	48.65	3113.60	75.15	4809.60		
1 Hyd. Jack with Rods		96.50		106.15	1.21	1.33
80 L.H., Daily Totals		$4020.50		$6167.75	$50.26	$77.10
Crew C-18	**Hr.**	**Daily**	**Hr.**	**Daily**	**Bare Costs**	**Incl. O&P**
.125 Labor Foreman (outside)	$39.60	$39.60	$60.95	$60.95	$37.82	$58.19
1 Laborer	37.60	300.80	57.85	462.80		
1 Concrete Cart, 10 C.F.		60.00		66.00	6.67	7.33
9 L.H., Daily Totals		$400.40		$589.75	$44.49	$65.53
Crew C-19	**Hr.**	**Daily**	**Hr.**	**Daily**	**Bare Costs**	**Incl. O&P**
.125 Labor Foreman (outside)	$39.60	$39.60	$60.95	$60.95	$37.82	$58.19
1 Laborer	37.60	300.80	57.85	462.80		
1 Concrete Cart, 18 C.F.		99.80		109.78	11.09	12.20
9 L.H., Daily Totals		$440.20		$633.53	$48.91	$70.39
Crew C-20	**Hr.**	**Daily**	**Hr.**	**Daily**	**Bare Costs**	**Incl. O&P**
1 Labor Foreman (outside)	$39.60	$316.80	$60.95	$487.60	$40.40	$61.70
5 Laborers	37.60	1504.00	57.85	2314.00		
1 Cement Finisher	45.00	360.00	66.65	533.20		
1 Equip. Oper. (medium)	50.60	404.80	76.75	614.00		
2 Gas Engine Vibrators		62.40		68.64		
1 Concrete Pump (Small)		720.00		792.00	12.23	13.45
64 L.H., Daily Totals		$3368.00		$4809.44	$52.63	$75.15

Crew No.	Bare Costs		Incl. Subs O&P		Cost Per Labor-Hour	
Crew C-21	**Hr.**	**Daily**	**Hr.**	**Daily**	**Bare Costs**	**Incl. O&P**
1 Labor Foreman (outside)	$39.60	$316.80	$60.95	$487.60	$40.40	$61.70
5 Laborers	37.60	1504.00	57.85	2314.00		
1 Cement Finisher	45.00	360.00	66.65	533.20		
1 Equip. Oper. (medium)	50.60	404.80	76.75	614.00		
2 Gas Engine Vibrators		62.40		68.64		
1 Concrete Conveyer		198.80		218.68	4.08	4.49
64 L.H., Daily Totals		$2846.80		$4236.12	$44.48	$66.19
Crew C-22	**Hr.**	**Daily**	**Hr.**	**Daily**	**Bare Costs**	**Incl. O&P**
1 Rodman Foreman (outside)	$54.55	$436.40	$85.20	$681.60	$52.74	$82.28
4 Rodmen (reinf.)	52.55	1681.60	82.10	2627.20		
.125 Equip. Oper. (crane)	51.70	51.70	78.45	78.45		
.125 Equip. Oper. (oiler)	45.20	45.20	68.55	68.55		
.125 Hyd. Crane, 25 Ton		92.08		101.28	2.19	2.41
42 L.H., Daily Totals		$2306.97		$3557.08	$54.93	$84.69
Crew C-23	**Hr.**	**Daily**	**Hr.**	**Daily**	**Bare Costs**	**Incl. O&P**
2 Skilled Worker Foremen (out)	$50.65	$810.40	$78.25	$1252.00	$49.01	$75.44
6 Skilled Workers	48.65	2335.20	75.15	3607.20		
1 Equip. Oper. (crane)	51.70	413.60	78.45	627.60		
1 Equip. Oper. (oiler)	45.20	361.60	68.55	548.40		
1 Lattice Boom Crane, 90 Ton		1511.00		1662.10	18.89	20.78
80 L.H., Daily Totals		$5431.80		$7697.30	$67.90	$96.22
Crew C-23A	**Hr.**	**Daily**	**Hr.**	**Daily**	**Bare Costs**	**Incl. O&P**
1 Labor Foreman (outside)	$39.60	$316.80	$60.95	$487.60	$42.34	$64.73
2 Laborers	37.60	601.60	57.85	925.60		
1 Equip. Oper. (crane)	51.70	413.60	78.45	627.60		
1 Equip. Oper. (oiler)	45.20	361.60	68.55	548.40		
1 Crawler Crane, 100 Ton		1667.00		1833.70		
3 Conc. Buckets, 8 C.Y.		608.40		669.24	56.88	62.57
40 L.H., Daily Totals		$3969.00		$5092.14	$99.22	$127.30
Crew C-24	**Hr.**	**Daily**	**Hr.**	**Daily**	**Bare Costs**	**Incl. O&P**
2 Skilled Worker Foremen (out)	$50.65	$810.40	$78.25	$1252.00	$49.01	$75.44
6 Skilled Workers	48.65	2335.20	75.15	3607.20		
1 Equip. Oper. (crane)	51.70	413.60	78.45	627.60		
1 Equip. Oper. (oiler)	45.20	361.60	68.55	548.40		
1 Lattice Boom Crane, 150 Ton		1813.00		1994.30	22.66	24.93
80 L.H., Daily Totals		$5733.80		$8029.50	$71.67	$100.37
Crew C-25	**Hr.**	**Daily**	**Hr.**	**Daily**	**Bare Costs**	**Incl. O&P**
2 Rodmen (reinf.)	$52.55	$840.80	$82.10	$1313.60	$41.30	$66.70
2 Rodmen Helpers	30.05	480.80	51.30	820.80		
32 L.H., Daily Totals		$1321.60		$2134.40	$41.30	$66.70
Crew C-27	**Hr.**	**Daily**	**Hr.**	**Daily**	**Bare Costs**	**Incl. O&P**
2 Cement Finishers	$45.00	$720.00	$66.65	$1066.40	$45.00	$66.65
1 Concrete Saw		167.40		184.14	10.46	11.51
16 L.H., Daily Totals		$887.40		$1250.54	$55.46	$78.16
Crew C-28	**Hr.**	**Daily**	**Hr.**	**Daily**	**Bare Costs**	**Incl. O&P**
1 Cement Finisher	$45.00	$360.00	$66.65	$533.20	$45.00	$66.65
1 Portable Air Compressor, Gas		17.90		19.69	2.24	2.46
8 L.H., Daily Totals		$377.90		$552.89	$47.24	$69.11
Crew C-29	**Hr.**	**Daily**	**Hr.**	**Daily**	**Bare Costs**	**Incl. O&P**
1 Laborer	$37.60	$300.80	$57.85	$462.80	$37.60	$57.85
1 Pressure Washer		70.20		77.22	8.78	9.65
8 L.H., Daily Totals		$371.00		$540.02	$46.38	$67.50

Crew No.	Bare Costs		Incl. Subs O&P		Cost Per Labor-Hour	
Crew C-30	**Hr.**	**Daily**	**Hr.**	**Daily**	**Bare Costs**	**Incl. O&P**
1 Laborer	$37.60	$300.80	$57.85	$462.80	$37.60	$57.85
1 Concrete Mixer, 10 C.F.		172.40		189.64	21.55	23.70
8 L.H., Daily Totals		$473.20		$652.44	$59.15	$81.56
Crew C-31	**Hr.**	**Daily**	**Hr.**	**Daily**	**Bare Costs**	**Incl. O&P**
1 Cement Finisher	$45.00	$360.00	$66.65	$533.20	$45.00	$66.65
1 Grout Pump		371.80		408.98	46.48	51.12
8 L.H., Daily Totals		$731.80		$942.18	$91.47	$117.77
Crew C-32	**Hr.**	**Daily**	**Hr.**	**Daily**	**Bare Costs**	**Incl. O&P**
1 Cement Finisher	$45.00	$360.00	$66.65	$533.20	$41.30	$62.25
1 Laborer	37.60	300.80	57.85	462.80		
1 Crack Chaser Saw, Gas, 6 H.P.		29.20		32.12		
1 Vacuum Pick-Up System		60.90		66.99	5.63	6.19
16 L.H., Daily Totals		$750.90		$1095.11	$46.93	$68.44
Crew D-1	**Hr.**	**Daily**	**Hr.**	**Daily**	**Bare Costs**	**Incl. O&P**
1 Bricklayer	$46.15	$369.20	$70.45	$563.60	$42.10	$64.28
1 Bricklayer Helper	38.05	304.40	58.10	464.80		
16 L.H., Daily Totals		$673.60		$1028.40	$42.10	$64.28
Crew D-2	**Hr.**	**Daily**	**Hr.**	**Daily**	**Bare Costs**	**Incl. O&P**
3 Bricklayers	$46.15	$1107.60	$70.45	$1690.80	$43.28	$66.12
2 Bricklayer Helpers	38.05	608.80	58.10	929.60		
.5 Carpenter	46.95	187.80	72.25	289.00		
44 L.H., Daily Totals		$1904.20		$2909.40	$43.28	$66.12
Crew D-3	**Hr.**	**Daily**	**Hr.**	**Daily**	**Bare Costs**	**Incl. O&P**
3 Bricklayers	$46.15	$1107.60	$70.45	$1690.80	$43.10	$65.83
2 Bricklayer Helpers	38.05	608.80	58.10	929.60		
.25 Carpenter	46.95	93.90	72.25	144.50		
42 L.H., Daily Totals		$1810.30		$2764.90	$43.10	$65.83
Crew D-4	**Hr.**	**Daily**	**Hr.**	**Daily**	**Bare Costs**	**Incl. O&P**
1 Bricklayer	$46.15	$369.20	$70.45	$563.60	$42.71	$65.10
2 Bricklayer Helpers	38.05	608.80	58.10	929.60		
1 Equip. Oper. (light)	48.60	388.80	73.75	590.00		
1 Grout Pump, 50 C.F./hr.		133.40		146.74	4.17	4.59
32 L.H., Daily Totals		$1500.20		$2229.94	$46.88	$69.69
Crew D-5	**Hr.**	**Daily**	**Hr.**	**Daily**	**Bare Costs**	**Incl. O&P**
1 Bricklayer	46.15	369.20	70.45	563.60	46.15	70.45
8 L.H., Daily Totals		$369.20		$563.60	$46.15	$70.45
Crew D-6	**Hr.**	**Daily**	**Hr.**	**Daily**	**Bare Costs**	**Incl. O&P**
3 Bricklayers	$46.15	$1107.60	$70.45	$1690.80	$42.29	$64.59
3 Bricklayer Helpers	38.05	913.20	58.10	1394.40		
.25 Carpenter	46.95	93.90	72.25	144.50		
50 L.H., Daily Totals		$2114.70		$3229.70	$42.29	$64.59
Crew D-7	**Hr.**	**Daily**	**Hr.**	**Daily**	**Bare Costs**	**Incl. O&P**
1 Tile Layer	$42.80	$342.40	$63.30	$506.40	$38.17	$56.45
1 Tile Layer Helper	33.55	268.40	49.60	396.80		
16 L.H., Daily Totals		$610.80		$903.20	$38.17	$56.45
Crew D-8	**Hr.**	**Daily**	**Hr.**	**Daily**	**Bare Costs**	**Incl. O&P**
3 Bricklayers	$46.15	$1107.60	$70.45	$1690.80	$42.91	$65.51
2 Bricklayer Helpers	38.05	608.80	58.10	929.60		
40 L.H., Daily Totals		$1716.40		$2620.40	$42.91	$65.51

Crew No.	Bare Costs		Incl. Subs O&P		Cost Per Labor-Hour	
Crew D-9	**Hr.**	**Daily**	**Hr.**	**Daily**	**Bare Costs**	**Incl. O&P**
3 Bricklayers	$46.15	$1107.60	$70.45	$1690.80	$42.10	$64.28
3 Bricklayer Helpers	38.05	913.20	58.10	1394.40		
48 L.H., Daily Totals		$2020.80		$3085.20	$42.10	$64.28
Crew D-10	**Hr.**	**Daily**	**Hr.**	**Daily**	**Bare Costs**	**Incl. O&P**
1 Bricklayer Foreman (outside)	$48.15	$385.20	$73.55	$588.40	$46.01	$70.14
1 Bricklayer	46.15	369.20	70.45	563.60		
1 Bricklayer Helper	38.05	304.40	58.10	464.80		
1 Equip. Oper. (crane)	51.70	413.60	78.45	627.60		
1 S.P. Crane, 4x4, 12 Ton		473.40		520.74	14.79	16.27
32 L.H., Daily Totals		$1945.80		$2765.14	$60.81	$86.41
Crew D-11	**Hr.**	**Daily**	**Hr.**	**Daily**	**Bare Costs**	**Incl. O&P**
1 Bricklayer Foreman (outside)	$48.15	$385.20	$73.55	$588.40	$44.12	$67.37
1 Bricklayer	46.15	369.20	70.45	563.60		
1 Bricklayer Helper	38.05	304.40	58.10	464.80		
24 L.H., Daily Totals		$1058.80		$1616.80	$44.12	$67.37
Crew D-12	**Hr.**	**Daily**	**Hr.**	**Daily**	**Bare Costs**	**Incl. O&P**
1 Bricklayer Foreman (outside)	$48.15	$385.20	$73.55	$588.40	$42.60	$65.05
1 Bricklayer	46.15	369.20	70.45	563.60		
2 Bricklayer Helpers	38.05	608.80	58.10	929.60		
32 L.H., Daily Totals		$1363.20		$2081.60	$42.60	$65.05
Crew D-13	**Hr.**	**Daily**	**Hr.**	**Daily**	**Bare Costs**	**Incl. O&P**
1 Bricklayer Foreman (outside)	$48.15	$385.20	$73.55	$588.40	$44.84	$68.48
1 Bricklayer	46.15	369.20	70.45	563.60		
2 Bricklayer Helpers	38.05	608.80	58.10	929.60		
1 Carpenter	46.95	375.60	72.25	578.00		
1 Equip. Oper. (crane)	51.70	413.60	78.45	627.60		
1 S.P. Crane, 4x4, 12 Ton		473.40		520.74	9.86	10.85
48 L.H., Daily Totals		$2625.80		$3807.94	$54.70	$79.33
Crew E-1	**Hr.**	**Daily**	**Hr.**	**Daily**	**Bare Costs**	**Incl. O&P**
1 Welder Foreman (outside)	$54.65	$437.20	$94.95	$759.60	$51.97	$86.72
1 Welder	52.65	421.20	91.45	731.60		
1 Equip. Oper. (light)	48.60	388.80	73.75	590.00		
1 Welder, Gas Engine, 300 amp		145.85		160.44	6.08	6.68
24 L.H., Daily Totals		$1393.05		$2241.64	$58.04	$93.40
Crew E-2	**Hr.**	**Daily**	**Hr.**	**Daily**	**Bare Costs**	**Incl. O&P**
1 Struc. Steel Foreman (outside)	$54.65	$437.20	$94.95	$759.60	$51.74	$86.82
4 Struc. Steel Workers	52.65	1684.80	91.45	2926.40		
1 Equip. Oper. (crane)	51.70	413.60	78.45	627.60		
1 Equip. Oper. (oiler)	45.20	361.60	68.55	548.40		
1 Lattice Boom Crane, 90 Ton		1511.00		1662.10	26.98	29.68
56 L.H., Daily Totals		$4408.20		$6524.10	$78.72	$116.50
Crew E-3	**Hr.**	**Daily**	**Hr.**	**Daily**	**Bare Costs**	**Incl. O&P**
1 Struc. Steel Foreman (outside)	$54.65	$437.20	$94.95	$759.60	$53.32	$92.62
1 Struc. Steel Worker	52.65	421.20	91.45	731.60		
1 Welder	52.65	421.20	91.45	731.60		
1 Welder, Gas Engine, 300 amp		145.85		160.44	6.08	6.68
24 L.H., Daily Totals		$1425.45		$2383.24	$59.39	$99.30

Crew No.	Bare Costs Hr.	Bare Costs Daily	Incl. Subs O&P Hr.	Incl. Subs O&P Daily	Cost Per Labor-Hour Bare Costs	Cost Per Labor-Hour Incl. O&P
Crew E-3A	**Hr.**	**Daily**	**Hr.**	**Daily**	**Bare Costs**	**Incl. O&P**
1 Struc. Steel Foreman (outside)	$54.65	$437.20	$94.95	$759.60	$53.32	$92.62
1 Struc. Steel Worker	52.65	421.20	91.45	731.60		
1 Welder	52.65	421.20	91.45	731.60		
1 Welder, Gas Engine, 300 amp		145.85		160.44		
1 Aerial Lift Truck, 40' Boom		301.40		331.54	18.64	20.50
24 L.H., Daily Totals		$1726.85		$2714.78	$71.95	$113.12
Crew E-4	**Hr.**	**Daily**	**Hr.**	**Daily**	**Bare Costs**	**Incl. O&P**
1 Struc. Steel Foreman (outside)	$54.65	$437.20	$94.95	$759.60	$53.15	$92.33
3 Struc. Steel Workers	52.65	1263.60	91.45	2194.80		
1 Welder, Gas Engine, 300 amp		145.85		160.44	4.56	5.01
32 L.H., Daily Totals		$1846.65		$3114.84	$57.71	$97.34
Crew E-5	**Hr.**	**Daily**	**Hr.**	**Daily**	**Bare Costs**	**Incl. O&P**
2 Struc. Steel Foremen (outside)	$54.65	$874.40	$94.95	$1519.20	$52.21	$88.56
5 Struc. Steel Workers	52.65	2106.00	91.45	3658.00		
1 Equip. Oper. (crane)	51.70	413.60	78.45	627.60		
1 Welder	52.65	421.20	91.45	731.60		
1 Equip. Oper. (oiler)	45.20	361.60	68.55	548.40		
1 Lattice Boom Crane, 90 Ton		1511.00		1662.10		
1 Welder, Gas Engine, 300 amp		145.85		160.44	20.71	22.78
80 L.H., Daily Totals		$5833.65		$8907.33	$72.92	$111.34
Crew E-6	**Hr.**	**Daily**	**Hr.**	**Daily**	**Bare Costs**	**Incl. O&P**
3 Struc. Steel Foremen (outside)	$54.65	$1311.60	$94.95	$2278.80	$52.25	$88.76
9 Struc. Steel Workers	52.65	3790.80	91.45	6584.40		
1 Equip. Oper. (crane)	51.70	413.60	78.45	627.60		
1 Welder	52.65	421.20	91.45	731.60		
1 Equip. Oper. (oiler)	45.20	361.60	68.55	548.40		
1 Equip. Oper. (light)	48.60	388.80	73.75	590.00		
1 Lattice Boom Crane, 90 Ton		1511.00		1662.10		
1 Welder, Gas Engine, 300 amp		145.85		160.44		
1 Air Compressor, 160 cfm		156.40		172.04		
2 Impact Wrenches		36.00		39.60	14.45	15.89
128 L.H., Daily Totals		$8536.85		$13394.98	$66.69	$104.65
Crew E-7	**Hr.**	**Daily**	**Hr.**	**Daily**	**Bare Costs**	**Incl. O&P**
1 Struc. Steel Foreman (outside)	$54.65	$437.20	$94.95	$759.60	$52.21	$88.56
4 Struc. Steel Workers	52.65	1684.80	91.45	2926.40		
1 Equip. Oper. (crane)	51.70	413.60	78.45	627.60		
1 Equip. Oper. (oiler)	45.20	361.60	68.55	548.40		
1 Welder Foreman (outside)	54.65	437.20	94.95	759.60		
2 Welders	52.65	842.40	91.45	1463.20		
1 Lattice Boom Crane, 90 Ton		1511.00		1662.10		
2 Welder, Gas Engine, 300 amp		291.70		320.87	22.53	24.79
80 L.H., Daily Totals		$5979.50		$9067.77	$74.74	$113.35
Crew E-8	**Hr.**	**Daily**	**Hr.**	**Daily**	**Bare Costs**	**Incl. O&P**
1 Struc. Steel Foreman (outside)	$54.65	$437.20	$94.95	$759.60	$52.00	$87.87
4 Struc. Steel Workers	52.65	1684.80	91.45	2926.40		
1 Welder Foreman (outside)	54.65	437.20	94.95	759.60		
4 Welders	52.65	1684.80	91.45	2926.40		
1 Equip. Oper. (crane)	51.70	413.60	78.45	627.60		
1 Equip. Oper. (oiler)	45.20	361.60	68.55	548.40		
1 Equip. Oper. (light)	48.60	388.80	73.75	590.00		
1 Lattice Boom Crane, 90 Ton		1511.00		1662.10		
4 Welder, Gas Engine, 300 amp		583.40		641.74	20.14	22.15
104 L.H., Daily Totals		$7502.40		$11441.84	$72.14	$110.02

Crew No.	Bare Costs Hr.	Bare Costs Daily	Incl. Subs O&P Hr.	Incl. Subs O&P Daily	Cost Per Labor-Hour Bare Costs	Cost Per Labor-Hour Incl. O&P
Crew E-9	**Hr.**	**Daily**	**Hr.**	**Daily**	**Bare Costs**	**Incl. O&P**
2 Struc. Steel Foremen (outside)	$54.65	$874.40	$94.95	$1519.20	$52.25	$88.76
5 Struc. Steel Workers	52.65	2106.00	91.45	3658.00		
1 Welder Foreman (outside)	54.65	437.20	94.95	759.60		
5 Welders	52.65	2106.00	91.45	3658.00		
1 Equip. Oper. (crane)	51.70	413.60	78.45	627.60		
1 Equip. Oper. (oiler)	45.20	361.60	68.55	548.40		
1 Equip. Oper. (light)	48.60	388.80	73.75	590.00		
1 Lattice Boom Crane, 90 Ton		1511.00		1662.10		
5 Welder, Gas Engine, 300 amp		729.25		802.17	17.50	19.25
128 L.H., Daily Totals		$8927.85		$13825.08	$69.75	$108.01
Crew E-10	**Hr.**	**Daily**	**Hr.**	**Daily**	**Bare Costs**	**Incl. O&P**
1 Welder Foreman (outside)	$54.65	$437.20	$94.95	$759.60	$53.65	$93.20
1 Welder	52.65	421.20	91.45	731.60		
1 Welder, Gas Engine, 300 amp		145.85		160.44		
1 Flatbed Truck, Gas, 3 Ton		303.20		333.52	28.07	30.87
16 L.H., Daily Totals		$1307.45		$1985.16	$81.72	$124.07
Crew E-11	**Hr.**	**Daily**	**Hr.**	**Daily**	**Bare Costs**	**Incl. O&P**
2 Painters, Struc. Steel	$41.45	$663.20	$72.95	$1167.20	$42.27	$69.38
1 Building Laborer	37.60	300.80	57.85	462.80		
1 Equip. Oper. (light)	48.60	388.80	73.75	590.00		
1 Air Compressor, 250 cfm		201.40		221.54		
1 Sandblaster, Portable, 3 C.F.		20.40		22.44		
1 Set Sand Blasting Accessories		14.05		15.46	7.37	8.11
32 L.H., Daily Totals		$1588.65		$2479.43	$49.65	$77.48
Crew E-11A	**Hr.**	**Daily**	**Hr.**	**Daily**	**Bare Costs**	**Incl. O&P**
2 Painters, Struc. Steel	$41.45	$663.20	$72.95	$1167.20	$42.27	$69.38
1 Building Laborer	37.60	300.80	57.85	462.80		
1 Equip. Oper. (light)	48.60	388.80	73.75	590.00		
1 Air Compressor, 250 cfm		201.40		221.54		
1 Sandblaster, Portable, 3 C.F.		20.40		22.44		
1 Set Sand Blasting Accessories		14.05		15.46		
1 Aerial Lift Truck, 60' Boom		435.60		479.16	20.98	23.08
32 L.H., Daily Totals		$2024.25		$2958.59	$63.26	$92.46
Crew E-11B	**Hr.**	**Daily**	**Hr.**	**Daily**	**Bare Costs**	**Incl. O&P**
2 Painters, Struc. Steel	$41.45	$663.20	$72.95	$1167.20	$40.17	$67.92
1 Building Laborer	37.60	300.80	57.85	462.80		
2 Paint Sprayer, 8 C.F.M.		99.80		109.78		
1 Aerial Lift Truck, 60' Boom		435.60		479.16	22.31	24.54
24 L.H., Daily Totals		$1499.40		$2218.94	$62.48	$92.46
Crew E-12	**Hr.**	**Daily**	**Hr.**	**Daily**	**Bare Costs**	**Incl. O&P**
1 Welder Foreman (outside)	$54.65	$437.20	$94.95	$759.60	$51.63	$84.35
1 Equip. Oper. (light)	48.60	388.80	73.75	590.00		
1 Welder, Gas Engine, 300 amp		145.85		160.44	9.12	10.03
16 L.H., Daily Totals		$971.85		$1510.04	$60.74	$94.38
Crew E-13	**Hr.**	**Daily**	**Hr.**	**Daily**	**Bare Costs**	**Incl. O&P**
1 Welder Foreman (outside)	$54.65	$437.20	$94.95	$759.60	$52.63	$87.88
.5 Equip. Oper. (light)	48.60	194.40	73.75	295.00		
1 Welder, Gas Engine, 300 amp		145.85		160.44	12.15	13.37
12 L.H., Daily Totals		$777.45		$1215.04	$64.79	$101.25
Crew E-14	**Hr.**	**Daily**	**Hr.**	**Daily**	**Bare Costs**	**Incl. O&P**
1 Welder Foreman (outside)	$54.65	$437.20	$94.95	$759.60	$54.65	$94.95
1 Welder, Gas Engine, 300 amp		145.85		160.44	18.23	20.05
8 L.H., Daily Totals		$583.05		$920.03	$72.88	$115.00

Crew No.	Bare Costs		Incl. Subs O&P		Cost Per Labor-Hour	
Crew E-16	**Hr.**	**Daily**	**Hr.**	**Daily**	**Bare Costs**	**Incl. O&P**
1 Welder Foreman (outside)	$54.65	$437.20	$94.95	$759.60	$53.65	$93.20
1 Welder	52.65	421.20	91.45	731.60		
1 Welder, Gas Engine, 300 amp		145.85		160.44	9.12	10.03
16 L.H., Daily Totals		$1004.25		$1651.64	$62.77	$103.23
Crew E-17	**Hr.**	**Daily**	**Hr.**	**Daily**	**Bare Costs**	**Incl. O&P**
1 Struc. Steel Foreman (outside)	$54.65	$437.20	$94.95	$759.60	$53.65	$93.20
1 Structural Steel Worker	52.65	421.20	91.45	731.60		
16 L.H., Daily Totals		$858.40		$1491.20	$53.65	$93.20
Crew E-18	**Hr.**	**Daily**	**Hr.**	**Daily**	**Bare Costs**	**Incl. O&P**
1 Struc. Steel Foreman (outside)	$54.65	$437.20	$94.95	$759.60	$52.64	$89.21
3 Structural Steel Workers	52.65	1263.60	91.45	2194.80		
1 Equipment Operator (med.)	50.60	404.80	76.75	614.00		
1 Lattice Boom Crane, 20 Ton		948.90		1043.79	23.72	26.09
40 L.H., Daily Totals		$3054.50		$4612.19	$76.36	$115.30
Crew E-19	**Hr.**	**Daily**	**Hr.**	**Daily**	**Bare Costs**	**Incl. O&P**
1 Struc. Steel Foreman (outside)	$54.65	$437.20	$94.95	$759.60	$51.97	$86.72
1 Structural Steel Worker	52.65	421.20	91.45	731.60		
1 Equip. Oper. (light)	48.60	388.80	73.75	590.00		
1 Lattice Boom Crane, 20 Ton		948.90		1043.79	39.54	43.49
24 L.H., Daily Totals		$2196.10		$3124.99	$91.50	$130.21
Crew E-20	**Hr.**	**Daily**	**Hr.**	**Daily**	**Bare Costs**	**Incl. O&P**
1 Struc. Steel Foreman (outside)	$54.65	$437.20	$94.95	$759.60	$51.85	$87.40
5 Structural Steel Workers	52.65	2106.00	91.45	3658.00		
1 Equip. Oper. (crane)	51.70	413.60	78.45	627.60		
1 Equip. Oper. (oiler)	45.20	361.60	68.55	548.40		
1 Lattice Boom Crane, 40 Ton		1164.00		1280.40	18.19	20.01
64 L.H., Daily Totals		$4482.40		$6874.00	$70.04	$107.41
Crew E-22	**Hr.**	**Daily**	**Hr.**	**Daily**	**Bare Costs**	**Incl. O&P**
1 Skilled Worker Foreman (out)	$50.65	$405.20	$78.25	$626.00	$49.32	$76.18
2 Skilled Workers	48.65	778.40	75.15	1202.40		
24 L.H., Daily Totals		$1183.60		$1828.40	$49.32	$76.18
Crew E-24	**Hr.**	**Daily**	**Hr.**	**Daily**	**Bare Costs**	**Incl. O&P**
3 Structural Steel Workers	$52.65	$1263.60	$91.45	$2194.80	$52.14	$87.78
1 Equipment Operator (med.)	50.60	404.80	76.75	614.00		
1 Hyd. Crane, 25 Ton		736.60		810.26	23.02	25.32
32 L.H., Daily Totals		$2405.00		$3619.06	$75.16	$113.10
Crew E-25	**Hr.**	**Daily**	**Hr.**	**Daily**	**Bare Costs**	**Incl. O&P**
1 Welder Foreman (outside)	$54.65	$437.20	$94.95	$759.60	$54.65	$94.95
1 Cutting Torch		11.40		12.54	1.43	1.57
8 L.H., Daily Totals		$448.60		$772.14	$56.08	$96.52
Crew F-3	**Hr.**	**Daily**	**Hr.**	**Daily**	**Bare Costs**	**Incl. O&P**
4 Carpenters	$46.95	$1502.40	$72.25	$2312.00	$47.90	$73.49
1 Equip. Oper. (crane)	51.70	413.60	78.45	627.60		
1 Hyd. Crane, 12 Ton		653.80		719.18	16.34	17.98
40 L.H., Daily Totals		$2569.80		$3658.78	$64.25	$91.47
Crew F-4	**Hr.**	**Daily**	**Hr.**	**Daily**	**Bare Costs**	**Incl. O&P**
4 Carpenters	$46.95	$1502.40	$72.25	$2312.00	$47.45	$72.67
1 Equip. Oper. (crane)	51.70	413.60	78.45	627.60		
1 Equip. Oper. (oiler)	45.20	361.60	68.55	548.40		
1 Hyd. Crane, 55 Ton		1128.00		1240.80	23.50	25.85
48 L.H., Daily Totals		$3405.60		$4728.80	$70.95	$98.52

Crew No.	Bare Costs		Incl. Subs O&P		Cost Per Labor-Hour	
Crew F-5	**Hr.**	**Daily**	**Hr.**	**Daily**	**Bare Costs**	**Incl. O&P**
1 Carpenter Foreman (outside)	$48.95	$391.60	$75.35	$602.80	$47.45	$73.03
3 Carpenters	46.95	1126.80	72.25	1734.00		
32 L.H., Daily Totals		$1518.40		$2336.80	$47.45	$73.03
Crew F-6	**Hr.**	**Daily**	**Hr.**	**Daily**	**Bare Costs**	**Incl. O&P**
2 Carpenters	$46.95	$751.20	$72.25	$1156.00	$44.16	$67.73
2 Building Laborers	37.60	601.60	57.85	925.60		
1 Equip. Oper. (crane)	51.70	413.60	78.45	627.60		
1 Hyd. Crane, 12 Ton		653.80		719.18	16.34	17.98
40 L.H., Daily Totals		$2420.20		$3428.38	$60.51	$85.71
Crew F-7	**Hr.**	**Daily**	**Hr.**	**Daily**	**Bare Costs**	**Incl. O&P**
2 Carpenters	$46.95	$751.20	$72.25	$1156.00	$42.27	$65.05
2 Building Laborers	37.60	601.60	57.85	925.60		
32 L.H., Daily Totals		$1352.80		$2081.60	$42.27	$65.05
Crew G-1	**Hr.**	**Daily**	**Hr.**	**Daily**	**Bare Costs**	**Incl. O&P**
1 Roofer Foreman (outside)	$42.10	$336.80	$71.85	$574.80	$37.51	$64.04
4 Roofers Composition	40.10	1283.20	68.45	2190.40		
2 Roofer Helpers	30.05	480.80	51.30	820.80		
1 Application Equipment		193.00		212.30		
1 Tar Kettle/Pot		169.80		186.78		
1 Crew Truck		206.40		227.04	10.16	11.18
56 L.H., Daily Totals		$2670.00		$4212.12	$47.68	$75.22
Crew G-2	**Hr.**	**Daily**	**Hr.**	**Daily**	**Bare Costs**	**Incl. O&P**
1 Plasterer	$42.95	$343.60	$64.40	$515.20	$39.55	$59.78
1 Plasterer Helper	38.10	304.80	57.10	456.80		
1 Building Laborer	37.60	300.80	57.85	462.80		
1 Grout Pump, 50 C.F./hr.		133.40		146.74	5.56	6.11
24 L.H., Daily Totals		$1082.60		$1581.54	$45.11	$65.90
Crew G-2A	**Hr.**	**Daily**	**Hr.**	**Daily**	**Bare Costs**	**Incl. O&P**
1 Roofer Composition	$40.10	$320.80	$68.45	$547.60	$35.92	$59.20
1 Roofer Helper	30.05	240.40	51.30	410.40		
1 Building Laborer	37.60	300.80	57.85	462.80		
1 Foam Spray Rig, Trailer-Mtd.		573.25		630.58		
1 Pickup Truck, 3/4 Ton		144.20		158.62	29.89	32.88
24 L.H., Daily Totals		$1579.45		$2209.99	$65.81	$92.08
Crew G-3	**Hr.**	**Daily**	**Hr.**	**Daily**	**Bare Costs**	**Incl. O&P**
2 Sheet Metal Workers	$55.95	$895.20	$85.50	$1368.00	$46.77	$71.67
2 Building Laborers	37.60	601.60	57.85	925.60		
32 L.H., Daily Totals		$1496.80		$2293.60	$46.77	$71.67
Crew G-4	**Hr.**	**Daily**	**Hr.**	**Daily**	**Bare Costs**	**Incl. O&P**
1 Labor Foreman (outside)	$39.60	$316.80	$60.95	$487.60	$38.27	$58.88
2 Building Laborers	37.60	601.60	57.85	925.60		
1 Flatbed Truck, Gas, 1.5 Ton		245.40		269.94		
1 Air Compressor, 160 cfm		156.40		172.04	16.74	18.42
24 L.H., Daily Totals		$1320.20		$1855.18	$55.01	$77.30
Crew G-5	**Hr.**	**Daily**	**Hr.**	**Daily**	**Bare Costs**	**Incl. O&P**
1 Roofer Foreman (outside)	$42.10	$336.80	$71.85	$574.80	$36.48	$62.27
2 Roofers Composition	40.10	641.60	68.45	1095.20		
2 Roofer Helpers	30.05	480.80	51.30	820.80		
1 Application Equipment		193.00		212.30	4.83	5.31
40 L.H., Daily Totals		$1652.20		$2703.10	$41.31	$67.58

Crew No.	Bare Costs		Incl. Subs O&P		Cost Per Labor-Hour	
Crew G-6A	**Hr.**	**Daily**	**Hr.**	**Daily**	**Bare Costs**	**Incl. O&P**
2 Roofers Composition	$40.10	$641.60	$68.45	$1095.20	$40.10	$68.45
1 Small Compressor, Electric		12.95		14.24		
2 Pneumatic Nailers		55.30		60.83	4.27	4.69
16 L.H., Daily Totals		$709.85		$1170.28	$44.37	$73.14

Crew G-7	Hr.	Daily	Hr.	Daily	Bare Costs	Incl. O&P
1 Carpenter	$46.95	$375.60	$72.25	$578.00	$46.95	$72.25
1 Small Compressor, Electric		12.95		14.24		
1 Pneumatic Nailer		27.65		30.41	5.08	5.58
8 L.H., Daily Totals		$416.20		$622.66	$52.02	$77.83

Crew H-1	Hr.	Daily	Hr.	Daily	Bare Costs	Incl. O&P
2 Glaziers	$45.10	$721.60	$68.70	$1099.20	$48.88	$80.08
2 Struc. Steel Workers	52.65	842.40	91.45	1463.20		
32 L.H., Daily Totals		$1564.00		$2562.40	$48.88	$80.08

Crew H-2	Hr.	Daily	Hr.	Daily	Bare Costs	Incl. O&P
2 Glaziers	$45.10	$721.60	$68.70	$1099.20	$42.60	$65.08
1 Building Laborer	37.60	300.80	57.85	462.80		
24 L.H., Daily Totals		$1022.40		$1562.00	$42.60	$65.08

Crew H-3	Hr.	Daily	Hr.	Daily	Bare Costs	Incl. O&P
1 Glazier	$45.10	$360.80	$68.70	$549.60	$40.27	$61.83
1 Helper	35.45	283.60	54.95	439.60		
16 L.H., Daily Totals		$644.40		$989.20	$40.27	$61.83

Crew H-4	Hr.	Daily	Hr.	Daily	Bare Costs	Incl. O&P
1 Carpenter	$46.95	$375.60	$72.25	$578.00	$43.90	$67.27
1 Carpenter Helper	35.45	283.60	54.95	439.60		
.5 Electrician	54.70	218.80	81.95	327.80		
20 L.H., Daily Totals		$878.00		$1345.40	$43.90	$67.27

Crew J-1	Hr.	Daily	Hr.	Daily	Bare Costs	Incl. O&P
3 Plasterers	$42.95	$1030.80	$64.40	$1545.60	$41.01	$61.48
2 Plasterer Helpers	38.10	609.60	57.10	913.60		
1 Mixing Machine, 6 C.F.		140.20		154.22	3.50	3.86
40 L.H., Daily Totals		$1780.60		$2613.42	$44.52	$65.34

Crew J-2	Hr.	Daily	Hr.	Daily	Bare Costs	Incl. O&P
3 Plasterers	$42.95	$1030.80	$64.40	$1545.60	$41.34	$61.80
2 Plasterer Helpers	38.10	609.60	57.10	913.60		
1 Lather	43.00	344.00	63.40	507.20		
1 Mixing Machine, 6 C.F.		140.20		154.22	2.92	3.21
48 L.H., Daily Totals		$2124.60		$3120.62	$44.26	$65.01

Crew J-3	Hr.	Daily	Hr.	Daily	Bare Costs	Incl. O&P
1 Terrazzo Worker	$42.95	$343.60	$63.50	$508.00	$39.23	$58.00
1 Terrazzo Helper	35.50	284.00	52.50	420.00		
1 Floor Grinder, 22" Path		115.55		127.11		
1 Terrazzo Mixer		188.20		207.02	18.98	20.88
16 L.H., Daily Totals		$931.35		$1262.13	$58.21	$78.88

Crew J-4	Hr.	Daily	Hr.	Daily	Bare Costs	Incl. O&P
2 Cement Finishers	$45.00	$720.00	$66.65	$1066.40	$42.53	$63.72
1 Laborer	37.60	300.80	57.85	462.80		
1 Floor Grinder, 22" Path		115.55		127.11		
1 Floor Edger, 7" Path		39.30		43.23		
1 Vacuum Pick-Up System		60.90		66.99	8.99	9.89
24 L.H., Daily Totals		$1236.55		$1766.53	$51.52	$73.61

Crew J-4A	Hr.	Daily	Hr.	Daily	Bare Costs	Incl. O&P
2 Cement Finishers	$45.00	$720.00	$66.65	$1066.40	$41.30	$62.25
2 Laborers	37.60	601.60	57.85	925.60		
1 Floor Grinder, 22" Path		115.55		127.11		
1 Floor Edger, 7" Path		39.30		43.23		
1 Vacuum Pick-Up System		60.90		66.99		
1 Floor Auto Scrubber		233.85		257.24	14.05	15.46
32 L.H., Daily Totals		$1771.20		$2486.56	$55.35	$77.70

Crew J-4B	Hr.	Daily	Hr.	Daily	Bare Costs	Incl. O&P
1 Laborer	$37.60	$300.80	$57.85	$462.80	$37.60	$57.85
1 Floor Auto Scrubber		233.85		257.24	29.23	32.15
8 L.H., Daily Totals		$534.65		$720.03	$66.83	$90.00

Crew J-6	Hr.	Daily	Hr.	Daily	Bare Costs	Incl. O&P
2 Painters	$40.35	$645.60	$60.80	$972.80	$41.73	$63.30
1 Building Laborer	37.60	300.80	57.85	462.80		
1 Equip. Oper. (light)	48.60	388.80	73.75	590.00		
1 Air Compressor, 250 cfm		201.40		221.54		
1 Sandblaster, Portable, 3 C.F.		20.40		22.44		
1 Set Sand Blasting Accessories		14.05		15.46	7.37	8.11
32 L.H., Daily Totals		$1571.05		$2285.03	$49.10	$71.41

Crew J-7	Hr.	Daily	Hr.	Daily	Bare Costs	Incl. O&P
2 Painters	$40.35	$645.60	$60.80	$972.80	$40.35	$60.80
1 Floor Belt Sander		14.20		15.62		
1 Floor Sanding Edger		12.60		13.86	1.68	1.84
16 L.H., Daily Totals		$672.40		$1002.28	$42.02	$62.64

Crew K-1	Hr.	Daily	Hr.	Daily	Bare Costs	Incl. O&P
1 Carpenter	$46.95	$375.60	$72.25	$578.00	$43.02	$65.67
1 Truck Driver (light)	39.10	312.80	59.10	472.80		
1 Flatbed Truck, Gas, 3 Ton		303.20		333.52	18.95	20.84
16 L.H., Daily Totals		$991.60		$1384.32	$61.98	$86.52

Crew K-2	Hr.	Daily	Hr.	Daily	Bare Costs	Incl. O&P
1 Struc. Steel Foreman (outside)	$54.65	$437.20	$94.95	$759.60	$48.80	$81.83
1 Struc. Steel Worker	52.65	421.20	91.45	731.60		
1 Truck Driver (light)	39.10	312.80	59.10	472.80		
1 Flatbed Truck, Gas, 3 Ton		303.20		333.52	12.63	13.90
24 L.H., Daily Totals		$1474.40		$2297.52	$61.43	$95.73

Crew L-1	Hr.	Daily	Hr.	Daily	Bare Costs	Incl. O&P
1 Electrician	$54.70	$437.60	$81.95	$655.60	$56.70	$85.30
1 Plumber	58.70	469.60	88.65	709.20		
16 L.H., Daily Totals		$907.20		$1364.80	$56.70	$85.30

Crew L-2	Hr.	Daily	Hr.	Daily	Bare Costs	Incl. O&P
1 Carpenter	$46.95	$375.60	$72.25	$578.00	$41.20	$63.60
1 Carpenter Helper	35.45	283.60	54.95	439.60		
16 L.H., Daily Totals		$659.20		$1017.60	$41.20	$63.60

Crew L-3	Hr.	Daily	Hr.	Daily	Bare Costs	Incl. O&P
1 Carpenter	$46.95	$375.60	$72.25	$578.00	$51.14	$77.99
.5 Electrician	54.70	218.80	81.95	327.80		
.5 Sheet Metal Worker	55.95	223.80	85.50	342.00		
16 L.H., Daily Totals		$818.20		$1247.80	$51.14	$77.99

Crew L-3A	Bare Costs Hr.	Bare Costs Daily	Incl. Subs O&P Hr.	Incl. Subs O&P Daily	Cost Per Labor-Hour Bare Costs	Cost Per Labor-Hour Incl. O&P
1 Carpenter Foreman (outside)	$48.95	$391.60	$75.35	$602.80	$51.28	$78.73
.5 Sheet Metal Worker	55.95	223.80	85.50	342.00		
12 L.H., Daily Totals		$615.40		$944.80	$51.28	$78.73

Crew L-4	Hr.	Daily	Hr.	Daily	Bare Costs	Incl. O&P
2 Skilled Workers	$48.65	$778.40	$75.15	$1202.40	$44.25	$68.42
1 Helper	35.45	283.60	54.95	439.60		
24 L.H., Daily Totals		$1062.00		$1642.00	$44.25	$68.42

Crew L-5	Hr.	Daily	Hr.	Daily	Bare Costs	Incl. O&P
1 Struc. Steel Foreman (outside)	$54.65	$437.20	$94.95	$759.60	$52.80	$90.09
5 Struc. Steel Workers	52.65	2106.00	91.45	3658.00		
1 Equip. Oper. (crane)	51.70	413.60	78.45	627.60		
1 Hyd. Crane, 25 Ton		736.60		810.26	13.15	14.47
56 L.H., Daily Totals		$3693.40		$5855.46	$65.95	$104.56

Crew L-5A	Hr.	Daily	Hr.	Daily	Bare Costs	Incl. O&P
1 Struc. Steel Foreman (outside)	$54.65	$437.20	$94.95	$759.60	$52.91	$89.08
2 Structural Steel Workers	52.65	842.40	91.45	1463.20		
1 Equip. Oper. (crane)	51.70	413.60	78.45	627.60		
1 S.P. Crane, 4x4, 25 Ton		599.40		659.34	18.73	20.60
32 L.H., Daily Totals		$2292.60		$3509.74	$71.64	$109.68

Crew L-5B	Hr.	Daily	Hr.	Daily	Bare Costs	Incl. O&P
1 Struc. Steel Foreman (outside)	$54.65	$437.20	$94.95	$759.60	$53.97	$85.46
2 Structural Steel Workers	52.65	842.40	91.45	1463.20		
2 Electricians	54.70	875.20	81.95	1311.20		
2 Steamfitters/Pipefitters	59.75	956.00	90.20	1443.20		
1 Equip. Oper. (crane)	51.70	413.60	78.45	627.60		
1 Equip. Oper. (oiler)	45.20	361.60	68.55	548.40		
1 Hyd. Crane, 80 Ton		1625.00		1787.50	22.57	24.83
72 L.H., Daily Totals		$5511.00		$7940.70	$76.54	$110.29

Crew L-6	Hr.	Daily	Hr.	Daily	Bare Costs	Incl. O&P
1 Plumber	$58.70	$469.60	$88.65	$709.20	$57.37	$86.42
.5 Electrician	54.70	218.80	81.95	327.80		
12 L.H., Daily Totals		$688.40		$1037.00	$57.37	$86.42

Crew L-7	Hr.	Daily	Hr.	Daily	Bare Costs	Incl. O&P
2 Carpenters	$46.95	$751.20	$72.25	$1156.00	$45.39	$69.52
1 Building Laborer	37.60	300.80	57.85	462.80		
.5 Electrician	54.70	218.80	81.95	327.80		
28 L.H., Daily Totals		$1270.80		$1946.60	$45.39	$69.52

Crew L-8	Hr.	Daily	Hr.	Daily	Bare Costs	Incl. O&P
2 Carpenters	$46.95	$751.20	$72.25	$1156.00	$49.30	$75.53
.5 Plumber	58.70	234.80	88.65	354.60		
20 L.H., Daily Totals		$986.00		$1510.60	$49.30	$75.53

Crew L-9	Hr.	Daily	Hr.	Daily	Bare Costs	Incl. O&P
1 Labor Foreman (inside)	$38.10	$304.80	$58.65	$469.20	$42.96	$68.17
2 Building Laborers	37.60	601.60	57.85	925.60		
1 Struc. Steel Worker	52.65	421.20	91.45	731.60		
.5 Electrician	54.70	218.80	81.95	327.80		
36 L.H., Daily Totals		$1546.40		$2454.20	$42.96	$68.17

Crew L-10	Bare Costs Hr.	Bare Costs Daily	Incl. Subs O&P Hr.	Incl. Subs O&P Daily	Cost Per Labor-Hour Bare Costs	Cost Per Labor-Hour Incl. O&P
1 Struc. Steel Foreman (outside)	$54.65	$437.20	$94.95	$759.60	$53.00	$88.28
1 Structural Steel Worker	52.65	421.20	91.45	731.60		
1 Equip. Oper. (crane)	51.70	413.60	78.45	627.60		
1 Hyd. Crane, 12 Ton		653.80		719.18	27.24	29.97
24 L.H., Daily Totals		$1925.80		$2837.98	$80.24	$118.25

Crew L-11	Hr.	Daily	Hr.	Daily	Bare Costs	Incl. O&P
2 Wreckers	$37.60	$601.60	$62.10	$993.60	$43.88	$69.10
1 Equip. Oper. (crane)	51.70	413.60	78.45	627.60		
1 Equip. Oper. (light)	48.60	388.80	73.75	590.00		
1 Hyd. Excavator, 2.5 C.Y.		1605.00		1765.50		
1 Loader, Skid Steer, 78 H.P.		318.60		350.46	60.11	66.12
32 L.H., Daily Totals		$3327.60		$4327.16	$103.99	$135.22

Crew M-1	Hr.	Daily	Hr.	Daily	Bare Costs	Incl. O&P
3 Elevator Constructors	$76.50	$1836.00	$114.20	$2740.80	$72.67	$108.49
1 Elevator Apprentice	61.20	489.60	91.35	730.80		
5 Hand Tools		46.00		50.60	1.44	1.58
32 L.H., Daily Totals		$2371.60		$3522.20	$74.11	$110.07

Crew M-3	Hr.	Daily	Hr.	Daily	Bare Costs	Incl. O&P
1 Electrician Foreman (outside)	$56.70	$453.60	$84.95	$679.60	$57.56	$86.48
1 Common Laborer	37.60	300.80	57.85	462.80		
.25 Equipment Operator (med.)	50.60	101.20	76.75	153.50		
1 Elevator Constructor	76.50	612.00	114.20	913.60		
1 Elevator Apprentice	61.20	489.60	91.35	730.80		
.25 S.P. Crane, 4x4, 20 Ton		138.65		152.51	4.08	4.49
34 L.H., Daily Totals		$2095.85		$3092.82	$61.64	$90.97

Crew M-4	Hr.	Daily	Hr.	Daily	Bare Costs	Incl. O&P
1 Electrician Foreman (outside)	$56.70	$453.60	$84.95	$679.60	$56.94	$85.58
1 Common Laborer	37.60	300.80	57.85	462.80		
.25 Equipment Operator, Crane	51.70	103.40	78.45	156.90		
.25 Equip. Oper. (oiler)	45.20	90.40	68.55	137.10		
1 Elevator Constructor	76.50	612.00	114.20	913.60		
1 Elevator Apprentice	61.20	489.60	91.35	730.80		
.25 S.P. Crane, 4x4, 40 Ton		175.75		193.32	4.88	5.37
36 L.H., Daily Totals		$2225.55		$3274.13	$61.82	$90.95

Crew Q-1	Hr.	Daily	Hr.	Daily	Bare Costs	Incl. O&P
1 Plumber	$58.70	$469.60	$88.65	$709.20	$52.83	$79.78
1 Plumber Apprentice	46.95	375.60	70.90	567.20		
16 L.H., Daily Totals		$845.20		$1276.40	$52.83	$79.78

Crew Q-1A	Hr.	Daily	Hr.	Daily	Bare Costs	Incl. O&P
.25 Plumber Foreman (outside)	$60.70	$121.40	$91.65	$183.30	$59.10	$89.25
1 Plumber	58.70	469.60	88.65	709.20		
10 L.H., Daily Totals		$591.00		$892.50	$59.10	$89.25

Crew Q-1C	Hr.	Daily	Hr.	Daily	Bare Costs	Incl. O&P
1 Plumber	$58.70	$469.60	$88.65	$709.20	$52.08	$78.77
1 Plumber Apprentice	46.95	375.60	70.90	567.20		
1 Equip. Oper. (medium)	50.60	404.80	76.75	614.00		
1 Trencher, Chain Type, 8' D		3402.00		3742.20	141.75	155.93
24 L.H., Daily Totals		$4652.00		$5632.60	$193.83	$234.69

Crew Q-2	Hr.	Daily	Hr.	Daily	Bare Costs	Incl. O&P
2 Plumbers	$58.70	$939.20	$88.65	$1418.40	$54.78	$82.73
1 Plumber Apprentice	46.95	375.60	70.90	567.20		
24 L.H., Daily Totals		$1314.80		$1985.60	$54.78	$82.73

Crew No.	Bare Costs		Incl. Subs O&P		Cost Per Labor-Hour	
Crew Q-3	**Hr.**	**Daily**	**Hr.**	**Daily**	**Bare Costs**	**Incl. O&P**
1 Plumber Foreman (inside)	$59.20	$473.60	$89.40	$715.20	$55.89	$84.40
2 Plumbers	58.70	939.20	88.65	1418.40		
1 Plumber Apprentice	46.95	375.60	70.90	567.20		
32 L.H., Daily Totals		$1788.40		$2700.80	$55.89	$84.40
Crew Q-4	**Hr.**	**Daily**	**Hr.**	**Daily**	**Bare Costs**	**Incl. O&P**
1 Plumber Foreman (inside)	$59.20	$473.60	$89.40	$715.20	$55.89	$84.40
1 Plumber	58.70	469.60	88.65	709.20		
1 Welder (plumber)	58.70	469.60	88.65	709.20		
1 Plumber Apprentice	46.95	375.60	70.90	567.20		
1 Welder, Electric, 300 amp		57.70		63.47	1.80	1.98
32 L.H., Daily Totals		$1846.10		$2764.27	$57.69	$86.38
Crew Q-5	**Hr.**	**Daily**	**Hr.**	**Daily**	**Bare Costs**	**Incl. O&P**
1 Steamfitter	$59.75	$478.00	$90.20	$721.60	$53.77	$81.20
1 Steamfitter Apprentice	47.80	382.40	72.20	577.60		
16 L.H., Daily Totals		$860.40		$1299.20	$53.77	$81.20
Crew Q-6	**Hr.**	**Daily**	**Hr.**	**Daily**	**Bare Costs**	**Incl. O&P**
2 Steamfitters	$59.75	$956.00	$90.20	$1443.20	$55.77	$84.20
1 Steamfitter Apprentice	47.80	382.40	72.20	577.60		
24 L.H., Daily Totals		$1338.40		$2020.80	$55.77	$84.20
Crew Q-7	**Hr.**	**Daily**	**Hr.**	**Daily**	**Bare Costs**	**Incl. O&P**
1 Steamfitter Foreman (inside)	$60.25	$482.00	$91.00	$728.00	$56.89	$85.90
2 Steamfitters	59.75	956.00	90.20	1443.20		
1 Steamfitter Apprentice	47.80	382.40	72.20	577.60		
32 L.H., Daily Totals		$1820.40		$2748.80	$56.89	$85.90
Crew Q-8	**Hr.**	**Daily**	**Hr.**	**Daily**	**Bare Costs**	**Incl. O&P**
1 Steamfitter Foreman (inside)	$60.25	$482.00	$91.00	$728.00	$56.89	$85.90
1 Steamfitter	59.75	478.00	90.20	721.60		
1 Welder (steamfitter)	59.75	478.00	90.20	721.60		
1 Steamfitter Apprentice	47.80	382.40	72.20	577.60		
1 Welder, Electric, 300 amp		57.70		63.47	1.80	1.98
32 L.H., Daily Totals		$1878.10		$2812.27	$58.69	$87.88
Crew Q-9	**Hr.**	**Daily**	**Hr.**	**Daily**	**Bare Costs**	**Incl. O&P**
1 Sheet Metal Worker	$55.95	$447.60	$85.50	$684.00	$50.35	$76.95
1 Sheet Metal Apprentice	44.75	358.00	68.40	547.20		
16 L.H., Daily Totals		$805.60		$1231.20	$50.35	$76.95
Crew Q-10	**Hr.**	**Daily**	**Hr.**	**Daily**	**Bare Costs**	**Incl. O&P**
2 Sheet Metal Workers	$55.95	$895.20	$85.50	$1368.00	$52.22	$79.80
1 Sheet Metal Apprentice	44.75	358.00	68.40	547.20		
24 L.H., Daily Totals		$1253.20		$1915.20	$52.22	$79.80
Crew Q-11	**Hr.**	**Daily**	**Hr.**	**Daily**	**Bare Costs**	**Incl. O&P**
1 Sheet Metal Foreman (inside)	$56.45	$451.60	$86.25	$690.00	$53.27	$81.41
2 Sheet Metal Workers	55.95	895.20	85.50	1368.00		
1 Sheet Metal Apprentice	44.75	358.00	68.40	547.20		
32 L.H., Daily Totals		$1704.80		$2605.20	$53.27	$81.41
Crew Q-12	**Hr.**	**Daily**	**Hr.**	**Daily**	**Bare Costs**	**Incl. O&P**
1 Sprinkler Installer	$56.15	$449.20	$84.95	$679.60	$50.52	$76.45
1 Sprinkler Apprentice	44.90	359.20	67.95	543.60		
16 L.H., Daily Totals		$808.40		$1223.20	$50.52	$76.45

Crew No.	Bare Costs		Incl. Subs O&P		Cost Per Labor-Hour	
Crew Q-13	**Hr.**	**Daily**	**Hr.**	**Daily**	**Bare Costs**	**Incl. O&P**
1 Sprinkler Foreman (inside)	$56.65	$453.20	$85.70	$685.60	$53.46	$80.89
2 Sprinkler Installers	56.15	898.40	84.95	1359.20		
1 Sprinkler Apprentice	44.90	359.20	67.95	543.60		
32 L.H., Daily Totals		$1710.80		$2588.40	$53.46	$80.89
Crew Q-14	**Hr.**	**Daily**	**Hr.**	**Daily**	**Bare Costs**	**Incl. O&P**
1 Asbestos Worker	$52.35	$418.80	$81.50	$652.00	$47.13	$73.38
1 Asbestos Apprentice	41.90	335.20	65.25	522.00		
16 L.H., Daily Totals		$754.00		$1174.00	$47.13	$73.38
Crew Q-15	**Hr.**	**Daily**	**Hr.**	**Daily**	**Bare Costs**	**Incl. O&P**
1 Plumber	$58.70	$469.60	$88.65	$709.20	$52.83	$79.78
1 Plumber Apprentice	46.95	375.60	70.90	567.20		
1 Welder, Electric, 300 amp		57.70		63.47	3.61	3.97
16 L.H., Daily Totals		$902.90		$1339.87	$56.43	$83.74
Crew Q-16	**Hr.**	**Daily**	**Hr.**	**Daily**	**Bare Costs**	**Incl. O&P**
2 Plumbers	$58.70	$939.20	$88.65	$1418.40	$54.78	$82.73
1 Plumber Apprentice	46.95	375.60	70.90	567.20		
1 Welder, Electric, 300 amp		57.70		63.47	2.40	2.64
24 L.H., Daily Totals		$1372.50		$2049.07	$57.19	$85.38
Crew Q-17	**Hr.**	**Daily**	**Hr.**	**Daily**	**Bare Costs**	**Incl. O&P**
1 Steamfitter	$59.75	$478.00	$90.20	$721.60	$53.77	$81.20
1 Steamfitter Apprentice	47.80	382.40	72.20	577.60		
1 Welder, Electric, 300 amp		57.70		63.47	3.61	3.97
16 L.H., Daily Totals		$918.10		$1362.67	$57.38	$85.17
Crew Q-17A	**Hr.**	**Daily**	**Hr.**	**Daily**	**Bare Costs**	**Incl. O&P**
1 Steamfitter	$59.75	$478.00	$90.20	$721.60	$53.08	$80.28
1 Steamfitter Apprentice	47.80	382.40	72.20	577.60		
1 Equip. Oper. (crane)	51.70	413.60	78.45	627.60		
1 Hyd. Crane, 12 Ton		653.80		719.18		
1 Welder, Electric, 300 amp		57.70		63.47	29.65	32.61
24 L.H., Daily Totals		$1985.50		$2709.45	$82.73	$112.89
Crew Q-18	**Hr.**	**Daily**	**Hr.**	**Daily**	**Bare Costs**	**Incl. O&P**
2 Steamfitters	$59.75	$956.00	$90.20	$1443.20	$55.77	$84.20
1 Steamfitter Apprentice	47.80	382.40	72.20	577.60		
1 Welder, Electric, 300 amp		57.70		63.47	2.40	2.64
24 L.H., Daily Totals		$1396.10		$2084.27	$58.17	$86.84
Crew Q-19	**Hr.**	**Daily**	**Hr.**	**Daily**	**Bare Costs**	**Incl. O&P**
1 Steamfitter	$59.75	$478.00	$90.20	$721.60	$54.08	$81.45
1 Steamfitter Apprentice	47.80	382.40	72.20	577.60		
1 Electrician	54.70	437.60	81.95	655.60		
24 L.H., Daily Totals		$1298.00		$1954.80	$54.08	$81.45
Crew Q-20	**Hr.**	**Daily**	**Hr.**	**Daily**	**Bare Costs**	**Incl. O&P**
1 Sheet Metal Worker	$55.95	$447.60	$85.50	$684.00	$51.22	$77.95
1 Sheet Metal Apprentice	44.75	358.00	68.40	547.20		
.5 Electrician	54.70	218.80	81.95	327.80		
20 L.H., Daily Totals		$1024.40		$1559.00	$51.22	$77.95
Crew Q-21	**Hr.**	**Daily**	**Hr.**	**Daily**	**Bare Costs**	**Incl. O&P**
2 Steamfitters	$59.75	$956.00	$90.20	$1443.20	$55.50	$83.64
1 Steamfitter Apprentice	47.80	382.40	72.20	577.60		
1 Electrician	54.70	437.60	81.95	655.60		
32 L.H., Daily Totals		$1776.00		$2676.40	$55.50	$83.64

Crew No.	Bare Costs		Incl. Subs O&P		Cost Per Labor-Hour	
Crew Q-22	**Hr.**	**Daily**	**Hr.**	**Daily**	**Bare Costs**	**Incl. O&P**
1 Plumber	$58.70	$469.60	$88.65	$709.20	$52.83	$79.78
1 Plumber Apprentice	46.95	375.60	70.90	567.20		
1 Hyd. Crane, 12 Ton		653.80		719.18	40.86	44.95
16 L.H., Daily Totals		$1499.00		$1995.58	$93.69	$124.72
Crew Q-22A	**Hr.**	**Daily**	**Hr.**	**Daily**	**Bare Costs**	**Incl. O&P**
1 Plumber	$58.70	$469.60	$88.65	$709.20	$48.74	$73.96
1 Plumber Apprentice	46.95	375.60	70.90	567.20		
1 Laborer	37.60	300.80	57.85	462.80		
1 Equip. Oper. (crane)	51.70	413.60	78.45	627.60		
1 Hyd. Crane, 12 Ton		653.80		719.18	20.43	22.47
32 L.H., Daily Totals		$2213.40		$3085.98	$69.17	$96.44
Crew Q-23	**Hr.**	**Daily**	**Hr.**	**Daily**	**Bare Costs**	**Incl. O&P**
1 Plumber Foreman (outside)	$60.70	$485.60	$91.65	$733.20	$56.67	$85.68
1 Plumber	58.70	469.60	88.65	709.20		
1 Equip. Oper. (medium)	50.60	404.80	76.75	614.00		
1 Lattice Boom Crane, 20 Ton		948.90		1043.79	39.54	43.49
24 L.H., Daily Totals		$2308.90		$3100.19	$96.20	$129.17
Crew R-1	**Hr.**	**Daily**	**Hr.**	**Daily**	**Bare Costs**	**Incl. O&P**
1 Electrician Foreman	$55.20	$441.60	$82.70	$661.60	$48.37	$73.08
3 Electricians	54.70	1312.80	81.95	1966.80		
2 Helpers	35.45	567.20	54.95	879.20		
48 L.H., Daily Totals		$2321.60		$3507.60	$48.37	$73.08
Crew R-1A	**Hr.**	**Daily**	**Hr.**	**Daily**	**Bare Costs**	**Incl. O&P**
1 Electrician	$54.70	$437.60	$81.95	$655.60	$45.08	$68.45
1 Helper	35.45	283.60	54.95	439.60		
16 L.H., Daily Totals		$721.20		$1095.20	$45.08	$68.45
Crew R-2	**Hr.**	**Daily**	**Hr.**	**Daily**	**Bare Costs**	**Incl. O&P**
1 Electrician Foreman	$55.20	$441.60	$82.70	$661.60	$48.84	$73.84
3 Electricians	54.70	1312.80	81.95	1966.80		
2 Helpers	35.45	567.20	54.95	879.20		
1 Equip. Oper. (crane)	51.70	413.60	78.45	627.60		
1 S.P. Crane, 4x4, 5 Ton		275.40		302.94	4.92	5.41
56 L.H., Daily Totals		$3010.60		$4438.14	$53.76	$79.25
Crew R-3	**Hr.**	**Daily**	**Hr.**	**Daily**	**Bare Costs**	**Incl. O&P**
1 Electrician Foreman	$55.20	$441.60	$82.70	$661.60	$54.30	$81.55
1 Electrician	54.70	437.60	81.95	655.60		
.5 Equip. Oper. (crane)	51.70	206.80	78.45	313.80		
.5 S.P. Crane, 4x4, 5 Ton		137.70		151.47	6.88	7.57
20 L.H., Daily Totals		$1223.70		$1782.47	$61.19	$89.12
Crew R-4	**Hr.**	**Daily**	**Hr.**	**Daily**	**Bare Costs**	**Incl. O&P**
1 Struc. Steel Foreman (outside)	$54.65	$437.20	$94.95	$759.60	$53.46	$90.25
3 Struc. Steel Workers	52.65	1263.60	91.45	2194.80		
1 Electrician	54.70	437.60	81.95	655.60		
1 Welder, Gas Engine, 300 amp		145.85		160.44	3.65	4.01
40 L.H., Daily Totals		$2284.25		$3770.43	$57.11	$94.26

Crew No.	Bare Costs		Incl. Subs O&P		Cost Per Labor-Hour	
Crew R-5	**Hr.**	**Daily**	**Hr.**	**Daily**	**Bare Costs**	**Incl. O&P**
1 Electrician Foreman	$55.20	$441.60	$82.70	$661.60	$47.75	$72.20
4 Electrician Linemen	54.70	1750.40	81.95	2622.40		
2 Electrician Operators	54.70	875.20	81.95	1311.20		
4 Electrician Groundmen	35.45	1134.40	54.95	1758.40		
1 Crew Truck		206.40		227.04		
1 Flatbed Truck, 20,000 GVW		248.60		273.46		
1 Pickup Truck, 3/4 Ton		144.20		158.62		
.2 Hyd. Crane, 55 Ton		225.60		248.16		
.2 Hyd. Crane, 12 Ton		130.76		143.84		
.2 Earth Auger, Truck-Mtd.		83.44		91.78		
1 Tractor w/Winch		427.00		469.70	16.66	18.32
88 L.H., Daily Totals		$5667.60		$7966.20	$64.40	$90.53
Crew R-6	**Hr.**	**Daily**	**Hr.**	**Daily**	**Bare Costs**	**Incl. O&P**
1 Electrician Foreman	$55.20	$441.60	$82.70	$661.60	$47.75	$72.20
4 Electrician Linemen	54.70	1750.40	81.95	2622.40		
2 Electrician Operators	54.70	875.20	81.95	1311.20		
4 Electrician Groundmen	35.45	1134.40	54.95	1758.40		
1 Crew Truck		206.40		227.04		
1 Flatbed Truck, 20,000 GVW		248.60		273.46		
1 Pickup Truck, 3/4 Ton		144.20		158.62		
.2 Hyd. Crane, 55 Ton		225.60		248.16		
.2 Hyd. Crane, 12 Ton		130.76		143.84		
.2 Earth Auger, Truck-Mtd.		83.44		91.78		
1 Tractor w/Winch		427.00		469.70		
3 Cable Trailers		598.20		658.02		
.5 Tensioning Rig		198.40		218.24		
.5 Cable Pulling Rig		1160.00		1276.00	38.89	42.78
88 L.H., Daily Totals		$7624.20		$10118.46	$86.64	$114.98
Crew R-7	**Hr.**	**Daily**	**Hr.**	**Daily**	**Bare Costs**	**Incl. O&P**
1 Electrician Foreman	$55.20	$441.60	$82.70	$661.60	$38.74	$59.58
5 Electrician Groundmen	35.45	1418.00	54.95	2198.00		
1 Crew Truck		206.40		227.04	4.30	4.73
48 L.H., Daily Totals		$2066.00		$3086.64	$43.04	$64.31
Crew R-8	**Hr.**	**Daily**	**Hr.**	**Daily**	**Bare Costs**	**Incl. O&P**
1 Electrician Foreman	$55.20	$441.60	$82.70	$661.60	$48.37	$73.08
3 Electrician Linemen	54.70	1312.80	81.95	1966.80		
2 Electrician Groundmen	35.45	567.20	54.95	879.20		
1 Pickup Truck, 3/4 Ton		144.20		158.62		
1 Crew Truck		206.40		227.04	7.30	8.03
48 L.H., Daily Totals		$2672.20		$3893.26	$55.67	$81.11
Crew R-9	**Hr.**	**Daily**	**Hr.**	**Daily**	**Bare Costs**	**Incl. O&P**
1 Electrician Foreman	$55.20	$441.60	$82.70	$661.60	$45.14	$68.54
1 Electrician Lineman	54.70	437.60	81.95	655.60		
2 Electrician Operators	54.70	875.20	81.95	1311.20		
4 Electrician Groundmen	35.45	1134.40	54.95	1758.40		
1 Pickup Truck, 3/4 Ton		144.20		158.62		
1 Crew Truck		206.40		227.04	5.48	6.03
64 L.H., Daily Totals		$3239.40		$4772.46	$50.62	$74.57
Crew R-10	**Hr.**	**Daily**	**Hr.**	**Daily**	**Bare Costs**	**Incl. O&P**
1 Electrician Foreman	$55.20	$441.60	$82.70	$661.60	$51.58	$77.58
4 Electrician Linemen	54.70	1750.40	81.95	2622.40		
1 Electrician Groundman	35.45	283.60	54.95	439.60		
1 Crew Truck		206.40		227.04		
3 Tram Cars		412.20		453.42	12.89	14.18
48 L.H., Daily Totals		$3094.20		$4404.06	$64.46	$91.75

Crew No.	Bare Costs		Incl. Subs O&P		Cost Per Labor-Hour	
Crew R-11	**Hr.**	**Daily**	**Hr.**	**Daily**	**Bare Costs**	**Incl. O&P**
1 Electrician Foreman	$55.20	$441.60	$82.70	$661.60	$51.90	$78.11
4 Electricians	54.70	1750.40	81.95	2622.40		
1 Equip. Oper. (crane)	51.70	413.60	78.45	627.60		
1 Common Laborer	37.60	300.80	57.85	462.80		
1 Crew Truck		206.40		227.04		
1 Hyd. Crane, 12 Ton		653.80		719.18	15.36	16.90
56 L.H., Daily Totals		$3766.60		$5320.62	$67.26	$95.01

Crew R-12	Hr.	Daily	Hr.	Daily	Bare Costs	Incl. O&P
1 Carpenter Foreman (inside)	$47.45	$379.60	$73.05	$584.40	$44.45	$69.24
4 Carpenters	46.95	1502.40	72.25	2312.00		
4 Common Laborers	37.60	1203.20	57.85	1851.20		
1 Equip. Oper. (medium)	50.60	404.80	76.75	614.00		
1 Steel Worker	52.65	421.20	91.45	731.60		
1 Dozer, 200 H.P.		1387.00		1525.70		
1 Pickup Truck, 3/4 Ton		144.20		158.62	17.40	19.14
88 L.H., Daily Totals		$5442.40		$7777.52	$61.85	$88.38

Crew R-13	Hr.	Daily	Hr.	Daily	Bare Costs	Incl. O&P
1 Electrician Foreman	$55.20	$441.60	$82.70	$661.60	$52.84	$79.37
3 Electricians	54.70	1312.80	81.95	1966.80		
.25 Equip. Oper. (crane)	51.70	103.40	78.45	156.90		
1 Equipment Oiler	45.20	361.60	68.55	548.40		
.25 Hydraulic Crane, 33 Ton		187.50		206.25	4.46	4.91
42 L.H., Daily Totals		$2406.90		$3539.95	$57.31	$84.28

Crew R-15	Hr.	Daily	Hr.	Daily	Bare Costs	Incl. O&P
1 Electrician Foreman	$55.20	$441.60	$82.70	$661.60	$53.77	$80.71
4 Electricians	54.70	1750.40	81.95	2622.40		
1 Equipment Oper. (light)	48.60	388.80	73.75	590.00		
1 Aerial Lift Truck, 40' Boom		301.40		331.54	6.28	6.91
48 L.H., Daily Totals		$2882.20		$4205.54	$60.05	$87.62

Crew R-18	Hr.	Daily	Hr.	Daily	Bare Costs	Incl. O&P
.25 Electrician Foreman	$55.20	$110.40	$82.70	$165.40	$42.89	$65.39
1 Electrician	54.70	437.60	81.95	655.60		
2 Helpers	35.45	567.20	54.95	879.20		
26 L.H., Daily Totals		$1115.20		$1700.20	$42.89	$65.39

Crew R-19	Hr.	Daily	Hr.	Daily	Bare Costs	Incl. O&P
.5 Electrician Foreman	$55.20	$220.80	$82.70	$330.80	$54.80	$82.10
2 Electricians	54.70	875.20	81.95	1311.20		
20 L.H., Daily Totals		$1096.00		$1642.00	$54.80	$82.10

Crew R-21	Hr.	Daily	Hr.	Daily	Bare Costs	Incl. O&P
1 Electrician Foreman	$55.20	$441.60	$82.70	$661.60	$54.72	$82.01
3 Electricians	54.70	1312.80	81.95	1966.80		
.1 Equip. Oper. (medium)	50.60	40.48	76.75	61.40		
.1 S.P. Crane, 4x4, 25 Ton		59.94		65.93	1.83	2.01
32.8 L.H., Daily Totals		$1854.82		$2755.73	$56.55	$84.02

Crew R-22	Hr.	Daily	Hr.	Daily	Bare Costs	Incl. O&P
.66 Electrician Foreman	$55.20	$291.46	$82.70	$436.66	$46.51	$70.47
2 Electricians	54.70	875.20	81.95	1311.20		
2 Helpers	35.45	567.20	54.95	879.20		
37.28 L.H., Daily Totals		$1733.86		$2627.06	$46.51	$70.47

Crew No.	Bare Costs		Incl. Subs O&P		Cost Per Labor-Hour	
Crew R-30	**Hr.**	**Daily**	**Hr.**	**Daily**	**Bare Costs**	**Incl. O&P**
.25 Electrician Foreman (outside)	$56.70	$113.40	$84.95	$169.90	$44.33	$67.35
1 Electrician	54.70	437.60	81.95	655.60		
2 Laborers, (Semi-Skilled)	37.60	601.60	57.85	925.60		
26 L.H., Daily Totals		$1152.60		$1751.10	$44.33	$67.35

Crew R-31	Hr.	Daily	Hr.	Daily	Bare Costs	Incl. O&P
1 Electrician	$54.70	$437.60	$81.95	$655.60	$54.70	$81.95
1 Core Drill, Electric, 2.5 H.P.		55.15		60.66	6.89	7.58
8 L.H., Daily Totals		$492.75		$716.26	$61.59	$89.53

Crew W-41E	Hr.	Daily	Hr.	Daily	Bare Costs	Incl. O&P
.5 Plumber Foreman (outside)	$60.70	$242.80	$91.65	$366.60	$50.66	$76.93
1 Plumber	58.70	469.60	88.65	709.20		
1 Laborer	37.60	300.80	57.85	462.80		
20 L.H., Daily Totals		$1013.20		$1538.60	$50.66	$76.93

Historical Cost Indexes

The table below lists both the RSMeans® historical cost index based on Jan. 1, 1993 = 100 as well as the computed value of an index based on Jan. 1, 2015 costs. Since the Jan. 1, 2015 figure is estimated, space is left to write in the actual index figures as they become available through either the quarterly *RSMeans Construction Cost Indexes* or as printed in the *Engineering News-Record.* To compute the actual index based on Jan. 1, 2015 = 100, divide the historical cost index for a particular year by the actual Jan. 1, 2015 construction cost index. Space has been left to advance the index figures as the year progresses.

Year	Historical Cost Index Jan. 1, 1993 = 100		Current Index Based on Jan. 1, 2015 = 100	
	Est.	Actual	Est.	Actual
Oct 2015*				
July 2015*				
April 2015*				
Jan 2015*	206.7		100.0	100.0
July 2014		204.9	99.1	
2013		201.2	97.3	
2012		194.6	94.1	
2011		191.2	92.5	
2010		183.5	88.8	
2009		180.1	87.1	
2008		180.4	87.3	
2007		169.4	82.0	
2006		162.0	78.4	
2005		151.6	73.3	
2004		143.7	69.5	
2003		132.0	63.9	
2002		128.7	62.3	
▼ 2001		125.1	60.5	

Year	Historical Cost Index Jan. 1, 1993 = 100	Current Index Based on Jan. 1, 2015 = 100	
	Actual	Est.	Actual
July 2000	120.9	58.5	
1999	117.6	56.9	
1998	115.1	55.7	
1997	112.8	54.6	
1996	110.2	53.3	
1995	107.6	52.1	
1994	104.4	50.5	
1993	101.7	49.2	
1992	99.4	48.1	
1991	96.8	46.8	
1990	94.3	45.6	
1989	92.1	44.6	
1988	89.9	43.5	
1987	87.7	42.4	
1986	84.2	40.7	
1985	82.6	40.0	
1984	82.0	39.7	
▼ 1983	80.2	38.8	

Year	Historical Cost Index Jan. 1, 1993 = 100	Current Index Based on Jan. 1, 2015 = 100	
	Actual	Est.	Actual
July 1982	76.1	36.8	
1981	70.0	33.9	
1980	62.9	30.4	
1979	57.8	28.0	
1978	53.5	25.9	
1977	49.5	23.9	
1976	46.9	22.7	
1975	44.8	21.7	
1974	41.4	20.0	
1973	37.7	18.2	
1972	34.8	16.8	
1971	32.1	15.5	
1970	28.7	13.9	
1969	26.9	13.0	
1968	24.9	12.0	
1967	23.5	11.4	
1966	22.7	11.0	
▼ 1965	21.7	10.5	

Adjustments to Costs

The "Historical Cost Index" can be used to convert national average building costs at a particular time to the approximate building costs for some other time.

Time Adjustment Using the Historical Cost Indexes:

$$\frac{\text{Index for Year A}}{\text{Index for Year B}} \times \text{Cost in Year B} = \text{Cost in Year A}$$

Example:

Estimate and compare construction costs for different years in the same city.

To estimate the national average construction cost of a building in 1970, knowing that it cost $900,000 in 2015:

INDEX in 1970 = 28.7

INDEX in 2015 = 206.7

$$\frac{\text{INDEX 1970}}{\text{INDEX 2015}} \times \text{Cost 2015} = \text{Cost 1970}$$

$$\frac{28.7}{206.7} \times \$900{,}000 = .139 \times \$900{,}000 = \$125{,}100$$

The construction cost of the building in 1970 is $125,100.

Note: The city cost indexes for Canada can be used to convert U.S. national averages to local costs in Canadian dollars.

Example:

To estimate and compare the cost of a building in Toronto, ON in 2015 with the known cost of $600,000 (US$) in New York, NY in 2015:

INDEX Toronto = 110.9

INDEX New York = 131.8

$$\frac{\text{INDEX Toronto}}{\text{INDEX New York}} \times \text{Cost New York} = \text{Cost Toronto}$$

$$\frac{110.9}{131.8} \times \$600{,}000 = .841 \times \$600{,}000 = \$504{,}600$$

The construction cost of the building in Toronto is $504,600 (CN$).

*Historical Cost Index updates and other resources are provided on the following website.
http://info.thegordiangroup.com/RSMeans.html

How to Use the City Cost Indexes

What you should know before you begin

RSMeans City Cost Indexes (CCI) are an extremely useful tool to use when you want to compare costs from city to city and region to region.

This publication contains average construction cost indexes for 731 U.S. and Canadian cities covering over 930 three-digit zip code locations, as listed directly under each city.

Keep in mind that a City Cost Index number is a percentage ratio of a specific city's cost to the national average cost of the same item at a stated time period.

In other words, these index figures represent relative construction factors (or, if you prefer, multipliers) for Material and Installation costs, as well as the weighted average for Total In Place costs for each CSI MasterFormat division. Installation costs include both labor and equipment rental costs. When estimating equipment rental rates only, for a specific location, use 01 54 33 EQUIPMENT RENTAL COSTS in the Reference Section at the back of the book.

The 30 City Average Index is the average of 30 major U.S. cities and serves as a National Average.

Index figures for both material and installation are based on the 30 major city average of 100 and represent the cost relationship as of July 1, 2014. The index for each division is computed from representative material and labor quantities for that division. The weighted average for each city is a weighted total of the components listed above it, but does not include relative productivity between trades or cities.

As changes occur in local material prices, labor rates, and equipment rental rates (including fuel costs), the impact of these changes should be accurately measured by the change in the City Cost Index for each particular city (as compared to the 30 City Average).

Therefore, if you know (or have estimated) building costs in one city today, you can easily convert those costs to expected building costs in another city.

In addition, by using the Historical Cost Index, you can easily convert National Average building costs at a particular time to the approximate building costs for some other time. The City Cost Indexes can then be applied to calculate the costs for a particular city.

Quick Calculations

Location Adjustment Using the City Cost Indexes:

$$\frac{\text{Index for City A}}{\text{Index for City B}} \times \text{Cost in City B} = \text{Cost in City A}$$

Time Adjustment for the National Average Using the Historical Cost Index:

$$\frac{\text{Index for Year A}}{\text{Index for Year B}} \times \text{Cost in Year B} = \text{Cost in Year A}$$

Adjustment from the National Average:

$$\frac{\text{Index for City A}}{100} \times \text{National Average Cost} = \text{Cost in City A}$$

Since each of the other RSMeans publications contains many different items, any *one* item multiplied by the particular city index may give incorrect results. However, the larger the number of items compiled, the closer the results should be to actual costs for that particular city.

The City Cost Indexes for Canadian cities are calculated using Canadian material and equipment prices and labor rates, in Canadian dollars. Therefore, indexes for Canadian cities can be used to convert U.S. National Average prices to local costs in Canadian dollars.

How to use this section

1. Compare costs from city to city.

In using the RSMeans Indexes, remember that an index number is not a fixed number but a ratio: It's a percentage ratio of a building component's cost at any stated time to the National Average cost of that same component at the same time period. Put in the form of an equation:

$$\frac{\text{Specific City Cost}}{\text{National Average Cost}} \times 100 = \text{City Index Number}$$

Therefore, when making cost comparisons between cities, do not subtract one city's index number from the index number of another city and read the result as a percentage difference. Instead, divide one city's index number by that of the other city. The resulting number may then be used as a multiplier to calculate cost differences from city to city.

The formula used to find cost differences between cities for the purpose of comparison is as follows:

$$\frac{\text{City A Index}}{\text{City B Index}} \times \text{City B Cost (Known)} = \text{City A Cost (Unknown)}$$

In addition, you can use RSMeans CCI to calculate and compare costs division by division between cities using the same basic formula. (Just be sure that you're comparing similar divisions.)

2. Compare a specific city's construction costs with the National Average.

When you're studying construction location feasibility, it's advisable to compare a prospective project's cost index with an index of the National Average cost.

For example, divide the weighted average index of construction costs of a specific city by that of the 30 City Average, which = 100.

$$\frac{\text{City Index}}{100} = \text{\% of National Average}$$

As a result, you get a ratio that indicates the relative cost of construction in that city in comparison with the National Average.

3. Convert U.S. National Average to actual costs in Canadian City.

$$\frac{\text{Index for Canadian City}}{100} \times \text{National Average Cost} = \text{Cost in Canadian City in \$ CAN}$$

4. Adjust construction cost data based on a National Average.

When you use a source of construction cost data which is based on a National Average (such as RSMeans cost data publications), it is necessary to adjust those costs to a specific location.

$$\frac{\text{City Index}}{100} \times \text{"Book" Cost Based on National Average Costs} = \text{City Cost (Unknown)}$$

5. When applying the City Cost Indexes to demolition projects, use the appropriate division installation index. For example, for removal of existing doors and windows, use Division 8 (Openings) index.

What you might like to know about how we developed the Indexes

The information presented in the CCI is organized according to the Construction Specifications Institute (CSI) MasterFormat 2014 classification system.

To create a reliable index, RSMeans researched the building type most often constructed in the United States and Canada. Because it was concluded that no one type of building completely represented the building construction industry, nine different types of buildings were combined to create a composite model.

The exact material, labor, and equipment quantities are based on detailed analyses of these nine building types, and then each quantity is weighted in proportion to expected usage. These various material items, labor hours, and equipment rental rates are thus combined to form a composite building representing as closely as possible the actual usage of materials, labor, and equipment used in the North American building construction industry.

The following structures were chosen to make up that composite model:

1. Factory, 1 story
2. Office, 2–4 story
3. Store, Retail
4. Town Hall, 2–3 story
5. High School, 2–3 story
6. Hospital, 4–8 story
7. Garage, Parking
8. Apartment, 1–3 story
9. Hotel/Motel, 2–3 story

For the purposes of ensuring the timeliness of the data, the components of the index for the composite model have been streamlined. They currently consist of:

- specific quantities of 66 commonly used construction materials;
- specific labor-hours for 21 building construction trades; and
- specific days of equipment rental for 6 types of construction equipment (normally used to install the 66 material items by the 21 trades.) Fuel costs and routine maintenance costs are included in the equipment cost.

A sophisticated computer program handles the updating of all costs for each city on a quarterly basis. Material and equipment price quotations are gathered quarterly from cities in the United States and Canada. These prices and the latest negotiated labor wage rates for 21 different building trades are used to compile the quarterly update of the City Cost Index.

The 30 major U.S. cities used to calculate the National Average are:

Atlanta, GA	Memphis, TN
Baltimore, MD	Milwaukee, WI
Boston, MA	Minneapolis, MN
Buffalo, NY	Nashville, TN
Chicago, IL	New Orleans, LA
Cincinnati, OH	New York, NY
Cleveland, OH	Philadelphia, PA
Columbus, OH	Phoenix, AZ
Dallas, TX	Pittsburgh, PA
Denver, CO	St. Louis, MO
Detroit, MI	San Antonio, TX
Houston, TX	San Diego, CA
Indianapolis, IN	San Francisco, CA
Kansas City, MO	Seattle, WA
Los Angeles, CA	Washington, DC

What the CCI does not indicate

The weighted average for each city is a total of the divisional components weighted to reflect typical usage, but it does not include the productivity variations between trades or cities.

In addition, the CCI does not take into consideration factors such as the following:

- managerial efficiency
- competitive conditions
- automation
- restrictive union practices
- unique local requirements
- regional variations due to specific building codes

City Cost Indexes

DIVISION		UNITED STATES			ALABAMA															
		30 CITY			ANNISTON			BIRMINGHAM			BUTLER			DECATUR			DOTHAN			
		AVERAGE			362			350 - 352			369			356			363			
		MAT.	INST.	TOTAL	MAT.	INST.	TOTAL	MAT.	INST.	TOTAL	MAT.	INST.	TOTAL	MAT.	INST.	TOTAL	MAT.	INST.	TOTAL	
015433	**CONTRACTOR EQUIPMENT**		100.0	100.0		101.0	101.0		101.1	101.1		98.2	98.2		101.0	101.0		98.2	98.2	
0241, 31 - 34	**SITE & INFRASTRUCTURE, DEMOLITION**	100.0	100.0	100.0	90.2	93.1	92.3	97.4	93.8	94.8	103.5	87.1	91.8	88.8	91.5	90.7	101.2	87.1	91.1	
0310	Concrete Forming & Accessories	100.0	100.0	100.0	90.1	47.1	53.0	93.1	75.9	78.2	86.4	42.6	48.6	94.6	45.4	52.1	95.5	42.0	49.4	
0320	Concrete Reinforcing	100.0	100.0	100.0	87.8	86.3	87.1	94.7	86.9	90.7	92.8	45.9	68.9	88.6	77.0	82.7	92.8	45.4	68.7	
0330	Cast-in-Place Concrete	100.0	100.0	100.0	101.8	49.7	80.4	110.0	74.6	95.4	99.3	54.8	81.0	103.4	64.1	87.2	99.3	46.3	77.5	
03	**CONCRETE**	100.0	100.0	100.0	102.1	57.0	79.9	102.2	78.4	90.5	103.0	49.3	76.6	98.3	59.3	79.1	102.3	46.0	74.7	
04	**MASONRY**	100.0	100.0	100.0	100.1	66.8	79.3	98.8	76.0	84.6	105.9	48.6	70.2	97.3	51.3	68.6	107.1	39.7	65.1	
05	**METALS**	100.0	100.0	100.0	104.0	91.5	100.1	104.0	94.1	101.0	102.8	76.6	94.8	106.2	86.1	100.0	102.9	75.3	94.4	
06	**WOOD, PLASTICS & COMPOSITES**	100.0	100.0	100.0	90.5	42.7	63.7	97.4	76.1	85.4	85.0	42.4	61.1	98.9	42.9	67.6	97.4	42.4	66.6	
07	**THERMAL & MOISTURE PROTECTION**	100.0	100.0	100.0	97.9	52.9	79.5	99.7	81.5	92.2	98.0	56.7	81.0	96.7	58.3	81.0	98.0	49.1	77.9	
08	**OPENINGS**	100.0	100.0	100.0	98.6	50.1	87.3	103.2	76.0	96.8	98.6	44.2	85.9	106.5	50.4	93.5	98.6	44.2	86.0	
0920	Plaster & Gypsum Board	100.0	100.0	100.0	90.8	41.4	57.4	94.8	75.8	81.9	87.9	41.0	56.2	95.8	41.6	59.2	98.6	41.0	59.7	
0950, 0980	Ceilings & Acoustic Treatment	100.0	100.0	100.0	80.4	41.4	54.7	89.8	75.8	80.6	80.4	41.0	54.5	87.2	41.6	57.2	80.4	41.0	54.5	
0960	Flooring	100.0	100.0	100.0	91.5	40.2	76.8	102.5	76.7	95.1	95.8	56.0	84.4	99.7	50.6	85.6	100.6	27.5	79.7	
0970, 0990	Wall Finishes & Painting/Coating	100.0	100.0	100.0	103.7	31.5	60.1	103.9	66.5	81.3	103.7	53.3	73.3	99.6	66.8	79.8	103.7	53.3	73.3	
09	**FINISHES**	100.0	100.0	100.0	87.0	43.0	62.8	95.4	74.8	84.1	89.9	45.7	65.5	91.9	47.7	67.6	92.5	40.1	63.6	
COVERS	**DIVS. 10 - 14, 25, 28, 41, 43, 44, 46**	100.0	100.0	100.0	100.0	68.3	93.6	100.0	88.4	97.7	100.0	41.4	88.2	100.0	43.2	88.5	100.0	41.5	88.2	
21, 22, 23	**FIRE SUPPRESSION, PLUMBING & HVAC**	100.0	100.0	100.0	100.0	55.4	82.0	100.0	70.4	88.1	97.3	33.3	71.5	100.0	40.8	76.1	97.3	33.2	71.5	
26, 27, 3370	**ELECTRICAL, COMMUNICATIONS & UTIL.**	100.0	100.0	100.0	93.1	57.3	74.2	99.0	61.6	79.3	95.0	40.2	66.0	94.2	64.9	78.7	93.8	57.4	74.6	
MF2014	**WEIGHTED AVERAGE**	100.0	100.0	100.0	98.6	61.8	82.6	100.6	76.7	90.2	98.9	49.4	77.3	99.8	57.9	81.5	99.0	49.4	77.4	

DIVISION		ALABAMA																	
		EVERGREEN			GADSDEN			HUNTSVILLE			JASPER			MOBILE			MONTGOMERY		
		364			359			357 - 358			355			365 - 366			360 - 361		
		MAT.	INST.	TOTAL	MAT.	INST.	TOTAL	MAT.	INST.	TOTAL	MAT.	INST.	TOTAL	MAT.	INST.	TOTAL	MAT.	INST.	TOTAL
015433	**CONTRACTOR EQUIPMENT**		98.2	98.2		101.0	101.0		101.0	101.0		101.0	101.0		98.2	98.2		98.2	98.2
0241, 31 - 34	**SITE & INFRASTRUCTURE, DEMOLITION**	104.0	87.4	92.2	94.9	92.7	93.3	88.6	92.7	91.5	94.6	92.4	93.1	96.4	88.1	90.5	96.5	88.1	90.5
0310	Concrete Forming & Accessories	83.0	44.2	49.5	86.9	42.5	48.6	94.6	69.9	73.3	92.1	34.6	42.5	94.5	54.0	59.6	94.1	44.1	51.0
0320	Concrete Reinforcing	92.9	45.9	69.0	94.0	86.8	90.3	88.6	81.6	85.1	88.6	85.6	87.1	90.6	83.6	87.0	95.8	86.1	90.9
0330	Cast-in-Place Concrete	99.3	57.3	82.1	103.4	66.7	88.3	100.7	69.5	87.9	114.5	44.1	85.6	104.1	66.4	88.6	105.2	56.5	85.2
03	**CONCRETE**	103.3	50.9	77.5	102.9	60.8	82.2	97.0	72.9	85.2	106.6	49.4	78.5	98.9	65.3	82.4	100.2	57.9	79.4
04	**MASONRY**	105.9	53.2	73.0	95.6	62.4	74.9	98.8	67.7	79.4	92.9	41.2	60.7	103.7	54.2	72.9	100.0	49.7	68.6
05	**METALS**	102.9	76.5	94.7	104.0	93.2	100.7	106.2	90.6	101.4	104.0	90.2	99.7	105.0	91.6	100.9	104.0	91.9	100.3
06	**WOOD, PLASTICS & COMPOSITES**	81.3	42.4	59.5	89.3	38.7	61.0	98.9	72.6	84.2	95.9	31.6	59.9	95.9	52.8	71.8	95.3	42.4	65.6
07	**THERMAL & MOISTURE PROTECTION**	97.9	56.9	81.1	96.8	71.7	86.5	96.6	77.6	88.8	96.8	45.4	75.7	97.6	70.1	86.3	97.9	66.6	85.0
08	**OPENINGS**	98.6	44.2	85.9	102.8	49.0	90.3	106.5	67.4	97.4	102.8	48.0	90.0	102.2	58.8	92.1	103.3	53.1	91.7
0920	Plaster & Gypsum Board	87.2	41.0	56.0	88.2	37.3	53.8	95.8	72.2	79.9	92.1	30.0	50.1	95.2	51.8	65.9	94.6	41.0	58.4
0950, 0980	Ceilings & Acoustic Treatment	80.4	41.0	54.5	83.8	37.3	53.2	89.0	72.2	78.0	83.8	30.0	48.4	85.6	51.8	63.4	88.0	41.0	57.1
0960	Flooring	93.9	56.0	83.1	95.5	76.7	90.1	99.7	76.7	93.1	97.8	40.2	81.3	100.5	58.5	88.5	97.3	58.5	86.2
0970, 0990	Wall Finishes & Painting/Coating	103.7	53.3	73.3	99.6	60.7	76.1	99.6	65.9	79.3	99.6	38.6	62.8	107.3	54.3	75.3	103.3	53.3	73.1
09	**FINISHES**	89.2	46.8	65.9	89.4	48.7	67.0	92.3	70.7	80.4	90.5	34.4	59.6	92.7	53.5	71.1	92.8	46.1	67.0
COVERS	**DIVS. 10 - 14, 25, 28, 41, 43, 44, 46**	100.0	42.9	88.5	100.0	79.8	95.9	100.0	85.3	97.0	100.0	39.8	87.8	100.0	83.1	96.6	100.0	79.6	95.9
21, 22, 23	**FIRE SUPPRESSION, PLUMBING & HVAC**	97.3	35.6	72.4	102.0	36.3	75.5	100.0	61.6	84.5	102.0	61.4	85.6	99.9	60.7	84.1	100.0	34.2	73.4
26, 27, 3370	**ELECTRICAL, COMMUNICATIONS & UTIL.**	92.5	40.2	64.8	94.3	61.6	77.0	95.3	64.9	79.3	93.8	61.3	76.6	95.7	59.1	76.3	96.3	60.9	77.6
MF2014	**WEIGHTED AVERAGE**	98.6	50.7	77.7	99.9	60.0	82.5	99.8	72.1	87.8	100.2	57.5	81.6	99.9	65.4	84.9	99.9	57.1	81.2

DIVISION		ALABAMA									ALASKA								
		PHENIX CITY			SELMA			TUSCALOOSA			ANCHORAGE			FAIRBANKS			JUNEAU		
		368			367			354			995 - 996			997			998		
		MAT.	INST.	TOTAL	MAT.	INST.	TOTAL	MAT.	INST.	TOTAL	MAT.	INST.	TOTAL	MAT.	INST.	TOTAL	MAT.	INST.	TOTAL
015433	**CONTRACTOR EQUIPMENT**		98.2	98.2		98.2	98.2		101.0	101.0		114.7	114.7		114.7	114.7		114.7	114.7
0241, 31 - 34	**SITE & INFRASTRUCTURE, DEMOLITION**	107.8	88.2	93.9	101.0	88.1	91.8	89.1	92.9	91.8	127.2	129.6	128.9	119.2	129.6	126.6	135.6	129.6	131.3
0310	Concrete Forming & Accessories	90.1	39.1	46.1	87.6	42.6	48.8	94.5	47.9	54.3	123.7	119.8	120.3	131.6	119.9	121.5	130.1	119.8	121.2
0320	Concrete Reinforcing	92.8	61.4	76.8	92.8	85.2	89.0	88.6	86.8	87.7	149.7	110.7	129.9	150.9	110.7	130.5	135.7	110.7	123.0
0330	Cast-in-Place Concrete	99.3	57.0	81.9	99.3	46.5	77.6	104.9	68.4	89.9	132.0	117.8	126.2	128.3	118.2	124.1	133.2	117.8	126.9
03	**CONCRETE**	106.2	51.3	79.2	101.7	53.7	78.1	99.0	63.8	81.7	138.6	116.7	127.9	123.8	116.9	120.4	136.0	116.7	126.5
04	**MASONRY**	105.9	41.1	65.5	109.6	38.7	65.4	97.6	65.5	77.6	184.6	125.3	147.6	189.5	125.3	149.5	176.6	125.3	144.6
05	**METALS**	102.8	81.9	96.4	102.8	89.3	98.7	105.3	93.2	101.6	113.6	103.4	110.5	117.6	103.5	113.3	116.3	103.4	112.3
06	**WOOD, PLASTICS & COMPOSITES**	90.2	35.4	59.5	87.0	42.4	62.0	98.9	44.4	68.4	129.2	118.7	123.4	136.1	118.7	126.4	129.4	118.7	123.4
07	**THERMAL & MOISTURE PROTECTION**	98.3	63.8	84.1	97.8	52.1	79.1	96.7	73.7	87.3	159.5	118.0	142.5	166.5	119.1	147.1	168.0	118.0	147.5
08	**OPENINGS**	98.6	43.5	85.8	98.6	53.1	88.0	106.5	58.8	95.4	132.7	115.6	128.7	128.9	115.7	125.8	129.8	115.6	126.5
0920	Plaster & Gypsum Board	92.2	33.9	52.8	90.1	41.0	56.9	95.8	43.1	60.2	140.9	119.1	126.2	164.8	119.1	133.9	148.1	119.1	128.5
0950, 0980	Ceilings & Acoustic Treatment	80.4	33.9	49.8	80.4	41.0	54.5	89.0	43.1	58.9	119.6	119.1	119.3	119.2	119.1	119.2	125.9	119.1	121.4
0960	Flooring	97.6	58.5	86.4	96.2	28.5	76.8	99.7	76.7	93.1	134.6	133.0	134.1	127.2	133.0	128.9	131.0	133.0	131.6
0970, 0990	Wall Finishes & Painting/Coating	103.7	53.3	73.3	103.7	53.3	73.3	99.6	45.9	67.2	136.8	116.3	124.4	133.5	121.2	126.1	132.1	116.3	122.6
09	**FINISHES**	91.4	42.2	64.3	90.0	40.2	62.6	92.3	51.2	69.7	134.3	122.2	127.7	132.4	122.8	127.1	134.4	122.2	127.7
COVERS	**DIVS. 10 - 14, 25, 28, 41, 43, 44, 46**	100.0	79.3	95.8	100.0	41.4	88.2	100.0	81.4	96.2	100.0	112.8	102.6	100.0	112.8	102.6	100.0	112.8	102.6
21, 22, 23	**FIRE SUPPRESSION, PLUMBING & HVAC**	97.3	34.5	72.0	97.3	33.6	71.6	100.0	34.0	73.4	100.3	105.0	102.2	100.2	108.0	103.4	100.3	105.0	102.2
26, 27, 3370	**ELECTRICAL, COMMUNICATIONS & UTIL.**	94.4	69.5	81.3	93.6	40.2	65.4	94.8	61.6	77.3	117.7	117.8	117.7	130.0	117.8	123.5	119.9	117.8	118.8
MF2014	**WEIGHTED AVERAGE**	99.5	54.6	79.9	98.7	49.9	77.4	99.8	61.2	83.0	121.1	115.6	118.7	121.0	116.4	119.0	121.3	115.6	118.8

DIVISION		ALASKA			ARIZONA															
		KETCHIKAN			CHAMBERS			FLAGSTAFF			GLOBE			KINGMAN			MESA/TEMPE			
		999			865			860			855			864			852			
		MAT.	INST.	TOTAL	MAT.	INST.	TOTAL	MAT.	INST.	TOTAL	MAT.	INST.	TOTAL	MAT.	INST.	TOTAL	MAT.	INST.	TOTAL	
015433	**CONTRACTOR EQUIPMENT**		114.7	114.7		93.0	93.0		93.0	93.0		92.2	92.2		93.0	93.0		92.2	92.2	
0241, 31 - 34	**SITE & INFRASTRUCTURE, DEMOLITION**	173.8	129.6	142.4	69.6	96.1	88.4	87.3	96.3	93.7	97.3	95.6	96.1	69.6	96.3	88.6	89.0	95.8	93.8	
0310	Concrete Forming & Accessories	122.9	119.8	120.2	98.8	58.1	63.7	104.4	65.1	70.5	99.0	58.1	63.7	97.0	65.1	69.5	102.2	69.3	73.9	
0320	Concrete Reinforcing	117.4	110.7	114.0	97.1	85.2	91.1	97.0	85.3	91.0	108.0	85.2	96.4	97.2	85.3	91.1	108.7	85.3	96.8	
0330	Cast-in-Place Concrete	259.2	117.8	201.2	90.7	73.2	83.5	90.8	73.4	83.7	94.3	72.0	85.1	90.4	73.4	83.4	95.0	72.3	85.7	
03	**CONCRETE**	208.7	116.7	163.5	95.0	68.6	82.0	114.1	71.8	93.3	110.0	68.2	89.5	94.6	71.8	83.4	101.8	73.3	87.8	
04	**MASONRY**	198.4	125.3	152.8	92.5	62.3	73.7	92.6	62.4	73.8	109.9	62.2	80.1	92.5	62.4	73.7	110.1	62.2	80.3	
05	**METALS**	117.8	103.4	113.4	96.2	75.6	89.8	96.7	76.3	90.4	93.3	76.0	88.0	96.8	76.3	90.5	93.6	76.8	88.4	
06	**WOOD, PLASTICS & COMPOSITES**	126.3	118.7	122.1	101.0	55.0	75.2	107.3	64.2	83.1	97.6	55.0	73.8	96.1	64.2	78.2	101.1	69.8	83.6	
07	**THERMAL & MOISTURE PROTECTION**	169.3	118.0	148.3	94.6	65.0	82.5	96.2	68.1	84.7	101.9	63.0	85.9	94.6	65.6	82.7	101.2	65.1	86.4	
08	**OPENINGS**	130.8	115.6	127.3	108.1	65.3	98.2	108.3	70.3	99.4	100.0	65.3	91.9	108.3	70.3	99.5	100.1	73.3	93.8	
0920	Plaster & Gypsum Board	152.7	119.1	130.0	90.2	53.7	65.5	93.6	63.2	73.0	96.6	53.7	67.6	82.9	63.2	69.6	98.7	68.9	78.6	
0950, 0980	Ceilings & Acoustic Treatment	112.7	119.1	116.9	99.7	53.7	69.5	100.5	63.2	76.0	88.9	53.7	65.8	100.5	63.2	76.0	88.9	68.9	75.8	
0960	Flooring	127.2	133.0	128.9	93.3	39.5	77.9	95.4	39.7	79.5	103.5	39.5	85.2	92.1	53.7	81.1	104.7	49.5	88.9	
0970, 0990	Wall Finishes & Painting/Coating	133.5	116.3	123.1	98.3	55.1	72.2	98.3	55.1	72.2	103.4	55.1	74.3	98.3	55.1	72.2	103.4	55.1	74.3	
09	**FINISHES**	133.4	122.2	127.2	93.7	52.9	71.2	96.6	58.4	75.6	97.1	53.0	72.8	92.5	60.7	75.0	96.8	63.6	78.5	
COVERS	**DIVS. 10 - 14, 25, 28, 41, 43, 44, 46**	100.0	112.8	102.6	100.0	82.3	96.4	100.0	83.3	96.6	100.0	82.4	96.4	100.0	83.3	96.6	100.0	84.0	96.8	
21, 22, 23	**FIRE SUPPRESSION, PLUMBING & HVAC**	98.4	105.0	101.1	97.0	78.9	89.7	100.2	79.0	91.6	95.2	78.9	88.7	97.0	79.0	89.7	100.0	79.0	91.5	
26, 27, 3370	**ELECTRICAL, COMMUNICATIONS & UTIL.**	129.9	117.8	123.5	104.5	70.9	86.7	103.4	61.3	81.1	97.5	61.2	78.3	104.5	61.3	81.7	94.2	61.3	76.8	
MF2014	**WEIGHTED AVERAGE**	132.0	115.6	124.8	97.6	71.5	86.2	101.3	71.8	88.4	98.8	70.0	86.3	97.6	72.0	86.4	98.6	72.8	87.4	

DIVISION		ARIZONA												ARKANSAS					
		PHOENIX			PRESCOTT			SHOW LOW			TUCSON			BATESVILLE			CAMDEN		
		850,853			863			859			856 - 857			725			717		
		MAT.	INST.	TOTAL	MAT.	INST.	TOTAL	MAT.	INST.	TOTAL	MAT.	INST.	TOTAL	MAT.	INST.	TOTAL	MAT.	INST.	TOTAL
015433	**CONTRACTOR EQUIPMENT**		92.7	92.7		93.0	93.0		92.2	92.2		92.2	92.2		89.1	89.1		89.1	89.1
0241, 31 - 34	**SITE & INFRASTRUCTURE, DEMOLITION**	89.4	96.0	94.1	76.0	96.0	90.2	99.2	95.6	96.6	85.2	95.8	92.7	77.1	84.9	82.6	78.3	84.4	82.7
0310	Concrete Forming & Accessories	103.1	68.3	73.1	100.4	52.5	59.1	106.3	69.1	74.2	102.5	68.1	72.9	85.9	44.9	50.5	83.7	30.7	38.0
0320	Concrete Reinforcing	107.0	85.4	96.0	97.0	85.2	91.0	108.7	85.2	96.8	89.9	85.3	87.5	87.4	67.5	77.3	93.6	67.4	80.3
0330	Cast-in-Place Concrete	95.1	72.4	85.8	90.7	73.1	83.5	94.3	72.1	85.2	97.8	72.3	87.3	77.4	45.3	64.2	82.6	38.3	64.4
03	**CONCRETE**	101.4	73.0	87.4	100.1	66.1	83.4	112.4	73.1	93.1	100.0	72.8	86.6	83.5	50.1	67.1	87.5	41.3	64.8
04	**MASONRY**	97.5	64.1	76.7	92.6	62.3	73.7	109.9	62.2	80.1	95.5	62.2	74.8	99.8	40.7	63.0	116.6	31.4	63.5
05	**METALS**	95.1	77.6	89.7	96.7	75.3	90.1	93.1	76.1	87.8	94.3	76.7	88.9	99.5	66.9	89.4	103.9	66.5	92.4
06	**WOOD, PLASTICS & COMPOSITES**	102.1	68.3	83.2	102.4	47.4	71.6	105.8	69.8	85.6	101.4	68.3	82.8	89.9	45.5	65.0	90.3	29.7	56.4
07	**THERMAL & MOISTURE PROTECTION**	100.9	67.0	87.0	95.1	64.2	82.4	102.1	65.0	86.9	102.4	64.2	86.7	98.2	43.9	76.0	94.2	35.7	70.2
08	**OPENINGS**	102.0	72.5	95.2	108.3	61.1	97.3	99.2	73.3	93.2	96.3	72.5	90.7	97.7	45.8	85.6	102.1	40.6	87.8
0920	Plaster & Gypsum Board	101.0	67.4	78.2	90.3	45.9	60.3	100.9	68.9	79.3	104.0	67.4	79.2	83.3	44.3	56.9	85.0	28.0	46.5
0950, 0980	Ceilings & Acoustic Treatment	96.5	67.4	77.3	98.8	45.9	64.0	88.9	68.9	75.8	89.8	67.4	75.0	83.2	44.3	57.6	83.3	28.0	47.0
0960	Flooring	104.9	51.7	89.7	94.1	39.5	78.5	106.2	45.0	88.7	95.2	44.2	80.6	96.7	59.1	85.9	99.6	39.6	82.4
0970, 0990	Wall Finishes & Painting/Coating	103.4	61.7	78.3	98.3	55.1	72.2	103.4	55.1	74.3	104.5	55.1	74.7	107.4	41.2	67.4	104.4	50.4	71.8
09	**FINISHES**	98.9	63.8	79.6	94.2	48.5	69.0	98.7	62.7	78.8	94.9	61.6	76.5	86.6	46.7	64.6	88.5	33.5	58.2
COVERS	**DIVS. 10 - 14, 25, 28, 41, 43, 44, 46**	100.0	83.9	96.7	100.0	81.4	96.3	100.0	84.0	96.8	100.0	83.9	96.7	100.0	39.5	87.8	100.0	35.6	87.0
21, 22, 23	**FIRE SUPPRESSION, PLUMBING & HVAC**	99.9	79.0	91.5	100.2	78.9	91.6	95.2	79.0	88.7	100.0	79.0	91.5	95.3	51.6	77.7	95.3	53.1	78.3
26, 27, 3370	**ELECTRICAL, COMMUNICATIONS & UTIL.**	101.0	66.7	82.9	103.0	61.2	80.9	94.6	61.2	77.0	96.5	61.3	77.9	96.5	63.5	79.0	96.8	62.1	78.5
MF2014	**WEIGHTED AVERAGE**	99.2	73.8	88.1	99.2	68.9	86.0	98.9	72.5	87.4	97.4	72.4	86.5	94.6	54.6	77.1	97.1	50.1	76.6

DIVISION		ARKANSAS																	
		FAYETTEVILLE			FORT SMITH			HARRISON			HOT SPRINGS			JONESBORO			LITTLE ROCK		
		727			729			726			719			724			720 - 722		
		MAT.	INST.	TOTAL	MAT.	INST.	TOTAL	MAT.	INST.	TOTAL	MAT.	INST.	TOTAL	MAT.	INST.	TOTAL	MAT.	INST.	TOTAL
015433	**CONTRACTOR EQUIPMENT**		89.1	89.1		89.1	89.1		89.1	89.1		89.1	89.1		108.3	108.3		89.1	89.1
0241, 31 - 34	**SITE & INFRASTRUCTURE, DEMOLITION**	76.4	86.3	83.5	81.8	86.7	85.3	82.1	84.9	84.1	81.4	85.8	84.5	103.2	101.1	101.7	93.1	86.8	88.6
0310	Concrete Forming & Accessories	81.2	38.5	44.4	101.3	57.3	63.4	90.6	44.8	51.1	81.1	35.7	41.9	89.6	48.2	53.9	96.4	60.0	65.0
0320	Concrete Reinforcing	87.4	71.3	79.2	88.4	72.0	80.1	87.0	63.7	75.1	91.8	67.5	79.4	84.5	70.1	77.2	93.4	68.3	80.6
0330	Cast-in-Place Concrete	77.4	49.4	65.9	88.5	70.3	81.0	85.8	43.4	68.4	84.5	38.8	65.7	84.3	55.0	72.2	87.3	70.3	80.3
03	**CONCRETE**	83.2	49.3	66.5	91.3	65.0	78.4	90.8	48.7	70.1	91.2	43.7	67.9	88.1	56.0	72.3	92.8	65.5	79.4
04	**MASONRY**	90.5	41.4	59.9	96.9	50.9	68.3	100.1	39.0	62.0	87.6	33.5	53.9	91.9	41.9	60.7	96.8	50.9	68.2
05	**METALS**	99.5	67.8	89.7	101.8	70.6	92.2	100.6	66.5	90.1	103.8	66.9	92.5	95.9	80.0	91.0	102.0	69.4	92.0
06	**WOOD, PLASTICS & COMPOSITES**	85.8	35.5	57.6	107.6	59.0	80.4	96.0	45.5	67.7	87.5	35.7	58.5	93.9	48.3	68.4	99.9	62.5	78.9
07	**THERMAL & MOISTURE PROTECTION**	99.0	45.3	77.0	99.4	57.5	82.2	98.5	42.6	75.6	94.4	37.9	71.2	103.3	49.1	81.1	95.7	57.8	80.2
08	**OPENINGS**	97.7	44.7	85.4	98.5	56.2	88.7	98.5	46.5	86.4	102.1	42.3	88.2	100.6	54.0	89.7	99.5	58.4	89.9
0920	Plaster & Gypsum Board	81.8	34.0	49.5	89.4	58.3	68.4	88.3	44.3	58.6	84.0	34.2	50.4	97.7	46.9	63.4	94.6	61.8	72.5
0950, 0980	Ceilings & Acoustic Treatment	83.2	34.0	50.9	84.8	58.3	67.4	84.8	44.3	58.2	83.3	34.2	51.0	86.2	46.9	60.3	88.0	61.8	70.8
0960	Flooring	94.0	59.1	84.0	102.9	60.8	90.9	98.8	59.1	87.5	98.5	59.1	87.2	68.9	53.4	64.5	101.0	60.8	89.5
0970, 0990	Wall Finishes & Painting/Coating	107.4	29.4	60.3	107.4	53.8	75.1	107.4	41.2	67.4	104.4	55.2	74.7	94.9	48.4	66.8	108.8	55.2	76.5
09	**FINISHES**	85.7	40.0	60.5	89.9	57.6	72.1	88.6	46.8	65.6	88.4	41.0	62.3	83.7	48.5	64.3	92.7	59.8	74.6
COVERS	**DIVS. 10 - 14, 25, 28, 41, 43, 44, 46**	100.0	48.7	89.6	100.0	78.8	95.7	100.0	49.1	89.7	100.0	36.5	87.2	100.0	45.2	88.9	100.0	79.2	95.8
21, 22, 23	**FIRE SUPPRESSION, PLUMBING & HVAC**	95.3	50.2	77.1	100.1	50.6	80.1	95.3	49.4	76.8	95.3	48.7	76.5	100.3	52.5	81.0	99.9	56.2	82.3
26, 27, 3370	**ELECTRICAL, COMMUNICATIONS & UTIL.**	90.1	51.2	69.5	93.7	64.5	78.3	95.0	37.1	64.4	99.0	67.8	82.5	100.2	63.5	80.8	100.7	70.2	84.6
MF2014	**WEIGHTED AVERAGE**	93.4	52.0	75.4	97.2	61.6	81.7	95.9	50.4	76.0	96.4	51.8	76.9	96.7	59.1	80.3	98.5	64.0	83.4

	DIVISION	ARKANSAS												CALIFORNIA					
		PINE BLUFF			RUSSELLVILLE			TEXARKANA			WEST MEMPHIS			ALHAMBRA			ANAHEIM		
		716			728			718			723			917 - 918			928		
		MAT.	INST.	TOTAL	MAT.	INST.	TOTAL	MAT.	INST.	TOTAL	MAT.	INST.	TOTAL	MAT.	INST.	TOTAL	MAT.	INST.	TOTAL
015433	CONTRACTOR EQUIPMENT		89.1	89.1		89.1	89.1		89.9	89.9		108.3	108.3		100.1	100.1		100.8	100.8
0241, 31 - 34	SITE & INFRASTRUCTURE, DEMOLITION	83.7	86.8	85.9	78.3	84.9	83.0	94.9	87.1	89.3	110.5	101.1	103.8	97.9	110.9	107.1	99.4	108.3	105.7
0310	Concrete Forming & Accessories	80.7	60.0	62.8	86.6	54.7	59.1	87.3	40.1	46.6	95.5	48.5	54.9	117.1	116.4	116.5	105.4	124.2	121.6
0320	Concrete Reinforcing	93.5	68.2	80.7	88.0	67.3	77.5	93.1	67.5	80.1	84.5	70.1	77.2	106.6	113.3	110.0	93.8	113.2	103.7
0330	Cast-in-Place Concrete	84.5	70.3	78.7	81.1	45.3	66.4	92.1	43.3	72.0	88.4	55.1	74.7	95.1	120.8	105.6	92.9	123.3	105.4
03	CONCRETE	92.1	65.5	79.0	86.7	54.4	70.8	89.8	47.2	68.9	95.5	56.2	76.2	101.1	116.5	108.7	101.8	120.8	111.2
04	MASONRY	125.0	50.9	78.8	96.1	37.3	59.5	102.4	33.1	59.2	79.4	41.9	56.0	122.7	121.1	121.7	80.0	118.8	104.2
05	METALS	104.6	69.4	93.8	99.5	66.5	89.3	96.8	66.9	87.6	95.0	80.3	90.5	85.3	100.9	90.1	102.7	101.1	102.2
06	WOOD, PLASTICS & COMPOSITES	87.0	62.5	73.3	91.3	59.0	73.2	95.5	42.0	65.5	100.1	48.3	71.1	100.0	112.1	106.7	101.9	122.6	113.5
07	THERMAL & MOISTURE PROTECTION	94.5	57.8	79.5	99.2	43.9	76.5	95.2	44.5	74.4	103.7	49.1	81.3	95.1	118.1	104.5	99.4	122.0	108.7
08	OPENINGS	102.1	58.4	91.9	97.7	55.7	87.9	107.3	46.5	93.2	100.6	54.0	89.7	91.8	114.8	97.2	104.2	120.6	108.0
0920	Plaster & Gypsum Board	83.6	61.8	68.9	83.3	58.3	66.4	86.6	40.7	55.6	100.2	46.9	64.2	97.4	112.5	107.6	107.1	123.2	118.0
0950, 0980	Ceilings & Acoustic Treatment	83.3	61.8	69.2	83.2	58.3	66.8	86.6	40.7	56.4	84.3	46.9	59.7	100.2	112.5	108.3	105.0	123.2	117.0
0960	Flooring	98.2	60.8	87.5	96.2	59.1	85.6	100.4	51.0	86.3	71.0	53.4	66.0	95.4	110.2	99.6	101.9	112.0	104.8
0970, 0990	Wall Finishes & Painting/Coating	104.4	55.2	74.7	107.4	35.0	63.7	104.4	30.3	59.7	94.9	50.4	68.0	100.5	111.5	107.1	98.9	106.8	103.7
09	FINISHES	88.3	59.8	72.6	86.7	54.1	68.7	90.7	40.4	62.9	85.0	48.8	65.0	99.8	114.0	107.6	100.5	120.1	111.3
COVERS	DIVS. 10 - 14, 25, 28, 41, 43, 44, 46	100.0	79.2	95.8	100.0	41.0	88.1	100.0	32.6	86.4	100.0	45.2	88.9	100.0	110.5	102.1	100.0	112.0	102.4
21, 22, 23	FIRE SUPPRESSION, PLUMBING & HVAC	100.1	52.9	81.1	95.3	51.5	77.6	100.1	53.5	81.3	95.6	65.8	83.6	95.2	115.6	103.4	100.0	118.6	107.5
26, 27, 3370	ELECTRICAL, COMMUNICATIONS & UTIL.	97.0	70.2	82.8	93.7	44.7	67.8	98.9	36.0	65.7	101.9	65.7	82.8	118.4	120.5	119.5	92.2	105.7	99.3
MF2014	WEIGHTED AVERAGE	99.4	63.3	83.6	94.6	53.7	76.8	98.2	49.2	76.8	96.1	62.4	81.4	98.3	114.8	105.5	99.4	114.9	106.2

	DIVISION	CALIFORNIA																	
		BAKERSFIELD			BERKELEY			EUREKA			FRESNO			INGLEWOOD			LONG BEACH		
		932 - 933			947			955			936 - 938			903 - 905			906 - 908		
		MAT.	INST.	TOTAL	MAT.	INST.	TOTAL	MAT.	INST.	TOTAL	MAT.	INST.	TOTAL	MAT.	INST.	TOTAL	MAT.	INST.	TOTAL
015433	CONTRACTOR EQUIPMENT		98.8	98.8		100.3	100.3		98.5	98.5		98.8	98.8		96.5	96.5		96.5	96.5
0241, 31 - 34	SITE & INFRASTRUCTURE, DEMOLITION	100.4	106.0	104.4	118.4	107.6	110.7	110.8	103.9	105.9	103.1	105.4	104.7	89.3	103.5	99.4	95.9	103.5	101.3
0310	Concrete Forming & Accessories	104.0	124.3	121.5	115.9	149.9	145.2	114.8	133.6	131.0	104.3	134.2	130.1	111.8	115.4	114.9	106.5	115.4	114.1
0320	Concrete Reinforcing	103.4	113.2	108.4	94.5	114.4	104.6	102.5	114.1	108.4	80.6	113.7	97.5	102.7	113.3	108.1	101.9	113.3	107.7
0330	Cast-in-Place Concrete	94.0	122.6	105.7	123.4	124.8	124.0	100.8	118.0	107.9	99.4	118.9	107.4	84.5	121.4	99.7	96.2	121.4	106.5
03	CONCRETE	97.9	120.6	109.1	107.3	132.7	119.8	112.8	123.2	117.9	98.6	123.8	111.0	90.9	116.3	103.4	99.8	116.3	107.9
04	MASONRY	100.2	118.3	111.5	113.4	134.3	126.5	104.0	132.6	121.8	103.6	122.0	115.1	76.3	121.2	104.3	85.3	121.2	107.7
05	METALS	105.3	100.9	104.0	106.9	100.5	104.9	102.6	98.9	101.4	105.6	100.5	104.1	94.4	101.5	96.6	94.3	101.5	96.5
06	WOOD, PLASTICS & COMPOSITES	96.0	122.7	111.0	111.4	154.9	135.8	116.6	138.5	128.9	108.1	138.5	125.1	103.8	110.7	107.6	96.9	110.7	104.6
07	THERMAL & MOISTURE PROTECTION	98.1	116.1	105.5	105.2	137.2	118.3	103.4	119.8	110.1	88.7	116.0	99.9	98.3	118.0	106.4	98.5	118.0	106.5
08	OPENINGS	97.8	116.1	102.0	93.0	140.6	104.1	103.5	113.5	105.8	98.1	124.7	104.3	88.8	114.1	94.7	88.8	114.1	94.7
0920	Plaster & Gypsum Board	101.9	123.2	116.3	111.6	156.0	141.6	113.4	139.5	131.0	100.6	139.5	126.9	103.8	111.0	108.7	99.9	111.0	107.4
0950, 0980	Ceilings & Acoustic Treatment	96.3	123.2	114.0	106.2	156.0	139.0	110.1	139.5	129.4	92.8	139.5	123.5	100.3	111.0	107.4	100.3	111.0	107.4
0960	Flooring	106.8	110.2	107.7	108.2	127.6	113.8	106.0	116.5	109.0	113.3	133.5	119.1	104.4	110.2	106.0	101.8	110.2	104.2
0970, 0990	Wall Finishes & Painting/Coating	111.4	102.0	105.7	102.5	143.9	127.5	100.5	46.2	67.7	129.0	107.2	115.8	100.9	111.5	107.3	100.9	111.5	107.3
09	FINISHES	100.0	120.5	111.3	105.4	147.2	128.4	106.1	123.6	115.8	102.1	133.2	119.2	102.7	113.2	108.5	101.8	113.2	108.1
COVERS	DIVS. 10 - 14, 25, 28, 41, 43, 44, 46	100.0	111.9	102.4	100.0	126.4	105.3	100.0	122.2	104.5	100.0	122.2	104.5	100.0	110.4	102.1	100.0	110.4	102.1
21, 22, 23	FIRE SUPPRESSION, PLUMBING & HVAC	100.1	117.1	107.0	95.3	147.8	116.5	95.2	110.6	101.4	100.2	111.3	104.7	94.8	115.6	103.2	94.8	115.6	103.2
26, 27, 3370	ELECTRICAL, COMMUNICATIONS & UTIL.	103.6	102.5	103.1	109.4	145.0	128.2	99.1	116.7	108.4	93.8	100.8	97.5	103.8	120.5	112.6	103.5	120.5	112.5
MF2014	WEIGHTED AVERAGE	100.7	113.6	106.3	102.7	135.1	116.8	102.1	116.4	108.4	100.1	115.3	106.7	94.8	114.1	103.2	96.2	114.1	104.0

	DIVISION	CALIFORNIA																	
		LOS ANGELES			MARYSVILLE			MODESTO			MOJAVE			OAKLAND			OXNARD		
		900 - 902			959			953			935			946			930		
		MAT.	INST.	TOTAL	MAT.	INST.	TOTAL	MAT.	INST.	TOTAL	MAT.	INST.	TOTAL	MAT.	INST.	TOTAL	MAT.	INST.	TOTAL
015433	CONTRACTOR EQUIPMENT		100.1	100.1		98.5	98.5		98.5	98.5		98.8	98.8		100.3	100.3		97.7	97.7
0241, 31 - 34	SITE & INFRASTRUCTURE, DEMOLITION	96.6	107.1	104.1	107.2	105.0	105.7	102.2	105.1	104.3	96.4	106.0	103.2	124.7	107.6	112.5	103.3	104.0	103.8
0310	Concrete Forming & Accessories	108.6	124.3	122.2	104.6	134.4	130.3	100.7	134.3	129.7	116.1	112.1	112.6	104.9	149.9	143.7	107.5	124.3	122.0
0320	Concrete Reinforcing	103.6	113.4	108.6	102.5	113.8	108.2	106.2	113.8	110.0	97.3	113.1	105.3	96.7	114.4	105.7	95.5	113.1	104.4
0330	Cast-in-Place Concrete	91.2	122.0	103.9	112.6	119.2	115.3	100.8	119.2	108.4	86.0	122.4	101.0	117.0	124.8	120.2	100.1	122.7	109.4
03	CONCRETE	95.9	120.6	108.0	113.9	124.0	118.9	104.7	124.0	114.2	91.0	115.1	102.8	106.8	132.7	119.5	99.2	120.6	109.7
04	MASONRY	90.5	122.6	110.5	104.9	123.1	116.2	102.9	123.1	115.5	101.9	118.3	112.1	120.6	134.3	129.2	104.8	116.5	112.1
05	METALS	101.1	102.9	101.7	102.0	101.8	102.0	99.1	101.7	99.9	102.8	100.3	102.0	101.4	100.4	101.1	100.7	100.7	100.7
06	WOOD, PLASTICS & COMPOSITES	103.0	122.6	113.9	103.0	138.5	122.9	98.5	138.5	120.9	107.4	106.8	107.1	98.8	154.9	130.2	102.1	122.7	113.6
07	THERMAL & MOISTURE PROTECTION	97.6	121.2	107.3	102.9	121.5	110.5	102.5	118.7	109.1	94.4	110.9	101.2	103.1	137.2	117.1	97.4	119.8	106.6
08	OPENINGS	95.4	120.7	101.3	102.8	125.3	108.0	101.6	125.4	107.1	93.2	107.5	96.5	93.1	140.6	104.1	95.7	120.6	101.5
0920	Plaster & Gypsum Board	103.6	123.2	116.9	105.0	139.5	128.3	107.5	139.5	129.1	110.8	106.8	108.1	105.7	156.0	139.7	104.7	123.2	117.2
0950, 0980	Ceilings & Acoustic Treatment	110.5	123.2	118.8	109.2	139.5	129.1	105.0	139.5	127.7	94.2	106.8	102.5	108.9	156.0	139.9	96.6	123.2	114.1
0960	Flooring	102.0	112.0	104.9	102.0	117.6	106.4	102.4	112.8	105.3	112.6	110.2	111.9	103.4	127.6	110.3	104.8	112.0	106.9
0970, 0990	Wall Finishes & Painting/Coating	100.0	111.5	106.9	100.5	115.8	109.7	100.5	117.0	110.5	111.1	102.0	105.6	102.5	143.9	127.5	111.1	101.1	105.1
09	FINISHES	104.2	120.6	113.3	103.1	131.5	118.8	102.2	130.9	118.0	101.8	109.9	106.2	104.3	147.2	127.9	99.4	119.6	110.5
COVERS	DIVS. 10 - 14, 25, 28, 41, 43, 44, 46	100.0	111.9	102.4	100.0	122.2	104.5	100.0	122.2	104.5	100.0	107.8	101.6	100.0	126.3	105.3	100.0	112.2	102.5
21, 22, 23	FIRE SUPPRESSION, PLUMBING & HVAC	99.9	118.6	107.5	95.2	109.0	100.8	100.0	111.3	104.6	95.3	115.1	103.3	100.1	147.8	119.4	100.1	118.6	107.6
26, 27, 3370	ELECTRICAL, COMMUNICATIONS & UTIL.	102.0	121.3	112.2	95.7	107.4	101.9	98.2	104.0	101.3	92.0	99.9	96.2	108.5	139.3	124.8	97.7	107.3	102.8
MF2014	WEIGHTED AVERAGE	99.2	117.5	107.2	101.3	115.9	107.7	100.8	115.7	107.3	96.8	109.8	102.5	103.1	134.3	116.7	99.6	114.4	106.0

	DIVISION	CALIFORNIA																	
		PALM SPRINGS			PALO ALTO			PASADENA			REDDING			RICHMOND			RIVERSIDE		
		922			943			910 - 912			960			948			925		
		MAT.	INST.	TOTAL	MAT.	INST.	TOTAL	MAT.	INST.	TOTAL	MAT.	INST.	TOTAL	MAT.	INST.	TOTAL	MAT.	INST.	TOTAL
015433	CONTRACTOR EQUIPMENT		99.6	99.6		100.3	100.3		100.1	100.1		98.5	98.5		100.3	100.3		99.6	99.6
0241, 31 - 34	SITE & INFRASTRUCTURE, DEMOLITION	90.9	106.3	101.8	114.4	107.6	109.6	94.9	110.9	106.3	125.1	105.0	110.8	123.8	107.6	112.3	97.9	106.3	103.9
0310	Concrete Forming & Accessories	101.9	115.3	113.4	103.0	146.3	140.3	106.2	115.3	114.1	107.8	142.7	137.9	118.7	149.7	145.5	105.8	124.2	121.7
0320	Concrete Reinforcing	107.8	113.1	110.5	94.5	114.4	104.6	107.5	113.3	110.5	118.7	113.8	116.2	94.5	114.3	104.6	104.7	113.1	109.0
0330	Cast-in-Place Concrete	88.6	123.2	102.8	104.4	124.8	112.8	90.2	120.8	102.8	118.0	119.2	118.5	120.0	124.8	122.0	96.3	123.3	107.4
03	CONCRETE	96.6	116.8	106.5	96.5	131.1	113.5	96.6	116.0	106.2	122.7	127.7	125.2	108.8	132.6	120.5	102.6	120.8	111.6
04	MASONRY	77.6	118.5	103.1	97.4	134.3	120.4	106.5	121.1	115.6	130.0	123.1	125.7	113.2	132.8	125.4	78.6	118.1	103.2
05	METALS	103.3	100.8	102.5	98.9	100.3	99.3	85.3	100.8	90.1	101.6	101.8	101.7	98.9	100.2	99.3	102.8	101.0	102.3
06	WOOD, PLASTICS & COMPOSITES	96.6	110.8	104.5	96.1	150.2	126.4	86.5	110.6	100.0	110.6	149.8	132.6	115.3	154.9	137.5	101.9	122.6	113.5
07	THERMAL & MOISTURE PROTECTION	99.1	118.7	107.2	102.7	137.5	117.0	94.9	117.4	104.1	121.6	122.1	121.8	103.3	135.9	116.7	99.6	121.1	108.4
08	OPENINGS	100.1	114.1	103.4	93.1	136.9	103.3	91.8	114.0	97.0	114.1	131.5	118.1	93.1	139.4	103.9	102.8	120.6	107.0
0920	Plaster & Gypsum Board	102.0	111.0	108.1	103.9	151.1	135.9	90.2	111.0	104.3	106.2	151.1	136.6	113.6	156.0	142.3	106.3	123.2	117.7
0950, 0980	Ceilings & Acoustic Treatment	101.7	111.0	107.8	107.1	151.1	136.0	100.2	111.0	107.3	131.5	151.1	144.4	107.1	156.0	139.3	109.2	123.2	118.4
0960	Flooring	104.6	107.2	105.3	102.4	127.6	109.6	90.8	110.2	96.3	98.0	117.6	103.6	109.9	127.6	115.0	106.0	112.0	107.7
0970, 0990	Wall Finishes & Painting/Coating	97.2	111.3	105.7	102.5	143.9	127.5	100.5	111.5	107.1	115.9	115.8	115.8	102.5	143.9	127.5	97.2	106.8	103.0
09	FINISHES	99.1	112.8	106.6	102.7	144.4	125.6	97.3	113.2	106.0	109.3	138.2	125.3	106.9	147.2	129.1	102.1	120.1	112.0
COVERS	DIVS. 10 - 14, 25, 28, 41, 43, 44, 46	100.0	110.7	102.2	100.0	125.9	105.2	100.0	110.3	102.1	100.0	123.4	104.7	100.0	126.3	105.3	100.0	112.0	102.4
21, 22, 23	FIRE SUPPRESSION, PLUMBING & HVAC	95.2	115.5	103.4	95.3	145.5	115.6	95.2	115.6	103.4	100.2	109.0	103.7	95.3	147.8	116.5	100.0	118.6	107.5
26, 27, 3370	ELECTRICAL, COMMUNICATIONS & UTIL.	95.3	105.4	100.7	108.4	150.2	130.5	115.1	120.5	118.0	99.2	107.4	103.5	109.0	132.7	121.5	91.9	105.4	99.1
MF2014	WEIGHTED AVERAGE	97.2	112.0	103.6	98.9	134.5	114.4	96.4	114.6	104.3	107.8	117.7	112.1	101.8	133.1	115.4	99.4	114.6	106.0

	DIVISION	CALIFORNIA																	
		SACRAMENTO			SALINAS			SAN BERNARDINO			SAN DIEGO			SAN FRANCISCO			SAN JOSE		
		942,956 - 958			939			923 - 924			919 - 921			940 - 941			951		
		MAT.	INST.	TOTAL	MAT.	INST.	TOTAL	MAT.	INST.	TOTAL	MAT.	INST.	TOTAL	MAT.	INST.	TOTAL	MAT.	INST.	TOTAL
015433	CONTRACTOR EQUIPMENT		99.9	99.9		98.8	98.8		99.6	99.6		100.1	100.1		110.7	110.7		99.3	99.3
0241, 31 - 34	SITE & INFRASTRUCTURE, DEMOLITION	100.6	113.3	109.6	116.3	105.7	108.8	77.7	106.3	98.0	101.7	104.0	103.4	127.0	113.6	117.5	133.5	100.4	109.9
0310	Concrete Forming & Accessories	103.3	137.0	132.4	111.0	137.4	133.7	109.6	115.2	114.4	105.6	113.3	112.2	104.6	150.9	144.6	107.0	149.8	143.9
0320	Concrete Reinforcing	89.4	113.8	101.8	96.0	114.2	105.2	104.7	113.0	108.9	104.2	113.1	108.7	110.1	115.0	112.6	93.5	114.5	104.1
0330	Cast-in-Place Concrete	99.0	120.3	107.7	98.9	119.6	107.4	66.6	123.2	89.8	98.9	107.8	102.6	120.0	126.4	122.6	117.1	124.2	120.0
03	CONCRETE	97.0	125.2	110.9	108.5	125.6	116.9	76.8	116.8	96.4	101.8	110.7	106.2	110.3	134.4	122.1	110.8	132.9	121.7
04	MASONRY	98.4	123.1	113.8	101.6	128.1	118.1	84.9	115.9	104.2	97.4	115.6	108.7	121.1	141.5	133.8	131.4	134.4	133.3
05	METALS	97.0	96.1	96.7	105.5	102.9	104.7	102.8	100.5	102.1	101.1	101.4	101.2	107.6	110.8	108.6	97.5	107.5	100.6
06	WOOD, PLASTICS & COMPOSITES	92.9	141.7	120.2	107.3	141.5	126.4	105.8	110.8	108.6	100.6	110.2	106.0	98.8	155.1	130.3	111.3	154.7	135.6
07	THERMAL & MOISTURE PROTECTION	110.8	121.3	115.1	95.0	126.1	107.7	98.3	117.7	106.3	99.3	107.1	102.5	104.9	141.9	120.1	98.9	139.3	115.4
08	OPENINGS	106.3	127.2	111.1	96.9	133.3	105.4	100.2	114.1	103.4	99.1	111.8	102.1	97.3	140.7	107.4	92.8	140.5	103.9
0920	Plaster & Gypsum Board	101.0	142.6	129.1	105.8	142.6	130.6	108.2	111.0	110.1	97.8	110.3	106.3	108.3	156.0	140.6	102.7	156.0	138.7
0950, 0980	Ceilings & Acoustic Treatment	107.1	142.6	130.4	94.2	142.6	126.0	105.0	111.0	109.0	107.7	110.3	109.4	117.1	156.0	142.7	103.3	156.0	138.0
0960	Flooring	102.6	117.6	106.9	107.1	121.1	111.1	107.8	110.2	108.5	98.5	112.0	102.4	103.4	127.6	110.3	95.4	127.6	104.6
0970, 0990	Wall Finishes & Painting/Coating	100.2	116.7	110.1	112.3	143.9	131.4	97.2	104.0	101.3	97.1	111.5	105.8	102.5	153.2	133.1	100.8	143.9	126.8
09	FINISHES	101.4	133.6	119.1	101.4	137.0	121.0	100.5	112.5	107.1	102.9	112.9	108.4	106.5	148.3	129.6	100.7	147.0	126.2
COVERS	DIVS. 10 - 14, 25, 28, 41, 43, 44, 46	100.0	123.0	104.6	100.0	122.6	104.6	100.0	108.1	101.6	100.0	109.7	102.0	100.0	126.8	105.4	100.0	125.9	105.2
21, 22, 23	FIRE SUPPRESSION, PLUMBING & HVAC	100.0	120.8	108.4	95.3	117.7	104.4	95.2	115.6	103.4	99.9	116.9	106.8	100.1	175.0	130.3	100.0	147.3	119.1
26, 27, 3370	ELECTRICAL, COMMUNICATIONS & UTIL.	103.5	108.7	106.2	93.2	120.9	107.8	95.3	103.4	99.6	101.2	98.5	99.8	108.6	157.8	134.6	101.1	155.9	130.1
MF2014	WEIGHTED AVERAGE	100.5	119.3	108.7	100.2	121.6	109.5	95.1	111.3	102.1	100.5	109.6	104.5	105.2	145.4	122.7	102.5	136.6	117.4

	DIVISION	CALIFORNIA																	
		SAN LUIS OBISPO			SAN MATEO			SAN RAFAEL			SANTA ANA			SANTA BARBARA			SANTA CRUZ		
		934			944			949			926 - 927			931			950		
		MAT.	INST.	TOTAL	MAT.	INST.	TOTAL	MAT.	INST.	TOTAL	MAT.	INST.	TOTAL	MAT.	INST.	TOTAL	MAT.	INST.	TOTAL
015433	CONTRACTOR EQUIPMENT		98.8	98.8		100.3	100.3		100.3	100.3		99.6	99.6		98.8	98.8		99.3	99.3
0241, 31 - 34	SITE & INFRASTRUCTURE, DEMOLITION	108.4	106.0	106.7	121.3	107.6	111.6	113.2	113.4	113.4	89.4	106.3	101.4	103.3	106.0	105.2	133.1	100.0	109.6
0310	Concrete Forming & Accessories	117.9	115.4	115.8	109.0	149.8	144.2	112.9	149.9	144.9	109.9	115.3	114.5	108.3	124.2	122.1	107.0	137.6	133.4
0320	Concrete Reinforcing	97.3	113.1	105.4	94.5	114.6	104.7	95.1	114.6	105.1	108.3	113.1	110.8	95.5	113.2	104.5	115.7	114.2	114.9
0330	Cast-in-Place Concrete	106.3	122.5	112.9	116.2	124.8	119.7	135.2	124.1	130.7	85.0	123.2	100.7	99.8	122.6	109.1	116.3	121.4	118.4
03	CONCRETE	106.9	116.6	111.7	105.4	132.7	118.8	126.0	132.4	129.1	94.2	116.8	105.3	99.0	120.6	109.6	113.9	126.5	120.1
04	MASONRY	103.5	118.3	112.7	112.9	137.3	128.1	93.2	137.4	120.7	74.6	118.8	102.1	102.2	119.1	112.7	135.3	128.2	130.9
05	METALS	103.5	100.6	102.6	98.7	100.6	99.3	100.1	98.8	99.7	102.9	100.8	102.2	101.2	100.9	101.1	104.5	106.3	105.0
06	WOOD, PLASTICS & COMPOSITES	109.9	110.9	110.5	104.0	154.9	132.5	101.8	154.7	131.4	107.8	110.8	109.5	102.1	122.7	113.6	111.3	141.6	128.3
07	THERMAL & MOISTURE PROTECTION	95.1	116.9	104.0	103.1	138.8	117.7	107.2	138.9	120.2	99.4	118.3	107.2	94.6	118.9	104.5	98.8	129.2	111.3
08	OPENINGS	95.1	109.7	98.5	93.1	139.4	103.8	103.8	139.3	112.1	99.5	114.1	102.9	96.6	120.6	102.2	94.0	133.4	103.2
0920	Plaster & Gypsum Board	112.2	111.0	111.4	108.9	156.0	140.8	111.5	156.0	141.6	109.6	111.0	110.6	104.7	123.2	117.2	110.1	142.6	132.0
0950, 0980	Ceilings & Acoustic Treatment	94.2	111.0	105.3	107.1	156.0	139.3	115.4	156.0	142.1	105.0	111.0	109.0	96.6	123.2	114.1	106.7	142.6	130.3
0960	Flooring	113.5	110.2	112.6	105.1	127.6	111.5	115.0	121.1	116.8	108.3	110.2	108.8	106.2	108.6	106.9	99.6	121.1	105.7
0970, 0990	Wall Finishes & Painting/Coating	111.1	101.7	105.5	102.5	143.9	127.5	98.9	142.6	125.3	97.2	106.8	103.0	111.1	101.1	105.1	101.0	143.9	126.9
09	FINISHES	103.2	112.3	108.2	104.7	147.2	128.1	107.6	145.8	128.6	102.0	112.8	107.9	100.0	119.0	110.5	103.7	137.1	122.1
COVERS	DIVS. 10 - 14, 25, 28, 41, 43, 44, 46	100.0	120.6	104.2	100.0	126.4	105.3	100.0	125.7	105.2	100.0	110.7	102.2	100.0	112.2	102.5	100.0	122.9	104.6
21, 22, 23	FIRE SUPPRESSION, PLUMBING & HVAC	95.3	115.6	103.5	95.3	142.2	114.2	95.3	169.5	125.2	95.2	114.3	102.9	100.1	118.6	107.6	100.0	117.8	107.2
26, 27, 3370	ELECTRICAL, COMMUNICATIONS & UTIL.	92.0	106.4	99.6	108.4	148.0	129.3	105.4	120.0	113.1	95.4	105.7	100.8	91.0	110.5	101.3	100.3	120.9	111.2
MF2014	WEIGHTED AVERAGE	99.4	112.0	104.9	100.9	134.6	115.6	103.5	136.7	118.0	96.9	111.8	103.4	98.9	115.1	106.0	104.5	121.8	112.0

City Cost Indexes

	DIVISION	CALIFORNIA															COLORADO		
		SANTA ROSA			STOCKTON			SUSANVILLE			VALLEJO			VAN NUYS			ALAMOSA		
		954			952			961			945			913 - 916			811		
		MAT.	INST.	TOTAL	MAT.	INST.	TOTAL	MAT.	INST.	TOTAL	MAT.	INST.	TOTAL	MAT.	INST.	TOTAL	MAT.	INST.	TOTAL
015433	CONTRACTOR EQUIPMENT		99.0	99.0		98.5	98.5		98.5	98.5		100.3	100.3		100.1	100.1		93.8	93.8
0241, 31 - 34	SITE & INFRASTRUCTURE, DEMOLITION	103.0	105.1	104.5	102.0	105.1	104.2	131.8	105.0	112.8	100.5	113.2	109.5	111.4	110.9	111.0	134.7	88.0	101.5
0310	Concrete Forming & Accessories	102.9	148.1	141.9	104.8	136.6	132.2	109.1	142.8	138.1	103.4	147.7	141.6	112.7	115.3	115.0	105.0	67.5	72.7
0320	Concrete Reinforcing	103.4	114.8	109.2	106.2	113.8	110.1	118.7	113.2	115.9	96.3	114.5	105.6	107.5	113.3	110.5	105.1	76.4	90.5
0330	Cast-in-Place Concrete	110.6	120.8	114.8	98.2	119.2	106.9	107.4	119.3	112.3	107.7	121.5	113.4	95.1	120.8	105.7	101.1	78.4	91.8
03	CONCRETE	113.7	130.9	122.1	103.7	125.0	114.2	125.2	127.7	126.4	102.2	130.5	116.1	110.2	116.0	113.0	113.5	73.3	93.8
04	MASONRY	103.1	137.2	124.3	102.8	123.1	115.5	127.9	112.7	118.4	71.6	133.1	109.9	122.7	121.1	121.7	126.0	71.9	92.3
05	METALS	103.2	105.0	103.8	99.3	101.8	100.0	100.6	101.1	100.7	100.1	98.1	99.5	84.4	100.8	89.5	98.2	80.2	92.7
06	WOOD, PLASTICS & COMPOSITES	98.0	154.5	129.6	104.3	141.5	125.1	112.5	149.8	133.4	90.6	154.7	126.5	95.0	110.6	103.7	98.0	67.5	80.9
07	THERMAL & MOISTURE PROTECTION	99.8	136.2	114.7	102.6	119.9	109.7	122.2	120.4	121.4	105.0	135.3	117.4	95.7	117.4	104.6	104.9	78.1	93.9
08	OPENINGS	101.0	140.4	110.2	101.6	127.1	107.5	113.9	131.5	118.0	105.6	140.5	113.7	91.6	114.0	96.8	98.3	74.1	92.7
0920	Plaster & Gypsum Board	103.9	156.0	139.2	107.5	142.6	131.2	107.1	151.1	136.9	105.8	156.0	139.8	95.3	111.0	105.9	79.0	66.3	70.4
0950, 0980	Ceilings & Acoustic Treatment	105.0	156.0	138.6	112.5	142.6	132.3	124.1	151.1	141.9	117.3	156.0	142.8	97.7	111.0	106.5	96.3	66.3	76.6
0960	Flooring	105.0	115.2	107.9	102.4	112.8	105.3	98.4	122.1	105.2	110.6	127.6	115.4	93.2	110.2	98.0	113.5	54.8	96.7
0970, 0990	Wall Finishes & Painting/Coating	97.2	142.6	124.6	100.5	117.0	110.5	115.9	115.8	115.8	99.8	142.6	125.6	100.5	111.5	107.1	114.1	39.8	69.3
09	FINISHES	101.1	143.6	124.5	103.8	132.7	119.7	108.9	139.0	125.5	104.6	145.8	127.3	99.3	113.2	107.0	102.5	61.7	80.0
COVERS	DIVS. 10 - 14, 25, 28, 41, 43, 44, 46	100.0	123.8	104.8	100.0	122.5	104.6	100.0	123.5	104.7	100.0	124.3	104.9	100.0	110.3	102.1	100.0	88.5	97.7
21, 22, 23	FIRE SUPPRESSION, PLUMBING & HVAC	95.2	167.3	124.3	100.0	111.3	104.6	95.4	109.0	100.9	100.1	127.5	111.1	95.2	115.6	103.4	95.2	72.0	85.9
26, 27, 3370	ELECTRICAL, COMMUNICATIONS & UTIL.	95.7	114.3	105.5	98.2	110.6	104.8	99.6	118.4	109.5	100.9	122.4	112.3	115.1	120.5	118.0	99.1	72.5	85.0
MF2014	WEIGHTED AVERAGE	100.8	134.7	115.6	100.9	117.2	108.0	106.8	118.1	111.7	100.1	127.2	111.9	99.1	114.6	105.9	102.2	73.7	89.8

	DIVISION	COLORADO																	
		BOULDER			COLORADO SPRINGS			DENVER			DURANGO			FORT COLLINS			FORT MORGAN		
		803			808 - 809			800 - 802			813			805			807		
		MAT.	INST.	TOTAL	MAT.	INST.	TOTAL	MAT.	INST.	TOTAL	MAT.	INST.	TOTAL	MAT.	INST.	TOTAL	MAT.	INST.	TOTAL
015433	CONTRACTOR EQUIPMENT		97.5	97.5		95.8	95.8		100.4	100.4		93.8	93.8		97.5	97.5		97.5	97.5
0241, 31 - 34	SITE & INFRASTRUCTURE, DEMOLITION	94.8	96.4	95.9	96.9	94.7	95.3	96.1	102.6	100.7	128.5	88.0	99.7	107.1	96.0	99.2	97.7	95.9	96.4
0310	Concrete Forming & Accessories	103.1	79.9	83.1	93.6	79.4	81.3	100.0	76.0	79.3	111.4	67.7	73.7	100.7	74.7	78.3	103.6	74.9	78.9
0320	Concrete Reinforcing	98.9	76.6	87.5	98.1	80.2	89.0	98.1	80.2	89.0	105.1	76.5	90.5	99.0	76.6	87.6	99.1	76.5	87.6
0330	Cast-in-Place Concrete	101.8	80.0	92.9	104.6	88.3	97.9	99.0	82.8	92.4	116.3	78.5	100.8	114.9	78.9	100.1	99.9	78.9	91.3
03	CONCRETE	103.1	79.6	91.5	106.3	82.9	94.8	100.7	79.5	90.3	116.1	73.4	95.1	113.5	76.7	95.4	101.5	76.8	89.4
04	MASONRY	91.0	72.1	79.2	91.3	81.9	85.4	93.3	72.2	80.1	113.6	71.9	87.6	108.5	75.7	88.1	104.9	72.2	84.5
05	METALS	96.0	83.5	92.2	99.1	85.7	95.0	101.6	85.6	96.7	98.2	80.4	92.7	97.2	80.4	92.0	95.7	80.3	91.0
06	WOOD, PLASTICS & COMPOSITES	102.2	83.0	91.4	91.7	77.3	83.6	99.8	77.2	87.2	107.6	67.5	85.1	99.6	77.2	87.0	102.2	77.2	88.2
07	THERMAL & MOISTURE PROTECTION	103.1	81.7	94.3	103.9	85.6	96.4	102.4	75.8	91.5	104.9	78.1	93.9	103.5	73.5	91.2	103.0	81.2	94.1
08	OPENINGS	98.2	82.5	94.5	102.4	80.5	97.3	103.0	80.4	97.8	105.3	74.1	98.0	98.2	79.4	93.8	98.1	79.4	93.8
0920	Plaster & Gypsum Board	105.1	82.7	89.9	87.9	76.7	80.3	100.7	76.8	84.5	92.5	66.3	74.8	99.1	76.7	84.0	105.1	76.7	85.9
0950, 0980	Ceilings & Acoustic Treatment	96.9	82.7	87.5	104.4	76.7	86.2	107.3	76.8	87.2	96.3	66.3	76.6	96.9	76.7	83.6	96.9	76.7	83.6
0960	Flooring	99.7	83.8	95.2	92.0	68.2	85.2	96.7	84.8	93.3	118.2	54.8	100.0	96.6	54.8	84.6	100.1	54.8	87.1
0970, 0990	Wall Finishes & Painting/Coating	101.7	66.6	80.5	101.4	40.5	64.7	101.7	75.7	86.0	114.1	39.8	69.3	101.7	40.2	64.6	101.7	53.6	72.6
09	FINISHES	101.7	79.4	89.4	98.8	72.8	84.5	102.4	77.3	88.6	105.0	61.7	81.1	100.5	67.4	82.3	101.8	68.9	83.7
COVERS	DIVS. 10 - 14, 25, 28, 41, 43, 44, 46	100.0	89.4	97.9	100.0	92.2	98.4	100.0	88.8	97.7	100.0	88.4	97.7	100.0	88.8	97.7	100.0	88.8	97.7
21, 22, 23	FIRE SUPPRESSION, PLUMBING & HVAC	95.3	77.3	88.1	100.2	88.2	95.4	100.0	79.9	91.9	95.2	83.1	90.3	100.1	77.2	90.9	95.3	77.2	88.0
26, 27, 3370	ELECTRICAL, COMMUNICATIONS & UTIL.	97.4	84.5	90.6	100.8	82.1	90.9	102.6	83.2	92.4	98.5	69.4	83.1	97.4	84.5	90.6	97.8	84.5	90.7
MF2014	WEIGHTED AVERAGE	97.8	81.3	90.6	100.4	84.0	93.2	100.8	81.7	92.5	102.7	75.7	90.9	101.3	79.0	91.6	98.3	79.0	89.9

	DIVISION	COLORADO																	
		GLENWOOD SPRINGS			GOLDEN			GRAND JUNCTION			GREELEY			MONTROSE			PUEBLO		
		816			804			815			806			814			810		
		MAT.	INST.	TOTAL	MAT.	INST.	TOTAL	MAT.	INST.	TOTAL	MAT.	INST.	TOTAL	MAT.	INST.	TOTAL	MAT.	INST.	TOTAL
015433	CONTRACTOR EQUIPMENT		96.7	96.7		97.5	97.5		96.7	96.7		97.5	97.5		95.2	95.2		93.8	93.8
0241, 31 - 34	SITE & INFRASTRUCTURE, DEMOLITION	143.2	95.1	109.0	107.3	96.2	99.4	128.4	94.7	104.5	94.0	95.4	95.0	137.1	91.2	104.5	120.6	91.0	99.6
0310	Concrete Forming & Accessories	102.0	75.0	78.7	96.1	74.8	77.7	110.2	74.4	79.3	98.6	78.4	81.2	101.4	74.8	78.4	107.3	79.4	83.3
0320	Concrete Reinforcing	104.0	76.5	90.0	99.1	76.4	87.6	104.3	76.3	90.1	98.9	75.1	86.8	103.9	76.4	89.9	100.6	80.2	90.2
0330	Cast-in-Place Concrete	101.1	77.9	91.6	100.0	78.9	91.3	111.9	77.1	97.6	96.1	59.1	80.9	101.1	77.8	91.5	100.4	87.8	95.2
03	CONCRETE	118.5	76.5	97.9	111.9	76.7	94.6	112.5	75.9	94.5	98.3	71.3	85.0	109.7	76.3	93.3	102.7	82.7	92.9
04	MASONRY	98.9	72.1	82.2	107.7	71.8	85.3	132.6	71.5	94.5	102.6	49.0	69.2	106.2	71.9	84.9	95.4	81.7	86.9
05	METALS	97.9	80.9	92.6	95.9	80.1	91.0	99.5	79.2	93.2	97.2	77.5	91.1	97.1	80.0	91.8	101.1	86.2	96.5
06	WOOD, PLASTICS & COMPOSITES	93.2	77.3	84.3	94.3	77.2	84.7	105.3	77.3	89.6	96.8	83.0	89.1	94.3	77.4	84.8	100.9	77.6	87.8
07	THERMAL & MOISTURE PROTECTION	104.8	79.2	94.3	103.9	75.4	92.2	104.0	68.5	89.5	102.9	66.4	87.9	105.0	79.2	94.4	103.5	83.6	95.3
08	OPENINGS	104.2	79.4	98.5	98.2	78.8	93.7	105.0	78.8	98.9	98.1	82.4	94.5	105.5	79.5	99.4	100.2	80.6	95.6
0920	Plaster & Gypsum Board	117.8	76.7	90.0	96.9	76.7	83.3	131.0	76.7	94.3	97.6	82.7	87.5	78.3	76.7	77.2	83.9	76.7	79.1
0950, 0980	Ceilings & Acoustic Treatment	95.5	76.7	83.2	96.9	76.7	83.6	95.5	76.7	83.2	96.9	82.7	87.5	96.3	76.7	83.4	104.7	76.7	86.3
0960	Flooring	112.8	50.4	95.0	94.7	54.8	83.3	117.5	54.8	99.6	95.7	54.8	84.0	115.8	45.4	95.7	114.6	84.8	106.1
0970, 0990	Wall Finishes & Painting/Coating	114.1	66.5	85.4	101.7	66.6	80.5	114.1	66.6	85.4	101.7	25.0	55.4	114.1	39.8	69.3	114.1	37.5	67.9
09	FINISHES	107.9	69.6	86.8	100.4	71.0	84.2	109.3	70.4	87.9	99.3	69.2	82.7	103.2	65.7	82.5	103.7	75.6	88.2
COVERS	DIVS. 10 - 14, 25, 28, 41, 43, 44, 46	100.0	88.8	97.7	100.0	88.8	97.7	100.0	88.8	97.7	100.0	89.4	97.9	100.0	89.1	97.8	100.0	92.7	98.5
21, 22, 23	FIRE SUPPRESSION, PLUMBING & HVAC	95.2	83.0	90.3	95.3	76.7	87.8	99.9	82.5	92.9	100.1	77.2	90.8	95.2	83.0	90.3	99.9	76.8	90.6
26, 27, 3370	ELECTRICAL, COMMUNICATIONS & UTIL.	95.7	69.4	81.8	97.8	84.5	90.7	98.2	53.8	74.7	97.4	84.5	90.6	98.2	56.4	76.1	99.1	73.0	85.3
MF2014	WEIGHTED AVERAGE	102.3	78.1	91.8	99.8	78.9	90.7	104.8	75.3	91.9	98.9	75.5	88.7	101.4	75.4	90.1	101.1	80.3	92.0

DIVISION		COLORADO			CONNECTICUT															
		SALIDA			BRIDGEPORT			BRISTOL			HARTFORD			MERIDEN			NEW BRITAIN		
		812			066			060			061			064			060		
		MAT.	INST.	TOTAL	MAT.	INST.	TOTAL	MAT.	INST.	TOTAL	MAT.	INST.	TOTAL	MAT.	INST.	TOTAL	MAT.	INST.	TOTAL
015433	**CONTRACTOR EQUIPMENT**		95.2	95.2		100.9	100.9		100.9	100.9		100.9	100.9		101.3	101.3		100.9	100.9
0241, 31 - 34	**SITE & INFRASTRUCTURE, DEMOLITION**	128.2	91.5	102.1	109.8	105.1	106.4	108.9	105.0	106.2	104.4	105.0	104.9	106.7	105.8	106.0	109.1	105.0	106.2
0310	Concrete Forming & Accessories	110.1	74.7	79.6	98.9	124.5	121.0	98.9	124.3	120.8	95.7	124.3	120.4	98.6	124.3	120.8	99.2	124.3	120.9
0320	Concrete Reinforcing	103.7	76.4	89.8	105.1	127.3	116.4	105.1	127.3	116.4	104.6	127.3	116.2	105.1	127.3	116.4	105.1	127.3	116.4
0330	Cast-in-Place Concrete	115.8	77.8	100.2	107.9	127.1	115.8	101.1	127.1	111.8	103.0	127.1	112.9	97.3	127.1	109.5	102.8	127.1	112.7
03	**CONCRETE**	111.2	76.3	94.1	106.6	125.7	116.0	103.4	125.6	114.3	104.0	125.6	114.6	101.6	125.6	113.4	104.2	125.6	114.7
04	**MASONRY**	133.9	71.9	95.3	111.8	134.0	125.6	102.9	134.0	122.3	107.7	134.0	124.1	102.5	134.0	122.1	105.2	134.0	123.1
05	**METALS**	96.7	79.9	91.6	100.5	125.5	108.2	100.5	125.4	108.1	105.6	125.4	111.7	97.4	125.4	106.0	96.5	125.4	105.4
06	**WOOD, PLASTICS & COMPOSITES**	101.9	77.4	88.2	98.8	123.4	112.6	98.8	123.4	112.6	92.3	123.4	109.7	98.8	123.4	112.6	98.8	123.4	112.6
07	**THERMAL & MOISTURE PROTECTION**	104.0	79.2	93.8	98.6	129.1	111.1	98.7	126.0	109.9	102.8	126.0	112.3	98.7	126.0	109.9	98.7	126.0	109.9
08	**OPENINGS**	98.4	79.5	94.0	101.2	132.1	108.4	101.2	132.1	108.4	100.3	132.1	107.7	103.6	132.1	110.2	101.2	132.1	108.4
0920	Plaster & Gypsum Board	78.6	76.7	77.3	101.7	123.5	116.4	101.7	123.5	116.4	97.4	123.5	115.0	103.2	123.5	116.9	101.7	123.5	116.4
0950, 0980	Ceilings & Acoustic Treatment	96.3	76.7	83.4	86.4	123.5	110.8	86.4	123.5	110.8	89.7	123.5	111.9	90.4	123.5	112.2	86.4	123.5	110.8
0960	Flooring	120.6	45.4	99.1	101.9	131.2	110.3	101.9	131.2	110.3	100.5	131.2	109.2	101.9	131.2	110.3	101.9	131.2	110.3
0970, 0990	Wall Finishes & Painting/Coating	114.1	41.7	70.4	104.9	122.9	115.8	104.9	122.9	115.8	105.9	122.9	116.2	104.9	122.9	115.8	104.9	122.9	115.8
09	**FINISHES**	103.6	65.9	82.8	97.2	125.0	112.6	97.3	125.0	112.6	96.3	125.0	112.1	98.4	125.0	113.1	97.3	125.0	112.6
COVERS	**DIVS. 10 - 14, 25, 28, 41, 43, 44, 46**	100.0	89.2	97.8	100.0	112.4	102.5	100.0	112.4	102.5	100.0	112.4	102.5	100.0	112.4	102.5	100.0	112.4	102.5
21, 22, 23	**FIRE SUPPRESSION, PLUMBING & HVAC**	95.2	72.0	85.8	100.0	116.4	106.6	100.0	116.4	106.6	100.0	116.4	106.6	95.2	116.4	103.8	100.0	116.4	106.6
26, 27, 3370	**ELECTRICAL, COMMUNICATIONS & UTIL.**	98.4	72.5	84.7	98.5	112.2	105.7	98.5	111.8	105.6	97.9	112.8	105.8	98.4	111.8	105.5	98.6	111.8	105.6
MF2014	**WEIGHTED AVERAGE**	101.9	75.3	90.3	101.3	120.8	109.8	100.5	120.6	109.3	101.4	120.8	109.8	98.9	120.7	108.4	100.1	120.6	109.1

DIVISION		CONNECTICUT																	
		NEW HAVEN			NEW LONDON			NORWALK			STAMFORD			WATERBURY			WILLIMANTIC		
		065			063			068			069			067			062		
		MAT.	INST.	TOTAL	MAT.	INST.	TOTAL	MAT.	INST.	TOTAL	MAT.	INST.	TOTAL	MAT.	INST.	TOTAL	MAT.	INST.	TOTAL
015433	**CONTRACTOR EQUIPMENT**		101.3	101.3		101.3	101.3		100.9	100.9		100.9	100.9		100.9	100.9		100.9	100.9
0241, 31 - 34	**SITE & INFRASTRUCTURE, DEMOLITION**	108.9	105.8	106.7	101.1	105.8	104.4	109.5	105.1	106.4	110.2	105.1	106.6	109.5	105.0	106.3	109.5	105.0	106.3
0310	Concrete Forming & Accessories	98.6	124.3	120.8	98.6	124.3	120.8	98.9	124.9	121.3	98.9	124.9	121.3	98.9	124.3	120.8	98.9	124.1	120.6
0320	Concrete Reinforcing	105.1	127.3	116.4	82.4	127.3	105.2	105.1	127.4	116.5	105.1	127.4	116.5	105.1	127.3	116.4	105.1	127.2	116.4
0330	Cast-in-Place Concrete	104.5	127.1	113.8	89.1	127.1	104.7	106.2	128.6	115.4	107.9	128.6	116.4	107.9	127.1	115.8	100.8	125.8	111.0
03	**CONCRETE**	118.7	125.6	122.1	91.4	125.6	108.2	105.8	126.4	115.9	106.6	126.4	116.3	106.6	125.6	115.9	103.3	125.0	114.0
04	**MASONRY**	103.2	134.0	122.4	101.5	134.0	121.7	102.7	135.4	123.1	103.4	135.4	123.3	103.4	134.0	122.5	102.7	134.0	122.2
05	**METALS**	96.8	125.4	105.6	96.5	125.4	105.4	100.5	126.1	108.3	100.5	126.1	108.3	100.5	125.4	108.1	100.2	125.2	107.9
06	**WOOD, PLASTICS & COMPOSITES**	98.8	123.4	112.6	98.8	123.4	112.6	98.8	123.4	112.6	98.8	123.4	112.6	98.8	123.4	112.6	98.8	123.4	112.6
07	**THERMAL & MOISTURE PROTECTION**	98.8	126.1	110.0	98.7	126.0	109.9	98.8	129.6	111.4	98.7	129.6	111.4	98.7	126.1	110.0	98.9	124.6	109.5
08	**OPENINGS**	101.2	132.1	108.4	104.0	132.1	110.6	101.2	132.1	108.4	101.2	132.1	108.4	101.2	132.1	108.4	104.0	132.1	110.6
0920	Plaster & Gypsum Board	101.7	123.5	116.4	101.7	123.5	116.4	101.7	123.5	116.4	101.7	123.5	116.4	101.7	123.5	116.4	101.7	123.5	116.4
0950, 0980	Ceilings & Acoustic Treatment	86.4	123.5	110.8	84.5	123.5	110.2	86.4	123.5	110.8	86.4	123.5	110.8	86.4	123.5	110.8	84.5	123.5	110.2
0960	Flooring	101.9	131.2	110.3	101.9	131.2	110.3	101.9	131.2	110.3	101.9	131.2	110.3	101.9	131.2	110.3	101.9	133.2	110.9
0970, 0990	Wall Finishes & Painting/Coating	104.9	122.9	115.8	104.9	122.9	115.8	104.9	122.9	115.8	104.9	122.9	115.8	104.9	122.9	115.8	104.9	122.9	115.8
09	**FINISHES**	97.3	125.0	112.6	96.4	125.0	112.2	97.3	125.0	112.6	97.4	125.0	112.6	97.2	125.0	112.5	97.0	125.4	112.6
COVERS	**DIVS. 10 - 14, 25, 28, 41, 43, 44, 46**	100.0	112.4	102.5	100.0	112.4	102.5	100.0	112.6	102.6	100.0	112.6	102.6	100.0	112.4	102.5	100.0	112.4	102.5
21, 22, 23	**FIRE SUPPRESSION, PLUMBING & HVAC**	100.0	116.4	106.6	95.2	116.4	103.8	100.0	116.4	106.6	100.0	116.4	106.6	100.0	116.4	106.6	100.0	116.1	106.5
26, 27, 3370	**ELECTRICAL, COMMUNICATIONS & UTIL.**	98.4	111.8	105.5	95.3	111.8	104.0	98.5	167.2	134.8	98.5	167.2	134.8	98.0	112.2	105.5	98.5	109.6	104.4
MF2014	**WEIGHTED AVERAGE**	101.6	120.7	109.9	97.1	120.7	107.4	100.8	128.7	113.0	100.9	128.7	113.0	100.9	120.7	109.5	100.8	120.2	109.2

DIVISION		D.C.			DELAWARE									FLORIDA					
		WASHINGTON			DOVER			NEWARK			WILMINGTON			DAYTONA BEACH			FORT LAUDERDALE		
		200 - 205			199			197			198			321			333		
		MAT.	INST.	TOTAL	MAT.	INST.	TOTAL	MAT.	INST.	TOTAL	MAT.	INST.	TOTAL	MAT.	INST.	TOTAL	MAT.	INST.	TOTAL
015433	**CONTRACTOR EQUIPMENT**		105.5	105.5		118.0	118.0		118.0	118.0		118.2	118.2		98.2	98.2		91.0	91.0
0241, 31 - 34	**SITE & INFRASTRUCTURE, DEMOLITION**	103.8	94.5	97.2	102.3	113.3	110.1	102.6	113.3	110.2	99.5	113.6	109.5	105.9	89.2	94.0	96.4	77.1	82.7
0310	Concrete Forming & Accessories	97.4	78.3	80.9	96.3	102.2	101.4	97.0	102.2	101.5	98.1	102.2	101.6	97.6	68.1	72.1	95.7	68.4	72.1
0320	Concrete Reinforcing	105.4	93.8	99.5	94.0	102.9	98.5	91.5	102.9	97.3	93.4	102.9	98.2	91.2	76.6	83.8	88.2	72.1	80.0
0330	Cast-in-Place Concrete	115.4	88.0	104.2	104.6	104.6	104.6	90.0	104.6	96.0	99.2	104.6	101.4	90.2	70.8	82.2	94.6	77.1	87.4
03	**CONCRETE**	107.7	85.9	97.0	102.2	104.0	103.1	94.6	104.0	99.2	99.6	104.0	101.8	91.6	71.8	81.9	94.5	73.3	84.0
04	**MASONRY**	98.2	80.0	86.9	107.5	98.1	101.6	103.2	98.1	100.0	108.5	98.1	102.0	99.1	65.7	78.3	102.8	68.3	81.3
05	**METALS**	99.9	108.6	102.6	102.0	116.7	106.5	103.5	116.7	107.5	102.0	116.7	106.5	104.7	91.1	100.5	102.0	89.6	98.2
06	**WOOD, PLASTICS & COMPOSITES**	95.3	76.2	84.6	94.0	101.9	98.4	95.8	101.9	99.2	92.0	101.9	97.6	96.3	69.2	81.1	84.3	67.2	74.7
07	**THERMAL & MOISTURE PROTECTION**	101.7	85.8	95.2	98.8	110.0	103.4	101.7	110.0	105.1	97.9	110.0	102.9	96.2	73.6	86.9	100.3	77.2	90.8
08	**OPENINGS**	99.9	87.9	97.1	91.3	110.9	95.9	91.6	110.9	96.1	89.9	110.9	94.8	98.6	67.4	91.3	97.5	65.5	90.1
0920	Plaster & Gypsum Board	108.0	75.5	86.0	97.1	101.8	100.3	99.5	101.8	101.1	101.7	101.8	101.8	95.6	68.7	77.4	105.4	66.6	79.2
0950, 0980	Ceilings & Acoustic Treatment	112.2	75.5	88.1	91.3	101.8	98.2	88.7	101.8	97.3	92.4	101.8	98.6	84.9	68.7	74.2	87.9	66.6	73.9
0960	Flooring	104.7	93.6	101.5	99.8	109.7	102.7	96.8	109.7	100.5	104.3	109.7	105.9	110.5	73.6	100.0	105.1	69.3	94.9
0970, 0990	Wall Finishes & Painting/Coating	108.0	82.2	92.4	97.2	106.4	102.8	97.3	106.4	102.8	95.0	106.4	101.9	104.5	74.1	86.2	98.9	70.2	81.6
09	**FINISHES**	100.8	80.5	89.6	96.2	103.6	100.3	95.0	103.6	99.7	98.7	103.6	101.4	96.9	69.3	81.7	94.9	68.1	80.1
COVERS	**DIVS. 10 - 14, 25, 28, 41, 43, 44, 46**	100.0	98.6	99.7	100.0	88.0	97.6	100.0	88.0	97.6	100.0	88.0	97.6	100.0	84.4	96.9	100.0	87.4	97.5
21, 22, 23	**FIRE SUPPRESSION, PLUMBING & HVAC**	100.1	92.1	96.9	100.0	117.4	107.0	100.1	117.4	107.1	100.1	117.4	107.1	99.9	76.1	90.3	100.0	66.5	86.5
26, 27, 3370	**ELECTRICAL, COMMUNICATIONS & UTIL.**	101.5	105.8	103.8	96.6	111.9	104.7	98.7	111.9	105.7	99.8	111.9	106.2	95.8	54.9	74.2	96.0	72.7	83.7
MF2014	**WEIGHTED AVERAGE**	101.1	91.9	97.1	99.3	109.2	103.6	98.8	109.2	103.3	99.3	109.2	103.6	99.0	72.9	87.6	98.6	72.6	87.2

DIVISION		FLORIDA																	
		FORT MYERS			GAINESVILLE			JACKSONVILLE			LAKELAND			MELBOURNE			MIAMI		
		339,341			326,344			320,322			338			329			330 - 332,340		
		MAT.	INST.	TOTAL	MAT.	INST.	TOTAL	MAT.	INST.	TOTAL	MAT.	INST.	TOTAL	MAT.	INST.	TOTAL	MAT.	INST.	TOTAL
015433	**CONTRACTOR EQUIPMENT**		98.2	98.2		98.2	98.2		98.2	98.2		98.2	98.2		98.2	98.2		91.0	91.0
0241, 31 - 34	**SITE & INFRASTRUCTURE, DEMOLITION**	107.3	88.3	93.8	113.9	88.3	95.7	106.0	88.6	93.6	109.2	88.7	94.6	112.7	88.5	95.5	99.5	76.9	83.4
0310	Concrete Forming & Accessories	91.5	74.8	77.1	92.7	54.3	59.6	97.4	54.7	60.5	88.0	75.3	77.1	93.8	69.8	73.1	100.7	68.7	73.1
0320	Concrete Reinforcing	89.3	92.7	91.0	96.8	64.8	80.5	91.2	64.9	77.8	91.5	93.6	92.5	92.3	76.6	84.3	94.3	72.2	83.0
0330	Cast-in-Place Concrete	98.7	68.2	86.2	103.6	62.6	86.7	91.1	68.3	81.8	100.9	69.6	88.1	108.7	73.1	94.1	95.5	78.1	88.3
03	**CONCRETE**	95.1	76.9	86.2	102.6	60.7	82.0	92.1	62.8	77.7	96.8	77.8	87.5	103.0	73.3	88.4	96.1	73.7	85.1
04	**MASONRY**	95.8	61.9	74.7	114.0	61.2	81.1	99.1	61.2	75.5	114.1	75.8	90.2	97.5	69.8	80.2	103.1	71.9	83.7
05	**METALS**	104.2	96.8	102.0	103.6	85.0	97.9	103.2	85.4	97.7	104.1	97.8	102.2	113.4	91.3	106.6	102.4	88.5	98.1
06	**WOOD, PLASTICS & COMPOSITES**	81.1	77.4	79.0	89.9	51.8	68.5	96.3	51.8	71.4	76.5	77.4	77.0	91.5	69.2	79.0	90.6	67.2	77.5
07	**THERMAL & MOISTURE PROTECTION**	100.1	80.0	91.9	96.5	61.7	82.2	96.4	62.3	82.4	100.1	84.6	93.7	96.6	76.0	88.2	101.6	71.6	89.3
08	**OPENINGS**	98.4	74.4	92.8	96.9	52.7	86.6	98.6	55.8	88.6	98.4	75.1	93.0	97.8	70.7	91.5	99.7	65.5	91.7
0920	Plaster & Gypsum Board	101.0	77.1	84.9	91.3	50.7	63.9	95.6	50.7	65.3	97.4	77.1	83.7	91.3	68.7	76.0	103.2	66.6	78.4
0950, 0980	Ceilings & Acoustic Treatment	82.7	77.1	79.0	79.5	50.7	60.6	84.9	50.7	62.5	82.7	77.1	79.0	83.3	68.7	73.7	92.2	66.6	75.4
0960	Flooring	102.3	55.1	88.8	108.1	43.1	89.5	110.5	64.5	97.4	100.3	56.5	87.8	108.3	73.6	98.4	107.3	73.0	97.5
0970, 0990	Wall Finishes & Painting/Coating	103.7	67.6	81.9	104.5	67.6	82.3	104.5	67.6	82.3	103.7	67.6	81.9	104.5	91.9	96.9	96.0	70.2	80.5
09	**FINISHES**	94.4	70.7	81.3	95.3	52.5	71.7	97.0	56.7	74.8	93.5	71.0	81.1	95.9	72.3	82.9	97.1	69.2	81.7
COVERS	**DIVS. 10 - 14, 25, 28, 41, 43, 44, 46**	100.0	72.6	94.5	100.0	82.8	96.5	100.0	81.0	96.2	100.0	72.6	94.5	100.0	85.8	97.1	100.0	88.0	97.6
21, 22, 23	**FIRE SUPPRESSION, PLUMBING & HVAC**	97.4	64.3	84.1	98.8	64.1	84.8	99.9	64.2	85.5	97.4	80.8	90.7	99.9	78.2	91.2	100.0	66.4	86.4
26, 27, 3370	**ELECTRICAL, COMMUNICATIONS & UTIL.**	98.1	62.8	79.5	96.1	71.8	83.3	95.5	62.2	77.9	96.3	61.6	77.9	97.0	67.8	81.6	100.1	74.9	86.8
MF2014	**WEIGHTED AVERAGE**	98.6	72.6	87.3	100.3	66.7	85.7	98.8	66.4	84.7	99.4	77.8	90.0	101.7	76.3	90.6	99.8	73.1	88.2

DIVISION		FLORIDA																	
		ORLANDO			PANAMA CITY			PENSACOLA			SARASOTA			ST. PETERSBURG			TALLAHASSEE		
		327 - 328,347			324			325			342			337			323		
		MAT.	INST.	TOTAL	MAT.	INST.	TOTAL	MAT.	INST.	TOTAL	MAT.	INST.	TOTAL	MAT.	INST.	TOTAL	MAT.	INST.	TOTAL
015433	**CONTRACTOR EQUIPMENT**		98.2	98.2		98.2	98.2		98.2	98.2		98.2	98.2		98.2	98.2		98.2	98.2
0241, 31 - 34	**SITE & INFRASTRUCTURE, DEMOLITION**	107.6	88.4	93.9	117.6	87.4	96.1	117.6	87.9	96.5	114.0	88.4	95.8	110.9	88.1	94.7	105.1	87.7	92.7
0310	Concrete Forming & Accessories	101.4	72.0	76.1	96.7	43.9	51.1	94.6	51.8	57.7	96.0	75.0	77.9	94.9	51.0	57.1	99.9	44.1	51.8
0320	Concrete Reinforcing	96.6	73.9	85.0	95.3	72.5	83.7	97.7	73.0	85.1	92.5	93.5	93.0	91.5	86.1	88.7	98.3	64.7	81.2
0330	Cast-in-Place Concrete	112.1	70.6	95.1	95.7	56.7	79.7	118.2	65.9	96.8	106.6	69.5	91.4	102.0	64.5	86.6	97.2	56.4	80.5
03	**CONCRETE**	103.4	73.0	88.4	100.8	55.4	78.5	110.7	62.2	86.9	100.2	77.5	89.1	98.5	63.8	81.4	97.3	54.0	76.0
04	**MASONRY**	100.6	65.7	78.8	103.9	47.3	68.6	124.7	54.5	80.9	99.8	75.8	84.8	157.9	49.0	90.0	103.6	53.8	72.6
05	**METALS**	102.4	89.6	98.5	104.4	86.5	98.9	105.6	87.9	100.2	105.2	97.5	102.8	105.0	92.9	101.3	102.2	84.6	96.8
06	**WOOD, PLASTICS & COMPOSITES**	95.8	75.2	84.2	95.1	41.8	65.2	92.7	51.5	69.7	95.7	77.4	85.4	85.5	50.1	65.7	95.0	41.3	64.9
07	**THERMAL & MOISTURE PROTECTION**	94.9	74.3	86.4	96.7	56.6	80.2	96.6	62.3	82.5	98.1	84.6	92.5	100.3	57.8	82.8	102.6	71.5	89.9
08	**OPENINGS**	101.4	69.2	93.9	96.5	46.5	84.9	96.5	56.9	87.3	99.7	74.2	93.8	98.4	60.9	89.6	100.2	47.3	87.9
0920	Plaster & Gypsum Board	99.7	74.9	82.9	94.5	40.4	57.9	97.4	50.5	65.7	97.8	77.1	83.8	103.5	49.0	66.7	108.1	40.0	62.0
0950, 0980	Ceilings & Acoustic Treatment	91.4	74.9	80.5	83.3	40.4	55.1	83.3	50.5	61.7	86.2	77.1	80.3	84.5	49.0	61.2	94.1	40.0	58.5
0960	Flooring	104.1	73.6	95.4	110.1	43.0	90.9	106.1	63.0	93.7	111.1	57.9	95.9	104.1	55.2	90.1	111.1	62.0	97.0
0970, 0990	Wall Finishes & Painting/Coating	103.4	70.2	83.4	104.5	64.9	80.6	104.5	67.6	82.3	109.5	67.6	84.2	103.7	60.8	77.8	99.5	67.6	80.3
09	**FINISHES**	98.2	72.4	84.0	97.5	44.8	68.5	96.4	54.8	73.5	99.9	71.2	84.1	95.9	51.7	71.6	100.7	48.5	72.0
COVERS	**DIVS. 10 - 14, 25, 28, 41, 43, 44, 46**	100.0	85.1	97.0	100.0	46.2	89.1	100.0	46.3	89.1	100.0	72.6	94.5	100.0	55.8	91.1	100.0	65.8	93.1
21, 22, 23	**FIRE SUPPRESSION, PLUMBING & HVAC**	99.9	56.6	82.4	99.9	52.3	80.7	99.9	52.5	80.8	99.9	64.8	85.7	100.0	58.4	83.2	100.0	38.9	75.4
26, 27, 3370	**ELECTRICAL, COMMUNICATIONS & UTIL.**	97.6	60.0	77.7	94.5	59.3	75.9	98.5	55.8	75.9	97.2	61.6	78.4	96.3	61.6	77.9	103.3	60.0	80.4
MF2014	**WEIGHTED AVERAGE**	100.5	70.0	87.2	100.2	57.8	81.7	102.7	61.2	84.6	100.8	74.3	89.2	102.6	63.3	85.5	100.8	56.9	81.7

DIVISION		FLORIDA						GEORGIA											
		TAMPA			WEST PALM BEACH			ALBANY			ATHENS			ATLANTA			AUGUSTA		
		335 - 336,346			334,349			317,398			306			300 - 303,399			308 - 309		
		MAT.	INST.	TOTAL	MAT.	INST.	TOTAL	MAT.	INST.	TOTAL	MAT.	INST.	TOTAL	MAT.	INST.	TOTAL	MAT.	INST.	TOTAL
015433	**CONTRACTOR EQUIPMENT**		98.2	98.2		91.0	91.0		91.9	91.9		94.2	94.2		94.7	94.7		94.2	94.2
0241, 31 - 34	**SITE & INFRASTRUCTURE, DEMOLITION**	111.4	88.6	95.2	93.2	77.1	81.8	98.8	78.7	84.5	100.9	93.4	95.6	97.6	94.8	95.7	94.5	93.2	93.6
0310	Concrete Forming & Accessories	97.7	75.6	78.6	99.2	68.1	72.4	90.4	43.6	50.0	92.9	45.9	52.4	96.7	73.5	76.7	94.2	65.5	69.4
0320	Concrete Reinforcing	88.2	93.6	91.0	90.7	71.8	81.1	90.6	80.2	85.3	95.3	77.9	86.5	94.6	81.2	87.8	95.7	71.8	83.6
0330	Cast-in-Place Concrete	99.7	69.7	87.4	90.0	74.3	83.5	92.4	54.5	76.8	107.8	55.6	86.4	107.8	70.8	92.6	101.9	49.9	80.5
03	**CONCRETE**	97.0	77.9	87.6	91.3	72.1	81.9	93.1	55.8	74.8	104.8	56.0	80.8	102.2	74.1	88.4	97.5	61.7	79.9
04	**MASONRY**	103.9	75.8	86.4	102.3	66.6	80.0	102.5	49.1	69.2	81.5	53.8	64.2	95.2	66.2	77.1	95.5	43.3	63.0
05	**METALS**	104.0	98.1	102.2	101.1	89.3	97.4	105.0	85.5	99.0	91.9	73.3	86.2	92.8	77.0	88.0	91.6	71.6	85.5
06	**WOOD, PLASTICS & COMPOSITES**	89.3	77.4	82.6	89.2	67.2	76.9	85.0	38.0	58.7	91.8	40.9	63.3	96.1	75.8	84.7	93.3	70.9	80.8
07	**THERMAL & MOISTURE PROTECTION**	100.5	84.6	94.0	100.1	70.2	87.8	95.9	60.3	81.3	94.1	52.8	77.2	94.0	72.0	85.0	93.7	57.0	78.7
08	**OPENINGS**	99.7	79.2	94.9	96.8	65.5	89.5	91.1	44.2	80.2	89.5	45.7	79.3	94.7	71.9	89.4	89.5	62.2	83.1
0920	Plaster & Gypsum Board	106.1	77.1	86.5	110.2	66.6	80.7	97.5	36.5	56.3	98.7	39.4	58.7	100.9	75.4	83.6	99.8	70.4	79.9
0950, 0980	Ceilings & Acoustic Treatment	87.9	77.1	80.8	82.7	66.6	72.1	84.0	36.5	52.8	97.2	39.4	59.2	97.2	75.4	82.9	98.1	70.4	79.9
0960	Flooring	105.1	56.5	91.2	106.8	66.4	95.3	110.9	47.8	92.9	96.4	53.9	84.2	97.7	65.2	88.4	96.6	46.6	82.3
0970, 0990	Wall Finishes & Painting/Coating	103.7	67.6	81.9	98.9	70.2	81.6	105.3	52.8	73.6	106.0	46.0	69.8	106.0	85.1	93.4	106.0	46.0	69.8
09	**FINISHES**	97.4	71.0	82.8	94.7	67.5	79.7	98.1	43.2	67.9	95.4	45.7	68.0	95.7	73.4	83.4	95.2	60.5	76.1
COVERS	**DIVS. 10 - 14, 25, 28, 41, 43, 44, 46**	100.0	85.1	97.0	100.0	87.4	97.5	100.0	80.2	96.0	100.0	77.3	95.4	100.0	85.6	97.1	100.0	78.5	95.7
21, 22, 23	**FIRE SUPPRESSION, PLUMBING & HVAC**	100.0	80.8	92.2	97.4	62.7	83.4	99.9	68.0	87.0	95.2	69.6	84.9	99.9	70.6	88.1	100.0	61.2	84.4
26, 27, 3370	**ELECTRICAL, COMMUNICATIONS & UTIL.**	96.0	61.6	77.8	97.2	72.7	84.3	96.9	58.7	76.7	99.7	69.5	83.8	99.0	71.9	84.7	100.4	61.8	80.0
MF2014	**WEIGHTED AVERAGE**	100.1	78.4	90.6	97.4	71.1	85.9	98.5	61.3	82.3	95.5	63.9	81.7	97.6	74.4	87.5	96.3	63.6	82.1

DIVISION		GEORGIA																	
		COLUMBUS			DALTON			GAINESVILLE			MACON			SAVANNAH			STATESBORO		
		318 - 319			307			305			310 - 312			313 - 314			304		
		MAT.	INST.	TOTAL	MAT.	INST.	TOTAL	MAT.	INST.	TOTAL	MAT.	INST.	TOTAL	MAT.	INST.	TOTAL	MAT.	INST.	TOTAL
015433	CONTRACTOR EQUIPMENT		91.9	91.9		106.1	106.1		94.2	94.2		102.3	102.3		92.8	92.8		93.4	93.4
0241, 31 - 34	SITE & INFRASTRUCTURE, DEMOLITION	98.7	78.7	84.5	101.1	97.8	98.8	100.8	93.3	95.5	99.8	93.5	95.3	100.6	80.1	86.1	102.0	77.5	84.6
0310	Concrete Forming & Accessories	90.3	54.3	59.3	85.7	46.7	52.1	96.4	42.9	50.2	89.9	51.9	57.1	92.1	50.0	55.8	80.3	51.4	55.4
0320	Concrete Reinforcing	90.9	80.6	85.7	94.8	73.9	84.2	95.1	77.8	86.3	92.1	80.3	86.1	98.1	71.9	84.7	94.4	41.8	67.6
0330	Cast-in-Place Concrete	92.1	53.9	76.4	104.7	50.2	82.3	113.3	52.9	88.5	90.9	65.6	80.5	100.2	55.2	81.7	107.6	59.7	87.9
03	CONCRETE	93.0	60.4	77.0	103.9	54.6	79.7	106.7	53.7	80.6	92.6	63.3	78.2	98.0	57.3	78.0	104.2	54.0	79.6
04	MASONRY	102.6	55.2	73.1	83.8	36.7	54.4	89.5	54.2	67.5	115.9	44.5	71.4	98.8	50.9	69.0	84.7	40.6	57.2
05	METALS	104.5	86.3	98.9	92.8	82.8	89.7	91.2	72.5	85.4	100.1	85.9	95.7	101.3	82.4	95.5	96.3	72.6	89.0
06	WOOD, PLASTICS & COMPOSITES	85.0	52.1	66.6	75.2	47.5	59.7	95.9	37.9	63.4	91.8	51.2	69.0	96.0	45.9	68.0	68.5	53.9	60.3
07	THERMAL & MOISTURE PROTECTION	95.9	63.5	82.6	96.1	51.5	77.8	94.1	54.8	78.0	94.4	62.2	81.2	95.0	56.6	79.2	94.7	51.1	76.8
08	OPENINGS	91.1	59.0	83.7	90.1	49.5	80.6	89.5	40.0	77.9	90.0	52.8	81.3	94.9	48.4	84.1	91.1	40.6	79.4
0920	Plaster & Gypsum Board	97.5	51.1	66.1	86.2	46.2	59.2	100.9	36.3	57.3	103.0	50.1	67.2	95.6	44.7	61.2	86.9	52.8	63.9
0950, 0980	Ceilings & Acoustic Treatment	84.0	51.1	62.4	109.3	46.2	67.8	97.2	36.3	57.2	79.2	50.1	60.1	91.5	44.7	60.8	105.7	52.8	71.0
0960	Flooring	110.9	50.4	93.6	96.8	46.9	82.5	97.6	46.6	83.0	87.9	47.8	76.5	108.6	46.4	90.8	115.6	44.9	95.4
0970, 0990	Wall Finishes & Painting/Coating	105.3	66.5	81.9	96.3	59.8	74.2	106.0	46.0	69.8	107.7	52.8	74.6	103.8	58.3	76.3	103.5	39.2	64.7
09	FINISHES	98.0	53.5	73.5	103.0	47.1	72.2	95.9	41.6	66.0	86.7	49.9	66.4	98.2	48.9	71.0	106.9	49.0	75.0
COVERS	DIVS. 10 - 14, 25, 28, 41, 43, 44, 46	100.0	81.8	96.3	100.0	22.8	84.4	100.0	39.4	87.8	100.0	80.2	96.0	100.0	78.0	95.6	100.0	43.5	88.6
21, 22, 23	FIRE SUPPRESSION, PLUMBING & HVAC	99.9	63.4	85.2	95.2	57.0	79.8	95.2	69.5	84.8	99.9	66.7	86.6	100.0	61.8	84.6	95.7	56.7	80.0
26, 27, 3370	ELECTRICAL, COMMUNICATIONS & UTIL.	97.1	69.6	82.5	109.7	67.7	87.5	99.7	69.5	83.8	96.1	61.2	77.7	101.7	57.4	78.3	99.9	57.4	77.4
MF2014	WEIGHTED AVERAGE	98.4	65.4	84.0	97.2	59.0	80.5	96.0	61.6	81.0	97.2	64.7	83.0	99.2	60.9	82.5	97.3	55.8	79.2

DIVISION		GEORGIA						HAWAII									IDAHO		
		VALDOSTA			WAYCROSS			HILO			HONOLULU			STATES & POSS., GUAM			BOISE		
		316			315			967			968			969			836 - 837		
		MAT.	INST.	TOTAL	MAT.	INST.	TOTAL	MAT.	INST.	TOTAL	MAT.	INST.	TOTAL	MAT.	INST.	TOTAL	MAT.	INST.	TOTAL
015433	CONTRACTOR EQUIPMENT		91.9	91.9		91.9	91.9		99.5	99.5		99.5	99.5		164.5	164.5		98.2	98.2
0241, 31 - 34	SITE & INFRASTRUCTURE, DEMOLITION	107.9	78.8	87.2	104.7	77.4	85.3	144.8	106.5	117.6	155.4	106.5	120.6	184.9	103.7	127.2	85.5	96.3	93.2
0310	Concrete Forming & Accessories	81.0	44.1	49.2	82.8	64.7	67.2	111.2	135.6	132.2	123.6	135.6	134.0	113.7	63.6	70.5	100.8	77.4	80.7
0320	Concrete Reinforcing	92.8	76.4	84.4	92.8	72.7	82.5	123.5	117.9	120.7	132.5	117.9	125.1	214.4	30.2	120.7	99.2	80.4	89.6
0330	Cast-in-Place Concrete	90.5	55.9	76.3	102.2	48.2	80.0	198.3	125.7	168.5	163.7	125.7	148.1	171.9	105.5	144.7	91.6	88.4	90.3
03	CONCRETE	98.1	55.8	77.3	101.2	61.4	81.6	158.1	127.7	143.2	150.8	127.7	139.5	159.9	72.3	116.9	99.1	81.9	90.7
04	MASONRY	108.9	50.5	72.5	109.7	39.3	65.8	150.5	128.3	136.7	150.6	128.3	136.7	213.9	43.5	107.7	121.4	84.2	98.2
05	METALS	104.1	84.0	97.9	103.1	77.7	95.3	108.1	107.3	107.8	120.7	107.3	116.5	139.2	76.2	119.8	102.2	80.9	95.6
06	WOOD, PLASTICS & COMPOSITES	73.4	37.6	53.4	75.0	72.4	73.5	114.5	139.2	128.3	134.4	139.2	137.1	127.1	67.4	93.7	93.1	75.9	83.4
07	THERMAL & MOISTURE PROTECTION	96.1	62.4	82.3	95.9	50.3	77.2	120.8	124.6	122.4	138.4	124.6	132.7	140.4	65.2	109.6	94.6	81.4	89.2
08	OPENINGS	87.3	42.5	76.9	87.4	57.8	80.5	111.3	132.8	116.3	121.8	132.8	124.4	116.6	54.1	102.1	99.1	71.2	92.6
0920	Plaster & Gypsum Board	90.4	36.2	53.8	90.4	72.0	78.0	109.7	140.1	130.3	151.3	140.1	143.7	219.9	55.5	108.8	92.4	75.1	80.7
0950, 0980	Ceilings & Acoustic Treatment	81.5	36.2	51.7	79.6	72.0	74.6	121.5	140.1	133.8	132.0	140.1	137.3	237.3	55.5	117.8	99.1	75.1	83.3
0960	Flooring	104.7	47.8	88.4	105.9	29.9	84.1	116.0	139.9	122.8	132.2	139.9	134.4	135.5	46.4	110.0	96.3	83.7	92.7
0970, 0990	Wall Finishes & Painting/Coating	105.3	52.2	73.2	105.3	46.0	69.5	110.4	143.8	130.6	119.4	143.8	134.2	115.9	35.9	67.6	103.2	39.5	64.7
09	FINISHES	95.6	43.6	67.0	95.2	57.4	74.4	113.7	138.6	127.4	128.4	138.6	134.0	188.8	58.9	117.2	96.0	74.8	84.3
COVERS	DIVS. 10 - 14, 25, 28, 41, 43, 44, 46	100.0	77.1	95.4	100.0	51.5	90.2	100.0	116.0	103.2	100.0	116.0	103.2	100.0	75.2	95.0	100.0	87.4	97.5
21, 22, 23	FIRE SUPPRESSION, PLUMBING & HVAC	99.9	70.0	87.9	96.9	57.1	80.8	100.2	107.7	103.2	100.3	107.7	103.3	102.6	37.4	76.3	100.0	71.8	88.6
26, 27, 3370	ELECTRICAL, COMMUNICATIONS & UTIL.	95.0	56.2	74.5	99.6	57.4	77.3	106.5	121.1	114.2	107.9	121.1	114.9	153.5	41.2	94.1	98.5	72.8	84.9
MF2014	WEIGHTED AVERAGE	98.5	61.4	82.3	98.3	59.4	81.4	114.8	120.2	117.2	119.5	120.2	119.8	136.2	58.0	102.1	100.1	78.6	90.7

DIVISION		IDAHO															ILLINOIS		
		COEUR D'ALENE			IDAHO FALLS			LEWISTON			POCATELLO			TWIN FALLS			BLOOMINGTON		
		838			834			835			832			833			617		
		MAT.	INST.	TOTAL	MAT.	INST.	TOTAL	MAT.	INST.	TOTAL	MAT.	INST.	TOTAL	MAT.	INST.	TOTAL	MAT.	INST.	TOTAL
015433	CONTRACTOR EQUIPMENT		92.8	92.8		98.2	98.2		92.8	92.8		98.2	98.2		98.2	98.2		101.6	101.6
0241, 31 - 34	SITE & INFRASTRUCTURE, DEMOLITION	83.1	91.7	89.2	83.1	96.1	92.4	89.9	92.5	91.7	85.8	96.3	93.3	92.1	97.2	95.8	97.4	98.4	98.1
0310	Concrete Forming & Accessories	112.6	81.1	85.5	94.7	78.1	80.4	117.9	82.2	87.1	101.0	77.1	80.4	102.0	54.7	61.2	84.3	117.2	112.7
0320	Concrete Reinforcing	106.2	96.5	101.3	101.0	80.0	90.3	106.2	96.8	101.4	99.6	80.2	89.7	101.3	79.9	90.4	94.6	110.7	102.8
0330	Cast-in-Place Concrete	99.0	87.2	94.1	87.2	74.3	81.9	102.9	86.0	95.9	94.1	88.3	91.7	96.6	63.9	83.1	102.8	114.8	107.7
03	CONCRETE	105.7	86.1	96.1	91.3	77.2	84.4	109.3	86.2	98.0	98.3	81.7	90.1	105.8	63.3	84.9	100.1	115.2	107.5
04	MASONRY	122.9	83.9	98.6	116.5	81.0	94.4	123.4	85.9	100.0	118.8	81.1	95.3	121.6	81.1	96.4	120.3	118.3	119.1
05	METALS	96.3	86.9	93.4	110.0	79.1	100.5	95.7	88.0	93.4	110.1	80.3	100.9	110.1	78.9	100.5	97.6	113.6	102.5
06	WOOD, PLASTICS & COMPOSITES	96.8	81.0	88.0	86.8	78.8	82.3	102.4	81.0	90.4	93.1	75.9	83.4	94.2	47.2	67.9	86.4	115.4	102.6
07	THERMAL & MOISTURE PROTECTION	148.3	80.8	120.6	94.3	71.7	85.0	148.5	81.4	121.0	94.7	73.7	86.1	95.4	74.7	86.9	98.4	112.8	104.3
08	OPENINGS	117.8	73.0	107.4	102.8	68.0	94.7	117.8	75.6	108.0	99.8	66.3	92.0	102.8	48.8	90.2	94.9	104.9	97.2
0920	Plaster & Gypsum Board	161.4	80.5	106.7	78.5	78.1	78.2	162.5	80.5	107.0	80.7	75.1	76.9	82.3	45.6	57.5	91.3	115.8	107.9
0950, 0980	Ceilings & Acoustic Treatment	129.0	80.5	97.1	97.2	78.1	84.6	129.0	80.5	97.1	104.7	75.1	85.2	99.7	45.6	64.1	88.0	115.8	106.3
0960	Flooring	138.3	45.9	111.9	96.4	43.1	81.2	141.2	93.8	127.7	99.6	83.7	95.1	100.7	43.1	84.3	93.8	118.7	100.9
0970, 0990	Wall Finishes & Painting/Coating	123.8	67.8	90.0	103.3	40.3	65.3	123.8	67.8	90.0	103.1	41.1	65.7	103.3	37.3	63.4	93.8	126.4	113.5
09	FINISHES	161.8	73.3	113.0	93.3	68.7	79.7	162.9	82.8	118.8	96.6	75.0	84.7	96.7	49.7	70.8	93.3	118.9	107.4
COVERS	DIVS. 10 - 14, 25, 28, 41, 43, 44, 46	100.0	87.4	97.5	100.0	47.7	89.4	100.0	87.6	97.5	100.0	87.4	97.5	100.0	44.2	88.7	100.0	105.2	101.0
21, 22, 23	FIRE SUPPRESSION, PLUMBING & HVAC	99.5	82.6	92.7	100.9	71.7	89.1	100.7	85.6	94.6	99.9	71.8	88.6	99.9	69.3	87.6	95.1	108.8	100.7
26, 27, 3370	ELECTRICAL, COMMUNICATIONS & UTIL.	91.0	77.6	83.9	90.8	70.4	80.0	88.9	80.7	84.5	96.2	70.4	82.6	92.3	60.1	75.3	94.2	94.6	94.4
MF2014	WEIGHTED AVERAGE	107.9	82.2	96.7	99.8	74.7	88.9	108.7	84.9	98.3	101.1	77.4	90.8	102.2	67.3	87.0	97.5	109.4	102.7

City Cost Indexes

	DIVISION	ILLINOIS																	
		CARBONDALE			CENTRALIA			CHAMPAIGN			CHICAGO			DECATUR			EAST ST. LOUIS		
		629			628			618 - 619			606 - 608			625			620 - 622		
		MAT.	INST.	TOTAL	MAT.	INST.	TOTAL	MAT.	INST.	TOTAL	MAT.	INST.	TOTAL	MAT.	INST.	TOTAL	MAT.	INST.	TOTAL
015433	**CONTRACTOR EQUIPMENT**		108.0	108.0		108.0	108.0		102.4	102.4		94.3	94.3		102.4	102.4		108.0	108.0
0241, 31 - 34	**SITE & INFRASTRUCTURE, DEMOLITION**	98.7	99.5	99.2	99.0	100.1	99.8	106.4	99.0	101.2	107.0	95.7	99.0	93.2	99.1	97.4	101.2	99.8	100.2
0310	Concrete Forming & Accessories	90.0	109.0	106.4	91.6	112.8	109.9	90.7	115.2	111.8	96.6	156.7	148.4	91.8	114.9	111.8	87.6	114.8	111.1
0320	Concrete Reinforcing	91.2	111.6	101.6	91.2	111.8	101.7	94.6	105.1	99.9	99.0	158.1	129.1	89.3	102.7	96.1	91.1	109.9	100.6
0330	Cast-in-Place Concrete	97.2	102.5	99.4	97.7	117.3	105.7	119.2	109.5	115.2	107.5	149.4	124.7	106.1	110.5	107.9	99.3	117.0	106.6
03	**CONCRETE**	90.1	107.9	98.9	90.6	114.8	102.5	113.1	111.3	112.2	102.6	153.2	127.5	101.3	111.1	106.1	91.7	115.2	103.2
04	**MASONRY**	83.6	108.6	99.2	83.6	116.7	104.2	145.3	117.7	128.1	101.1	156.6	135.7	79.6	114.7	101.5	83.9	116.6	104.3
05	**METALS**	96.2	119.6	103.4	96.2	120.8	103.8	97.6	108.7	101.0	94.3	134.3	106.6	99.9	108.1	102.4	97.3	119.4	104.1
06	**WOOD, PLASTICS & COMPOSITES**	91.7	106.1	99.8	94.1	109.7	102.8	93.5	113.9	104.9	104.0	155.9	133.1	91.9	113.9	104.2	89.2	112.1	102.0
07	**THERMAL & MOISTURE PROTECTION**	97.1	101.5	98.9	97.2	111.7	103.1	99.0	113.2	104.9	98.0	145.1	117.3	103.2	109.3	105.7	97.2	110.2	102.5
08	**OPENINGS**	89.4	114.2	95.2	89.4	116.2	95.7	95.5	111.2	99.2	105.3	158.7	117.8	100.9	110.6	103.2	89.5	116.9	95.9
0920	Plaster & Gypsum Board	97.2	106.2	103.3	98.3	110.0	106.2	93.7	114.2	107.6	92.7	157.6	136.6	100.0	114.2	109.6	96.1	112.4	107.2
0950, 0980	Ceilings & Acoustic Treatment	91.4	106.2	101.2	91.4	110.0	103.6	88.0	114.2	105.2	99.2	157.6	137.6	98.1	114.2	108.7	91.4	112.4	105.2
0960	Flooring	122.2	120.2	121.6	123.2	115.7	121.0	96.7	120.2	103.4	96.9	148.0	111.5	109.1	116.7	111.3	121.2	115.7	119.6
0970, 0990	Wall Finishes & Painting/Coating	112.6	97.4	103.5	112.6	106.6	109.0	93.8	110.0	103.6	91.9	153.0	128.8	102.5	106.5	104.9	112.6	103.5	107.1
09	**FINISHES**	101.8	108.6	105.5	102.2	112.6	107.9	95.2	115.9	106.6	98.0	155.6	129.7	101.8	115.0	109.1	101.4	113.8	108.2
COVERS	**DIVS. 10 - 14, 25, 28, 41, 43, 44, 46**	100.0	102.6	100.5	100.0	104.1	100.8	100.0	104.4	100.9	100.0	124.5	105.0	100.0	104.3	100.9	100.0	104.4	100.9
21, 22, 23	**FIRE SUPPRESSION, PLUMBING & HVAC**	95.1	106.1	99.5	95.1	95.3	95.2	95.1	105.5	99.3	99.8	133.9	113.6	99.9	98.5	99.4	99.9	98.8	99.4
26, 27, 3370	**ELECTRICAL, COMMUNICATIONS & UTIL.**	95.4	107.7	101.9	96.8	107.7	102.6	97.4	94.9	96.1	98.3	133.9	117.1	99.5	90.4	94.7	96.4	103.5	100.1
MF2014	**WEIGHTED AVERAGE**	94.7	107.9	100.5	95.0	108.5	100.9	101.0	107.6	103.9	99.8	139.7	117.2	99.2	104.9	101.7	96.4	108.7	101.8

	DIVISION	ILLINOIS																	
		EFFINGHAM			GALESBURG			JOLIET			KANKAKEE			LA SALLE			NORTH SUBURBAN		
		624			614			604			609			613			600 - 603		
		MAT.	INST.	TOTAL	MAT.	INST.	TOTAL	MAT.	INST.	TOTAL	MAT.	INST.	TOTAL	MAT.	INST.	TOTAL	MAT.	INST.	TOTAL
015433	**CONTRACTOR EQUIPMENT**		102.4	102.4		101.6	101.6		92.5	92.5		92.5	92.5		101.6	101.6		92.5	92.5
0241, 31 - 34	**SITE & INFRASTRUCTURE, DEMOLITION**	97.6	98.9	98.5	99.9	98.3	98.8	106.9	94.9	98.4	100.5	94.6	96.3	99.2	99.2	99.2	106.1	94.9	98.1
0310	Concrete Forming & Accessories	96.3	114.0	111.6	90.6	116.6	113.1	98.4	159.6	151.2	91.8	143.2	136.2	104.6	125.3	122.4	97.8	154.0	146.3
0320	Concrete Reinforcing	92.2	99.9	96.1	94.1	110.6	102.5	99.0	150.8	125.4	99.8	147.4	124.0	94.3	144.7	119.9	99.0	156.5	128.3
0330	Cast-in-Place Concrete	105.7	108.4	106.8	105.9	106.2	106.0	107.4	146.5	123.5	100.1	132.5	113.4	105.7	122.5	112.6	107.5	141.2	121.3
03	**CONCRETE**	102.1	109.4	105.7	103.2	112.0	107.5	102.7	152.1	126.9	96.7	139.3	117.7	104.1	127.9	115.8	102.7	148.8	125.4
04	**MASONRY**	88.2	108.5	100.8	120.5	118.0	118.9	104.3	148.5	131.8	100.6	141.2	125.9	120.5	125.1	123.4	101.1	144.1	127.9
05	**METALS**	97.0	105.6	99.7	97.6	113.0	102.4	92.2	129.1	103.6	92.2	126.6	102.8	97.7	131.6	108.1	93.4	131.4	105.1
06	**WOOD, PLASTICS & COMPOSITES**	94.2	113.9	105.2	93.3	115.5	105.7	105.7	161.0	136.7	97.8	142.2	122.7	108.8	123.7	117.1	104.0	155.7	132.9
07	**THERMAL & MOISTURE PROTECTION**	102.7	106.6	104.3	98.5	107.6	102.2	97.8	141.6	115.8	97.0	135.7	112.9	98.7	117.8	106.5	98.2	139.3	115.1
08	**OPENINGS**	94.9	109.9	98.4	94.9	111.0	98.6	102.8	159.6	116.0	95.4	148.4	107.7	94.9	129.5	102.9	102.9	158.2	115.8
0920	Plaster & Gypsum Board	99.7	114.2	109.5	93.7	115.9	108.7	89.9	162.9	139.2	86.7	143.4	125.0	100.1	124.3	116.5	92.7	157.3	136.4
0950, 0980	Ceilings & Acoustic Treatment	91.4	114.2	106.4	88.0	115.9	106.3	99.2	162.9	141.0	99.2	143.4	128.3	88.0	124.3	111.9	99.2	157.3	137.4
0960	Flooring	110.1	120.2	113.0	96.6	118.7	102.9	96.5	140.3	109.1	93.7	131.9	104.6	102.8	122.6	108.5	96.9	140.3	109.3
0970, 0990	Wall Finishes & Painting/Coating	102.5	104.2	103.5	93.8	95.1	94.5	90.1	152.5	127.8	90.1	126.4	112.0	93.8	126.4	113.5	91.9	152.5	128.4
09	**FINISHES**	100.6	115.2	108.7	94.6	115.4	106.1	97.4	156.9	130.2	95.7	137.9	119.0	97.3	124.3	112.2	97.9	152.7	128.1
COVERS	**DIVS. 10 - 14, 25, 28, 41, 43, 44, 46**	100.0	70.8	94.1	100.0	105.1	101.0	100.0	124.5	105.0	100.0	120.8	104.2	100.0	102.5	100.5	100.0	122.3	104.5
21, 22, 23	**FIRE SUPPRESSION, PLUMBING & HVAC**	95.2	102.8	98.2	95.1	105.3	99.2	99.9	131.2	112.5	95.1	129.0	108.7	95.1	123.4	106.6	99.8	128.6	111.4
26, 27, 3370	**ELECTRICAL, COMMUNICATIONS & UTIL.**	97.3	107.6	102.8	95.1	86.6	90.6	97.5	137.9	118.9	92.3	137.1	116.0	92.2	137.1	116.0	97.3	128.5	113.8
MF2014	**WEIGHTED AVERAGE**	97.3	106.0	101.1	98.2	106.7	101.9	99.2	138.4	116.3	95.6	131.7	111.3	98.4	124.4	109.7	99.3	135.1	114.9

	DIVISION	ILLINOIS																	
		PEORIA			QUINCY			ROCK ISLAND			ROCKFORD			SOUTH SUBURBAN			SPRINGFIELD		
		615 - 616			623			612			610 - 611			605			626 - 627		
		MAT.	INST.	TOTAL	MAT.	INST.	TOTAL	MAT.	INST.	TOTAL	MAT.	INST.	TOTAL	MAT.	INST.	TOTAL	MAT.	INST.	TOTAL
015433	**CONTRACTOR EQUIPMENT**		101.6	101.6		102.4	102.4		101.6	101.6		101.6	101.6		92.5	92.5		102.4	102.4
0241, 31 - 34	**SITE & INFRASTRUCTURE, DEMOLITION**	100.4	98.3	98.9	96.5	98.7	98.0	98.0	97.4	97.5	99.8	99.7	99.7	106.1	94.9	98.1	99.0	99.1	99.1
0310	Concrete Forming & Accessories	93.7	117.4	114.2	94.2	111.8	109.4	92.3	104.2	102.5	98.1	131.3	126.7	97.8	154.0	146.3	92.6	115.3	112.1
0320	Concrete Reinforcing	91.7	110.8	101.4	91.8	105.3	98.7	94.1	103.7	99.0	86.7	136.9	112.2	99.0	156.5	128.3	94.2	105.1	99.7
0330	Cast-in-Place Concrete	102.7	114.1	107.3	105.9	102.9	104.7	103.6	97.3	101.0	105.1	126.4	113.8	107.5	141.2	121.3	100.8	109.2	104.2
03	**CONCRETE**	100.2	115.0	107.5	101.7	107.6	104.6	101.0	102.0	101.5	100.9	130.4	115.4	102.7	148.8	125.4	99.1	111.3	105.1
04	**MASONRY**	120.1	118.1	118.8	111.7	104.1	106.9	120.3	97.0	105.8	94.1	135.8	120.0	101.1	144.1	127.9	91.0	117.8	107.7
05	**METALS**	100.4	113.8	104.5	97.1	108.0	100.5	97.7	108.4	101.0	100.4	128.2	108.9	93.4	131.4	105.1	97.5	109.3	101.1
06	**WOOD, PLASTICS & COMPOSITES**	101.2	115.5	109.2	91.8	113.9	104.2	95.0	104.4	100.3	101.1	128.3	116.4	104.0	155.7	132.9	89.2	113.9	103.0
07	**THERMAL & MOISTURE PROTECTION**	99.3	112.4	104.7	102.7	103.5	103.0	98.5	98.3	98.4	101.8	129.0	113.0	98.2	139.3	115.1	104.4	112.2	107.6
08	**OPENINGS**	101.3	116.0	104.7	95.8	111.4	99.4	94.9	103.2	96.8	101.3	132.1	108.5	102.9	158.2	115.8	103.0	111.2	104.9
0920	Plaster & Gypsum Board	97.1	115.9	109.8	98.3	114.2	109.0	93.7	104.5	101.0	97.1	129.1	118.8	92.7	157.3	136.4	98.7	114.2	109.2
0950, 0980	Ceilings & Acoustic Treatment	93.0	115.9	108.1	91.4	114.2	106.4	88.0	104.5	98.8	93.0	129.1	116.7	99.2	157.3	137.4	102.1	114.2	110.1
0960	Flooring	99.9	118.7	105.3	109.1	108.1	108.8	97.7	105.6	100.0	99.9	122.6	106.4	96.9	140.3	109.3	113.5	107.8	111.9
0970, 0990	Wall Finishes & Painting/Coating	93.8	126.4	113.5	102.5	106.5	104.9	93.8	95.1	94.5	93.8	133.1	117.5	91.9	152.5	128.4	101.1	106.5	104.3
09	**FINISHES**	97.2	118.9	109.2	100.1	111.8	106.6	94.9	103.7	99.7	97.2	130.0	115.3	97.9	152.7	128.1	104.9	113.5	109.6
COVERS	**DIVS. 10 - 14, 25, 28, 41, 43, 44, 46**	100.0	105.2	101.0	100.0	71.2	94.2	100.0	99.0	99.8	100.0	114.3	102.9	100.0	122.3	104.5	100.0	104.5	100.9
21, 22, 23	**FIRE SUPPRESSION, PLUMBING & HVAC**	99.9	104.9	101.9	95.2	101.0	97.5	95.1	99.9	97.1	100.0	116.5	106.7	99.8	128.6	111.4	99.9	103.3	101.3
26, 27, 3370	**ELECTRICAL, COMMUNICATIONS & UTIL.**	96.1	96.7	96.4	94.7	80.8	87.3	87.3	95.5	91.6	96.4	131.1	114.7	97.3	128.5	113.8	102.4	92.0	96.9
MF2014	**WEIGHTED AVERAGE**	100.5	109.3	104.4	98.1	101.0	99.3	97.2	100.5	98.6	99.5	124.8	110.5	99.3	135.1	114.9	100.0	106.5	102.8

City Cost Indexes

	DIVISION	INDIANA																	
		ANDERSON			BLOOMINGTON			COLUMBUS			EVANSVILLE			FORT WAYNE			GARY		
		460			474			472			476 - 477			467 - 468			463 - 464		
		MAT.	INST.	TOTAL	MAT.	INST.	TOTAL	MAT.	INST.	TOTAL	MAT.	INST.	TOTAL	MAT.	INST.	TOTAL	MAT.	INST.	TOTAL
015433	**CONTRACTOR EQUIPMENT**		97.0	97.0		86.5	86.5		86.5	86.5		116.0	116.0		97.0	97.0		97.0	97.0
0241, 31 - 34	**SITE & INFRASTRUCTURE, DEMOLITION**	93.9	96.1	95.5	85.8	94.4	91.9	82.2	94.3	90.8	91.0	123.9	114.4	94.8	96.0	95.7	94.5	99.6	98.1
0310	Concrete Forming & Accessories	97.8	81.5	83.7	101.0	80.7	83.5	95.0	78.6	80.8	94.4	82.7	84.3	96.1	75.2	78.1	97.9	116.3	113.8
0320	Concrete Reinforcing	95.8	82.8	89.2	86.8	80.9	83.8	87.2	80.9	84.0	95.1	78.1	86.4	95.8	75.5	85.5	95.8	110.8	103.5
0330	Cast-in-Place Concrete	109.4	80.2	97.4	103.6	78.2	93.1	103.1	71.2	90.0	99.0	87.9	94.4	116.3	82.8	102.5	114.3	113.9	114.1
03	**CONCRETE**	100.8	81.7	91.5	105.0	79.7	92.6	104.2	76.3	90.5	105.3	83.8	94.7	104.0	78.4	91.4	103.2	114.2	108.6
04	**MASONRY**	94.1	80.1	85.4	95.6	76.5	83.7	95.4	76.5	83.6	91.2	82.5	85.8	98.5	77.9	85.6	95.6	114.6	107.5
05	**METALS**	92.8	89.3	91.7	95.8	77.6	90.2	95.9	77.0	90.0	89.3	85.0	87.9	92.8	85.9	90.7	92.8	108.7	97.7
06	**WOOD, PLASTICS & COMPOSITES**	100.2	81.9	89.9	112.9	80.9	95.0	107.7	78.2	91.2	93.5	82.0	87.0	99.9	74.7	85.8	97.7	115.7	107.8
07	**THERMAL & MOISTURE PROTECTION**	107.7	75.9	94.7	95.1	78.3	88.2	94.7	78.1	87.9	99.4	83.7	93.0	107.5	77.7	95.3	106.3	108.6	107.2
08	**OPENINGS**	98.5	82.1	94.7	105.8	81.1	100.1	101.5	79.6	96.4	99.1	80.8	94.9	98.5	74.2	92.9	98.5	120.3	103.6
0920	Plaster & Gypsum Board	102.5	81.6	88.4	97.5	81.0	86.4	95.0	78.2	83.6	93.4	80.8	84.9	101.8	74.2	83.1	96.1	116.4	109.8
0950, 0980	Ceilings & Acoustic Treatment	87.9	81.6	83.7	80.9	81.0	81.0	80.9	78.2	79.1	85.5	80.8	82.4	87.9	74.2	78.9	87.9	116.4	106.6
0960	Flooring	100.6	85.0	96.2	105.5	75.0	96.8	100.8	75.0	93.4	100.3	80.7	94.7	100.6	78.8	94.4	100.6	123.4	107.1
0970, 0990	Wall Finishes & Painting/Coating	105.6	69.4	83.7	95.3	82.4	87.5	95.3	82.4	87.5	101.2	84.9	91.4	105.6	73.9	86.5	105.6	123.9	116.7
09	**FINISHES**	95.6	81.0	87.6	94.8	79.8	86.5	93.0	78.2	84.9	93.8	82.6	87.6	95.4	75.7	84.5	94.7	118.7	107.9
COVERS	**DIVS. 10 - 14, 25, 28, 41, 43, 44, 46**	100.0	91.8	98.3	100.0	90.8	98.1	100.0	90.5	98.1	100.0	95.9	99.2	100.0	91.7	98.3	100.0	106.1	101.2
21, 22, 23	**FIRE SUPPRESSION, PLUMBING & HVAC**	100.0	78.8	91.5	99.7	78.9	91.3	94.9	78.8	88.4	99.9	79.6	91.7	100.0	72.1	88.7	100.0	106.6	102.7
26, 27, 3370	**ELECTRICAL, COMMUNICATIONS & UTIL.**	87.0	88.4	87.7	98.9	88.0	93.2	98.1	87.9	92.7	95.1	88.3	91.5	87.7	79.2	83.2	98.7	107.6	103.4
MF2014	**WEIGHTED AVERAGE**	97.0	83.8	91.3	99.3	81.8	91.6	97.2	80.9	90.1	97.1	86.9	92.6	97.6	79.1	89.5	98.3	110.4	103.6

	DIVISION	INDIANA																	
		INDIANAPOLIS			KOKOMO			LAFAYETTE			LAWRENCEBURG			MUNCIE			NEW ALBANY		
		461 - 462			469			479			470			473			471		
		MAT.	INST.	TOTAL	MAT.	INST.	TOTAL	MAT.	INST.	TOTAL	MAT.	INST.	TOTAL	MAT.	INST.	TOTAL	MAT.	INST.	TOTAL
015433	**CONTRACTOR EQUIPMENT**		93.7	93.7		97.0	97.0		86.5	86.5		104.4	104.4		95.0	95.0		94.5	94.5
0241, 31 - 34	**SITE & INFRASTRUCTURE, DEMOLITION**	93.5	99.3	97.7	90.3	96.0	94.4	83.1	94.3	91.0	81.2	110.4	102.0	85.6	94.6	92.0	78.2	96.8	91.4
0310	Concrete Forming & Accessories	98.6	85.6	87.4	101.2	76.6	80.0	92.6	82.6	84.0	91.5	75.4	77.6	92.6	81.0	82.6	89.5	73.3	75.5
0320	Concrete Reinforcing	99.4	83.0	91.1	86.7	81.1	83.9	86.8	82.7	84.7	86.1	74.2	80.1	96.0	82.7	89.3	87.4	77.1	82.2
0330	Cast-in-Place Concrete	101.9	86.4	95.5	108.3	82.7	97.8	103.7	81.8	94.7	97.0	74.8	87.9	108.8	78.6	96.4	100.1	74.0	89.4
03	**CONCRETE**	100.9	85.1	93.1	97.5	80.1	89.0	104.5	82.1	93.5	97.4	75.5	86.6	103.5	81.0	92.4	102.8	74.6	89.0
04	**MASONRY**	95.8	80.3	86.2	93.7	79.1	84.6	101.2	80.3	88.2	79.9	74.7	76.7	97.7	80.1	86.7	86.9	68.3	75.3
05	**METALS**	93.5	80.5	89.5	89.3	88.3	89.0	94.3	78.4	89.4	91.0	83.5	88.7	97.6	89.2	95.0	92.9	81.0	89.3
06	**WOOD, PLASTICS & COMPOSITES**	99.7	86.2	92.2	103.4	75.0	87.5	104.7	83.1	92.6	91.9	75.2	82.6	106.3	81.5	92.4	94.1	74.3	83.0
07	**THERMAL & MOISTURE PROTECTION**	100.7	80.8	92.5	107.3	75.7	94.3	94.7	80.6	88.9	100.1	76.7	90.5	97.7	76.8	89.2	86.8	69.6	79.7
08	**OPENINGS**	105.7	84.5	100.8	93.3	77.9	89.8	99.9	82.8	95.9	101.4	75.3	95.3	99.0	81.9	95.0	98.7	76.3	93.5
0920	Plaster & Gypsum Board	97.0	85.8	89.5	107.5	74.5	85.2	92.2	83.3	86.1	71.5	75.0	73.9	93.4	81.6	85.4	91.6	73.8	79.6
0950, 0980	Ceilings & Acoustic Treatment	93.8	85.8	88.6	87.9	74.5	79.1	76.7	83.3	81.0	89.7	75.0	80.0	81.7	81.6	81.6	85.5	73.8	77.8
0960	Flooring	101.4	85.0	96.7	104.6	93.3	101.3	99.7	88.8	96.6	74.1	85.0	77.2	99.9	85.0	95.6	97.9	62.8	87.9
0970, 0990	Wall Finishes & Painting/Coating	103.6	82.4	90.8	105.6	71.7	85.2	95.3	93.1	94.0	96.2	71.6	81.4	95.3	69.4	79.7	101.2	82.7	90.0
09	**FINISHES**	96.1	85.4	90.2	97.3	78.8	87.1	91.4	85.0	87.9	83.7	77.2	80.1	92.4	80.6	85.9	93.1	72.7	81.8
COVERS	**DIVS. 10 - 14, 25, 28, 41, 43, 44, 46**	100.0	93.3	98.7	100.0	91.1	98.2	100.0	91.1	98.2	100.0	43.6	88.6	100.0	90.8	98.1	100.0	43.1	88.5
21, 22, 23	**FIRE SUPPRESSION, PLUMBING & HVAC**	99.9	79.4	91.6	95.2	79.0	88.7	94.9	79.4	88.6	95.8	74.1	87.1	99.7	78.7	91.2	95.2	75.2	87.1
26, 27, 3370	**ELECTRICAL, COMMUNICATIONS & UTIL.**	101.0	88.4	94.4	91.4	78.7	84.7	97.6	82.5	89.6	93.0	73.5	82.7	90.9	79.6	84.9	93.6	75.8	84.2
MF2014	**WEIGHTED AVERAGE**	99.1	84.8	92.9	94.8	81.5	89.0	96.9	82.7	90.7	94.0	77.7	86.9	97.9	82.3	91.1	95.1	75.4	86.5

	DIVISION	INDIANA									IOWA								
		SOUTH BEND			TERRE HAUTE			WASHINGTON			BURLINGTON			CARROLL			CEDAR RAPIDS		
		465 - 466			478			475			526			514			522 - 524		
		MAT.	INST.	TOTAL	MAT.	INST.	TOTAL	MAT.	INST.	TOTAL	MAT.	INST.	TOTAL	MAT.	INST.	TOTAL	MAT.	INST.	TOTAL
015433	**CONTRACTOR EQUIPMENT**		105.3	105.3		116.0	116.0		116.0	116.0		99.4	99.4		99.4	99.4		96.1	96.1
0241, 31 - 34	**SITE & INFRASTRUCTURE, DEMOLITION**	96.7	96.2	96.3	92.9	124.1	115.1	92.4	121.3	113.0	98.3	96.5	97.0	87.5	96.5	93.9	100.0	95.1	96.5
0310	Concrete Forming & Accessories	99.5	81.4	83.9	95.4	82.2	84.0	96.2	78.8	81.2	94.8	75.5	78.2	82.6	50.4	54.8	100.5	82.3	84.8
0320	Concrete Reinforcing	97.3	79.9	88.4	95.1	82.9	88.9	87.9	48.8	68.0	91.0	82.7	86.8	91.8	78.7	85.1	91.7	84.2	87.9
0330	Cast-in-Place Concrete	106.6	79.8	95.6	95.9	85.6	91.6	104.3	84.8	96.3	111.7	55.1	88.4	111.7	59.7	90.3	111.9	79.7	98.7
03	**CONCRETE**	98.3	81.8	90.2	108.4	83.7	96.2	114.3	75.3	95.1	102.6	70.8	86.9	101.3	60.2	81.1	102.7	82.3	92.7
04	**MASONRY**	105.7	77.8	88.3	98.9	79.6	86.9	91.3	80.8	84.8	102.3	64.9	79.0	104.2	74.1	85.4	108.2	78.0	89.4
05	**METALS**	92.8	99.9	95.0	90.0	87.4	89.2	84.6	68.2	79.5	88.8	93.0	90.1	88.9	88.7	88.8	91.3	92.2	91.5
06	**WOOD, PLASTICS & COMPOSITES**	94.8	81.6	87.4	95.7	82.1	88.1	96.1	78.9	86.5	95.1	76.6	84.8	81.6	44.9	61.1	101.9	82.6	91.1
07	**THERMAL & MOISTURE PROTECTION**	101.9	80.6	93.2	99.5	80.8	91.8	99.6	82.2	92.5	102.9	73.9	91.0	103.2	69.4	89.3	103.9	80.9	94.5
08	**OPENINGS**	96.0	80.5	92.4	99.7	82.2	95.6	96.4	67.6	89.7	94.8	70.4	89.1	99.3	53.4	88.6	100.3	81.9	96.0
0920	Plaster & Gypsum Board	90.8	81.3	84.4	93.4	81.0	85.0	93.4	77.7	82.8	102.3	75.9	84.5	97.7	43.2	60.9	106.9	82.3	90.3
0950, 0980	Ceilings & Acoustic Treatment	88.9	81.3	83.9	85.5	81.0	82.6	79.6	77.7	78.3	98.7	75.9	83.7	98.7	43.2	62.3	101.2	82.3	88.8
0960	Flooring	99.3	91.1	96.9	100.3	85.0	95.9	101.2	76.4	94.1	107.9	37.8	87.8	102.2	33.2	82.5	124.0	78.3	110.9
0970, 0990	Wall Finishes & Painting/Coating	99.2	87.1	91.9	101.2	83.1	90.3	101.2	83.4	90.5	106.3	78.5	89.5	106.3	78.7	89.6	107.6	72.2	86.3
09	**FINISHES**	95.4	83.9	89.1	93.8	82.8	87.7	93.1	79.6	85.6	102.5	68.9	84.0	98.7	48.3	71.0	108.4	80.6	93.1
COVERS	**DIVS. 10 - 14, 25, 28, 41, 43, 44, 46**	100.0	92.8	98.5	100.0	93.7	98.7	100.0	95.3	99.0	100.0	86.8	97.3	100.0	65.1	92.9	100.0	91.8	98.3
21, 22, 23	**FIRE SUPPRESSION, PLUMBING & HVAC**	99.9	77.1	90.7	99.9	79.6	91.7	95.2	77.6	88.1	95.4	75.2	87.3	95.4	72.6	86.2	100.2	80.5	92.2
26, 27, 3370	**ELECTRICAL, COMMUNICATIONS & UTIL.**	99.6	86.6	92.7	93.3	86.7	89.8	93.8	84.2	88.8	100.6	67.7	83.2	101.3	81.3	90.7	98.1	80.6	88.8
MF2014	**WEIGHTED AVERAGE**	98.0	84.4	92.1	97.9	86.5	92.9	95.8	81.8	89.7	97.2	75.3	87.6	97.0	71.2	85.8	100.0	83.2	92.7

	DIVISION	IOWA																	
		COUNCIL BLUFFS			CRESTON			DAVENPORT			DECORAH			DES MOINES			DUBUQUE		
		515			508			527 - 528			521			500 - 503,509			520		
		MAT.	INST.	TOTAL	MAT.	INST.	TOTAL	MAT.	INST.	TOTAL	MAT.	INST.	TOTAL	MAT.	INST.	TOTAL	MAT.	INST.	TOTAL
015433	CONTRACTOR EQUIPMENT		95.5	95.5		99.4	99.4		99.4	99.4		99.4	99.4		101.1	101.1		94.9	94.9
0241, 31 - 34	SITE & INFRASTRUCTURE, DEMOLITION	103.8	91.5	95.0	93.9	95.5	95.1	98.7	98.6	98.6	96.9	95.5	95.9	102.9	99.8	100.7	97.9	92.4	94.0
0310	Concrete Forming & Accessories	82.0	71.2	72.7	78.4	65.8	67.5	100.0	91.3	92.5	92.3	45.4	51.8	95.4	81.8	83.6	83.3	74.9	76.1
0320	Concrete Reinforcing	93.6	79.0	86.1	89.3	82.4	85.8	91.7	97.0	94.4	91.0	76.5	83.6	98.2	86.7	92.3	90.4	84.1	87.2
0330	Cast-in-Place Concrete	116.4	73.6	98.8	115.2	64.1	94.2	107.8	90.1	100.5	108.6	58.0	87.8	101.4	87.5	95.7	109.6	98.6	105.1
03	CONCRETE	105.0	74.3	89.9	102.9	69.3	86.4	100.7	92.5	96.7	100.5	57.0	79.1	100.0	85.4	92.8	99.4	85.5	92.6
04	MASONRY	109.8	75.0	88.1	108.1	81.9	91.7	105.3	86.8	93.7	124.0	71.4	91.2	101.0	81.2	88.7	109.2	73.9	87.2
05	METALS	96.3	89.3	94.1	93.8	90.5	92.8	91.3	102.7	94.8	89.0	85.8	88.0	99.9	96.5	98.8	89.8	92.0	90.5
06	WOOD, PLASTICS & COMPOSITES	80.4	70.9	75.1	74.1	61.6	67.1	101.9	91.0	95.8	92.2	37.1	61.3	91.8	81.1	85.8	82.1	73.5	77.3
07	THERMAL & MOISTURE PROTECTION	103.3	67.1	88.5	104.2	77.6	93.3	103.4	87.5	96.8	103.1	53.4	82.7	98.5	79.4	90.6	103.6	73.9	91.4
08	OPENINGS	99.3	75.9	93.9	108.4	63.9	98.1	100.3	90.8	98.1	97.8	48.6	86.4	102.7	86.5	98.9	99.3	79.3	94.7
0920	Plaster & Gypsum Board	97.7	70.3	79.2	93.6	60.5	71.2	106.9	90.8	96.0	101.2	35.2	56.6	90.4	80.6	83.8	97.7	72.9	81.0
0950, 0980	Ceilings & Acoustic Treatment	98.7	70.3	80.0	90.8	60.5	70.9	101.2	90.8	94.4	98.7	35.2	57.0	94.2	80.6	85.2	98.7	72.9	81.8
0960	Flooring	100.8	80.7	95.0	97.4	33.2	79.1	110.3	97.0	106.5	107.6	47.2	90.3	102.5	89.6	98.8	114.9	77.0	104.1
0970, 0990	Wall Finishes & Painting/Coating	102.2	65.8	80.2	101.0	78.7	87.5	106.3	95.5	99.8	106.3	32.9	62.0	96.2	88.6	91.6	106.7	63.8	80.8
09	FINISHES	99.4	72.1	84.4	94.2	60.2	75.4	104.3	92.7	97.9	102.2	42.6	69.4	97.1	83.7	89.7	103.7	73.5	87.1
COVERS	DIVS. 10 - 14, 25, 28, 41, 43, 44, 46	100.0	88.6	97.7	100.0	69.5	93.8	100.0	94.5	98.9	100.0	82.9	96.5	100.0	92.1	98.4	100.0	90.2	98.0
21, 22, 23	FIRE SUPPRESSION, PLUMBING & HVAC	100.2	74.2	89.7	95.2	79.7	89.0	100.2	92.8	97.2	95.4	72.9	86.3	99.9	79.7	91.7	100.2	75.3	90.1
26, 27, 3370	ELECTRICAL, COMMUNICATIONS & UTIL.	103.6	81.5	91.9	93.8	81.3	87.2	95.6	91.0	93.2	98.1	44.3	69.7	105.4	81.3	92.7	102.1	78.2	89.4
MF2014	WEIGHTED AVERAGE	100.7	78.1	90.9	98.2	77.3	89.1	99.0	93.1	96.4	98.0	64.2	83.3	100.5	85.2	93.8	99.1	80.2	90.9

	DIVISION	IOWA																	
		FORT DODGE			MASON CITY			OTTUMWA			SHENANDOAH			SIBLEY			SIOUX CITY		
		505			504			525			516			512			510 - 511		
		MAT.	INST.	TOTAL	MAT.	INST.	TOTAL	MAT.	INST.	TOTAL	MAT.	INST.	TOTAL	MAT.	INST.	TOTAL	MAT.	INST.	TOTAL
015433	CONTRACTOR EQUIPMENT		99.4	99.4		99.4	99.4		94.9	94.9		95.5	95.5		99.4	99.4		99.4	99.4
0241, 31 - 34	SITE & INFRASTRUCTURE, DEMOLITION	102.2	94.4	96.6	102.3	95.4	97.4	98.1	90.5	92.7	102.2	90.5	93.9	108.1	94.2	98.2	109.9	95.0	99.3
0310	Concrete Forming & Accessories	78.9	45.1	49.7	83.1	45.5	50.6	90.3	73.7	75.9	83.6	56.5	60.2	84.0	37.9	44.2	100.5	66.4	71.1
0320	Concrete Reinforcing	89.3	67.2	78.1	89.2	81.8	85.4	91.0	86.1	88.5	93.6	68.4	80.7	93.6	65.3	79.2	91.7	77.9	84.7
0330	Cast-in-Place Concrete	108.2	44.2	81.9	108.2	57.5	87.4	112.4	53.7	88.3	112.5	59.8	90.9	110.3	44.7	83.4	111.0	55.5	88.2
03	CONCRETE	98.2	50.4	74.7	98.5	57.9	78.5	102.1	69.9	86.3	102.4	61.0	82.1	101.4	47.0	74.6	102.0	65.8	84.2
04	MASONRY	106.9	37.9	63.9	120.5	69.8	88.9	105.8	57.8	75.9	109.4	75.0	88.0	128.6	38.1	72.2	102.2	54.6	72.6
05	METALS	93.9	81.3	90.1	94.0	88.7	92.4	88.8	91.7	89.7	95.3	83.3	91.6	89.1	79.7	86.2	91.3	87.4	90.1
06	WOOD, PLASTICS & COMPOSITES	74.5	45.2	58.1	78.4	37.1	55.3	89.3	80.9	84.6	82.1	52.8	65.7	82.9	35.9	56.6	101.9	66.6	82.1
07	THERMAL & MOISTURE PROTECTION	103.5	57.1	84.5	103.0	63.5	86.8	103.7	64.6	87.7	102.6	65.2	87.2	102.9	48.3	80.5	103.4	64.3	87.4
08	OPENINGS	101.9	49.5	89.7	93.5	50.1	83.4	99.3	77.7	94.3	90.3	54.7	82.0	95.9	43.8	83.8	100.3	65.9	92.3
0920	Plaster & Gypsum Board	93.6	43.5	59.8	93.6	35.2	54.1	98.8	80.6	86.5	97.7	51.6	66.5	97.7	33.9	54.6	106.9	65.7	79.0
0950, 0980	Ceilings & Acoustic Treatment	90.8	43.5	59.7	90.8	35.2	54.2	98.7	80.6	86.8	98.7	51.6	67.7	98.7	33.9	56.1	101.2	65.7	77.9
0960	Flooring	98.9	47.2	84.1	100.9	47.2	85.6	117.9	50.2	98.5	101.5	34.3	82.3	103.1	34.0	83.3	110.3	53.7	94.1
0970, 0990	Wall Finishes & Painting/Coating	101.0	61.7	77.3	101.0	30.1	58.2	106.7	78.7	89.8	102.2	65.8	80.2	106.3	61.7	79.4	106.3	63.5	80.5
09	FINISHES	96.1	46.1	68.5	96.7	42.0	66.5	104.9	70.4	85.9	99.5	52.3	73.4	101.9	38.0	66.7	105.8	63.6	82.5
COVERS	DIVS. 10 - 14, 25, 28, 41, 43, 44, 46	100.0	81.7	96.3	100.0	86.5	97.3	100.0	82.4	96.4	100.0	66.1	93.2	100.0	80.8	96.1	100.0	88.2	97.6
21, 22, 23	FIRE SUPPRESSION, PLUMBING & HVAC	95.2	65.5	83.3	95.2	70.5	85.2	95.4	70.9	85.5	95.4	73.5	86.6	95.4	68.0	84.3	100.2	75.8	90.4
26, 27, 3370	ELECTRICAL, COMMUNICATIONS & UTIL.	100.3	41.9	69.4	99.4	55.6	76.3	100.4	72.0	85.4	98.1	81.5	89.3	98.1	41.4	68.1	98.1	73.0	84.8
MF2014	WEIGHTED AVERAGE	97.9	58.2	80.6	97.6	65.8	83.8	97.9	73.7	87.3	97.6	71.2	86.1	98.3	56.4	80.0	99.7	72.6	87.9

	DIVISION	IOWA						KANSAS											
		SPENCER			WATERLOO			BELLEVILLE			COLBY			DODGE CITY			EMPORIA		
		513			506 - 507			669			677			678			668		
		MAT.	INST.	TOTAL	MAT.	INST.	TOTAL	MAT.	INST.	TOTAL	MAT.	INST.	TOTAL	MAT.	INST.	TOTAL	MAT.	INST.	TOTAL
015433	CONTRACTOR EQUIPMENT		99.4	99.4		99.4	99.4		103.4	103.4		103.4	103.4		103.4	103.4		101.6	101.6
0241, 31 - 34	SITE & INFRASTRUCTURE, DEMOLITION	108.1	94.2	98.2	107.6	95.3	98.8	111.0	95.0	99.7	112.3	95.2	100.1	114.9	94.9	100.7	103.0	92.5	95.5
0310	Concrete Forming & Accessories	90.2	37.9	45.1	94.1	55.1	60.5	96.2	54.1	59.8	99.5	58.3	64.0	93.1	58.2	63.0	87.2	65.3	68.3
0320	Concrete Reinforcing	93.6	67.1	80.1	89.9	83.9	86.8	96.6	55.1	75.5	98.9	55.2	76.7	96.5	55.1	75.4	95.3	55.7	75.2
0330	Cast-in-Place Concrete	110.3	44.7	83.4	115.8	59.2	92.6	124.2	55.7	96.1	127.1	55.8	97.8	129.3	55.5	99.0	120.1	53.6	92.8
03	CONCRETE	101.8	47.3	75.0	104.4	63.1	84.1	119.3	56.2	88.3	119.9	58.1	89.5	121.3	57.9	90.2	111.5	60.5	86.5
04	MASONRY	128.6	38.1	72.2	107.7	73.1	86.1	99.2	58.7	73.9	108.1	59.9	78.1	118.9	58.9	81.5	105.5	68.5	82.4
05	METALS	89.0	80.5	86.4	96.4	91.1	94.7	95.6	77.3	90.0	96.0	77.7	90.4	97.4	76.7	91.1	95.3	78.9	90.3
06	WOOD, PLASTICS & COMPOSITES	89.2	35.9	59.3	91.6	48.0	67.2	97.9	52.1	72.3	103.6	57.9	78.0	95.8	57.9	74.5	88.7	66.6	76.3
07	THERMAL & MOISTURE PROTECTION	103.8	48.2	81.0	103.3	71.3	90.2	94.8	60.7	80.8	99.1	61.6	83.7	99.1	60.7	83.4	93.1	75.9	86.0
08	OPENINGS	107.5	44.4	92.9	94.4	58.2	86.0	98.6	48.4	86.9	103.2	51.5	91.2	103.1	51.5	91.1	96.3	56.3	87.0
0920	Plaster & Gypsum Board	98.8	33.9	54.9	101.9	46.5	64.4	97.2	50.6	65.7	99.7	56.5	70.5	94.0	56.5	68.7	94.4	65.5	74.9
0950, 0980	Ceilings & Acoustic Treatment	98.7	33.9	56.1	94.2	46.5	62.8	86.3	50.6	62.8	84.0	56.5	65.9	84.0	56.5	65.9	86.3	65.5	72.7
0960	Flooring	105.7	34.0	85.2	106.2	64.9	94.4	103.0	39.2	84.8	103.1	39.2	84.9	99.6	39.2	82.3	98.0	36.4	80.4
0970, 0990	Wall Finishes & Painting/Coating	106.3	61.7	79.4	101.0	77.8	87.0	100.0	38.9	63.1	104.7	38.9	65.0	104.7	38.9	65.0	100.0	38.9	63.1
09	FINISHES	102.8	38.0	67.1	100.3	57.6	76.8	96.9	49.3	70.6	96.6	52.7	72.4	94.9	52.7	71.6	94.0	57.3	73.8
COVERS	DIVS. 10 - 14, 25, 28, 41, 43, 44, 46	100.0	80.8	96.1	100.0	87.8	97.5	100.0	40.8	88.0	100.0	41.4	88.2	100.0	41.4	88.2	100.0	40.5	88.0
21, 22, 23	FIRE SUPPRESSION, PLUMBING & HVAC	95.4	68.0	84.3	100.0	79.8	91.9	95.2	71.6	85.7	95.2	69.0	84.6	100.0	69.0	87.5	95.2	73.4	86.4
26, 27, 3370	ELECTRICAL, COMMUNICATIONS & UTIL.	99.9	41.4	69.0	95.8	55.7	74.6	96.5	67.7	81.3	100.0	64.5	81.2	96.8	73.8	84.6	93.9	71.4	82.0
MF2014	WEIGHTED AVERAGE	99.9	56.6	81.0	99.5	71.8	87.4	99.6	64.9	84.4	101.1	65.0	85.4	102.7	66.0	86.7	97.9	69.2	85.4

City Cost Indexes

	DIVISION	KANSAS																	
		FORT SCOTT			HAYS			HUTCHINSON			INDEPENDENCE			KANSAS CITY			LIBERAL		
		667			676			675			673			660 - 662			679		
		MAT.	INST.	TOTAL	MAT.	INST.	TOTAL	MAT.	INST.	TOTAL	MAT.	INST.	TOTAL	MAT.	INST.	TOTAL	MAT.	INST.	TOTAL
015433	CONTRACTOR EQUIPMENT		102.5	102.5		103.4	103.4		103.4	103.4		103.4	103.4		99.9	99.9		103.4	103.4
0241, 31 - 34	SITE & INFRASTRUCTURE, DEMOLITION	99.8	93.4	95.2	117.7	95.2	101.7	96.2	95.0	95.4	116.0	95.1	101.1	94.4	93.1	93.5	117.3	95.0	101.5
0310	Concrete Forming & Accessories	104.2	76.8	80.6	97.2	58.3	63.7	87.9	58.2	62.2	107.9	66.3	72.0	100.9	94.2	95.1	93.5	58.2	63.0
0320	Concrete Reinforcing	94.6	93.3	94.0	96.5	55.2	75.5	96.5	55.1	75.4	95.9	62.7	79.0	91.8	97.2	94.6	97.8	55.1	76.1
0330	Cast-in-Place Concrete	111.4	54.5	88.0	100.3	55.8	82.0	92.9	53.1	76.6	129.9	53.4	98.5	95.4	97.4	96.3	100.3	53.1	80.9
03	CONCRETE	106.6	72.9	90.1	110.4	58.1	84.7	92.4	57.1	75.0	122.6	62.2	92.9	98.4	96.2	97.3	112.5	57.1	85.2
04	MASONRY	106.7	60.3	77.8	118.0	59.9	81.8	107.9	58.6	77.2	105.3	64.5	79.9	107.8	99.4	102.6	116.6	58.6	80.4
05	METALS	95.3	92.8	94.5	95.6	77.7	90.1	95.4	76.7	89.6	95.3	80.5	90.8	102.9	100.1	102.1	95.9	76.7	90.0
06	WOOD, PLASTICS & COMPOSITES	108.7	83.5	94.6	100.6	57.9	76.7	90.9	57.9	72.4	113.7	68.0	88.1	104.4	94.5	98.8	96.4	57.9	74.8
07	THERMAL & MOISTURE PROTECTION	94.0	80.8	88.6	99.4	61.6	83.9	98.0	60.6	82.7	99.1	80.5	91.5	93.7	98.0	95.5	99.5	60.6	83.6
08	OPENINGS	96.3	78.2	92.1	103.1	51.5	91.1	103.0	51.5	91.0	100.7	58.8	90.9	97.7	89.0	95.7	103.1	51.5	91.1
0920	Plaster & Gypsum Board	99.7	82.9	88.4	97.3	56.5	69.7	93.0	56.5	68.3	106.9	67.0	79.9	92.9	94.2	93.8	94.8	56.5	68.9
0950, 0980	Ceilings & Acoustic Treatment	86.3	82.9	84.1	84.0	56.5	65.9	84.0	56.5	65.9	84.0	67.0	72.8	86.3	94.2	91.5	84.0	56.5	65.9
0960	Flooring	113.7	39.4	92.5	102.0	39.2	84.0	96.8	39.2	80.3	107.3	39.2	87.8	91.4	98.7	93.5	99.8	39.2	82.5
0970, 0990	Wall Finishes & Painting/Coating	101.8	38.9	63.8	104.7	38.9	65.0	104.7	38.9	65.0	104.7	38.9	65.0	108.5	67.7	83.9	104.7	38.9	65.0
09	FINISHES	99.6	66.9	81.6	96.4	52.7	72.3	92.4	52.7	70.5	98.9	58.7	76.8	94.3	91.0	92.5	95.7	52.7	72.0
COVERS	DIVS. 10 - 14, 25, 28, 41, 43, 44, 46	100.0	46.2	89.1	100.0	41.4	88.2	100.0	41.4	88.2	100.0	42.5	88.4	100.0	61.9	92.3	100.0	41.4	88.2
21, 22, 23	FIRE SUPPRESSION, PLUMBING & HVAC	95.2	68.3	84.3	95.2	69.0	84.6	95.2	69.0	84.6	95.2	70.1	85.1	99.9	96.4	98.5	95.2	67.2	83.9
26, 27, 3370	ELECTRICAL, COMMUNICATIONS & UTIL.	93.2	71.4	81.6	98.9	67.7	82.4	93.9	67.7	80.0	96.1	71.4	83.0	98.7	97.2	97.9	96.8	73.8	84.6
MF2014	WEIGHTED AVERAGE	97.9	73.0	87.0	100.5	65.4	85.2	96.5	65.0	82.8	100.9	69.2	87.1	99.5	94.8	97.5	100.4	65.5	85.2

	DIVISION	KANSAS									KENTUCKY								
		SALINA			TOPEKA			WICHITA			ASHLAND			BOWLING GREEN			CAMPTON		
		674			664 - 666			670 - 672			411 - 412			421 - 422			413 - 414		
		MAT.	INST.	TOTAL	MAT.	INST.	TOTAL	MAT.	INST.	TOTAL	MAT.	INST.	TOTAL	MAT.	INST.	TOTAL	MAT.	INST.	TOTAL
015433	CONTRACTOR EQUIPMENT		103.4	103.4		101.6	101.6		103.4	103.4		97.3	97.3		94.5	94.5		101.1	101.1
0241, 31 - 34	SITE & INFRASTRUCTURE, DEMOLITION	105.2	95.2	98.1	98.2	91.5	93.4	101.8	93.8	96.1	112.3	86.3	93.8	78.6	97.1	91.8	87.0	98.3	95.0
0310	Concrete Forming & Accessories	89.7	54.9	59.7	98.4	40.9	48.8	95.3	51.0	57.1	87.9	105.2	102.8	85.7	83.0	83.4	89.4	82.6	83.6
0320	Concrete Reinforcing	95.9	59.4	77.3	94.9	99.5	97.2	95.3	77.3	86.1	88.8	108.7	98.9	86.2	91.3	88.8	87.0	101.8	94.5
0330	Cast-in-Place Concrete	112.3	53.7	88.2	100.4	45.5	77.9	105.3	51.7	83.3	91.1	101.1	95.2	90.4	95.5	92.5	100.8	73.6	89.6
03	CONCRETE	107.3	56.7	82.5	99.4	55.1	77.7	101.6	57.4	79.9	98.6	104.8	101.6	96.8	88.9	92.9	100.7	83.2	92.1
04	MASONRY	134.8	59.9	88.2	102.5	55.9	73.4	106.3	50.2	71.4	97.8	103.7	101.5	99.7	80.2	87.5	96.9	69.0	79.5
05	METALS	97.3	80.6	92.1	99.4	96.9	98.6	99.4	84.2	94.7	92.1	110.3	97.7	93.6	87.3	91.7	92.9	92.0	92.6
06	WOOD, PLASTICS & COMPOSITES	92.4	52.1	69.8	97.1	37.3	63.6	95.1	51.3	70.6	76.0	105.2	92.3	88.6	83.1	85.5	87.1	89.0	88.2
07	THERMAL & MOISTURE PROTECTION	98.6	61.4	83.3	96.7	67.4	84.7	97.2	56.1	80.4	91.0	97.5	93.7	86.6	80.2	84.0	99.7	67.5	86.5
08	OPENINGS	103.1	48.5	90.4	100.8	55.4	90.2	105.3	56.2	93.9	97.9	99.6	98.3	98.7	79.3	94.2	100.2	87.6	97.3
0920	Plaster & Gypsum Board	93.0	50.6	64.3	96.8	35.3	55.2	91.8	49.8	63.4	60.9	105.4	91.0	87.4	82.9	84.3	87.4	88.1	87.9
0950, 0980	Ceilings & Acoustic Treatment	84.0	50.6	62.0	92.0	35.3	54.7	88.8	49.8	63.2	80.8	105.4	97.0	85.5	82.9	83.8	85.5	88.1	87.2
0960	Flooring	98.2	61.9	87.8	103.4	73.7	94.9	105.4	63.0	93.3	79.4	102.4	86.0	95.8	86.0	93.0	98.0	41.2	81.7
0970, 0990	Wall Finishes & Painting/Coating	104.7	38.9	65.0	101.8	52.5	72.1	103.6	48.0	70.0	103.1	101.2	102.0	101.2	72.7	84.0	101.2	65.4	79.6
09	FINISHES	93.6	53.6	71.6	98.5	45.9	69.5	97.7	52.3	72.7	80.9	104.8	94.1	91.8	82.8	86.8	92.6	74.0	82.4
COVERS	DIVS. 10 - 14, 25, 28, 41, 43, 44, 46	100.0	86.0	97.2	100.0	47.2	89.3	100.0	83.0	96.6	100.0	94.1	98.8	100.0	62.3	92.4	100.0	55.8	91.1
21, 22, 23	FIRE SUPPRESSION, PLUMBING & HVAC	100.0	69.1	87.5	100.0	69.1	87.5	99.8	64.8	85.7	95.0	92.5	94.0	99.9	82.7	93.0	95.2	79.8	89.0
26, 27, 3370	ELECTRICAL, COMMUNICATIONS & UTIL.	96.6	73.8	84.5	97.5	71.8	83.9	100.9	73.8	86.5	91.4	97.3	94.5	94.0	81.5	87.4	91.4	66.1	78.0
MF2014	WEIGHTED AVERAGE	101.5	67.6	86.7	99.5	66.0	84.9	100.7	66.0	85.6	94.6	99.3	96.6	96.2	83.9	90.8	95.8	78.6	88.3

	DIVISION	KENTUCKY																	
		CORBIN			COVINGTON			ELIZABETHTOWN			FRANKFORT			HAZARD			HENDERSON		
		407 - 409			410			427			406			417 - 418			424		
		MAT.	INST.	TOTAL	MAT.	INST.	TOTAL	MAT.	INST.	TOTAL	MAT.	INST.	TOTAL	MAT.	INST.	TOTAL	MAT.	INST.	TOTAL
015433	CONTRACTOR EQUIPMENT		101.1	101.1		104.4	104.4		94.5	94.5		101.1	101.1		101.1	101.1		116.0	116.0
0241, 31 - 34	SITE & INFRASTRUCTURE, DEMOLITION	87.9	98.1	95.1	82.6	112.3	103.7	73.1	97.2	90.2	88.9	99.2	96.2	84.8	99.6	95.3	81.1	124.0	111.6
0310	Concrete Forming & Accessories	84.5	70.6	72.5	85.0	82.0	82.4	80.7	80.5	80.5	94.2	71.8	74.8	86.1	83.5	83.8	92.5	81.5	83.0
0320	Concrete Reinforcing	86.0	69.8	77.7	85.7	91.1	88.5	86.6	90.8	88.8	97.7	90.8	94.2	87.4	107.6	97.7	86.3	84.8	85.6
0330	Cast-in-Place Concrete	93.6	58.6	79.2	96.5	97.6	96.9	81.8	73.0	78.2	89.4	69.1	81.1	97.0	82.5	91.0	79.9	90.6	84.3
03	CONCRETE	93.3	66.8	80.3	98.9	89.7	94.4	88.6	80.0	84.4	93.8	74.8	84.5	97.3	87.7	92.6	93.9	85.4	89.7
04	MASONRY	94.4	61.9	74.1	111.8	101.5	105.4	83.5	75.4	78.5	94.6	79.4	85.1	94.8	71.5	80.3	103.6	92.9	96.9
05	METALS	90.4	78.2	86.6	91.0	96.9	92.8	92.8	87.2	91.1	93.0	87.8	91.4	92.9	94.2	93.3	84.4	86.9	85.1
06	WOOD, PLASTICS & COMPOSITES	76.9	75.9	76.4	85.0	71.7	77.6	83.7	82.0	82.7	91.4	68.7	78.7	84.1	89.0	86.8	91.3	79.0	84.4
07	THERMAL & MOISTURE PROTECTION	99.3	63.8	84.7	100.3	92.3	97.0	86.3	76.9	82.5	101.0	75.1	90.4	99.7	69.5	87.3	99.0	91.3	95.9
08	OPENINGS	97.9	65.6	90.4	102.4	81.7	97.6	98.7	83.2	95.1	104.5	76.7	98.0	100.6	79.9	95.8	96.8	79.9	92.9
0920	Plaster & Gypsum Board	92.3	74.6	80.3	68.8	71.4	70.6	86.7	81.7	83.3	96.5	67.2	76.7	86.7	88.1	87.6	90.2	77.8	81.8
0950, 0980	Ceilings & Acoustic Treatment	80.4	74.6	76.6	88.8	71.4	77.4	85.5	81.7	83.0	89.6	67.2	74.8	85.5	88.1	87.2	79.6	77.8	78.4
0960	Flooring	97.2	41.2	81.2	71.8	88.6	76.6	93.4	86.0	91.3	106.0	54.4	91.2	96.4	42.0	80.9	99.5	86.0	95.6
0970, 0990	Wall Finishes & Painting/Coating	107.2	49.4	72.3	96.2	83.2	88.3	101.2	76.2	86.1	109.1	79.8	91.4	101.2	65.4	79.6	101.2	96.0	98.1
09	FINISHES	90.7	63.6	75.8	82.7	82.0	82.4	90.6	80.7	85.1	97.6	68.8	81.7	91.9	74.8	82.5	91.4	83.8	87.2
COVERS	DIVS. 10 - 14, 25, 28, 41, 43, 44, 46	100.0	48.4	89.6	100.0	98.7	99.7	100.0	84.2	96.8	100.0	65.5	93.0	100.0	56.6	91.2	100.0	65.4	93.0
21, 22, 23	FIRE SUPPRESSION, PLUMBING & HVAC	95.2	71.4	85.6	95.9	92.4	94.5	95.5	80.1	89.3	100.1	82.3	92.9	95.2	80.7	89.4	95.5	77.6	88.3
26, 27, 3370	ELECTRICAL, COMMUNICATIONS & UTIL.	91.4	80.1	85.4	95.2	79.3	86.8	91.2	80.2	85.4	101.3	87.1	93.8	91.4	58.4	73.9	93.4	80.3	86.5
MF2014	WEIGHTED AVERAGE	94.0	71.8	84.3	95.9	91.4	93.9	92.8	82.0	88.1	98.1	80.7	90.5	95.2	78.7	88.0	93.7	86.1	90.4

City Cost Indexes

	DIVISION	KENTUCKY																	
		LEXINGTON			LOUISVILLE			OWENSBORO			PADUCAH			PIKEVILLE			SOMERSET		
		403 - 405			400 - 402			423			420			415 - 416			425 - 426		
		MAT.	INST.	TOTAL	MAT.	INST.	TOTAL	MAT.	INST.	TOTAL	MAT.	INST.	TOTAL	MAT.	INST.	TOTAL	MAT.	INST.	TOTAL
015433	CONTRACTOR EQUIPMENT		101.1	101.1		94.5	94.5		116.0	116.0		116.0	116.0		97.3	97.3		101.1	101.1
0241, 31 - 34	SITE & INFRASTRUCTURE, DEMOLITION	90.1	100.6	97.5	84.8	97.3	93.7	91.0	124.5	114.8	83.7	123.4	111.9	123.2	85.3	96.3	77.9	98.8	92.8
0310	Concrete Forming & Accessories	96.4	73.0	76.2	95.8	82.0	83.9	90.9	85.1	85.9	88.9	79.9	81.1	96.8	88.9	90.0	87.0	75.9	77.4
0320	Concrete Reinforcing	94.4	92.2	93.3	96.8	92.6	94.7	86.3	91.8	89.1	86.8	86.2	86.5	89.3	108.4	99.0	86.6	90.8	88.7
0330	Cast-in-Place Concrete	95.8	92.7	94.5	96.4	73.9	87.2	93.2	91.1	92.4	85.2	86.5	85.7	100.1	96.7	98.7	79.9	98.3	87.5
03	CONCRETE	96.5	83.7	90.2	97.1	81.3	89.4	106.3	88.5	97.5	98.6	83.5	91.2	112.7	96.0	104.5	83.6	86.6	85.1
04	MASONRY	93.7	73.3	81.0	91.6	78.8	83.6	95.9	91.5	93.2	98.8	88.4	92.3	95.3	96.8	96.2	89.8	76.7	81.6
05	METALS	92.8	89.0	91.6	93.9	88.2	92.1	85.8	90.6	87.3	82.9	86.7	84.1	92.0	108.8	97.2	92.8	87.6	91.2
06	WOOD, PLASTICS & COMPOSITES	92.0	68.7	79.0	90.9	83.1	86.5	89.2	83.3	85.9	86.8	79.1	82.5	85.5	88.3	87.1	84.6	75.9	79.7
07	THERMAL & MOISTURE PROTECTION	99.5	84.1	93.2	96.3	78.2	88.9	99.4	92.8	96.7	99.1	78.8	90.8	91.7	79.5	86.7	99.0	71.2	87.6
08	OPENINGS	98.1	73.9	92.5	93.5	84.3	91.4	96.8	85.7	94.2	96.0	77.5	91.7	98.6	87.8	96.1	99.5	77.7	94.4
0920	Plaster & Gypsum Board	101.9	67.2	78.4	98.0	82.9	87.8	88.7	82.2	84.3	87.7	77.9	81.1	63.8	88.1	80.2	86.7	74.6	78.5
0950, 0980	Ceilings & Acoustic Treatment	84.5	67.2	73.1	88.9	82.9	84.9	79.6	82.2	81.3	79.6	77.9	78.5	80.8	88.1	85.6	85.5	74.6	78.3
0960	Flooring	101.9	70.3	92.9	98.6	86.0	95.0	98.9	86.0	95.2	97.8	59.4	86.8	83.4	102.4	88.8	96.7	41.2	80.8
0970, 0990	Wall Finishes & Painting/Coating	107.2	81.0	91.4	104.7	76.2	87.5	101.2	96.0	98.1	101.2	79.7	88.2	103.1	81.2	89.9	101.2	76.2	86.1
09	FINISHES	94.4	72.3	82.2	94.9	82.3	88.0	91.5	86.0	88.5	90.7	76.1	82.7	83.3	90.9	87.5	91.2	69.4	79.2
COVERS	DIVS. 10 - 14, 25, 28, 41, 43, 44, 46	100.0	86.5	97.3	100.0	84.9	96.9	100.0	102.9	100.6	100.0	61.8	92.3	100.0	56.3	91.2	100.0	58.5	91.6
21, 22, 23	FIRE SUPPRESSION, PLUMBING & HVAC	100.0	79.5	91.7	100.0	80.9	92.3	99.9	78.6	91.4	95.5	81.1	89.7	95.0	87.2	91.8	95.5	77.4	88.2
26, 27, 3370	ELECTRICAL, COMMUNICATIONS & UTIL.	94.2	81.5	87.5	99.4	81.5	89.9	93.5	81.4	87.1	95.8	79.9	87.4	94.4	73.4	83.3	91.8	80.2	85.6
MF2014	WEIGHTED AVERAGE	96.6	81.4	90.0	96.6	83.3	90.8	96.3	88.9	93.1	93.9	84.5	89.8	96.9	88.6	93.3	93.2	79.9	87.4

	DIVISION	LOUISIANA																	
		ALEXANDRIA			BATON ROUGE			HAMMOND			LAFAYETTE			LAKE CHARLES			MONROE		
		713 - 714			707 - 708			704			705			706			712		
		MAT.	INST.	TOTAL	MAT.	INST.	TOTAL	MAT.	INST.	TOTAL	MAT.	INST.	TOTAL	MAT.	INST.	TOTAL	MAT.	INST.	TOTAL
015433	CONTRACTOR EQUIPMENT		89.9	89.9		89.5	89.5		90.1	90.1		90.1	90.1		89.5	89.5		89.9	89.9
0241, 31 - 34	SITE & INFRASTRUCTURE, DEMOLITION	101.2	87.1	91.1	104.0	87.2	92.0	102.1	87.8	91.9	103.2	88.1	92.5	103.9	87.0	91.9	101.2	87.0	91.1
0310	Concrete Forming & Accessories	82.9	43.4	48.9	97.6	60.8	65.8	78.4	46.2	50.7	96.0	54.1	59.9	96.8	56.7	62.2	82.4	43.3	48.7
0320	Concrete Reinforcing	94.9	54.3	74.2	101.1	58.9	79.6	97.9	59.3	78.3	99.3	58.9	78.7	99.3	59.1	78.9	93.8	54.0	73.6
0330	Cast-in-Place Concrete	96.2	49.9	77.2	97.2	58.8	81.4	95.8	44.0	74.5	95.3	47.9	75.8	100.3	67.3	86.8	96.2	56.8	80.0
03	CONCRETE	95.5	48.8	72.6	98.9	60.4	80.0	95.2	48.9	72.5	96.4	53.7	75.4	98.8	61.5	80.5	95.3	51.1	73.6
04	MASONRY	121.3	53.4	79.0	94.9	54.1	69.5	95.7	51.2	68.0	95.7	52.4	68.7	95.0	58.8	72.5	115.6	47.9	73.4
05	METALS	95.4	70.4	87.7	100.6	72.1	91.8	95.6	71.2	88.1	94.8	72.1	87.8	94.8	72.6	87.9	95.4	69.9	87.6
06	WOOD, PLASTICS & COMPOSITES	90.5	41.7	63.2	99.0	63.8	79.3	84.2	47.0	63.3	105.7	55.7	77.7	103.8	57.7	78.0	89.8	42.2	63.1
07	THERMAL & MOISTURE PROTECTION	95.6	62.9	82.2	96.2	65.4	83.5	96.5	61.4	82.1	97.1	63.8	83.4	96.1	66.8	84.1	95.6	60.8	81.3
08	OPENINGS	109.1	44.7	94.1	97.9	59.1	88.9	93.0	51.7	83.4	96.7	53.2	86.6	96.7	54.9	87.0	109.1	47.6	94.8
0920	Plaster & Gypsum Board	84.2	40.4	54.6	96.7	63.0	74.0	97.5	45.7	62.5	105.7	54.7	71.2	105.7	56.8	72.6	83.8	40.9	54.8
0950, 0980	Ceilings & Acoustic Treatment	84.1	40.4	55.4	91.3	63.0	72.7	94.1	45.7	62.3	92.4	54.7	67.6	93.3	56.8	69.3	84.1	40.9	55.7
0960	Flooring	98.5	65.7	89.1	101.8	65.7	91.5	96.3	65.7	87.5	104.9	65.7	93.7	104.9	72.9	95.7	98.1	57.1	86.4
0970, 0990	Wall Finishes & Painting/Coating	104.4	64.4	80.3	94.7	46.3	65.5	99.4	48.3	68.6	99.4	57.9	74.4	99.4	50.0	69.6	104.4	48.1	70.4
09	FINISHES	89.8	48.9	67.2	94.7	60.2	75.7	94.8	49.7	70.0	98.1	56.3	75.0	98.3	58.6	76.4	89.6	45.5	65.3
COVERS	DIVS. 10 - 14, 25, 28, 41, 43, 44, 46	100.0	50.0	89.9	100.0	80.8	96.1	100.0	45.4	89.0	100.0	79.4	95.8	100.0	80.3	96.0	100.0	45.7	89.0
21, 22, 23	FIRE SUPPRESSION, PLUMBING & HVAC	100.1	55.3	82.1	100.0	59.5	83.7	95.3	42.8	74.1	100.1	61.1	84.4	100.1	62.0	84.7	100.1	54.3	81.6
26, 27, 3370	ELECTRICAL, COMMUNICATIONS & UTIL.	95.7	57.9	75.7	100.8	61.6	80.1	96.7	56.0	75.2	97.9	67.4	81.8	97.4	65.0	80.3	97.7	59.6	77.6
MF2014	WEIGHTED AVERAGE	99.4	57.2	81.0	99.1	63.7	83.7	95.8	54.5	77.8	97.9	63.0	82.7	98.1	65.0	83.7	99.3	56.5	80.6

	DIVISION	LOUISIANA									MAINE								
		NEW ORLEANS			SHREVEPORT			THIBODAUX			AUGUSTA			BANGOR			BATH		
		700 - 701			710 - 711			703			043			044			045		
		MAT.	INST.	TOTAL	MAT.	INST.	TOTAL	MAT.	INST.	TOTAL	MAT.	INST.	TOTAL	MAT.	INST.	TOTAL	MAT.	INST.	TOTAL
015433	CONTRACTOR EQUIPMENT		90.4	90.4		89.9	89.9		90.1	90.1		100.9	100.9		100.9	100.9		100.9	100.9
0241, 31 - 34	SITE & INFRASTRUCTURE, DEMOLITION	103.4	91.1	94.7	105.5	87.1	92.4	104.4	88.1	92.8	91.6	101.1	98.4	93.3	101.3	99.0	91.0	101.1	98.2
0310	Concrete Forming & Accessories	95.1	65.5	69.5	98.5	44.2	51.7	90.4	64.6	68.1	95.9	94.8	94.9	93.1	96.6	96.1	88.9	94.9	94.1
0320	Concrete Reinforcing	100.2	60.1	79.8	98.3	54.4	76.0	97.9	59.1	78.2	96.1	106.7	101.5	87.7	108.1	98.1	86.8	107.9	97.5
0330	Cast-in-Place Concrete	98.0	69.4	86.3	101.1	53.6	81.6	102.9	51.6	81.8	101.5	58.8	84.0	82.2	110.6	93.9	82.2	58.9	72.7
03	CONCRETE	99.0	66.3	82.9	100.4	50.5	75.9	100.5	59.7	80.4	103.1	84.1	93.8	94.1	102.9	98.4	94.1	84.4	89.4
04	MASONRY	97.7	59.8	74.1	107.1	50.5	71.8	120.5	48.8	75.8	107.4	63.6	80.1	118.0	103.9	109.2	125.6	86.1	101.0
05	METALS	110.3	73.3	98.9	99.5	70.6	90.6	95.6	71.6	88.2	106.0	86.6	100.1	96.1	89.3	94.0	94.6	88.4	92.7
06	WOOD, PLASTICS & COMPOSITES	97.9	66.3	80.2	101.7	42.7	68.7	92.3	69.7	79.6	91.1	105.3	99.0	90.1	96.4	93.6	84.6	105.3	96.2
07	THERMAL & MOISTURE PROTECTION	94.9	68.8	84.2	95.3	59.8	80.8	96.4	61.8	82.2	100.1	65.0	85.7	97.8	84.4	92.3	97.8	71.8	87.1
08	OPENINGS	97.8	65.4	90.2	108.0	45.3	93.4	97.6	64.1	89.8	106.3	87.9	102.0	102.8	83.0	98.2	102.8	87.8	99.3
0920	Plaster & Gypsum Board	101.4	65.7	77.2	94.0	41.5	58.5	99.7	69.2	79.0	100.6	104.8	103.5	99.4	95.7	96.9	95.5	104.8	101.8
0950, 0980	Ceilings & Acoustic Treatment	98.2	65.7	76.8	88.1	41.5	57.4	94.1	69.2	77.7	97.4	104.8	102.3	86.2	95.7	92.5	84.3	104.8	97.8
0960	Flooring	105.7	65.7	94.2	101.9	61.3	90.3	102.3	44.1	85.7	103.9	55.3	90.0	97.1	114.3	102.0	95.3	55.3	83.9
0970, 0990	Wall Finishes & Painting/Coating	101.9	64.1	79.1	100.4	44.6	66.7	100.7	49.6	69.9	112.7	63.3	82.9	105.0	44.1	68.3	105.0	44.1	68.3
09	FINISHES	101.2	65.1	81.3	94.5	46.3	67.9	97.2	59.8	76.6	100.2	85.8	92.3	95.7	94.8	95.2	94.2	83.7	88.4
COVERS	DIVS. 10 - 14, 25, 28, 41, 43, 44, 46	100.0	83.1	96.6	100.0	78.1	95.6	100.0	81.4	96.2	100.0	100.5	100.1	100.0	109.8	102.0	100.0	100.5	100.1
21, 22, 23	FIRE SUPPRESSION, PLUMBING & HVAC	100.0	65.1	85.9	99.9	58.6	83.3	95.3	61.7	81.8	99.9	64.2	85.5	100.1	74.3	89.7	95.4	64.3	82.8
26, 27, 3370	ELECTRICAL, COMMUNICATIONS & UTIL.	102.1	73.3	86.9	103.5	66.6	84.0	95.3	73.3	83.7	100.9	80.1	89.9	97.2	75.8	85.9	95.2	80.1	87.2
MF2014	WEIGHTED AVERAGE	101.4	69.5	87.5	101.0	59.6	83.0	98.2	65.4	83.9	102.1	79.5	92.3	99.0	89.5	94.9	97.5	81.9	90.7

DIVISION		MAINE																	
		HOULTON			KITTERY			LEWISTON			MACHIAS			PORTLAND			ROCKLAND		
		047			039			042			046			040 - 041			048		
		MAT.	INST.	TOTAL	MAT.	INST.	TOTAL	MAT.	INST.	TOTAL	MAT.	INST.	TOTAL	MAT.	INST.	TOTAL	MAT.	INST.	TOTAL
015433	CONTRACTOR EQUIPMENT		100.9	100.9		100.9	100.9		100.9	100.9		100.9	100.9		100.9	100.9		100.9	100.9
0241, 31 - 34	SITE & INFRASTRUCTURE, DEMOLITION	93.0	102.2	99.5	84.8	101.2	96.5	90.9	101.3	98.3	92.3	102.2	99.4	90.1	101.3	98.1	88.7	102.2	98.3
0310	Concrete Forming & Accessories	96.7	101.6	101.0	89.2	95.5	94.6	98.5	96.7	96.9	93.9	101.6	100.5	98.5	96.7	96.9	94.9	101.6	100.7
0320	Concrete Reinforcing	87.7	106.5	97.2	84.1	107.9	96.2	107.8	108.1	107.9	87.7	106.5	97.2	102.2	108.1	105.2	87.7	106.5	97.2
0330	Cast-in-Place Concrete	82.3	71.7	77.9	81.2	59.8	72.4	83.9	110.7	94.9	82.2	71.1	77.7	96.7	110.7	102.4	83.9	71.1	78.7
03	CONCRETE	95.1	91.5	93.4	89.7	85.0	87.4	94.7	102.9	98.7	94.6	91.3	93.0	101.9	102.9	102.4	92.0	91.3	91.6
04	MASONRY	101.3	76.3	85.7	113.3	87.1	97.0	101.9	103.9	103.2	101.3	76.3	85.7	112.8	103.9	107.2	95.2	76.3	83.4
05	METALS	94.8	86.6	92.3	85.5	88.6	86.4	99.5	89.3	96.4	94.8	86.5	92.3	107.9	89.3	102.2	94.7	86.5	92.2
06	WOOD, PLASTICS & COMPOSITES	94.0	105.3	100.3	89.2	105.3	98.2	96.0	96.4	96.2	91.0	105.3	99.0	93.7	96.4	95.2	91.9	105.3	99.4
07	THERMAL & MOISTURE PROTECTION	97.9	72.6	87.5	99.2	71.1	87.7	97.6	84.4	92.2	97.8	71.6	87.1	100.1	84.4	93.7	97.5	71.7	86.9
08	OPENINGS	102.9	86.1	99.0	102.0	87.8	98.7	106.2	83.0	100.8	102.9	86.1	99.0	104.5	83.0	99.5	102.8	86.1	98.9
0920	Plaster & Gypsum Board	101.6	104.8	103.8	100.0	104.8	103.3	104.9	95.7	98.7	100.1	104.8	103.3	102.5	95.7	97.9	100.1	104.8	103.3
0950, 0980	Ceilings & Acoustic Treatment	84.3	104.8	97.8	94.2	104.8	101.2	95.4	95.7	95.6	84.3	104.8	97.8	99.1	95.7	96.9	84.3	104.8	97.8
0960	Flooring	98.3	51.7	85.0	99.8	57.7	87.7	100.0	114.3	104.1	97.5	51.7	84.4	98.6	114.3	103.1	97.9	51.7	84.7
0970, 0990	Wall Finishes & Painting/Coating	105.0	137.1	124.4	96.4	38.4	61.4	105.0	44.1	68.3	105.0	137.1	124.4	102.0	44.1	67.1	105.0	137.1	124.4
09	FINISHES	95.9	97.8	97.0	96.5	83.7	89.5	98.9	94.8	96.6	95.4	97.8	96.7	98.8	94.8	96.6	95.2	97.8	96.6
COVERS	DIVS. 10 - 14, 25, 28, 41, 43, 44, 46	100.0	106.4	101.3	100.0	100.8	100.2	100.0	109.8	102.0	100.0	106.4	101.3	100.0	109.8	102.0	100.0	106.4	101.3
21, 22, 23	FIRE SUPPRESSION, PLUMBING & HVAC	95.4	73.2	86.4	95.3	76.6	87.8	100.1	74.3	89.7	95.4	73.2	86.4	100.0	74.3	89.6	95.4	73.2	86.4
26, 27, 3370	ELECTRICAL, COMMUNICATIONS & UTIL.	99.2	80.1	89.1	97.0	80.1	88.1	99.3	80.1	89.1	99.2	80.1	89.1	98.9	80.1	89.0	99.1	80.1	89.1
MF2014	WEIGHTED AVERAGE	97.2	85.6	92.1	95.2	84.7	90.6	99.6	90.1	95.5	97.1	85.5	92.0	102.0	90.1	96.8	96.3	85.5	91.6

DIVISION		MAINE			MARYLAND														
		WATERVILLE			ANNAPOLIS			BALTIMORE			COLLEGE PARK			CUMBERLAND			EASTON		
		049			214			210 - 212			207 - 208			215			216		
		MAT.	INST.	TOTAL	MAT.	INST.	TOTAL	MAT.	INST.	TOTAL	MAT.	INST.	TOTAL	MAT.	INST.	TOTAL	MAT.	INST.	TOTAL
015433	CONTRACTOR EQUIPMENT		100.9	100.9		99.8	99.8		103.4	103.4		105.5	105.5		99.8	99.8		99.8	99.8
0241, 31 - 34	SITE & INFRASTRUCTURE, DEMOLITION	92.8	101.1	98.7	103.0	91.5	94.9	101.5	95.6	97.3	100.5	94.5	96.2	94.3	91.8	92.5	101.4	88.6	92.3
0310	Concrete Forming & Accessories	88.4	94.8	93.9	97.2	73.6	76.8	101.4	71.1	75.3	84.3	70.2	72.2	91.4	81.0	82.4	89.2	70.2	72.8
0320	Concrete Reinforcing	87.7	106.7	97.3	100.3	80.7	90.3	105.6	80.7	92.9	105.4	78.1	91.5	86.6	72.3	79.3	86.0	79.0	82.4
0330	Cast-in-Place Concrete	82.3	58.8	72.6	110.0	75.7	95.9	107.5	76.4	94.7	116.5	77.5	100.4	91.7	85.6	89.2	101.8	48.1	79.8
03	CONCRETE	95.6	84.1	90.0	104.8	76.6	90.9	104.8	75.8	90.6	107.5	75.5	91.8	90.2	81.8	86.1	98.0	65.2	81.9
04	MASONRY	112.2	63.6	81.9	103.5	71.6	83.6	100.1	71.7	82.4	111.8	69.0	85.1	99.1	83.8	89.5	113.4	42.9	69.4
05	METALS	94.8	86.6	92.3	104.0	93.4	100.8	100.6	93.9	98.5	86.8	97.4	90.0	98.5	90.5	96.1	98.8	87.2	95.2
06	WOOD, PLASTICS & COMPOSITES	84.0	105.3	95.9	92.9	75.4	83.1	99.6	72.1	84.2	80.0	70.0	74.4	84.9	80.1	82.2	82.7	77.4	79.7
07	THERMAL & MOISTURE PROTECTION	97.9	65.0	84.4	100.2	78.8	91.4	101.9	78.6	92.3	101.9	78.7	92.4	100.2	78.9	91.5	100.3	59.0	83.4
08	OPENINGS	102.9	87.9	99.4	105.8	80.3	99.9	99.0	78.5	94.2	94.3	74.8	89.7	99.7	78.0	94.7	98.0	71.4	91.8
0920	Plaster & Gypsum Board	95.5	104.8	101.8	100.5	74.9	83.2	102.3	71.3	81.4	99.0	69.0	78.8	102.3	79.8	87.1	102.3	77.0	85.2
0950, 0980	Ceilings & Acoustic Treatment	84.3	104.8	97.8	91.3	74.9	80.6	94.0	71.3	79.1	102.2	69.0	80.4	93.3	79.8	84.4	93.3	77.0	82.6
0960	Flooring	95.0	55.3	83.6	99.8	77.2	93.3	100.7	77.2	94.0	97.8	79.7	92.6	95.5	90.8	94.2	94.7	51.7	82.4
0970, 0990	Wall Finishes & Painting/Coating	105.0	63.3	79.8	92.1	79.2	84.3	97.9	79.2	86.6	108.0	77.6	89.7	96.5	73.4	82.5	96.5	77.6	85.1
09	FINISHES	94.3	85.8	89.6	94.8	74.3	83.5	100.1	72.3	84.8	95.3	71.9	82.4	96.5	81.9	88.5	96.7	69.1	81.5
COVERS	DIVS. 10 - 14, 25, 28, 41, 43, 44, 46	100.0	100.5	100.1	100.0	86.2	97.2	100.0	86.2	97.2	100.0	85.5	97.1	100.0	90.5	98.1	100.0	75.3	95.0
21, 22, 23	FIRE SUPPRESSION, PLUMBING & HVAC	95.4	64.2	82.8	100.1	80.5	92.2	100.0	80.6	92.2	95.4	83.6	90.7	95.2	73.0	86.3	95.2	68.3	84.4
26, 27, 3370	ELECTRICAL, COMMUNICATIONS & UTIL.	99.2	80.1	89.1	99.5	91.9	95.5	102.0	91.9	96.6	101.1	99.7	100.4	98.3	81.8	89.6	97.8	65.2	80.5
MF2014	WEIGHTED AVERAGE	97.5	79.5	89.7	101.5	82.0	93.0	100.8	81.9	92.6	97.2	83.4	91.2	96.8	81.8	90.2	98.3	68.6	85.3

DIVISION		MARYLAND															MASSACHUSETTS		
		ELKTON			HAGERSTOWN			SALISBURY			SILVER SPRING			WALDORF			BOSTON		
		219			217			218			209			206			020 - 022, 024		
		MAT.	INST.	TOTAL	MAT.	INST.	TOTAL	MAT.	INST.	TOTAL	MAT.	INST.	TOTAL	MAT.	INST.	TOTAL	MAT.	INST.	TOTAL
015433	CONTRACTOR EQUIPMENT		99.8	99.8		99.8	99.8		99.8	99.8		98.1	98.1		98.1	98.1		106.5	106.5
0241, 31 - 34	SITE & INFRASTRUCTURE, DEMOLITION	88.4	89.5	89.2	92.6	92.3	92.4	101.3	88.6	92.3	89.0	87.6	88.0	95.4	87.3	89.7	98.9	109.1	106.2
0310	Concrete Forming & Accessories	95.4	81.0	82.9	90.3	75.1	77.2	103.9	51.7	58.9	92.8	70.4	73.5	100.2	68.7	73.0	102.7	143.8	138.1
0320	Concrete Reinforcing	86.0	105.3	95.8	86.6	72.3	79.3	86.0	63.5	74.6	104.1	77.9	90.8	104.8	78.0	91.1	108.9	155.8	132.8
0330	Cast-in-Place Concrete	82.5	75.2	79.5	87.4	85.7	86.7	101.8	46.7	79.2	119.3	79.3	102.8	133.6	76.5	110.2	104.0	151.2	123.4
03	CONCRETE	83.3	84.3	83.8	86.8	79.2	83.0	99.0	53.7	76.8	105.9	76.1	91.2	116.5	74.4	95.8	105.7	147.5	126.2
04	MASONRY	98.3	58.9	73.8	105.3	83.8	91.9	113.1	47.3	72.1	110.9	72.3	86.8	95.4	67.2	77.9	106.6	164.2	142.5
05	METALS	98.8	101.3	99.6	98.7	90.7	96.2	98.8	81.2	93.4	91.0	93.8	91.9	91.0	93.9	91.9	100.2	132.4	110.1
06	WOOD, PLASTICS & COMPOSITES	90.0	86.8	88.3	84.0	71.9	77.2	100.6	55.1	75.1	86.5	69.2	76.8	94.0	69.2	80.1	98.6	143.3	123.6
07	THERMAL & MOISTURE PROTECTION	99.9	73.5	89.1	100.0	76.8	90.5	100.6	63.7	85.4	105.4	84.9	97.0	105.8	82.7	96.4	103.4	151.9	123.3
08	OPENINGS	98.0	81.3	94.1	98.0	73.0	92.2	98.2	61.7	89.7	85.7	74.3	83.0	86.2	74.3	83.5	101.4	146.2	111.8
0920	Plaster & Gypsum Board	105.2	86.7	92.7	102.3	71.3	81.4	111.6	54.0	72.6	105.1	69.0	80.7	108.3	69.0	81.8	107.5	144.1	132.2
0950, 0980	Ceilings & Acoustic Treatment	93.3	86.7	89.0	94.3	71.3	79.2	93.3	54.0	67.5	112.2	69.0	83.8	112.2	69.0	83.8	106.1	144.1	131.1
0960	Flooring	97.0	58.5	86.0	95.2	90.8	93.9	100.7	64.7	90.4	104.1	79.7	97.2	107.7	79.0	99.5	97.9	182.7	122.2
0970, 0990	Wall Finishes & Painting/Coating	96.5	77.6	85.1	96.5	77.6	85.1	96.5	77.6	85.1	115.8	77.6	92.8	115.8	77.6	92.8	104.4	155.9	135.4
09	FINISHES	96.9	76.9	85.9	96.4	77.5	86.0	99.8	56.3	75.8	96.9	71.4	82.8	98.6	70.7	83.2	104.2	152.7	130.9
COVERS	DIVS. 10 - 14, 25, 28, 41, 43, 44, 46	100.0	59.9	91.9	100.0	89.6	97.9	100.0	38.7	87.6	100.0	84.7	96.9	100.0	83.0	96.6	100.0	119.4	103.9
21, 22, 23	FIRE SUPPRESSION, PLUMBING & HVAC	95.2	80.5	89.3	100.0	86.2	94.4	95.2	59.9	81.0	95.4	84.8	91.2	95.4	82.3	90.1	100.1	132.4	113.1
26, 27, 3370	ELECTRICAL, COMMUNICATIONS & UTIL.	99.6	91.9	95.5	98.1	81.8	89.5	96.5	67.1	81.0	98.4	99.7	99.1	95.7	99.7	97.8	100.8	135.4	119.1
MF2014	WEIGHTED AVERAGE	95.9	81.9	89.8	97.6	83.3	91.4	98.7	62.1	82.7	96.5	83.2	90.7	97.1	81.7	90.4	101.6	139.5	118.1

DIVISION		MASSACHUSETTS																	
		BROCKTON			BUZZARDS BAY			FALL RIVER			FITCHBURG			FRAMINGHAM			GREENFIELD		
		023			025			027			014			017			013		
		MAT.	INST.	TOTAL	MAT.	INST.	TOTAL	MAT.	INST.	TOTAL	MAT.	INST.	TOTAL	MAT.	INST.	TOTAL	MAT.	INST.	TOTAL
015433	CONTRACTOR EQUIPMENT		102.7	102.7		102.7	102.7		103.7	103.7		100.9	100.9		101.9	101.9		100.9	100.9
0241, 31 - 34	SITE & INFRASTRUCTURE, DEMOLITION	94.5	105.7	102.4	84.8	105.6	99.6	93.5	105.8	102.2	86.3	105.6	100.0	83.0	105.3	98.9	90.0	104.3	100.1
0310	Concrete Forming & Accessories	101.6	138.7	133.6	99.2	138.5	133.1	101.6	138.8	133.6	93.8	130.8	125.7	101.2	138.8	133.6	92.1	113.2	110.3
0320	Concrete Reinforcing	106.4	155.6	131.4	85.3	160.1	123.4	106.4	160.2	133.7	83.9	151.4	118.3	83.9	155.6	120.4	87.3	124.4	106.2
0330	Cast-in-Place Concrete	97.7	151.5	119.8	81.2	151.5	110.1	94.5	152.0	118.1	85.9	150.5	112.4	85.9	148.3	111.5	88.3	131.2	105.9
03	CONCRETE	101.8	145.2	123.1	86.2	145.9	115.5	100.3	146.2	122.9	85.9	140.4	112.7	88.7	144.1	116.0	89.6	120.9	104.9
04	MASONRY	101.7	160.0	138.0	94.4	160.0	135.3	102.6	159.9	138.3	101.0	159.0	137.2	107.2	160.1	140.1	105.4	136.0	124.4
05	METALS	97.4	130.0	107.5	92.3	131.7	104.4	97.4	132.1	108.1	95.5	125.5	104.7	95.5	130.0	106.2	97.8	110.4	101.7
06	WOOD, PLASTICS & COMPOSITES	99.1	138.4	121.1	96.1	138.4	119.8	99.1	138.7	121.3	93.7	128.3	113.1	100.1	138.2	121.4	91.5	110.9	102.4
07	THERMAL & MOISTURE PROTECTION	100.9	148.7	120.5	100.1	147.0	119.4	100.8	146.3	119.5	99.2	141.5	116.5	99.3	149.1	119.7	99.2	121.9	108.5
08	OPENINGS	102.2	143.5	111.9	97.9	139.2	107.5	102.2	139.3	110.9	104.6	137.0	112.2	94.9	143.4	106.2	104.7	115.0	107.1
0920	Plaster & Gypsum Board	93.7	139.0	124.3	89.5	139.0	123.0	93.7	139.0	124.3	100.3	128.6	119.4	102.8	139.0	127.3	101.0	110.7	107.5
0950, 0980	Ceilings & Acoustic Treatment	102.2	139.0	126.4	88.2	139.0	121.6	102.2	139.0	126.4	89.0	128.6	115.0	89.0	139.0	121.9	97.4	110.7	106.1
0960	Flooring	98.6	182.7	122.7	96.7	182.7	121.3	97.7	182.7	122.0	98.1	182.7	122.3	99.7	182.7	123.5	97.3	154.0	113.5
0970, 0990	Wall Finishes & Painting/Coating	99.9	155.1	133.2	99.9	155.1	133.2	99.9	155.1	133.2	99.5	155.1	133.1	100.5	155.1	133.5	99.5	115.8	109.4
09	FINISHES	98.9	149.1	126.6	94.2	149.1	124.4	98.7	149.3	126.6	94.6	143.1	121.3	95.3	148.9	124.8	96.7	121.0	110.0
COVERS	DIVS. 10 - 14, 25, 28, 41, 43, 44, 46	100.0	117.9	103.6	100.0	117.9	103.6	100.0	118.5	103.7	100.0	109.0	101.8	100.0	117.5	103.5	100.0	104.6	100.9
21, 22, 23	FIRE SUPPRESSION, PLUMBING & HVAC	100.1	111.1	104.6	95.4	111.1	101.7	100.1	111.2	104.6	96.0	113.6	103.1	96.0	126.1	108.1	96.0	102.8	98.7
26, 27, 3370	ELECTRICAL, COMMUNICATIONS & UTIL.	99.3	100.1	99.7	96.3	100.1	98.3	99.2	100.1	99.7	100.1	105.1	102.8	96.6	128.9	113.7	100.1	98.9	99.5
MF2014	WEIGHTED AVERAGE	99.9	128.1	112.2	94.4	128.1	109.1	99.7	128.3	112.2	96.5	126.5	109.6	95.7	135.1	112.9	97.7	112.3	104.1

DIVISION		MASSACHUSETTS																	
		HYANNIS			LAWRENCE			LOWELL			NEW BEDFORD			PITTSFIELD			SPRINGFIELD		
		026			019			018			027			012			010 - 011		
		MAT.	INST.	TOTAL	MAT.	INST.	TOTAL	MAT.	INST.	TOTAL	MAT.	INST.	TOTAL	MAT.	INST.	TOTAL	MAT.	INST.	TOTAL
015433	CONTRACTOR EQUIPMENT		102.7	102.7		102.7	102.7		100.9	100.9		103.7	103.7		100.9	100.9		100.9	100.9
0241, 31 - 34	SITE & INFRASTRUCTURE, DEMOLITION	90.9	105.6	101.3	95.2	105.7	102.7	94.2	105.6	102.3	92.1	105.8	101.8	95.2	104.2	101.6	94.6	104.4	101.6
0310	Concrete Forming & Accessories	93.5	138.5	132.3	102.4	139.0	134.0	99.0	139.0	133.5	101.6	138.8	133.6	99.0	112.2	110.4	99.3	113.6	111.6
0320	Concrete Reinforcing	85.3	160.1	123.4	103.8	150.4	127.5	104.6	150.4	127.9	106.4	160.2	133.7	86.6	122.3	104.8	104.6	124.4	114.7
0330	Cast-in-Place Concrete	89.1	151.5	114.7	99.4	148.9	119.7	90.3	148.9	114.4	83.3	152.0	111.5	98.5	129.7	111.3	94.0	131.7	109.5
03	CONCRETE	92.4	145.9	118.7	103.7	143.5	123.2	95.1	143.3	118.8	95.0	146.2	120.2	96.2	119.5	107.6	96.9	121.2	108.8
04	MASONRY	100.6	160.0	137.6	112.3	160.7	142.5	100.2	160.0	137.5	100.6	159.9	137.6	100.9	133.2	121.0	100.5	136.9	123.2
05	METALS	93.7	131.7	105.4	98.2	128.3	107.5	98.2	126.0	106.7	97.4	132.1	108.1	98.0	109.5	101.5	100.9	110.4	103.8
06	WOOD, PLASTICS & COMPOSITES	89.3	138.4	116.8	100.6	138.4	121.8	99.8	138.4	121.4	99.1	138.7	121.3	99.8	110.9	106.0	99.8	110.9	106.0
07	THERMAL & MOISTURE PROTECTION	100.4	147.0	119.5	99.7	148.5	119.7	99.5	148.6	119.7	100.7	146.3	119.4	99.6	120.7	108.2	99.5	122.2	108.8
08	OPENINGS	98.5	139.2	108.0	99.0	142.1	109.0	105.9	142.1	114.4	102.2	139.3	110.9	105.9	114.4	107.9	105.9	115.0	108.0
0920	Plaster & Gypsum Board	85.2	139.0	121.6	105.6	139.0	128.2	105.6	139.0	128.2	93.7	139.0	124.3	105.6	110.7	109.0	105.6	110.7	109.0
0950, 0980	Ceilings & Acoustic Treatment	93.8	139.0	123.5	99.3	139.0	125.4	99.3	139.0	125.4	102.2	139.0	126.4	99.3	110.7	106.8	99.3	110.7	106.8
0960	Flooring	94.4	182.7	119.6	100.2	182.7	123.8	100.2	182.7	123.8	97.7	182.7	122.0	100.6	154.0	115.9	99.7	154.0	115.3
0970, 0990	Wall Finishes & Painting/Coating	99.9	155.1	133.2	99.6	155.1	133.1	99.5	155.1	133.1	99.9	155.1	133.2	99.5	115.8	109.4	100.9	115.8	109.9
09	FINISHES	94.4	149.1	124.6	98.8	149.1	126.5	98.7	149.1	126.5	98.6	149.3	126.5	98.8	120.3	110.6	98.7	121.2	111.1
COVERS	DIVS. 10 - 14, 25, 28, 41, 43, 44, 46	100.0	117.9	103.6	100.0	118.1	103.6	100.0	118.1	103.6	100.0	118.5	103.7	100.0	103.7	100.7	100.0	104.9	101.0
21, 22, 23	FIRE SUPPRESSION, PLUMBING & HVAC	100.1	111.1	104.6	100.1	124.8	110.0	100.1	125.7	110.4	100.1	111.2	104.6	100.1	101.4	100.6	100.1	103.3	101.4
26, 27, 3370	ELECTRICAL, COMMUNICATIONS & UTIL.	96.8	100.1	98.5	99.1	128.9	114.9	99.6	128.9	115.1	100.0	100.1	100.1	99.6	98.9	99.2	99.7	98.9	99.2
MF2014	WEIGHTED AVERAGE	97.0	128.1	110.6	100.3	134.6	115.3	99.5	134.5	114.8	99.1	128.3	111.8	99.7	111.3	104.8	100.2	112.6	105.6

DIVISION		MASSACHUSETTS			MICHIGAN														
		WORCESTER			ANN ARBOR			BATTLE CREEK			BAY CITY			DEARBORN			DETROIT		
		015 - 016			481			490			487			481			482		
		MAT.	INST.	TOTAL	MAT.	INST.	TOTAL	MAT.	INST.	TOTAL	MAT.	INST.	TOTAL	MAT.	INST.	TOTAL	MAT.	INST.	TOTAL
015433	CONTRACTOR EQUIPMENT		100.9	100.9		110.7	110.7		102.8	102.8		110.7	110.7		110.7	110.7		98.8	98.8
0241, 31 - 34	SITE & INFRASTRUCTURE, DEMOLITION	94.6	105.6	102.4	82.3	98.8	94.0	93.3	87.5	89.2	73.9	97.8	90.9	82.1	98.9	94.1	94.7	100.6	98.9
0310	Concrete Forming & Accessories	99.6	130.7	126.5	98.3	111.6	109.8	97.6	87.6	89.0	98.4	87.1	88.6	98.2	115.8	113.3	100.8	115.8	113.7
0320	Concrete Reinforcing	104.6	150.6	128.0	92.3	120.1	106.5	91.4	93.3	92.4	92.3	119.2	106.0	92.3	120.2	106.5	94.9	120.1	107.8
0330	Cast-in-Place Concrete	93.5	150.5	116.9	88.8	107.4	96.4	99.0	100.4	99.6	85.0	90.6	87.3	86.8	109.9	96.3	94.1	109.9	100.6
03	CONCRETE	96.6	140.2	118.0	93.5	112.0	102.6	98.1	92.5	95.4	91.7	95.2	93.4	92.6	114.7	103.5	96.6	113.6	105.0
04	MASONRY	100.0	159.0	136.8	104.4	108.4	106.9	105.2	87.2	94.0	104.0	87.0	93.4	104.3	111.6	108.9	97.9	111.6	106.5
05	METALS	101.0	125.2	108.4	94.2	118.8	101.8	98.3	87.6	95.0	94.8	115.4	101.2	94.3	119.1	101.9	94.9	102.1	97.1
06	WOOD, PLASTICS & COMPOSITES	100.3	128.3	116.0	97.5	112.8	106.0	96.5	87.2	91.3	97.5	86.5	91.3	97.5	116.8	108.3	103.0	116.8	110.7
07	THERMAL & MOISTURE PROTECTION	99.5	141.5	116.7	100.4	108.2	103.6	94.6	84.8	90.6	98.2	92.4	95.8	98.9	115.7	105.8	97.7	115.7	105.1
08	OPENINGS	105.9	136.7	113.1	99.4	110.4	101.9	96.6	82.3	93.3	99.4	93.9	98.1	99.4	112.6	102.5	99.2	112.9	102.4
0920	Plaster & Gypsum Board	105.6	128.6	121.2	103.8	112.4	109.6	93.7	82.9	86.4	103.8	85.3	91.3	103.8	116.6	112.4	100.5	116.6	111.3
0950, 0980	Ceilings & Acoustic Treatment	99.3	128.6	118.5	88.7	112.4	104.3	96.3	82.9	87.5	89.7	85.3	86.8	88.7	116.6	107.0	91.2	116.6	107.9
0960	Flooring	100.2	179.8	123.0	97.5	117.6	103.2	105.7	91.9	101.7	97.5	81.5	92.9	96.9	113.9	101.8	95.9	113.9	101.1
0970, 0990	Wall Finishes & Painting/Coating	99.5	155.1	133.1	96.2	98.7	97.7	107.9	81.9	92.2	96.2	82.0	87.6	96.2	100.8	99.0	97.3	100.8	99.4
09	FINISHES	98.7	142.5	122.9	92.8	111.6	103.1	98.0	87.6	92.3	92.6	84.8	88.3	92.6	114.3	104.5	93.9	114.3	105.1
COVERS	DIVS. 10 - 14, 25, 28, 41, 43, 44, 46	100.0	109.0	101.8	100.0	105.8	101.2	100.0	96.4	99.3	100.0	94.1	98.8	100.0	107.1	101.4	100.0	107.1	101.4
21, 22, 23	FIRE SUPPRESSION, PLUMBING & HVAC	100.1	113.6	105.5	100.0	98.8	99.5	100.0	87.4	94.9	100.0	83.6	93.4	100.0	109.2	103.7	100.0	110.2	104.1
26, 27, 3370	ELECTRICAL, COMMUNICATIONS & UTIL.	99.7	105.1	102.5	96.3	108.8	102.9	94.4	83.7	88.8	95.2	90.3	92.6	96.3	107.0	101.9	98.2	107.0	102.9
MF2014	WEIGHTED AVERAGE	100.2	126.4	111.6	97.1	107.5	101.7	98.3	87.6	93.7	96.6	91.7	94.5	97.0	111.0	103.1	97.8	109.6	102.9

City Cost Indexes

DIVISION		MICHIGAN																	
		FLINT			GAYLORD			GRAND RAPIDS			IRON MOUNTAIN			JACKSON			KALAMAZOO		
		484 - 485			497			493,495			498 - 499			492			491		
		MAT.	INST.	TOTAL	MAT.	INST.	TOTAL	MAT.	INST.	TOTAL	MAT.	INST.	TOTAL	MAT.	INST.	TOTAL	MAT.	INST.	TOTAL
015433	CONTRACTOR EQUIPMENT		110.7	110.7		104.7	104.7		102.8	102.8		93.7	93.7		104.7	104.7		102.8	102.8
0241, 31 - 34	SITE & INFRASTRUCTURE, DEMOLITION	71.7	98.0	90.4	88.5	85.0	86.0	93.2	87.6	89.2	96.1	93.7	94.4	109.4	87.3	93.7	93.6	87.5	89.3
0310	Concrete Forming & Accessories	101.3	89.9	91.5	95.5	77.3	79.8	97.1	84.6	86.3	87.8	85.0	85.4	92.5	86.7	87.5	97.6	87.3	88.7
0320	Concrete Reinforcing	92.3	119.6	106.2	85.2	118.6	102.2	97.8	93.2	95.5	85.0	99.8	92.5	82.8	119.4	101.4	91.4	93.4	92.4
0330	Cast-in-Place Concrete	89.4	92.4	90.6	98.7	86.1	93.5	104.1	97.6	101.4	116.8	85.4	103.9	98.6	92.2	96.0	101.0	98.8	100.1
03	CONCRETE	94.0	97.1	95.5	94.6	88.9	91.8	101.6	90.2	96.0	104.3	87.9	96.2	89.3	95.4	92.3	101.6	91.8	96.8
04	MASONRY	104.5	96.5	99.5	116.2	79.1	93.1	102.0	87.4	92.9	101.2	86.1	91.8	95.1	91.6	92.9	103.8	89.1	94.6
05	METALS	94.3	116.3	101.0	99.7	109.9	102.8	95.3	87.1	92.8	99.0	89.6	96.1	99.9	114.1	104.2	98.3	86.5	94.7
06	WOOD, PLASTICS & COMPOSITES	100.9	88.2	93.8	89.4	76.8	82.4	97.7	83.2	89.6	85.0	85.1	85.0	88.1	84.7	86.2	96.5	87.2	91.3
07	THERMAL & MOISTURE PROTECTION	98.2	96.3	97.4	93.2	74.5	85.5	97.4	76.6	88.9	96.7	81.4	90.4	92.5	92.6	92.5	94.6	85.1	90.7
08	OPENINGS	99.4	95.6	98.5	97.5	75.1	92.3	103.4	84.7	99.0	104.6	76.0	98.0	96.5	93.1	95.7	96.6	81.7	93.1
0920	Plaster & Gypsum Board	105.2	87.1	93.0	93.4	74.9	80.9	95.4	78.8	84.2	51.7	85.2	74.3	91.7	82.9	85.8	93.7	82.9	86.4
0950, 0980	Ceilings & Acoustic Treatment	88.7	87.1	87.6	95.4	74.9	81.9	104.3	78.8	87.5	94.4	85.2	88.3	95.4	82.9	87.2	96.3	82.9	87.5
0960	Flooring	97.5	95.8	97.0	97.8	92.4	96.2	105.5	85.7	99.8	122.4	93.8	114.2	96.5	82.7	92.6	105.7	82.7	99.1
0970, 0990	Wall Finishes & Painting/Coating	96.2	85.4	89.7	103.8	82.0	90.6	109.7	80.8	92.3	126.1	72.1	93.5	103.8	96.8	99.6	107.9	81.9	92.2
09	FINISHES	92.1	89.8	90.9	97.4	79.1	87.3	101.6	84.2	92.0	100.0	85.3	91.9	98.3	86.2	91.6	98.0	85.6	91.2
COVERS	DIVS. 10 - 14, 25, 28, 41, 43, 44, 46	100.0	95.5	99.1	100.0	88.6	97.7	100.0	96.4	99.3	100.0	87.2	97.4	100.0	100.9	100.2	100.0	96.4	99.3
21, 22, 23	FIRE SUPPRESSION, PLUMBING & HVAC	100.0	89.6	95.8	95.6	78.4	88.6	100.0	83.0	93.1	95.5	85.3	91.4	95.6	88.4	92.7	100.0	80.8	92.3
26, 27, 3370	ELECTRICAL, COMMUNICATIONS & UTIL.	96.3	96.0	96.1	92.4	79.1	85.4	99.7	79.1	88.8	98.7	84.3	91.1	96.3	108.8	102.9	94.3	81.8	87.7
MF2014	WEIGHTED AVERAGE	96.8	96.0	96.4	97.2	83.6	91.3	99.7	85.1	93.3	99.3	86.2	93.6	96.5	95.1	95.9	98.7	85.7	93.0

DIVISION		MICHIGAN															MINNESOTA		
		LANSING			MUSKEGON			ROYAL OAK			SAGINAW			TRAVERSE CITY			BEMIDJI		
		488 - 489			494			480,483			486			496			566		
		MAT.	INST.	TOTAL	MAT.	INST.	TOTAL	MAT.	INST.	TOTAL	MAT.	INST.	TOTAL	MAT.	INST.	TOTAL	MAT.	INST.	TOTAL
015433	CONTRACTOR EQUIPMENT		110.7	110.7		102.8	102.8		96.2	96.2		110.7	110.7		93.7	93.7		98.2	98.2
0241, 31 - 34	SITE & INFRASTRUCTURE, DEMOLITION	95.6	98.0	97.3	91.2	87.5	88.5	86.5	98.4	95.0	74.9	97.8	91.2	82.4	93.2	90.0	95.4	97.5	96.9
0310	Concrete Forming & Accessories	98.2	89.7	90.9	98.0	84.1	86.0	94.1	112.6	110.1	98.3	87.6	89.1	87.8	75.2	76.9	86.1	89.6	89.1
0320	Concrete Reinforcing	94.3	119.4	107.1	92.1	93.3	92.7	83.7	119.9	102.1	92.3	119.2	106.0	86.3	92.5	89.5	93.7	106.6	100.2
0330	Cast-in-Place Concrete	103.5	92.1	98.8	98.6	94.8	97.1	77.8	106.3	89.5	87.7	90.5	88.9	91.2	78.2	85.9	106.0	103.0	104.8
03	CONCRETE	100.8	96.9	98.9	96.4	89.1	92.8	80.5	110.8	95.4	93.0	95.4	94.2	85.8	79.8	82.9	97.6	98.5	98.1
04	MASONRY	106.6	94.2	98.9	102.2	88.5	93.7	98.2	112.0	106.8	106.0	87.0	94.1	99.2	82.4	88.7	102.5	103.7	103.3
05	METALS	93.9	115.6	100.6	96.0	87.2	93.3	97.3	98.0	97.5	94.3	115.2	100.7	99.0	88.2	95.7	92.1	120.0	100.7
06	WOOD, PLASTICS & COMPOSITES	96.3	88.8	92.1	93.2	83.4	87.7	93.0	113.1	104.3	93.8	87.6	90.3	85.0	75.1	79.4	71.3	85.5	79.3
07	THERMAL & MOISTURE PROTECTION	98.6	89.6	94.9	93.6	78.9	87.6	96.8	111.2	102.7	99.0	92.6	96.4	95.8	71.7	85.9	104.7	95.4	100.9
08	OPENINGS	103.2	95.1	101.3	95.8	84.5	93.2	99.3	109.4	101.7	97.4	94.5	96.7	104.6	70.6	96.7	100.2	107.2	101.8
0920	Plaster & Gypsum Board	96.9	87.7	90.7	75.7	79.0	77.9	101.3	112.7	109.0	103.8	86.4	92.1	51.7	74.9	67.3	102.2	85.3	90.8
0950, 0980	Ceilings & Acoustic Treatment	94.0	87.7	89.8	97.1	79.0	85.2	88.0	112.7	104.3	88.7	86.4	87.2	94.4	74.9	81.6	122.8	85.3	98.2
0960	Flooring	106.6	85.7	100.6	104.5	85.7	99.1	94.8	113.9	100.3	97.5	81.5	92.9	122.4	94.9	114.5	104.4	120.8	109.1
0970, 0990	Wall Finishes & Painting/Coating	108.0	80.8	91.6	106.3	80.8	90.9	97.6	96.8	97.1	96.2	82.0	87.6	126.1	44.7	76.9	102.2	95.9	98.4
09	FINISHES	98.3	87.6	92.4	94.7	83.2	88.4	91.7	111.5	102.6	92.4	85.5	88.6	99.0	74.1	85.3	106.2	95.2	100.1
COVERS	DIVS. 10 - 14, 25, 28, 41, 43, 44, 46	100.0	100.6	100.1	100.0	96.0	99.2	100.0	103.0	100.6	100.0	94.3	98.8	100.0	84.6	96.9	100.0	96.9	99.4
21, 22, 23	FIRE SUPPRESSION, PLUMBING & HVAC	99.9	88.6	95.3	99.9	82.3	92.8	95.6	107.2	100.3	100.0	83.1	93.2	95.5	77.7	88.3	95.6	83.4	90.6
26, 27, 3370	ELECTRICAL, COMMUNICATIONS & UTIL.	98.6	93.3	95.8	94.8	79.1	86.5	98.4	104.4	101.6	93.9	90.5	92.1	94.0	79.1	86.1	104.3	104.6	104.5
MF2014	WEIGHTED AVERAGE	99.3	94.7	97.3	97.2	84.8	91.8	94.9	106.9	100.1	96.4	91.8	94.4	96.3	80.1	89.3	98.2	98.2	98.2

DIVISION		MINNESOTA																	
		BRAINERD			DETROIT LAKES			DULUTH			MANKATO			MINNEAPOLIS			ROCHESTER		
		564			565			556 - 558			560			553 - 555			559		
		MAT.	INST.	TOTAL	MAT.	INST.	TOTAL	MAT.	INST.	TOTAL	MAT.	INST.	TOTAL	MAT.	INST.	TOTAL	MAT.	INST.	TOTAL
015433	CONTRACTOR EQUIPMENT		100.8	100.8		98.2	98.2		99.1	99.1		100.8	100.8		102.6	102.6		99.1	99.1
0241, 31 - 34	SITE & INFRASTRUCTURE, DEMOLITION	96.9	102.2	100.7	93.6	97.9	96.6	100.6	99.4	99.7	93.8	102.9	100.3	97.9	104.7	102.8	97.4	99.0	98.5
0310	Concrete Forming & Accessories	87.7	91.6	91.1	83.1	91.6	90.4	98.7	109.9	108.4	95.6	99.1	98.6	100.7	128.6	124.8	99.3	106.5	105.5
0320	Concrete Reinforcing	92.5	106.7	99.7	93.7	106.6	100.2	97.4	107.1	102.3	92.4	116.2	104.5	92.2	116.6	104.6	91.7	116.4	104.3
0330	Cast-in-Place Concrete	115.2	106.7	111.7	102.9	106.1	104.2	114.8	106.0	111.2	106.2	107.3	106.6	109.6	117.4	112.8	112.3	99.4	107.0
03	CONCRETE	101.7	100.7	101.2	95.2	100.4	97.8	106.5	108.7	107.6	97.2	106.1	101.5	103.4	122.8	112.9	102.5	106.8	104.6
04	MASONRY	126.4	112.0	117.4	126.3	110.1	116.2	107.6	117.6	113.9	114.5	110.4	112.0	109.0	126.7	120.1	104.5	114.1	110.5
05	METALS	93.2	120.0	101.5	92.1	120.0	100.6	101.7	122.4	108.0	93.1	125.5	103.0	99.1	128.9	108.3	100.4	127.3	108.7
06	WOOD, PLASTICS & COMPOSITES	89.4	85.3	87.1	68.4	85.5	78.0	98.2	109.3	104.4	98.8	96.4	97.5	106.7	127.4	118.3	103.2	105.3	104.4
07	THERMAL & MOISTURE PROTECTION	103.4	107.7	105.2	104.6	108.3	106.1	101.6	114.9	107.1	103.8	99.0	101.8	102.8	127.4	112.9	104.9	100.9	103.2
08	OPENINGS	86.9	107.1	91.6	100.2	107.2	101.8	102.5	116.5	105.8	91.5	115.5	97.1	97.1	133.6	105.6	96.4	121.5	102.2
0920	Plaster & Gypsum Board	86.9	85.3	85.8	101.1	85.3	90.5	90.2	110.0	103.6	91.5	96.8	95.1	95.8	128.6	117.9	97.6	105.8	103.2
0950, 0980	Ceilings & Acoustic Treatment	59.3	85.3	76.4	122.8	85.3	98.2	96.1	110.0	105.3	59.3	96.8	84.0	98.2	128.6	118.2	93.2	105.8	101.5
0960	Flooring	103.3	120.8	108.3	103.2	120.8	108.2	101.9	122.1	107.6	105.0	120.8	109.5	97.9	120.8	104.4	102.9	120.8	108.0
0970, 0990	Wall Finishes & Painting/Coating	96.0	95.9	95.9	102.2	95.9	98.4	101.9	110.3	107.0	108.3	104.9	106.3	98.2	126.1	115.1	97.2	104.9	101.9
09	FINISHES	89.4	96.4	93.2	105.5	96.5	100.6	98.3	112.5	106.1	91.1	103.4	97.9	98.6	127.6	114.6	97.4	109.1	103.8
COVERS	DIVS. 10 - 14, 25, 28, 41, 43, 44, 46	100.0	98.3	99.7	100.0	98.6	99.7	100.0	97.6	99.5	100.0	98.6	99.7	100.0	109.7	102.0	100.0	102.8	100.6
21, 22, 23	FIRE SUPPRESSION, PLUMBING & HVAC	94.8	86.1	91.3	95.6	86.0	91.7	99.8	96.5	98.5	94.8	87.7	91.9	99.9	115.6	106.2	99.9	96.2	98.4
26, 27, 3370	ELECTRICAL, COMMUNICATIONS & UTIL.	101.8	102.4	102.1	104.0	67.5	84.7	98.4	102.4	100.6	108.6	89.2	98.3	101.2	111.4	106.6	98.6	89.2	93.6
MF2014	WEIGHTED AVERAGE	96.9	100.5	98.5	98.9	95.1	97.3	101.3	107.3	103.9	97.2	101.3	99.0	100.4	120.0	108.9	99.9	104.7	102.0

DIVISION		MINNESOTA															MISSISSIPPI		
		SAINT PAUL			ST. CLOUD			THIEF RIVER FALLS			WILLMAR			WINDOM			BILOXI		
		550 - 551			563			567			562			561			395		
		MAT.	INST.	TOTAL	MAT.	INST.	TOTAL	MAT.	INST.	TOTAL	MAT.	INST.	TOTAL	MAT.	INST.	TOTAL	MAT.	INST.	TOTAL
015433	**CONTRACTOR EQUIPMENT**		99.1	99.1		100.8	100.8		98.2	98.2		100.8	100.8		100.8	100.8		98.8	98.8
0241, 31 - 34	**SITE & INFRASTRUCTURE, DEMOLITION**	95.2	100.1	98.7	92.5	104.3	100.9	94.4	97.5	96.6	91.8	102.0	99.1	86.0	101.0	96.6	100.6	88.4	91.9
0310	Concrete Forming & Accessories	93.9	123.0	119.0	84.9	117.7	113.2	86.8	89.0	88.7	84.7	91.1	90.2	89.1	84.8	85.4	94.7	52.9	58.7
0320	Concrete Reinforcing	98.9	116.6	107.9	92.6	116.4	104.7	94.0	106.5	100.4	92.2	116.0	104.3	92.2	115.3	104.0	95.0	58.3	76.4
0330	Cast-in-Place Concrete	124.0	116.7	121.0	101.7	114.8	107.1	105.1	88.1	98.1	103.3	85.3	95.9	89.4	66.3	79.9	108.4	54.0	86.0
03	**CONCRETE**	111.2	120.1	115.5	92.9	116.9	104.7	96.3	93.1	94.7	92.9	94.9	93.9	83.6	85.5	84.5	101.6	55.9	79.2
04	**MASONRY**	111.7	126.7	121.1	111.1	119.9	116.6	102.4	104.5	103.7	115.4	112.0	113.3	125.8	92.1	104.8	102.7	44.3	66.3
05	**METALS**	101.1	128.6	109.6	93.9	126.8	104.0	92.2	119.4	100.6	93.0	124.5	102.7	92.9	122.4	102.0	94.9	81.7	90.8
06	**WOOD, PLASTICS & COMPOSITES**	93.4	119.9	108.2	86.7	114.9	102.5	72.3	85.5	79.7	86.4	85.5	85.9	90.7	82.8	86.3	99.1	56.1	75.0
07	**THERMAL & MOISTURE PROTECTION**	104.1	126.4	113.3	103.6	117.5	109.3	105.6	98.4	102.7	103.4	108.6	105.5	103.4	89.6	97.7	97.4	56.1	80.4
08	**OPENINGS**	99.8	129.5	106.7	91.5	126.7	99.7	100.2	107.2	101.8	89.0	91.9	89.7	92.7	90.4	92.2	99.6	57.7	89.8
0920	Plaster & Gypsum Board	88.9	120.9	110.6	86.9	115.8	106.4	101.8	85.3	90.7	86.9	85.6	86.0	86.9	82.8	84.1	101.7	55.2	70.3
0950, 0980	Ceilings & Acoustic Treatment	95.6	120.9	112.3	59.3	115.8	96.4	122.8	85.3	98.2	59.3	85.6	76.6	59.3	82.8	74.7	87.3	55.2	66.2
0960	Flooring	103.7	120.8	108.6	100.0	120.8	105.9	104.1	120.8	108.8	101.6	120.8	107.1	103.9	120.8	108.8	101.4	55.1	88.1
0970, 0990	Wall Finishes & Painting/Coating	105.3	126.1	117.8	108.3	126.1	119.1	102.2	95.9	98.4	102.2	95.9	98.4	102.2	104.9	103.8	99.5	43.4	65.7
09	**FINISHES**	97.9	123.1	111.8	88.9	119.2	105.6	106.0	95.2	100.0	88.9	94.2	91.8	89.1	91.3	90.3	93.3	52.4	70.8
COVERS	**DIVS. 10 - 14, 25, 28, 41, 43, 44, 46**	100.0	108.6	101.7	100.0	104.3	100.9	100.0	96.8	99.3	100.0	98.2	99.6	100.0	93.7	98.7	100.0	55.4	91.0
21, 22, 23	**FIRE SUPPRESSION, PLUMBING & HVAC**	99.9	113.8	105.5	99.6	113.8	105.3	95.6	83.1	90.5	94.8	106.8	99.6	94.8	84.1	90.5	100.0	49.1	79.5
26, 27, 3370	**ELECTRICAL, COMMUNICATIONS & UTIL.**	97.1	111.4	104.7	101.8	111.4	106.9	101.3	67.5	83.5	101.8	85.2	93.0	108.6	89.2	98.3	103.8	56.9	79.0
MF2014	**WEIGHTED AVERAGE**	101.4	117.9	108.6	96.9	116.0	105.2	97.8	92.4	95.4	95.5	101.3	98.0	95.9	92.3	94.3	99.2	58.1	81.3

DIVISION		MISSISSIPPI																	
		CLARKSDALE			COLUMBUS			GREENVILLE			GREENWOOD			JACKSON			LAUREL		
		386			397			387			389			390 - 392			394		
		MAT.	INST.	TOTAL	MAT.	INST.	TOTAL	MAT.	INST.	TOTAL	MAT.	INST.	TOTAL	MAT.	INST.	TOTAL	MAT.	INST.	TOTAL
015433	**CONTRACTOR EQUIPMENT**		98.8	98.8		98.8	98.8		98.8	98.8		98.8	98.8		98.8	98.8		98.8	98.8
0241, 31 - 34	**SITE & INFRASTRUCTURE, DEMOLITION**	98.3	88.2	91.1	98.9	88.4	91.5	103.9	88.4	92.9	101.0	87.9	91.7	97.6	88.4	91.1	104.4	88.3	92.9
0310	Concrete Forming & Accessories	85.0	49.2	54.2	83.3	51.2	55.6	81.7	63.7	66.1	94.2	49.2	55.4	93.6	57.5	62.5	83.3	52.9	57.1
0320	Concrete Reinforcing	99.8	41.6	70.1	101.8	42.5	71.6	100.3	60.3	79.9	99.8	52.5	75.7	100.2	56.1	77.7	102.4	39.1	70.2
0330	Cast-in-Place Concrete	105.6	53.5	84.2	110.5	58.6	89.2	108.7	55.6	86.9	113.5	53.2	88.7	98.7	56.8	81.5	108.1	55.2	86.4
03	**CONCRETE**	98.7	50.9	75.2	103.2	53.7	78.9	104.2	61.6	83.3	105.2	52.7	79.4	97.7	58.5	78.5	106.0	52.7	79.8
04	**MASONRY**	100.6	45.0	66.0	131.6	51.6	81.7	149.8	49.5	87.3	101.3	44.9	66.1	109.2	49.5	72.0	127.1	48.1	77.9
05	**METALS**	95.4	73.4	88.6	91.9	74.2	86.5	96.4	82.5	92.1	95.4	74.8	89.0	101.8	80.8	95.4	92.0	73.3	86.3
06	**WOOD, PLASTICS & COMPOSITES**	82.1	51.2	64.8	83.9	52.2	66.1	78.7	68.1	72.8	95.0	51.2	70.5	96.3	59.6	75.8	85.1	54.6	68.0
07	**THERMAL & MOISTURE PROTECTION**	95.7	47.2	75.8	97.3	51.5	78.5	96.0	60.0	81.2	96.1	47.1	76.0	97.5	59.2	81.8	97.4	53.1	79.3
08	**OPENINGS**	97.0	49.4	85.9	98.9	49.5	87.4	97.0	61.4	88.7	97.0	50.7	86.2	103.7	55.6	92.5	95.6	50.3	85.0
0920	Plaster & Gypsum Board	90.2	50.2	63.1	91.3	51.2	64.2	89.8	67.6	74.8	100.9	50.2	66.6	95.0	58.8	70.5	91.3	53.7	65.8
0950, 0980	Ceilings & Acoustic Treatment	82.9	50.2	61.4	82.0	51.2	61.8	85.7	67.6	73.8	82.9	50.2	61.4	89.7	58.8	69.4	82.0	53.7	63.4
0960	Flooring	105.7	52.8	90.6	95.5	59.3	85.1	104.1	52.8	89.4	111.1	52.8	94.4	99.3	55.1	86.7	94.2	52.8	82.4
0970, 0990	Wall Finishes & Painting/Coating	104.4	46.1	69.2	99.5	53.0	71.4	104.4	65.9	81.2	104.4	46.1	69.2	96.5	65.9	78.0	99.5	60.9	76.2
09	**FINISHES**	94.0	49.9	69.7	89.1	53.2	69.3	94.6	62.9	77.1	97.4	49.9	71.2	93.2	58.3	74.0	89.2	54.3	70.0
COVERS	**DIVS. 10 - 14, 25, 28, 41, 43, 44, 46**	100.0	55.7	91.1	100.0	56.8	91.3	100.0	58.5	91.6	100.0	55.7	91.1	100.0	57.5	91.4	100.0	37.5	87.4
21, 22, 23	**FIRE SUPPRESSION, PLUMBING & HVAC**	97.9	41.3	75.1	97.4	37.5	73.3	100.0	47.6	78.8	97.9	39.3	74.3	100.1	57.0	82.7	97.5	45.8	76.6
26, 27, 3370	**ELECTRICAL, COMMUNICATIONS & UTIL.**	97.1	46.4	70.3	100.8	49.1	73.5	97.1	62.7	78.9	97.1	43.4	68.8	105.4	62.7	82.8	102.6	56.9	78.5
MF2014	**WEIGHTED AVERAGE**	97.3	52.6	77.8	98.8	53.9	79.2	101.0	61.7	83.9	98.4	52.2	78.3	100.7	62.2	83.9	98.9	55.8	80.1

DIVISION		MISSISSIPPI									MISSOURI								
		MCCOMB			MERIDIAN			TUPELO			BOWLING GREEN			CAPE GIRARDEAU			CHILLICOTHE		
		396			393			388			633			637			646		
		MAT.	INST.	TOTAL	MAT.	INST.	TOTAL	MAT.	INST.	TOTAL	MAT.	INST.	TOTAL	MAT.	INST.	TOTAL	MAT.	INST.	TOTAL
015433	**CONTRACTOR EQUIPMENT**		98.8	98.8		98.8	98.8		98.8	98.8		106.9	106.9		106.9	106.9		101.9	101.9
0241, 31 - 34	**SITE & INFRASTRUCTURE, DEMOLITION**	92.2	88.0	89.2	96.2	88.7	90.8	95.6	88.1	90.3	89.6	95.6	93.9	91.3	95.4	94.2	105.2	94.6	97.7
0310	Concrete Forming & Accessories	83.3	50.9	55.4	80.4	63.3	65.7	82.2	51.2	55.5	92.3	87.3	88.0	85.2	85.1	85.1	89.0	93.0	92.5
0320	Concrete Reinforcing	103.0	41.0	71.4	101.8	56.1	78.5	97.6	52.5	74.7	101.4	102.8	102.1	102.7	88.7	95.5	96.8	106.5	101.8
0330	Cast-in-Place Concrete	96.0	53.0	78.3	102.7	59.2	84.8	105.6	54.9	84.8	91.2	84.7	88.5	90.2	86.0	88.5	103.0	85.8	95.9
03	**CONCRETE**	92.9	51.3	72.5	97.2	61.9	79.8	98.2	54.2	76.6	95.8	90.5	93.2	94.9	87.3	91.2	105.6	93.6	99.7
04	**MASONRY**	133.0	44.4	77.7	102.4	53.5	71.9	138.4	47.9	82.0	117.6	98.7	105.8	113.9	80.0	92.8	102.9	98.8	100.3
05	**METALS**	92.2	71.6	85.8	93.0	81.0	89.3	95.3	75.1	89.1	94.6	114.4	100.7	95.6	107.0	99.1	95.1	108.5	99.2
06	**WOOD, PLASTICS & COMPOSITES**	83.9	54.0	67.2	81.2	65.1	72.2	79.3	52.2	64.1	90.8	86.0	88.1	83.5	85.5	84.6	93.6	93.3	93.4
07	**THERMAL & MOISTURE PROTECTION**	96.9	46.4	76.2	97.0	61.6	82.5	95.7	52.9	78.2	100.3	98.9	99.7	100.2	84.2	93.6	98.3	94.5	96.8
08	**OPENINGS**	99.0	51.0	87.8	98.9	64.1	90.8	97.0	50.6	86.2	97.7	97.5	97.6	97.7	82.9	94.2	95.0	95.7	95.2
0920	Plaster & Gypsum Board	91.3	53.0	65.4	91.3	64.5	73.2	89.8	51.2	63.7	95.2	85.6	88.7	94.2	85.1	88.0	102.0	92.9	95.8
0950, 0980	Ceilings & Acoustic Treatment	82.0	53.0	63.0	83.9	64.5	71.2	82.9	51.2	62.0	92.6	85.6	88.0	92.6	85.1	87.6	89.1	92.9	91.6
0960	Flooring	95.5	52.8	83.3	94.2	55.1	83.0	104.4	52.8	89.6	96.4	96.8	96.5	93.3	84.0	90.6	97.3	102.5	98.8
0970, 0990	Wall Finishes & Painting/Coating	99.5	42.3	65.0	99.5	76.1	85.4	104.4	58.7	76.8	97.1	102.8	100.5	97.1	73.2	82.7	87.9	106.3	99.0
09	**FINISHES**	88.5	51.0	67.8	88.8	63.7	75.0	93.5	52.6	71.0	97.8	89.7	93.4	96.7	83.0	89.2	100.2	96.1	97.9
COVERS	**DIVS. 10 - 14, 25, 28, 41, 43, 44, 46**	100.0	59.1	91.7	100.0	59.5	91.8	100.0	56.8	91.3	100.0	88.2	97.6	100.0	93.5	98.7	100.0	88.5	97.7
21, 22, 23	**FIRE SUPPRESSION, PLUMBING & HVAC**	97.4	34.6	72.1	100.0	59.1	83.5	98.1	43.2	76.0	95.2	96.2	95.6	99.9	96.8	98.7	95.3	97.0	96.0
26, 27, 3370	**ELECTRICAL, COMMUNICATIONS & UTIL.**	99.2	49.4	72.9	102.6	57.8	78.9	96.9	49.3	71.7	96.9	82.6	89.3	96.9	101.5	99.3	92.4	78.6	85.1
MF2014	**WEIGHTED AVERAGE**	97.4	51.7	77.5	97.6	64.1	83.0	98.8	54.9	79.7	97.3	94.4	96.0	98.2	92.4	95.7	97.6	94.5	96.3

DIVISION		MISSOURI																	
		COLUMBIA			FLAT RIVER			HANNIBAL			HARRISONVILLE			JEFFERSON CITY			JOPLIN		
		652			636			634			647			650 - 651			648		
		MAT.	INST.	TOTAL	MAT.	INST.	TOTAL	MAT.	INST.	TOTAL	MAT.	INST.	TOTAL	MAT.	INST.	TOTAL	MAT.	INST.	TOTAL
015433	**CONTRACTOR EQUIPMENT**		108.0	108.0		106.9	106.9		106.9	106.9		101.9	101.9		108.0	108.0		105.7	105.7
0241, 31 - 34	**SITE & INFRASTRUCTURE, DEMOLITION**	102.8	97.2	98.8	92.2	95.6	94.6	87.6	95.7	93.4	96.6	95.7	95.9	102.3	97.2	98.6	105.6	99.7	101.4
0310	Concrete Forming & Accessories	86.9	81.8	82.5	98.3	84.3	86.2	90.5	79.0	80.6	86.2	98.7	97.0	98.3	81.8	84.1	101.3	75.3	78.9
0320	Concrete Reinforcing	96.3	119.0	107.9	102.7	107.7	105.3	100.9	97.7	99.3	96.4	115.3	106.0	98.5	111.9	105.3	99.9	98.5	99.2
0330	Cast-in-Place Concrete	96.1	87.8	92.7	94.2	93.8	94.0	86.3	86.0	86.2	105.5	103.9	104.8	102.0	87.0	95.9	111.5	77.9	97.7
03	**CONCRETE**	93.2	92.2	92.7	98.8	93.2	96.0	92.1	86.3	89.2	101.8	104.0	102.9	98.9	90.6	94.8	105.2	81.4	93.5
04	**MASONRY**	154.1	90.0	114.2	114.5	81.4	93.9	109.0	101.3	104.2	97.4	103.4	101.2	112.0	90.0	98.3	96.5	85.9	89.9
05	**METALS**	98.4	121.2	105.4	94.5	115.6	101.0	94.6	111.9	99.9	95.6	113.2	101.0	98.0	118.3	104.2	98.2	100.5	98.9
06	**WOOD, PLASTICS & COMPOSITES**	91.7	78.7	84.4	99.1	82.6	89.9	88.9	73.8	80.4	89.9	97.5	94.2	100.0	78.7	88.1	106.0	74.2	88.2
07	**THERMAL & MOISTURE PROTECTION**	96.2	87.5	92.6	100.5	93.8	97.7	100.1	97.0	98.8	97.6	103.2	99.9	101.8	87.6	96.0	97.7	78.9	90.0
08	**OPENINGS**	99.5	95.0	98.4	97.7	97.2	97.6	97.7	82.3	94.1	95.1	102.8	96.9	99.1	93.1	97.7	96.2	78.5	92.1
0920	Plaster & Gypsum Board	87.2	78.0	81.0	101.3	82.1	88.3	94.5	73.0	79.9	97.3	97.3	97.3	91.3	78.0	82.3	108.2	73.2	84.5
0950, 0980	Ceilings & Acoustic Treatment	93.2	78.0	83.2	92.6	82.1	85.7	92.6	73.0	79.7	89.1	97.3	94.5	97.2	78.0	84.6	90.0	73.2	78.9
0960	Flooring	97.6	98.6	97.9	99.4	84.0	95.0	95.8	96.8	96.1	92.6	102.5	95.4	104.4	98.6	102.7	123.8	74.9	109.8
0970, 0990	Wall Finishes & Painting/Coating	105.2	77.1	88.2	97.1	78.4	85.8	97.1	102.8	100.5	92.5	103.8	99.3	102.3	77.1	87.1	87.5	74.9	79.9
09	**FINISHES**	92.8	83.0	87.4	99.7	82.2	90.1	97.4	83.1	89.5	97.8	99.5	98.7	97.2	83.0	89.4	107.0	75.3	89.6
COVERS	**DIVS. 10 - 14, 25, 28, 41, 43, 44, 46**	100.0	96.2	99.2	100.0	91.8	98.3	100.0	87.8	97.5	100.0	90.6	98.1	100.0	96.2	99.2	100.0	87.9	97.6
21, 22, 23	**FIRE SUPPRESSION, PLUMBING & HVAC**	99.8	96.4	98.4	95.2	97.8	96.2	95.2	97.5	96.1	95.2	100.0	97.2	99.8	98.1	99.1	100.1	72.1	88.8
26, 27, 3370	**ELECTRICAL, COMMUNICATIONS & UTIL.**	94.8	86.5	90.4	101.1	101.5	101.3	95.7	82.6	88.8	98.8	99.6	99.2	100.4	86.5	93.0	90.4	77.5	83.6
MF2014	**WEIGHTED AVERAGE**	100.2	93.9	97.5	98.1	95.0	96.8	96.2	92.4	94.6	97.1	101.5	99.1	99.9	93.7	97.2	99.5	81.7	91.8

DIVISION		MISSOURI																	
		KANSAS CITY			KIRKSVILLE			POPLAR BLUFF			ROLLA			SEDALIA			SIKESTON		
		640 - 641			635			639			654 - 655			653			638		
		MAT.	INST.	TOTAL	MAT.	INST.	TOTAL	MAT.	INST.	TOTAL	MAT.	INST.	TOTAL	MAT.	INST.	TOTAL	MAT.	INST.	TOTAL
015433	**CONTRACTOR EQUIPMENT**		103.3	103.3		99.1	99.1		101.5	101.5		108.0	108.0		99.9	99.9		101.5	101.5
0241, 31 - 34	**SITE & INFRASTRUCTURE, DEMOLITION**	98.9	98.2	98.4	91.8	91.4	91.5	79.1	95.4	90.7	101.5	97.1	98.3	99.0	92.8	94.6	82.4	95.4	91.7
0310	Concrete Forming & Accessories	100.1	105.4	104.6	83.2	80.6	81.0	83.3	85.0	84.7	95.0	91.9	92.3	92.4	81.4	82.9	84.3	84.9	84.9
0320	Concrete Reinforcing	94.9	119.8	107.6	101.6	98.8	100.2	104.7	94.7	99.6	96.7	94.7	95.7	95.3	106.2	100.8	104.0	98.4	101.2
0330	Cast-in-Place Concrete	103.7	106.6	104.9	94.1	83.8	89.9	72.4	86.5	78.2	98.3	96.2	97.5	102.7	83.8	94.9	77.4	86.5	81.1
03	**CONCRETE**	101.1	108.7	104.8	110.8	86.0	98.6	85.9	88.1	86.9	95.2	95.0	95.1	109.8	87.7	99.0	89.5	88.7	89.1
04	**MASONRY**	99.5	107.5	104.5	121.0	88.9	101.0	112.5	80.1	92.3	126.3	87.9	102.4	132.9	88.3	105.1	112.2	80.1	92.2
05	**METALS**	105.9	115.8	108.9	94.2	103.8	97.2	94.8	101.7	96.9	97.8	110.5	101.7	96.5	108.0	100.1	95.1	103.2	97.6
06	**WOOD, PLASTICS & COMPOSITES**	105.3	105.0	105.1	77.0	78.7	77.9	76.1	85.5	81.4	100.3	92.8	96.1	93.1	78.7	85.1	77.7	85.5	82.1
07	**THERMAL & MOISTURE PROTECTION**	97.8	106.8	101.5	106.8	88.2	99.2	105.2	85.9	97.3	96.4	93.4	95.1	102.5	93.1	98.7	105.3	84.5	96.8
08	**OPENINGS**	102.5	109.2	104.0	102.7	85.8	98.8	103.7	84.5	99.2	99.5	88.3	96.9	104.5	87.4	100.5	103.7	85.5	99.5
0920	Plaster & Gypsum Board	105.2	105.0	105.0	89.8	78.0	81.8	90.2	85.1	86.7	89.3	92.5	91.5	82.5	78.0	79.5	92.0	85.1	87.3
0950, 0980	Ceilings & Acoustic Treatment	96.7	105.0	102.1	90.9	78.0	82.4	92.6	85.1	87.6	93.2	92.5	92.7	93.2	78.0	83.2	92.6	85.1	87.6
0960	Flooring	98.0	109.0	101.2	74.4	95.8	80.5	88.3	82.0	86.5	101.2	95.8	99.6	78.8	95.8	83.7	88.7	82.0	86.8
0970, 0990	Wall Finishes & Painting/Coating	92.5	112.2	104.4	92.5	82.1	86.2	92.0	73.2	80.7	105.2	100.6	102.5	105.2	106.3	105.9	92.0	73.2	80.7
09	**FINISHES**	102.0	106.6	104.5	96.5	82.7	88.9	96.5	82.7	88.9	94.3	93.4	93.8	91.9	85.4	88.3	97.1	82.7	89.2
COVERS	**DIVS. 10 - 14, 25, 28, 41, 43, 44, 46**	100.0	100.6	100.1	100.0	87.1	97.4	100.0	93.6	98.7	100.0	89.0	97.8	100.0	90.2	98.0	100.0	94.2	98.8
21, 22, 23	**FIRE SUPPRESSION, PLUMBING & HVAC**	100.0	104.3	101.7	95.2	97.3	96.1	95.2	96.3	95.6	95.0	98.2	96.3	95.0	95.9	95.4	95.2	96.3	95.6
26, 27, 3370	**ELECTRICAL, COMMUNICATIONS & UTIL.**	100.3	100.1	100.2	95.8	82.5	88.8	96.1	101.5	99.0	93.2	86.5	89.6	94.7	128.8	112.7	95.3	101.5	98.6
MF2014	**WEIGHTED AVERAGE**	101.4	105.7	103.3	99.5	89.9	95.3	96.2	92.0	94.4	97.9	94.6	96.5	100.2	97.3	98.9	96.7	92.2	94.8

DIVISION		MISSOURI									MONTANA								
		SPRINGFIELD			ST. JOSEPH			ST. LOUIS			BILLINGS			BUTTE			GREAT FALLS		
		656 - 658			644 - 645			630 - 631			590 - 591			597			594		
		MAT.	INST.	TOTAL	MAT.	INST.	TOTAL	MAT.	INST.	TOTAL	MAT.	INST.	TOTAL	MAT.	INST.	TOTAL	MAT.	INST.	TOTAL
015433	**CONTRACTOR EQUIPMENT**		102.5	102.5		101.9	101.9		107.9	107.9		98.5	98.5		98.2	98.2		98.2	98.2
0241, 31 - 34	**SITE & INFRASTRUCTURE, DEMOLITION**	101.6	95.1	97.0	100.5	93.9	95.8	91.7	98.9	96.8	95.0	96.3	95.9	102.0	96.1	97.8	105.7	96.2	98.9
0310	Concrete Forming & Accessories	101.4	79.1	82.2	100.0	89.6	91.0	98.4	104.8	103.9	98.4	71.5	75.2	85.5	71.8	73.7	98.4	71.5	75.2
0320	Concrete Reinforcing	92.7	118.7	105.9	93.8	115.0	104.6	93.9	112.7	103.4	90.2	81.7	85.9	97.8	81.9	89.7	90.2	81.8	85.9
0330	Cast-in-Place Concrete	104.3	79.6	94.2	103.6	103.3	103.5	90.2	105.0	96.3	129.3	67.6	103.9	142.3	69.0	112.2	150.4	63.7	114.8
03	**CONCRETE**	105.8	87.4	96.8	100.9	99.7	100.3	94.4	107.1	100.7	109.8	72.9	91.7	113.8	73.5	94.0	119.5	71.5	96.0
04	**MASONRY**	102.9	89.8	94.8	99.3	96.1	97.3	95.4	111.9	105.7	128.0	73.7	94.2	123.5	76.6	94.3	128.0	75.4	95.2
05	**METALS**	102.8	108.8	104.7	101.9	112.4	105.2	100.3	120.2	106.4	105.8	90.0	100.9	100.1	90.1	97.1	103.2	90.0	99.2
06	**WOOD, PLASTICS & COMPOSITES**	101.8	77.1	88.0	106.0	87.6	95.7	99.0	103.0	101.2	98.6	70.5	82.8	85.7	70.8	77.4	99.9	70.5	83.4
07	**THERMAL & MOISTURE PROTECTION**	100.7	82.2	93.2	98.2	95.6	97.1	100.2	106.9	103.0	106.5	69.9	91.5	106.3	71.0	91.8	106.9	68.5	91.2
08	**OPENINGS**	107.2	84.3	101.9	100.6	98.6	100.1	97.6	110.9	100.7	97.3	68.7	90.6	95.6	70.0	89.6	98.5	68.0	91.4
0920	Plaster & Gypsum Board	90.4	76.3	80.9	109.7	87.0	94.4	102.2	103.1	102.8	98.0	69.8	78.9	98.7	70.2	79.4	107.3	69.8	82.0
0950, 0980	Ceilings & Acoustic Treatment	93.2	76.3	82.1	95.8	87.0	90.0	97.6	103.1	101.2	81.5	69.8	73.8	88.3	70.2	76.4	90.0	69.8	76.7
0960	Flooring	99.9	74.9	92.8	101.5	107.3	103.2	99.2	97.6	98.7	106.6	70.2	96.2	104.4	74.2	95.7	111.5	74.2	100.8
0970, 0990	Wall Finishes & Painting/Coating	99.0	74.9	84.4	87.9	86.7	87.2	97.1	106.0	102.5	105.2	86.6	94.0	103.4	51.2	71.9	103.4	86.6	93.3
09	**FINISHES**	96.7	78.0	86.4	103.2	91.7	96.9	100.8	103.1	102.1	96.6	72.6	83.4	97.2	69.7	82.0	101.1	73.3	85.8
COVERS	**DIVS. 10 - 14, 25, 28, 41, 43, 44, 46**	100.0	94.8	99.0	100.0	96.9	99.4	100.0	102.4	100.5	100.0	94.7	98.9	100.0	94.8	98.9	100.0	94.7	98.9
21, 22, 23	**FIRE SUPPRESSION, PLUMBING & HVAC**	99.8	74.6	89.6	100.1	89.0	95.6	100.0	107.4	103.0	100.1	74.7	89.9	100.2	75.6	90.3	100.2	70.7	88.3
26, 27, 3370	**ELECTRICAL, COMMUNICATIONS & UTIL.**	99.2	69.7	83.6	98.8	78.6	88.2	99.2	108.3	104.0	98.5	72.2	84.6	105.5	71.7	87.7	97.7	71.9	84.1
MF2014	**WEIGHTED AVERAGE**	101.7	83.7	93.8	100.6	93.5	97.5	98.7	107.8	102.7	102.7	77.1	91.5	102.6	77.3	91.6	104.1	76.2	91.9

DIVISION		MONTANA																	
		HAVRE			HELENA			KALISPELL			MILES CITY			MISSOULA			WOLF POINT		
		595			596			599			593			598			592		
		MAT.	INST.	TOTAL	MAT.	INST.	TOTAL	MAT.	INST.	TOTAL	MAT.	INST.	TOTAL	MAT.	INST.	TOTAL	MAT.	INST.	TOTAL
015433	**CONTRACTOR EQUIPMENT**		98.2	98.2		98.2	98.2		98.2	98.2		98.2	98.2		98.2	98.2		98.2	98.2
0241, 31 - 34	**SITE & INFRASTRUCTURE, DEMOLITION**	109.2	95.8	99.7	98.5	95.8	96.6	92.2	96.0	94.9	98.5	95.4	96.3	85.1	95.8	92.7	115.5	95.6	101.3
0310	Concrete Forming & Accessories	78.4	69.0	70.3	101.5	65.2	70.2	88.7	71.4	73.8	96.8	67.6	71.6	88.7	71.4	73.8	89.2	67.9	70.8
0320	Concrete Reinforcing	98.6	81.7	90.0	102.5	81.7	91.9	100.4	83.4	91.8	98.2	81.8	89.8	99.5	83.4	91.3	99.6	81.8	90.5
0330	Cast-in-Place Concrete	153.3	60.4	115.1	115.0	62.9	93.6	123.6	62.7	98.6	135.3	60.3	104.5	104.9	64.9	88.4	151.5	60.6	114.2
03	**CONCRETE**	122.7	69.3	96.5	106.7	68.4	87.9	102.6	71.5	87.3	110.4	68.6	89.9	90.0	72.2	81.3	126.5	68.9	98.2
04	**MASONRY**	124.5	69.8	90.4	120.1	75.7	92.4	122.2	77.0	94.0	129.7	65.3	89.6	147.9	76.6	103.4	131.0	66.0	90.5
05	**METALS**	96.3	89.6	94.2	102.1	88.8	98.0	96.1	90.6	94.4	95.4	89.9	93.7	96.6	90.7	94.8	95.5	89.2	93.6
06	**WOOD, PLASTICS & COMPOSITES**	77.4	70.5	73.5	102.9	62.3	80.2	89.1	70.5	78.7	96.9	70.5	82.1	89.1	70.5	78.7	88.5	70.5	78.4
07	**THERMAL & MOISTURE PROTECTION**	106.6	68.0	90.8	103.9	70.7	90.3	105.9	72.8	92.3	106.2	65.7	89.6	105.5	79.1	94.7	107.1	66.0	90.3
08	**OPENINGS**	95.6	68.0	89.2	99.8	65.2	91.7	95.6	68.5	89.3	95.1	68.0	88.8	95.6	68.3	89.2	95.1	67.4	88.7
0920	Plaster & Gypsum Board	94.8	69.8	77.9	96.0	61.4	72.6	98.7	69.8	79.2	106.7	69.8	81.8	98.7	69.8	79.2	102.1	69.8	80.3
0950, 0980	Ceilings & Acoustic Treatment	88.3	69.8	76.2	90.8	61.4	71.5	88.3	69.8	76.2	87.5	69.8	75.9	88.3	69.8	76.2	87.5	69.8	75.9
0960	Flooring	101.8	74.2	93.9	106.7	59.1	93.1	106.2	74.2	97.1	111.3	70.2	99.5	106.2	74.2	97.1	107.6	70.2	96.9
0970, 0990	Wall Finishes & Painting/Coating	103.4	48.6	70.4	99.7	47.2	68.0	103.4	47.2	69.5	103.4	47.2	69.5	103.4	62.0	78.4	103.4	48.6	70.4
09	**FINISHES**	96.5	67.7	80.6	98.6	61.6	78.2	97.2	69.1	81.7	100.0	65.7	81.1	96.7	70.7	82.3	99.6	66.0	81.1
COVERS	**DIVS. 10 - 14, 25, 28, 41, 43, 44, 46**	100.0	70.8	94.1	100.0	94.0	98.8	100.0	93.9	98.8	100.0	90.4	98.1	100.0	93.7	98.7	100.0	90.6	98.1
21, 22, 23	**FIRE SUPPRESSION, PLUMBING & HVAC**	95.4	67.8	84.3	100.1	70.1	88.0	95.4	69.2	84.8	95.4	69.6	85.0	100.2	69.2	87.7	95.4	70.0	85.1
26, 27, 3370	**ELECTRICAL, COMMUNICATIONS & UTIL.**	97.7	71.2	83.7	104.6	71.3	87.0	102.2	69.3	84.8	97.7	76.9	86.7	103.2	67.7	84.5	97.7	76.9	86.7
MF2014	**WEIGHTED AVERAGE**	101.2	73.1	89.0	102.4	73.8	89.9	99.0	75.3	88.7	100.1	74.0	88.7	99.9	75.5	89.3	102.3	74.2	90.0

DIVISION		NEBRASKA																	
		ALLIANCE			COLUMBUS			GRAND ISLAND			HASTINGS			LINCOLN			MCCOOK		
		693			686			688			689			683 - 685			690		
		MAT.	INST.	TOTAL	MAT.	INST.	TOTAL	MAT.	INST.	TOTAL	MAT.	INST.	TOTAL	MAT.	INST.	TOTAL	MAT.	INST.	TOTAL
015433	**CONTRACTOR EQUIPMENT**		97.8	97.8		101.6	101.6		101.6	101.6		101.6	101.6		101.6	101.6		101.6	101.6
0241, 31 - 34	**SITE & INFRASTRUCTURE, DEMOLITION**	100.5	98.8	99.3	100.2	92.3	94.6	105.0	93.1	96.5	103.7	92.3	95.6	92.7	93.0	92.9	103.8	92.2	95.6
0310	Concrete Forming & Accessories	87.3	56.4	60.7	95.5	67.4	71.2	95.1	72.4	75.5	98.2	71.9	75.5	95.7	68.2	71.9	93.0	56.1	61.2
0320	Concrete Reinforcing	108.8	85.1	96.7	98.3	76.5	87.2	97.7	76.4	86.9	97.7	77.1	87.2	96.7	76.0	86.1	101.8	76.2	88.8
0330	Cast-in-Place Concrete	113.7	60.0	91.6	115.7	60.0	92.8	122.5	67.2	99.8	122.5	63.2	98.2	92.4	69.6	83.1	122.6	59.3	96.6
03	**CONCRETE**	123.9	63.5	94.2	108.8	67.3	88.4	113.7	72.1	93.3	113.9	70.6	92.6	95.6	71.0	83.5	112.9	62.0	87.9
04	**MASONRY**	115.6	74.1	89.7	122.7	77.8	94.7	115.4	76.1	90.9	124.9	88.0	101.9	104.8	67.2	81.4	110.7	73.9	87.8
05	**METALS**	101.5	78.7	94.5	96.1	84.6	92.6	97.8	86.3	94.3	98.7	85.8	94.7	99.9	85.5	95.5	96.3	83.6	92.4
06	**WOOD, PLASTICS & COMPOSITES**	87.0	52.2	67.5	101.2	66.7	81.9	100.4	72.2	84.6	104.1	72.2	86.3	98.4	66.7	80.7	95.7	52.2	71.4
07	**THERMAL & MOISTURE PROTECTION**	104.3	67.8	89.4	100.8	69.8	88.1	100.9	72.8	89.4	100.9	83.4	93.8	98.1	69.9	86.5	98.9	66.6	85.6
08	**OPENINGS**	94.5	58.6	86.1	95.4	66.1	88.6	95.5	69.7	89.5	95.5	69.1	89.3	107.8	63.0	97.4	95.0	56.6	86.1
0920	Plaster & Gypsum Board	82.7	50.7	61.1	90.9	65.6	73.8	90.1	71.3	77.4	91.9	71.3	78.0	96.2	65.6	75.5	93.7	50.7	64.7
0950, 0980	Ceilings & Acoustic Treatment	91.9	50.7	64.8	80.2	65.6	70.6	80.2	71.3	74.3	80.2	71.3	74.3	90.4	65.6	74.1	88.0	50.7	63.5
0960	Flooring	99.9	82.1	94.8	92.9	100.4	95.0	92.7	107.2	96.8	93.9	100.4	95.7	101.0	83.6	96.0	97.6	82.1	93.1
0970, 0990	Wall Finishes & Painting/Coating	168.5	54.8	99.9	85.5	63.7	72.4	85.5	67.6	74.7	85.5	63.7	72.4	103.8	77.9	88.1	93.8	47.7	66.0
09	**FINISHES**	98.4	59.5	77.0	89.9	71.9	80.0	90.0	77.1	82.9	90.6	75.2	82.1	96.9	71.7	83.0	95.6	58.6	75.2
COVERS	**DIVS. 10 - 14, 25, 28, 41, 43, 44, 46**	100.0	65.3	93.0	100.0	86.1	97.2	100.0	88.7	97.7	100.0	86.7	97.3	100.0	88.1	97.6	100.0	64.5	92.8
21, 22, 23	**FIRE SUPPRESSION, PLUMBING & HVAC**	95.3	70.5	85.3	95.2	76.0	87.4	100.0	78.0	91.1	95.2	76.0	87.5	99.9	78.0	91.0	95.1	75.9	87.4
26, 27, 3370	**ELECTRICAL, COMMUNICATIONS & UTIL.**	92.3	69.4	80.2	93.6	82.3	87.6	92.2	67.0	78.9	91.6	83.6	87.4	105.4	67.0	85.1	94.4	69.4	81.2
MF2014	**WEIGHTED AVERAGE**	101.0	70.5	87.7	98.4	77.1	89.1	100.0	77.1	90.0	99.4	79.8	90.9	100.5	74.9	89.4	98.7	71.1	86.7

DIVISION		NEBRASKA												NEVADA					
		NORFOLK			NORTH PLATTE			OMAHA			VALENTINE			CARSON CITY			ELKO		
		687			691			680 - 681			692			897			898		
		MAT.	INST.	TOTAL	MAT.	INST.	TOTAL	MAT.	INST.	TOTAL	MAT.	INST.	TOTAL	MAT.	INST.	TOTAL	MAT.	INST.	TOTAL
015433	**CONTRACTOR EQUIPMENT**		92.8	92.8		101.6	101.6		92.8	92.8		95.9	95.9		98.2	98.2		98.2	98.2
0241, 31 - 34	**SITE & INFRASTRUCTURE, DEMOLITION**	83.6	91.5	89.2	105.2	92.3	96.0	90.0	92.1	91.5	88.6	96.5	94.2	84.5	99.0	94.8	66.9	97.6	88.8
0310	Concrete Forming & Accessories	81.7	71.1	72.5	95.4	71.5	74.8	94.3	72.4	75.4	83.7	55.8	59.6	103.3	92.8	94.2	109.2	81.3	85.2
0320	Concrete Reinforcing	98.4	66.5	82.2	101.2	77.1	88.9	99.7	76.6	87.9	101.8	66.1	83.7	103.2	118.4	111.0	102.7	113.2	108.1
0330	Cast-in-Place Concrete	116.7	62.4	94.4	122.6	64.5	98.7	101.1	75.5	90.6	108.4	57.2	87.4	100.4	85.7	94.4	95.1	80.2	89.0
03	**CONCRETE**	107.3	67.5	87.8	113.0	70.8	92.3	100.1	74.5	87.5	110.2	58.9	85.0	100.9	95.2	98.1	95.4	87.2	91.4
04	**MASONRY**	129.4	79.1	98.0	98.6	87.9	92.0	106.8	79.0	89.4	110.9	73.9	87.8	117.3	94.6	103.2	115.0	71.9	88.1
05	**METALS**	99.6	73.5	91.6	95.5	85.4	92.4	99.9	78.9	93.4	107.6	73.4	97.1	94.0	104.7	97.3	96.9	99.4	97.7
06	**WOOD, PLASTICS & COMPOSITES**	84.0	71.7	77.1	97.8	72.2	83.5	94.1	72.0	81.7	81.7	51.7	64.9	92.1	94.4	93.4	100.6	81.8	90.1
07	**THERMAL & MOISTURE PROTECTION**	100.8	73.6	89.6	98.9	82.8	92.3	95.5	78.7	88.6	99.6	67.6	86.5	103.5	92.3	99.0	99.9	76.1	90.1
08	**OPENINGS**	97.2	66.5	90.0	94.3	69.1	88.4	100.0	71.5	93.3	96.9	55.6	87.3	99.5	107.9	101.4	100.9	85.9	97.4
0920	Plaster & Gypsum Board	90.8	71.3	77.6	93.7	71.3	78.6	100.5	71.6	81.0	95.6	50.7	65.3	100.9	94.1	96.3	105.3	81.2	89.0
0950, 0980	Ceilings & Acoustic Treatment	92.0	71.3	78.4	88.0	71.3	77.0	93.0	71.6	78.9	104.9	50.7	69.3	94.0	94.1	94.1	94.4	81.2	85.7
0960	Flooring	117.4	100.4	112.5	98.5	100.4	99.1	113.4	81.9	104.4	127.4	79.8	113.8	101.9	107.9	103.6	105.4	51.1	89.9
0970, 0990	Wall Finishes & Painting/Coating	151.7	63.7	98.6	93.8	63.7	75.6	126.2	67.2	90.6	167.2	66.1	106.1	101.2	80.8	88.9	102.6	95.9	98.5
09	**FINISHES**	108.8	75.2	90.3	95.9	75.2	84.5	106.0	73.3	88.0	118.2	59.7	86.0	97.9	94.3	95.9	97.6	77.2	86.3
COVERS	**DIVS. 10 - 14, 25, 28, 41, 43, 44, 46**	100.0	85.5	97.1	100.0	66.8	93.3	100.0	87.6	97.5	100.0	62.5	92.4	100.0	89.8	97.9	100.0	62.8	92.5
21, 22, 23	**FIRE SUPPRESSION, PLUMBING & HVAC**	95.0	75.8	87.2	99.9	76.0	90.2	99.9	76.8	90.6	94.8	75.7	87.1	100.0	80.1	92.0	97.7	79.6	90.4
26, 27, 3370	**ELECTRICAL, COMMUNICATIONS & UTIL.**	92.4	82.3	87.1	92.6	77.1	84.4	99.1	82.3	90.2	89.7	86.6	88.0	103.2	96.7	99.7	100.7	96.6	98.6
MF2014	**WEIGHTED AVERAGE**	100.0	76.7	89.8	99.0	78.3	90.0	100.2	78.6	90.8	101.2	72.5	88.7	99.7	93.5	97.0	98.3	84.9	92.4

City Cost Indexes

DIVISION		NEVADA									NEW HAMPSHIRE								
		ELY			LAS VEGAS			RENO			CHARLESTON			CLAREMONT			CONCORD		
		893			889 - 891			894 - 895			036			037			032 - 033		
		MAT.	INST.	TOTAL	MAT.	INST.	TOTAL	MAT.	INST.	TOTAL	MAT.	INST.	TOTAL	MAT.	INST.	TOTAL	MAT.	INST.	TOTAL
015433	CONTRACTOR EQUIPMENT		98.2	98.2		98.2	98.2		98.2	98.2		100.9	100.9		100.9	100.9		100.9	100.9
0241, 31 - 34	SITE & INFRASTRUCTURE, DEMOLITION	72.2	99.1	91.3	75.4	101.0	93.6	72.2	99.0	91.3	86.9	99.6	96.0	80.9	99.6	94.2	93.7	103.9	100.9
0310	Concrete Forming & Accessories	102.3	105.7	105.2	103.5	109.6	108.8	98.7	92.7	93.5	86.8	81.1	81.9	92.6	81.1	82.7	95.5	93.6	93.8
0320	Concrete Reinforcing	101.5	111.6	106.7	93.7	119.9	107.1	96.0	119.9	108.1	84.1	93.5	88.9	84.1	93.5	88.9	95.3	94.1	94.7
0330	Cast-in-Place Concrete	102.1	105.4	103.4	98.9	107.8	102.5	108.0	85.7	98.8	97.1	70.5	86.2	89.2	70.5	81.6	112.2	90.5	103.3
03	CONCRETE	103.4	106.4	104.9	98.7	110.6	104.6	103.1	95.4	99.3	99.1	80.0	89.7	91.1	80.0	85.6	106.6	92.5	99.6
04	MASONRY	119.7	102.5	109.0	108.3	104.0	105.6	114.3	94.6	102.0	97.5	84.7	89.5	96.9	84.7	89.3	110.9	104.2	106.7
05	METALS	96.9	103.3	98.9	104.5	108.6	105.8	98.4	105.1	100.5	92.8	90.1	92.0	92.8	90.1	92.0	98.8	93.8	97.3
06	WOOD, PLASTICS & COMPOSITES	91.3	106.8	100.0	89.8	107.9	99.9	85.9	94.4	90.6	87.4	90.2	89.0	93.6	90.2	91.7	95.4	92.4	93.7
07	THERMAL & MOISTURE PROTECTION	100.3	95.7	98.4	113.8	102.6	109.2	99.9	92.3	96.8	98.8	72.3	88.0	98.7	72.3	87.9	101.9	91.5	97.6
08	OPENINGS	100.8	98.8	100.3	99.9	115.6	103.5	98.7	102.3	99.6	102.0	76.4	96.0	103.3	76.4	97.0	104.9	87.2	100.8
0920	Plaster & Gypsum Board	101.0	107.0	105.1	97.5	108.1	104.7	92.0	94.1	93.4	100.0	89.3	92.8	101.0	89.3	93.1	103.9	91.6	95.6
0950, 0980	Ceilings & Acoustic Treatment	94.4	107.0	102.7	102.7	108.1	106.3	99.4	94.1	95.9	94.2	89.3	91.0	94.2	89.3	91.0	99.9	91.6	94.4
0960	Flooring	103.0	57.2	89.9	94.6	107.9	98.4	99.9	107.9	102.2	98.4	32.1	79.4	100.6	32.1	81.0	102.7	116.9	106.8
0970, 0990	Wall Finishes & Painting/Coating	102.6	121.0	113.7	105.4	121.0	114.8	102.6	80.8	89.4	96.4	45.7	65.8	96.4	46.0	65.9	95.8	95.4	95.6
09	FINISHES	96.8	98.9	98.0	95.8	110.4	103.9	95.6	94.3	94.9	95.3	69.7	81.2	95.6	69.8	81.4	97.4	97.9	97.7
COVERS	DIVS. 10 - 14, 25, 28, 41, 43, 44, 46	100.0	63.6	92.6	100.0	104.4	100.9	100.0	89.8	97.9	100.0	92.2	98.4	100.0	92.2	98.4	100.0	104.3	100.9
21, 22, 23	FIRE SUPPRESSION, PLUMBING & HVAC	97.7	104.5	100.5	100.1	104.4	101.8	100.0	80.1	92.0	95.3	39.6	72.8	95.3	39.7	72.9	99.9	84.4	93.7
26, 27, 3370	ELECTRICAL, COMMUNICATIONS & UTIL.	101.0	109.1	105.3	105.4	119.9	113.1	101.4	96.7	98.9	98.5	52.9	74.4	98.5	52.9	74.4	99.5	83.6	91.1
MF2014	WEIGHTED AVERAGE	99.4	102.2	100.6	100.9	108.7	104.3	99.6	93.3	96.8	96.7	69.5	84.8	95.9	69.5	84.4	101.1	92.5	97.4

DIVISION		NEW HAMPSHIRE															NEW JERSEY		
		KEENE			LITTLETON			MANCHESTER			NASHUA			PORTSMOUTH			ATLANTIC CITY		
		034			035			031			030			038			082,084		
		MAT.	INST.	TOTAL	MAT.	INST.	TOTAL	MAT.	INST.	TOTAL	MAT.	INST.	TOTAL	MAT.	INST.	TOTAL	MAT.	INST.	TOTAL
015433	CONTRACTOR EQUIPMENT		100.9	100.9		100.9	100.9		100.9	100.9		100.9	100.9		100.9	100.9		99.1	99.1
0241, 31 - 34	SITE & INFRASTRUCTURE, DEMOLITION	94.0	99.9	98.2	81.0	99.9	94.4	93.6	103.9	100.9	95.7	103.9	101.5	89.2	104.0	99.7	96.4	105.2	102.7
0310	Concrete Forming & Accessories	91.3	83.0	84.2	102.9	83.0	85.8	96.9	94.0	94.4	99.3	94.0	94.8	88.2	94.3	93.4	109.0	127.5	125.0
0320	Concrete Reinforcing	84.1	93.6	88.9	84.8	93.6	89.3	103.7	94.1	98.8	105.1	94.1	99.5	84.1	94.2	89.2	76.3	118.1	97.6
0330	Cast-in-Place Concrete	97.6	72.8	87.4	87.6	72.7	81.5	110.6	113.1	111.7	92.4	113.1	100.9	87.6	113.2	98.1	86.1	133.0	105.3
03	CONCRETE	98.7	81.6	90.3	90.5	81.6	86.1	107.2	100.4	103.9	98.9	100.4	99.7	90.6	100.6	95.5	93.1	126.5	109.5
04	MASONRY	101.3	88.3	93.2	107.9	88.3	95.7	102.9	104.2	103.7	101.4	104.2	103.1	97.2	104.2	101.5	111.8	132.0	124.4
05	METALS	93.5	90.7	92.6	93.5	90.7	92.6	101.3	94.3	99.2	98.8	94.3	97.5	95.0	94.9	94.9	95.6	106.1	98.8
06	WOOD, PLASTICS & COMPOSITES	92.0	90.2	91.0	103.5	90.2	96.0	95.3	92.4	93.7	101.9	92.4	96.6	88.7	92.4	90.8	109.6	127.2	119.4
07	THERMAL & MOISTURE PROTECTION	99.2	75.3	89.4	98.8	73.8	88.5	101.5	95.0	98.8	99.6	95.0	97.7	99.3	115.0	105.7	105.9	124.2	113.4
08	OPENINGS	100.4	81.2	96.0	104.4	77.1	98.0	105.2	87.2	101.0	105.4	87.2	101.2	106.2	84.0	101.0	101.7	124.4	107.0
0920	Plaster & Gypsum Board	100.3	89.3	92.9	114.6	89.3	97.5	104.7	91.6	95.8	110.0	91.6	97.5	100.0	91.6	94.3	105.4	127.5	120.3
0950, 0980	Ceilings & Acoustic Treatment	94.2	89.3	91.0	94.2	89.3	91.0	101.7	91.6	95.0	105.1	91.6	96.2	95.1	91.6	92.8	81.3	127.5	111.6
0960	Flooring	100.2	52.6	86.6	109.9	32.1	87.6	98.6	116.9	103.8	103.7	116.9	107.4	98.5	116.9	103.8	105.4	159.9	121.0
0970, 0990	Wall Finishes & Painting/Coating	96.4	45.7	65.8	96.4	59.9	74.4	100.7	95.4	97.5	96.4	95.4	95.8	96.4	95.4	95.8	93.9	130.2	115.8
09	FINISHES	97.0	74.5	84.6	100.2	72.2	84.8	98.6	97.9	98.2	101.7	97.9	99.6	96.2	97.9	97.1	96.0	134.5	117.2
COVERS	DIVS. 10 - 14, 25, 28, 41, 43, 44, 46	100.0	93.4	98.7	100.0	98.6	99.7	100.0	104.3	100.9	100.0	104.3	100.9	100.0	104.3	100.9	100.0	111.5	102.3
21, 22, 23	FIRE SUPPRESSION, PLUMBING & HVAC	95.3	43.3	74.3	95.3	63.8	82.6	99.9	84.4	93.7	100.1	84.4	93.8	100.1	84.4	93.8	99.7	123.6	109.3
26, 27, 3370	ELECTRICAL, COMMUNICATIONS & UTIL.	98.5	63.2	79.9	99.8	55.9	76.6	99.8	83.6	91.2	101.2	83.6	91.9	99.1	83.6	90.9	91.1	138.1	116.0
MF2014	WEIGHTED AVERAGE	97.2	73.3	86.7	97.1	76.3	88.0	101.4	93.8	98.1	100.5	93.8	97.6	97.9	94.3	96.4	98.2	124.8	109.8

DIVISION		NEW JERSEY																	
		CAMDEN			DOVER			ELIZABETH			HACKENSACK			JERSEY CITY			LONG BRANCH		
		081			078			072			076			073			077		
		MAT.	INST.	TOTAL	MAT.	INST.	TOTAL	MAT.	INST.	TOTAL	MAT.	INST.	TOTAL	MAT.	INST.	TOTAL	MAT.	INST.	TOTAL
015433	CONTRACTOR EQUIPMENT		99.1	99.1		100.9	100.9		100.9	100.9		100.9	100.9		99.1	99.1		98.7	98.7
0241, 31 - 34	SITE & INFRASTRUCTURE, DEMOLITION	97.8	105.5	103.3	102.0	106.3	105.1	106.1	106.3	106.3	103.0	106.3	105.3	93.4	106.3	102.5	97.5	106.0	103.6
0310	Concrete Forming & Accessories	100.2	127.5	123.8	97.8	128.3	124.1	110.2	128.4	125.9	97.8	128.3	124.1	101.8	128.3	124.7	102.4	128.0	124.5
0320	Concrete Reinforcing	100.1	118.3	109.4	77.2	137.2	107.7	77.2	137.2	107.7	77.2	137.2	107.7	100.1	137.2	119.0	77.2	137.1	107.7
0330	Cast-in-Place Concrete	83.5	132.9	103.8	101.5	127.7	112.2	87.1	131.4	105.3	99.2	131.4	112.4	79.2	127.7	99.1	88.0	132.9	106.4
03	CONCRETE	94.0	126.5	110.0	98.8	128.7	113.5	94.5	130.1	112.0	97.0	130.0	113.2	92.1	128.6	110.0	96.2	130.2	112.9
04	MASONRY	101.3	132.0	120.4	93.0	132.5	117.6	108.7	132.5	123.5	96.9	132.5	119.1	87.0	132.5	115.4	101.3	132.0	120.5
05	METALS	101.1	106.1	102.6	93.3	116.2	100.4	94.8	116.2	101.4	93.4	116.1	100.4	98.9	113.9	103.5	93.4	113.7	99.7
06	WOOD, PLASTICS & COMPOSITES	98.4	127.2	114.5	96.2	127.2	113.5	111.7	127.2	120.4	96.2	127.2	113.5	97.1	127.2	114.0	98.2	127.1	114.4
07	THERMAL & MOISTURE PROTECTION	105.8	123.3	113.0	100.6	132.8	113.8	100.8	133.3	114.1	100.4	125.5	110.7	100.2	132.8	113.5	100.3	124.4	110.2
08	OPENINGS	104.1	124.5	108.8	105.5	127.7	110.6	103.6	127.7	109.2	102.9	127.7	108.7	101.6	127.7	107.7	97.8	127.6	104.7
0920	Plaster & Gypsum Board	101.5	127.5	119.0	100.9	127.5	118.9	107.7	127.5	121.0	100.9	127.5	118.9	104.3	127.5	120.0	102.7	127.5	119.4
0950, 0980	Ceilings & Acoustic Treatment	91.5	127.5	115.1	83.8	127.5	112.5	85.7	127.5	113.1	83.8	127.5	112.5	94.9	127.5	116.3	83.8	127.5	112.5
0960	Flooring	101.5	159.9	118.2	93.6	178.1	117.8	98.5	178.1	121.3	93.6	178.1	117.8	94.6	178.1	118.5	94.8	178.1	118.6
0970, 0990	Wall Finishes & Painting/Coating	93.9	130.2	115.8	98.4	132.5	119.0	98.4	132.5	119.0	98.4	132.5	119.0	98.5	132.5	119.0	98.5	130.2	117.6
09	FINISHES	96.4	134.5	117.4	93.7	136.9	117.5	96.9	136.9	118.9	93.5	136.9	117.4	96.6	136.9	118.8	94.5	137.5	118.2
COVERS	DIVS. 10 - 14, 25, 28, 41, 43, 44, 46	100.0	111.5	102.3	100.0	118.1	103.7	100.0	118.1	103.7	100.0	118.1	103.7	100.0	118.1	103.7	100.0	111.3	102.3
21, 22, 23	FIRE SUPPRESSION, PLUMBING & HVAC	100.0	123.6	109.5	99.7	127.4	110.9	100.1	125.5	110.3	99.7	127.4	110.9	100.1	127.4	111.1	99.7	127.1	110.8
26, 27, 3370	ELECTRICAL, COMMUNICATIONS & UTIL.	96.0	138.1	118.2	92.8	137.7	116.5	93.4	137.7	116.8	92.8	140.1	117.8	97.6	140.1	120.1	92.5	130.8	112.7
MF2014	WEIGHTED AVERAGE	99.5	124.8	110.5	97.9	127.8	110.9	98.8	127.6	111.3	97.6	128.1	110.9	97.9	127.9	111.0	97.1	126.3	109.8

DIVISION		NEW JERSEY																	
		NEW BRUNSWICK			NEWARK			PATERSON			POINT PLEASANT			SUMMIT			TRENTON		
		088 - 089			070 - 071			074 - 075			087			079			085 - 086		
		MAT.	INST.	TOTAL	MAT.	INST.	TOTAL	MAT.	INST.	TOTAL	MAT.	INST.	TOTAL	MAT.	INST.	TOTAL	MAT.	INST.	TOTAL
015433	CONTRACTOR EQUIPMENT		98.7	98.7		100.9	100.9		100.9	100.9		98.7	98.7		100.9	100.9		98.7	98.7
0241, 31 - 34	SITE & INFRASTRUCTURE, DEMOLITION	109.2	106.0	107.0	107.8	106.3	106.8	105.0	106.3	105.9	110.9	106.0	107.4	103.7	106.3	105.6	95.8	106.0	103.0
0310	Concrete Forming & Accessories	103.3	128.3	124.8	97.6	128.4	124.2	99.9	128.2	124.3	97.7	127.9	123.7	100.6	128.4	124.5	98.6	127.7	123.7
0320	Concrete Reinforcing	77.2	137.2	107.7	99.8	137.2	118.8	100.1	137.2	119.0	77.2	137.1	107.7	77.2	137.2	107.7	99.8	112.4	106.2
0330	Cast-in-Place Concrete	106.3	133.2	117.4	108.8	131.5	118.1	100.8	131.4	113.4	106.3	132.8	117.2	84.3	131.4	103.6	101.8	132.8	114.5
03	CONCRETE	110.4	130.5	120.3	105.8	130.1	117.7	102.2	130.0	115.9	110.1	130.1	119.9	91.5	130.1	110.5	102.5	125.5	113.8
04	MASONRY	109.3	132.5	123.7	99.6	132.5	120.1	93.6	132.5	117.9	97.0	132.0	118.8	95.5	132.5	118.6	103.9	132.0	121.4
05	METALS	95.6	113.8	101.2	100.8	116.3	105.5	94.2	116.1	100.9	95.6	113.5	101.1	93.3	116.2	100.4	100.8	105.2	102.1
06	WOOD, PLASTICS & COMPOSITES	103.0	127.1	116.5	93.9	127.2	112.5	98.8	127.2	114.7	95.9	127.1	113.4	100.1	127.2	115.3	95.6	127.1	113.3
07	THERMAL & MOISTURE PROTECTION	106.1	132.0	116.7	101.1	133.3	114.3	100.7	125.5	110.9	106.2	124.4	113.7	101.0	133.3	114.2	105.5	124.3	113.2
08	OPENINGS	96.1	127.6	103.5	104.4	127.7	109.8	108.3	127.7	112.8	98.2	129.2	105.4	110.2	127.7	114.2	106.1	122.2	109.8
0920	Plaster & Gypsum Board	102.6	127.5	119.4	101.1	127.5	118.9	104.3	127.5	120.0	98.3	127.5	118.0	102.7	127.5	119.4	98.7	127.5	118.1
0950, 0980	Ceilings & Acoustic Treatment	81.3	127.5	111.6	97.6	127.5	117.2	94.9	127.5	116.3	81.3	127.5	111.6	83.8	127.5	112.5	93.1	127.5	115.7
0960	Flooring	103.0	178.1	124.5	95.7	178.1	119.2	94.6	178.1	118.5	100.5	159.9	117.5	94.9	178.1	118.7	102.1	172.2	122.1
0970, 0990	Wall Finishes & Painting/Coating	93.9	132.5	117.2	99.8	132.5	119.5	98.4	132.5	119.0	93.9	130.2	115.8	98.4	132.5	119.0	99.1	130.2	117.8
09	FINISHES	96.0	136.8	118.5	95.6	136.9	118.3	96.7	136.9	118.9	94.7	134.5	116.6	94.6	136.9	117.9	96.6	136.5	118.6
COVERS	DIVS. 10 - 14, 25, 28, 41, 43, 44, 46	100.0	118.0	103.6	100.0	118.1	103.7	100.0	118.1	103.7	100.0	108.8	101.8	100.0	118.1	103.7	100.0	111.3	102.3
21, 22, 23	FIRE SUPPRESSION, PLUMBING & HVAC	99.7	127.4	110.8	100.1	127.4	111.1	100.1	127.4	111.1	99.7	127.0	110.7	99.7	125.5	110.1	100.1	126.9	110.9
26, 27, 3370	ELECTRICAL, COMMUNICATIONS & UTIL.	91.8	136.8	115.6	101.2	140.1	121.8	97.6	137.7	118.8	91.1	130.8	112.1	93.4	137.7	116.8	99.3	135.4	118.4
MF2014	WEIGHTED AVERAGE	99.7	127.6	111.9	101.2	128.4	113.0	99.6	127.8	111.9	99.2	125.9	110.8	97.9	127.6	110.8	100.9	125.1	111.5

DIVISION		NEW JERSEY			NEW MEXICO														
		VINELAND			ALBUQUERQUE			CARRIZOZO			CLOVIS			FARMINGTON			GALLUP		
		080,083			870 - 872			883			881			874			873		
		MAT.	INST.	TOTAL	MAT.	INST.	TOTAL	MAT.	INST.	TOTAL	MAT.	INST.	TOTAL	MAT.	INST.	TOTAL	MAT.	INST.	TOTAL
015433	CONTRACTOR EQUIPMENT		99.1	99.1		110.1	110.1		110.1	110.1		110.1	110.1		110.1	110.1		110.1	110.1
0241, 31 - 34	SITE & INFRASTRUCTURE, DEMOLITION	100.7	105.5	104.1	87.6	103.3	98.7	105.8	103.3	104.0	94.1	103.3	100.6	94.1	103.3	100.6	102.1	103.3	102.9
0310	Concrete Forming & Accessories	95.2	127.6	123.1	101.7	64.6	69.7	99.2	64.6	69.3	99.2	64.5	69.2	101.7	64.6	69.7	101.7	64.6	69.7
0320	Concrete Reinforcing	76.3	116.1	96.6	98.4	68.9	83.4	107.0	68.9	87.6	108.2	68.9	88.2	107.6	68.9	87.9	103.0	68.9	85.6
0330	Cast-in-Place Concrete	92.8	133.0	109.3	98.9	70.4	87.2	95.8	70.4	85.4	95.7	70.3	85.3	99.8	70.4	87.7	93.9	70.4	84.3
03	CONCRETE	97.9	126.2	111.8	100.2	68.5	84.6	117.0	68.5	93.2	105.8	68.4	87.4	103.9	68.5	86.5	110.0	68.5	89.6
04	MASONRY	99.4	132.0	119.7	105.0	58.7	76.1	103.4	58.7	75.5	103.4	58.7	75.5	113.7	58.7	79.4	99.4	58.7	74.0
05	METALS	95.5	105.5	98.6	105.2	86.8	99.5	99.5	86.8	95.6	99.2	86.7	95.3	102.9	86.8	97.9	102.0	86.8	97.3
06	WOOD, PLASTICS & COMPOSITES	92.8	127.2	112.1	97.5	65.5	79.6	93.1	65.5	77.6	93.1	65.5	77.6	97.7	65.5	79.7	97.7	65.5	79.7
07	THERMAL & MOISTURE PROTECTION	105.7	124.2	113.3	94.8	73.5	86.0	100.3	73.5	89.3	99.2	73.5	88.7	94.9	73.5	86.2	95.9	73.5	86.7
08	OPENINGS	97.7	124.1	103.8	101.2	67.7	93.4	98.6	67.7	91.4	98.7	67.7	91.5	103.6	67.7	95.3	103.6	67.7	95.3
0920	Plaster & Gypsum Board	96.9	127.5	117.5	95.8	64.2	74.4	78.6	64.2	68.8	78.6	64.2	68.8	89.4	64.2	72.3	89.4	64.2	72.3
0950, 0980	Ceilings & Acoustic Treatment	81.3	127.5	111.6	95.5	64.2	74.9	96.3	64.2	75.2	96.3	64.2	75.2	92.6	64.2	73.9	92.6	64.2	73.9
0960	Flooring	99.6	159.9	116.8	101.9	66.0	91.6	101.2	66.0	91.2	101.2	66.0	91.2	103.7	66.0	92.9	103.7	66.0	92.9
0970, 0990	Wall Finishes & Painting/Coating	93.9	130.2	115.8	113.2	66.6	85.1	103.3	66.6	81.1	103.3	66.6	81.1	107.1	66.6	82.7	107.1	66.6	82.7
09	FINISHES	93.5	134.5	116.1	95.9	64.8	78.8	96.6	64.8	79.1	95.4	64.8	78.5	94.8	64.8	78.3	96.0	64.8	78.8
COVERS	DIVS. 10 - 14, 25, 28, 41, 43, 44, 46	100.0	111.5	102.3	100.0	82.9	96.5	100.0	82.9	96.5	100.0	82.9	96.5	100.0	82.9	96.5	100.0	82.9	96.5
21, 22, 23	FIRE SUPPRESSION, PLUMBING & HVAC	99.7	123.6	109.3	100.1	69.0	87.6	97.2	69.0	85.8	97.2	68.7	85.7	100.0	69.0	87.5	97.1	69.0	85.8
26, 27, 3370	ELECTRICAL, COMMUNICATIONS & UTIL.	91.1	138.1	116.0	91.2	71.2	80.6	92.8	71.2	81.4	90.2	71.2	80.2	88.8	71.2	79.5	88.0	71.2	79.1
MF2014	WEIGHTED AVERAGE	97.5	124.7	109.4	99.6	72.6	87.8	100.3	72.6	88.2	98.3	72.5	87.1	100.1	72.6	88.1	99.5	72.6	87.8

DIVISION		NEW MEXICO																	
		LAS CRUCES			LAS VEGAS			ROSWELL			SANTA FE			SOCORRO			TRUTH/CONSEQUENCES		
		880			877			882			875			878			879		
		MAT.	INST.	TOTAL	MAT.	INST.	TOTAL	MAT.	INST.	TOTAL	MAT.	INST.	TOTAL	MAT.	INST.	TOTAL	MAT.	INST.	TOTAL
015433	CONTRACTOR EQUIPMENT		86.0	86.0		110.1	110.1		110.1	110.1		110.1	110.1		110.1	110.1		86.0	86.0
0241, 31 - 34	SITE & INFRASTRUCTURE, DEMOLITION	94.0	82.9	86.1	93.6	103.3	100.5	96.3	103.3	101.3	98.8	103.3	102.0	89.8	103.3	99.4	108.3	82.9	90.3
0310	Concrete Forming & Accessories	95.7	63.4	67.8	101.7	64.6	69.7	99.2	64.6	69.3	100.4	64.6	69.5	101.7	64.6	69.7	99.3	63.4	68.3
0320	Concrete Reinforcing	104.6	68.7	86.4	104.7	68.9	86.5	108.2	68.9	88.2	103.8	68.9	86.1	106.8	68.9	87.5	100.8	68.7	84.5
0330	Cast-in-Place Concrete	90.4	62.7	79.0	97.1	70.4	86.1	95.8	70.4	85.3	105.3	70.4	91.0	95.1	70.4	85.0	104.2	62.7	87.1
03	CONCRETE	84.9	65.0	75.1	101.2	68.5	85.1	106.5	68.5	87.8	104.0	68.5	86.5	100.2	68.5	84.6	94.1	65.0	79.8
04	MASONRY	99.2	58.3	73.7	99.7	58.7	74.1	114.3	58.7	79.6	104.1	58.7	75.8	99.6	58.7	74.1	97.2	58.3	73.0
05	METALS	98.1	80.3	92.6	101.7	86.8	97.1	100.4	86.8	96.2	99.0	86.8	95.2	102.0	86.8	97.3	101.6	80.3	95.0
06	WOOD, PLASTICS & COMPOSITES	82.5	64.4	72.4	97.7	65.5	79.7	93.1	65.5	77.6	99.3	65.5	80.4	97.7	65.5	79.7	89.1	64.4	75.3
07	THERMAL & MOISTURE PROTECTION	86.0	68.6	78.8	94.5	73.5	85.9	99.3	73.5	88.7	96.9	73.5	87.3	94.5	73.5	85.9	83.3	68.6	77.2
08	OPENINGS	91.5	67.1	85.8	99.9	67.7	92.4	98.5	67.7	91.4	102.0	67.7	94.0	99.8	67.7	92.3	93.1	67.1	87.0
0920	Plaster & Gypsum Board	77.5	64.2	68.5	89.4	64.2	72.3	78.6	64.2	68.8	98.9	64.2	75.4	89.4	64.2	72.3	91.1	64.2	72.9
0950, 0980	Ceilings & Acoustic Treatment	84.2	64.2	71.0	92.6	64.2	73.9	96.3	64.2	75.2	93.0	64.2	74.0	92.6	64.2	73.9	82.4	64.2	70.4
0960	Flooring	131.2	66.0	112.5	103.7	66.0	92.9	101.2	66.0	91.2	110.8	66.0	98.0	103.7	66.0	92.9	135.3	66.0	115.4
0970, 0990	Wall Finishes & Painting/Coating	90.7	66.6	76.2	107.1	66.6	82.7	103.3	66.6	81.1	111.8	66.6	84.5	107.1	66.6	82.7	97.7	66.6	79.0
09	FINISHES	105.2	63.9	82.5	94.7	64.8	78.2	95.5	64.8	78.6	99.7	64.8	80.5	94.6	64.8	78.2	107.5	63.9	83.5
COVERS	DIVS. 10 - 14, 25, 28, 41, 43, 44, 46	100.0	80.2	96.0	100.0	82.9	96.5	100.0	82.9	96.5	100.0	82.9	96.5	100.0	82.9	96.5	100.0	80.2	96.0
21, 22, 23	FIRE SUPPRESSION, PLUMBING & HVAC	100.3	68.7	87.6	97.1	69.0	85.8	99.9	69.0	87.5	100.0	69.0	87.5	97.1	69.0	85.8	97.1	68.7	85.6
26, 27, 3370	ELECTRICAL, COMMUNICATIONS & UTIL.	92.1	71.2	81.0	90.6	71.2	80.4	91.7	71.2	80.9	102.9	71.2	86.2	88.5	71.2	79.4	92.3	71.2	81.2
MF2014	WEIGHTED AVERAGE	96.1	69.4	84.4	98.0	72.6	86.9	100.0	72.6	88.0	100.8	72.6	88.5	97.6	72.6	86.7	97.5	69.4	85.2

	DIVISION	NEW MEXICO			NEW YORK														
		TUCUMCARI			ALBANY			BINGHAMTON			BRONX			BROOKLYN			BUFFALO		
		884			120 - 122			137 - 139			104			112			140 - 142		
		MAT.	INST.	TOTAL	MAT.	INST.	TOTAL	MAT.	INST.	TOTAL	MAT.	INST.	TOTAL	MAT.	INST.	TOTAL	MAT.	INST.	TOTAL
015433	CONTRACTOR EQUIPMENT		110.1	110.1		112.9	112.9		114.1	114.1		110.6	110.6		113.3	113.3		97.1	97.1
0241, 31 - 34	SITE & INFRASTRUCTURE, DEMOLITION	93.8	103.3	100.5	83.8	106.3	99.8	95.9	94.1	94.6	108.8	120.8	117.3	120.3	127.6	125.5	98.3	98.2	98.2
0310	Concrete Forming & Accessories	99.2	64.5	69.2	100.1	100.6	100.5	100.8	88.5	90.2	98.4	175.3	164.8	107.2	182.9	172.5	97.3	116.5	113.9
0320	Concrete Reinforcing	106.0	68.9	87.1	104.0	103.8	103.9	93.7	97.8	95.8	103.9	184.4	144.9	95.1	205.9	151.5	97.8	102.1	100.0
0330	Cast-in-Place Concrete	95.7	70.3	85.3	91.9	111.3	99.9	102.7	127.9	113.1	95.9	173.4	127.7	104.8	172.1	132.4	106.7	119.7	112.0
03	CONCRETE	105.0	68.4	87.0	99.0	105.4	102.1	95.8	105.2	100.4	96.1	174.8	134.8	107.5	181.3	143.7	103.0	114.1	108.4
04	MASONRY	114.6	58.7	79.8	101.0	112.6	108.2	109.7	127.8	121.0	92.7	177.5	145.5	120.2	177.4	155.9	105.1	121.0	115.0
05	METALS	99.2	86.7	95.3	104.1	110.3	106.0	96.3	118.9	103.3	99.9	151.6	115.8	104.4	150.1	118.4	99.6	95.4	98.3
06	WOOD, PLASTICS & COMPOSITES	93.1	65.5	77.6	98.7	97.9	98.3	104.9	84.7	93.6	94.4	174.9	139.5	106.9	185.1	150.6	99.4	116.4	108.9
07	THERMAL & MOISTURE PROTECTION	99.1	73.5	88.6	105.9	105.8	105.9	107.3	103.6	105.8	108.5	163.2	131.0	108.4	164.1	131.2	101.8	110.6	105.4
08	OPENINGS	98.5	67.7	91.3	102.6	94.0	100.6	93.0	85.0	91.1	87.9	175.5	108.3	90.4	180.0	111.2	97.7	103.9	99.1
0920	Plaster & Gypsum Board	78.6	64.2	68.8	97.2	97.6	97.5	107.4	83.8	91.4	99.2	176.9	151.7	102.8	187.6	160.2	98.3	116.7	110.7
0950, 0980	Ceilings & Acoustic Treatment	96.3	64.2	75.2	92.4	97.6	95.8	91.1	83.8	86.3	83.8	176.9	145.0	87.1	187.6	153.2	102.0	116.7	111.6
0960	Flooring	101.2	66.0	91.2	97.1	113.7	101.9	106.8	103.6	105.9	98.3	186.8	123.6	113.9	186.8	134.8	101.3	121.2	107.0
0970, 0990	Wall Finishes & Painting/Coating	103.3	66.6	81.1	103.8	94.3	98.0	93.3	98.8	96.6	102.9	157.4	135.8	123.4	157.4	143.9	98.8	112.4	107.0
09	FINISHES	95.4	64.8	78.5	94.7	101.9	98.7	94.9	91.1	92.8	94.0	175.9	139.1	108.4	181.9	148.9	100.6	117.9	110.2
COVERS	DIVS. 10 - 14, 25, 28, 41, 43, 44, 46	100.0	82.9	96.5	100.0	99.0	99.8	100.0	96.2	99.2	100.0	135.0	107.1	100.0	135.4	107.1	100.0	105.6	101.1
21, 22, 23	FIRE SUPPRESSION, PLUMBING & HVAC	97.2	68.7	85.7	100.0	102.8	101.1	100.5	90.2	96.3	100.2	165.5	126.5	99.7	165.4	126.2	100.0	96.7	98.7
26, 27, 3370	ELECTRICAL, COMMUNICATIONS & UTIL.	92.8	71.2	81.4	98.7	104.1	101.6	99.9	105.4	102.8	97.0	181.9	141.9	99.7	181.9	143.1	100.2	102.7	101.5
MF2014	WEIGHTED AVERAGE	99.0	72.5	87.4	100.1	104.7	102.1	98.5	101.4	99.8	97.7	166.1	127.5	102.8	168.5	131.4	100.3	106.2	102.9

	DIVISION	NEW YORK																	
		ELMIRA			FAR ROCKAWAY			FLUSHING			GLENS FALLS			HICKSVILLE			JAMAICA		
		148 - 149			116			113			128			115,117,118			114		
		MAT.	INST.	TOTAL	MAT.	INST.	TOTAL	MAT.	INST.	TOTAL	MAT.	INST.	TOTAL	MAT.	INST.	TOTAL	MAT.	INST.	TOTAL
015433	CONTRACTOR EQUIPMENT		116.0	116.0		113.3	113.3		113.3	113.3		112.9	112.9		113.3	113.3		113.3	113.3
0241, 31 - 34	SITE & INFRASTRUCTURE, DEMOLITION	97.1	94.1	95.0	123.4	127.6	126.4	123.4	127.6	126.4	73.6	105.8	96.5	113.3	126.3	122.6	117.7	127.6	124.7
0310	Concrete Forming & Accessories	81.3	93.2	91.6	93.5	175.1	163.9	97.3	175.1	164.4	85.7	90.3	89.7	90.0	154.5	145.6	97.3	175.1	164.4
0320	Concrete Reinforcing	97.4	95.7	96.5	95.1	205.9	151.5	96.7	205.9	152.3	95.4	94.4	94.9	95.1	210.2	153.7	95.1	205.9	151.5
0330	Cast-in-Place Concrete	93.7	103.2	97.6	113.3	172.1	137.5	113.3	172.1	137.5	85.3	106.7	94.1	96.3	165.5	124.7	104.8	172.1	132.4
03	CONCRETE	90.8	98.6	94.6	113.6	177.8	145.1	114.1	177.8	145.4	88.3	97.5	92.9	99.4	167.1	132.6	106.8	177.8	141.7
04	MASONRY	103.5	102.1	102.6	124.1	177.4	157.4	118.1	177.4	155.1	100.7	105.6	103.7	114.6	166.8	147.1	122.2	177.4	156.6
05	METALS	96.7	117.3	103.0	104.4	150.1	118.4	104.4	150.1	118.4	97.7	106.1	100.3	105.8	149.1	119.2	104.4	150.1	118.4
06	WOOD, PLASTICS & COMPOSITES	85.0	92.2	89.0	90.1	174.6	137.4	94.8	174.6	139.5	87.1	87.6	87.3	86.6	152.6	123.6	94.8	174.6	139.5
07	THERMAL & MOISTURE PROTECTION	103.9	94.1	99.9	108.3	163.0	130.7	108.3	163.0	130.7	98.8	97.2	98.2	107.9	156.3	127.8	108.1	163.0	130.6
08	OPENINGS	99.4	88.9	97.0	89.0	174.4	108.8	89.0	174.4	108.8	93.3	86.2	91.7	89.0	163.4	106.3	89.0	174.4	108.8
0920	Plaster & Gypsum Board	97.8	91.7	93.7	91.6	176.9	149.2	94.1	176.9	150.0	89.1	87.0	87.7	91.2	154.2	133.8	94.1	176.9	150.0
0950, 0980	Ceilings & Acoustic Treatment	95.4	91.7	93.0	76.2	176.9	142.4	76.2	176.9	142.4	82.1	87.0	85.3	75.2	154.2	127.2	76.2	176.9	142.4
0960	Flooring	94.6	103.6	97.1	109.2	186.8	131.4	110.6	186.8	132.4	86.2	111.3	93.3	108.3	185.1	130.2	110.6	186.8	132.4
0970, 0990	Wall Finishes & Painting/Coating	98.9	90.0	93.5	123.4	157.4	143.9	123.4	157.4	143.9	101.1	87.4	92.8	123.4	157.4	143.9	123.4	157.4	143.9
09	FINISHES	95.5	94.7	95.0	103.5	175.7	143.3	104.3	175.7	143.6	86.1	93.1	90.0	102.1	159.7	133.8	103.8	175.7	143.4
COVERS	DIVS. 10 - 14, 25, 28, 41, 43, 44, 46	100.0	99.4	99.9	100.0	134.2	106.9	100.0	134.2	106.9	100.0	96.2	99.2	100.0	128.2	105.7	100.0	134.2	106.9
21, 22, 23	FIRE SUPPRESSION, PLUMBING & HVAC	95.5	91.5	93.9	95.0	165.3	123.4	95.0	165.3	123.4	95.5	95.7	95.6	99.7	154.4	121.8	95.0	165.3	123.4
26, 27, 3370	ELECTRICAL, COMMUNICATIONS & UTIL.	96.5	96.2	96.3	107.0	181.9	146.6	107.0	181.9	146.6	93.6	100.5	97.3	99.0	143.4	122.5	97.9	181.9	142.3
MF2014	WEIGHTED AVERAGE	96.7	97.4	97.0	102.6	166.8	130.6	102.5	166.8	130.5	94.0	98.6	96.0	100.9	153.3	123.7	100.8	166.8	129.6

	DIVISION	NEW YORK																	
		JAMESTOWN			KINGSTON			LONG ISLAND CITY			MONTICELLO			MOUNT VERNON			NEW ROCHELLE		
		147			124			111			127			105			108		
		MAT.	INST.	TOTAL	MAT.	INST.	TOTAL	MAT.	INST.	TOTAL	MAT.	INST.	TOTAL	MAT.	INST.	TOTAL	MAT.	INST.	TOTAL
015433	CONTRACTOR EQUIPMENT		93.9	93.9		113.3	113.3		113.3	113.3		113.3	113.3		110.6	110.6		110.6	110.6
0241, 31 - 34	SITE & INFRASTRUCTURE, DEMOLITION	98.4	94.2	95.4	140.6	122.9	128.0	121.3	127.6	125.8	135.5	122.7	126.4	115.2	118.5	117.6	114.6	118.5	117.4
0310	Concrete Forming & Accessories	81.3	86.9	86.1	87.1	104.4	102.0	101.6	175.1	165.0	94.7	102.9	101.8	88.8	139.1	132.1	103.6	139.0	134.2
0320	Concrete Reinforcing	97.6	98.9	98.2	95.8	140.6	118.6	95.1	205.9	151.5	95.1	140.1	118.0	102.9	183.3	143.8	103.0	183.3	143.9
0330	Cast-in-Place Concrete	97.2	102.1	99.2	115.2	135.1	123.4	108.1	172.1	134.4	107.8	124.8	114.8	107.1	141.3	121.1	107.0	141.3	121.1
03	CONCRETE	93.8	94.3	94.0	110.8	121.5	116.0	109.9	177.8	143.3	105.5	117.2	111.2	105.4	146.9	125.8	105.0	146.9	125.6
04	MASONRY	111.8	99.9	104.4	116.0	141.3	131.8	116.9	177.4	154.6	108.6	128.7	121.1	98.1	147.3	128.8	98.1	147.3	128.8
05	METALS	94.0	92.1	93.5	106.1	117.2	109.5	104.4	150.1	118.4	106.0	116.2	109.2	99.6	135.6	110.7	99.9	135.6	110.9
06	WOOD, PLASTICS & COMPOSITES	83.5	84.0	83.8	88.4	96.4	92.9	100.9	174.6	142.2	96.1	96.4	96.3	84.6	136.2	113.5	101.4	136.2	120.9
07	THERMAL & MOISTURE PROTECTION	103.4	93.5	99.3	122.3	137.2	128.4	108.3	163.0	130.7	122.0	132.1	126.1	109.4	144.1	123.6	109.5	144.1	123.7
08	OPENINGS	99.2	85.6	96.1	95.9	117.3	100.9	89.0	174.4	108.8	91.1	117.3	97.2	87.9	148.6	102.0	88.0	148.6	102.1
0920	Plaster & Gypsum Board	88.5	83.2	84.9	91.4	96.3	94.7	98.7	176.9	151.5	92.1	96.3	94.9	94.9	137.0	123.3	106.3	137.0	127.0
0950, 0980	Ceilings & Acoustic Treatment	92.0	83.2	86.2	72.9	96.3	88.3	76.2	176.9	142.4	72.9	96.3	88.3	82.1	137.0	118.2	82.1	137.0	118.2
0960	Flooring	97.7	103.6	99.4	103.8	72.5	94.9	112.1	186.8	133.5	106.1	72.5	96.5	90.2	186.8	117.8	97.1	186.8	122.8
0970, 0990	Wall Finishes & Painting/Coating	100.2	94.2	96.6	134.4	116.8	123.8	123.4	157.4	143.9	134.4	116.8	123.8	101.1	157.4	135.1	101.1	157.4	135.1
09	FINISHES	94.6	89.9	92.0	101.0	97.1	98.8	105.1	175.7	144.0	101.4	96.3	98.6	91.2	149.2	123.2	94.6	149.2	124.7
COVERS	DIVS. 10 - 14, 25, 28, 41, 43, 44, 46	100.0	98.6	99.7	100.0	111.9	102.4	100.0	134.2	106.9	100.0	110.8	102.2	100.0	125.2	105.1	100.0	121.9	104.4
21, 22, 23	FIRE SUPPRESSION, PLUMBING & HVAC	95.4	87.2	92.1	95.4	122.3	106.3	99.7	165.3	126.2	95.4	116.7	104.0	95.5	134.1	111.1	95.5	134.1	111.1
26, 27, 3370	ELECTRICAL, COMMUNICATIONS & UTIL.	95.4	96.3	95.9	95.1	111.0	103.5	98.5	181.9	142.6	95.1	111.0	103.5	95.2	155.2	126.9	95.2	155.2	126.9
MF2014	WEIGHTED AVERAGE	96.8	92.5	94.9	102.6	118.5	109.6	102.4	166.8	130.5	101.1	115.1	107.2	97.5	141.4	116.6	97.9	141.3	116.8

DIVISION		NEW YORK																	
		NEW YORK			NIAGARA FALLS			PLATTSBURGH			POUGHKEEPSIE			QUEENS			RIVERHEAD		
		100 - 102			143			129			125 - 126			110			119		
		MAT.	INST.	TOTAL	MAT.	INST.	TOTAL	MAT.	INST.	TOTAL	MAT.	INST.	TOTAL	MAT.	INST.	TOTAL	MAT.	INST.	TOTAL
015433	CONTRACTOR EQUIPMENT		111.1	111.1		93.9	93.9		98.9	98.9		113.3	113.3		113.3	113.3		113.3	113.3
0241, 31 - 34	SITE & INFRASTRUCTURE, DEMOLITION	117.3	121.5	120.2	100.5	95.5	97.0	106.1	103.0	103.9	136.5	123.5	127.2	116.6	127.6	124.4	114.5	125.9	122.6
0310	Concrete Forming & Accessories	102.6	183.3	172.2	81.3	114.4	109.8	91.1	95.4	94.9	87.1	165.0	154.3	90.1	175.1	163.4	94.6	153.4	145.3
0320	Concrete Reinforcing	109.9	210.4	161.0	96.3	102.5	99.5	99.9	102.7	101.3	95.8	141.0	118.8	96.7	205.9	152.3	96.9	184.2	141.3
0330	Cast-in-Place Concrete	107.8	177.7	136.5	100.6	120.3	108.7	104.4	104.4	104.4	111.5	137.9	122.4	99.6	172.1	129.4	97.9	164.5	125.3
03	CONCRETE	106.2	184.2	144.5	95.9	113.3	104.4	102.5	99.3	101.0	107.9	149.5	128.3	102.6	177.8	139.6	100.1	161.4	130.2
04	MASONRY	102.5	177.5	149.2	119.9	127.5	124.6	95.0	101.6	99.1	108.6	144.5	131.0	111.1	177.4	152.4	119.9	166.3	148.8
05	METALS	113.1	151.9	125.0	96.7	93.8	95.8	102.0	91.2	98.7	106.1	119.8	110.3	104.4	150.1	118.4	106.3	138.4	116.2
06	WOOD, PLASTICS & COMPOSITES	98.5	185.4	147.2	83.4	110.3	98.5	93.8	93.5	93.6	88.4	174.6	136.7	86.7	174.6	135.9	91.8	152.6	125.9
07	THERMAL & MOISTURE PROTECTION	108.6	164.5	131.5	103.5	112.9	107.4	116.3	100.5	109.8	122.3	146.8	132.3	107.9	163.0	130.5	108.8	156.1	128.2
08	OPENINGS	93.7	180.9	114.0	99.2	100.7	99.6	101.6	90.8	99.1	95.9	159.5	110.7	89.0	174.4	108.8	89.0	157.5	104.9
0920	Plaster & Gypsum Board	106.0	187.6	161.2	88.5	110.3	103.3	109.4	92.6	98.0	91.4	176.9	149.2	91.2	176.9	149.1	92.5	154.2	134.2
0950, 0980	Ceilings & Acoustic Treatment	102.2	187.6	158.4	92.0	110.3	104.1	98.1	92.6	94.5	72.9	176.9	141.2	76.2	176.9	142.4	76.1	154.2	127.5
0960	Flooring	99.5	186.8	124.5	97.7	121.2	104.4	108.9	113.7	110.3	103.8	167.2	122.0	108.3	186.8	130.7	109.2	142.7	118.8
0970, 0990	Wall Finishes & Painting/Coating	102.9	157.4	135.8	100.2	112.4	107.6	130.0	91.4	106.7	134.4	117.2	124.0	123.4	157.4	143.9	123.4	157.4	143.9
09	FINISHES	99.9	182.1	145.2	94.7	115.9	106.4	97.9	98.0	98.0	100.8	163.0	135.1	102.5	175.7	142.8	102.7	151.3	129.5
COVERS	DIVS. 10 - 14, 25, 28, 41, 43, 44, 46	100.0	136.1	107.3	100.0	107.1	101.4	100.0	98.0	99.6	100.0	121.6	104.4	100.0	134.2	106.9	100.0	127.6	105.6
21, 22, 23	FIRE SUPPRESSION, PLUMBING & HVAC	100.1	165.5	126.5	95.4	100.0	97.2	95.4	96.6	95.9	95.4	127.0	108.1	99.7	165.3	126.2	99.9	151.4	120.7
26, 27, 3370	ELECTRICAL, COMMUNICATIONS & UTIL.	104.7	181.9	145.5	94.0	100.3	97.3	91.6	90.3	90.9	95.1	119.9	108.2	99.0	181.9	142.8	100.6	133.6	118.1
MF2014	WEIGHTED AVERAGE	103.4	168.7	131.8	97.8	106.4	101.5	99.1	96.7	98.0	101.8	136.7	117.0	100.9	166.8	129.6	101.6	148.2	121.9

DIVISION		NEW YORK																	
		ROCHESTER			SCHENECTADY			STATEN ISLAND			SUFFERN			SYRACUSE			UTICA		
		144 - 146			123			103			109			130 - 132			133 - 135		
		MAT.	INST.	TOTAL	MAT.	INST.	TOTAL	MAT.	INST.	TOTAL	MAT.	INST.	TOTAL	MAT.	INST.	TOTAL	MAT.	INST.	TOTAL
015433	CONTRACTOR EQUIPMENT		116.7	116.7		112.9	112.9		110.6	110.6		110.6	110.6		112.9	112.9		112.9	112.9
0241, 31 - 34	SITE & INFRASTRUCTURE, DEMOLITION	85.9	109.9	102.9	84.2	106.3	99.9	119.9	120.8	120.6	111.4	116.3	114.9	94.8	105.6	102.4	73.5	104.3	95.4
0310	Concrete Forming & Accessories	98.7	98.2	98.3	103.0	100.6	100.9	88.3	183.3	170.2	96.9	135.2	129.9	99.8	91.0	92.2	101.0	87.9	89.7
0320	Concrete Reinforcing	100.1	95.7	97.9	94.4	103.8	99.2	103.9	210.4	158.1	103.0	140.9	122.3	94.7	96.4	95.6	94.7	95.7	95.2
0330	Cast-in-Place Concrete	94.2	104.4	98.4	102.9	111.3	106.4	107.1	173.5	134.4	103.8	138.2	117.9	95.4	105.9	99.7	87.3	104.7	94.4
03	CONCRETE	99.3	100.6	99.9	102.9	105.4	104.1	107.1	182.7	144.2	102.0	136.3	118.9	98.9	97.8	98.4	96.9	95.9	96.4
04	MASONRY	107.0	104.9	105.7	97.7	112.6	107.0	104.7	177.5	150.1	97.6	142.6	125.7	101.5	105.2	103.8	93.1	103.9	99.8
05	METALS	104.1	107.3	105.1	101.9	110.3	104.5	97.8	151.8	114.4	97.8	119.4	104.5	99.8	105.6	101.6	97.7	105.1	100.0
06	WOOD, PLASTICS & COMPOSITES	97.6	97.6	97.6	107.4	97.9	102.1	83.3	185.4	140.4	94.0	136.2	117.6	101.3	88.5	94.2	101.3	84.7	92.0
07	THERMAL & MOISTURE PROTECTION	103.5	101.7	102.7	100.3	105.8	102.6	108.9	164.3	131.6	109.3	142.1	122.8	102.3	97.6	100.4	90.6	97.6	93.5
08	OPENINGS	105.3	92.2	102.2	99.4	94.0	98.1	87.9	181.2	109.6	88.0	139.0	99.8	95.0	85.6	92.8	98.0	83.4	94.6
0920	Plaster & Gypsum Board	107.2	97.4	100.6	98.8	97.6	98.0	95.0	187.6	157.6	98.5	137.0	124.5	98.0	88.0	91.2	98.0	84.0	88.5
0950, 0980	Ceilings & Acoustic Treatment	100.4	97.4	98.5	88.7	97.6	94.6	83.8	187.6	152.0	82.1	137.0	118.2	91.1	88.0	89.0	91.1	84.0	86.4
0960	Flooring	93.8	113.3	99.4	92.9	113.7	98.9	94.2	186.8	120.7	93.5	185.1	119.7	94.6	102.1	96.8	92.1	102.2	95.0
0970, 0990	Wall Finishes & Painting/Coating	100.3	99.1	99.6	101.1	94.3	97.0	102.9	157.4	135.8	101.1	124.9	115.5	98.5	99.8	99.3	91.3	99.8	96.4
09	FINISHES	100.4	101.3	100.9	91.6	101.9	97.3	93.2	182.1	142.2	92.3	140.0	118.6	94.0	93.5	93.7	92.1	91.1	91.6
COVERS	DIVS. 10 - 14, 25, 28, 41, 43, 44, 46	100.0	99.8	100.0	100.0	99.0	99.8	100.0	136.1	107.3	100.0	123.6	104.8	100.0	96.5	99.3	100.0	96.2	99.2
21, 22, 23	FIRE SUPPRESSION, PLUMBING & HVAC	100.1	90.3	96.1	100.2	102.8	101.2	100.2	165.5	126.5	95.5	123.3	106.7	100.2	91.9	96.9	100.2	92.2	97.0
26, 27, 3370	ELECTRICAL, COMMUNICATIONS & UTIL.	98.7	94.6	96.6	98.1	104.1	101.3	97.0	181.9	141.9	102.8	119.9	111.8	100.0	102.2	101.1	98.1	102.2	100.3
MF2014	WEIGHTED AVERAGE	101.1	99.1	100.2	99.3	104.7	101.7	99.2	168.4	129.4	97.6	129.0	111.3	98.9	98.0	98.5	97.1	97.1	97.1

DIVISION		NEW YORK									NORTH CAROLINA								
		WATERTOWN			WHITE PLAINS			YONKERS			ASHEVILLE			CHARLOTTE			DURHAM		
		136			106			107			287 - 288			281 - 282			277		
		MAT.	INST.	TOTAL	MAT.	INST.	TOTAL	MAT.	INST.	TOTAL	MAT.	INST.	TOTAL	MAT.	INST.	TOTAL	MAT.	INST.	TOTAL
015433	CONTRACTOR EQUIPMENT		112.9	112.9		110.6	110.6		110.6	110.6		96.3	96.3		96.3	96.3		101.7	101.7
0241, 31 - 34	SITE & INFRASTRUCTURE, DEMOLITION	80.9	105.7	98.5	108.5	118.5	115.6	116.7	118.5	118.0	102.4	76.9	84.3	105.6	76.9	85.2	100.4	85.7	89.9
0310	Concrete Forming & Accessories	85.9	94.3	93.1	101.9	139.1	134.0	102.2	143.9	138.2	96.0	41.3	48.8	99.7	42.8	50.6	100.2	44.3	52.0
0320	Concrete Reinforcing	95.3	96.5	95.9	103.0	183.3	143.9	107.1	183.3	145.9	93.0	63.0	77.7	98.8	58.5	78.3	91.2	57.9	74.2
0330	Cast-in-Place Concrete	101.5	107.6	104.1	95.1	141.3	114.1	106.3	141.4	120.7	115.4	51.3	89.1	120.2	49.8	91.3	101.6	47.1	79.2
03	CONCRETE	109.3	99.9	104.7	95.7	146.9	120.9	105.2	149.1	126.7	106.0	50.6	78.8	108.6	49.9	79.8	98.7	49.5	74.5
04	MASONRY	94.2	107.9	102.8	97.1	147.3	128.4	102.1	147.3	130.3	93.6	43.7	62.5	101.9	51.2	70.3	86.4	37.9	56.2
05	METALS	97.8	105.6	100.2	99.3	135.6	110.5	108.9	135.7	117.1	103.1	82.3	96.7	104.0	80.9	96.9	121.4	79.9	108.6
06	WOOD, PLASTICS & COMPOSITES	83.0	92.2	88.1	99.5	136.2	120.0	99.3	142.7	123.6	97.9	40.3	65.6	103.0	41.9	68.8	96.7	44.7	67.6
07	THERMAL & MOISTURE PROTECTION	90.9	100.1	94.7	109.2	144.1	123.5	109.5	144.8	124.0	106.7	43.7	80.8	100.8	46.6	78.6	106.9	45.7	81.8
08	OPENINGS	98.0	89.4	96.0	88.0	148.6	102.1	91.1	151.7	105.2	97.1	44.1	84.8	102.5	45.0	89.1	105.8	47.1	92.2
0920	Plaster & Gypsum Board	88.7	91.7	90.8	101.3	137.0	125.4	105.6	143.6	131.3	100.1	38.3	58.3	100.0	39.9	59.4	105.9	42.8	63.3
0950, 0980	Ceilings & Acoustic Treatment	91.1	91.7	91.5	82.1	137.0	118.2	100.5	143.6	128.9	85.7	38.3	54.5	88.8	39.9	56.7	88.1	42.8	58.3
0960	Flooring	86.0	102.2	90.6	95.6	186.8	121.7	95.2	186.8	121.4	102.2	42.7	85.2	101.6	43.4	84.9	103.9	42.7	86.4
0970, 0990	Wall Finishes & Painting/Coating	91.3	96.2	94.3	101.1	157.4	135.1	101.1	157.4	135.1	113.6	40.3	69.3	113.8	49.4	74.9	105.4	37.4	64.3
09	FINISHES	89.7	95.7	93.0	92.9	149.2	123.9	98.0	153.0	128.3	95.6	40.7	65.3	95.5	42.9	66.5	96.0	42.9	66.7
COVERS	DIVS. 10 - 14, 25, 28, 41, 43, 44, 46	100.0	97.5	99.5	100.0	125.2	105.1	100.0	126.3	105.3	100.0	77.7	95.5	100.0	78.1	95.6	100.0	72.1	94.4
21, 22, 23	FIRE SUPPRESSION, PLUMBING & HVAC	100.2	85.9	94.4	100.3	134.1	114.0	100.3	134.1	114.0	100.4	53.4	81.4	100.0	54.1	81.5	100.5	52.9	81.3
26, 27, 3370	ELECTRICAL, COMMUNICATIONS & UTIL.	100.0	90.4	94.9	95.2	155.2	126.9	102.9	163.8	135.1	102.2	55.7	77.6	101.2	58.5	78.6	96.1	55.6	74.7
MF2014	WEIGHTED AVERAGE	98.5	96.3	97.5	97.6	141.4	116.7	102.1	143.7	120.2	100.8	55.3	80.9	101.9	56.8	82.2	102.9	55.3	82.1

City Cost Indexes

	DIVISION	NORTH CAROLINA																	
		ELIZABETH CITY 279			FAYETTEVILLE 283			GASTONIA 280			GREENSBORO 270,272 - 274			HICKORY 286			KINSTON 285		
		MAT.	INST.	TOTAL	MAT.	INST.	TOTAL	MAT.	INST.	TOTAL	MAT.	INST.	TOTAL	MAT.	INST.	TOTAL	MAT.	INST.	TOTAL
015433	CONTRACTOR EQUIPMENT		106.0	106.0		101.7	101.7		96.3	96.3		101.7	101.7		101.7	101.7		101.7	101.7
0241, 31 - 34	SITE & INFRASTRUCTURE, DEMOLITION	104.6	87.4	92.4	101.7	85.6	90.3	102.3	76.9	84.2	100.3	85.7	89.9	101.2	85.5	90.1	100.3	85.6	89.8
0310	Concrete Forming & Accessories	85.3	42.5	48.4	95.6	60.3	65.2	103.1	38.7	47.5	99.9	44.4	52.1	92.0	37.1	44.6	88.2	42.1	48.4
0320	Concrete Reinforcing	89.2	45.9	67.2	96.8	58.0	77.0	93.5	56.5	74.7	90.1	58.0	73.8	93.0	56.1	74.2	92.5	45.8	68.7
0330	Cast-in-Place Concrete	101.8	47.0	79.3	121.0	48.3	91.2	112.8	52.7	88.1	100.8	47.6	79.0	115.4	48.1	87.7	111.4	44.8	84.0
03	CONCRETE	98.7	46.4	73.0	108.1	57.1	83.1	104.5	48.7	77.1	98.1	49.8	74.4	105.7	46.3	76.5	102.4	45.5	74.4
04	MASONRY	98.6	48.0	67.1	97.3	38.9	60.9	98.3	50.9	68.7	82.5	41.4	56.9	82.3	43.7	58.2	89.1	48.1	63.6
05	METALS	107.1	75.8	97.5	124.2	80.0	110.6	103.9	80.0	96.5	113.7	80.0	103.4	103.2	78.7	95.7	102.0	74.8	93.6
06	WOOD, PLASTICS & COMPOSITES	80.1	43.6	59.7	97.1	66.8	80.1	107.0	36.6	67.6	96.4	44.9	67.6	92.1	35.2	60.2	88.6	42.9	63.0
07	THERMAL & MOISTURE PROTECTION	106.2	43.3	80.4	106.3	45.9	81.5	106.9	45.2	81.6	106.7	42.8	80.5	107.0	41.7	80.2	106.8	42.8	80.6
08	OPENINGS	102.6	38.4	87.7	97.2	59.1	88.4	100.9	41.0	87.0	105.8	47.2	92.2	97.2	36.8	83.1	97.3	43.1	84.7
0920	Plaster & Gypsum Board	98.6	41.0	59.7	104.7	65.6	78.3	107.0	34.5	58.0	107.5	43.0	63.9	100.1	33.1	54.8	99.8	41.0	60.1
0950, 0980	Ceilings & Acoustic Treatment	88.1	41.0	57.1	86.5	65.6	72.7	89.0	34.5	53.2	88.1	43.0	58.4	85.7	33.1	51.1	89.0	41.0	57.4
0960	Flooring	95.7	23.2	75.0	102.4	42.7	85.3	105.4	42.7	87.4	103.9	39.7	85.6	102.1	32.8	82.3	99.6	22.6	77.6
0970, 0990	Wall Finishes & Painting/Coating	105.4	42.0	67.1	113.6	33.2	65.0	113.6	40.3	69.3	105.4	31.6	60.9	113.6	40.3	69.3	113.6	39.2	68.7
09	FINISHES	93.0	38.7	63.0	96.5	55.3	73.8	98.1	38.6	65.3	96.3	41.9	66.3	95.8	35.4	62.5	95.5	38.0	63.8
COVERS	DIVS. 10 - 14, 25, 28, 41, 43, 44, 46	100.0	79.4	95.8	100.0	74.3	94.8	100.0	77.3	95.4	100.0	75.9	95.1	100.0	77.1	95.4	100.0	71.6	94.3
21, 22, 23	FIRE SUPPRESSION, PLUMBING & HVAC	95.6	51.1	77.7	100.2	52.7	81.0	100.4	52.4	81.0	100.4	53.1	81.3	95.6	52.2	78.1	95.6	51.2	77.7
26, 27, 3370	ELECTRICAL, COMMUNICATIONS & UTIL.	95.9	34.5	63.4	101.4	50.7	74.6	101.6	57.4	78.2	95.2	55.7	74.3	99.6	57.4	77.3	99.4	46.1	71.2
MF2014	WEIGHTED AVERAGE	99.3	51.6	78.5	104.5	58.1	84.2	101.6	55.1	81.3	101.3	55.7	81.4	98.8	53.9	79.2	98.4	52.7	78.5

	DIVISION	NORTH CAROLINA															NORTH DAKOTA		
		MURPHY 289			RALEIGH 275 - 276			ROCKY MOUNT 278			WILMINGTON 284			WINSTON-SALEM 271			BISMARCK 585		
		MAT.	INST.	TOTAL	MAT.	INST.	TOTAL	MAT.	INST.	TOTAL	MAT.	INST.	TOTAL	MAT.	INST.	TOTAL	MAT.	INST.	TOTAL
015433	CONTRACTOR EQUIPMENT		96.3	96.3		101.7	101.7		101.7	101.7		96.3	96.3		101.7	101.7		98.2	98.2
0241, 31 - 34	SITE & INFRASTRUCTURE, DEMOLITION	103.5	76.7	84.5	101.3	85.7	90.2	102.6	85.7	90.6	103.7	77.1	84.8	100.6	85.7	90.0	101.4	96.9	98.2
0310	Concrete Forming & Accessories	103.7	39.3	48.2	98.9	46.6	53.8	91.9	44.2	50.8	97.5	48.8	55.5	101.9	48.0	55.4	104.9	40.8	49.6
0320	Concrete Reinforcing	92.6	45.4	68.6	96.2	56.0	75.7	89.2	55.8	72.2	93.7	57.9	75.5	90.1	57.9	73.7	99.9	92.7	96.2
0330	Cast-in-Place Concrete	119.4	44.4	88.6	106.7	51.4	84.0	99.6	47.7	78.3	115.0	49.5	88.1	103.4	50.0	81.4	102.6	47.5	79.9
03	CONCRETE	109.2	44.1	77.2	101.7	51.7	77.1	99.5	49.3	74.9	105.9	52.4	79.6	99.4	52.2	76.2	101.5	54.2	78.3
04	MASONRY	85.3	41.2	57.8	86.6	42.2	58.9	76.7	39.5	53.5	82.8	41.7	57.2	82.7	38.6	55.2	114.6	56.4	78.3
05	METALS	100.9	74.2	92.7	105.7	79.2	97.5	106.3	78.4	97.7	102.7	79.9	95.7	110.8	79.9	101.3	103.4	88.4	98.8
06	WOOD, PLASTICS & COMPOSITES	107.8	39.7	69.7	95.1	47.5	68.5	87.4	44.7	63.5	100.1	49.0	71.5	96.4	49.4	70.1	95.3	34.7	61.4
07	THERMAL & MOISTURE PROTECTION	106.9	40.8	79.8	100.5	47.0	78.5	106.6	40.8	79.7	106.7	46.9	82.2	106.7	43.5	80.8	112.4	51.3	87.4
08	OPENINGS	97.1	38.9	83.5	104.8	47.4	91.4	101.8	41.9	87.9	97.3	51.0	86.5	105.8	49.7	92.8	108.0	50.1	94.5
0920	Plaster & Gypsum Board	106.1	37.6	59.8	99.8	45.7	63.3	100.5	42.8	61.5	102.4	47.2	65.1	107.5	47.6	67.0	98.8	33.0	54.3
0950, 0980	Ceilings & Acoustic Treatment	85.7	37.6	54.1	88.8	45.7	60.5	85.6	42.8	57.5	86.5	47.2	60.7	88.1	47.6	61.5	112.9	33.0	60.4
0960	Flooring	105.7	24.1	82.4	101.2	42.7	84.5	99.4	21.8	77.2	102.9	44.4	86.2	103.9	42.7	86.4	99.4	72.1	91.6
0970, 0990	Wall Finishes & Painting/Coating	113.6	39.4	68.8	104.2	36.9	63.5	105.4	38.5	65.0	113.6	38.7	68.4	105.4	36.7	64.0	101.9	29.4	58.1
09	FINISHES	97.7	36.1	63.7	95.5	44.7	67.5	93.9	39.3	63.8	96.3	46.8	69.0	96.3	45.7	68.4	102.7	42.8	69.7
COVERS	DIVS. 10 - 14, 25, 28, 41, 43, 44, 46	100.0	76.9	95.3	100.0	72.6	94.5	100.0	72.5	94.5	100.0	73.8	94.7	100.0	78.7	95.7	100.0	82.5	96.5
21, 22, 23	FIRE SUPPRESSION, PLUMBING & HVAC	95.6	50.9	77.6	100.0	52.5	80.8	95.6	52.7	78.3	100.4	54.9	82.0	100.4	53.3	81.4	100.1	71.9	88.7
26, 27, 3370	ELECTRICAL, COMMUNICATIONS & UTIL.	103.2	29.3	64.1	98.3	40.4	67.7	97.9	39.7	67.1	102.4	50.7	75.1	95.2	55.7	74.3	102.5	71.6	86.2
MF2014	WEIGHTED AVERAGE	99.6	48.3	77.2	100.5	54.1	80.3	98.5	52.3	78.3	100.3	55.9	81.0	101.0	56.5	81.6	103.1	66.2	87.0

	DIVISION	NORTH DAKOTA																	
		DEVILS LAKE 583			DICKINSON 586			FARGO 580 - 581			GRAND FORKS 582			JAMESTOWN 584			MINOT 587		
		MAT.	INST.	TOTAL	MAT.	INST.	TOTAL	MAT.	INST.	TOTAL	MAT.	INST.	TOTAL	MAT.	INST.	TOTAL	MAT.	INST.	TOTAL
015433	CONTRACTOR EQUIPMENT		98.2	98.2		98.2	98.2		98.2	98.2		98.2	98.2		98.2	98.2		98.2	98.2
0241, 31 - 34	SITE & INFRASTRUCTURE, DEMOLITION	105.7	94.4	97.7	113.6	92.9	98.9	102.6	96.9	98.5	109.6	92.9	97.7	104.7	92.9	96.3	107.1	96.9	99.8
0310	Concrete Forming & Accessories	101.2	35.4	44.4	90.8	34.9	42.5	99.3	41.5	49.4	94.5	34.2	42.5	92.3	34.1	42.1	90.5	67.0	70.2
0320	Concrete Reinforcing	100.2	93.1	96.6	101.1	92.8	96.9	96.9	92.8	94.8	98.7	92.9	95.7	100.8	91.3	95.9	102.1	93.2	97.6
0330	Cast-in-Place Concrete	126.3	46.0	93.3	114.3	44.4	85.6	113.6	49.2	87.2	114.3	44.1	85.5	124.7	44.1	91.6	114.3	46.5	86.4
03	CONCRETE	110.6	51.3	81.5	109.6	50.2	80.4	105.9	55.1	80.9	106.8	49.8	78.8	109.1	48.7	79.5	105.5	65.6	85.9
04	MASONRY	121.7	64.9	86.3	124.3	60.4	84.5	114.2	56.4	78.2	115.9	64.4	83.8	135.1	32.8	71.3	114.2	65.4	83.8
05	METALS	99.9	87.4	96.0	99.8	82.5	94.5	103.0	88.7	98.6	99.8	82.1	94.4	99.8	63.7	88.7	100.1	89.3	96.8
06	WOOD, PLASTICS & COMPOSITES	95.2	32.0	59.8	83.1	32.0	54.5	92.9	35.2	60.6	87.4	32.0	56.4	85.0	32.0	55.3	82.8	70.1	75.7
07	THERMAL & MOISTURE PROTECTION	107.6	49.8	83.9	108.1	48.5	83.7	107.4	51.6	84.5	107.8	49.8	84.0	107.4	41.5	80.4	107.5	57.4	87.0
08	OPENINGS	100.7	42.7	87.2	100.7	42.7	87.2	100.5	50.3	88.9	100.7	42.7	87.2	100.7	32.2	84.8	100.9	69.3	93.5
0920	Plaster & Gypsum Board	118.4	30.2	58.8	109.1	30.2	55.8	96.6	33.5	54.0	110.5	30.2	56.2	110.2	30.2	56.1	109.1	69.4	82.3
0950, 0980	Ceilings & Acoustic Treatment	114.3	30.2	59.0	114.3	30.2	59.0	112.1	33.5	60.5	114.3	30.2	59.0	114.3	30.2	59.0	114.3	69.4	84.8
0960	Flooring	107.9	35.9	87.3	101.4	35.9	82.7	102.8	72.1	94.0	103.3	35.9	84.0	102.1	35.9	83.2	101.1	89.3	97.8
0970, 0990	Wall Finishes & Painting/Coating	104.4	22.5	54.9	104.4	31.5	60.4	101.2	70.7	82.8	104.4	28.2	58.4	104.4	22.5	54.9	104.4	27.9	58.2
09	FINISHES	108.2	32.2	66.3	105.9	33.2	65.8	103.8	47.7	72.9	106.1	32.8	65.7	105.3	32.2	65.0	105.0	67.0	84.1
COVERS	DIVS. 10 - 14, 25, 28, 41, 43, 44, 46	100.0	32.4	86.3	100.0	32.6	86.4	100.0	82.6	96.5	100.0	32.5	86.4	100.0	80.6	96.1	100.0	86.4	97.3
21, 22, 23	FIRE SUPPRESSION, PLUMBING & HVAC	95.6	74.0	86.9	95.6	66.2	83.7	100.1	77.0	90.8	100.4	33.3	73.3	95.6	35.0	71.2	100.4	60.2	84.2
26, 27, 3370	ELECTRICAL, COMMUNICATIONS & UTIL.	98.3	35.8	65.3	107.7	72.5	89.1	102.4	68.4	84.4	102.1	53.5	76.4	98.3	35.7	65.2	105.5	75.7	89.8
MF2014	WEIGHTED AVERAGE	102.0	58.6	83.1	102.8	60.9	84.6	102.7	67.6	87.4	102.7	51.6	80.4	102.1	45.2	77.3	102.7	71.3	89.0

DIVISION		NORTH DAKOTA			OHIO														
		WILLISTON			AKRON			ATHENS			CANTON			CHILLICOTHE			CINCINNATI		
		588			442 - 443			457			446 - 447			456			451 - 452		
		MAT.	INST.	TOTAL	MAT.	INST.	TOTAL	MAT.	INST.	TOTAL	MAT.	INST.	TOTAL	MAT.	INST.	TOTAL	MAT.	INST.	TOTAL
015433	CONTRACTOR EQUIPMENT		98.2	98.2		94.4	94.4		90.4	90.4		94.4	94.4		99.7	99.7		99.5	99.5
0241, 31 - 34	SITE & INFRASTRUCTURE, DEMOLITION	107.5	92.9	97.1	98.1	101.6	100.6	111.0	91.6	97.2	98.2	101.0	100.2	97.0	102.6	101.0	94.1	102.4	100.0
0310	Concrete Forming & Accessories	96.3	34.9	43.3	99.0	94.3	95.0	95.0	84.3	85.8	99.0	83.7	85.8	97.4	92.3	93.0	99.3	79.2	82.0
0320	Concrete Reinforcing	103.1	92.8	97.9	99.5	92.1	95.7	93.1	87.0	90.0	99.5	75.8	87.5	90.1	80.4	85.1	95.5	79.2	87.2
0330	Cast-in-Place Concrete	114.3	44.4	85.6	94.0	98.0	95.7	111.5	96.5	105.4	94.9	95.0	95.0	101.2	99.3	100.4	93.2	92.4	92.9
03	CONCRETE	106.9	50.2	79.1	97.3	94.4	95.9	107.6	88.6	98.3	97.7	85.7	91.8	101.9	92.4	97.2	96.3	84.0	90.3
04	MASONRY	108.8	60.4	78.7	93.3	96.2	95.1	86.4	90.4	88.9	94.0	84.3	87.9	94.0	100.5	98.1	93.7	85.0	88.2
05	METALS	100.0	82.5	94.6	95.4	81.6	91.1	103.0	80.0	95.9	95.4	74.5	88.9	94.9	85.7	92.1	97.2	84.6	93.3
06	WOOD, PLASTICS & COMPOSITES	88.8	32.0	57.0	99.1	93.6	96.0	85.7	84.9	85.3	99.5	82.8	90.2	98.4	89.6	93.5	100.9	76.8	87.4
07	THERMAL & MOISTURE PROTECTION	107.7	48.5	83.5	110.8	96.5	104.9	99.4	94.8	97.5	112.0	91.4	103.5	101.2	97.4	99.7	99.2	88.7	94.9
08	OPENINGS	100.8	42.7	87.3	110.7	93.3	106.7	101.7	82.3	97.2	104.3	77.9	98.2	93.8	83.5	91.4	102.0	77.4	96.3
0920	Plaster & Gypsum Board	110.5	30.2	56.2	96.9	93.2	94.4	91.4	84.1	86.5	98.0	82.1	87.2	93.3	89.5	90.7	94.9	76.3	82.3
0950, 0980	Ceilings & Acoustic Treatment	114.3	30.2	59.0	93.0	93.2	93.1	103.3	84.1	90.7	93.0	82.1	85.8	97.9	89.5	92.4	98.7	76.3	84.0
0960	Flooring	104.1	35.9	84.6	96.9	93.4	95.9	121.3	99.0	114.9	97.1	81.7	92.7	98.3	97.7	98.1	99.3	90.7	96.8
0970, 0990	Wall Finishes & Painting/Coating	104.4	31.5	60.4	96.3	106.3	102.3	100.5	99.1	99.6	96.3	83.2	88.4	97.8	92.9	94.9	97.8	83.9	89.4
09	FINISHES	106.2	33.2	66.0	98.0	95.3	96.5	101.8	89.0	94.8	98.2	82.9	89.8	98.8	93.3	95.8	99.2	81.1	89.3
COVERS	DIVS. 10 - 14, 25, 28, 41, 43, 44, 46	100.0	32.6	86.4	100.0	98.2	99.6	100.0	52.0	90.3	100.0	95.4	99.1	100.0	92.6	98.5	100.0	89.3	97.8
21, 22, 23	FIRE SUPPRESSION, PLUMBING & HVAC	95.6	66.2	83.7	100.1	94.6	97.9	95.3	51.3	77.6	100.1	81.4	92.5	95.8	95.2	95.6	100.0	83.3	93.3
26, 27, 3370	ELECTRICAL, COMMUNICATIONS & UTIL.	102.5	72.5	86.6	99.3	93.9	96.5	95.5	98.1	96.9	98.5	92.7	95.5	95.0	85.6	90.0	93.9	79.4	86.3
MF2014	WEIGHTED AVERAGE	101.2	60.9	83.7	99.9	94.2	97.4	99.6	80.4	91.2	99.3	85.6	93.3	96.8	93.0	95.1	98.3	84.5	92.2

DIVISION		OHIO																	
		CLEVELAND			COLUMBUS			DAYTON			HAMILTON			LIMA			LORAIN		
		441			430 - 432			453 - 454			450			458			440		
		MAT.	INST.	TOTAL	MAT.	INST.	TOTAL	MAT.	INST.	TOTAL	MAT.	INST.	TOTAL	MAT.	INST.	TOTAL	MAT.	INST.	TOTAL
015433	CONTRACTOR EQUIPMENT		94.7	94.7		93.6	93.6		94.7	94.7		99.7	99.7		92.9	92.9		94.4	94.4
0241, 31 - 34	SITE & INFRASTRUCTURE, DEMOLITION	98.0	102.2	101.0	95.5	98.3	97.5	92.9	101.8	99.2	92.8	102.1	99.4	104.6	91.6	95.4	97.5	102.9	101.4
0310	Concrete Forming & Accessories	99.1	99.6	99.5	100.3	84.5	86.7	99.3	79.0	81.8	99.4	79.4	82.1	95.0	86.9	88.1	99.1	86.8	88.5
0320	Concrete Reinforcing	100.0	92.5	96.2	102.7	82.6	92.5	95.5	81.3	88.3	95.5	79.2	87.2	93.1	81.5	87.2	99.5	92.4	95.9
0330	Cast-in-Place Concrete	92.2	106.5	98.1	94.1	93.7	93.9	86.7	86.4	86.6	92.9	92.7	92.8	102.4	97.5	100.4	89.5	104.1	95.5
03	CONCRETE	96.5	99.8	98.2	97.9	87.1	92.6	93.3	81.8	87.6	96.2	84.2	90.3	100.4	89.3	94.9	95.2	93.2	94.2
04	MASONRY	97.8	105.6	102.7	96.6	94.0	95.0	93.2	83.5	87.1	93.5	85.5	88.5	117.7	86.4	98.2	89.9	104.6	99.1
05	METALS	96.9	84.4	93.1	96.9	80.7	91.9	96.4	78.6	91.0	96.5	84.5	92.8	103.0	81.5	96.4	96.0	82.8	91.9
06	WOOD, PLASTICS & COMPOSITES	98.2	97.1	97.6	97.1	82.3	88.8	102.2	77.1	88.1	100.9	76.8	87.4	85.6	86.3	86.0	99.1	80.9	88.9
07	THERMAL & MOISTURE PROTECTION	109.5	108.6	109.1	100.9	94.9	98.4	104.7	87.6	97.7	101.4	88.8	96.2	99.0	95.0	97.3	111.9	103.0	108.2
08	OPENINGS	100.6	95.2	99.3	102.4	80.0	97.2	101.3	78.1	95.9	98.9	77.4	93.9	101.7	79.8	96.6	104.3	86.4	100.2
0920	Plaster & Gypsum Board	96.2	96.8	96.6	93.5	81.8	85.6	94.9	76.6	82.6	94.9	76.3	82.3	91.4	85.5	87.4	96.9	80.1	85.5
0950, 0980	Ceilings & Acoustic Treatment	91.3	96.8	94.9	97.7	81.8	87.3	99.7	76.6	84.5	98.7	76.3	84.0	102.4	85.5	91.3	93.0	80.1	84.5
0960	Flooring	96.7	105.0	99.1	92.0	90.3	91.5	102.0	80.7	95.9	99.3	90.7	96.8	120.3	92.2	112.3	97.1	105.0	99.3
0970, 0990	Wall Finishes & Painting/Coating	96.3	105.1	101.6	95.5	92.9	94.0	97.8	85.0	90.1	97.8	84.7	89.9	100.5	82.7	89.8	96.3	105.1	101.6
09	FINISHES	97.5	100.9	99.4	94.1	85.9	89.6	100.2	79.3	88.7	99.1	81.4	89.3	100.8	87.2	93.3	98.0	91.1	94.2
COVERS	DIVS. 10 - 14, 25, 28, 41, 43, 44, 46	100.0	102.5	100.5	100.0	93.3	98.6	100.0	89.1	97.8	100.0	89.5	97.9	100.0	95.2	99.0	100.0	100.3	100.1
21, 22, 23	FIRE SUPPRESSION, PLUMBING & HVAC	100.0	100.5	100.2	100.1	92.0	96.8	100.9	86.6	95.2	100.6	83.6	93.7	95.3	89.0	92.8	100.1	92.1	96.9
26, 27, 3370	ELECTRICAL, COMMUNICATIONS & UTIL.	98.8	105.0	102.1	97.6	87.8	92.4	92.6	83.0	87.5	93.0	83.3	87.9	95.8	80.3	87.6	98.7	88.0	93.0
MF2014	WEIGHTED AVERAGE	99.0	100.3	99.6	98.6	89.2	94.5	98.0	84.4	92.1	97.9	85.1	92.3	100.0	86.8	94.3	98.9	93.0	96.3

DIVISION		OHIO																	
		MANSFIELD			MARION			SPRINGFIELD			STEUBENVILLE			TOLEDO			YOUNGSTOWN		
		448 - 449			433			455			439			434 - 436			444 - 445		
		MAT.	INST.	TOTAL	MAT.	INST.	TOTAL	MAT.	INST.	TOTAL	MAT.	INST.	TOTAL	MAT.	INST.	TOTAL	MAT.	INST.	TOTAL
015433	CONTRACTOR EQUIPMENT		94.4	94.4		93.2	93.2		94.7	94.7		98.1	98.1		96.0	96.0		94.4	94.4
0241, 31 - 34	SITE & INFRASTRUCTURE, DEMOLITION	93.9	101.5	99.3	91.8	97.1	95.6	93.2	100.5	98.4	133.1	106.9	114.5	94.8	99.0	97.8	98.0	102.0	100.8
0310	Concrete Forming & Accessories	89.1	83.7	84.5	96.7	81.2	83.3	99.3	83.5	85.7	98.3	89.4	90.7	100.3	97.1	97.6	99.0	87.5	89.1
0320	Concrete Reinforcing	90.6	76.4	83.4	94.7	82.5	88.5	95.5	81.3	88.3	92.3	85.5	88.8	102.7	85.4	93.9	99.5	85.6	92.4
0330	Cast-in-Place Concrete	87.1	94.7	90.2	85.9	92.7	88.7	89.1	86.3	87.9	93.3	94.1	93.6	94.1	100.0	96.5	93.1	96.9	94.7
03	CONCRETE	89.7	85.7	87.7	89.9	85.1	87.6	94.4	83.8	89.2	93.8	89.7	91.8	97.9	95.5	96.7	96.9	89.8	93.4
04	MASONRY	92.4	97.0	95.3	98.8	95.5	96.7	93.4	83.5	87.2	86.2	94.2	91.2	104.8	99.9	101.7	93.6	93.6	93.6
05	METALS	96.2	75.5	89.8	96.0	78.7	90.6	96.4	78.4	90.9	92.5	79.5	88.5	96.7	85.5	93.3	95.4	78.7	90.3
06	WOOD, PLASTICS & COMPOSITES	86.7	80.9	83.4	92.8	79.8	85.5	103.6	83.6	92.4	88.4	88.3	88.3	97.1	96.9	97.0	99.1	86.1	91.8
07	THERMAL & MOISTURE PROTECTION	111.3	94.4	104.4	100.5	85.2	94.2	104.6	88.2	97.9	112.8	96.4	106.1	102.5	102.7	102.6	112.1	94.3	104.8
08	OPENINGS	104.9	76.0	98.2	96.6	76.5	91.9	99.3	79.0	94.6	97.0	83.7	94.0	99.7	91.3	97.7	104.3	85.1	99.8
0920	Plaster & Gypsum Board	90.7	80.1	83.5	91.4	79.3	83.2	94.9	83.3	87.1	90.7	87.5	88.6	93.5	96.9	95.8	96.9	85.5	89.2
0950, 0980	Ceilings & Acoustic Treatment	93.9	80.1	84.8	97.7	79.3	85.6	99.7	83.3	88.9	94.8	87.5	90.0	97.7	96.9	97.2	93.0	85.5	88.1
0960	Flooring	92.5	107.5	96.8	90.9	107.5	95.6	102.0	80.7	95.9	118.6	100.4	113.4	91.2	99.7	93.6	97.1	93.9	96.2
0970, 0990	Wall Finishes & Painting/Coating	96.3	88.5	91.6	95.5	47.5	66.5	97.8	85.0	90.1	107.8	103.1	104.9	95.6	102.0	99.4	96.3	92.6	94.1
09	FINISHES	95.8	88.2	91.6	93.1	82.6	87.3	100.2	83.2	90.8	110.0	92.6	100.4	93.8	98.1	96.2	98.1	88.8	93.0
COVERS	DIVS. 10 - 14, 25, 28, 41, 43, 44, 46	100.0	96.2	99.2	100.0	54.4	90.8	100.0	89.8	97.9	100.0	95.0	99.0	100.0	98.7	99.7	100.0	96.4	99.3
21, 22, 23	FIRE SUPPRESSION, PLUMBING & HVAC	95.3	88.9	92.7	95.3	91.0	93.6	100.9	81.3	93.0	95.7	92.0	94.2	100.1	100.4	100.2	100.1	87.1	94.8
26, 27, 3370	ELECTRICAL, COMMUNICATIONS & UTIL.	96.2	79.0	87.1	91.9	79.0	85.1	92.6	87.8	90.1	86.9	114.1	101.3	97.6	104.7	101.4	98.7	86.0	92.0
MF2014	WEIGHTED AVERAGE	96.7	87.3	92.6	95.1	85.3	90.8	98.0	84.7	92.2	96.8	94.9	96.0	98.7	98.1	98.4	99.2	89.0	94.7

	DIVISION	OHIO			OKLAHOMA															
		ZANESVILLE			ARDMORE			CLINTON			DURANT			ENID			GUYMON			
		437 - 438			734			736			747			737			739			
		MAT.	INST.	TOTAL	MAT.	INST.	TOTAL	MAT.	INST.	TOTAL	MAT.	INST.	TOTAL	MAT.	INST.	TOTAL	MAT.	INST.	TOTAL	
015433	CONTRACTOR EQUIPMENT		93.2	93.2		82.7	82.7		81.8	81.8		81.8	81.8		81.8	81.8		81.8	81.8	
0241, 31 - 34	SITE & INFRASTRUCTURE, DEMOLITION	94.5	98.7	97.5	96.2	91.6	92.9	97.6	90.2	92.3	95.5	89.9	91.5	99.3	90.2	92.8	101.6	90.0	93.3	
0310	Concrete Forming & Accessories	93.7	81.3	83.0	94.0	40.4	47.8	92.5	46.6	53.0	85.5	44.1	49.8	96.2	34.1	42.7	99.8	44.9	52.4	
0320	Concrete Reinforcing	94.1	85.1	89.5	90.8	79.8	85.2	91.3	79.8	85.5	95.8	80.2	87.9	90.7	79.8	85.2	91.3	79.8	85.5	
0330	Cast-in-Place Concrete	90.5	91.7	91.0	97.5	43.4	75.3	94.3	45.1	74.1	91.7	42.8	71.6	94.3	46.4	74.6	94.3	43.1	73.3	
03	CONCRETE	93.5	85.3	89.5	95.2	49.3	72.7	94.7	52.7	74.1	92.3	50.8	71.9	95.3	47.6	71.8	98.2	51.2	75.1	
04	MASONRY	95.5	85.9	89.5	103.4	57.1	74.5	129.7	57.1	84.4	95.6	62.3	74.9	110.1	57.1	77.0	105.9	53.7	73.3	
05	METALS	97.4	80.4	92.1	102.5	70.1	92.5	102.6	70.1	92.6	93.5	70.7	86.5	104.0	70.0	93.6	103.2	69.8	92.9	
06	WOOD, PLASTICS & COMPOSITES	88.5	79.8	83.6	97.4	37.9	64.0	96.4	46.2	68.3	88.1	43.0	62.9	100.1	29.3	60.5	104.0	46.0	71.5	
07	THERMAL & MOISTURE PROTECTION	100.6	92.6	97.3	108.9	60.6	89.1	109.1	61.5	89.6	99.7	61.8	84.2	109.2	59.9	89.0	109.5	57.3	88.1	
08	OPENINGS	96.6	79.3	92.5	102.9	49.0	90.3	102.9	53.3	91.4	96.3	52.0	86.0	102.9	43.9	89.1	103.0	49.2	90.5	
0920	Plaster & Gypsum Board	88.2	79.3	82.2	89.8	36.6	53.8	89.4	45.2	59.6	79.4	41.9	54.1	90.5	27.8	48.1	90.7	44.9	59.8	
0950, 0980	Ceilings & Acoustic Treatment	97.7	79.3	85.6	85.7	36.6	53.4	85.7	45.2	59.1	82.4	41.9	55.8	85.7	27.8	47.6	86.5	44.9	59.2	
0960	Flooring	89.3	90.3	89.6	107.0	43.2	88.7	105.8	41.1	87.3	102.8	61.9	91.1	107.7	41.1	88.6	109.3	24.4	85.0	
0970, 0990	Wall Finishes & Painting/Coating	95.5	92.9	94.0	104.7	51.3	72.5	104.7	51.3	72.5	103.9	51.3	72.1	104.7	51.3	72.5	104.7	33.0	61.5	
09	FINISHES	92.4	83.6	87.5	92.9	40.1	63.8	92.7	44.7	66.3	90.2	46.6	66.2	93.5	34.6	61.1	94.5	38.9	63.9	
COVERS	DIVS. 10 - 14, 25, 28, 41, 43, 44, 46	100.0	88.5	97.7	100.0	77.6	95.5	100.0	78.6	95.7	100.0	77.9	95.5	100.0	76.7	95.3	100.0	77.4	95.4	
21, 22, 23	FIRE SUPPRESSION, PLUMBING & HVAC	95.3	89.9	93.1	95.6	65.8	83.6	95.6	65.9	83.6	95.5	65.4	83.4	100.3	65.8	86.4	95.6	63.8	82.7	
26, 27, 3370	ELECTRICAL, COMMUNICATIONS & UTIL.	92.3	83.4	87.5	93.6	70.9	81.6	94.6	70.9	82.1	96.5	70.9	83.0	94.6	70.9	82.1	96.2	61.5	77.9	
MF2014	WEIGHTED AVERAGE	95.6	86.5	91.6	98.3	61.9	82.4	99.6	63.1	83.7	95.1	63.5	81.3	100.2	60.5	82.9	99.4	59.7	82.1	

	DIVISION	OKLAHOMA																	
		LAWTON			MCALESTER			MIAMI			MUSKOGEE			OKLAHOMA CITY			PONCA CITY		
		735			745			743			744			730 - 731			746		
		MAT.	INST.	TOTAL	MAT.	INST.	TOTAL	MAT.	INST.	TOTAL	MAT.	INST.	TOTAL	MAT.	INST.	TOTAL	MAT.	INST.	TOTAL
015433	CONTRACTOR EQUIPMENT		82.7	82.7		81.8	81.8		89.9	89.9		89.9	89.9		83.0	83.0		81.8	81.8
0241, 31 - 34	SITE & INFRASTRUCTURE, DEMOLITION	95.7	91.6	92.8	89.1	90.1	89.9	90.5	87.3	88.2	90.8	87.1	88.2	95.1	92.0	92.9	95.9	90.1	91.8
0310	Concrete Forming & Accessories	99.8	44.3	52.0	83.5	43.6	49.1	96.9	66.2	70.4	101.4	33.4	42.7	97.7	57.9	63.4	92.2	46.3	52.6
0320	Concrete Reinforcing	91.0	79.8	85.3	95.5	79.8	87.5	94.0	79.9	86.8	94.9	79.3	87.0	96.2	79.9	87.9	94.9	79.9	87.2
0330	Cast-in-Place Concrete	91.2	46.4	72.8	80.4	45.2	65.9	84.3	47.5	69.2	85.3	45.8	69.1	92.9	46.9	74.0	94.2	44.7	73.8
03	CONCRETE	91.5	52.1	72.1	82.9	51.3	67.4	87.6	63.0	75.5	89.4	47.7	68.9	94.9	58.3	76.9	94.4	52.4	73.8
04	MASONRY	105.8	57.1	75.4	114.3	56.1	78.1	98.4	56.3	72.1	116.9	45.2	72.2	109.8	56.4	76.5	90.9	56.1	69.2
05	METALS	108.0	70.1	96.4	93.4	70.0	86.2	93.4	82.1	89.9	94.8	80.4	90.4	100.7	70.2	91.3	93.4	70.3	86.3
06	WOOD, PLASTICS & COMPOSITES	103.0	43.1	69.5	85.6	43.0	61.7	101.0	72.9	85.3	105.6	30.7	63.6	97.0	61.8	77.3	96.3	46.0	68.1
07	THERMAL & MOISTURE PROTECTION	108.9	61.3	89.4	99.4	60.6	83.5	99.8	64.3	85.2	99.9	48.6	78.8	101.1	62.9	85.4	99.9	66.2	86.1
08	OPENINGS	104.6	51.4	92.2	96.2	51.8	85.9	96.2	68.3	89.7	96.2	42.4	83.7	104.0	61.6	94.2	96.2	53.1	86.2
0920	Plaster & Gypsum Board	92.4	42.0	58.3	78.4	41.9	53.7	84.8	72.6	76.5	86.9	29.0	47.8	98.0	61.2	73.1	83.7	44.9	57.5
0950, 0980	Ceilings & Acoustic Treatment	93.2	42.0	59.6	82.4	41.9	55.8	82.4	72.6	75.9	90.8	29.0	50.2	96.4	61.2	73.3	82.4	44.9	57.8
0960	Flooring	109.7	41.1	90.1	101.8	41.1	84.4	108.8	61.9	95.4	111.3	39.8	90.8	106.2	41.1	87.6	105.9	41.1	87.4
0970, 0990	Wall Finishes & Painting/Coating	104.7	51.3	72.5	103.9	37.8	64.0	103.9	77.5	87.9	103.9	34.2	61.8	106.4	51.3	73.2	103.9	51.3	72.1
09	FINISHES	95.5	42.8	66.5	89.2	41.0	62.6	92.0	67.3	78.4	94.9	33.0	60.8	97.6	53.7	73.4	91.8	44.3	65.6
COVERS	DIVS. 10 - 14, 25, 28, 41, 43, 44, 46	100.0	78.2	95.6	100.0	77.9	95.5	100.0	81.6	96.3	100.0	76.1	95.2	100.0	80.0	96.0	100.0	78.2	95.6
21, 22, 23	FIRE SUPPRESSION, PLUMBING & HVAC	100.3	65.9	86.4	95.5	61.6	81.8	95.5	61.9	81.9	100.3	60.6	84.3	100.1	65.5	86.2	95.5	61.8	81.9
26, 27, 3370	ELECTRICAL, COMMUNICATIONS & UTIL.	96.2	70.9	82.8	94.9	64.5	78.8	96.3	64.6	79.5	94.5	52.4	72.2	101.7	70.9	85.4	94.5	70.4	81.7
MF2014	WEIGHTED AVERAGE	100.7	62.8	84.2	94.5	60.4	79.7	94.8	67.5	82.9	97.4	55.7	79.2	100.3	65.7	85.2	95.1	62.1	80.7

	DIVISION	OKLAHOMA												OREGON					
		POTEAU			SHAWNEE			TULSA			WOODWARD			BEND			EUGENE		
		749			748			740 - 741			738			977			974		
		MAT.	INST.	TOTAL	MAT.	INST.	TOTAL	MAT.	INST.	TOTAL	MAT.	INST.	TOTAL	MAT.	INST.	TOTAL	MAT.	INST.	TOTAL
015433	CONTRACTOR EQUIPMENT		89.1	89.1		81.8	81.8		89.9	89.9		81.8	81.8		98.8	98.8		98.8	98.8
0241, 31 - 34	SITE & INFRASTRUCTURE, DEMOLITION	77.2	85.8	83.3	99.0	90.1	92.7	97.1	87.5	90.3	97.9	90.2	92.4	105.4	102.9	103.6	96.1	102.9	100.9
0310	Concrete Forming & Accessories	90.1	41.0	47.8	85.4	43.7	49.5	101.5	40.3	48.7	92.6	46.8	53.1	110.1	98.8	100.4	106.5	98.7	99.8
0320	Concrete Reinforcing	95.9	79.9	87.8	94.9	79.8	87.2	95.1	79.8	87.3	90.7	79.9	85.2	93.3	99.6	96.5	97.4	99.6	98.5
0330	Cast-in-Place Concrete	84.3	44.9	68.1	97.2	42.9	74.9	92.9	46.2	73.7	94.3	45.3	74.2	104.4	102.7	103.7	101.1	102.6	101.7
03	CONCRETE	89.9	50.9	70.7	96.0	50.6	73.7	94.7	51.0	73.2	95.0	52.8	74.3	107.1	100.0	103.6	98.8	100.0	99.4
04	MASONRY	98.7	56.2	72.2	115.6	56.1	78.5	99.3	60.0	74.8	98.6	57.1	72.7	104.1	103.3	103.6	101.1	103.3	102.5
05	METALS	93.4	81.7	89.8	93.3	69.9	86.1	98.1	81.4	93.0	102.7	70.5	92.8	91.9	96.4	93.3	92.6	96.2	93.7
06	WOOD, PLASTICS & COMPOSITES	93.0	39.1	62.8	88.0	43.0	62.8	104.8	36.6	66.6	96.5	46.0	68.2	101.7	98.4	99.9	97.6	98.4	98.1
07	THERMAL & MOISTURE PROTECTION	99.9	60.5	83.7	99.9	59.3	83.2	99.8	62.2	84.4	109.1	66.6	91.7	107.9	94.9	102.6	107.2	91.6	100.8
08	OPENINGS	96.2	49.9	85.5	96.2	51.8	85.9	97.8	47.5	86.1	102.9	53.1	91.3	96.9	102.3	98.1	97.1	102.3	98.4
0920	Plaster & Gypsum Board	82.3	37.8	52.2	79.4	41.9	54.1	86.9	35.2	51.9	89.6	44.9	59.4	107.4	98.2	101.2	106.0	98.2	100.7
0950, 0980	Ceilings & Acoustic Treatment	82.4	37.8	53.1	82.4	41.9	55.8	90.8	35.2	54.3	86.5	44.9	59.2	91.3	98.2	95.8	92.3	98.2	96.1
0960	Flooring	105.2	61.9	92.8	102.8	32.7	82.8	110.0	42.8	90.8	105.8	43.2	87.9	110.5	103.4	108.5	108.9	103.4	107.3
0970, 0990	Wall Finishes & Painting/Coating	103.9	51.3	72.1	103.9	34.1	61.8	103.9	40.5	65.6	104.7	51.3	72.5	106.4	74.8	87.3	106.4	74.8	87.3
09	FINISHES	89.9	44.3	64.8	90.4	39.2	62.2	94.7	38.9	63.9	93.0	44.9	66.5	102.7	96.8	99.5	101.2	96.8	98.8
COVERS	DIVS. 10 - 14, 25, 28, 41, 43, 44, 46	100.0	77.7	95.5	100.0	77.9	95.5	100.0	78.8	95.7	100.0	78.5	95.7	100.0	99.6	99.9	100.0	99.6	99.9
21, 22, 23	FIRE SUPPRESSION, PLUMBING & HVAC	95.5	61.8	81.9	95.5	65.4	83.4	100.3	63.5	85.4	95.6	65.9	83.6	95.1	100.8	97.4	99.9	100.8	100.2
26, 27, 3370	ELECTRICAL, COMMUNICATIONS & UTIL.	94.6	64.6	78.7	96.6	70.9	83.0	96.5	64.6	79.6	96.1	70.9	82.8	101.6	96.1	98.7	100.1	96.1	98.0
MF2014	WEIGHTED AVERAGE	94.3	61.4	80.0	96.5	61.8	81.4	98.1	61.6	82.2	98.3	63.4	83.1	98.9	99.4	99.1	98.6	99.3	98.9

	DIVISION	OREGON																	
		KLAMATH FALLS			MEDFORD			PENDLETON			PORTLAND			SALEM			VALE		
		976			975			978			970 - 972			973			979		
		MAT.	INST.	TOTAL	MAT.	INST.	TOTAL	MAT.	INST.	TOTAL	MAT.	INST.	TOTAL	MAT.	INST.	TOTAL	MAT.	INST.	TOTAL
015433	CONTRACTOR EQUIPMENT		98.8	98.8		98.8	98.8		96.2	96.2		98.8	98.8		98.8	98.8		96.2	96.2
0241, 31 - 34	SITE & INFRASTRUCTURE, DEMOLITION	109.2	102.9	104.7	103.4	102.9	103.0	102.5	96.4	98.2	98.5	102.9	101.6	91.9	102.9	99.7	90.5	96.4	94.7
0310	Concrete Forming & Accessories	102.9	98.6	99.2	102.0	98.6	99.1	103.3	99.0	99.6	107.7	98.9	100.1	106.0	98.8	99.8	109.9	98.7	100.2
0320	Concrete Reinforcing	93.3	99.6	96.5	94.9	99.6	97.3	92.6	99.7	96.2	98.1	99.7	98.9	103.6	99.6	101.6	90.4	99.6	95.1
0330	Cast-in-Place Concrete	104.4	102.6	103.7	104.4	102.6	103.7	105.2	103.9	104.7	103.9	102.7	103.4	97.9	102.7	99.9	82.9	103.8	91.5
03	CONCRETE	109.7	99.9	104.9	104.6	99.9	102.3	91.4	100.6	95.9	100.3	100.1	100.2	98.2	100.0	99.1	76.9	100.4	88.5
04	MASONRY	117.5	103.3	108.6	98.2	103.3	101.4	107.8	103.4	105.0	102.5	103.3	103.0	109.3	103.3	105.6	106.1	103.4	104.4
05	METALS	91.9	96.2	93.2	92.2	96.2	93.4	98.1	96.9	97.7	93.5	96.5	94.4	98.7	96.4	98.0	98.0	96.6	97.5
06	WOOD, PLASTICS & COMPOSITES	92.6	98.4	95.9	91.5	98.4	95.3	94.7	98.5	96.8	98.6	98.4	98.5	95.5	98.4	97.1	103.2	98.5	100.6
07	THERMAL & MOISTURE PROTECTION	108.2	92.9	101.9	107.8	92.9	101.7	101.1	93.5	98.0	107.1	96.7	102.9	104.9	94.9	100.8	100.5	94.4	98.0
08	OPENINGS	96.9	102.3	98.1	99.7	102.3	100.3	93.2	102.4	95.4	95.0	102.3	96.7	98.7	102.3	99.5	93.2	94.1	93.4
0920	Plaster & Gypsum Board	102.9	98.2	99.7	102.2	98.2	99.5	90.2	98.2	95.6	105.5	98.2	100.5	102.1	98.2	99.4	95.9	98.2	97.4
0950, 0980	Ceilings & Acoustic Treatment	98.9	98.2	98.4	105.4	98.2	100.7	64.4	98.2	86.6	94.2	98.2	96.8	99.3	98.2	98.5	64.4	98.2	86.6
0960	Flooring	107.6	103.4	106.4	107.1	103.4	106.1	74.7	103.4	82.9	106.4	103.4	105.5	107.2	103.4	106.1	76.7	103.4	84.4
0970, 0990	Wall Finishes & Painting/Coating	106.4	70.4	84.6	106.4	70.4	84.6	96.6	72.8	82.2	106.2	72.8	86.0	104.6	74.8	86.6	96.6	74.8	83.4
09	FINISHES	103.4	96.3	99.5	103.8	96.3	99.6	72.3	96.7	85.7	100.8	96.6	98.5	100.0	96.8	98.2	72.7	96.9	86.0
COVERS	DIVS. 10 - 14, 25, 28, 41, 43, 44, 46	100.0	99.5	99.9	100.0	99.5	99.9	100.0	90.1	98.0	100.0	99.6	99.9	100.0	99.6	99.9	100.0	99.9	100.0
21, 22, 23	FIRE SUPPRESSION, PLUMBING & HVAC	95.1	100.7	97.4	99.9	100.7	100.2	97.0	113.2	103.5	99.9	100.8	100.2	99.9	100.8	100.2	97.0	98.0	97.4
26, 27, 3370	ELECTRICAL, COMMUNICATIONS & UTIL.	100.2	80.8	89.9	103.9	80.8	91.6	92.3	97.1	94.8	100.4	103.2	101.9	108.0	96.1	101.7	92.3	97.0	94.8
MF2014	WEIGHTED AVERAGE	99.8	97.1	98.6	100.0	97.1	98.8	94.8	101.4	97.7	98.8	100.4	99.5	100.5	99.4	100.0	92.9	98.1	95.2

	DIVISION	PENNSYLVANIA																	
		ALLENTOWN			ALTOONA			BEDFORD			BRADFORD			BUTLER			CHAMBERSBURG		
		181			166			155			167			160			172		
		MAT.	INST.	TOTAL	MAT.	INST.	TOTAL	MAT.	INST.	TOTAL	MAT.	INST.	TOTAL	MAT.	INST.	TOTAL	MAT.	INST.	TOTAL
015433	CONTRACTOR EQUIPMENT		112.9	112.9		112.9	112.9		109.6	109.6		112.9	112.9		112.9	112.9		112.1	112.1
0241, 31 - 34	SITE & INFRASTRUCTURE, DEMOLITION	93.6	104.5	101.3	96.8	104.4	102.2	101.4	100.1	100.5	92.4	103.5	100.3	88.1	105.8	100.7	89.2	101.2	97.8
0310	Concrete Forming & Accessories	99.2	113.0	111.1	84.2	80.4	80.9	84.1	80.8	81.3	86.4	81.9	82.5	85.7	96.0	94.6	90.2	80.0	81.4
0320	Concrete Reinforcing	94.7	108.1	101.5	91.8	102.7	97.3	93.1	80.9	86.9	93.7	103.2	98.5	92.4	109.1	100.9	91.7	103.0	97.4
0330	Cast-in-Place Concrete	86.5	104.9	94.0	96.3	86.1	92.1	106.7	70.1	91.6	92.0	92.6	92.3	85.1	96.7	89.9	91.1	69.8	82.3
03	CONCRETE	93.6	110.1	101.7	89.1	88.1	88.6	102.1	78.5	90.5	95.4	91.0	93.2	81.3	99.9	90.4	101.9	82.2	92.3
04	MASONRY	97.4	101.0	99.7	100.5	64.5	78.0	114.1	87.0	97.2	97.6	88.4	91.9	102.6	98.9	100.3	102.6	87.4	93.1
05	METALS	100.0	123.0	107.1	93.9	116.8	101.0	97.6	104.0	99.6	97.9	116.5	103.6	93.6	121.9	102.3	97.5	114.6	102.7
06	WOOD, PLASTICS & COMPOSITES	100.7	115.8	109.1	79.2	83.2	81.4	84.0	80.7	82.2	85.5	80.1	82.5	80.6	95.5	89.0	88.0	80.8	84.0
07	THERMAL & MOISTURE PROTECTION	102.3	119.1	109.2	101.4	91.0	97.1	103.0	87.7	96.7	102.3	91.7	97.9	101.1	100.7	100.9	99.3	72.5	88.3
08	OPENINGS	95.0	115.0	99.6	88.7	89.3	88.8	98.0	81.6	94.2	95.2	91.0	94.2	88.7	104.8	92.4	94.1	82.8	91.4
0920	Plaster & Gypsum Board	95.9	116.0	109.5	87.7	82.5	84.2	92.3	80.0	84.0	88.5	79.2	82.2	87.7	95.2	92.7	96.7	80.0	85.4
0950, 0980	Ceilings & Acoustic Treatment	82.8	116.0	104.6	87.0	82.5	84.0	95.0	80.0	85.1	85.4	79.2	81.3	87.9	95.2	92.7	84.0	80.0	81.4
0960	Flooring	94.6	95.9	95.0	88.2	97.5	90.9	93.1	88.2	91.7	89.4	105.3	94.0	89.0	81.0	86.7	97.9	46.2	83.1
0970, 0990	Wall Finishes & Painting/Coating	98.5	70.5	81.6	94.0	109.2	103.2	100.2	81.6	89.0	98.5	94.6	96.1	94.0	109.2	103.2	105.9	81.6	91.2
09	FINISHES	91.8	105.5	99.4	90.0	85.8	87.7	95.0	80.8	87.2	89.9	85.5	87.5	89.9	93.6	91.9	92.1	73.7	82.0
COVERS	DIVS. 10 - 14, 25, 28, 41, 43, 44, 46	100.0	105.6	101.1	100.0	95.4	99.1	100.0	98.3	99.7	100.0	99.2	99.8	100.0	102.4	100.5	100.0	96.1	99.2
21, 22, 23	FIRE SUPPRESSION, PLUMBING & HVAC	100.2	112.1	105.0	99.7	82.5	92.8	95.2	87.3	92.0	95.5	90.7	93.6	95.0	93.9	94.6	95.5	87.1	92.1
26, 27, 3370	ELECTRICAL, COMMUNICATIONS & UTIL.	99.1	98.4	98.8	89.3	112.2	101.4	94.3	112.2	103.8	92.7	112.2	103.0	89.9	111.2	101.1	89.6	86.3	87.8
MF2014	WEIGHTED AVERAGE	97.9	108.6	102.5	94.6	91.9	93.4	98.2	91.3	95.2	95.7	96.4	96.0	92.5	102.0	96.6	96.2	88.0	92.6

	DIVISION	PENNSYLVANIA																	
		DOYLESTOWN			DUBOIS			ERIE			GREENSBURG			HARRISBURG			HAZLETON		
		189			158			164 - 165			156			170 - 171			182		
		MAT.	INST.	TOTAL	MAT.	INST.	TOTAL	MAT.	INST.	TOTAL	MAT.	INST.	TOTAL	MAT.	INST.	TOTAL	MAT.	INST.	TOTAL
015433	CONTRACTOR EQUIPMENT		93.9	93.9		109.6	109.6		112.9	112.9		109.6	109.6		112.1	112.1		112.9	112.9
0241, 31 - 34	SITE & INFRASTRUCTURE, DEMOLITION	106.9	90.2	95.0	106.1	100.5	102.1	93.7	105.2	101.9	97.8	102.9	101.4	89.6	103.5	99.5	86.9	104.7	99.6
0310	Concrete Forming & Accessories	83.2	128.2	122.0	83.5	84.6	84.4	98.5	88.1	89.5	90.7	95.9	95.2	100.4	87.9	89.6	80.8	88.7	87.6
0320	Concrete Reinforcing	91.6	130.9	111.6	92.6	103.5	98.1	93.7	103.3	98.6	92.6	109.1	101.0	101.4	106.2	103.9	91.9	107.1	99.6
0330	Cast-in-Place Concrete	81.7	86.7	83.8	102.9	92.9	98.8	94.7	81.8	89.4	99.0	96.3	97.9	93.2	95.8	94.3	81.7	93.1	86.4
03	CONCRETE	89.2	114.2	101.5	103.6	92.2	98.0	88.2	90.0	89.1	97.2	99.6	98.4	97.7	95.4	96.6	86.3	95.0	90.6
04	MASONRY	100.7	128.1	117.7	114.4	91.1	99.9	89.6	92.5	91.4	124.8	98.9	108.7	102.5	89.6	94.5	110.2	95.7	101.2
05	METALS	97.5	123.8	105.6	97.6	116.1	103.3	94.1	117.0	101.2	97.5	120.6	104.6	104.0	121.4	109.4	99.7	120.6	106.1
06	WOOD, PLASTICS & COMPOSITES	81.0	130.7	108.8	82.9	83.1	83.0	96.9	86.6	91.1	90.9	95.4	93.4	95.8	86.7	90.7	79.6	86.6	83.5
07	THERMAL & MOISTURE PROTECTION	100.1	131.4	112.9	103.2	96.5	100.5	101.9	92.6	98.1	102.8	100.7	102.0	102.5	110.2	105.7	101.8	106.6	103.8
08	OPENINGS	97.2	137.9	106.7	98.0	92.7	96.7	88.8	92.2	89.6	97.9	104.7	99.5	100.6	94.0	99.0	95.6	92.3	94.8
0920	Plaster & Gypsum Board	86.2	131.4	116.7	91.2	82.5	85.3	95.9	86.0	89.2	93.5	95.2	94.6	100.6	86.0	90.7	86.6	86.0	86.2
0950, 0980	Ceilings & Acoustic Treatment	82.0	131.4	114.5	95.0	82.5	86.7	82.8	86.0	84.9	94.2	95.2	94.8	92.6	86.0	88.3	83.7	86.0	85.2
0960	Flooring	79.4	134.3	95.1	92.9	105.3	96.4	93.2	91.8	92.8	96.2	68.0	88.2	103.2	91.7	99.9	87.0	94.6	89.2
0970, 0990	Wall Finishes & Painting/Coating	98.0	69.0	80.5	100.2	106.6	104.1	103.5	94.6	98.1	100.2	106.6	104.1	106.4	89.6	96.3	98.5	106.4	103.2
09	FINISHES	83.2	122.4	104.8	95.2	89.5	92.1	92.2	88.8	90.4	95.6	91.7	93.4	96.9	88.0	92.0	88.1	89.5	88.9
COVERS	DIVS. 10 - 14, 25, 28, 41, 43, 44, 46	100.0	70.6	94.1	100.0	99.1	99.8	100.0	100.8	100.2	100.0	102.2	100.4	100.0	97.4	99.5	100.0	101.0	100.2
21, 22, 23	FIRE SUPPRESSION, PLUMBING & HVAC	95.0	127.4	108.1	95.2	88.4	92.4	99.7	92.9	97.0	95.2	91.1	93.5	100.1	92.1	96.9	95.5	99.2	97.0
26, 27, 3370	ELECTRICAL, COMMUNICATIONS & UTIL.	92.1	128.1	111.1	94.9	112.2	104.0	91.1	96.7	94.0	94.9	112.2	104.1	96.8	88.3	92.3	93.6	91.7	92.6
MF2014	WEIGHTED AVERAGE	94.9	120.6	106.1	98.5	96.8	97.8	94.5	95.8	95.1	98.2	100.9	99.4	99.8	95.6	98.0	95.4	98.3	96.6

DIVISION		PENNSYLVANIA																	
		INDIANA			JOHNSTOWN			KITTANNING			LANCASTER			LEHIGH VALLEY			MONTROSE		
		157			159			162			175 - 176			180			188		
		MAT.	INST.	TOTAL	MAT.	INST.	TOTAL	MAT.	INST.	TOTAL	MAT.	INST.	TOTAL	MAT.	INST.	TOTAL	MAT.	INST.	TOTAL
015433	**CONTRACTOR EQUIPMENT**		109.6	109.6		109.6	109.6		112.9	112.9		112.1	112.1		112.9	112.9		112.9	112.9
0241, 31 - 34	**SITE & INFRASTRUCTURE, DEMOLITION**	95.9	101.2	99.7	101.9	102.2	102.1	90.7	105.7	101.4	81.6	103.5	97.2	90.7	104.3	100.4	89.4	102.1	98.4
0310	Concrete Forming & Accessories	84.7	86.3	86.1	83.5	84.8	84.6	85.7	95.9	94.5	92.3	87.5	88.2	92.8	112.7	110.0	81.8	88.5	87.6
0320	Concrete Reinforcing	91.8	109.2	100.6	93.1	108.9	101.2	92.4	109.2	101.0	91.4	106.1	98.9	91.9	108.1	100.1	96.3	106.3	101.4
0330	Cast-in-Place Concrete	97.1	95.9	96.6	107.6	92.4	101.4	88.4	96.5	91.7	77.5	97.9	85.9	88.4	104.7	95.1	86.7	90.9	88.4
03	**CONCRETE**	94.7	95.2	94.9	102.9	93.2	98.2	83.8	99.7	91.6	89.8	96.0	92.8	92.6	109.8	101.1	91.3	93.7	92.5
04	**MASONRY**	110.1	100.1	103.9	110.9	90.6	98.3	105.4	100.1	102.1	108.2	90.1	97.0	97.3	101.0	99.6	97.3	95.1	95.9
05	**METALS**	97.7	120.1	104.6	97.6	118.9	104.2	93.7	121.9	102.4	97.5	120.8	104.7	99.7	122.0	106.5	98.0	114.1	102.9
06	**WOOD, PLASTICS & COMPOSITES**	84.8	83.1	83.8	82.9	83.1	83.0	80.6	95.5	89.0	90.8	86.7	88.5	92.2	115.8	105.4	80.4	88.4	84.9
07	**THERMAL & MOISTURE PROTECTION**	102.7	98.2	100.9	103.0	95.1	99.7	101.1	101.0	101.1	98.8	97.1	98.1	102.2	102.8	102.4	101.8	91.7	97.7
08	**OPENINGS**	98.0	94.4	97.1	98.0	90.9	96.3	88.7	101.2	91.6	94.1	98.6	95.1	95.6	115.0	100.1	92.2	93.1	92.4
0920	Plaster & Gypsum Board	92.7	82.5	85.8	91.0	82.5	85.2	87.7	95.2	92.7	98.5	86.0	90.1	89.5	116.0	107.4	87.1	87.8	87.6
0950, 0980	Ceilings & Acoustic Treatment	95.0	82.5	86.7	94.2	82.5	86.5	87.9	95.2	92.7	84.0	86.0	85.3	83.7	116.0	105.0	85.4	87.8	87.0
0960	Flooring	93.7	105.3	97.0	92.9	98.2	94.4	89.0	105.3	93.6	98.8	91.7	96.8	91.9	95.9	93.0	87.6	57.9	79.1
0970, 0990	Wall Finishes & Painting/Coating	100.2	106.6	104.1	100.2	109.2	105.6	94.0	106.6	101.6	105.9	57.1	76.5	98.5	66.4	79.1	98.5	106.4	103.2
09	**FINISHES**	94.8	90.0	92.1	94.6	88.8	91.4	90.1	97.4	94.1	92.0	84.5	87.9	90.2	104.1	97.9	88.9	84.7	86.6
COVERS	**DIVS. 10 - 14, 25, 28, 41, 43, 44, 46**	100.0	100.6	100.1	100.0	99.3	99.9	100.0	102.2	100.5	100.0	97.4	99.5	100.0	106.0	101.2	100.0	101.2	100.2
21, 22, 23	**FIRE SUPPRESSION, PLUMBING & HVAC**	95.2	90.8	93.4	95.2	88.7	92.5	95.0	96.9	95.8	95.5	92.3	94.2	95.5	112.0	102.2	95.5	98.4	96.7
26, 27, 3370	**ELECTRICAL, COMMUNICATIONS & UTIL.**	94.9	112.2	104.1	94.9	112.2	104.0	89.3	112.2	101.4	91.0	44.0	66.2	93.6	143.4	119.9	92.7	97.6	95.3
MF2014	**WEIGHTED AVERAGE**	97.1	99.3	98.0	98.1	97.2	97.7	92.9	103.3	97.4	95.1	89.0	92.4	95.8	114.0	103.7	94.7	96.9	95.7

DIVISION		PENNSYLVANIA																	
		NEW CASTLE			NORRISTOWN			OIL CITY			PHILADELPHIA			PITTSBURGH			POTTSVILLE		
		161			194			163			190 - 191			150 - 152			179		
		MAT.	INST.	TOTAL	MAT.	INST.	TOTAL	MAT.	INST.	TOTAL	MAT.	INST.	TOTAL	MAT.	INST.	TOTAL	MAT.	INST.	TOTAL
015433	**CONTRACTOR EQUIPMENT**		112.9	112.9		98.8	98.8		112.9	112.9		98.2	98.2		110.8	110.8		112.1	112.1
0241, 31 - 34	**SITE & INFRASTRUCTURE, DEMOLITION**	88.5	105.8	100.8	95.8	100.6	99.2	87.1	103.7	98.9	101.6	100.0	100.5	101.3	104.9	103.8	84.4	103.4	97.9
0310	Concrete Forming & Accessories	85.7	95.6	94.3	83.2	129.7	123.3	85.7	83.4	83.7	98.9	140.1	134.4	98.5	96.7	97.0	83.3	89.7	88.8
0320	Concrete Reinforcing	91.3	92.5	91.9	90.2	137.9	114.5	92.4	92.3	92.4	100.6	137.9	119.6	93.6	109.4	101.6	90.7	102.7	96.8
0330	Cast-in-Place Concrete	85.9	96.5	90.2	85.0	127.1	102.3	83.4	95.5	88.4	98.0	132.2	112.1	102.9	96.5	100.3	82.6	100.9	90.1
03	**CONCRETE**	81.6	96.4	88.9	89.2	130.1	109.3	80.1	90.5	85.2	99.4	136.4	117.6	100.5	100.1	100.3	93.3	97.3	95.3
04	**MASONRY**	101.9	97.7	99.3	110.4	124.9	119.4	101.8	95.8	98.0	96.7	130.7	117.9	103.0	102.1	102.4	102.1	92.2	95.9
05	**METALS**	93.7	113.6	99.8	99.2	130.0	108.7	93.7	111.5	99.2	102.0	130.2	110.7	99.0	121.3	105.9	97.7	118.8	104.2
06	**WOOD, PLASTICS & COMPOSITES**	80.6	95.9	89.2	79.9	130.6	108.3	80.6	80.1	80.3	98.6	141.9	122.9	100.6	95.8	97.9	80.2	88.1	84.6
07	**THERMAL & MOISTURE PROTECTION**	101.1	98.0	99.8	101.4	131.1	113.5	101.0	94.4	98.3	102.8	135.2	116.1	103.1	101.6	102.4	98.9	106.7	102.1
08	**OPENINGS**	88.7	96.6	90.5	86.9	140.1	99.2	88.7	80.3	86.7	98.6	146.2	109.7	101.8	105.0	102.5	94.1	91.8	93.6
0920	Plaster & Gypsum Board	87.7	95.6	93.0	85.6	131.4	116.5	87.7	79.2	82.0	96.6	143.1	128.0	99.8	95.6	96.9	93.5	87.5	89.5
0950, 0980	Ceilings & Acoustic Treatment	87.9	95.6	92.9	84.5	131.4	115.3	87.9	79.2	82.2	94.0	143.1	126.2	95.0	95.6	95.4	84.0	87.5	86.3
0960	Flooring	89.0	57.5	80.0	92.9	139.2	106.1	89.0	105.3	93.6	98.4	139.2	110.1	99.5	106.6	101.5	95.2	91.7	94.2
0970, 0990	Wall Finishes & Painting/Coating	94.0	109.2	103.2	98.6	147.8	128.3	94.0	106.6	101.6	99.8	154.7	133.0	100.2	119.2	111.7	105.9	106.4	106.2
09	**FINISHES**	90.0	89.6	89.8	90.7	133.0	114.0	89.8	87.8	88.7	98.9	141.8	122.5	97.7	100.2	99.1	90.5	91.3	90.9
COVERS	**DIVS. 10 - 14, 25, 28, 41, 43, 44, 46**	100.0	102.4	100.5	100.0	117.5	103.5	100.0	100.6	100.1	100.0	120.8	104.2	100.0	102.2	100.4	100.0	98.7	99.7
21, 22, 23	**FIRE SUPPRESSION, PLUMBING & HVAC**	95.0	93.6	94.4	95.2	127.3	108.2	95.0	92.9	94.2	100.0	131.8	112.9	99.9	100.4	100.1	95.5	98.6	96.8
26, 27, 3370	**ELECTRICAL, COMMUNICATIONS & UTIL.**	89.9	98.1	94.2	93.0	148.4	122.3	91.8	111.1	102.0	97.1	148.4	124.2	97.2	112.2	105.1	89.2	95.5	92.5
MF2014	**WEIGHTED AVERAGE**	92.5	97.9	94.8	95.0	129.5	110.0	92.4	96.9	94.4	99.7	133.6	114.5	99.9	104.6	102.0	94.9	98.5	96.5

DIVISION		PENNSYLVANIA																	
		READING			SCRANTON			STATE COLLEGE			STROUDSBURG			SUNBURY			UNIONTOWN		
		195 - 196			184 - 185			168			183			178			154		
		MAT.	INST.	TOTAL	MAT.	INST.	TOTAL	MAT.	INST.	TOTAL	MAT.	INST.	TOTAL	MAT.	INST.	TOTAL	MAT.	INST.	TOTAL
015433	**CONTRACTOR EQUIPMENT**		118.0	118.0		112.9	112.9		112.1	112.1		112.9	112.9		112.9	112.9		109.6	109.6
0241, 31 - 34	**SITE & INFRASTRUCTURE, DEMOLITION**	100.3	112.7	109.1	94.1	104.8	101.7	84.4	103.1	97.7	88.5	102.2	98.2	96.6	104.5	102.2	96.5	102.9	101.0
0310	Concrete Forming & Accessories	98.8	90.2	91.4	99.3	88.6	90.0	84.4	80.3	80.8	87.2	89.4	89.1	96.1	88.5	89.6	77.6	96.0	93.5
0320	Concrete Reinforcing	91.5	104.7	98.2	94.7	106.6	100.8	93.0	103.4	98.3	95.0	111.6	103.4	93.2	106.2	99.8	92.6	109.1	101.0
0330	Cast-in-Place Concrete	76.2	97.3	84.9	90.3	93.1	91.5	87.1	65.6	78.2	85.1	72.6	80.0	90.2	95.5	92.4	97.1	96.4	96.8
03	**CONCRETE**	88.2	96.7	92.4	95.4	94.9	95.1	95.5	81.0	88.4	90.1	88.9	89.5	96.7	95.5	96.1	94.4	99.7	97.0
04	**MASONRY**	99.5	93.0	95.4	97.7	95.7	96.4	103.1	77.9	87.4	95.1	100.4	98.4	102.3	89.7	94.5	127.0	98.9	109.5
05	**METALS**	99.5	121.2	106.2	102.1	120.5	107.8	97.7	117.6	103.8	99.7	115.9	104.7	97.4	120.0	104.4	97.4	120.6	104.5
06	**WOOD, PLASTICS & COMPOSITES**	98.5	88.1	92.6	100.7	86.2	92.6	87.6	83.2	85.1	86.4	88.4	87.5	88.9	88.1	88.5	76.7	95.4	87.2
07	**THERMAL & MOISTURE PROTECTION**	101.8	112.1	106.0	102.2	95.6	99.5	101.5	90.6	97.0	102.0	85.3	95.2	100.0	104.9	102.0	102.6	100.7	101.8
08	**OPENINGS**	91.3	98.9	93.1	95.0	91.9	94.3	92.0	89.3	91.4	95.6	88.2	93.9	94.2	92.8	93.9	97.9	104.7	99.5
0920	Plaster & Gypsum Board	96.8	87.5	90.5	98.0	85.6	89.6	89.6	82.5	84.8	88.0	87.8	87.9	92.7	87.5	89.2	89.2	95.2	93.2
0950, 0980	Ceilings & Acoustic Treatment	76.9	87.5	83.9	91.1	85.6	87.5	82.9	82.5	82.6	82.0	87.8	85.8	80.7	87.5	85.2	94.2	95.2	94.8
0960	Flooring	96.8	91.7	95.4	94.6	107.0	98.2	92.3	97.5	93.8	89.9	51.3	78.9	95.9	87.7	93.5	90.4	105.3	94.7
0970, 0990	Wall Finishes & Painting/Coating	97.3	106.4	102.8	98.5	106.4	103.2	98.5	109.2	105.0	98.5	64.6	78.0	105.9	94.3	98.9	100.2	109.2	105.6
09	**FINISHES**	92.0	90.8	91.3	93.9	93.1	93.4	89.4	85.8	87.4	88.9	80.0	84.0	91.3	88.7	89.9	93.2	97.5	95.6
COVERS	**DIVS. 10 - 14, 25, 28, 41, 43, 44, 46**	100.0	100.3	100.1	100.0	101.0	100.2	100.0	93.1	98.6	100.0	62.4	92.4	100.0	97.9	99.6	100.0	102.2	100.4
21, 22, 23	**FIRE SUPPRESSION, PLUMBING & HVAC**	100.1	108.8	103.7	100.2	99.2	99.8	95.5	85.9	91.6	95.5	100.0	97.3	95.5	92.0	94.1	95.2	91.1	93.5
26, 27, 3370	**ELECTRICAL, COMMUNICATIONS & UTIL.**	99.6	95.5	97.5	99.2	97.7	98.4	91.9	112.2	102.6	93.6	143.3	119.9	89.5	90.8	90.2	92.0	112.2	102.7
MF2014	**WEIGHTED AVERAGE**	97.1	102.1	99.3	98.6	99.2	98.9	95.3	92.8	94.2	95.3	101.4	97.9	95.8	95.8	95.8	97.3	101.7	99.2

	DIVISION	PENNSYLVANIA																	
		WASHINGTON 153			WELLSBORO 169			WESTCHESTER 193			WILKES-BARRE 186 - 187			WILLIAMSPORT 177			YORK 173 - 174		
		MAT.	INST.	TOTAL	MAT.	INST.	TOTAL	MAT.	INST.	TOTAL	MAT.	INST.	TOTAL	MAT.	INST.	TOTAL	MAT.	INST.	TOTAL
015433	CONTRACTOR EQUIPMENT		109.6	109.6		112.9	112.9		98.8	98.8		112.9	112.9		112.9	112.9		112.1	112.1
0241, 31 - 34	SITE & INFRASTRUCTURE, DEMOLITION	96.6	102.9	101.1	95.9	102.0	100.3	101.6	97.8	98.9	86.5	104.7	99.5	87.9	103.4	98.9	85.2	103.5	98.2
0310	Concrete Forming & Accessories	84.8	96.2	94.6	85.8	86.7	86.6	89.8	128.2	122.9	89.9	88.9	89.1	92.5	61.0	65.3	86.9	88.1	87.9
0320	Concrete Reinforcing	92.6	109.3	101.1	93.0	106.2	99.7	89.3	113.6	101.7	93.7	107.1	100.6	92.5	64.6	78.3	93.2	106.2	99.8
0330	Cast-in-Place Concrete	97.1	96.4	96.8	91.2	88.9	90.3	94.1	126.4	107.4	81.7	93.0	86.4	76.4	74.2	75.5	83.0	98.1	89.2
03	CONCRETE	94.9	99.8	97.3	97.8	92.2	95.1	96.9	124.5	110.5	87.2	95.1	91.1	84.7	68.2	76.6	94.5	96.3	95.4
04	MASONRY	109.4	100.5	103.9	103.6	89.7	94.9	104.7	124.9	117.3	110.6	95.3	101.0	93.9	93.7	93.8	103.6	90.1	95.2
05	METALS	97.3	120.9	104.6	97.8	114.2	102.9	99.2	117.2	104.8	97.9	121.0	105.0	97.5	100.4	98.4	99.1	121.4	105.9
06	WOOD, PLASTICS & COMPOSITES	84.9	95.4	90.8	84.9	88.1	86.7	87.1	130.6	111.5	88.9	86.6	87.6	85.2	51.0	66.1	84.1	86.7	85.5
07	THERMAL & MOISTURE PROTECTION	102.7	101.1	102.1	102.5	89.6	97.2	101.7	129.9	113.3	101.8	106.4	103.7	99.4	102.4	100.6	99.0	110.4	103.7
08	OPENINGS	97.9	104.7	99.5	95.1	93.0	94.6	86.9	126.1	96.0	92.2	98.8	93.8	94.2	53.5	84.7	94.1	94.0	94.1
0920	Plaster & Gypsum Board	92.4	95.2	94.3	87.9	87.5	87.6	87.0	131.4	117.0	88.8	86.0	86.9	93.5	49.3	63.6	94.3	86.0	88.7
0950, 0980	Ceilings & Acoustic Treatment	94.2	95.2	94.8	82.9	87.5	85.9	84.5	131.4	115.3	85.4	86.0	85.8	84.0	49.3	61.2	83.1	86.0	85.0
0960	Flooring	93.8	105.3	97.1	89.1	50.8	78.2	95.7	139.2	108.1	90.7	94.6	91.8	94.8	49.7	81.9	96.4	91.7	95.1
0970, 0990	Wall Finishes & Painting/Coating	100.2	109.2	105.6	98.5	106.4	103.2	98.6	147.8	128.3	98.5	106.4	103.2	105.9	106.4	106.2	105.9	89.6	96.1
09	FINISHES	94.6	97.8	96.4	89.5	82.2	85.5	92.1	131.0	113.6	89.9	90.8	90.4	91.1	62.3	75.2	90.8	88.1	89.3
COVERS	DIVS. 10 - 14, 25, 28, 41, 43, 44, 46	100.0	102.2	100.4	100.0	97.9	99.6	100.0	117.3	103.5	100.0	100.8	100.2	100.0	94.6	98.9	100.0	97.6	99.5
21, 22, 23	FIRE SUPPRESSION, PLUMBING & HVAC	95.2	97.2	96.0	95.5	91.1	93.7	95.2	125.8	107.6	95.5	99.0	96.9	95.5	93.1	94.5	100.2	92.4	97.1
26, 27, 3370	ELECTRICAL, COMMUNICATIONS & UTIL.	94.3	112.2	103.8	92.7	86.5	89.5	92.9	114.9	104.5	93.6	91.7	92.6	90.0	74.6	81.8	91.0	88.3	89.5
MF2014	WEIGHTED AVERAGE	97.0	103.2	99.7	96.3	92.6	94.7	95.9	121.5	107.1	95.1	98.7	96.6	93.8	83.0	89.1	96.7	95.9	96.3

	DIVISION	PUERTO RICO			RHODE ISLAND						SOUTH CAROLINA								
		SAN JUAN 009			NEWPORT 028			PROVIDENCE 029			AIKEN 298			BEAUFORT 299			CHARLESTON 294		
		MAT.	INST.	TOTAL	MAT.	INST.	TOTAL	MAT.	INST.	TOTAL	MAT.	INST.	TOTAL	MAT.	INST.	TOTAL	MAT.	INST.	TOTAL
015433	CONTRACTOR EQUIPMENT		90.5	90.5		102.4	102.4		102.4	102.4		101.3	101.3		101.3	101.3		101.3	101.3
0241, 31 - 34	SITE & INFRASTRUCTURE, DEMOLITION	133.8	91.2	103.5	89.2	105.0	100.4	91.4	105.0	101.1	118.9	87.4	96.5	114.3	85.6	93.9	99.7	86.3	90.2
0310	Concrete Forming & Accessories	92.4	17.8	28.1	101.5	122.4	119.5	99.8	122.4	119.3	97.5	67.4	71.5	96.4	37.9	46.0	95.4	63.0	67.5
0320	Concrete Reinforcing	188.3	12.7	98.9	106.4	149.2	128.2	102.3	149.2	126.2	93.7	65.2	79.2	92.8	25.8	58.7	92.7	59.3	75.7
0330	Cast-in-Place Concrete	103.8	30.4	73.7	79.5	124.1	97.8	95.5	124.1	107.2	79.2	69.8	75.4	79.2	47.2	66.1	92.9	49.5	75.1
03	CONCRETE	108.7	22.2	66.2	93.2	127.5	110.0	100.4	127.5	113.7	102.7	68.9	86.1	99.9	40.9	70.9	94.4	58.9	77.0
04	MASONRY	90.1	16.9	44.5	95.4	132.5	118.5	101.3	132.5	120.7	79.6	61.4	68.3	93.9	33.3	56.1	95.1	40.9	61.4
05	METALS	118.4	34.6	92.6	97.4	125.6	106.1	103.1	125.6	110.0	101.8	83.8	96.3	101.8	68.2	91.5	103.8	80.4	96.6
06	WOOD, PLASTICS & COMPOSITES	94.4	17.5	51.3	99.0	120.9	111.3	100.1	120.9	111.7	97.0	68.6	81.1	95.3	37.8	63.1	94.0	68.4	79.7
07	THERMAL & MOISTURE PROTECTION	129.1	21.6	85.0	100.6	121.5	109.2	100.5	121.5	109.1	102.7	66.4	87.8	102.4	38.8	76.3	101.6	47.3	79.3
08	OPENINGS	152.0	15.4	120.2	102.2	128.2	108.3	107.3	128.2	112.2	99.0	64.8	91.1	99.0	36.7	84.5	102.9	63.3	93.7
0920	Plaster & Gypsum Board	159.8	14.9	61.9	92.9	120.9	111.9	94.5	120.9	112.4	104.0	67.4	79.3	107.4	35.7	59.0	109.0	67.3	80.8
0950, 0980	Ceilings & Acoustic Treatment	225.7	14.9	87.2	93.2	120.9	111.4	90.1	120.9	110.4	86.4	67.4	73.9	89.8	35.7	54.3	89.8	67.3	75.0
0960	Flooring	224.3	16.6	164.9	97.7	141.5	110.2	98.0	141.5	110.4	105.4	68.4	94.8	106.8	51.2	90.9	106.5	58.2	92.7
0970, 0990	Wall Finishes & Painting/Coating	213.5	16.8	94.8	99.9	129.2	117.6	97.7	129.2	116.7	109.5	70.3	85.8	109.5	34.7	64.3	109.5	66.9	83.8
09	FINISHES	210.6	17.6	104.2	96.6	126.9	113.3	93.7	126.9	112.0	98.3	68.0	81.6	99.5	39.8	66.6	97.8	63.5	78.9
COVERS	DIVS. 10 - 14, 25, 28, 41, 43, 44, 46	100.0	17.5	83.3	100.0	108.6	101.7	100.0	108.6	101.7	100.0	71.8	94.3	100.0	70.4	94.0	100.0	69.0	93.7
21, 22, 23	FIRE SUPPRESSION, PLUMBING & HVAC	103.3	14.0	67.3	100.1	110.6	104.3	99.9	110.6	104.2	95.7	63.1	82.5	95.7	36.5	71.8	100.5	53.7	81.6
26, 27, 3370	ELECTRICAL, COMMUNICATIONS & UTIL.	125.9	13.2	66.3	100.0	99.0	99.5	99.6	99.0	99.2	96.9	65.9	80.5	100.7	33.6	65.2	99.0	88.8	93.6
MF2014	WEIGHTED AVERAGE	122.7	24.4	79.8	98.4	117.5	106.7	100.6	117.5	108.0	98.6	69.1	85.7	99.3	44.9	75.6	99.9	65.2	84.8

	DIVISION	SOUTH CAROLINA															SOUTH DAKOTA		
		COLUMBIA 290 - 292			FLORENCE 295			GREENVILLE 296			ROCK HILL 297			SPARTANBURG 293			ABERDEEN 574		
		MAT.	INST.	TOTAL	MAT.	INST.	TOTAL	MAT.	INST.	TOTAL	MAT.	INST.	TOTAL	MAT.	INST.	TOTAL	MAT.	INST.	TOTAL
015433	CONTRACTOR EQUIPMENT		101.3	101.3		101.3	101.3		101.3	101.3		101.3	101.3		101.3	101.3		98.2	98.2
0241, 31 - 34	SITE & INFRASTRUCTURE, DEMOLITION	99.8	86.3	90.2	108.8	86.3	92.8	104.1	85.9	91.2	101.7	85.0	89.8	103.9	86.0	91.1	99.2	93.7	95.3
0310	Concrete Forming & Accessories	94.4	45.1	51.9	83.2	45.3	50.5	95.0	45.1	52.0	93.2	38.7	46.2	98.0	45.3	52.5	94.8	36.6	44.6
0320	Concrete Reinforcing	95.8	59.1	77.1	92.3	59.3	75.5	92.2	44.5	67.9	93.0	44.1	68.1	92.2	57.0	74.3	96.1	38.1	66.6
0330	Cast-in-Place Concrete	95.8	50.1	77.0	79.2	49.4	66.9	79.2	49.3	66.9	79.2	43.9	64.7	79.2	49.4	66.9	105.6	43.5	80.1
03	CONCRETE	96.2	51.1	74.1	94.2	51.0	73.0	93.0	48.2	71.0	90.9	43.3	67.5	93.2	50.6	72.3	102.8	40.7	72.3
04	MASONRY	91.5	37.6	57.9	79.8	40.9	55.6	77.4	40.9	54.7	101.6	34.6	59.8	79.8	40.9	55.6	115.9	55.4	78.2
05	METALS	100.9	79.6	94.3	102.6	80.0	95.6	102.6	74.5	93.9	101.8	71.2	92.4	102.6	79.1	95.4	95.9	63.6	86.0
06	WOOD, PLASTICS & COMPOSITES	98.0	44.8	68.2	79.7	44.8	60.2	93.7	44.8	66.4	92.0	38.9	62.3	98.0	44.8	68.2	99.6	36.0	64.0
07	THERMAL & MOISTURE PROTECTION	97.2	43.3	75.1	101.9	44.8	78.5	101.8	44.8	78.4	101.6	39.7	76.3	101.8	44.8	78.5	99.2	47.7	78.1
08	OPENINGS	104.6	50.5	92.0	99.1	50.5	87.8	99.0	47.0	86.9	99.0	39.4	85.2	99.0	50.0	87.6	99.1	36.7	84.6
0920	Plaster & Gypsum Board	102.3	42.9	62.2	97.1	42.9	60.5	102.6	42.9	62.3	101.9	36.8	57.9	105.5	42.9	63.2	103.7	34.3	56.8
0950, 0980	Ceilings & Acoustic Treatment	91.3	42.9	59.5	87.3	42.9	58.1	86.4	42.9	57.8	86.4	36.8	53.8	86.4	42.9	57.8	93.2	34.3	54.5
0960	Flooring	100.5	43.6	84.2	98.4	43.6	82.7	104.3	57.2	90.8	103.3	44.4	86.5	105.6	57.2	91.8	108.4	50.3	91.8
0970, 0990	Wall Finishes & Painting/Coating	106.2	66.9	82.5	109.5	66.9	83.8	109.5	66.9	83.8	109.5	39.2	67.0	109.5	66.9	83.8	102.6	37.1	63.1
09	FINISHES	95.9	46.6	68.7	94.4	47.1	68.3	96.3	49.3	70.4	95.7	39.4	64.6	97.1	49.3	70.8	100.3	38.8	66.4
COVERS	DIVS. 10 - 14, 25, 28, 41, 43, 44, 46	100.0	66.3	93.2	100.0	66.3	93.2	100.0	66.3	93.2	100.0	64.6	92.9	100.0	66.3	93.2	100.0	40.5	88.0
21, 22, 23	FIRE SUPPRESSION, PLUMBING & HVAC	100.0	53.0	81.1	100.5	53.1	81.4	100.5	53.0	81.3	95.7	44.2	74.9	100.5	53.1	81.3	100.1	39.8	75.8
26, 27, 3370	ELECTRICAL, COMMUNICATIONS & UTIL.	99.0	59.7	78.3	96.8	59.7	77.2	99.1	57.4	77.0	99.1	56.6	76.6	99.1	57.4	77.1	101.7	51.1	75.0
MF2014	WEIGHTED AVERAGE	99.3	56.5	80.6	98.2	56.9	80.2	98.3	55.8	79.8	97.8	50.3	77.1	98.6	56.7	80.3	100.5	49.6	78.3

DIVISION		SOUTH DAKOTA																	
		MITCHELL			MOBRIDGE			PIERRE			RAPID CITY			SIOUX FALLS			WATERTOWN		
		573			576			575			577			570 - 571			572		
		MAT.	INST.	TOTAL	MAT.	INST.	TOTAL	MAT.	INST.	TOTAL	MAT.	INST.	TOTAL	MAT.	INST.	TOTAL	MAT.	INST.	TOTAL
015433	CONTRACTOR EQUIPMENT		98.2	98.2		98.2	98.2		98.2	98.2		98.2	98.2		99.2	99.2		98.2	98.2
0241, 31 - 34	SITE & INFRASTRUCTURE, DEMOLITION	96.0	93.7	94.3	95.9	93.7	94.3	100.5	93.7	95.7	97.7	93.8	94.9	94.3	95.4	95.0	95.8	93.7	94.3
0310	Concrete Forming & Accessories	94.0	37.0	44.8	84.9	36.8	43.4	97.7	38.4	46.6	102.2	36.7	45.7	98.8	40.5	48.5	81.6	36.6	42.8
0320	Concrete Reinforcing	95.5	46.3	70.5	98.0	38.1	67.5	99.0	71.1	84.8	89.9	71.3	80.4	98.0	71.2	84.3	92.9	38.2	65.1
0330	Cast-in-Place Concrete	102.6	42.3	77.8	102.6	43.5	78.3	98.5	42.2	75.4	101.8	42.8	77.6	92.3	42.9	72.0	102.6	45.9	79.3
03	CONCRETE	100.5	41.8	71.7	100.3	40.7	71.0	99.0	47.3	73.6	99.8	46.7	73.7	95.9	48.4	72.6	99.2	41.5	70.9
04	MASONRY	103.9	53.3	72.4	113.4	55.4	77.2	117.4	53.5	77.6	113.5	56.4	77.9	105.1	53.6	73.0	141.1	57.4	88.9
05	METALS	95.0	63.8	85.4	95.0	63.6	85.3	98.4	79.1	92.4	97.8	79.7	92.2	98.2	79.3	92.4	95.0	64.1	85.5
06	WOOD, PLASTICS & COMPOSITES	98.5	36.5	63.8	87.6	36.2	58.8	104.5	36.8	66.6	103.5	33.3	64.2	96.8	39.3	64.6	83.8	36.0	57.0
07	THERMAL & MOISTURE PROTECTION	98.9	45.6	77.1	99.0	47.8	78.0	101.8	45.9	78.9	99.6	47.8	78.3	101.9	48.2	79.9	98.8	48.3	78.1
08	OPENINGS	97.7	37.3	83.6	100.4	36.3	85.5	105.7	46.7	92.0	103.1	44.8	89.6	106.5	48.1	92.9	97.7	36.5	83.4
0920	Plaster & Gypsum Board	102.1	34.8	56.7	96.2	34.6	54.5	100.1	35.1	56.2	103.2	31.6	54.8	90.7	37.7	54.9	93.9	34.3	53.6
0950, 0980	Ceilings & Acoustic Treatment	89.9	34.8	53.7	93.2	34.6	54.7	93.3	35.1	55.0	95.7	31.6	53.6	90.6	37.7	55.8	89.9	34.3	53.3
0960	Flooring	107.9	50.3	91.4	103.5	50.3	88.3	107.4	35.8	86.9	107.6	78.6	99.3	103.8	74.9	95.5	102.1	50.3	87.3
0970, 0990	Wall Finishes & Painting/Coating	102.6	40.6	65.2	102.6	41.7	65.8	105.7	44.6	68.8	102.6	44.6	67.6	101.7	44.6	67.2	102.6	37.1	63.1
09	FINISHES	99.1	39.5	66.2	97.7	39.4	65.6	100.8	37.3	65.8	100.5	43.9	69.3	97.1	46.5	69.2	96.2	38.8	64.6
COVERS	DIVS. 10 - 14, 25, 28, 41, 43, 44, 46	100.0	37.6	87.4	100.0	40.5	88.0	100.0	75.8	95.1	100.0	75.7	95.1	100.0	76.2	95.2	100.0	40.5	88.0
21, 22, 23	FIRE SUPPRESSION, PLUMBING & HVAC	95.3	39.4	72.8	95.3	39.9	73.0	100.0	64.5	85.7	100.1	65.0	85.9	100.0	38.6	75.2	95.3	39.9	73.0
26, 27, 3370	ELECTRICAL, COMMUNICATIONS & UTIL.	99.9	44.0	70.3	101.7	44.8	71.7	104.8	51.8	76.8	98.0	51.8	73.6	101.3	75.1	87.5	99.0	44.8	70.4
MF2014	WEIGHTED AVERAGE	97.8	48.4	76.3	98.5	48.8	76.8	101.7	58.4	82.8	100.4	59.5	82.6	100.0	57.7	81.5	99.0	49.1	77.2

DIVISION		TENNESSEE																	
		CHATTANOOGA			COLUMBIA			COOKEVILLE			JACKSON			JOHNSON CITY			KNOXVILLE		
		373 - 374			384			385			383			376			377 - 379		
		MAT.	INST.	TOTAL	MAT.	INST.	TOTAL	MAT.	INST.	TOTAL	MAT.	INST.	TOTAL	MAT.	INST.	TOTAL	MAT.	INST.	TOTAL
015433	CONTRACTOR EQUIPMENT		103.8	103.8		98.2	98.2		98.2	98.2		104.5	104.5		97.7	97.7		97.7	97.7
0241, 31 - 34	SITE & INFRASTRUCTURE, DEMOLITION	104.5	97.9	99.8	89.1	86.9	87.5	94.5	86.4	88.8	97.7	96.8	97.1	111.1	86.6	93.7	90.7	87.3	88.3
0310	Concrete Forming & Accessories	97.1	56.4	62.0	82.1	62.0	64.8	82.2	35.4	41.8	89.0	44.9	51.0	83.7	38.9	45.1	95.6	61.0	65.7
0320	Concrete Reinforcing	91.2	67.4	79.1	87.4	61.2	74.1	87.4	61.2	74.1	87.4	62.0	74.4	91.7	60.3	75.7	91.2	62.3	76.5
0330	Cast-in-Place Concrete	101.3	62.7	85.5	94.2	51.6	76.7	106.7	43.5	80.7	104.2	49.3	81.6	81.6	60.2	72.8	95.2	65.7	83.1
03	CONCRETE	96.0	62.3	79.4	96.7	59.8	78.6	106.9	45.2	76.6	97.7	51.6	75.1	103.3	52.3	78.3	93.4	64.4	79.1
04	MASONRY	109.7	51.7	73.5	127.7	54.6	82.1	121.9	43.8	73.2	127.9	47.2	77.6	125.7	44.8	75.3	86.6	55.3	67.1
05	METALS	98.7	89.2	95.8	96.4	86.0	93.2	96.5	85.9	93.2	98.8	85.9	94.8	95.9	85.2	92.6	99.3	87.0	95.5
06	WOOD, PLASTICS & COMPOSITES	104.0	57.3	77.8	70.1	64.4	66.9	70.3	33.7	49.8	85.4	45.4	63.0	76.9	36.9	54.5	91.0	60.6	73.9
07	THERMAL & MOISTURE PROTECTION	99.4	59.9	83.2	93.6	59.2	79.5	94.0	45.7	74.2	96.0	54.1	78.8	94.5	52.6	77.3	92.5	61.6	79.8
08	OPENINGS	99.7	57.6	89.9	92.6	57.9	84.6	92.7	42.2	80.9	100.3	51.6	89.0	96.2	45.1	84.3	93.3	58.6	85.2
0920	Plaster & Gypsum Board	82.8	56.5	65.0	87.7	63.7	71.5	87.7	32.1	50.1	89.3	44.1	58.8	102.4	35.5	57.2	109.0	59.8	75.8
0950, 0980	Ceilings & Acoustic Treatment	96.6	56.5	70.2	77.0	63.7	68.3	77.0	32.1	47.5	86.2	44.1	58.5	93.0	35.5	55.2	93.8	59.8	71.5
0960	Flooring	99.1	58.9	87.7	89.3	20.5	69.7	89.4	58.6	80.6	87.1	40.8	73.9	96.7	40.6	80.7	101.3	55.8	88.3
0970, 0990	Wall Finishes & Painting/Coating	102.5	70.7	83.3	94.0	33.4	57.4	94.0	36.5	59.3	95.8	49.6	67.9	99.6	42.1	64.9	99.6	81.2	88.5
09	FINISHES	97.2	58.2	75.7	89.8	52.0	69.0	90.3	38.4	61.7	89.7	43.7	64.4	101.9	38.1	66.7	94.5	62.0	76.6
COVERS	DIVS. 10 - 14, 25, 28, 41, 43, 44, 46	100.0	41.8	88.3	100.0	47.9	89.5	100.0	41.0	88.1	100.0	43.0	88.5	100.0	75.5	95.1	100.0	81.7	96.3
21, 22, 23	FIRE SUPPRESSION, PLUMBING & HVAC	100.2	62.6	85.0	97.4	78.1	89.6	97.4	72.6	87.4	100.1	68.0	87.1	99.9	59.1	83.4	99.9	67.9	87.0
26, 27, 3370	ELECTRICAL, COMMUNICATIONS & UTIL.	102.3	69.2	84.8	94.2	57.4	74.7	95.9	61.6	77.8	101.0	58.5	78.5	92.5	46.3	68.1	98.1	56.0	75.8
MF2014	WEIGHTED AVERAGE	100.0	66.2	85.2	96.9	66.1	83.5	98.1	59.1	81.1	99.9	61.2	83.0	99.8	56.5	80.9	96.6	66.9	83.7

DIVISION		TENNESSEE									TEXAS								
		MCKENZIE			MEMPHIS			NASHVILLE			ABILENE			AMARILLO			AUSTIN		
		382			375,380 - 381			370 - 372			795 - 796			790 - 791			786 - 787		
		MAT.	INST.	TOTAL	MAT.	INST.	TOTAL	MAT.	INST.	TOTAL	MAT.	INST.	TOTAL	MAT.	INST.	TOTAL	MAT.	INST.	TOTAL
015433	CONTRACTOR EQUIPMENT		98.2	98.2		102.8	102.8		103.7	103.7		89.9	89.9		89.9	89.9		89.4	89.4
0241, 31 - 34	SITE & INFRASTRUCTURE, DEMOLITION	94.3	86.5	88.7	94.1	93.7	93.8	98.9	97.5	97.9	99.4	88.7	91.8	97.3	88.6	91.1	102.4	88.4	92.4
0310	Concrete Forming & Accessories	90.0	38.1	45.2	94.3	64.9	69.0	98.9	65.8	70.3	99.0	65.5	70.1	97.9	57.1	62.7	98.6	59.1	64.5
0320	Concrete Reinforcing	87.5	61.9	74.5	99.7	68.7	83.9	96.0	67.7	81.6	92.9	52.5	72.3	99.0	51.3	74.7	96.5	48.5	72.1
0330	Cast-in-Place Concrete	104.4	57.3	85.1	94.1	61.3	80.6	90.3	65.9	80.3	96.6	63.7	83.1	94.1	63.7	81.6	93.4	66.3	82.3
03	CONCRETE	105.5	51.3	78.9	94.1	65.8	80.2	92.4	67.6	80.2	95.0	63.0	79.3	96.7	59.0	78.2	95.7	60.2	78.3
04	MASONRY	126.2	46.2	76.3	99.8	62.7	76.7	93.8	59.0	72.1	104.2	63.6	78.9	109.7	63.6	81.0	107.6	56.5	75.7
05	METALS	96.5	86.1	93.3	100.1	90.6	97.2	100.4	89.9	97.2	105.3	69.8	94.4	101.6	69.2	91.7	100.5	66.3	90.0
06	WOOD, PLASTICS & COMPOSITES	79.1	36.3	55.1	95.4	67.0	79.5	101.6	66.6	82.0	101.1	68.5	82.8	100.8	57.1	76.4	94.4	59.9	75.0
07	THERMAL & MOISTURE PROTECTION	94.0	48.7	75.4	95.5	63.4	82.4	95.7	61.5	81.7	100.0	67.7	86.7	97.9	65.3	84.5	104.2	64.8	88.1
08	OPENINGS	92.7	44.1	81.4	100.1	66.0	92.2	98.5	65.9	90.9	97.1	63.3	89.2	101.1	56.8	90.8	104.3	55.6	92.9
0920	Plaster & Gypsum Board	90.5	34.8	52.8	95.6	66.3	75.8	95.8	66.0	75.6	84.0	68.0	73.2	92.1	56.3	67.9	86.7	59.2	68.1
0950, 0980	Ceilings & Acoustic Treatment	77.0	34.8	49.2	93.9	66.3	75.8	95.5	66.0	76.1	88.9	68.0	75.2	98.0	56.3	70.6	87.9	59.2	69.0
0960	Flooring	92.1	39.3	77.0	99.4	36.1	81.3	97.7	64.2	88.2	110.9	76.5	101.1	106.5	76.5	97.9	105.7	64.1	93.8
0970, 0990	Wall Finishes & Painting/Coating	94.0	49.6	67.2	97.0	55.7	72.1	100.5	72.0	83.3	106.6	56.1	76.1	100.8	56.1	73.8	104.3	48.8	70.8
09	FINISHES	91.3	38.2	62.1	97.3	58.5	75.9	100.2	66.0	81.4	93.7	67.1	79.0	96.8	60.3	76.7	95.6	58.7	75.2
COVERS	DIVS. 10 - 14, 25, 28, 41, 43, 44, 46	100.0	27.4	85.3	100.0	82.4	96.4	100.0	82.9	96.5	100.0	82.2	96.4	100.0	68.6	93.7	100.0	80.5	96.1
21, 22, 23	FIRE SUPPRESSION, PLUMBING & HVAC	97.4	68.3	85.6	100.0	73.9	89.5	100.0	84.0	93.5	100.3	49.0	79.6	100.0	54.9	81.8	100.1	60.1	84.0
26, 27, 3370	ELECTRICAL, COMMUNICATIONS & UTIL.	95.6	62.9	78.3	101.7	65.3	82.4	97.6	63.1	79.4	99.3	48.7	72.5	101.1	64.1	81.6	96.8	63.9	79.4
MF2014	WEIGHTED AVERAGE	98.2	59.2	81.2	99.0	71.3	86.9	98.4	74.1	87.8	99.7	62.2	83.3	100.2	63.2	84.1	100.0	63.6	84.1

DIVISION		TEXAS																	
		BEAUMONT			BROWNWOOD			BRYAN			CHILDRESS			CORPUS CHRISTI			DALLAS		
		776 - 777			768			778			792			783 - 784			752 - 753		
		MAT.	INST.	TOTAL	MAT.	INST.	TOTAL	MAT.	INST.	TOTAL	MAT.	INST.	TOTAL	MAT.	INST.	TOTAL	MAT.	INST.	TOTAL
015433	CONTRACTOR EQUIPMENT		91.5	91.5		89.9	89.9		91.5	91.5		89.9	89.9		96.6	96.6		99.1	99.1
0241, 31 - 34	SITE & INFRASTRUCTURE, DEMOLITION	90.0	89.9	89.9	105.2	89.6	94.1	81.8	90.5	88.0	109.7	88.0	94.3	140.1	84.4	100.5	104.3	89.8	94.0
0310	Concrete Forming & Accessories	101.8	61.9	67.4	99.0	60.5	65.8	81.0	67.1	69.0	97.3	65.5	69.8	100.2	58.1	63.9	101.2	65.8	70.7
0320	Concrete Reinforcing	95.9	65.2	80.3	92.9	52.1	72.1	98.3	48.6	73.0	93.0	52.1	72.2	88.0	48.5	67.9	100.2	52.5	75.9
0330	Cast-in-Place Concrete	89.5	64.7	79.3	101.4	63.6	85.9	71.7	65.4	69.1	99.1	63.7	84.5	109.8	64.4	91.2	98.0	64.6	84.3
03	CONCRETE	96.3	64.2	80.5	101.2	60.7	81.3	80.9	63.7	72.5	104.0	62.9	83.8	100.1	59.9	80.4	99.0	64.1	81.9
04	MASONRY	106.8	63.9	80.0	141.8	56.6	88.7	146.3	60.4	92.7	108.5	56.6	76.2	91.0	56.6	69.6	102.4	56.7	73.9
05	METALS	95.2	75.5	89.1	99.9	69.1	90.5	94.8	70.1	87.2	102.7	69.2	92.4	94.7	77.9	89.5	100.4	80.0	94.1
06	WOOD, PLASTICS & COMPOSITES	106.1	61.8	81.3	99.5	61.9	78.4	73.5	68.5	70.7	100.4	68.5	82.6	116.0	58.4	83.7	104.1	68.6	84.2
07	THERMAL & MOISTURE PROTECTION	101.9	66.7	87.5	98.2	65.2	84.7	93.9	66.3	82.5	100.5	64.8	85.9	104.5	66.2	88.8	93.0	66.4	82.1
08	OPENINGS	94.0	62.3	86.7	94.7	59.7	86.6	97.2	61.6	88.9	94.2	63.3	87.1	109.1	54.7	96.5	99.5	63.5	91.1
0920	Plaster & Gypsum Board	102.3	61.2	74.5	87.8	61.2	69.8	90.3	68.0	75.2	83.6	68.0	73.1	98.4	57.5	70.7	96.7	68.0	77.3
0950, 0980	Ceilings & Acoustic Treatment	98.2	61.2	73.9	80.6	61.2	67.8	91.2	68.0	76.0	87.3	68.0	74.6	92.7	57.5	69.6	97.8	68.0	78.2
0960	Flooring	112.4	73.0	101.1	95.3	64.1	86.3	85.8	64.1	79.6	109.1	64.1	96.2	118.7	64.1	103.1	102.8	64.1	91.7
0970, 0990	Wall Finishes & Painting/Coating	95.1	59.4	73.6	104.2	56.1	75.1	92.5	62.2	74.2	106.6	56.1	76.1	120.1	56.5	81.7	107.0	56.1	76.3
09	FINISHES	93.7	63.4	77.0	87.7	60.7	72.8	82.3	66.2	73.4	94.0	64.6	77.8	105.9	58.7	79.9	100.0	64.7	80.5
COVERS	DIVS. 10 - 14, 25, 28, 41, 43, 44, 46	100.0	83.4	96.7	100.0	81.4	96.3	100.0	82.7	96.5	100.0	82.2	96.4	100.0	82.6	96.5	100.0	82.5	96.5
21, 22, 23	FIRE SUPPRESSION, PLUMBING & HVAC	100.2	64.1	85.6	95.4	50.0	77.1	95.4	66.5	83.8	95.5	54.9	79.1	100.2	58.9	83.5	100.0	62.4	84.9
26, 27, 3370	ELECTRICAL, COMMUNICATIONS & UTIL.	95.2	70.6	82.2	93.4	42.8	66.7	93.4	67.4	79.7	99.2	64.1	80.7	92.8	60.0	75.5	95.2	64.6	79.0
MF2014	WEIGHTED AVERAGE	97.4	68.6	84.9	98.9	59.4	81.7	94.9	68.2	83.3	99.3	64.3	84.0	100.9	63.5	84.6	99.5	67.4	85.5

DIVISION		TEXAS																	
		DEL RIO			DENTON			EASTLAND			EL PASO			FORT WORTH			GALVESTON		
		788			762			764			798 - 799,885			760 - 761			775		
		MAT.	INST.	TOTAL	MAT.	INST.	TOTAL	MAT.	INST.	TOTAL	MAT.	INST.	TOTAL	MAT.	INST.	TOTAL	MAT.	INST.	TOTAL
015433	CONTRACTOR EQUIPMENT		89.4	89.4		95.8	95.8		89.9	89.9		89.9	89.9		89.9	89.9		100.5	100.5
0241, 31 - 34	SITE & INFRASTRUCTURE, DEMOLITION	121.8	88.4	98.0	105.5	81.3	88.3	108.0	87.6	93.5	101.7	87.6	91.7	102.1	88.7	92.6	107.2	88.4	93.8
0310	Concrete Forming & Accessories	96.5	57.8	63.1	106.2	65.3	70.9	99.8	65.3	70.0	97.8	63.4	68.1	98.0	65.6	70.1	89.8	64.6	68.1
0320	Concrete Reinforcing	88.7	48.4	68.2	94.3	52.1	72.8	93.1	50.9	71.6	99.2	52.0	75.2	98.4	52.4	75.0	97.8	62.4	79.8
0330	Cast-in-Place Concrete	118.2	63.2	95.6	78.5	64.6	72.8	107.2	63.6	89.3	88.4	63.7	78.3	95.0	63.7	82.1	95.2	66.4	83.4
03	CONCRETE	121.3	58.6	90.5	80.0	63.9	72.1	105.9	62.6	84.6	94.0	61.9	78.3	97.0	63.1	80.4	97.5	66.4	82.2
04	MASONRY	106.9	56.5	75.5	152.1	56.7	92.6	106.8	56.6	75.5	99.0	57.7	73.3	104.7	56.6	74.7	102.5	60.4	76.3
05	METALS	94.4	66.2	85.7	99.5	80.6	93.7	99.7	68.5	90.1	98.5	67.4	89.0	101.5	69.5	91.7	96.3	88.3	93.9
06	WOOD, PLASTICS & COMPOSITES	96.5	58.3	75.1	111.3	68.6	87.4	106.1	68.5	85.1	91.3	65.7	76.9	97.3	68.5	81.2	89.4	64.7	75.6
07	THERMAL & MOISTURE PROTECTION	101.3	65.3	86.5	96.0	66.7	84.0	98.6	65.9	85.1	99.8	64.4	85.3	96.0	65.9	83.6	93.0	67.1	82.4
08	OPENINGS	101.8	54.7	90.8	112.2	63.4	100.8	70.4	63.0	68.7	93.3	58.1	85.1	99.4	63.4	91.0	101.6	63.2	92.7
0920	Plaster & Gypsum Board	94.7	57.5	69.6	92.1	68.0	75.8	87.8	68.0	74.4	95.9	65.1	75.1	91.8	68.0	75.7	96.8	64.0	74.7
0950, 0980	Ceilings & Acoustic Treatment	89.1	57.5	68.3	84.7	68.0	73.8	80.6	68.0	72.3	91.4	65.1	74.1	90.5	68.0	75.7	94.6	64.0	74.5
0960	Flooring	100.7	64.1	90.2	90.4	64.1	82.9	121.4	64.1	105.0	110.4	64.1	97.1	117.9	64.1	102.5	100.1	64.1	89.8
0970, 0990	Wall Finishes & Painting/Coating	106.4	48.8	71.6	115.6	56.1	79.7	105.8	56.1	75.8	106.2	53.7	74.5	103.2	56.1	74.7	103.4	60.0	77.2
09	FINISHES	98.2	57.7	75.9	85.8	64.7	74.2	95.8	64.6	78.6	97.3	63.0	78.4	98.4	64.6	79.8	91.4	63.7	76.2
COVERS	DIVS. 10 - 14, 25, 28, 41, 43, 44, 46	100.0	80.5	96.1	100.0	82.5	96.5	100.0	82.2	96.4	100.0	81.0	96.2	100.0	82.3	96.4	100.0	84.5	96.9
21, 22, 23	FIRE SUPPRESSION, PLUMBING & HVAC	95.4	57.3	80.0	95.4	42.6	74.1	95.4	49.9	77.1	100.0	50.3	80.0	100.0	58.4	83.2	95.4	66.7	83.8
26, 27, 3370	ELECTRICAL, COMMUNICATIONS & UTIL.	94.9	68.8	81.1	95.9	64.6	79.3	93.3	64.5	78.1	99.1	57.4	77.0	94.7	64.5	78.8	95.0	67.9	80.7
MF2014	WEIGHTED AVERAGE	100.8	63.2	84.4	98.9	62.5	83.0	95.9	63.2	81.6	98.0	61.7	82.2	99.3	65.3	84.5	97.0	70.0	85.2

DIVISION		TEXAS																	
		GIDDINGS			GREENVILLE			HOUSTON			HUNTSVILLE			LAREDO			LONGVIEW		
		789			754			770 - 772			773			780			756		
		MAT.	INST.	TOTAL	MAT.	INST.	TOTAL	MAT.	INST.	TOTAL	MAT.	INST.	TOTAL	MAT.	INST.	TOTAL	MAT.	INST.	TOTAL
015433	CONTRACTOR EQUIPMENT		89.4	89.4		96.7	96.7		100.4	100.4		91.5	91.5		89.4	89.4		91.4	91.4
0241, 31 - 34	SITE & INFRASTRUCTURE, DEMOLITION	107.8	88.4	94.0	97.9	85.4	89.0	106.0	88.2	93.4	96.6	89.9	91.8	102.1	88.4	92.3	95.8	92.3	93.3
0310	Concrete Forming & Accessories	94.1	57.9	62.9	92.3	65.3	69.0	91.8	64.6	68.4	87.8	61.6	65.2	96.6	58.0	63.3	88.4	64.9	68.1
0320	Concrete Reinforcing	89.2	45.9	67.2	100.7	52.1	76.0	97.5	61.4	79.1	98.5	50.6	74.1	88.7	48.8	68.4	99.6	52.0	75.4
0330	Cast-in-Place Concrete	100.2	63.2	85.0	89.3	63.9	78.9	92.3	66.4	81.7	98.6	64.6	84.6	84.6	63.3	75.8	103.8	62.8	86.9
03	CONCRETE	97.6	58.2	78.2	91.6	63.6	77.8	95.2	66.2	80.9	103.9	61.3	83.0	92.5	58.8	75.9	107.5	62.3	85.3
04	MASONRY	115.7	56.5	78.8	160.0	56.6	95.6	102.3	66.5	80.0	145.1	58.9	91.4	100.2	63.5	77.3	155.7	56.6	93.9
05	METALS	93.9	65.3	85.1	97.7	79.3	92.1	99.2	88.2	95.8	94.7	69.2	86.8	97.0	66.9	87.8	91.1	67.8	83.9
06	WOOD, PLASTICS & COMPOSITES	95.6	58.3	74.7	93.5	68.6	79.5	91.7	64.7	76.6	82.0	61.8	70.7	96.5	58.3	75.1	87.2	68.4	76.7
07	THERMAL & MOISTURE PROTECTION	102.0	64.5	86.6	93.0	66.0	81.9	92.6	67.8	82.4	94.9	65.9	83.0	100.2	66.5	86.4	94.5	65.2	82.5
08	OPENINGS	100.8	54.0	89.9	97.7	63.4	89.7	104.3	62.8	94.6	97.2	58.5	88.2	101.6	54.7	90.7	87.8	63.3	82.1
0920	Plaster & Gypsum Board	93.7	57.5	69.2	90.3	68.0	75.3	99.1	64.0	75.4	94.2	61.2	71.9	95.8	57.5	69.9	89.1	68.0	74.9
0950, 0980	Ceilings & Acoustic Treatment	89.1	57.5	68.3	93.6	68.0	76.8	99.6	64.0	76.2	91.2	61.2	71.5	93.3	57.5	69.8	90.3	68.0	75.7
0960	Flooring	101.1	64.1	90.5	98.6	64.1	88.7	101.2	64.1	90.6	89.4	64.1	82.1	100.5	64.1	90.1	103.5	64.1	92.2
0970, 0990	Wall Finishes & Painting/Coating	106.4	48.8	71.6	107.0	56.1	76.3	103.4	62.2	78.5	92.5	59.4	72.5	106.4	53.2	74.3	96.8	56.1	72.2
09	FINISHES	97.1	57.7	75.4	96.6	64.6	79.0	98.9	64.0	79.7	84.7	61.6	72.0	97.7	58.2	75.9	101.7	64.5	81.2
COVERS	DIVS. 10 - 14, 25, 28, 41, 43, 44, 46	100.0	80.8	96.1	100.0	82.3	96.4	100.0	84.5	96.9	100.0	81.0	96.2	100.0	80.5	96.1	100.0	82.0	96.4
21, 22, 23	FIRE SUPPRESSION, PLUMBING & HVAC	95.4	64.6	83.0	95.3	49.0	76.6	100.1	66.7	86.6	95.4	65.9	83.5	100.2	64.1	85.6	95.3	33.6	70.4
26, 27, 3370	ELECTRICAL, COMMUNICATIONS & UTIL.	91.8	63.9	77.1	92.0	64.6	77.5	96.9	67.5	81.4	93.4	68.0	80.0	95.1	60.9	77.0	92.2	55.6	72.9
MF2014	WEIGHTED AVERAGE	97.7	64.0	83.0	98.8	64.0	83.6	99.4	70.5	86.8	98.0	66.7	84.4	98.3	64.5	83.6	98.5	58.7	81.2

DIVISION		TEXAS																	
		LUBBOCK			LUFKIN			MCALLEN			MCKINNEY			MIDLAND			ODESSA		
		793 - 794			759			785			750			797			797		
		MAT.	INST.	TOTAL	MAT.	INST.	TOTAL	MAT.	INST.	TOTAL	MAT.	INST.	TOTAL	MAT.	INST.	TOTAL	MAT.	INST.	TOTAL
015433	CONTRACTOR EQUIPMENT		98.5	98.5		91.4	91.4		96.8	96.8		96.7	96.7		98.5	98.5		89.9	89.9
0241, 31 - 34	SITE & INFRASTRUCTURE, DEMOLITION	123.8	86.1	97.0	91.2	94.0	93.1	144.1	84.4	101.7	94.6	85.3	88.0	126.7	86.1	97.8	99.7	88.0	91.4
0310	Concrete Forming & Accessories	97.9	57.2	62.8	91.7	61.7	65.9	101.2	57.5	63.5	91.3	65.3	68.8	101.9	65.4	70.5	98.9	65.4	70.0
0320	Concrete Reinforcing	94.1	52.5	72.9	101.3	64.9	82.8	88.2	48.4	67.9	100.7	50.9	75.3	95.1	52.1	73.2	92.9	52.1	72.1
0330	Cast-in-Place Concrete	96.8	64.7	83.6	92.9	64.0	81.0	119.3	64.1	96.6	83.8	63.8	75.6	102.9	64.6	87.2	96.6	63.6	83.1
03	CONCRETE	93.7	60.4	77.4	99.8	63.6	82.0	107.9	59.5	84.1	87.0	63.3	75.3	98.2	64.0	81.4	95.0	62.8	79.2
04	MASONRY	103.6	64.0	78.9	119.4	58.8	81.6	106.8	56.4	75.4	172.2	56.6	100.2	121.5	56.7	81.1	104.2	56.6	74.6
05	METALS	109.0	81.3	100.5	97.9	72.4	90.0	94.3	77.5	89.2	97.6	78.8	91.8	107.2	80.7	99.0	104.6	69.1	93.7
06	WOOD, PLASTICS & COMPOSITES	101.1	57.3	76.5	95.3	61.8	76.5	114.6	58.4	83.1	92.3	68.6	79.0	106.0	68.6	85.1	101.1	68.5	82.8
07	THERMAL & MOISTURE PROTECTION	89.3	66.2	79.8	94.3	65.6	82.5	104.8	59.8	86.3	92.8	66.0	81.8	89.5	65.4	79.7	100.0	65.9	86.0
08	OPENINGS	108.5	57.2	96.6	66.7	63.0	65.9	105.7	54.7	93.9	97.7	63.0	89.6	107.5	63.4	97.2	97.1	63.3	89.2
0920	Plaster & Gypsum Board	84.4	56.3	65.4	88.0	61.2	69.9	99.4	57.5	71.1	90.0	68.0	75.1	85.9	68.0	73.8	84.0	68.0	73.2
0950, 0980	Ceilings & Acoustic Treatment	90.6	56.3	68.1	84.4	61.2	69.1	93.3	57.5	69.8	93.6	68.0	76.8	88.1	68.0	74.9	88.9	68.0	75.2
0960	Flooring	104.4	64.1	92.8	138.2	64.1	117.0	118.1	63.8	102.6	98.2	64.1	88.4	105.7	64.1	93.8	110.9	64.1	97.5
0970, 0990	Wall Finishes & Painting/Coating	119.0	56.1	81.0	96.8	56.1	72.2	120.1	48.8	77.1	107.0	56.1	76.3	119.0	56.1	81.0	106.6	56.1	76.1
09	FINISHES	96.4	58.0	75.2	110.2	61.2	83.2	106.4	57.8	79.6	96.2	64.6	78.8	96.8	64.7	79.1	93.7	64.6	77.7
COVERS	DIVS. 10 - 14, 25, 28, 41, 43, 44, 46	100.0	81.3	96.2	100.0	80.9	96.1	100.0	80.7	96.1	100.0	82.3	96.4	100.0	81.2	96.2	100.0	80.9	96.1
21, 22, 23	FIRE SUPPRESSION, PLUMBING & HVAC	99.8	50.7	80.0	95.3	64.4	82.8	95.4	58.7	80.6	95.3	45.4	75.2	95.0	49.1	76.5	100.3	49.1	79.6
26, 27, 3370	ELECTRICAL, COMMUNICATIONS & UTIL.	97.9	64.2	80.1	93.5	70.6	81.4	92.6	36.3	62.8	92.1	64.6	77.5	97.9	64.2	80.1	99.4	64.1	80.7
MF2014	WEIGHTED AVERAGE	101.6	63.6	85.0	95.6	67.8	83.5	101.0	59.8	83.0	98.7	63.1	83.2	101.5	64.2	85.2	99.6	63.1	83.7

DIVISION		TEXAS																	
		PALESTINE			SAN ANGELO			SAN ANTONIO			TEMPLE			TEXARKANA			TYLER		
		758			769			781 - 782			765			755			757		
		MAT.	INST.	TOTAL	MAT.	INST.	TOTAL	MAT.	INST.	TOTAL	MAT.	INST.	TOTAL	MAT.	INST.	TOTAL	MAT.	INST.	TOTAL
015433	CONTRACTOR EQUIPMENT		91.4	91.4		89.9	89.9		91.9	91.9		89.9	89.9		91.4	91.4		91.4	91.4
0241, 31 - 34	SITE & INFRASTRUCTURE, DEMOLITION	96.6	92.6	93.8	101.5	89.7	93.1	100.9	92.0	94.5	89.8	89.1	89.3	85.9	92.0	90.2	95.1	92.4	93.2
0310	Concrete Forming & Accessories	82.7	65.4	67.8	99.3	58.7	64.3	95.0	58.1	63.2	102.8	57.8	64.0	99.0	57.4	63.1	93.6	65.4	69.3
0320	Concrete Reinforcing	98.8	52.1	75.1	92.7	52.1	72.1	93.6	49.8	71.3	92.9	49.1	70.6	98.7	52.3	75.1	99.6	52.4	75.6
0330	Cast-in-Place Concrete	84.9	63.0	75.9	95.7	64.8	83.0	85.6	65.2	77.2	78.5	63.1	72.2	85.5	62.7	76.2	101.9	63.0	85.9
03	CONCRETE	101.6	62.6	82.4	96.7	60.3	78.8	91.7	59.7	76.0	82.9	58.7	71.0	92.8	59.0	76.2	107.1	62.7	85.3
04	MASONRY	113.8	56.6	78.2	138.0	58.7	88.5	96.8	65.3	77.2	151.0	56.4	92.1	176.5	56.6	101.8	166.0	56.6	97.8
05	METALS	97.6	68.3	88.6	100.1	69.1	90.6	96.9	68.6	88.2	99.8	66.5	89.6	91.0	67.8	83.9	97.5	68.4	88.5
06	WOOD, PLASTICS & COMPOSITES	85.3	68.4	75.8	99.8	58.3	76.5	94.7	57.4	73.8	108.9	58.3	80.6	99.7	58.2	76.5	97.0	68.4	81.0
07	THERMAL & MOISTURE PROTECTION	94.8	65.3	82.7	98.0	66.0	84.9	96.4	68.2	84.8	97.7	64.8	84.2	94.1	64.2	81.9	94.6	65.3	82.6
08	OPENINGS	66.7	63.3	65.9	94.7	57.8	86.1	102.7	54.4	91.5	67.1	57.0	64.7	87.7	57.8	80.8	66.6	63.4	65.9
0920	Plaster & Gypsum Board	85.2	68.0	73.6	87.8	57.5	67.3	98.0	56.5	70.0	87.8	57.5	67.3	93.4	57.5	69.1	88.0	68.0	74.5
0950, 0980	Ceilings & Acoustic Treatment	84.4	68.0	73.6	80.6	57.5	65.4	102.7	56.5	72.3	80.6	57.5	65.4	90.3	57.5	68.7	84.4	68.0	73.6
0960	Flooring	129.9	64.1	111.1	95.3	64.1	86.4	104.1	64.1	92.6	123.2	64.1	106.3	111.1	64.1	97.6	140.1	64.1	118.4
0970, 0990	Wall Finishes & Painting/Coating	96.8	56.1	72.2	104.2	56.1	75.1	106.3	53.2	74.2	105.8	48.8	71.4	96.8	56.1	72.2	96.8	56.1	72.2
09	FINISHES	108.0	64.5	84.0	87.5	59.1	71.8	103.4	58.1	78.4	95.1	57.7	74.5	104.0	58.5	78.9	111.1	64.5	85.4
COVERS	DIVS. 10 - 14, 25, 28, 41, 43, 44, 46	100.0	82.0	96.4	100.0	81.0	96.2	100.0	81.2	96.2	100.0	81.1	96.2	100.0	80.9	96.1	100.0	82.0	96.4
21, 22, 23	FIRE SUPPRESSION, PLUMBING & HVAC	95.3	59.6	80.9	95.4	50.1	77.1	100.0	63.2	85.1	95.4	54.7	79.0	95.3	37.0	71.8	95.3	62.4	82.0
26, 27, 3370	ELECTRICAL, COMMUNICATIONS & UTIL.	89.8	55.6	71.7	97.4	44.8	69.6	96.4	60.9	77.6	94.5	59.8	76.2	93.4	60.1	75.8	92.2	55.5	72.8
MF2014	WEIGHTED AVERAGE	95.0	64.4	81.7	98.5	59.6	81.5	98.5	65.1	84.0	94.7	61.7	80.3	98.0	58.4	80.7	98.5	65.0	83.9

DIVISION		TEXAS															UTAH		
		VICTORIA			WACO			WAXAHACKIE			WHARTON			WICHITA FALLS			LOGAN		
		779			766 - 767			751			774			763			843		
		MAT.	INST.	TOTAL	MAT.	INST.	TOTAL	MAT.	INST.	TOTAL	MAT.	INST.	TOTAL	MAT.	INST.	TOTAL	MAT.	INST.	TOTAL
015433	CONTRACTOR EQUIPMENT		99.3	99.3		89.9	89.9		96.7	96.7		100.5	100.5		89.9	89.9		97.5	97.5
0241, 31 - 34	SITE & INFRASTRUCTURE, DEMOLITION	111.8	85.7	93.2	98.6	88.7	91.6	96.0	85.4	88.4	117.1	87.8	96.2	99.3	88.7	91.8	96.0	95.7	95.8
0310	Concrete Forming & Accessories	90.0	59.0	63.3	101.2	65.5	70.4	91.3	65.5	69.0	84.9	61.7	64.9	101.2	65.5	70.4	104.7	58.1	64.5
0320	Concrete Reinforcing	93.9	48.5	70.8	92.5	48.5	70.1	100.7	50.9	75.4	97.7	50.6	73.7	92.5	51.3	71.5	99.6	81.3	90.3
0330	Cast-in-Place Concrete	106.9	65.6	89.9	84.9	66.8	77.4	88.4	63.9	78.3	109.8	65.5	91.6	90.7	63.7	79.6	87.3	73.4	81.6
03	CONCRETE	104.9	60.9	83.3	89.7	63.3	76.7	90.6	63.4	77.2	109.1	62.5	86.2	92.4	62.8	77.9	106.0	68.3	87.5
04	MASONRY	120.0	58.9	81.9	105.2	56.6	74.9	160.7	56.6	95.8	103.6	58.9	75.8	105.7	63.6	79.5	105.8	63.2	79.3
05	METALS	94.9	80.7	90.5	102.2	67.8	91.6	97.7	79.0	92.0	96.3	81.6	91.8	102.2	69.8	92.2	102.2	79.0	95.0
06	WOOD, PLASTICS & COMPOSITES	92.5	58.4	73.4	107.1	68.5	85.5	92.3	68.6	79.0	83.1	62.0	71.3	107.1	68.5	85.5	85.3	54.9	68.3
07	THERMAL & MOISTURE PROTECTION	96.8	66.9	84.6	98.5	65.8	85.1	92.9	66.0	81.9	93.4	66.6	82.4	98.5	67.7	85.9	97.4	67.9	85.3
08	OPENINGS	101.4	56.1	90.8	77.7	62.4	74.1	97.7	63.0	89.6	101.6	58.6	91.6	77.7	63.3	74.4	94.3	56.5	85.5
0920	Plaster & Gypsum Board	93.8	57.5	69.3	88.2	68.0	74.6	90.4	68.0	75.3	92.5	61.2	71.3	88.2	68.0	74.6	78.9	53.5	61.7
0950, 0980	Ceilings & Acoustic Treatment	95.4	57.5	70.5	82.2	68.0	72.9	95.3	68.0	77.4	94.6	61.2	72.6	82.2	68.0	72.9	97.2	53.5	68.5
0960	Flooring	99.5	64.1	89.4	122.4	64.1	105.8	98.2	64.1	88.4	97.7	64.1	88.1	123.4	76.5	110.0	101.2	47.7	85.9
0970, 0990	Wall Finishes & Painting/Coating	103.6	59.4	76.9	105.8	48.8	71.4	107.0	56.1	76.3	103.4	59.4	76.8	108.5	56.1	76.8	103.3	60.7	77.5
09	FINISHES	89.4	59.6	73.0	95.8	63.8	78.1	96.8	64.6	79.1	90.8	61.7	74.8	96.3	67.1	80.2	96.2	55.3	73.6
COVERS	DIVS. 10 - 14, 25, 28, 41, 43, 44, 46	100.0	80.8	96.1	100.0	82.3	96.4	100.0	82.3	96.4	100.0	81.3	96.2	100.0	69.8	93.9	100.0	85.1	97.0
21, 22, 23	FIRE SUPPRESSION, PLUMBING & HVAC	95.4	65.9	83.5	100.2	54.7	81.8	95.3	58.5	80.5	95.4	65.0	83.1	100.2	54.2	81.6	99.9	69.5	87.6
26, 27, 3370	ELECTRICAL, COMMUNICATIONS & UTIL.	99.8	57.6	77.5	97.6	63.6	79.6	92.1	64.6	77.5	98.6	68.0	82.4	99.3	60.1	78.6	97.5	73.0	84.6
MF2014	WEIGHTED AVERAGE	99.0	65.6	84.4	96.5	64.1	82.4	98.6	65.9	84.4	98.9	67.7	85.3	97.1	64.4	82.8	99.8	70.1	86.9

	DIVISION	UTAH												VERMONT					
		OGDEN			PRICE			PROVO			SALT LAKE CITY			BELLOWS FALLS			BENNINGTON		
		842,844			845			846 - 847			840 - 841			051			052		
		MAT.	INST.	TOTAL	MAT.	INST.	TOTAL	MAT.	INST.	TOTAL	MAT.	INST.	TOTAL	MAT.	INST.	TOTAL	MAT.	INST.	TOTAL
015433	CONTRACTOR EQUIPMENT		97.5	97.5		96.6	96.6		96.6	96.6		97.5	97.5		100.9	100.9		100.9	100.9
0241, 31 - 34	SITE & INFRASTRUCTURE, DEMOLITION	84.9	95.7	92.6	93.3	93.8	93.7	92.3	94.1	93.6	84.5	95.6	92.4	88.2	100.5	96.9	87.5	100.5	96.7
0310	Concrete Forming & Accessories	104.7	58.1	64.5	107.1	48.3	56.4	106.3	58.0	64.7	107.1	58.0	64.8	98.3	85.5	87.3	95.9	106.6	105.2
0320	Concrete Reinforcing	99.2	81.3	90.1	106.8	81.1	93.7	107.7	81.3	94.3	101.5	81.3	91.2	83.2	85.7	84.5	83.2	85.7	84.5
0330	Cast-in-Place Concrete	88.7	73.4	82.4	87.4	59.5	75.9	87.4	73.3	81.6	96.9	73.3	87.2	87.0	110.7	96.7	87.0	110.7	96.7
03	CONCRETE	95.9	68.3	82.3	107.3	59.0	83.6	105.9	68.2	87.4	114.8	68.2	91.9	91.0	94.2	92.6	90.8	103.6	97.1
04	MASONRY	99.7	63.2	77.0	110.9	61.3	80.0	111.1	63.2	81.2	113.1	63.2	82.0	103.6	95.2	98.4	112.4	95.2	101.6
05	METALS	102.7	79.0	95.4	99.4	76.6	92.4	100.3	78.9	93.7	106.2	78.9	97.8	94.8	88.3	92.8	94.7	88.2	92.7
06	WOOD, PLASTICS & COMPOSITES	85.3	54.9	68.3	88.6	44.3	63.8	86.8	54.9	68.9	87.2	54.9	69.1	103.7	83.5	92.4	100.6	112.4	107.2
07	THERMAL & MOISTURE PROTECTION	96.4	67.9	84.7	99.0	59.9	82.9	99.0	67.9	86.2	103.4	67.9	88.9	97.0	82.7	91.1	97.0	81.1	90.4
08	OPENINGS	94.3	56.5	85.5	98.2	48.9	86.7	98.2	56.5	88.5	96.2	56.5	86.9	105.0	86.5	100.7	105.0	102.3	104.3
0920	Plaster & Gypsum Board	78.9	53.5	61.7	81.7	42.5	55.2	79.2	53.5	61.8	90.5	53.5	65.5	106.0	82.4	90.1	104.2	112.2	109.6
0950, 0980	Ceilings & Acoustic Treatment	97.2	53.5	68.5	97.2	42.5	61.3	97.2	53.5	68.5	91.5	53.5	66.5	91.5	82.4	85.6	91.5	112.2	105.1
0960	Flooring	99.0	47.7	84.3	102.3	35.5	83.2	102.0	47.7	86.5	103.2	47.7	87.3	100.6	104.6	101.8	99.7	104.6	101.1
0970, 0990	Wall Finishes & Painting/Coating	103.3	60.7	77.5	103.3	37.5	63.6	103.3	62.6	78.7	106.7	62.6	80.1	97.4	105.8	102.5	97.4	105.8	102.5
09	FINISHES	94.3	55.3	72.8	97.3	43.5	67.7	96.8	55.5	74.0	96.6	55.5	74.0	95.9	90.7	93.0	95.4	107.8	102.2
COVERS	DIVS. 10 - 14, 25, 28, 41, 43, 44, 46	100.0	85.1	97.0	100.0	45.8	89.0	100.0	85.0	97.0	100.0	85.0	97.0	100.0	97.6	99.5	100.0	100.7	100.1
21, 22, 23	FIRE SUPPRESSION, PLUMBING & HVAC	99.9	69.5	87.6	97.5	67.7	85.5	99.9	69.4	87.6	100.1	69.4	87.7	95.3	96.4	95.7	95.3	96.4	95.7
26, 27, 3370	ELECTRICAL, COMMUNICATIONS & UTIL.	97.9	73.0	84.7	103.5	73.0	87.4	98.4	59.8	78.0	100.5	59.8	79.0	101.5	89.3	95.1	101.5	61.2	80.2
MF2014	WEIGHTED AVERAGE	98.1	70.1	85.9	100.2	64.6	84.7	100.3	68.2	86.3	102.2	68.3	87.4	97.2	93.0	95.3	97.5	93.5	95.8

	DIVISION	VERMONT																	
		BRATTLEBORO			BURLINGTON			GUILDHALL			MONTPELIER			RUTLAND			ST. JOHNSBURY		
		053			054			059			056			057			058		
		MAT.	INST.	TOTAL	MAT.	INST.	TOTAL	MAT.	INST.	TOTAL	MAT.	INST.	TOTAL	MAT.	INST.	TOTAL	MAT.	INST.	TOTAL
015433	CONTRACTOR EQUIPMENT		100.9	100.9		100.9	100.9		100.9	100.9		100.9	100.9		100.9	100.9		100.9	100.9
0241, 31 - 34	SITE & INFRASTRUCTURE, DEMOLITION	88.9	100.5	97.1	93.0	100.4	98.2	87.2	99.0	95.6	91.5	100.1	97.6	91.7	100.1	97.7	87.3	99.0	95.6
0310	Concrete Forming & Accessories	98.6	85.4	87.2	98.0	87.6	89.0	96.1	78.3	80.7	97.9	84.1	86.0	98.9	87.6	89.1	94.4	78.3	80.5
0320	Concrete Reinforcing	82.3	85.7	84.0	101.1	85.6	93.2	83.9	85.6	84.8	94.1	85.6	89.8	103.9	85.6	94.6	82.3	85.6	84.0
0330	Cast-in-Place Concrete	89.7	110.7	98.3	100.9	109.6	104.5	84.3	101.0	91.2	100.9	109.6	104.5	85.2	109.6	95.2	84.3	101.0	91.2
03	CONCRETE	93.0	94.1	93.6	101.2	94.7	98.0	88.6	87.6	88.1	100.1	93.1	96.7	94.3	94.7	94.5	88.2	87.6	87.9
04	MASONRY	111.5	95.2	101.3	115.7	93.7	102.0	111.8	78.2	90.8	109.6	93.7	99.7	93.9	93.7	93.8	138.8	78.2	101.0
05	METALS	94.7	88.2	92.7	103.5	87.5	98.6	94.8	87.3	92.5	99.6	87.4	95.9	100.6	87.4	96.5	94.8	87.3	92.5
06	WOOD, PLASTICS & COMPOSITES	104.0	83.5	92.5	99.9	88.2	93.3	100.0	83.5	90.8	97.7	83.5	89.7	104.2	88.2	95.2	94.8	83.5	88.5
07	THERMAL & MOISTURE PROTECTION	97.1	78.0	89.3	103.8	80.4	94.2	96.8	70.7	86.1	103.3	79.9	93.7	97.2	80.4	90.3	96.7	70.7	86.1
08	OPENINGS	105.0	86.5	100.7	108.4	85.0	103.0	105.0	82.4	99.7	105.8	82.4	100.3	108.4	85.0	102.9	105.0	82.4	99.7
0920	Plaster & Gypsum Board	106.0	82.4	90.1	109.7	87.2	94.5	112.4	82.4	92.2	108.9	82.4	91.0	106.6	87.2	93.5	113.8	82.4	92.6
0950, 0980	Ceilings & Acoustic Treatment	91.5	82.4	85.6	101.0	87.2	91.9	91.5	82.4	85.6	95.1	82.4	86.8	95.9	87.2	90.2	91.5	82.4	85.6
0960	Flooring	100.7	104.6	101.8	102.4	104.6	103.1	103.6	104.6	103.9	103.2	104.6	103.6	100.6	104.6	101.8	106.9	104.6	106.3
0970, 0990	Wall Finishes & Painting/Coating	97.4	105.8	102.5	102.9	85.2	92.2	97.4	85.2	90.0	101.7	85.2	91.7	97.4	85.2	90.0	97.4	85.2	90.0
09	FINISHES	96.0	90.7	93.1	99.7	90.8	94.8	97.5	84.2	90.2	97.9	88.0	92.5	97.1	90.8	93.6	98.7	84.2	90.7
COVERS	DIVS. 10 - 14, 25, 28, 41, 43, 44, 46	100.0	97.6	99.5	100.0	97.5	99.5	100.0	91.8	98.3	100.0	97.0	99.4	100.0	97.5	99.5	100.0	91.8	98.3
21, 22, 23	FIRE SUPPRESSION, PLUMBING & HVAC	95.3	96.4	95.7	99.9	71.8	88.6	95.3	64.0	82.7	95.1	71.8	85.7	100.1	71.8	88.7	95.3	64.0	82.7
26, 27, 3370	ELECTRICAL, COMMUNICATIONS & UTIL.	101.5	89.3	95.1	102.2	61.2	80.5	101.5	61.2	80.2	99.7	61.2	79.3	101.6	61.2	80.2	101.5	61.2	80.2
MF2014	WEIGHTED AVERAGE	97.8	92.8	95.6	102.4	83.6	94.2	97.4	77.8	88.8	99.5	82.8	92.2	99.7	83.6	92.7	98.6	77.8	89.6

	DIVISION	VERMONT			VIRGINIA														
		WHITE RIVER JCT.			ALEXANDRIA			ARLINGTON			BRISTOL			CHARLOTTESVILLE			CULPEPER		
		050			223			222			242			229			227		
		MAT.	INST.	TOTAL	MAT.	INST.	TOTAL	MAT.	INST.	TOTAL	MAT.	INST.	TOTAL	MAT.	INST.	TOTAL	MAT.	INST.	TOTAL
015433	CONTRACTOR EQUIPMENT		100.9	100.9		103.0	103.0		101.7	101.7		101.7	101.7		106.1	106.1		101.7	101.7
0241, 31 - 34	SITE & INFRASTRUCTURE, DEMOLITION	91.5	99.2	97.0	114.3	89.8	96.9	124.5	87.7	98.3	108.8	86.1	92.6	113.5	88.1	95.4	111.8	87.6	94.6
0310	Concrete Forming & Accessories	93.1	78.9	80.8	92.2	72.8	75.5	91.2	72.0	74.7	87.1	42.5	48.6	85.4	48.7	53.7	82.5	71.4	72.9
0320	Concrete Reinforcing	83.2	85.6	84.4	83.0	87.0	85.1	93.7	84.4	88.9	93.7	68.4	80.8	93.1	72.8	82.7	93.7	84.2	88.9
0330	Cast-in-Place Concrete	89.7	101.9	94.7	105.1	81.8	95.6	102.3	80.3	93.3	101.8	48.5	79.9	105.9	55.5	85.2	104.8	69.2	90.2
03	CONCRETE	94.7	88.2	91.5	102.0	79.6	91.0	106.7	78.3	92.8	102.9	51.2	77.5	103.5	57.2	80.8	100.6	74.1	87.6
04	MASONRY	124.5	79.7	96.6	91.3	72.8	79.8	106.3	70.7	84.1	95.4	51.4	68.0	120.3	53.2	78.5	107.5	70.7	84.6
05	METALS	94.8	87.3	92.5	105.7	97.0	103.0	104.3	97.2	102.1	103.1	85.7	97.8	103.4	90.2	99.3	103.5	95.3	101.0
06	WOOD, PLASTICS & COMPOSITES	97.4	83.5	89.6	95.0	71.0	81.6	91.5	71.0	80.0	83.7	40.0	59.2	82.2	45.6	61.7	80.8	71.0	75.3
07	THERMAL & MOISTURE PROTECTION	97.1	71.4	86.6	102.7	81.5	94.0	104.7	80.6	94.8	104.2	60.1	86.1	103.8	67.7	89.0	104.0	78.1	93.4
08	OPENINGS	105.0	82.4	99.7	100.9	75.2	94.9	99.0	75.2	93.5	102.0	46.5	89.1	100.2	52.0	89.0	100.5	75.2	94.6
0920	Plaster & Gypsum Board	102.8	82.4	89.0	107.4	69.9	82.1	104.2	69.9	81.0	99.7	38.0	58.0	99.7	43.0	61.4	99.9	69.9	79.6
0950, 0980	Ceilings & Acoustic Treatment	91.5	82.4	85.6	91.4	69.9	77.3	89.8	69.9	76.7	88.9	38.0	55.4	88.9	43.0	58.8	89.8	69.9	76.7
0960	Flooring	98.6	104.6	100.4	106.0	82.9	99.4	104.5	81.9	98.0	101.0	66.5	91.1	99.7	66.5	90.2	99.7	81.9	94.6
0970, 0990	Wall Finishes & Painting/Coating	97.4	85.2	90.0	124.0	80.0	97.5	124.0	80.0	97.5	109.6	54.3	76.2	109.6	76.1	89.3	124.0	76.1	95.1
09	FINISHES	95.2	84.6	89.3	100.1	74.6	86.0	100.0	73.8	85.6	96.5	47.2	69.3	96.2	53.5	72.6	96.9	73.2	83.9
COVERS	DIVS. 10 - 14, 25, 28, 41, 43, 44, 46	100.0	92.3	98.4	100.0	88.6	97.7	100.0	87.6	97.5	100.0	67.8	93.5	100.0	79.3	95.8	100.0	87.6	97.5
21, 22, 23	FIRE SUPPRESSION, PLUMBING & HVAC	95.3	64.8	83.0	100.3	86.9	94.9	100.3	85.4	94.3	95.6	50.5	77.4	95.6	69.3	85.0	95.6	72.9	86.4
26, 27, 3370	ELECTRICAL, COMMUNICATIONS & UTIL.	101.5	61.2	80.2	97.1	102.4	99.9	94.7	99.2	97.1	96.8	35.4	64.3	96.7	71.9	83.6	99.3	99.2	99.2
MF2014	WEIGHTED AVERAGE	98.5	78.3	89.7	101.1	85.5	94.3	101.9	84.0	94.1	99.4	54.9	80.0	100.5	67.1	86.0	99.9	80.5	91.4

DIVISION		VIRGINIA																	
		FAIRFAX			FARMVILLE			FREDERICKSBURG			GRUNDY			HARRISONBURG			LYNCHBURG		
		220 - 221			239			224 - 225			246			228			245		
		MAT.	INST.	TOTAL	MAT.	INST.	TOTAL	MAT.	INST.	TOTAL	MAT.	INST.	TOTAL	MAT.	INST.	TOTAL	MAT.	INST.	TOTAL
015433	CONTRACTOR EQUIPMENT		101.7	101.7		106.1	106.1		101.7	101.7		101.7	101.7		101.7	101.7		101.7	101.7
0241, 31 - 34	SITE & INFRASTRUCTURE, DEMOLITION	123.1	87.8	98.0	109.2	87.6	93.8	111.4	87.7	94.5	106.6	84.8	91.1	120.1	86.3	96.1	107.5	86.3	92.4
0310	Concrete Forming & Accessories	85.4	72.0	73.9	98.0	43.7	51.2	85.4	69.4	71.6	90.3	36.4	43.8	81.4	42.6	47.9	87.1	60.3	64.0
0320	Concrete Reinforcing	93.7	87.0	90.3	90.9	49.9	70.0	94.4	87.0	90.6	92.4	50.4	71.0	93.7	64.9	79.0	93.1	68.5	80.5
0330	Cast-in-Place Concrete	102.3	80.6	93.4	103.1	53.5	82.7	103.9	69.3	89.7	101.8	47.5	79.5	102.3	58.0	84.1	101.8	57.7	83.7
03	CONCRETE	106.3	78.8	92.8	101.2	49.9	76.0	100.3	73.8	87.3	101.6	44.4	73.5	104.1	53.8	79.4	101.5	62.3	82.3
04	MASONRY	106.2	71.4	84.5	103.5	44.9	66.9	106.6	70.7	84.2	96.5	47.3	65.8	104.2	50.3	70.6	112.0	53.2	75.4
05	METALS	103.6	97.0	101.5	101.0	77.1	93.6	103.5	96.2	101.3	103.1	69.9	92.9	103.4	83.9	97.4	103.3	86.3	98.1
06	WOOD, PLASTICS & COMPOSITES	83.7	71.0	76.6	96.4	44.1	67.1	83.7	68.1	75.0	86.7	35.1	57.8	79.8	39.0	56.9	83.7	62.6	71.9
07	THERMAL & MOISTURE PROTECTION	104.6	73.7	91.9	104.3	49.3	81.8	104.0	78.5	93.6	104.2	45.3	80.0	104.4	63.9	87.8	104.0	63.9	87.6
08	OPENINGS	99.0	75.2	93.5	100.7	43.0	87.3	100.2	73.6	94.0	102.0	34.4	86.3	100.5	49.4	88.6	100.5	58.8	90.8
0920	Plaster & Gypsum Board	99.9	69.9	79.6	108.6	41.5	63.2	99.9	66.9	77.6	99.7	32.9	54.5	99.7	36.9	57.3	99.7	61.3	73.7
0950, 0980	Ceilings & Acoustic Treatment	89.8	69.9	76.7	84.8	41.5	56.3	89.8	66.9	74.8	88.9	32.9	52.1	88.9	36.9	54.8	88.9	61.3	70.8
0960	Flooring	101.4	81.9	95.8	106.5	60.3	93.3	101.4	81.9	95.8	102.3	33.7	82.6	99.4	74.0	92.1	101.0	66.5	91.1
0970, 0990	Wall Finishes & Painting/Coating	124.0	80.0	97.5	112.8	38.7	68.1	124.0	76.1	95.1	109.6	35.1	64.6	124.0	54.3	81.9	109.6	54.3	76.2
09	FINISHES	98.5	74.0	85.0	98.0	46.0	69.4	97.4	71.5	83.1	96.7	35.1	62.8	97.3	47.8	70.1	96.3	60.2	76.4
COVERS	DIVS. 10 - 14, 25, 28, 41, 43, 44, 46	100.0	87.8	97.5	100.0	45.7	89.0	100.0	83.4	96.6	100.0	43.7	88.6	100.0	67.6	93.5	100.0	70.9	94.1
21, 22, 23	FIRE SUPPRESSION, PLUMBING & HVAC	95.6	85.7	91.6	95.5	43.4	74.5	95.6	85.1	91.3	95.6	62.1	82.1	95.6	66.9	84.0	95.6	69.0	84.8
26, 27, 3370	ELECTRICAL, COMMUNICATIONS & UTIL.	97.9	99.2	98.6	90.1	48.2	68.0	94.9	99.2	97.2	96.8	44.6	69.2	96.9	96.4	96.7	97.9	52.7	74.0
MF2014	WEIGHTED AVERAGE	100.7	84.0	93.4	98.7	52.4	78.5	99.5	82.7	92.1	99.3	52.5	78.9	100.2	67.2	85.8	99.9	65.6	85.0

DIVISION		VIRGINIA																	
		NEWPORT NEWS			NORFOLK			PETERSBURG			PORTSMOUTH			PULASKI			RICHMOND		
		236			233 - 235			238			237			243			230 - 232		
		MAT.	INST.	TOTAL	MAT.	INST.	TOTAL	MAT.	INST.	TOTAL	MAT.	INST.	TOTAL	MAT.	INST.	TOTAL	MAT.	INST.	TOTAL
015433	CONTRACTOR EQUIPMENT		106.1	106.1		106.7	106.7		106.1	106.1		106.0	106.0		101.7	101.7		106.1	106.1
0241, 31 - 34	SITE & INFRASTRUCTURE, DEMOLITION	108.2	88.8	94.4	108.9	89.8	95.3	111.7	89.0	95.6	106.7	88.6	93.8	105.9	85.5	91.4	103.6	89.0	93.2
0310	Concrete Forming & Accessories	97.3	64.1	68.7	97.9	64.3	68.9	90.6	57.2	61.8	86.4	53.1	57.7	90.3	38.7	45.8	97.6	57.2	62.8
0320	Concrete Reinforcing	90.6	72.0	81.2	94.7	72.1	83.2	90.3	73.0	81.5	90.3	72.0	81.0	92.4	64.6	78.3	97.5	73.0	85.0
0330	Cast-in-Place Concrete	100.2	66.1	86.2	105.2	66.1	89.1	106.3	53.7	84.7	99.2	66.0	85.6	101.8	48.2	79.8	98.1	53.7	79.8
03	CONCRETE	98.4	67.6	83.3	101.5	67.7	84.9	103.6	60.4	82.4	97.1	62.6	80.2	101.6	48.7	75.6	98.5	60.4	79.8
04	MASONRY	97.8	53.9	70.5	103.8	53.9	72.7	111.9	52.6	75.0	103.8	53.9	72.7	91.5	48.2	64.5	102.5	52.6	71.4
05	METALS	103.3	89.8	99.2	103.1	89.9	99.1	101.1	90.7	97.9	102.2	89.1	98.2	103.2	82.5	96.8	107.1	90.7	102.0
06	WOOD, PLASTICS & COMPOSITES	95.3	66.2	79.0	93.2	66.2	78.1	86.4	58.6	70.8	82.9	51.5	65.3	86.7	36.8	58.7	95.0	58.6	74.6
07	THERMAL & MOISTURE PROTECTION	104.3	68.4	89.6	100.9	68.4	87.5	104.3	68.3	89.5	104.3	66.8	88.9	104.2	51.1	82.4	102.2	68.3	88.3
08	OPENINGS	101.1	62.2	92.1	100.4	62.2	91.5	100.4	59.0	90.8	101.2	54.1	90.2	102.0	42.2	88.1	104.8	59.0	94.1
0920	Plaster & Gypsum Board	109.4	64.3	78.9	105.5	64.3	77.6	101.3	56.4	71.0	101.6	49.1	66.1	99.7	34.7	55.7	104.2	56.4	71.9
0950, 0980	Ceilings & Acoustic Treatment	88.2	64.3	72.5	88.9	64.3	72.7	85.7	56.4	66.4	88.2	49.1	62.5	88.9	34.7	53.3	90.6	56.4	68.1
0960	Flooring	106.5	66.5	95.1	105.6	66.5	94.4	102.5	71.2	93.6	99.5	66.5	90.0	102.3	66.5	92.0	103.4	71.2	94.2
0970, 0990	Wall Finishes & Painting/Coating	112.8	42.0	70.1	113.0	76.1	90.7	112.8	76.1	90.6	112.8	76.1	90.6	109.6	54.3	76.2	110.0	76.1	89.5
09	FINISHES	98.7	62.2	78.6	98.2	66.0	80.5	96.3	61.2	76.9	95.6	56.7	74.2	96.7	43.7	67.5	98.5	61.2	77.9
COVERS	DIVS. 10 - 14, 25, 28, 41, 43, 44, 46	100.0	82.2	96.4	100.0	82.2	96.4	100.0	79.6	95.9	100.0	72.0	94.3	100.0	44.0	88.7	100.0	79.6	95.9
21, 22, 23	FIRE SUPPRESSION, PLUMBING & HVAC	100.3	63.6	85.5	100.0	64.3	85.6	95.5	67.5	84.2	100.3	64.3	85.8	95.6	62.2	82.1	99.9	67.5	86.8
26, 27, 3370	ELECTRICAL, COMMUNICATIONS & UTIL.	92.8	63.5	77.3	95.5	58.3	75.8	93.0	71.9	81.8	91.2	58.3	73.8	96.8	54.4	74.4	96.8	71.9	83.6
MF2014	WEIGHTED AVERAGE	100.0	68.2	86.1	100.5	68.2	86.4	99.4	68.7	86.0	99.4	65.3	84.5	99.0	57.3	80.9	101.3	68.7	87.1

DIVISION		VIRGINIA									WASHINGTON								
		ROANOKE			STAUNTON			WINCHESTER			CLARKSTON			EVERETT			OLYMPIA		
		240 - 241			244			226			994			982			985		
		MAT.	INST.	TOTAL	MAT.	INST.	TOTAL	MAT.	INST.	TOTAL	MAT.	INST.	TOTAL	MAT.	INST.	TOTAL	MAT.	INST.	TOTAL
015433	CONTRACTOR EQUIPMENT		101.7	101.7		106.1	106.1		101.7	101.7		91.2	91.2		101.6	101.6		101.6	101.6
0241, 31 - 34	SITE & INFRASTRUCTURE, DEMOLITION	106.4	86.3	92.1	109.9	87.7	94.1	118.7	87.6	96.6	99.2	90.1	92.7	91.5	108.4	103.5	93.4	108.4	104.1
0310	Concrete Forming & Accessories	96.8	60.5	65.5	89.9	50.0	55.5	83.8	69.5	71.4	112.7	67.4	73.6	113.4	100.4	102.2	99.0	100.3	100.1
0320	Concrete Reinforcing	93.4	68.5	80.7	93.1	48.9	70.6	93.1	86.4	89.7	112.0	87.2	99.4	104.2	98.8	101.5	110.8	98.7	104.6
0330	Cast-in-Place Concrete	115.7	57.7	91.9	105.9	51.5	83.6	102.3	69.3	88.7	96.3	81.9	90.4	103.5	106.2	104.6	98.9	106.2	101.9
03	CONCRETE	105.9	62.4	84.5	102.9	52.0	77.9	103.5	73.7	88.9	107.2	76.3	92.0	98.7	101.6	100.1	98.9	101.6	100.2
04	MASONRY	97.7	53.2	70.0	106.9	51.4	72.3	101.7	59.7	75.5	102.3	81.5	89.3	110.6	100.8	104.5	103.0	99.1	100.6
05	METALS	105.5	86.5	99.7	103.4	80.5	96.3	103.5	94.7	100.8	88.3	81.6	86.2	101.8	92.4	98.9	101.2	92.1	98.4
06	WOOD, PLASTICS & COMPOSITES	95.9	62.6	77.3	86.7	49.9	66.1	82.2	68.1	74.3	113.2	63.8	85.5	111.7	100.0	105.2	91.1	100.0	96.1
07	THERMAL & MOISTURE PROTECTION	103.8	66.8	88.7	103.8	52.0	82.6	104.5	75.6	92.7	143.8	77.2	116.5	103.7	101.7	102.9	103.5	96.8	100.7
08	OPENINGS	100.9	58.8	91.1	100.5	48.1	88.3	102.1	73.7	95.5	119.0	68.7	107.3	101.3	99.8	101.0	106.9	99.9	105.3
0920	Plaster & Gypsum Board	107.4	61.3	76.2	99.7	47.5	64.4	99.9	66.9	77.6	145.7	62.5	89.5	109.4	99.9	103.0	101.3	99.9	100.4
0950, 0980	Ceilings & Acoustic Treatment	91.4	61.3	71.6	88.9	47.5	61.7	89.8	66.9	74.8	97.6	62.5	74.5	99.8	99.9	99.9	98.1	99.9	99.3
0960	Flooring	106.0	66.5	94.7	101.9	39.5	84.0	100.7	81.9	95.4	91.6	47.1	78.9	113.8	91.6	107.4	103.8	91.7	100.3
0970, 0990	Wall Finishes & Painting/Coating	109.6	54.3	76.2	109.6	34.5	64.3	124.0	87.2	101.8	100.0	58.5	75.0	95.4	91.5	93.0	99.0	91.5	94.5
09	FINISHES	99.0	61.1	78.1	96.6	46.6	69.0	97.9	72.7	84.0	110.3	61.3	83.3	104.7	97.6	100.8	98.9	97.6	98.2
COVERS	DIVS. 10 - 14, 25, 28, 41, 43, 44, 46	100.0	70.9	94.1	100.0	70.9	94.1	100.0	87.3	97.4	100.0	75.4	95.0	100.0	99.0	99.8	100.0	100.5	100.1
21, 22, 23	FIRE SUPPRESSION, PLUMBING & HVAC	100.3	64.8	86.0	95.6	58.9	80.8	95.6	87.7	92.4	95.6	83.4	90.7	100.1	99.1	99.7	100.0	99.1	99.7
26, 27, 3370	ELECTRICAL, COMMUNICATIONS & UTIL.	96.8	54.4	74.4	95.6	74.9	84.7	95.3	99.2	97.4	92.4	94.8	93.7	104.9	98.5	101.5	104.4	99.1	101.6
MF2014	WEIGHTED AVERAGE	101.5	65.1	85.6	99.7	61.8	83.2	100.0	82.2	92.2	101.5	80.2	92.2	101.6	99.6	100.7	101.2	99.4	100.4

	DIVISION	WASHINGTON																	
		RICHLAND			SEATTLE			SPOKANE			TACOMA			VANCOUVER			WENATCHEE		
		993			980 - 981,987			990 - 992			983 - 984			986			988		
		MAT.	INST.	TOTAL	MAT.	INST.	TOTAL	MAT.	INST.	TOTAL	MAT.	INST.	TOTAL	MAT.	INST.	TOTAL	MAT.	INST.	TOTAL
015433	CONTRACTOR EQUIPMENT		91.2	91.2		101.6	101.6		91.2	91.2		101.6	101.6		96.7	96.7		101.6	101.6
0241, 31 - 34	SITE & INFRASTRUCTURE, DEMOLITION	101.8	91.0	94.1	95.1	106.7	103.4	101.1	91.0	93.9	94.5	108.4	104.4	104.2	96.1	98.4	103.2	107.1	106.0
0310	Concrete Forming & Accessories	112.8	78.5	83.2	108.3	100.9	102.0	118.2	78.1	83.6	104.2	100.4	100.9	104.9	90.5	92.5	105.8	77.6	81.5
0320	Concrete Reinforcing	107.4	87.4	97.2	105.1	98.9	101.9	108.2	87.3	97.6	103.1	98.8	100.9	103.9	98.4	101.1	103.9	87.7	95.6
0330	Cast-in-Place Concrete	96.5	84.4	91.5	100.9	106.4	103.2	100.3	84.2	93.7	106.4	106.2	106.3	118.8	98.4	110.4	108.6	77.3	95.7
03	CONCRETE	106.8	82.2	94.7	99.6	102.1	100.8	109.1	82.0	95.8	100.4	101.6	101.0	110.7	94.6	102.8	108.4	79.6	94.2
04	MASONRY	104.3	84.3	91.8	108.2	100.9	103.6	104.9	84.3	92.0	106.2	100.9	102.9	106.7	95.6	99.8	108.6	91.4	97.9
05	METALS	88.7	84.9	87.5	102.5	94.2	99.9	90.9	84.5	89.0	103.6	92.3	100.1	101.0	91.8	98.2	101.0	86.2	96.4
06	WOOD, PLASTICS & COMPOSITES	113.4	76.1	92.5	104.9	100.0	102.2	122.4	76.1	96.5	101.0	100.0	100.4	92.9	90.2	91.4	102.5	75.8	87.5
07	THERMAL & MOISTURE PROTECTION	145.0	78.8	117.8	102.1	101.8	102.0	141.4	79.5	116.0	103.5	98.5	101.4	104.0	91.6	98.9	104.1	81.5	94.8
08	OPENINGS	121.4	71.3	109.7	100.1	99.9	100.0	121.9	71.8	110.3	102.0	99.9	101.5	98.3	92.8	97.0	101.5	71.0	94.4
0920	Plaster & Gypsum Board	145.7	75.2	98.1	103.6	99.9	101.1	137.7	75.2	95.5	107.8	99.9	102.5	105.7	90.1	95.1	110.2	74.9	86.4
0950, 0980	Ceilings & Acoustic Treatment	104.2	75.2	85.1	110.2	99.9	103.5	99.5	75.2	83.5	103.1	99.9	101.0	100.2	90.1	93.5	95.4	74.9	82.0
0960	Flooring	92.0	80.1	88.6	108.3	106.8	107.8	91.2	85.3	89.5	107.2	91.7	102.8	113.5	102.2	110.2	109.9	62.6	96.4
0970, 0990	Wall Finishes & Painting/Coating	100.0	67.6	80.4	96.3	91.5	93.4	100.2	70.9	82.5	95.4	91.5	93.0	97.6	70.8	81.4	95.4	72.6	81.6
09	FINISHES	112.0	76.7	92.6	105.0	100.6	102.6	109.8	78.1	92.3	103.4	97.6	100.2	101.5	90.6	95.5	103.8	73.5	87.1
COVERS	DIVS. 10 - 14, 25, 28, 41, 43, 44, 46	100.0	85.4	97.0	100.0	100.5	100.1	100.0	85.3	97.0	100.0	100.5	100.1	100.0	60.7	92.1	100.0	77.2	95.4
21, 22, 23	FIRE SUPPRESSION, PLUMBING & HVAC	100.5	107.6	103.3	100.0	115.1	106.1	100.4	84.3	93.9	100.2	99.2	99.8	100.3	87.0	94.9	95.4	86.7	91.9
26, 27, 3370	ELECTRICAL, COMMUNICATIONS & UTIL.	89.7	94.8	92.4	102.6	107.0	104.9	88.0	79.0	83.3	104.7	99.1	101.7	109.7	101.0	105.1	105.5	98.4	101.8
MF2014	WEIGHTED AVERAGE	103.1	89.3	97.1	101.4	104.7	102.8	103.4	82.3	94.2	101.8	99.7	100.9	102.7	92.1	98.1	101.6	86.6	95.0

	DIVISION	WASHINGTON			WEST VIRGINIA														
		YAKIMA			BECKLEY			BLUEFIELD			BUCKHANNON			CHARLESTON			CLARKSBURG		
		989			258 - 259			247 - 248			262			250 - 253			263 - 264		
		MAT.	INST.	TOTAL	MAT.	INST.	TOTAL	MAT.	INST.	TOTAL	MAT.	INST.	TOTAL	MAT.	INST.	TOTAL	MAT.	INST.	TOTAL
015433	CONTRACTOR EQUIPMENT		101.6	101.6		101.7	101.7		101.7	101.7		101.7	101.7		101.7	101.7		101.7	101.7
0241, 31 - 34	SITE & INFRASTRUCTURE, DEMOLITION	96.8	107.8	104.7	100.5	89.9	92.9	100.6	89.9	92.9	106.8	89.9	94.8	101.5	90.7	93.8	107.4	89.9	95.0
0310	Concrete Forming & Accessories	104.6	95.8	97.0	83.5	93.6	92.2	87.0	93.6	92.7	86.4	94.3	93.2	96.7	94.1	94.4	83.8	94.2	92.8
0320	Concrete Reinforcing	103.5	87.4	95.3	92.7	89.0	90.8	92.0	84.1	88.0	92.6	84.2	88.3	101.0	89.1	94.9	92.6	84.1	88.3
0330	Cast-in-Place Concrete	113.6	83.9	101.4	98.1	103.0	100.1	99.6	102.9	101.0	99.3	101.8	100.3	96.8	103.0	99.3	108.8	100.0	105.2
03	CONCRETE	105.3	89.9	97.7	97.6	96.4	97.0	97.4	95.5	96.5	100.4	95.4	98.0	99.1	96.7	97.9	104.4	94.8	99.7
04	MASONRY	99.8	83.1	89.4	98.9	98.9	98.9	93.3	98.9	96.8	104.5	98.9	101.0	95.8	97.9	97.1	108.4	98.9	102.4
05	METALS	101.6	86.3	96.9	100.9	98.1	100.0	103.4	96.3	101.2	103.6	97.1	101.6	99.6	98.9	99.4	103.6	96.9	101.5
06	WOOD, PLASTICS & COMPOSITES	101.4	100.0	100.6	83.5	93.4	89.1	85.6	93.4	90.0	84.9	93.4	89.7	98.4	93.4	95.6	81.5	93.4	88.2
07	THERMAL & MOISTURE PROTECTION	103.6	82.7	95.0	102.9	90.6	97.9	104.0	90.6	98.5	104.2	91.7	99.1	99.4	90.4	95.7	104.1	90.6	98.6
08	OPENINGS	101.5	81.0	96.7	101.3	85.8	97.7	102.5	84.6	98.3	102.5	84.6	98.4	103.6	85.8	99.4	102.5	85.1	98.5
0920	Plaster & Gypsum Board	107.4	99.9	102.4	97.0	93.0	94.3	98.9	93.0	94.9	99.3	93.0	95.1	100.3	93.0	95.4	97.2	93.0	94.4
0950, 0980	Ceilings & Acoustic Treatment	97.3	99.9	99.0	83.9	93.0	89.9	87.3	93.0	91.1	88.9	93.0	91.6	92.2	93.0	92.7	88.9	93.0	91.6
0960	Flooring	108.1	58.0	93.8	102.4	118.0	106.8	98.9	118.0	104.3	98.6	118.0	104.1	106.4	118.0	109.7	97.6	118.0	103.4
0970, 0990	Wall Finishes & Painting/Coating	95.4	70.9	80.6	107.4	91.5	97.8	109.6	94.6	100.6	109.6	94.3	100.3	106.9	95.9	100.2	109.6	94.3	100.3
09	FINISHES	102.5	86.6	93.7	94.7	98.6	96.9	94.8	98.9	97.1	95.7	98.9	97.5	99.1	98.8	99.0	95.0	98.9	97.2
COVERS	DIVS. 10 - 14, 25, 28, 41, 43, 44, 46	100.0	97.6	99.5	100.0	56.9	91.3	100.0	56.9	91.3	100.0	101.5	100.3	100.0	101.2	100.2	100.0	101.5	100.3
21, 22, 23	FIRE SUPPRESSION, PLUMBING & HVAC	100.2	107.2	103.0	95.5	93.5	94.7	95.6	92.8	94.4	95.6	95.1	95.4	100.0	93.4	97.4	95.6	95.0	95.3
26, 27, 3370	ELECTRICAL, COMMUNICATIONS & UTIL.	107.8	94.9	101.0	93.9	91.1	92.4	95.7	91.1	93.3	97.1	96.1	96.6	99.1	91.1	94.8	97.1	96.1	96.6
MF2014	WEIGHTED AVERAGE	102.0	94.1	98.6	97.9	93.3	95.9	98.3	92.9	96.0	99.6	95.5	97.8	99.9	94.8	97.6	100.1	95.4	98.1

	DIVISION	WEST VIRGINIA																	
		GASSAWAY			HUNTINGTON			LEWISBURG			MARTINSBURG			MORGANTOWN			PARKERSBURG		
		266			255 - 257			249			254			265			261		
		MAT.	INST.	TOTAL	MAT.	INST.	TOTAL	MAT.	INST.	TOTAL	MAT.	INST.	TOTAL	MAT.	INST.	TOTAL	MAT.	INST.	TOTAL
015433	CONTRACTOR EQUIPMENT		101.7	101.7		101.7	101.7		101.7	101.7		101.7	101.7		101.7	101.7		101.7	101.7
0241, 31 - 34	SITE & INFRASTRUCTURE, DEMOLITION	104.2	89.9	94.0	105.1	90.9	95.0	116.3	89.8	97.5	104.2	90.2	94.3	101.6	90.5	93.7	109.8	90.6	96.2
0310	Concrete Forming & Accessories	85.8	92.2	91.4	95.3	95.9	95.8	84.2	93.5	92.2	83.6	82.8	82.9	84.2	93.9	92.6	88.6	90.3	90.1
0320	Concrete Reinforcing	92.6	84.0	88.3	94.1	92.9	93.5	92.6	83.9	88.2	92.7	80.6	86.6	92.6	84.1	88.3	92.0	83.9	87.9
0330	Cast-in-Place Concrete	104.0	103.5	103.7	106.9	102.5	105.1	99.7	102.9	101.0	102.8	96.7	100.3	99.3	101.6	100.2	101.5	96.4	99.4
03	CONCRETE	100.8	95.1	98.0	102.8	98.0	100.4	107.1	95.4	101.4	101.2	87.9	94.7	97.2	95.2	96.2	102.5	91.8	97.2
04	MASONRY	109.2	95.9	100.9	96.4	100.8	99.1	95.9	95.9	95.9	99.7	89.5	93.3	126.5	98.9	109.3	82.9	91.7	88.4
05	METALS	103.5	96.8	101.4	103.3	100.4	102.4	103.5	96.0	101.2	101.3	94.2	99.1	103.6	96.6	101.5	104.2	96.6	101.9
06	WOOD, PLASTICS & COMPOSITES	84.0	90.7	87.8	94.7	95.3	95.1	81.8	93.4	88.3	83.5	81.6	82.5	81.8	93.4	88.3	85.0	89.1	87.3
07	THERMAL & MOISTURE PROTECTION	104.0	90.6	98.5	103.0	90.8	98.0	104.8	89.8	98.7	103.1	79.3	93.4	104.0	90.6	98.5	103.9	89.0	97.8
08	OPENINGS	100.7	83.2	96.6	100.5	87.6	97.5	102.5	84.6	98.4	103.3	73.4	96.3	103.8	85.1	99.4	101.4	82.3	96.9
0920	Plaster & Gypsum Board	98.6	90.2	92.9	104.8	95.0	98.1	97.2	93.0	94.4	97.6	80.9	86.3	97.2	93.0	94.4	99.7	88.5	92.1
0950, 0980	Ceilings & Acoustic Treatment	88.9	90.2	89.8	86.4	95.0	92.0	88.9	93.0	91.6	86.4	80.9	82.8	88.9	93.0	91.6	88.9	88.5	88.7
0960	Flooring	98.4	118.0	104.0	109.8	121.7	113.2	97.7	118.0	103.5	102.4	118.0	106.8	97.7	118.0	103.5	101.5	118.0	106.2
0970, 0990	Wall Finishes & Painting/Coating	109.6	95.9	101.3	107.4	94.6	99.7	109.6	75.4	88.9	107.4	41.4	67.5	109.6	94.3	100.3	109.6	94.6	100.6
09	FINISHES	95.2	97.5	96.5	98.5	100.8	99.8	96.1	96.8	96.5	95.6	83.1	88.7	94.6	98.9	97.0	96.6	95.9	96.2
COVERS	DIVS. 10 - 14, 25, 28, 41, 43, 44, 46	100.0	101.2	100.2	100.0	101.8	100.4	100.0	56.9	91.3	100.0	97.7	99.5	100.0	65.8	93.1	100.0	100.5	100.1
21, 22, 23	FIRE SUPPRESSION, PLUMBING & HVAC	95.6	93.9	94.9	100.3	90.5	96.4	95.6	93.5	94.7	95.5	86.0	91.7	95.6	95.0	95.3	100.3	88.3	95.4
26, 27, 3370	ELECTRICAL, COMMUNICATIONS & UTIL.	97.1	91.1	93.9	97.8	95.3	96.5	93.0	91.1	92.0	100.1	81.8	90.4	97.3	96.1	96.7	97.2	92.7	94.8
MF2014	WEIGHTED AVERAGE	99.5	93.9	97.1	100.7	95.7	98.5	99.8	92.4	96.6	99.3	86.4	93.7	100.2	94.4	97.6	100.1	91.8	96.5

		WEST VIRGINIA									WISCONSIN								
	DIVISION	PETERSBURG			ROMNEY			WHEELING			BELOIT			EAU CLAIRE			GREEN BAY		
		268			267			260			535			547			541 - 543		
		MAT.	INST.	TOTAL	MAT.	INST.	TOTAL	MAT.	INST.	TOTAL	MAT.	INST.	TOTAL	MAT.	INST.	TOTAL	MAT.	INST.	TOTAL
015433	CONTRACTOR EQUIPMENT		101.7	101.7		101.7	101.7		101.7	101.7		102.3	102.3		100.4	100.4		98.2	98.2
0241, 31 - 34	SITE & INFRASTRUCTURE, DEMOLITION	100.9	90.5	93.5	103.6	90.5	94.3	110.6	90.4	96.2	94.7	108.3	104.3	96.1	102.9	101.0	99.7	98.9	99.2
0310	Concrete Forming & Accessories	87.4	91.6	91.0	83.4	91.8	90.6	90.3	93.2	92.8	100.2	100.3	100.3	97.1	99.4	99.1	105.5	98.7	99.7
0320	Concrete Reinforcing	92.0	82.5	87.2	92.6	81.0	86.7	91.4	84.0	87.7	92.3	116.0	104.3	90.4	99.2	94.9	88.7	92.4	90.6
0330	Cast-in-Place Concrete	99.3	99.1	99.2	104.0	99.1	102.0	101.5	100.1	100.9	100.6	100.9	100.7	104.0	99.2	102.0	107.6	99.7	104.3
03	CONCRETE	97.3	93.0	95.2	100.6	92.8	96.8	102.5	94.4	98.5	98.8	103.5	101.1	99.7	99.4	99.6	102.8	98.0	100.5
04	MASONRY	99.5	98.9	99.1	96.2	98.9	97.9	107.6	93.9	99.1	100.6	104.3	102.9	94.7	102.1	99.3	126.7	101.4	110.9
05	METALS	103.7	95.8	101.2	103.7	95.5	101.2	104.4	96.8	102.1	93.1	107.7	97.6	95.0	101.4	97.0	97.4	98.2	97.6
06	WOOD, PLASTICS & COMPOSITES	85.9	90.7	88.6	80.9	90.7	86.4	86.7	93.4	90.5	99.9	99.2	99.5	105.1	99.3	101.9	110.0	99.3	104.0
07	THERMAL & MOISTURE PROTECTION	104.0	86.0	96.6	104.1	83.4	95.6	104.3	90.0	98.4	102.1	96.8	99.9	102.4	86.7	95.9	104.3	86.0	96.8
08	OPENINGS	103.8	82.8	98.9	103.7	82.4	98.8	102.2	85.1	98.2	102.5	108.8	103.9	103.7	96.2	102.0	100.1	96.0	99.2
0920	Plaster & Gypsum Board	99.3	90.2	93.2	96.8	90.2	92.4	99.7	93.0	95.2	94.5	99.6	97.9	102.9	99.6	100.6	98.7	99.6	99.3
0950, 0980	Ceilings & Acoustic Treatment	88.9	90.2	89.8	88.9	90.2	89.8	88.9	93.0	91.6	88.1	99.6	95.6	90.4	99.6	96.4	81.9	99.6	93.5
0960	Flooring	99.4	118.0	104.7	97.5	118.0	103.4	102.3	118.0	106.8	99.8	118.1	105.1	91.4	114.1	97.9	110.2	114.1	111.3
0970, 0990	Wall Finishes & Painting/Coating	109.6	94.3	100.3	109.6	94.3	100.3	109.6	94.3	100.3	98.3	97.2	97.7	94.5	86.4	89.6	105.4	83.2	92.0
09	FINISHES	95.4	97.3	96.4	94.7	97.3	96.1	96.9	98.2	97.6	97.4	103.1	100.5	94.6	101.2	98.3	99.0	100.7	99.9
COVERS	DIVS. 10 - 14, 25, 28, 41, 43, 44, 46	100.0	65.5	93.0	100.0	101.2	100.2	100.0	95.4	99.1	100.0	96.6	99.3	100.0	96.8	99.3	100.0	96.5	99.3
21, 22, 23	FIRE SUPPRESSION, PLUMBING & HVAC	95.6	93.7	94.8	95.6	93.4	94.7	100.3	93.6	97.6	100.1	98.6	99.5	100.1	87.6	95.0	100.3	85.1	94.2
26, 27, 3370	ELECTRICAL, COMMUNICATIONS & UTIL.	100.5	81.8	90.6	99.8	81.8	90.3	94.4	96.1	95.3	101.0	86.7	93.4	103.8	84.9	93.8	98.2	82.0	89.6
MF2014	WEIGHTED AVERAGE	99.3	91.3	95.8	99.4	92.2	96.3	101.2	94.3	98.2	98.9	100.7	99.7	99.3	95.3	97.5	101.2	93.4	97.8

		WISCONSIN																	
	DIVISION	KENOSHA			LA CROSSE			LANCASTER			MADISON			MILWAUKEE			NEW RICHMOND		
		531			546			538			537			530,532			540		
		MAT.	INST.	TOTAL	MAT.	INST.	TOTAL	MAT.	INST.	TOTAL	MAT.	INST.	TOTAL	MAT.	INST.	TOTAL	MAT.	INST.	TOTAL
015433	CONTRACTOR EQUIPMENT		100.2	100.2		100.4	100.4		102.3	102.3		102.3	102.3		90.5	90.5		100.8	100.8
0241, 31 - 34	SITE & INFRASTRUCTURE, DEMOLITION	100.3	105.2	103.8	90.1	102.8	99.1	93.9	108.2	104.1	92.4	108.3	103.7	94.6	98.5	97.4	94.4	103.3	100.8
0310	Concrete Forming & Accessories	108.0	100.5	101.6	84.1	99.1	97.1	99.5	97.9	98.2	102.0	100.1	100.4	104.1	118.0	116.1	92.4	100.9	99.7
0320	Concrete Reinforcing	92.1	97.9	95.0	90.1	91.5	90.9	93.5	91.3	92.3	97.9	91.6	94.7	95.7	98.2	97.0	87.8	99.0	93.5
0330	Cast-in-Place Concrete	109.4	107.3	108.5	93.5	95.9	94.5	100.0	100.9	100.4	91.1	100.8	95.1	99.8	109.6	103.8	108.3	82.5	97.7
03	CONCRETE	103.5	102.4	103.0	90.8	96.8	93.7	98.5	97.8	98.1	95.7	98.8	97.2	99.7	110.5	105.0	97.5	94.3	95.9
04	MASONRY	97.9	114.1	108.0	93.8	105.9	101.4	100.7	104.3	102.9	101.5	104.3	103.2	104.9	120.6	114.7	121.6	102.3	109.6
05	METALS	93.9	101.2	96.2	94.9	98.2	95.9	90.6	95.0	92.0	91.9	97.2	93.6	97.5	94.8	96.7	95.1	99.7	96.5
06	WOOD, PLASTICS & COMPOSITES	105.0	96.4	100.2	90.1	99.3	95.3	99.1	99.2	99.2	96.7	99.2	98.1	104.7	118.5	112.5	94.0	102.5	98.7
07	THERMAL & MOISTURE PROTECTION	102.2	108.7	104.9	101.8	90.2	97.1	101.9	82.5	93.9	102.8	105.1	103.8	100.8	114.0	106.2	103.6	92.2	98.9
08	OPENINGS	96.9	101.1	97.9	103.7	85.0	99.4	97.9	85.8	95.1	107.9	100.5	106.2	101.0	113.2	103.8	89.1	97.5	91.1
0920	Plaster & Gypsum Board	85.3	96.7	93.0	97.7	99.6	99.0	93.3	99.6	97.5	99.6	99.6	99.6	101.0	119.3	113.4	87.7	103.0	98.1
0950, 0980	Ceilings & Acoustic Treatment	88.1	96.7	93.7	89.5	99.6	96.1	83.1	99.6	93.9	94.0	99.6	97.6	95.9	119.3	111.3	58.5	103.0	87.8
0960	Flooring	118.8	110.4	116.4	85.6	114.1	93.7	99.5	110.5	102.7	103.2	110.5	105.3	98.2	114.0	102.7	103.2	114.1	106.3
0970, 0990	Wall Finishes & Painting/Coating	108.3	115.5	112.6	94.5	77.9	84.5	98.3	66.4	79.1	102.4	97.2	99.3	94.4	122.9	111.6	108.3	86.4	95.1
09	FINISHES	102.5	103.6	103.1	91.6	100.3	96.4	96.0	96.5	96.3	98.6	101.9	100.4	100.9	118.6	110.7	90.0	101.3	96.2
COVERS	DIVS. 10 - 14, 25, 28, 41, 43, 44, 46	100.0	101.1	100.2	100.0	96.7	99.3	100.0	49.7	89.8	100.0	99.7	99.9	100.0	104.6	100.9	100.0	96.1	99.2
21, 22, 23	FIRE SUPPRESSION, PLUMBING & HVAC	100.3	97.0	98.9	100.1	87.5	95.0	95.3	87.3	92.1	100.0	95.6	98.2	100.0	103.5	101.4	94.8	86.9	91.6
26, 27, 3370	ELECTRICAL, COMMUNICATIONS & UTIL.	101.5	98.7	100.1	104.2	85.1	94.1	100.8	85.1	92.5	102.1	92.1	96.8	100.8	99.5	100.1	101.9	85.1	93.0
MF2014	WEIGHTED AVERAGE	99.5	102.2	100.7	97.8	94.5	96.3	96.7	92.5	94.8	99.1	99.0	99.1	100.0	107.2	103.1	96.8	94.6	95.9

		WISCONSIN																	
	DIVISION	OSHKOSH			PORTAGE			RACINE			RHINELANDER			SUPERIOR			WAUSAU		
		549			539			534			545			548			544		
		MAT.	INST.	TOTAL	MAT.	INST.	TOTAL	MAT.	INST.	TOTAL	MAT.	INST.	TOTAL	MAT.	INST.	TOTAL	MAT.	INST.	TOTAL
015433	CONTRACTOR EQUIPMENT		98.2	98.2		102.3	102.3		102.3	102.3		98.2	98.2		100.8	100.8		98.2	98.2
0241, 31 - 34	SITE & INFRASTRUCTURE, DEMOLITION	91.5	98.9	96.8	85.0	108.0	101.4	94.5	109.0	104.8	103.3	98.8	100.1	91.3	103.3	99.9	87.4	98.9	95.6
0310	Concrete Forming & Accessories	88.6	98.1	96.8	91.7	98.9	97.9	100.5	100.5	100.5	86.2	97.8	96.2	90.5	91.9	91.7	88.1	98.7	97.2
0320	Concrete Reinforcing	88.8	92.3	90.6	93.5	91.4	92.5	92.3	97.9	95.1	89.0	90.9	90.0	87.8	91.4	89.6	89.0	91.3	90.2
0330	Cast-in-Place Concrete	99.7	99.4	99.6	85.9	101.8	92.4	98.7	107.0	102.1	113.1	99.3	107.4	102.0	97.9	100.3	93.0	99.6	95.7
03	CONCRETE	92.4	97.6	94.9	86.4	98.5	92.4	97.9	102.3	100.1	104.7	97.2	101.0	91.8	94.2	93.0	87.3	97.8	92.5
04	MASONRY	107.9	101.4	103.8	99.6	104.3	102.5	100.6	114.0	109.0	125.3	101.4	110.4	120.8	104.5	110.6	107.4	101.4	103.6
05	METALS	95.4	96.7	95.8	91.3	96.0	92.7	94.8	101.2	96.8	95.3	96.7	95.7	96.1	98.2	96.8	95.1	97.7	95.9
06	WOOD, PLASTICS & COMPOSITES	89.8	99.3	95.1	89.2	99.2	94.8	100.2	96.4	98.1	87.3	99.3	94.0	92.3	90.1	91.0	89.2	99.3	94.9
07	THERMAL & MOISTURE PROTECTION	103.4	85.7	96.1	101.4	92.6	97.7	102.1	108.3	104.7	104.2	83.5	95.7	103.3	92.2	98.7	103.3	84.4	95.5
08	OPENINGS	95.9	95.5	95.8	98.1	92.3	96.8	102.5	101.1	102.2	96.0	87.1	93.9	88.6	91.1	89.2	96.1	95.7	96.0
0920	Plaster & Gypsum Board	86.2	99.6	95.2	86.4	99.6	95.3	94.5	96.7	96.0	86.2	99.6	95.2	87.6	90.2	89.4	86.2	99.6	95.2
0950, 0980	Ceilings & Acoustic Treatment	81.9	99.6	93.5	85.6	99.6	94.8	88.1	96.7	93.7	81.9	99.6	93.5	59.3	90.2	79.6	81.9	99.6	93.5
0960	Flooring	101.6	114.1	105.2	96.2	118.1	102.5	99.8	110.4	102.8	100.9	114.1	104.7	104.4	122.4	109.6	101.5	114.1	105.1
0970, 0990	Wall Finishes & Painting/Coating	102.2	101.4	101.7	98.3	66.4	79.1	98.3	115.5	108.7	102.2	60.4	76.9	96.0	102.8	100.1	102.2	83.2	90.7
09	FINISHES	94.0	100.9	97.8	94.1	98.6	96.6	97.4	103.6	100.8	94.8	97.0	96.0	89.3	98.6	94.4	93.7	100.7	97.5
COVERS	DIVS. 10 - 14, 25, 28, 41, 43, 44, 46	100.0	64.6	92.8	100.0	63.4	92.6	100.0	101.1	100.2	100.0	55.0	90.9	100.0	94.1	98.8	100.0	96.5	99.3
21, 22, 23	FIRE SUPPRESSION, PLUMBING & HVAC	95.6	85.0	91.3	95.3	95.3	95.3	100.1	97.0	98.8	95.6	86.7	92.0	94.8	89.4	92.6	95.6	86.4	91.9
26, 27, 3370	ELECTRICAL, COMMUNICATIONS & UTIL.	102.5	80.4	90.9	104.8	92.1	98.1	100.8	98.6	99.6	101.8	81.8	91.3	107.1	99.0	102.8	103.7	81.8	92.1
MF2014	WEIGHTED AVERAGE	96.9	91.9	94.7	95.3	96.6	95.9	99.1	102.4	100.5	99.3	91.2	95.8	96.6	96.3	96.5	96.2	93.5	95.0

	DIVISION	WYOMING																	
		CASPER			CHEYENNE			NEWCASTLE			RAWLINS			RIVERTON			ROCK SPRINGS		
		826			820			827			823			825			829 - 831		
		MAT.	INST.	TOTAL	MAT.	INST.	TOTAL	MAT.	INST.	TOTAL	MAT.	INST.	TOTAL	MAT.	INST.	TOTAL	MAT.	INST.	TOTAL
015433	CONTRACTOR EQUIPMENT		98.2	98.2		98.2	98.2		98.2	98.2		98.2	98.2		98.2	98.2		98.2	98.2
0241, 31 - 34	SITE & INFRASTRUCTURE, DEMOLITION	98.6	94.4	95.6	92.5	94.4	93.8	85.1	94.0	91.5	98.4	94.0	95.3	92.1	94.0	93.5	89.2	94.0	92.6
0310	Concrete Forming & Accessories	102.5	50.1	57.3	104.4	53.5	60.5	94.8	62.2	66.7	99.2	62.2	67.3	93.7	62.1	66.5	101.5	62.1	67.5
0320	Concrete Reinforcing	105.7	70.6	87.8	98.0	70.6	84.1	105.5	70.8	87.9	105.2	70.8	87.7	106.1	70.6	88.0	106.1	70.6	88.0
0330	Cast-in-Place Concrete	102.8	72.0	90.2	95.0	72.1	85.6	95.9	61.6	81.8	96.0	61.6	81.9	96.0	60.0	81.2	95.9	59.9	81.1
03	CONCRETE	101.7	62.1	82.3	98.6	63.7	81.4	99.0	63.9	81.8	111.8	63.9	88.3	106.7	63.3	85.4	99.5	63.2	81.7
04	MASONRY	108.2	51.9	73.1	103.7	51.9	71.4	100.3	54.8	71.9	100.3	54.8	71.9	100.3	50.5	69.3	160.4	50.4	91.8
05	METALS	99.3	72.1	90.9	101.7	72.1	92.6	97.8	72.0	89.9	97.9	72.0	89.9	98.0	71.6	89.8	98.7	71.1	90.2
06	WOOD, PLASTICS & COMPOSITES	97.2	49.7	70.6	96.5	54.3	72.9	86.4	66.9	75.5	90.4	66.9	77.2	85.3	66.9	75.0	95.4	66.9	79.4
07	THERMAL & MOISTURE PROTECTION	105.9	55.1	85.1	100.1	55.6	81.8	101.8	55.8	82.9	103.2	55.8	83.7	102.7	59.9	85.1	101.9	59.9	84.7
08	OPENINGS	108.0	53.3	95.3	108.3	55.8	96.1	112.6	61.7	100.8	112.3	61.7	100.5	112.5	61.7	100.7	113.0	61.3	101.0
0920	Plaster & Gypsum Board	97.4	48.1	64.1	89.3	52.9	64.7	85.2	65.8	72.1	85.5	65.8	72.2	85.2	65.8	72.1	97.3	65.8	76.0
0950, 0980	Ceilings & Acoustic Treatment	99.2	48.1	65.6	91.0	52.9	66.0	92.9	65.8	75.1	92.9	65.8	75.1	92.9	65.8	75.1	92.9	65.8	75.1
0960	Flooring	112.0	56.4	96.1	109.9	56.4	94.6	103.5	56.4	90.0	106.2	56.4	92.0	103.0	56.4	89.7	108.4	56.4	93.5
0970, 0990	Wall Finishes & Painting/Coating	104.7	43.1	67.5	111.2	43.1	70.0	107.1	55.4	75.9	107.1	55.4	75.9	107.1	55.4	75.9	107.1	55.4	75.9
09	FINISHES	103.2	49.6	73.6	100.0	52.3	73.7	95.0	60.8	76.1	97.2	60.8	77.1	95.7	60.8	76.5	98.1	60.8	77.5
COVERS	DIVS. 10 - 14, 25, 28, 41, 43, 44, 46	100.0	87.6	97.5	100.0	88.2	97.6	100.0	89.1	97.8	100.0	89.1	97.8	100.0	81.4	96.2	100.0	80.8	96.1
21, 22, 23	FIRE SUPPRESSION, PLUMBING & HVAC	99.9	64.1	85.4	99.9	64.1	85.4	97.4	63.4	83.7	97.4	63.4	83.7	97.4	63.4	83.7	99.8	63.3	85.1
26, 27, 3370	ELECTRICAL, COMMUNICATIONS & UTIL.	103.3	64.1	82.6	105.8	68.6	86.1	104.5	69.0	85.7	104.5	69.0	85.7	104.5	62.9	82.5	102.4	62.9	81.5
MF2014	WEIGHTED AVERAGE	102.0	63.9	85.4	101.5	65.2	85.7	99.9	66.9	85.5	101.8	66.9	86.6	101.0	65.4	85.5	103.7	65.3	87.0

	DIVISION	WYOMING												CANADA					
		SHERIDAN			WHEATLAND			WORLAND			YELLOWSTONE NAT'L PA			BARRIE, ONTARIO			BATHURST, NEW BRUNSWICK		
		828			822			824			821								
		MAT.	INST.	TOTAL	MAT.	INST.	TOTAL	MAT.	INST.	TOTAL	MAT.	INST.	TOTAL	MAT.	INST.	TOTAL	MAT.	INST.	TOTAL
015433	CONTRACTOR EQUIPMENT		98.2	98.2		98.2	98.2		98.2	98.2		98.2	98.2		102.1	102.1		102.2	102.2
0241, 31 - 34	SITE & INFRASTRUCTURE, DEMOLITION	92.4	94.4	93.8	89.6	94.0	92.8	87.1	94.0	92.0	87.2	94.2	92.2	118.5	100.4	105.6	103.4	96.6	98.6
0310	Concrete Forming & Accessories	102.2	50.6	57.7	96.8	52.7	58.8	96.9	62.0	66.8	96.9	63.1	67.8	124.4	89.4	94.2	104.4	63.7	69.3
0320	Concrete Reinforcing	106.1	70.9	88.2	105.5	70.8	87.8	106.1	70.6	88.0	107.9	70.6	88.9	173.2	83.8	127.7	137.0	56.5	96.0
0330	Cast-in-Place Concrete	99.2	72.0	88.1	100.2	61.5	84.3	95.9	59.9	81.1	95.9	61.6	81.8	165.6	89.0	134.2	131.6	61.9	103.0
03	CONCRETE	106.9	62.4	85.0	103.7	59.7	82.1	99.2	63.2	81.6	99.5	64.3	82.2	150.3	88.4	119.9	124.6	62.4	94.1
04	MASONRY	100.6	56.3	73.0	100.7	54.7	72.0	100.3	50.4	69.2	100.4	53.4	71.1	171.5	98.0	125.7	160.8	64.9	101.1
05	METALS	101.6	72.4	92.6	97.8	71.7	89.8	98.0	71.5	89.8	98.6	71.5	90.3	110.2	90.2	104.0	109.0	71.5	97.5
06	WOOD, PLASTICS & COMPOSITES	97.9	50.4	71.3	88.3	54.3	69.2	88.3	66.9	76.3	88.3	66.9	76.3	119.8	88.2	102.1	101.8	64.0	80.7
07	THERMAL & MOISTURE PROTECTION	102.9	56.3	83.8	102.1	55.8	83.1	101.9	57.5	83.7	101.4	58.8	83.9	111.8	89.1	102.5	106.0	62.2	88.1
08	OPENINGS	113.2	52.7	99.1	111.1	55.8	98.3	112.8	61.7	100.9	105.6	61.7	95.4	93.9	86.5	92.2	87.9	55.9	80.5
0920	Plaster & Gypsum Board	109.4	48.8	68.5	85.2	52.8	63.3	85.2	65.8	72.1	85.4	65.8	72.1	151.1	87.7	108.2	137.6	62.7	87.0
0950, 0980	Ceilings & Acoustic Treatment	95.7	48.8	64.9	92.9	52.8	66.6	92.9	65.8	75.1	93.7	65.8	75.4	91.5	87.7	89.0	106.4	62.7	77.7
0960	Flooring	107.5	56.4	92.9	105.0	56.4	91.1	105.0	56.4	91.1	105.0	56.4	91.1	127.2	98.9	119.1	107.9	47.2	90.6
0970, 0990	Wall Finishes & Painting/Coating	109.4	43.1	69.4	107.1	34.4	63.3	107.1	55.4	75.9	107.1	55.4	75.9	116.2	89.9	100.3	117.6	51.6	77.8
09	FINISHES	102.8	50.0	73.7	95.8	51.0	71.1	95.5	60.8	76.3	95.7	61.5	76.8	115.0	91.5	102.1	111.3	59.7	82.8
COVERS	DIVS. 10 - 14, 25, 28, 41, 43, 44, 46	100.0	87.7	97.5	100.0	80.3	96.0	100.0	81.4	96.2	100.0	82.4	96.4	139.2	74.4	126.1	131.1	65.3	117.8
21, 22, 23	FIRE SUPPRESSION, PLUMBING & HVAC	97.4	64.1	83.9	97.4	63.3	83.6	97.4	63.3	83.6	97.4	64.9	84.3	101.8	99.3	100.8	102.0	69.8	89.0
26, 27, 3370	ELECTRICAL, COMMUNICATIONS & UTIL.	107.4	63.0	83.9	104.5	68.6	85.5	104.5	62.9	82.5	103.2	62.9	81.9	116.8	90.5	102.9	112.7	61.8	85.8
MF2014	WEIGHTED AVERAGE	102.6	64.3	85.9	100.4	64.3	84.7	100.0	65.3	84.9	99.3	66.3	84.9	117.7	93.0	106.9	111.5	67.3	92.2

	DIVISION	CANADA																	
		BRANDON, MANITOBA			BRANTFORD, ONTARIO			BRIDGEWATER, NOVA SCOTIA			CALGARY, ALBERTA			CAP-DE-LA-MADELEINE, QUEBEC			CHARLESBOURG, QUEBEC		
		MAT.	INST.	TOTAL	MAT.	INST.	TOTAL	MAT.	INST.	TOTAL	MAT.	INST.	TOTAL	MAT.	INST.	TOTAL	MAT.	INST.	TOTAL
015433	CONTRACTOR EQUIPMENT		104.5	104.5		102.1	102.1		101.9	101.9		107.1	107.1		102.8	102.8		102.8	102.8
0241, 31 - 34	SITE & INFRASTRUCTURE, DEMOLITION	133.0	99.3	109.1	118.0	100.7	105.7	102.6	98.2	99.5	124.0	105.4	110.7	98.4	99.9	99.4	98.4	99.9	99.4
0310	Concrete Forming & Accessories	146.5	72.5	82.7	124.9	96.9	100.7	97.1	71.6	75.1	124.4	97.4	101.1	130.7	88.0	93.8	130.7	88.0	93.8
0320	Concrete Reinforcing	185.9	53.2	118.4	165.4	82.5	123.2	140.6	46.7	92.8	135.6	71.9	103.2	140.6	78.8	109.2	140.6	78.8	109.2
0330	Cast-in-Place Concrete	133.4	76.7	110.1	153.6	110.3	135.8	157.8	72.2	122.7	189.0	107.3	155.5	124.0	97.8	113.2	124.0	97.8	113.2
03	CONCRETE	145.6	71.0	109.0	141.8	98.8	120.7	139.4	67.9	104.3	159.1	96.1	128.2	124.3	89.8	107.4	124.3	89.8	107.4
04	MASONRY	233.5	67.6	130.1	168.0	102.6	127.2	163.4	71.9	106.4	196.3	91.4	130.9	163.9	87.1	116.0	163.9	87.1	116.0
05	METALS	125.6	76.8	110.6	109.1	90.9	103.5	108.2	74.2	97.7	137.0	89.3	122.3	107.4	87.6	101.3	107.4	87.6	101.3
06	WOOD, PLASTICS & COMPOSITES	154.2	73.5	109.0	123.7	95.7	108.1	92.9	71.1	80.7	102.3	97.2	99.4	135.3	88.0	108.8	135.3	88.0	108.8
07	THERMAL & MOISTURE PROTECTION	135.0	71.8	109.1	113.8	94.7	106.0	108.6	69.9	92.8	117.8	94.6	108.3	107.4	89.9	100.3	107.4	89.9	100.3
08	OPENINGS	105.5	64.8	96.1	92.4	92.7	92.5	86.1	64.9	81.1	89.9	86.2	89.1	93.6	81.1	90.7	93.6	81.1	90.7
0920	Plaster & Gypsum Board	119.3	72.2	87.4	130.1	95.5	106.7	130.7	70.0	89.7	139.4	96.6	110.5	157.0	87.4	109.9	157.0	87.4	109.9
0950, 0980	Ceilings & Acoustic Treatment	109.6	72.2	85.0	96.5	95.5	95.8	96.5	70.0	79.1	138.8	96.6	111.0	96.5	87.4	90.5	96.5	87.4	90.5
0960	Flooring	138.6	70.2	119.1	120.5	98.9	114.3	102.7	67.2	92.5	122.3	92.6	113.8	120.5	99.5	114.5	120.5	99.5	114.5
0970, 0990	Wall Finishes & Painting/Coating	126.5	58.7	85.6	118.3	98.6	106.4	118.3	64.1	85.6	127.4	112.4	118.3	118.3	91.7	102.2	118.3	91.7	102.2
09	FINISHES	122.5	71.0	94.1	112.4	97.9	104.4	106.6	70.3	86.6	125.5	98.7	110.7	115.3	90.8	101.8	115.3	90.8	101.8
COVERS	DIVS. 10 - 14, 25, 28, 41, 43, 44, 46	131.1	67.6	118.3	131.1	76.6	120.1	131.1	66.6	118.1	131.1	96.9	124.2	131.1	85.2	121.9	131.1	85.2	121.9
21, 22, 23	FIRE SUPPRESSION, PLUMBING & HVAC	102.1	84.1	94.9	102.0	102.3	102.1	102.0	84.4	94.9	101.2	92.9	97.8	102.3	91.0	97.8	102.3	91.0	97.8
26, 27, 3370	ELECTRICAL, COMMUNICATIONS & UTIL.	118.6	69.6	92.7	111.5	89.9	100.1	117.1	64.4	89.2	113.5	101.3	107.1	111.2	72.6	90.8	111.2	72.6	90.8
MF2014	WEIGHTED AVERAGE	125.2	75.8	103.7	114.9	96.9	107.1	112.9	74.6	96.2	123.5	95.7	111.4	112.4	87.6	101.6	112.4	87.6	101.6

DIVISION		CANADA																	
		CHARLOTTETOWN, PRINCE EDWARD ISLAND			CHICOUTIMI, QUEBEC			CORNER BROOK, NEWFOUNDLAND			CORNWALL, ONTARIO			DALHOUSIE, NEW BRUNSWICK			DARTMOUTH, NOVA SCOTIA		
		MAT.	INST.	TOTAL	MAT.	INST.	TOTAL	MAT.	INST.	TOTAL	MAT.	INST.	TOTAL	MAT.	INST.	TOTAL	MAT.	INST.	TOTAL
015433	CONTRACTOR EQUIPMENT		101.8	101.8		102.8	102.8		103.3	103.3		102.1	102.1		102.2	102.2		101.9	101.9
0241, 31 - 34	SITE & INFRASTRUCTURE, DEMOLITION	121.3	95.7	103.1	100.7	99.1	99.6	137.7	97.1	108.9	116.1	100.2	104.8	100.9	96.6	97.8	124.8	98.2	105.9
0310	Concrete Forming & Accessories	110.4	58.2	65.4	133.2	93.1	98.6	121.8	61.7	69.9	122.6	89.6	94.1	103.5	63.9	69.3	111.8	71.6	77.2
0320	Concrete Reinforcing	141.5	46.8	93.3	103.3	95.9	99.5	170.3	48.6	108.4	165.4	82.2	123.1	142.7	56.6	98.9	178.2	46.7	111.3
0330	Cast-in-Place Concrete	166.8	61.2	123.4	124.0	99.8	114.0	155.3	70.6	120.5	138.2	100.6	122.7	131.8	62.0	103.2	149.8	72.2	118.0
03	CONCRETE	147.1	57.9	103.3	114.6	95.9	105.4	182.2	63.2	123.7	134.3	92.2	113.6	132.5	62.5	98.1	162.1	67.9	115.8
04	MASONRY	178.4	62.6	106.3	162.7	95.0	120.5	229.3	64.1	126.4	166.8	94.0	121.4	160.2	64.9	100.8	244.8	71.9	137.1
05	METALS	133.9	67.6	113.5	108.8	92.2	103.7	126.0	72.9	109.7	109.1	89.6	103.1	104.4	71.8	94.3	125.9	74.2	110.0
06	WOOD, PLASTICS & COMPOSITES	97.8	57.8	75.4	137.4	93.6	112.9	128.3	60.8	90.5	122.1	89.0	103.6	100.2	64.0	80.0	116.1	71.1	90.9
07	THERMAL & MOISTURE PROTECTION	121.3	60.4	96.3	106.5	96.4	102.3	139.8	61.8	107.8	113.6	89.3	103.6	110.8	62.2	90.9	136.8	69.9	109.4
08	OPENINGS	88.2	50.0	79.3	92.2	80.6	89.5	112.4	57.1	99.5	93.6	86.2	91.9	89.1	55.9	81.4	95.3	64.9	88.2
0920	Plaster & Gypsum Board	131.6	56.3	80.7	159.4	93.2	114.6	149.0	59.3	88.4	189.4	88.6	121.3	141.2	62.7	88.2	144.6	70.0	94.2
0950, 0980	Ceilings & Acoustic Treatment	120.2	56.3	78.2	105.6	93.2	97.4	110.5	59.3	76.9	99.0	88.6	92.2	98.9	62.7	75.1	117.2	70.0	86.2
0960	Flooring	111.7	62.9	97.8	123.6	99.5	116.8	121.0	57.2	102.8	120.5	97.5	113.9	106.5	72.1	96.7	116.1	67.2	102.1
0970, 0990	Wall Finishes & Painting/Coating	125.6	42.9	75.7	117.6	106.4	110.8	126.4	61.8	87.4	118.3	91.9	102.3	120.7	51.6	79.0	126.4	64.1	88.8
09	FINISHES	117.1	57.8	84.4	118.4	96.5	106.4	122.0	60.9	88.4	121.0	91.6	104.8	110.6	64.5	85.2	120.1	70.3	92.6
COVERS	DIVS. 10 - 14, 25, 28, 41, 43, 44, 46	131.1	64.8	117.7	131.1	86.5	122.1	131.1	65.6	117.9	131.1	74.0	119.6	131.1	65.3	117.8	131.1	66.6	118.1
21, 22, 23	FIRE SUPPRESSION, PLUMBING & HVAC	102.3	63.8	86.8	102.0	92.3	98.1	102.1	71.8	89.9	102.3	100.1	101.4	102.0	69.8	89.0	102.1	84.4	95.0
26, 27, 3370	ELECTRICAL, COMMUNICATIONS & UTIL.	110.6	52.3	79.8	109.7	86.8	97.6	116.0	58.3	85.5	112.4	90.9	101.0	114.5	58.4	84.8	120.7	64.4	90.9
MF2014	WEIGHTED AVERAGE	120.0	62.8	95.1	111.4	92.9	103.3	129.6	67.6	102.6	115.0	93.3	105.5	111.9	67.5	92.5	126.1	74.6	103.7

DIVISION		CANADA																	
		EDMONTON, ALBERTA			FORT MCMURRAY, ALBERTA			FREDERICTON, NEW BRUNSWICK			GATINEAU, QUEBEC			GRANBY, QUEBEC			HALIFAX, NOVA SCOTIA		
		MAT.	INST.	TOTAL	MAT.	INST.	TOTAL	MAT.	INST.	TOTAL	MAT.	INST.	TOTAL	MAT.	INST.	TOTAL	MAT.	INST.	TOTAL
015433	CONTRACTOR EQUIPMENT		107.1	107.1		104.4	104.4		102.2	102.2		102.8	102.8		102.8	102.8		101.9	101.9
0241, 31 - 34	SITE & INFRASTRUCTURE, DEMOLITION	144.1	105.4	116.6	123.1	101.3	107.6	105.0	96.6	99.1	98.2	99.8	99.4	98.7	99.8	99.5	109.2	98.3	101.5
0310	Concrete Forming & Accessories	123.2	97.4	100.9	122.8	92.3	96.5	120.3	64.2	71.9	130.7	87.8	93.7	130.7	87.8	93.7	108.4	78.8	82.9
0320	Concrete Reinforcing	134.2	71.9	102.5	152.9	71.8	111.7	132.8	56.7	94.1	148.7	78.8	113.2	148.7	78.8	113.2	144.3	62.7	102.7
0330	Cast-in-Place Concrete	183.9	107.3	152.4	203.9	105.1	163.3	128.3	62.0	101.1	122.3	97.8	112.2	126.4	97.7	114.6	149.4	81.9	121.7
03	CONCRETE	156.4	96.1	126.8	163.6	93.1	129.0	126.6	62.7	95.2	124.8	89.8	107.6	126.7	89.7	108.5	136.4	77.3	107.4
04	MASONRY	187.2	91.4	127.5	207.9	89.1	133.8	182.6	66.5	110.2	163.7	87.1	116.0	164.1	87.1	116.1	185.9	84.7	122.8
05	METALS	136.9	89.3	122.3	135.4	88.9	121.1	129.1	72.3	111.6	107.4	87.4	101.2	107.4	87.4	101.2	134.9	81.5	118.4
06	WOOD, PLASTICS & COMPOSITES	105.2	97.2	100.7	117.6	91.8	103.2	116.7	64.0	87.2	135.3	88.0	108.8	135.3	88.0	108.8	97.1	77.8	86.3
07	THERMAL & MOISTURE PROTECTION	122.7	94.6	111.2	121.9	91.9	109.6	116.0	63.2	94.4	107.4	89.9	100.3	107.4	88.3	99.6	119.0	79.4	102.8
08	OPENINGS	89.2	86.2	88.5	93.6	83.3	91.2	89.3	54.8	81.2	93.6	76.3	89.6	93.6	76.3	89.6	93.5	72.0	88.5
0920	Plaster & Gypsum Board	134.3	96.6	108.8	128.2	91.1	103.1	137.4	62.7	86.9	126.6	87.4	100.1	129.0	87.4	100.8	127.0	77.0	93.2
0950, 0980	Ceilings & Acoustic Treatment	139.5	96.6	111.3	104.9	91.1	95.8	116.1	62.7	81.0	96.5	87.4	90.5	96.5	87.4	90.5	118.7	77.0	91.3
0960	Flooring	122.5	92.6	114.0	120.5	92.6	112.5	116.8	75.3	104.9	120.5	99.5	114.5	120.5	99.5	114.5	109.2	88.3	103.3
0970, 0990	Wall Finishes & Painting/Coating	122.4	112.4	116.3	118.4	95.5	104.5	123.1	65.8	88.5	118.3	91.7	102.2	118.3	91.7	102.2	124.9	87.4	102.2
09	FINISHES	126.6	98.7	111.2	115.3	93.0	103.0	117.7	66.7	89.6	111.2	90.8	100.0	111.5	90.8	100.1	115.6	81.8	97.0
COVERS	DIVS. 10 - 14, 25, 28, 41, 43, 44, 46	131.1	96.9	124.2	131.1	95.2	123.9	131.1	65.3	117.8	131.1	85.2	121.9	131.1	85.2	121.9	131.1	68.5	118.5
21, 22, 23	FIRE SUPPRESSION, PLUMBING & HVAC	101.1	92.9	97.8	102.4	99.0	101.0	102.3	79.3	93.0	102.3	91.0	97.8	102.0	91.0	97.6	100.6	77.6	91.3
26, 27, 3370	ELECTRICAL, COMMUNICATIONS & UTIL.	110.6	101.3	105.7	105.8	85.4	95.1	116.1	77.1	95.5	111.2	72.6	90.8	111.8	72.6	91.1	117.0	81.2	98.1
MF2014	WEIGHTED AVERAGE	123.1	95.7	111.2	123.6	92.8	110.2	117.4	72.5	97.9	112.1	87.4	101.3	112.3	87.4	101.4	119.6	80.9	102.7

DIVISION		CANADA																	
		HAMILTON, ONTARIO			HULL, QUEBEC			JOLIETTE, QUEBEC			KAMLOOPS, BRITISH COLUMBIA			KINGSTON, ONTARIO			KITCHENER, ONTARIO		
		MAT.	INST.	TOTAL	MAT.	INST.	TOTAL	MAT.	INST.	TOTAL	MAT.	INST.	TOTAL	MAT.	INST.	TOTAL	MAT.	INST.	TOTAL
015433	CONTRACTOR EQUIPMENT		108.9	108.9		102.8	102.8		102.8	102.8		106.1	106.1		104.4	104.4		104.3	104.3
0241, 31 - 34	SITE & INFRASTRUCTURE, DEMOLITION	115.5	112.4	113.3	98.2	99.8	99.4	98.8	99.9	99.6	120.5	103.3	108.3	116.1	104.0	107.5	99.6	104.8	103.3
0310	Concrete Forming & Accessories	120.9	94.7	98.3	130.7	87.8	93.7	130.7	88.0	93.8	123.4	90.8	95.3	122.8	89.6	94.2	113.0	87.2	90.8
0320	Concrete Reinforcing	146.4	92.0	118.7	148.7	78.8	113.2	140.6	78.8	109.2	110.3	76.8	93.2	165.4	82.2	123.1	100.8	91.9	96.3
0330	Cast-in-Place Concrete	143.5	101.0	126.1	122.3	97.8	112.2	127.4	97.8	115.2	110.3	101.6	106.7	138.2	100.6	122.7	129.0	93.6	114.4
03	CONCRETE	132.8	96.3	114.8	124.8	89.8	107.6	125.9	89.8	108.2	132.9	92.0	112.8	136.2	92.2	114.6	110.8	90.4	100.8
04	MASONRY	183.0	101.2	132.0	163.7	87.1	116.0	164.1	87.1	116.1	170.7	95.0	123.5	173.5	94.1	124.0	150.8	97.6	117.6
05	METALS	123.4	92.5	113.9	107.4	87.4	101.2	107.4	87.6	101.3	109.8	87.5	102.9	110.8	89.5	104.3	117.4	92.2	109.6
06	WOOD, PLASTICS & COMPOSITES	100.4	93.9	96.8	135.3	88.0	108.8	135.3	88.0	108.8	105.2	89.4	96.4	122.1	89.2	103.7	109.7	85.5	96.1
07	THERMAL & MOISTURE PROTECTION	118.2	95.7	109.0	107.4	89.9	100.2	107.4	89.9	100.2	123.3	86.8	108.3	113.6	90.4	104.1	108.6	92.8	102.1
08	OPENINGS	91.3	91.6	91.4	93.6	76.3	89.6	93.6	81.1	90.7	90.2	85.7	89.2	93.6	85.9	91.8	84.7	85.3	84.8
0920	Plaster & Gypsum Board	149.5	93.6	111.7	126.6	87.4	100.1	157.0	87.4	109.9	113.6	88.5	96.6	192.8	88.7	122.4	120.2	84.9	96.3
0950, 0980	Ceilings & Acoustic Treatment	122.1	93.6	103.4	96.5	87.4	90.5	96.5	87.4	90.5	96.5	88.5	91.2	112.4	88.7	96.8	98.6	84.9	89.6
0960	Flooring	123.1	102.3	117.2	120.5	99.5	114.5	120.5	99.5	114.5	119.6	55.6	101.3	120.5	97.5	113.9	110.5	102.3	108.1
0970, 0990	Wall Finishes & Painting/Coating	121.6	102.6	110.2	118.3	91.7	102.2	118.3	91.7	102.2	118.3	84.0	97.6	118.3	84.9	98.1	115.3	92.6	101.6
09	FINISHES	122.0	97.0	108.2	111.2	90.8	100.0	115.3	90.8	101.8	111.7	84.2	96.6	124.4	90.9	105.9	106.4	90.2	97.5
COVERS	DIVS. 10 - 14, 25, 28, 41, 43, 44, 46	131.1	98.8	124.6	131.1	85.2	121.9	131.1	85.2	121.9	131.1	95.6	124.0	131.1	74.0	119.6	131.1	97.0	124.2
21, 22, 23	FIRE SUPPRESSION, PLUMBING & HVAC	101.3	89.2	96.4	102.0	91.0	97.6	102.0	91.0	97.6	102.0	94.4	98.9	102.3	100.3	101.5	100.4	87.7	95.3
26, 27, 3370	ELECTRICAL, COMMUNICATIONS & UTIL.	108.3	100.5	104.2	112.9	72.6	91.6	111.8	72.6	91.1	115.1	81.9	97.6	112.4	89.6	100.3	111.5	98.0	104.4
MF2014	WEIGHTED AVERAGE	117.0	96.7	108.2	112.2	87.4	101.4	112.5	87.6	101.7	114.5	90.6	104.1	116.0	93.4	106.2	109.6	92.9	102.3

		CANADA																	
	DIVISION	LAVAL, QUEBEC			LETHBRIDGE, ALBERTA			LLOYDMINSTER, ALBERTA			LONDON, ONTARIO			MEDICINE HAT, ALBERTA			MONCTON, NEW BRUNSWICK		
		MAT.	INST.	TOTAL	MAT.	INST.	TOTAL	MAT.	INST.	TOTAL	MAT.	INST.	TOTAL	MAT.	INST.	TOTAL	MAT.	INST.	TOTAL
015433	CONTRACTOR EQUIPMENT		102.8	102.8		104.4	104.4		104.4	104.4		104.4	104.4		104.4	104.4		102.2	102.2
0241, 31 - 34	SITE & INFRASTRUCTURE, DEMOLITION	98.7	99.8	99.5	115.6	101.9	105.9	115.4	101.3	105.4	113.1	104.9	107.3	114.2	101.4	105.1	102.8	96.8	98.5
0310	Concrete Forming & Accessories	130.8	87.8	93.8	124.1	92.4	96.8	122.3	82.6	88.1	119.1	88.5	92.7	124.1	82.6	88.3	104.4	64.5	69.9
0320	Concrete Reinforcing	148.7	78.8	113.2	152.9	71.8	111.7	152.9	71.8	111.7	132.8	91.9	112.0	152.9	71.8	111.7	137.0	61.5	98.6
0330	Cast-in-Place Concrete	126.4	97.8	114.6	153.0	105.1	133.3	141.9	101.4	125.3	147.4	98.6	127.4	141.9	101.4	125.3	126.8	78.7	107.0
03	CONCRETE	126.7	89.8	108.6	139.5	93.2	116.7	134.1	87.5	111.3	132.4	92.7	112.9	134.3	87.5	111.3	122.3	69.7	96.5
04	MASONRY	164.0	87.1	116.1	181.7	89.1	124.0	163.4	82.5	113.0	182.6	98.7	130.3	163.4	82.5	113.0	160.5	65.0	101.0
05	METALS	107.4	87.4	101.3	128.8	88.9	116.5	110.0	88.8	103.4	124.9	91.8	114.7	110.0	88.7	103.4	109.0	80.4	100.2
06	WOOD, PLASTICS & COMPOSITES	135.5	88.0	108.9	121.3	91.8	104.8	117.6	82.0	97.7	108.2	86.6	96.1	121.3	82.0	99.3	101.8	64.0	80.7
07	THERMAL & MOISTURE PROTECTION	107.9	89.9	100.5	119.1	91.9	108.0	116.0	87.5	104.3	119.1	93.5	108.6	122.3	87.5	108.0	110.2	65.3	91.8
08	OPENINGS	93.6	76.3	89.6	93.6	83.3	91.2	93.6	77.9	89.9	86.4	86.6	86.4	93.6	77.9	89.9	87.9	61.0	81.6
0920	Plaster & Gypsum Board	129.3	87.4	100.9	119.1	91.1	100.2	114.4	81.0	91.8	152.4	86.1	107.6	117.2	81.0	92.8	137.6	62.7	87.0
0950, 0980	Ceilings & Acoustic Treatment	96.5	87.4	90.5	104.1	91.1	95.5	96.5	81.0	86.3	121.1	86.1	98.1	96.5	81.0	86.3	106.4	62.7	77.7
0960	Flooring	120.5	99.5	114.5	120.5	92.6	112.5	120.5	92.6	112.5	117.2	102.3	113.0	120.5	92.6	112.5	107.9	72.1	97.7
0970, 0990	Wall Finishes & Painting/Coating	118.3	91.7	102.2	118.2	104.2	109.8	118.4	81.2	96.0	122.0	99.4	108.3	118.2	81.2	95.9	117.6	51.6	77.8
09	FINISHES	111.5	90.8	100.1	112.9	94.0	102.5	110.7	84.0	96.0	120.5	91.9	104.7	110.9	84.0	96.1	111.3	64.5	85.5
COVERS	DIVS. 10 - 14, 25, 28, 41, 43, 44, 46	131.1	85.2	121.9	131.1	95.2	123.9	131.1	91.9	123.2	131.1	97.5	124.3	131.1	91.9	123.2	131.1	65.3	117.8
21, 22, 23	FIRE SUPPRESSION, PLUMBING & HVAC	100.4	91.0	96.6	102.2	95.6	99.6	102.3	95.6	99.6	101.4	86.6	95.4	102.0	92.3	98.1	102.0	70.1	89.1
26, 27, 3370	ELECTRICAL, COMMUNICATIONS & UTIL.	112.9	72.6	91.6	107.1	85.4	95.7	104.9	85.4	94.6	107.0	97.5	102.0	104.9	85.4	94.6	117.2	61.8	88.0
MF2014	WEIGHTED AVERAGE	112.1	87.4	101.3	118.4	92.2	107.0	113.4	88.9	102.8	116.4	93.3	106.3	113.6	88.2	102.5	111.8	70.1	93.6

		CANADA																	
	DIVISION	MONTREAL, QUEBEC			MOOSE JAW, SASKATCHEWAN			NEW GLASGOW, NOVA SCOTIA			NEWCASTLE, NEW BRUNSWICK			NORTH BAY, ONTARIO			OSHAWA, ONTARIO		
		MAT.	INST.	TOTAL	MAT.	INST.	TOTAL	MAT.	INST.	TOTAL	MAT.	INST.	TOTAL	MAT.	INST.	TOTAL	MAT.	INST.	TOTAL
015433	CONTRACTOR EQUIPMENT		104.5	104.5		100.6	100.6		101.9	101.9		102.2	102.2		102.1	102.1		104.3	104.3
0241, 31 - 34	SITE & INFRASTRUCTURE, DEMOLITION	112.1	99.4	103.1	115.5	95.9	101.6	118.5	98.2	104.1	103.4	96.6	98.6	133.9	99.8	109.7	110.7	103.9	105.9
0310	Concrete Forming & Accessories	124.3	93.4	97.7	107.1	60.5	66.9	111.7	71.6	77.1	104.4	63.9	69.5	147.4	87.0	95.3	118.4	89.1	93.1
0320	Concrete Reinforcing	134.0	96.0	114.7	107.9	61.9	84.5	170.3	46.7	107.4	137.0	56.6	96.1	201.7	81.8	140.7	159.6	84.4	121.3
0330	Cast-in-Place Concrete	161.7	101.2	136.8	137.7	71.1	110.4	149.8	72.2	118.0	131.6	62.0	103.0	144.3	87.1	120.8	149.1	87.9	124.0
03	CONCRETE	139.7	96.5	118.4	117.9	65.1	92.0	160.9	67.9	115.2	124.6	62.5	94.1	163.4	86.3	125.5	135.6	88.0	112.2
04	MASONRY	167.9	95.1	122.5	162.0	64.4	101.1	228.9	71.9	131.1	160.8	64.9	101.1	235.3	89.9	144.7	153.8	94.4	116.8
05	METALS	128.0	92.7	117.1	106.3	73.9	96.3	123.6	74.2	108.4	109.0	71.8	97.5	124.6	89.2	113.7	108.6	91.4	103.3
06	WOOD, PLASTICS & COMPOSITES	114.4	93.9	102.9	102.7	59.1	78.3	116.1	71.1	90.9	101.8	64.0	80.7	157.4	87.6	118.3	116.7	88.2	100.7
07	THERMAL & MOISTURE PROTECTION	116.5	96.9	108.5	106.8	64.1	89.3	136.8	69.9	109.4	110.2	62.2	90.5	143.5	85.6	119.8	109.4	87.4	100.4
08	OPENINGS	90.1	82.4	88.3	89.3	56.8	81.7	95.3	64.9	88.2	87.9	55.9	80.5	104.2	83.7	99.4	90.4	87.9	89.8
0920	Plaster & Gypsum Board	135.1	93.2	106.8	110.7	57.8	74.9	142.7	70.0	93.6	137.6	62.7	87.0	140.3	87.1	104.4	124.7	87.7	99.7
0950, 0980	Ceilings & Acoustic Treatment	126.1	93.2	104.4	96.5	57.8	71.0	109.6	70.0	83.6	106.4	62.7	77.7	109.6	87.1	94.8	95.3	87.7	90.3
0960	Flooring	122.0	99.5	115.6	110.6	63.3	97.1	116.1	67.2	102.1	107.9	72.1	97.7	138.6	97.5	126.9	113.7	104.9	111.2
0970, 0990	Wall Finishes & Painting/Coating	120.8	106.4	112.1	118.3	66.8	87.2	126.4	64.1	88.8	117.6	51.6	77.8	126.4	91.2	105.1	115.3	106.7	110.1
09	FINISHES	119.9	96.7	107.1	107.1	61.4	81.9	118.2	70.3	91.8	111.3	64.5	85.5	125.1	89.8	105.7	107.9	93.9	100.2
COVERS	DIVS. 10 - 14, 25, 28, 41, 43, 44, 46	131.1	87.1	122.3	131.1	64.6	117.7	131.1	66.6	118.1	131.1	65.3	117.8	131.1	72.7	119.3	131.1	97.2	124.3
21, 22, 23	FIRE SUPPRESSION, PLUMBING & HVAC	101.2	92.4	97.7	102.3	76.1	91.8	102.1	84.4	95.0	102.0	69.8	89.0	102.1	98.1	100.5	100.4	100.7	100.5
26, 27, 3370	ELECTRICAL, COMMUNICATIONS & UTIL.	112.6	86.8	99.0	113.9	62.8	86.9	116.3	64.4	88.9	112.0	61.8	85.5	116.4	90.8	102.9	112.6	90.5	100.9
MF2014	WEIGHTED AVERAGE	117.8	93.2	107.1	110.7	69.5	92.7	124.1	74.6	102.5	111.6	67.9	92.5	127.3	91.1	111.5	112.3	94.3	104.4

		CANADA																	
	DIVISION	OTTAWA, ONTARIO			OWEN SOUND, ONTARIO			PETERBOROUGH, ONTARIO			PORTAGE LA PRAIRIE, MANITOBA			PRINCE ALBERT, SASKATCHEWAN			PRINCE GEORGE, BRITISH COLUMBIA		
		MAT.	INST.	TOTAL	MAT.	INST.	TOTAL	MAT.	INST.	TOTAL	MAT.	INST.	TOTAL	MAT.	INST.	TOTAL	MAT.	INST.	TOTAL
015433	CONTRACTOR EQUIPMENT		104.3	104.3		102.1	102.1		102.1	102.1		104.5	104.5		100.6	100.6		106.1	106.1
0241, 31 - 34	SITE & INFRASTRUCTURE, DEMOLITION	108.2	104.6	105.7	118.5	100.3	105.5	118.0	100.2	105.3	116.6	99.3	104.3	110.9	96.1	100.3	124.0	103.3	109.3
0310	Concrete Forming & Accessories	121.2	91.4	95.5	124.4	85.5	90.9	124.9	88.1	93.1	124.3	72.0	79.2	107.1	60.3	66.8	112.7	85.5	89.3
0320	Concrete Reinforcing	138.9	91.9	115.0	173.2	83.8	127.7	165.4	82.3	123.1	152.9	53.2	102.2	112.7	61.8	86.8	110.3	76.8	93.2
0330	Cast-in-Place Concrete	145.6	99.9	126.8	165.6	83.1	131.7	153.6	88.7	127.0	141.9	76.1	114.9	124.8	71.0	102.7	138.2	101.6	123.2
03	CONCRETE	132.6	94.4	113.8	150.3	84.6	118.1	141.8	87.4	115.1	127.8	70.6	99.7	112.6	65.0	89.2	145.4	89.7	118.0
04	MASONRY	168.0	98.8	124.9	171.5	95.6	124.2	168.0	96.7	123.5	166.5	66.6	104.2	161.1	64.4	100.8	172.9	95.0	124.3
05	METALS	127.9	91.7	116.8	110.2	90.0	104.0	109.1	89.7	103.1	110.0	76.7	99.7	106.3	73.7	96.3	109.8	87.6	102.9
06	WOOD, PLASTICS & COMPOSITES	109.4	90.6	98.9	119.8	84.3	99.9	123.7	86.3	102.8	121.1	73.5	94.4	102.7	59.1	78.3	105.2	82.1	92.3
07	THERMAL & MOISTURE PROTECTION	124.4	93.8	111.8	111.8	86.4	101.4	113.8	91.2	104.5	107.3	71.3	92.5	106.7	63.0	88.8	117.2	86.1	104.4
08	OPENINGS	93.9	88.7	92.7	93.9	83.1	91.4	92.4	85.5	90.8	93.6	64.8	86.9	88.2	56.8	80.9	90.2	81.8	88.2
0920	Plaster & Gypsum Board	168.3	90.2	115.5	151.1	83.7	105.6	130.1	85.8	100.1	114.2	72.2	85.8	110.7	57.8	74.9	113.6	81.0	91.6
0950, 0980	Ceilings & Acoustic Treatment	125.6	90.2	102.4	91.5	83.7	86.4	96.5	85.8	89.4	96.5	72.2	80.5	96.5	57.8	71.0	96.5	81.0	86.3
0960	Flooring	112.3	97.5	108.1	127.2	98.9	119.1	120.5	97.5	113.9	120.5	70.2	106.1	110.6	63.3	97.1	116.0	76.2	104.6
0970, 0990	Wall Finishes & Painting/Coating	123.2	94.2	105.7	116.2	89.9	100.3	118.3	93.5	103.3	118.4	58.7	82.4	118.3	57.0	81.3	118.3	84.0	97.6
09	FINISHES	122.6	92.9	106.2	115.0	88.6	100.5	112.4	90.5	100.3	110.5	70.8	88.6	107.1	60.3	81.3	110.6	83.4	95.6
COVERS	DIVS. 10 - 14, 25, 28, 41, 43, 44, 46	131.1	95.7	124.0	139.2	73.2	125.9	131.1	74.2	119.6	131.1	67.2	118.2	131.1	64.6	117.7	131.1	94.8	123.8
21, 22, 23	FIRE SUPPRESSION, PLUMBING & HVAC	101.4	87.2	95.7	101.8	98.1	100.3	102.0	101.7	101.9	102.0	83.6	94.5	102.3	68.6	88.7	102.0	94.4	98.9
26, 27, 3370	ELECTRICAL, COMMUNICATIONS & UTIL.	107.3	98.7	102.8	118.2	89.6	103.1	111.5	90.4	100.4	113.6	60.1	85.3	113.9	62.8	86.9	111.7	81.9	96.0
MF2014	WEIGHTED AVERAGE	117.3	94.0	107.2	117.8	91.2	106.2	114.9	93.1	105.4	113.4	74.2	96.3	109.9	67.7	91.5	115.4	89.9	104.3

	DIVISION	CANADA																	
		QUEBEC, QUEBEC			RED DEER, ALBERTA			REGINA, SASKATCHEWAN			RIMOUSKI, QUEBEC			ROUYN-NORANDA, QUEBEC			SAINT HYACINTHE, QUEBEC		
		MAT.	INST.	TOTAL	MAT.	INST.	TOTAL	MAT.	INST.	TOTAL	MAT.	INST.	TOTAL	MAT.	INST.	TOTAL	MAT.	INST.	TOTAL
015433	CONTRACTOR EQUIPMENT		104.9	104.9		104.4	104.4		100.6	100.6		102.8	102.8		102.8	102.8		102.8	102.8
0241, 31 - 34	SITE & INFRASTRUCTURE, DEMOLITION	110.8	99.5	102.8	114.2	101.4	105.1	124.4	98.0	105.6	98.7	99.1	99.0	98.2	99.8	99.4	98.7	99.8	99.5
0310	Concrete Forming & Accessories	125.6	93.7	98.1	139.6	82.6	90.4	117.6	90.9	94.6	130.7	93.1	98.3	130.7	87.8	93.7	130.7	87.8	93.7
0320	Concrete Reinforcing	138.3	96.0	116.8	152.9	71.8	111.7	138.0	84.8	111.0	106.0	95.9	100.9	148.7	78.8	113.2	148.7	78.8	113.2
0330	Cast-in-Place Concrete	141.8	101.6	125.2	141.9	101.4	125.3	171.5	95.7	140.4	128.6	99.8	116.8	122.3	97.8	112.2	126.4	97.8	114.6
03	CONCRETE	131.0	96.7	114.1	135.3	87.5	111.8	148.9	91.4	120.6	121.1	95.9	108.7	124.8	89.8	107.6	126.7	89.8	108.5
04	MASONRY	163.0	95.1	120.7	163.4	82.5	113.0	183.5	96.7	129.4	163.6	95.0	120.9	163.7	87.1	116.0	164.0	87.1	116.1
05	METALS	128.0	93.0	117.2	110.0	88.7	103.4	129.3	86.4	116.1	106.9	92.2	102.4	107.4	87.4	101.2	107.4	87.4	101.2
06	WOOD, PLASTICS & COMPOSITES	115.8	93.9	103.6	121.3	82.0	99.3	101.9	91.9	96.3	135.3	93.6	112.0	135.3	88.0	108.8	135.3	88.0	108.8
07	THERMAL & MOISTURE PROTECTION	118.4	97.0	109.6	132.2	87.5	113.9	123.2	86.0	107.9	107.4	96.4	102.9	107.4	89.9	100.2	107.7	89.9	100.4
08	OPENINGS	93.0	90.1	92.3	93.6	77.9	89.9	91.9	80.7	89.3	93.2	80.6	90.2	93.6	76.3	89.6	93.6	76.3	89.6
0920	Plaster & Gypsum Board	142.1	93.2	109.0	117.2	81.0	92.8	130.5	91.6	104.2	156.8	93.2	113.8	126.4	87.4	100.0	128.7	87.4	100.8
0950, 0980	Ceilings & Acoustic Treatment	126.1	93.2	104.4	96.5	81.0	86.3	125.4	91.6	103.2	95.7	93.2	94.0	95.7	87.4	90.2	95.7	87.4	90.2
0960	Flooring	125.8	99.5	118.3	123.0	92.6	114.3	120.0	65.5	104.4	120.5	99.5	114.5	120.5	99.5	114.5	120.5	99.5	114.5
0970, 0990	Wall Finishes & Painting/Coating	127.8	106.4	114.9	118.2	81.2	95.9	123.9	92.9	105.2	118.3	106.4	111.1	118.3	91.7	102.2	118.3	91.7	102.2
09	FINISHES	122.2	96.8	108.2	111.7	84.0	96.4	122.7	87.2	103.1	115.1	96.5	104.9	111.0	90.8	99.9	111.3	90.8	100.0
COVERS	DIVS. 10 - 14, 25, 28, 41, 43, 44, 46	131.1	87.3	122.3	131.1	91.9	123.2	131.1	72.4	119.3	131.1	86.5	122.1	131.1	85.2	121.9	131.1	85.2	121.9
21, 22, 23	FIRE SUPPRESSION, PLUMBING & HVAC	101.2	92.4	97.7	102.0	92.3	98.1	100.7	84.8	94.3	102.0	92.3	98.1	102.0	91.0	97.6	98.8	91.0	95.7
26, 27, 3370	ELECTRICAL, COMMUNICATIONS & UTIL.	111.4	86.8	98.4	104.9	85.4	94.6	114.4	92.4	102.8	111.8	86.8	98.6	111.8	72.6	91.1	112.5	72.6	91.4
MF2014	WEIGHTED AVERAGE	117.1	93.6	106.8	114.1	88.2	102.8	120.7	89.0	106.9	111.8	92.9	103.6	112.0	87.4	101.3	111.6	87.4	101.1

	DIVISION	CANADA																	
		SAINT JOHN, NEW BRUNSWICK			SARNIA, ONTARIO			SASKATOON, SASKATCHEWAN			SAULT STE MARIE, ONTARIO			SHERBROOKE, QUEBEC			SOREL, QUEBEC		
		MAT.	INST.	TOTAL	MAT.	INST.	TOTAL	MAT.	INST.	TOTAL	MAT.	INST.	TOTAL	MAT.	INST.	TOTAL	MAT.	INST.	TOTAL
015433	CONTRACTOR EQUIPMENT		102.2	102.2		102.1	102.1		100.6	100.6		102.1	102.1		102.8	102.8		102.8	102.8
0241, 31 - 34	SITE & INFRASTRUCTURE, DEMOLITION	103.5	98.0	99.6	116.5	100.3	105.0	111.7	98.0	102.0	106.7	99.8	101.8	98.7	99.8	99.5	98.8	99.9	99.6
0310	Concrete Forming & Accessories	124.6	67.0	74.9	123.6	95.3	99.1	107.4	90.9	93.2	112.8	87.9	91.3	130.7	87.8	93.7	130.7	88.0	93.8
0320	Concrete Reinforcing	137.0	62.0	98.8	117.4	83.6	100.2	114.5	84.8	99.4	106.1	82.1	93.9	148.7	78.8	113.2	140.6	78.8	109.2
0330	Cast-in-Place Concrete	129.7	80.1	109.3	141.7	102.1	125.5	135.4	95.7	119.1	127.4	87.2	110.9	126.4	97.8	114.6	127.4	97.8	115.2
03	CONCRETE	125.2	71.4	98.8	128.6	95.4	112.3	119.6	91.4	105.7	113.1	86.8	100.2	126.7	89.8	108.5	125.9	89.8	108.2
04	MASONRY	180.8	73.7	114.0	179.1	99.5	129.5	168.8	96.7	123.8	164.5	94.4	120.8	164.1	87.1	116.1	164.1	87.1	116.1
05	METALS	108.9	81.7	100.5	109.1	90.1	103.2	104.5	86.4	99.0	108.3	90.2	102.7	107.4	87.4	101.2	107.4	87.6	101.3
06	WOOD, PLASTICS & COMPOSITES	125.7	65.4	91.9	122.6	94.7	107.0	100.5	91.9	95.7	109.6	88.3	97.6	135.3	88.0	108.8	135.3	88.0	108.8
07	THERMAL & MOISTURE PROTECTION	110.5	70.9	94.3	113.9	94.7	106.0	107.1	86.0	98.4	112.7	88.4	102.7	107.4	89.9	100.3	107.4	89.9	100.2
08	OPENINGS	87.8	60.6	81.5	94.7	89.4	93.4	88.9	80.7	87.0	86.5	85.6	86.3	93.6	76.3	89.6	93.6	81.1	90.7
0920	Plaster & Gypsum Board	152.7	64.2	92.9	150.6	94.4	112.6	123.5	91.6	101.9	119.6	87.8	98.1	128.7	87.4	100.8	156.8	87.4	109.9
0950, 0980	Ceilings & Acoustic Treatment	111.5	64.2	80.4	100.7	94.4	96.6	117.4	91.6	100.4	96.5	87.8	90.8	95.7	87.4	90.2	95.7	87.4	90.2
0960	Flooring	119.5	72.1	105.9	120.5	106.6	116.5	111.0	65.5	98.0	113.4	100.8	109.8	120.5	99.5	114.5	120.5	99.5	114.5
0970, 0990	Wall Finishes & Painting/Coating	117.6	80.2	95.0	118.3	105.7	110.7	120.7	92.9	103.9	118.3	97.9	106.0	118.3	91.7	102.2	118.3	91.7	102.2
09	FINISHES	117.9	69.0	91.0	116.1	98.9	106.6	114.3	87.2	99.3	108.1	91.5	99.0	111.3	90.8	100.0	115.1	90.8	101.7
COVERS	DIVS. 10 - 14, 25, 28, 41, 43, 44, 46	131.1	66.3	118.0	131.1	75.7	119.9	131.1	74.1	119.6	131.1	96.4	124.1	131.1	85.2	121.9	131.1	85.2	121.9
21, 22, 23	FIRE SUPPRESSION, PLUMBING & HVAC	102.0	79.8	93.0	102.0	107.8	104.3	100.5	84.8	94.2	102.0	94.8	99.1	102.3	91.0	97.8	102.0	91.0	97.6
26, 27, 3370	ELECTRICAL, COMMUNICATIONS & UTIL.	120.3	88.2	103.3	114.2	92.7	102.8	116.2	92.4	103.6	113.0	90.8	101.3	111.8	72.6	91.1	111.8	72.6	91.1
MF2014	WEIGHTED AVERAGE	114.0	77.9	98.3	114.8	97.5	107.3	111.1	89.1	101.5	110.3	92.1	102.3	112.4	87.4	101.5	112.5	87.6	101.7

	DIVISION	CANADA																	
		ST CATHARINES, ONTARIO			ST JEROME, QUEBEC			ST JOHNS, NEWFOUNDLAND			SUDBURY, ONTARIO			SUMMERSIDE, PRINCE EDWARD ISLAND			SYDNEY, NOVA SCOTIA		
		MAT.	INST.	TOTAL	MAT.	INST.	TOTAL	MAT.	INST.	TOTAL	MAT.	INST.	TOTAL	MAT.	INST.	TOTAL	MAT.	INST.	TOTAL
015433	CONTRACTOR EQUIPMENT		102.1	102.1		102.8	102.8		103.3	103.3		102.1	102.1		101.8	101.8		101.9	101.9
0241, 31 - 34	SITE & INFRASTRUCTURE, DEMOLITION	100.0	101.4	101.0	98.2	99.8	99.4	123.4	99.9	106.7	100.1	100.9	100.7	128.4	95.7	105.2	114.3	98.2	102.8
0310	Concrete Forming & Accessories	110.9	94.2	96.5	130.7	87.8	93.7	123.9	84.8	90.1	107.0	89.0	91.5	112.0	58.3	65.7	111.7	71.6	77.1
0320	Concrete Reinforcing	101.7	92.0	96.7	148.7	78.8	113.2	158.9	75.5	116.4	102.5	91.3	96.8	168.1	46.8	106.4	170.3	46.7	107.4
0330	Cast-in-Place Concrete	123.2	100.6	113.9	122.3	97.8	112.2	167.5	96.4	138.3	124.2	97.0	113.0	142.3	61.2	109.0	115.6	72.2	97.8
03	CONCRETE	108.0	95.9	102.1	124.8	89.8	107.6	152.6	87.2	120.5	108.4	92.2	100.4	170.1	57.9	115.0	144.7	67.9	106.9
04	MASONRY	150.3	100.7	119.4	163.7	87.1	116.0	187.7	92.4	128.3	150.4	96.6	116.8	228.1	62.6	125.0	226.2	71.9	130.1
05	METALS	107.7	92.2	102.9	107.4	87.4	101.2	136.3	84.3	120.3	107.7	91.5	102.7	123.6	67.6	106.4	123.6	74.2	108.4
06	WOOD, PLASTICS & COMPOSITES	107.4	93.9	99.8	135.3	88.0	108.8	114.7	82.8	96.8	103.6	88.3	95.0	116.4	57.8	83.6	116.1	71.1	90.9
07	THERMAL & MOISTURE PROTECTION	108.6	96.5	103.6	107.4	89.9	100.2	127.1	91.4	112.5	108.0	91.7	101.3	136.1	61.1	105.4	136.8	69.9	109.4
08	OPENINGS	84.1	90.2	85.5	93.6	76.3	89.6	93.7	74.5	89.2	84.8	85.6	85.0	108.0	50.0	94.5	95.3	64.9	88.2
0920	Plaster & Gypsum Board	108.5	93.6	98.4	126.4	87.4	100.0	143.9	82.1	102.1	112.9	87.8	95.9	143.1	56.3	84.4	142.7	70.0	93.6
0950, 0980	Ceilings & Acoustic Treatment	95.3	93.6	94.1	95.7	87.4	90.2	120.6	82.1	95.3	90.3	87.8	88.6	109.6	56.3	74.6	109.6	70.0	83.6
0960	Flooring	109.3	98.9	106.3	120.5	99.5	114.5	116.0	59.3	99.8	107.3	100.8	105.5	116.2	62.9	100.9	116.1	67.2	102.1
0970, 0990	Wall Finishes & Painting/Coating	115.3	116.0	115.7	118.3	91.7	102.2	120.4	95.9	105.6	115.3	93.6	102.2	126.4	42.9	76.0	126.4	64.1	88.8
09	FINISHES	103.8	97.8	100.4	111.0	90.8	99.9	122.0	81.4	99.6	102.7	91.6	96.5	119.2	57.8	85.4	118.2	70.3	91.8
COVERS	DIVS. 10 - 14, 25, 28, 41, 43, 44, 46	131.1	75.2	119.8	131.1	85.2	121.9	131.1	72.6	119.3	131.1	97.1	124.3	131.1	64.8	117.7	131.1	66.6	118.1
21, 22, 23	FIRE SUPPRESSION, PLUMBING & HVAC	100.4	87.8	95.3	102.0	91.0	97.6	100.8	82.1	93.3	101.9	87.8	96.2	102.1	63.8	86.7	102.1	84.4	95.0
26, 27, 3370	ELECTRICAL, COMMUNICATIONS & UTIL.	112.9	98.8	105.5	112.4	72.6	91.3	114.4	83.2	97.9	110.3	99.2	104.4	115.4	52.3	82.0	116.3	64.4	88.9
MF2014	WEIGHTED AVERAGE	107.6	94.5	101.9	112.1	87.4	101.3	122.7	85.2	106.4	107.7	93.0	101.3	126.7	62.8	98.8	122.1	74.6	101.4

	DIVISION	CANADA																	
		THUNDER BAY, ONTARIO			TIMMINS, ONTARIO			TORONTO, ONTARIO			TROIS RIVIERES, QUEBEC			TRURO, NOVA SCOTIA			VANCOUVER, BRITISH COLUMBIA		
		MAT.	INST.	TOTAL	MAT.	INST.	TOTAL	MAT.	INST.	TOTAL	MAT.	INST.	TOTAL	MAT.	INST.	TOTAL	MAT.	INST.	TOTAL
015433	CONTRACTOR EQUIPMENT		102.1	102.1		102.1	102.1		104.4	104.4		102.8	102.8		101.9	101.9		112.7	112.7
0241, 31 - 34	SITE & INFRASTRUCTURE, DEMOLITION	104.8	101.3	102.3	118.0	99.8	105.1	133.8	105.6	113.7	114.8	99.9	104.2	102.8	98.2	99.5	118.2	107.6	110.6
0310	Concrete Forming & Accessories	118.3	92.2	95.8	124.9	87.0	92.2	124.3	100.8	104.0	155.5	88.0	97.3	97.1	71.6	75.1	121.1	91.0	95.1
0320	Concrete Reinforcing	90.9	90.8	90.9	165.4	81.8	122.9	146.0	94.4	119.7	170.3	78.8	123.8	140.6	46.7	92.8	134.1	79.2	106.2
0330	Cast-in-Place Concrete	135.6	99.7	120.8	153.6	87.1	126.3	136.8	111.8	126.5	119.7	97.8	110.7	159.5	72.2	123.7	143.5	97.9	124.8
03	CONCRETE	115.8	94.5	105.3	141.8	86.3	114.5	129.7	103.1	116.7	147.4	89.8	119.1	140.2	67.9	104.7	137.1	91.4	114.6
04	MASONRY	151.0	100.6	119.6	168.0	89.9	119.3	182.0	110.9	137.7	230.7	87.1	141.2	163.6	71.9	106.5	166.7	91.4	119.7
05	METALS	107.5	91.1	102.5	109.1	89.2	103.0	132.5	93.9	120.7	122.7	87.6	111.9	108.2	74.2	97.7	139.3	90.3	124.2
06	WOOD, PLASTICS & COMPOSITES	116.7	90.9	102.2	123.7	87.6	103.5	115.1	98.9	106.0	173.0	88.0	125.4	92.9	71.1	80.7	104.8	91.5	97.4
07	THERMAL & MOISTURE PROTECTION	108.8	93.8	102.7	113.8	85.6	102.2	123.9	102.8	115.2	135.2	89.9	116.6	108.6	69.9	92.8	125.2	86.9	109.5
08	OPENINGS	83.3	88.0	84.4	92.4	83.7	90.4	90.5	96.8	92.0	105.5	81.1	99.8	86.1	64.9	81.1	88.9	86.9	88.4
0920	Plaster & Gypsum Board	139.8	90.5	106.5	130.1	87.1	101.1	133.0	98.8	109.9	173.4	87.4	115.3	130.7	70.0	89.7	129.7	90.6	103.3
0950, 0980	Ceilings & Acoustic Treatment	90.3	90.5	90.4	96.5	87.1	90.3	114.9	98.8	104.3	108.8	87.4	94.7	96.5	70.0	79.1	132.2	90.6	104.8
0960	Flooring	113.7	105.9	111.4	120.5	97.5	113.9	117.0	108.4	114.6	138.6	99.5	127.5	102.7	67.2	92.5	124.6	96.0	116.4
0970, 0990	Wall Finishes & Painting/Coating	115.3	94.6	102.8	118.3	91.2	101.9	120.4	106.7	112.1	126.4	91.7	105.4	118.3	64.1	85.6	121.7	95.6	106.0
09	FINISHES	108.5	95.1	101.1	112.4	89.8	99.9	116.9	103.0	109.2	128.7	90.8	107.8	106.6	70.3	86.6	123.8	92.7	106.6
COVERS	DIVS. 10 - 14, 25, 28, 41, 43, 44, 46	131.1	75.2	119.8	131.1	72.7	119.3	131.1	101.2	125.1	131.1	85.2	121.9	131.1	66.6	118.1	131.1	94.9	123.8
21, 22, 23	FIRE SUPPRESSION, PLUMBING & HVAC	100.4	88.0	95.4	102.0	98.1	100.4	101.0	97.3	99.5	102.1	91.0	97.7	102.0	84.4	94.9	101.2	80.5	92.8
26, 27, 3370	ELECTRICAL, COMMUNICATIONS & UTIL.	111.5	97.0	103.8	113.0	90.8	101.3	108.8	101.0	104.7	116.6	72.6	93.4	111.5	64.4	86.6	109.0	83.3	95.4
MF2014	WEIGHTED AVERAGE	108.8	93.4	102.1	115.1	91.1	104.6	118.3	101.3	110.9	124.8	87.6	108.6	112.5	74.6	96.0	119.5	89.1	106.2

	DIVISION	CANADA																	
		VICTORIA, BRITISH COLUMBIA			WHITEHORSE, YUKON			WINDSOR, ONTARIO			WINNIPEG, MANITOBA			YARMOUTH, NOVA SCOTIA			YELLOWKNIFE, NWT		
		MAT.	INST.	TOTAL	MAT.	INST.	TOTAL	MAT.	INST.	TOTAL	MAT.	INST.	TOTAL	MAT.	INST.	TOTAL	MAT.	INST.	TOTAL
015433	CONTRACTOR EQUIPMENT		109.3	109.3		102.1	102.1		102.1	102.1		106.5	106.5		101.9	101.9		101.9	101.9
0241, 31 - 34	SITE & INFRASTRUCTURE, DEMOLITION	123.0	107.1	111.7	142.3	96.7	109.9	95.9	101.3	99.7	118.2	100.9	105.9	118.3	98.2	104.0	154.1	101.0	116.4
0310	Concrete Forming & Accessories	112.7	90.4	93.5	129.4	59.6	69.2	118.3	90.5	94.3	128.5	69.3	77.4	111.7	71.6	77.1	132.7	80.2	87.4
0320	Concrete Reinforcing	112.0	79.1	95.2	160.8	60.0	109.5	99.6	91.9	95.7	132.9	57.9	94.7	170.3	46.7	107.4	151.4	62.7	106.2
0330	Cast-in-Place Concrete	140.3	97.2	122.6	209.3	70.3	152.2	126.2	101.9	116.2	163.9	75.0	127.4	148.3	72.2	117.0	211.2	88.1	160.6
03	CONCRETE	150.1	90.7	120.9	182.4	64.1	124.3	109.6	94.7	102.3	144.4	69.7	107.7	160.1	67.9	114.8	182.0	80.0	131.9
04	MASONRY	171.7	91.3	121.6	260.3	62.8	137.2	150.5	99.9	119.0	176.6	70.8	110.7	228.8	71.9	131.0	260.3	74.8	144.7
05	METALS	107.0	86.3	100.6	135.3	73.6	116.3	107.6	91.7	102.7	141.7	75.0	121.2	123.6	74.2	108.4	137.9	78.7	119.7
06	WOOD, PLASTICS & COMPOSITES	104.4	91.4	97.1	117.6	58.3	84.4	116.7	89.0	101.2	113.6	70.2	89.3	116.1	71.1	90.9	124.0	81.9	100.4
07	THERMAL & MOISTURE PROTECTION	116.9	86.7	104.5	139.1	62.0	107.5	108.6	93.9	102.6	120.8	71.8	100.7	136.8	69.9	109.4	151.5	78.4	121.5
08	OPENINGS	90.5	83.2	88.8	107.6	55.5	95.4	83.1	87.5	84.1	85.5	62.6	80.2	95.3	64.9	88.2	102.2	68.9	94.4
0920	Plaster & Gypsum Board	115.8	90.6	98.8	150.5	56.8	87.2	125.7	88.6	100.6	129.5	68.6	88.3	142.7	70.0	93.6	163.7	81.1	107.9
0950, 0980	Ceilings & Acoustic Treatment	98.1	90.6	93.2	144.6	56.8	86.9	90.3	88.6	89.2	136.2	68.6	91.8	109.6	70.0	83.6	138.9	81.1	100.9
0960	Flooring	116.4	76.2	104.9	141.0	61.3	118.2	113.7	103.1	110.6	117.0	75.6	105.2	116.1	67.2	102.1	135.5	92.5	123.2
0970, 0990	Wall Finishes & Painting/Coating	120.7	95.6	105.5	143.1	56.0	90.5	115.3	95.5	103.3	124.5	56.5	83.4	126.4	64.1	88.8	135.3	80.3	102.1
09	FINISHES	112.1	89.3	99.5	140.9	59.3	95.9	106.3	93.4	99.2	122.3	69.5	93.2	118.2	70.3	91.8	137.5	82.0	106.9
COVERS	DIVS. 10 - 14, 25, 28, 41, 43, 44, 46	131.1	71.5	119.1	131.1	63.9	117.6	131.1	74.7	119.7	131.1	68.7	118.5	131.1	66.6	118.1	131.1	67.1	118.2
21, 22, 23	FIRE SUPPRESSION, PLUMBING & HVAC	102.0	80.5	93.3	102.5	74.8	91.3	100.4	88.0	95.4	101.1	66.9	87.3	102.1	84.4	95.0	102.7	94.0	99.2
26, 27, 3370	ELECTRICAL, COMMUNICATIONS & UTIL.	113.1	82.7	97.0	133.5	62.2	95.8	116.0	99.0	107.0	113.3	68.0	89.4	116.3	64.4	88.9	130.5	85.2	106.5
MF2014	WEIGHTED AVERAGE	115.7	87.2	103.3	135.3	68.5	106.2	108.2	93.4	101.7	121.0	71.8	99.5	124.0	74.6	102.5	135.3	84.2	113.0

Location Factors

Costs shown in RSMeans cost data publications are based on national averages for materials and installation. To adjust these costs to a specific location, simply multiply the base cost by the factor and divide by 100 for that city. The data is arranged alphabetically by state and postal zip code numbers. For a city not listed, use the factor for a nearby city with similar economic characteristics.

STATE/ZIP	CITY	MAT.	INST.	TOTAL
ALABAMA				
350-352	Birmingham	100.6	76.7	90.2
354	Tuscaloosa	99.8	61.2	83.0
355	Jasper	100.2	57.5	81.6
356	Decatur	99.8	57.9	81.5
357-358	Huntsville	99.8	72.1	87.8
359	Gadsden	99.9	60.0	82.5
360-361	Montgomery	99.9	57.1	81.2
362	Anniston	98.6	61.8	82.6
363	Dothan	99.0	49.4	77.4
364	Evergreen	98.6	50.7	77.7
365-366	Mobile	99.9	65.4	84.9
367	Selma	98.7	49.9	77.4
368	Phenix City	99.5	54.6	79.9
369	Butler	98.9	49.4	77.3
ALASKA				
995-996	Anchorage	121.1	115.6	118.7
997	Fairbanks	121.0	116.4	119.0
998	Juneau	121.3	115.6	118.8
999	Ketchikan	132.0	115.6	124.8
ARIZONA				
850,853	Phoenix	99.2	73.8	88.1
851,852	Mesa/Tempe	98.6	72.8	87.4
855	Globe	98.8	70.0	86.3
856-857	Tucson	97.4	72.4	86.5
859	Show Low	98.9	72.5	87.4
860	Flagstaff	101.3	71.8	88.4
863	Prescott	99.2	68.9	86.0
864	Kingman	97.6	72.0	86.4
865	Chambers	97.6	71.5	86.2
ARKANSAS				
716	Pine Bluff	99.4	63.3	83.6
717	Camden	97.1	50.1	76.6
718	Texarkana	98.2	49.2	76.8
719	Hot Springs	96.4	51.8	76.9
720-722	Little Rock	98.5	64.0	83.4
723	West Memphis	96.1	62.4	81.4
724	Jonesboro	96.7	59.1	80.3
725	Batesville	94.6	54.6	77.1
726	Harrison	95.9	50.4	76.0
727	Fayetteville	93.4	52.0	75.4
728	Russellville	94.6	53.7	76.8
729	Fort Smith	97.2	61.6	81.7
CALIFORNIA				
900-902	Los Angeles	99.2	117.5	107.2
903-905	Inglewood	94.8	114.1	103.2
906-908	Long Beach	96.2	114.1	104.0
910-912	Pasadena	96.4	114.6	104.3
913-916	Van Nuys	99.1	114.6	105.9
917-918	Alhambra	98.3	114.8	105.5
919-921	San Diego	100.5	109.6	104.5
922	Palm Springs	97.2	112.0	103.6
923-924	San Bernardino	95.1	111.3	102.1
925	Riverside	99.4	114.6	106.0
926-927	Santa Ana	96.9	111.8	103.4
928	Anaheim	99.4	114.9	106.2
930	Oxnard	99.6	114.4	106.0
931	Santa Barbara	98.9	115.1	106.0
932-933	Bakersfield	100.7	113.6	106.3
934	San Luis Obispo	99.4	112.0	104.9
935	Mojave	96.8	109.8	102.5
936-938	Fresno	100.1	115.3	106.7
939	Salinas	100.2	121.6	109.5
940-941	San Francisco	105.2	145.4	122.7
942,956-958	Sacramento	100.5	119.3	108.7
943	Palo Alto	98.9	134.5	114.4
944	San Mateo	100.9	134.6	115.6
945	Vallejo	100.1	127.2	111.9
946	Oakland	103.1	134.3	116.7
947	Berkeley	102.7	135.1	116.8
948	Richmond	101.8	133.1	115.4
949	San Rafael	103.5	136.7	118.0
950	Santa Cruz	104.5	121.8	112.0

STATE/ZIP	CITY	MAT.	INST.	TOTAL
CALIFORNIA (CONT'D)				
951	San Jose	102.5	136.6	117.4
952	Stockton	100.9	117.2	108.0
953	Modesto	100.8	115.7	107.3
954	Santa Rosa	100.8	134.7	115.6
955	Eureka	102.1	116.4	108.4
959	Marysville	101.3	115.9	107.7
960	Redding	107.8	117.7	112.1
961	Susanville	106.8	118.1	111.7
COLORADO				
800-802	Denver	100.8	81.7	92.5
803	Boulder	97.8	81.3	90.6
804	Golden	99.8	78.9	90.7
805	Fort Collins	101.3	79.0	91.6
806	Greeley	98.9	75.5	88.7
807	Fort Morgan	98.3	79.0	89.9
808-809	Colorado Springs	100.4	84.0	93.2
810	Pueblo	101.1	80.3	92.0
811	Alamosa	102.2	73.7	89.8
812	Salida	101.9	75.3	90.3
813	Durango	102.7	75.7	90.9
814	Montrose	101.4	75.4	90.1
815	Grand Junction	104.8	75.3	91.9
816	Glenwood Springs	102.3	78.1	91.8
CONNECTICUT				
060	New Britain	100.1	120.6	109.1
061	Hartford	101.4	120.8	109.8
062	Willimantic	100.8	120.2	109.2
063	New London	97.1	120.7	107.4
064	Meriden	98.9	120.7	108.4
065	New Haven	101.6	120.7	109.9
066	Bridgeport	101.3	120.8	109.8
067	Waterbury	100.9	120.7	109.5
068	Norwalk	100.8	128.7	113.0
069	Stamford	100.9	128.7	113.0
D.C.				
200-205	Washington	101.1	91.9	97.1
DELAWARE				
197	Newark	98.8	109.2	103.3
198	Wilmington	99.3	109.2	103.6
199	Dover	99.3	109.2	103.6
FLORIDA				
320,322	Jacksonville	98.8	66.4	84.7
321	Daytona Beach	99.0	72.9	87.6
323	Tallahassee	100.8	56.9	81.7
324	Panama City	100.2	57.8	81.7
325	Pensacola	102.7	61.2	84.6
326,344	Gainesville	100.3	66.7	85.7
327-328,347	Orlando	100.5	70.0	87.2
329	Melbourne	101.7	76.3	90.6
330-332,340	Miami	99.8	73.1	88.2
333	Fort Lauderdale	98.6	72.6	87.2
334,349	West Palm Beach	97.4	71.1	85.9
335-336,346	Tampa	100.1	78.4	90.6
337	St. Petersburg	102.6	63.3	85.5
338	Lakeland	99.4	77.8	90.0
339,341	Fort Myers	98.6	72.6	87.3
342	Sarasota	100.8	74.3	89.2
GEORGIA				
300-303,399	Atlanta	97.6	74.4	87.5
304	Statesboro	97.3	55.8	79.2
305	Gainesville	96.0	61.6	81.0
306	Athens	95.5	63.9	81.7
307	Dalton	97.2	59.0	80.5
308-309	Augusta	96.3	63.6	82.1
310-312	Macon	97.2	64.7	83.0
313-314	Savannah	99.2	60.9	82.5
315	Waycross	98.3	59.4	81.4
316	Valdosta	98.5	61.4	82.3
317,398	Albany	98.5	61.3	82.3
318-319	Columbus	98.4	65.4	84.0

Location Factors

STATE/ZIP	CITY	MAT.	INST.	TOTAL
HAWAII				
967	Hilo	114.8	120.2	117.2
968	Honolulu	119.5	120.2	119.8
STATES & POSS.				
969	Guam	136.2	58.0	102.1
IDAHO				
832	Pocatello	101.1	77.4	90.8
833	Twin Falls	102.2	67.3	87.0
834	Idaho Falls	99.8	74.7	88.9
835	Lewiston	108.7	84.9	98.3
836-837	Boise	100.1	78.6	90.7
838	Coeur d'Alene	107.9	82.2	96.7
ILLINOIS				
600-603	North Suburban	99.3	135.1	114.9
604	Joliet	99.2	138.4	116.3
605	South Suburban	99.3	135.1	114.9
606-608	Chicago	99.8	139.7	117.2
609	Kankakee	95.6	131.7	111.3
610-611	Rockford	99.5	124.8	110.5
612	Rock Island	97.2	100.5	98.6
613	La Salle	98.4	124.4	109.7
614	Galesburg	98.2	106.7	101.9
615-616	Peoria	100.5	109.3	104.4
617	Bloomington	97.5	109.4	102.7
618-619	Champaign	101.0	107.6	103.9
620-622	East St. Louis	96.4	108.7	101.8
623	Quincy	98.1	101.0	99.3
624	Effingham	97.3	106.0	101.1
625	Decatur	99.2	104.9	101.7
626-627	Springfield	100.0	106.5	102.8
628	Centralia	95.0	108.5	100.9
629	Carbondale	94.7	107.9	100.5
INDIANA				
460	Anderson	97.0	83.8	91.3
461-462	Indianapolis	99.1	84.8	92.9
463-464	Gary	98.3	110.4	103.6
465-466	South Bend	98.0	84.4	92.1
467-468	Fort Wayne	97.6	79.1	89.5
469	Kokomo	94.8	81.5	89.0
470	Lawrenceburg	94.0	77.7	86.9
471	New Albany	95.1	75.4	86.5
472	Columbus	97.2	80.9	90.1
473	Muncie	97.9	82.3	91.1
474	Bloomington	99.3	81.8	91.6
475	Washington	95.8	81.8	89.7
476-477	Evansville	97.1	86.9	92.6
478	Terre Haute	97.9	86.5	92.9
479	Lafayette	96.9	82.7	90.7
IOWA				
500-503,509	Des Moines	100.5	85.2	93.8
504	Mason City	97.6	65.8	83.8
505	Fort Dodge	97.9	58.2	80.6
506-507	Waterloo	99.5	71.8	87.4
508	Creston	98.2	77.3	89.1
510-511	Sioux City	99.7	72.6	87.9
512	Sibley	98.3	56.4	80.0
513	Spencer	99.9	56.6	81.0
514	Carroll	97.0	71.2	85.8
515	Council Bluffs	100.7	78.1	90.9
516	Shenandoah	97.6	71.2	86.1
520	Dubuque	99.1	80.2	90.9
521	Decorah	98.0	64.2	83.3
522-524	Cedar Rapids	100.0	83.2	92.7
525	Ottumwa	97.9	73.7	87.3
526	Burlington	97.2	75.3	87.6
527-528	Davenport	99.0	93.1	96.4
KANSAS				
660-662	Kansas City	99.5	94.8	97.5
664-666	Topeka	99.5	66.0	84.9
667	Fort Scott	97.9	73.0	87.0
668	Emporia	97.9	69.2	85.4
669	Belleville	99.6	64.9	84.4
670-672	Wichita	100.7	66.0	85.6
673	Independence	100.9	69.2	87.1
674	Salina	101.5	67.6	86.7
675	Hutchinson	96.5	65.0	82.8
676	Hays	100.5	65.4	85.2
677	Colby	101.1	65.0	85.4

STATE/ZIP	CITY	MAT.	INST.	TOTAL
KANSAS (CONT'D)				
678	Dodge City	102.7	66.0	86.7
679	Liberal	100.4	65.5	85.2
KENTUCKY				
400-402	Louisville	96.6	83.3	90.8
403-405	Lexington	96.6	81.4	90.0
406	Frankfort	98.1	80.7	90.5
407-409	Corbin	94.0	71.8	84.3
410	Covington	95.9	91.4	93.9
411-412	Ashland	94.6	99.3	96.6
413-414	Campton	95.8	78.6	88.3
415-416	Pikeville	96.9	88.6	93.3
417-418	Hazard	95.2	78.7	88.0
420	Paducah	93.9	84.5	89.8
421-422	Bowling Green	96.2	83.9	90.8
423	Owensboro	96.3	88.9	93.1
424	Henderson	93.7	86.1	90.4
425-426	Somerset	93.2	79.9	87.4
427	Elizabethtown	92.8	82.0	88.1
LOUISIANA				
700-701	New Orleans	101.4	69.5	87.5
703	Thibodaux	98.2	65.4	83.9
704	Hammond	95.8	54.5	77.8
705	Lafayette	97.9	63.0	82.7
706	Lake Charles	98.1	65.0	83.7
707-708	Baton Rouge	99.1	63.7	83.7
710-711	Shreveport	101.0	59.6	83.0
712	Monroe	99.3	56.5	80.6
713-714	Alexandria	99.4	57.2	81.0
MAINE				
039	Kittery	95.2	84.7	90.6
040-041	Portland	102.0	90.1	96.8
042	Lewiston	99.6	90.1	95.5
043	Augusta	102.1	79.5	92.3
044	Bangor	99.0	89.5	94.9
045	Bath	97.5	81.9	90.7
046	Machias	97.1	85.5	92.0
047	Houlton	97.2	85.6	92.1
048	Rockland	96.3	85.5	91.6
049	Waterville	97.5	79.5	89.7
MARYLAND				
206	Waldorf	97.1	81.7	90.4
207-208	College Park	97.2	83.4	91.2
209	Silver Spring	96.5	83.2	90.7
210-212	Baltimore	100.8	81.9	92.6
214	Annapolis	101.5	82.0	93.0
215	Cumberland	96.8	81.8	90.2
216	Easton	98.3	68.6	85.3
217	Hagerstown	97.6	83.3	91.4
218	Salisbury	98.7	62.1	82.7
219	Elkton	95.9	81.9	89.8
MASSACHUSETTS				
010-011	Springfield	100.2	112.6	105.6
012	Pittsfield	99.7	111.3	104.8
013	Greenfield	97.7	112.3	104.1
014	Fitchburg	96.5	126.5	109.6
015-016	Worcester	100.2	126.4	111.6
017	Framingham	95.7	135.1	112.9
018	Lowell	99.5	134.5	114.8
019	Lawrence	100.3	134.6	115.3
020-022, 024	Boston	101.6	139.5	118.1
023	Brockton	99.9	128.1	112.2
025	Buzzards Bay	94.4	128.1	109.1
026	Hyannis	97.0	128.1	110.6
027	New Bedford	99.1	128.3	111.8
MICHIGAN				
480,483	Royal Oak	94.9	106.9	100.1
481	Ann Arbor	97.1	107.5	101.7
482	Detroit	97.8	109.6	102.9
484-485	Flint	96.8	96.0	96.4
486	Saginaw	96.4	91.8	94.4
487	Bay City	96.6	91.7	94.5
488-489	Lansing	99.3	94.7	97.3
490	Battle Creek	98.3	87.6	93.7
491	Kalamazoo	98.7	85.7	93.0
492	Jackson	96.5	95.1	95.9
493,495	Grand Rapids	99.7	85.1	93.3
494	Muskegon	97.2	84.8	91.8

STATE/ZIP	CITY	MAT.	INST.	TOTAL
MICHIGAN (CONT'D)				
496	Traverse City	96.3	80.1	89.3
497	Gaylord	97.2	83.6	91.3
498-499	Iron mountain	99.3	86.2	93.6
MINNESOTA				
550-551	Saint Paul	101.4	117.9	108.6
553-555	Minneapolis	100.4	120.0	108.9
556-558	Duluth	101.3	107.3	103.9
559	Rochester	99.9	104.7	102.0
560	Mankato	97.2	101.3	99.0
561	Windom	95.9	92.3	94.3
562	Willmar	95.5	101.3	98.0
563	St. Cloud	96.9	116.0	105.2
564	Brainerd	96.9	100.5	98.5
565	Detroit Lakes	98.9	95.1	97.3
566	Bemidji	98.2	98.2	98.2
567	Thief River Falls	97.8	92.4	95.4
MISSISSIPPI				
386	Clarksdale	97.3	52.6	77.8
387	Greenville	101.0	61.7	83.9
388	Tupelo	98.8	54.9	79.7
389	Greenwood	98.4	52.2	78.3
390-392	Jackson	100.7	62.2	83.9
393	Meridian	97.6	64.1	83.0
394	Laurel	98.9	55.8	80.1
395	Biloxi	99.2	58.1	81.3
396	Mccomb	97.4	51.7	77.5
397	Columbus	98.8	53.9	79.2
MISSOURI				
630-631	St. Louis	98.7	107.8	102.7
633	Bowling Green	97.3	94.4	96.0
634	Hannibal	96.2	92.4	94.6
635	Kirksville	99.5	89.9	95.3
636	Flat River	98.1	95.0	96.8
637	Cape Girardeau	98.2	92.4	95.7
638	Sikeston	96.7	92.2	94.8
639	Poplar Bluff	96.2	92.0	94.4
640-641	Kansas City	101.4	105.7	103.3
644-645	St. Joseph	100.6	93.5	97.5
646	Chillicothe	97.6	94.5	96.3
647	Harrisonville	97.1	101.5	99.1
648	Joplin	99.5	81.7	91.8
650-651	Jefferson City	99.9	93.7	97.2
652	Columbia	100.2	93.9	97.5
653	Sedalia	100.2	97.3	98.9
654-655	Rolla	97.9	94.6	96.5
656-658	Springfield	101.7	83.7	93.8
MONTANA				
590-591	Billings	102.7	77.1	91.5
592	Wolf Point	102.3	74.2	90.0
593	Miles City	100.1	74.0	88.7
594	Great Falls	104.1	76.2	91.9
595	Havre	101.2	73.1	89.0
596	Helena	102.4	73.8	89.9
597	Butte	102.6	77.3	91.6
598	Missoula	99.9	75.5	89.3
599	Kalispell	99.0	75.3	88.7
NEBRASKA				
680-681	Omaha	100.2	78.6	90.8
683-685	Lincoln	100.5	74.9	89.4
686	Columbus	98.4	77.1	89.1
687	Norfolk	100.0	76.7	89.8
688	Grand Island	100.0	77.1	90.0
689	Hastings	99.4	79.8	90.9
690	Mccook	98.7	71.1	86.7
691	North Platte	99.0	78.3	90.0
692	Valentine	101.2	72.5	88.7
693	Alliance	101.0	70.5	87.7
NEVADA				
889-891	Las Vegas	100.9	108.7	104.3
893	Ely	99.4	102.2	100.6
894-895	Reno	99.6	93.3	96.8
897	Carson City	99.7	93.5	97.0
898	Elko	98.3	84.9	92.4
NEW HAMPSHIRE				
030	Nashua	100.5	93.8	97.6
031	Manchester	101.4	93.8	98.1

STATE/ZIP	CITY	MAT.	INST.	TOTAL
NEW HAMPSHIRE (CONT'D)				
032-033	Concord	101.1	92.5	97.4
034	Keene	97.2	73.3	86.7
035	Littleton	97.1	76.3	88.0
036	Charleston	96.7	69.5	84.8
037	Claremont	95.9	69.5	84.4
038	Portsmouth	97.9	94.3	96.4
NEW JERSEY				
070-071	Newark	101.2	128.4	113.0
072	Elizabeth	98.8	127.6	111.3
073	Jersey City	97.9	127.9	111.0
074-075	Paterson	99.6	127.8	111.9
076	Hackensack	97.6	128.1	110.9
077	Long Branch	97.1	126.3	109.8
078	Dover	97.9	127.8	110.9
079	Summit	97.9	127.6	110.8
080,083	Vineland	97.5	124.7	109.4
081	Camden	99.5	124.8	110.5
082,084	Atlantic City	98.2	124.8	109.8
085-086	Trenton	100.9	125.1	111.5
087	Point Pleasant	99.2	125.9	110.8
088-089	New Brunswick	99.7	127.6	111.9
NEW MEXICO				
870-872	Albuquerque	99.6	72.6	87.8
873	Gallup	99.5	72.6	87.8
874	Farmington	100.1	72.6	88.1
875	Santa Fe	100.8	72.6	88.5
877	Las Vegas	98.0	72.6	86.9
878	Socorro	97.6	72.6	86.7
879	Truth/Consequences	97.5	69.4	85.2
880	Las Cruces	96.1	69.4	84.4
881	Clovis	98.3	72.5	87.1
882	Roswell	100.0	72.6	88.0
883	Carrizozo	100.3	72.6	88.2
884	Tucumcari	99.0	72.5	87.4
NEW YORK				
100-102	New York	103.4	168.7	131.8
103	Staten Island	99.2	168.4	129.4
104	Bronx	97.7	166.1	127.5
105	Mount Vernon	97.5	141.4	116.6
106	White Plains	97.6	141.4	116.7
107	Yonkers	102.1	143.7	120.2
108	New Rochelle	97.9	141.3	116.8
109	Suffern	97.6	129.0	111.3
110	Queens	100.9	166.8	129.6
111	Long Island City	102.4	166.8	130.5
112	Brooklyn	102.8	168.5	131.4
113	Flushing	102.5	166.8	130.5
114	Jamaica	100.8	166.8	129.6
115,117,118	Hicksville	100.9	153.3	123.7
116	Far Rockaway	102.6	166.8	130.6
119	Riverhead	101.6	148.2	121.9
120-122	Albany	100.1	104.7	102.1
123	Schenectady	99.3	104.7	101.7
124	Kingston	102.6	118.5	109.6
125-126	Poughkeepsie	101.8	136.7	117.0
127	Monticello	101.1	115.1	107.2
128	Glens Falls	94.0	98.6	96.0
129	Plattsburgh	99.1	96.7	98.0
130-132	Syracuse	98.9	98.0	98.5
133-135	Utica	97.1	97.1	97.1
136	Watertown	98.5	96.3	97.5
137-139	Binghamton	98.5	101.4	99.8
140-142	Buffalo	100.3	106.2	102.9
143	Niagara Falls	97.8	106.4	101.5
144-146	Rochester	101.1	99.1	100.2
147	Jamestown	96.8	92.5	94.9
148-149	Elmira	96.7	97.4	97.0
NORTH CAROLINA				
270,272-274	Greensboro	101.3	55.7	81.4
271	Winston-Salem	101.0	56.5	81.6
275-276	Raleigh	100.5	54.1	80.3
277	Durham	102.9	55.3	82.1
278	Rocky Mount	98.5	52.3	78.3
279	Elizabeth City	99.3	51.6	78.5
280	Gastonia	101.6	55.1	81.3
281-282	Charlotte	101.9	56.8	82.2
283	Fayetteville	104.5	58.1	84.2
284	Wilmington	100.3	55.9	81.0
285	Kinston	98.4	52.7	78.5

Location Factors

STATE/ZIP	CITY	MAT.	INST.	TOTAL
NORTH CAROLINA (CONT'D)				
286	Hickory	98.8	53.9	79.2
287-288	Asheville	100.8	55.3	80.9
289	Murphy	99.6	48.3	77.2
NORTH DAKOTA				
580-581	Fargo	102.7	67.6	87.4
582	Grand Forks	102.7	51.6	80.4
583	Devils Lake	102.0	58.6	83.1
584	Jamestown	102.1	45.2	77.3
585	Bismarck	103.1	66.2	87.0
586	Dickinson	102.8	60.9	84.6
587	Minot	102.7	71.3	89.0
588	Williston	101.2	60.9	83.7
OHIO				
430-432	Columbus	98.6	89.2	94.5
433	Marion	95.1	85.3	90.8
434-436	Toledo	98.7	98.1	98.4
437-438	Zanesville	95.6	86.5	91.6
439	Steubenville	96.8	94.9	96.0
440	Lorain	98.9	93.0	96.3
441	Cleveland	99.0	100.3	99.6
442-443	Akron	99.9	94.2	97.4
444-445	Youngstown	99.2	89.0	94.7
446-447	Canton	99.3	85.6	93.3
448-449	Mansfield	96.7	87.3	92.6
450	Hamilton	97.9	85.1	92.3
451-452	Cincinnati	98.3	84.5	92.2
453-454	Dayton	98.0	84.4	92.1
455	Springfield	98.0	84.7	92.2
456	Chillicothe	96.8	93.0	95.1
457	Athens	99.6	80.4	91.2
458	Lima	100.0	86.8	94.3
OKLAHOMA				
730-731	Oklahoma City	100.3	65.7	85.2
734	Ardmore	98.3	61.9	82.4
735	Lawton	100.7	62.8	84.2
736	Clinton	99.6	63.1	83.7
737	Enid	100.2	60.5	82.9
738	Woodward	98.3	63.4	83.1
739	Guymon	99.4	59.7	82.1
740-741	Tulsa	98.1	61.6	82.2
743	Miami	94.8	67.5	82.9
744	Muskogee	97.4	55.7	79.2
745	Mcalester	94.5	60.4	79.7
746	Ponca City	95.1	62.1	80.7
747	Durant	95.1	63.5	81.3
748	Shawnee	96.5	61.8	81.4
749	Poteau	94.3	61.4	80.0
OREGON				
970-972	Portland	98.8	100.4	99.5
973	Salem	100.5	99.4	100.0
974	Eugene	98.6	99.3	98.9
975	Medford	100.0	97.1	98.8
976	Klamath Falls	99.8	97.1	98.6
977	Bend	98.9	99.4	99.1
978	Pendleton	94.8	101.4	97.7
979	Vale	92.9	98.1	95.2
PENNSYLVANIA				
150-152	Pittsburgh	99.9	104.6	102.0
153	Washington	97.0	103.2	99.7
154	Uniontown	97.3	101.7	99.2
155	Bedford	98.2	91.3	95.2
156	Greensburg	98.2	100.9	99.4
157	Indiana	97.1	99.3	98.0
158	Dubois	98.5	96.8	97.8
159	Johnstown	98.1	97.2	97.7
160	Butler	92.5	102.0	96.6
161	New Castle	92.5	97.9	94.8
162	Kittanning	92.9	103.3	97.4
163	Oil City	92.4	96.9	94.4
164-165	Erie	94.5	95.8	95.1
166	Altoona	94.6	91.9	93.4
167	Bradford	95.7	96.4	96.0
168	State College	95.3	92.8	94.2
169	Wellsboro	96.3	92.6	94.7
170-171	Harrisburg	99.8	95.6	98.0
172	Chambersburg	96.2	88.0	92.6
173-174	York	96.7	95.9	96.3
175-176	Lancaster	95.1	89.0	92.4

STATE/ZIP	CITY	MAT.	INST.	TOTAL
PENNSYLVANIA (CONT'D)				
177	Williamsport	93.8	83.0	89.1
178	Sunbury	95.8	95.8	95.8
179	Pottsville	94.9	98.5	96.5
180	Lehigh Valley	95.8	114.0	103.7
181	Allentown	97.9	108.6	102.5
182	Hazleton	95.4	98.3	96.6
183	Stroudsburg	95.3	101.4	97.9
184-185	Scranton	98.6	99.2	98.9
186-187	Wilkes-Barre	95.1	98.7	96.6
188	Montrose	94.7	96.9	95.7
189	Doylestown	94.9	120.6	106.1
190-191	Philadelphia	99.7	133.6	114.5
193	Westchester	95.9	121.5	107.1
194	Norristown	95.0	129.5	110.0
195-196	Reading	97.1	102.1	99.3
PUERTO RICO				
009	San Juan	122.7	24.4	79.8
RHODE ISLAND				
028	Newport	98.4	117.5	106.7
029	Providence	100.6	117.5	108.0
SOUTH CAROLINA				
290-292	Columbia	99.3	56.5	80.6
293	Spartanburg	98.6	56.7	80.3
294	Charleston	99.9	65.2	84.8
295	Florence	98.2	56.9	80.2
296	Greenville	98.3	55.8	79.8
297	Rock Hill	97.8	50.3	77.1
298	Aiken	98.6	69.1	85.7
299	Beaufort	99.3	44.9	75.6
SOUTH DAKOTA				
570-571	Sioux Falls	100.0	57.7	81.5
572	Watertown	99.0	49.1	77.2
573	Mitchell	97.8	48.4	76.3
574	Aberdeen	100.5	49.6	78.3
575	Pierre	101.7	58.4	82.8
576	Mobridge	98.5	48.8	76.8
577	Rapid City	100.4	59.5	82.6
TENNESSEE				
370-372	Nashville	98.4	74.1	87.8
373-374	Chattanooga	100.0	66.2	85.2
375,380-381	Memphis	99.0	71.3	86.9
376	Johnson City	99.8	56.5	80.9
377-379	Knoxville	96.6	66.9	83.7
382	Mckenzie	98.2	59.2	81.2
383	Jackson	99.9	61.2	83.0
384	Columbia	96.9	66.1	83.5
385	Cookeville	98.1	59.1	81.1
TEXAS				
750	Mckinney	98.7	63.1	83.2
751	Waxahackie	98.6	65.9	84.4
752-753	Dallas	99.5	67.4	85.5
754	Greenville	98.8	64.0	83.6
755	Texarkana	98.0	58.4	80.7
756	Longview	98.5	58.7	81.2
757	Tyler	98.5	65.0	83.9
758	Palestine	95.0	64.4	81.7
759	Lufkin	95.6	67.8	83.5
760-761	Fort Worth	99.3	65.3	84.5
762	Denton	98.9	62.5	83.0
763	Wichita Falls	97.1	64.4	82.8
764	Eastland	95.9	63.2	81.6
765	Temple	94.7	61.7	80.3
766-767	Waco	96.5	64.1	82.4
768	Brownwood	98.9	59.4	81.7
769	San Angelo	98.5	59.6	81.5
770-772	Houston	99.4	70.5	86.8
773	Huntsville	98.0	66.7	84.4
774	Wharton	98.9	67.7	85.3
775	Galveston	97.0	70.0	85.2
776-777	Beaumont	97.4	68.6	84.9
778	Bryan	94.9	68.2	83.3
779	Victoria	99.0	65.6	84.4
780	Laredo	98.3	64.5	83.6
781-782	San Antonio	98.5	65.1	84.0
783-784	Corpus Christi	100.9	63.5	84.6
785	Mc Allen	101.0	59.8	83.0
786-787	Austin	100.0	63.6	84.1

STATE/ZIP	CITY	MAT.	INST.	TOTAL
TEXAS (CONT'D)				
788	Del Rio	100.8	63.2	84.4
789	Giddings	97.7	64.0	83.0
790-791	Amarillo	100.2	63.2	84.1
792	Childress	99.3	64.3	84.0
793-794	Lubbock	101.6	63.6	85.0
795-796	Abilene	99.7	62.2	83.3
797	Midland	101.5	64.2	85.2
798-799,885	El Paso	98.0	61.7	82.2
UTAH				
840-841	Salt Lake City	102.2	68.3	87.4
842,844	Ogden	98.1	70.1	85.9
843	Logan	99.8	70.1	86.9
845	Price	100.2	64.6	84.7
846-847	Provo	100.3	68.2	86.3
VERMONT				
050	White River Jct.	98.5	78.3	89.7
051	Bellows Falls	97.2	93.0	95.3
052	Bennington	97.5	93.5	95.8
053	Brattleboro	97.8	92.8	95.6
054	Burlington	102.4	83.6	94.2
056	Montpelier	99.5	82.8	92.2
057	Rutland	99.7	83.6	92.7
058	St. Johnsbury	98.6	77.8	89.6
059	Guildhall	97.4	77.8	88.8
VIRGINIA				
220-221	Fairfax	100.7	84.0	93.4
222	Arlington	101.9	84.0	94.1
223	Alexandria	101.1	85.5	94.3
224-225	Fredericksburg	99.5	82.7	92.1
226	Winchester	100.0	82.2	92.2
227	Culpeper	99.9	80.5	91.4
228	Harrisonburg	100.2	67.2	85.8
229	Charlottesville	100.5	67.1	86.0
230-232	Richmond	101.3	68.7	87.1
233-235	Norfolk	100.5	68.2	86.4
236	Newport News	100.0	68.2	86.1
237	Portsmouth	99.4	65.3	84.5
238	Petersburg	99.4	68.7	86.0
239	Farmville	98.7	52.4	78.5
240-241	Roanoke	101.5	65.1	85.6
242	Bristol	99.4	54.9	80.0
243	Pulaski	99.0	57.3	80.9
244	Staunton	99.7	61.8	83.2
245	Lynchburg	99.9	65.6	85.0
246	Grundy	99.3	52.5	78.9
WASHINGTON				
980-981,987	Seattle	101.4	104.7	102.8
982	Everett	101.6	99.6	100.7
983-984	Tacoma	101.8	99.7	100.9
985	Olympia	101.2	99.4	100.4
986	Vancouver	102.7	92.1	98.1
988	Wenatchee	101.6	86.6	95.0
989	Yakima	102.0	94.1	98.6
990-992	Spokane	103.4	82.3	94.2
993	Richland	103.1	89.3	97.1
994	Clarkston	101.5	80.2	92.2
WEST VIRGINIA				
247-248	Bluefield	98.3	92.9	96.0
249	Lewisburg	99.8	92.4	96.6
250-253	Charleston	99.9	94.8	97.6
254	Martinsburg	99.3	86.4	93.7
255-257	Huntington	100.7	95.7	98.5
258-259	Beckley	97.9	93.3	95.9
260	Wheeling	101.2	94.3	98.2
261	Parkersburg	100.1	91.8	96.5
262	Buckhannon	99.6	95.5	97.8
263-264	Clarksburg	100.1	95.4	98.1
265	Morgantown	100.2	94.4	97.6
266	Gassaway	99.5	93.9	97.1
267	Romney	99.4	92.2	96.3
268	Petersburg	99.3	91.3	95.8
WISCONSIN				
530,532	Milwaukee	100.0	107.2	103.1
531	Kenosha	99.5	102.2	100.7
534	Racine	99.1	102.4	100.5
535	Beloit	98.9	100.7	99.7
537	Madison	99.1	99.0	99.1

STATE/ZIP	CITY	MAT.	INST.	TOTAL
WISCONSIN (CONT'D)				
538	Lancaster	96.7	92.5	94.8
539	Portage	95.3	96.6	95.9
540	New Richmond	96.8	94.6	95.9
541-543	Green Bay	101.2	93.4	97.8
544	Wausau	96.2	93.5	95.0
545	Rhinelander	99.3	91.2	95.8
546	La Crosse	97.8	94.5	96.3
547	Eau Claire	99.3	95.3	97.5
548	Superior	96.6	96.3	96.5
549	Oshkosh	96.9	91.9	94.7
WYOMING				
820	Cheyenne	101.5	65.2	85.7
821	Yellowstone Nat'l Park	99.3	66.3	84.9
822	Wheatland	100.4	64.3	84.7
823	Rawlins	101.8	66.9	86.6
824	Worland	100.0	65.3	84.9
825	Riverton	101.0	65.4	85.5
826	Casper	102.0	63.9	85.4
827	Newcastle	99.9	66.9	85.5
828	Sheridan	102.6	64.3	85.9
829-831	Rock Springs	103.7	65.3	87.0
CANADIAN FACTORS (reflect Canadian currency)				
ALBERTA				
	Calgary	123.5	95.7	111.4
	Edmonton	123.1	95.7	111.2
	Fort McMurray	123.6	92.8	110.2
	Lethbridge	118.4	92.2	107.0
	Lloydminster	113.4	88.9	102.8
	Medicine Hat	113.6	88.2	102.5
	Red Deer	114.1	88.2	102.8
BRITISH COLUMBIA				
	Kamloops	114.5	90.6	104.1
	Prince George	115.4	89.9	104.3
	Vancouver	119.5	89.1	106.2
	Victoria	115.7	87.2	103.3
MANITOBA				
	Brandon	125.2	75.8	103.7
	Portage la Prairie	113.4	74.2	96.3
	Winnipeg	121.0	71.8	99.5
NEW BRUNSWICK				
	Bathurst	111.5	67.3	92.2
	Dalhousie	111.9	67.5	92.5
	Fredericton	117.4	72.5	97.9
	Moncton	111.8	70.1	93.6
	Newcastle	111.6	67.9	92.5
	St. John	114.0	77.9	98.3
NEWFOUNDLAND				
	Corner Brook	129.6	67.6	102.6
	St Johns	122.7	85.2	106.4
NORTHWEST TERRITORIES				
	Yellowknife	135.3	84.2	113.0
NOVA SCOTIA				
	Bridgewater	112.9	74.6	96.2
	Dartmouth	126.1	74.6	103.7
	Halifax	119.6	80.9	102.7
	New Glasgow	124.1	74.6	102.5
	Sydney	122.1	74.6	101.4
	Truro	112.5	74.6	96.0
	Yarmouth	124.0	74.6	102.5
ONTARIO				
	Barrie	117.7	93.0	106.9
	Brantford	114.9	96.9	107.1
	Cornwall	115.0	93.3	105.5
	Hamilton	117.0	96.7	108.2
	Kingston	116.0	93.4	106.2
	Kitchener	109.6	92.9	102.3
	London	116.4	93.3	106.3
	North Bay	127.3	91.1	111.5
	Oshawa	112.3	94.3	104.4
	Ottawa	117.3	94.0	107.2
	Owen Sound	117.8	91.2	106.2
	Peterborough	114.9	93.1	105.4
	Sarnia	114.8	97.5	107.3
	Sault Ste Marie	110.3	92.1	102.3

Location Factors

STATE/ZIP	CITY	MAT.	INST.	TOTAL
ONTARIO (CONT'D)				
	St. Catharines	107.6	94.5	101.9
	Sudbury	107.7	93.0	101.3
	Thunder Bay	108.8	93.4	102.1
	Timmins	115.1	91.1	104.6
	Toronto	118.3	101.3	110.9
	Windsor	108.2	93.4	101.7
PRINCE EDWARD ISLAND				
	Charlottetown	120.0	62.8	95.1
	Summerside	126.7	62.8	98.8
QUEBEC				
	Cap-de-la-Madeleine	112.4	87.6	101.6
	Charlesbourg	112.4	87.6	101.6
	Chicoutimi	111.4	92.9	103.3
	Gatineau	112.1	87.4	101.3
	Granby	112.3	87.4	101.4
	Hull	112.2	87.4	101.4
	Joliette	112.5	87.6	101.7
	Laval	112.1	87.4	101.3
	Montreal	117.8	93.2	107.1
	Quebec	117.1	93.6	106.8
	Rimouski	111.8	92.9	103.6
	Rouyn-Noranda	112.0	87.4	101.3
	Saint Hyacinthe	111.6	87.4	101.1
	Sherbrooke	112.4	87.4	101.5
	Sorel	112.5	87.6	101.7
	St Jerome	112.1	87.4	101.3
	Trois Rivieres	124.8	87.6	108.6
SASKATCHEWAN				
	Moose Jaw	110.7	69.5	92.7
	Prince Albert	109.9	67.7	91.5
	Regina	120.7	89.0	106.9
	Saskatoon	111.1	89.1	101.5
YUKON				
	Whitehorse	135.3	68.5	106.2

R011105-05 Tips for Accurate Estimating

1. Use pre-printed or columnar forms for orderly sequence of dimensions and locations and for recording telephone quotations.
2. Use only the front side of each paper or form except for certain pre-printed summary forms.
3. Be consistent in listing dimensions: For example, length x width x height. This helps in rechecking to ensure that the total length of partitions is appropriate for the building area.
4. Use printed (rather than measured) dimensions where given.
5. Add up multiple printed dimensions for a single entry where possible.
6. Measure all other dimensions carefully.
7. Use each set of dimensions to calculate multiple related quantities.
8. Convert foot and inch measurements to decimal feet when listing. Memorize decimal equivalents to .01 parts of a foot (1/8″ equals approximately .01′).
9. Do not "round off" quantities until the final summary.
10. Mark drawings with different colors as items are taken off.
11. Keep similar items together, different items separate.
12. Identify location and drawing numbers to aid in future checking for completeness.
13. Measure or list everything on the drawings or mentioned in the specifications.
14. It may be necessary to list items not called for to make the job complete.
15. Be alert for: Notes on plans such as N.T.S. (not to scale); changes in scale throughout the drawings; reduced size drawings; discrepancies between the specifications and the drawings.
16. Develop a consistent pattern of performing an estimate. For example:
    a. Start the quantity takeoff at the lower floor and move to the next higher floor.
    b. Proceed from the main section of the building to the wings.
    c. Proceed from south to north or vice versa, clockwise or counterclockwise.
    d. Take off floor plan quantities first, elevations next, then detail drawings.
17. List all gross dimensions that can be either used again for different quantities, or used as a rough check of other quantities for verification (exterior perimeter, gross floor area, individual floor areas, etc.).
18. Utilize design symmetry or repetition (repetitive floors, repetitive wings, symmetrical design around a center line, similar room layouts, etc.). Note: Extreme caution is needed here so as not to omit or duplicate an area.
19. Do not convert units until the final total is obtained. For instance, when estimating concrete work, keep all units to the nearest cubic foot, then summarize and convert to cubic yards.
20. When figuring alternatives, it is best to total all items involved in the basic system, then total all items involved in the alternates. Therefore, you work with positive numbers in all cases. When adds and deducts are used, it is often confusing whether to add or subtract a portion of an item; especially on a complicated or involved alternate.

R011105-10 Unit Gross Area Requirements

The figures in the table below indicate typical ranges in square feet as a function of the "occupant" unit. This table is best used in the preliminary design stages to help determine the probable size requirement for the total project.

Building Type	Unit	Gross Area in S.F. 1/4	Median	3/4
Apartments	Unit	660	860	1,100
Auditorium & Play Theaters	Seat	18	25	38
Bowling Alleys	Lane		940	
Churches & Synagogues	Seat	20	28	39
Dormitories	Bed	200	230	275
Fraternity & Sorority Houses	Bed	220	315	370
Garages, Parking	Car	325	355	385
Hospitals	Bed	685	850	1,075
Hotels	Rental Unit	475	600	710
Housing for the elderly	Unit	515	635	755
Housing, Public	Unit	700	875	1,030
Ice Skating Rinks	Total	27,000	30,000	36,000
Motels	Rental Unit	360	465	620
Nursing Homes	Bed	290	350	450
Restaurants	Seat	23	29	39
Schools, Elementary	Pupil	65	77	90
Junior High & Middle	↓	85	110	129
Senior High	↓	102	130	145
Vocational	↓	110	135	195
Shooting Ranges	Point		450	
Theaters & Movies	Seat		15	

R011105-50 Metric Conversion Factors

Description: This table is primarily for converting customary U.S. units in the left hand column to SI metric units in the right hand column. In addition, conversion factors for some commonly encountered Canadian and non-SI metric units are included.

	If You Know		Multiply By		To Find
Length	Inches	x	25.4[a]	=	Millimeters
	Feet	x	0.3048[a]	=	Meters
	Yards	x	0.9144[a]	=	Meters
	Miles (statute)	x	1.609	=	Kilometers
Area	Square inches	x	645.2	=	Square millimeters
	Square feet	x	0.0929	=	Square meters
	Square yards	x	0.8361	=	Square meters
Volume (Capacity)	Cubic inches	x	16,387	=	Cubic millimeters
	Cubic feet	x	0.02832	=	Cubic meters
	Cubic yards	x	0.7646	=	Cubic meters
	Gallons (U.S. liquids)[b]	x	0.003785	=	Cubic meters[c]
	Gallons (Canadian liquid)[b]	x	0.004546	=	Cubic meters[c]
	Ounces (U.S. liquid)[b]	x	29.57	=	Milliliters[c, d]
	Quarts (U.S. liquid)[b]	x	0.9464	=	Liters[c, d]
	Gallons (U.S. liquid)[b]	x	3.785	=	Liters[c, d]
Force	Kilograms force[d]	x	9.807	=	Newtons
	Pounds force	x	4.448	=	Newtons
	Pounds force	x	0.4536	=	Kilograms force[d]
	Kips	x	4448	=	Newtons
	Kips	x	453.6	=	Kilograms force[d]
Pressure, Stress, Strength (Force per unit area)	Kilograms force per square centimeter[d]	x	0.09807	=	Megapascals
	Pounds force per square inch (psi)	x	0.006895	=	Megapascals
	Kips per square inch	x	6.895	=	Megapascals
	Pounds force per square inch (psi)	x	0.07031	=	Kilograms force per square centimeter[d]
	Pounds force per square foot	x	47.88	=	Pascals
	Pounds force per square foot	x	4.882	=	Kilograms force per square meter[d]
Bending Moment Or Torque	Inch-pounds force	x	0.01152	=	Meter-kilograms force[d]
	Inch-pounds force	x	0.1130	=	Newton-meters
	Foot-pounds force	x	0.1383	=	Meter-kilograms force[d]
	Foot-pounds force	x	1.356	=	Newton-meters
	Meter-kilograms force[d]	x	9.807	=	Newton-meters
Mass	Ounces (avoirdupois)	x	28.35	=	Grams
	Pounds (avoirdupois)	x	0.4536	=	Kilograms
	Tons (metric)	x	1000	=	Kilograms
	Tons, short (2000 pounds)	x	907.2	=	Kilograms
	Tons, short (2000 pounds)	x	0.9072	=	Megagrams[e]
Mass per Unit Volume	Pounds mass per cubic foot	x	16.02	=	Kilograms per cubic meter
	Pounds mass per cubic yard	x	0.5933	=	Kilograms per cubic meter
	Pounds mass per gallon (U.S. liquid)[b]	x	119.8	=	Kilograms per cubic meter
	Pounds mass per gallon (Canadian liquid)[b]	x	99.78	=	Kilograms per cubic meter
Temperature	Degrees Fahrenheit		(F-32)/1.8	=	Degrees Celsius
	Degrees Fahrenheit		(F+459.67)/1.8	=	Degrees Kelvin
	Degrees Celsius		C+273.15	=	Degrees Kelvin

[a]The factor given is exact
[b]One U.S. gallon = 0.8327 Canadian gallon
[c]1 liter = 1000 milliliters = 1000 cubic centimeters
1 cubic decimeter = 0.001 cubic meter
[d]Metric but not SI unit
[e]Called "tonne" in England and "metric ton" in other metric countries

R011105-60 Weights and Measures

Measures of Length
1 Mile = 1760 Yards = 5280 Feet
1 Yard = 3 Feet = 36 inches
1 Foot = 12 Inches
1 Mil = 0.001 Inch
1 Fathom = 2 Yards = 6 Feet
1 Rod = 5.5 Yards = 16.5 Feet
1 Hand = 4 Inches
1 Span = 9 Inches
1 Micro-inch = One Millionth Inch or 0.000001 Inch
1 Micron = One Millionth Meter + 0.00003937 Inch

Surveyor's Measure
1 Mile = 8 Furlongs = 80 Chains
1 Furlong = 10 Chains = 220 Yards
1 Chain = 4 Rods = 22 Yards = 66 Feet = 100 Links
1 Link = 7.92 Inches

Square Measure
1 Square Mile = 640 Acres = 6400 Square Chains
1 Acre = 10 Square Chains = 4840 Square Yards = 43,560 Sq. Ft.
1 Square Chain = 16 Square Rods = 484 Square Yards = 4356 Sq. Ft.
1 Square Rod = 30.25 Square Yards = 272.25 Square Feet = 625 Square Lines
1 Square Yard = 9 Square Feet
1 Square Foot = 144 Square Inches
An Acre equals a Square 208.7 Feet per Side

Cubic Measure
1 Cubic Yard = 27 Cubic Feet
1 Cubic Foot = 1728 Cubic Inches
1 Cord of Wood = 4 x 4 x 8 Feet = 128 Cubic Feet
1 Perch of Masonry = 16½ x 1½ x 1 Foot = 24.75 Cubic Feet

Avoirdupois or Commercial Weight
1 Gross or Long Ton = 2240 Pounds
1 Net or Short ton = 2000 Pounds
1 Pound = 16 Ounces = 7000 Grains
1 Ounce = 16 Drachms = 437.5 Grains
1 Stone = 14 Pounds

Shipping Measure
For Measuring Internal Capacity of a Vessel:
1 Register Ton = 100 Cubic Feet

For Measurement of Cargo:
Approximately 40 Cubic Feet of Merchandise is considered a Shipping Ton, unless that bulk would weigh more than 2000 Pounds, in which case Freight Charge may be based upon weight.

40 Cubic Feet = 32.143 U.S. Bushels = 31.16 Imp. Bushels

Liquid Measure
1 Imperial Gallon = 1.2009 U.S. Gallon = 277.42 Cu. In.
1 Cubic Foot = 7.48 U.S. Gallons

General Requirements — R0111 Summary of Work

R011110-30 Engineering Fees

Typical **Structural Engineering Fees** based on type of construction and total project size. These fees are included in Architectural Fees.

Type of Construction	Total Project Size (in thousands of dollars)			
	$500	$500-$1,000	$1,000-$5,000	Over $5000
Industrial buildings, factories & warehouses	Technical payroll times 2.0 to 2.5	1.60%	1.25%	1.00%
Hotels, apartments, offices, dormitories, hospitals, public buildings, food stores	↓	2.00%	1.70%	1.20%
Museums, banks, churches and cathedrals	↓	2.00%	1.75%	1.25%
Thin shells, prestressed concrete, earthquake resistive	↓	2.00%	1.75%	1.50%
Parking ramps, auditoriums, stadiums, convention halls, hangars & boiler houses	↓	2.50%	2.00%	1.75%
Special buildings, major alterations, underpinning & future expansion	↓	Add to above 0.5%	Add to above 0.5%	Add to above 0.5%

For complex reinforced concrete or unusually complicated structures, add 20% to 50%.

Typical **Mechanical and Electrical Engineering Fees** are based on the size of the subcontract. The fee structure for both are shown below. These fees are included in Architectural Fees.

Type of Construction	Subcontract Size							
	$25,000	$50,000	$100,000	$225,000	$350,000	$500,000	$750,000	$1,000,000
Simple structures	6.4%	5.7%	4.8%	4.5%	4.4%	4.3%	4.2%	4.1%
Intermediate structures	8.0	7.3	6.5	5.6	5.1	5.0	4.9	4.8
Complex structures	10.1	9.0	9.0	8.0	7.5	7.5	7.0	7.0

For renovations, add 15% to 25% to applicable fee.

General Requirements — R0121 Allowances

R012153-10 Repair and Remodeling

Cost figures are based on new construction utilizing the most cost-effective combination of labor, equipment and material with the work scheduled in proper sequence to allow the various trades to accomplish their work in an efficient manner.

The costs for repair and remodeling work must be modified due to the following factors that may be present in any given repair and remodeling project.

1. Equipment usage curtailment due to the physical limitations of the project, with only hand-operated equipment being used.
2. Increased requirement for shoring and bracing to hold up the building while structural changes are being made and to allow for temporary storage of construction materials on above-grade floors.
3. Material handling becomes more costly due to having to move within the confines of an enclosed building. For multi-story construction, low capacity elevators and stairwells may be the only access to the upper floors.
4. Large amounts of cutting and patching and attempting to match the existing construction is required. It is often more economical to remove entire walls rather than create many new door and window openings. This sort of trade-off has to be carefully analyzed.
5. Cost of protection of completed work is increased since the usual sequence of construction usually cannot be accomplished.
6. Economies of scale usually associated with new construction may not be present. If small quantities of components must be custom fabricated due to job requirements, unit costs will naturally increase. Also, if only small work areas are available at a given time, job scheduling between trades becomes difficult and subcontractor quotations may reflect the excessive start-up and shut-down phases of the job.
7. Work may have to be done on other than normal shifts and may have to be done around an existing production facility which has to stay in production during the course of the repair and remodeling.
8. Dust and noise protection of adjoining non-construction areas can involve substantial special protection and alter usual construction methods.
9. Job may be delayed due to unexpected conditions discovered during demolition or removal. These delays ultimately increase construction costs.
10. Piping and ductwork runs may not be as simple as for new construction. Wiring may have to be snaked through walls and floors.
11. Matching "existing construction" may be impossible because materials may no longer be manufactured. Substitutions may be expensive.
12. Weather protection of existing structure requires additional temporary structures to protect building at openings.
13. On small projects, because of local conditions, it may be necessary to pay a tradesman for a minimum of four hours for a task that is completed in one hour.

All of the above areas can contribute to increased costs for a repair and remodeling project. Each of the above factors should be considered in the planning, bidding and construction stage in order to minimize the increased costs associated with repair and remodeling jobs.

R012909-80 Sales Tax by State

State sales tax on materials is tabulated below (5 states have no sales tax). Many states allow local jurisdictions, such as a county or city, to levy additional sales tax.

Some projects may be sales tax exempt, particularly those constructed with public funds.

State	Tax (%)	State	Tax (%)	State	Tax (%)	State	Tax (%)
Alabama	4	Illinois	6.25	Montana	0	Rhode Island	7
Alaska	0	Indiana	7	Nebraska	5.5	South Carolina	6
Arizona	5.6	Iowa	6	Nevada	6.85	South Dakota	4
Arkansas	6	Kansas	6.3	New Hampshire	0	Tennessee	7
California	7.5	Kentucky	6	New Jersey	7	Texas	6.25
Colorado	2.9	Louisiana	4	New Mexico	5.13	Utah	4.7
Connecticut	6.35	Maine	5	New York	4	Vermont	6
Delaware	0	Maryland	6	North Carolina	4.75	Virginia	5
District of Columbia	6	Massachusetts	6.25	North Dakota	5	Washington	6.5
Florida	6	Michigan	6	Ohio	5.5	West Virginia	6
Georgia	4	Minnesota	6.88	Oklahoma	4.5	Wisconsin	5
Hawaii	4	Mississippi	7	Oregon	0	Wyoming	4
Idaho	6	Missouri	4.23	Pennsylvania	6	Average	5.04 %

Sales Tax by Province (Canada)

GST - a value-added tax, which the government imposes on most goods and services provided in or imported into Canada. PST - a retail sales tax, which three of the provinces impose on the price of most goods and some services. QST - a value-added tax, similar to the federal GST, which Quebec imposes. HST - Five provinces have combined their retail sales tax with the federal GST into one harmonized tax.

Province	PST (%)	QST (%)	GST(%)	HST(%)
Alberta	0	0	5	0
British Columbia	7	0	5	0
Manitoba	8	0	5	0
New Brunswick	0	0	0	13
Newfoundland	0	0	0	13
Northwest Territories	0	0	5	0
Nova Scotia	0	0	0	15
Ontario	0	0	0	13
Prince Edward Island	0	0	0	14
Quebec	0	9.975	5	0
Saskatchewan	5	0	5	0
Yukon	0	0	5	0

R012909-85 Unemployment Taxes and Social Security Taxes

State unemployment tax rates vary not only from state to state, but also with the experience rating of the contractor. The federal unemployment tax rate is 6.0% of the first $7,000 of wages. This is reduced by a credit of up to 5.4% for timely payment to the state. The minimum federal unemployment tax is 0.6% after all credits.

Social security (FICA) for 2015 is estimated at time of publication to be 7.65% of wages up to $117,000.

R012909-90 Overtime

One way to improve the completion date of a project or eliminate negative float from a schedule is to compress activity duration times. This can be achieved by increasing the crew size or working overtime with the proposed crew.

To determine the costs of working overtime to compress activity duration times, consider the following examples. Below is an overtime efficiency and cost chart based on a five, six, or seven day week with an eight through twelve hour day. Payroll percentage increases for time and one half and double time are shown for the various working days.

Days per Week	Hours per Day	Production Efficiency					Payroll Cost Factors	
		1st Week	2nd Week	3rd Week	4th Week	Average 4 Weeks	@ 1-1/2 Times	@ 2 Times
5	8	100%	100%	100%	100%	100 %	1.000	1.000
	9	100	100	95	90	96	1.056	1.111
	10	100	95	90	85	93	1.100	1.200
	11	95	90	75	65	81	1.136	1.273
	12	90	85	70	60	76	1.167	1.333
6	8	100	100	95	90	96	1.083	1.167
	9	100	95	90	85	93	1.130	1.259
	10	95	90	85	80	88	1.167	1.333
	11	95	85	70	65	79	1.197	1.394
	12	90	80	65	60	74	1.222	1.444
7	8	100	95	85	75	89	1.143	1.286
	9	95	90	80	70	84	1.183	1.365
	10	90	85	75	65	79	1.214	1.429
	11	85	80	65	60	73	1.240	1.481
	12	85	75	60	55	69	1.262	1.524

R013113-40 Builder's Risk Insurance

Builder's Risk Insurance is insurance on a building during construction. Premiums are paid by the owner or the contractor. Blasting, collapse and underground insurance would raise total insurance costs above those listed. Floater policy for materials delivered to the job runs $.75 to $1.25 per $100 value. Contractor equipment insurance runs $.50 to $1.50 per $100 value. Insurance for miscellaneous tools to $1,500 value runs from $3.00 to $7.50 per $100 value.

Tabulated below are New England Builder's Risk insurance rates in dollars per $100 value for $1,000 deductible. For $25,000 deductible, rates can be reduced 13% to 34%. On contracts over $1,000,000, rates may be lower than those tabulated. Policies are written annually for the total completed value in place. For "all risk" insurance (excluding flood, earthquake and certain other perils) add $.025 to total rates below.

Coverage	Frame Construction (Class 1)		Brick Construction (Class 4)		Fire Resistive (Class 6)	
	Range	Average	Range	Average	Range	Average
Fire Insurance	$.350 to $ .850	$.600	$.158 to $.189	$.174	$.052 to $.080	$.070
Extended Coverage	.115 to .200	.158	.080 to .105	.101	.081 to .105	.100
Vandalism	.012 to .016	.014	.008 to .011	.011	.008 to .011	.010
Total Annual Rate	$.477 to $1.066	$.772	$.246 to $.305	$.286	$.141 to $.196	$.180

R013113-50 General Contractor's Overhead

There are two distinct types of overhead on a construction project: Project overhead and main office overhead. Project overhead includes those costs at a construction site not directly associated with the installation of construction materials. Examples of project overhead costs include the following:

1. Superintendent
2. Construction office and storage trailers
3. Temporary sanitary facilities
4. Temporary utilities
5. Security fencing
6. Photographs
7. Cleanup
8. Performance and payment bonds

The above project overhead items are also referred to as general requirements and therefore are estimated in Division 1. Division 1 is the first division listed in the CSI MasterFormat but it is usually the last division estimated. The sum of the costs in Divisions 1 through 49 is referred to as the sum of the direct costs.

All construction projects also include indirect costs. The primary components of indirect costs are the contractor's main office overhead and profit. The amount of the main office overhead expense varies depending on the following:

1. Owner's compensation
2. Project managers' and estimators' wages
3. Clerical support wages
4. Office rent and utilities
5. Corporate legal and accounting costs
6. Advertising
7. Automobile expenses
8. Association dues
9. Travel and entertainment expenses

These costs are usually calculated as a percentage of annual sales volume. This percentage can range from 35% for a small contractor doing less than $500,000 to 5% for a large contractor with sales in excess of $100 million.

R013113-60 Workers' Compensation Insurance Rates by Trade

The table below tabulates the national averages for workers' compensation insurance rates by trade and type of building. The average "Insurance Rate" is multiplied by the "% of Building Cost" for each trade. This produces the "Workers' Compensation" cost by % of total labor cost, to be added for each trade by building type to determine the weighted average workers' compensation rate for the building types analyzed.

Trade	Insurance Rate (% Labor Cost) Range			Insurance Rate Average	% of Building Cost Office Bldgs.	% of Building Cost Schools & Apts.	% of Building Cost Mfg.	Workers' Compensation Office Bldgs.	Workers' Compensation Schools & Apts.	Workers' Compensation Mfg.
Excavation, Grading, etc.	3.4 %	to	21.2%	9.7%	4.8%	4.9%	4.5%	0.47%	0.48%	0.44%
Piles & Foundations	6.1	to	28.3	14.3	7.1	5.2	8.7	1.02	0.74	1.24
Concrete	4.2	to	35.8	12.9	5.0	14.8	3.7	0.65	1.91	0.48
Masonry	4.9	to	36.4	13.7	6.9	7.5	1.9	0.95	1.03	0.26
Structural Steel	6.1	to	101.1	31.7	10.7	3.9	17.6	3.39	1.24	5.58
Miscellaneous & Ornamental Metals	4.3	to	31.5	12.2	2.8	4.0	3.6	0.34	0.49	0.44
Carpentry & Millwork	5.0	to	36.8	14.9	3.7	4.0	0.5	0.55	0.60	0.07
Metal or Composition Siding	6.1	to	69.2	18.4	2.3	0.3	4.3	0.42	0.06	0.79
Roofing	6.1	to	100.3	31.7	2.3	2.6	3.1	0.73	0.82	0.98
Doors & Hardware	4.1	to	36.8	11.3	0.9	1.4	0.4	0.10	0.16	0.05
Sash & Glazing	4.8	to	30.7	13.3	3.5	4.0	1.0	0.47	0.53	0.13
Lath & Plaster	3.1	to	36.2	10.9	3.3	6.9	0.8	0.36	0.75	0.09
Tile, Marble & Floors	3	to	24.8	8.9	2.6	3.0	0.5	0.23	0.27	0.04
Acoustical Ceilings	3.8	to	33.8	8.4	2.4	0.2	0.3	0.20	0.02	0.03
Painting	4.5	to	36.1	11.7	1.5	1.6	1.6	0.18	0.19	0.19
Interior Partitions	5.0	to	36.8	14.9	3.9	4.3	4.4	0.58	0.64	0.66
Miscellaneous Items	2.5	to	132.3	13.4	5.2	3.7	9.7	0.70	0.50	1.30
Elevators	1.3	to	12.8	5.3	2.1	1.1	2.2	0.11	0.06	0.12
Sprinklers	2.9	to	18.5	7.3	0.5	—	2.0	0.04	—	0.15
Plumbing	2.2	to	19.3	7.0	4.9	7.2	5.2	0.34	0.50	0.36
Heat., Vent., Air Conditioning	3.8	to	18.1	8.8	13.5	11.0	12.9	1.19	0.97	1.14
Electrical	2.4	to	13.5	5.8	10.1	8.4	11.1	0.59	0.49	0.64
Total	1.3 %	to	132.3%	—	100.0%	100.0%	100.0%	13.61%	12.45%	15.18%
				Overall Weighted Average	13.75%					

Workers' Compensation Insurance Rates by States

The table below lists the weighted average workers' compensation base rate for each state with a factor comparing this with the national average of 13.5%.

State	Weighted Average	Factor	State	Weighted Average	Factor	State	Weighted Average	Factor
Alabama	20.6%	153	Kentucky	12.5%	93	North Dakota	8.1%	60
Alaska	12.5	93	Louisiana	19.6	145	Ohio	8.1	60
Arizona	13.8	102	Maine	11.4	84	Oklahoma	10.3	76
Arkansas	7.7	57	Maryland	13.5	100	Oregon	11.3	84
California	27.1	201	Massachusetts	11.5	85	Pennsylvania	18.3	136
Colorado	7.3	54	Michigan	13.5	100	Rhode Island	14.7	109
Connecticut	25.9	192	Minnesota	22.8	169	South Carolina	16.9	125
Delaware	14.0	104	Mississippi	14.0	104	South Dakota	14.1	104
District of Columbia	10.4	77	Missouri	13.9	103	Tennessee	12.2	90
Florida	10.8	80	Montana	8.3	61	Texas	9.3	69
Georgia	29.5	219	Nebraska	17.0	126	Utah	8.8	65
Hawaii	8.7	64	Nevada	9.2	68	Vermont	13.3	99
Idaho	10.5	78	New Hampshire	18.8	139	Virginia	8.7	64
Illinois	26.9	199	New Jersey	15.7	116	Washington	9.5	70
Indiana	5.0	37	New Mexico	15.3	113	West Virginia	7.4	55
Iowa	15.5	115	New York	19.0	141	Wisconsin	13.2	98
Kansas	9.2	68	North Carolina	17.3	128	Wyoming	6.5	48
			Weighted Average for U.S. is	13.7% of payroll = 100%				

Rates in the following table are the base or manual costs per $100 of payroll for workers' compensation in each state. Rates are usually applied to straight time wages only and not to premium time wages and bonuses.

The weighted average skilled worker rate for 35 trades is 13.5%. For bidding purposes, apply the full value of workers' compensation directly to total labor costs, or if labor is 38%, materials 42% and overhead and profit 20% of total cost, carry 38/80 × 13.5% =6.4% of cost (before overhead and profit) into overhead. Rates vary not only from state to state but also with the experience rating of the contractor.

Rates are the most current available at the time of publication.

R013113-80 Performance Bond

This table shows the cost of a Performance Bond for a construction job scheduled to be completed in 12 months. Add 1% of the premium cost per month for jobs requiring more than 12 months to complete. The rates are "standard" rates offered to contractors that the bonding company considers financially sound and capable of doing the work. Preferred rates are offered by some bonding companies based upon financial strength of the contractor. Actual rates vary from contractor to contractor and from bonding company to bonding company. Contractors should prequalify through a bonding agency before submitting a bid on a contract that requires a bond.

Contract Amount	Building Construction Class B Projects	Highways & Bridges Class A New Construction	Highways & Bridges Class A-1 Highway Resurfacing
First $ 100,000 bid	$25.00 per M	$15.00 per M	$9.40 per M
Next 400,000 bid	$ 2,500 plus $15.00 per M	$ 1,500 plus $10.00 per M	$ 940 plus $7.20 per M
Next 2,000,000 bid	8,500 plus 10.00 per M	5,500 plus 7.00 per M	3,820 plus 5.00 per M
Next 2,500,000 bid	28,500 plus 7.50 per M	19,500 plus 5.50 per M	15,820 plus 4.50 per M
Next 2,500,000 bid	47,250 plus 7.00 per M	33,250 plus 5.00 per M	28,320 plus 4.50 per M
Over 7,500,000 bid	64,750 plus 6.00 per M	45,750 plus 4.50 per M	39,570 plus 4.00 per M

General Requirements R0151 Temporary Utilities

R015113-65 Temporary Power Equipment

Cost data for the temporary equipment was developed utilizing the following information.

1) Re-usable material-services, transformers, equipment and cords are based on new purchase and prorated to three projects.
2) PVC feeder includes trench and backfill.
3) Connections include disconnects and fuses.
4) Labor units include an allowance for removal.
5) No utility company charges or fees are included.
6) Concrete pads or vaults are not included.
7) Utility company conduits not included.

R015423-10 Steel Tubular Scaffolding

On new construction, tubular scaffolding is efficient up to 60′ high or five stories. Above this it is usually better to use a hung scaffolding if construction permits. Swing scaffolding operations may interfere with tenants. In this case, the tubular is more practical at all heights.

In repairing or cleaning the front of an existing building the cost of tubular scaffolding per S.F. of building front increases as the height increases above the first tier. The first tier cost is relatively high due to leveling and alignment.

The minimum efficient crew for erecting and dismantling is three workers. They can set up and remove 18 frame sections per day up to 5 stories high. For 6 to 12 stories high, a crew of four is most efficient. Use two or more on top and two on the bottom for handing up or hoisting. They can also set up and remove 18 frame sections per day. At 7' horizontal spacing, this will run about 800 S.F. per day of erecting and dismantling. Time for placing and removing planks must be added to the above. A crew of three can place and remove 72 planks per day up to 5 stories. For over 5 stories, a crew of four can place and remove 80 planks per day.

The table below shows the number of pieces required to erect tubular steel scaffolding for 1000 S.F. of building frontage. This area is made up of a scaffolding system that is 12 frames (11 bays) long by 2 frames high.

For jobs under twenty-five frames, add 50% to rental cost. Rental rates will be lower for jobs over three months duration. Large quantities for long periods can reduce rental rates by 20%.

Description of Component	Number of Pieces for 1000 S.F. of Building Front	Unit
5′ Wide Standard Frame, 6′-4″ High	24	Ea.
Leveling Jack & Plate	24	
Cross Brace	44	
Side Arm Bracket, 21″	12	
Guardrail Post	12	
Guardrail, 7′ section	22	
Stairway Section	2	
Stairway Starter Bar	1	
Stairway Inside Handrail	2	
Stairway Outside Handrail	2	
Walk-Thru Frame Guardrail	2	↓

Scaffolding is often used as falsework over 15′ high during construction of cast-in-place concrete beams and slabs. Two foot wide scaffolding is generally used for heavy beam construction. The span between frames depends upon the load to be carried with a maximum span of 5′.

Heavy duty shoring frames with a capacity of 10,000#/leg can be spaced up to 10′ O.C. depending upon form support design and loading.

Scaffolding used as horizontal shoring requires less than half the material required with conventional shoring.

On new construction, erection is done by carpenters.

Rolling towers supporting horizontal shores can reduce labor and speed the job. For maintenance work, catwalks with spans up to 70′ can be supported by the rolling towers.

R015433-10 Contractor Equipment

Rental Rates shown elsewhere in the book pertain to late model high quality machines in excellent working condition, rented from equipment dealers. Rental rates from contractors may be substantially lower than the rental rates from equipment dealers depending upon economic conditions; for older, less productive machines, reduce rates by a maximum of 15%. Any overtime must be added to the base rates. For shift work, rates are lower. Usual rule of thumb is 150% of one shift rate for two shifts; 200% for three shifts.

For periods of less than one week, operated equipment is usually more economical to rent than renting bare equipment and hiring an operator.

Costs to move equipment to a job site (mobilization) or from a job site (demobilization) are not included in rental rates, nor in any Equipment costs on any Unit Price line items or crew listings. These costs can be found elsewhere. If a piece of equipment is already at a job site, it is not appropriate to utilize mob/demob costs in an estimate again.

Rental rates vary throughout the country with larger cities generally having lower rates. Lease plans for new equipment are available for periods in excess of six months with a percentage of payments applying toward purchase.

Rental rates can also be treated as reimbursement costs for contractor-owned equipment. Owned equipment costs include depreciation, loan payments, interest, taxes, insurance, storage, and major repairs.

Monthly rental rates vary from 2% to 5% of the cost of the equipment depending on the anticipated life of the equipment and its wearing parts. Weekly rates are about 1/3 the monthly rates and daily rental rates about 1/3 the weekly rate.

The hourly operating costs for each piece of equipment include costs to the user such as fuel, oil, lubrication, normal expendables for the equipment, and a percentage of mechanic's wages chargeable to maintenance. The hourly operating costs listed do not include the operator's wages.

The daily cost for equipment used in the standard crews is figured by dividing the weekly rate by five, then adding eight times the hourly operating cost to give the total daily equipment cost, not including the operator. This figure is in the right hand column of the Equipment listings under Equipment Cost/Day.

Pile Driving rates shown for pile hammer and extractor do not include leads, crane, boiler or compressor. Vibratory pile driving requires an added field specialist during set-up and pile driving operation for the electric model. The hydraulic model requires a field specialist for set-up only. Up to 125 reuses of sheet piling are possible using vibratory drivers. For normal conditions, crane capacity for hammer type and size are as follows.

Crane Capacity	Hammer Type and Size		
	Air or Steam	Diesel	Vibratory
25 ton	to 8,750 ft.-lb.		70 H.P.
40 ton	15,000 ft.-lb.	to 32,000 ft.-lb.	170 H.P.
60 ton	25,000 ft.-lb.		300 H.P.
100 ton		112,000 ft.-lb.	

Cranes should be specified for the job by size, building and site characteristics, availability, performance characteristics, and duration of time required.

Backhoes & Shovels rent for about the same as equivalent size cranes but maintenance and operating expense is higher. Crane operators rate must be adjusted for high boom heights. Average adjustments: for 150′ boom add 2% per hour; over 185′, add 4% per hour; over 210′, add 6% per hour; over 250′, add 8% per hour and over 295′, add 12% per hour.

Tower Cranes of the climbing or static type have jibs from 50′ to 200′ and capacities at maximum reach range from 4,000 to 14,000 pounds. Lifting capacities increase up to maximum load as the hook radius decreases.

Typical rental rates, based on purchase price are about 2% to 3% per month.

Erection and dismantling runs between 500 and 2000 labor hours. Climbing operation takes 10 labor hours per 20′ climb. Crane dead time is about 5 hours per 40′ climb. If crane is bolted to side of the building add cost of ties and extra mast sections. Climbing cranes have from 80′ to 180′ of mast while static cranes have 80′ to 800′ of mast.

Truck Cranes can be converted to tower cranes by using tower attachments. Mast heights over 400′ have been used.

A single 100′ high material **Hoist and Tower** can be erected and dismantled in about 400 labor hours; a double 100′ high hoist and tower in about 600 labor hours. Erection times for additional heights are 3 and 4 labor hours per vertical foot respectively up to 150′, and 4 to 5 labor hours per vertical foot over 150′ high. A 40′ high portable Buck hoist takes about 160 labor hours to erect and dismantle. Additional heights take 2 labor hours per vertical foot to 80′ and 3 labor hours per vertical foot for the next 100′. Most material hoists do not meet local code requirements for carrying personnel.

A 150′ high **Personnel Hoist** requires about 500 to 800 labor hours to erect and dismantle. Budget erection time at 5 labor hours per vertical foot for all trades. Local code requirements or labor scarcity requiring overtime can add up to 50% to any of the above erection costs.

Earthmoving Equipment: The selection of earthmoving equipment depends upon the type and quantity of material, moisture content, haul distance, haul road, time available, and equipment available. Short haul cut and fill operations may require dozers only, while another operation may require excavators, a fleet of trucks, and spreading and compaction equipment. Stockpiled material and granular material are easily excavated with front end loaders. Scrapers are most economically used with hauls between 300′ and 1-1/2 miles if adequate haul roads can be maintained. Shovels are often used for blasted rock and any material where a vertical face of 8′ or more can be excavated. Special conditions may dictate the use of draglines, clamshells, or backhoes. Spreading and compaction equipment must be matched to the soil characteristics, the compaction required and the rate the fill is being supplied.

R015436-50 Mobilization

Costs to move rented construction equipment to a job site from an equipment dealer's or contractor's yard (mobilization), or to move the equipment off the job site (demobilization), are not included in the rental or operating rates, nor in the equipment cost on a unit price line or in a crew listing. These costs can be found consolidated in the Mobilization section of the data and elsewhere in particular site work sections. If a piece of equipment is already on the job site, it is not appropriate to include mob/demob costs in a new estimate that requires use of that equipment. The following table identifies approximate sizes of rented construction equipment that would be hauled on a towed trailer. Because this listing is not all-encompassing, the user can infer as to what size trailer might be required for a piece of equipment not listed.

3-ton Trailer	20-ton Trailer	40-ton Trailer	50-ton Trailer
20 H.P. Excavator	110 H.P. Excavator	200 H.P. Excavator	270 H.P. Excavator
50 H.P. Skid Steer	165 H.P. Dozer	300 H.P. Dozer	Small Crawler Crane
35 H.P. Roller	150 H.P. Roller	400 H.P. Scraper	500 H.P. Scraper
40 H.P. Trencher	Backhoe	450 H.P. Art. Dump Truck	500 H.P. Art. Dump Truck

R024119-10 Demolition Defined

Whole Building Demolition - Demolition of the whole building with no concern for any particular building element, component, or material type being demolished. This type of demolition is accomplished with large pieces of construction equipment that break up the structure, load it into trucks and haul it to a disposal site, but disposal or dump fees are not included. Demolition of below-grade foundation elements, such as footings, foundation walls, grade beams, slabs on grade, etc., is not included. Certain mechanical equipment containing flammable liquids or ozone-depleting refrigerants, electric lighting elements, communication equipment components, and other building elements may contain hazardous waste, and must be removed, either selectively or carefully, as hazardous waste before the building can be demolished.

Foundation Demolition - Demolition of below-grade foundation footings, foundation walls, grade beams, and slabs on grade. This type of demolition is accomplished by hand or pneumatic hand tools, and does not include saw cutting, or handling, loading, hauling, or disposal of the debris.

Gutting - Removal of building interior finishes and electrical/mechanical systems down to the load-bearing and sub-floor elements of the rough building frame, with no concern for any particular building element, component, or material type being demolished. This type of demolition is accomplished by hand or pneumatic hand tools, and includes loading into trucks, but not hauling, disposal or dump fees, scaffolding, or shoring. Certain mechanical equipment containing flammable liquids or ozone-depleting refrigerants, electric lighting elements, communication equipment components, and other building elements may contain hazardous waste, and must be removed, either selectively or carefully, as hazardous waste, before the building is gutted.

Selective Demolition - Demolition of a selected building element, component, or finish, with some concern for surrounding or adjacent elements, components, or finishes (see the first Subdivision (s) at the beginning of appropriate Divisions). This type of demolition is accomplished by hand or pneumatic hand tools, and does not include handling, loading, storing, hauling, or disposal of the debris, scaffolding, or shoring. "Gutting" methods may be used in order to save time, but damage that is caused to surrounding or adjacent elements, components, or finishes may have to be repaired at a later time.

Careful Removal - Removal of a piece of service equipment, building element or component, or material type, with great concern for both the removed item and surrounding or adjacent elements, components or finishes. The purpose of careful removal may be to protect the removed item for later re-use, preserve a higher salvage value of the removed item, or replace an item while taking care to protect surrounding or adjacent elements, components, connections, or finishes from cosmetic and/or structural damage. An approximation of the time required to perform this type of removal is 1/3 to 1/2 the time it would take to install a new item of like kind (see Reference Number R260105-30). This type of removal is accomplished by hand or pneumatic hand tools, and does not include loading, hauling, or storing the removed item, scaffolding, shoring, or lifting equipment.

Cutout Demolition - Demolition of a small quantity of floor, wall, roof, or other assembly, with concern for the appearance and structural integrity of the surrounding materials. This type of demolition is accomplished by hand or pneumatic hand tools, and does not include saw cutting, handling, loading, hauling, or disposal of debris, scaffolding, or shoring.

Rubbish Handling - Work activities that involve handling, loading or hauling of debris. Generally, the cost of rubbish handling must be added to the cost of all types of demolition, with the exception of whole building demolition.

Minor Site Demolition - Demolition of site elements outside the footprint of a building. This type of demolition is accomplished by hand or pneumatic hand tools, or with larger pieces of construction equipment, and may include loading a removed item onto a truck (check the Crew for equipment used). It does not include saw cutting, hauling or disposal of debris, and, sometimes, handling or loading.

R024119-20 Dumpsters

Dumpster rental costs on construction sites are presented in two ways.

The cost per week rental includes the delivery of the dumpster; its pulling or emptying once per week, and its final removal. The assumption is made that the dumpster contractor could choose to empty a dumpster by simply bringing in an empty unit and removing the full one. These costs also include the disposal of the materials in the dumpster.

The Alternate Pricing can be used when actual planned conditions are not approximated by the weekly numbers. For example, these lines can be used when a dumpster is needed for 4 weeks and will need to be emptied 2 or 3 times per week. Conversely the Alternate Pricing lines can be used when a dumpster will be rented for several weeks or months but needs to be emptied only a few times over this period.

R053100-10 Decking Descriptions

General - All Deck Products

Steel deck is made by cold forming structural grade sheet steel into a repeating pattern of parallel ribs. The strength and stiffness of the panels are the result of the ribs and the material properties of the steel. Deck lengths can be varied to suit job conditions, but because of shipping considerations, are usually less than 40 feet. Standard deck width varies with the product used but full sheets are usually 12″, 18″, 24″, 30″, or 36″. Deck is typically furnished in a standard width with the ends cut square. Any cutting for width, such as at openings or for angular fit, is done at the job site.

Deck is typically attached to the building frame with arc puddle welds, self-drilling screws, or powder or pneumatically driven pins. Sheet to sheet fastening is done with screws, button punching (crimping), or welds.

Composite Floor Deck

After installation and adequate fastening, floor deck serves several purposes. It (a) acts as a working platform, (b) stabilizes the frame, (c) serves as a concrete form for the slab, and (d) reinforces the slab to carry the design loads applied during the life of the building. Composite decks are distinguished by the presence of shear connector devices as part of the deck. These devices are designed to mechanically lock the concrete and deck together so that the concrete and the deck work together to carry subsequent floor loads. These shear connector devices can be rolled-in embossments, lugs, holes, or wires welded to the panels. The deck profile can also be used to interlock concrete and steel.

Composite deck finishes are either galvanized (zinc coated) or phosphatized/painted. Galvanized deck has a zinc coating on both the top and bottom surfaces. The phosphatized/painted deck has a bare (phosphatized) top surface that will come into contact with the concrete. This bare top surface can be expected to develop rust before the concrete is placed. The bottom side of the deck has a primer coat of paint.

Composite floor deck is normally installed so the panel ends do not overlap on the supporting beams. Shear lugs or panel profile shape often prevent a tight metal to metal fit if the panel ends overlap; the air gap caused by overlapping will prevent proper fusion with the structural steel supports when the panel end laps are shear stud welded.

Adequate end bearing of the deck must be obtained as shown on the drawings. If bearing is actually less in the field than shown on the drawings, further investigation is required.

Roof Deck

Roof deck is not designed to act compositely with other materials. Roof deck acts alone in transferring horizontal and vertical loads into the building frame. Roof deck rib openings are usually narrower than floor deck rib openings. This provides adequate support of rigid thermal insulation board.

Roof deck is typically installed to endlap approximately 2″ over supports. However, it can be butted (or lapped more than 2″) to solve field fit problems. Since designers frequently use the installed deck system as part of the horizontal bracing system (the deck as a diaphragm), any fastening substitution or change should be approved by the designer. Continuous perimeter support of the deck is necessary to limit edge deflection in the finished roof and may be required for diaphragm shear transfer.

Standard roof deck finishes are galvanized or primer painted. The standard factory applied paint for roof deck is a primer paint and is not intended to weather for extended periods of time. Field painting or touching up of abrasions and deterioration of the primer coat or other protective finishes is the responsibility of the contractor.

Cellular Deck

Cellular deck is made by attaching a bottom steel sheet to a roof deck or composite floor deck panel. Cellular deck can be used in the same manner as floor deck. Electrical, telephone, and data wires are easily run through the chase created between the deck panel and the bottom sheet.

When used as part of the electrical distribution system, the cellular deck must be installed so that the ribs line up and create a smooth cell transition at abutting ends. The joint that occurs at butting cell ends must be taped or otherwise sealed to prevent wet concrete from seeping into the cell. Cell interiors must be free of welding burrs, or other sharp intrusions, to prevent damage to wires.

When used as a roof deck, the bottom flat plate is usually left exposed to view. Care must be maintained during erection to keep good alignment and prevent damage.

Cellular deck is sometimes used with the flat plate on the top side to provide a flat working surface. Installation of the deck for this purpose requires special methods for attachment to the frame because the flat plate, now on the top, can prevent direct access to the deck material that is bearing on the structural steel. It may be advisable to treat the flat top surface to prevent slipping.

Cellular deck is always furnished galvanized or painted over galvanized.

Form Deck

Form deck can be any floor or roof deck product used as a concrete form. Connections to the frame are by the same methods used to anchor floor and roof deck. Welding washers are recommended when welding deck that is less than 20 gauge thickness.

Form deck is furnished galvanized, prime painted, or uncoated. Galvanized deck must be used for those roof deck systems where form deck is used to carry a lightweight insulating concrete fill.

R078413-30 Firestopping

Firestopping is the sealing of structural, mechanical, electrical and other penetrations through fire-rated assemblies. The basic components of firestop systems are safing insulation and firestop sealant on both sides of wall penetrations and the top side of floor penetrations.

Pipe penetrations are assumed to be through concrete, grout, or joint compound and can be sleeved or unsleeved. Costs for the penetrations and sleeves are not included. An annular space of 1″ is assumed. Escutcheons are not included.

Metallic pipe is assumed to be copper, aluminum, cast iron or similar metallic material. Insulated metallic pipe is assumed to be covered with a thermal insulating jacket of varying thickness and materials.

Non-metallic pipe is assumed to be PVC, CPVC, FR Polypropylene or similar plastic piping material. Intumescent firestop sealant or wrap strips are included. Collars on both sides of wall penetrations and a sheet metal plate on the underside of floor penetrations are included.

Ductwork is assumed to be sheet metal, stainless steel or similar metallic material. Duct penetrations are assumed to be through concrete, grout or joint compound. Costs for penetrations and sleeves are not included. An annular space of 1/2″ is assumed.

Multi-trade openings include costs for sheet metal forms, firestop mortar, wrap strips, collars and sealants as necessary.

Structural penetrations joints are assumed to be 1/2″ or less. CMU walls are assumed to be within 1-1/2″ of metal deck. Drywall walls are assumed to be tight to the underside of metal decking.

Metal panel, glass or curtain wall systems include a spandrel area of 5′ filled with mineral wool foil-faced insulation. Fasteners and stiffeners are included.

R220102-20 Labor Adjustment Factors

Labor Adjustment Factors are provided for Divisions 21, 22, and 23 to assist the mechanical estimator account for the various complexities and special conditions of any particular project. While a single percentage has been entered on each line of Division 22 01 02.20, it should be understood that these are just suggested midpoints of ranges of values commonly used by mechanical estimators. They may be increased or decreased depending on the severity of the special conditions.

The group for "existing occupied buildings", has been the subject of requests for explanation. Actually there are two stages to this group: buildings that are existing and "finished" but unoccupied, and those that also are occupied. Buildings that are "finished" may result in higher labor costs due to the workers having to be more careful not to damage finished walls, ceilings, floors, etc. and may necessitate special protective coverings and barriers. Also corridor bends and doorways may not accommodate long pieces of pipe or larger pieces of equipment. Work above an already hung ceiling can be very time consuming. The addition of occupants may force the work to be done on premium time (nights and/or weekends), eliminate the possible use of some preferred tools such as pneumatic drivers, powder charged drivers etc. The estimator should evaluate the access to the work area and just how the work is going to be accomplished to arrive at an increase in labor costs over "normal" new construction productivity.

R224000-10 Water Consumption Rates

Fixture Type	Water Supply Fixture Unit Value		
	Hot Water	Cold Water	Combined
Bathtub	1.0	1.0	1.4
Clothes Washer	1.0	1.0	1.4
Dishwasher	1.4	0.0	1.4
Kitchen Sink	1.0	1.0	1.4
Laundry Tub	1.0	1.0	1.4
Lavatory	0.5	0.5	0.7
Shower Stall	1.0	1.0	1.4
Water Closet (tank type)	0.0	2.2	2.2
Full Bath Group w/bathtub or shower stall	1.5	2.7	3.6
Half Bath Group w/W.C. and Lavatory	0.5	2.5	2.6
Kitchen Group w/Dishwasher and Sink	1.9	1.0	2.5
Laundry Group w/Clothes Washer and Laundry Tub	1.8	1.8	2.5
Hose bibb (sillcock)	0.0	2.5	2.5

Notes:
Typically, WSFU = 1GPM

Supply loads in the building water-distribution system shall be determined by total load on the pipe being sized, in terms of water supply fixture units (WSFU) and gallons per minute (GPM) flow rates. For fixtures not listed, choose a WSFU value of a fixture with similar flow characteristics. Water Fixture Supply Units determined the required water supply to fixtures and their service systems. Fixture units are equal to (1) cubic foot of water drained in a 1-1/4″ pipe per minute. It is not a flow rate unit but a design factor.

Excerpted from the 2012 *International Plumbing Code*, Copyright 2011. Washington, D.C.: International Code Council. Reproduced with permission.

R224000-20 Fixture Demands in Gallons per Fixture per Hour

Table below is based on 140°F final temperature except for dishwashers in public places (*) where 180°F water is mandatory.

Supply Systems for Flush Tanks			Supply Systems for Flushometer Valves		
Load	Demand		Load	Demand	
WSFU	GPM	CU. FT.	WSFU	GPM	CU. FT.
1.0	3.0	0.041040	-	-	-
2.0	5.0	0.068400	-	-	-
3.0	6.5	0.868920	-	-	-
4.0	8.0	1.069440	-	-	-
5.0	9.4	1.256592	5.0	15	2.0052
6.0	10.7	1.430376	6.0	17.4	2.326032
7.0	11.8	1.577424	7.0	19.8	2.646364
8.0	12.8	1.711104	8.0	22.2	2.967696
9.0	13.7	1.831416	9.0	24.6	3.288528
10.0	14.6	1.951728	10.0	27	3.60936
11.0	15.4	2.058672	11.0	27.8	3.716304
12.0	16.0	2.138880	12.0	28.6	3.823248
13.0	16.5	2.205720	13.0	29.4	3.930192
14.0	17.0	2.272560	14.0	30.2	4.037136
15.0	17.5	2.339400	15.0	31	4.14408
16.0	18.0	2.906240	16.0	31.8	4.241024
17.0	18.4	2.459712	17.0	32.6	4.357968
18.0	18.8	2.513184	18.0	33.4	4.464912
19.0	19.2	2.566656	19.0	34.2	4.571856
20.0	19.6	2.620218	20.0	35.0	4.678800
25.0	21.5	2.874120	25.0	38.0	5.079840
30.0	23.3	3.114744	30.0	42.0	5.611356
35.0	24.9	3.328632	35.0	44.0	5.881920
40.0	26.3	3.515784	40.0	46.0	6.149280
45.0	27.7	3.702936	45.0	48.0	6.416640
50.0	29.1	3.890088	50.0	50.0	6.684000

Notes:
When designing a plumbing system that utilizes fixtures other than, or in addition to, water closets, use the data provided in the Supply Systems for Flush Tanks section of the above table.

To obtain the probable maximum demand, multiply the total demands for the fixtures (gal./fixture/hour) by the demand factor. The heater should have a heating capacity in gallons per hour equal to this maximum. The storage tank should have a capacity in gallons equal to the probable maximum demand multiplied by the storage capacity factor.

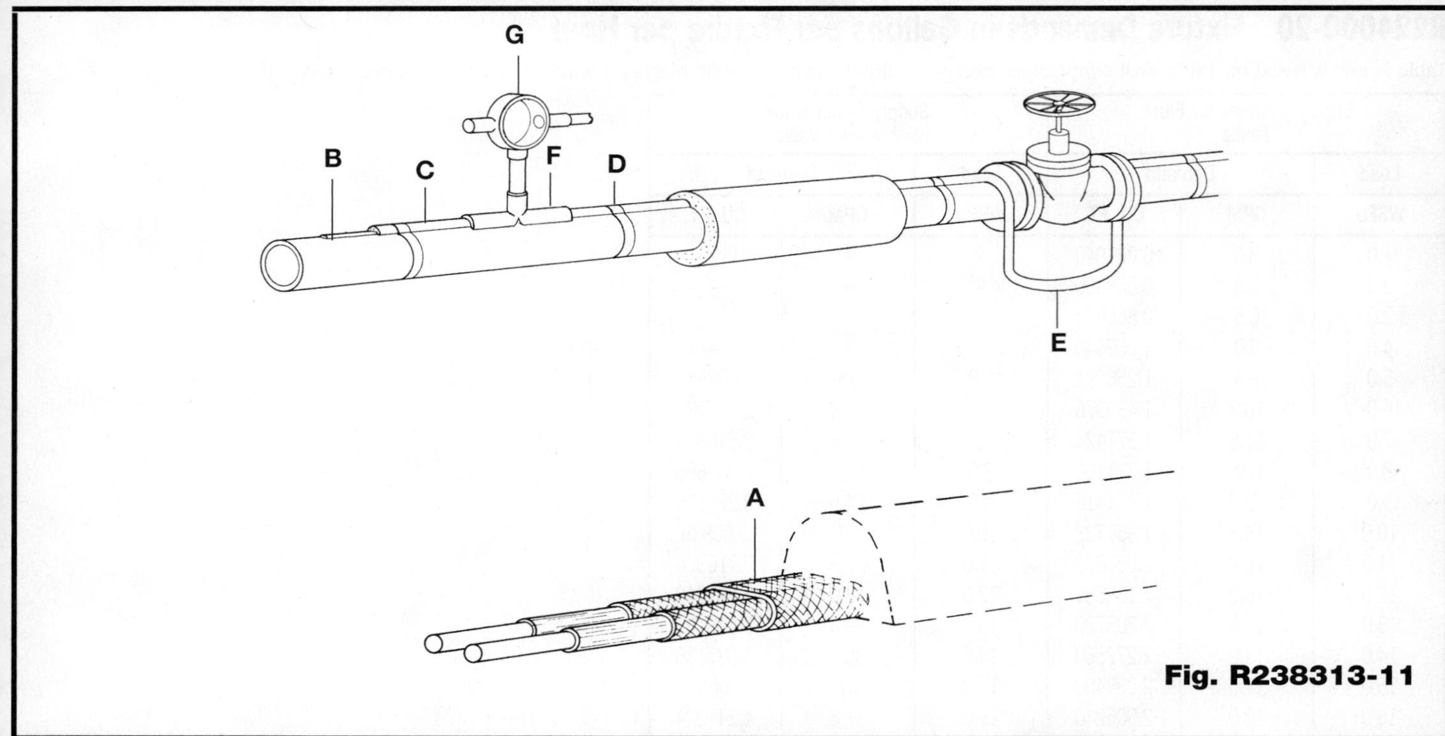

Fig. R238313-11

R238313-10 Heat Trace Systems

Before you can determine the cost of a HEAT TRACE installation, the method of attachment must be established. There are (4) common methods:

1. Cable is simply attached to the pipe with polyester tape every 12′.
2. Cable is attached with a continuous cover of 2″ wide aluminum tape.
3. Cable is attached with factory extruded heat transfer cement and covered with metallic raceway with clips every 10′.
4. Cable is attached between layers of pipe insulation using either clips or polyester tape.

Example: Components for method 3 must include:

A. Heat trace cable by voltage and watts per linear foot.
B. Heat transfer cement, 1 gallon per 60 linear feet of cover.
C. Metallic raceway by size and type.
D. Raceway clips by size of pipe.

When taking off linear foot lengths of cable add the following for each valve in the system. (E)

In all of the above methods each component of the system must be priced individually.

SCREWED OR WELDED VALVE:

Size		Length
1/2″	=	6″
3/4″	=	9″
1″	=	1′ -0″
1-1/2″	=	1′ -6″
2″	=	2′
2-1/2″	=	2′ -6″
3″	=	2′ -6″
4″	=	4′ -0″
6″	=	7′ -0″
8″	=	9′ -6″
10″	=	12′ -6″
12″	=	15′ -0″
14″	=	18′ -0″
16″	=	21′ -6″
18″	=	25′ -6″
20″	=	28′ -6″
24″	=	34′ -0″
30″	=	40′ -0″

FLANGED VALVE:

Size		Length
1/2″	=	1′ -0″
3/4″	=	1′ -6″
1″	=	2′ -0″
1-1/2″	=	2′ -6″
2″	=	2′ -6″
2-1/2″	=	3′ -0″
3″	=	3′ -6″
4″	=	4′ -0″
6″	=	8′ -0″
8″	=	11′ -0″
10″	=	14′ -0″
12″	=	16′ -6″
14″	=	19′ -6″
16″	=	23′ -0″
18″	=	27′ -0″
20″	=	30′ -0″
24″	=	36′ -0″
30″	=	42′ -0″

BUTTERFLY VALVES:

Size		Length
1/2″	=	0′
3/4″	=	0′
1″	=	1′ -0″
1-1/2″	=	1′ -6″
2″	=	2′ -0″
2-1/2″	=	2′ -6″
3″	=	2′ -6″
4″	=	3′ -0″
6″	=	3′ -6″
8″	=	4′ -0″
10″	=	4′ -0″
12″	=	5′ -0″
14″	=	5′ -6″
16″	=	6′ -0″
18″	=	6′ -6″
20″	=	7′ -0″
24″	=	8′ -0″
30″	=	10′ -0″

R238313-10 Heat Trace Systems (cont.)

Add the following quantities of heat transfer cement to linear foot totals for each valve:

Nominal Valve Size	Gallons of Cement per Valve
1/2″	0.14
3/4″	0.21
1″	0.29
1-1/2″	0.36
2″	0.43
2-1/2″	0.70
3″	0.71
4″	1.00
6″	1.43
8″	1.48
10″	1.50
12″	1.60
14″	1.75
16″	2.00
18″	2.25
20″	2.50
24″	3.00
30″	3.75

The following must be added to the list of components to accurately price HEAT TRACE systems:

1. Expediter fitting and clamp fasteners (F)
2. Junction box and nipple connected to expediter fitting (G)
3. Field installed terminal blocks within junction box
4. Ground lugs
5. Piping from power source to expediter fitting
6. Controls
7. Thermostats
8. Branch wiring
9. Cable splices
10. End of cable terminations
11. Branch piping fittings and boxes

Deduct the following percentages from labor if cable lengths in the same area exceed:

150′ to 250′	10%	351′ to 500′	20%
251′ to 350′	15%	Over 500′	25%

Add the following percentages to labor for elevated installations:

15′ to 20′ high	10%	31′ to 35′ high	40%
21′ to 25′ high	20%	36′ to 40′ high	50%
26′ to 30′ high	30%	Over 40′ high	60%

R238313-20 Spiral-Wrapped Heat Trace Cable (Pitch Table)

In order to increase the amount of heat, occasionally heat trace cable is wrapped in a spiral fashion around a pipe; increasing the number of feet of heater cable per linear foot of pipe.

Engineers first determine the heat loss per foot of pipe (based on the insulating material, its thickness, and the temperature differential across it). A ratio is then calculated by the formula:

$$\text{Feet of Heat Trace per Foot of Pipe} = \frac{\text{Watts/Foot of Heat Loss}}{\text{Watts/Foot of the Cable}}$$

The linear distance between wraps (pitch) is then taken from a chart or table. Generally, the pitch is listed on a drawing leaving the estimator to calculate the total length of heat tape required. An approximation may be taken from this table.

Feet of Heat Trace Per Foot of Pipe																
	Nominal Pipe Size in Inches															
Pitch In Inches	1	1¼	1½	2	2½	3	4	6	8	10	12	14	16	18	20	24
3.5	1.80															
4	1.65															
5	1.46	1.60	1.80													
6	1.34	1.45	1.55	1.75												
7	1.25	1.35	1.43	1.57	1.75											
8	1.20	1.28	1.34	1.45	1.60	1.80										
9	1.16	1.23	1.28	1.37	1.51	1.68										
10	1.13	1.19	1.24	1.32	1.44	1.57	1.82									
15	1.06	1.08	1.10	1.15	1.21	1.29	1.42	1.78								
20	1.04	1.05	1.06	1.08	1.13	1.17	1.25	1.49	1.73							
25		1.04	1.04	1.06	1.08	1.11	1.17	1.33	1.51	1.72						
30				1.04	1.05	1.07	1.12	1.24	1.37	1.54	1.70	1.80				
35					1.06	1.06	1.09	1.17	1.28	1.42	1.54	1.64	1.78			
40						1.05	1.07	1.14	1.22	1.33	1.44	1.52	1.64	1.75		
50							1.05	1.09	1.15	1.22	1.29	1.35	1.44	1.53	1.64	1.83
60								1.06	1.11	1.16	1.21	1.25	1.31	1.39	1.46	1.62
70								1.05	1.08	1.12	1.17	1.19	1.24	1.30	1.35	1.47
80									1.06	1.09	1.13	1.15	1.19	1.24	1.28	1.38
90									1.04	1.06	1.10	1.13	1.16	1.19	1.23	1.32
100										1.05	1.08	1.10	1.13	1.15	1.19	1.23

Note: Common practice would normally limit the lower end of the table to 5% of additional heat and above 80% an engineer would likely opt for two (2) parallel cables.

R260105-30 Electrical Demolition (Removal for Replacement)

The purpose of this reference number is to provide a guide to users for electrical "removal for replacement" by applying the rule of thumb: 1/3 of new installation time (typical range from 20% to 50%) for removal. Remember to use reasonable judgment when applying the suggested percentage factor. For example:

Contractors have been requested to remove an existing fluorescent lighting fixture and replace with a new fixture utilizing energy saver lamps and electronic ballast:

In order to fully understand the extent of the project, contractors should visit the job site and estimate the time to perform the renovation work in accordance with applicable national, state and local regulation and codes.

The contractor may need to add extra labor hours to his estimate if he discovers unknown concealed conditions such as: contaminated asbestos ceiling, broken acoustical ceiling tile and need to repair, patch and touch-up paint all the damaged or disturbed areas, tasks normally assigned to general contractors. In addition, the owner could request that the contractors salvage the materials removed and turn over the materials to the owner or dispose of the materials to a reclamation station. The normal removal item is 0.5 labor hour for a lighting fixture and 1.5 labor-hours for new installation time. Revise the estimate times from 2 labor-hours work up to a minimum 4 labor-hours work for just fluorescent lighting fixture.

For removal of large concentrations of lighting fixtures in the same area, apply an "economy of scale" to reduce estimating labor hours.

Figure R260110-10 Typical Overhead Service Entrance

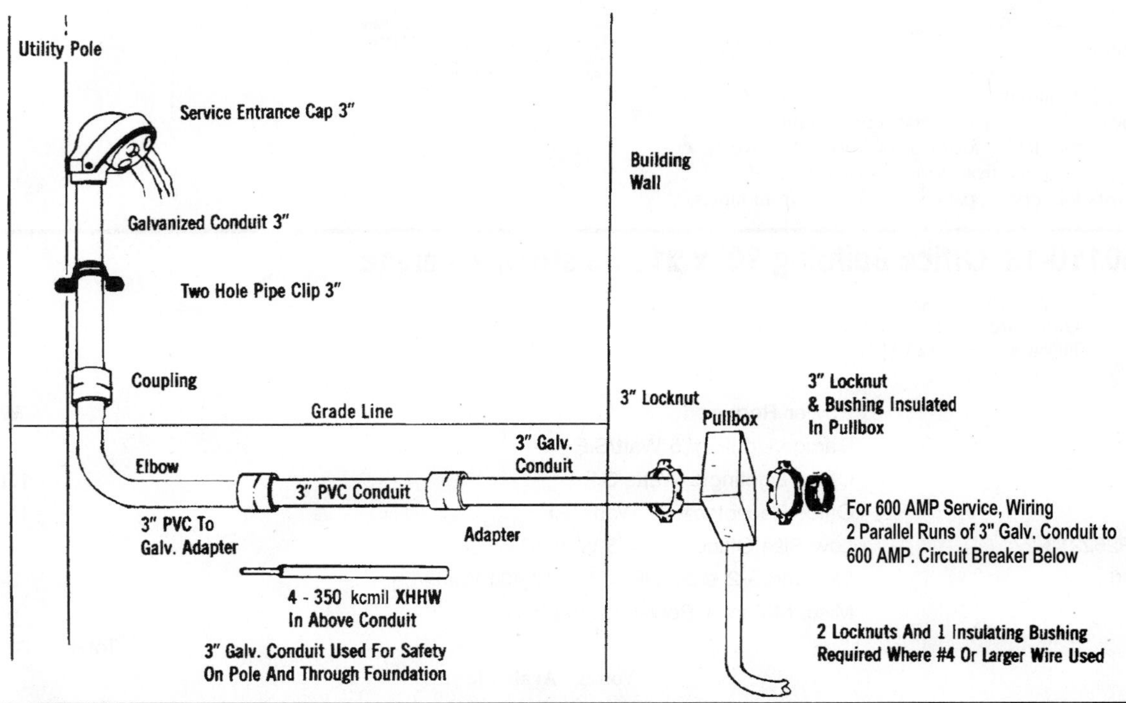

Figure R260110-11 Typical Commercial Electric System

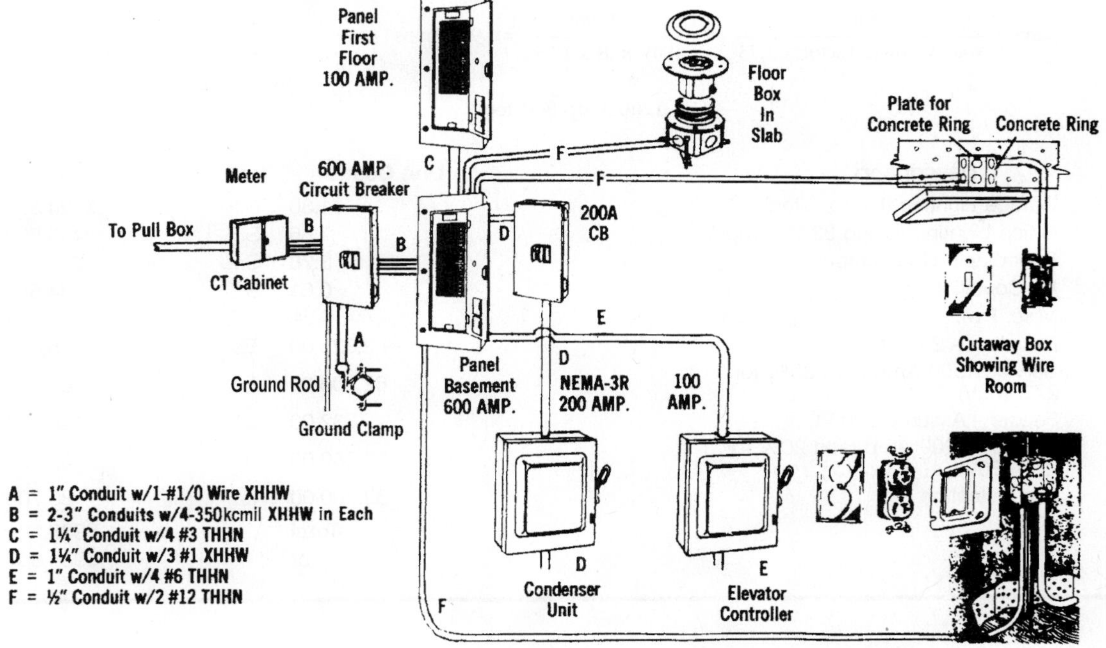

Figure R260110-12 Preliminary Procedure

1. Determine building size and use.
2. Develop total load in watts.
   a. Lighting
   b. Receptacle
   c. Air Conditioning
   d. Elevator
   e. Other power requirements
3. Determine best voltage available from utility company.
4. Determine cost from tables for loads (a) thru (e) above.
5. Determine size of service from formulas (R260110-17).
6. Determine costs for service, panels, and feeders from tables.

Figure R260110-13 Office Building 90′ x 210′, 3 story, w/garage

Garage Area = 18,900 S.F.
Office Area = 56,700 S.F.
Elevator = 2 @ 125 FPM

Tables	Power Required	Watts	
R260110-14	Garage Lighting .5 Watt/S.F.	9,450	
	Office Lighting 3 Watts/S.F.	170,100	
R260110-14	Office Receptacles 2 Watts/S.F.	113,400	
R260110-14, R262213-27	Low Rise Office A.C. 4.3 Watts/S.F.	243,810	
R260110-15, 16	Elevators - 2 @ 20 HP = 2 @ 17,404 Watts/Ea.	34,808	
R260110-14	Misc. Motors + Power 1.2 Watts/S.F.	68,040	
	Total	639,608	Watts

Voltage Available

277/480V, 3 Phase, 4 Wire

Formula

R260110-17

$$\text{Amperes} = \frac{\text{Watts}}{\text{Volts x Power Factor x 1.73}} = \frac{639{,}608}{480\text{V x .8 x 1.73}} = 963 \text{ Amps}$$

Use 1200 Amp Service

System	Description	Unit Cost	Unit	Total
D5020 210 0200	Garage Lighting (Interpolated)	$1.30	S.F.	$ 24,570
D5020 210 0540	Office Lighting (Using 32 W lamps)	7.86	S.F.	445,662
D5020 115 0880	Receptacle-Undercarpet	3.78	S.F.	214,326
D5020 140 0280	Air Conditioning	0.61	S.F.	34,587
D5020 135 0320	Misc. Pwr.	0.32	S.F.	18,144
D5020 145 2120	Elevators - 2 @ 20HP	2,825.00	Ea.	5,650
D5010 120 0480	Service-1200 Amp (add 25% for 277/480V)	62,900.00	Ea.	62,900
D5010 230 0480	Feeder - Assume 200 Ft.	332.00	L.F.	66,400
D5010 240 0320	Panels - 1200 Amp (add 20% for 277/480V)	32,350.00	Ea.	32,350
D5030 910 0400	Fire Detection	33,700.00	Ea.	33,700
		Total		$938,289
		or		$938,300

Table R260110-14 Nominal Watts Per S.F. for Electric Systems for Various Building Types

Type Construction	1. Lighting	2. Devices	3. HVAC	4. Misc.	5. Elevator	Total Watts
Apartment, luxury high rise	2	2.2	3	1		
Apartment, low rise	2	2	3	1		
Auditorium	2.5	1	3.3	.8		
Bank, branch office	3	2.1	5.7	1.4		
Bank, main office	2.5	1.5	5.7	1.4		
Church	1.8	.8	3.3	.8		
College, science building	3	3	5.3	1.3		
College, library	2.5	.8	5.7	1.4		
College, physical education center	2	1	4.5	1.1		
Department store	2.5	.9	4	1		
Dormitory, college	1.5	1.2	4	1		
Drive-in donut shop	3	4	6.8	1.7		
Garage, commercial	.5	.5	0	.5		
Hospital, general	2	4.5	5	1.3		
Hospital, pediatric	3	3.8	5	1.3		
Hotel, airport	2	1	5	1.3		
Housing for the elderly	2	1.2	4	1		
Manufacturing, food processing	3	1	4.5	1.1		
Manufacturing, apparel	2	1	4.5	1.1		
Manufacturing, tools	4	1	4.5	1.1		
Medical clinic	2.5	1.5	3.2	1		
Nursing home	2	1.6	4	1		
Office building, hi rise	3	2	4.7	1.2		
Office building, low rise	3	2	4.3	1.2		
Radio-TV studio	3.8	2.2	7.6	1.9		
Restaurant	2.5	2	6.8	1.7		
Retail store	2.5	.9	5.5	1.4		
School, elementary	3	1.9	5.3	1.3		
School, junior high	3	1.5	5.3	1.3		
School, senior high	2.3	1.7	5.3	1.3		
Supermarket	3	1	4	1		
Telephone exchange	1	.6	4.5	1.1		
Theater	2.5	1	3.3	.8		
Town Hall	2	1.9	5.3	1.3		
U.S. Post Office	3	2	5	1.3		
Warehouse, grocery	1	.6	0	.5		

Rule of Thumb: 1 KVA = 1 HP (Single Phase)

Three Phase:

Watts = 1.73 x Volts x Current x Power Factor x Efficiency

$$\text{Horsepower} = \frac{\text{Volts x Current x 1.73 x Power Factor}}{\text{746 Watts}}$$

See R260120 for an additional reference

Table R260110-15 Horsepower Requirements for Elevators with 3 Phase Motors

Type	Maximum Travel Height in Ft.	Travel Speeds in FPM	Capacity of Cars in Lbs.		
			1200	1500	1800
Hydraulic	70	70	10	15	15
		85	15	15	15
		100	15	15	20
		110	20	20	20
		125	20	20	20
		150	25	25	25
		175	25	30	30
		200	30	30	40
Geared Traction	300	200			
		350			
			2000	2500	3000
Hydraulic	70	70	15	20	20
		85	20	20	25
		100	20	25	30
		110	20	25	30
		125	25	30	40
		150	30	40	50
		175	40	50	50
		200	40	50	60
Geared Traction	300	200	10	10	15
		350	15	15	23
			3500	4000	4500
Hydraulic	70	70	20	25	30
		85	25	30	30
		100	30	40	40
		110	40	40	50
		125	40	50	50
		150	50	50	60
		175	60		
		200	60		
Geared Traction	300	200	15		23
		350	23		35

The power factor of electric motors varies from 80% to 90% in larger size motors. The efficiency likewise varies from 80% on a small motor to 90% on a large motor.

Table R260110-16 Watts per Motor

90% Power Factor & Efficiency @ 200 or 460V			
HP	Watts	HP	Watts
10	9024	30	25784
15	13537	40	33519
20	17404	50	41899
25	21916	60	49634

Table R260110-17 Electrical Formulas

Ohm's Law

Ohm's Law is a method of explaining the relation existing between voltage, current, and resistance in an electrical circuit. It is practically the basis of all electrical calculations. The term "electromotive force" is often used to designate pressure in volts. This formula can be expressed in various forms.

To find the current in amperes:

$$\text{Current} = \frac{\text{Voltage}}{\text{Resistance}} \quad \text{or} \quad \text{Amperes} = \frac{\text{Volts}}{\text{Ohms}} \quad \text{or} \quad I = \frac{E}{R}$$

The flow of current in amperes through any circuit is equal to the voltage or electromotive force divided by the resistance of that circuit.

To find the pressure or voltage:

Voltage = Current x Resistance or Volts = Amperes x Ohms or E = I x R

The voltage required to force a current through a circuit is equal to the resistance of the circuit multiplied by the current.

To find the resistance:

$$\text{Resistance} = \frac{\text{Voltage}}{\text{Current}} \quad \text{or} \quad \text{Ohms} = \frac{\text{Volts}}{\text{Amperes}} \quad \text{or} \quad R = \frac{E}{I}$$

The resistance of a circuit is equal to the voltage divided by the current flowing through that circuit.

Power Formulas

One horsepower = 746 watts One kilowatt = 1000 watts

The power factor of electric motors varies from 80% to 90% in the larger size motors.

Single-Phase Alternating Current Circuits

Power in Watts = Volts x Amperes x Power Factor

To find current in amperes:

$$\text{Current} = \frac{\text{Watts}}{\text{Volts x Power Factor}} \quad \text{or}$$

$$\text{Amperes} = \frac{\text{Watts}}{\text{Volts x Power Factor}} \quad \text{or} \quad I = \frac{W}{E \text{ x } PF}$$

To find current of a motor, single phase:

$$\text{Current} = \frac{\text{Horsepower x 746}}{\text{Volts x Power Factor x Efficiency}} \quad \text{or}$$

$$I = \frac{HP \text{ x } 746}{E \text{ x } PF \text{ x Eff.}}$$

To find horsepower of a motor, single phase:

$$\text{Horsepower} = \frac{\text{Volts x Current x Power Factor x Efficiency}}{\text{746 Watts}}$$

$$HP = \frac{E \text{ x } I \text{ x } PF \text{ x Eff.}}{746}$$

To find power in watts of a motor, single phase:

Watts = Volts x Current x Power Factor x Efficiency or

Watts = E x I x PF x Eff.

To find single phase kVA:

$$\text{1 Phase kVA} = \frac{\text{Volts x Amps}}{1000}$$

Three-Phase Alternating Current Circuits

Power in Watts = Volts x Amperes x Power Factor x 1.73

To find current in amperes in each wire:

$$\text{Current} = \frac{\text{Watts}}{\text{Voltage x Power Factor x 1.73}} \quad \text{or}$$

$$\text{Amperes} = \frac{\text{Watts}}{\text{Volts x Power Factor x 1.73}} \quad \text{or} \quad I = \frac{W}{E \text{ x } PF \text{ x } 1.73}$$

To find current of a motor, 3 phase:

$$\text{Current} = \frac{\text{Horsepower x 746}}{\text{Volts x Power Factor x Efficiency x 1.73}} \quad \text{or}$$

$$I = \frac{HP \text{ x } 746}{E \text{ x } PF \text{ x Eff. x } 1.73}$$

To find horsepower of a motor, 3 phase:

$$\text{Horsepower} = \frac{\text{Volts x Current x 1.73 x Power Factor}}{\text{746 Watts}}$$

$$HP = \frac{E \text{ x } I \text{ x } 1.73 \text{ x } PF}{746}$$

To find power in watts of a motor, 3 phase:

Watts = Volts x Current x 1.73 x Power Factor x Efficiency or

Watts = E x I x 1.73 x PF x Eff.

To find 3 phase kVA:

$$\text{3 phase kVA} = \frac{\text{Volts x Amps x 1.73}}{1000} \quad \text{or}$$

$$\text{kVA} = \frac{V \text{ x } A \text{ x } 1.73}{1000}$$

Power Factor (PF) is the percentage ratio of the measured watts (effective power) to the volt-amperes (apparent watts)

$$\text{Power Factor} = \frac{\text{Watts}}{\text{Volts x Amperes}} \text{ x } 100\%$$

A Conceptual Estimate of the costs for a building when final drawings are not available can be quickly figured by using **Table R260110-18 Cost Per S.F. for Electrical Systems for Various Building Types.** The following definitions apply to this table.

1. **Service and Distribution:** This system includes the incoming primary feeder from the power company, main building transformer, metering arrangement, switchboards, distribution panel boards, stepdown transformers, and power and lighting panels. Items marked (*) include the cost of the primary feeder and transformer. In all other projects the cost of the primary feeder and transformer is paid for by the local power company.
2. **Lighting:** Includes all interior fixtures for decor, illumination, exit and emergency lighting. Fixtures for exterior building lighting are included, but parking area lighting is not included unless mentioned. See also Section D5020 for detailed analysis of lighting requirements and costs.
3. **Devices:** Includes all outlet boxes, receptacles, switches for lighting control, dimmers and cover plates.
4. **Equipment Connections:** Includes all materials and equipment for making connections for heating, ventilating and air conditioning, food service and other motorized items requiring connections.
5. **Basic Materials:** This category includes all disconnect power switches not part of service equipment, raceways for wires, pull boxes, junction boxes, supports, fittings, grounding materials, wireways, busways, and wire and cable systems.
6. **Special Systems:** Includes installed equipment only for the particular system such as fire detection and alarm, sound, emergency generator and others as listed in the table.

Table R260110-18 Cost per S.F. for Electric Systems for Various Building Types

	1.	2.	3.	4.	5.	6. Special Systems		
Type Construction	Service & Distrib.	Lighting	Devices	Equipment Connections	Basic Materials	Fire Alarm & Detection	Telephone	Television Distribution
Apartment, Low Rise	$0.81	$0.88	$0.22	$0.29	$ 2.21	$ 0.52	0.81	$0.22
Apartment, High Rise	2.44	1.35	0.41	0.54	4.20	0.81	1.35	0.27
Assisted Living	4.18	1.30	0.29	0.58	2.74	1.15	0.72	0.72
Auditorium	2.45	4.23	0.45	0.67	7.13	1.34	1.11	
Bank	4.17	2.98	0.30	0.89	6.56	3.58	2.39	
Church	1.08	3.78	0.36	0.90	6.48	1.80	0.72	
College, Classroom	3.74	3.30	0.44	0.66	5.94	0.88	0.88	0.88
College, Laboratory	6.69	7.14	0.89	0.89	13.38	2.68	2.23	
Community Center	1.77	2.76	0.39	0.79	5.91	2.56	0.79	0.39
Department Store	1.09	3.03	0.12	0.48	4.12	0.73	0.48	
Dormitory, Low Rise	0.93	1.87	0.62	0.62	5.13	1.24	1.4	0.31
Factory	1.35	5.57	0.17	0.51	5.74	1.35	0.51	0.17
Funeral Home	1.58	1.79	0.21	0.53	3.38	1.27	1.37	
Garage, Commercial	3.41	0.93	0.16	0.31	2.79	5.89	0.78	
Garage, Parking	0.21	0.58	0.12	0.04	1.78	0.08	0.04	
Gymnasium	1.04	2.81	0.30	0.59	5.92	1.63	0.44	
Hospital, Low Rise	2.64	3.95	1.76	1.76	14.94	3.08	2.2	0.88
Hospital, High Rise	4.66	3.81	1.69	1.27	16.95	2.12	1.69	0.42
Housing, Elderly	4.18	1.30	0.29	0.58	2.74	1.15	0.72	0.72
Library	1.69	5.30	0.48	1.45	9.88	2.17	0.72	
Medical Office	2.41	1.41	0.60	0.40	4.62	1.61	$1.61	0.4
Motel	1.48	1.03	0.57	0.11	3.77	1.26	1.1	0.23
Nursing Home	2.34	2.56	0.43	0.64	5.54	1.28	1.49	1.49
Office, Low Rise	3.10	2.24	0.34	0.69	5.33	1.89	1.20	0.17
Office, High Rise	0.73	2.32	0.29	0.44	5.37	1.45	1.16	0.15
Post Office	2.98	1.58	0.18	0.53	5.95	3.68	1.58	
Restaurant	8.70	3.17	0.26	1.06	6.59	4.22	0.53	
Restaurant, Fast Food	6.22	3.67	0.28	1.13	8.76	5.65	0.57	
Retail Store	1.83	2.32	0.12	0.49	4.16	1.10	0.37	
School, Elementary	1.65	2.93	0.37	0.55	6.04	2.19	0.91	0.37
School, Junior High	1.47	2.95	0.55	0.74	5.89	0.92	1.1	0.37
School, High	1.38	3.46	0.46	0.69	7.14	1.15	1.38	1.38
Supermarket	3.13	2.71	0.14	0.43	4.56	1.42	0.28	
Telephone Exchange	6.79	5.34	0.97	2.43	14.55	11.16	5.82	
Theater (Movie)	2.79	5.07	0.25	0.76	5.83	3.55	1.01	
Town Hall	1.63	3.45	0.61	1.22	7.72	1.63	1.63	
Warehouse	0.32	1.21	0.24	0.16	2.89	1.93	0.4	

*Includes cost of primary feeder and transformer. Cont'd on next page.

COST ASSUMPTIONS:

Each of the projects analyzed in Table R260110-18 was bid within the last 10 years in the northeastern part of the United States. Bid prices have been adjusted to Jan. 1 levels. The list of projects is by no means all-inclusive, yet by carefully examining the various systems for a particular building type, certain cost relationships will emerge. The use of Square Foot/Cubic Costs, in the Reference Section, along with the Project Size Modifier, should produce a budget S.F. cost for the electrical portion of a job that is consistent with the amount of design information normally available at the conceptual estimate stage.

Table R260110-18 Cost per S.F. for Electric Systems for Various Building Types (cont.)

	6. Special Systems, (cont.)						
Type Construction	Intercom Systems	Sound Systems	Video Monitoring	Internet	Emergency Generator	Access Control	Master Clock Sys.
Apartment, Low Rise	$0.66			0.29	0.15	0.29	
Apartment, High Rise	0.95			0.54	0.14	0.54	
Assisted Living	1.73	$0.14	$0.43		$0.14	0.29	
Auditorium		2.45	0.89		1.34	$0.22	
Bank			2.98	2.68	0.60	2.68	
Church		2.34			0.18	0.36	
College, Classroom	0.88	1.10		1.76	0.44	0.44	0.66
College, Laboratory	2.23			4.46	0.45	1.34	2.23
Community Center		1.97		0.99	0.39	0.99	
Department Store			0.85	0.36		0.85	
Dormitory, Low Rise	1.40			1.55		0.47	
Factory			0.51		0.17	0.51	0.34
Funeral Home				0.11	0.11	0.21	
Garage, Commercial					0.47	0.78	
Garage, Parking			0.41		0.04	0.83	
Gymnasium		1.33			0.15	0.59	
Hospital, Low Rise	2.64	0.44	0.88	0.44	6.59	1.32	0.44
Hospital, High Rise	1.27	0.42	0.42	0.42	5.51	1.27	0.42
Housing, Elderly	1.73	0.14	0.43		0.14	0.29	
Library		0.48		0.96	0.48	0.48	
Medical Office	2.41		1	1.81	0.80	1	
Motel				1.26	0.11	0.46	
Nursing Home	3.62	0.21	0.85		0.43	0.43	
Office, Low Rise			0.34	1.38	0.17	0.34	
Office, High Rise			0.44	1.31	0.44	0.44	
Post Office				0.53	0.18	0.35	
Restaurant				0.53	0.53	0.79	
Restaurant, Fast Food				0.57	0.57	0.85	
Retail Store		0.12	0.73	0.12	0.12	0.73	
School, Elementary	1.1	0.91		0.37		0.55	0.37
School, Junior High	1.29	0.92		0.55	0.37	0.55	0.74
School, High	1.38	1.15	0.92	$0.46	0.69	0.69	0.69
Supermarket		0.14	0.57	0.14		0.71	
Telephone Exchange					0.49	0.97	
Theater (Movie)		4.56		0.25	0.25	1.01	
Town Hall				1.83	0.20	0.41	
Warehouse			0.4	0.08		0.4	

*Includes cost of primary feeder and transformer. Cont'd on next page.

General: Variations in the following square foot costs are due to the type of structural systems of the buildings, geographical location, local electrical codes, designer's preference for specific materials and equipment, and the owner's particular requirements.

Table R260110-18 Cost per S.F. for Total Electric Systems for Various Building Types (cont.)

Type Construction	Additional Information	Total Floor Area in Square Feet	Total Cost per Square Foot for Total Electric Systems
Apartment, Low Rise	Stories = 3, Height = 10′, Perimeter = 715′	40,200	7.37
Apartment, High Rise	Stories = 15, Height = 10′, Perimeter = 350′	115,000	13.54
Assisted Living	Stories = 1, Height = 10′, Perimeter = 400′	10,000	14.41
Auditorium	Stories = 1, Height = 24′, Perimeter = 747′	28,000	22.28
Bank	Stories = 1, Height = 14′, Perimeter = 168′	2,700	29.82
Church	Stories = 1, Height = 24′, Perimeter = 562′	17,700	18.00
College, Classroom	Stories = 2, Height = 12′, Perimeter = 630′	50,000	22.00
College, Laboratory	Stories = 1, Height = 12′, Perimeter = 550′	27,500	44.61
Community Center	Stories = 1, Height = 12′, Perimeter = 453′	10,000	19.70
Department Store	Stories = 1, Height = 14′, Perimeter = 1039′	85,800	12.12
Dormitory, Low Rise	Stories = 3, Height = 12′, Perimeter = 1008′	63,000	15.55
Factory	Stories = 1, Height = 20′, Perimeter = 730′	30,000	16.88
Funeral Home	Stories = 1, Height = 12′, Perimeter = 424′	10,000	10.55
Garage, Commercial	Stories = 1, Height = 12′, Perimeter = 153′	1,400	15.51
Garage, Parking	Stories = 5, Height = 10′, Perimeter = 723′	145,000	4.14
Gymnasium	Stories = 1, Height = 25′, Perimeter = 660′	22,000	14.80
Hospital, Low Rise	Stories = 3, Height = 12′, Perimeter = 566′	55,000	43.93
Hospital, High Rise	Stories = 6, Height = 12′, Perimeter = 866′	200,000	42.36
Housing, Elderly	Stories = 1, Height = 10′, Perimeter = 400′	10,000	14.41
Library	Stories = 2, Height = 14′, Perimeter = 435′	22,000	24.10
Medical Office	Stories = 2, Height = 10′, Perimeter = 240′	7,000	20.09
Motel	Stories = 3, Height = 9′, Perimeter = 606′	49,000	11.42
Nursing Home	Stories = 2, Height = 10′, Perimeter = 384′	21,000	21.32
Office, Low Rise	Stories = 3, Height = 12′, Perimeter = 360′	20,000	17.20
Office, High Rise	Stories = 15, Height = 10′, Perimeter = 634′	311,200	14.50
Post Office	Stories = 1, Height = 14′, Perimeter = 486′	13,000	17.50
Restaurant	Stories = 1, Height = 12′, Perimeter = 174′	2,900	26.38
Restaurant, Fast Food	Stories = 1, Height = 10′, Perimeter = 260′	4,000	28.26
Retail Store	Stories = 1, Height = 14′, Perimeter = 360′	8,000	12.23
School, Elementary	Stories = 1, Height = 15′, Perimeter = 1325′	39,500	18.29
School, Junior High	Stories = 2, Height = 15′, Perimeter = 1700′	110,000	18.41
School, High	Stories = 2, Height = 15′, Perimeter = 2411′	158,300	23.04
Supermarket	Stories = 1, Height = 18′, Perimeter = 584′	30,600	14.24
Telephone Exchange	Stories = 1, Height = 12′, Perimeter = 286′	5,000	48.51
Theater (Movie)	Stories = 1, Height = 20′, Perimeter = 513′	14,000	25.35
Town Hall	Stories = 3, Height = 12′, Perimeter = 356′	20,000	20.32
Warehouse	Stories = 1, Height = 24′, Perimeter = 700′	30,000	8.04

*Includes cost of primary feeder and transformer.

Table R260120-10 ASHRAE Climate Zone Map

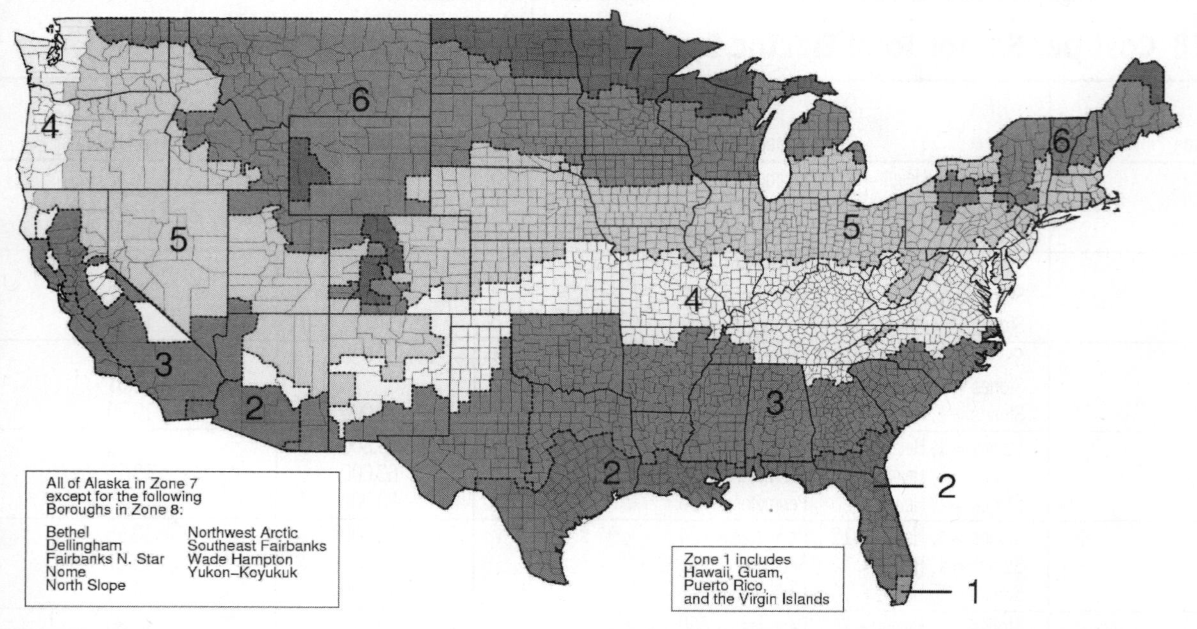

Table R260120-20 Watts per Square Foot for Lighting and Misc. Power by Building Type & Space

Building / Space	Percent	Lighting	Misc
Apartment, Low: Area = 20,000 SF; Floors = 2; Heating = DX Coils; Cooling = DX Coils			
Room	85%	0.50	0.10
Storage	8%	1.19	0.00
Laundry	7%	1.28	0.15
Weighted Average		**0.61**	**0.10**
Apartment, Mid: Area = 40,000 SF; Floors = 5; Heating = DX Coils; Cooling = DX Coils			
Room	71%	0.50	0.30
Corridor	16%	0.57	0.00
Storage	7%	1.19	0.00
Laundry	6%	1.28	0.15
Average		**0.61**	**0.22**
Apartment, High: Area = 72,000 SF; Floors = 8; Heating = DX Coils; Cooling = DX Coils			
Room	71%	0.50	0.30
Corridor	16%	0.57	0.00
Storage	7%	1.19	0.00
Laundry	6%	1.28	0.15
Weighted Average		**0.61**	**0.22**
Assisted Living: Area = 10,000 SF; Floors = 1; Heating = Furnace; Cooling = DX Coils			
Room	70%	1.46	0.25
Dining	15%	1.85	0.10
Food Prep	10%	1.19	1.00
Care	5%	1.34	1.00
Weighted Average		**1.49**	**0.34**

Building / Space	Percent	Lighting	Misc
Auditorum: Area = 70,000 SF; Floors = 1; Heating = Hot Water Coils; Cooling = Chilled Water Coils			
Theater	70%	0.90	
Lobby	15%	1.50	
Office	5%	1.30	
Corridor	5%	0.60	
MEP	5%	0.70	
Weighted Average		**0.99**	
Bank: Area = 25,000 SF; Floors = 2; Heating = Furnace; Cooling = DX Coils			
Bank	60%	2.33	0.45
Corridor	10%	0.57	0.00
Office	10%	1.49	0.75
Lobby	5%	1.52	0.25
Restrooms	5%	0.77	0.10
Conference	4%	0.92	0.10
MEP	4%	0.81	0.10
Production	2%	1.64	0.70
Weighted Average		**1.82**	**0.38**
Community Center: Area = 35,000 SF; Floors = 1; Heating = Furnace; Cooling = DX Coils			
Convention	50%	1.27	
Dining	15%	1.85	
Food Prep	15%	1.19	
Storage	10%	1.19	
Corridor	5%	0.57	
Restrooms	5%	0.77	
Weighted Average		**1.28**	
Convention Center: Area = 150,000 SF; Floors = 2; Heating = Hot Water Coils; Cooling = Chilled Water Coils			
Convention	65%	1.27	0.25
Conference	20%	0.92	0.10
Dining	5%	1.85	0.10
Corridor	4%	0.57	0.00
Food Prep	3%	1.19	1.00
Restrooms	3%	0.77	0.10
Weighted Average		**1.18**	**0.22**

Building / Space	Percent	Lighting	Misc
Day Care: Area = 2,500 SF; Floors = 1; Heating = Furnace; Cooling = DX Coils			
Classroom	45%	1.45	
Corridor	15%	0.57	
Library	15%	1.67	
Food Prep	10%	1.19	
Restrooms	7%	0.77	
Office	5%	1.49	
Storage	3%	1.19	
Weighted Average		**1.27**	
Fitness Center: Area = 20,000 SF; Floors = 1; Heating = Furnace; Cooling = DX Coils			
Exercise	80%	1.13	0.10
Corridor	5%	0.57	0.00
MEP	5%	0.81	0.10
Office	5%	1.24	0.75
Restrooms	7%	0.77	0.10
Weighted Average		**1.07**	**0.13**
Hospital: Area = 250,000 SF; Floors = 4; Heating = Hot Water Coils; Cooling = Chilled Water Coils			
Medical	60%	1.34	1.00
Laboratory	15%	3.19	1.00
Corridor	10%	0.57	0.00
Laundry	5%	1.28	0.15
MEP	5%	0.81	0.10
Restrooms	5%	0.77	0.10
Weighted Average		**1.48**	**0.77**
Hotel: Area = 180,000 SF; Floors = 10; Heating = Furnace; Cooling = DX Coils			
Room	75%	1.46	
Meeting Center	15%	1.27	
Corridor	8%	0.57	
MEP	2%	0.81	
Weighted Average		**1.35**	

Steps:

1. Determine climate zone using R260120-10
2. Use R260120-20 to determine lighting and miscellaneous power for appropriate building type
3. Use corresponding building type and climate zone in R260120-30 to determine other systems

*Values for R260120-20, & -30 were derived by performing an energy analysis using the eQUEST® software from James J. Hirsch & Associates. eQUEST is a trademark of James J. Hirsch

**Additional systems may be necessary

Table R260120-20 Watts per Square Foot for Lighting and Misc. Power by Building Type & Space (cont)

Building / Space	Percent	Lighting	Misc
Manufacturing, General: Area = 100,000 SF; Floors = 10; Heating = Furnace; Cooling = DX Coils			
Work	60%	2.66	1.00
Storage	25%	1.19	0.00
Office	10%	1.24	0.75
Restrooms	3%	0.77	0.10
MEP	2%	0.81	0.10
Weighted Average		**2.06**	**0.68**
Manufacturing, High Tech: Area = 200,000 SF; Floors = 1; Heating = Furnace; Cooling = DX Coils			
Work	45%	3.19	1.00
Office	27%	1.24	0.75
Corridor	18%	0.57	0.00
Dining	3%	1.85	0.10
Conference	2%	0.92	0.10
Server	2%	1.24	5.00
MEP	2%	0.81	0.10
Food Prep	1%	1.19	1.00
Weighted Average		**2.00**	**0.77**
Medical Clinc: Area = 20,000 SF; Floors = 1; Heating = Furnace; Cooling = DX Coils			
Clinic	50%	1.40	
General Office	15%	1.30	
Private Office	10%	1.30	
Laboratory	10%	1.30	
Storage	5%	0.60	
Restrooms	5%	0.60	
MEP	5%	0.70	
Weighted Average		**1.25**	
Motel: Area = 10,000 SF; Floors = 1; Heating = DX Coils; Cooling = DX Coils			
Room	95%	1.46	0.25
Office	3%	1.24	0.75
MEP	2%	0.81	0.10
Weighted Average		**1.44**	**0.26**
Museum: Area = 20,000 SF; Floors = 1; Heating = Hot Water Coils; Cooling = Chilled Water Coils			
Exhibit Area	65%	1.03	0.25
Theater	15%	1.02	0.25
Lobby	5%	1.77	0.25
MEP	5%	0.81	0.10
Office	5%	1.49	0.75
Restrooms	5%	0.77	0.10
Weighted Average		**1.06**	**0.26**
Office, Low: Area = 25,000 SF; Floors = 2; Heating = Furnace; Cooling = DX Coils			
Office	70%	1.30	0.75
Corridor	10%	0.60	0.00
Lobby	5%	1.10	0.25
Restrooms	5%	0.60	0.10
Conference	4%	1.60	0.10
MEP	4%	0.70	0.10
Production	2%	1.50	0.70
Weighted Average		**1.18**	**0.56**
Office, Mid: Area = 125,000 SF; Floors = 4; Heating = Hot Water Coils; Cooling = Chilled Water Coils			
Office Open	40%	1.24	0.75
Office Private	30%	1.49	0.75
Corridor	10%	0.57	0.00
Lobby	5%	1.52	0.25
Restrooms	5%	0.77	0.10
Conference	4%	0.92	0.10
MEP	4%	0.81	0.10
Production	2%	1.64	0.70
Weighted Average		**1.22**	**0.56**
Office, High: Area = 250,000 SF; Floors = 8; Heating = Hot Water Coils; Cooling = Chilled Water Coils			
Office Open	45%	1.24	0.75
Office Private	25%	1.49	0.75
Corridor	10%	0.57	0.00
Lobby	5%	1.52	0.25
Conference	4%	0.92	0.10
Production	2%	1.64	0.70
Restrooms	5%	0.77	0.10
MEP	4%	0.81	0.10
Weighted Average		**1.20**	**0.56**

Building / Space	Percent	Lighting	Misc
Restaurant, Bar/Lounge: Area = 4,000 SF; Floors = 1; Heating = Furnace; Cooling = DX Coils			
Lounge	35%	1.85	0.10
Dining	30%	1.85	0.10
Food Prep	25%	1.19	1.00
MEP	5%	0.81	0.10
Restrooms	5%	0.77	0.10
Weighted Average		**1.58**	**0.33**
Restaurant, Full: Area = 5,000 SF; Floors = 1; Heating = Furnace; Cooling = DX Coils			
Dining	60%	1.85	0.10
Food Prep	30%	1.19	1.00
Restrooms	5%	0.77	0.10
MEP	5%	0.81	0.10
Weighted Average		**1.55**	**0.37**
Restaurant, Quick: Area = 2,500 SF; Floors = 1; Heating = Furnace; Cooling = DX Coils			
Dining	50%	1.85	0.10
Food Prep	35%	1.19	1.00
Restrooms	10%	0.77	0.10
MEP	5%	0.81	0.10
Weighted Average		**1.46**	**0.42**
Store, Conveniece: Area = 3,000 SF; Floors = 1; Heating = Furnace; Cooling = DX Coils			
Retail	80%	2.38	0.25
Office	10%	1.49	0.75
Restrooms	10%	0.77	0.10
Weighted Average		**2.13**	**0.29**
Store, Department: Area = 120,000 SF; Floors = 2; Heating = Hot Water Coils; Cooling = Chilled Water Coils			
Retail	55%	2.38	0.25
Exhibit Area	15%	1.03	0.25
Storage	16%	1.19	0.00
Office	5%	1.24	0.75
Smoking Lounge	2%	1.85	0.10
Restrooms	2%	0.77	0.10
MEP	5%	0.81	0.10
Weighted Average		**1.81**	**0.22**
Store, Large: Area = 70,000 SF; Floors = 1; Heating = Furnace; Cooling = DX Coils			
Retail	65%	2.38	0.25
Storage	20%	1.19	0.00
Office	5%	1.24	0.75
Dining	3%	1.85	0.10
Food Prep	1%	1.19	1.00
Lounge	2%	1.85	0.10
Restrooms	2%	0.77	0.10
MEP	2%	0.81	0.10
Weighted Average		**1.98**	**0.22**
Store, Small: Area = 5,000 SF; Floors = 1; Heating = Furnace; Cooling = DX Coils			
Retail	58%	2.38	0.25
Exhibit Area	15%	1.03	0.25
Storage	15%	1.19	0.00
Office	5%	1.24	0.75
Restrooms	5%	0.77	0.10
MEP	2%	0.81	0.10
Weighted Average		**1.83**	**0.23**
Store, Strip Mall: Area = 15,000 SF; Floors = 1; Heating = Furnace; Cooling = DX Coils			
Retail	38%	2.38	0.25
Atrium	20%	1.55	0.13
Exhibit Area	15%	1.03	0.25
Storage	15%	1.19	0.00
Office	5%	1.24	0.75
Restrooms	5%	0.77	0.10
MEP	2%	0.81	0.10
Weighted Average		**1.66**	**0.20**

Building / Space	Percent	Lighting	Misc
Store, Warehouse: Area = 150,000 SF; Floors = 1; Heating = Furnace; Cooling = DX Coils			
Retail	75%	2.38	0.25
Storage	20%	1.19	0.00
Office	4%	1.24	0.75
Restrooms	1%	0.77	0.10
Weighted Average		**2.08**	**0.22**
School, Elementary: Area = 35,000 SF; Floors = 1; Heating = Furnace; Cooling = DX Coils			
Classroom	50%	1.45	0.50
Corridor	15%	0.57	0.00
Gym	10%	1.13	0.10
Dining	5%	1.85	0.10
Food Prep	5%	1.19	1.00
Library	5%	1.78	0.25
Auditorum	5%	1.62	0.25
Restrooms	5%	0.77	0.10
Weighted Average		**1.28**	**0.35**
School, Middle: Area = 75,000 SF; Floors = 1; Heating = Furnace; Cooling = DX Coils			
Classroom	50%	1.45	0.50
Corridor	14%	0.57	0.00
Gym	9%	1.13	0.10
Auditorum	7%	1.62	0.25
Dining	6%	1.85	0.10
Food Prep	5%	1.19	1.00
Library	5%	1.78	0.25
Restrooms	4%	0.77	0.10
Weighted Average		**1.31**	**0.35**
School, High: Area = 150,000 SF; Floors = 2; Heating = Furnace; Cooling = DX Coils			
Classroom	45%	1.45	0.50
Corridor	10%	0.57	0.00
Gym	10%	1.13	0.10
Dining	10%	1.85	0.10
Food Prep	5%	1.19	1.00
Library	5%	1.78	0.25
Auditorum	10%	1.62	0.25
Restrooms	5%	0.77	0.10
Weighted Average		**1.36**	**0.34**
School, College: Area = 250,000 SF; Floors = 4; Heating = Furnace; Cooling = DX Coils			
Classroom	50%	1.45	0.50
Office	15%	1.49	0.75
Corridor	12%	0.57	0.00
Gym	10%	1.13	0.10
Library	5%	1.78	0.25
Dining	3%	1.85	0.10
Restrooms	3%	0.77	0.10
Food Prep	2%	1.19	1.00
Weighted Average		**1.32**	**0.41**
Service Station: Area = 2,000 SF; Floors = 1; Heating = Furnace; Cooling = DX Coils			
Workshop	70%	1.31	0.50
Office	10%	1.49	0.75
Restrooms	10%	0.77	0.10
Retail	10%	2.38	0.25
Weighted Average		**1.38**	**0.46**
Storage, High Bay: Area = 200,000 SF; Floors = 1; Heating = Furnace; Cooling = DX Coils			
Storage	85%	1.19	0.00
Office	10%	1.24	0.75
Restrooms	3%	0.77	0.10
MEP	2%	0.81	0.10
Weighted Average		**1.17**	**0.08**
Storage, Low Bay: Area = 70,000 SF; Floors = 1; Heating = Furnace; Cooling = DX Coils			
Storage	85%	1.19	0.00
Office	10%	1.24	0.75
Restrooms	3%	0.77	0.10
MEP	2%	0.81	0.10
Weighted Average		**1.17**	**0.08**

Table R260120-30 Watts per S.F. for Various Systems by Building Type & Climate Zone

Building / Zone	Heating	Cooling	Heat rejection	Auxiliary (pumps)	Vent fan	Supp heat pump
Apartment, Low						
Zone 1	0.42	3.65	0.00	0.00	0.12	1.47
Zone 2	0.63	3.53	0.00	0.00	0.12	4.01
Zone 3	0.64	3.45	0.00	0.00	0.13	4.63
Zone 4	0.65	3.09	0.00	0.00	0.13	7.41
Zone 5	0.67	2.88	0.00	0.00	0.12	10.30
Zone 6	0.69	2.66	0.00	0.00	0.13	10.68
Zone 7	0.85	2.11	0.00	0.00	0.12	11.76
Average	**0.65**	**3.05**	**0.00**	**0.00**	**0.13**	**7.18**
Apartment, Mid						
Zone 1	0.00	1.90	0.00	0.00	0.11	0.00
Zone 2	0.43	1.82	0.00	0.00	0.11	2.12
Zone 3	0.48	1.69	0.00	0.00	0.11	2.32
Zone 4	0.67	1.67	0.00	0.00	0.12	4.25
Zone 5	0.70	1.54	0.00	0.00	0.12	4.79
Zone 6	0.72	1.45	0.00	0.00	0.12	5.07
Zone 7	0.93	1.39	0.00	0.00	0.12	5.11
Average	**0.56**	**1.63**	**0.00**	**0.00**	**0.12**	**3.38**
Apartment, High						
Zone 1	0.00	2.11	0.00	0.00	0.14	0.00
Zone 2	0.00	2.03	0.00	0.00	0.13	0.00
Zone 3	0.48	1.88	0.00	0.00	0.14	1.75
Zone 4	0.57	1.77	0.00	0.00	0.14	4.22
Zone 5	0.66	1.64	0.00	0.00	0.14	5.40
Zone 6	0.71	1.63	0.00	0.00	0.15	5.64
Zone 7	0.91	1.18	0.00	0.00	0.14	5.75
Average	**0.48**	**1.75**	**0.00**	**0.00**	**0.14**	**3.25**
Assisted Living						
Zone 1	0.00	2.65	0.00	0.04	0.31	0.00
Zone 2	0.00	2.63	0.00	0.04	0.31	0.00
Zone 3	0.00	2.57	0.00	0.04	0.31	0.00
Zone 4	0.00	2.50	0.00	0.04	0.31	0.00
Zone 5	0.00	2.42	0.00	0.04	0.30	0.00
Zone 6	0.00	1.99	0.00	0.04	0.32	0.00
Zone 7	0.00	1.90	0.00	0.04	0.30	0.00
Average	**0.00**	**2.38**	**0.00**	**0.04**	**0.31**	**0.00**
Auditorium						
Zone 1	0.00	2.32	0.34	0.37	1.03	0.00
Zone 2	0.00	2.32	0.34	0.38	1.03	0.00
Zone 3	0.00	2.29	0.33	0.38	1.04	0.00
Zone 4	0.00	2.26	0.29	0.35	1.05	0.00
Zone 5	0.00	2.20	0.34	0.39	1.04	0.00
Zone 6	0.00	1.71	0.14	0.34	1.05	0.00
Zone 7	0.00	1.38	0.23	0.36	1.04	0.00
Average	**0.00**	**2.07**	**0.29**	**0.37**	**1.04**	**0.00**
Bank						
Zone 1	0.00	3.32	0.00	0.00	0.60	0.00
Zone 2	0.00	3.28	0.00	0.00	0.61	0.00
Zone 3	0.00	3.28	0.00	0.00	0.61	0.00
Zone 4	0.00	3.21	0.00	0.00	0.60	0.00
Zone 5	0.00	3.06	0.00	0.00	0.59	0.00
Zone 6	0.00	2.82	0.00	0.00	0.61	0.00
Zone 7	0.00	2.66	0.00	0.00	0.60	0.00
Average	**0.00**	**3.09**	**0.00**	**0.00**	**0.60**	**0.00**

Building / Zone	Heating	Cooling	Heat rejection	Auxiliary (pumps)	Vent fan	Supp heat pump
Community Center						
Zone 1	0.00	2.70	0.00	0.00	0.44	0.00
Zone 2	0.00	2.66	0.00	0.00	0.44	0.00
Zone 3	0.00	2.66	0.00	0.00	0.45	0.00
Zone 4	0.00	2.44	0.00	0.00	0.45	0.00
Zone 5	0.00	2.42	0.00	0.00	0.44	0.00
Zone 6	0.00	2.16	0.00	0.00	0.44	0.00
Zone 7	0.00	1.91	0.00	0.00	0.44	0.00
Average	**0.00**	**2.42**	**0.00**	**0.00**	**0.45**	**0.00**
Convention Center						
Zone 1	0.00	1.46	0.25	0.27	0.31	0.00
Zone 2	0.00	1.44	0.25	0.27	0.29	0.00
Zone 3	0.00	1.43	0.24	0.26	0.30	0.00
Zone 4	0.00	1.42	0.23	0.26	0.28	0.00
Zone 5	0.00	1.30	0.29	0.31	0.22	0.00
Zone 6	0.00	1.17	0.12	0.29	0.16	0.00
Zone 7	0.00	0.90	0.20	0.31	0.16	0.00
Average	**0.00**	**1.30**	**0.22**	**0.28**	**0.25**	**0.00**
Day Care						
Zone 1	0.00	3.47	0.00	0.02	0.29	0.00
Zone 2	0.00	3.42	0.00	0.02	0.28	0.00
Zone 3	0.00	3.40	0.00	0.02	0.29	0.00
Zone 4	0.00	3.29	0.00	0.02	0.29	0.00
Zone 5	0.00	3.08	0.00	0.02	0.29	0.00
Zone 6	0.00	2.29	0.00	0.02	0.29	0.00
Zone 7	0.00	2.12	0.00	0.02	0.24	0.00
Average	**0.05**	**3.01**	**0.05**	**0.07**	**0.29**	**0.05**
Fitness Center						
Zone 1	0.00	4.97	0.00	0.00	0.59	0.00
Zone 2	0.00	4.94	0.00	0.00	0.59	0.00
Zone 3	0.00	4.80	0.00	0.00	0.60	0.00
Zone 4	0.00	4.76	0.00	0.00	0.60	0.00
Zone 5	0.00	4.75	0.00	0.00	0.59	0.00
Zone 6	0.00	4.13	0.00	0.00	0.60	0.00
Zone 7	0.00	3.27	0.00	0.00	0.60	0.00
Average	**0.00**	**4.52**	**0.00**	**0.00**	**0.59**	**0.00**
Hospital						
Zone 1	0.00	1.54	0.23	0.26	0.80	0.00
Zone 2	0.00	1.47	0.22	0.25	0.80	0.00
Zone 3	0.00	1.44	0.22	0.25	0.81	0.00
Zone 4	0.00	1.42	0.22	0.25	0.81	0.00
Zone 5	0.00	1.37	0.23	0.27	0.81	0.00
Zone 6	0.00	1.26	0.15	0.25	0.82	0.00
Zone 7	0.00	1.07	0.19	0.26	0.82	0.00
Average	**0.00**	**1.36**	**0.21**	**0.26**	**0.81**	**0.00**
Hotel						
Zone 1	0.00	1.77	0.00	0.01	0.26	0.00
Zone 2	0.00	1.64	0.00	0.01	0.26	0.00
Zone 3	0.00	1.62	0.00	0.01	0.27	0.00
Zone 4	0.00	1.62	0.00	0.01	0.27	0.00
Zone 5	0.00	1.51	0.00	0.01	0.27	0.00
Zone 6	0.00	1.25	0.00	0.01	0.28	0.00
Zone 7	0.00	1.23	0.00	0.01	0.28	0.00
Average	**0.00**	**1.52**	**0.00**	**0.01**	**0.27**	**0.00**

Table R260120-30 Watts per S.F. for Various Systems by Building Type & Climate Zone (cont.)

Building / Zone	Heating	Cooling	Heat rejection	Auxiliary (pumps)	Vent fan	Supp heat pump
Manufacturing, General						
Zone 1	0.00	1.46	0.00	0.00	0.13	0.00
Zone 2	0.00	1.45	0.00	0.00	0.13	0.00
Zone 3	0.00	1.43	0.00	0.00	0.13	0.00
Zone 4	0.00	1.39	0.00	0.00	0.13	0.00
Zone 5	0.00	1.39	0.00	0.00	0.13	0.00
Zone 6	0.00	1.21	0.00	0.00	0.13	0.00
Zone 7	0.00	1.00	0.00	0.00	0.13	0.00
Average	**0.00**	**1.33**	**0.00**	**0.00**	**0.13**	**0.00**
Manufacturing, High Tech						
Zone 1	0.00	2.15	0.00	0.00	0.17	0.00
Zone 2	0.00	2.12	0.00	0.00	0.17	0.00
Zone 3	0.00	2.04	0.00	0.00	0.17	0.00
Zone 4	0.00	2.01	0.00	0.00	0.17	0.00
Zone 5	0.00	1.88	0.00	0.00	0.17	0.00
Zone 6	0.00	1.63	0.00	0.00	0.17	0.00
Zone 7	0.00	1.50	0.00	0.00	0.17	0.00
Average	**0.00**	**1.90**	**0.00**	**0.00**	**0.17**	**0.00**
Medial Clinic						
Zone 1	0.00	3.15	0.00	0.00	0.54	0.00
Zone 2	0.00	3.09	0.00	0.00	0.53	0.00
Zone 3	0.00	3.07	0.00	0.00	0.54	0.00
Zone 4	0.00	3.02	0.00	0.00	0.54	0.00
Zone 5	0.00	2.99	0.00	0.00	0.54	0.00
Zone 6	0.00	2.58	0.00	0.00	0.55	0.00
Zone 7	0.00	2.45	0.00	0.00	0.52	0.00
Average	**0.00**	**2.91**	**0.00**	**0.00**	**0.54**	**0.00**
Motel						
Zone 1	0.97	2.68	0.00	0.00	0.19	2.15
Zone 2	1.19	2.52	0.00	0.00	0.18	4.43
Zone 3	1.19	2.52	0.00	0.00	0.19	4.81
Zone 4	1.19	2.42	0.00	0.00	0.21	7.39
Zone 5	1.28	2.26	0.00	0.00	0.21	7.82
Zone 6	1.29	2.07	0.00	0.00	0.23	8.89
Zone 7	1.39	1.97	0.00	0.00	0.23	9.51
Average	**1.21**	**2.35**	**0.00**	**0.00**	**0.21**	**6.43**
Museum						
Zone 1	0.00	3.77	0.00	0.11	0.67	0.00
Zone 2	0.00	3.71	0.00	0.12	0.77	0.00
Zone 3	0.00	3.59	0.00	0.12	0.72	0.00
Zone 4	0.00	3.47	0.00	0.14	0.74	0.00
Zone 5	0.00	3.42	0.00	0.16	0.51	0.00
Zone 6	0.00	2.39	0.00	0.15	0.23	0.00
Zone 7	0.00	1.69	0.00	0.17	0.24	0.00
Average	**0.00**	**3.15**	**0.00**	**0.14**	**0.55**	**0.00**
Office, Low						
Zone 1	0.00	2.90	0.00	0.02	0.29	0.00
Zone 2	0.00	2.90	0.00	0.02	0.30	0.00
Zone 3	0.00	2.81	0.00	0.02	0.30	0.00
Zone 4	0.00	2.70	0.00	0.02	0.29	0.00
Zone 5	0.00	2.66	0.00	0.02	0.29	0.00
Zone 6	0.00	2.04	0.00	0.02	0.30	0.00
Zone 7	0.00	1.82	0.00	0.02	0.29	0.00
Average	**0.00**	**2.55**	**0.00**	**0.02**	**0.29**	**0.00**

Building / Zone	Heating	Cooling	Heat rejection	Auxiliary (pumps)	Vent fan	Supp heat pump
Office, Mid						
Zone 1	0.00	1.61	0.31	0.33	0.45	0.00
Zone 2	0.00	1.60	0.29	0.32	0.46	0.00
Zone 3	0.00	1.59	0.28	0.34	0.42	0.00
Zone 4	0.00	1.56	0.27	0.33	0.43	0.00
Zone 5	0.00	1.55	0.34	0.36	0.37	0.00
Zone 6	0.00	1.29	0.16	0.34	0.31	0.00
Zone 7	0.00	1.10	0.22	0.35	0.32	0.00
Average	**0.00**	**1.47**	**0.27**	**0.34**	**0.40**	**0.00**
Office, High						
Zone 1	0.00	1.70	0.30	0.33	0.44	0.00
Zone 2	0.00	1.59	0.29	0.32	0.41	0.00
Zone 3	0.00	1.59	0.32	0.34	0.42	0.00
Zone 4	0.00	1.58	0.28	0.35	0.42	0.00
Zone 5	0.00	1.58	0.32	0.36	0.41	0.00
Zone 6	0.00	1.11	0.16	0.29	0.33	0.00
Zone 7	0.00	0.86	0.23	0.31	0.37	0.00
Average	**0.00**	**1.43**	**0.27**	**0.33**	**0.40**	**0.00**
Restaurant, Bar/Lounge						
Zone 1	0.00	3.07	0.00	0.06	0.24	0.00
Zone 2	0.00	2.87	0.00	0.06	0.25	0.00
Zone 3	0.00	2.85	0.00	0.06	0.24	0.00
Zone 4	0.00	2.83	0.00	0.06	0.24	0.00
Zone 5	0.00	2.80	0.00	0.06	0.24	0.00
Zone 6	0.00	1.88	0.00	0.06	0.24	0.00
Zone 7	0.00	1.87	0.00	0.06	0.23	0.00
Average	**0.00**	**2.59**	**0.00**	**0.06**	**0.24**	**0.00**
Restaurant, Full						
Zone 1	0.00	3.37	0.00	0.05	0.38	0.00
Zone 2	0.00	3.30	0.00	0.05	0.39	0.00
Zone 3	0.00	3.25	0.00	0.05	0.39	0.00
Zone 4	0.00	3.20	0.00	0.05	0.39	0.00
Zone 5	0.00	3.03	0.00	0.05	0.38	0.00
Zone 6	0.00	2.35	0.00	0.05	0.39	0.00
Zone 7	0.00	2.33	0.00	0.05	0.38	0.00
Average	**0.00**	**2.98**	**0.00**	**0.05**	**0.38**	**0.00**
Restaurant, Quick						
Zone 1	0.00	3.78	0.00	0.08	0.48	0.00
Zone 2	0.00	3.72	0.00	0.08	0.50	0.00
Zone 3	0.00	3.64	0.00	0.08	0.50	0.00
Zone 4	0.00	3.60	0.00	0.08	0.50	0.00
Zone 5	0.00	3.42	0.00	0.08	0.49	0.00
Zone 6	0.00	2.74	0.00	0.08	0.51	0.00
Zone 7	0.00	2.69	0.00	0.08	0.49	0.00
Average	**0.00**	**3.37**	**0.00**	**0.08**	**0.50**	**0.00**
Store, Convenience						
Zone 1	0.00	4.08	0.00	0.02	0.29	0.00
Zone 2	0.00	3.66	0.00	0.02	0.30	0.00
Zone 3	0.00	3.60	0.00	0.02	0.30	0.00
Zone 4	0.00	3.46	0.00	0.02	0.29	0.00
Zone 5	0.00	3.28	0.00	0.02	0.29	0.00
Zone 6	0.00	2.66	0.00	0.02	0.27	0.00
Zone 7	0.00	2.50	0.00	0.02	0.24	0.00
Average	**0.00**	**3.32**	**0.00**	**0.02**	**0.28**	**0.00**

Table R260120-30 Watts per S.F. for Various Systems by Building Type & Climate Zone (cont.)

Building / Zone	Heating	Cooling	Heat rejection	Auxiliary (pumps)	Vent fan	Supp heat pump
Store, Department						
Zone 1	0.00	1.65	0.20	0.23	0.63	0.00
Zone 2	0.00	1.61	0.20	0.23	0.63	0.00
Zone 3	0.00	1.59	0.19	0.22	0.64	0.00
Zone 4	0.00	1.52	0.19	0.23	0.64	0.00
Zone 5	0.00	1.51	0.21	0.24	0.63	0.00
Zone 6	0.00	1.32	0.13	0.21	0.63	0.00
Zone 7	0.00	0.98	0.17	0.23	0.63	0.00
Average	**0.00**	**1.45**	**0.18**	**0.23**	**0.63**	**0.00**
Store, Large						
Zone 1	0.00	2.68	0.00	0.00	0.16	0.00
Zone 2	0.00	2.61	0.00	0.00	0.17	0.00
Zone 3	0.00	2.61	0.00	0.00	0.17	0.00
Zone 4	0.00	2.56	0.00	0.00	0.16	0.00
Zone 5	0.00	2.53	0.00	0.00	0.16	0.00
Zone 6	0.00	1.84	0.00	0.00	0.16	0.00
Zone 7	0.00	1.50	0.00	0.00	0.16	0.00
Average	**0.00**	**2.33**	**0.00**	**0.00**	**0.16**	**0.00**
Store, Small						
Zone 1	0.00	3.60	0.00	0.01	0.25	0.00
Zone 2	0.00	3.51	0.00	0.01	0.26	0.00
Zone 3	0.00	3.38	0.00	0.01	0.25	0.00
Zone 4	0.00	3.32	0.00	0.01	0.25	0.00
Zone 5	0.00	3.23	0.00	0.01	0.24	0.00
Zone 6	0.00	2.49	0.00	0.01	0.23	0.00
Zone 7	0.00	2.17	0.00	0.01	0.21	0.00
Average	**0.00**	**3.10**	**0.00**	**0.01**	**0.24**	**0.00**
Store, Strip Mall						
Zone 1	0.00	3.36	0.00	0.02	0.25	0.00
Zone 2	0.00	3.30	0.00	0.02	0.25	0.00
Zone 3	0.00	3.14	0.00	0.02	0.25	0.00
Zone 4	0.00	3.14	0.00	0.02	0.25	0.00
Zone 5	0.00	2.24	0.00	0.02	0.24	0.00
Zone 6	0.00	1.96	0.00	0.02	0.25	0.00
Zone 7	0.00	1.64	0.00	0.02	0.24	0.00
Average	**0.00**	**2.52**	**0.00**	**0.02**	**0.25**	**0.00**
Store, Warehouse						
Zone 1	0.00	2.99	0.00	0.00	0.21	0.00
Zone 2	0.00	2.97	0.00	0.00	0.21	0.00
Zone 3	0.00	2.94	0.00	0.00	0.21	0.00
Zone 4	0.00	2.83	0.00	0.00	0.21	0.00
Zone 5	0.00	2.70	0.00	0.00	0.21	0.00
Zone 6	0.00	2.10	0.00	0.00	0.21	0.00
Zone 7	0.00	1.78	0.00	0.00	0.21	0.00
Average	**0.00**	**2.61**	**0.00**	**0.00**	**0.21**	**0.00**
School, Elementary						
Zone 1	0.00	2.73	0.00	0.01	0.21	0.00
Zone 2	0.00	2.69	0.00	0.01	0.21	0.00
Zone 3	0.00	2.63	0.00	0.01	0.21	0.00
Zone 4	0.00	2.61	0.00	0.01	0.20	0.00
Zone 5	0.00	2.51	0.00	0.01	0.20	0.00
Zone 6	0.00	1.71	0.00	0.01	0.20	0.00
Zone 7	0.00	1.64	0.00	0.01	0.20	0.00
Average	**0.00**	**2.36**	**0.00**	**0.01**	**0.20**	**0.00**

Building / Zone	Heating	Cooling	Heat rejection	Auxiliary (pumps)	Vent fan	Supp heat pump
School, Middle						
Zone 1	0.00	2.48	0.00	0.00	0.17	0.00
Zone 2	0.00	2.45	0.00	0.00	0.17	0.00
Zone 3	0.00	2.41	0.00	0.00	0.18	0.00
Zone 4	0.00	2.35	0.00	0.00	0.17	0.00
Zone 5	0.00	2.26	0.00	0.00	0.17	0.00
Zone 6	0.00	1.61	0.00	0.00	0.17	0.00
Zone 7	0.00	1.49	0.00	0.00	0.17	0.00
Average	**0.00**	**2.15**	**0.00**	**0.00**	**0.17**	**0.00**
School, High						
Zone 1	0.00	3.24	0.00	0.01	0.22	0.00
Zone 2	0.00	3.08	0.00	0.01	0.22	0.00
Zone 3	0.00	3.05	0.00	0.01	0.23	0.00
Zone 4	0.00	3.00	0.00	0.01	0.23	0.00
Zone 5	0.00	2.94	0.00	0.01	0.23	0.00
Zone 6	0.00	1.98	0.00	0.01	0.23	0.00
Zone 7	0.00	1.81	0.00	0.01	0.23	0.00
Average	**0.00**	**2.73**	**0.00**	**0.01**	**0.23**	**0.00**
School, College						
Zone 1	0.00	2.92	0.00	0.00	0.20	0.00
Zone 2	0.00	2.91	0.00	0.00	0.20	0.00
Zone 3	0.00	2.78	0.00	0.00	0.20	0.00
Zone 4	0.00	2.72	0.00	0.00	0.20	0.00
Zone 5	0.00	2.43	0.00	0.00	0.20	0.00
Zone 6	0.00	1.77	0.00	0.00	0.20	0.00
Zone 7	0.00	1.51	0.00	0.00	0.19	0.00
Average	**0.00**	**2.43**	**0.00**	**0.00**	**0.20**	**0.00**
Service Station						
Zone 1	0.00	10.42	0.00	0.03	0.47	0.00
Zone 2	0.00	9.01	0.00	0.03	0.46	0.00
Zone 3	0.00	7.82	0.00	0.03	0.49	0.00
Zone 4	0.00	7.02	0.00	0.03	0.47	0.00
Zone 5	0.00	6.13	0.00	0.03	0.46	0.00
Zone 6	0.00	4.32	0.00	0.03	0.45	0.00
Zone 7	0.00	4.18	0.00	0.03	0.40	0.00
Average	**0.00**	**6.99**	**0.00**	**0.03**	**0.46**	**0.00**
Storage, High Bay						
Zone 1	0.00	1.98	0.00	0.00	0.17	0.00
Zone 2	0.00	1.97	0.00	0.00	0.18	0.00
Zone 3	0.00	1.90	0.00	0.00	0.18	0.00
Zone 4	0.00	1.89	0.00	0.00	0.18	0.00
Zone 5	0.00	1.85	0.00	0.00	0.17	0.00
Zone 6	0.00	1.29	0.00	0.00	0.17	0.00
Zone 7	0.00	1.10	0.00	0.00	0.17	0.00
Average	**0.00**	**1.71**	**0.00**	**0.00**	**0.17**	**0.00**
Storage, Low Bay						
Zone 1	0.00	1.87	0.00	0.00	0.16	0.00
Zone 2	0.00	1.85	0.00	0.00	0.16	0.00
Zone 3	0.00	1.82	0.00	0.00	0.16	0.00
Zone 4	0.00	1.77	0.00	0.00	0.16	0.00
Zone 5	0.00	1.76	0.00	0.00	0.16	0.00
Zone 6	0.00	1.20	0.00	0.00	0.16	0.00
Zone 7	0.00	1.00	0.00	0.00	0.16	0.00
Average	**0.00**	**1.61**	**0.00**	**0.00**	**0.16**	**0.00**

R260519-20 Armored Cable

Armored Cable – Quantities are taken off in the same manner as wire.

Bx Type Cable – Productivities are based on an average run of 50′ before terminating at a box fixture, etc. Each 50′ section includes field preparation of (2) ends with hacksaw, identification and tagging of wire. Set up is open coil type without reels attaching cable to snake and pulling across a suspended ceiling or open face wood or steel studding, price does not include drilling of studs.

Cable in Tray – Productivities are based on an average run of 100 L.F. with set up of pulling equipment for (2) 90° bends, attaching cable to pull-in means, identification and tagging of wires, set up of reels. Wire termination to breakers equipment, etc., are not included.

Job Conditions – Productivities are based on new construction to a height of 15′ using rolling staging in an unobstructed area. Material staging is assumed to be within 100′ of work being performed.

R260519-80 Undercarpet Systems

Takeoff Procedure for Power Systems: List components for each fitting type, tap, splice, and bend on your quantity takeoff sheet. Each component must be priced separately. Start at the power supply transition fittings and survey each circuit for the components needed. List the quantities of each component under a specific circuit number. Use the floor plan layout scale to get cable footage.

Reading across the list, combine the totals of each component in each circuit and list the total quantity in the last column. Calculate approximately 5% for scrap for items such as cable, top shield, tape, and spray adhesive. Also provide for final variations that may occur on-site.

Suggested guidelines are:

1. Equal amounts of cable and top shield should be priced.
2. For each roll of cable, price a set of cable splices.
3. For every 1 ft. of cable, price 2-1/2 ft. of hold-down tape.
4. For every 3 rolls of hold-down tape, price 1 can of spray adhesive.

Adjust final figures wherever possible to accommodate standard packaging of the product. This information is available from the distributor.

Each transition fitting requires:

1. 1 base
2. 1 cover
3. 1 transition block

Each floor fitting requires:

1. 1 frame/base kit
2. 1 transition block
3. 2 covers (duplex/blank)

Each tap requires:

1. 1 tap connector for each conductor
2. 1 pair insulating patches
3. 2 top shield connectors

Each splice requires:

1. 1 splice connector for each conductor
2. 1 pair insulating patches
3. 3 top shield connectors

Each cable bend requires:

1. 2 top shield connectors

Each cable dead end (outside of transition block) requires:

1. 1 pair insulating patches

Labor does not include:

1. Patching or leveling uneven floors.
2. Filling in holes or removing projections from concrete slabs.
3. Sealing porous floors.
4. Sweeping and vacuuming floors.
5. Removal of existing carpeting.
6. Carpet square cut-outs.
7. Installation of carpet squares.

Takeoff Procedures for Telephone Systems: After reviewing floor plans identify each transition. Number or letter each cable run from that fitting.

Start at the transition fitting and survey each circuit for the components needed. List the cable type, terminations, cable length, and floor fitting type under the specific circuit number. Use the floor plan layout scale to get the cable footage. Add some extra length (next higher increment of 5 feet) to preconnectorized cable.

Transition fittings require:

1. 1 base plate
2. 1 cover
3. 1 transition block

Floor fittings require:

1. 1 frame/base kit
2. 2 covers
3. Modular jacks

Reading across the list, combine the list of components in each circuit and list the total quantity in the last column. Calculate the necessary scrap factors for such items as tape, bottom shield and spray adhesive. Also provide for final variations that may occur on-site.

Adjust final figures whenever possible to accommodate standard packaging. Check that items such as transition fittings, floor boxes, and floor fittings that are to utilize both power and telephone have been priced as combination fittings, so as to avoid duplication.

Make sure to include marking of floors and drilling of fasteners if fittings specified are not the adhesive type.

Labor does not include:

1. Conduit or raceways before transition of floor boxes
2. Telephone cable before transition boxes
3. Terminations before transition boxes
4. Floor preparation as described in power section

Be sure to include all cable folds when pricing labor.

Takeoff Procedure for Data Systems: Start at the transition fittings and take off quantities in the same manner as the telephone system, keeping in mind that data cable does not require top or bottom shields.

The data cable is simply cross-taped on the cable run to the floor fitting.

Data cable can be purchased in either bulk form in which case coaxial connector material and labor must be priced, or in preconnectorized cut lengths.

Data cable cannot be folded and must be notched at 1 inch intervals. A count of all turns must be added to the labor portion of the estimate. (Note: Some manufacturers have prenotched cable.)

Notching required:

1. 90 degree turn requires 8 notches per side.
2. 180 degree turn requires 16 notches per side.

Floor boxes, transition boxes, and fittings are the same as described in the power and telephone procedures.

Since undercarpet systems require special hand tools, be sure to include this cost in proportion to number of crews involved in the installation.

Job Conditions: Productivity is based on new construction in an unobstructed area. Staging area is assumed to be within 200′ of work being performed.

R260519-90 Wire

Wire quantities are taken off by either measuring each cable run or by extending the conduit and raceway quantities times the number of conductors in the raceway. Ten percent should be added for waste and tie-ins. Keep in mind that the unit of measure of wire is C.L.F. not L.F. as in raceways so the formula would read:

$$\frac{\text{(L.F. Raceway x No. of Conductors) x 1.10}}{100} = \text{C.L.F.}$$

Price per C.L.F. of wire includes:
1. Set up wire coils or spools on racks
2. Attaching wire to pull in means
3. Measuring and cutting wire
4. Pulling wire into a raceway
5. Identifying and tagging

Price does not include:
1. Connections to breakers, panelboards or equipment
2. Splices

Job Conditions: Productivity is based on new construction to a height of 15′ using rolling staging in an unobstructed area. Material staging is assumed to be within 100′ of work being performed.

Economy of Scale: If more than three wires at a time are being pulled, deduct the following percentages from the labor of that grouping:

4-5 wires	25%
6-10 wires	30%
11-15 wires	35%
over 15	40%

If a wire pull is less than 100′ in length and is interrupted several times by boxes, lighting outlets, etc., it may be necessary to add the following lengths to each wire being pulled:

Junction box to junction box	2 L.F.
Lighting panel to junction box	6 L.F.
Distribution panel to sub panel	8 L.F.
Switchboard to distribution panel	12 L.F.
Switchboard to motor control center	20 L.F.
Switchboard to cable tray	40 L.F.

Measure of Drops and Riser: It is important when taking off wire quantities to include the wire for drops to electrical equipment. If heights of electrical equipment are not clearly stated, use the following guide:

	Bottom A.F.F.	Top A.F.F.	Inside Cabinet
Safety switch to 100A	5′	6′	2′
Safety switch 400 to 600A	4′	6′	3′
100A panel 12 to 30 circuit	4′	6′	3′
42 circuit panel	3′	6′	4′
Switch box	3′	3′6″	1′
Switchgear	0′	8′	8′
Motor control centers	0′	8′	8′
Transformers - wall mount	4′	8′	2′
Transformers - floor mount	0′	12′	4′

R260519-92 Minimum Copper and Aluminum Wire Size Allowed for Various Types of Insulation

Minimum Wire Sizes									
	Copper		Aluminum			Copper		Aluminum	
Amperes	THW THWN or XHHW	THHN XHHW *	THW XHHW	THHN XHHW *	Amperes	THW THWN or XHHW	THHN XHHW *	THW XHHW	THHN XHHW *
15A	#14	#14	#12	#12	195	3/0	2/0	250kcmil	4/0
20	#12	#12	#10	#10	200	3/0	3/0	250kcmil	4/0
25	#10	#10	#10	#10	205	4/0	3/0	250kcmil	4/0
30	#10	#10	# 8	# 8	225	4/0	3/0	300kcmil	250kcmil
40	# 8	# 8	# 8	# 8	230	4/0	4/0	300kcmil	250kcmil
45	# 8	# 8	# 6	# 8	250	250kcmil	4/0	350kcmil	300kcmil
50	# 8	# 8	# 6	# 6	255	250kcmil	4/0	400kcmil	300kcmil
55	# 6	# 8	# 4	# 6	260	300kcmil	4/0	400kcmil	350kcmil
60	# 6	# 6	# 4	# 6	270	300kcmil	250kcmil	400kcmil	350kcmil
65	# 6	# 6	# 4	# 4	280	300kcmil	250kcmil	500kcmil	350kcmil
75	# 4	# 6	# 3	# 4	285	300kcmil	250kcmil	500kcmil	400kcmil
85	# 4	# 4	# 2	# 3	290	350kcmil	250kcmil	500kcmil	400kcmil
90	# 3	# 4	# 2	# 2	305	350kcmil	300kcmil	500kcmil	400kcmil
95	# 3	# 4	# 1	# 2	310	350kcmil	300kcmil	500kcmil	500kcmil
100	# 3	# 3	# 1	# 2	320	400kcmil	300kcmil	600kcmil	500kcmil
110	# 2	# 3	1/0	# 1	335	400kcmil	350kcmil	600kcmil	500kcmil
115	# 2	# 2	1/0	# 1	340	500kcmil	350kcmil	600kcmil	500kcmil
120	# 1	# 2	1/0	1/0	350	500kcmil	350kcmil	700kcmil	500kcmil
130	# 1	# 2	2/0	1/0	375	500kcmil	400kcmil	700kcmil	600kcmil
135	1/0	# 1	2/0	1/0	380	500kcmil	400kcmil	750kcmil	600kcmil
150	1/0	# 1	3/0	2/0	385	600kcmil	500kcmil	750kcmil	600kcmil
155	2/0	1/0	3/0	3/0	420	600kcmil	500kcmil		700kcmil
170	2/0	1/0	4/0	3/0	430		500kcmil		750kcmil
175	2/0	2/0	4/0	3/0	435		600kcmil		750kcmil
180	3/0	2/0	4/0	4/0	475		600kcmil		

*Dry Locations Only

Notes:
1. Size #14 to 4/0 is in AWG units (American Wire Gauge).
2. Size 250 to 750 is in kcmil units (Thousand Circular Mils).
3. Use next higher ampere value if exact value is not listed in table.
4. For loads that operate continuously increase ampere value by 25% to obtain proper wire size.
5. Table R260519-92 has been written for estimating purpose only, based on ambient temperature of 30°C (86° F); for ambient temperature other than 30°C (86° F), ampacity correction factors will be applied.

R260519-93 Metric Equivalent, Wire

U.S. vs. European Wire – Approximate Equivalents			
United States		European	
Size AWG or kcmil	Area Cir. Mils.(cmil) mm^2	Size mm^2	Area Cir. Mils.
18	1620/.82	.75	1480
16	2580/1.30	1.0	1974
14	4110/2.08	1.5	2961
12	6530/3.30	2.5	4935
10	10,380/5.25	4	7896
8	16,510/8.36	6	11,844
6	26,240/13.29	10	19,740
4	41,740/21.14	16	31,584
3	52,620/26.65	25	49,350
2	66,360/33.61	–	–
1	83,690/42.39	35	69,090
1/0	105,600/53.49	50	98,700
2/0	133,100/67.42	–	–
3/0	167,800/85.00	70	138,180
4/0	211,600/107.19	95	187,530
250	250,000/126.64	120	236,880
300	300,000/151.97	150	296,100
350	350,000/177.30	–	–
400	400,000/202.63	185	365,190
500	500,000/253.29	240	473,760
600	600,000/303.95	300	592,200
700	700,000/354.60	–	–
750	750,000/379.93	–	–

R260519-94 Size Required and Weight (Lbs./1000 L.F.) of Aluminum and Copper THW Wire by Ampere Load

Amperes	Copper Size	Aluminum Size	Copper Weight	Aluminum Weight
15	14	12	24	11
20	12	10	33	17
30	10	8	48	39
45	8	6	77	52
65	6	4	112	72
85	4	2	167	101
100	3	1	205	136
115	2	1/0	252	162
130	1	2/0	324	194
150	1/0	3/0	397	233
175	2/0	4/0	491	282
200	3/0	250	608	347
230	4/0	300	753	403
255	250	400	899	512
285	300	500	1068	620
310	350	500	1233	620
335	400	600	1396	772
380	500	750	1732	951

R260526-80 Grounding

Grounding When taking off grounding systems, identify separately the type and size of wire.

Example:

Bare copper & size
Bare aluminum & size
Insulated copper & size
Insulated aluminum & size

Count the number of ground rods and their size.

Example:

1. 8′ grounding rod – 5/8″ dia. 20 Ea.
2. 10′ grounding rod – 5/8″ dia. 12 Ea.
3. 15′ grounding rod – 3/4″ dia. 4 Ea.

Count the number of connections; the size of the largest wire will determine the productivity.

Example:

Braze a #2 wire to a #4/0 cable
The 4/0 cable will determine the L.H. and cost to be used.

Include individual connections to:

1. Ground rods
2. Building steel
3. Equipment
4. Raceways

Price does not include:

1. Excavation
2. Backfill
3. Sleeves or raceways used to protect grounding wires
4. Wall penetrations
5. Floor cutting
6. Core drilling

Job Conditions: Productivity is based on a ground floor area, using cable reels in an unobstructed area. Material staging area assumed to be within 100′ of work being performed.

R260533-20 Conduit To 15′ High

List conduit by quantity, size, and type. Do not deduct for lengths occupied by fittings, since this will be allowance for scrap. Example:

A. Aluminum — size
B. Rigid galvanized — size
C. Steel intermediate (IMC) — size
D. Rigid steel, plastic-coated 20 Mil. — size
E. Rigid steel, plastic-coated 40 Mil. — size
F. Electric metallic tubing (EMT) — size
G. PVC Schedule 40 — size

Types (A) thru (E) listed above contain the following per 100 L.F.:

1. (11) Threaded couplings
2. (11) Beam-type hangers
3. (2) Factory sweeps
4. (2) Fiber bushings
5. (4) Locknuts
6. (2) Field threaded pipe terminations
7. (2) Removal of concentric knockouts

Type (F) contains per 100 L.F.:

1. (11) Set screw couplings
2. (11) Beam clamps
3. (2) Field bends on 1/2″ and 3/4″ diameter
4. (2) Factory sweeps for 1″ and above
5. (2) Set screw steel connectors
6. (2) Removal of concentric knockouts

Type (G) contains per 100 L.F.:

1. (11) Field cemented couplings
2. (34) Beam clamps
3. (2) Factory sweeps
4. (2) Adapters
5. (2) Locknuts
6. (2) Removal of concentric knockouts

Labor-hours for all conduit to 15′ include:

1. Unloading by hand
2. Hauling by hand to an area up to 200′ from loading dock
3. Setup of rolling staging
4. Installation of conduit and fittings as described in Conduit models (A) thru (G)

Not included in the material and labor are:

1. Staging rental or purchase
2. Structural modifications
3. Wire
4. Junction boxes
5. Fittings in excess of those described in conduit models (A) thru (G)
6. Painting of conduit

Fittings

Only those fittings listed above are included in the linear foot totals, although they should be listed separately from conduit lengths, without prices, to ensure proper quantities for material procurement.

If the fittings required exceed the quantities included in the model conduit runs, then material and labor costs must be added to the difference. If actual needs per 100 L.F. of conduit are: (2) sweeps, (4) LBs and (1) field bend, then, (4) LBs and (1) field bend must be priced additionally.

R260533-21 Hangers

It is sometimes desirable to substitute an alternate style of hanger if the support being used is not the type described in the conduit models.

One approach is the substitution method:

1. Find the cost of the type hanger described in the conduit model.
2. Calculate the cost of the desired type hanger (it may be necessary to calculate individual components such as drilling, expansion shields, etc.).
3. Calculate the cost difference (delta) between the two types of hangers.
4. Multiply the cost delta by the number of hangers in the model.
5. Divide the total delta cost for hangers in the model by the length of the model to find the delta cost per L.F. for that model.
6. Modify the given unit costs per L.F. for the model by the delta cost per L.F.

Another approach to hanger configurations would be to start with the conduit only and add all the supports and any other items as separate lines. This procedure is most useful if the project involves racking many runs of conduit on a single hanger, for instance, a trapeze type hanger.

Example: Five (5) 2″ RGS conduits, 50 L.F. each, are to be run on trapeze hangers from one pull box to another. The run includes one 90° bend.

1. List the hangers' components to create an assembly cost for each 2′-wide trapeze.
2. List the components for the 50′ conduit run, noting that 6 trapeze supports will be required.

Job Conditions: Productivities are based on new construction to 15′ high, using scaffolding in an unobstructed area. Material storage is assumed to be within 100′ of work being performed.

Add to labor for elevated installations:

15′ to 20′ High–10%	30′ to 35 High–30%
20′ to 25′ High–20%	35′ to 40′ High–35%
25′ to 30′ High–25%	Over 40′ High–40%

Add these percentages to the L.F. labor cost, but not to fittings. Add these percentages only to quantities exceeding the different height levels, rather than the total conduit quantities.

Linear foot price for labor does not include penetrations in walls or floors and must be added to the estimate

R260533-22 Conductors in Conduit

Table below lists maximum number of conductors for various sized conduit using THW, TW or THWN insulations.

Copper Wire Size	Conduit Size																										
	½″			¾″			1″			1-¼″			1-½″			2″			2-½″			3″		3-½″		4″	
	TW	THW	THWN	TW	THW	THWN	TW	THW	THWN	TW	THW	THWN	TW	THW	THWN	TW	THW	THWN	TW	THW	THWN	THW	THWN	THW	THWN	THW	THWN
#14	9	6	13	15	10	24	25	16	39	44	29	69	60	40	94	99	65	154	142	93		143		192			
#12	7	4	10	12	8	18	19	13	29	35	24	51	47	32	70	78	53	114	111	76	164	117		157			
#10	5	4	6	9	6	11	15	11	18	26	19	32	36	26	44	60	43	73	85	61	104	95	160	127		163	
#8	2	1	3	4	3	5	7	5	9	12	10	16	17	13	22	28	22	36	40	32	51	49	79	66	106	85	136
#6		1	1		2	4		4	6		7	11		10	15		16	26		23	37	36	57	48	76	62	98
#4		1	1		1	2		3	4		5	7		7	9		12	16		17	22	27	35	36	47	47	60
#3		1	1		1	1		2	3		4	6		6	8		10	13		15	19	23	29	31	39	40	51
#2		1	1		1	1		2	3		4	5		5	7		9	11		13	16	20	25	27	33	34	43
#1					1	1		1	1		3	3		4	5		6	8		9	12	14	18	19	25	25	32
1/0					1	1		1	1		2	3		3	4		5	7		8	10	12	15	16	21	21	27
2/0					1	1		1	1		1	2		3	3		5	6		7	8	10	13	14	17	18	22
3/0					1	1		1	1		1	1		2	3		4	5		6	7	9	11	12	14	15	18
4/0						1		1	1		1	1		1	2		3	4		5	6	7	9	10	12	13	15
250 kcmil								1	1		1	1		1	1		2	3		4	4	6	7	8	10	10	12
300								1	1		1	1		1	1		2	3		3	4	5	6	7	8	9	11
350									1		1	1		1	1		1	2		3	3	4	5	6	7	8	9
400											1	1		1	1		1	1		2	3	4	5	5	6	7	8
500											1	1		1	1		1	1		1	2	3	4	4	5	6	7
600												1		1	1		1	1		1	1	3	3	4	4	5	5
700														1	1		1	1		1	1	2	3	3	4	4	5
750 ▼														1	1		1	1		1	1	2	2	3	3	4	4

R260533-25 Conduit Weight Comparisons (Lbs. per 100 ft.) with Maximum Cable Fill*

Type	1/2″	3/4″	1″	1-1/4″	1-1/2″	2″	2-1/2″	3″	3-1/2″	4″	5″	6″
Rigid Galvanized Steel (RGS)	104	140	235	358	455	721	1022	1451	1749	2148	3083	4343
Intermediate Steel (IMC)	84	113	186	293	379	611	883	1263	1501	1830		
Electrical Metallic Tubing (EMT)	54	116	183	296	368	445	641	930	1215	1540		

*Conduit & Heaviest Conductor Combination

R260533-30 Labor for Couplings and Fittings

The labor included in the unit price lines for couplings is for their installation separate from a fitting. However, the labor included in the unit price lines for fittings covers their complete installation, which may include the labor installation of the coupling(s). If coupling(s) are required to complete a fitting installation, only the material for the required number of couplings needs to be added to the appropriate unit price lines.

R260533-60 Wireway

When "taking off" Wireway, list by size and type.

Example:
1. Screw cover, unflanged + size
2. Screw cover, flanged + size
3. Hinged cover, flanged + size
4. Hinged cover, unflanged + size

Each 10′ length on Wireway contains:
1. 10′ of cover either screw or hinged type
2. (1) Coupling or flange gasket
3. (1) Wall type mount

All fittings must be priced separately.

Substitution of hanger types is done the same as described in R260533-21, "HANGERS," keeping in mind that the wireway model is based on 10′ sections instead of a 100′ conduit run.

Labor-hours for wireway include:
1. Unloading by hand
2. Hauling by hand up to 100′ from loading dock
3. Measuring and marking
4. Mounting wall bracket using (2) anchor type lead fasteners
5. Installing wireway on brackets, to 15′ high (For higher elevations use factors in R260533-20)

Job Conditions: Productivity is based on new construction, to a height of 15′ using rolling staging in an unobstructed area.

Material staging area is assumed to be within 100′ of work being performed.

R260533-65 Outlet Boxes

Outlet boxes should be included on the same takeoff sheet as branch piping or devices to better explain what is included in each circuit.

Each unit price in this section is a stand alone item and contains no other component unless specified. For example, to estimate a duplex outlet, components that must be added are:
1. 4″ square box
2. 4″ plaster ring
3. Duplex receptacle
4. Device cover

The method of mounting outlet boxes is (2) plastic shield fasteners.

Outlet boxes plastic, labor-hours include:
1. Marking box location on wood studding
2. Mounting box

Economy of Scale – For large concentrations of plastic boxes in the same area deduct the following percentages from labor-hour totals:

1	to	10	0%
11	to	25	20%
26	to	50	25%
51	to	100	30%
	over	100	35%

Note: It is important to understand that these percentages are not used on the total job quantities, but only areas where concentrations exceed the levels specified.

R260533-70 Pull Boxes and Cabinets

List cabinets and pull boxes by NEMA type and size.

Example:	TYPE	SIZE
	NEMA 1	6″W x 6″H x 4″D
	NEMA 3R	6″W x 6″H x 4″D

Labor-hours for wall mount (indoor or outdoor) installations include:
1. Unloading and uncrating
2. Handling of enclosures up to 200′ from loading dock using a dolly or pipe rollers
3. Measuring and marking
4. Drilling (4) anchor type lead fasteners using a hammer drill
5. Mounting and leveling boxes

Note: A plywood backboard is not included.

Labor-hours for ceiling mounting include:
1. Unloading and uncrating
2. Handling boxes up to 100′ from loading dock
3. Measuring and marking
4. Drilling (4) anchor type lead fasteners using a hammer drill
5. Installing and leveling boxes to a height of 15′ using rolling staging

Labor-hours for free standing cabinets include:
1. Unloading and uncrating
2. Handling of cabinets up to 200′ from loading dock using a dolly or pipe rollers
3. Marking of floor
4. Drilling (4) anchor type lead fasteners using a hammer drill
5. Leveling and shimming

Labor-hours for telephone cabinets include:
1. Unloading and uncrating
2. Handling cabinets up to 200′ using a dolly or pipe rollers
3. Measuring and marking
4. Mounting and leveling, using (4) lead anchor type fasteners

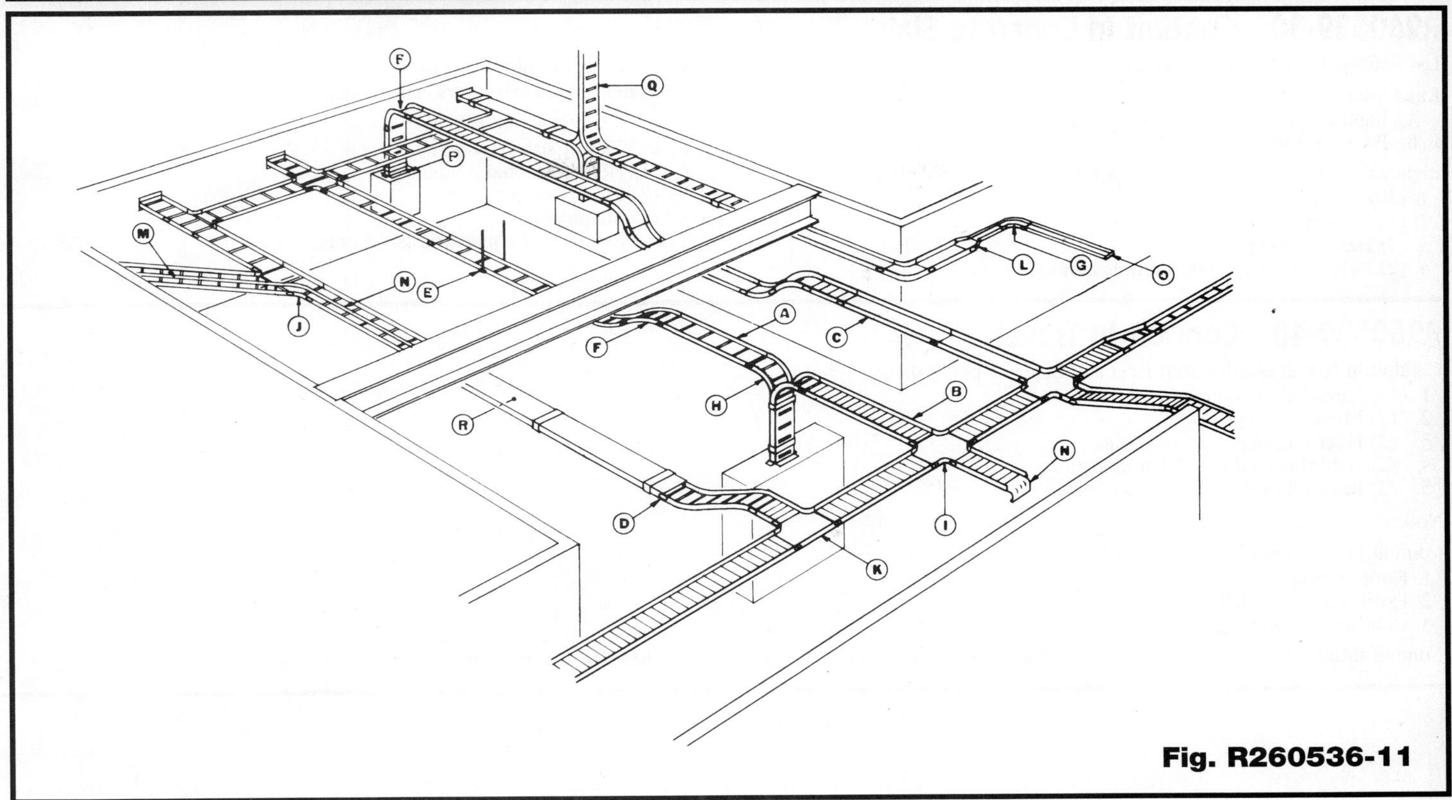

Fig. R260536-11

R260536-10 Cable Tray

Cable Tray - When taking off cable tray it is important to identify separately the different types and sizes involved in the system being estimated. (Fig. R260536-11)

A. – Ladder Type, galvanized or aluminum
B. – Trough Type, galvanized or aluminum
C. – Solid Bottom, galvanized or aluminum

The unit of measure is calculated in linear feet; do not deduct from this footage any length occupied by fittings, this will be the only allowance for scrap. Be sure to include all vertical drops to panels, switch gear, etc.

Hangers – Included in the linear footage of cable tray is

D. – 1 – Pair of connector plates per 12 L.F.
E. – 1 – Pair clamp type hangers and 4′ of 3/8″ threaded rod per 12 L.F.

Not included are structural supports, which must be priced in addition to the hangers.

Fittings – Identify separately the different types of fittings

1.) Ladder Type, galvanized or aluminum
2.) Trough Type, galvanized or aluminum
3.) Solid Bottom Type, galvanized or aluminum

The configuration, radius and rung spacing must also be listed. The unit of measure is "Ea."

F. – Elbow, vertical Ea.
G. – Elbow, horizontal Ea.
H. – Tee, vertical Ea.
I. – Cross, horizontal Ea.
J. – Wye, horizontal Ea.
K. – Tee, horizontal Ea.
L. – Reducing fitting Ea.

Depending on the use of the system other examples of units which must be included are:

M. – Divider strip L.F.
N. – Drop-outs Ea.
O. – End caps Ea.
P. – Panel connectors Ea.

Wire and cable are not included and should be taken off separately, see Unit Price sections.

Job Conditions – Unit prices are based on a new installation to a work plane of 15′ using rolling staging.

Add to labor for elevated installations:

15′ to 20′ High	10%
20′ to 25′ High	20%
25′ to 30′ High	25%
30′ to 35′ High	30%
35′ to 40′ High	35%
Over 40′ High	40%

Add these percentages for L.F. totals but not to fittings. Add percentages to only those quantities that fall in the different elevations, in other words, if the total quantity of cable tray is 200′ but only 75′ is above 15′ then the 10% is added to the 75′ only.

Linear foot costs do not include penetrations through walls and floors which must be added to the estimate.

Cable Tray Covers

Covers – Cable tray covers are taken off in the same manner as the tray itself, making distinctions as to the type of cover. (Fig. R260536-11)

Q. – Vented, galvanized or aluminum
R. – Solid, galvanized or aluminum

Cover configurations are taken off separately noting type, specific radius and widths.

Note: Care should be taken to identify from plans and specifications exactly what is being covered. In many systems only vertical fittings are covered to retain wire and cable.

R260539-30 Conduit In Concrete Slab

List conduit by quantity, size and type.

Example:

A. Rigid galvanized steel + size
B. P.V.C. + size

Rigid galvanized steel (A) contains per 100 L.F.:

1. (20) Ties to slab reinforcing
2. (11) Threaded steel couplings
3. (2) Factory sweeps
4. (2) Field threaded conduit terminations
5. (2) Fiber bushings + locknuts
6. (2) Removal of concentric knockouts

P.V.C. (B) contains per 100 L.F.:

1. (20) Ties to slab reinforcing
2. (11) Field cemented couplings
3. (2) Factory sweeps
4. (2) Adapters
5. (2) Removal of concentric knockouts

R260539-40 Conduit In Trench

Conduit in trench is galvanized steel and contains per 100 L.F.:

1. (11) Threaded couplings
2. (2) Factory sweeps
3. (2) Fiber bushings + (4) locknuts
4. (2) Field threaded conduit terminations
5. (2) Removal of concentric knockouts

Note:

Conduit in Unit Price sections do not include:

1. Floor cutting
2. Excavation or backfill
3. Grouting or patching

Conduit fittings in excess of those listed in the above Conduit model must be added. (Refer to R260533-20 for Procedure example.)

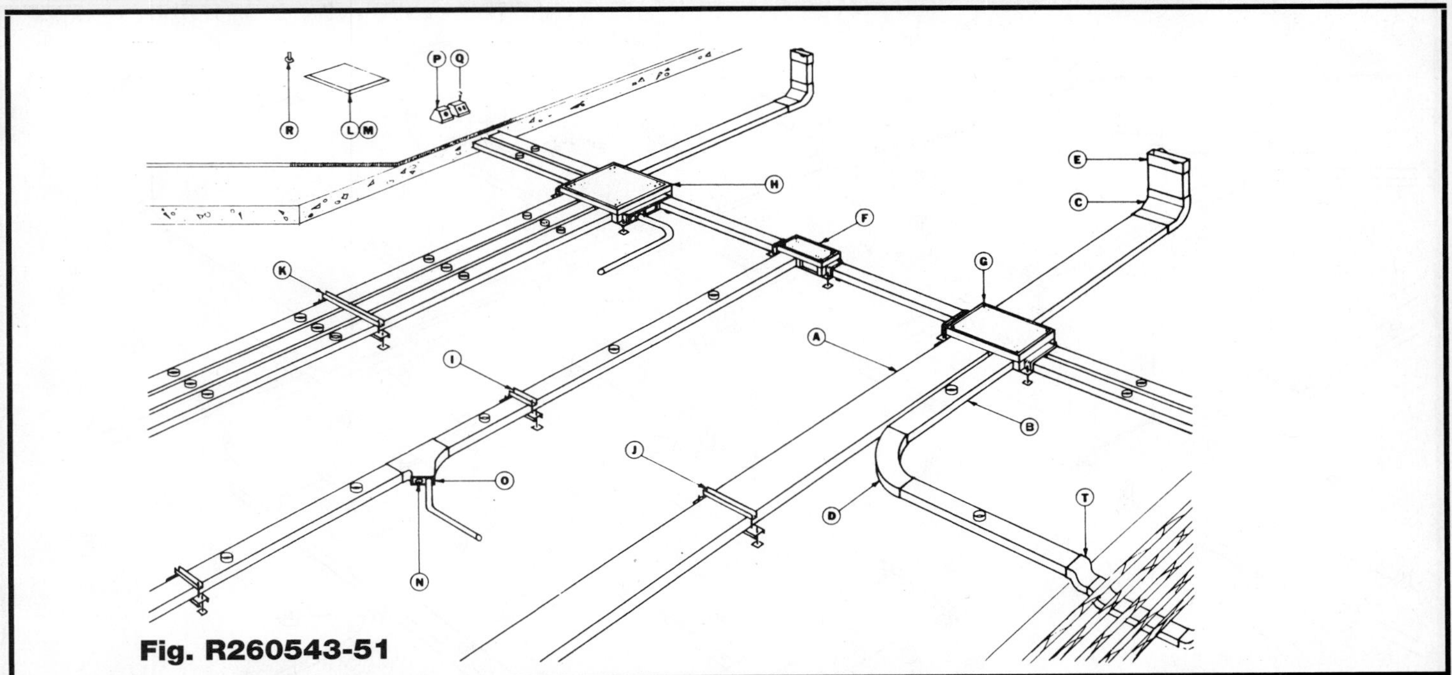

Fig. R260543-51

R260543-50 Underfloor Duct

When pricing Underfloor Duct it is important to identify and list each component, since costs vary significantly from one type of fitting to another. Do not deduct boxes or fittings from linear foot totals; this will be your allowance for scrap.

The first step is to identify the system as either:

FIG. R260543-51 Single Level

FIG. R260543-52 Dual Level

Single Level System

Include on your "takeoff sheet" the following unit price items, making sure to distinquish between Standard and Super duct:

A. Feeder duct (blank) in L.F.
B. Distribution duct (Inserts 2′ on center) in L.F.
C. Elbows (Vertical) Ea.
D. Elbows (Horizontal) Ea.
E. Cabinet connector Ea.
F. Single duct junction box Ea.
G. Double duct junction box Ea.
H. Triple duct junction box Ea.
I. Support, single cell Ea.
J. Support, double cell Ea.
K. Support, triple cell Ea.
L. Carpet pan Ea.
M. Terrazzo pan Ea.
N. Insert to conduit adapter Ea.
O. Conduit adapter Ea.
P. Low tension outlet Ea.
Q. High tension outlet Ea.
R. Galvanized nipple Ea.
S. Wire per C.L.F.
T. Offset (Duct type) Ea.

Dual Level System + Labor
see next page

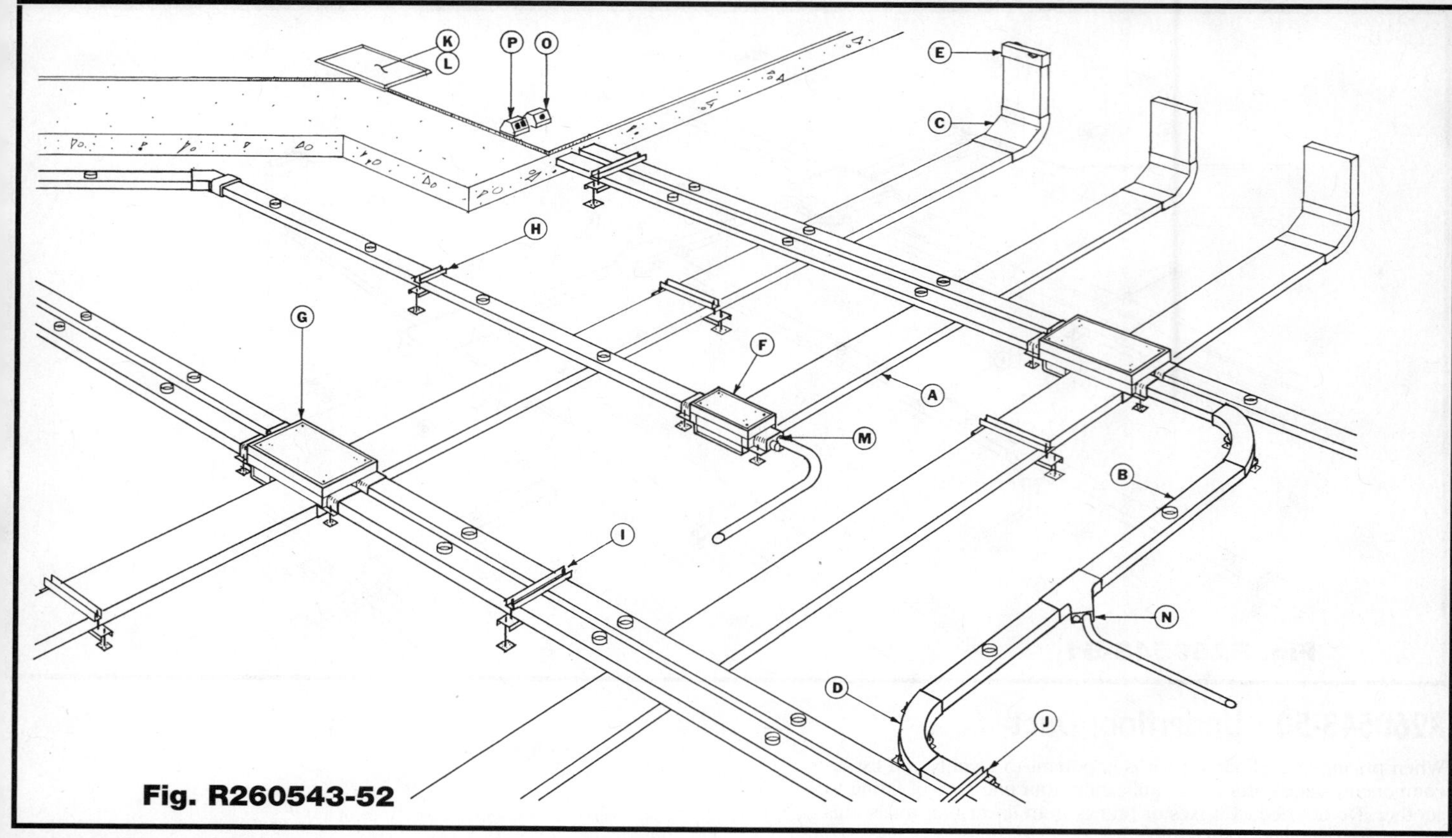

Fig. R260543-52

R260543-50 Underfloor Duct (cont.)

Dual Level

Include the following when "taking off" Dual Level systems:

Distinguish between Standard and Super duct.

A. Feeder duct (blank) in L.F.
B. Distribution duct (Inserts 2′ on center) in L.F.
C. Elbows (Vertical) Ea.
D. Elbows (Horizontal) Ea.
E. Cabinet connector Ea.
F. Single duct, 2 level, junction box Ea.
G. Double duct, 2 level, junction box Ea.
H. Support, single cell Ea.
I. Support, double cell Ea.
J. Support, triple cell Ea.
K. Carpet pan Ea.
L. Terrazzo pan Ea.
M. Insert to conduit adapter Ea.
N. Conduit adapter Ea.
O. Low tension outlet Ea.
P. High tension outlet Ea.
Q. Wire per C.L.F.

Note: Make sure to include risers in linear foot totals. High tension outlets include box, receptacle, covers and related mounting hardware.

Labor-hours for both Single and Dual Level systems include:

1. Unloading and uncrating
2. Hauling up to 200′ from loading dock
3. Measuring and marking
4. Setting raceway and fittings in slab or on grade
5. Leveling raceway and fittings

Labor-hours do not include:

1. Floor cutting
2. Excavation or backfill
3. Concrete pour
4. Grouting or patching
5. Wire or wire pulls
6. Additional outlets after concrete is poured
7. Piping to or from Underfloor Duct

Note: Installation is based on installing up to 150′ of duct. If quantities exceed this, deduct the following percentages:

1. 150′ to 250′ - 10%
2. 250′ to 350′ - 15%
3. 350′ to 500′ - 20%
4. over 500′ - 25%

Deduct these percentages from labor only.

Deduct these percentages from straight sections only.

Do not deduct from fittings or junction boxes.

Job Conditions: Productivity is based on new construction.

Underfloor duct to be installed on first three floors.

Material staging area within 100′ of work being performed.

Area unobstructed and duct not subject to physical damage.

R260580-75 Motor Connections

Motor connections should be listed by size and type of motor.
Included in the material and labor cost is:

1. (2) Flex connectors
2. 18″ of flexible metallic wireway
3. Wire identification and termination
4. Test for rotation
5. (2) or (3) Conductors

Price does not include:

1. Mounting of motor
2. Disconnect Switch
3. Motor Starter
4. Controls
5. Conduit or wire ahead of flex

Note: When "Taking off" Motor connections, it is advisable to list connections on the same quantity sheet as Motors, Motor Starters and controls.

R260913-80 Switchboard Instruments

Switchboard instruments are added to the price of switchboards according to job specifications. This equipment is usually included when ordering "Gear" from the manufacturer and will arrive factory installed.

Included in the labor cost is:

1. Internal wiring connections
2. Wire identification
3. Wire tagging

Transition sections include:

1. Uncrating
2. Hauling sections up to 100′ from loading dock
3. Positioning sections
4. Leveling sections
5. Bolting enclosures
6. Bolting vertical bus bars

Price does not include:

1. Equipment pads
2. Steel channels embedded or grouted in concrete
3. Special knockouts
4. Rigging

Job Conditions: Productivity is based on new construction, equipment to be installed on the first floor within 200′ of the loading dock.

R262213-10 Electric Circuit Voltages

General: The following method provides the user with a simple non-technical means of obtaining comparative costs of wiring circuits. The circuits considered serve the electrical loads of motors, electric heating, lighting and transformers, for example, that require low voltage 60 Hertz alternating current.

The method used here is suitable only for obtaining estimated costs. It is **not** intended to be used as a substitute for electrical engineering design applications.

Conduit and wire circuits can represent from twenty to thirty percent of the total building electrical cost. By following the described steps and using the tables the user can translate the various types of electric circuits into estimated costs.

Wire Size: Wire size is a function of the electric load which is usually listed in one of the following units:

1. Amperes (A)
2. Watts (W)
3. Kilowatts (kW)
4. Volt amperes (VA)
5. Kilovolt amperes (kVA)
6. Horsepower (HP)

These units of electric load must be converted to amperes in order to obtain the size of wire necessary to carry the load. To convert electric load units to amperes one must have an understanding of the voltage classification of the power source and the voltage characteristics of the electrical equipment or load to be energized. The seven A.C. circuits commonly used are illustrated in Figures R262213-11 thru R262213-17 showing the tranformer load voltage and the point of use voltage at the point on the circuit where the load is connected. The difference between the source and point of use voltages is attributed to the circuit voltage drop and is considered to be approximately 4%.

Motor Voltages: Motor voltages are listed by their point of use voltage and not the power source voltage.

For example: 460 volts instead of 480 volts
200 instead of 208 volts
115 volts instead of 120 volts

Lighting and Heating Voltages: Lighting and heating equipment voltages are listed by the power source voltage and not the point of wire voltage.

For example: 480, 277, 120 volt lighting
480 volt heating or air conditioning unit
208 volt heating unit

Transformer Voltages: Transformer primary (input) and secondary (output) voltages are listed by the power source voltage.

For example: Single phase 10 kVA
Primary 240/480 volts
Secondary 120/240 volts

In this case, the primary voltage may be 240 volts with a 120 volts secondary or may be 480 volts with either a 120V or a 240V secondary.

For example: Three phase 10 kVA
Primary 480 volts
Secondary 208Y/120 volts

In this case the transformer is suitable for connection to a circuit with a 3 phase 3 wire or 3 phase 4 wire circuit with a 480 voltage. This application will provide a secondary circuit of 3 phase 4 wire with 208 volts between phase wires and 120 volts between any phase wire and the neutral (white) wire.

R262213-11

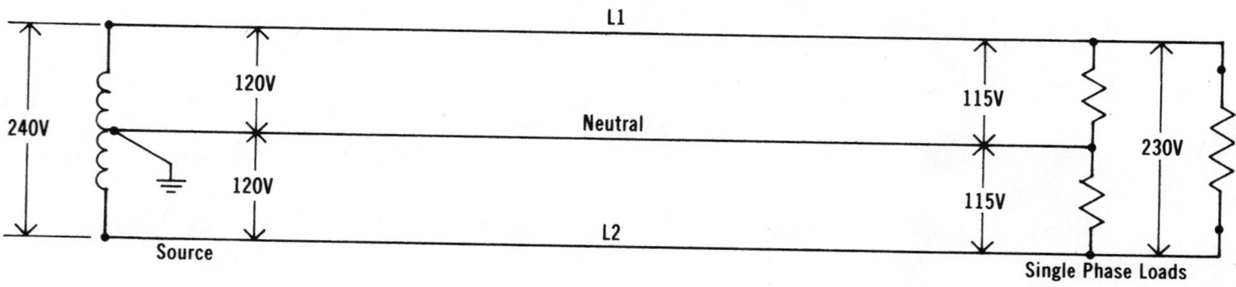

R262213-12

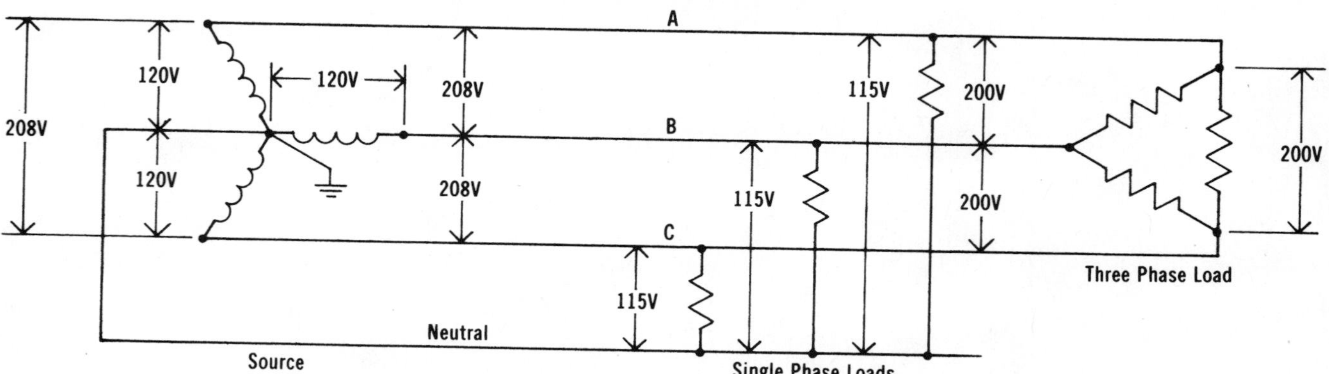

R262213-13 Electric Circuit Voltages (cont.)

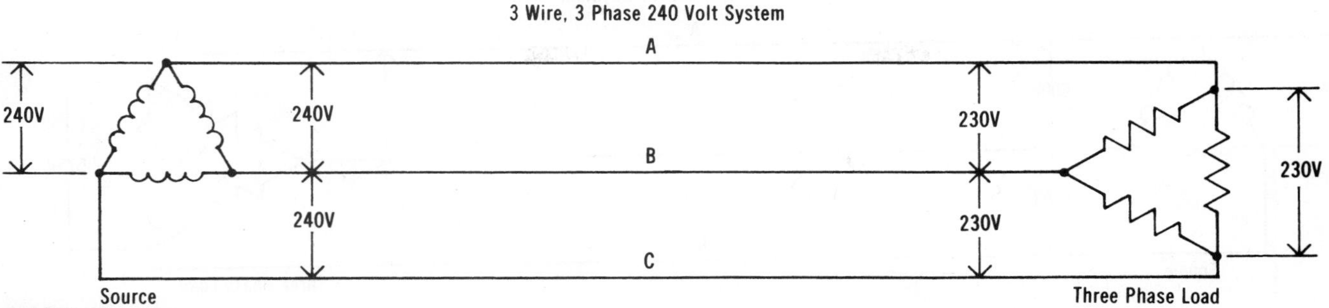

R262213-14

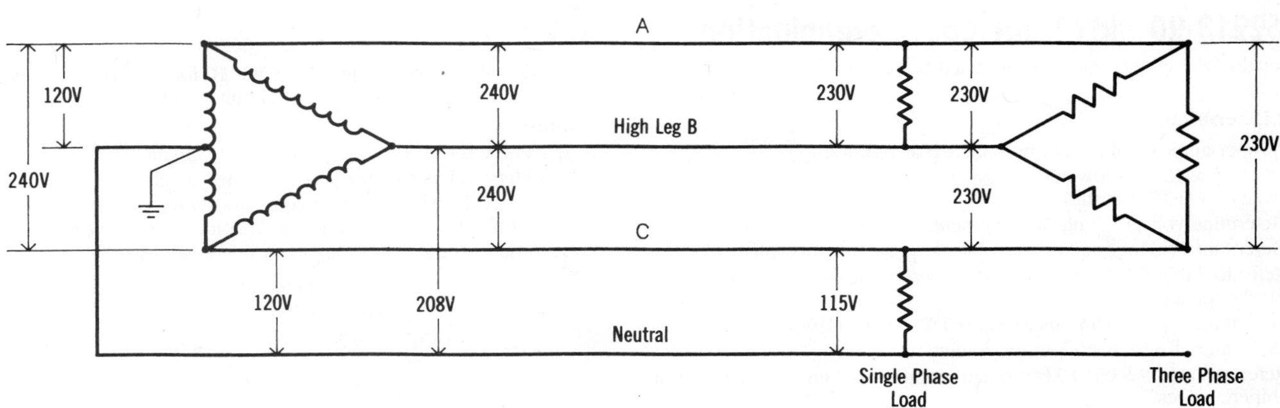

R262213-15

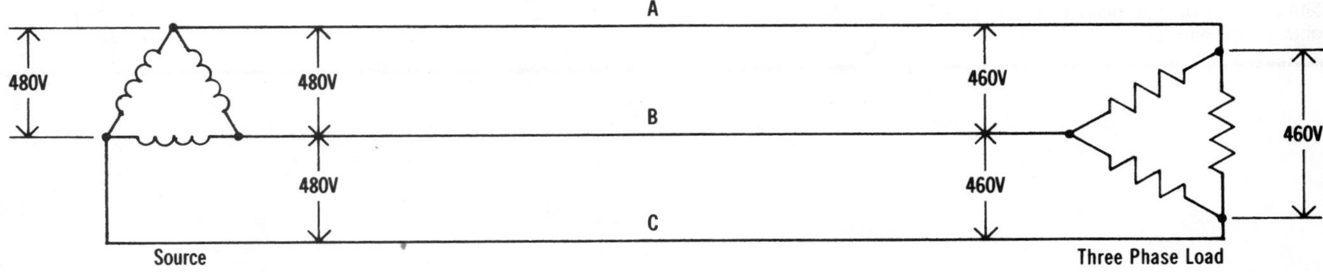

R262213-16

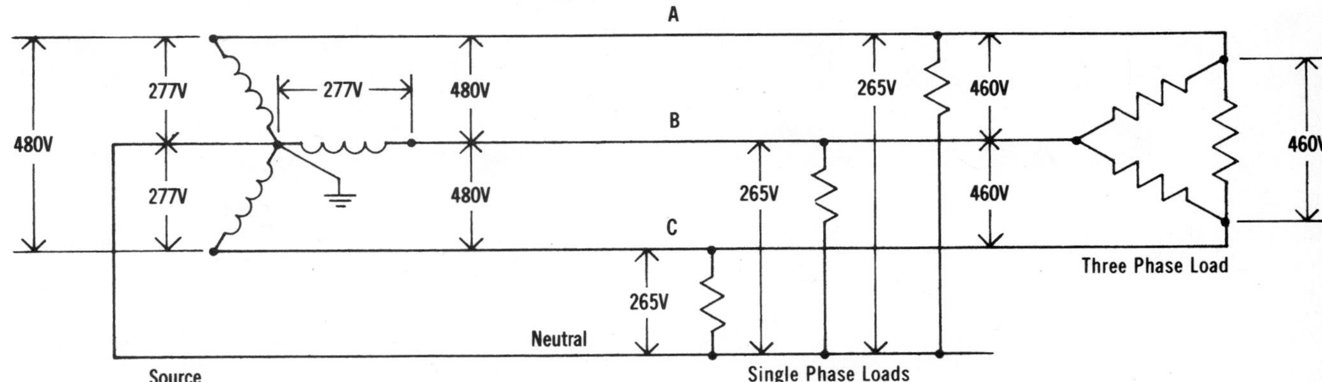

R262213-17 Electric Circuit Voltages (cont.)

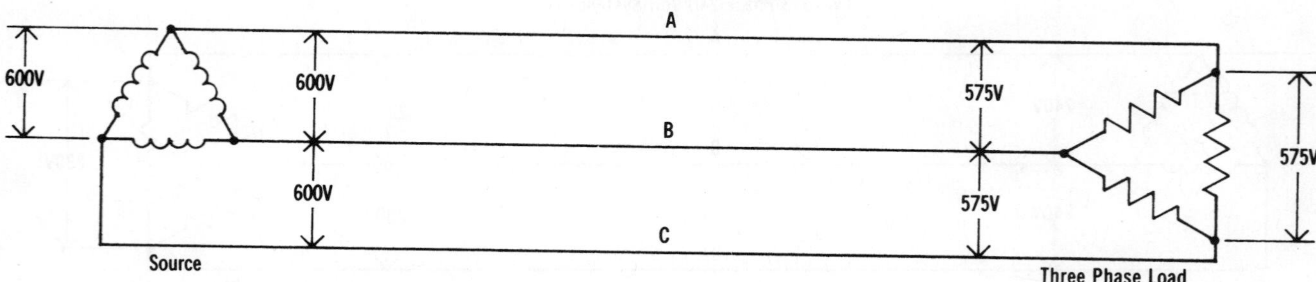

R262213-20 kW Value/Cost Determination

General: Lighting and electric heating loads are expressed in watts and kilowatts.

Cost Determination:

The proper ampere values can be obtained as follows:

1. Convert watts to kilowatts (watts 1000 ÷ kilowatts)
2. Determine voltage rating of equipment.
3. Determine whether equipment is single phase or three phase.
4. Refer to Table R262213-21 to find ampere value from kW, Ton and Btu/hr. values.
5. Determine type of wire insulation – TW, THW, THWN.
6. Determine if wire is copper or aluminum.
7. Refer to Table R260519-92 to obtain copper or aluminum wire size from ampere values.
8. Next refer to Table R260533-22 for the proper conduit size to accommodate the number and size of wires in each particular case.
9. Next refer to unit cost data for the per linear foot cost of the conduit.
10. Next refer to unit cost data for the per linear foot cost of the wire. Multiply cost of wire per L.F. x number of wires in the circuits to obtain total wire cost per L.F.
11. Add values obtained in Step 9 and 10 for total cost per linear foot for conduit and wire x length of circuit = Total Cost.

Notes:

1. 1 Phase refers to single phase, 2 wire circuits.
2. 3 Phase refers to three phase, 3 wire circuits.
3. For circuits which operate continuously for 3 hours or more, multiply the ampere values by 1.25 for a given kw requirement.
4. For kW ratings not listed, add ampere values.

   For example: Find the ampere value of 9 kW at 208 volt, single phase.

   4 kW = 19.2A
   5 kW = 24.0A
   9 kW = 43.2A
5. "Length of Circuit" refers to the one way distance of the run, not to the total sum of wire lengths.

R262213-21 Ampere Values as Determined by kW Requirements, BTU/HR or Ton, Voltage and Phase Values

			Ampere Values						
			120V	208V		240V		277V	480V
kW	Ton	BTU/HR	1 Phase	1 Phase	3 Phase	1 Phase	3 Phase	1 Phase	3 Phase
0.5	.1422	1,707	4.2A	2.4A	1.4A	2.1A	1.2A	1.8A	0.6A
0.75	.2133	2,560	6.2	3.6	2.1	3.1	1.9	2.7	.9
1.0	.2844	3,413	8.3	4.9	2.8	4.2	2.4	3.6	1.2
1.25	.3555	4,266	10.4	6.0	3.5	5.2	3.0	4.5	1.5
1.5	.4266	5,120	12.5	7.2	4.2	6.3	3.1	5.4	1.8
2.0	.5688	6,826	16.6	9.7	5.6	8.3	4.8	7.2	2.4
2.5	.7110	8,533	20.8	12.0	7.0	10.4	6.1	9.1	3.1
3.0	.8532	10,239	25.0	14.4	8.4	12.5	7.2	10.8	3.6
4.0	1.1376	13,652	33.4	19.2	11.1	16.7	9.6	14.4	4.8
5.0	1.4220	17,065	41.6	24.0	13.9	20.8	12.1	18.1	6.1
7.5	2.1331	25,598	62.4	36.0	20.8	31.2	18.8	27.0	9.0
10.0	2.8441	34,130	83.2	48.0	27.7	41.6	24.0	36.5	12.0
12.5	3.5552	42,663	104.2	60.1	35.0	52.1	30.0	45.1	15.0
15.0	4.2662	51,195	124.8	72.0	41.6	62.4	37.6	54.0	18.0
20.0	5.6883	68,260	166.4	96.0	55.4	83.2	48.0	73.0	24.0
25.0	7.1104	85,325	208.4	120.2	70.0	104.2	60.0	90.2	30.0
30.0	8.5325	102,390		144.0	83.2	124.8	75.2	108.0	36.0
35.0	9.9545	119,455		168.0	97.1	145.6	87.3	126.0	42.1
40.0	11.3766	136,520		192.0	110.8	166.4	96.0	146.0	48.0
45.0	12.7987	153,585			124.8	187.5	112.8	162.0	54.0
50.0	14.2208	170,650			140.0	208.4	120.0	180.4	60.0
60.0	17.0650	204,780			166.4		150.4	216.0	72.0
70.0	19.9091	238,910			194.2		174.6		84.2
80.0	22.7533	273,040			221.6		192.0		96.0
90.0	25.5975	307,170					225.6		108.0
100.0	28.4416	341,300							120.0

R262213-25 kVA Value/Cost Determination

General: Control transformers are listed in VA. Step-down and power transformers are listed in kVA.

Cost Determination:
1. Convert VA to kVA. Volt amperes (VA) ÷ 1000 = Kilovolt amperes (kVA).
2. Determine voltage rating of equipment.
3. Determine whether equipment is single phase or three phase.
4. Refer to Table R262213-26 to find ampere value from kVA value.
5. Determine type of wire insulation – TW, THW, THWN.
6. Determine if wire is copper or aluminum.
7. Refer to Table R260519-92 to obtain copper or aluminum wire size from ampere values.
8. Next refer to Table R260533-22 for the proper conduit size to accommodate the number and size of wires in each particular case.
9. Next refer to unit price data for the per linear foot cost of the conduit.
10. Next refer to unit price data for the per linear foot cost of the wire. Multiply cost of wire per L.F. x number of wires in the circuits to obtain total wire cost.
11. Add values obtained in Step 9 and 10 for total cost per linear foot for conduit and wire x length of circuit = Total Cost.

Example: A transformer rated 10 kVA 480 volts primary, 240 volts secondary, 3 phase has the capacity to furnish the following:
1. Primary amperes = 10 kVA x 1.20 = 12 amperes (from Table R262213-26)
2. Secondary amperes = 10 kVA x 2.40 = 24 amperes (from Table R262213-26)

Note: Transformers can deliver generally 125% of their rated kVA. For instance, a 10 kVA rated transformer can safely deliver 12.5 kVA.

R262213-26 Multiplier Values for kVA to Amperes Determined by Voltage and Phase Values

Volts	Multiplier for Circuits	
	2 Wire, 1 Phase	3 Wire, 3 Phase
115	8.70	
120	8.30	
230	4.30	2.51
240	4.16	2.40
200	5.00	2.89
208	4.80	2.77
265	3.77	2.18
277	3.60	2.08
460	2.17	1.26
480	2.08	1.20
575	1.74	1.00
600	1.66	0.96

R262213-60 Oil Filled Transformers

Transformers in this section include:
1. Rigging (as required)
2. Rental of crane and operator
3. Setting of oil filled transformer
4. (4) Anchor bolts, nuts and washers in concrete pad

Price does not include:
1. Primary and secondary terminations
2. Transformer pad
3. Equipment grounding
4. Cable
5. Conduit locknuts or bushings

Transformers in Unit Price sections for dry type, back-boost and isolating transformers include:
1. Unloading and uncrating
2. Hauling transformer to within 200′ of loading dock
3. Setting in place
4. Wall mounting hardware
5. Testing

Price does not include:
1. Structural supports
2. Suspension systems
3. Welding or fabrication
4. Primary & secondary terminations

Add the following percentages to the labor for ceiling mounted transformers:

10' to 15′	= + 15%
15′ to 25′	= + 30%
Over 25′	= + 35%

Job Conditions: Productivities are based on new construction. Installation is assumed to be on the first floor, in an obstructed area to a height of 10′. Material staging area is within 100′ of final transformer location.

R262416-50 Load Centers and Panelboards

When pricing Load Centers list panels by size and type. List Breakers in a separate column of the "Quantity Sheet," and define by phase and ampere rating.

Material and Labor prices include breakers; for example: a 100A, 3-Wire, 102/240V, 18 circuit panel w/ main breaker, as described in the unit cost section, contains 18 single pole 20A breakers.

If you do not choose to include a full panel of single pole breakers, use the following method to adjust material and labor costs.

Example: In an 18 circuit panel only 16 single pole breakers are desired, requiring that the cost of 2 breakers be subtracted from the panel cost.

1. Go to the appropriate unit cost section of the book to find the unit prices of the given circuit breaker type.
2. Modify those costs as follows:
   Bare material price x 0.50
   Bare labor cost x 0.60.
3. Multiply those modified bare costs by 2 (breakers in this example).
4. Subtract the modified costs for the 2 breakers from the given cost of the panel.

Labor-hours for Load Center installation includes:

1. Unloading, uncrating, and handling enclosures 200′ from unloading area
2. Measuring and marking
3. Drilling (4) lead anchor type fasteners using a hammer drill
4. Mounting and leveling panel to a height of 6′
5. Preparation and termination of feeder cable to lugs or main breaker
6. Branch circuit identification
7. Lacing using tie wraps
8. Testing and load balancing
9. Marking panel directory

Not included in the material and labor are:

1. Modifications to enclosure
2. Structural supports
3. Additional lugs
4. Plywood backboards
5. Painting or lettering

Note: Knockouts are included in the price of terminating pipe runs and need not be added to the Load Center costs.

Job Conditions: Productivity is based on new construction to a height of 6′, in an unobstructed area. Material staging area is assumed to be within 100′ of work being performed.

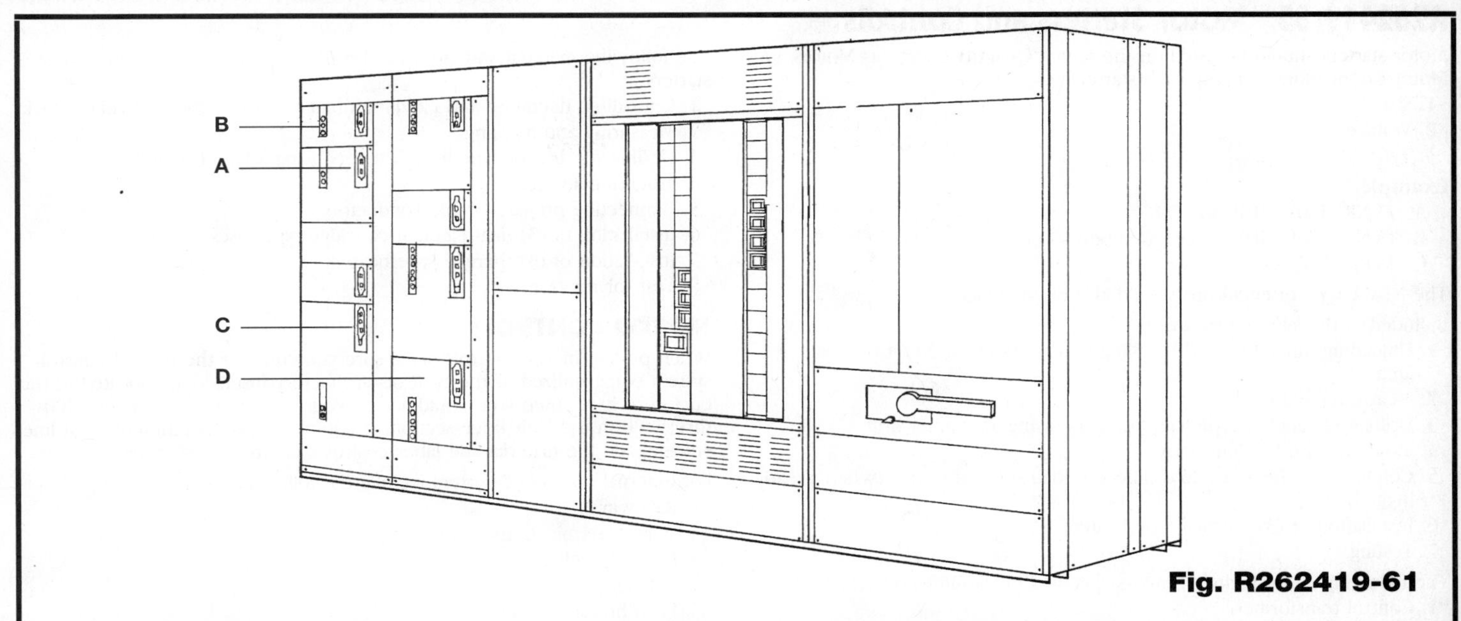

Fig. R262419-61

R262419-60 Motor Control Centers

When taking off Motor Control Centers, list the size, type and height of structures.

Example:

1. 600A, 22,000 RMS, 72″ high
2. 600A, back to back, 72″ high

Next take off individual starters; the number of structures can also be determined by adding the height in inches of starters divided by the height of the structure, and list on the same quantity sheet as the structures. Identify starters by Type, Horsepower rating, Size and Height in inches.

Example:

A. Class l, Type B, FVNR starter, 25 H.P., 18″ high.
   Add to the list with Starters, factory installed controls.

Example:

B. Pilot lights
C. Push buttons
D. Auxiliary contacts

Identify starters and structures as either copper or aluminum and by the NEMA type of enclosure.

When pricing starters and structures, be sure to add or deduct adjustments, using lines in the unit price section.

Included in the cost of Motor Control Structures are:

1. Uncrating
2. Hauling to location within 100′ of loading dock
3. Setting structures
4. Leveling
5. Aligning
6. Bolting together structure frames
7. Bolting horizontal bus bars

Labor-hours do not include:

1. Equipment pad
2. Steel channels embedded or grouted in concrete
3. Pull boxes
4. Special knockouts
5. Main switchboard section
6. Transition section
7. Instrumentation
8. External control wiring
9. Conduit or wire

Material for Starters includes:

1. Circuit breaker or fused disconnect
2. Magnetic motor starter
3. Control transformer
4. Control fuse and fuse block

Labor-hours for Starters include:

1. Handling
2. Installing starter within structure
3. Internal wiring connections
4. Lacing within enclosure
5. Testing
6. Phasing

Job Conditions: Productivity is based on new construction. Motor control location assumed to be on first floor in an unobstructed area. Material staging area within 100′ of final location.

Note: Additional labor-hours must be added if M.C.C. is to be installed on other than the first floor, or if rigging is required.

R262419-65 Motor Starters and Controls

Motor starters should be listed on the same "Quantity Sheet" as Motors and Motor Connections. Identify each starter by:

1. Size
2. Voltage
3. Type

Example:

A. FVNR, 480V, 2HP, Size 00
B. FVNR, 480V, 5HP, Size 0, Combination type
C. FVR, 480V, Size 2

The NEMA type of enclosure should also be identified.

Included in the labor-hours are:

1. Unloading, uncrating and handling of starters up to 200′ from loading area
2. Measuring and marking
3. Drilling (4) anchor type lead fasteners, using a hammer drill
4. Mounting and leveling starter
5. Connecting wire or cable to line and load sides of starter (when already lugged)
6. Installation of (3) thermal type heaters
7. Testing

The following is not included unless specified in the unit price description.

1. Control transformer
2. Controls, either factory or field installed
3. Conduit and wire to or from starter
4. Plywood backboard
5. Cable terminations

The following material and labor has been included for Combination type starters:

1. Unloading, uncrating and handling of starter up to 200′ of loading dock
2. Measuring and marking
3. Drilling (4) anchor type lead fasteners using a hammer drill
4. Mounting and leveling
5. Connecting prepared cable conductors
6. Installation of (3) dual element cartridge type fuses
7. Installation of (3) thermal type heaters
8. Test for rotation

MOTOR CONTROLS

When pricing motor controls make sure you consider the type of control system being utilized. If the controls are factory installed and located in the enclosure itself, then you would add to your starter price the items 5200 thru 5800 in the Unit Price section. If control voltage is different from line voltage, add the material and labor cost of a control transformer.

For external control of starters include the following items:

1. Raceways
2. Wire & terminations
3. Control enclosures
4. Fittings
5. Push button stations
6. Indicators

Job Conditions: Productivity is based on new construction to a height of 10′.

Material staging area is assumed to be within 100′ of work being performed.

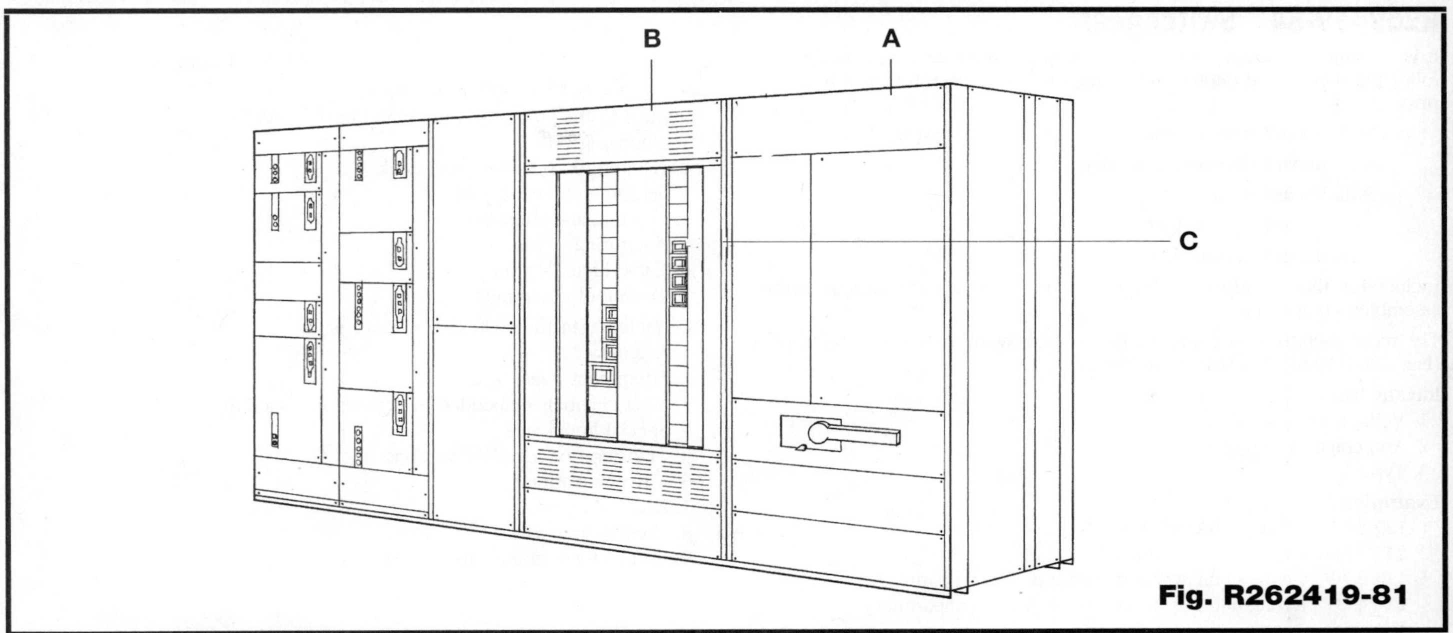

Fig. R262419-81

R262419-80 Distribution Section

After "Taking off" the Switchboard section, include on the same "Quantity sheet" the Distribution section; identify by:

1. Voltage
2. Ampere rating
3. Type

Example:

(Fig. R262419-81) (B) Distribution Section

Included in the labor costs of the Distribution section is:

1. Uncrating
2. Hauling to 200′ of loading dock
3. Setting of distribution panel
4. Leveling & shimming
5. Anchoring of equipment to pad or floor
6. Bolting of horizontal bus bars between
7. Testing of equipment

Not included in the Distribution section is:

1. Breakers (C)
2. Equipment pads
3. Steel channels embedded or grouted in concrete
4. Pull boxes
5. Special knockouts
6. Transition section
7. Conduit or wire

R262419-82 Feeder Section

List quantities on the same sheet as the Distribution section. Identify breakers by: (Fig. R262419-81)

1. Frame type
2. Number of poles
3. Ampere rating

Installation includes:

1. Handling
2. Placing breakers in distribution panel
3. Preparing wire or cable
4. Lacing wire
5. Marking each phase with colored tape
6. Marking panel legend
7. Testing
8. Balancing

R262419-84 Switchgear

It is recommended that "Switchgear" or those items contained in the following sections be quoted from equipment manufacturers as a package price.

Switchboard instruments

Switchboard distribution sections

Switchboard feeder sections

Switchboard Service Disc.

Switchboard In-plant Dist.

Included in these sections are the most common types and sizes of factory assembled equipment.

The recommended procedure for low voltage switchgear would be to price (Fig. R262419-81) (A) Main Switchboard

Identify by:
1. Voltage
2. Amperage
3. Type

Example:
1. 120/208V, 4-wire, 600A, nonfused
2. 277/480V, 4-wire, 600A, nonfused
3. 120/208V, 4-wire, 400A w/fused switch & CT compartment
4. 277/480V, 4-wire, 400A, w/fused switch & CT compartment
5. 120/208V, 4-wire, 800A, w/pressure switch & CT compartment
6. 277/480V, 4-wire, 800A, w/molded CB & CT compartment

Included in the labor costs for Switchboards are:
1. Uncrating
2. Hauling to 200′ of loading dock
3. Setting equipment
4. Leveling and shimming
5. Anchoring
6. Cable identification
7. Testing of equipment

Not included in the Switchboard price is:
1. Rigging
2. Equipment pads
3. Steel channels embedded or grouted in concrete
4. Special knockouts
5. Transition or Auxiliary sections
6. Instrumentation
7. External control
8. Conduit and wire
9. Conductor terminations

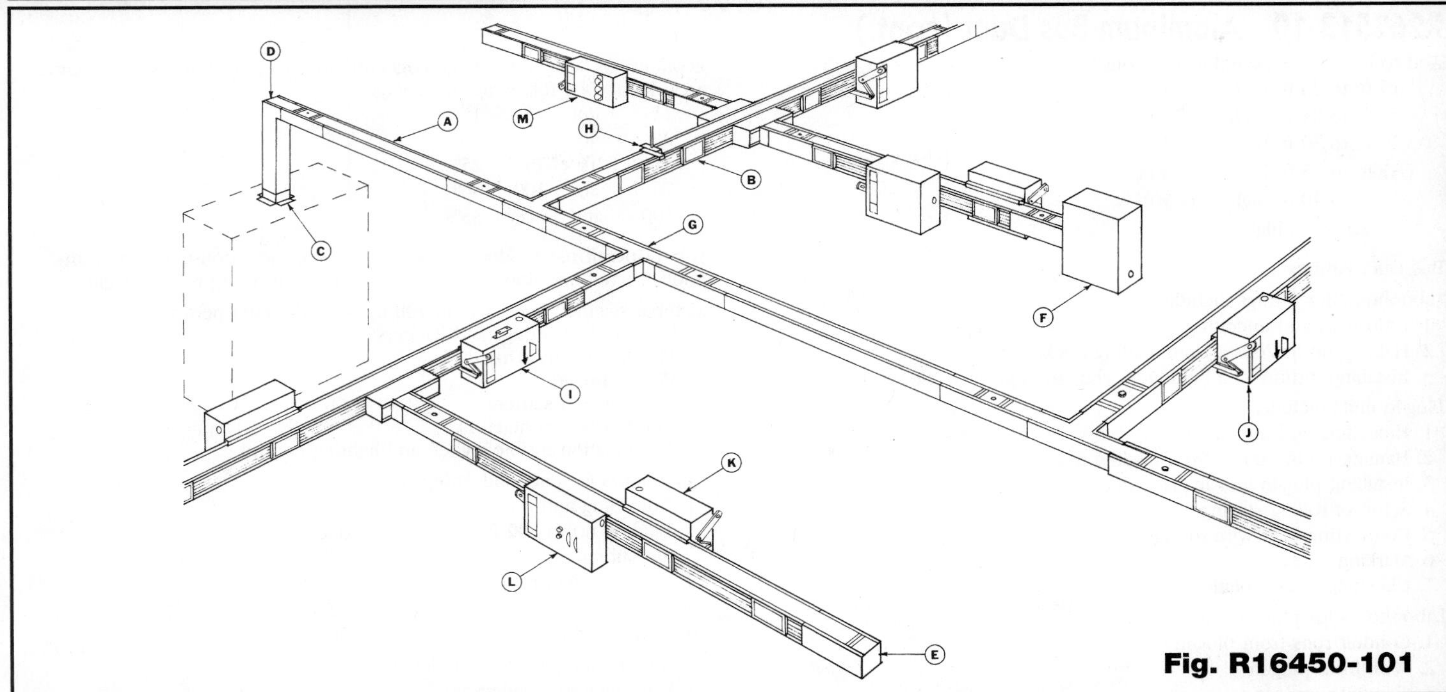

Fig. R16450-101

R262513-10 Aluminum Bus Duct

When taking off bus duct identify the system as either:

1. Aluminum
2. Copper

List straight lengths by type and size

A. Feeder — 800 A
B. Plug in — 800 A

Do not measure thru fittings as you would on conduit, since there is no allowance for scrap in bus duct systems.

If upon taking off linear foot quantities of bus duct you find your quantities are not divisible evenly by 10 ft., then the remainder must be priced as a special item and quoted from the manufactuer. Do not use the bare material cost per L.F. for these special items. You can, however, safely use the bare labor cost per L.F. for the entire length.

Identify fittings by type and ampere rating.

Example:

C. Switchboard stub 800 A

D. Elbows 800 A

E. End box 800 A

F. Cable tap box 800 A

G. Tee Fittings 800 A

H. Hangers

Plug-in Units – List separately plug-in units and identify by type and ampere rating

I. Plug-in switches 600 Volt 3 phase 60 A

J. Plug-in molded case C.B. 60 A

K. Combination starter FVNR NEMA 1

L. Combination contactor & fused switch NEMA 1

M. Combination fusible switch & lighting control 60 A

Labor-hours for feeder and plug-in sections include:

1. Unloading and uncrating
2. Hauling up to 200 ft. from loading dock
3. Measuring and marking
4. Setup of rolling staging
5. Installing hangers
6. Hanging and bolting sections
7. Aligning and leveling
8. Testing

Labor-hours do not include:

1. Modifications to existing structure for hanger supports
2. Threaded rod in excess of 2 ft.
3. Welding
4. Penetrations thru walls
5. Staging rental

Deduct the following percentages from labor only:

150 ft. to 250 ft. — 10%
251 ft. to 350 ft. — 15%
351 ft. to 500 ft. — 20%
Over 500 ft. — 25%

Deduct percentage only if runs are contained in the same area.

Example: If the job entails running 100 ft. in 5 different locations do not deduct 20%, but if the duct is being run in 1 area and the quantity is 500 ft. then you would deduct 20%.

Deduct only from straight lengths, not fittings or plug-in units.

R262513-10 Aluminum Bus Duct (cont.)

Add to labor for elevated installations:

15 ft. to 20 ft. high	10%
21 ft. to 25 ft. high	20%
26 ft. to 30 ft. high	30%
31 ft. to 35 ft. high	40%
36 ft. to 40 ft. high	50%
Over 40 ft. high	60%

Bus Duct Fittings:

Labor-hours for fittings include:
1. Unloading and uncrating
2. Hauling up to 200 ft. from loading dock
3. Installing, fitting, and bolting all ends to in-place sections

Plug-in units include:
1. Unloading and uncrating
2. Hauling up to 200 ft. from loading dock
3. Installing plug-in into in-place duct
4. Setup of rolling staging
5. Connecting load wire to lugs
6. Marking wire
7. Checking phase rotation

Labor-hours for plug-ins do not include:
1. Conduit runs from plug-in
2. Wire from plug-in
3. Conduit termination

Economy of Scale – For large concentrations of plug-in units in the same area deduct the following percentages:

11	to	25	15%
26	to	50	20%
51	to	75	25%
76	to	100	30%
100	and	over	35%

Job Conditions: Productivities are based on new construction in an unobstructed first floor area to a height of 15 ft. using rolling staging.

Material staging area is within 100 ft. of work being performed.
Add to the duct fittings and hangers:
- Plug-in switches (fused)
- Plug-in breakers
- Combination starters
- Combination contactors
- Combination fusible switch and lighting control

Labor-hours for duct and fittings include:
1. Unloading and uncrating
2. Hauling up to 200 ft. from loading dock
3. Measuring and marking
4. Installing duct runs
5. Leveling
6. Sound testing

Labor-hours for plug-ins include:
1. Unloading and uncrating
2. Hauling up to 200 ft. from loading dock
3. Installing plug-ins
4. Preparing wire
5. Wire connections and marking

R262716-40 Standard Electrical Enclosure Types

NEMA Enclosures

Electrical enclosures serve two basic purposes; they protect people from accidental contact with enclosed electrical devices and connections, and they protect the enclosed devices and connections from specified external conditions. The National Electrical Manufacturers Association (NEMA) has established the following standards. Because these descriptions are not intended to be complete representations of NEMA listings, consultation of NEMA literature is advised for detailed information.

The following definitions and descriptions pertain to NONHAZARDOUS locations.

NEMA Type 1: General purpose enclosures intended for use indoors, primarily to prevent accidental contact of personnel with the enclosed equipment in areas that do not involve unusual conditions.

NEMA Type 2: Dripproof indoor enclosures intended to protect the enclosed equipment against dripping noncorrosive liquids and falling dirt.

NEMA Type 3: Dustproof, raintight and sleet-resistant (ice-resistant) enclosures intended for use outdoors to protect the enclosed equipment against wind-blown dust, rain, sleet, and external ice formation.

NEMA Type 3R: Rainproof and sleet-resistant (ice-resistant) enclosures which are intended for use outdoors to protect the enclosed equipment against rain. These enclosures are constructed so that the accumulation and melting of sleet (ice) will not damage the enclosure and its internal mechanisms.

NEMA Type 3S: Enclosures intended for outdoor use to provide limited protection against wind-blown dust, rain, and sleet (ice) and to allow operation of external mechanisms when ice-laden.

NEMA Type 4: Watertight and dust-tight enclosures intended for use indoors and out – to protect the enclosed equipment against splashing water, see page of water, falling or hose-directed water, and severe external condensation.

NEMA Type 4X: Watertight, dust-tight, and corrosion-resistant indoor and outdoor enclosures featuring the same provisions as Type 4 enclosures, plus corrosion resistance.

NEMA Type 5: Indoor enclosures intended primarily to provide limited protection against dust and falling dirt.

NEMA Type 6: Enclosures intended for indoor and outdoor use – primarily to provide limited protection against the entry of water during occasional temporary submersion at a limited depth.

NEMA Type 6R: Enclosures intended for indoor and outdoor use – primarily to provide limited protection against the entry of water during prolonged submersion at a limited depth.

NEMA Type 11: Enclosures intended for indoor use – primarily to provide, by means of oil immersion, limited protection to enclosed equipment against the corrosive effects of liquids and gases.

NEMA Type 12: Dust-tight and driptight indoor enclosures intended for use indoors in industrial locations to protect the enclosed equipment against fibers, flyings, lint, dust, and dirt, as well as light splashing, see page, dripping, and external condensation of noncorrosive liquids.

NEMA Type 13: Oil-tight and dust-tight indoor enclosures intended primarily to house pilot devices, such as limit switches, foot switches, push buttons, selector switches, and pilot lights, and to protect these devices against lint and dust, see page, external condensation, and sprayed water, oil, and noncorrosive coolant.

The following definitions and descriptions pertain to HAZARDOUS, or CLASSIFIED, locations:

NEMA Type 7: Enclosures intended to use in indoor locations classified as Class 1, Groups A, B, C, or D, as defined in the National Electrical Code.

NEMA Type 9: Enclosures intended for use in indoor locations classified as Class 2, Groups E, F, or G, as defined in the National Electrical Code.

R262726-90 Wiring Devices

Wiring devices should be priced on a separate takeoff form which includes boxes, covers, conduit and wire.

Labor-hours for devices include:

1. Stripping of wire
2. Attaching wire to device using terminators on the device itself, lugs, set screws etc.
3. Mounting of device in box

Labor-hours do not include:

1. Conduit
2. Wire
3. Boxes
4. Plates

Economy of Scale – for large concentrations of devices in the same area deduct the following percentages from labor-hours:

1	to	10	0%
11	to	25	20%
26	to	50	25%
51	to	100	30%
	over	100	35%

R262726-90 Wiring Devices (cont.)

NEMA No.	15 R	20 R	30 R	50 R	60 R
1 125V 2 Pole, 2 Wire	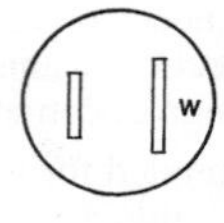 W				
2 250V 2 Pole, 2 Wire		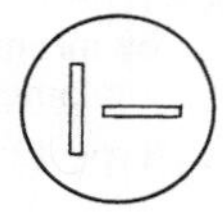	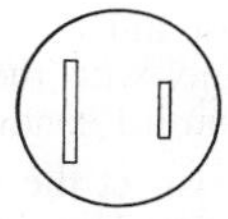		
5 125V 2 Pole, 3 Wire	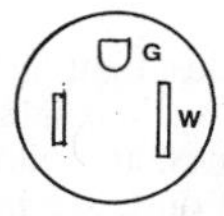 G W	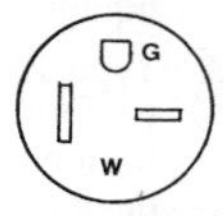 G W	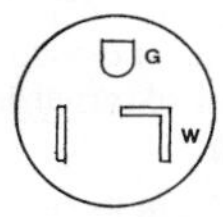 G W	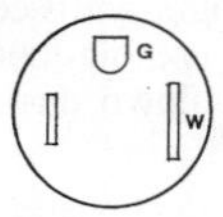 G W	
6 250V 2 Pole, 3 Wire	 G	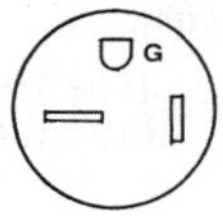 G	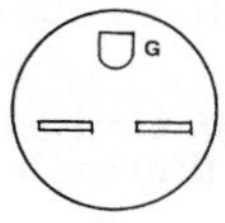 G	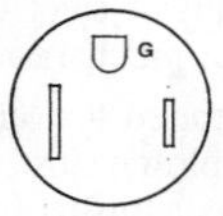 G	
7 277V, AC 2 Pole, 3 Wire	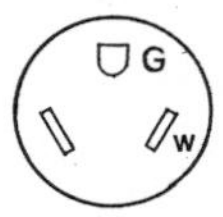 G W	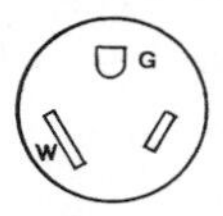 G W	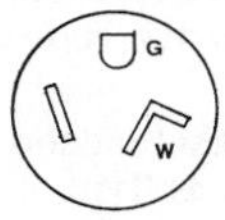 G W	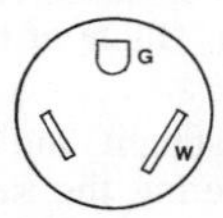 G W	
10 125/250V 3 Pole, 3 Wire		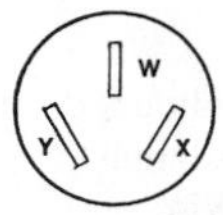 W Y X	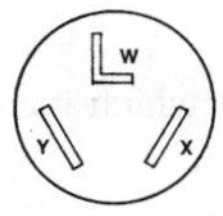 W Y X	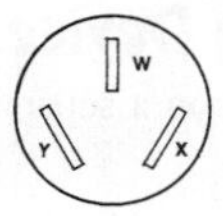 W Y X	
11 3 Phase 250V 3 Pole, 3 Wire	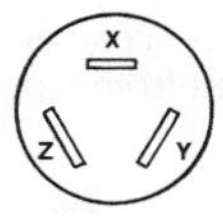 X Z Y	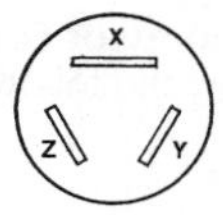 X Z Y	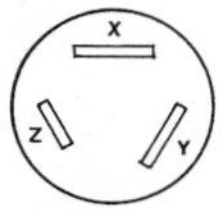 X Z Y	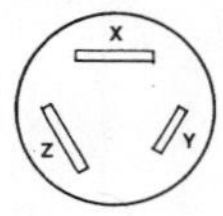 X Z Y	
14 125/250V 3 Pole, 4 Wire	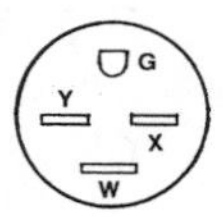 G Y X W	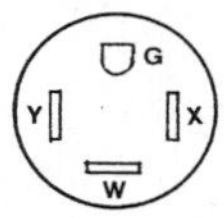 G Y X W	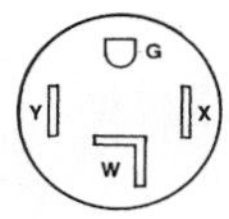 G Y X W	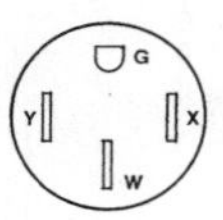 G Y X W	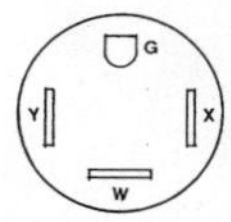 G Y X W
15 3 Phase 250V 3 Pole, 4 Wire	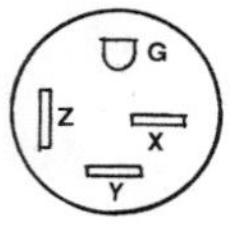 G Z X Y	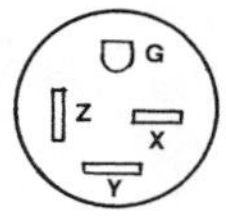 G Z X Y	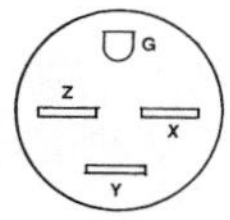 G Z X Y	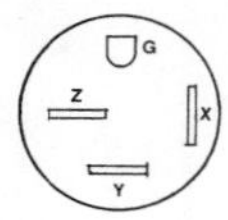 G Z X Y	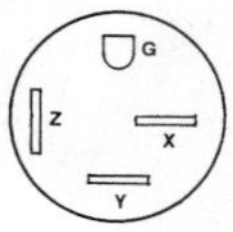 G Z X Y
18 3 Phase 208Y/120V 4 Pole, 4 Wire	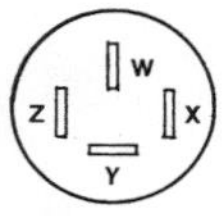 W Z X Y	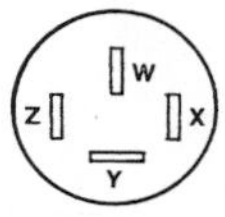 W Z X Y	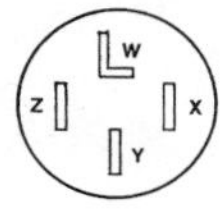 W Z X Y	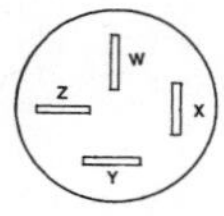 W Z X Y	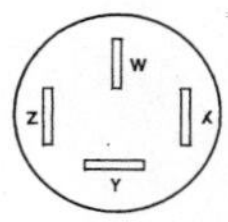 W Z X Y

R262726-90 Wiring Devices (cont.)

NEMA No.	15 R	20 R	30 R	NEMA No.	15 R	20 R	30 R
L 1 125V 2 Pole, 2 Wire				L 13 3 Phase 600V 3 Pole, 3 Wire			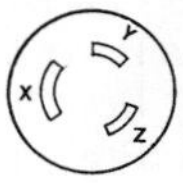
L 2 250V 2 Pole, 2 Wire		15A		L 14 125/250V 3 Pole, 4 Wire			
L 5 125 V 2 Pole, 3 Wire				L 15 3 Phase 250 V 3 Pole, 4 Wire			
L 6 250 V 2 Pole, 3 Wire				L 16 3 Phase 480V 3 Pole, 4 Wire			
L 7 227 V, AC 2 Pole, 3 Wire				L 17 3 Phase 600 V 3 Pole, 4 Wire			
L 8 480 V 2 Pole, 3 Wire				L 18 3 Phase 208Y/120V 4 Pole, 4 Wire			
L 9 600 V 2 Pole, 3 Wire				L 19 3 Phase 480Y/277V 4 Pole, 4 Wire			
L 10 125 /250V 3 Pole, 3 Wire				L 20 3 Phase 600Y/347V 4 Pole, 4 Wire		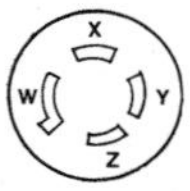	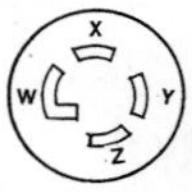
L 11 3 Phase 250 V 3 Pole, 3 Wire				L 21 3 Phase 208Y/120V 4 Pole, 5 Wire			
L 12 3 Phase 480 V 3 Pole, 3 Wire				L 22 3 Phase 480Y/277V 4 Pole, 5 Wire			
				L 23 3 Phase 600Y/347V 4 Pole, 5 Wire		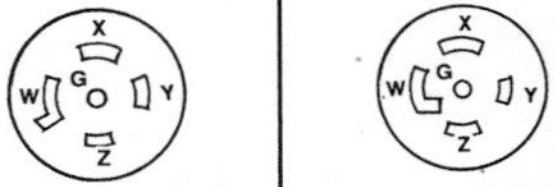	

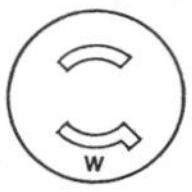

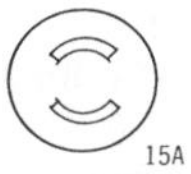

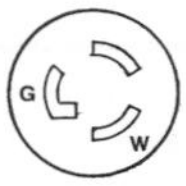

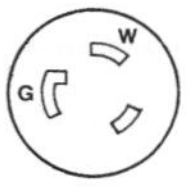

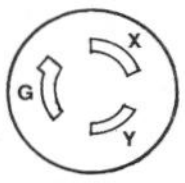

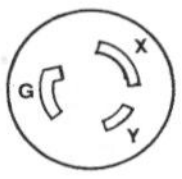

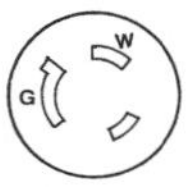

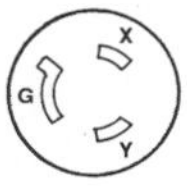

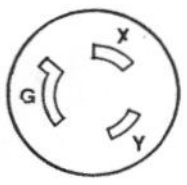

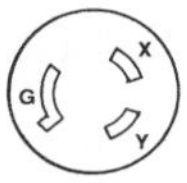

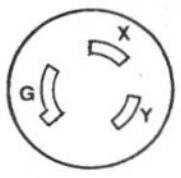

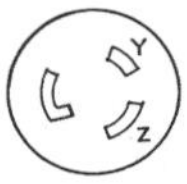

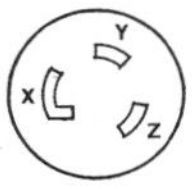

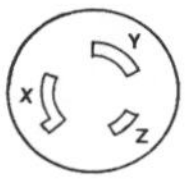

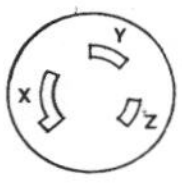

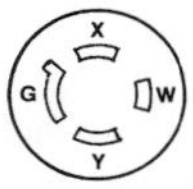

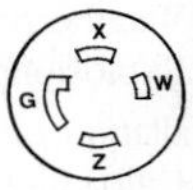

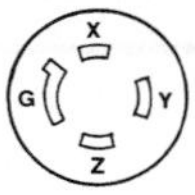

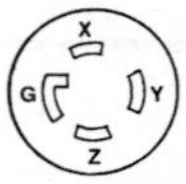

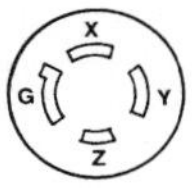

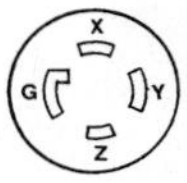

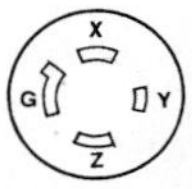

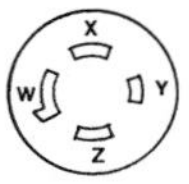

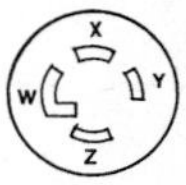

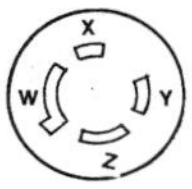

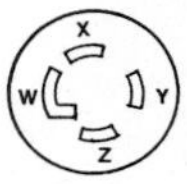

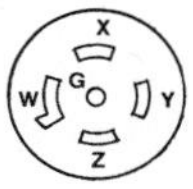

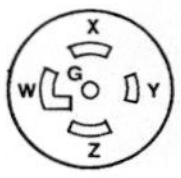

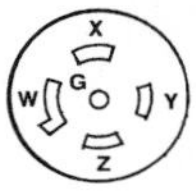

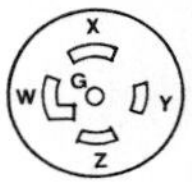

R262816-80 Safety Switches

List each Safety Switch by type, ampere rating, voltage, single or three phase, fused or nonfused.

Example:

A. General duty, 240V, 3-pole, fused
B. Heavy Duty, 600V, 3-pole, nonfused
C. Heavy Duty, 240V, 2-pole, fused
D. Heavy Duty, 600V, 3-pole, fused

Also include NEMA enclosure type and identify as:

1. Indoor
2. Weatherproof
3. Explosionproof

Installation of Safety Switches includes:

1. Unloading and uncrating
2. Handling disconnects up to 200′ from loading dock
3. Measuring and marking location
4. Drilling (4) anchor type lead fasteners, using a hammer drill
5. Mounting and leveling Safety Switch
6. Installing (3) fuses
7. Phasing and tagging line and load wires

Price does not include:

1. Modifications to enclosure
2. Plywood backboard
3. Conduit or wire
4. Fuses
5. Termination of wires

Job Conditions: Productivities are based on new construction to an installed height of 6′ above finished floor.

Material staging area is assumed to be within 100′ of work in progress.

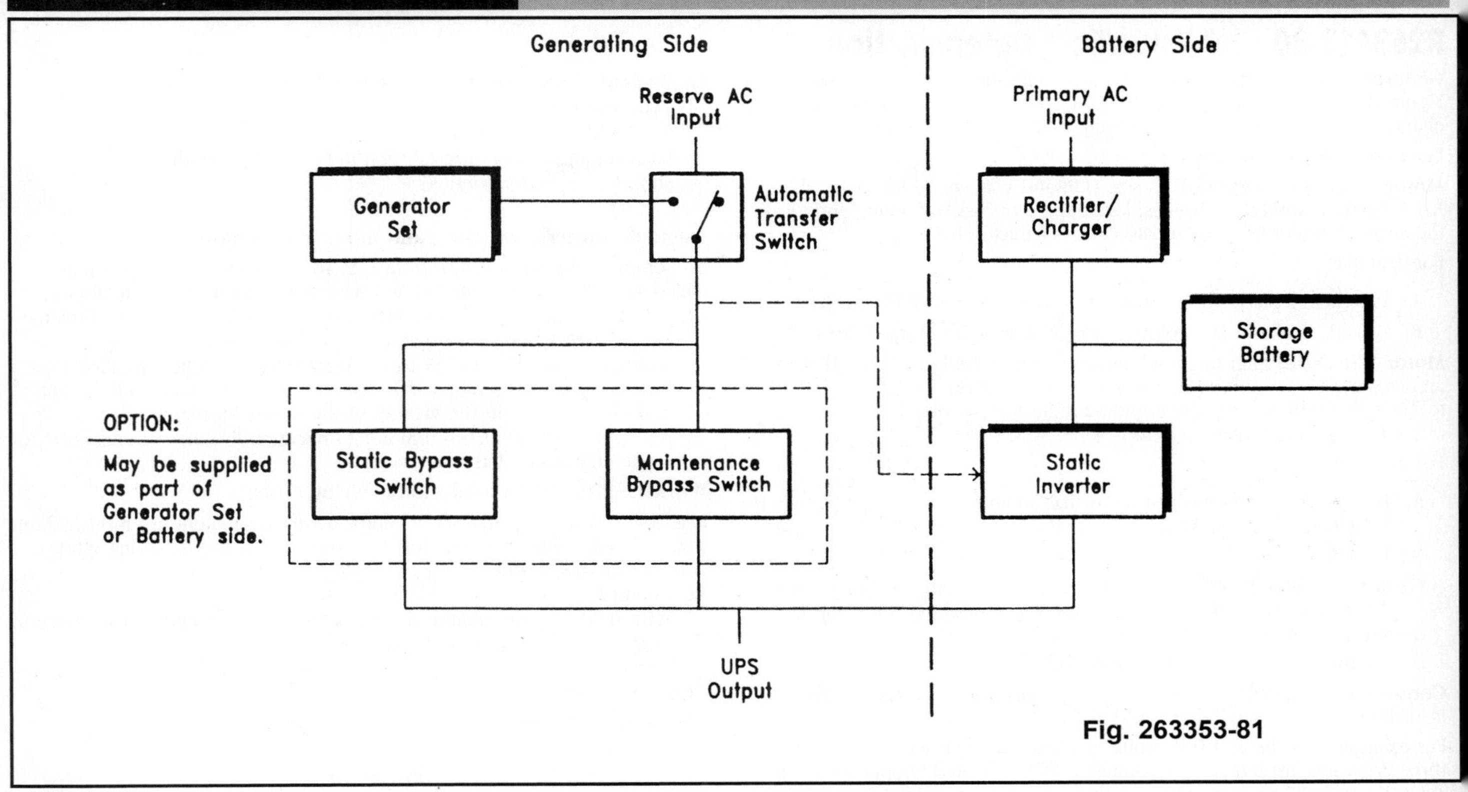

Fig. 263353-81

R263353-80 Uninterruptible Power Supply Systems

General: Uninterruptible Power Supply (UPS) Systems are used to provide power for legally required standby systems. They are also designed to protect computers and provide optional additional coverage for any or all loads in a facility. Figure R263353-81 shows a typical configuration for a UPS System with generator set.

Cost Determination:

It is recommended that UPS System material costs be obtained from equipment manufacturers as a package. Installation costs should be obtained from a vendor or contractor.

The recommended procedure for pricing UPS Systems would be to identify:

1. Frequency – by Hertz (Hz.)
   For example: 50, 60, and 400/415 Hz.
2. Apparent power and real power
   For example: 200 kVA/170 kW
3. Input/Output voltage For example: 120V, 208V or 240V
4. Phase – either single or three-phase
5. Options and accessories – For extended run time more batteries would be required. Other accessories include battery cabinets, battery racks, remote control panel, power distribution unit (PDU), and warranty enhancement plans.

Note:

1. Larger systems can be configured by paralleling two or more standard small size single modules.
   For example: two 15 kVA modules, combined together and configured to 30 kVA.
2. Maximum input current during battery recharge is typically 15% higher than normal current.
3. UPS Systems weights vary depending on options purchased and power rating in kVA.

R263413-30 HP Value/Cost Determination

General: Motors can be powered by any of the seven systems shown in Figure R262213-11 thru Figure R262213-17 provided the motor voltage characteristics are compatible with the power system characteristics.

Cost Determination:

Motor Amperes for the various size H.P. and voltage are listed in Table R263413-31. To find the amperes, locate the required H.P. rating and locate the amperes under the appropriate circuit characteristics.

For example:

A. 100 H.P., 3 phase, 460 volt motor = 124 amperes (Table R263413-31)

B. 10 H.P., 3 phase, 200 volt motor = 32.2 amperes (Table R263413-31)

Motor Wire Size: After the amperes are found in Table R263413-31 the amperes must be increased 25% to compensate for power losses. Next refer to Table R260519-92. Find the appropriate insulation column for copper or aluminum wire to determine the proper wire size.

For example:

A. 100 H.P., 3 phase, 460 volt motor has an ampere value of 124 amperes from Table R263413-31

B. 124A x 1.25 = 155 amperes

C. Refer to Table R260519-92 for THW or THWN wire insulations to find the proper wire size. For a 155 ampere load using copper wire a size 2/0 wire is needed.

D. For the 3 phase motor three wires of 2/0 size are required.

Conduit Size: To obtain the proper conduit size for the wires and type of insulation used, refer to Table R260533-22.

For example: For the 100 H.P., 460V, 3 phase motor, it was determined that three 2/0 wires are required. Assuming THWN insulated copper wire, use Table R260533-22 to determine that three 2/0 wires require 1-1/2″ conduit.

Material Cost of the conduit and wire system depends on:

1. Wire size required
2. Copper or aluminum wire
3. Wire insulation type selected
4. Steel or plastic conduit
5. Type of conduit raceway selected.

Labor Cost of the conduit and wire system depends on:

1. Type and size of conduit
2. Type and size of wires installed
3. Location and height of installation in building or depth of trench
4. Support system for conduit.

Magnetic starters, switches, and motor connection:

To complete the cost picture from H.P. to Costs additional items must be added to the cost of the conduit and wire system to arrive at a total cost.

1. Assembly Table D5020-160 Magnetic Starters Installed Cost lists the various size starters for single phase and three phase motors.
2. Assembly Table D5020-165 Heavy Duty Safety Switches Installed Cost lists safety switches required at the beginning of a motor circuit and also one required in the vicinity of the motor location.
3. Assembly Table D5020-170 Motor Connection lists the various costs for single and three phase motors.

Worksheet to obtain total motor wiring costs:

It is assumed that the motors or motor driven equipment are furnished and installed under other sections for this estimate and the following work is done under this section:

1. Conduit
2. Wire (add 10% for additional wire beyond conduit ends for connections to switches, boxes, starters, etc.)
3. Starters
4. Safety switches
5. Motor connections

R263413-31 Ampere Values Determined by Horsepower, Voltage and Phase Values

H.P.	Amperes							
	Single Phase			Three Phase				
	115V	208V	230V	200V	208V	230V	460V	575V
1/6	4.4A	2.4A	2.2A					
1/4	5.8	3.2	2.9					
1/3	7.2	4.0	3.6					
1/2	9.8	5.4	4.9	2.5A	2.4A	2.2A	1.1A	0.9A
3/4	13.8	7.6	6.9	3.7	3.5	3.2	1.6	1.3
1	16	8.8	8	4.8	4.6	4.2	2.1	1.7
1-1/2	20	11	10	6.9	6.6	6.0	3.0	2.4
2	24	13.2	12	7.8	7.5	6.8	3.4	2.7
3	34	18.7	17	11.0	10.6	9.6	4.8	3.9
5	56	30.8	28	17.5	16.7	15.2	7.6	6.1
7-1/2	80	44	40	25.3	24.2	22	11	9
10	100	55	50	32.2	30.8	28	14	11
15				48.3	46.2	42	21	17
20				62.1	59.4	54	27	22
25				78.2	74.8	68	34	27
30				92	88	80	40	32
40				120	114	104	52	41
50				150	143	130	65	52
60				177	169	154	77	62
75				221	211	192	96	77
100				285	273	248	124	99
125				359	343	312	156	125
150				414	396	360	180	144
200				552	528	480	240	192
250							302	242
300							361	289
350							414	336
400							477	382

Figure R263413-32

Worksheet for Motor Circuits					
				Cost	
Item	Type	Size	Quantity	Unit	Total
Wire					
Conduit					
Switch					
Starter					
Switch					
Motor Connection					
Other					
Total Cost					

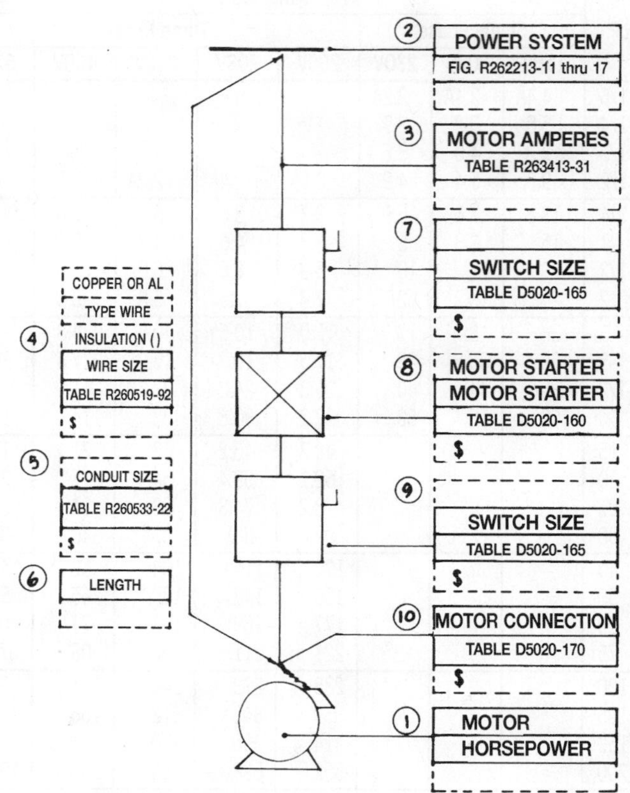

R263413-33 Maximum Horsepower for Starter Size by Voltage

Starter	Maximum HP (3φ)			
Size	208V	240V	480V	600V
00	1½	1½	2	2
0	3	3	5	5
1	7½	7½	10	10
2	10	15	25	25
3	25	30	50	50
4	40	50	100	100
5		100	200	200
6		200	300	300
7		300	600	600
8		450	900	900
8L		700	1500	1500

Electrical — R2636 Transfer Switches

R263623-60 Automatic Transfer Switches

When taking off Automatic transfer switches identify by voltage, amperage and number of poles.

Example: Automatic transfer switch, 480V, 3 phase, 30A, NEMA 1 enclosure

Labor-hours for transfer switches include:

1. Unloading, uncrating and handling switches up to 200′ from loading dock
2. Measuring and marking
3. Drilling (4) lead type anchors, using a hammer drill
4. Mounting and leveling
5. Circuit identification
6. Testing and load balancing

Labor-hours do not include:

1. Modifications in enclosure
2. Structural supports
3. Additional lugs
4. Plywood backboard
5. Painting or lettering
6. Conduit runs to or from transfer switch
7. Wire
8. Termination of wires

Electrical — R2651 Interior Lighting

R265113-40 Interior Lighting Fixtures

When taking off interior lighting fixtures, it is advisable to set up your quantity worksheet to conform to the lighting schedule as it appears on the print. Include the alpha-numeric code plus the symbol on your work sheet.

Take off a particular section or floor of the building and count each type of fixture before going on to another type. It would also be advantageous to include on the same worksheet the pipe, wire, fittings and circuit number associated with each type of lighting fixture. This will help you identify the costs associated with any particular lighting system and in turn make material purchases more specific as to when and how much to order under the classification of lighting.

By taking off lighting first you can get a complete "WALK THRU" of the job. This will become helpful when doing other phases of the project.

Materials for a recessed fixture include:

1. Fixture
2. Lamps
3. 6′ of jack chain
4. (2) S hooks
5. (2) Wire nuts

Labor for interior recessed fixtures include:

1. Unloading by hand
2. Hauling by hand to an area up to 200′ from loading dock
3. Uncrating
4. Layout
5. Installing fixture
6. Attaching jack chain & S hooks
7. Connecting circuit power
8. Reassembling fixture
9. Installing lamps
10. Testing

Material for surface mounted fixtures includes:

1. Fixture
2. Lamps
3. Either (4) lead type anchors, (4) toggle bolts, or (4) ceiling grid clips
4. (2) Wire nuts

Material for pendent mounted fixtures includes:

1. Fixture
2. Lamps
3. (2) Wire nuts
4. Rigid pendents as required by type of fixtures
5. Canopies as required by type of fixture

Labor hours include the following for both surface and pendent fixtures:

1. Unloading by hand
2. Hauling by hand to an area up to 200′ from loading dock
3. Uncrating
4. Layout and marking
5. Drilling (4) holes for either lead anchors or toggle bolts using a hammer drill
6. Installing fixture
7. Leveling fixture
8. Connecting circuit power
9. Installing lamps
10. Testing

Labor for surface or pendent fixtures does not include:

1. Conduit
2. Boxes or covers
3. Connectors
4. Fixture whips
5. Special support
6. Switching
7. Wire

Economy of Scale: For large concentrations of lighting fixtures in the same area deduct the following percentages from labor:

25 to 50	fixtures	15%
51 to 75	fixtures	20%
76 to 100	fixtures	25%
101 and over		30%

Job Conditions: Productivity is based on new construction in a unobstructed first floor location, using rolling staging to 15′ high.

Material staging is assumed to be within 100′ of work being performed.

Add the following percentages to labor for elevated installations:

15′ to 20′	high	10%
21′ to 25′	high	20%
26′ to 30′	high	30%
31′ to 35′	high	40%
36′ to 40′	high	50%
41′ and over		60%

R265723-10 For Other than Regular Cool White (CW) Lamps

Multiply Material Costs as Follows:					
Regular Lamps	Cool white deluxe (CWX)	x 1.35	Energy Saving Lamps	Cool white (CW/ES)	x 1.35
	Warm white deluxe (WWX)	x 1.35		Cool white deluxe (CWX/ES)	x 1.65
	Warm white (WW)	x 1.30		Warm white (WW/ES)	x 1.55
	Natural (N)	x 2.05		Warm white deluxe (WWX/ES)	x 1.65

R271323-40 Fiber Optics

Fiber optic systems use optical fiber such as plastic, glass, or fused silica, a transparent material, to transmit radiant power (i.e., light) for control, communication, and signaling applications. The types of fiber optic cables can be nonconductive, conductive, or composite. The composite cables contain fiber optics and current-carrying electrical conductors. The configuration for one of the fiber optic systems is as follows:

The transceiver module acts as transmitting and receiving equipment in a common house, which converts electrical energy to light energy or vice versa.

Pricing the fiber optic system is not an easy task. The performance of the whole system will affect the cost significantly. New specialized tools and techniques decrease the installing cost tremendously. In the fiber optic section of Means Electrical Cost Data, a benchmark for labor-hours and material costs is set up so that users can adjust their costs according to unique project conditions.

Units for Measure: Fiber optic cable is measured in hundred linear feet (C.L.F.) or industry units of measure - meter (m) or kilometer (km). The connectors are counted as units (EA.)

Material Units: Generally, the material costs include only the cable. All the accessories shall be priced separately.

Labor Units: The following procedures are generally included for the installation of fiber optic cables:

- Receiving
- Material handling
- Setting up pulling equipment
- Measuring and cutting cable
- Pulling cable

These additional items are listed and extended: Terminations

Takeoff Procedure: Cable should be taken off by type, size, number of fibers, and number of terminations required. List the lengths of each type of cable on the takeoff sheets. Total and add 10% for waste. Transfer the figures to a cost analysis sheet and extend.

R271513-75 High Performance Cable

There are several categories used to describe high performance cable. The following information includes a description of categories CAT 3, 5, 5e, 6, and 7, and details classifications of frequency and specific standards. The category standards have evolved under the sponsorship of organizations such as the Telecommunication Industry Association (TIA), the Electronic Industries Alliance (EIA), the American National Standards Institute (ANSI), the International Organization for Standardization (ISO), and the International Electrotechnical Commission (IEC), all of which have catered to the increasing complexities of modern network technology. For network cabling, users must comply with national or international standards. A breakdown of these categories is as follows:

Category 3: Designed to handle frequencies up to 16 MHz.

Category 5: (TIA/EIA 568A) Designed to handle frequencies up to 100 MHz.

Category 5e: Additional transmission performance to exceed Category 5.

Category 6: Development by TIA and other international groups to handle frequencies of 250 MHz.

Category 7: Development to handle a frequency range up to 600 MHz.

R337116-60 Average Transmission Line Material Requirements (Per Mile)

Terrain:		Flat				Rolling				Mountain			
Item		**69kV**	**161kV**	**161kV**	**500kV**	**69kV**	**161kV**	**161kV**	**500kV**	**69kV**	**161kV**	**161kV**	**500kV**
Pole Type	**Unit**	**Wood**	**Wood**	**Steel**	**Steel**	**Wood**	**Wood**	**Steel**	**Steel**	**Wood**	**Wood**	**Steel**	**Steel**
Conductor: 397,500 – Cir. Mil., 26/7–ACSR													
Structures	Ea.	12[1]				9[2]				7[2]			
Poles	Ea.	12				18				14			
Crossarms	Ea.	24[3]				9[4]				7[4]			
Conductor	Ft.	15,990				15,990				15,990			
Insulators[5]	Ea.	180				135				105			
Ground Wire	Ft.	5,330				10,660				10,660			
Conductor: 636,000 – Cir. Mil., 26/7–ACSR													
Structures	Ea.	13[1]	11[6]			10[2]	9[2]	6[9]		8[2]	8[2]	6[9]	
Excavation	C.Y.	–	–			–	–	120		–	–	120	
Concrete	C.Y.	–	–			–	–	10		–	–	10	
Steel Towers	Tons	–	–			–	–	32		–	–	32	
Poles	Ea.	13	11			20	18	–		16	16	–	
Crossarms	Ea.	26[3]	33[7]			10[4]	9[8]	–		8[4]	8[8]	–	
Conductor	Ft.	15,990	15,990			15,990	15,990	15,990		15,990	15,990	15,990	
Insulators[5]	Ea.	195	165			150	297	297		120	264	297	
Ground Wire	Ft.	5330	5330			10,660	10,660	10,660		10,660	10,660	10,660	
Conductor: 954,000 – Cir. Mil., 45/7–ACSR													
Structures	Ea.	14[1]	12[6]		4[11]	10[2]	9[2]	6[9]	4[11]	8[2]	8[2]	6[9]	4[13]
Excavation	C.Y.	–	–		200	–	–	125	214	–	–	125	233
Concrete	C.Y.	–	–		20	–	–	10	21	–	–	10	21
Steel Towers	Tons	–	–		57	–	–	33	57	–	–	33	63
Poles	Ea.	14	12		–	20	18	–	–	16	16	–	–
Crossarms	Ea.	28[10]	36[7]		–	10[4]	9[8]	–	–	8[4]	8[8]	–	–
Conductor	Ft.	15,990	15,990		47,970[12]	15,990	15,990	15,990	47,970[12]	15,990	15,990	15,990	47,970[12]
Insulators[5]	Ea.	210	180		288	150	297	297	288	120	264	297	576
Ground Wire	Ft.	5330	5330		10,660	10,660	10,660	10,660	10,660	10,660	10,660	10,660	10,660
Conductor: 1,351,500 – Cir. Mil., 45/7–ACSR													
Structures	Ea.			8[14]				8[14]					
Excavation	C.Y.			220				220					
Concrete	C.Y.			28				28					
Steel Towers	Ton			46				46					
Conductor	Ft.			31,680[15]				31,680[15]					
Insulators	Ea.			528				528					
Ground Wire	Ft.			10,660				10,660					

1. Single pole two-arm suspension type construction
2. Two-pole wood H-frame construction
3. 4¾″ x 5¾″ x 8′ and 4¾″ x 5¾″ x 10′ wood crossarm
4. 6″ x 8″ x 26′-0″ wood crossarm
5. 5¾″ x 10″ disc insulator
6. Single pole construction with 3 fiberglass crossarms (5 fog- type insulators per phase)
7. 7′-0″ fiberglass crossarms
8. 6″ x 10″ x 35′-0″ wood crossarm
9. Laced steel tower, single circuit construction
10. 5″ x 7″ x 8′ and 5″ x 7″ x 10′ wood crossarm
11. Laced steel tower, single circuit 500-kV construction
12. Bundled conductor (3 sub-conductors per phase)
13. Laced steel tower, single circuit restrained phases (500-kV)
14. Laced steel tower, double circuit construction
15. Both sides of double circuit strung

Note: To allow for sagging, a mile (5280 Ft.) of transmission line uses 5330 Ft. of conductor per wire (called a wire mile).

R337119-30 Concrete for Conduit Encasement

Table below lists C.Y. of concrete for 100 L.F. of trench. Conduits separation center to center should meet 7.5″ (N.E.C.).

Number of Conduits	1	2	3	4	6	8	9	Number of Conduits
Trench Dimension	11.5″ x 11.5″	11.5″ x 19″	11.5″ x 27″	19″ x 19″	19″ x 27″	19″ x 38″	27″ x 27″	Trench Dimension
Conduit Diameter 2.0″	3.29	5.39	7.64	8.83	12.51	17.66	17.72	Conduit Diameter 2.0″
2.5″	3.23	5.29	7.49	8.62	12.19	17.23	17.25	2.5″
3.0″	3.15	5.13	7.24	8.29	11.71	16.59	16.52	3.0″
3.5″	3.08	4.97	7.02	7.99	11.26	15.98	15.84	3.5″
4.0″	2.99	4.80	6.76	7.65	10.74	15.30	15.07	4.0″
5.0″	2.78	4.37	6.11	6.78	9.44	13.57	13.12	5.0″
6.0″	2.52	3.84	5.33	5.74	7.87	11.48	10.77	6.0″
	●	● ●	● ● ●	● ● ● ●	● ● ● ● ● ●	● ● ● ● ● ● ● ●	● ● ● ● ● ● ● ● ●	

Change Orders

Change Order Considerations

A change order is a written document, usually prepared by the design professional, and signed by the owner, the architect/engineer, and the contractor. A change order states the agreement of the parties to: an addition, deletion, or revision in the work; an adjustment in the contract sum, if any; or an adjustment in the contract time, if any. Change orders, or "extras" in the construction process occur after execution of the construction contract and impact architects/ engineers, contractors, and owners.

Change orders that are properly recognized and managed can ensure orderly, professional, and profitable progress for all who are involved in the project. There are many causes for change orders and change order requests. In all cases, change orders or change order requests should be addressed promptly and in a precise and prescribed manner. The following paragraphs include information regarding change order pricing and procedures.

The Causes of Change Orders

Reasons for issuing change orders include:

- Unforeseen field conditions that require a change in the work
- Correction of design discrepancies, errors, or omissions in the contract documents
- Owner-requested changes, either by design criteria, scope of work, or project objectives
- Completion date changes for reasons unrelated to the construction process
- Changes in building code interpretations, or other public authority requirements that require a change in the work
- Changes in availability of existing or new materials and products

Procedures

Properly written contract documents must include the correct change order procedures for all parties—owners, design professionals and contractors—to follow in order to avoid costly delays and litigation.

Being "in the right" is not always a sufficient or acceptable defense. The contract provisions requiring notification and documentation must be adhered to within a defined or reasonable time frame.

The appropriate method of handling change orders is by a written proposal and acceptance by all parties involved. Prior to starting work on a project, all parties should identify their authorized agents who may sign and accept change orders, as well as any limits placed on their authority.

Time may be a critical factor when the need for a change arises. For such cases, the contractor might be directed to proceed on a "time and materials" basis, rather than wait for all paperwork to be processed—a delay that could impede progress. In this situation, the contractor must still follow the prescribed change order procedures including, but not limited to, notification and documentation.

Lack of documentation can be very costly, especially if legal judgments are to be made, and if certain field personnel are no longer available. For time and material change orders, the contractor should keep accurate daily records of all labor and material allocated to the change.

Owners or awarding authorities who do considerable and continual building construction (such as the federal government) realize the inevitability of change orders for numerous reasons, both predictable and unpredictable. As a result, the federal government, the American Institute of Architects (AIA), the Engineers Joint Contract Documents Committee (EJCDC) and other contractor, legal, and technical organizations have developed standards and procedures to be followed by all parties to achieve contract continuance and timely completion, while being financially fair to all concerned.

Pricing Change Orders

When pricing change orders, regardless of their cause, the most significant factor is when the change occurs. The need for a change may be perceived in the field or requested by the architect/engineer *before* any of the actual installation has begun, or may evolve or appear *during* construction when the item of work in question is partially installed. In the latter cases, the original sequence of construction is disrupted, along with all contiguous and supporting systems. Change orders cause the greatest impact when they occur *after* the installation has been completed and must be uncovered, or even replaced. Post-completion changes may be caused by necessary design changes, product failure, or changes in the owner's requirements that are not discovered until the building or the systems begin to function.

Specified procedures of notification and record keeping must be adhered to and enforced regardless of the stage of construction: *before, during,* or *after* installation. Some bidding documents anticipate change orders by requiring that unit prices including overhead and profit percentages—for additional as well as deductible changes—be listed. Generally these unit prices do not fully take into account the ripple effect, or impact on other trades, and should be used for general guidance only.

When pricing change orders, it is important to classify the time frame in which the change occurs. There are two basic time frames for change orders: *pre-installation change orders*, which occur before the start of construction, and *post-installation change orders*, which involve reworking after the original installation. Change orders that occur between these stages may be priced according to the extent of work completed using a combination of techniques developed for pricing *pre-* and *post-installation* changes.

Factors To Consider When Pricing Change Orders

As an estimator begins to prepare a change order, the following questions should be reviewed to determine their impact on the final price.

General

- *Is the change order work* pre-installation *or* post-installation?

  Change order work costs vary according to how much of the installation has been completed. Once workers have the project scoped in their minds, even though they have not started, it can be difficult to refocus. Consequently they may spend more than the normal amount of time understanding the change. Also, modifications to work in place, such as trimming or refitting, usually take more time than was initially estimated. The greater the amount of work in place, the more reluctant workers are to change it. Psychologically they may resent the change and as a result the rework takes longer than normal. Post-installation change order estimates must include demolition of existing work as required to accomplish the change. If the work is performed at a later time, additional obstacles, such as building finishes, may be present which must be protected. Regardless of whether the change occurs pre-installation or post-installation, attempt to isolate the identifiable factors and price them separately. For example, add shipping costs that may be required pre-installation or any demolition required post-installation. Then analyze the potential impact on productivity of psychological and/or learning curve factors and adjust the output rates accordingly. One approach is to break down the typical workday into segments and quantify the impact on each segment.

Change Order Installation Efficiency

The labor-hours expressed (for new construction) are based on average installation time, using an efficiency level. For change order situations, adjustments to this efficiency level should reflect the daily labor-hour allocation for that particular occurrence.

- *Will the change substantially delay the original completion date?*

  A significant change in the project may cause the original completion date to be extended. The extended schedule may subject the contractor to new wage rates dictated by relevant labor contracts. Project supervision and other project overhead must also be extended beyond the original completion date. The schedule extension may also put installation into a new weather season. For example, underground piping scheduled for October installation was delayed until January. As a result, frost penetrated the trench area, thereby changing the degree of difficulty of the task. Changes and delays may have a ripple effect throughout the project. This effect must be analyzed and negotiated with the owner.

- *What is the net effect of a deduct change order?*

  In most cases, change orders resulting in a deduction or credit reflect only bare costs. The contractor may retain the overhead and profit based on the original bid.

Materials

- *Will you have to pay more or less for the new material, required by the change order, than you paid for the original purchase?*

  The same material prices or discounts will usually apply to materials purchased for change orders as new construction. In some instances, however, the contractor may forfeit the advantages of competitive pricing for change orders. Consider the following example:

  A contractor purchased over $20,000 worth of fan coil units for an installation, and obtained the maximum discount. Some time later it was determined the project required an additional matching unit. The contractor has to purchase this unit from the original supplier to ensure a match. The supplier at this time may not discount the unit because of the small quantity, and the fact that he is no longer in a competitive situation. The impact of quantity on purchase can add between 0% and 25% to material prices and/or subcontractor quotes.

- *If materials have been ordered or delivered to the job site, will they be subject to a cancellation charge or restocking fee?*

  Check with the supplier to determine if ordered materials are subject to a cancellation charge. Delivered materials not used as result of a change order may be subject to a restocking fee if returned to the supplier. Common restocking charges run between 20% and 40%. Also, delivery charges to return the goods to the supplier must be added.

Labor

- *How efficient is the existing crew at the actual installation?*

  Is the same crew that performed the initial work going to do the change order? Possibly the change consists of the installation of a unit identical to one already installed; therefore, the change should take less time. Be sure to consider this potential productivity increase and modify the productivity rates accordingly.

- *If the crew size is increased, what impact will that have on supervision requirements?*

  Under most bargaining agreements or management practices, there is a point at which a working foreman is replaced by a nonworking foreman. This replacement increases project overhead by adding a nonproductive worker. If additional workers are added to accelerate the project or to perform changes while maintaining the schedule, be sure to add additional supervision time if warranted. Calculate the hours involved and the additional cost directly if possible.

- *What are the other impacts of increased crew size?*

  The larger the crew, the greater the potential for productivity to decrease. Some of the factors that cause this productivity loss are: overcrowding (producing restrictive conditions in the working space), and possibly a shortage of any special tools and equipment required. Such factors affect not only the crew working on the elements directly involved in the change order, but other crews whose movement may also be hampered. As the crew increases, check its basic composition for changes by the addition or deletion of apprentices or nonworking foreman, and quantify the potential effects of equipment shortages or other logistical factors.

- *As new crews, unfamiliar with the project, are brought onto the site, how long will it take them to become oriented to the project requirements?*

  The orientation time for a new crew to become 100% effective varies with the site and type of project. Orientation is easiest at a new construction site, and most difficult at existing, very restrictive renovation sites. The type of work also affects orientation time. When all elements of the work are exposed, such as concrete or masonry work, orientation is decreased. When the work is concealed or less visible, such as existing electrical systems, orientation takes longer. Usually orientation can be accomplished in one day or less. Costs for added orientation should be itemized and added to the total estimated cost.

- *How much actual production can be gained by working overtime?*

  Short term overtime can be used effectively to accomplish more work in a day. However, as overtime is scheduled to run beyond several weeks, studies have shown marked decreases in output. The following chart shows the effect of long term overtime on worker efficiency. If the anticipated change requires extended overtime to keep the job on schedule, these factors can be used as a guide to predict the impact on time and cost. Add project overhead, particularly supervision, that may also be incurred.

Days per Week	Hours per Day	Production Efficiency					Payroll Cost Factors	
		1 Week	2 Weeks	3 Weeks	4 Weeks	Average 4 Weeks	@ 1-1/2 Times	@ 2 Times
5	8	100%	100%	100%	100%	100%	100%	100%
	9	100	100	95	90	96.25	105.6	111.1
	10	100	95	90	85	91.25	110.0	120.0
	11	95	90	75	65	81.25	113.6	127.3
	12	90	85	70	60	76.25	116.7	133.3
6	8	100	100	95	90	96.25	108.3	116.7
	9	100	95	90	85	92.50	113.0	125.9
	10	95	90	85	80	87.50	116.7	133.3
	11	95	85	70	65	78.75	119.7	139.4
	12	90	80	65	60	73.75	122.2	144.4
7	8	100	95	85	75	88.75	114.3	128.6
	9	95	90	80	70	83.75	118.3	136.5
	10	90	85	75	65	78.75	121.4	142.9
	11	85	80	65	60	72.50	124.0	148.1
	12	85	75	60	55	68.75	126.2	152.4

Effects of Overtime

Caution: Under many labor agreements, Sundays and holidays are paid at a higher premium than the normal overtime rate.

The use of long-term overtime is counterproductive on almost any construction job; that is, the longer the period of overtime, the lower the actual production rate. Numerous studies have been conducted, and while they have resulted in slightly different numbers, all reach the same conclusion. The figure above tabulates the effects of overtime work on efficiency.

As illustrated, there can be a difference between the *actual* payroll cost per hour and the *effective* cost per hour for overtime work. This is due to the reduced production efficiency with the increase in weekly hours beyond 40. This difference between actual and effective cost results from overtime work over a prolonged period. Short-term overtime work does not result in as great a reduction in efficiency and, in such cases, effective cost may not vary significantly from the actual payroll cost. As the total hours per week are increased on a regular basis, more time is lost due to fatigue, lowered morale, and an increased accident rate.

As an example, assume a project where workers are working 6 days a week, 10 hours per day. From the figure above (based on productivity studies), the average effective productive hours over a 4-week period are:

$$0.875 \times 60 = 52.5$$

Depending upon the locale and day of week, overtime hours may be paid at time and a half or double time. For time and a half, the overall (average) *actual* payroll cost (including regular and overtime hours) is determined as follows:

$$\frac{40 \text{ reg. hrs.} + (20 \text{ overtime hrs.} \times 1.5)}{60 \text{ hrs.}} = 1.167$$

Based on 60 hours, the payroll cost per hour will be 116.7% of the normal rate at 40 hours per week. However, because the effective production (efficiency) for 60 hours is reduced to the equivalent of 52.5 hours, the effective cost of overtime is calculated as follows:

For time and a half:

$$\frac{40 \text{ reg. hrs.} + (20 \text{ overtime hrs.} \times 1.5)}{52.5 \text{ hrs.}} = 1.33$$

Installed cost will be 133% of the normal rate (for labor).

Thus, when figuring overtime, the actual cost per unit of work will be higher than the apparent overtime payroll dollar increase, due to the reduced productivity of the longer workweek. These efficiency calculations are true only for those cost factors determined by hours worked. Costs that are applied weekly or monthly, such as equipment rentals, will not be similarly affected.

Equipment

- *What equipment is required to complete the change order?*

  Change orders may require extending the rental period of equipment already on the job site, or the addition of special equipment brought in to accomplish the change work. In either case, the additional rental charges and operator labor charges must be added.

Summary

The preceding considerations and others you deem appropriate should be analyzed and applied to a change order estimate. The impact of each should be quantified and listed on the estimate to form an audit trail.

Change orders that are properly identified, documented, and managed help to ensure the orderly, professional and profitable progress of the work. They also minimize potential claims or disputes at the end of the project.

Estimating Tips

- The cost figures in this section were derived from approximately 11,000 projects contained in the RSMeans database of completed construction projects. They include the contractor's overhead and profit, but do not generally include architectural fees or land costs. The figures have been adjusted to January of the current year. New projects are added to our files each year, and outdated projects are discarded. For this reason, certain costs may not show a uniform annual progression. In no case are all subdivisions of a project listed.
- These projects were located throughout the U.S. and reflect a tremendous variation in square foot (S.F.) and cubic foot (C.F.) costs. This is due to differences, not only in labor and material costs, but also in individual owners' requirements. For instance, a bank in a large city would have different features than one in a rural area. This is true of all the different types of buildings analyzed. Therefore, caution should be exercised when using these square foot costs. For example, for courthouses, costs in the database are local courthouse costs and will not apply to the larger, more elaborate federal courthouses. As a general rule, the projects in the 1/4 column do not include any site work or equipment, while the projects in the 3/4 column may include both equipment and site work. The median figures do not generally include site work.
- None of the figures "go with" any others. All individual cost items were computed and tabulated separately. Thus, the sum of the median figures for plumbing, HVAC, and electrical will not normally total up to the total mechanical and electrical costs arrived at by separate analysis and tabulation of the projects.
- Each building was analyzed as to total and component costs and percentages. The figures were arranged in ascending order with the results tabulated as shown. The 1/4 column shows that 25% of the projects had lower costs and 75% had higher. The 3/4 column shows that 75% of the projects had lower costs and 25% had higher. The median column shows that 50% of the projects had lower costs and 50% had higher.
- There are two times when square foot costs are useful. The first is in the conceptual stage when no details are available. Then, square foot costs make a useful starting point. The second is after the bids are in and the costs can be worked back into their appropriate categories for information purposes. As soon as details become available in the project design, the square foot approach should be discontinued and the project priced as to its particular components. When more precision is required, or for estimating the replacement cost of specific buildings, the current edition of *RSMeans Square Foot Costs* should be used.
- In using the figures in this section, it is recommended that the median column be used for preliminary figures if no additional information is available. The median figures, when multiplied by the total city construction cost index figures (see City Cost Indexes) and then multiplied by the project size modifier at the end of this section, should present a fairly accurate base figure, which would then have to be adjusted in view of the estimator's experience, local economic conditions, code requirements, and the owner's particular requirements. There is no need to factor the percentage figures, as these should remain constant from city to city. All tabulations mentioning air conditioning had at least partial air conditioning.
- The editors of this book would greatly appreciate receiving cost figures on one or more of your recent projects, which would then be included in the averages for next year. All cost figures received will be kept confidential, except that they will be averaged with other similar projects to arrive at square foot cost figures for next year's book. See the last page of the book for details and the discount available for submitting one or more of your projects. ■

50 17 00 | S.F. Costs

		50 17 00 | S.F. Costs	UNIT	UNIT COSTS 1/4	UNIT COSTS MEDIAN	UNIT COSTS 3/4	% OF TOTAL 1/4	% OF TOTAL MEDIAN	% OF TOTAL 3/4	
01	0010	**APARTMENTS Low Rise (1 to 3 story)**	S.F.	74.50	94	125				01
	0020	Total project cost	C.F.	6.70	8.85	10.95				
	0100	Site work	S.F.	5.45	8.70	15.30	6.05%	10.55%	13.95%	
	0500	Masonry		1.47	3.62	5.95	1.54%	3.92%	6.50%	
	1500	Finishes		7.90	10.85	13.45	9.05%	10.75%	12.85%	
	1800	Equipment		2.44	3.70	5.50	2.71%	3.99%	5.95%	
	2720	Plumbing		5.80	7.45	9.50	6.65%	8.95%	10.05%	
	2770	Heating, ventilating, air conditioning		3.70	4.56	6.70	4.20%	5.60%	7.60%	
	2900	Electrical		4.33	5.75	7.80	5.20%	6.65%	8.35%	
	3100	Total: Mechanical & Electrical		15.40	19.95	24.50	16.05%	18.20%	23%	
	9000	Per apartment unit, total cost	Apt.	69,500	106,000	156,500				
	9500	Total: Mechanical & Electrical	"	13,100	20,700	27,000				
02	0010	**APARTMENTS Mid Rise (4 to 7 story)**	S.F.	99	119	147				02
	0020	Total project costs	C.F.	7.70	10.65	14.55				
	0100	Site work	S.F.	3.95	8	15.45	5.25%	6.70%	9.20%	
	0500	Masonry		4.13	9.05	12.40	5.10%	7.25%	10.50%	
	1500	Finishes		12.95	17.35	20.50	10.70%	13.50%	17.70%	
	1800	Equipment		2.77	4.30	5.65	2.54%	3.47%	4.31%	
	2500	Conveying equipment		2.24	2.76	3.33	2.05%	2.27%	2.69%	
	2720	Plumbing		5.80	9.30	9.85	5.70%	7.20%	8.95%	
	2900	Electrical		6.50	8.85	10.75	6.35%	7.20%	8.95%	
	3100	Total: Mechanical & Electrical		21	26	31.50	18.25%	21%	23%	
	9000	Per apartment unit, total cost	Apt.	112,000	132,000	218,500				
	9500	Total: Mechanical & Electrical	"	21,100	24,400	25,600				
03	0010	**APARTMENTS High Rise (8 to 24 story)**	S.F.	112	129	155				03
	0020	Total project costs	C.F.	10.90	12.65	14.95				
	0100	Site work	S.F.	4.07	6.60	9.20	2.58%	4.84%	6.15%	
	0500	Masonry		6.50	11.80	14.65	4.74%	9.65%	11.05%	
	1500	Finishes		12.45	15.55	18.35	9.75%	11.80%	13.70%	
	1800	Equipment		3.61	4.44	5.85	2.78%	3.49%	4.35%	
	2500	Conveying equipment		2.55	3.87	5.25	2.23%	2.78%	3.37%	
	2720	Plumbing		7.15	9.75	11.95	6.80%	7.20%	10.45%	
	2900	Electrical		7.70	9.75	13.15	6.45%	7.65%	8.80%	
	3100	Total: Mechanical & Electrical		23	29.50	35.50	17.95%	22.50%	24.50%	
	9000	Per apartment unit, total cost	Apt.	116,500	128,500	178,000				
	9500	Total: Mechanical & Electrical	"	23,200	28,800	30,500				
04	0010	**AUDITORIUMS**	S.F.	117	161	234				04
	0020	Total project costs	C.F.	7.30	10.15	14.55				
	2720	Plumbing	S.F.	6.90	10.20	12.20	5.85%	7.20%	8.70%	
	2900	Electrical		9.40	13.55	22.50	6.85%	9.50%	11.40%	
	3100	Total: Mechanical & Electrical		62.50	83	101	24.50%	27.50%	31%	
05	0010	**AUTOMOTIVE SALES**	S.F.	86.50	119	146				05
	0020	Total project costs	C.F.	5.70	6.80	8.85				
	2720	Plumbing	S.F.	3.93	6.85	7.45	2.89%	6.05%	6.50%	
	2770	Heating, ventilating, air conditioning		6.05	9.25	10	4.61%	10%	10.35%	
	2900	Electrical		6.95	10.90	16	7.25%	8.80%	12.15%	
	3100	Total: Mechanical & Electrical		19.35	31	37.50	17.30%	20.50%	22%	
06	0010	**BANKS**	S.F.	170	211	268				06
	0020	Total project costs	C.F.	12.10	16.45	21.50				
	0100	Site work	S.F.	19.40	30.50	43	7.90%	12.95%	17%	
	0500	Masonry		7.70	16.50	29.50	3.36%	6.90%	10.05%	
	1500	Finishes		15.15	23	28.50	5.85%	8.65%	11.70%	
	1800	Equipment		6.35	14	28.50	1.34%	5.55%	10.50%	
	2720	Plumbing		5.30	7.55	11.05	2.82%	3.90%	4.93%	
	2770	Heating, ventilating, air conditioning		10.05	13.45	17.90	4.86%	7.15%	8.50%	
	2900	Electrical		16.05	21.50	28	8.25%	10.20%	12.20%	
	3100	Total: Mechanical & Electrical		38.50	51.50	62	16.55%	19.45%	24%	
	3500	See also division 11 22 00								

50 17 00 | S.F. Costs

		50 17 00 | S.F. Costs	UNIT	UNIT COSTS			% OF TOTAL			
				1/4	MEDIAN	3/4	1/4	MEDIAN	3/4	
13	0010	**CHURCHES**	S.F.	114	145	190				13
	0020	Total project costs	C.F.	7.05	8.90	11.75				
	1800	Equipment	S.F.	1.22	3.19	6.80	.83%	2.04%	4.30%	
	2720	Plumbing		4.43	6.20	9.15	3.51%	4.96%	6.25%	
	2770	Heating, ventilating, air conditioning		10.35	13.50	19.15	7.50%	10%	12%	
	2900	Electrical		9.60	13.20	18	7.30%	8.80%	10.95%	
	3100	Total: Mechanical & Electrical		29.50	39.50	53	18.30%	22%	25%	
	3500	See also division 11 91 00								
15	0010	**CLUBS, COUNTRY**	S.F.	122	147	185				15
	0020	Total project costs	C.F.	9.85	12	16.55				
	2720	Plumbing	S.F.	7.40	10.95	25	5.60%	7.90%	10%	
	2900	Electrical		9.60	13.15	17.15	7%	8.95%	11%	
	3100	Total: Mechanical & Electrical		51	64	67.50	19%	26.50%	29.50%	
17	0010	**CLUBS, SOCIAL Fraternal**	S.F.	103	141	188				17
	0020	Total project costs	C.F.	6.10	9.25	11				
	2720	Plumbing	S.F.	6.15	7.65	11.55	5.60%	6.90%	8.55%	
	2770	Heating, ventilating, air conditioning		7.15	10.70	13.75	8.20%	9.25%	14.40%	
	2900	Electrical		7.15	12.05	13.75	5.95%	9.30%	10.55%	
	3100	Total: Mechanical & Electrical		38	41	52	21%	23%	23.50%	
18	0010	**CLUBS, Y.M.C.A.**	S.F.	134	167	252				18
	0020	Total project costs	C.F.	5.65	9.45	14.05				
	2720	Plumbing	S.F.	7.75	15.40	17.25	5.65%	7.60%	10.85%	
	2900	Electrical		10.05	13	22.50	6.45%	7.95%	8.90%	
	3100	Total: Mechanical & Electrical		41	46.50	76	20.50%	21.50%	28.50%	
19	0010	**COLLEGES Classrooms & Administration**	S.F.	118	167	216				19
	0020	Total project costs	C.F.	7.10	12.50	20				
	0500	Masonry	S.F.	9.30	17.15	21	5.10%	8.05%	10.50%	
	2720	Plumbing		5.30	12.90	24	5.10%	6.60%	8.95%	
	2900	Electrical		10.05	15.85	22	7.70%	9.85%	12%	
	3100	Total: Mechanical & Electrical		42.50	58.50	77.50	23%	28%	31.50%	
21	0010	**COLLEGES Science, Engineering, Laboratories**	S.F.	230	269	310				21
	0020	Total project costs	C.F.	13.20	19.25	22				
	1800	Equipment	S.F.	6.25	29	31.50	2%	6.45%	12.65%	
	2900	Electrical		18.95	27	41.50	7.10%	9.40%	12.10%	
	3100	Total: Mechanical & Electrical		70.50	83.50	129	28.50%	31.50%	41%	
	3500	See also division 11 53 00								
23	0010	**COLLEGES Student Unions**	S.F.	147	199	241				23
	0020	Total project costs	C.F.	8.20	10.70	13.25				
	3100	Total: Mechanical & Electrical	S.F.	55	59.50	70.50	23.50%	26%	29%	
25	0010	**COMMUNITY CENTERS**	S.F.	122	150	203				25
	0020	Total project costs	C.F.	7.90	11.30	14.65				
	1800	Equipment	S.F.	2.45	4.80	7.90	1.47%	3.01%	5.30%	
	2720	Plumbing		5.75	10.05	13.70	4.85%	7%	8.95%	
	2770	Heating, ventilating, air conditioning		8.65	13.35	19.15	6.80%	10.35%	12.90%	
	2900	Electrical		10.25	13.55	19.90	7.15%	8.90%	10.40%	
	3100	Total: Mechanical & Electrical		34	43	61.50	18.80%	23%	30%	
28	0010	**COURT HOUSES**	S.F.	175	206	284				28
	0020	Total project costs	C.F.	13.40	16	20				
	2720	Plumbing	S.F.	8.30	11.60	13.15	5.95%	7.45%	8.20%	
	2900	Electrical		18.55	21	30.50	9.05%	10.65%	12.15%	
	3100	Total: Mechanical & Electrical		52	68	74.50	22.50%	26.50%	30%	
30	0010	**DEPARTMENT STORES**	S.F.	64.50	87.50	110				30
	0020	Total project costs	C.F.	3.46	4.48	6.10				
	2720	Plumbing	S.F.	2.01	2.54	3.85	1.82%	4.21%	5.90%	
	2770	Heating, ventilating, air conditioning		4.88	9.05	13.65	8.20%	9.10%	14.80%	

	50 17 00 | S.F. Costs		UNIT	UNIT COSTS 1/4	UNIT COSTS MEDIAN	UNIT COSTS 3/4	% OF TOTAL 1/4	% OF TOTAL MEDIAN	% OF TOTAL 3/4	
30	2900	Electrical	S.F.	7.40	10.15	12	9.05%	12.15%	14.95%	30
	3100	Total: Mechanical & Electrical		13.05	16.65	29	13.20%	21.50%	50%	
31	0010	**DORMITORIES Low Rise (1 to 3 story)**	S.F.	122	170	212				31
	0020	Total project costs	C.F.	6.90	11.20	16.75				
	2720	Plumbing	S.F.	7.35	9.85	12.45	8.05%	9%	9.65%	
	2770	Heating, ventilating, air conditioning		7.80	9.35	12.45	4.61%	8.05%	10%	
	2900	Electrical		8.10	12.35	16.90	6.40%	8.65%	9.50%	
	3100	Total: Mechanical & Electrical		42	45	70.50	22%	25%	27%	
	9000	Per bed, total cost	Bed	52,000	57,500	123,500				
32	0010	**DORMITORIES Mid Rise (4 to 8 story)**	S.F.	150	196	242				32
	0020	Total project costs	C.F.	16.55	18.20	22				
	2900	Electrical	S.F.	15.95	18.15	24.50	8.20%	10.20%	11.95%	
	3100	Total: Mechanical & Electrical	"	44.50	89	90.50	25%	30.50%	35.50%	
	9000	Per bed, total cost	Bed	21,400	48,800	282,500				
34	0010	**FACTORIES**	S.F.	57	85	131				34
	0020	Total project costs	C.F.	3.65	5.45	9.05				
	0100	Site work	S.F.	6.50	11.85	18.75	6.95%	11.45%	17.95%	
	2720	Plumbing		3.07	5.70	9.45	3.73%	6.05%	8.10%	
	2770	Heating, ventilating, air conditioning		5.95	8.55	11.55	5.25%	8.45%	11.35%	
	2900	Electrical		7.05	11.20	17.05	8.10%	10.50%	14.20%	
	3100	Total: Mechanical & Electrical		20	32.50	41	21%	28.50%	35.50%	
36	0010	**FIRE STATIONS**	S.F.	113	156	213				36
	0020	Total project costs	C.F.	6.60	9.05	12.05				
	0500	Masonry	S.F.	15.40	30	38	7.50%	10.95%	15.55%	
	1140	Roofing		3.68	9.95	11.35	1.90%	4.94%	5.05%	
	1580	Painting		2.86	4.27	4.37	1.37%	1.57%	2.07%	
	1800	Equipment		1.40	2.69	4.97	.62%	1.63%	3.42%	
	2720	Plumbing		6.30	10.05	14.25	5.85%	7.35%	9.45%	
	2770	Heating, ventilating, air conditioning		6.25	10.15	15.65	5.15%	7.40%	9.40%	
	2900	Electrical		8.15	14.35	19.35	6.90%	8.60%	10.60%	
	3100	Total: Mechanical & Electrical		41.50	53	60	18.40%	23%	27%	
37	0010	**FRATERNITY HOUSES & Sorority Houses**	S.F.	113	145	199				37
	0020	Total project costs	C.F.	11.25	11.70	14.10				
	2720	Plumbing	S.F.	8.55	9.75	17.90	6.80%	8%	10.85%	
	2900	Electrical		7.45	16.10	19.70	6.60%	9.90%	10.65%	
38	0010	**FUNERAL HOMES**	S.F.	119	162	295				38
	0020	Total project costs	C.F.	12.15	13.55	26				
	2900	Electrical	S.F.	5.25	9.65	10.55	3.58%	4.44%	5.95%	
39	0010	**GARAGES, COMMERCIAL (Service)**	S.F.	67.50	104	144				39
	0020	Total project costs	C.F.	4.43	6.55	9.50				
	1800	Equipment	S.F.	3.80	8.55	13.30	2.21%	4.62%	6.80%	
	2720	Plumbing		4.67	7.20	13.10	5.45%	7.85%	10.65%	
	2730	Heating & ventilating		4.60	7.95	10.15	5.25%	6.85%	8.20%	
	2900	Electrical		6.40	9.75	14.10	7.15%	9.25%	10.85%	
	3100	Total: Mechanical & Electrical		12.10	27	40	12.35%	17.40%	26%	
40	0010	**GARAGES, MUNICIPAL (Repair)**	S.F.	105	135	192				40
	0020	Total project costs	C.F.	6.20	7.85	13.50				
	0500	Masonry	S.F.	5.20	18.20	28	4.03%	9.15%	12.50%	
	2720	Plumbing		4.44	8.55	16.05	3.59%	6.70%	7.95%	
	2730	Heating & ventilating		7.60	11	21	6.15%	7.45%	13.50%	
	2900	Electrical		7.65	12.35	22.50	6.90%	9.40%	13%	
	3100	Total: Mechanical & Electrical		34.50	56.50	68	21.50%	25.50%	35.50%	
41	0010	**GARAGES, PARKING**	S.F.	39.50	56	96.50				41
	0020	Total project costs	C.F.	3.62	4.92	7.15				

50 17 | Square Foot Costs

	50 17 00 | S.F. Costs		UNIT	UNIT COSTS 1/4	UNIT COSTS MEDIAN	UNIT COSTS 3/4	% OF TOTAL 1/4	% OF TOTAL MEDIAN	% OF TOTAL 3/4	
41	2720	Plumbing	S.F.	.66	1.69	2.61	1.72%	2.70%	3.85%	41
	2900	Electrical		2.12	2.80	4.08	4.52%	5.40%	6.35%	
	3100	Total: Mechanical & Electrical		4.11	6.05	7.50	7%	8.90%	11.05%	
	3200									
	9000	Per car, total cost	Car	16,300	20,400	26,000				
43	0010	**GYMNASIUMS**	S.F.	108	143	196				43
	0020	Total project costs	C.F.	5.35	7.25	8.90				
	1800	Equipment	S.F.	2.55	4.78	8.60	1.76%	3.26%	6.70%	
	2720	Plumbing		5.90	7.90	10.40	4.65%	6.40%	7.75%	
	2770	Heating, ventilating, air conditioning		6.40	11.10	22.50	5.15%	9.05%	11.10%	
	2900	Electrical		8.50	11.25	14.95	6.75%	8.50%	10.30%	
	3100	Total: Mechanical & Electrical		29	40.50	49.50	19.75%	23.50%	29%	
	3500	See also division 11 66 00								
46	0010	**HOSPITALS**	S.F.	206	258	355				46
	0020	Total project costs	C.F.	15.60	19.40	28				
	1800	Equipment	S.F.	5.20	10.05	17.30	.80%	2.53%	4.80%	
	2720	Plumbing		17.70	25	32	7.60%	9.10%	10.85%	
	2770	Heating, ventilating, air conditioning		26	33.50	46.50	7.80%	12.95%	16.65%	
	2900	Electrical		22.50	30.50	45.50	10%	11.75%	14.10%	
	3100	Total: Mechanical & Electrical		66.50	92.50	137	28%	33.50%	37%	
	9000	Per bed or person, total cost	Bed	238,000	328,500	378,500				
	9900	See also division 11 71 00								
48	0010	**HOUSING For the Elderly**	S.F.	101	128	157				48
	0020	Total project costs	C.F.	7.20	10	12.80				
	0100	Site work	S.F.	7.05	10.95	16.05	5.05%	7.90%	12.10%	
	0500	Masonry		1.91	11.55	16.85	1.30%	6.05%	11%	
	1800	Equipment		2.45	3.37	5.35	1.88%	3.23%	4.43%	
	2510	Conveying systems		2.29	3.31	4.49	1.78%	2.20%	2.81%	
	2720	Plumbing		7.50	9.60	12.10	8.15%	9.55%	10.50%	
	2730	Heating, ventilating, air conditioning		3.86	5.45	8.15	3.30%	5.60%	7.25%	
	2900	Electrical		7.55	10.25	13.10	7.30%	8.50%	10.25%	
	3100	Total: Mechanical & Electrical		26	32	41	18.10%	22.50%	29%	
	9000	Per rental unit, total cost	Unit	94,000	110,000	122,500				
	9500	Total: Mechanical & Electrical	"	21,000	24,100	28,100				
50	0010	**HOUSING Public (Low Rise)**	S.F.	85	118	154				50
	0020	Total project costs	C.F.	6.75	9.45	11.75				
	0100	Site work	S.F.	10.85	15.60	25.50	8.35%	11.75%	16.50%	
	1800	Equipment		2.31	3.77	5.75	2.26%	3.03%	4.24%	
	2720	Plumbing		6.15	8.10	10.25	7.15%	9.05%	11.60%	
	2730	Heating, ventilating, air conditioning		3.08	6	6.55	4.26%	6.05%	6.45%	
	2900	Electrical		5.15	7.65	10.65	5.10%	6.55%	8.25%	
	3100	Total: Mechanical & Electrical		24.50	31.50	35	14.50%	17.55%	26.50%	
	9000	Per apartment, total cost	Apt.	93,500	106,500	133,500				
	9500	Total: Mechanical & Electrical	"	19,900	24,600	27,200				
51	0010	**ICE SKATING RINKS**	S.F.	72.50	170	187				51
	0020	Total project costs	C.F.	5.35	5.45	6.30				
	2720	Plumbing	S.F.	2.72	5.10	5.20	3.12%	3.23%	5.65%	
	2900	Electrical		7.80	11.95	12.65	6.30%	10.15%	15.05%	
52	0010	**JAILS**	S.F.	189	286	370				52
	0020	Total project costs	C.F.	19.95	28	32.50				
	1800	Equipment	S.F.	8.65	25.50	43.50	2.80%	6.95%	9.80%	
	2720	Plumbing		21.50	28.50	37.50	7%	8.90%	13.35%	
	2770	Heating, ventilating, air conditioning		19.95	26.50	51.50	7.50%	9.45%	17.75%	
	2900	Electrical		23.50	31.50	40	9.40%	11.70%	15.25%	
	3100	Total: Mechanical & Electrical		62	110	131	28%	31%	36%	
53	0010	**LIBRARIES**	S.F.	143	190	248				53
	0020	Total project costs	C.F.	9.55	12	15.30				

50 17 00 | S.F. Costs

		50 17 00 | S.F. Costs	UNIT	UNIT COSTS 1/4	UNIT COSTS MEDIAN	UNIT COSTS 3/4	% OF TOTAL 1/4	% OF TOTAL MEDIAN	% OF TOTAL 3/4	
53	0500	Masonry	S.F.	9.50	19.35	32.50	5.60%	7.15%	10.95%	53
	1800	Equipment		1.91	5.15	7.75	.28%	1.39%	4.07%	
	2720	Plumbing		5.15	7.30	10.20	3.38%	4.60%	5.70%	
	2770	Heating, ventilating, air conditioning		11.45	19.35	23.50	7.80%	10.95%	12.80%	
	2900	Electrical		14.40	19	25.50	8.40%	10.55%	12%	
	3100	Total: Mechanical & Electrical	↓	47	57	67	21%	24%	28%	
54	0010	**LIVING, ASSISTED**	S.F.	129	154	180				54
	0020	Total project costs	C.F.	10.90	12.70	14.45				
	0500	Masonry	S.F.	3.75	4.53	5.55	2.36%	3.16%	3.86%	
	1800	Equipment		2.87	3.42	4.38	2.12%	2.45%	2.87%	
	2720	Plumbing		10.85	14.50	15.05	6.05%	8.15%	10.60%	
	2770	Heating, ventilating, air conditioning		12.85	13.45	14.75	7.95%	9.35%	9.70%	
	2900	Electrical		12.70	14.25	16.50	9%	9.95%	10.65%	
	3100	Total: Mechanical & Electrical	↓	35	41	48	24%	28.50%	31.50%	
55	0010	**MEDICAL CLINICS**	S.F.	132	162	207				55
	0020	Total project costs	C.F.	9.65	12.50	16.60				
	1800	Equipment	S.F.	1.68	7.50	11.65	1.05%	2.94%	6.35%	
	2720	Plumbing		8.75	12.30	16.50	6.15%	8.40%	10.10%	
	2770	Heating, ventilating, air conditioning		10.45	13.70	20	6.65%	8.85%	11.35%	
	2900	Electrical		11.30	16.10	21	8.10%	10%	12.20%	
	3100	Total: Mechanical & Electrical	↓	36	49	67.50	22%	27%	33.50%	
	3500	See also division 11 71 00								
57	0010	**MEDICAL OFFICES**	S.F.	124	154	190				57
	0020	Total project costs	C.F.	9.25	12.50	16.95				
	1800	Equipment	S.F.	4.09	8.10	11.35	.66%	4.87%	6.50%	
	2720	Plumbing		6.85	10.55	14.25	5.60%	6.80%	8.50%	
	2770	Heating, ventilating, air conditioning		8.25	11.95	15.75	6.10%	8%	9.70%	
	2900	Electrical		9.90	14.50	20.50	7.50%	9.80%	11.70%	
	3100	Total: Mechanical & Electrical	↓	27	39.50	56	19.30%	22.50%	27.50%	
59	0010	**MOTELS**	S.F.	78	113	148				59
	0020	Total project costs	C.F.	6.95	9.30	15.25				
	2720	Plumbing	S.F.	7.90	10.10	12.05	9.45%	10.60%	12.55%	
	2770	Heating, ventilating, air conditioning		4.83	7.20	12.90	5.60%	5.60%	10%	
	2900	Electrical		7.40	9.35	11.65	7.45%	9.05%	10.45%	
	3100	Total: Mechanical & Electrical	↓	22.50	31.50	54	18.50%	24%	25.50%	
	5000									
	9000	Per rental unit, total cost	Unit	39,800	75,500	82,000				
	9500	Total: Mechanical & Electrical	"	7,750	11,700	13,600				
60	0010	**NURSING HOMES**	S.F.	123	158	197				60
	0020	Total project costs	C.F.	9.65	12.05	16.45				
	1800	Equipment	S.F.	3.34	5.10	8.55	2%	3.62%	4.99%	
	2720	Plumbing		10.50	15.90	19.20	8.75%	10.10%	12.70%	
	2770	Heating, ventilating, air conditioning		11.05	16.80	22.50	9.70%	11.45%	11.80%	
	2900	Electrical		12.15	15.20	21	9.40%	10.60%	12.50%	
	3100	Total: Mechanical & Electrical	↓	29	40.50	68	26%	28%	30%	
	9000	Per bed or person, total cost	Bed	54,500	68,000	88,000				
61	0010	**OFFICES Low Rise (1 to 4 story)**	S.F.	103	135	175				61
	0020	Total project costs	C.F.	7.35	10.15	13.40				
	0100	Site work	S.F.	8.30	14.45	21.50	5.95%	9.70%	13.55%	
	0500	Masonry		4	8.05	14.70	2.61%	5.40%	8.45%	
	1800	Equipment		.93	2.15	5.85	.57%	1.50%	3.42%	
	2720	Plumbing		3.69	5.70	8.35	3.66%	4.50%	6.10%	
	2770	Heating, ventilating, air conditioning		8.15	11.40	16.65	7.20%	10.30%	11.70%	
	2900	Electrical		8.45	12.10	17.20	7.45%	9.65%	11.40%	
	3100	Total: Mechanical & Electrical	↓	23.50	32.50	48.50	18.20%	22%	27%	
62	0010	**OFFICES Mid Rise (5 to 10 story)**	S.F.	109	132	180				62
	0020	Total project costs	C.F.	7.75	9.90	14				

	50 17 00 | S.F. Costs		UNIT	UNIT COSTS 1/4	UNIT COSTS MEDIAN	UNIT COSTS 3/4	% OF TOTAL 1/4	% OF TOTAL MEDIAN	% OF TOTAL 3/4	
62	2720	Plumbing	S.F.	3.30	5.10	7.35	2.83%	3.74%	4.50%	62
	2770	Heating, ventilating, air conditioning	↓	8.30	11.85	18.95	7.65%	9.40%	11%	
	2900	Electrical		8.10	10.40	15.70	6.35%	7.80%	10%	
	3100	Total: Mechanical & Electrical	↓	21	27	51.50	18.95%	21%	27.50%	
63	0010	**OFFICES High Rise (11 to 20 story)**	S.F.	134	169	208				63
	0020	Total project costs	C.F.	9.40	11.75	16.85				
	2900	Electrical	S.F.	8.15	9.95	14.80	5.80%	7.85%	10.50%	
	3100	Total: Mechanical & Electrical	↓	26.50	35.50	59.50	16.90%	23.50%	34%	
64	0010	**POLICE STATIONS**	S.F.	161	212	270				64
	0020	Total project costs	C.F.	12.85	15.70	21.50				
	0500	Masonry	S.F.	16.10	26.50	33.50	7.80%	9.10%	11.35%	
	1800	Equipment		2.31	11.25	17.80	.98%	3.35%	6.70%	
	2720	Plumbing		9	17.95	22.50	5.65%	6.90%	10.75%	
	2770	Heating, ventilating, air conditioning		13.20	18.70	26.50	5.85%	10.55%	11.70%	
	2900	Electrical		17.60	25.50	33	9.80%	11.70%	14.50%	
	3100	Total: Mechanical & Electrical	↓	67	70.50	94	28.50%	31.50%	32.50%	
65	0010	**POST OFFICES**	S.F.	127	157	200				65
	0020	Total project costs	C.F.	7.65	9.70	11				
	2720	Plumbing	S.F.	5.15	7.10	8.95	4.24%	5.30%	5.60%	
	2770	Heating, ventilating, air conditioning		8	11.05	12.30	6.65%	7.15%	9.35%	
	2900	Electrical		10.50	14.75	17.50	7.25%	9%	11%	
	3100	Total: Mechanical & Electrical	↓	30.50	39.50	45	16.25%	18.80%	22%	
66	0010	**POWER PLANTS**	S.F.	880	1,175	2,150				66
	0020	Total project costs	C.F.	24.50	53	113				
	2900	Electrical	S.F.	62.50	132	197	9.30%	12.75%	21.50%	
	8100	Total: Mechanical & Electrical	↓	155	505	1,125	32.50%	32.50%	52.50%	
67	0010	**RELIGIOUS EDUCATION**	S.F.	103	137	170				67
	0020	Total project costs	C.F.	5.70	8.15	10.20				
	2720	Plumbing	S.F.	4.28	6.05	8.60	4.40%	5.30%	7.10%	
	2770	Heating, ventilating, air conditioning		10.80	12.25	17.30	10.05%	11.45%	12.35%	
	2900	Electrical		8.15	11.65	17.10	7.70%	9.05%	10.35%	
	3100	Total: Mechanical & Electrical	↓	35.50	45	53	22%	23.50%	26%	
69	0010	**RESEARCH Laboratories & Facilities**	S.F.	157	222	325				69
	0020	Total project costs	C.F.	11.40	22.50	27				
	1800	Equipment	S.F.	6.80	13.40	32.50	1.22%	4.80%	9.35%	
	2720	Plumbing		11.20	19.70	31.50	6.15%	8.30%	10.80%	
	2770	Heating, ventilating, air conditioning		13.80	46.50	55	7.25%	16.50%	17.50%	
	2900	Electrical		18.45	29.50	48.50	9.45%	11.15%	14.60%	
	3100	Total: Mechanical & Electrical	↓	59	104	147	29.50%	36%	41%	
70	0010	**RESTAURANTS**	S.F.	149	192	249				70
	0020	Total project costs	C.F.	12.50	16.40	21.50				
	1800	Equipment	S.F.	8.80	23.50	35.50	6.10%	13%	15.65%	
	2720	Plumbing		11.80	14.30	18.75	6.10%	8.15%	9%	
	2770	Heating, ventilating, air conditioning		14.95	21	25	9.20%	12%	12.40%	
	2900	Electrical		15.70	19.40	25	8.35%	10.55%	11.55%	
	3100	Total: Mechanical & Electrical	↓	48.50	52	67.50	21%	25%	29.50%	
	9000	Per seat unit, total cost	Seat	5,450	7,275	8,600				
	9500	Total: Mechanical & Electrical	"	1,375	1,825	2,150				
72	0010	**RETAIL STORES**	S.F.	69.50	93.50	124				72
	0020	Total project costs	C.F.	4.71	6.70	9.35				
	2720	Plumbing	S.F.	2.52	4.20	7.15	3.26%	4.60%	6.80%	
	2770	Heating, ventilating, air conditioning		5.45	7.45	11.20	6.75%	8.75%	10.15%	
	2900	Electrical		6.25	8.55	12.35	7.25%	9.90%	11.60%	
	3100	Total: Mechanical & Electrical	↓	16.65	21.50	28.50	17.05%	21%	23.50%	
74	0010	**SCHOOLS Elementary**	S.F.	114	141	172				74
	0020	Total project costs	C.F.	7.40	9.50	12.25				
	0500	Masonry	S.F.	8	17.30	25.50	4.89%	10.50%	14%	
	1800	Equipment	↓	2.63	4.96	9.40	1.83%	3.13%	4.61%	

	50 17 00 | S.F. Costs		UNIT	UNIT COSTS 1/4	UNIT COSTS MEDIAN	UNIT COSTS 3/4	% OF TOTAL 1/4	% OF TOTAL MEDIAN	% OF TOTAL 3/4	
74	2720	Plumbing	S.F.	6.50	9.20	12.30	5.70%	7.15%	9.35%	74
	2730	Heating, ventilating, air conditioning		9.80	15.55	22.50	8.15%	10.80%	14.90%	
	2900	Electrical		10.70	14.25	18.20	8.45%	10.05%	11.85%	
	3100	Total: Mechanical & Electrical		39	47.50	60	25%	27.50%	30%	
	9000	Per pupil, total cost	Ea.	13,000	19,300	43,300				
	9500	Total: Mechanical & Electrical	"	3,675	4,650	11,700				
76	0010	**SCHOOLS Junior High & Middle**	S.F.	117	145	177				76
	0020	Total project costs	C.F.	7.40	9.60	10.75				
	0500	Masonry	S.F.	12.55	18.80	23	7.55%	11.10%	14.30%	
	1800	Equipment		3.20	6	8.80	1.80%	3.03%	4.80%	
	2720	Plumbing		6.80	8.40	10.40	5.30%	6.80%	7.25%	
	2770	Heating, ventilating, air conditioning		13.60	16.55	29	8.90%	11.55%	14.20%	
	2900	Electrical		11.90	14.70	18.60	8.05%	9.55%	10.60%	
	3100	Total: Mechanical & Electrical		37.50	48	59.50	23%	25.50%	29.50%	
	9000	Per pupil, total cost	Ea.	14,800	19,400	26,100				
78	0010	**SCHOOLS Senior High**	S.F.	123	150	188				78
	0020	Total project costs	C.F.	7.35	10.30	17.40				
	1800	Equipment	S.F.	3.24	7.55	10.45	1.88%	2.67%	4.30%	
	2720	Plumbing		6.80	10.25	18.70	5.60%	6.90%	8.30%	
	2770	Heating, ventilating, air conditioning		11.75	15.95	24.50	8.95%	11.60%	15%	
	2900	Electrical		12.30	16.40	24	8.70%	10.35%	12.50%	
	3100	Total: Mechanical & Electrical		40.50	48	77.50	24%	26.50%	28.50%	
	9000	Per pupil, total cost	Ea.	11,500	23,300	29,200				
80	0010	**SCHOOLS Vocational**	S.F.	98.50	143	177				80
	0020	Total project costs	C.F.	6.10	8.80	12.15				
	0500	Masonry	S.F.	4.33	14.25	22	3.20%	4.61%	10.95%	
	1800	Equipment		3.08	7.65	10.60	1.24%	3.10%	4.26%	
	2720	Plumbing		6.30	9.35	13.80	5.40%	6.90%	8.55%	
	2770	Heating, ventilating, air conditioning		8.80	16.40	27.50	8.60%	11.90%	14.65%	
	2900	Electrical		10.25	14	19.15	8.45%	11%	13.20%	
	3100	Total: Mechanical & Electrical		38.50	67.50	86.50	27.50%	29.50%	31%	
	9000	Per pupil, total cost	Ea.	13,700	36,700	54,500				
83	0010	**SPORTS ARENAS**	S.F.	86	115	177				83
	0020	Total project costs	C.F.	4.67	8.35	10.80				
	2720	Plumbing	S.F.	4.31	7.55	15.95	4.35%	6.35%	9.40%	
	2770	Heating, ventilating, air conditioning		10.75	12.70	17.60	8.80%	10.20%	13.55%	
	2900	Electrical		7.95	12.15	15.70	7.65%	9.75%	11.90%	
85	0010	**SUPERMARKETS**	S.F.	79.50	92	108				85
	0020	Total project costs	C.F.	4.42	5.35	8.10				
	2720	Plumbing	S.F.	4.44	5.60	6.50	5.40%	6%	7.45%	
	2770	Heating, ventilating, air conditioning		6.55	8.65	10.55	8.60%	8.65%	9.60%	
	2900	Electrical		9.30	11.35	13.50	10.40%	12.45%	13.60%	
	3100	Total: Mechanical & Electrical		25.50	27	36.50	23.50%	27.50%	28.50%	
86	0010	**SWIMMING POOLS**	S.F.	129	300	460				86
	0020	Total project costs	C.F.	10.30	12.85	14				
	2720	Plumbing	S.F.	11.90	13.60	18.25	4.80%	9.70%	20.50%	
	2900	Electrical		9.75	15.65	34.50	6.05%	6.95%	7.60%	
	3100	Total: Mechanical & Electrical		61	80.50	106	11.15%	14.10%	23.50%	
87	0010	**TELEPHONE EXCHANGES**	S.F.	160	243	315				87
	0020	Total project costs	C.F.	10.65	17.05	23.50				
	2720	Plumbing	S.F.	7.20	11.15	16.30	4.52%	5.80%	6.90%	
	2770	Heating, ventilating, air conditioning		16.75	33.50	42	11.80%	16.05%	18.40%	
	2900	Electrical		17.40	27.50	49	10.90%	14%	17.85%	
	3100	Total: Mechanical & Electrical		51.50	97.50	138	29.50%	33.50%	44.50%	
91	0010	**THEATERS**	S.F.	102	138	203				91
	0020	Total project costs	C.F.	4.96	7.35	10.80				

	50 17 00 | S.F. Costs		UNIT	UNIT COSTS 1/4	MEDIAN	3/4	% OF TOTAL 1/4	MEDIAN	3/4	
91	2720	Plumbing	S.F.	3.34	3.88	15.85	2.92%	4.70%	6.80%	91
	2770	Heating, ventilating, air conditioning		10.45	12.65	15.65	8%	12.25%	13.40%	
	2900	Electrical		9.40	12.70	26	8.05%	9.35%	12.25%	
	3100	Total: Mechanical & Electrical	▼	24	32.50	38.50	21.50%	25.50%	27.50%	
94	0010	**TOWN HALLS City Halls & Municipal Buildings**	S.F.	112	152	199				94
	0020	Total project costs	C.F.	8.10	12.60	18.45				
	2720	Plumbing	S.F.	5	9.35	16.45	4.31%	5.95%	7.95%	
	2770	Heating, ventilating, air conditioning		9.05	17.95	26.50	7.05%	9.05%	13.45%	
	2900	Electrical		11.40	15.70	21.50	8.05%	9.45%	11.65%	
	3100	Total: Mechanical & Electrical	▼	39.50	50	76.50	22%	26.50%	31%	
97	0010	**WAREHOUSES & Storage Buildings**	S.F.	44.50	63.50	95				97
	0020	Total project costs	C.F.	2.34	3.67	6.05				
	0100	Site work	S.F.	4.62	9.15	13.80	6.05%	12.95%	19.55%	
	0500	Masonry		2.54	6.35	13.70	3.60%	7.35%	12%	
	1800	Equipment		.68	1.55	8.70	.72%	1.69%	5.55%	
	2720	Plumbing		1.49	2.68	5	2.90%	4.80%	6.55%	
	2730	Heating, ventilating, air conditioning		1.70	4.80	6.45	2.41%	5%	8.90%	
	2900	Electrical		2.65	4.98	8.20	5.15%	7.20%	10.05%	
	3100	Total: Mechanical & Electrical	▼	7.40	11.35	22.50	13.30%	18.90%	26%	
99	0010	**WAREHOUSE & OFFICES Combination**	S.F.	55	73.50	101				99
	0020	Total project costs	C.F.	2.82	4.09	6.05				
	1800	Equipment	S.F.	.95	1.84	2.74	.42%	1.19%	2.40%	
	2720	Plumbing		2.13	3.70	5.50	3.74%	4.76%	6.30%	
	2770	Heating, ventilating, air conditioning		3.36	5.25	7.35	5%	5.65%	10.05%	
	2900	Electrical		3.75	5.55	8.65	5.85%	8%	10%	
	3100	Total: Mechanical & Electrical	▼	10.45	15.85	25	14.55%	19.95%	24.50%	

Square Foot Project Size Modifier

One factor that affects the S.F. cost of a particular building is the size. In general, for buildings built to the same specifications in the same locality, the larger building will have the lower S.F. cost. This is due mainly to the decreasing contribution of the exterior walls, plus the economy of scale usually achievable in larger buildings. The area conversion scale shown below will give a factor to convert costs for the typical size building to an adjusted cost for the particular project.

The square foot base size table lists the median costs, most typical project size in our accumulated data, and the range in size of the projects.

The size factor for your project is determined by dividing your project area in S.F. by the typical project size for the particular building type. With this factor, enter the area conversion scale at the appropriate size factor and determine the appropriate cost multiplier for your building size.

Example: Determine the cost per S.F. for a 100,000 S.F. Mid-rise apartment building.

$$\frac{\text{Proposed building area} = 100{,}000 \text{ S.F.}}{\text{Typical size from below} = 50{,}000 \text{ S.F.}} = 2.00$$

Enter area conversion scale at 2.0, intersect curve, read horizontally the appropriate cost multiplier of .94. Size adjusted cost becomes .94 × $119.00 = $112.00 based on national average costs.

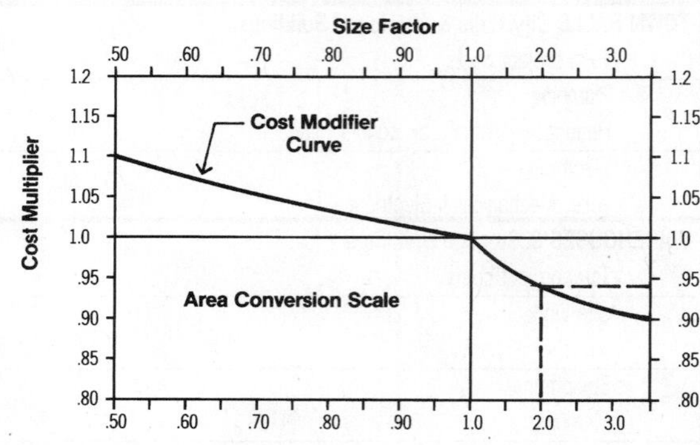

Note: For size factors less than .50, the cost multiplier is 1.1
For size factors greater than 3.5, the cost multiplier is .90

Square Foot Base Size

Building Type	Median Cost per S.F.	Typical Size Gross S.F.	Typical Range Gross S.F.	Building Type	Median Cost per S.F.	Typical Size Gross S.F.	Typical Range Gross S.F.
Apartments, Low Rise	$ 94.00	21,000	9,700 - 37,200	Jails	$ 286.00	40,000	5,500 - 145,000
Apartments, Mid Rise	119.00	50,000	32,000 - 100,000	Libraries	190.00	12,000	7,000 - 31,000
Apartments, High Rise	129.00	145,000	95,000 - 600,000	Living, Assisted	154.00	32,300	23,500 - 50,300
Auditoriums	161.00	25,000	7,600 - 39,000	Medical Clinics	162.00	7,200	4,200 - 15,700
Auto Sales	119.00	20,000	10,800 - 28,600	Medical Offices	154.00	6,000	4,000 - 15,000
Banks	211.00	4,200	2,500 - 7,500	Motels	113.00	40,000	15,800 - 120,000
Churches	145.00	17,000	2,000 - 42,000	Nursing Homes	158.00	23,000	15,000 - 37,000
Clubs, Country	147.00	6,500	4,500 - 15,000	Offices, Low Rise	135.00	20,000	5,000 - 80,000
Clubs, Social	141.00	10,000	6,000 - 13,500	Offices, Mid Rise	132.00	120,000	20,000 - 300,000
Clubs, YMCA	167.00	28,300	12,800 - 39,400	Offices, High Rise	169.00	260,000	120,000 - 800,000
Colleges (Class)	167.00	50,000	15,000 - 150,000	Police Stations	212.00	10,500	4,000 - 19,000
Colleges (Science Lab)	269.00	45,600	16,600 - 80,000	Post Offices	157.00	12,400	6,800 - 30,000
College (Student Union)	199.00	33,400	16,000 - 85,000	Power Plants	1175.00	7,500	1,000 - 20,000
Community Center	150.00	9,400	5,300 - 16,700	Religious Education	137.00	9,000	6,000 - 12,000
Court Houses	206.00	32,400	17,800 - 106,000	Research	222.00	19,000	6,300 - 45,000
Dept. Stores	87.50	90,000	44,000 - 122,000	Restaurants	192.00	4,400	2,800 - 6,000
Dormitories, Low Rise	170.00	25,000	10,000 - 95,000	Retail Stores	93.50	7,200	4,000 - 17,600
Dormitories, Mid Rise	196.00	85,000	20,000 - 200,000	Schools, Elementary	141.00	41,000	24,500 - 55,000
Factories	85.00	26,400	12,900 - 50,000	Schools, Jr. High	145.00	92,000	52,000 - 119,000
Fire Stations	156.00	5,800	4,000 - 8,700	Schools, Sr. High	150.00	101,000	50,500 - 175,000
Fraternity Houses	145.00	12,500	8,200 - 14,800	Schools, Vocational	143.00	37,000	20,500 - 82,000
Funeral Homes	162.00	10,000	4,000 - 20,000	Sports Arenas	115.00	15,000	5,000 - 40,000
Garages, Commercial	104.00	9,300	5,000 - 13,600	Supermarkets	92.00	44,000	12,000 - 60,000
Garages, Municipal	135.00	8,300	4,500 - 12,600	Swimming Pools	300.00	20,000	10,000 - 32,000
Garages, Parking	56.00	163,000	76,400 - 225,300	Telephone Exchange	243.00	4,500	1,200 - 10,600
Gymnasiums	143.00	19,200	11,600 - 41,000	Theaters	138.00	10,500	8,800 - 17,500
Hospitals	258.00	55,000	27,200 - 125,000	Town Halls	152.00	10,800	4,800 - 23,400
House (Elderly)	128.00	37,000	21,000 - 66,000	Warehouses	63.50	25,000	8,000 - 72,000
Housing (Public)	118.00	36,000	14,400 - 74,400	Warehouse & Office	73.50	25,000	8,000 - 72,000
Ice Rinks	170.00	29,000	27,200 - 33,600				

Abbreviations

A	Area Square Feet; Ampere
AAFES	Army and Air Force Exchange Service
ABS	Acrylonitrile Butadiene Stryrene; Asbestos Bonded Steel
A.C., AC	Alternating Current; Air-Conditioning; Asbestos Cement; Plywood Grade A & C
ACI	American Concrete Institute
ACR	Air Conditioning Refrigeration
ADA	Americans with Disabilities Act
AD	Plywood, Grade A & D
Addit.	Additional
Adj.	Adjustable
af	Audio-frequency
AFUE	Annual Fuel Utilization Efficiency
AGA	American Gas Association
Agg.	Aggregate
A.H., Ah	Ampere Hours
A hr.	Ampere-hour
A.H.U., AHU	Air Handling Unit
A.I.A.	American Institute of Architects
AIC	Ampere Interrupting Capacity
Allow.	Allowance
alt., alt	Alternate
Alum.	Aluminum
a.m.	Ante Meridiem
Amp.	Ampere
Anod.	Anodized
ANSI	American National Standards Institute
APA	American Plywood Association
Approx.	Approximate
Apt.	Apartment
Asb.	Asbestos
A.S.B.C.	American Standard Building Code
Asbe.	Asbestos Worker
ASCE.	American Society of Civil Engineers
A.S.H.R.A.E.	American Society of Heating, Refrig. & AC Engineers
ASME	American Society of Mechanical Engineers
ASTM	American Society for Testing and Materials
Attchmt.	Attachment
Avg., Ave.	Average
AWG	American Wire Gauge
AWWA	American Water Works Assoc.
Bbl.	Barrel
B&B, BB	Grade B and Better; Balled & Burlapped
B&S	Bell and Spigot
B.&W.	Black and White
b.c.c.	Body-centered Cubic
B.C.Y.	Bank Cubic Yards
BE	Bevel End
B.F.	Board Feet
Bg. cem.	Bag of Cement
BHP	Boiler Horsepower; Brake Horsepower
B.I.	Black Iron
bidir.	bidirectional
Bit., Bitum.	Bituminous
Bit., Conc.	Bituminous Concrete
Bk.	Backed
Bkrs.	Breakers
Bldg., bldg	Building
Blk.	Block
Bm.	Beam
Boil.	Boilermaker
bpm	Blows per Minute
BR	Bedroom
Brg.	Bearing
Brhe.	Bricklayer Helper
Bric.	Bricklayer
Brk., brk	Brick
brkt	Bracket
Brng.	Bearing
Brs.	Brass
Brz.	Bronze
Bsn.	Basin
Btr.	Better
Btu	British Thermal Unit
BTUH	BTU per Hour
Bu.	bushels
BUR	Built-up Roofing
BX	Interlocked Armored Cable
°C	degree centegrade
c	Conductivity, Copper Sweat
C	Hundred; Centigrade
C/C	Center to Center, Cedar on Cedar
C-C	Center to Center
Cab	Cabinet
Cair.	Air Tool Laborer
Cal.	caliper
Calc	Calculated
Cap.	Capacity
Carp.	Carpenter
C.B.	Circuit Breaker
C.C.A.	Chromate Copper Arsenate
C.C.F.	Hundred Cubic Feet
cd	Candela
cd/sf	Candela per Square Foot
CD	Grade of Plywood Face & Back
CDX	Plywood, Grade C & D, exterior glue
Cefi.	Cement Finisher
Cem.	Cement
CF	Hundred Feet
C.F.	Cubic Feet
CFM	Cubic Feet per Minute
CFRP	Carbon Fiber Reinforced Plastic
c.g.	Center of Gravity
CHW	Chilled Water; Commercial Hot Water
C.I., CI	Cast Iron
C.I.P., CIP	Cast in Place
Circ.	Circuit
C.L.	Carload Lot
CL	Chain Link
Clab.	Common Laborer
Clam	Common maintenance laborer
C.L.F.	Hundred Linear Feet
CLF	Current Limiting Fuse
CLP	Cross Linked Polyethylene
cm	Centimeter
CMP	Corr. Metal Pipe
CMU	Concrete Masonry Unit
CN	Change Notice
Col.	Column
CO_2	Carbon Dioxide
Comb.	Combination
comm.	Commercial, Communication
Compr.	Compressor
Conc.	Concrete
Cont., cont	Continuous; Continued, Container
Corr.	Corrugated
Cos	Cosine
Cot	Cotangent
Cov.	Cover
C/P	Cedar on Paneling
CPA	Control Point Adjustment
Cplg.	Coupling
CPM	Critical Path Method
CPVC	Chlorinated Polyvinyl Chloride
C.Pr.	Hundred Pair
CRC	Cold Rolled Channel
Creos.	Creosote
Crpt.	Carpet & Linoleum Layer
CRT	Cathode-ray Tube
CS	Carbon Steel, Constant Shear Bar Joist
Csc	Cosecant
C.S.F.	Hundred Square Feet
CSI	Construction Specifications Institute
CT	Current Transformer
CTS	Copper Tube Size
Cu	Copper, Cubic
Cu. Ft.	Cubic Foot
cw	Continuous Wave
C.W.	Cool White; Cold Water
Cwt.	100 Pounds
C.W.X.	Cool White Deluxe
C.Y.	Cubic Yard (27 cubic feet)
C.Y./Hr.	Cubic Yard per Hour
Cyl.	Cylinder
d	Penny (nail size)
D	Deep; Depth; Discharge
Dis., Disch.	Discharge
Db	Decibel
Dbl.	Double
DC	Direct Current
DDC	Direct Digital Control
Demob.	Demobilization
d.f.t.	Dry Film Thickness
d.f.u.	Drainage Fixture Units
D.H.	Double Hung
DHW	Domestic Hot Water
DI	Ductile Iron
Diag.	Diagonal
Diam., Dia	Diameter
Distrib.	Distribution
Div.	Division
Dk.	Deck
D.L.	Dead Load; Diesel
DLH	Deep Long Span Bar Joist
dlx	Deluxe
Do.	Ditto
DOP	Dioctyl Phthalate Penetration Test (Air Filters)
Dp., dp	Depth
D.P.S.T.	Double Pole, Single Throw
Dr.	Drive
DR	Dimension Ratio
Drink.	Drinking
D.S.	Double Strength
D.S.A.	Double Strength A Grade
D.S.B.	Double Strength B Grade
Dty.	Duty
DWV	Drain Waste Vent
DX	Deluxe White, Direct Expansion
dyn	Dyne
e	Eccentricity
E	Equipment Only; East; emissivity
Ea.	Each
EB	Encased Burial
Econ.	Economy
E.C.Y	Embankment Cubic Yards
EDP	Electronic Data Processing
EIFS	Exterior Insulation Finish System
E.D.R.	Equiv. Direct Radiation
Eq.	Equation
EL	elevation
Elec.	Electrician; Electrical
Elev.	Elevator; Elevating
EMT	Electrical Metallic Conduit; Thin Wall Conduit
Eng.	Engine, Engineered
EPDM	Ethylene Propylene Diene Monomer
EPS	Expanded Polystyrene
Eqhv.	Equip. Oper., Heavy
Eqlt.	Equip. Oper., Light
Eqmd.	Equip. Oper., Medium
Eqmm.	Equip. Oper., Master Mechanic
Eqol.	Equip. Oper., Oilers
Equip.	Equipment
ERW	Electric Resistance Welded

Abbreviations

E.S.	Energy Saver
Est.	Estimated
esu	Electrostatic Units
E.W.	Each Way
EWT	Entering Water Temperature
Excav.	Excavation
excl	Excluding
Exp., exp	Expansion, Exposure
Ext., ext	Exterior; Extension
Extru.	Extrusion
f.	Fiber stress
F	Fahrenheit; Female; Fill
Fab., fab	Fabricated; fabric
FBGS	Fiberglass
F.C.	Footcandles
f.c.c.	Face-centered Cubic
f'c.	Compressive Stress in Concrete; Extreme Compressive Stress
F.E.	Front End
FEP	Fluorinated Ethylene Propylene (Teflon)
F.G.	Flat Grain
F.H.A.	Federal Housing Administration
Fig.	Figure
Fin.	Finished
FIPS	Female Iron Pipe Size
Fixt.	Fixture
FJP	Finger jointed and primed
Fl. Oz.	Fluid Ounces
Flr.	Floor
FM	Frequency Modulation; Factory Mutual
Fmg.	Framing
FM/UL	Factory Mutual/Underwriters Labs
Fdn.	Foundation
FNPT	Female National Pipe Thread
Fori.	Foreman, Inside
Foro.	Foreman, Outside
Fount.	Fountain
fpm	Feet per Minute
FPT	Female Pipe Thread
Fr	Frame
F.R.	Fire Rating
FRK	Foil Reinforced Kraft
FSK	Foil/scrim/kraft
FRP	Fiberglass Reinforced Plastic
FS	Forged Steel
FSC	Cast Body; Cast Switch Box
Ft., ft	Foot; Feet
Ftng.	Fitting
Ftg.	Footing
Ft lb.	Foot Pound
Furn.	Furniture
FVNR	Full Voltage Non-Reversing
FVR	Full Voltage Reversing
FXM	Female by Male
Fy.	Minimum Yield Stress of Steel
g	Gram
G	Gauss
Ga.	Gauge
Gal., gal.	Gallon
gpm, GPM	Gallon per Minute
Galv., galv	Galvanized
GC/MS	Gas Chromatograph/Mass Spectrometer
Gen.	General
GFI	Ground Fault Interrupter
GFRC	Glass Fiber Reinforced Concrete
Glaz.	Glazier
GPD	Gallons per Day
gpf	Gallon per flush
GPH	Gallons per Hour
GPM	Gallons per Minute
GR	Grade
Gran.	Granular
Grnd.	Ground
GVW	Gross Vehicle Weight
GWB	Gypsum wall board
H	High Henry
HC	High Capacity
H.D., HD	Heavy Duty; High Density
H.D.O.	High Density Overlaid
HDPE	High density polyethelene plastic
Hdr.	Header
Hdwe.	Hardware
H.I.D., HID	High Intensity Discharge
Help.	Helper Average
HEPA	High Efficiency Particulate Air Filter
Hg	Mercury
HIC	High Interrupting Capacity
HM	Hollow Metal
HMWPE	high molecular weight polyethylene
HO	High Output
Horiz.	Horizontal
H.P., HP	Horsepower; High Pressure
H.P.F.	High Power Factor
Hr.	Hour
Hrs./Day	Hours per Day
HSC	High Short Circuit
Ht.	Height
Htg.	Heating
Htrs.	Heaters
HVAC	Heating, Ventilation & Air-Conditioning
Hvy.	Heavy
HW	Hot Water
Hyd.; Hydr.	Hydraulic
Hz	Hertz (cycles)
I.	Moment of Inertia
IBC	International Building Code
I.C.	Interrupting Capacity
ID	Inside Diameter
I.D.	Inside Dimension; Identification
I.F.	Inside Frosted
I.M.C.	Intermediate Metal Conduit
In.	Inch
Incan.	Incandescent
Incl.	Included; Including
Int.	Interior
Inst.	Installation
Insul., insul	Insulation/Insulated
I.P.	Iron Pipe
I.P.S., IPS	Iron Pipe Size
IPT	Iron Pipe Threaded
I.W.	Indirect Waste
J	Joule
J.I.C.	Joint Industrial Council
K	Thousand; Thousand Pounds; Heavy Wall Copper Tubing, Kelvin
K.A.H.	Thousand Amp. Hours
kcmil	Thousand Circular Mils
KD	Knock Down
K.D.A.T.	Kiln Dried After Treatment
kg	Kilogram
kG	Kilogauss
kgf	Kilogram Force
kHz	Kilohertz
Kip	1000 Pounds
KJ	Kiljoule
K.L.	Effective Length Factor
K.L.F.	Kips per Linear Foot
Km	Kilometer
KO	Knock Out
K.S.F.	Kips per Square Foot
K.S.I.	Kips per Square Inch
kV	Kilovolt
kVA	Kilovolt Ampere
kVAR	Kilovar (Reactance)
KW	Kilowatt
KWh	Kilowatt-hour
L	Labor Only; Length; Long; Medium Wall Copper Tubing
Lab.	Labor
lat	Latitude
Lath.	Lather
Lav.	Lavatory
lb.; #	Pound
L.B., LB	Load Bearing; L Conduit Body
L. & E.	Labor & Equipment
lb./hr.	Pounds per Hour
lb./L.F.	Pounds per Linear Foot
lbf/sq.in.	Pound-force per Square Inch
L.C.L.	Less than Carload Lot
L.C.Y.	Loose Cubic Yard
Ld.	Load
LE	Lead Equivalent
LED	Light Emitting Diode
L.F.	Linear Foot
L.F. Nose	Linear Foot of Stair Nosing
L.F. Rsr	Linear Foot of Stair Riser
Lg.	Long; Length; Large
L & H	Light and Heat
LH	Long Span Bar Joist
L.H.	Labor Hours
L.L., LL	Live Load
L.L.D.	Lamp Lumen Depreciation
lm	Lumen
lm/sf	Lumen per Square Foot
lm/W	Lumen per Watt
LOA	Length Over All
log	Logarithm
L-O-L	Lateralolet
long.	longitude
L.P., LP	Liquefied Petroleum; Low Pressure
L.P.F.	Low Power Factor
LR	Long Radius
L.S.	Lump Sum
Lt.	Light
Lt. Ga.	Light Gauge
L.T.L.	Less than Truckload Lot
Lt. Wt.	Lightweight
L.V.	Low Voltage
M	Thousand; Material; Male; Light Wall Copper Tubing
M^2CA	Meters Squared Contact Area
m/hr.; M.H.	Man-hour
mA	Milliampere
Mach.	Machine
Mag. Str.	Magnetic Starter
Maint.	Maintenance
Marb.	Marble Setter
Mat; Mat'l.	Material
Max.	Maximum
MBF	Thousand Board Feet
MBH	Thousand BTU's per hr.
MC	Metal Clad Cable
MCC	Motor Control Center
M.C.F.	Thousand Cubic Feet
MCFM	Thousand Cubic Feet per Minute
M.C.M.	Thousand Circular Mils
MCP	Motor Circuit Protector
MD	Medium Duty
MDF	Medium-density fibreboard
M.D.O.	Medium Density Overlaid
Med.	Medium
MF	Thousand Feet
M.F.B.M.	Thousand Feet Board Measure
Mfg.	Manufacturing
Mfrs.	Manufacturers
mg	Milligram
MGD	Million Gallons per Day
MGPH	Thousand Gallons per Hour
MH, M.H.	Manhole; Metal Halide; Man-Hour
MHz	Megahertz
Mi.	Mile
MI	Malleable Iron; Mineral Insulated
MIPS	Male Iron Pipe Size
mj	Mechanical Joint
m	Meter
mm	Millimeter
Mill.	Millwright
Min., min.	Minimum, minute

Misc. Miscellaneous
ml Milliliter, Mainline
M.L.F. Thousand Linear Feet
Mo. Month
Mobil. Mobilization
Mog. Mogul Base
MPH Miles per Hour
MPT Male Pipe Thread
MRGWB Moisture Resistant Gypsum Wallboard
MRT Mile Round Trip
ms Millisecond
M.S.F. Thousand Square Feet
Mstz. Mosaic & Terrazzo Worker
M.S.Y. Thousand Square Yards
Mtd., mtd., mtd Mounted
Mthe. Mosaic & Terrazzo Helper
Mtng. Mounting
Mult. Multi; Multiply
M.V.A. Million Volt Amperes
M.V.A.R. Million Volt Amperes Reactance
MV Megavolt
MW Megawatt
MXM Male by Male
MYD Thousand Yards
N Natural; North
nA Nanoampere
NA Not Available; Not Applicable
N.B.C. National Building Code
NC Normally Closed
NEMA National Electrical Manufacturers Assoc.
NEHB Bolted Circuit Breaker to 600V.
NFPA National Fire Protection Association
NLB Non-Load-Bearing
NM Non-Metallic Cable
nm Nanometer
No. Number
NO Normally Open
N.O.C. Not Otherwise Classified
Nose. Nosing
NPT National Pipe Thread
NQOD Combination Plug-on/Bolt on Circuit Breaker to 240V.
N.R.C., NRC Noise Reduction Coefficient/ Nuclear Regulator Commission
N.R.S. Non Rising Stem
ns Nanosecond
nW Nanowatt
OB Opposing Blade
OC On Center
OD Outside Diameter
O.D. Outside Dimension
ODS Overhead Distribution System
O.G. Ogee
O.H. Overhead
O&P Overhead and Profit
Oper. Operator
Opng. Opening
Orna. Ornamental
OSB Oriented Strand Board
OS&Y Outside Screw and Yoke
OSHA Occupational Safety and Health Act
Ovhd. Overhead
OWG Oil, Water or Gas
Oz. Ounce
P. Pole; Applied Load; Projection
p. Page
Pape. Paperhanger
P.A.P.R. Powered Air Purifying Respirator
PAR Parabolic Reflector
P.B., PB Push Button
Pc., Pcs. Piece, Pieces
P.C. Portland Cement; Power Connector
P.C.F. Pounds per Cubic Foot
PCM Phase Contrast Microscopy
PDCA Painting and Decorating Contractors of America
P.E., PE Professional Engineer; Porcelain Enamel; Polyethylene; Plain End
P.E.C.I. Porcelain Enamel on Cast Iron
Perf. Perforated
PEX Cross linked polyethylene
Ph. Phase
P.I. Pressure Injected
Pile. Pile Driver
Pkg. Package
Pl. Plate
Plah. Plasterer Helper
Plas. Plasterer
plf Pounds Per Linear Foot
Pluh. Plumbers Helper
Plum. Plumber
Ply. Plywood
p.m. Post Meridiem
Pntd. Painted
Pord. Painter, Ordinary
pp Pages
PP, PPL Polypropylene
P.P.M. Parts per Million
Pr. Pair
P.E.S.B. Pre-engineered Steel Building
Prefab. Prefabricated
Prefin. Prefinished
Prop. Propelled
PSF, psf Pounds per Square Foot
PSI, psi Pounds per Square Inch
PSIG Pounds per Square Inch Gauge
PSP Plastic Sewer Pipe
Pspr. Painter, Spray
Psst. Painter, Structural Steel
P.T. Potential Transformer
P. & T. Pressure & Temperature
Ptd. Painted
Ptns. Partitions
Pu Ultimate Load
PVC Polyvinyl Chloride
Pvmt. Pavement
PRV Pressure Relief Valve
Pwr. Power
Q Quantity Heat Flow
Qt. Quart
Quan., Qty. Quantity
Q.C. Quick Coupling
r Radius of Gyration
R Resistance
R.C.P. Reinforced Concrete Pipe
Rect. Rectangle
recpt. receptacle
Reg. Regular
Reinf. Reinforced
Req'd. Required
Res. Resistant
Resi. Residential
RF Radio Frequency
RFID Radio-frequency identification
Rgh. Rough
RGS Rigid Galvanized Steel
RHW Rubber, Heat & Water Resistant; Residential Hot Water
rms Root Mean Square
Rnd. Round
Rodm. Rodman
Rofc. Roofer, Composition
Rofp. Roofer, Precast
Rohe. Roofer Helpers (Composition)
Rots. Roofer, Tile & Slate
R.O.W. Right of Way
RPM Revolutions per Minute
R.S. Rapid Start
Rsr Riser
RT Round Trip
S. Suction; Single Entrance; South
SBS Styrene Butadiere Styrene
SC Screw Cover
SCFM Standard Cubic Feet per Minute
Scaf. Scaffold
Sch., Sched. Schedule
S.C.R. Modular Brick
S.D. Sound Deadening
SDR Standard Dimension Ratio
S.E. Surfaced Edge
Sel. Select
SER, SEU Service Entrance Cable
S.F. Square Foot
S.F.C.A. Square Foot Contact Area
S.F. Flr. Square Foot of Floor
S.F.G. Square Foot of Ground
S.F. Hor. Square Foot Horizontal
SFR Square Feet of Radiation
S.F. Shlf. Square Foot of Shelf
S4S Surface 4 Sides
Shee. Sheet Metal Worker
Sin. Sine
Skwk. Skilled Worker
SL Saran Lined
S.L. Slimline
Sldr. Solder
SLH Super Long Span Bar Joist
S.N. Solid Neutral
SO Stranded with oil resistant inside insulation
S-O-L Socketolet
sp Standpipe
S.P. Static Pressure; Single Pole; Self-Propelled
Spri. Sprinkler Installer
spwg Static Pressure Water Gauge
S.P.D.T. Single Pole, Double Throw
SPF Spruce Pine Fir; Sprayed Polyurethane Foam
S.P.S.T. Single Pole, Single Throw
SPT Standard Pipe Thread
Sq. Square; 100 Square Feet
Sq. Hd. Square Head
Sq. In. Square Inch
S.S. Single Strength; Stainless Steel
S.S.B. Single Strength B Grade
sst, ss Stainless Steel
Sswk. Structural Steel Worker
Sswl. Structural Steel Welder
St.; Stl. Steel
STC Sound Transmission Coefficient
Std. Standard
Stg. Staging
STK Select Tight Knot
STP Standard Temperature & Pressure
Stpi. Steamfitter, Pipefitter
Str. Strength; Starter; Straight
Strd. Stranded
Struct. Structural
Sty. Story
Subj. Subject
Subs. Subcontractors
Surf. Surface
Sw. Switch
Swbd. Switchboard
S.Y. Square Yard
Syn. Synthetic
S.Y.P. Southern Yellow Pine
Sys. System
t. Thickness
T Temperature; Ton
Tan Tangent
T.C. Terra Cotta
T & C Threaded and Coupled
T.D. Temperature Difference
Tdd Telecommunications Device for the Deaf
T.E.M. Transmission Electron Microscopy
temp Temperature, Tempered, Temporary
TFFN Nylon Jacketed Wire

Abbreviations

TFE	Tetrafluoroethylene (Teflon)
T. & G.	Tongue & Groove; Tar & Gravel
Th.,Thk.	Thick
Thn.	Thin
Thrded	Threaded
Tilf.	Tile Layer, Floor
Tilh.	Tile Layer, Helper
THHN	Nylon Jacketed Wire
THW.	Insulated Strand Wire
THWN	Nylon Jacketed Wire
T.L.,TL	Truckload
T.M.	Track Mounted
Tot.	Total
T-O-L	Threadolet
tmpd	Tempered
TPO	Thermoplastic Polyolefin
T.S.	Trigger Start
Tr.	Trade
Transf.	Transformer
Trhv.	Truck Driver, Heavy
Trlr	Trailer
Trlt.	Truck Driver, Light
TTY	Teletypewriter
TV	Television
T.W.	Thermoplastic Water Resistant Wire
UCI	Uniform Construction Index
UF	Underground Feeder
UGND	Underground Feeder
UHF	Ultra High Frequency
U.I.	United Inch
U.L., UL	Underwriters Laboratory
Uld.	unloading
Unfin.	Unfinished
UPS	Uninterruptible Power Supply
URD	Underground Residential Distribution
US	United States
USGBC	U.S. Green Building Council
USP	United States Primed
UTMCD	Uniform Traffic Manual For Control Devices
UTP	Unshielded Twisted Pair
V	Volt
VA	Volt Amperes
VAT	Vinyl Asbestos Tile
V.C.T.	Vinyl Composition Tile
VAV	Variable Air Volume
VC	Veneer Core
VDC	Volts Direct Current
Vent.	Ventilation
Vert.	Vertical
V.F.	Vinyl Faced
V.G.	Vertical Grain
VHF	Very High Frequency
VHO	Very High Output
Vib.	Vibrating
VLF	Vertical Linear Foot
VOC	Volitile Organic Compound
Vol.	Volume
VRP	Vinyl Reinforced Polyester
W	Wire; Watt; Wide; West
w/	With
W.C., WC	Water Column; Water Closet
W.F.	Wide Flange
W.G.	Water Gauge
Wldg.	Welding
W. Mile	Wire Mile
W-O-L	Weldolet
W.R.	Water Resistant
Wrck.	Wrecker
W.S.P.	Water, Steam, Petroleum
WT., Wt.	Weight
WWF	Welded Wire Fabric
XFER	Transfer
XFMR	Transformer
XHD	Extra Heavy Duty
XHHW, XLPE	Cross-Linked Polyethylene Wire Insulation
XLP	Cross-linked Polyethylene
Xport	Transport
Y	Wye
yd	Yard
yr	Year
Δ	Delta
%	Percent
~	Approximately
Ø	Phase; diameter
@	At
#	Pound; Number
<	Less Than
>	Greater Than
Z	zone

D

E

F

G

M

N

O

S

T

U

V

W

X

Notes

Notes

Notes

Notes

Notes

Notes

Notes

Other RSMeans Products & Services

RSMeans—
a tradition of excellence in construction cost information and services since 1942

For more information visit the RSMeans website at **www.rsmeans.com**

Unit prices according to the latest MasterFormat!

Book Selection Guide

The following table provides definitive information on the content of each cost data publication. The number of lines of data provided in each unit price or assemblies division, as well as the number of crews, is listed for each book. The presence of other elements such as reference tables, square foot models, equipment rental costs, historical cost indexes, and city cost indexes, is also indicated. You can use the table to help select the RSMeans book that has the quantity and type of information you most need in your work.

Unit Cost Divisions	Building Construction	Mechanical	Electrical	Commercial Renovation	Square Foot	Site Work Landsc.	Green Building	Interior	Concrete Masonry	Open Shop	Heavy Construction	Light Commercial	Facilities Construction	Plumbing	Residential
1	573	393	410	506		509	191	312	456	572	501	246	1042	403	178
2	777	279	85	733		993	178	399	213	776	734	481	1221	286	257
3	1690	340	230	1085		1471	987	355	2032	1689	1685	481	1788	316	388
4	957	21	0	922		720	179	612	1155	925	612	529	1172	0	441
5	1893	158	155	1090		841	1799	1095	720	1893	1037	979	1909	204	721
6	2452	18	18	2110		110	589	1528	281	2448	123	2140	2124	22	2660
7	1584	215	128	1623		581	757	532	524	1581	26	1317	1685	227	1037
8	2082	81	45	2609		265	1123	1753	105	2077	0	2193	2879	0	1526
9	1971	72	26	1767		313	449	2056	390	1911	15	1659	2220	54	1435
10	1026	17	10	642		215	27	842	158	1026	29	511	1118	235	224
11	1087	208	165	546		124	54	932	28	1071	0	230	1107	169	110
12	544	0	2	319		217	137	1654	14	532	0	352	1704	23	310
13	719	140	137	242		343	111	248	66	695	244	83	746	94	79
14	273	36	0	221		0	0	257	0	273	0	12	293	16	6
21	90	0	16	37		0	0	250	0	87	0	68	426	431	220
22	1157	7544	150	1190		1568	1067	838	20	1121	1677	857	7440	9347	712
23	1177	6971	580	928		158	902	776	38	1101	111	878	5187	1918	469
26	1354	454	10081	1021		793	611	1136	55	1349	562	1298	10030	399	617
27	72	0	297	34		13	0	71	0	72	39	52	279	0	22
28	96	58	136	71		0	21	78	0	99	0	40	151	44	25
31	1498	735	610	803		3217	289	7	1208	1443	3256	600	1560	660	611
32	822	53	8	886		4426	356	405	291	793	1841	420	1701	165	468
33	530	1079	538	252		2178	41	0	237	522	2159	128	1699	1286	154
34	107	0	47	4		190	0	0	31	62	193	0	128	0	0
35	18	0	0	0		327	0	0	0	18	442	0	84	0	0
41	60	0	0	32		7	0	22	0	60	30	0	67	14	0
44	75	79	0	0		0	0	0	0	0	0	0	75	75	0
46	23	16	0	0		274	261	0	0	23	264	0	33	33	0
48	12	0	25	0		0	25	0	0	12	17	12	12	0	12
Totals	24719	18967	13899	19673		19853	10154	16158	8022	24231	15597	15566	49880	16421	12682

Assem Div	Building Construction	Mechanical	Elec-trical	Commercial Renovation	Square Foot	Site Work Landscape	Assemblies	Green Building	Interior	Concrete Masonry	Heavy Construction	Light Commercial	Facilities Construction	Plumbing	Asm Div	Residential
A		15	0	188	150	577	598	0	0	536	571	154	24	0	1	376
B		0	0	848	2498	0	5658	56	329	1975	368	2089	174	0	2	211
C		0	0	647	926	0	1304	0	1629	146	0	816	250	0	3	588
D		1067	941	712	1859	72	2538	330	825	0	0	1345	1105	1088	4	851
E		0	0	86	260	0	300	0	5	0	0	257	5	0	5	392
F		0	0	0	114	0	114	0	0	0	0	114	3	0	6	357
G		527	447	318	249	3365	729	0	0	535	1350	205	293	677	7	307
															8	760
															9	80
															10	0
															11	0
															12	0
Totals		1609	1388	2799	6056	4014	11241	386	2788	3192	2289	4980	1854	1765		3922

Reference Section	Building Construction Costs	Mechanical	Electrical	Commercial Renovation	Square Foot	Site Work Landscape	Assem.	Green Building	Interior	Concrete Masonry	Open Shop	Heavy Construction	Light Commercial	Facilities Construction	Plumbing	Resi.
Reference Tables	yes	yes	yes	yes	no	yes	yes	yes	yes	yes	yes	yes	yes	yes	yes	yes
Models					111			25					50			28
Crews	571	571	571	550		571		571	571	571	548	571	548	550	571	548
Equipment Rental Costs	yes	yes	yes	yes		yes		yes	yes	yes	yes	yes	yes	yes	yes	yes
Historical Cost Indexes	yes	yes	yes	yes	yes	yes	yes	yes	yes	yes	yes	yes	yes	yes	yes	no
City Cost Indexes	yes	yes	yes	yes	yes	yes	yes	yes	yes	yes	yes	yes	yes	yes	yes	yes

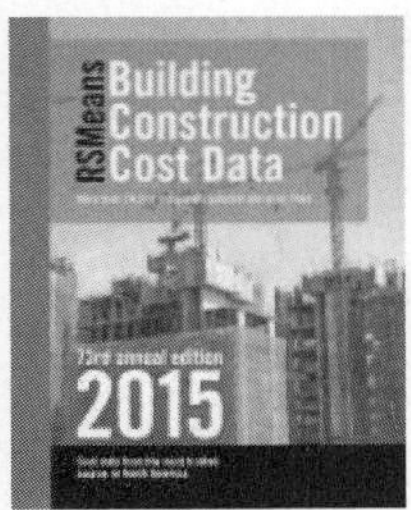

RSMeans Building Construction Cost Data 2015

Offers you unchallenged unit price reliability in an easy-to-use format. Whether used for verifying complete, finished estimates or for periodic checks, it supplies more cost facts better and faster than any comparable source. More than 24,700 unit prices have been updated for 2015. The City Cost Indexes and Location Factors cover more than 930 areas, for indexing to any project location in North America. Order and get *RSMeans Quarterly Update Service* FREE.

$206.95 | Available Sept. 2014 | Catalog no. 60015

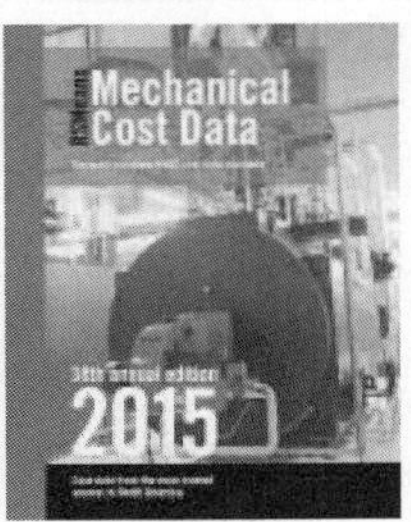

RSMeans Mechanical Cost Data 2015

Total unit and systems price guidance for mechanical construction . . . materials, parts, fittings, and complete labor cost information. Includes prices for piping, heating, air conditioning, ventilation, and all related construction.

Plus new 2015 unit costs for:

- Thousands of installed HVAC/controls, sub-assemblies and assemblies
- "On-site" Location Factors for more than 930 cities and towns in the U.S. and Canada
- Crews, labor, and equipment

$203.95 | Available Oct. 2014 | Catalog no. 60025

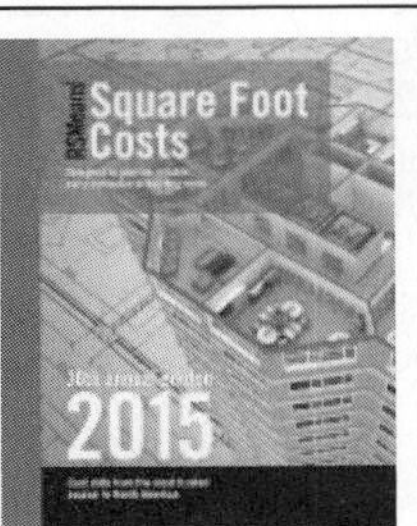

RSMeans Square Foot Costs 2015

Accurate and Easy-to-Use

- **Updated price information** based on nationwide figures from suppliers, estimators, labor experts, and contractors.
- Green building models
- Realistic graphics, offering true-to-life illustrations of building projects
- Extensive information on using square foot cost data, including sample estimates and alternate pricing methods

$218.95 | Available Oct. 2014 | Catalog no. 60055

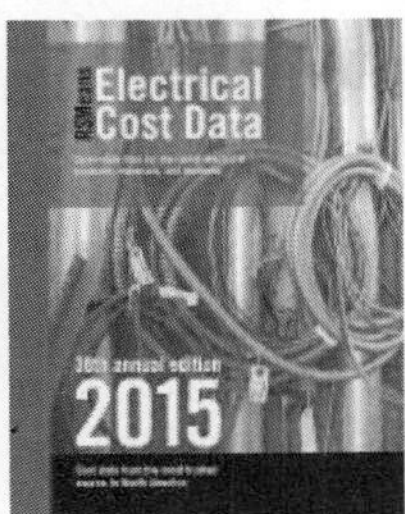

RSMeans Electrical Cost Data 2015

Pricing information for every part of electrical cost planning. More than 13,800 unit and systems costs with design tables; clear specifications and drawings; engineering guides; illustrated estimating procedures; complete labor-hour and materials costs for better scheduling and procurement; and the latest electrical products and construction methods.

- A variety of special electrical systems, including cathodic protection
- Costs for maintenance, demolition, HVAC/mechanical, specialties, equipment, and more

$208.95 | Available Oct. 2014 | Catalog no. 60035

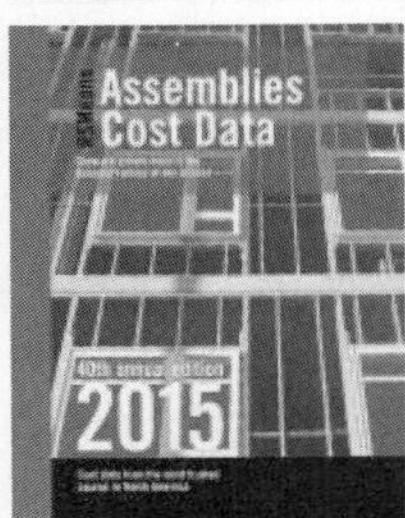

RSMeans Assemblies Cost Data 2015

RSMeans Assemblies Cost Data takes the guesswork out of preliminary or conceptual estimates. Now you don't have to try to calculate the assembled cost by working up individual component costs. We've done all the work for you.

Presents detailed illustrations, descriptions, specifications, and costs for every conceivable building assembly—over 350 types in all—arranged in the easy-to-use UNIFORMAT II system. Each illustrated "assembled" cost includes a complete grouping of materials and associated installation costs, including the installing contractor's overhead and profit.

$334.95 | Available Sept. 2014 | Catalog no. 60065

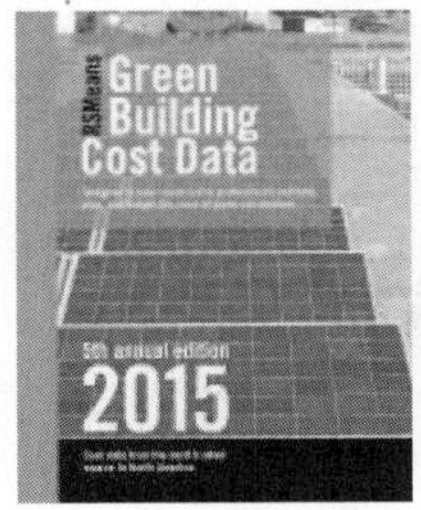

RSMeans Green Building Cost Data 2015

Estimate, plan, and budget the costs of green building for both new commercial construction and renovation work with this fifth edition of *RSMeans Green Building Cost Data*. More than 10,000 unit costs for a wide array of green building products plus assemblies costs. Easily identified cross references to LEED and Green Globes building rating systems criteria.

$170.95 | Available Nov. 2014 | Catalog no. 60555

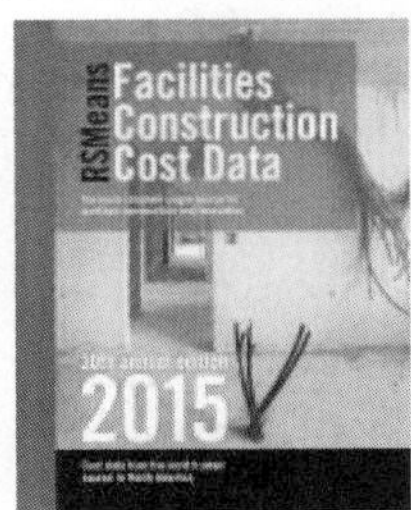

RSMeans Facilities Construction Cost Data 2015

For the maintenance and construction of commercial, industrial, municipal, and institutional properties. Costs are shown for new and remodeling construction and are broken down into materials, labor, equipment, and overhead and profit. Special emphasis is given to sections on mechanical, electrical, furnishings, site work, building maintenance, finish work, and demolition.

More than 49,800 unit costs, plus assemblies costs and a comprehensive Reference Section are included.

$524.95 | Available Nov. 2014 | Catalog no. 60205

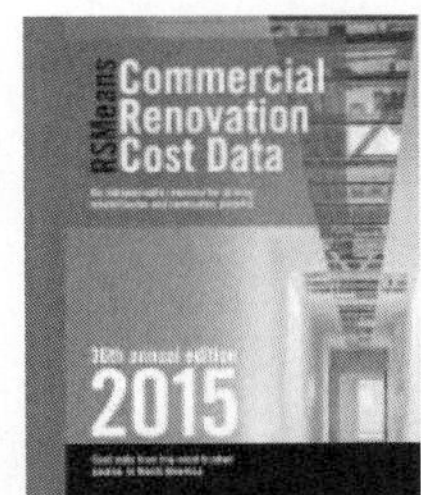

RSMeans Commercial Renovation Cost Data 2015

Commercial/Multi-family Residential

Use this valuable tool to estimate commercial and multi-family residential renovation and remodeling.

Includes: Updated costs for hundreds of unique methods, materials, and conditions that only come up in repair and remodeling, PLUS:

- Unit costs for more than 19,600 construction components
- Installed costs for more than 2,700 assemblies
- More than 930 "on-site" localization factors for the U.S. and Canada

$166.95 | Available Oct. 2014 | Catalog no. 60045

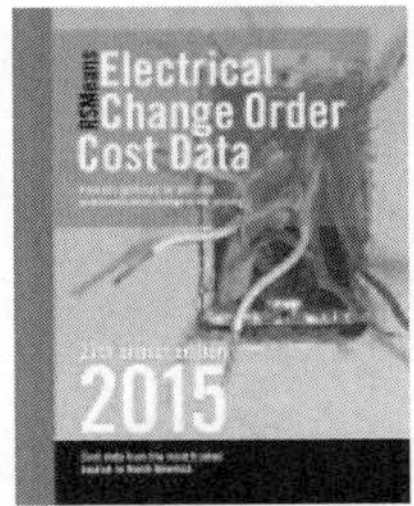

RSMeans Electrical Change Order Cost Data 2015

RSMeans Electrical Change Order Cost Data provides you with electrical unit prices exclusively for pricing change orders—based on the recent, direct experience of contractors and suppliers. Analyze and check your own change order estimates against the experience others have had doing the same work. It also covers productivity analysis, and change order cost justifications. With useful information for calculating the effects of change orders and dealing with their administration.

$199.95 | Available Dec. 2014 | Catalog no. 60235

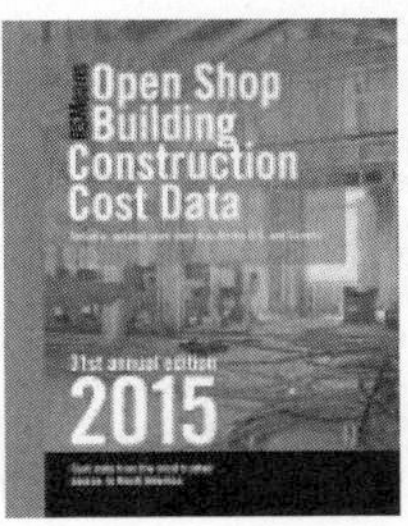

RSMeans Open Shop Building Construction Cost Data 2015

The latest costs for accurate budgeting and estimating of new commercial and residential construction . . . renovation work . . . change orders . . . cost engineering.

RSMeans Open Shop "BCCD" will assist you to:

- Develop benchmark prices for change orders.
- Plug gaps in preliminary estimates and budgets.
- Estimate complex projects.
- Substantiate invoices on contracts.
- Price ADA-related renovations.

$175.95 | Available Dec. 2014 | Catalog no. 60155

For more information visit the RSMeans website at **www.rsmeans.com**

RSMeans Residential Cost Data 2015

Contains square foot costs for 28 basic home models with the look of today, plus hundreds of custom additions and modifications you can quote right off the page. Includes more than 3,900 costs for 89 residential systems. Complete with blank estimating forms, sample estimates, and step-by-step instructions.

Contains line items for cultured stone and brick, PVC trim, lumber, and TPO roofing.

$148.95 | Available Oct. 2014 | Catalog no. 60175

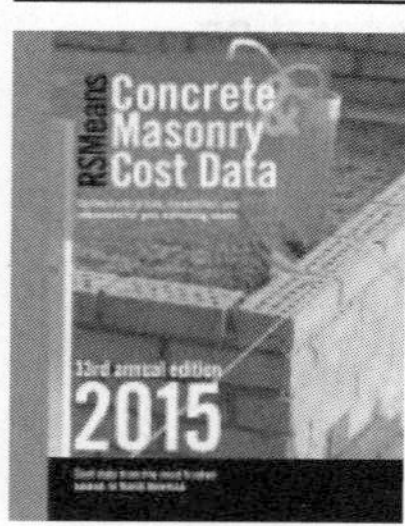

RSMeans Facilities Maintenance & Repair Cost Data 2015

RSMeans Facilities Maintenance & Repair Cost Data gives you a complete system to manage and plan your facility repair and maintenance costs and budget efficiently. Guidelines for auditing a facility and developing an annual maintenance plan. Budgeting is included, along with reference tables on cost and management, and information on frequency and productivity of maintenance operations.

The only nationally recognized source of maintenance and repair costs. Developed in cooperation with the Civil Engineering Research Laboratory (CERL) of the Army Corps of Engineers.

$461.95 | Available Nov. 2014 | Catalog no. 60305

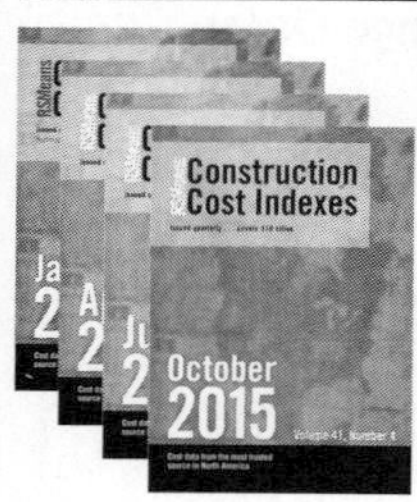

RSMeans Concrete & Masonry Cost Data 2015

Provides you with cost facts for virtually all concrete/masonry estimating needs, from complicated formwork to various sizes and face finishes of brick and block—all in great detail. The comprehensive Unit Price Section contains more than 8,000 selected entries. Also contains an Assemblies [Cost] Section, and a detailed Reference Section that supplements the cost data.

$188.95 | Available Dec. 2014 | Catalog no. 60115

RSMeans Construction Cost Indexes 2015

What materials and labor costs will change unexpectedly this year? By how much?

- Breakdowns for 318 major cities
- National averages for 30 key cities
- Expanded five major city indexes
- Historical construction cost indexes

$359.95 per year (subscription) | Catalog no. 50145
$94.95 individual quarters | Catalog no. 60145 A,B,C,D

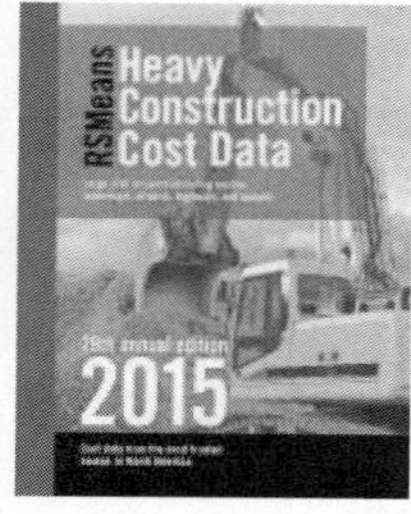

RSMeans Heavy Construction Cost Data 2015

A comprehensive guide to heavy construction costs. Includes costs for highly specialized projects such as tunnels, dams, highways, airports, and waterways. Information on labor rates, equipment, and materials costs is included. Features unit price costs, systems costs, and numerous reference tables for costs and design.

$204.95 | Available Dec. 2014 | Catalog no. 60165

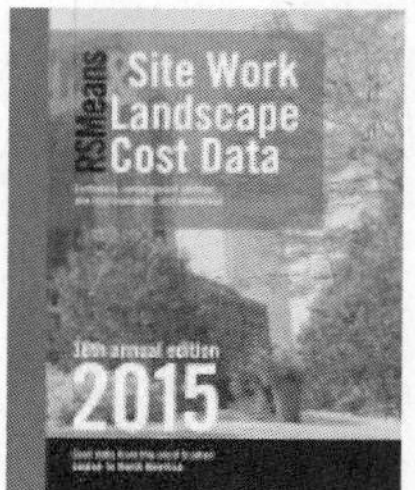

RSMeans Site Work & Landscape Cost Data 2015

Includes unit and assemblies costs for earthwork, sewerage, piped utilities, site improvements, drainage, paving, trees and shrubs, street openings/repairs, underground tanks, and more. Contains more than 60 types of assemblies costs for accurate conceptual estimates.

Includes:

- Estimating for infrastructure improvements
- Environmentally-oriented construction
- ADA-mandated handicapped access
- Hazardous waste line items

$199.95 | Available Dec. 2014 | Catalog no. 60285

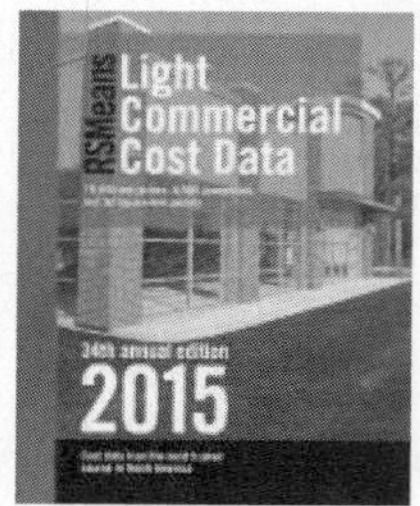

RSMeans Light Commercial Cost Data 2015

Specifically addresses the light commercial market, which is a specialized niche in the construction industry. Aids you, the owner/designer/contractor, in preparing all types of estimates—from budgets to detailed bids. Includes new advances in methods and materials.

Assemblies Section allows you to evaluate alternatives in the early stages of design/planning.

More than 15,500 unit costs ensure that you have the prices you need, when you need them.

$153.95 | Available Nov. 2014 | Catalog no. 60185

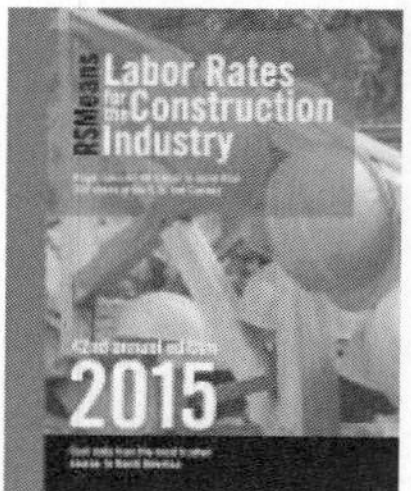

RSMeans Labor Rates for the Construction Industry 2015

Complete information for estimating labor costs, making comparisons, and negotiating wage rates by trade for more than 300 U.S. and Canadian cities. With 46 construction trades in each city, and historical wage rates included for comparison. Each city chart lists the county and is alphabetically arranged with handy visual flip tabs for quick reference.

$450.95 | Available Dec. 2014 | Catalog no. 60125

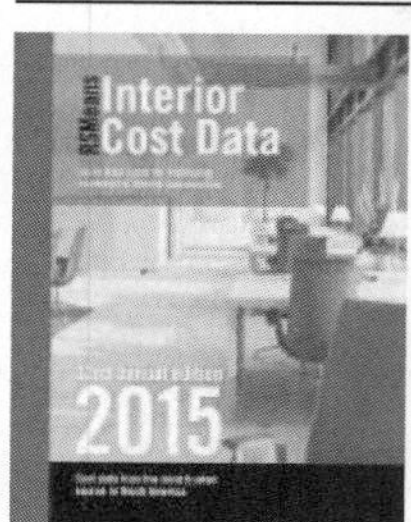

RSMeans Interior Cost Data 2015

Provides you with prices and guidance needed to make accurate interior work estimates. Contains costs on materials, equipment, hardware, custom installations, furnishings, and labor for new and remodel commercial and industrial interior construction, including updated information on office furnishings, and reference information.

$208.95 | Available Nov. 2014 | Catalog no. 60095

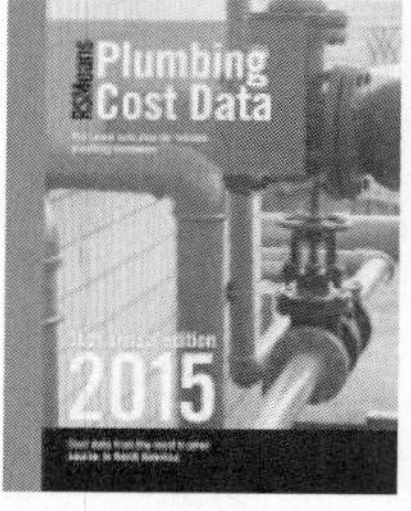

RSMeans Plumbing Cost Data 2015

Comprehensive unit prices and assemblies for plumbing, irrigation systems, commercial and residential fire protection, point-of-use water heaters, and the latest approved materials. This publication and its companion, *RSMeans Mechanical Cost Data*, provide full-range cost estimating coverage for all the mechanical trades.

Contains updated costs for potable water, radiant heat systems, and high efficiency fixtures.

$206.95 | Available Oct. 2014 | Catalog no. 60215

RSMeans Online and RSMeans CostWorks CD Comparison Matrix

Both the RSMeans CostWorks CD package and the RSMeans Online estimating tool provide the same comprehensive, up-to-date, and localized cost information to improve planning, estimating, and budgeting with RSMeans—the industry's most-trusted source of construction cost data.

In addition to key differences noted in the comparison matrix below, both products let you do the following:

- Quickly locate costs in the searchable RSMeans database.
- Instantly adjust the data to your specific location for accurate costs anywhere in the U.S. and Canada.
- Create reliable, updated cost estimates in minutes.
- Manage, share, and export your estimates to MS Excel® with ease.
- Access a wide range of reference materials, including crew rates and national average labor rates.

Both the CD and Online options are available in a convenient one-year subscription!

Use the handy comparison matrix below to help you decide the best product for your cost estimating needs!

Why us?

For more than 70 years, RSMeans has provided the quality cost data construction professionals everywhere depend on to estimate with confidence, save time, improve decision-making, and increase profits.

Improve planning and decision-making

Back your estimates with complete, accurate, and up-to-date cost data for informed business decisions.

- Verify construction costs from third parties.
- Check validity of subcontractor proposals.
- Evaluate material and assembly alternatives.

Increase profits

Use RSMeans to estimate projects quickly and accurately, so you can gain an edge over your competition.

- Create accurate and competitive bids.
- Minimize the risk of cost overruns.
- Reduce variability.
- Gain control over costs.

The following matrix summarizes the key comparison points between RSMeans' two electronic cost data products.

	Question	CostWorks CD	RSMeans Online.com
1	How is the data presented?	• The data is organized by book title and sold at comparable book cost. • Several packaged bundles of five titles are available at a cost savings. • Available in a convenient one-year subscription	• The data is organized by book title and sold at comparable book cost. • Several packaged bundles of five titles are available at a cost savings. • RSMeans Online also offers a Complete Library title with all online data at a cost savings. • Available in a convenient one-year subscription
2	Where is the data stored?	With a typical install, the data is read off of your hard drive or off of the CD.	All data is stored on our secure server.
3	How are updates delivered?	Updates are available on a quarterly basis for download from the Costworks HomePort website.	Data is updated automatically on a quarterly basis, and users are notified when updates occur.
4	Where are my projects and estimates stored?	Upon installation, a "My CostWorks Projects" folder is created within the "My Documents" directory. All projects are defaulted to save here unless otherwise specified.	Estimates are stored on our secure server.
5	Does my IT team need to assist me in installing and running the program?	Yes, depending on the level of your permission. See #11 for system requirements.	While the product is entirely Web-based, you may need IT support with permissions for certain tasks. See #11 for system requirements.
6	Can users collaborate on projects/estimates?	Yes, you can copy project folders between computers.	You can create shared folders that can be viewed, copied, and edited by anyone within your account workgroup.

Comparison matrix continued

7	Can I create and use my own data with the RSMeans database?	Custom lines cannot be added to the CD database. With the Estimator tool, however, you can create and add custom cost lines to any estimate. You can also save custom lines as an Estimator template for use in creating new estimates.	You can create • a custom line within an estimate • a custom cost book consisting of custom unit (not assembly) cost lines, no more than once per quarter **Note** that custom data is searchable using the Include My Custom Data feature.
8	Can I enter my own trade/ labor rates and markups?	Although you cannot adjust the trade labor rates, you can add custom markups for each report template using the Estimator tool.	For each quarterly update, you can create one custom unit cost book. You can edit the labor rates by individual trade and the crew equipment costs, and also edit the associated material and equipment markups.
9	What are my reporting options?	The Estimator tool lets you create editable custom reports and report templates. Reports can also be exported to MS Excel®.	You can generate pre-formatted estimate summaries and cost lists in both PDF and Excel formats.
10	How do I find specific cost data?	You can search by CSI line number and by keyword within a division.	You can search by CSI line number and by keyword within a division. There are several additional search options, including keyword and advanced search using special characters.
11	What are the system requirements?	• Operating system – Windows XP or later • PC requirements - 256 MB RAM (recommended), 1024 x 768 High Color Monitor, 800MHz Computer (recommended), 24X CD-ROM, 250 MB Hard Disk Space	• Resolution – Recommended resolution is 1024 x 768 or higher • Compatibility - Internet Explorer version 8.x, 9.x and Mozilla Firefox version 10 and higher • iPad - iOS 5 and higher

2015 RSMeans Seminar Schedule

Note: call for exact dates and details.

Location	Dates
Seattle, WA	January and August
Dallas/Ft. Worth, TX	January
Austin, TX	February
Anchorage, AK	March and September
Las Vegas, NV	March and October
Washington, DC	April and September
Phoenix, AZ	April
Kansas City, MO	April
Toronto	May
Denver, CO	May
San Francisco, CA	June
Bethesda, MD	June
Columbus, GA	June
El Segundo, CA	August

Location	Dates
Jacksonville FL	September
Dallas, TX	September
Charleston, SC	October
Houston, TX	October
Atlanta, GA	November
Baltimore, MD	November
Orlando, FL	November
San Diego, CA	December
San Antonio, TX	December
Raleigh, NC	December

1-800-334-3509, Press 1

Professional Development

For more information visit the RSMeans website at www.rsmeans.com

eLearning Training Sessions

Learn how to use *RSMeans Online*® or *RSMeans CostWorks*® CD from the convenience of your home or office. Our eLearning training sessions let you join a training conference call and share the instructors' desktops, so you can view the presentation and step-by-step instructions on your own computer screen. The live webinars are held from 9 a.m. to 4 p.m. eastern standard time, with a one-hour break for lunch.

For these sessions, you must have a computer with high speed Internet access and a compatible Web browser. Learn more at www.rsmeansonline.com or call for a schedule: 1-800-334-3509 and press 1. Webinars are generally held on selected Wednesdays each month.

RSMeans Online®
$299 per person

RSMeans CostWorks® CD
$299 per person

RSMeans Online™ Training

Construction estimating is vital to the decision-making process at each state of every project. RSMeansOnline works the way you do. It's systematic, flexible and intuitive. In this one day class you will see how you can estimate any phase of any project faster and better.

Some of what you'll learn:
- Customizing RSMeansOnline
- Making the most of RSMeans "Circle Reference" numbers
- How to integrate your cost data
- Generate reports, exporting estimates to MS Excel, sharing, collaborating and more

Also available: RSMeans Online® training webinar

Maintenance & Repair Estimating for Facilities

This *two-day* course teaches attendees how to plan, budget, and estimate the cost of ongoing and preventive maintenance and repair for existing buildings and grounds.

Some of what you'll learn:
- The most financially favorable maintenance, repair, and replacement scheduling and estimating
- Auditing and value engineering facilities
- Preventive planning and facilities upgrading
- Determining both in-house and contract-out service costs
- Annual, asset-protecting M&R plan

Who should attend: facility managers, maintenance supervisors, buildings and grounds superintendents, plant managers, planners, estimators, and others involved in facilities planning and budgeting.

Facilities Construction Estimating

In this *two-day* course, professionals working in facilities management can get help with their daily challenges to establish budgets for all phases of a project.

Some of what you'll learn:
- Determining the full scope of a project
- Identifying the scope of risks and opportunities
- Creative solutions to estimating issues
- Organizing estimates for presentation and discussion
- Special techniques for repair/remodel and maintenance projects
- Negotiating project change orders

Who should attend: facility managers, engineers, contractors, facility tradespeople, planners, and project managers.

Practical Project Management for Construction Professionals

In this *two-day* course, acquire the essential knowledge and develop the skills to effectively and efficiently execute the day-to-day responsibilities of the construction project manager.

Covers:
- General conditions of the construction contract
- Contract modifications: change orders and construction change directives
- Negotiations with subcontractors and vendors
- Effective writing: notification and communications
- Dispute resolution: claims and liens

Who should attend: architects, engineers, owner's representatives, project managers.

Construction Cost Estimating: Concepts and Practice

This *one-day* introductory course to improve estimating skills and effectiveness starts with the details of interpreting bid documents, ending with the summary of the estimate and bid submission.

Some of what you'll learn:
- Using the plans and specifications for creating estimates
- The takeoff process—deriving all tasks with correct quantities
- Developing pricing, using various sources; how subcontractor pricing fits in
- Summarizing the estimate to arrive at the final number
- Formulas for area and cubic measure, adding waste and adjusting productivity to specific projects
- Evaluating subcontractors' proposals and prices
- Adding Insurance and bonds
- Understanding how labor costs are calculated
- Submitting bids and proposals

Who should attend: project managers, architects, engineers, owner's representatives, contractors, and anyone who's responsible for budgeting or estimating construction projects.

Mechanical & Electrical Estimating

This *two-day* course teaches attendees how to prepare more accurate and complete mechanical/electrical estimates, avoiding the pitfalls of omission and double-counting, while understanding the composition and rationale within the RSMeans mechanical/electrical database.

Some of what you'll learn:
- The unique way mechanical and electrical systems are interrelated
- M&E estimates—conceptual, planning, budgeting, and bidding stages
- Order of magnitude, square foot, assemblies, and unit price estimating
- Comparative cost analysis of equipment and design alternatives

Who should attend: architects, engineers, facilities managers, mechanical and electrical contractors, and others who need a highly reliable method for developing, understanding, and evaluating mechanical and electrical contracts.

Unit Price Estimating

This interactive *two-day* seminar teaches attendees how to interpret project information and process it into final, detailed estimates with the greatest accuracy level.

The most important credential an estimator can take to the job is the ability to visualize construction and estimate accurately.

Some of what you'll learn:

- Interpreting the design in terms of cost
- The most detailed, time-tested methodology for accurate pricing
- Key cost drivers—material, labor, equipment, staging, and subcontracts
- Understanding direct and indirect costs for accurate job cost accounting and change order management

Who should attend: corporate and government estimators and purchasers, architects, engineers, and others who need to produce accurate project estimates.

Conceptual Estimating Using RSMeans CostWorks® CD

This *two-day* class uses the leading industry data and a powerful software package to develop highly accurate conceptual estimates for your construction projects. All attendees must bring a laptop computer loaded with the current year *Square Foot Costs* and the *Assemblies Cost Data* CostWorks titles.

Some of what you'll learn:

- Introduction to conceptual estimating
- Types of conceptual estimates
- Helpful hints
- Order of magnitude estimating
- Square foot estimating
- Assemblies estimating

Who should attend: architects, engineers, contractors, construction estimators, owner's representatives, and anyone looking for an electronic method for performing square foot estimating.

RSMeans CostWorks® CD Training

This one-day course helps users become more familiar with the functionality of *RSMeans CostWork*s program. Each menu, icon, screen, and function found in the program is explained in depth. Time is devoted to hands-on estimating exercises.

Some of what you'll learn:

- Searching the database using all navigation methods
- Exporting RSMeans data to your preferred spreadsheet format
- Viewing crews, assembly components, and much more
- Automatically regionalizing the database

This training session requires you to bring a laptop computer to class.

When you register for this course you will receive an outline for your laptop requirements.

Also offering web training for RSMeans CostWorks CD!

Assessing Scope of Work for Facility Construction Estimating

This *two-day* practical training program addresses the vital importance of understanding the SCOPE of projects in order to produce accurate cost estimates for facility repair and remodeling.

Some of what you'll learn:

- Discussions of site visits, plans/specs, record drawings of facilities, and site-specific lists
- Review of CSI divisions, including means, methods, materials, and the challenges of scoping each topic
- Exercises in SCOPE identification and SCOPE writing for accurate estimating of projects
- Hands-on exercises that require SCOPE, take-off, and pricing

Who should attend: corporate and government estimators, planners, facility managers, and others who need to produce accurate project estimates.

Facilities Estimating Using RSMeans CostWorks® CD

Combines hands-on skill building with best estimating practices and real-life problems. Brings you up-to-date with key concepts, and provides tips, pointers, and guidelines to save time and avoid cost oversights and errors.

Some of what you'll learn:

- Estimating process concepts
- Customizing and adapting RSMeans cost data
- Establishing scope of work to account for all known variables
- Budget estimating: when, why, and how
- Site visits: what to look for—what you can't afford to overlook
- How to estimate repair and remodeling variables

This training session requires you to bring a laptop computer to class.

Who should attend: facility managers, architects, engineers, contractors, facility tradespeople, planners, project managers and anyone involved with JOC, SABRE, or IDIQ.

Unit Price Estimating Using RSMeans CostWorks® CD

Step-by-step instruction and practice problems to identify and track key cost drivers—material, labor, equipment, staging, and subcontractors—for each specific task. Learn the most detailed, time-tested methodology for accurately "pricing" these variables, their impact on each other and on total cost.

Some of what you'll learn:

- Unit price cost estimating
- Order of magnitude, square foot, and assemblies estimating
- Quantity takeoff
- Direct and indirect construction costs
- Development of contractor's bill rates
- How to use *RSMeans Building Construction Cost Data*

This training session requires you to bring a laptop computer to class.

Who should attend: architects, engineers, corporate and government estimators, facility managers, and government procurement staff.

Registration Information

Register early . . . Save up to $100! Register 30 days before the start date of a seminar and save $100 off your total fee. *Note: This discount can be applied only once per order. It cannot be applied to team discount registrations or any other special offer.*

How to register Register by phone today! The RSMeans toll-free number for making reservations is **1-800-334-3509, Press 1.**

Two-day seminar registration fee - $935. One-day *RSMeans CostWorks*® training registration fee - $375. To register by mail, complete the registration form and return, with your full fee, to: RSMeans Seminars, 700 Longwater Drive, Norwell, MA 02061.

Government pricing All federal government employees save off the regular seminar price. Other promotional discounts cannot be combined with the government discount.

Team discount program for two to four seminar registrations. Call for pricing: 1-800-334-3509, Press 1

Multiple course discounts When signing up for two or more courses, call for pricing.

Refund policy Cancellations will be accepted up to ten business days prior to the seminar start. There are no refunds for cancellations received later than ten working days prior to the first day of the seminar. A $150 processing fee will be applied for all cancellations. Written notice of cancellation is required. Substitutions can be made at any time before the session starts. **No-shows are subject to the full seminar fee.**

AACE approved courses Many seminars described and offered here have been approved for 14 hours (1.4 recertification credits) of credit by the AACE International Certification Board toward meeting the continuing education requirements for recertification as a Certified Cost Engineer/Certified Cost Consultant.

AIA Continuing Education We are registered with the AIA Continuing Education System (AIA/CES) and are committed to developing quality learning activities in accordance with the CES criteria. Many seminars meet the AIA/CES criteria for Quality Level 2. AIA members may receive (14) learning units (LUs) for each two-day RSMeans course.

Daily course schedule The first day of each seminar session begins at 8:30 a.m. and ends at 4:30 p.m. The second day begins at 8:00 a.m. and ends at 4:00 p.m. Participants are urged to bring a hand-held calculator, since many actual problems will be worked out in each session.

Continental breakfast Your registration includes the cost of a continental breakfast, and a morning and afternoon refreshment break. These informal segments allow you to discuss topics of mutual interest with other seminar attendees. (You are free to make your own lunch and dinner arrangements.)

Hotel/transportation arrangements RSMeans arranges to hold a block of rooms at most host hotels. To take advantage of special group rates when making your reservation, be sure to mention that you are attending the RSMeans seminar. You are, of course, free to stay at the lodging place of your choice. **(Hotel reservations and transportation arrangements should be made directly by seminar attendees.)**

Important Class sizes are limited, so please register as soon as possible.

Note: *Pricing subject to change.*

Registration Form

ADDS-1000

Call 1-800-334-3509, Press 1 to register or FAX this form 1-800-632-6732. Visit our website: www.rsmeans.com

Please register the following people for the RSMeans construction seminars as shown here. We understand that we must make our own hotel reservations if overnight stays are necessary.

☐ Full payment of $________________ enclosed.

☐ Bill me.

Please print name of registrant(s).

(To appear on certificate of completion)

____

____

____

____

P.O. #: ________________

GOVERNMENT AGENCIES MUST SUPPLY PURCHASE ORDER NUMBER OR TRAINING FORM.

Firm name ________________

Address ________________

City/State/Zip ________________

Telephone no. __________ Fax no. __________

E-mail address ________________

Charge registration(s) to: ☐ MasterCard ☐ VISA ☐ American Express

Account no. __________ Exp. date __________

Cardholder's signature ________________

Seminar name ________________

Seminar City ________________

Please mail check to: RSMeans Seminars, 700 Longwater Drive, Norwell, MA 02061 USA

For more information visit the RSMeans website at **www.rsmeans.com**

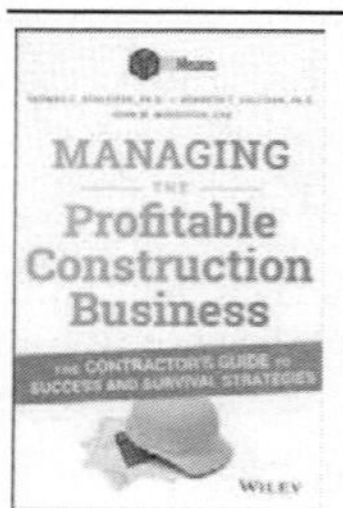

Managing the Profitable Construction Business: The Contractor's Guide to Success and Survival Strategies, 2nd Ed.

by Thomas C. Scheifler, Ph.D.

Learn from a team of construction business veterans led by Thomas C. Schleifer, who is commonly referred to as a construction business "turnaround" expert due to the number of construction companies he has rescued from financial distress. His financial acumen, combined with his practical, hands-on experience, has made him a sought-after private consultant.

$60.00 | 288 pages, hardover | Catalog no. 67370

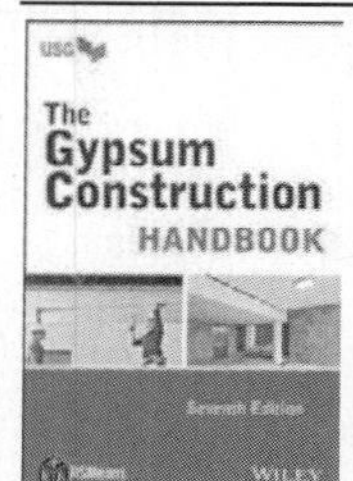

The Gypsum Construction Handbook, 7th Edition

by USG

The tried-and-true Gypsum Construction Handbook is a systematic guide to selecting and using gypsum drywall, veneer plaster, tile backers, ceilings, and conventional plaster building materials. A widely respected training text for aspiring architects and engineers, the book provides detailed product information and efficient installation methodology.

The 7th edition features updates in gypsum products, including ultralight panels, glass-mat panels, paper-faced plastic bead, and ultralightweight joint compound, and modern specialty acoustical and ceiling product guidelines. This comprehensive reference also incorporates the latest in sustainable products.

$40.00 | 576 pages, softcover | Catalog no. 67357B

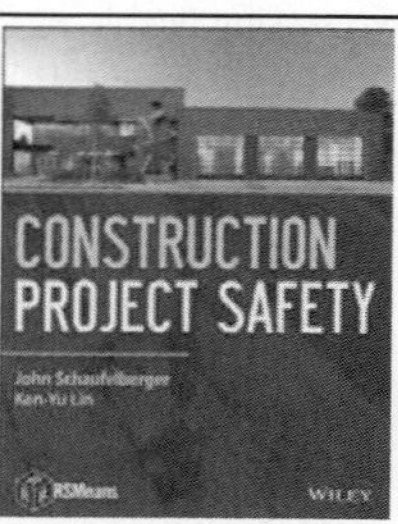

Construction Project Safety

by John Schaufelberger, Ken-Yu Lin

This essential introduction to construction safety for construction management personnel takes a project-based approach to present potential hazards in construction and their mitigation or prevention.

Beginning with an introduction to Accident Prevention Programs and OSHA compliance requirements, the book integrates safety instruction into the building process by following a building project from site construction through interior finish. Reinforcing this applied approach are photographs, drawings, contract documentation, and an online 3D BIM model to help visualize the on-site scenarios.

$85.00 | 320 pages, hardover | Catalog no. 67368

Solar Energy: Technologies & Project Delivery for Buildings

by Andy Walker

An authoritative reference on the design of solar energy systems in building projects, with applications, operating principles, and simple tools for the construction, engineering, and design professionals, the book simplifies the solar design and engineering process and provides sample documentation for the complete design of a solar energy system for buildings.

$85.00 | 320 pages, hardover | Catalog no. 67365

Project Control: Integrating Cost & Schedule in Construction

by Wayne J. Del Pico

Written by a seasoned professional in the field, *Project Control: Integrating Cost and Schedule in Construction* fills a void in the area of project control as applied in the construction industry today. It demonstrates how productivity models for an individual project are created, monitored, and controlled, and how corrective actions are implemented as deviations from the baseline occur.

$65.00 | 240 pages, softcover | Catalog no. 67366

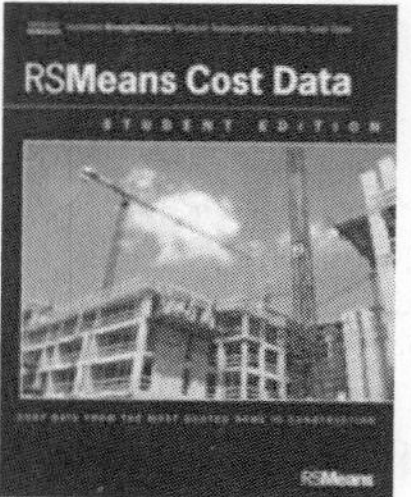

RSMeans Cost Data, Student Edition

Provides a thorough introduction to cost estimating in a print and online package. Clear explanations and example-driven. The ideal reference for students and new professionals. Features include:

- Commercial and residential construction cost data in print and online formats
- Complete how-to guidance on the essentials of cost estimating
- A supplemental website with plans, problem sets, and a full sample estimate

$110.00 | 512 pages, softcover | Catalog no. 67363

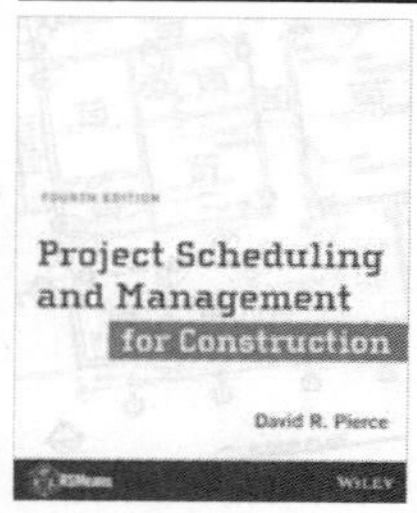

Project Scheduling and Management for Construction, 4th Edition

by David R. Pierce

First published in 1988 by RSMeans, the new edition of *Project Scheduling and Management for Construction* has been substantially revised for both professionals and students enrolled in construction management and civil engineering programs. While retaining its emphasis on developing practical, professional-level scheduling skills, the new edition is a relatable, real world case study.

$95.00 | 272 pages, softcover | Catalog no. 67367

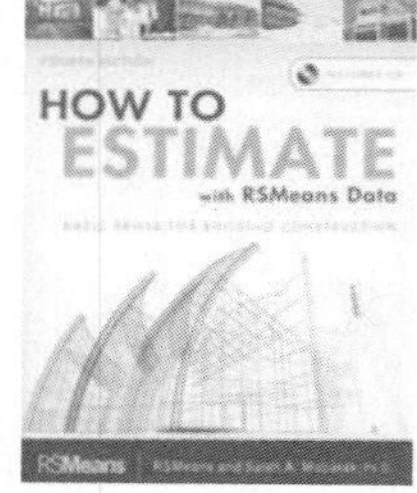

How to Estimate with Means Data & CostWorks, 4th Edition

by RSMeans and Saleh A. Mubarak, Ph.D.

This step-by-step guide takes you through all the major construction items with extensive coverage of site work, concrete and masonry, wood and metal framing, doors and windows, and other divisions. The only construction cost estimating handbook that uses the most popular source of construction data, RSMeans, this indispensible guide features access to the instructional version of *CostWorks* in electronic form, enabling you to practice techniques to solve real-world estimating problems.

$75.00 | 320 pages, softcover | Includes CostWorks CD | Catalog no. 67324C

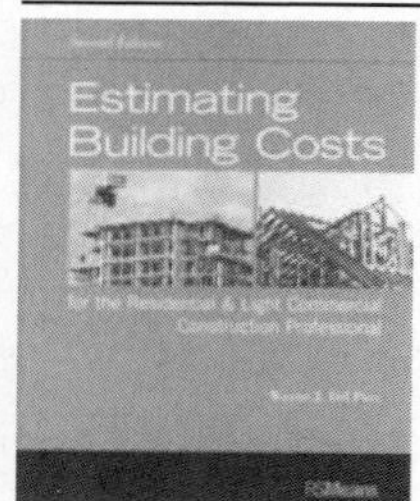

Estimating Building Costs, 2nd Edition

For the Residential & Light Commercial Construction Professional

by Wayne J. Del Pico

Explains, in detail, how to put together a reliable estimate that can be used not only for budgeting, but also for developing a schedule, managing a project, dealing with contingencies, and, ultimately, making a profit.

Completely revised and updated to reflect the CSI MasterFormat 2010 system.

$75.00 | 528 pages, softcover | Catalog no. 67343A

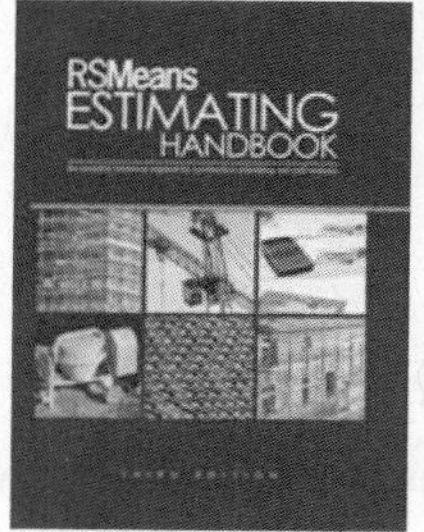

RSMeans Estimating Handbook, 3rd Edition

Widely used in the industry for tasks ranging from routine estimates to special cost analysis projects. This handbook will help construction professionals:

- Evaluate architectural plans and specifications
- Prepare accurate quantity takeoffs
- Compare design alternatives and costs
- Perform value engineering
- Double-check estimates and quotes
- Estimate change orders

$110.00 | Over 976 pages, hardcover | Catalog No. 67276B

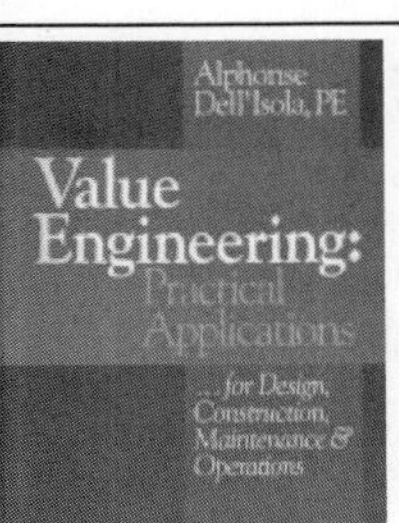

Value Engineering: Practical Applications

For Design, Construction, Maintenance & Operations

by Alphonse Dell'Isola, PE

A tool for immediate application—for engineers, architects, facility managers, owners, and contractors. Includes making the case for VE—the management briefing; integrating VE into planning, budgeting, and design; conducting life cycle costing; using VE methodology in design review and consultant selection; and case studies.

$90.00 | Over 427 pages, illustrated, softcover | Catalog no. 67319A

Risk Management for Design and Construction

Introduces risk as a central pillar of project management and shows how a project manager can prepare to deal with uncertainty. Includes:

- Integrated cost and schedule risk analysis
- An introduction to a ready-to-use system of analyzing a project's risks and tools to proactively manage risks
- A methodology that was developed and used by the Washington State Department of Transportation
- Case studies and examples on the proper application of principles
- Combining value analysis with risk analysis

$125.00 | Over 288 pages, softcover | Catalog no. 67359

How Your House Works, 2nd Edition

by Charlie Wing

Knowledge of your home's systems helps you control repair and construction costs and makes sure the correct elements are being installed or replaced. This book uncovers the mysteries behind just about every major appliance and building element in your house. See-through, cross-section drawings in full color show you exactly how these things should be put together and how they function, including what to check if they don't work. It just might save you having to call in a professional.

$22.95 | 208 pages, softcover | Catalog no. 67351A

Construction Business Management

by Nick Ganaway

Only 43% of construction firms stay in business after four years. Make sure your company thrives with valuable guidance from a pro with 25 years of success as a commercial contractor. Find out what it takes to build all aspects of a business that is profitable, enjoyable, and enduring. With a bonus chapter on retail construction.

$60.00 | 201 pages, softcover | Catalog no. 67352

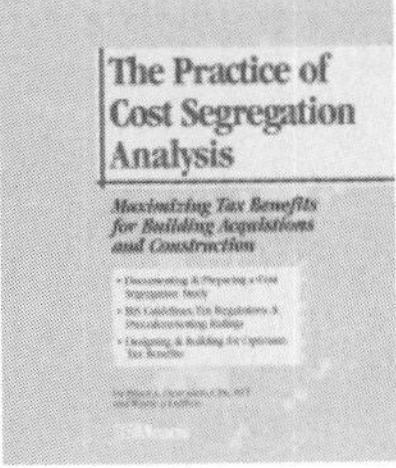

The Practice of Cost Segregation Analysis

by Bruce A. Desrosiers and Wayne J. Del Pico

This expert guide walks you through the practice of cost segregation analysis, which enables property owners to defer taxes and benefit from "accelerated cost recovery" through depreciation deductions on assets that are properly identified and classified.

With a glossary of terms, sample cost segregation estimates for various building types, key information resources, and updates via a dedicated website, this book is a critical resource for anyone involved in cost segregation analysis.

$105.00 | Over 240 pages | Catalog no. 67345

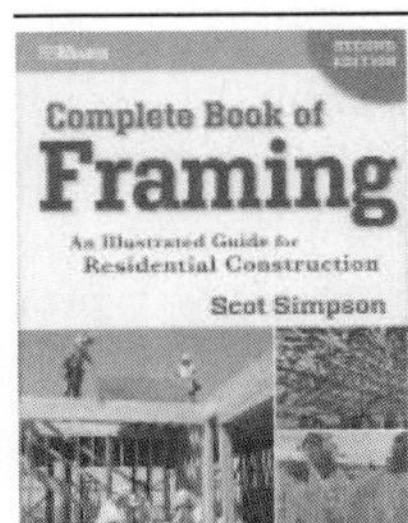

Complete Book of Framing, 2nd Edition

by Scot Simpson

This updated, easy-to-learn guide to rough carpentry and framing is written by an expert with more than thirty years of framing experience. Starting with the basics, this book begins with types of lumber, nails, and what tools are needed, followed by detailed, fully illustrated steps for framing each building element. Framer-Friendly Tips throughout the book show how to get a task done right—and more easily.

$29.95 | 368 pages, softcover | Catalog no. 67353A

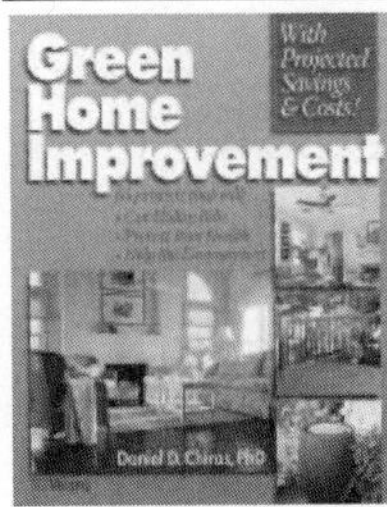

Green Home Improvement

by Daniel D. Chiras, PhD

With energy costs rising and environmental awareness increasing, people are looking to make their homes greener. This book, with 65 projects and actual costs and projected savings, helps homeowners prioritize their green improvements.

Projects range from simple water savers that cost only a few dollars, to bigger-ticket items such as HVAC systems. With color photos and cost estimates, each project compares options and describes the work involved, the benefits, and the savings.

$34.95 | 320 pages, illustrated, softcover | Catalog no. 67355

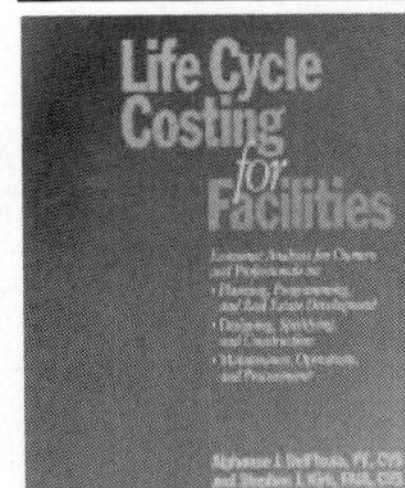

Life Cycle Costing for Facilities

by Alphonse Dell'Isola and Dr. Stephen Kirk

Guidance for achieving higher quality design and construction projects at lower costs! Cost-cutting efforts often sacrifice quality to yield the cheapest product. Life cycle costing enables building designers and owners to achieve both. The authors of this book show how LCC can work for a variety of projects—from roads to HVAC upgrades to different types of buildings.

$110.00 | 396 pages, hardcover | Catalog no. 67341

Project Scheduling and Management for Construction, 3rd Edition

by David R. Pierce, Jr.

A comprehensive, yet easy-to-follow guide to construction project scheduling and control—from vital project management principles through the latest scheduling, tracking, and controlling techniques. The author is a leading authority on scheduling, with years of field and teaching experience at leading academic institutions. Spend a few hours with this book and come away with a solid understanding of this essential management topic.

$64.95 | Over 286 pages, illustrated, hardcover | Catalog no. 67247B

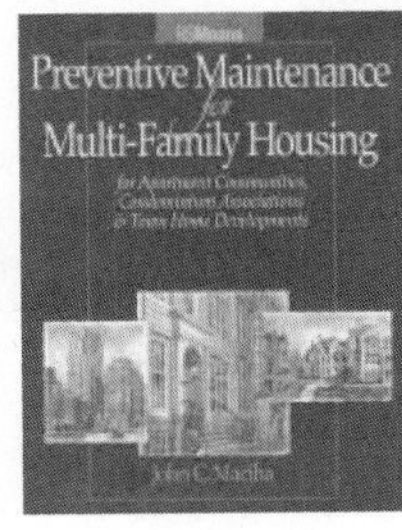

Preventive Maintenance for Multi-Family Housing

by John C. Maciha

Prepared by one of the nation's leading experts on multi-family housing.

This complete PM system for apartment and condominium communities features expert guidance, checklists for buildings and grounds maintenance tasks and their frequencies, and a dedicated website featuring customizable electronic forms. A must-have for anyone involved with multi-family housing maintenance and upkeep.

$95.00 | 290 pages | Catalog no. 67346

Green Building: Project Planning & Cost Estimating, 3rd Edition

Since the widely read first edition of this book, green building has gone from a growing trend to a major force in design and construction.

This new edition has been updated with the latest in green building technologies, design concepts, standards, and costs. Full-color with all new case studies—plus a new chapter on commercial real estate.

$110.00 | 480 pages, softcover | Catalog no. 67338B

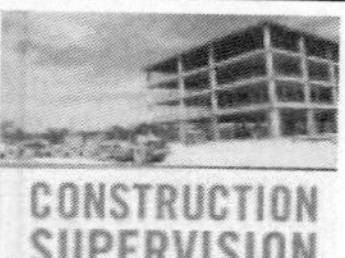

Construction Supervision

Inspires excellence with proven tactics and techniques applied by thousands of construction supervisors. Leadership guidelines carve out a practical blueprint for motivating work performance and increasing productivity through effective communication. Features:

- A unique focus on field supervision and crew management
- Coverage of supervision from the foreman to the superintendent level
- An overview of technical skills whose mastery will build confidence and success for the supervisor
- A detailed view of "soft" management and communication skills

$95.00 | 464 pages, softcover | Catalog no. 67358

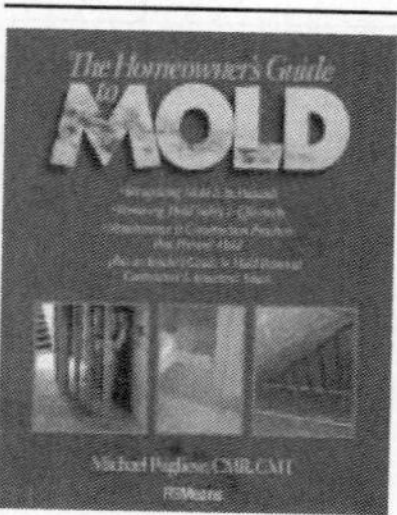

The Homeowner's Guide to Mold

By Michael Pugliese

Mold, whether caused by leaks, humidity or flooding, is a real health and financial issue—for homeowners and contractors. This full-color book explains:

- Construction and maintenance practices to prevent mold
- How to inspect for and remove mold
- Mold remediation procedures and costs
- What to do after a flood
- How to deal with insurance companies

$21.95 | 144 pages, softcover | Catalog no. 67344

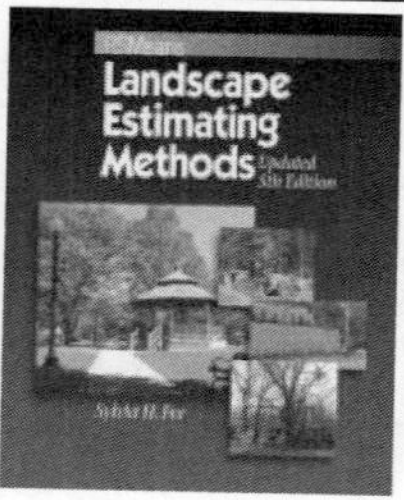

Landscape Estimating Methods, 5th Edition

Answers questions about preparing competitive landscape construction estimates, with up-to-date cost estimates and the new MasterFormat classification system. Expanded and revised to address the latest materials and methods, including new coverage on approaches to green building. Includes:

- Step-by-step explanation of the estimating process
- Sample forms and worksheets that save time and prevent errors

$75.00 | 336 pages, softcover | Catalog no. 67295C

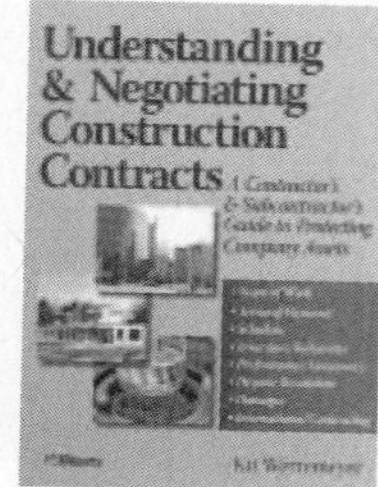

Understanding & Negotiating Construction Contracts

by Kit Werremeyer

Take advantage of the author's 30 years' experience in small-to-large (including international) construction projects. Learn how to identify, understand, and evaluate high risk terms and conditions typically found in all construction contracts—then, negotiate to lower or eliminate the risk, improve terms of payment, and reduce exposure to claims and disputes. The author avoids "legalese" and gives real-life examples from actual projects.

$80.00 | 320 pages, softcover | Catalog no. 67350

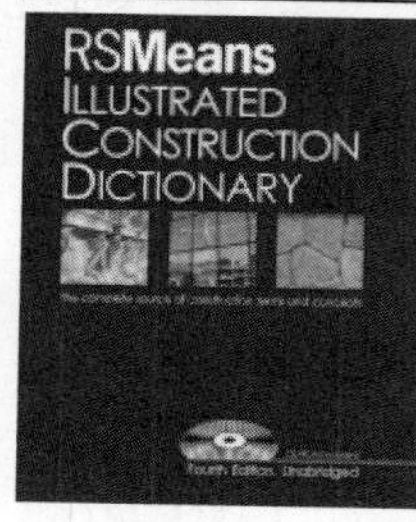

RSMeans Illustrated Construction Dictionary Unabridged 4th Edition, with CD-ROM

Long regarded as the industry's finest, *Means Illustrated Construction Dictionary* is now even better. With nearly 20,000 terms and more than 1,400 illustrations and photos, it is the clear choice for the most comprehensive and current information. The companion CD-ROM that comes with this new edition adds extra features, such as larger graphics and expanded definitions.

$110.00 | Over 880 pages, illust., hardcover | Catalog no. 67292B

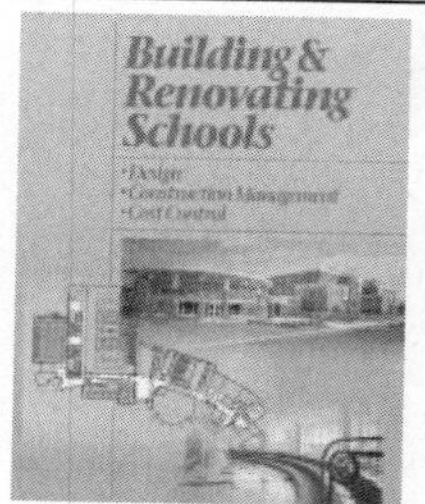

Building & Renovating Schools

This all-inclusive guide covers every step of the school construction process—from initial planning, needs assessment, and design, right through moving into the new facility. A must-have resource for anyone concerned with new school construction or renovation. With square foot cost models for elementary, middle, and high school facilities, and real-life case studies of recently completed school projects.

The contributors to this book—architects, construction project managers, contractors, and estimators who specialize in school construction—provide start-to-finish, expert guidance on the process.

$110.00 | Over 412 pages, hardcover | Catalog no. 67342

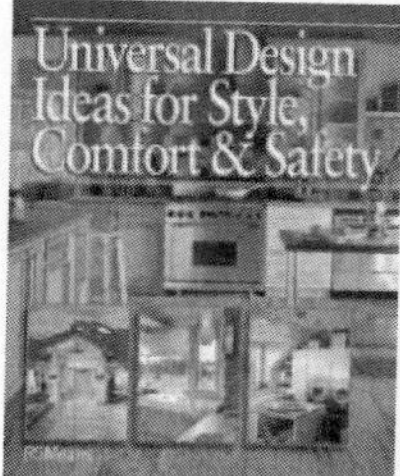

Universal Design Ideas for Style, Comfort & Safety

by RSMeans and Lexicon Consulting, Inc.

Incorporating universal design when building or remodeling helps people of any age and physical ability more fully and safely enjoy their living spaces. This book shows how universal design can be artfully blended into the most attractive homes. It discusses specialized products like adjustable countertops and chair lifts, as well as simple ways to enhance a home's safety and comfort. With color photos and expert guidance, every area of the home is covered. Includes budget estimates that give an idea how much projects will cost.

$21.95 | 160 pages, illustrated, softcover | Catalog no. 67354

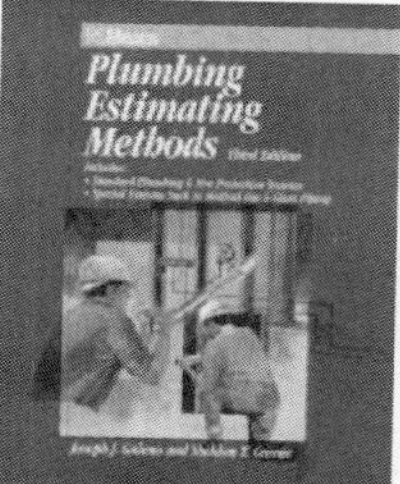

Plumbing Estimating Methods, 3rd Edition

by Joseph J. Galeno and Sheldon T. Greene

Updated and revised! This practical guide walks you through a plumbing estimate, from basic materials and installation methods through change order analysis. *Plumbing Estimating Methods* covers residential, commercial, industrial, and medical systems, and features sample takeoff and estimate forms and detailed illustrations of systems and components.

$65.00 | 381+ pages, softcover | Catalog no. 67283B

2015 Order Form

ORDER TOLL FREE 1-800-334-3509
OR FAX 1-800-632-6732

For more information visit the RSMeans website at **www.rsmeans.com**

Qty.	Book no.	COST ESTIMATING BOOKS	Unit Price	Total
	60065	Assemblies Cost Data 2015	$334.95	
	60015	Building Construction Cost Data 2015	206.95	
	60045	Commercial Renovaion Cost Data 2015	166.95	
	60115	Concrete & Masonry Cost Data 2015	188.95	
	50145	Construction Cost Indexes 2015 (subscription)	359.95	
	60145A	Construction Cost Index–January 2015	94.95	
	60145B	Construction Cost Index–April 2015	94.95	
	60145C	Construction Cost Index–July 2015	94.95	
	60145D	Construction Cost Index–October 2015	94.95	
	60345	Contr. Pricing Guide: Resid. R & R Costs 2015	39.95	
	60235	Electrical Change Order Cost Data 2015	199.95	
	60035	Electrical Cost Data 2015	208.95	
	60205	Facilities Construction Cost Data 2015	524.95	
	60305	Facilities Maintenance & Repair Cost Data 2015	461.95	
	60555	Green Building Cost Data 2015	170.95	
	60165	Heavy Construction Cost Data 2015	204.95	
	60095	Interior Cost Data 2015	208.95	
	60125	Labor Rates for the Const. Industry 2015	450.95	
	60185	Light Commercial Cost Data 2015	153.95	
	60025	Mechanical Cost Data 2015	203.95	
	60155	Open Shop Building Const. Cost Data 2015	175.95	
	60215	Plumbing Cost Data 2015	206.95	
	60175	Residential Cost Data 2015	148.95	
	60285	Site Work & Landscape Cost Data 2015	199.95	
	60055	Square Foot Costs 2015	218.95	
	62014	Yardsticks for Costing (2014)	189.95	
	62015	Yardsticks for Costing (2015)	201.95	
		REFERENCE BOOKS		
	67329	Bldrs Essentials: Best Bus. Practices for Bldrs	45.00	
	67307	Bldrs Essentials: Plan Reading & Takeoff	50.00	
	67342	Building & Renovating Schools	110.00	
	67261A	The Building Prof. Guide to Contract Documents	75.00	
	67353A	Complete Book of Framing, 2nd Ed.	29.95	
	67146	Concrete Repair & Maintenance Illustrated	75.00	
	67352	Construction Business Management	60.00	
	67364	Construction Law	115.00	
	67358	Construction Supervision	95.00	
	67363	Cost Data Student Edition	110.00	
	67314	Cost Planning & Est. for Facil. Maint.	95.00	
	67230B	Electrical Estimating Methods, 3rd Ed.	75.00	
	67343A	Est. Bldg. Costs for Resi. & Lt. Comm., 2nd Ed.	75.00	
	67276B	Estimating Handbook, 3rd Ed.	110.00	
	67318	Facilities Operations & Engineering Reference	115.00	
	67338B	Green Building: Proj. Planning & Cost Est., 3rd Ed.	110.00	
	67355	Green Home Improvement	34.95	
	67357B	The Gypsum Construction Handbook	40.00	
	67148	Heavy Construction Handbook	110.00	
	67308E	Home Improvement Costs–Int. Projects, 9th Ed.	24.95	
	67309E	Home Improvement Costs–Ext. Projects, 9th Ed.	24.95	
	67344	Homeowner's Guide to Mold	21.95	
	67324C	How to Est.w/Means Data & CostWorks, 4th Ed.	75.00	

Qty.	Book no.	REFERENCE BOOKS (Cont.)	Unit Price	Total
	67351A	How Your House Works, 2nd Ed.	22.95	
	67282A	Illustrated Const. Dictionary, Condensed, 2nd Ed.	65.00	
	67292B	Illustrated Const. Dictionary, w/CD-ROM, 4th Ed.	110.00	
	67362	Illustrated Const. Dictionary, Student Ed.	60.00	
	67348	Job Order Contracting	130.00	
	67295C	Landscape Estimating Methods, 5th Ed.	75.00	
	67341	Life Cycle Costing for Facilities	110.00	
	67370	Managing the Profitable Construction Business	60.00	
	67294B	Mechanical Estimating Methods, 4th Ed.	75.00	
	67345	Practice of Cost Segregation Analysis	105.00	
	67366	Project Control: Integrating Cost & Sched.	95.00	
	67346	Preventive Maint. for Multi-Family Housing	95.00	
	67367	Project Sched. and Manag. for Constr., 4th Ed.	95.00	
	67322B	Resi. & Light Commercial Const. Stds., 3rd Ed.	65.00	
	67359	Risk Management for Design and Construction	125.00	
	67365	Solar Energy: Technology & Project Delivery	85.00	
	67145B	Sq. Ft. & Assem. Estimating Methods, 3rd Ed.	75.00	
	67321	Total Productive Facilities Management	85.00	
	67350	Understanding and Negotiating Const. Contracts	80.00	
	67303B	Unit Price Estimating Methods, 4th Ed.	75.00	
	67354	Universal Design	21.95	
	67319A	Value Engineering: Practical Applications	90.00	

MA residents add 6.25% state sales tax	
Shipping & Handling**	
Total (U.S. Funds)*	

Prices are subject to change and are for U.S. delivery only. *Canadian customers may call for current prices. **Shipping & handling charges: Add 8% of total order for check and credit card payments. Add 8% of total order for invoiced orders.

Send order to: ADDV-1000

Name (please print) ____________________

Company ____________________

☐ Company

☐ Home Address ____________________

____

City/State/Zip ____________________

Phone # ____________________ P.O. # ____________________

(Must accompany all orders being billed)

Mail to: RSMeans, 700 Longwater Drive, Norwell, MA 02061

RSMeans Project Cost Report

By filling out this report, your project data will contribute to the database that supports the RSMeans® Project Cost Square Foot Data. When you fill out this form, RSMeans will provide a $50 discount off one of the RSMeans products advertised in the preceding pages. Please complete the form including all items where you have cost data, and all the items marked (☑).

$50.00 Discount per product for each report you submit

Project Description (NEW construction only, please)

☑ Building Use (Office School . . .) ____________

☑ Address (City, State) ____________

____

☑ Total Building Area (SF) ____________

☑ Ground Floor (SF) ____________

☑ Frame (Wood, Steel . . .) ____________

☑ Exterior Wall (Brick, Tilt-up . . .) ____________

☑ Basement: (checkone) ☐ Full ☐ Partial ☐ None

☑ Number of Stories ____________

☑ Floor-to-Floor Height ____________

☑ Volume (C.F.) ____________

% Air Conditioned ____________ Tons ____________

Total Project Cost $ ____________

Owner ____________

Architect ____________

General Contractor ____________

☑ Bid Date ____________

☑ Typical Bay Size ____________

☑ Occupant Capacity ____________

☑ Labor Force: _____ % Union _____ % Non-Union

☑ Project Description (Circle one number in each line.)

1. Economy 2. Average 3. Custom 4. Luxury

1. Square 2. Rectangular 3. Irregular 4. Very Irregular

Comments ____________

____

A	☑	General Conditions	$
B	☑	Site Work	$
C	☑	Concrete	$
D	☑	Masonry	$
E	☑	Metals	$
F	☑	Wood & Plastics	$
G	☑	Thermal & Moisture Protection	$
GR		Roofing & Flashing	$
H	☑	Doors and Windows	$
J	☑	Finishes	$
JP		Painting & Wall Covering	$

K	☑	Specialties	$
L	☑	Equipment	$
M	☑	Furnishings	$
N	☑	Special Construction	$
P	☑	Conveying Systems	$
Q	☑	Mechanical	$
QP		Plumbing	$
QB		HVAC	$
R	☑	Electrical	$
S	☑	Mech./Elec. Combined	$

Please specify the RSMeans product you wish to receive. Complete the address information.

Product Name ____________

Product Number ____________

Your Name ____________

Title ____________

Company ____________

☐ Company

☐ Home Street Address ____________

City, State, Zip ____________

Email Address ____________

Return by mail or fax 888-492-6770.

Method of Payment:

Credit Card # ____________

Expiration Date ____________

Check ____________

Purchase Order ____________

Phone Number ____________
(if you would prefer a sales call)

RSMeans
Square Foot Costs Department
700 Longwater Drive
Norwell, MA 02061
www.RSMeans.com